COMMISSION OF THE EUROPEAN COMMUNITIES

AGREP

PERMANENT INVENTORY
OF
AGRICULTURAL RESEARCH
PROJECTS IN THE
EUROPEAN COMMUNITIES

VOL. I MAIN LIST: RESEARCH PROJECTS

(containing data on approx. 21,000 research projects)

MAY 1980

SPRINGER-SCIENCE+BUSINESS MEDIA, B.V.

This volume has a Library of Congress Cataloging in Publication classification.

ISBN 978-94-009-8266-6   ISBN 978-94-009-8264-2 (eBook)
DOI 10.1007/978-94-009-8264-2

Publication arranged by:
Commission of the European Communities,
Directorate-General Information Market and Innovation,
Luxembourg

Data processing by
I/S Datacentralen af 1959,
Retortvej 6-8,
2500 Valby,
Denmark

EUR 5895

LEGAL NOTICE
Neither the Commission of the European Communities nor any person acting on behalf of
the Commission is responsible for the use which might be made of the following information.

Photocomposition by Special-Trykkeriet Viborg a-s

Table of Contents

INTRODUCTION

The Permanent Inventory of AGricultural REsearch Projects (AGREP) is an information system on current projects in agriculture and related fields in the Member States of the European Communities (Belgium, Denmark, Federal Republic of Germany, France, Ireland, Italy, Luxembourg, the Netherlands and the United Kingdom). Its creation is based upon a regulation of the Council of Ministers of the European Communities concerning the coordination of agricultural research (Council Regulation No 1728/74 of 27 June, 1974).

AGREP covers agriculture in an extensive sense of the word, i.e. including fields like fisheries and forestry, land use and development, conservation of nature, veterinary medicine, food science and technology, as well as agricultural economics and rural sociology. A detailed scope description is given in the subject classification facets "fields of research" and "subject areas" as presented on the introductory pages of the two printed volumes.

The data base is fully computerized. It is available in printed form and on magnetic tape. On-line access is offered by EURONET/DIANE (DIrect Access Network for Europe). The data base is updated annually.

The 1980 issue of AGREP contains about 21.000 title citations of agricultural research projects.

AGREP is a cooperative system. The national focal points (NFPs) collect input and provide it in an agreed format to the Commission which sponsors the management of the system and the provision of output to the Member Countries. The Commission is advised by the Standing Committee on Agricultural Research (SCAR) and its sub-group, the AGREP Advisory Board. Data processing and technical assistance are carried out by a contractor to the Commission.

The AGREP publication consists of two volumes:

Vol. I contains the Main List with the titles (both in the original language and in English) of the research projects, complemented by some basic information on each project, i.e. name(s) of researcher(s), country and research organisation (in the Main List given as codes), and availability of publications.

Vol. II contains various indexes, i.e.

– a Subject Index which permits the identification of research projects relevant to subject areas,

– the List of Research Organisations ordered by country and within each country by the hierarchical organisation code (see also explanatory note, vol. I, page IX). The names of research organisations are complemented by postal adresses, – the List of Scientists containing the name(s) of scientist(s) involved in the research projects referred to.

Further information may be obtained from the

AGREP Secretariat
Mr. G. Trevisan
Commission of the European Communities
Bâtiment Jean Monnet
LUXEMBOURG – KIRCHBERG

or by contacting one of the National Focal Points listed on the following page.

NATIONAL FOCAL POINTS OF EC MEMBER COUNTRIES PROVIDING INPUT TO AGREP.

BELGIUM
Ministère de l'Agriculture
Mr. J. Gooris
Administration de la Recherche Agronomique
Chaussée d'Ixelles 29-31
1050 – BRUXELLES
Telephone: 32-2-5123910

F.R. of GERMANY
Zentralstelle für Agrardokumentation
und -information
Mr. E. Müller
Villichgasse 17
5300 BONN 2
Telephone: (0228) 357097

FRANCE
Ministère de l'Agriculture et du
Développement Rural
Institut National de la Recherche Agronomique
Mme. G. Michel
149, Rue de Grenelle
75341 PARIS CEDEX 07
Telephone: 33-1-5503200

DENMARK
Landbrugets Samråd for Forskning og Forsøg
Mr. O. Kjeldsen Rasmussen
Vesterbrogade 4A^{IV}
1620 COPENHAGEN V
Telephone 45-1-127561

IRELAND
The Agricultural Institute
Headquarters
Mr. V. Reilly
19, Sandymount Avenue
DUBLIN 4
Telephone 353-1-688188
Telex 30459

ITALY
Ministero dell' Agricoltura e delle Foreste
Direzione Generale della Produzione Agricola
Divisione II Informazione Divulgazione
Mr. A. D'Ambrosio
Via XX Settembre
ROMA
Telephone: 39-6-4759580
Telex: 61251

Consiglio Nazionale delle Ricerche
Mr. E. Galante
Piazzale Aldo Moro, 7
ROMA
Telephone: 39-6-4953337
Telex: 610076

THE NETHERLANDS
Nationale Raad Landbouwkundig Onderzoek – TNO
Mr. H. Wansink
Office: Adelheidstraat 84,
Post: Juliana van Stolberglaan 148,
Postbus 297
2501 BD's-GRAVENHAGE
Telephone: 31-70-471021

UNITED KINGDOM
Agricultural Research Council
Mr. W. S. Wise
160, Great Portland Str.
LONDON W1N 6DT
Telephone: 44-1-5806655

COVERAGE

In the table below is indicated the degree of coverage within the different subject areas as defined by the Standing Committee for Agricultural Research (SCAR.).

By coverage is meant the percentage of all current research projects on subject areas A-H in a country, which are registered in AGREP. The figures should be viewed in conjunction with the notes following the table.

It should be noted, that the percentages are estimates and therefore subject to considerable uncertainty.

Percentage coverage of SCAR-recommended subject areas for the 1980 AGREP-inventory.

Subject Area	BE	DE	DK	GB	IE	IT	NL
A. Natural resources							
– Soil science and management of land, water, fauna and flora	90	85	80	40	95	85	80
B. Plant production	90						
– Crop husbandry, breeding and protection		85	95	75	99	85	100
– forestry		85	40	10	100	90	100
C. Animal production	90						
– Livestock husbandry, breeding and veterinary medicine		80	95	75	99	85	100
– Fisheries and fishery products		85	0	75	80	85	75
D. Agricultural engineering and building	90	85	95	85	100	85	100
E. Food and nutrition	85	90	50	85	99	80	75
F. Economic and social aspects	90						
– Management, economy and marketing studies		85	95	20	90	85	100
– Social studies		85	50	–	99	85	80
G. Research related to developing countries (on fields A-F)	90	80	50	5	100	85	90
H. General research methodology and service institutions (related to field A-F)	90	–	95	–	100	85	90

Notes regarding type of research organisation and financing in contributing countries.

Belgium
Both publicly financed and executed and privately financed and executed research projects are included in AGREP. The coverage indicated is the one for research carried out at public institutes – for privately executed research the coverage is considerably less.

Denmark
Mainly research carried out at public and semi-public institutions is included in AGREP.

Federal Republic of Germany
Research projects of public, semi-public and private institutions are included in AGREP. The percentage of coverage indicated relates mainly to public and semi-public organisations. For privately executed research the coverage is considerably less.

Ireland
Research carried out at public institutions – whether it is publicly financed research or done under contract for private industry – has been included in AGREP. Research carried out in private industry has not been included.

Italy
Research projects sponsored by:
– the public Administrations (governmental organizations in the strict sense – e.g. MAF, CNR, CNEN, universities and other research and public bodies),
– the public enterprises which are funded by the Ministry for Public Investment,
– private firms and other bodies,
  are included.

The Netherlands
Both public and private research is included in AGREP. The coverage indicated is the one for public research. Some public research on fishery products, foods and nutrition as well as research carried out at universities other than the Agricultural University and Faculty of Veterinary Medicine is not yet included in AGREP. For privately executed research the coverage is considerably less.

United Kingdom
Only public research is included in AGREP.

EXPLANATORY NOTES

MAIN LIST: Research projects (Vol. I)

Each entry of the main list contains input data collected per research project as listed below.

The entries are ordered
1) according to "Fields of Research" subdivided by "subject areas" (commodities) (table D of the classification scheme, see Table of Contents)
2) within each group according to the AGREP code, i.e. country, research institute and year the project was started. In cases where more than one D-classification code has been assigned to a project, the reference numbers of these projects are listed under the related chapter headings initiated by "see also".

Example of a typical entry:

```
                2)                                    3)
1)  3086 │ Prummel, J.              │NL 010103/62/0534
    Bemesting van grove tuinbouwgewassen op
4)  landbouwgronden│ │Manuring of vegetables in agriculture.   5)

6)  Publications│
```

1) A computer-assigned sequential number (reference number) at the beginning of each entry.
2) Scientists involved. Up to 6 names of research workers involved; where more than one name is given, the first name is that of the project leader.

3) **AGREP code**

A 14-character alpha-numeric code composed of
a) a 2-character alphabetical country code

BE	Belgium
DE	Federal Republic of Germany
DK	Denmark
FR	France
GB	United Kingdom
IE	Ireland
IT	Italy
NL	Netherlands
XE	The Commission of the European Communities

b) a 6-digit Research Organisation Code (see following page)
c) a 2-digit code indicating the year when the project started (e.g. 75)
For the Federal Republic of Germany only: The code indicates the year in which the project became known to the Focal Point.
d) a 4-digit current number assigned by the research organisations or by the National Focal Point.
4) **Title of the research project in the original language** except English
5) **English title** (original or translated) in bold face
6) **Publications**
"Publications" means that documents relevant to the research project are available.

RESEARCH ORGANISATION CODES

The organisation code is a six digit, three-level hierarchical numeric code, assigned by the National Focal Points. With the exception of the Federal Republic of Germany, the code should be interpreted in the following way:

Digit 1-2: Highest level   (e.g. ministry)
Digit 3-4: Intermediate level   (e.g. university)
Digit 5-6: Lowest level   (e.g. institute)

For the Federal Republic of Germany only, the code should be interpreted as follows:

Digit 1-3: Highest level   (e.g. university)
Digit 4-5: Intermediate level   (e.g. institute)
Digit 6:   Lowest level   (e.g. department)
Note: Digit 1 may also be a code for the type of the organisation (e.g. Federal)

This three-level hierarchical code makes it possible to identify the different units in the chain of responsibility, e.g. AA.02.03.04:

AA   Country Code
    02.   Faculté des Sciences Agronomiques de
                            l'Université Catholique
        03.   Département de Phytotechnique Tropicale
                            et Subtropicale
            04.   Laboratoir de Phytotechnique Tropicale et
                            Subtropicale

In the list of Research Organisations, each organisation code is followed by the name of the corresponding organisation, while the address in general is given only if it is different from the next higher level.

Example: AA.06.00.00. Faculté de Médecine
                            Vétérinaire de l'Université
                            45, Rue des Vétérinaires

                            AA.06.01.00 Chaire d'Anatomie
                            AA.06.02.00 Chaire de Bactériologie
                            AA.06.03.00 Chaire de Génétique
                            AA.06.04.00 Chaire d'Histologie
                            AA.06.06.00 Chaire d'Inspection des Denrées Alimentaires

Main List – Fields of Research

(Fields of Research are subdivided by Subject Areas, see Vol. II).

D 1100 – Soil science

See also 1490, 2406, 2908, 2916, 2951, 2997, 3336, 3846, 4406, 5279, 6269

1 De Boodt, M.; Verdonck, O.　　BE 030013/63/0004 R
Onderzoek van de optimalisatie van de substraten gebruikt in de tuinbouw, het opmaken van een bodemgeschiktheidskaart voor de sierplantenteelt in België, en automatisatie van irrigatiesystemen. **Research on optimalisation of substrates used in horticulture and mapping of soils nutable for horticultural purpuses in Belgium and automatisation of irrigation systems.** Publications.

2 Van Assche, C.; Van Wambeke, E.; Kesavan, R.
　　BE 040203/76/0016R
Onderzoek naar de persistentie van systemische fungiciden in de bodem. **Study on the persistency of systemic fungicides in the soil.**

3 Seutin, E.; Haquenne, W.; Meeus, P.　　BE 080700/78/0024 R
Recherches sur l'évolution des pesticides dans le sol. **Study of pesticides residues behaviour in soil.**

4 Vanderstappen, R.; Hoenig, M.; Ledent, G.
　　BE 100000/69/0002 R
Inventaires et études relatives à la pollution par éléments minéraux de l'environnement rural. **Inventory and study of pollution by mineral elements of the rural surrounding.** Publications.

5 De Borger, R.; Jacobs, T.; Monseur, X.; Neirinckx, G.; Hofman, M.; Meeus, K.　　BE 100000/76/0032 R
Studie van de problemen van verontreiniging ontstaan door het uitspreiden van dierlijke afval. **Study of the problems concerning pollution due to the spreading of animal wastes.** Publications.

6 Kickuth, R.; Aldag, R.　　DE 132034/70/0002
Trocknung und Vererdung von industriellen und kommunalen Schlämmen 1968. **Drying and bringing of industrial and communal sludges into soils.**

7 Meiwes, K.J.; Khanna, P.K.; Ulrich, B.
　　DE 132601/75/0001
Schwefelformen und –verhalten in Waldökosystemen. **Sulphur forms and behaviour in forest ecosystems.**

8 Jungk, A.; Hendriks, L.　　DE 138060/75/0002
Veränderungen des Phosphatgehaltes im wurzelnahen Boden lebender Pflanzen. **Changes of the phosphate content in the soil near the roots of living plants.**

9 Krause, R.　　DE 201070/75/0007
Bestimmung von Stoff-, Werkzeug– und Systemparametern; Ermittlung von Gesetzmässigkeiten von Werkzeugwirkungen beim Einbringen und Mischen von Stoffen mit Boden. **Determination of parameters of materials, tools and systems; tool effect in incorporating and mixing different matters with soil.**

10 Schöllhorn, J.　　DE 501202/78/0001 N
Prüfung der Möglichkeiten zur Bestandes–, Narben– und Strukturverbesserung auf Moorböden des Donauriedes. **Trials on possibilities to improve plant association, sod and soil structure on bog in the "Donau–Ried" near Ulm.**

11 Schöllhorn, J.　　DE 501202/78/0002 N
Wirkung verschiedener Güllebehandlungen auf Narbe und

Ertrag von Dauergrünland. **Effects of different treatments with semi–liquid manure on sod and yield of permanent grassland.**

12 Wolf, R.　　DE 501251/78/0001 N
Umgestaltung von Feuchtgebieten; Bericht über drei im Landkreis Ludwigsburg 1977/78 durchgeführte Projekte. **Reorganization of wetlands; report on three projects realized in 1977 and 1978 in the district of Ludwigsburg.** Publications.

13 Schlenker, G.; Mühlhäusser, G.; Müller, S.
　　DE 501502/70/0002 R
Standortliche Voraussetzungen für das Vorkommen von Werteichen in Baden–Württemberg 1970. **Locational conditions of the occurrence of valuable oaks in Baden–Wuerttemberg.**

14 Süss, A.; Eben, C.　　DE 502050/72/0016
Der Abbau verschiedener Pestizide im Boden. **Degradation of different pesticides in soil.**

15 Krämer, F.　　DE 508302/70/0002 R
Schütthöhe und Tieflockerung als Standortfaktor auf Löss–Rohböden und ihr Einfluss auf das Pflanzenwachstum 1957. **Piling height and deep loosening as site factors on uncultivated loess soils and influence on plant growth.**

16 Günther, K.-H.　　DE 508302/72/0003
Über den Einfluss von im Niederschlagswasser mitgeführten Luftverunreinigungen auf die Bodenazidität, insbesondere im Wald. **Influence of air–polluted rainwater to the acidity of soils, especially in the forests.**

17 Werner, W.; Wienschierz, H.　　DE 903000/78/0003 N
Phosphat–Wirkung von Klärschlämmen aus der 3. Reinigungsstufe. **Investigations on the phosphorus–effectiveness of sludges from chemical sewage cleaning plants.** Publications.

18 Hansen, L.　　DK 010105/77/8021
Kortlægning af vandbevægelsens og afstrømningens indflydelse på kvælstofbalancen under markforhold. **Survey of the influence of water movement and flow–off on the nitrogen balance under field conditions.**

19 Hansen, l.　　DK 010105/78/9008
Transport af vand og kvælstof til dræn og undergrund i relation til vandbalance og kvælstofbalance under markforhold. **Transport of water and nitrogen to drains and subsoil in relation to water balance and nitrogen balance under field conditions.**

20 Helweg, A.　　DK 010117/77/8135
Indflydelse af jordbundens temperatur og fugtighed på pesticiders nedbrydning i jord, for bedre udnyttelse af udenlandske forskningsresultater. **Influence of soil temperature and moisture on the degradation of pesticides in soil, for better utilisation of foreign research results.**

21 Olesen, S.E.　　DK 010900/75/0005
Vinderosionsundersøgelser. **Wind erosion investigations.**

22 Pedersen, B.; Willems, M.; Jørgensen, S.S.
　　DK 030106/73/0001
Tilførsel af spildevandsslam og slamaske til jord. **The addition of waste water and ash sludge to the soil.**

23 Haarløv, N.; Andersen, C.   DK 030109/76/0007
Jordbundsfaunaens sammensætning i dansk agerjord og dens andel i omsætningen af tilførte organiske og uorganiske gødningsmidler under vekslende driftsformer. **The composition of soil fauna on Danish arable land and its part in the conversion of added organic and inorganic fertilizer under changing working methods.**

24 Aslyng, H.C.; Jensen, S.E.; Hansen, S.
     DK 030143/79/0001 N
Analytisk–statistisk behandling af klimadata.
Mathematical–statistical analysis of climatic data.

25 Burford   GB 010503/00/0010
Soil porosity as affected by cultivation in relation to plant growth and gas exchange.

26 Goss   GB 010503/77/0012
Effects of soil physical conditions on root growth with special reference to cultivation.

27 Smith   GB 010802/00/0009
Persistence in soil of paraquat applied repeatedly to vegetation cover or bare soil.

28 Atkinson   GB 011008/78/0040
Root growth water and nutrient use and requirements in intensive plum orchards.

29 Atkinson   GB 011008/78/0046
Soil studies in orchards.

30 Rose   GB 011102/75/0063
Transport processes in porous materials.

31 Bunt   GB 011106/00/0008
Loamless potting substrates.

32 Brown; Weir, G.   GB 012003/00/0052
Description, classification and identification of soil clays.

33 Brown; Rayner   GB 012003/00/0053
Interstratification, interlayering and other imperfections in clay minerals and soil clays.

34 Weir; Catt   GB 012003/00/0058
Soils of fields used for experiments.

35 Catt; Weir   GB 012003/00/0059
Soils developed in superficial deposits (loess, boulder clay and stratified clay vale drift) in S and E England.

36 Bloomfeld; Pruden   GB 012003/00/0066
Oxidation of pyrite in soil.

37 Bloomfied; Pruden   GB 012003/00/0069
Organic and inorganic constituents of soil solutions, in relation to process of podzolisation.

38 Brown; Hughes   GB 012003/00/0072
The mineralogy of soils of the humid tropics.

39 Newman; Wright   GB 012003/00/0073
Effects of water regime and physical stress on the structure of soils.

40 Jenkinson   GB 012003/00/0074
Effects of mechanical disturbance on the decomposition of organic matter in soil.

41 Cooke; Williams   GB 012003/75/0038
Effects of organic matter and other constituents on the structure and physical properties of soils.

42 North   GB 012008/00/0013
Quantification of soil stability by ultrasonic dispersion methods.

43 Brown, N.   GB 012008/00/0014 R
Changes in soil physical properties caused by tillage and traffic.

44 Brown, N.   GB 012008/00/0016 R
Root development in relation to soil structure.

45 Towner   GB 012008/00/0017
Mechanical strength of soil and deformation properties of swelling soils.

46 Floate   GB 030304/77/0011
Interactions between acidity aluminium and phosphorus availability in hill soils.

47 Mitchill; Logan, B.   GB 030401/00/0001
Scottish soil types: chemical and physical characterization in relation to development.

48 Wilson; Bain   GB 030401/00/0003
Soil mineralogy: relationship with soil type and soil properties.

49 Wilson   GB 030401/00/0004
Minerals: alteration during weathering and soil development.

50 Mitchill   GB 030401/00/0005
Soil colloids: nature, origin and behaviour of inorganic, organic and organomineral complexes.

51 Paterson; Mitchell   GB 030401/00/0006
Surface characteristics of soil particles.

52 Mackenzie   GB 030401/00/0007
Mineral and organic soils: development of chemical and instrumental methods of examination.

53 Bracewell   GB 030401/00/0008
Mineral and organic soils: characterization by products of thermal decomposition.

54 McHardy   GB 030401/00/0009
Mineral and biological materials: structure and composition by electron–optical and electron probe methods.

55 Boggie   GB 030401/00/0010
Organic soils: moisture retention and root development.

56 Robertson, R.; Boggie; Williams   GB 030401/00/0011 R
Organic soils: site capability and amelioration.

57 Robertson, R.   GB 030401/00/0012 R
Scottish peat deposits: survey, classification and characterization.

58 Berrow   GB 030402/00/0001

Distribution and location of trace elements in soils, effect of soil parent material and drainage conditions.

59 Berrow   GB 030402/00/0004
Geochemical distribution and pedological behaviour of trace elements.

60 Stein   GB 030402/00/0005
Develop techniques for the determination of trace elements: direct reading methods and computer processing.

61 Farmer; Russell   GB 030402/00/0007
Characterise soil minerals and study surface properties and weathering by infra–red spectroscopy.

62 Farmer; Russell   GB 030402/00/0008
Characterization of soil organic matter by infrared and ultraviolet methods.

63 Cheshire   GB 030403/00/0007
Characterization of soil humic substances by means of their paramagnetic properties.

64 Bache   GB 030406/00/0004
Soil acidity: aluminium solubility and cation exchange equilibria in different soil types.

65 Bache; Ireland   GB 030406/00/0005
Anion sorption: kinetics and eqilibria of phosphate reactions in relation to soil composition.

66 Bache; Buchan   GB 030406/00/0012
Soil physical conditions and crop growth.

67 Bound; Edmonds   GB 030406/77/0014
Electrochemical studies on soil–nutrient–plant relationships.

68 Birse; Roberson J   GB 030408/00/0002
Plant communities and their relation to genetic soil types.

69 Romans   GB 030408/00/0004
Studies of soil structure and genesis.

70 Soane; Kenworthy   GB 030503/00/0009
Soil strength characteristics at high strain rates.

71 Soane; Campbell   GB 030503/00/0010
Intransient mechanical properties of agricultural soils.

72 Soane; Campbell   GB 030503/00/0012
Develop method of measuring size, density and strength of soil clods.

73 Soane   GB 030503/00/0013
Effect of temperature fluctuations on clod strength characteristics.

74 McConaghy   GB 040402/00/0008
Characteristics of soils derived from materials of basic igneous origin.

75 McAllister   GB 040402/00/0013
Relation between cation contents of soils and herbage.

76 Adams   GB 040402/00/0022
Release of nitrogen from the decomposition of soil organic matter.

77 Benians   GB 040402/76/0019
Ion activities in soil solutions and extracts, in relation to plant nutrition, ion exchange and soil classification.

78 Gracey   GB 041101/75/0017
Studies on the residual value of phosphate applied as slurry and the depletion of soil phosphate reserves.

79 Seaby   GB 041502/00/0002
The effect of lime in different forms on mycorrhizal formation and peat and litter breakdown.

80 Researcher not indicated   GB 050125/00/0001 R
Soils: land drainage.

81 Researcher not indicated   GB 050125/00/0002 R
Soils: soil management.

82 Researcher not indicated   GB 050125/00/0003 R
Soils: soil organic matter.

83 Researcher not indicated   GB 050125/00/0004 R
Soils: chemical fertility, long term manurial trials, leaching of nutrients.

84 Researcher not indicated   GB 050165/00/0001
Soils: land drainage.

85 Researcher not indicated   GB 050165/00/0002
Soils: soil management.

86 Researcher not indicated   GB 050165/00/0003
Soils: soil organic matter.

87 Researcher not indicated   GB 050165/00/0004
Soils: irrigation techniques.

88 Researcher not indicated   GB 050165/00/0005
Soils: chemical fertility, long term manurial trials, leaching of nutrients.

89 Witney   GB 060109/00/0006
Effects of cultivation on soil physical properties and crop growth.

90 Purves   GB 060115/00/0003
Trace elements in relation to pollution.

91 Farr   GB 060219/00/0011
Studies of plant root environment and ionic diffusion in soil and nutrient availability to plants.

92 McGregor   GB 060306/00/0006
Development and evaluation of field and laboratory methods for examination of soil physical properties.

93 Flegg, A.M.; MacDermod, C.; Pyne, J.P.; Max, M.
     IE 040101/76/9099 N
The clare phosphate project : an investigation into the occurrence and reserves of rock phosphate in the basal namurian shales of county clare in the republic of ireland.

94 Kiely, J.   IE 060400/72/0715 R
Detailed reconnaisance survey of east Cork. Publications.

95 Grubb, L.; Daly, P.J.   IE 060706/77/1369 R
An evaluation of the potential of reclaimed blanket peat under
an intensive lowland sheep system. Publications.

96 Bazzoffi, P.; Zanchi, C.; Torri, D.   IT 020100/79/0008 N
Studio dell'energia cinetica delle gocce di pioggia. **Study of
raindrop kinetic energy.** Publications.

97 Giardina, M.C.   IT 060200/74/0075
Isolamento, studio tassonomico e fisiologico di specie di
Arthrobacter provenienti da suoli pedologicamente diversi.
**Isolation, taxonomy and physiology of Arthrobacter species
from soils different in pedology.**

Soil in general (B 1100)

See also 1, 5, 6, 7, 8, 9, 338, 1175, 1176, 1332, 1432, 1435, 1498,
1807, 1816, 2109, 2189, 2332, 2424, 2728, 2873, 2883, 3251,
3392, 4008, 4066, 4826, 5005, 5019, 5149, 5187, 5458, 5687,
5818, 5853, 5855, 5944, 5965, 5997, 7824, 8233, 8371, 10119,
10130, 10159, 10172, 10322, 11966, 20312, 20313, 20314, 20315,
20316

98 Hanotiaux, G.; Bollinne, A.   BE 010016/72/0001 R
Etude de l'érosion en région limoneuse. **Study of erosion in the
loam area of Belgium.** Publications.

99 De Boodt, M.; De Wulf, F.   BE 030013/72/0005
Studie van de fysische bodemdegradatie. **Study of the physical
soil degradation.**

100 Van Assche, C.; Vanachter, A.; Van Wambeke, E.
    BE 040203/76/0014 R
Studie van de nevenwerkingen (bodemkarakteristieken en
plant kwaliteitseigenschappen) ten gevolge van chemische
grondontsmetting. **Study on side effects (soil characteristics and
plant quality) due to chemical soil disinfection.**

101 D'Hoore, J.; Gombeer, R.; Sougnez, N.; Gulinck, H.;
Ceusters, A.M.   BE 040800/77/0001
Landbouwkundige en Ecologische Interpretatie van
Teledetektie gegevens. **Agricultural and Ecological
Interpretation of remote sensing data.**

102 Raimond, Y.; Destain, J.P.   BE 080100/59/0001
Evolution des propriétés physiques et chimiques du sol sous
l'influence de divers systèmes d'exploitation culturale.
**Evolution of soil physical and chemical properties as related to
the different cropping systems.** Publications.

103 Raimond, Y.; Destain, J.P.   BE 080100/67/0002
Influence des cultures sans labour et en semis directs sur les
propriétés physiques et chimiques du sol. **Soil physical and
chemical properties as related to reduced cultivations and direct
drilling.** Publications.

104 Istas, J.; De Temmerman, L.; Baeten, H.;
Vanderstappen, R.; Dupire, S.; Raekelboom, E.
    BE 100000/75/0033 R
Studie van de bestaande verontreinigingsgraad van grond en
planten door zware metalen en fluor. **Study of the existing
degree of pollution of soil and plants by heavy metals and
fluorine.** Publications.

105 Jacobs, T.   BE 100000/76/0030

Bodemverontreiniging door atmosferische neerslag. **Soil
pollution by atmospheric fall–out.**

106 Rogister, J.   BE 130000/73/0008 R
Corrélations entre les caractères pédologiques des stations les
types de végétation et l'accroissement des arbres. **Correlations
between the pedological caracteristics of the growthsites, the
vegetation types and the tree growth.** Publications.

107 Huyh, A.; Baeyens, L.; Berben, J.; de Jamblinne de
Meux, A.   BE 140000/77/0015 R
Herstelmogelijkheden van verontreinigde en verlaten
gronden. **Reutilization possibilities of polluted and fallow land.**
Publications.

108 Blume, H.–P.; Drewes, H.   DE 105050/73/0003
Abbau und Bewegung von Pestiziden im Boden. **Degradation
and action of pesticides in soil.**

109 Blume, H.–P.; Alaily, F.   DE 105050/77/0002
Zur Genese der Gley–Podsole. **Genesis of Aquods.**

110 Blume, H.–P.; Petermann, T.; Litz, N.
    DE 105050/78/0002 N
Bodenkartierung mit Nutzungsbewertung eines
Sahara–Ausschnittes – in Fezzan –. **Soil mapping and land use
suitability survey of a part of the central Sahara – Fezzan –.**

111 Pahlke, K.   DE 105250/78/0001 N
Auswirkungen der Blausandmelioration auf junger, schluffiger
Seemarsch. **Effects of 'Blausand'– amelioration on young, silty
sea marsh soils.**

112 Gracanin, Z.   DE 108150/73/0003
Aktuelle Veränderungen der Böden, der Landschaft und der
Bodenkultur im Kaiserstuhl. **Actual changes in soils, landscape
and soil use in Kaiserstuhl.**

113 Mückenhausen, E.   DE 111050/70/0001
Fortschritte der Bodensystematik 1968. **Progress in soil
taxonomy.**

114 Mückenhausen, E.; Matuschka, W.   DE 111050/70/0006
Entstehung und Eigenschaften der Böden an der
Afrika–Strasse zwischen Addis–Abeba und Kenia 1969.
**Formation and characteristics of soils by the African Highway
between Addis–Abeba and Kenya.**

115 Zakosek, H.; Mückenhausen, E.; Schröder, D.
    DE 111050/75/0003
Perkolationsversuche mit Bodensäulen. **Percolation
experiments with soil columns.**

116 Zakosek, H.; Plass, W.; Stephan, S.; Stöhr, W.
    DE 111050/75/0008
Verbreitung, Genese und Systematik der Smonica in
Rheinhessen. **Distribution, genesis and systematic of the
smonica in Rhenish Hesse.**

117 Mückenhausen, E.   DE 111050/77/0005
Beitrag zu den fossilen Böden – Paläoböden – der Eifel/
Westdeutschland. **Contribution to the fossil – paleosols – of the
Eifel Mountains/West–Germany.** Publications.

118 Zakosek, H.; Steinegger, U.; Stephan, S.; Wiechmann,
H.; Beckmann, H.   DE 111050/77/0006

Paläoböden und Gesteinszersatz im Zusammenhang mit den Vertikalbewegungen des Rheinischen Schildes. **Paleosols and rock detritus as connected with the vertical movement of the Rhenish shield.**

119 Schröder, D.; Brunnacker, K.; Wiechmann, H.; Stephan, S.; Zakosek, H.　　　　DE 111050/77/0007
Untersuchung der Auenböden in der Niederrheinischen Bucht. **Research on the alluvial soils of the Lower Rhine basin.**

120 Stephan, S.　　　　DE 111050/78/0001 N
Mikromorphologische und tonmineralogische Untersuchung von Böden und vulkanischen Ablagerungen aus Ecuador. **Micromorphological and clay–mineralogical investigation of soils and volcanic deposits from Ecuador.**

121 Boeker, P.; Franken, H.　　　　DE 111251/74/0002
Erforschung der Bodenfruchtbarkeit in Abhängigkeit von Boden, Klima, Pflanze und Stickstoffdüngung. **Research on soil fertility in dependence on soil type, climatic conditions, plants and nitrogen fertilization.**

122 Boeker, P.; Franken, H.; Oestrich, A.
　　　　DE 111251/74/0014
Handelsdünger– und – spezifische – Stallmistwirkungen auf Boden und Pflanze. **Effects of commercial fertilizers and specific manures on soil and plant.**

123 Boeker, P.; Franken, H.; Oestrich, A.
　　　　DE 111251/74/0015
Die Wirkung verschiedener Kali– und Stickstofformen auf Boden und Pflanze. **The effects of different types of potash and nitrogen on soil and plant.**

124 Zöttl, H.W.; Keilen, K.; Stahr, K.　DE 126050/78/0003 N
Untersuchungen zur Bodenentwicklung und den Standortseigenschaften im südlichen Hochschwarzwald. **Investigations of soil development and forest site properties in southern Black Forest.** Publications.

125 Moll, W.; Blum, W.E.　　　　DE 126050/78/0006 N
Bodenkarte 1 : 1 Mill. "Lake Victoria" – Kenya, Uganda, Tansania – mit Erläuterungsband. **Soil map 1 : 1 mill. "Lake Victoria" – Kenya, Uganda, Tansania – and commentary volume.**

126 Schönhals, E.; Poetsch, T.; Altemüller, H.–J.
　　　　DE 129020/73/0002
Vergleichende mineralogische und mikromorphologische Untersuchungen zum Problem der Genese und Systematik der Locker– braunerden. **Comparative mineralogical and micromorphological studies of problems in genesis and systems of loose brown soils.**

127 Harrach, T.; Djafari, H.　　　　DE 129020/74/0002
Bodenschätzung in der Flurbereinigung auf der Grundlage der Reichsbodenschätzung. **Soil taxation within land consolidation on the basis of soil taxation during the period of the German Reich.**

128 Breburda, J.　　　　DE 129382/71/0002 R
Bodenerosion und Bodenerhaltung im östlichen Mediterrangebiet 1971. **Soil erosion and soil preservation in the eastern Mediterranean Region.**

129 Breburda, J.　　　　DE 129382/71/0003 R
Bodenkundliche Probleme in der Sowjetunion 1971. **Soil scientific problems in Soviet Russia.**

130 Kowald, R.; Mollenhauer, K.; Vogel, C.
　　　　DE 129450/72/0002
Landwirtschaftliche Verwertung von Abwasserschlamm auf schweren Böden. **Agricultural utilization of sewage sludge on heavy soils.** Publications.

131 Kowald, R.; Vogel, C.　　　　DE 129450/74/0002
Die Filterwirkung verschiedener Böden bei Anwendung von Siedlungsabfällen. **Filtering efficiency of different soils when domestic refuse is applied.**

132 Jung, L.; Brechtel, R.　　　　DE 129551/70/0003
Messungen von Oberflächenabfluss und Bodenabtrag auf verschiedenen Böden. **Measurement of top–soil run–off and soil erosion on different soils.**

133 Atanasiu, N.; Westphal, A.　　　　DE 129552/70/0001
Untersuchungen über die Aufnahme durch 15N markierte N–Formen auf deutschen und tropischen Böden. **Investigations on the uptake of different forms of nitrogen marked with 15N by German and tropical soils.**

134 Atanasiu, N.; Westphal, A.; Alkämper, J.
　　　　DE 129552/70/0008
Untersuchungen über die Wirkung von verschiedenen N–,P–,K–Düngerformen auf tropischen Böden. **Investigations of the effect of different N–,P–,K–fertilizer forms on tropical soils.**

135 Atanasiu, N.; Westphal, A.　　　　DE 129552/71/0002
Standortuntersuchungen – Merkmale der Bodenfruchtbarkeit und des Leistungspotentials – im ägäischen Küstengebiet. **Investigations on the soil fertility and the capacity of different locations in the Aegean maritime provinces.**

136 Nitz, H.–J.; Meyer, G.–U.　　DE 132511/72/0002 R
Untersuchungen zur Dynamik der Landwirtschaftsformationen an regionalen Beispielen aus den Marsch–, Geest– und Jungmoränengebieten Schleswig–Holsteins. **Exemplified by marsh, geest, and recent–moraine areas in Sleswich–Holstein.**

137 Ulrich, B.; Mayer, R.; Khanna, P.K.; Prenzel, J.
　　　　DE 132600/72/0001
Modellbildung der Bioelementkreisläufe in Waldökosystemen. **Modelling of bioelement cycles in forest ecosystems.**

138 Ulrich, B.; Beese, F.　　　　DE 132603/75/0001
Stickstoffumwandlung bei stationärem und nichtstationärem Transport durch den Boden. **Nitrogen transformation during transport through the soil in steady and transient state.**

139 Beese, F.; Ulrich, B.　　　　DE 132603/77/0001
Die horizontale und vertikale Variabilität des scheinbaren Diffusionskoeffizienten einiger repräsentativer Böden und die Bedeutung dieses bodenphysikalischen Parameters für die modellmässige Beschreibung des Lösungstransports. **Horizontal and vertical variability of the apparent diffusioncoefficient of some representative soils and the influence of this parameter on modelling the solute transport.**

140 Richter, J.　　　　DE 138030/75/0001
Mathematische Modelle für Bodenprozesse. **Mathematical**

D 1100 - Soil science

models for soil processes.

141 Ottow, J.C.G.; Benckiser, G.   DE 144321/75/0002
Denitrifikation mit relativ persistenten organischen
H–Donatoren. **Denitrification with relatively persistent organic
H–donators.**

142 Hurle, K.; Kibler, E.   DE 144540/75/0002
Der gegenseitige Einfluss von Pestiziden auf ihren Abbau im
Boden. **The mutual influence of pesticides on their degradation
in the soil.**

143 Schroeder, D.; Kneib, W.; Lamp, J.; Mutert, E.
    DE 148051/74/0003
Anwendung taxonometrischer Verfahren zur Herstellung und
Auswertung schleswigholsteinischer Bodenkarten. **Application
of taxonometric methods to the production and interpretation of
soil maps in Sleswick–Holstein.**

144 Schroeder, D.; Mutert, E.   DE 148051/74/0004
Untersuchung des Stoffhaushaltes einer ostholsteinischen
Landschaftseinheit. **Investigations of pedosystems in a unit of
landscape in East Holstein.**

145 Schroeder, D.; Kneib, W.   DE 148051/74/0005
Untersuchungen zur regionalen Gruppierung von Böden in
topischer und chorischer Dimension. **Investigations on regional
grouping of soils in topic and choric dimensions.**

146 Brümmer, G.; Gerth, J.   DE 148051/75/0001
Belastung und Belastbarkeit von Böden und Sedimenten mit
Schwermetallen. **Current pollution and limits of pollution with
heavy metals in soils and sediments.**

147 Schroeder, D.; Friesel, P.   DE 148051/75/0003
Teilprozesse der Podsolierung. **Partial processes of
podzolization.**

148 Schroeder, D.; Kneib, W.   DE 148051/78/0001 N
Pedogenese und Pedofunktion. **Pedogenesis and pedofunction.**

149 Becher, H.H.   DE 161020/75/0002
Feststofftransport im Boden in Abhängigkeit vom
Saugspannungsgradienten; Entkalkung und Tonverlagerung in
Abhängigkeit von Saugspannungsgradienten und
Perkolationsmedium. **Translocation of suspended material in
soils depending on suction gradient; Decarbonisation and clay
migration as influenced by suction gradient and percolation
media.**

150 Schwertmann, U.; Taylor, R.M.; Cornwell, S.; Murad,
E.; Fitzpatrick, R.W.   DE 161020/77/0002
Eigenschaften pedogener Goethite; Al–Substitution in
Goethiten und Hämatiten – Synthese – ; Fe–oxid–Synthese in
Gegenwart organischer Stoffe. **Properties of pedogenic
goethites. Al substitution in goethites and hematites – synthesis
– . Fe–oxide synthesis in the presence of organic compounds.**

151 Schwertmann, U.; Kämpf, N.   DE 161020/78/0003 N
Pedosequenzen im Staate Rio Grande do Sul, Brasilien.
Pedosequences in the state of Rio Grande do Sul, Brasil.

152 Schwertmann, U.; Schulze, D.   DE 161020/78/0004 N
Eigenschaften von Bodengeothiten. **Properties of soil goethites.**

153 Amberger, A.; Bosch, M.   DE 161040/75/0001
Wirkung verschiedener N–Dünger auf Boden und Pflanze in
einem 50–jährigen Feldversuch auf Ackerbraunerde –
Weihenstephan –. **Effect of different nitrogen fertilizers on soil
and plant in a 50 years field experiment on brown earth –
Weihenstephan –.**

154 Amberger, A.; Vilsmeier, K.   DE 161040/78/0001 N
Ammonifikations– und Nitrifikations–Studien in
verschiedenen Böden unter dem Einfluss von
Nitrifikations–Hemmstoffen. **Ammonification and nitrification
studies on different soils as influenced by nitrification
inhibitors.** Publications.

155 Spatz, G.; Popp, T.; Voigtländer, G.   DE 161255/77/0002
Die Dynamik von Landnutzung und Bodenerosion in den
Bayerischen Alpen, in den letzten 60 Jahren. **Dynamics of
land use and soil erosion in the Bavarian Alps during the last six
decades.**

156 Burberg, P.–H.   DE 164350/72/0004 R
Flächenbilanz, Aufstockungsbilanz und
Tragfähigkeitsberechnung als Grundlage für regionale
Entwicklungspläne der Landwirtschaft 1971. **Area balance,
increase balance and calculation of capacity as basis of regional
development plans on agriculture.**

157 Stühmeier, K.; El–Bassam, N.   DE 201010/78/0001 N
Mathematische Darstellung der Schadstoffverlagerung in
Böden. **Mathematical description of the migration of toxic
elements in soils.**

158 Tietjen, C.; El–Bassam, N.; Esser, J.   DE 201040/75/0009
Dynamik der Schwermetalle im System
Boden–Wasser–Pflanze. **Dynamics of heavy metals in the
soil–water–plant system.**

159 Zach, M.   DE 201040/78/0007 N
Korrektur der Bodenbearbeitung zur Verbesserung des
Feldaufgangs auf leichten Böden und zur Verminderung der
Erosion. **Corrections of the soil tillage in order to improve the
field germination on light soils and to reduce erosion.**

160 Schmidt–Lorenz, R.   DE 202010/77/0002
Oxidreiche rote und braune Böden – Aufklärung ihrer
Mikromorphogenese und Dynamik. **Red and brown soils rich
in oxide – clarification of their micromorphogenesis and
dynamics.**

161 Kloke, A.; Schoenhard, G.   DE 215060/71/4001 N
Untersuchungen über die Belastbarkeit des Bodens mit
Pflanzennährstoffen. **Studies on the capacity of soil towards
plant nutrients.**

162 Geike, F.   DE 215060/78/0005 N
Beeinflussung bestimmter Parameter des Bodens durch
Umweltchemikalien. **Effect of environmental pollutants on
certain soil parameters.**

163 Hoffmann, G.; Scholl, W.; Schmid, R.
    DE 501010/78/0001 N
Schwermetalle in Ackerböden von Baden–Württemberg.
**Contents of heavy metals in arable soils of
Baden–Wuerttemberg.**

164 Pfulb, K.; Scholl, W.; Völkel, R.   DE 501012/75/0001

Einfluss verschieden hoher Gaben Müllklärschlammkompost auf den Gehalt von Rebböden an verschiedenen Schwermetallen. **The influence of different doses of refuse sewage–sludge on the content of different heavy metals in vineyard soils.**

165 Diez, T.; Beck, T.; Bucher, R.; Sommer, G.
　　　　　　　　　　　　　　　　DE 502055/73/0027
N–Dynamik des Bodens in Abhängigkeit von Witterung, Vegetation und Düngung. **N–dynamics in soils depending on climatic conditions, vegetation and fertilization.**

166 Schmid, G.; Schuch, M.; Haisch, A.; Jordan, F.; Weigelt, R. 　　　　　　　　　　　　DE 502055/75/0001
Abbau der Abwärme von Kernkraftwerken in landwirtschaftlich genutzten Böden – Versuch Gundremmingen. **Decomposition of waste heat from nuclear power stations in agricultural soils – Experiment at Gundremmingen.** Publications.

167 Pfrogner, J.; Graf, J.　　　　　DE 502055/77/0007
Bestandsaufnahme und Wertung der landwirtschaftlich nutzbaren Flächen in Bayern – EDV–technische Aufbereitung und Auswertung. **Data collection on the quality Bavarian arable land as basis for regional planning.**

168 Diez, T.; Hege, U.　　　　　DE 502055/78/0002 N
Massnahmen zur Minderung der Bodenerosion. **Means to reduce soil erosion.**

169 Tepe, W.; Leidenfrost, E.　　　DE 506110/70/0001
Die Bestimmung der Nährstoffleistung der Böden. **The determination of the nutrition effect in soils.**

170 Tepe, W.; Leidenfrost, E.　　　DE 506110/78/0001 N
Die Bestimmung der Aktivität der Ionen im Boden. **The determination of the activity of the ions in the soil.** Publications.

171 Puffe, D.; Zerr, W.　　　　　DE 506156/78/0002 N
Hydrologische Untersuchungen an verschiedenartigen Böden im Fuldatal und chemische Untersuchungen des Grundwassers bei intensiver Nutzung. **Hydrological studies on different soils in Fulda valley and chemical studies on groundwater in intensive utilization.**

172 Merkel, D.　　　　　　　　DE 507051/78/0001 N
Vergleichende Prüfung verschiedener Methoden zur Bestimmung von Schwermetallen in Böden durch Erhebungsuntersuchungen an Grünlandstandorten im Oker– und Allertal. **Comparative studies on different methods for the estimation of heavy metals in soils by means of inquiries on meadows in the valleys of Oker and Aller.** Publications.

173 Kuntze, H.; Hagemann, P.　　　DE 507103/72/0001
Bodenentwicklung und Bewertung von Sandmischkulturen. **Soil development and valuation of sand–peat–mixed cultures.**

174 Ernst, P.　　　　　　　　DE 508301/78/0004 N
Einfluss steigender Güllegaben auf Ertrag, Inhaltsstoffe, Pflanzenbestand, Stickstoffwirkung und Nährstoffbelastung des Bodens. **Influence of increasing use of semi–liquid manure on yield, constituents, plant stock, nitrogen effect and nutrient load of soil.**

175 Werner, W.　　　　　　　DE 903000/78/0005 N
Dünger–P–Umwandlung im Boden und Mobilität der Umwandlungsprodukte in verschiedenen Bodentypen. **Transformation of fertilizer phosphorus and mobility of the transformation products in different soil types.**

176 Hansen, L.; Jessen, T.　　　　DK 010103/59/3201
Forskellige drænings–, gødnings–, kalknings– og jordbehandlingsmetoder ved op– og videredyrkning af lavbundsjord. **Different methods of drainage, fertilizer use, liming and soil tillage in the reclamation and cultivation of lowland soils.**

177 Jensen, J.　　　　　　　DK 010117/71/4604
Udvikling og afprøvning af analysemetoder til jord. **Development and testing of methods for soil analysis.**

178 Jensen, J.　　　　　　　DK 010117/78/0006 R
Kvælstofmineralisering i jord. **Nitrogen mineralisation in soils.**

179 Møberg, J.P.; Msaye, B.M.; Kilasara, J.P.
　　　　　　　　　　　　　　　DK 030106/77/0004 N
Undersøgelser af jordbundsudviklingen i to sammenhængende catenaer i Uluguru–Mimdu området, Tanzania. **Properties of soils in two catenaes forming a continuum in the Uluguru–Mindu area in Tanzania.**

180 Møberg, J.P.　　　　　　DK 030106/79/0002 N
En fysisk–kemisk undersøgelse af en leret bakkeøjord fra Vestjylland. **A physico–chemical investigation of a clayey soil formed in morainic material from the Saale glaciation.**

181 Begon, J.C.　　　　　　　FR 010103/66/6353
Pédogénèse des sols de boulbène des vallées d'Aquitaine (thèse de doctorat d'État). **Pedogenesis of boulbene soils in aquitaine valley.**

182 Bétremieux, R.; Isambert, M.; Gobillot, T.
　　　　　　　　　　　　　　　FR 010103/68/6327
Chateaudun 1/100.000e. **Pedologic map of france : 1/100,000, 77. sheet of chateaudun.**

183 Simon, G.　　　　　　　FR 010103/69/6349
Arrière–action d'une fertilisation soufrée répétée. **After effect of repeated sulphur fertilisation.**

184 Begon, J.C.; Hardy, R.; Roque, J.　FR 010103/69/6354
Etude de potentialités des sols. **Study of soil potentialities.**

185 Baize, D.; Roque, J.; Hardy, R.; Wilbert, J.; Isambert, M.　　　　　　　　　　　　FR 010103/70/6355
Evolution et comportement des sols lessivés. **Evolution and behaviour of leached soils.**

186 Bétremieux, R.; Lefebvre, E.　　　FR 010103/72/6328
Etude expérimentale des processus pédologiques en conditions hydromorphes. **Experimental study on pedologic processes in hydromorphic conditions.**

187 Simon, G.　　　　　　　FR 010103/72/6350
Arrière–effet d'un enfouissement de résidus de féveroles. **After effect of ploughing of horse bean residue.**

188 Terce, M.; Calvet, R.　　　　FR 010103/72/6351
Etude des caractéristiques physico–chimiques du comportement des herbicides dans les sols. **Study of physico–chemical characteristics and behaviour of herbicides in soil.**

189 Renaud, J.　　　　　　　　FR 010112/70/3026
Effets de l'utilisation de certains appareils sur le sol et les végétaux. **Utilization of different machines and effects on soil and plants.** Publications.

190 Anstett, A.　　　　　　　　FR 010115/60/6223
Transformation d'un sol agricole en sol maraîcher. **Transformation of an agricultural soil into a market garden soil.**

191 Levy, G.; Becker, M.　　　　FR 010303/68/4151
Les sols à hydromorphie temporaire : comportement de diverses essences, influence du drainage sur le comportement de la forêt, l'évolution des sols et de la flore. **Low permeability soils ; behaviour of several forest species ; influence of drainage on the behaviour of forest and on the evolution of soils and ground vegetation.**

192 Morlat, R.; Salette, J.　　　FR 010403/74/6025
Etude pédologique des vignobles de la vallée de la Loire. **Pedological study of the vine areas in the loire valley.**

193 Remoue, M.; Lemaitre, Cl.　　FR 010405/70/0230
Comparaison de méthodes de culture du sol. **Comparison of tillage methods.**

194 Moulinier, H.　　　　　　　FR 010502/71/0513
Evolution des sols de serre. **Evolution of greenhouse–soils.**

195 Arvieu, J.C.; Bellenand–Mayeur, P.　FR 010502/71/0520
Remanence des pesticides dans les sols. **Pesticide remanence in soils.**

196 Arvieu, J.C.; Bellenand–Mayeur, P.; Mars, S.
　　　　　　　　　　　　　　　FR 010502/74/6011
Contribution au dosage de l'aldicarbe et de ses produits de dégradation dans le sol et la plante. **Contribution to aldicarb and its degradations produce determination in soil and plant.**

197 Faure, A.　　　　　　　　　FR 010601/69/6264
Analyse de la portance des sols. **Analysis of the bearing capacity of soils.**

198 Cornillon, P.　　　　　　　FR 010601/73/0536
Influence de la température du sol sur la croissance et le développement de la tomate. **Effect of soil temperature on tomato growth and development.** Publications.

199 Monnier, G.　　　　　　　　FR 010601/73/6267
Recherches de caractéristiques et mécaniques optimales pour les substrats artificiels de culture. **Search for optima physical and mechanical characteristics for artificial culture substrates.**

200 Guennelon, R.　　　　　　　FR 010601/74/6262
Propriétés adsorbantes de milieux artificiels (pollution, décontamination). **Adsorbing properties of artifical substrates (pollution and decontamination).**

201 Durand, J.H.　　　　　　　FR 010704/70/0548
Synthèse régionale qui se traduit par la carte des aptitudes des terres de la région. **Regional synthesis and mapping of soil potentialities.**

202 Loiseau, P. – Merle, P.; Niqueux, M; Arnaud
　　　　　　　　　　　　　　　FR 010802/73/6108

Lutte contre les érosions et reverdissement des sols récemment dénudés par l'homme en zone alpine. **Fight against erosions and returfing of soils lovely bared by man in alpine belt.**

203 Lienard, G.; Wolfer; Dejou, J.; Montard, F.X.; Larrère
　　　　　　　　　　　　　　　FR 010802/74/6281
Examen du système agraire dans la rétion N.E. des Monts–Dore (Puy–de–Dôme). **Assessment of the agricultural system in the N.E. area of Mont–Dore.**

204 Loiseau, P.; de Montard, F. X.; Robelin, M.; Mingeau, M.　　　　　　　　　　　　　FR 010802/76/6095
Flux gazeux et bilan de masse dans les systèmes sol–peuplement prairial. **Gaseous flux and mass balance in soil–sward systems.**

205 Clair, A.　　　　　　　　　FR 011001/73/6283
Cartographie géologique des formations superficielles. **Geological mapping of surface formations.**

206 Concaret, J.; Vermi, P.; Perrey, C.　FR 011001/73/6284
Carte pédologique au 1/100.000 (feuille Beaune). **Pedological map of france (1/100 000) sheet of beaune.**

207 Libois, A.　　　　　　　　　FR 011001/76/6137
Etude de la fertilisation en sols drainés. **Study of fertilization in drained soils.**

208 Fournier, J.C.　　　　　　　FR 011002/70/0593
Étude en modèle de la biodégradabilité d'un résidu d'herbicide dans le sol: la 2,6 – dichlorobenzamide. **Modilisation of the study on biodegradability of an herbicide residue in the soil: 2.6 dichlorobenzamid.**

209 Servat, E.　　　　　　　　　FR 011211/60/0619
Cartographie et aménagement. **Soil maps and land reclamation.** Publications.

210 Bonfils, P.　　　　　　　　　FR 011211/64/6306
Carte pédologique de Brive (1/100.000). **Pedologic map of France, sheet og Brive, 1:100.000.**

211 Bornand, M.; Legros, J.P.; Moinereau, J.
　　　　　　　　　　　　　　　FR 011211/67/6309
Cartographie pédologique au 1/100.000 éme. Feuille de Privas. **Pedologic map of france: 1/100,000. Sheet of privas.**

212 Favrot, J.C.　　　　　　　　FR 011211/69/6316
Etablissement de la carte pédologique de France au 1/100.000 et carte des secteurs naturels d'aménagement, feuille de moulins. **Pedologic map of france : 1/100,000 and map of natural areas of development.**

213 Callot, G.　　　　　　　　　FR 011211/70/0608
Cartographie – 1/100.000 Angoulême. **Mapping – 1/100.000 Angoulême.** Publications.

214 Bonfils, P.　　　　　　　　　FR 011211/70/0618
Caractérisation Andosols et bruns andiques sur basalte. **Andosoils and brown soils on basalt.** Publications.

215 Bonfils, P.　　　　　　　　　FR 011211/70/6307
Caractérisation des andosols et des sols andiques sur roches basaltiques. **Characterization of andosols on basalt rocks.**

216 Callot, G.; Legros, J.P.　　　FR 011211/71/0609

Cartographie du climat et des sols en zone viticole. **Mapping climate and soils in a vineyard country.** Publications.

217 Legros, J–P.; Maury, C.   FR 011211/71/0612
Méthodologie de la collecte et du stockage de l'information en cartographie pédologique. **Method of collecting and storing data in soil mapping.** Publications.

218 Bornand, M.   FR 011211/71/0614
Paléosols rubéfiés des terrasses rhodaniennes. **Reddish paleosol of terraces of the Rhone valley.** Publications.

219 Callot, G.   FR 011211/71/6311
Dynamique et genèse des sols formés sur roches carbonatées. **Dynamic and genesis of soils on carbonate rocks.**

220 Favrot, J.C.; Bouzigues, R.   FR 011211/72/0616
Etude pédologique en vue du drainage des terres agricoles du Pays d'Ouche (Eure). **Soil survey before drainage in "Pays d'Ouche".** Publications.

221 Legros, J.P.; Bornand, M.   FR 011211/72/6319
Cartographie pédologique de la France au 1/100.000. Feuille de St–Etienne. **Pedologic map of france : 1/100,000. Sheet of Saint–Etienne.**

222 Naert, B.   FR 011211/72/6320
Carte pédologique au 1/100.000 éme. Feuille de Millau. **Pedologic map of france : 1/100,000. Sheet of Millau.**

223 Naert, B.   FR 011211/72/6321
Télédétection et méthodologie cartographique appliquées à la pédologie et à l'aménagement du territoire en paysage calcaire. **Remote sensing and cartographic methodology in pedology and national development in calcareous landscape.**

224 Dupuis, M.; Favrot, J.P.   FR 011211/74/6313
Incidence des facteurs pédologiques sur l'économie et les modalités de la fertilisation. **Effect of pedologic factors on economy and ways and means of fertilization.**

225 Dupuis, M.; Bonfils, P.   FR 011211/74/6314
Essai de révision de "schéma directeur" de l'analyse des sols en pédologie. **Tentative revision of the leading plan for soil analysis in pedology.**

226 Bornand, M.   FR 011211/75/6310
Rôle de la contrainte "sol" dans les choix et orientations des aménagements régionaux. L'exemple du couloir rhodanien. **The soil as a parameter in choices and orientation of regional developments. For example the rhone valley.**

227 Bosc, M.; Blanchet, R.; Marty, J.R.   FR 011401/72/6194
Alimentation et fertilisation phospho–potassique selon les sols et les conditions culturales. **Phospho–potassic alimentation and fertilisation according to soils and to cultural conditions.**

228 de Crecy, J.   FR 011602/76/6037
Potentialités agricoles du département de la Guadeloupe. **Land capability (Guadeloupe. French West Indies).**

229 Recamier, A.; Fourbet, J.F.   FR 011801/60/3008
Etude de divers niveaux de réduction de travail du sol en rotation céréalière. **Study of different level of tillage reduction in cereal cropping.** Publications.

230 Fourbet, J.P.; Recamier, A.; Morel, R.; Riviere, J.   FR 011801/75/9067
Pouvoir épurateur d'un sol soumis à une pollution accidentelle d'hydrocarbures. **Purifying power of the soil in respect of an accidental hydrocarbon pollution.**

231 Cassin, J.; Ciccoli, H.; Nicoli, M.   FR 012202/67/0217
Définition de la meilleure méthode d'entretien du sol dans les vergers d'agrumes. **Determing the best fertilization level in citrus orchards.** Publications.

232 Durand, R.   FR 012210/71/6277
Carte pédologique de France (1/100.000) Reims. **Pedological map (1/100 000) reims.**

233 Soignet, G.   FR 012216/00/0510
Traitement de l'information au laboratoire. **Data processing in soil–analysis.**

234 Soignet, G.   FR 012216/68/0510
Traitement de l'information au laboratoire. **Data processing in soil–analysis.**

235 Cie Sielski, H.; Soignet, G.   FR 012216/72/0512
Détermination des éléments totaux dans les sols et les roches. **Determination of total elements in soils and rocks.**

236 Ciesielski, H.; Soignet, G.   FR 012216/74/6255
Analyse totale des sols et des roches. Mise au point de techniques analytiques. **Bulk analysis of soils and rocks. Perfecting of analytical procedures.**

237 Lefevre, G.; Desmet; Hiroux, G.; Arnoux; Lefebvre, R.   FR 012222/54/6044
Effets à long terme sur le sol et les productions de fumures minérales différenciées (phosphatées ou potassiques.). **Long terme effects on soil and vegetal productions from different mineral fertilization.**

238 Lefevre, P.   FR 012222/65/0504
Classification, Pédologie, Cartographie des sols de la plaine maritime de Picardie. **Classification, Pedology, mapping of the soils of the maritime plain of Picardy.** Publications.

239 Lefevre, G.; Graffin, Ph.   FR 012222/72/0501
Pollution. Utilisation d'eaux résiduaires de sucrerie et de conserverie. **Pollution. Spreading sugar refinery, cannery, and effluents on soil.** Publications.

240 Lefevre, P.   FR 012222/72/6038
Caractérisation agro–géologique de la Région Picardie et des départements de la Somme et de l'Oise. **Agro–geological characterisation of the Picardy Region and of the Somme, Oise departments.**

241 Hebert, J.; Maucorps, J.   FR 012226/60/0605
Cartographie et utilisation des sols de l'Aisne. **Soil mapping and land utilisation in the district Aisne.** Publications.

242 Marin–Lafleche, A.   FR 012226/70/0606
Recherche de critères pratiques d'évaluation de l'aptitude culturale. **Research of practical criteria for evaluating cropping possibilities.**

243 Chrétien, J.   FR 012227/67/6256
Carte pédologique 1/100.000 de Dijon. **Pedological mapping of**

the Dijon area (1/100 000).

244 Baize, D.   FR 012227/68/0665
Etude des sols développés dans la couverture des plateaux
Jurassiques de Bourgogne. **Studies on soil formation on
Jurassic platform in Burgundy.** Publications.

245 Chrétien, J.   FR 012227/69/0666
Influence de la forme et de la nature des sables sur les
propriétés physiques des sols. **Effect of shape and nature of
sand on soil physical properties.** Publications.

246 Marin–Lafleche, A.   FR 012503/71/6146
Détermination des aptitudes agronomiques d'un périmètre
possédant une cartographie des sols à grande échelle.
**Determination of the agricultural suitability of a soil large scale
surveyed area.**

247 Remy, J.C.; Badia, J.; Marin–Lafleche, A.
                    FR 012503/74/6148
Interprétation automatique des analyses de terre.
Interpretation of soil analysis by a programming system.

248 Marin–Lafleche, A.   FR 012503/74/6150
Valeur agronomique des déchets urbains. **Evaluation of
agronomic value of urban wastes.**

249 Researcher not indicated   GB 050126/00/0002 R
Farm waste: effects on soil and crops.

250 Voss   GB 060306/00/0002 R
**Effects of slurry application on soil condition and plant
composition.**

251 Gardiner, M.; Finch, T.F.   IE 060200/56/0117 N
Soil survey of Meath. Publications.

252 Kiely, P.V.   IE 060200/66/0385 R
**Mineralogical and chemical composition of Irish soils and
parent materials.** Publications.

253 Kiely, P.V.; Fleming, G.A.   IE 060200/66/0387 R
**Geochemical survey–mapping the distribution of trace and
major elements in the soils of the country.** Publications.

254 Diamond, J.J.; Ryan, M.J.   IE 060200/71/0361 N
Soil productivity experiments: classification of grassland sites.
Publications.

255 Diamond, J.J.   IE 060200/71/0362
Soil survey – County waterford. Publications.

256 Finch, T.F.   IE 060200/74/0830 N
Soil survey of Tipperary (north riding). Publications.

257 Tunney, H.   IE 060200/78/1388 N
Slurry soil injection experiment. Publications.

258 Hammond, R.F.   IE 060201/68/0418 R
**Peat soil classification and correlation the determination of
morphological and physical parameters, utilising a field and
laboratory programme, to provide the necessary data for a
classification system suitable for use in this country.**
Publications.

259 Hammond, R.F.   IE 060201/70/0420 R

**Soil survey– Co. Offaly. (1) the systematic classification and
mapping of the different soils within the county.** Publications.

260 Bulfin, M.   IE 060300/72/0724 R
Soil and land use survey of Co. Wicklow. Publications.

261 Conry, M.J.   IE 060500/69/0262 R
Soil survey of Co. Laois. Publications.

262 Conry, M.J.; Hammond, R.F.; Bulfin, M.
                    IE 060500/74/0998 R
**Soil type distribution and land use evaluation of Slieve Blooms –
problems of naturally handicapped areas in relation to
reclamation (for grassland) forestry and amenity purposes.**
Publications.

263 Walsh, M.   IE 060700/72/0462 R
Soils of Co. Galway. Publications.

264 Walsh, M.   IE 060700/74/0914 R
Iron–pan podzol soils in the west of Ireland. Publications.

265 Walsh, M.   IE 060706/77/1372
**Soil survey of the 'benefitting land' in the catchment areas of the
rivers boyle mayo–roscommon) and bonet north leitrim).**

266 Healy, B.; Bracken, J.J.   IE 120403/78/9165 N
Baseline studies of Aughinish island on the river Shannon.
Publications.

267 Piovanelli, C.; Gregori, E.; Arcara, P.G.; Miclaus, N.
                    IT 020100/75/0002 R
Modificazioni indotte sull'attività biofisica e sulla sostanza
organica di un suolo tendenzialmente argilloso, da diverse
tecniche di coltivazione e di diserbo del mais. **Modifications on
biophisical activity and organic matter on clayey soil due to
different tillage techniques and weed control in maize.**

268 Zanchi, C.; Chisci, G.   IT 020100/75/0006
Comportamento idrologico ed erosione, in funzione del
pascolamento bovino, di suoli prevalentemente argillosi.
**Hydrology and erosion in relation to cattle grazing on
loam–clay soil.**

269 Ronchetti, G.; Arcara, P.G.; Rodolfi, G.
                    IT 020100/75/0008
Studi pedologici di dettaglio nel bacino idrografico del
T.Prescudin (Pordenone). **Detailed pedological studies of the
watershed of the Prescudin river (Pordenone).** Publications.

270 Lulli, L.; Lorenzoni, P.; Rodolfi, G.; Arretini, A.;
Sfalanga, M.; Frascati, F.   IT 020100/75/0009 R
Analisi geomorfologiche e pedologiche di due bacini
idrografici della Toscana attrezzati per lo studio dell'erosione.
**Geomorphological and pedological analysis of two Tuscany's
fitted up for erosional studies watersheds.**

271 Lulli, L.   IT 020100/76/0001 N
Studio dei suoli su materiali vulcanici in ambiente
mediterraneo. **Soil study on volcanic materials in
mediterranean areas.**

272 Chisci, G.C.   IT 020100/77/0625 R
Studio dell'erodibilità di alcuni suoli italiani in laboratorio a
mezzo simulatori di pioggia. **A laboratory study of the erosion
of certain Italian soils under simulated rainfall.**

D 1100 – Soil science

273 Rodolfi, G.; Canuti, P.; Arretini, A.; Bezoari, G.;
Focardi, P.; Monti, S.   IT 020100/78/0004 R
Studio di due versanti, attrezzati per il controllo dell'erosione
di massa, in situazioni diverse e caratterizzate da un diverso
uso del suolo. **Study on two fitted up for mass erosion control
slopes, distinct environmental situations, characterized by a
different land use.** Publications.

274 Panicucci, M.; Bazzoffi, P.; Ronchetti, G.; Miclaus, N.;
Bidini, D.   IT 020100/78/0005 R
Indagini sulla quantità e qualità dell'interrimento in serbatoi
artificiali, in relazione alle caratteristiche bacinali. **Study on
quality and quantity of silting up in artificial reservoirs in
relation to the basins characteristics.**

275 Magaldi, D.; Bidini, D.; Lulli, L.; Arretini, A.; Rodolfi,
G.; Ferrari, G.A.   IT 020100/78/0006 R
Cartografia e valutazione dei suoli su sedimenti quaternari
della pianura lucchese. **Cartography and evaluation of soils
from quaternary sediments in Lucca (Italy) plain.**

276 Magaldi, D.; Frascati, F.; Bazzoffi, P.; Gregori, E.;
Bidini, D.; Lorenzoni, P.   IT 020100/79/0002 N
Messa a punto di metodologie multidisciplinari per la
valutazione del territorio italiano. **Appronty to study of
multidisciplinary methods for land evaluation in Italy.**

277 Rodolfi, G.; Lulli, L.; Tellini, M.   IT 020100/79/0005 N
Studio geomorfologico, pedologico, e delle caratteristiche
idrografiche di un ambiente montano interappenninico. **Study
on the geomorphological, pedological and hydrographic
features of a inter–Appennines montainous environment.**

278 Raglione, M.; Ronchetti, G.; Rodolfi, G.; Sfalanga, M.;
Lulli, L.; Gregori, E.   IT 020100/79/0006 N
Rilevamento ed analisi delle caratteristiche idrologiche,
geomorfologiche e pedologiche del bacino idrografico del
fiume Tacina (Calabria). **Survey and analises of some
hydrolocic geomorpholical and pedological characteristics of the
Tacina River Basin.** Publications.

279 Zanchi, C.; Falciai, M.; Giacomin, A.
     IT 020100/79/0009 N
Studio delle modificazioni indotte dal drenaggio tubato sulle
caratteristiche fisiche e produttive del suolo in zone di pianura.
**Influence of tile drain on physical soil characteristics on flat
lands.**

280 Piovanelli, C.; Tellini, M.; Bidini, D.
     IT 020100/79/0010 N
Effetti dei trattamenti geosterilizzanti con bromuro di metile,
vapam e vapore su alcune cultivar di garofano; persistenza dei
residui nel terreno e modificazioni indotte sull'attività
biologica. **Soil disinfestations with methyl bromide, vapam and
steam on some carnations ipersistence of residues into soil and
modifications of soil biological activity.** Publications.

281 Chisci, G.; Torri, D.; D'Egidio, G.; Zanchi, C.; Sfalanga,
M.   IT 020100/79/0011 N
Studio delle correlazioni tra lunghezza e pendenza di un
versante ed entità delle perdite di suolo per erosione
idrometeorica su substrati argillosi. **Relationship among length,
slop and soil erosion losses in clayey soils.**

282 Bidini, D.; Miclaus, N.; Lulli, L.; Gregori, E.
     IT 020100/79/0012 N
Studio dei processi pedogenetici di alcuni suoli forestali
provenienti da substrati vulcanici in relazione alla loro fertilità.
**Study of pedogenetic processes in some forest volcanic soils in
relation to their fertility.** Publications.

283 Mecella, G.; Costantini, A.; Del Vecchio, C.; Di Blasi, N.

     IT 020200/69/0001
Classificazione dei terreni in rapporto alla redditività.
Land–classification in relation to yield. Publications.

284 Tombesi, L.; Mecella, G.; Di Blasi, N.; Pierandrei, F.;
Biondi, A.; Favola, G.   IT 020200/78/0003 N
Studio dei terreni dell'Alta Valle del Tevere. **A study of some
soils of Tiber High Valley.** Publications.

285 Marizza, L.   IT 020200/79/0003 N
Indagini sulla dinamica dei concimi azotati in terreni
ferrettizati nelle normali condizioni di campagna in presenza di
irrigazione. **Field investigations on N–NO$_3$ dynamics in
"ferretti" soils under irrigation.**

286 Leandri, A.; Imbroglini, G.   IT 020300/69/0002
Fatteri pedoclimatici in una clorosi nervale non infettiva della
vite nel Lazio. **Pedoclimatic factors on a non–infectious vein
chlorosis of grapevines occurring in Latium.** Publications.

287 Lopez, G.; Colucci, R.   IT 020500/73/0002
Carta della potenzialità dei terreni pugliesi. **Soil potentiality
map of Apulia.**

288 Venezian Scarascia, M.E.; Losavio, N.   IT 020500/75/0001
Sistemazione, lavorazione e messa a coltura di terreni della
bassa Murgia barese. **Land preparation, tillage and cultivation
on the lower Murgia Barese.**

289 Lopez, G.; Colucci, R.   IT 020500/75/0006
Studio agropedologico di dettaglio di una azienda della bassa
Murgia sud–orientale. **Detailed agro–pedological study of a
farm in the lower South–Eastern Murgia.**

290 Lanza, F.; Lopez, G.; Rizzo, V.; Spallacci, P.; Onofrii,
M.   IT 020500/75/0008
Significato teorico e pratico della capacità idrica di campo e sue
variazioni nel tempo. **Theoretical and practical field water
capacity and its daily variation.** Publications.

291 Boschi, V.; Chisci, G.; Spallacci, P.   IT 020500/75/0012
Caratteristiche idrologiche ed entità dell'erosione su argille
scagliose sistemate a fosse livellari parallele tra loro ed
investite con differenti colture: collina emiliana. **Hydrology
and amount of erosion on scaly clays with parallel contour
ditches frown to different crops: Emilia hills.** Publications.

292 Spallacci, P.; Montorsi, M.; Boschi, V.   IT 020500/75/0013
Limiti dello spandimento dei liquami di suini su terreni diversi.
(Studio condotto in cassoni lisimetrici). **Limits of pig wastes
spreading on different soil types (studies made on lysimeters).**
Publications.

293 Boschi, V.   IT 020500/77/0621 R
Erosione su argille scagliose sistemate a fosse livellari. **Erosive
impact on the scaly clay lining of catchment basins.**

294 Lavezzi, A.; Egger, E.; Giulivo, C.   IT 021300/79/0004 N
Tecniche di lavorazione e diserbo del vigneto. **Vineyard tilling**

D 1100 – Soil science

and weeding techniques.

295 Ciancio, O.   IT 021700/77/0626 R
Effetti del tipo di bosco sull'entità dell'erosione. **Type of
wooded area : its influence on the grade of erosion.**

296 Marzi, V.   IT 040101/77/0639 R
Tecniche colturali ed erosione del suolo. **Cultural techniques
and soil erosion.**

297 Catizone, P.   IT 040201/77/6794
Inquinamento del terreno da diserbanti in funzione della
diversa caratteristica del terreno. **Soil pollution by weedkillers
depending on the type of soil.**

298 Rossi, N.; Lanzoni, L.; Lugo, P.; Diegoli, A.
                                        IT 040202/75/0001
Fertilità chimica e produzione del terreno sottoposto a
differenti trattamenti di concimazione e di irrigazione.
**Relationship betwen soil fertility and yield in fertili sation and
irrigation essays.**

299 Indelicato, S.   IT 040311/77/0634 R
Valutazione della prevedibile erosione. **Assessment of
foreseeable erosion.**

300 Cantiani, M.   IT 040504/78/1041 N
Inventario per la cubatura dei soprassuoli boschivi con metodo
fotogrammetrico. **Inventory by photogrammetric methods of
the cubic capacity of wooded top–soil.**

301 Goldberg, F.L.   IT 040603/74/0565
Effetti dell'inquinamento del terreno sulla produzione
vegetale. **Effects of soil pollution on plant production.**

302 Basso, F.   IT 040701/77/0619 R
Valutazione dei fenomeni erosivi e del comportamento delle
colture. **Evaluation of erosive agents and the behaviour of
cultures.**

303 Carravetta, R.   IT 040710/77/0624 R
Dinamica dell'erosione idrometeorica. **The dynamics of
hydrometric erosion.**

304 Barone, L.   IT 040725/79/0001 N
Indagine sul riscaldamento del suolo, con utilizzazione diretta
dell'energia solare, per la disinfestazione dei terreni.
**Investigation on soil heating, with direct utilization of solar
energy, for soil disinfestation.** Publications.

305 Benini, G.   IT 040807/77/0620 R
Ricerca sui parametri caratterizzanti l'attività erosiva dei bacini
montani. **Research on parameters defining the erosive action of
mountain basins.**

306 Ballatore, G.P.   IT 040904/73/0137
Ricerca sulla difesa del suolo. **Research on soil protection.**

307 Lo Cascio, B.   IT 040904/74/0576
Ricerca sulla difesa del suolo. **Research on soil protection.**

308 Lo Cascio, B.   IT 040904/77/0636 R
Effetti dei sistemi di lavorazione del suolo sull'erosione. **The
effects of cultural techniques on soil erosion.**

309 Melisenda, I.   IT 040915/73/2146

Ricerca sull'erosione idrica dei terreni a coltura altamente
meccanizzata. **Research on water erosion of soils under highly
mechanized cultivation.**

310 Melisenda, I.   IT 040915/77/0640 R
Erosione idrica di terreni sottoposti a sistemi di lavorazione
diversi. **Water erosion of soils under different cultural
techniques.**

311 Tafuri, F.; Businelli, M.; Giusquiani, P.L.
                                        IT 041004/78/0001
Necessità di concimazioni fosfatiche e loro dosaggio in base
alle isoterme di adsorbimento. **Phosphate requirements of soils
evaluated by absorption isotherms.**

312 Giovagnotti, C.; Calandra, R.   IT 041014/75/0001
Studio dei suoli relativi alla comunita montana del m. Subasio.
**Pedological researches into the mountain community of M.
Subasio (Italy).** Publications.

313 Levi Minzi, R.   IT 041103/77/0214 R
Ulteriori indagini sull'inquinamento del terreno e dei prodotti
agrari da parte dei metalli pesanti. **Further research on ground
and crop pollution due to heavy metals.**

314 Grossi, P.   IT 041107/74/0567
Valutazione dell'influenza di alcuni interventi sistematori sulla
fertilità e sul potere regimante del terreno. **Evaluation of the
influence of certain management works on soil fertility and its
capacity to regulate water downflow.**

315 Sapetti, C.   IT 041202/72/0556
Comportamento dei microelementi durante la trasformazione
dei minerali nel corso della pedogenesi. **Behaviour of
microelements during the transformation of minerals during
soil genesis.** Publications.

316 Aru, A.   IT 042401/77/0618
Studio pedologico su alcune aree modello dell'Italia
meridionale ed insulare. **Pedology: a study on model areas in
southern Italy and the Italian isles.**

317 Sequi, P.   IT 060400/72/0007
Attività dei residui nel terreno (erbicidi, insetticidi, fungicidi,
ecc.). **Residue activity in soil (herbicides, testicides, fungicides,
etc.).** Publications.

318 Sequi, P.   IT 060400/73/0006
Problemi metodologici nell'analisi fisica e chimica del terreno.
Methods of soil analysis (physical and chemical properties).
Publications.

319 Cervelli, S.   IT 060400/77/0774
Interazione tra pesticidi e fertilizzanti. **Pesticide and fertilizer
interaction.**

320 Sequi, P.   IT 060400/77/0867
Attività dei prodotti residui (insetticidi, fungicidi, erbicidi
ecc.). **Activity of residues (insecticides, fungicides, herbicides).**

321 Malquori, A.   IT 061200/71/0146
Interazione terreno–erbicidi. **Soil–herbicides interactions.**
Publications.

322 Malquori, A.   IT 061200/77/0775
Caratterizzazione suoli e interazione terreni–erbicidi.

D 1100 — Soil science

Characterisation of soils; soil and herbicide interaction.

323 Sanesi, G.　　　　　IT 062300/77/0644
Metodi di rilevamento della potenzialità e delle limitazioni d'uso nel suolo. **Survey methods for assessing the potentialities and limitations of soil exploitation.**

324 Evangelisti, L.　　　　　IT 062300/77/0778
Rilevamento pedologico di aree sperimentali. **Pedological survey of experimental areas.**

325 Mancini, F.　　　　　IT 062300/77/0970
Genesi del suolo. **Soil genesis.**

326 Mancini, F.　　　　　IT 062300/77/0971
Classificazione del suolo. **Soil classification.**

327 Mancini, F.　　　　　IT 062300/77/0972
Cartografia del suolo. **Soil mapping.**

328 Baldoni, R.　　　　　IT 062400/77/0950
Aspetti della fertilizzazione del terreno nell'azienda produttrice di carne bovina con l'utilizzazione di foraggi di mais e sorgo. **Aspects of soil fertilisation in beef cattle farms using maize and sorgnum fodder.**

329 Govi, M.　　　　　IT 063600/77/0627
Classificazione e cartografia dei suoli: quantificazione dell'erosione dei versanti. **Classification and mapping of soils : quantification of slope erosion.**

330 Salandin, R. and Giordano, A.; Bertolino, D.; Assandri, G.; Amprimo, G.; Chiavacci, R.; Penon, A.
　　　　　IT 120100/79/0001 N
Carta della capacità d'uso dei suoli e delle loro limitazioni di una superficie omogenea comprendente il comprensorio di Ivrea e parte dei comprensori di Torino e Vercelli. Scala 1:100.000. **Land use capability and limitation map of a homogeneous area including the Ivrea district and part of the districs of Turin and Vercelli, (Piedmont, Italy). Scale 1:100.000.**

331 Giordano, A.　　　　　IT 121900/77/0629
Erosione dei versanti in relazione alle caratteristiche ambientali mediante simulatori di pioggia. **Hill–side erosion related to the type of environment using simulated rainfalls.**

332 Grootenhuis, J.A.　　　　　NL 010103/44/0497
Betekenis van vruchtopvolging voor akkerbouw op klei– en zavelgronden. **The importance of crop rotation to agriculture on clay and silty clay soils.** Publications.

333 Ris, J.　　　　　NL 010103/47/0507 R
Fluctuaties van de vruchtbaarheid van de grond ontstaan onder invloed van het weer, als oorzaak van verschillen in jaarlijkse opbrengsten. **Soil fertility fluctuations originated from weather influences, as a cause of differences in yields.** Publications.

334 Ferrari, Th.J.　　　　　NL 010103/51/0500
Analyse van de bodemvruchtbaarheid van de Gelderse Vallei. **Soil fertility analysis of the "Gelderse Vallei".**

335 Venekamp, J.T.N.　　　　　NL 010103/57/0502 R
Toepassing van wiskundig–statistische kennis in het bodemvruchtbaarheidsonderzoek. **Application of mathematical statistics in soil fertility research.** Publications.

336 Ris, J.　　　　　NL 010103/61/0528
Beschrijving en verklaring van veranderingen in de vruchtbaarheid van de grond onder invloed van het weer. **Description and explanation of changes of soil fertility as affected by meteorological fluctuations.**

337 Sissingh, H.A.　　　　　NL 010103/63/0556
Incidenteel chemisch onderzoek als begeleiding van toegepast onderzoek. **Analytical chemical investigations on behalf of the applied research projects.**

338 Ferrari, Th.J.　　　　　NL 010103/69/0510 R
Factoren die in het gematigde klimaat van West–Europa de opbrengsten van de akkerbouwgewassen bepalen. **Factors determining field–crop yields in the temperate climate of West–Europe.** Publications.

339 Kolenbrander, G.J.; Ferrari, Th.J.　　　NL 010103/69/2588
Simulatie van de stikstofhuishouding van de grond. **Simulation of the soil nitrogen transformations.** Publications.

340 Knottnerus, D.J.C.　　　　　NL 010103/71/3154
Bestrijding van het stuiven van grond en andere materialen om economische en hygiënische redenen. **Control measures against the wind erosion of soil and other materials for reasons of economy and health (including testing of anti–erosion materials).** Publications.

341 Knottnerus, D.J.C.　　　　　NL 010103/73/5586
Indringing van de vorst in de grond onder op verschillende manieren gezaaide winterrogge (i.v.m. aardappelopslag). **Investigation on the depth of frost penetration into the soil under winter rye sown in different ways, and its effect on the incidence of volunteer potatoes.** Publications.

342 Sluijsmans, C.M.J.　　　　　NL 010103/75/6252 R
Onderzoek naar landbouwkundig en ecologisch aanvaardbare maximale hoeveelheden dierlijke mest op cultuurgrond. **Investigation to establish acceptable maximum applications of animal manure on farm lands from an agricultural and ecological point of view.**

343 Willigen, P. de　　　　　NL 010103/75/6586 R
Toepassing van de simulatietechniek bij de bestudering van fysische, chemische en biologische processen in de grond. **The use of simulation techniques in the study of physical, chemical and biological processes in soils.**

344 Sluijsmans, C.J.M.　　　　　NL 010103/79/8713 N
Vergelijkend bodemvruchtbaarheidsonderzoek bij gangbaar en alternatief geleide bedrijfsvoering. **Comparative soil fertility research on conventional versus alternative ("biological") systems of husbandry.**

345 Lynden, K.R. van　　　　　NL 010109/68/1439
Onderzoek naar het verband tussen de bodemgesteldheid en de groei van de fijnspar als onderdeel van groeiplaatsonderzoek. **Study on the relation between soil condition and growth of spruce as a part of habitat research.**

346 Frissel, M.J.　　　　　NL 010110/67/7284
Migration of 90Sr, 137 Cs and Pu in soils. Verification of a computer model on the behaviour of these radiocontaminants in soils of Western Europe.

D 1100 – Soil science

347 Sinnaeve, J.   NL 010110/76/7286
Vastlegging en vermindering van de beschikbaarheid van radioactieve metalen in gronden. **Immobilization and reduction of the availability of radio contaminants in soils.**

348 Cate, J.A.M. ten   NL 010119/50/1421
Datering van afzettingen en bodemprofielen en karakterisering van het milieu, met behulp van pollenanalyse. **Dating deposits and profiles and establishing the environment by pollen analysis.**

349 Cate, J.A.M. ten   NL 010119/66/1420
Het vervaardigen van een geomorfologische kaart van Nederland. **The preparation of a geomorphological map of the Netherlands.**

350 Smet, L.A.H. de   NL 010119/67/1411 R
Heroriëntatie van bodembestanddelen o.a. slemp, in samenhang met de landbouwkundige problemen in het proefgebied "Noordelijke Zeeklei". **The rearrangement of the soil constituents, such as "slemp" in relation to agricultural problems.** Publications.

351 Jongerius, A.   NL 010119/67/7795 R
Het gebruik van elektronische apparatuur (Quantimet) voor het analyseren van bodemkenmerken aan slijpplaten en van diverse andere in het landbouwkundig onderzoek gebruikte parameters. **The use of electronic epuipment (Quantimet) for analysis of soil characteristics and several other in agricultural research used parameters.**

352 Edelman–Vlam, A.W.   NL 010119/68/1412
Historisch–geografisch onderzoek van oude bouwlanden. **The historical geography of old earth soils.**

353 Sluis, P. van der   NL 010119/68/2608 R
Bewortelingsonderzoek akkerbouwgewassen in relatie tot de kaarteenheden. **Study on the rootability of agricultural crops in relation with mapping units.** Publications.

354 Sluijs, P. van der   NL 010119/69/5999
Bodemkundig onderzoek t.b.v. de ruilverkaveling van de Lopikerwaard. **Soil studies for the re–allotment of the Lopikerwaard.** Publications.

355 Dam, J.G.C. van   NL 010119/70/3235
Bodemgeschiktheidsonderzoek voor de teelt van waspeen. **Study on the relation between soil conditions and suitability for cultivation of carrots.**

356 Breeuwsma, A.   NL 010119/70/3237
De mineralogische en chemische karakterisering van de Nederlandse afzettingen. **The mineralogical and chemical characterization of the sediments in the Netherlands.**

357 Bouma, J.   NL 010119/70/3241 R
Onderzoek naar de fysisch–hydrologische eigenschappen van de grond in kassen en in de polder Heerhugowaard. **Study on the physical–hydrological properties of soils in greenhouses in the polder Heerhugowaard.** Publications.

358 Jongerius, A.   NL 010119/70/4789
Begeleiding van micro–morfologische studies, uitgevoerd door buitenlandse collega's. **Supervision of micromorphological studies conducted by visiting foreign colleagues.**

359 Jongerius, A.   NL 010119/71/4790 R
Ontwikkeling sub–microscopische technieken voor bodemonderzoek. **Development of sub–microscopic techniques for soil investigations.**

360 Poelman, J.N.B.   NL 010119/72/4798
Onderzoek naar de landschappelijke ontstaanswijze van het gebied tussen de benedenloop van de Roer en de Maas. **Genesis of the landscape of the area between the lower course of the rivers Roer and Maas.** Publications.

361 Dam, J.G.C. van   NL 010119/73/4800
Het verband tussen bodemgesteldheid en de aanlegs – en onderhoudskosten en de bespeelbaarheid van sportvelden. **Relations between soil conditions and the construction costs, maintenance costs and quality of the surface of sprotsfields.**

362 Ismaël, S.   NL 010119/73/5965
Porositeitskarakterisering van Nederlandse gronden met behulp van de Quantimet. **Porosity investigation of soils in the Netherlands by means of Quantimet.** Publications.

363 Pape, J.C.; Bannink, J.F.   NL 010119/73/5967
Vegetatiekartering ten behoeve van de bodemgeschiktheid voor de houtteelt. **Vegetation mapping on behalf of the soil suitability for forestry.**

364 Van Wallenburg, C.; Vos, G.A.   NL 010119/73/5968
Karakterisering van de rijpingstoestand van diverse gronden. **Characterization of the ripening status of various soils.**

365 Dam, J.G.C. van; Hulshof, J.A.   NL 010119/73/5971
Bodemgeschiktheidsonderzoek voor de teelt van witloof. **Study on soil conditions for chicory.**

366 Poelman, J.N.B.; Teunissen van Manen, T.C.
    NL 010119/73/5976
Samenstellen gezamenlijk kaartblad, Venlo/Krefeld. **Composition of a joint soilmap Venlo/Krefeld.**

367 Breeuwsma, A.   NL 010119/73/5977
Onderzoek naar het optreden van bodemverontreiniging bij vuilstortplaatsen. **Research on soil pollution by refuse tips.**

368 Pape, J.C.   NL 010119/73/5985
Ontwikkeling van methoden ter bepaling en afweging van ruimtelijke aanspraken in een landschappelijk en natuurwetenschappelijk waardevol gebied (Midden–Brabant) – inventarisatie bodem en landschapselementen, bodemgeschiktheidsonderzoek. **Development of methods for determination and evaluation of multiple land–use in areas with high landscape and nature values – survey of soil and landscape elements, soil suitability studies.** Publications.

369 Sluijs, P. van der; Hoekstra, C.   NL 010119/73/5994
De invloed van de dichtheid van de grond op eigenschappen in graslanden met een landbouwkundig gebruik. **The influence of soilcompaction on properties of grasslands used for agricultural purposes.**

370 Smet, L.A.H. de; Domhof, J.   NL 010119/73/5997 R
Bodemgeschiktheidsonderzoek t.b.v. de moderne weidebouw en groenvoedergewassen. **Research into soil condition and suitability for modern dairy farming and forage crops.**

371 Clingeborg, A.E.   NL 010119/74/5979 R

D 1100 – Soil science

De bodemkundige opbouw van het "woudgebied" in relatie tot zijn omgeving. **The pedological construction of the forest area in relation to the environment.**

372 Lynden, K.R. van; Waenink, A.W.; Vis, T.
NL 010119/74/5981
De opstelling van bodemgeschiktheidsclassificatie voor de bosbouw. **Development of a soil suitability classification for forestry.**

373 Lynden, K.R. van; Vis, T.; Waenink, A.W.
NL 010119/74/5982
Standaard bodemgeschiktheid voor bos. **Development of a soil suitability standard for forestry.**

374 Bakker, H. de
NL 010119/74/5995 R
Herdruk systeem voor bodemclassificatie voor Nederland, de hogere niveaus. **Reprint system of soil classification for the Netherlands.**

375 Dam, J.G.C. van; Wopereis, F.A.
NL 010119/74/7546
Onderzoek naar het verband tussen bodemgesteldheid en de geschiktheid voor bomen in het stadsmilieu. **Research into the connection between soil constitution and suitability for trees in the urban environment.** Publications.

376 Pape, J.C.; Stuurman, F.J.; Bannink, J.F.
NL 010119/75/6907
Gebruiksmogelijkheden van de bodemkaart bij beoordeling van de milieugradient. **The possibilities for using the soil map for judging environmental gradients.** Publications.

377 Marsman, B.A.; Bie, S.W.; Gruijter, J.J. de
NL 010119/75/6909
Vergelijking van methoden van bodemkartering op grond van hun resultaten. **Studies of soil survey methods from their results.**

378 Wallenburg, C. van; Kleinsman, W.B.
NL 010119/75/6912
Onderzoek naar enkele geologische en cultuurhistorische aspecten in het ruilverkavelingsgebied Schagerkogge. **Research into some geological and cultural–historical aspects in the Schagerkogge reallotment area.** Publications.

379 Beekman, A.G.; Holst, A.F. van   NL 010119/76/7548
Relatie standaardreeks en schattingsklassen in het ruilverkavelingsgebied Laren (t.b.v. de ontwikkeling van meer betrouwbare methoden voor de vaststelling van de uitruilwaarde van gronden in ruilverkavelingsgebieden). **Compilation of standard series of soil profiles in the Laren Reallotment area (for development of more reliable methods for determination of the value of soils in reallotments areas).**

380 Smet, L.A.H. de   NL 010119/76/7552 R
Onderzoek naar de gevolgen van bouwplanvernauwing op de bodemgesteldheid ter bepaling van de toelaatbare teelt–frequentie op verschillende gronden. **Research into the relations between soil crop rotation.** Publications.

381 Poelman, J.N.B.   NL 010119/76/7555 R
Stevigheid van met gras begroeide bovengrond bij zandgronden. **Research into the firmness of grass–stabilised sandy soils.**

382 Lynden, K.R. van   NL 010119/76/7787

Onderzoek naar de relatie tussen groeiplaats en groei van de "Robusta" populier. **Research on growth–site relationships of "Robusta" poplar.**

383 Lynden, K.R. van   NL 010119/77/7788 R
Toetsing van het bodemgeschiktheidsbeoordelingssysteem voor de bosbouw. **Testing-system of judging soil suitability for forestry.** Publications.

384 Veer, A.A. de   NL 010119/77/7792
Relatie bodem–boombegroeiing in de Gelderse Vallei en Achterhoek. **Relations soil – tree vegetation in the Gelderland valley and Achterhoek.**

385 Cate, J.A.M. ten   NL 010119/77/7794 R
Onderzoek naar de morfologie en genese van "niet–dalvormige" laagten t.b.v. de geomorfologische kaart van Nederland, schaal 1 : 50 000. **Research into the morphology and genesis of non–valley type depressions for the geomorphological map of the Netherlands, scale 1 :50.000.**

386 Smet, L.A.H. de   NL 010119/77/7797
Bodemkundige opname van proefvelden t.b.v. periodieke profielbemonstering bij teeltkundig en teelttechnisch onderzoek. **Soil survey of experimental fields to be used for periodical soil sampling for plant research.**

387 Wallenburg, C. van   NL 010119/77/7799 R
Veldcriteria voor toepassing van de zee– en rivierkleigronden legenda op de kaartbladen 44 Oost en 44 West. **Fieldcriteria for the use of the marine and riverclay legend on the sheets 44 East and 44 West.**

388 Wallenburg, C. van   NL 010119/77/7800 R
Rijpingstoestand en voortgang van de rijping bij grienden en rietgorzen in de Biesbosch en Het Eiland van Dordrecht. **Ripening condition and progress of ripening in holms and reedmarsh in the Biesbosch and the Eiland van Dordrecht.**

389 Bie, S.W.   NL 010119/77/7804 R
Automatische kartografische verwerking bodemkarteringsgegevens ruilverkaveling Eemland. **Automated cartographic handling of soil map data Eemland reallotment plan.**

390 Haans, J.C.F.M.   NL 010119/77/7807
Inbreng en bewerking gegevens Structuurschema Landinrichting in INFA–systeem. **Input and compilation of the data of the Structure Scheme Landinrichting in the INFA system.**

391 Bijlsma, S.   NL 010119/78/8308
Morfologie en genese fluvioglacide terreinvorming. **Morphology and genesis of fluvioglacial forms.** Publications.

392 Cate, J.A.M. ten   NL 010119/78/8309
Morfologie en genese van gordeldekzand–gebieden. **Morphology and genesis of circular cover sand areas.**

393 Veer, A.A. de   NL 010119/78/8310 N
Onderzoek naar de relatie tussen bodemgesteldheid en landschapsfysiognomie. **Research on the relationship between soil conditions and landscape physiognomy.**

394 Pape, J.C.   NL 010119/79/9077 N
Bodemkundig onderzoek van de natuurterreinen Empese en

Tondense hei, Koolmandsijk en Nijkamps heide. **Soil survey of the nature reserves Empese en Tondense hei, Koolmansdijk and Nijkamps heide.**

395 Nicolaï, P.   NL 010207/77/7855 R
Onderzoek naar de invloed van het bodemprofiel op de opbrengst en kwaliteit van akkerbouw– en vollegrondsgroentegewassen. **Investigations on the influence of the soil profile on yield and quality of arable and vegetable crops.**

396 Delver, P.; Pouwer, A.   NL 010212/74/6014
Stip in appel (invloed van bodem en bemesting). **Bitter pit in apples (effects of soil and fertilization).** Publications.

397 Voerman, S.   NL 010301/71/3490
Kwantitatieve bepaling van bestrijdingsmiddelen in de bodem. **Quantitative determination of pesticides in soil.** Publications.

398 Houx, N.W.H.   NL 010301/78/7888
Omzettingsprodukten van bestrijdingsmiddelen in bodem en water. **Conversion products of pesticides in soil and water.**

399 Burg, J. van den; Schoenfeld, P.H.   NL 010601/74/6011
Invloed van bos op de produktiviteit van voormalige heidegronden. **Influence of forest stands on productivity of former heathlands.**

400 Drift, J. van der   NL 010602/75/7329 N
Bodem–eutrofiëring in bossen en natuurterreinen. **Soil eutrophication in forests and nature areas.**

401 Janssen, B.H.   NL 020007/71/4966
Bodemvruchtbaarheidsbepalingen voor de (humide) tropen. **Assessment of soil fertility in the (humid) tropics.** Publications.

402 Breemen, N. van   NL 020008/63/4255
Genese van kattekleigronden in de kustvlakte van Thailand. **Genesis of acid sulphate soils in the coastal plain of Thailand.** Publications.

403 Plas, L. van der; Rogaar, H.   NL 020008/64/4911
Verwering en bodemvorming onder alpiene omstandigheden. **Weathering and soil formation in an Alpine environment.** Publications.

404 Havinga, A.J.   NL 020008/66/4917
Organische stof in de bodem (pollencorrosie; karakterisering en verweringspotentie van humus (fracties)). **Organic matter in soils (Corrosion of pollen; Characterisation and wheathering power of humus (fractions)).** Publications.

405 Miedema, R.   NL 020008/66/4921
Regionaal bodemkundige studies aan loëss en dekzanden. **Regional soils studies in loess and lower sand landscapes.** Publications.

406 Plas, L. van der   NL 020008/67/4915
Thermodynamisch en mineralogisch onderzoek t.b.v. de bodemkunde. **Thermodynamic and mineralogical investigations pertaining to soil processes and soil material.** Publications.

407 Begheijn, L.T.   NL 020008/67/4916
Ontwikkeling van methodieken en technieken t.b.v. de bodemkunde. **Development of methods and techniques useful for soil science.** Publications.

408 Havinga, A.J.   NL 020008/67/4920
Regionaal bodemkundige studies in alluviale gronden (m.u.v. katteklei) (Nederlands rivierkleigebied en Biesbosch). **Regional soils studies: alluvial soils, catclays not included (Dutch rivier clay region and Biesbosch).** Publications.

409 Bennema, J.   NL 020008/68/4922
Regionaal bodemkundig onderzoek aan rode en gele tropische en subtrppische gronden (Suriname, Brazilië, Sierra Leone, Italië,enz. doch m.u.v. Kenya). **Regional soil research on red and yellow tropical soils (Surinam, Brazil, Sierra Leone, Italy, etc. but Kenya not included).** Publications.

410 Westeringh, W. van de   NL 020008/70/4909
Oude cultuurgronden: ontstaan, karakteristieken en betekenis. **Old arable land soils; their genesis, characteristics and importance/interest.** Publications.

411 Rogaar, H.   NL 020008/70/4918
Biologische rijping in alluviale gronden. **Biological ripening in alluvial soils.** Publications.

412 Bennema, J.   NL 020008/70/6779
Bodemkundig onderzoek in de wildparken in Oost–Afrika. **Pedological research in game reserves of East–Afrika.**

413 Bennema, J.   NL 020008/71/4912 R
Land–evaluatie: ontwikkeling van methoden en toepassingen. **Land evaluation: development of methods and application.** Publications.

414 Brinkman, R.   NL 020008/71/4924
Bodemvorming in periodiek natte, zure gronden: regionaal, analytisch en experimenteel studie–object. **Soil formation in periodically wet, acid, soils: a regional, analytical and experimental study.** Publications.

415 Dolleman, H.   NL 020008/71/8845 N
Grond als industriële grondstof. **Soil as an industrial commodity.**

416 Meester, T. de   NL 020008/72/4913
Bodemkundige aspecten van erosie en bodembescherming (Spanje, Middellandse Zeegebied o.a.). **Soil factors affecting erosion and soil conservation (Spain, Mediterranean a.o.).** Publications.

417 Meester, T. de   NL 020008/73/4910 R
Kartering, landevaluatie en bodemkundig onderzoek in South Nyanza en Kisii (Kenya). **Survey, land evaluation and soil research in South Nyanza and Kisii (Kenya).** Publications.

418 Buurman, P.   NL 020008/75/6365
Ontwikkeling van "red yellow podzolic soils" in Indonesia, in verband met landbouwkundig gebruik. **The genesis of "red yellow podzolic soils" in Indonesia, with emphasis on agricultural suitability.** Publications.

419 Boerma, P.N.   NL 020008/76/6780
Morfogenese van het Zuidlimburgse vereffeningsvak. **Morphogenesis of the Southlimburg planation surface.** Publications.

420 Slager, S.   NL 020008/77/7541

Onderzoek naar de vegetatie–ontwikkeling in een aantal Alno–Padionbossen met van elkaar afwijkende voorgeschiedenis. **A study of vegetation development in major poplar afforestations of different origins.**

421 Ernte, P.J.   NL 040007/60/4081
Geologisch–sedimentologisch onderzoek naar de opbouw van de bodem in Oostelijk Flevoland. **Sedimentological and geological survey of the Holocene and Pleistocene in Oostelijk Flevoland.**

422 Glopper, R.J. de   NL 040007/60/4085
Onderzoek naar de opgetreden inklinking en het opstellen van prognoses voor de te verwachten inklinking in verschillende gebieden. **Investigation of appeared subsidence and prediction of subsidence in various areas.**

423 Glopper, R.J. de   NL 040007/60/4086
Onderzoek naar veranderingen in de bodemgesteldheid in de kuststrook van Friesland en Groningen. **Soil survey in the coastal area of Friesland and Groningen.**

424 Glopper, R.J. de   NL 040007/60/4087
De sedimentatie–omstandigheden langs de Fries–Groningse kust. **Sedimentation conditions along the coast of Friesland and Groningen.**

425 Glopper, R.J. de   NL 040007/60/4089
De chemische en fysische rijping van in Friesland en Groningen gelegen gronden. **Chemical and physical ripening of sediments in Friesland and Groningen.**

426 Benning, G.   NL 040007/68/8484
Onderzoek naar de geschiktheid van bodem en waterbeheersing ter bepaling van de juiste bestemming en inrichting van Flevoland (behalve Almere gebied). **Research on the soil suitability and water management in relation to land use planning of Flevoland (excluded Almere).**

427 Ente, P.J.   NL 040007/71/4079
De opbouw van het Holoceen en de diepte van het Pleistoceen in het Waddengebied. **Geological survey of the Holocene and the depth of the Pleistocene in the Wadden–area.**

428 Andriesse, J.P.   NL 040012/77/7903
De karakteristieken en klassifikatie van de belangrijkste gronden in Azië in gebruik voor de rubbercultuur. **The characteristics and classification of the main soils in Asia in use for rubber cultivation.**

429 Scholte Ubing, D.W.; Somers, J.A.; Ruiter, M.A. de
              NL 050401/72/6236
Verontreiniging en het reinigen van het water– en bodemmilieu. **Pollution and self–purification of water and soil.**

Soil composition – general (B 1110)

See also 14, 15, 158, 171, 826, 876, 925, 929, 976, 984, 1008, 1493, 1788, 2052, 2088, 3373, 3594, 4488, 5133, 5259, 5263, 5264, 5441, 5480, 5501, 5581, 5820, 6246, 7784, 10066, 10112, 10169, 10176, 19782, 20308, 20310

430 Cottenie, A.; Van de Maele, F.; D'Haese, A.
              BE 030008/72/0002 R
Studie van de waardevermindering van het landbouwproduktiemilieu (grond, water, plant). **Study of the**

depreciation of the agricultural production environment (soil, water, plant). Publications.

431 Baert, L.; Vanderdeelen, J.; Biermans, V.
              BE 030009/69/0002 R
Fosfaatretrogradatie in kalk– en zandgronden. **Phosphate retrogradation in calcareous soils and sandy soils.** Publications.

432 Uytterhoeven, J.; Stul, M.; Van Leemput, L.; Schoonheydt, R.   BE 040101/70/0001 R
Adsorptie van organisch materiaal aan kleimineralen. **Adsorption of organic material on clays.** Publications.

433 Livens, J.; Verstraeten, M.J.; Vlassak, K.
              BE 040201/58/0003 R
Mobilisatie en immobilisatie van stikstof in de bodem. **Mobilisation and immobilisation of nitrogen in soil.** Publications.

434 Livens, J.; Vlassak, K.; Verstraeten, L.
              BE 040201/76/0004 R
Neveneffekten van pesticiden op biologische processen van de N–cyclus in de bodem. **Side–effects of pesticides on biological processes of the N–cycle in soil.**

435 Scheys, I.; Vandamme, J.; Lamberts, D.; Appelmans, F.; De Forche, F.; De la Kéthulle, A.   BE 040202/00/0008 R
Studie van de fysische en chemische factoren van de bodemgeschiktheid voor de groenteteelt. **Research of the physical and chemical factors of soil suitability for horticultural crops.** Publications.

436 Van Onsem, J.G.; Verdonck, O.   BE 070600/68/0069 R
Studie van nieuwe substraten voor de sierplantenteelt. **Research on new substrates for horticultural purposes.** Publications.

437 Mayaudon, J.; Batistic, L.; Nenquin–Bellinck, C.
              BE 140000/74/0045 R
Tests enzymatiques pour la fertilité biologique des sols horticoles et recherche de la dégradation microbiologique des pesticides. **Research of enzymatical test of biological fertility by horticultural soils and microbiological degradation of pesticides.** Publications.

438 Blume, H.–P.; Chacon, J.; Marschner, H.
              DE 105050/75/0003
Phosphatstatus typischer Böden Malaysias und Möglichkeiten seiner Verbesserung. **Phosphate status of typical soils in Malaysia and possibilities of improvement.**

439 Blume, H.–P.; Horn, R.; Meshref, H.; Drewes, H.
              DE 105050/77/0003
Wasser– und Nährstoffdynamik Berliner Böden unter Forst–, Acker– und Rieselwiesen–Nutzung. **Water and nutrient metabolism in forest, agricultural and wastewater soils of Berlin.**

440 Kick, H.; Peters, F.   DE 111100/74/0006
Lysimeterversuch über Nährstoffbilanzen in tunesischen Böden unter besonderer Berücksichtigung des Stickstoffes. **Lysimetric experiments on the nutrient balance in Tunesien soils with particular regard to nitrogen.**

441 Boeker, P.; Franken, H.   DE 111251/74/0001
Der Einfluss verschiedener Gründüngungspflanzen auf den

D 1100 - Soil science

Strohabbau im Boden. **The influence of divers green manuring plants on straw decomposition in soil.**

442 Hädrich, F.   DE 126050/73/0009
Untersuchungen zum Wasserhaushalt einer ZweischichtPararendzina unter jungen Kiefernbeständen – Pinus silv. –, in alluvialen Ablagerungen des Rheines bei Hartheim – südl. Oberrheingebiet –. **Investigations on soil–water relationships in a two–layer pararendzina under young pine stands – Pinus silv. – in alluvial deposits of the Rhine near Hartheim in the southern Upper–Rhine region.**

443 Preusse, H.–U.; Haschemzadeh, A.   DE 129020/77/0001
Bodenchemische und tonmineralogische Untersuchungen an typischen Böden West–Aserbeidschans – Mahabad–Ebene –. **Soil chemical and clay mineralogical analysis of typical soils in West–Aserbeidschan – Mahabad–plain –.**

444 Boguslawski, E.von   DE 129120/75/0001
Biologisch–dynamische Düngung und konventionelle Verfahren der Bodenfruchtbarkeit. **Biodynamic fertilization and conventional methods of soil fertility.**

445 Atanasiu, N.; Thiagalingam, K.; Westphal, A.;
Suwannarat, C.   DE 129552/75/0004
P–Fixierung und P–Freisetzung von tropischen Böden. **P–fixing capacity and P–release in tropical soils.**

446 Meyer, B.; Hugenroth, P.   DE 132033/70/0001
Bodenkundliche Untersuchungen der im Raum Göttingen durchgeführten vorgeschichtlichen Grabungen 1968. **Soil scientific studies during archaeological excavations in Goettingen area.**

447 Meyer, B.; Gebhardt, H.   DE 132033/70/0002
Bodenkundliche Untersuchungen über die Ursachen des Humusabbaus und der Aufhellung niedersächsischer Schwarzerden 1968. **Soil scientific investigations on the causes of humus degradation and lightening of black earths in Lower Saxony.**

448 Timmermann, F.; Wessolek, G.   DE 132063/73/0001
Die Nährstoffverlagerung in einer südniedersächsischen Lössparabraunerde in Abhängigkeit von Düngerart, Düngermenge und Pflanzenbewuchs. **Nutrient movement in a typical grey brown podsolic soil in the South of Lower Saxony depending on kind and amount of fertilizer and on vegetation.** Publications.

449 Timmermann, F.; Benaguid, T.   DE 132063/75/0001
Nitratentwicklung in landwirtschaftlich genutzten Böden und Ertragsbildung in Abhängigkeit von Stickstoffdüngung, Fruchtart und Witterung. **Nitrate development in agricultural soils and yield formation in dependence upon nitrogen fertilization, crops and weather conditions.**

450 Welte, E.; Timmermann, F.   DE 132065/75/0002
Untersuchungen über die Nährstoffbelastung von Grundwasser und Oberflächengewässer aus Boden und Düngung. **Investigations into leaching and nutrient runoff due to soil management and fertilization.** Publications.

451 Beug, H.–J.; Henrion, I.   DE 132092/75/0001
Geschichte der Harzvermoorungen. **History of peatbogs in the Harz Mountains.**

452 Ulrich, B.; Hetsch, W.   DE 132600/77/0002
Auswirkungen unterschiedlicher Waldbrandintensität auf den Nährstoffhaushalt des Bodens. **The effect of forest fire intensity on soil nutrient status.**

453 Mayer, R.; Ulrich, B.   DE 132601/74/0001
Umweltrelevante Spurenstoffe in der Bodendecke. Q/I–Beziehungen und Bindungsformen von Spurenstoffen in Böden. **Trace elements of environmental relevance in the soil covering. Quantity/Intensity relations and binding forms of trace elements in soils.**

454 Ulrich, B.; Prenzel, J.   DE 132601/78/0001 N
Physikalisch–chemische Beschreibung multipler Kationenaustauschgleichgewichte in Böden. **Physico–chemical desription of multiple cation exchange equilibria in soils.**

455 Reichenbach, H.von; Beyme, B.   DE 138030/74/0002
Die Oxidation von Gitter–Fe und ihr Einfluss auf die Austauschereigenschaften von Biotiten. **Cation exchange of biotites as influenced by the oxidation of lattice–structure ferrous iron.**

456 Ottow, J.C.G.; Demerdash, M.E.–H.
                                      DE 144321/78/0001 N
Potentielle und aktuelle Denitrifikationsintensität unterschiedlicher Standorte. **Potential and actual denitrification rate of different soils.**

457 Schweikle, V.   DE 144322/74/0004
Orientierung von Tonpartikeln in Bodenaggregaten. **Orientation of clay particles in soil aggregates.**

458 Schlichting, E.; Reinfelder, H. DE 144322/78/0001 N
Vergleichende Untersuchungen zur Prognose des N–Düngebedarfs von Böden unterschiedlicher Standorte. **Comparative investigations for prognosticating the N fertilizer requirement of soils from different sites.**

459 Brümmer, G.; Herms, U.   DE 148051/75/0002
Untersuchungen zur Schwermetall–Löslichkeit in Abhängigkeit vom Redox–Potential, pH–Wert und Stoffbestand von Böden und Sedimenten. **Solubility of heavy metals in relation to redox potential, pH value and composition of soils and sediments.**

460 Baumann, H.; Thies, A.E.   DE 148550/78/0003 N
Fernerkundung von Erosion und Versalzung über Satellit. **Remote sensing of erosion and salinity by satellite.** Publications.

461 Buch, M.–W.von   DE 202010/77/0001
Degradierte Böden unter Nadelholzbeständen im subtropischen Bereich und vergleichende Untersuchungen zu Vulkanascheböden. **Degraded soils under coniferous stands in subtropics and comparative studies on volcanic ash soils.**

462 Franken, W.; Franken, M.   DE 404010/75/0004
Wasserbilanz und Nährstoffkreislauf im Einzugsbebiet eines Quellbaches im tropischen Regenwald bei Manaus, Brasilien. **Water balance and nutrient cycling in the catchment area of a river source in the tropical rain forest near Manaus, Brazil.**

463 Völkel, R.; Scholl, W.; Pfulb, K.   DE 501010/75/0001
Einfluss der Acidität des Bodens auf die Aufnahme von aus kompost. Siedlungsabfällen stammenden Schwermetallen –

D 1100 – Soil science

komb. Gefäss– und Freilandversuch –. **The influence of the soil–acidity on the uptake of heavy metals from refuse sewage–sludge – combined pot and field experiments –.**

464 Kannenberg, J.   DE 501106/74/0001
Untersuchungen zur Verlagerung von Nährstoffen, insbesondere Nitraten bei Weinbergsböden des Oberrheintales. **Studies on the translocation of nutrients spec. nitrates in vineyard soils in the Upper Rhine Valley.**

465 Schuch, M.   DE 502050/70/0001
Untersuchungen über das Quellen und Schwinden des Bodens von kultivierten und unberührten Hochmooren in verschiedenen Bodentiefen in Abhängigkeit von Niederschlag, Bodenwasserhaushalt und Bodentemperatur 1958. **Investigations of swelling and shrinking of soils in cultivated and in virgin marshland at different depths of soil as affected by precipitation, water content and temperature of soil.**

466 Schuch, M.; Schmeidl, H.   DE 502050/70/0005
Klima– und Bodentemperaturbeobachtungen in Hochmooren 1958. **Observation on climatic conditions and on temperatures of soils in high marshes.**

467 Sommer, G.   DE 502055/72/0001
Die Dynamik der Bodenphosphate unter dem Einfluss von Düngung und Entzug. **The dynamics of soil phosphates under the influence of fertilization plant utilization.** Publications.

468 Sommer, G.   DE 502055/73/0017
Die N–Dynamik von Ackerböden unter dem Einfluss von Düngung und Entzug. **The N–dynamics in arable soils under the influence of fertilization and plant uptake.**

469 Beck, T.   DE 502055/74/0003 R
Der Verlauf der mikrobiellen N–Umformung im Boden. **Disposal of sludge from sewage plants on pasture land affecting soil fertility, yield and quality of feed.**

470 Krämer, F.   DE 508302/70/0001 N
Die Bewirtschaftung von Löss–Rohböden im Hinblick auf die Entwicklung der nachhaltigen Bodenfruchtbarkeit. **Cultivation of uncultivated loess soils with regard to development of persistent soil fertility.**

471 Sunkel, R.   DE 508302/70/0008 N
Ermittlung von Kennwerten zur Beurteilung der physikalischen und chemischen Eigenschaften rekultivierter Lössböden. **Determination of characteristics for estimation of physical and chemical properties of recultvated loess soils.**

472 Madsen, H.B.   DK 030801/79/5166 N
Danske jordes pedologiske udvikling set i forhold til terræn, udgangsmateriale og klima og dettes indflydelse på afgrødernes vandforsyning. **The influence of ground, original material and climate on the development af danish soils and their capacity of water supply to crops.**

473 Trocme, S.; Boniface, R. Mme   FR 010103/46/6215
Influence à court et à long terme de bilans négatifs, nuls ou positifs en éléments fertilisants (P,K,Mg). **Effects over short or long term periods of null, negative or positive balance in mineral elements.**

474 Calvet, R.; Vallet, (Melle).   FR 010103/72/0649
Phénomènes de déplacement des produits herbicides dans des milieux poreux. **Herbicide movement in porous media.**

475 Chaussidon, J.   FR 010103/72/0653
Etude des propriétés physico–chimiques de l'eau présente dans le système eau–argile à une teneur en eau supérieure à 30%. **Studies of physico–chemical properties of water in clay–water systems when water contents is higher than 30%.** Publications.

476 Nys, C.; Levy, G.   FR 010303/71/4161
Evolution de la fertilité des sols sous les reboisements de résineux. **Evolution of soil fertility in afforestations with conifers.**

477 Arvieu, J.C.   FR 010502/72/6252
Evolution d'un insecticide–nématicide (temik) et d'un fongicide (benlate) dans les conditions du sol. **Evolution of a nematicide–insecticide (temik) and of a fungicide (benlate) under soil conditions.**

478 Arvieu, J.C.; Mars, S.; Bellon, Nicole; Ferry, Geneviève   FR 010502/73/6019
Etude des processus physico–chimiques de la dynamique des produits pesticides dans les sols (essentiellement produits fongicides ou nématicides de traitement du sol : aldicarbe, bénomyl, bromure de méthyle). **Physico chemical processes of the behavior of pesticides in soils.**

479 Blanc, D. Mme; Moulinier, H.; Morisot, A.   FR 010502/74/6009
Dynamique des nitrates dans les sols horticoles. **Dynamics of nitrate nitrogen horticultural soils.**

480 André, J.P.   FR 010502/74/6253
Etude des propriétés physicochimiques des substrats de culture végétale. **On physico–chemical properties of plant culture substrates.**

481 Moulinier, H.; Sandra Garrigos, I.   FR 010502/75/6007
Evolution de diverses matières organiques incorporées á une terre en présence de nitrates. **Fate of some organic matters mixed with soil containing nitrates.**

482 Monnier, G.   FR 010601/72/0527
Evaluation et contrôle des facteurs physiques de la production de la tomate en plein champ. **Evaluation and control of physical characteristics of the soil for field tomato production.**

483 Arnoux, M.; Defrance, H.; Golinsky, P.   FR 010615/76/9039
Méthodes naturelles de contrôle de la fatigue des sols à Pêcher. **Natural methods for evaluating the dregree of exhaustion of soils from peach tree orchards.**

484 Juste, C.; Lineres–Solda; Delas, J.; Gomez, A.; Lasserre; Dureau, P. Mme   FR 010704/63/6083
Rôle de la matière organique dans la formation et les propriétés des sols du sud ouest atlantique. **Influence of organic matter on formation and characteristics of acid soils in the western south.**

485 Juste, C.; Solda; Lubet; Lasserre   FR 010704/63/6089
Influence du chaulage sur la productivité et l'évolution des sols acides du sud–ouest consacrés à la monoculture du maïs. **Effect of liming on the productivity and the evolution of acid soils in monoculture of corn in the western south.**

D 1100 – Soil science

486 Soyer, J.P.; Chignon; Menet, M.; Lubet
FR 010704/67/6078
Fertilisation d'entretien phosphopotassique et magnésienne en culture céréalière à rotation simple (monoculture maïs ou rotation blé–maïs). **Fertility maintenance in major nutrients (p, k, mg) on cereals with simple sequences of annual crops (continuous corn or wheatcorn rotation).**

487 Gachon, L.; Robelin, M.
FR 010802/52/0573
Bilans hydriques et chimiques sous diverses conditions écologiques (sol–climat–fumures–systeme cultural). **Water and chemical balances under various ecological conditions. (Soil, climate, manures, cultural practice).** Publications.

488 Conesa, A.; Delphin; Specty, R.
FR 010901/76/6131
Conséquences de l'exportation systématique des pailles sur la fertilité d'un sol. **Effects of straws carrying away on soil fertility.**

489 Grosman, R.
FR 011001/71/6291
Evaluation du potentiel chimique des sols régionaux. **Evaluation of chemical potential of regional soils.**

490 Libois, A.; Grosmann, R.
FR 011001/72/6135
Fixation du phosphore dans différents types de sols du Centre–Est. **Phosphorus fixation in different types of soils in the Centre–Est area.**

491 Decau
FR 011401/66/6208
Influence, à long terme, de la fertilisation azotée et de l'enfouissement ou de l'enlèvement des pailles sur la production de matière sèche et l'évolution de la fertilité d'un sol argilo–calcaire en rotation céréalière. **Long term effect of straw ploughing in or removing and of nitrogen fertilization on dry matter production and fertility of a clay calcareous cropped with cereal rotation.**

492 Marty, J.R.; Charpenteau; Hutter, W.; Rellier, J.P.; Decau; Maertens, C.
FR 011401/69/6207
Fertilité du sol dans différentes rotations culturales avec ou sans irrigation. **Soil Fertility in different irrigated and non–irrigated agricultural systems.**

493 Langlet, A.; Begon, J. C.; Turc, L.; Hentgen, A.; Jamagne, M.
FR 011401/76/6211
Potentialités du milieu sol – climat pour la production de paille de céréales. **Possibilities of soils and climates for production of cereal straw.**

494 Sebillotte, M.; Boiffin, J.
FR 011701/63/6165
Etude des rotations culturales – Evolution de la fertilité du sol sous l'influence de différentes conditions de rotation et d'entretien du sol – Répercussions sur l'élaboration du rendement du blé, du colza et du maïs. **Study on crop rotation – Soil fertility evolution under different conditions of crop rotation and soil management – Consequences on yield elaboration of wheat winter rape and maize.**

495 Sebillotte, M.; Boiffin, J.; Caneill,
FR 011701/74/6167
Evolution de la fertilité des terres des exploitations agricoles d'une petite région agricole. **Study on land fertility evolution in farms of a given agricultural area.**

496 Manichon, H.; Sebillotte, M.
FR 011701/76/6170
Analyse de l'évolution du profil cultural au champ et des répercussions sur la végétation. Elaboration d'un modèle prédictif. **Analysis of soil profile evolution and repercussions on crops. Elaboration of a predictive model.**

497 Coppenet, M.; Duval, L.
FR 012208/73/6177
Bilans des éléments fertilisants au niveau des exploitations d'élevage. Modifications chimiques des sols des élevages intensifs et partiellement hors sol. **Fertilizers balance of cattle farms Chemical variations of soils.**

498 Maurice, J.
FR 012208/74/6182
Etude de la dynamique du bore dans les sols. **Dynamics of boron in soils.**

499 Remy; Studer, R.
FR 012209/71/6134
Influence du type de sol sur la prévision de la fumure azotée du blé d'hiver. **Influence of soil type on winter wheat nitrogen requirement.**

500 Duval, Y.; Masclet, A.
FR 012219/64/6184
Efficacité comparée de deux formes d'engrais phosphatés dans un sol d'alluvions modernes limono–argilo–calcaire. **Comparative efficiency of two forms of phosphorus fertilizer on young alluvial soil calcareous loamy clay.**

501 Hebert, J.; Remy, J.C.
FR 012503/58/6154
Description de l'évolution de l'azote dans différents sols au cours d'une succession d'années. **Description of nitrogen evolution in some soils during a series of years.**

502 Remy, J.C.
FR 012503/71/6159
Recherche des causes de moindre efficacité des engrais azotés en solution. **Studies on the lower efficiency of the nitrogen fertilizer solution.**

503 Remy, J.C.
FR 012503/74/6157
Devenir des engrais nitriques en sol de limon au moyen de 15 N. **The fate of nitric nitrogen fertilizers 15 N–labelled in loam soil.**

504 Remy, J.C.
FR 012503/76/6155
Détermination de l'azote total et du rapport isotopique par couplage four – spectromètre de masse. **Simultaneous determination of total nitrogen and isotopic ratio by dry combustion – mass spectrometry couplage.**

505 Remy, J.C.; Maucorps, J.
FR 012503/76/6156
Etude des variations de la teneur isotopique naturelle en 15 N dans le profil de sol et sous l'influence de la culture. **Study of natural isotopic abundance 15 N in the soil profile under crop conditions.**

506 Remy, J.C.; Hebert, J.
FR 012503/76/6158
Contribution des résidus de récolte à la nutrition azotée des cultures. **Crop residues contribution to nitrogen nutrition.**

507 Hebert, J.; Machet, J.M.
FR 012503/76/6164
Transfert de l'azote du profil cultural vers la nappe de la craie. **Transfer of nitrogen from soil profile down to ground water in the chalk.**

508 Mercer; Hill
GB 010501/79/0007 N
Movement of water nitrate, and other ions in soil and chalk.

509 Bloomfeld; Pruden
GB 012003/00/0067 R
Reactions between minor elements and soil organic matter and sewage sludge.

D 1100 – Soil science

510 Blagden, P.   IE 060200/67/0369 R
Soil productivity experiment – potassium experiments.
Publications.

511 McGrath, D.   IE 060200/69/0400
The carbohydrates of peat. Publications.

512 O'Sullivan, M.   IE 060200/72/0388 R
Delineation of trace element deficient areas throughout the country. Publications.

513 Jelley, R.M.; Gardiner, M.; Burke, W.
    IE 060300/70/0126 R
Determination of physical properties of soils mapped by the National Soil Survey. Publications.

514 Arcara, P.G.; Rodolfi, G.; Bidini, D.
    IT 020100/79/0001 N
Caratterizzazione biologica e biochimica di suoli provenienti da substrati carbonatici in differenti condizioni ambientali. **Biological and biochemical characterization of soils on limestone in different climatic conditions.**

515 Zanchi, C.; Arcara, G.; Tellini, M.; Piovanelli, C.; Bidini, A.   IT 020100/79/0004 N
Studio dell'influenza del condizionamento con poliidrossidi di ferro e di alluminio e di differenti colture, nei confronti del ruscellamento superficiale, dell'erosione e dell'asportazione degli elementi nutritivi. **Influence of a soil conditioner and of different crops on runoff, soil losses and nutritive losses.** Publications.

516 Gregori, E.; Miclaus, N.   IT 020100/79/0007 N
Influenza di alcuni fertilizzanti azotati sulla nodulazione e sulla resa produttiva di leguminose coltivate su regosuoli. **Influence of some nitrogen fertilizers on nodulation and yield of legumes in euric and calcaric regosoils (FAO 1970).**

517 Nigro, C.; Scandella, P.   IT 020200/77/0001
Fertilità biologica del terreno. **Biological fertility of soil.**
Publications.

518 Izza, C.   IT 020200/78/0001 N
Determinazione della disponibilità in elementi nutritivi (Fosforo in particolare) mediante curve isoterme di assorbimento. **Sorption–desorption isotherm of Phosphorus in some soils of the Tiber high valley.** Publications.

519 De Rossi, C.; Bevilacqua, M.; Santroni, R.; Fraticelli, A.; Piccolo, A.   IT 020200/78/0002 N
Indagine sui metodi di determinazione del fosforo assimilabile in terreni con basso rapporto silice/sesquiossidi. **Studies on the methods for determining available phosphorus in soils with low silica/sesquioxides ratio.** Publications.

520 Romanin, M.   IT 020200/78/0005 N
Controllo delle perdite di azoto nella concimazione di pieno campo. Concimazione con ammoniaca anidra. **Direct field measurement of ammonia losses. Fertilization with anhydrous ammonia.** Publications.

521 Marizza, L.; Romanin, M.   IT 020200/78/0006 N
Fertilizzazione fluida. Prove di concimazione a livello aziendale. **Field tests with fluid fertilization.**

522 Spallacci, P.; Lanza, F.   IT 020500/67/0004
Fertilità residua da letamazione a lungo termine e concimazione chimica annuale sul mais in successione a se stesso e su colture avvicendate. **The residual effect of long–term manuring and chemical fertilizer dressing upon continuous corn and rotated crops.** Publications.

523 Boschi, V.; Spallacci, P.; Montorsi, M.   IT 020500/70/0002
Possibilità e limiti dell'utilizzazione agricola dei liquami di suini e bovini. **Possibilities and limits in the agricultural use of pig and cow slurry.** Publications.

524 Lopez, G.; Colucci, R.; Rizzo, V.; Tritto, F.
    IT 020500/72/0004
Variazioni del contenuto in N,P,K del terreno in funzione del volume di adacquamento e delle concimazioni. **Influence of irrigation and of fertilizers on N,P,K content in soil.**

525 Spallacci, P.; Boschi, V.   IT 020500/73/0008
Rilievo pedo–agronomico della provincia di Parma. **Chemical and physical study of soils in the province of Parma.**

526 Rizzo, V.; Di Bari, V.; Perniola, M.; Convertini, G.; Lopez, G.; Colucci, R.   IT 020500/79/0001 N
Interramento stoppie: influenza sul terreno e sulla produzione vegetale. **Influence of wheat stubble plowed under on soil conditions and crop production.**

527 Lanza, F.; Colucci, R.; Lopez, G.; Perniola, M.
    IT 020500/79/0003 N
Influenza di differenti modalità di interventi agronomici sulla fertilità dei terreni. **Influence of different agronomical practices on soil fertility.**

528 Dellacecca, V.   IT 040101/77/0165 R
Ricerca internazionale sull'influenza dell'interramento della paglia e della concimazione azotata a dosi crescenti. Nel guadro dell'accordo di collaborazione Italia–Germania. **International research on the response to straw ploughing–in and of nitrogen fertilizing in increasing doses (within the framework of the Italian–German collaboration agreement.**

529 Rossi, N.   IT 040202/77/0288 R
Analisi chimica del terreno, studi di correlazione e di calibrazione. **Chemical soil analysis. A study on correlation and calibration.**

530 Silva, S.; Beghi, B.; Chimenti, I.   IT 040406/79/0002 N
Distribuzione di radionuclidi applicati a terreni fra colloidi organici ed inorganici. **Distribution of radionuclides applied to the soils between organic and inorganic colloids.**

531 Basile, G.   IT 040700/77/0790
Adsorbimento e persistenza delle atrazine nel suolo. **Adsorption and persistence of atrazine in the soil.**

532 Postiglione, L.   IT 040701/77/0267 R
Ordinamenti colturali ed evoluzione della fertilità. **Cultural lay–out and fertility evolution.**

533 Eschena, T.   IT 040705/77/0170 R
La presenza e la genesi dell'inogolite negli andosuoli d'Italia. **Presence and genesis of inogolite in Italian andosoils.**

534 Tafuri, F.; Scarponi, L.; Giusquiani, P.L.
    IT 041004/79/0001 N

D 1100 — Soil science

Perdite di azoto per volatilizzazione da terreni trattati con urea in presenza di sostanza organica. **The volatilisation losses of nitrogen from soils treated with urea in the presence of organic materials.**

535 Rambelli, A. IT 041607/77/0279 R
Degradazione e mineralizzazione della sostanza organica di origine vegetale. Nel quadro dell'accordo di collaborazione Italia–Francia. **Degradation and mineralisation of organic matter of vegetal origin. Within the framework of the Franco–Italian collaboration agreement.**

536 Sequi, P. IT 060400/73/0003
Sostanze cementanti degli aggregati del terreno. **Cementing substances of soil aggregates.** Publications.

537 Sequi, P. IT 060400/74/0002
Interazioni tra argille e sostanza organica e formazione del terreno. **Clay–organic matter interactions and soil formation.**

538 Giovannini, G. IT 060400/77/0630
Parametri chimici correlati con la erodibilità del terreno. **Chemical parameters related to soil liability to erosion.**

539 Petruzzelli, G. IT 060400/77/0780
Inquinamento metalli pesanti ed adsorbimento delle piante. **Heavy metal pollution and plant adsorption.**

540 Mannipieri, P. IT 060400/77/0783
Attività enzimatiche dei suoli. **Soils : their enzymatic activity.**

541 Malquori, A. IT 061200/74/0144
Diffusione di fosfati semplici e condensati attraverso il terreno. **Diffusion of ortho– and poly–phosphates in the soil.** Publications.

542 Malquori, A. IT 061200/74/0145
Influenza di metalli considerati tossici sulle caratteristiche biochimiche del suolo. **Influence of toxic metals on biological characteristics of the soil.** Publications.

543 Florenzano, G. IT 061300/77/0777
Indicatori biologici di suoli inquinati. **Biological indicators of soil pollution.**

544 Ceruti, A. IT 061400/70/0155
Biochimismo dei funghi in relazione agli alti polimeri che si possono trovare nel terreno. **Fungal biochemistry concerning high polymers of the soil.** Publications.

545 Tafuri, F. IT 061600/74/0026
Adsorbimento di alcuni erbicidi su argille e su carbone attivato. **Adsorption of some herbicides on clays and activated carbon.** Publications.

546 Tafuri, F. IT 061600/77/0781
Erbicidi nel terreno e loro presenza nei vegetali. **The presence of herbicides in soils and plants.**

547 Tafuri, F. IT 061600/77/0873
Persistenza nel terreno dell'atrazina e dei suoi metaboliti. **Persistence of atrazine and its metabolites in the soil.**

548 Giardini, L. IT 062900/70/0138
Studio sulla persistenza di azione dell'atrazina in terreno franco–limoso. **Study on persistency of atrazine in a silt loam soil.**

549 Senni, L. IT 121900/77/0809
Denitrificazione biologica ad alto rendimento. **Highly efficacious biological denitration.**

550 Boekel, P.; Prummel, J. NL 010103/62/0520 R
Binding van spoorelementen aan (componenten van) de organische stof. **Adsorption and chelation of trace elements by soil organic matter.** Publications.

551 Boekel, P. NL 010103/74/5589 R
De stevigheid, gladheid en doorlatendheid van grassportvelden. **Firmness, slip resistance and permeability of sports fields.** Publications.

552 Smet, L.A.H. de; Soesbergen, G.A. van
NL 010119/71/3973
Invloed van bodemkundige factoren en beworteling op opbrengst en kwaliteit van consumptie–aardappelen. **Influence of the properties of soils and root development upon yield and quality of ware potatoes.**

553 Breeuwsma, A.; Boers, J.A.; Sjardijn, R.C.; Zwijnen, R.
NL 010119/73/5978 R
De beïnvloeding van de granulaire samenstelling van ijzerrijke gronden door de bepalingsmethode. **The influence of iron oxides on particle size distribution analysis.**

554 Gruijter, J.J. de; Holst, A.F. van NL 010119/74/5983
Proefproject standaard textuur–verlopen in zandgronden. **The determination of texture–profiles in sandy soil for representation on soil maps.**

555 Wallenburg, C. van; Buck, J. de NL 010119/75/6914
Bepaling van A–cijfers van geoxydeerd en niet–geoxydeerd veen ter beoordeling van de 'klinkgevoeligheid'. **A–coefficients for oxidised and unoxidised peat soils.**

556 Dam, J.G.C. van NL 010119/77/7796 R
Onderzoek naar de groei van sierconiferen afkomstig van verschillende bodemeenheden na verplanting. **Research on the conifer growth for trees originating on different soil types after transplantation to a research plot.**

557 Pape, J.C. NL 010119/77/8304
Onderzoek naar het verband bodemvruchtbaarheid en vegetatie. **Research into the connection between vegetation and soil fertility.**

558 Luten, W.; Roozeboom, L. NL 010208/69/5923
Hantering van het grondbewerkingswerktuig t.b.v. de grasland; exploitatie – de techniek van herinzaai van grasland. **The handling of soil tillage implements on grassland farming; the technique of re–seeding grassland.** Publications.

559 Burg, J. van den NL 010601/70/3111
Toepassing van gezuiverd effluent in bosgebieden. **Application of renovated effluent in forest areas.** Publications.

560 Burg, J.van den; Faber, P.J. NL 010601/72/3533
Onderzoek naar de mogelijkheden en de methoden voor het bergen van vaste en vloeibare mest in bossen. **Storage of solid and fluid manure in forests.** Publications.

561 Burg, J. van den NL 010601/79/8981 N

De invloed van de volle boom–oogstmethode op de bodemvruchtbaarheid en de groei van naaldhoutsoorten op pleistocene zandgronden. **Impact of full–tree logging on soil fertility and growth of trees on pleistocene sandy soils.**

562 Janse, A.R.P. NL 020007/72/4955
Spectrale karakteristieken gemeten aan bodemoppervlakten. **Distribution of radiant temperature on natural surfaces.** Publications.

563 Bruggenwert, M.G.M. NL 020007/76/8507
Overzicht van de experimentele resultaten op het gebied van kationen omwisseling in de bodem en aan bodemcomponenten. **Survey of experimental information on cation exchange in soil systems.** Publications.

564 Linde, A.J. van der NL 020016/75/4254
Adsorptie van polymeren aan bodem–belangrijke oxiden. **Adsorption of polymers on soil–important oxides.**

565 Ente, P.J. NL 040007/68/4080
De chemische rijping van de bodem in Zuidelijk Flevoland. **Chemical ripening of the soils in Zuidelijk Flevoland.**

566 Rosmalen, H.A. van NL 040012/77/7902
Bepaling van uitwisselbare kationen en CEC in kalkhoudende gronden. **Determination of exchangeable cations and CEC in lime containing soils.**

567 Wormer, T.M. NL 040012/77/8940 N
De relatie tussen bodemvruchtbaarheid en het voorkomen van Orobanche. **The relationship between soil fertility and Orobranche development.**

568 Pol, F. van der NL 040012/78/8942 N
Optimalisering van bestaande analyses en aanpassings/invoering nieuwe analyses t.b.v. service laboratorium OBA. **Improvement of soil analytical methods on behalf of the Services Laboratory.** Publications.

Soil composition – inorganic (B 1111)

See also 16, 169, 746, 764, 767, 832, 835, 837, 840, 841, 851, 1067, 1100, 1103, 1401, 1421, 1443, 1444, 1747, 1762, 2030, 3699, 5461, 5489, 5492, 5659, 5858, 6581, 7826, 10335

569 Tonnard, V.; Marcoen, J. BE 010005/72/0001
Etude de l'altération des minéraux dans le cadre de la pédogénèse. **Study of weathering of minerals in soil genesis.** Publications.

570 Fripiat, J. BE 020102/49/0001 N
Caractérisation des propriétés de surface d'oxydes et silicates naturels et synthétiques finement divisés. **Caracterization of the surface properties of finely divided natural and synthetic oxides and silicates.** Publications.

571 De Borger, R. BE 100000/71/0016 R
Scheikunde studie van de invloed van bekalkingsproeven in een beukenbos op leemgrond. **Chemical study of the influence of liming a loam soil beech forest.**

572 Zakosek, H.; Müller, W. DE 111050/78/0004 N
Ein– und Austrag von Nitraten in Weinbergsanlagen der Mosel. **Imput and output of nitrates in vineyards of the Mosel valley.**

573 Gewehr, H.; Beckmann, H. DE 111050/78/0005 N
Mineralogische Untersuchung der Sand– und der Tonfraktion niederrheinischer Auenböden. **Mineralogical investigation of the sand and clay fraction from soils on lower Rhine alluvial deposits.**

574 Schröder, D.; Schegiewal, A. DE 111050/78/0006 N
Zur P– und K–Versorgung verschiedener Bodentypen bei einheitlicher Bewirtschaftung. **The P and K supply in various soil types under equal management.**

575 Zakosek, H.; Wiechmann, H. DE 111053/78/0001 N
Untersuchungen zur Verteilung, Bindung und Mobilität umweltrelevanter Spurenelemente in den wichtigsten Auenböden der Niederrheinischen Bucht. **Distribution, sorption and mobility of heavy metals in the most important alluvial soils of the Lower Rhine valley.**

576 Kick, H.; Bürger, H.; Sommer, K. DE 111100/75/0001
Vorkommen von Schwermetallen in landwirtschaftlich genutzten Böden Nordrhein–Westfalens und in den Pflanzen der einzelnen Standorte. **Occurrence of heavy metals in agricultural soils of Northrhine–Westphalia and in the plants of the individual sites.**

577 Schaffer, G.; Diestel, H.; Friedel, B DE 114301/77/0001
Salzverlagerung im Boden mit Diskontinuitäten. **Salt displacement in soils with discontinuities.**

578 Schaffer, G.; Schleiff, U. DE 114301/77/0002
Salztoleranz und Wasserverbrauch in Abhängigkeit von der Giesswasserqualität – Salzgehalt – und der Kaliumversorgung. **Salt tolerance and water use in dependence on irrigation water quality – salt content – and potassium supply.**

579 Zöttl, H.W.; Gudmundsson, T.; Keilen, K.; Stahr, K. DE 126050/78/0001 N
Elementselektive Verwitterung und Verlagerung in Böden auf Bärhaldegranit und ihre Bilanzierung. **Differences in weathering and translocation between several elements in soils formed from Bärhalde–granite including element balancing.** Publications.

580 Moll, W.; Trüby, P. DE 126050/78/0007 N
Modellversuche zur Ermittlung der SchwermetallSorptionsfähigkeit und –Belastbarkeit verschiedener forstlich genutzter Böden. **Model experiments for the determination of heavy metal absorption capacity and the tolerance limits of heavy metals in various forest soils.**

581 Mengel, K.; Keerthisinghe, D.G. DE 129047/78/0002 N
Phosphatverfügbarkeit verschiedener Böden. **Phosphate availability of different types of soil.**

582 Schönhals, E.; Tributh, H. DE 129382/70/0010
Tonmineralogische Untersuchungen zur Klärung der Genese rumänischer Steppenböden und ihrer Übergangsformen. **Clay mineralogy and genesis of some loess–derived soils of Rumania.**

583 Wohlrab, B.; Ditter, P. DE 129450/75/0007
Wirkungen verschiedener naturräumlicher Ausstattung und verschiedener Flächennutzung auf den Schwermetallgehalt der Böden und den Schwermetalleintrag in Gewässer. **Effects of various types of natural spaces and of land use on the heavy metal content of soils and on the transport of heavy metals into**

D 1100 - Soil science

waters.

584 Ulrich, B.; Zaher; Khanna, P.K.   DE 132601/77/0001
Manganverhalten in sauren Böden. Manganformen, Q/I
Beziehungen, Transportvorgänge. **Behavior of manganese in
acid soils. Forms of manganese, Q/I relationships, transport
processes.**

585 Ottow, J.C.G.; Makboul, H.   DE 144321/75/0001
Charakterisierung von Pelosolen durch ihren Enzymhaushalt.
Characterization of clay soils by soil enzyme activity.

586 Ottow, J.C.G.; Munch, J.C.   DE 144321/78/0002 N
Einfluss des Aktivitätsgrades – Feo/Fed – pedogener Eisen
'III'-Oxide auf das Ausmass der bakteriellen Eisenreduktion
in hydromorphen Böden. **Role of the – Feo/Fed – ratio of
pedogenic iron 'III'–oxide in the degree of bacterial
iron–reduction.**

587 Schlichting, E.; Schramm, A.   DE 144322/74/0001
Untersuchungen über die Bor–Sorption durch Tonminerale.
Investigations into the boron sorption by clay minerals.

588 Schlichting, E.; Bredemeier, J.   DE 144322/74/0006
Verteilung umweltrelevanter Spurenstoffe in Bodenprofilen
und Bodenlandschaften SW–Deutschlands. **Distribution of
trace of environmental relevance in soil profiles and soil
landscapes in Southwest Germany.**

589 Schlichting, E.; Himmelhan, C.   DE 144322/74/0008
Schwermetall–Umsetzungen in Müllkompost–gedüngten
Böden. **Heavy metal reactions in soils fertilized with domestic
wastes.**

590 Schlichting, E.; Metzger, F.   DE 144322/74/0009
P– und Schwermetall–Umsetzungen in Klärschlammgedüngten
Böden. **Phosphorus and heavy metal reactions in soils fertilized
with sewage sludge.**

591 Schlichting, E.; Müller, D.   DE 144322/77/0001
Austrag umweltrelevanter Haupt– und Spurenstoffe aus
naturnahen Ökochoren des schwäbischen Keuperberglandes
und Albvorlandes. **Export from natural ecochores of the
Swabian keuper hills and Alb forelands of macro– und
micro–elements relevant for environment.**

592 Schlichting, E.; Hauffe, H.   DE 144322/77/0002
Untersuchungen über die Phosphat– und
Schwermetall–Verteilung in unterschiedlich benutzten und
besiedelten Moränelandschaften. **Investigations on the
phosphate and heavy metal distribution in moraine landscapes
in South Germany with different landuse and urbanisation.**

593 Schlichting, E.; Thoma, G.   DE 144322/77/0003
Umweltrelevante Unterschiede zwischen "konventionellen"
und "biologischen" Produktionsverfahren in der
Landwirtschaft. Hier: Mineralisierung, Aufnahme und
Auswaschung von Stickstoff in unterschiedlich
bewirtschafteten Böden verschiedener Standorte. **Difference
between conventional and biological production systems in
agriculture relevant to environment, here: mineralization
uptake and leaching of nitrogen from soils of different sites and
different management.**

594 Schwertmann, U.; Deller, B.   DE 161020/74/0001
P–Haushalt von landwirtschaftlich genutzten Böden. **P status**

in agricultural soils.

595 Fischer, W.R.   DE 161020/74/0002
Hg, Pb, Cd und Zn in Süsswassersedimenten. **Hg, Pb, Cd, and
Zn in limnetic underwater sediments.**

596 Schwertmann, U.   DE 161020/74/0006
Spurenelement in Bodensequenzen. **Trace elements in
pedosequences.**

597 Schwertmann, U.; Fischer, W.R.   DE 161020/75/0005
Verhalten von Spurenelementen – Cu, Zn, Cr, Mn und wenn
möglich Pb und Cd – in mit Siedlungsabfällen gedüngtem
Boden. **Behaviour of trace elements – Cu, Zn, Cr, Mn and
possibly Pb and Cd – in soils fertilized with composted waste
and sewage.**

598 Niederbudde, E.A.   DE 161020/75/0006
Umwandlung von Tonmineralen in Böden aus illithaltigen
Sedimenten jungtertiärer Hochflächen. **Transformation of clay
minerals in soils from illitic sediments of younger tertiary
plateaus.**

599 Niederbudde, E.A.; Kussmaul, H.   DE 161020/77/0001
Tonmineraleigenschaften, Tonmineralumwandlungen und
Tonmineralverlagerungen in Parabraunerden unter Acker und
Laubwald aus Löss in Unterfranken. **Clay mineral properties,
transformations and migration in grey brown podzolic soils
below arable and deciduous forest land derived from loess in
Lower Franconia.**

600 Fischer, W.R.   DE 161020/77/0003
Bildung von Eisenoxiden in Gegenwart von organischen
Substanzen des Bodens. **Formation of iron oxides in the
presence of soil organic substances.**

601 Schweisfurth, R.; Arnold, M.   DE 167150/77/0003
Mikrobiologische und geochemische Untersuchungen über die
Anreicherung von Schwermetallen in Manganoxiden.
**Microbiological and geochemical experiments on the
concentration of heavy metals in manganese oxides.**
Publications.

602 Rietz, E.   DE 201010/77/0001
Wechselwirkung anorganischer und organischer
Bodenbestandteile im Verlauf von deren Umwandlung und
Transport. **Interrelations between inorganic and organic
components in soil during their transformation and transfer.**

603 Tietjen, C.; El–Bassam, N.; Keppel, N.; Esser, J.
   DE 201040/74/5002
Kontamination von Boden– und Grundwasser durch
Schwermetalle aus Siedlungsabfällen. **Contamination of soils
and ground water with heavy metals from domestic wastes.**

604 Schoenhard, D.   DE 215060/77/0005
Vergleich der Wirkung von Kalk und Lewatit bei der Bindung
von Schwermetallen im Boden. **Comparison of effects of lime
and lewatit in binding of heavy metals on soil.**

605 Hoffmann, G.; Wiechens, E.   DE 501010/78/0003 N
Vergleichende Untersuchungen zur Methodik der Bestimmung
des pflanzenaufnehmbaren Phosphors in Böden. **Comparative
investigations of methods to determine the plant–available
phosphor in soils.** Publications.

606 Schlichting, E.; Müller, S.　　　　DE 501502/77/0003
Austrag umweltrelevanter Haupt- und Spurenstoffe aus
naturnahen Ökochoren des schwäbischen Keuperberglandes
und Albvorlandes. **Output of main and trace elements from
undisturbed forest sites of the Keuper and Jura region of South
West Germany.**

607 Diez, T.　　　　DE 502055/74/0008
Ermittlung des Nährstoffpotentials und des
Nährstoffnachlieferungsvermögens ackerbaulich genutzter
Böden des Tertiär- hügellandes. **Determination of nutrient
potential and nutrient activation in cultivated soils in tertiary
hilly country.**

608 Teuteberg, W.; Grunwaldt, H.–S.; Patzke, W.
　　　　DE 511052/77/0001
Nitratuntersuchungen an Ackerstandorten
Schleswig–Holsteins. **Nitrate content of fields in
Schleswig–Holstein.** Publications.

609 Munk, H.; Lueg, F.　　　　DE 901000/71/0001 N
Umwandlung von Düngephosphaten im Boden und Methoden
zur Kennzeichnung der Pflanzenverfügbarkeit. **Fertilizer
transformation in soils and methods for indicating plant
availability.**

610 Munk, H.; Bärmann, C.　　　　DE 901000/78/0001 N
Einfluss kalkhaltiger Mineraldünger auf den Reaktionszustand
und das Ertragspotential von Ackerböden. **Influence of
lime–containing fertilizers on reaction and yield potential of
soils.**

611 Werner, W.　　　　DE 903000/78/0001 N
Untersuchungen zur P–Dynamik tropischer Böden – Latosole
–. **Investigations on the phosphorus dynamics of tropical soils –
latosols –.**

612 Werner, W.; Solle, A.　　　　DE 903000/78/0004 N
Kennzeichnung des P–Versorgungszustandes hoch gedüngter
Böden durch verschiedene chemische
Bodenuntersuchungsmethoden. **Evaluation of the P–status of
highly fertilized soils by different chemical soil tests.**

613 Jensen, J.　　　　DK 010117/79/0001 N
Anvendelse af Dolomitkalk som jordbrugskalk. **The use of
dolomite as a liming amendment in soil.**

614 Rasmussen, K.　　　　DK 030106/71/0003
Calciumcarbonats omsætninger i stærkt sur jord. **Calcium
carbonate turnover in strongly acidified soil.**

615 Borggaard, O.K.　　　　DK 030106/73/0002
Ekstraktion af jordbundens plantenæringsstoffer med EDTA.
Extraction of plant nutrients from the soil with EDTA.

616 Petersen, L.; Rasmussen, K.　　　　DK 030106/76/0006
Studier over lerfraktionens mineralogiske sammensætning i to
fluvio–glaciale sedimenter fra Østgrønland. **Studies of the
mineralogical composition of the clay fraction from two
fluvio–glacial sediments in eastern Greenland.**

617 Møberg, J.P.　　　　DK 030106/79/0004 N
Undersøgelse af den mineralogiske sammensætning af
forskellige jorder i Tanzania. **Determination of the
mineralogical composition of various soils in Tanzania.**

618 Larsen, S.　　　　DK 030148/76/0008
Faktorer, der påvirker jord som fosforkilde for planter.
Factors affecting soil as a source of phosphorus for plants.

619 Hovmand, M.F.　　　　DK 030202/77/8145
Bestemmelse af spormetalbalance for landbrugsjord.
Determination of trace element balance for agricultural soil.

620 Chassin, P.　　　　FR 010103/70/0652
Etude des propriétés d'adsorption des silicates pour des
matières organiques polaires de faible poids moléculaire.
**Studies on adsorption of polar organic matter with low
molecular weights by silicates.** Publications.

621 Henin, S.; Besson, H.　　　　FR 010103/70/6334
Formation et transformation des hydrocarbonates et des
phyllites. **Formation and transformation of hydrocarbonates
and phyllites.**

622 Chassin, P.　　　　FR 010103/71/0651
Etude de l'hydratation de la montmorillonite en présence de
matières organiques adsorbées. **Studies on montmorillonite
hydration when organic matter is adsorbed.** Publications.

623 Mamy, J.; Calvet, R.　　　　FR 010103/71/0657
Etude des propriétés diélectriques de l'eau adsorbée sur les
silicates phylliteux. Etats de l'eau. **Studies on dielectric
properties of water adsorbed on phyllite silicates. State of
water.** Publications.

624 Boniface, Mme; Trocme, S.　　　　FR 010103/72/0646
Pertes par drainage des éléments minéraux en lysimètres.
Losses by leaching of mineral elements in lysimeters.

625 Calvet, R.; Vallet, (Melle).　　　　FR 010103/72/0648
Adsorption des herbicides sur les constituants minéraux des
sols. **Herbicide adsorption by mineral constituents of soils.**

626 Boniface, M.; Trocme, M.　　　　FR 010103/72/6217
Bilan des éléments minéraux en sols cultivés ou non, évalué
par lysimétrie. **Lysimeter studies on mineral balance in bare
and cultivated soils.**

627 Calvet, R.; Terce, M.　　　　FR 010103/72/6330
Phénomènes d'adsorption des herbicides sur les argiles.
Herbicide adsorption phenomenon on clays.

628 Mamy, J.　　　　FR 010103/72/6340
Etude des modifications structurales des micas au cours de
leurs altérations, et structure des courbes d'eau adsorbées.
**Study of structural changes of micas during weathering and
structure of adsorbed water layers.**

629 Mamy, J.　　　　FR 010103/73/6341
Fixation du potassium et transformations structurales dans la
montmorillonite. **Potassium fixation and structural
transformations in montmorillonite.**

630 Pedro, G.; Tessier, D.　　　　FR 010103/73/6342
Comportement et organisation des matériaux argileux en
fonction de leur humidité. **Behaviour and organization of clay
material as related to water content.**

631 Baize, D.　　　　FR 010103/73/6352
Evolution pédologique des matériaux argileux. **Pedologic
evolution of clay minerals.**

632 Pedro, G.; Chauvel; Berrier   FR 010103/74/6343
Étude des causes de l'activation et de l'inactivation des particules argileuses dans les sols. **Study on causes for activation and inactivation of clay particles in soil.**

633 Robert, M.; Veneau, G.   FR 010103/74/6345
Variations du complexe adsorbant des minéraux argileux 2/1 au cours de l'altération. **Variations of adsorbed complex in clay minerals 2/1 during weathering.**

634 Arvieu, J.C.   FR 010502/67/0521
Cinétique des réactions des engrais phosphatés en sol calcaire. Physico–chimie et assimilabilité des produits formés. **Kinetic of reactions of phosphate fertilizers in calcareous soils. Physico–chemical properties and availability of phosphoric compounds.** Publications.

635 Arvieu, J.C.; Bouvier, O.   FR 010502/67/6251
Etude cinétique des transformations chimiques subies par les engrais phosphatés dans les conditions des sols calcaires. **Kinetics of chemical changes underwent by phosphate fertilizers under calcareous soils conditions.**

636 Gouny, M.   FR 010601/70/0528
Soil salinity and field crops behaviour.

637 Souty, N.   FR 010601/74/6268
Etude du comportement et des transformations du carbonate de calcium sous l'influence de l'évolution de la matière organique. **Behaviour and transformation of calcium carbonate under the influence of organic matter evolution.**

638 Delas, J.; Molot, P.; Juste, C.   FR 010704/59/6070
Problèmes posés par les sols viticoles acides. **Problems in vineyards with acid soils.**

639 Gachon, L.; Triboi, E.   FR 010802/60/6103
Etude comparative de systèmes culturaux en sol argilo–calcaire de Limagne. **Comparative study of cultural systems in calcareous clay–soil of Limagne.**

640 Gachon, L.   FR 010802/65/0567
Diagnostic et entretien de la fertilité phosphorique des sols. **Diagnosis and maintenance of phosphoric level in soils.**

641 Conesa, A.; Mettauer, H.   FR 010901/75/6125
Etude de la fertilité phosphatée des sols de l'est de la France. **Study of soil phosphate fertility.**

642 Grosman, R.; Lefebvre, J.M.; Libois, A.; Concaret, J.   FR 011001/75/6290
Essai en pots sur interaction soufre et divers éléments. **Pot experimentation on the interactions between sulphur and various elements.**

643 Servant, J.   FR 011211/64/0615
Contribution à l'étude des sols salins et sols salins à alcali. **Contribution to the study of saline and alkaline soils.** Publications.

644 Dupuis, M.   FR 011211/69/0617
Répartition granulométrique des Carbonates (Ca,Mg) dans les sols calci–magnesiens. **Size particles repartition of carbonates (Ca, Mg) in calcareous magnesium soils.** Publications.

645 Dupuis, M.   FR 011211/69/6315
Répartition granulométrique des carbonates dans les sols calcimagnésiques. **Particle size distribution of carbonates in calcium–magnesium rich soils.**

646 Servant, J.   FR 011211/70/6323
Contributiona l'étude pédologique des terrains halomorphes en zone tempérée et méditerranéene. L'exemple des sols sales du Sud et du Sud–Ouest de la France. **Pedological study of halomorphic soils in temperate and mediterranean zone. Example of saline soils in the south and the southwest of france.**

647 Bonfils, P.   FR 011211/73/6305
Extractions comparées et dosages de S 02, AL 203, Fe 203 Ti 02 dans les roches et dans les sols. **Comparative extraction and titration of s 02, al203, Fe 203, Ti 02 in rocks and soils.**

648 Servant, J.   FR 011211/73/6324
Influence des données pédologiques et en particulier de la salinité sur la culture de la canne de provence (pâte à papier) en Camargue. **Influence of pedological data in particular salinity on giant reed, arundo donax (papermaking) in camargue.**

649 Servant, J.; Concaret, J.   FR 011211/74/6325
Expérimentations sur la mise en valeur des terrains sales en Camargue. **Saline soil reclamation in camargue.**

650 Bosc, M.; Blanchet, R.   FR 011401/70/0633
Diffusion et convection des ions vers la rhizosphère, et répercussions sur l'absorption. **Ion diffusion and convection to the rhizosphere, effects on absorption.**

651 Fourbet, J.F.; Huet, Ph.   FR 011801/58/9074
Recherche de la teneur minimum d'un sol en p et k nécessitant des apports de ces éléments dans diverses situations culturales en sol de limon argileux. **Research on minimum amount of P and K in a clay silt soil, with several tilling situations, requering the use of P and K manure.**

652 Maurice, J.   FR 012208/66/0627
Géochimie du bore. **Boron geochemistry.** Publications.

653 Duval, L.   FR 012208/71/0624
Etude du molybdène des sols et des roches. **Studies on molybdenum in soils and rocks.**

654 Durand, R.   FR 012210/69/6278
Contribution à l'étude géochimique de l'alteration des roches carbonatées et de la pédogénèse en milieu calcimagnesique. **Contribution to a geochemical study of carbonated rocks weathering and of pedogenesis in a calcium–magnesium rich medium.**

655 Dutil, P.   FR 012210/74/6274
Evolution des phosphates en sols de craie. **Evolution of phosphates in chalky soils.**

656 Hiroux, G.; Desmet   FR 012222/73/6045
Etude in situ de la répartition de l'azote minéral de profils de sols sous blé. **In situ study of mineral N distribution of soil profils bearing wheat.**

657 Dowdell   GB 010503/00/0006 R
Denitrification and leaching of nitrogen fertilizers.

658 Burford; Mercer GB 010503/77/0013 R
The fate of fertilizer nitrogen applied to cereals on shallow soil overlying chalk , especially leaching loss.

659 Dowdell GB 010503/79/0016 N
Effects of soil management on soil nitrogen.

660 Jones GB 011201/00/0001 R
Form of occurence of N.K.S. and trace elements in soil and availibility to forage plants.

661 Mattingly; Talibudeen GB 012003/00/0011 R
Chemistry of soil phosphates.

662 Talibudn; Goulding GB 012003/00/0013 R
Nutrient cations relations in crops and soils, the calorimetry of their sorption and release.

663 Ashworth GB 012003/00/0017 R
Sorption isotherms and calorimetry of sorption of ammonia from gas and aqueous phases.

664 Addiscott; Ashworth GB 012003/00/0028 R
Efficiency of nitrogen fertilisers.

665 Bolton GB 012003/00/0032 R
Compare magnesium reserves in soil types and test their availability to crops.

666 Bolton GB 012003/00/0033 R
Effects of liming on soil composition, crop growth and nutrient uptake.

667 Newman; Talibureen GB 012003/00/0054 R
Contribution of soil minerals containing potassium to the fraction of potassium active in soils.

668 Williams GB 012003/00/0077 R
Distribution of elements within soils and the association of the elements with various soil components.

669 Briggs GB 012005/00/0047 R
Factors influencing the effectiveness of nitrification inhibitors applied to soil.

670 Draycott; Durrant GB 012301/00/0003 R
Soil analysis and phosphorus response.

671 Williams, B. GB 030401/00/0016 R
Nitrogen mineralization: factors controlling release of nitrogen immobilised in peat and humus.

672 Goodman GB 030402/00/0003 R
Forms of occurence of trace elements in soils and the mechanism of their movement towards the plant root.

673 Anderson GB 030403/00/0003 R
Nitrogenous constituents of soils, peat and leaf litter: relationships with co–occurring macromolecules.

674 Anderson GB 030403/00/0017 R
The nature and properties of organically bound phosphate in soils.

675 Williams; Craigmyle GB 030406/00/0001 R
Inorganic soil phosphorus and sulphur evaluation of available forms and effects of fertilizers.

676 Sinclair GB 030406/00/0011 R
Soil magnesium and potassium: distribution, solubility and availability in different soil series.

677 McLaren GB 060115/00/0001 R
Soil and plant sulphur.

678 Robinson GB 060220/00/0006 R
Effect of waste management practices on nitrogen transformation by the soil microflora.

679 Voss GB 060306/00/0007 R
Survey of trace element levels in soil.

680 Flegg, A.M.; Bell, A. IE 040101/79/9214 N
The identification and evaluation of native sources of industrial talc for use in the fertilizer industry.

681 Magaldi, D.; Raglione, M.; Lulli, L.; Rodolfi, G.
 IT 020100/78/0003 R
Studio dei suoli soggetti a limitazioni d'uso determinate da forti concentrazioni di carbonati. **Study of soils with strong limitations by heavy carbonatic accumulations.**

682 Spallacci, P. IT 020500/70/0001
Studio della lisciviazione e del bilancio dei principi nutritivi nella monocoltura di mais da granella. **Leaching and fertility balance of a soil under continuous grain–corn.**

683 Lopez, G.; Colucci, R. IT 020500/75/0005
Stato del fosforo nei terreni. **Phosphorus forms in soils.**

684 Biancardi, E.; Leoni, O. IT 021100/79/0014 N
Dinamica del sodio nel rapporto terreno–bietola. **Sodium dynamics in the soil : sugar beet ration.**

685 Casalicchio, G.; Rosciglione, L.; Vianello, G.; Del Monte, M. IT 040202/78/0002
I microelementi nei suoli della regione Emilia – Romagna. **Microelements in the soils of the Emilia–Romagna region.**

686 Eschena, T. IT 040705/74/0548
Interazione fra H–argille e sali basici di alluminio. **Interactions between H–clays and aluminium basic salts.**

687 Violente, P.; Violante, A. IT 040705/75/0001
Influenza delle argilla sulla cristallizzazione di AL (OH)$_3$ in presenza di anioni organici. **Clay activity on AL (OH)$_3$ crystallization, in presence of organic anions.** Publications.

688 Eschena, T.; Palmieri, F. IT 040705/76/0001
Interazioni minerali argillosi e molecole organiche. **Interactions between clay minerals and organic molecules.** Publications.

689 Marano, B.; Palmieri, F.; Palmieri, G.
 IT 040705/78/0002 N
Concimazione minerale e movimento dei soluti in suolo limo–argilloso. **Mineral fertilization action on solute movement in clay soils.**

690 Violante, P. IT 040705/78/1120 N
I rapporti di superficie fra minerali argillosi e idrossidi di alluminio. **Surface relation of clay minerals to aluminium**

hydroxide.

691 Fierotti, G.　　　　　IT 040904/77/0628 R
Classificazione e cartografia a varia scala dei suoli della collina argillosa siciliana. **Classification and various scale mapping of soils in the Sicilian clay hills.**

692 Tafuri, F.; Businelli, M.　　　IT 041004/79/0002 N
I microelementi assimilabili nei terreni umbri. **Assimilable microelements in Umbrian soils.**

693 Riffaldi, R.　　　　　IT 041103/74/0611
Interazioni fra cadmio e frazioni umiche del terreno. **Interactions between cadmium and humic fractions of soil.**

694 Veniale, F.　　　　　IT 041804/77/0650
Mineralogia dei costituenti argillosi e parametri geotecnici. **Mineralogy of clay components and geotechnical parameters.**

695 Sequi, P.　　　　　IT 060400/77/0865
Complessi organo–minerali del terreno. **Mineral–organic complexes in the soil.**

696 Sequi, P.　　　　　IT 060400/77/0868
Fertilità del terreno ed attività enzimatiche. **Soil fertility and enzymatic activity.**

697 Malquori, A.　　　　　IT 061200/74/0142
Smectiti nei terreni italiani. **Smectites in italian soils.**

698 Malquori, A.　　　　　IT 061200/77/0944
Interazione terreno–erbicidi: Valutazione di residui di erbicidi nel suolo e loro assorbimento su materiali inorganici amorfi e cristallini. **Soil–herbicides interaction: evaluation of herbicide residues present in the soil, their absorption by inorganic amorphous and cristalline substances.**

699 Malquori, A.　　　　　IT 061200/77/0945
Interazioni terreno–erbicidi: 1) Influenza di carboni attivi sulla eliminazione di residui di erbicidi nel terreno. **Soil – herbicides interaction: 1) Influence of activated charcoals on the elimination of herbicide residues from the soil.**

700 Malquori, A.　　　　　IT 061200/77/0946
Attacco biologico di alluminio silicati ed interazioni fra i colloidi minerali ed alcuni composti organici presenti nel suolo. **Biological impact of aluminium silicates, interactions of mineral colloids and other organic compounds present in the soil.**

701 Sissingh, H.A.　　　　NL 010103/59/0476
Ontwikkeling en toetsing van grondonderzoek op basis van de meest oplosbare fractie van het fosfaat in de grond. **Development and evaluation of a soil extraction method for the determination of the actual available fraction of soil phosphorus.** Publications.

702 Sissingh, H.A.　　　　NL 010103/62/0541
Chemisch onderzoek over het in oplossing gaan en fixeren van fosfaat in grond. **Chemical investigations on the solubility and fixation of phosphates in the soil.** Publications.

703 Kolenbrander, G.J.　　　NL 010103/70/3153 R
De invloed van de grondwaterstand op de verliezen van stikstof en enkele andere plantevoedende stoffen in lysimeters. **Lysimeter investigations. Influence of depth of water table on losses of nitrogen and some other nutrient elements.**

704 Groot, A.J. de　　　　NL 010103/71/3004
Zware metalen in fluviatiele en mariene ecosystemen. **Heavy metals in fluvial and marine ecosystems.** Publications.

705 Prummel, J.　　　　　NL 010103/71/3312
Veranderingen in de fosfaattoestand van bouwland op lange termijn. **Long–term changes in the phosphorus status of arable land.** Publications.

706 Goor, B.J. van　　　　NL 010103/77/7689 R
Inventarisatie zware metalen in nederlandse gronden en gewassen. **Survey of heavy metals in Dutch soils and crops.**

707 Veen, J.A. van　　　　NL 010110/70/7937
Experimentele analyse en computer simulatie van de stikstofcyclus in de atmosfeer–bodem–plant relatie. **Experimental analysis and computer simulation of nitrogen cycling in atmosphere–soil–plant interrelations.** Publications.

708 Poelstra, P.　　　　　NL 010110/75/7285
A survey of the concentration of Pu, Am and radioactive metals in soils of the Rhine river delta. Publications.

709 Poelman, J.N.B.; Kanters, H.L.　　NL 010119/73/5970
Onderzoek naar het voorkomen en de verbreiding van kalkrijke Maasafzettingen. **Occurrence and distribution of calcareous deposits from the Maas.**

710 Breeuwsma, A.　　　　NL 010119/75/7547
Pyrietbepaling in veengronden. **Determination of the pyrite content of peat soils.** Publications.

711 Breeuwsma, A.　　　　NL 010119/77/7789 R
Chemische bepalingen ten behoeve van de bodemkartering. **Chemical determinations for soil survey purposes.** Publications.

712 Poelman, J.N.B.　　　　NL 010119/77/7803 R
Fosfaatkartering Wijk bij Duurstede. **Mapping of the occurence of phosphate near Wijk bij Duurstede.**

713 Kamphorst, A.　　　　NL 020007/72/4953
Zoutbeweging in relatie tot de vochtbeweging in het bodemprofiel. **Salt transport as related to the transport of water in the soil.** Publications.

714 Haan, F.A.M. de; Beek, J.; Riemsdijk, W.H. van
　　　　　　　　　　NL 020007/73/4957
Accumulatie van fosfaat in de bodem. **Accumulation of phosphates in soil.** Publications.

715 Haan, F.A.M. de; Harmsen, K.; Lexmond, Th.; Keizer, M.G.　　　　　　　　NL 020007/73/4958
Gedrag van zware metalen in de bodem. **Behaviour of heavy metals in soil.** Publications.

716 Diest, A. van　　　　　NL 020007/73/4965
Relaties tussen macro– en microelementen in grond en gewas. **Relationships between macro– and micronutrients in soils and crops.** Publications.

717 Janssen, B.H.　　　　　NL 020007/74/4967
Humusopbouw en –afbraak in verband met permanente plantenteelt in de humide tropen. **Humus formation and decay in relation to continuous growing of annual crops in the tropics.**

D 1100 – Soil science

Publications.

718 Bruggenwert, M.G.M. NL 020007/74/6348
Adsorptie van A1–ionen aan kleimineralen. **Adsorption of A1–ions onto clay–minerals.** Publications.

719 Bruggenwert, M.C.M. NL 020007/76/6773
Invloed van Al op de adsorptie van kat– en anionen in de bodem. **Influence of Al on the adsorption of positive and negative ions in the soil.**

720 Lexmond, Th. NL 020007/76/6774
Gedrag in de bodem van Cu uit varkensdrijfmest. **Mobility in soil of Cu from pig manure.**

721 Eyk, D. van der NL 020007/76/6778
Relaties tussen fosfaatfixatie en bemestingsregiem op enkele gronden in Z.W.–Kenya. **Relations between phosphate fixation and fertilizing measures on some South Western Kenyan soils.**

722 Vergouwe, A.A. NL 020008/77/7568
Mineralogisch onderzoek van typische mineralen die voorkomen in zoute gronden. **Mineralogical investigation of minerals in saline soils.**

723 Leffelaar, P.A. NL 020054/78/8858 N
Het ontstaan van plaatselijke anaerobie in niet met water verzadigde gronden in verband met denitrificatie. **The development of partial anaerobiosis in non water saturated soils with respect to denitrification.**

724 Ente, P.J. NL 040007/72/4337
Het voorkomen van zware metalen in water, bodem en organismen in het IJsselmeergebied. **The presence of heavy metals in water, soil and organisms in the IJsselmeer areas.**

Soil composition – organic (B 1112)

See also 10, 169, 585, 600, 602, 607, 618, 637, 654, 674, 683, 687, 688, 693, 695, 700, 711, 834, 1067, 1081, 1104, 2609, 2988, 4972, 5062, 5172, 5298, 5302, 5332, 5604

725 De Borger, R.; Pussemier, L. BE 100000/72/0015 R
Studie van de adsorptie van pesticiden aan de organische stof van de bodem. **Study of the adsorption of pesticides by soil organic matter.** Publications.

726 Zakosek, H.; Schröder, D.; Bäll, H. DE 111050/77/0002
Beeinflussung der Zelluloseabbauaktivität durch Herbizidanwendung im Weinbau. **Influence of herbicide application on the cellulose decomposition activity in vineyards.**

727 Schröder, D. DE 111050/78/0002 N
Der Einfluss agrochemischer Substanzen auf den Stroh– und Zelluloseabbau im Boden. **The influence of agrochemical compounds on the breakdown of straw and cellulose in soil.**

728 Sauerbeck, D. DE 111101/72/0007
Radiometrische Untersuchungen zur Humusbilanz von Böden. **Radiometric investigations into the turnover of soil organic matter.** Publications.

729 Sauerbeck, D.; Gonzalez, M.A. DE 111101/78/0001 N
Umsetzung C–14–markierter Pflanzenrückstände in tropischen Böden von Costa Rica. **Turnover of C–14–labelled plant residues in tropical soils of Costa Rica.** Publications.

730 Kickuth, R.; Aldag, R. DE 132034/72/0002 R
Die ökochemische Funktion von – proteinogenen und nichtproteinogenen – Aminoverbindungen in unbelasteten und belasteten Böden und Gewässern 1972. **The ecochemical function of – proteinogenous and non–proteinogenous – amino compounds in unstressed and stressed soils and waters.**

731 Babel, U. DE 144000/74/0001
Morphologie und Dynamik von Humusprofilen in Waldbeständen mit stark vermindertem Zuwachs. **Morphology and dynamics of humic soil profiles in forest stand with strong reduction in growth.**

732 Ottow, J.C.G.; Schütt, C. DE 144320/74/0003
Mikrobiologische Untersuchungen an Manganknollen aus dem Meer. **Microbiological investigations on ferromanganese nodules from the sea bottom.**

733 Niederbudde, E.A. DE 161020/75/0007
Diskriminanzanalytische Trennung von Bodenformen sowie von Böden und Sedimenten mittels Tonmineraleigenschaften und Nährstoffen. **Discriminant analytical separation of soil forms as well as of soils and sediments by clay mineral properties and nutrients.**

734 Niederbudde, E.A.; Fischer, W.R.; Pfanneberg, T.
  DE 161020/77/0005
Wirkung von Tonmineralen unterschiedlichen NH4–Fixierungsvermögens auf die N–Mineralisierung und N–Immobilisierung beim Abbau von organischer Substanz im Boden und Zufuhr von NH4. **Effects of clay minerals of different NH4 fixing capacity on N–mineralization and N–immobilization in decomposition of organic matter in soil and in urea supply.**

735 Rietz, E. DE 201010/71/4001
Charakterisierung von polymeren organischen Bodensubstanzen durch Bestimmung der Teilchengewichtsverteilung aus Sedimentations– und Diffusions–Geschwindigkeitsmessungen. **Characterization of polymeric organic substances in soil by determination of distribution of weight of particles from measurement of sedimentation and diffusion speed.**

736 Salfeld, J.–C. DE 201010/71/4002
Methodik der Humusanalyse. **Methods of humus analysis.**

737 Altemüller, H.–J. DE 201010/71/4011
Polarisations– und phasenoptische Merkmale toniger und organischer Bodenanteile und ihre Bedeutung für die Gefügebildung und Gefügeeigenschaften. **Characteristics of polarization and phase optics of clayey and organic parts of soil and their importance to formation and properties of soil structure.**

738 Ellwardt, P.–C. DE 201010/74/0004
Organische Bodensubstanz und Umweltschutz: Analytik von cancerogenen polycyclischen aromatischen Kohlenwasserstoffen in mit Müll–Klärschlamm–Kompost behandelten Böden und in den darauf gezogenen Nutzpflanzen. **Organic matter in soil and environment protection: analytics of cancerogenic, polycyclic aromatic hydrocarbons in soils treated with sewage sludge composts and in plants grown on them.**

739 Salfeld, J.–C. DE 201010/77/0004
Humusdynamik in Ackerböden. **Humus dynamics in arable soils.**

740 Salfeld, J.–C. DE 201010/77/0005
Stoffumsetzungen im System der organischen Stoffe eines Bodens. **Transformation of substances in system of organic matters in soil.**

741 Haider, K. DE 201010/77/0006
Einfluss von neugebildeten Huminstoffen auf die Austauschbarkeit von Natriumionen in natron–alkalischen Böden. **Influence of new–formed humic substances on the interchangeability of Na ions in sodium carbonate–alkaline soils.**

742 Haider, K. DE 201010/77/0007
Biochemischer Abbau von Pflanzenbestandteilen und Umwandlung in die organische Substanz des Bodens. **Biochemical decomposition of plant components and transformation into organic matter of soil.**

743 Haider, K. DE 201010/77/0008
Abbau von Ligninen und technisch abgewandelten Ligninen durch isolierte Organismen und im Boden. **Breakdown of lignins and technically modified lignins by isolated organisms and in soil.**

744 Haider, K. DE 201010/77/0010
Mikrobieller Abbau chlorierter Cycloalkane und Aromate unter anaeroben und aeroben Bedingungen. **Microbial breakdown of chlorinated cycloalkanes and aromatics under anaerobic and aerobic conditions.**

745 Haider, K. DE 201010/77/0011
Bindung und Komplexierung chlorierter Phenole und Anilin–Derivate in der organischen Bodensubstanz. **Binding and complex of chlorinated phenols and anilinederivatives in organic matter of soil.**

746 Söchtig, H. DE 201010/77/0012
Kreislauf von markiertem Stickstoff im Ökosystem Boden ohne und mit zusätzlicher organischer Düngung. **Cycle of labelled nitrogen in ecosystem of soil without and with additional organic fertilization.**

747 Harms, H. DE 201010/77/0013
Wirkung reduzierter Stickstoffverbindungen –NH+4, N–Lignin auf die Denovo–Synthese von Purin–Nukleotiden. **Effects of reduced nitrogen compounds –NH+4, N–lignin on the denovo–synthesis of purine–nucleotides.**

748 Douglas; Söchtig, H.; Flaig, W. DE 201010/77/0015
Einfluss niedermolekularer Verbindungen aus der organischen Bodensubstanz auf Enzyme insbesondere im Stickstoffkreislauf. **Influence of low–molecular compounds from organic matter in soil on enzymes especially in nitrogen cycle.**

749 Tietjen, C. DE 201040/77/0001
Humuswirtschaft bei verstärktem Getreidebau. **Humus balance in intensified cereal growing.**

750 Beinhauer, R. DE 301060/75/0004
Untersuchungen zum Temperatur– und Strahlungshaushalt von brachfallenden Niedermoorböden zum Zwecke der Wiederaufforstung oder Neukulturen. **Research on temperature and radiation of fallow fen soils for the purpose of forest or other recultivation.**

751 Edwards, C.A.; Bauchhenss, J. DE 502055/77/0004
Der Einfluss von organischem Material im Boden auf die Bodenfauna. Eventuelle Beeinträchtigung des Pflanzenbaues durch Massenvermehrungen. **The role of organic matter in pest and disease problems in agriculture. Possible impair of plant production by mass propagation.**

752 Kuntze, H.; Schwaar, J.; Eggelsmann, R. DE 507103/75/0001
Regeneration von teilabgetorftem Hochmoor. **Regeneration of partial cut off peatlands.**

753 Werner, W.; Wienschierz, H. DE 903000/78/0002 N
Sofortwirksamkeit und Nachwirkung von Güllephosphaten. **Immediate and long term phosphorus effectiveness of liquid manures.**

754 Jørgensen, S.S.; Pedersen, B.; Willems, M. DK 030106/79/0003 N
Spektroskopisk og kemisk karakterisering af nogle danske jordtypers humusfraktion. **Spectroscopic and chemical characterisation of organic material from Danish soil types.**

755 Sørensen, H. DK 030146/74/0019
Humusstoffernes dannelse og biostabilitet. **The formation and biostability of humus.**

756 Robert, M.; Razzaghe, K.; Vicente, M.A. FR 010103/72/6346
Action des composés organiques dans la pédogénèse. **Action of organic compounds in pedogenesis.**

757 Lemaire, F.; Salette, J.; Morlat, R. FR 010403/76/6026
Entretien du taux de matière organique dans les sols supportant des cultures pérennes. **Organic matter levels in perennial crops soils.**

758 Peyriere, J.; Brun, R.; Rico, F. FR 010616/73/9086
Influence et évolution de la matière organique dans les sols de serre. **Influence and evolution of organic matter in glasshouse soils.**

759 Delas, J.; Molot, P. FR 010704/60/6071
Entretien humique des sols viticoles. **Conservation of soil organic matter in vineyards.**

760 Juste, C.; Dureau, P. Mme; Soyer, J.P.; Solda; Menet, M.; Lubet FR 010704/63/6084
Etude de l'influence de différents systèmes de culture sur l'évolution de la matière organique des sols du sud–ouest atlantique. **Influence of cropping systems on the evolution of soil organic matter in the western south.**

761 Juste, C.; Delas, J. FR 010704/67/0553
Etude des complexes organo–minéraux des sols. **Study on organo–mineral compounds of soils.** Publications.

762 Durand, J.H. FR 010704/73/6094
Migration du calcaire en milieu calcaire. **Lime migration in calcareous environment.**

763 Fournier, J.C. FR 011002/75/6297

Etude de la formation et de la dégradation de divers métabolites des phénylurées substituées dans le sol. **Study on formation and degradation of some metabolites of substituted phenylurea in soil.**

764 Dumas, Y.; Clairon, M.; Sobesky, O. Mme
FR 011602/76/6036
La matière organique dans les sols argileux des Antilles. **Organic matter in Caribbean Clay soils.**

765 Morel, R.; Chabouis, Mme   FR 011801/00/0621
Etude de la dynamique de l'azote et du carbone organique du sol dans le dispositif Deherain. Essai de longue durée. **Studies on organic nitrogen and carbon dynamics in the soil (Deherain long term experiment).** Publications.

766 Studer, R.   FR 012209/74/6133
Bilan de l'acide phosphorique en sols calcaires – Applications à la rationalisation de la fumure phosphatée en rotations essentiellement céréaliéres. **Phosphorus balance in superficial calcareous soils.**

767 Muller, J.   FR 012210/71/6271
Matière organique en sol de craie. **Organic matter in chalky soils.**

768 Hebert, J.; Remy, J. C.   FR 012226/58/0600
Observations sur l'évolution de l'azote en plein champ. **Observations on nitrogen evolution in the field.**

769 Remy, J. C.   FR 012226/60/0603
Action de l'azote sur la conservation de la matière organique. **Effect of nitrogen on the conservation of organic matter.**

770 Hebert, J.; Remy, J.C.   FR 012226/72/0601
Pertes par drainage d'éléments eutrophisants. **Leaching of eutrophic elements.**

771 Whitehead   GB 011201/00/0008 R
Soil organic matter in relation to soil physical conditions and productivity.

772 Johnston; Mattingly   GB 012003/00/0006 R
Soil organic matter in relation to organic manure and cropping.

773 Jenkinson   GB 012003/00/0064 R
Age, stability and turnover of organic matter.

774 North   GB 012008/00/0012 R
Complex formation between high molecular weight polysaccharides and clay domains.

775 Vaughan; Anderson H   GB 030403/00/0001 R
Chemical and biochemical investigations of organic material of microbial origin.

776 Linehan   GB 030403/00/0004 R
Nature,distribution and properties of humic soil substances.

777 Cheshire   GB 030403/00/0005 R
The synthesis and degradation of polysaccharides and related constituents of soil organic matter.

778 Vaughn; Ord   GB 030403/00/0011 R
The effects of organic constituents of soil on biochemical processes in plants.

779 Sparling; Darbyshire   GB 030405/77/0012 R
Microbial degradation of soil organic matter.

780 Scott; Craigmyle   GB 030406/00/0002 R
Organic phosphorus and sulphur in relation to soil type and nutrient supply.

781 Swift   GB 060114/00/0002 R
Influence of soil organic matter on soil structure and plant nutrition.

782 Gregori, E.; Miclaus, N.; Bidini, D.   IT 020100/78/0002 R
Ciclo della sostanza organica e valutazione della fertilità stazionale mediante indici biologici, bio–chimici e fisici in suoli sotto foresta di faggio (Fagus Sylvatica). **Turn–over of organic matter and evaluation of site fertility by biological, bio–chemical and physical indexes in beech–forest soils (Fagus Sylvatica).**

783 Nigro, C.; Cavallari, L.; Marini Bettolo, G.; Castagnola, M.; Liberti, A.   IT 020200/77/0003 N
Bilancio umico azotato del terreno. **Balance of humus and nitrogen of soil.** Publications.

784 Perniola, M.   IT 020500/74/0601
I costituenti la sostanza organica del terreno. Approfondimento delle conoscenze sui componenti la frazione lipidica. **Components of soil organic matter. Further studies on components of the lipid fraction.** Publications.

785 Perniola, M.; Convertini, G.; Ferri, D.
IT 020500/76/0006 N
Sostanze fenoliche e carboidrati nella sostanza organica del terreno.. **Phenolic substances and carbohydrates in the soil organic matter..**

786 Perniola, M.   IT 020500/78/1084 N
I costituenti della sostanza organica del terreno, approfondimento delle conoscenze sui componenti della frazione lipidica. **Organic matter components of the soil; further investigations on the lipid fraction components.**

787 Senesi, N.   IT 040104/78/1111 N
Componenti fisiologicamente attivi dei composti umici del terreno. **Physiologically active components of the soil humic compounds.**

788 Toderi, G.; Giordani, G.   IT 040201/66/0001
Effetti della letamazione e dell'interramento degli stocchi di mais e della paglia di grano in una successione colturale mais–frumento. **Effects of manuring and ploughing under corn stalks and wheat straw in a corn–wheat rotation.** Publications.

789 Cavazza, L.   IT 040201/77/0140 R
Influenza dell'interramento della paglia sull'azione dello azoto e sulla fertilità del terreno. **Influence of straw ploughing–isa on nitrogen action and on soil fertility.**

790 Casalicchio, G.   IT 040202/74/0528
Ricerche sui costituenti della sostanza organica del suolo, con particolare riguardo alla frazione lipidica. **Research on the components of soil organic matter with particular emphasis on lipid fraction.**

791 Ramunni, A.U.; Scialdone, R.   IT 040705/74/0001

D 1100 - Soil science

Struttura ed attività chimica della sostanza organica nel suolo. **Chemical structure and activity of organic matter in soil.** Publications.

792 Dell'aendla, G.; Caeco, E.   IT 040802/75/0001
Separazione di frazioni umiche biologicamente attive. **Isolation and characterization of humic fractions with biological activity.** Publications.

793 Petronici, C.   IT 040901/74/0602
Ricerche sulla sostanza organica del terreno, frazionamento e studio di particolari costituenti. **Research on soil organic matter, soil fractioning and study on special components.**

794 Patronici, C.   IT 040901/78/1085 N
Le interazioni dei fumiganti col terreno, effetto del bromuro di metile su alcune proprietà della sostanza organica. **Interaction of fumigants and the soil; methyl bromide action on certain properties of organic matter.**

795 Lippi Boncambi, C.   IT 041014/73/1663
Origine, classificazione e proprietà fisico–chimiche dei suoli organici dell'Italia centrale. **Origin, classification and physical and chemical properties of organic soils in central Italy.** Publications.

796 Riffaldi, R.   IT 041103/74/0610
Ulteriori ricerche sulla formazione e l'evoluzione della sostanza organica nei terreni formatisi su rocce calcaree. **Further research on the formation and evolution of organic matter in soils formed on limestone.**

797 Riffaldi, R.   IT 041103/78/1096 N
Il ruolo della sostanza organica sulla disponibilità del fosforo nel terreno. **The part played by organic matter in the availability of phosphorus in the soil.**

798 Arduino, E.   IT 041202/73/1183
Evoluzione dell'azoto in profili di terreni forestali come indice di alterazione della sostanza organica. **Evolution of nitrogen in forest soil profiles as an indication of the alteration of organic substance.**

799 Testini, C.   IT 041305/73/0318
L'assorbimento degli erbicidi S–triacinici da parte della frazione organica del terreno. **Apsorption of S–triazine herbicides by the organic fraction of the soil.**

800 Testini, C.   IT 041305/77/0306 R
Ricerche in spettroscopia I ed NMR su frazioni umiche e non umiche del terreno. **Spectrographic I and NMR studies on soils with and without humus.**

801 Sequi, P.   IT 060400/71/0001
Complessi tra sostanza organica del terreno e metalli e loro influenza sulla nutrizione delle piante. **Organo–metallic complexes in soil and their influence on plant nutrition.** Publications.

802 Lande Cremer, L.C.N. de la   NL 010103/47/0467
Invloed van organische bemesting op het gehalte van de grond aan organische stof. **Influence of organic manuring on the organic matter content of the soil.** Publications.

803 Bruins, E.H.   NL 010103/56/0450
Ontwikkeling van een microbiologische bepalingsmethode voor bestendige en aantastbare organische stof in de grond. **Development of a microbiological determination method for the persistent and decomposable fractions of the organic matter in the soil.**

804 Jager, G.   NL 010103/60/0452
De invloed van organische bemesting op de afbraak van bestendige humus. **The influence of organic manuring on the decomposition of stabilized soil organic matter (humus).**

805 Haan, S. de   NL 010103/63/3568 R
Hoeveelheid en hoedanigheid van de organische stof in verschillende korrelgroottefracties van de grond. **Quantity and quality of organic matter in different particle size fractions of the soil.**

806 Riem Vis, F.   NL 010103/74/5590
Betekenis van het humusgehalte voor de kwaliteit van de toplaag en van de grasmat van sportvelden. Kwantitatieve benadering van de beheersing van het humusgehalte. **Humus content as a measure of the quality of the top layer and the turf of sports fields. Quantitative approximation for the regulation of the humus content.** Publications.

807 Gerritse, R.G.   NL 010103/74/5593
Verplaatsing en mineralisatie van organische fosforverbindingen uit stalmest en gier in de grond. **Transport and mineralization of organic phosphorus compounds from animal wastes in the soil.** Publications.

808 Dijk, H. van   NL 010103/78/7967
Organische stikstofverbindingen in mest en in grond en hun mineralisatiesnelheid. **Organic nitrogen compounds in manure and in soil and their mineralisation rate.**

809 Gerritse, R.G.   NL 010103/78/7968
Biogeochemische aspekten van organische complexen van zware metalen in relatie tot het bodemmilieu. **Environmental biogeochemical aspects of organic heavy metal complexes.**

810 Jongerius, A.; Reijmerink, A.; Heintsberger, G.; Boersma, D.H.; Schoonderbeek, D.   NL 010119/75/6916 R
Kartering van humusvormen. **Research into humus formation.**

811 Breeuwsma, A.   NL 010119/77/7790
Bepaling van het lutumgehalte in moerig materiaal. **Determination of clay percentage in peaty soil materials.**

812 Linden, M.J.H.A. van de   NL 010602/64/4539
Bestendigheid van organische stikstofcomplexen in bosgrond. **Stability of organic nitrogen complexes in forest soil.**

813 Minderman, G.   NL 010602/77/8588 N
Karakterisering van organische stof. **Charaterization of organic matter.**

814 Haan, F.A.M. de   NL 020007/73/4960
Omzetting en afbraak van organische stoffen in de bodem. **Transformation and decomposition of organic compounds in soil.** Publications.

815 Halma, G.   NL 020008/76/6782
Verbetering van methodes voor het karakteriseren van organische stof in de bodem. **Improvement of methods for characterization of soil organic matter.** Publications.

D 1100 – Soil science

816 Antheunisse, J.   NL 020033/74/6340
De afbraak van cocosvezel in de grond. **The decomposition of coconut fibre in soil.**

817 Pol, P. van der   NL 040012/78/8945 N
Fosfaatbijdrage uit organische stof in P–arme gronden. **P–contribution from organic matter in P–deficient soils.**

Soil composition – soil air, soil water (B 1113)

See also 20, 578, 603, 628, 630, 657, 658, 703, 713, 723, 873, 875, 878, 880, 881, 887, 888, 917, 922, 923, 957, 958, 1009, 1095, 1179, 1180, 1332, 1403, 1404, 1421, 1428, 1429, 1431, 1443, 1444, 1496, 1504, 1523, 1543, 1551, 5298, 5421, 20116

818 Sine, L.; Calembert, J.; Bentz, A.; Dendas, J.; Ben Harrath, A.; Fraakinet, M.   BE 010009/59/0003 R
Etude de l'évapo–transpiration potentielle et actuelle en application agricole. **Study of potential and actual evapotranspiration in agricultural application.** Publications.

819 Schröder, D.; Beckmann, H.   DE 111050/78/0003 N
Einfluss von Temperatur und Feuchtigkeit auf den Verlust an gasförmigem Stickstoff im Boden. **The effect of temperature and humidity on the loss of gaseous nitrogen from the soil.**

820 Kick, H.; Prömse, M.   DE 111100/77/0001
Untersuchungen über die Ackerböden, Belastung von Grundund Oberflächenwasser durch Böden und Düngung. **Studies of the charge on surface– and groundwater by soils and fertilization.**

821 Wohlrab, B.; Sokollek, V.   DE 129450/78/0001 N
Untersuchungen über den Einfluss der Brache auf den Wasserertrag von Einzugsgebieten und auf die Bodenerosion. **Investigations into the influence of fallow land on water yield of catchment area and on soil erosion.**

822 Kickuth, R.   DE 132034/72/0001 R
Bindung der Nährstofffracht in kommunalen Abwässern durch induzierte Flockenbildung im Wurzelraum geeigneter höherer Pflanzen 1972. **Binding of nutrient load in communal sewages by induced flake in roots of suitable higher plants.**

823 Ulrich, B.; Ploeg, R.R. van der; Beese, F.
DE 132600/77/0007
Entwicklung von Modellen zur Beschreibung des Wasser– und Lösungstransports in forstlich und landwirtschaftlich genutzten Böden. **Construction of deterministic models for the water and solute transport in silvicultural and agricultural soils.** Publications.

824 Fischer, W.R.   DE 161020/73/0003
Charakterisierung von Unterwasserböden. **Characterization of underwater soils.**

825 Schwertmann, U.; Vogl, W.; Becher, H.H.
DE 161020/78/0001 N
Bodenwasser- und Nährstoffdynamik von 4 flächendeckenden Bodenformen der Hallertau – Bayern –. **Dynamics of soil water and nutrients of 4 soil types covering most of the area of the Hallertau – Bavaria –.**

826 Heger, K.   DE 301010/75/0008
Entwicklung eines Modells zur Beschreibung der Wasserverteilung in der zur Versorgung von Kulturen benötigten Bodenschicht. **Development of a model for describing the water distribution in the soil layer supplying the cultures.**

827 Häckel, H.   DE 301100/77/0007
Bodenfeuchteklimatologie. **Soil moisture climatology.**

828 Schuch, M.; Schmeidl, H.   DE 502050/70/0004
Vergleichende Wasserhaushalts– und Klimabeobachtungen auf kultivierten, forstlich genutzten und unberührten Hochmooren in Südbayern 1958. **Comparative observations on water content and climatic conditions of cultivated marshland used for forestry and of virgin marshland in South Bavaria.** Publications.

829 Schuch, M.; Jordan, F.   DE 502050/70/0009
Vergleichende Wasserhaushaltsbeobachtungen verschieden meliorierter staunasser Böden 1967. **Comparative observations on the water content of varying meliorated moisture–retaining soils.** Publications.

830 Kuntze, H.; Burghardt, W.; Weetjen, N.
DE 507103/75/0002
Kombinierte Maulwurf– und Rohr–Dränung von ackerbaulich genutzten Alluvialböden zur Staunässebeseitigung. **Combined amelioration of alluvial soils for the removal of stagnation.**

831 Prost, R.   FR 010103/68/0659
Etude des propriétés physico–chimiques de l'eau adsorbée sur les argiles. **Studies on physico–chemical properties of water adsorbed by clays.** Publications.

832 Prost, R.   FR 010103/68/6344
Etude de l'eau adsorbée sur les argiles. **Study on adsorbed water on clays.**

833 Mamy, J.; Le Renard, J.   FR 010103/70/0658
Etude par diffraction des rayons X de la structure des couches d'eau adsorbées par les micas. **Studies of the structure of water layers absorbed by micas, using X. ray diffraction.** Publications.

834 Chassin, P.   FR 010103/70/6331
Adsorption compétitive entre l'eau et les matières organiques sur l'argile. **Competitive adsorption on clay between water and organic matter.**

835 Gachon, L.; Dejou, J.; Robelin, M.; Morizet, J.
FR 010802/72/6120
Bilans hydriques et minéraux en différentes conditions de sols sous climat semi–continental de limagne. Potentialités des terres noires. **Moisture and mineral balances in different soils conditions under semi–continental climate of Limagne.**

836 Clair, A.; Perrey, C.; Concaret, J.; Grosman, R.
FR 011001/74/6282
Evolution de la qualité chimique des eaux percolant à travers le sol. **Evolution of the chemical quality of waters percolating through soil.**

837 Bosc, M.; Blanchet, R.   FR 011401/74/6192
Influence de l'humidité du sol sur la concentration en K de la phase liquide et sur les mouvements d'ions potassium vers les racines. **Influence of moisture content of soil on K concentration of the soil solution and on movements of potassium ions to roots.**

838 Drew; Saker GB 010502/79/0020 N
Factors contributing to waterlogging damage and alleviation in cereals.

839 Dowdell; Burford GB 010503/00/0005 R
Composition of the soil atmosphere.

840 Cannell; Ellis GB 010503/78/0015 R
Effect of reduced cultivation on hydrology, drainage requirements and leaching of major nutrients.

841 Colbourn; Dowdell GB 010503/79/0017 N
Denitrification and soil aeration status.

842 Rowse; Goodman GB 011802/00/0018 R
Develop and test models for the distribution of water in cropped and uncropped soil.

843 Rowse; Drew GB 011802/00/0019 R
Effect in ADAS experiments of ley treatments of soil on available water capacity and moisture release.

844 Williams GB 012003/00/0025 R
Composition of drainage waters related to cropping and fertilising, implications for water pollution.

845 Currie; Pritchard GB 012008/00/0009 R
Gas and water vapour movements in soil and their dependence on structure.

846 Youngs; Towner GB 012008/00/0020 R
Physical phenomena of ground water movement.

847 Smith GB 060114/00/0001 R
Soil and root aeration.

848 Sherwood, M. IE 060200/74/0826 R
Movement of fertiliser nitrate nitrogen through soils and the effects of meteorological and soil factors on the process. Publications.

849 Panicucci, M. IT 020100/73/0001 R
Correlazione tra coefficienti di deflusso ed utilizzazione del suolo in alcuni corsi d'acqua italiani. **Relationship between runoff coefficient and land use in some italian rivers.**

850 Natali, S. IT 041104/77/0247 R
Relazione fra contenuto idrico del suolo, potenziale idrico della pianta e traspirazione su pesco. **Correlation between the water content of the soil, the water potential of the plant and transpiration in the peach-tree.**

851 Raats, P.A.C. NL 010103/77/7691 R
Fysisch-mathematische beschrijving van transport- en afvoerverschijnselen in de grond i.v.m. mogelijke beinvloeding van het milieu van het wortelstelsel en de kwaliteit van het grond- en oppervlaktewater. **Physical-mathematical description of accumulation, depletion and transport processes in soil in connection with possible influences upon the environment of plant roots and the quality of ground and surface water.**

852 Poelstra, P. NL 010110/72/7289
The contamination of vegetation, surface water and deep ground water (drinking water) by heavy metals unintentionally released into soils. Publications.

853 Poelman, J.N.B.; Hoekstra, C.; Jókövi, P.; Krabbenborg, J.A. NL 010119/75/6910 R
pF-waarden en volumegewichten van Nederlandse gronden. **pF-curves and bulk density for soils in the Netherlands.**

854 Bouma, J.; Jongerius, A. NL 010119/75/6915
Het voorspellen van het hydrologisch gedrag van de grond op basis van morfometrische kenmerken. **The prediction of soil hydrologic properties using morphological features.** Publications.

855 Bouma, J.; Dekker, L.W. NL 010119/75/6917
Vochtleverantie aan het gewas door het bodemprofiel bij kleigronden. **Moisture delivery to crops by clay soils.** Publications.

856 Sluijs, P. van der; Houben, J.M.M.Th. NL 010119/76/7550
Onderzoek K-relatie in zandgronden. **Research into K-relations in sandy soils.** Publications.

857 Heesen, H.C. van; Krabbenborg, A.J. NL 010119/76/7556
Vochtleverantie van zandgronden. **Research into moisture supply in sandy soils.**

858 Stakman, W.P. NL 010501/67/9047 N
Ontwikkeling van bepalingsmethoden voor bodemfysische eigenschappen. **Determination methods of soil-physical characteristics.**

859 Hoeks, J. NL 010501/73/5210 R
Biochemische processen bij afvalgassen in de bodem. **Biochemical processes resulting from waste gases in soil.** Publications.

860 Bakker, J.W. NL 010501/73/5223 R
Gastransport in de waterfase van de grond. **Gas transport in the fluid phase of the soil.**

861 Koorevaar, P. NL 020007/76/6775
Meting van de fysische eigenschappen van de bodem wat betreft het watertransport. **Measurement of physical soil properties concerning watertransport.**

862 Muller, A.; Schelhaas, R.M. NL 040012/69/3815
Invloed van luchtdroging op vochtkarakteristiek van vulkanische asgronden. **Influence of air drying on the moisture characteristics of volcanic ash soils.** Publications.

863 Galli de Paratesi, S.; Gillot, J. XE 060101/76/0001
Soil moisture and heat budget. Evaluation in selected zones of agricultural and environmental interest. (Tellus project). Publications.

Soil composition – other (B 1119)

See also 9035

864 Kapol, F. DE 301020/75/0003
Temperaturentwicklung in Bodensubstraten verschiedenartiger Pflanzengefässe. **Development of temperature in soil substrates of various containers for plants.**

865 Geiger, K.; Schottdorf, W.   DE 502153/75/0002
Verwendung von Müllklärschlammkompost im Weinbau.
Utilization of refuse sewage sludge compost in viticulture.

Soil structure (B 1120)

See also 435, 442, 460, 480, 482, 515, 525, 552, 556, 558, 616,
645, 646, 678, 698, 699, 713, 718, 737, 781, 821, 860, 1009,
1099, 1403, 1414, 1482, 1768, 2684, 2685, 2811, 2952, 3260,
3670, 3671, 4862, 5200, 5298, 5426, 5441, 5480, 5492, 5820,
8398, 10073, 10086, 17190

866 Sine, L.; Noirfalise, A.; Brull, A.; Calembert, J.; Gaspar,
S.; Bentz, A.   BE 010009/59/0001 R
Etude des propriétés hydrodynamiques de substrats. **Study of
the hydrodynamic characteristics of soils.** Publications.

867 Sine, L.; Servotte, G.; Bentz, A.; Pliez, A.; Ben Harrath,
A.   BE 010009/67/0002 R
Etude des transferts d'eau dans des sols non saturés en
particulier la relation entre la teneur en eau, les tensions et les
flux d'humidité. **Study of dynamics of water in unsaturated soils
especially the relation of water content, conductivity to suction
and wetness.** Publications.

868 De Boodt, M.; Gabriels, D.; Callebaout, F.
         BE 030013/72/0001 R
Fundamenteel onderzoek op de lucht– en warmte economie in
natuurlijke en landbouwgronden. **Fundamental research on the
air– and heat economy in natural and structured soils.**
Publications.

869 Blume, H.–P.; Horn, R.   DE 105050/78/0001 N
Welche Bedeutung haben einige im Boden auftretende
Aggregatformen im Hinblick auf die Verdichtbarkeit unter
statischer und dynamischer Belastung. – Ein Beitrag zur
Klärung der bodenphysikalischen Bildungsbedingungen und
deren ökologische Auswirkungen anhand von Labor– und
Freilanduntersuchungen. **What is the importance of some forms
of soil aggregates in respect of the compressibility by static and
dynamic surcharge. – A contribution to elucidate the soil
aggregation processes by soil physical aspects as well as their
ecological effects, investigated by soil cores taken in the open air
and by artificial soils.**

870 Mückenhausen, E.; Kick, M.C.; Gärtel, W.; Zakosek, H.
         DE 111050/75/0012
Die Wanderung der Kationen und Anionen in
Weinbergsböden in Abhängigkeit von Textur,
bodentypologischer Entwicklung und Ausgangsgestein. **The
movement of cations and anions in vineyard soils depending on
texture, typological soil development and parent material.**

871 Brinkmann, W.; Flake, E.   DE 111850/77/0001
Untersuchungen zur Gestaltung des Keimbettes. **Investigations
on preparation of seed–bed.**

872 Stahr, K.   DE 126050/73/0010 R
Die Bedeutung pleistozäner Deckschichten für die
Bodenbildung und die Standorteigenschaften im
Südschwarzwald. **Importance of periglacial slope sediments on
soil genesis and site properties in southern Black Forest.**

873 Ehlers, W.   DE 132181/73/0001
Wasserleitfähigkeit in bearbeiteten und unbearbeiteten
Ackerböden, Effekt einer Pflugsohle. **Water conductivity in
tilled and untilled soils, possible effects of a plough pan.**

874 Weihe, K.von; Neugebohrn, L.   DE 135053/77/0001
Untersuchungen zur Bodenfestigkeit der Hamburgischen
Hochwasserschutzanlagen in Abhängigkeit von der Vegetation
un der Bewirtschaftungsform. **Analysis of soil consolidation of
flood protective plants in Hamburg in dependence on vegetation
and system of management.**

875 Shawki, E.; Hartge, K.H.   DE 138030/78/0001 N
Einfluss von Korngrössen auf die Benetzbarkeit von Böden
und ihre Bestimmung. **Influence of grain sizes on wettability of
soils and on the modus of their determination.**

876 Hartge, K.H.; Reichenbach, H.von   DE 138030/78/0002 N
Der Einfluss von Mineralbestand, Aggregatform,
Kationenbelag und Feuchte auf die Stabilität von
Aggregatpackung aus Tonen. **Effects of minerals, forms of
aggregate, cations, and moisture content on the stability of
packings of aggregates of clay.**

877 Schlichting, E.; Ahmed, H.A.   DE 144322/74/0005
Filterfunktion von Böden in Verdichtungsgebieten. **Filter
function of soils in conurbations.**

878 Geisler, G.   DE 148100/75/0001 N
Einfluss von Wachstumsfaktoren im Boden, insbesondere
Porengrösse, Porengrössenverteilung, Bodenluft,
Wassergehalt und Nährstoffangebot auf Wurzel– und
Sprosswachstum von Kulturpflanzen. **The influence of growth
factors in the soil, esp. volume of pores, distribution of pore
sizes, soil air, soil water and fertilizer, on root and shoot growth
of cultivated plants.**

879 Laatsch, W.   DE 160030/73/0005
Mechanik des Hangabtrags im Spitzingseegebiet, Bayerische
Alpen. **Mechanism of slope erosion in the Spitzing Lake area,
Bavarian Alps.**

880 Löffler, H.; Pospischil, L.   DE 160311/73/0007 N
Bodenphysikalische und bodenmechanische Untersuchungen
von Waldböden im bayerischen Alpenraum. **Research on
physical and mechanic factors of forest soils in Bavarian Alpine
regions.**

881 Becher, H.H.   DE 161020/73/0001
Eindringwiderstand von Bodenaggregaten in Abhängigkeit
vom Wassergehalt. **Penetration resistance of soil aggregates
depending on water content.**

882 Becher, H.H.; Vogl, W.   DE 161020/75/0001
Bodenphysikalische Eigenschaften ausgewählter Lössprofile
des tertiären Hügellandes, nördlich Freising. **Soil physical
properties of selected loess profiles of the tertiary rolling
country in the North–East of Munich.**

883 Knittel, H.   DE 161250/77/0008
Versuche zur Bodenphysik des Saatbettes. **Trials on soil
physics of seed bed.**

884 Estler, H.; Schönhammer, H.   DE 161525/70/0001 R
Technische Ausrüstung und Verfahren der
Minimalbodenbearbeitung. Einfluss der MB auf die
physikalischen Bodeneigenschaften. Messmethoden für die
Ermittlung des Bodenbearbeitungseffektes. Einfluss der

Minimalbestelltechnik auf die Pflanzenentwicklung – Beeinflussung der Bodenstruktur durch Grossmaschinen – 1970. **Technology and method for minimum cultivation of soil. Influence of minimum cultivation of soil on physical properties of soil. Measuring methods for the determination of the effect of soil cultivation technology on the growth plants – soil structure as affected by big machines –.**

885 Altemüller, H.–J.　　　　DE 201010/71/4017
Gefügekundliche Untersuchung von Verschlämmungsprozessen intensiv genutzter Böden im Hinblick auf Abtragungsvor gänge und den Transport kleiner Teilchen im Bodeninnern 1971. **english title not indicated.**

886 Zach, M.; Dambroth, M.　　　　DE 201040/71/5034 R
Entwicklung von kulturartenspezifischen Bodenbearbeitungssystemen 1976. **Development of crops–specific systems of soil cultivation.**

887 Sommer, C.; Zach, M.　　　　DE 201040/78/0009 N
Modellversuche über den Einfluss von Bodendichte und Bodenwasserpotential auf die Pflanzenentwicklung. **Model trials on the influence of soil density and soil–water–potential with regard to plant growth.**

888 Sommer, C.　　　　DE 201040/78/0010 N
Erprobung der Doppel–Energie–Gamma Methode zur simultanen Messung von Bodendichte und Bodenfeuchte am Gefässversuch und im Feld. **Testing the double–energy–gamma method for measuring soil density and soil moisture at pots and in the field, simultaneously.**

889 Müller, S.; Mühlhäusser, G.　　　　DE 501502/74/0002
Untersuchungen über Verdichtungserscheinungen auf schluffreichen Waldböden, die vermutlich nach dem Einsatz schwerer Brennungsmaschinen auftreten. **Investigations on compaction phenomena in silty forest soils caused probably by using heavy burning machinery.**

890 Süss, A.; Borchert, H.　　　　DE 502055/73/0016
Auswirkungen hoher Gaben behandelter Klärschlämme auf die physikalischen Bodeneigenschaften. **Influences of high amounts of hygienized sewage sludge on physical soil characteristics.**

891 Borchert, H.　　　　DE 502055/73/0024
Methodenvergleiche zur Erfassung der Bodendichte. **Comparison of methods for determination of bulk density.**

892 Borchert, H.; Wendland, E.　　　　DE 502055/73/0025
Untersuchung der Zusammenhänge zwischen Bodenstruktur und Pflanzenkrankheiten in Sonderkulturen. **Investigations on the relationship between soil structure and disease of plants in special cultures.**

893 Kern, H.　　　　DE 502055/78/0001 N
Prüfung von Alternativen zum Pflugeinsatz auf tonreichen Böden. **Testing of soil cultivation on heavy clayey soils without plough.**

894 Frenz, F.–W.; Lechl, P.　　　　DE 502104/71/0001 R
Vergleich von Weisstorf und Schwarztorf zur Verbesserung des Gewächshausbodens. Ausgangsboden: verwitterter Lösslehm 1971. **Comparison of white and black peat for improvement of soil in greenhouses. Starter soil: weathered loess loam.**

895 Hansen, R.; Müssel, H.　　　　DE 502107/77/0001
Bodenstrukturverbesserung für Pflanzungen der hohen Bartiris. **Soil structure improvement for plantations of the tall bearded iris.**

896 Sunkel, R.　　　　DE 508302/72/0002
Messung von Gefügeänderungen beim Schütthöhenversuch Neurath. **Measurement of structural changes in recultivated soils.**

897 Munk, H.; Bärmann, C.　　　　DE 901000/78/0003 N
Einfluss kalksilicatischer Schlacken auf Reaktionszustand, Gefüge und Ertragspotential von Ackerböden. **Influence of calcium alkaline silicates on reaction, texture and yield potential of soils.**

898 Grimme, H.; Nemeth, K.　　　　DE 902001/75/0001
K–Dynamik schwerer Böden in Abhängigkeit vom Tonmineralbestand. **K dynamics of fine textured soils as a function of clay mineralogy.**

899 Hansen, L.; Rasmussen, K.J.　　　　DK 010105/68/3401
Udvikling og tilpasning af jordfysiske målemetoder til bedømmelse af jordstruktur. **Development and adaptation of soil physical methods of measurement for evaluating soil structure.**

900 Olesen, S.E.　　　　DK 010900/75/0006
Sætningsundersøgelser på tørvejord. **Investigations of settling in peat soils.**

901 Møberg, J.P.　　　　DK 030106/75/0035 N
Jordbehandlingens indflydelse på oxidhydroxidernes egenskaber. **The influence of tilling and mulching on the properties of oxyhydroxides.**

902 Petersen, L.　　　　DK 030106/77/0008
Studier over luvisols. **Studies of alluvial soils.**

903 Jensen, J.R.　　　　DK 030143/76/0013 N
Studier over iontransport i porøse materialer. **Ion transport in porous media.**

904 Calvet, R.; Le Renard, J.　　　　FR 010103/74/6329
Transfert des herbicides dans les milieux poreux. **Herbicide transfert in porous medium.**

905 Gras, R.　　　　FR 010502/74/6005
Etude des propriétés physiques des sols de serre. **Study about physical properties of greenhouse soils.**

906 Fies, M. Mme　　　　FR 010601/68/0541
Analyse de la porosité des sols en tant que propriété texturale. **Studies on soil porosity as a textural property. Publications.**

907 Faure, A.　　　　FR 010601/68/6265
Mécanisme de compactage des sols, rôle de la texture. **Mechanics of soil compacting, effect of texture.**

908 Fies, J.C.　　　　FR 010601/68/6266
Analyse de la porosité des sols en tant que propriété texturale. **Analysis of soil porosity as a textural property.**

909 Stengel, P.　　　　FR 010601/74/6269
Caractérisation et déterminisme de l'état structural du sol. **Characterization and determinism of the soil structural state.**

910 Menet, M.; Juste, C.   FR 010704/75/6091
Lutte contre l'érosion éolienne dans les sols sableux cultivés
des landes de Gascogne. **Control of the wind erosion in sandy
soils.**

911 Dejou, J.; Guyot, J.; Robert, M.   FR 010802/71/6280
Etude d'une béidellite pure localisée dans les diaclases
traversant l'arène d'un granite au sud de l'Indre. **Study of a
near pure beidellite located in the diaclasis through a granitic
arena in the SOUTH of INDRE AREA.**

912 Dejou, J.; Guyot, J.; Robert, M.   FR 010802/72/6279
Evolution superficielle des roches cristallines et
cristallophylliennes dans les régions tempérées humides.
**Surface evolution of crystalline and foliated crystalline rocks in
wet temperate zones.**

913 Guyot, J.; Concaret, J.   FR 011001/74/6289
Aptitude à la fissuration d'agrégats terreux de petite taille.
Fissuration capability of small–sized soil aggregates.

914 Bornand, M.   FR 011211/71/6308
Etude des sols sur matériaux fluvio–glaciaires rhodaniens,
genèse et évolution chronologique. **Study of red soils on rhone
fluvio–glacial material, genesis and chronologic evolution.**

915 Legros, J.P.   FR 011211/72/6318
Etude des sols de montagne sur roches cristallines et
cristallophylliennes. **Study of mountain soils on crystalline and
foliated crystalline rocks.**

916 Hutter, W.; Marty, J.R.   FR 011401/74/6202
Travail du sol dans les systèmes de culture. **Tillage in Cropping
Systems.**

917 de Crecy, J.   FR 011602/72/6035
Comportement structural des sols en relation avec les
alternances hydriques et les travaux culturaux. **Climatic
relationships to soil structure in tropical zone (French West
Indies).**

918 Chrétien, J.   FR 012227/69/6257
Forme des sables et propriétés physiques des sols. **Shapes of
sands as related to physical properties of soils.**

919 Stafford; Tanner   GB 011605/00/0002 R
Soil dynamic studies for cultivation implements.

920 Audsley; Wheeler   GB 011611/78/0017 N
System studies of the effect of soil compaction on crop yield.

921 Rayner   GB 012003/00/0051 R
Structure and properties of micas.

922 Poulovassilis   GB 012008/00/0018 R
Hysteresis in soil–water properties.

923 Youngs; Poulovassilis   GB 012008/00/0019 R
Physical phenomena of water movement in unsaturated soil.

924 Jelley, R.M.; Burke, W.; Power, R.
                        IE 060300/71/0124 R
**Soil structure and moisture research with tillage crops,
especially sugar beet.** Publications.

925 Clear, T.; Joyce, P.M.; Gardiner, J.J.

                           IE 120107/64/9091 N
**Sociological and technological aspects of afforestation on
drumlin soils.** Publications.

926 Imperiale, G.   IT 012500/77/0633
Caratteristiche geomorfologiche di piccoli bacini
dell'appennino ligure. **Geomorphological characteristics of
small Apennine basins in Liguria.**

927 Arcara, P.G.; D'Egidio, G.; Torri, D.; Stiattesi, M.
                           IT 020100/75/0001 R
Determinare in campo la variabilità dei principali parametri
fisici e fisico–meccanici dei più diffusi suoli italiani. **Field
determination of the variability of the principal
phico–mechanical parameters in the principal italian soils.**

928 Zanchi, C.; Torri, D.   IT 020100/78/0007
Studio dell'erodibilità in campo su due diversi tipi di suolo.
Field study of the erodibility of two different types of soils.

929 Polemio, M.   IT 040104/77/0264 R
La conducibilità idraulica di terreni a diversa mineralogia in
relazione ai composti organici naturali e di sintesi. **Water
movement in soils of differing mineralogical composition and
the use of organic manures and fertilizers.**

930 Puglisi, S.   IT 040110/77/0272
Ricerche sulla sistemazione idraulico–forestale di aree
appenniniche in argille azzurre a morfologia calanchiva.
**Hydrological and forestal planning applied to blue clay
Appennine creeks.**

931 Toderi, G.; Stefanelli, G.   IT 040201/78/0003
Ricerche sulle lavorazioni del terreno a differenti profondità e
con vari attrezzi. **Researches on tilling: depth and tools.**

932 Radaelli, L.   IT 040304/77/0276 R
L'influenza della composizione delle acque sulle proprietà
fisiche del terreno e sulla nutrizione minerale di piante ortive.
**Influence of water composition on soil physical properties and
on the mineral nutrition of marketgarden plants.**

933 Ferrari, G.   IT 040802/77/0796
Effetto di diserbanti sulla struttura del suolo e meccanismi di
assorbimento naturale. **Impact of weedkillers on soil structure
and mechanisms of natural absorption.**

934 Sarcinelli, S.   IT 040913/73/2156
Ricerche sulla evoluzione della struttura del terreno in
funzione delle lavorazioni meccaniche e degli avvicendamenti
colturali. **Research on the evolution of the structure of soils with
respect to mechanized cultivation and crop rotation.**
Publications.

935 Sequi, P.   IT 060400/72/0005
Influenza delle pratiche agronomiche sulla struttura del
terreno. **The influence of agronomic practices on soil structure.**
Publications.

936 Sequi, P.   IT 060400/77/0866
Struttura e stabilità di struttura degli aggregati. **Structure and
structural stability of aggregates.**

937 Tedeschi, P.   IT 060700/77/0997
Studio delle variazioni per piani orizzontali di proprietà fisiche
del terreno come conseguenza della presenza delle piante e

delle tecniche colturali. **A study of the horizontal plane variations of the physical properties of the soil due to cultural techniques and to the presence of plants.**

938 Ouwerkerk, C. van   NL 010103/57/0443
Invloed van vochtgehalte en dichtheid op de reologische eigenschappen van de grond. **Influence of moisture content and density on the rheological properties of soil.** Publications.

939 Boekel, P.   NL 010103/60/0513 R
Vergelijkend structuuronderzoek op Friese en Groningse klei– en zavelgronden. **A comparative study on soil structure of clayey and silty soils in the northern part of the Netherlands.** Publications.

940 Boekel, P.; Pelgrum, A.   NL 010103/62/0463
Invloed van de kalktoestand op de structuur van kleigronden. **The effect of lime status on the structure of clay soils.** Publications.

941 Boekel, P.; Zwiers, J.S.   NL 010103/66/0570 R
Invloed van bodemstructuur op de groei van gewassen. **Effect of soil structure on crop growth.** Publications.

942 Boekel, P.   NL 010103/67/0941 R
Onderzoek naar de bodemstructuur op klei– en zavelgronden in Zuidwest–Nederland en op lössgronden in Limburg. **A study on soil structure of clayey and silty soils in the southern part of the Netherlands.** Publications.

943 Groot, A.J. de   NL 010103/68/2593
Onderzoek naar de mengende werking van grondbewerkingswerktuigen door toepassing van merktechnieken en activeringsanalyse. **Evaluation of soil–displacement characteristics of tillage systems using labeling techniques and activation analysis.**

944 Ouwerkerk, C. van   NL 010103/69/0518
Invloed van verschillende factoren op de structuur van het zaaibed en de kieming en de eerste groei van het gewas. **Effects of various factors on seed–bed structure and on emergence and first growth of the crop.** Publications.

945 Boekel, P.   NL 010103/70/3005 R
Structuurverandering onder invloed van bouwplan, vruchtopvolging en grondontsmetting. **Changes in soil structure as affected by cropping system, crop rotation and soil disinfectants.** Publications.

946 Ouwerkerk, C. van   NL 010103/70/3011 R
Ontwikkeling van eenvoudige meetmethoden voor de stevigheid van de toplaag van grassportvelden. **Development of simple methods for measuring the firmness of the top layer of sports fields.** Publications.

947 Boekel, P.   NL 010103/71/3155 R
Behoud en verbetering van de structuur van kleigronden door organische bemesting. **Preservation and improvement of the structure of clay soils by means of organic manures.** Publications.

948 Ouwerkerk, C. van   NL 010103/71/3157
De betekenis van verschillende grondbewerkingssystemen voor het behoud of de verbetering van het producerend vermogen van de grond. **The importance of different soil tillage systems for preservation or improvement of the productive**

capacity of the soil. Publications.

949 Ouwerkerk, C. van   NL 010103/71/3158 R
Invloed van de reologische eigenschappen van de bovengrond op de bewerkbaarheid, de berijdbaarheid en de geschiktheid voor recreatief gebruik. **Influence of rheological properties of the top soil on workability, trafficability and suitability for recreative purposes.** Publications.

950 Raats, P.A.C.   NL 010103/77/7690
Onderzoek naar het verband tussen krachten en vervormingen bij grond. **Analysis of the relationship between stresses and strains in soils.**

951 Boekel, P.   NL 010103/79/8711 N
Bewerkbaarheid van de grond in het voorjaar in relatie met vochtgehalte en vochttransport. **Workability of soils in spring in relation to water content and water flow.** Publications.

952 Boekel, P.   NL 010103/79/8712 N
Het effect van prikrollen van grassportvelden. **The effect of spiking of athletic turfs.** Publications.

953 Smet, L.A.H. de   NL 010119/67/1450
Onderzoek naar de fysische bodemvruchtbaarheid in verband met de toenemende mechanisatie. **Study on physical soil fertility in relation to increasing mechanization.** Publications.

954 Jongerius, A.   NL 010119/68/3239 R
Onderzoek naar de reologische eigenschappen van klei– en zavelgronden in (toekomstige) ruilverkavelingsgebieden. **Study on the rheological properties of silty clay loam and loam soils in (planned) re–allotment areas.** Publications.

955 Dam, J.G.C. van; Hulshof, J.A.   NL 010119/73/5973
Verbetering van ondiep bewortelbare zandgronden voor de teelt van asperges door middel van diepe grondbewerking met een mengrotor. **The improvement of shallow rooting sandy soils for growing asparagus by deep cultivation with a rotovator.**

956 Smet, L.A.H. de   NL 010119/76/7553 N
De invloed van de mechanisatie en grondbewerking op bodem en beworteling en opbrengst van het gewas. **The influence of mechanised tillage on soil, rooting and vegetation production.**

957 Bouma, J.   NL 010119/78/9073 N
Bodemkundige bijdrage aan het factorenanalyse–onderzoek naar de aardappelopbrengst in het zuidwestelijk zeekleigebied. **Soil research as a part of the factor analysis research on potato yields in south western marine clay soils in the Netherlands.**

958 Bouma, J.   NL 010119/79/9076 N
Het fysisch gedrag van kleigronden in het vroege voorjaar. **The physical behaviour of clayey soils in the early spring.**

959 Lumkes, L.M.   NL 010207/71/3268 R
Onderzoek naar de toepassing van nieuwe grondbewerkingssystemen. **Research on the application of new soil tillage systems.** Publications.

960 Loon, C.D. van; Houwing, J.F.   NL 010207/75/6259
Het effect van de grondbewerking op de opkomst en groei van aardappelen. **Effect of soil tillage on emergence and growth of potatoes.** Publications.

961 Kromwijk, P.A.M.   NL 010207/75/6261 R

Kiembedbereiding voor suikerbieten. **Seedbed preparation for sugar–beets.** Publications.

962 Lumkes, L.M.; Hoekstra, O.; Lamers, J.G.
NL 010207/77/7623
Onderzoek naar de relatie vruchtwisseling, hoofdgrondbewerking en bodemstructuur. **Research on the relation between soil structure, crop–rotation and soil tillage methods.**

963 Lumkes, L.M. NL 010207/77/7624 N
Onderzoek naar wijzen van bodemstabilisatie (bouwvoorconditie) op stuifgevoelige en slempgevoelige gronden. **Applied research on soil degradation, especially on control of erosion.**

964 Boels, D. NL 010501/73/5230
Effect van belasten op de dichtheid van de grond. **Effect of pressure on soil density.** Publications.

965 Wind, G.P. NL 010501/73/5235 R
Hydraulisch modelonderzoek naar datum en tijdsduur van de bewerkbaarheid van kleibouwland in het voorjaar. **Investigation with the aid of a hydraulic analog into the date and duration of soil workability in spring.** Publications.

966 Miedema, R. NL 020008/74/4919
Structuurstabiliteit in oude alluviale gronden en löss. **Structure stability in old alluvial soils and loëss.** Publications.

967 Boone, F.R.; Kroesbergen, B.; Boers, A.
NL 020020/59/4437
Het karakteriseren van de structuur van de grond alsdoor de grondbewerking te beinvloeden factor voor de plantenteelt. **The characterization of soil structure changes induced by tillage as a plant growth determining factor.** Publications.

968 Boone, F.R.; Kouwenhoven, J.K. NL 020020/59/4439
Rationalisatie van grond–bewerkingssystemen. **Rationalization of tillage systems.** Publications.

969 Boone, F.R.; Kroesbergen, B.; Boers, A.
NL 020020/59/4982
Gewasreacties op grondbewerking onder praktijkomstandigheden. **Crop reactions in practice due to tillage.** Publications.

970 Koolen, A.J.; Rijpma, P.J. NL 020020/59/6382
Het meten en beschrijven van de mechanische eigenschappen van grond. **Measuring and describing soil mechanical properties.** Publications.

971 Kouwenhoven, J.K.; Terpstra, R. NL 020020/64/4440
Ontwikkeling van ruggenteeltsystemen voor diverse gewassen in Nederland en in de tropen. **Development of ridge cultivation systems for various crops in the Netherlands and in the tropical zone.** Publications.

972 Kouwenhoven, J.K.; Terpstra, R. NL 020020/64/4444
Het effect van grondverplaatsing en –vervorming door werktuigen en trekkers op gewasgroei en gewasopbrengst. **Tillage implements effects and tillage systems.** Publications.

973 Koolen, A.J. NL 020020/68/4436
Proces analyse in de grondbewerking. Studie van de basisprocessen, die bij het mechanisch grondtransport

optreden. **Process–analysis in soil tillage. Study of the basic processes occurring in mechanical soil handling.** Publications.

974 Kouwenhoven, J.K. NL 020020/70/4438
Grondbewerking in (humide) tropische gebieden. **Soil tillage in the (humid) tropical zone.** Publications.

975 Boone, F.R.; Kroesbergen, B.; Boers, A.
NL 020020/70/4983
De invloed van de grondbewerking op de luchthuishouding en mechanische weerstand van de grond (i.v.m. de ontwikkelingsmogelijlheden van gewassen). **The influence of tillage on soil aeration and mechanical resistance in relation to plant development.** Publications.

976 Boone, F.R. NL 020020/71/3973
Invloed van bodemkundige factoren (o.a. vochtvoorziening) en beworteling op opbrengst en kwaliteit van consumptie–aardappelen. **Influence of the properties of soils (a.o. watersupply) and root development upon yield and quality of ware potatoes.**

977 Boone, F.R. NL 020020/75/7683 R
Grondbewerking als onkruidbestrijdingsmaatregel. **Tillage as a weed control measure.**

978 Hoogmoed, W.B. NL 020020/76/7321
Ontwikkeling van kriteria en methoden ter verbetering van de efficiëntie van de uitvoering van grondbewerkingen, speciaal gericht op aride en semi–aride streken. **Development of criteria and methods for improving the efficiency of soil management and tillage operations, with special references to arid and semi–arid regions.** Publications.

979 Glopper, R.J. de NL 040007/70/8473
Onderzoek naar opgetreden zettingen en het opstellen van prognoses voor de te verwachten zettingen onder opgebrachte belastingen. **Investigations of subsidence and the prediction of subsidence to be expected under overburden.**

Bio–communities in the soil (B 1130)

See also 97, 434, 514, 516, 517, 535, 542, 544, 549, 586, 601, 669, 696, 700, 707, 727, 746, 751, 763, 779, 782, 785, 816, 1005, 1012, 1768, 3899, 4348, 5139, 5281, 5660, 5748, 7789, 7808, 7820, 8099, 8148, 8184, 8378, 8398, 8911, 10113, 10125, 10168

980 Bonnier, C.; Saive, R.; Cornet, D. BE 010023/50/0001
Etude de la fertilité biologique des sols spécialement sur la symbiose Rhizobium–Légumineuses et l'influence de la culture moderne sur la fertilité biologique des sols. **Studies of the soil biological fertility especially on Leguminous–Rhizobium symbiosis and influence of modern tillage on soil biological fertility.** Publications.

981 Bonnier, C.; Brackel, J.; Saive, R. BE 010023/76/0002
L'influence de la culture moderne sur la fertilité biologique des sols. **Influence of modern tillage on soil biological fertility.**

982 Baert, L.; Van Cleemput, O.; Biermans, V.
BE 030009/69/0001
Stikstofverlies in de bodem. **Loss of nitrogen in soils.** Publications.

983 Gillard, A.; Pelerents, C.; Heungens, A.; Van Daele, E.
BE 030024/78/0009 R

D 1100 – Soil science

Dierlijke bio indicatoren van de bodemvruchtbaarheid. **Animal bio indicators of soil fertility.**

984 Livens, J.; Vlassak, K.; Reynders, L.; Fayez, X.
BE 040201/70/0002 R
Onderzoek over de biologische stikstoffixatie in de ecosystemen (kultuurgewassen, soja) in modelsystemen in microsites. **Research on biological nitrogen fixation in ecosystems (crops, soja, vetch clover) in modelsystems and in microsites (rhizosphere). Publications.**

985 Van Assche, C.; Coosemans, J.   BE 040203/76/0015 R
Studie van de invloed van systemische nematiciden (aldicarb) op vrijlevende nematoden en op de bodem microfauna. **Study on the influence of systemic nematicides (aldicarb) on freeliving nematodes and soil microfauna.**

986 Van Assche, C.   BE 040203/76/0018
Invloed van zware metalen op bodem microflora. **Study on the influence of heavy metals on the soil microflora. Publications.**

987 Van Assche, C.; Geypens, M.   BE 040203/77/0020
Biologische bodemanalyse. **Biological soilanalysis.**

988 Huge, P.   BE 080100/69/0006 N
Influence de divers traitements culturaux (pesticides, engrais, lisier) sur les groupes fonctionnels de microorganisme du sol. **Influence of various cropping systems (pesticides, fertilizers, slurry) on functional groups of soil microorganisms. Publications.**

989 Jadot, R.   BE 080600/76/0009
Equilibre biologique des sols de monoculture céréalière. **Biological balance in the soil of cereal monoculture.**

990 Blume, H.–P.; Dümmler, H.   DE 105050/70/0003
Bodengesellschaften West–Berlins 1970. **Soil communities in West Berlin.**

991 Blume, H.–P.; Hellriegel, T.   DE 105050/74/0002
Beeinflussung der Böden verschiedener Berliner StrassenÖkosysteme durch Blei– und Cadmiumimmissionen. **Consequences of Pb and Cd immissions on the soils in various Berlin street ecosystems.**

992 Dutzler–Franz, G.   DE 111052/77/0003
Der Einfluss von Klärschlamm auf die mikrobielle Aktivität verschiedener Bodentypen. **Influence of sewage sludge on microbial activity in different soil types.**

993 Heumann, W.   DE 120101/77/0001
Konjugation bei Rhizobium zur genetischen Untersuchung der symbiontischen N2–Fixierung. **Conjugation in Rhizobium. Genetic investigation of the symbiotic dinitrogen fixation. Publications.**

994 Heumann, W.; Kamberger, W.   DE 120101/77/0002
Vergleichende Charakterisierung der Lipopolysaccharide – LPS – von infektiösen und nichtinfektiösen Bodenbakterien der Gattung Rhizobium. **Comparative characterization of the lipopolysaccharide – LPS – of infective and noninfective Rhizobia.**

995 Pühler, A.   DE 120101/77/0004
Molekulare Genetik der N2–Fixierung bei freilebenden und symbiontischen Bakterien. **Molecular genetics of the dinitrogen fixation in free living and symbiotic bacteria.**

996 Ahrens, E.; Cengel, M.   DE 129080/75/0003
Die mikrobielle Dynamik in Böden extremer Standorte, insbesondere von solchen der westlichen und mittleren Türkei. **The microbial dynamics in extreme soils, particularly in those of West and Central Turkey.**

997 Küster, E.; Filip, Z.   DE 129080/78/0003 N
Infrarotspektroskopische und elektronenmikroskopische Untersuchungen an Bodenmikroorganismen. **Infrared spectroscopic and electron microscopic studies on soil microorganisms.**

998 Graff, O.   DE 201020/73/4001
Beeinflussung pflanzlicher Inhaltsstoffe durch tierische Tätigkeit im Wurzelbereich. **Plant components as affected by activity of animals in roots.**

999 Borkott, H.   DE 201020/77/0002
Indikatororganismen für die Kontrolle von Bodenbelastungen bei Schwemmistanwendung. **Indicator organisms for the control of soil strain by use of liquid manure.**

1000 Zadrazil, F.; Grabbe, K.   DE 201020/77/0003
Substratansprüche saprophytischer Basidiomyceten. **Substrate demand of saprophytic Basidiomycetes.**

1001 Domsch, K.H.   DE 201020/77/0004
Erarbeitung von Beurteilungskriterien für Pestizid–Nebenwirkungen auf Bodenmikroorganismen und –prozesse. **Establishment of criteria for valuation of side–effects of pesticides on micro–organisms and processes in soil.**

1002 Domsch, K.H.   DE 201020/77/0005
Auswertung vorhandener Informationen über PestizidNebenwirkungen für eine Datenbank der EG. **Evaluation of given informations on side–effects of pesticides for a data bank of the EC.**

1003 Anderson, J.P.E.; Reber, H.; Martens, R.
DE 201020/77/0006
Mikrobiologische und bodenkundliche Grundlagen für den Pestizidabbau im Boden. **Microbiological and pedological conditions of breakdown of pesticides in soil.**

1004 Brunnert, H.   DE 201020/77/0007
Freisetzung, Aufnahme, Transport und Verbleib von Schwermetallionen durch und in Mikroorganismen. **Release, intake, transport and retention of heavy metal ions by and in micro–organisms.**

1005 Jagnow, G.; Heinemeyer, O.   DE 201020/78/0002 N
Beeinflussung der mikrobiellen Denitrifikation durch Pflanzenschutzmittel. **The influence of pesticides on microbial denitrification.**

1006 Domsch, K.H.; Anderson, J.P.E.
DE 201020/78/0003 N
Bestimmung der mikrobiellen Biomasse in Böden. **Measurement of microbial biomass in soils.**

1007 Thielemann, R.   DE 215190/75/0005
Einfluss einer wiederholten Nematizidbehandlung auf das Oekosystem des Bodens. **Influence of repeated treatment with nematicide on the ecosystem of the soil.**

1008 Domsch, K.H.; Anderson, J.P.E.; Hoyningen–Huene, J.von   DE 301030/77/0003
Bedeutung von bodenchemischen und bodenklimatischen Stresssituationen für die Biomasse des Bodens. **Effectiveness of chemical and climatological soil stress situations on the biomass of soil.**

1009 Domsch, K.H.; Anderson, J.P.E.; Hoyningen–Huene, J.von   DE 301030/77/0004
Beziehungen zwischen bodenphysikalischen Parametern und der Ausbildung von Mikroorganismen–Populationen. **Relations between physical soil properties and the development of microbial populations.**

1010 Spatz, G.; Bauchhenss, J.   DE 502055/77/0002
Freihaltung der Kulturlandschaft mit Schafen im Voralpengebiet. Der Einfluss von Schafbeweidung auf die Bodenfauna. **Maintenance works on cultural landscapes by means of sheep management. The effect of sheep management on Collembola– and Oribatid mite faunae.**

1011 Bauchhenss, J.   DE 502055/77/0003
Auswirkungen des Abflämmens auf die Collembola– und Oribatidenfauna einer Brachfläche im Hochspessart. **The effect of burning over on Collembola and Oribatid mite faunae in the upper Spessart mountains.**

1012 Rieder, J.B.; Thalmann, H.; Bauchhenss, J.
                   DE 502058/78/0009 N
Einfluss der Güllebelüftung auf die Pflanzensoziologie und die Bodenfauna des Dauergrünlandes sowie die Nährstoffwanderung im Boden. **Influence of aeration of liquid manure on the composition of plant communities and soil–animals and on the nutrient migration in soil.**

1013 Brauns, A.; Guttmann, R.   DE 507800/78/0001 N
Beiträge zur Entwicklung der Bodenfauna vom rekultivierten Schutthalden. **Contributions to the growth of soil fauna on recultivated spoil banks.** Publications.

1014 Munk, H.; Lueg, F.   DE 901000/78/0002 N
Einfluss von Kalk- und Silicatzufuhr auf die biologische Aktivität von Ackerböden. **Influence of calcium alkaline silicates on biological activities of soils.**

1015 Nissen, T.V.   DK 010117/22/4605
Jordbundsmikroorganismer og deres virksomhed med hensyn til plantenæringsstoffernes kredsløb og jordbundens sundhed. **Soil microorganisms and their activity in relation to plant nutrient cycles and soil health.**

1016 Nissen, T.V.; Eiland, F.   DK 010117/76/9019 R
Metodestudier over mikrobiologisk aktivitet i jordbunden – bestemmelse af iltforbrug, biomasse og tælling af mikroorganismer. **Method studies of micro biological activity in the soil – determination of oxygen consumption, biomass and counts of microorganisms.**

1017 Jensen, V.   DK 030108/77/0007
Biologisk kvælstofbinding i danske landbrugsjorder. **Biological nitrogen fixation in Danish farm land soils.**

1018 Madsen, P.P.   DK 030300/78/9010
Kortlægning af nitratreduktionsveje for biologisk nitratreduktion i rodzonen i danske jorde under anvendelse af kvælstof 15 som sporstof. **Survey of nitrate reduction pathways for biological nitrate reduction in the root zone of Danish soils with the use of nitrogen 15 as tracer.**

1019 Simon, G.Mme   FR 010103/63/0662
Pesticides et Microflore du sol. **Pesticides and soil microflora.** Publications.

1020 Simon, S.   FR 010103/73/6348
Pesticides et populations microbiennes des sols. **Pesticides and microbial population of soil.**

1021 Joannes, H.   FR 010312/76/8284
Dynamique de populations de vers de terre (3). **Dynamics of populations of earth–worms.**

1022 Goulas, J.P.   FR 010703/72/0557
Dégradation des matériaux végétaux par les champignons supérieurs. **Decay of plant materials by higher fungi.**

1023 Lagacherie   FR 011002/67/6301
Symbiose rhizobium japonicum soja. Fixation biologique de l'azote. **Rhizobium japonicum, soybean symbiosis, biological nitrogen fixation.**

1024 Obaton, M.   FR 011002/70/0598
Symbiose en pays chaud. **Symbiosis in tropical countries.** Publications.

1025 Fournier, J.C.   FR 011002/70/6299
Etude en modele de la biodegradabilite d'un residu d'herbicide dans le sol: La 2,6 dichlorobenzamide. **Study on model of biodegradability of an herbicide residue in soil: The 2,6 dichlorobenzamide.**

1026 Amarger, n.   FR 011002/71/6292
Fixation symbiotique = relation entre rhizobium et légumineuse dans le processus d'infection. **Symbiotic fixation, relations between rhizobium and legue during infection process.**

1027 Catroux, G.; Germon, J.C.   FR 011002/71/6296
Utilisation du sol comme système épurateur. **Soil as a purifying system.**

1028 Soulas, G.   FR 011002/72/0590
Etude dans des conditions contrôlées des conséquences de l'adsorption des herbicides par les constituants physico–chimiques du sol, sur leur métabolisme par les microorganismes. **Studies on the consequences of Herbicide adsorption by soil constituents: its effects on their degradation by microorganisms.**

1029 Soulas, G.; Fournier, J.C.   FR 011002/72/0592
Métabolisme par des microorganismes isolés du sol, d'un herbicide: le Benzophenuron. **Metabolism of benzophenuron by microorganisms isolated from the soil.**

1030 Fournier, J.C.   FR 011002/73/6298
Ecologie de la degradation microbienne des pesticides. **Ecology of microbial degradation of pesticides.**

1031 Pussard, M.; Pons, R.   FR 011007/64/5001
Etude monographique d'amibes libres du sol. **Monographs of soil free living Amoebas.**

1032 Bouche, M.   FR 011007/65/1743

Rôle des vers de terre dans la prairie. **Role of earthworms in grassland.** Publications.

1033 Bouche, M.; Hadjibiros, K.   FR 011007/65/5002
Dynamique et niveaux des populations de lombriciens.
Dynamic and level of earthworm populations.

1034 Athias, C.   FR 011007/67/1742
Les gamasidés dans la biologie des sols. **Gamasides in soil biology.** Publications.

1035 Bouche, M.   FR 011007/67/5003
Lombriciens de France – Taxonomie et mésologie.
Earthworms from France – Taxonomy and environmental factor limits.

1036 Bouche, M.; Ferriere, G.; Kretzschmar, A.; Rouelle, J.
    FR 011007/70/5004
Quantification du rôle écologique et agronomique des lombriciens. **Quantification of ecological and agronomical function of the earthworms.**

1037 Bouche, M.; Fayolle, L.   FR 011007/71/5005
Méthodologie des essais de pesticides et de polluants sur les lombriciens. **Methodology to test pesticid or pollutant effects on earthworms.**

1038 Bouche, M.; Joannes, H.   FR 011007/72/5006
Lombriciens belgo–luxembourgeois; taxonomie et mésologie.
Earthworms from Belgium and Luxembourg; taxonomy and environmental factor limits.

1039 Kretzschmar, A.; Joannes, H.   FR 011007/72/5007
Action mécanique des lombriciens dans les sols. **Mechanical role of earthworms in the soil.**

1040 de Guiran, G.; Bonnet, L.; Stawiecki, K.
    FR 011007/72/5008
Survie des meloidogynes dans le sol. **Survival of Meloidogyne in the soil.**

1041 Rouelle, J.   FR 011007/73/5009
Lombriciens et dispersion des germes. **Lumbricid as a factor of germ dispersal.**

1042 Rouelle, J.; Loquet   FR 011007/73/5010
Action des lombriciens sur l'activité des microorganismes du cycle du carbone. **Lumbricid microorganism relationships in carbon cycle.**

1043 Rouelle, J.   FR 011007/73/5011
Action des lombriciens sur l'activité des microorganismes du cycle de l'azote. **Lumbricid – microorganism relationships in nitrogene cycle.**

1044 de Guiran, G.; Bonnet, L.; Stawiecki, K.
    FR 011007/74/5012
Influence des apports organiques sur les nématodes. **Effect of organic manures on soil nematods.**

1045 Ferriere, G.; Bouche, M.   FR 011007/75/5014
Ecophysiologie des lombriciens. **Earthworm ecophysiology.**

1046 Pussard, M.; Pons, R.   FR 011007/75/5016
Systématique des genres Acanthamoeba et Naegleria.
Systematic studies of genus Acanthamoeba and Naegleria.

1047 Bouche, M.; Fayolle, L.   FR 011007/76/5017
Effet du lisier de porcs sur les lombriciens. **Pig slurry effect on earthworms.**

1048 Pussard, M.; Pons, R.   FR 011007/77/5019
Etude des amibes mycophages. **Studies on the mycophage Amoeba.**

1049 Kermarrec, A.   FR 011605/72/5270
Analyse des communautés d'Helminthes Telluriques en Guadeloupe. **Nematode communities in tropical soils and mangroves ; mathematical and ècophysiological approach.**

1050 Kermarrec, A.   FR 011605/77/5276
Biologie des sols de mangrove en Guadeloupe. **Soil biology and physiology of mangroves.**

1051 Bouché, M.; Ricou, G.   FR 012219/72/1719
Fonctionnement et productivité d'écosystèmes prairiaux naturels. **Functioning and productivity of natural grassland ecosystems.** Publications.

1052 Masclet, A.; Nagy, Mme; Ricou, G. Mme
    FR 012219/76/6187
Biodégradation d'hydrocarbures par épandage sur le sol.
Hydrocarbons degradation by sludge farming.

1053 Marsh; Davies   GB 010803/00/0001 R
Effects of herbicides and their metabolites on microbial populations and their activities in the soil.

1054 Wingfield; Greaves   GB 010803/78/0006 R
Interactions between herbicides and the physiology and population dynamics of model microbial ecosystems.

1055 Stringer; Wright   GB 011514/00/0005 R
Ecology of earthworms.

1056 Jenkinson; Powlson   GB 012003/00/0065 R
Effect of fumigation on soil metabolism.

1057 Jenkinson; Brookes   GB 012003/79/0079 N
Measurement and turn over of nitrogen, phosphorus and carbon in the soil biomass.

1058 Currie; Pritchard   GB 012008/00/0008 R
Field measurement of root and soil respiration.

1059 Skinner   GB 012010/00/0001 R
Anaerobic soil bacteria.

1060 Darbyshire   GB 030405/00/0010 R
Investigation of soil protozoan populations.

1061 Darbyshire; Sparlin   GB 030405/77/0013 R
Interrelationships of soil actinomycetes,bacteria and protozoa with plant roots.

1062 Darbyshir   GB 030405/77/0014 R
Asymbiotic nitrogen fixation by soil microbes in the rhizosphere of agricultural plants and in peat.

1063 Arcara, G.   IT 020100/77/0785
Microbiologia dei suoli marginali, studio dei processi di azoto–fissazione. **Marginal soils microbiology, a study of**

D 1100 – Soil science

nitrogen fixation processes.

1064 Gregori, E.; Arcara, P.G.   IT 020100/78/0008
Studio dei rapporti tra vegetazione e microflora del suolo in
relazione a diverse tecniche colturali. **Study of the relationship
between vegetation and soil microflora according to different
agricoltural practices.**

1065 Lanza, G.   IT 021600/76/0002 N
Studi preliminari sulla microflora dei terreni agrumetati.
Preliminary studies on soil microflora of citrus orchards.

1066 Materassi, R.   IT 040513/73/0257
Ricerche ecologiche, sistematiche e fisiologico–biochimiche
sugli arthrobacter del terreno. **Ecological, systematic,
physiological and biochemical research on the Arthrobacter of
the soil.** Publications.

1067 Galli Fossati, E.   IT 040611/74/0556
Degradazione microbica degli acidi biliari e dei composti
organici di sintesi. Anticrittogamici ed erbicidi. **Microbial
degradation of biliary acids and of synthetic organic
compounds. Fungicides and herbicides.**

1068 Locci, R.   IT 040612/73/1817
Ricerche ecologiche su funghi fitopatogeni della rizosfera.
**Ecological research on phytopathogenic fungi of the
rhizosphere.**

1069 D'Errico, F.P.   IT 040725/75/0002 R
Un forno rotativo per la disinfestazione dei terreni a fiamma
diretta. **A rotative oven for soil disinfestation by direct flame.**
Publications.

1070 Santini, L.   IT 041106/77/0293 R
Studio delle caratteristiche della
micromammalofauna–rodentia insectivora– di alcuni terreni
agrari e forestali dell'Italia centrale. **A study of the
characteristics of the insectivorous small mammalian fauna –
Rodentia – in some cultivated and forestal areas of Central
Italy.**

1071 Rambelli, A.S.   IT 041602/72/0415
Ricerche sulla microflora fungina dei terreni a savana ed a
foresta nella zona di Lamto: Costa d'Avorio. In collaborazione
con il Laboratorio di Zoologia – Scuola Normale Superiore di
Parigi. Prof. M. Lamotte. **Research on the fungal microflora of
savanna and forest lands in the area of Lamto, Ivory Coast. In
collaboration with the Zoological Laboratory – Ecole Normale
Superieure of Paris, Prof. M. Lamotte.** Publications.

1072 Giardina, M.C.   IT 060200/72/0080
Degradazione della prometrina
(2–metiltio–4–6–isopropilammino triazine) da parte di un
ceppo di Sarcina isolato dal suolo. **Degradation of prometryne
by a Sarcina strain isolated from soil.** Publications.

1073 Cacciari, I.   IT 060200/77/0776
Interazioni pesticidi e batteri del suolo. **Pesticide and soil
bacteria interaction.**

1074 Veri, G.   IT 060200/77/0983 R
Effetti della microflora del suolo. **Action of the soil
micro–flora.**

1075 Florenzano, G.   IT 061300/77/0969

Azotofissazione e sue applicazioni. **Nitrogen fixation and its
applications.**

1076 Ceruti, A.   IT 061400/70/0156
Sintesi di micorrize. **Synthesis of mycorrhizae.** Publications.

1077 Ceruti, A.   IT 061400/70/0158
Individuazione dei funghi micorrizogeni. **Characterization of
mycorrhizal fungi.** Publications.

1078 Ceruti, A.   IT 061400/70/0159
Nutrizione dei funghi micorrizogeni. **Nutrition of mycorrhizal
fungi.** Publications.

1079 Ceruti, A.   IT 061400/70/0160
Micoflora dei terreni forestali ed agrari. **Mycoflora of forestal
and cultivated soils.** Publications.

1080 Ceruti, A.   IT 061400/77/0916
Micoflora dei terreni agrari e forestali. **The mycoflora in
agricultural and forest lands.**

1081 Ceruti, A.   IT 061400/77/0921
Biochimismo dei funghi in relazione agli alti polimeri che si
possono trovare nel terreno. **Fungi biochemistry related to the
high polymers eventually present in the soil.**

1082 Foschi, S.   IT 061800/77/1016
Ricerca sulla difesa dei microrganismi fungini terricoli.
Research on the resistance of soil fungus microorganisms.

1083 Nuti, M.P.   IT 061900/73/0011
Ricerche sull'integrità biologica del terreno agrario: Influenza
dei microrganismi sui condizionatori del suolo. **Researches on
biological integrity of agricultural soil: Effects of
microorganisms on soil conditioners.** Publications.

1084 De Bertoldi, M.   IT 061900/74/0012
Ricerche sull'integrità biologica del terreno agrario:
Meccanismi di resistenza ad alcuni fungicidi da parte di funghi
terricoli. **Researches on biological integrity of agricultural soil:
Drug–resistence to some fungicides of agricultural use in some
soil fungi.**

1085 Picci, G.   IT 061900/77/0881
Indagine biologiche sull'azotofissazione. **Biological research on
nitrogen fixation.**

1086 Picci, G.   IT 061900/77/0882 R
Studio degli effetti genotossici dei fungicidi agricoli. **Study on
the genetic toxic effects of fungicides used in agriculture.**

1087 Senni, L.   IT 121900/77/0810
La microbiologia del suolo e la crescita delle piante con
l'applicazione di humus ricavato da rifiuti urbani e fanghi. **Soil
microbiology and plant growth applying humus obtained from
urban wastes and mud.**

1088 Jager, G.   NL 010103/59/0448
Rizosfeer dynamiek; afgifte van organische stoffen door de
wortels, specificiteit van micro–organismen t.o.v. bepaalde
gewassen mede in verband met vruchtwisseling en voorkomen
van wortelparasieten. **Rhizosphere dynamics, root
"excretions", the specificity of soil bacteria towards some crops
– also in relation to the crop rotation.** Publications.

D 1100 – Soil science

1089 Lebbink, G. NL 010103/72/3560 R
Neveneffecten van grondontsmetting met natrium
N–methyldithiocarbamaat (SMDC, Metam, Vapam), in het
bijzonder t.a.v. de microflora van de grond. **Side–effects of soil
disinfection with sodium N–methyldithiocarbamate (SMDC,
Metam, Vapam), particularly with respect to the soil
micro–flora.** Publications.

1090 Jager, G.; Hoopen, A. ten NL 010103/75/6248
Evaluatie van bepalingsmethoden voor de antifytopathogene
potentiaal in grond t.o.v. Rhizoctonia solani. **Evaluation of
methods to determine the antagonism in soil with regard to
Rhizoctonia solani.**

1091 Dijk, H. van NL 010103/77/7217
Meetmethoden en criteria voor het effect van
bestrijdingsmiddelen op de microflora en de enzymactiviteit in
de grond. **Measuring methods and criteria for the effect of
pesticides on the microflora and enzyme activity of the soil.**

1092 Dijk, H. van NL 010103/77/7218 R
Beoordeling van bestrijdingsmiddelen op schadelijke gevolgen
voor het milieu, inzonderheid de bodembiosfeer. **Assessment
of pesticides regarding detrimental effects on the environment,
particularly the soil biosphere.**

1093 Gunst, J.H. de NL 010602/59/4547
Kwalitatieve en kwantitatieve inventarisatie van de
bodemfauna. **Qualitative and quantitative composition of the
soil fauna.**

1094 Eijsackers, H.J.P.; Heymans, G. NL 010602/71/5598
Effecten van bestrijdingsmiddelen en andere toxische stoffen
op de oeco–systemen in de grond. **Effects of pesticides and
other toxic agents on oeco systems in soil.**

1095 Eijsackers, H.J.P. NL 010602/74/7331 N
De invloed van de grondwaterstand op de regenwormenstand.
The effect of ground water level on earthworm populations.

1096 Eijsackers, H.J.P. NL 010602/75/7332 N
Wqrmeninventarisatie ten dienste van natuurbeheer,
milieubeheer en agrarish beheer. **Earthworm sampling for
nature, environmental and agricultural management.**

1097 Doelman, P. NL 010602/77/8783 N
Invloed van zware metalen op microbiologische
bodemprocessen. **Effects of heavy metals on microbiological
soil processes.**

1098 Haaker, H. NL 020006/70/4816
Het mechanisme van de aërobe N_2–reductie en de aërobe en
anaërobe NH_3–vorming in azotobacter en andere
micro–organismen. **The mechanism of the aerobic N_2–reduction
and the aerobic and anaerobic NH_3–formation in Azotobacter
vinelandii and other micro–organisms.** Publications.

1099 Brussaard, L. NL 020008/78/8647 N
Een onderzoek naar het functioneren van populaties van de
mestkever Typhaeus typhaeus (Linnaeus) en hun bijdrage aan
de vorming van poriënstelsels in enkele habitats op
zandgronden. **A study of the population of the dung beetle
Typhaeus typhaeus (Linnaeus) and their contribution to the
formation of poresystems in some sandy habitats.**

1100 Mulder, E.G. NL 020033/65/4701

De microbiologische aspecten van mangaan en ijzer in de
bodem en in oppervlaktewater. **The effect of micro–organisms
on the oxydation of manganese and iron in soil and water.**

1101 Mulder, E.G. NL 020033/69/4705
Factoren, die de efficiëntie van de stikstofbinding bij
Azotobacter bepalen. **Factors, affecting the efficiency of
nitrogen fixation by Azotobacter.** Publications.

1102 Crombach, W.H.J. NL 020033/74/6267
Discrepantie tussen genotype en fenotype bij Arthrobacter.
**Discrepancy between genotype and phenotype of
Arthrobacter.**

1103 Arkesteyn, G.J.M.W. NL 020033/76/4388
Microbiologische oxidatie van pyriet bij de vorming van
kattekleien. **Microbial oxidation of pyrite related to the
formation of acid sulphate soils.**

1104 Huntjes, J.L.M. NL 020033/78/8802 N
De afbraak en humificatie van organische verbindingen in
grond onder aërobe omstandigheden. **The decomposition and
humification of organic compounds in soil under aerobic
conditions.**

1105 Egeraat, A.W.M. van NL 020033/78/8803 N
Stikstofverbinding door vrijlevende rhizobia. **Nitrogen fixation
by free–living rhizobia.**

1106 Eenkhoorn, W. NL 040007/68/4122
Ontwikkeling van bacteriën en aaltjes in de bodem van
Zuidelijk Flevoland. **Development of bacteria and eelworms in
the soil of Zuidelijk Flevoland.**

Other subjects related to soil (B 1190)

See also 2, 3, 434, 2601

1107 Martens, P.; Deleu, R.; Copin, A. BE 010021/72/0001 R
Etude sur l'altération du milieu agricole par les pesticides.
**Studies on the deterioration of agricultural environment by
pesticides.** Publications.

1108 Rixhon, L.; Frankinet, M. BE 080800/67/0003
Etude du travail du sol. **Study of soil tillage.**

1109 Sine, L.; Van Bladel, P.; Nangniot, P.; Cloos, P.;
Morcale, A.; Copin, A. BE 140000/78/0054 R
Essais de prévisions logiques du comportement des résidus de
pesticides dans le sol. **Research of logical pronostication from
evolution of pesticides in soil.** Publications.

D 1300 – Land and water management

See also 4, 223, 226, 228, 269, 270, 277, 278, 300, 649, 849,
1834, 4101, 5463, 5467, 5908, 13806, 19570, 19816

1110 De Backer, L.; Persoons, E.; Bastin, G.; Bazier, G.
BE 020400/72/0004 R
Etablissement de modèles hydrogéologiques et de ressources
en eau. **Water resources and hydrogeological models
establishment.** Publications.

1111 Persoons, E.; Bazier, G.; Brossel, P.
BE 020400/77/0005 R
Etude du bassin de la Dyle. **Dyle bassin study.** Publications.

D 1300 – Land and water management

1112 Istas, J.; Neirinckx, G.   BE 100000/68/0004 R
Bepaling en bestrijding van watervervuiling door koolwaterstoffen. **Determination and prevention of water pollution by hydrocarbons.** Publications.

1113 De Borger, R.; Vanderstappen, R.; Meeus, K.; Guns, M.; Ledent, G.   BE 100000/71/0010 R
Inventaire et étude de la pollution des eaux et des sols en Belgique. **Inventory and study of water and soil pollution in Belgium.** Publications.

1114 Decleire, M.; De Cat, W.   BE 100000/72/0005 R
Mise au point de biotests rapides pour la détection d'herbicides et d'insecticides dans les eaux. **The detection of herbicides and insecticides in water by rapid bioessays.** Publications.

1115 Monseur, X.; Dourte, P.; Termonia, M.; Walravens, J.
     BE 100000/73/0013 R
Recherches sur les micro–polluants organiques de l'eau et sur les substances polluantes organiques volatiles du lisier dans les eaux de drainage et les eaux souterraines. **Research on the organic micro–pollutants in water and on volatil organic polluants of manure in drain–water and underground water.** Publications.

1116 Timmermans, J.   BE 130000/00/0004 R
Contrôle de la végétation aquatique à l'aide d'herbicides dans les eaux piscicoles. **Control of aquatic weeds by herbicides in inland fish waters.** Publications.

1117 Timmermans, J.; Gérard, D.   BE 130000/38/0001 R
Biologie générale et productivité piscicole des eaux douces, y compris l'étude de l'altération du milieu. **General biology and fish production of inland waters, pollution problems included.** Publications.

1118 Pflug, W.; Bauer, H.J.   DE 101100/71/0005 R
Landschaftsökologische Luftbildinterpretation zur Bewertung des Naturpotentials 1969. **Landscape ecological interpretation of aerial photos for assessment of nature potential.**

1119 Schulze, E.; Droege, H.P.   DE 111253/75/0005
Bodenbearbeitung in ariden und humiden Gebieten Guatemalas. **Tillage in arid and humid regions of Guatemala.**

1120 Ellenberg, H.; Schmidt, W.   DE 132093/71/0003
Die Vegetationsentwicklung eines Brachlandes unter dem Einfluss unterschiedlicher Nutzung. **Development of the vegetation of fallow land under different utilization.** Publications.

1121 Ewald, U.   DE 142300/73/0002
Land– und forstwirtschaftliche Erschliessung des Orinokodeltas und seiner Randgebiete, Venezuela. **Reclamation of land for agriculture and forestry in the Lower Orinoco areas, Venezuela.**

1122 Schwille, F.   DE 218000/72/0006 R
Der Gang des Grundwassers im Gebiet der Bundesrepublik Deutschland 1972–1978. **The course of groundwater in the Federal Republic of Germany.**

1123 Dancau, B.   DE 502050/72/0010 R
Untersuchungen zur Rekultivierung im Bereich der Bayer. Braunkohlen–Industrie AG – BBI – Schwandorf 1965.

Investigations on recultivation in the area of the Bavarian brown coal mining industry at Schwandorf.

1124 Callot, G.; Rochon.   FR 011211/74/6322
Aménagement des terres et des parcours dans les Alpes du Sud. **Development of lands and pastures in the southern alps.**

1125 Langlet, A.   FR 011401/74/6214
Evaluation des aptitudes culturales et des potentialités de production agricole à l'échelle d'un territoire. **Evaluation of agricultural aptitudes and capabilities of an area.**

1126 Langlet, A. – Flamand, A.; Chassany, J. P.; de Montard, F. X.; Labouesse, F.; Theriez, M.   FR 011401/76/6213
Potentialités et utilisation des terrains de parcours des Causses. **Capabilities and utilization of dry range land.**

1127 Damour, L.; Snegaroff, J.; Chevallier, C.; Lesage, B.
     FR 012201/76/9055
Incidence de la mise en valeur agricole des marais de l'Ouest sur la qualité des eaux. **Influence of the reclaiming of west marshes in France on water quality.**

1128 Damour, L.; Jeannin, B.; Garreau, J.; Chevallier, C.; Raichon, C.   FR 012201/77/9059
Conditions et conséquences de la maîtrise de l'eau dans les marais côtiers du Centre–Ouest. **Conditions and consequences of water management and control in the coast–line marsh–lands of central–western France.**

1129 Pelletier, J.; Druart, J.C.; Cherubino, M.T.
     FR 012218/57/5555
Etude qualitative et quantitative du phytoplancton du Léman. **Qualitative and quantitative study of phytoplancton in lake of Geneva.**

1130 Balvay, G.; Laurent, M. Mme   FR 012218/67/5573
Evolution de la biocénose zooplanctonique du lac d'Annecy. **Evolution of zooplankton biocenosis in lake Annecy.**

1131 Pelletier, J.P.; Moille, J.P.; Cherubino, M.T.
     FR 012218/68/5552
Evaluation de la production primaire du Léman. **Primary production assessment in lake of Geneva.**

1132 Balvay, G.   FR 012218/70/5570
Apports de phosphore dans la zone épilimnique par les larves de Chaoborus. **Phosphorus increase in epilimnion by Chaoborus larvae.**

1133 Pelletier, J.P.; Moille, J.P.; Cherubino, M.T.
     FR 012218/72/5553
Evaluation de la production primaire des lacs de Nantua, d'Annecy et du Bourget. **Primary production assessment in subalpine lakes.**

1134 Barroin, G.; Orand, A.; Colon, M.   FR 012218/72/5564
Inactivation des sédiments. **Sediments inactivation.**

1135 Feuillade, J.; Feuillade, M.; Blanc, P.
     FR 012218/73/5557
Etude de l'écologie d'Oscillatoria rubescens D.C. **Oscillatoria rubescens'ecology.**

1136 Balvay, G.; Pelletier, J.; Druart, C.; Laurent, M. Mme
     FR 012218/74/5572

D 1300 – Land and water management

Le zooplancton du lac de Paladru et son évolution à la suite du soutirage des eaux profondes. **Zooplancton of lac Paladru and evolution after the drawing off of the hypolimnic waters.**

1137 Feuillade, M.; Feuillade, J.   FR 012218/75/5558
Valeur des dosages de pigments pour l'estimation de la biomasse des cyanophycées. **Value of pigment measurements to estimate cyanophycean biomass.**

1138 Pelletier, J.P.; Orand, A.   FR 012218/76/5554
Etude de la distribution du phytoplancton dans le Léman par mesure "in situ" de la chlorophylle. **Study of phytoplancton distribution in lake of Geneva by "in situ" chlorophyll measurement.**

1139 Feuillade, J.; Feuillade, M.   FR 012218/76/5556
Etude de la physiologie d'Oscillatoria rubescens D.C. **Physiological studies on Oscillatoria rubescens D.C.**

1140 Barroin, G.; Colon, M.   FR 012218/76/5562
Incinération des sédiments. **Sediments Burning.**

1141 Barroin, G.; Orand, A.; Colon, M.   FR 012218/76/5563
Charrue subaquatique. **Subaquatic plow.**

1142 Feuillade, J.; Feuillade, M.; Blanc, P.
                                        FR 012218/77/5560
Etude de l'action d'une oxygénation hypolimnique sur une biocénose à Oscillatoria rubescens dominante. **Action of a deep oxygenation on an Oscillatoria rubescens population.**

1143 Barroin, G.; Orand, A.; Colon, M.   FR 012218/77/5561
Destratification épilimnique. **Epilimnetic Destratification.**

1144 Dubois, J.P.; Pelletier, J.P.   FR 012218/77/5565
Etude méthodologique des techniques d'évaluation de la production du phytoplancton. Application aux eaux turbides. **Methodology of phytoplancton productivity measurments applied to turbid waters.**

1145 Masclet, A.   FR 012219/74/6188
Aménagement écologique – Remblais de dragage de la Basse Seine. **Filling dredging of Basse Seine river.**

1146 Smith   GB 010802/00/0012 R
Effect of repeated application of glyphosate on fertility of soils or growth of cereals at Begbroke Hill.

1147 Scholefield   GB 011201/00/0009 R
Factors determining the extent and consequences of poaching on grassland.

1148 Gibson; Stephens   GB 040801/00/0006 R
Causes of oxygen depletion of Lough Neagh water.

1149 Stephens   GB 040801/00/0007 R
The nature of forms of phosphorous,other than orthophosphate,available for algal growth in Lough Neagh.

1150 Smith   GB 040801/00/0009 R
The effects of a major drainage scheme on the flow and water quality of River.Main.

1151 Foy   GB 040801/76/0010 R
Monitoring of algal toxins in enriched lakes.

1152 Foy; Stevens   GB 040801/78/0011 R
Phosphate removal in lakes.

1153 Strange   GB 040901/00/0002 R
Effect of drainage on the fish and invertebrate fauna of the River Camowen.

1154 Parker   GB 040902/76/0004 R
The distribution of benthic organisms in Belfast Lough in relation to pollution.

1155 Parker   GB 040902/78/0005 R
The effects of pollution upon a benthic ecosystem in Lough Foyle.

1156 Whittle; Hardy   GB 051002/00/0028 R
Hydrocarbon pollution in the marine environment.

1157 Holding   GB 060113/00/0006 R
Nitrogen and phosphorus compounds in relation to pollution.

1158 Sutherland   GB 060215/79/0011 N
Shelter belts for the outer hebrides.

1159 MacLeod   GB 060320/00/0002 R
North Argyll land development group.

1160 Gleeson, T.   IE 060300/64/0147 R
Surface treatments to reduce treading damage. Publications.

1161 Gleeson, T.   IE 060300/64/0149 R
Effect of lime fertilizer and sward management on susceptibility of pasture to treading damage. Publications.

1162 Galvin, L.   IE 060300/68/0131
Evaluation of plastic land drainage pipes. Publications.

1163 Galvin, L.   IE 060300/68/0132
Drainage and reclamation of peatland. Publications.

1164 Gleeson, T.   IE 060300/68/0146 R
Effect of traffic and treading on grass production and soil characters. Publications.

1165 Burke, W.   IE 060300/70/0123 R
Hydrology of forest on blanket peat. Publications.

1166 Galvin, L.   IE 060300/70/0135 R
Plastic pipe specification (standardisation, drainage). Publications.

1167 Jelley, R.M.; Burke, W.; McEntee, M.; O'Callaghan, T.F.; Murphy, R.F.   IE 060300/71/0125 R
Recording of soil moisture deficit and response to irrigation. Publications.

1168 Galvin, L.   IE 060300/71/0134
Land grading and planing. Publications.

1169 Galvin, L.   IE 060300/71/0141 R
Economics of land drainage. Publications.

1170 Gleeson, T.   IE 060300/71/0145 R
Soil physical problems of intensive pasture usage. Publications.

1171 Mulqueen, J.   IE 060701/65/0653 N

D 1300 – Land and water management

Development of the methodology of mole drainage. Publications.

1172 Mulqueen, J.   IE 060701/65/0655 R
Development of the methodology of gravel drains. Publications.

1173 Mulqueen, J.   IE 060701/67/0654 R
Deep ploughing and subsoil raising on bog and mountain land. Publications.

1174 Mulqueen, J.   IE 060701/72/0658 N
Development of reclamation and drainage methods. Publications.

1175 Chisci, G.; Torri, D.; Lulli, L.; Sfalanga, M.; Zanchi, C.; Tellini, M.   IT 020100/71/0001 R
Studio dei modelli deterministici per la prevenzione del deflusso e dell'erosione in piccoli bacini. **Study of deterministic models to predict runoff and prevent erosion in small basins.**

1176 Panicucci, M.   IT 020100/76/0002 R
Studio sulla distribuzione delle precipitazioni in funzione dell'orientamento delle pendici e sua influenza sui fenomeni erosivi e sul deflusso. **Study on precipitation distribution in relation to the slop orientation and its influence on erosion phenomena and runoff.**

1177 Lulli, L.   IT 020100/77/0637 R
Dinamica evolutiva di un'ambiente. **The dynamics of environmental evolution.**

1178 Mecella, G.; Di Blasi, N.; Biondi, A.; Saiella, M.
IT 020200/78/0004 N
Studio di un bacino idrografico. **A study of an hydrographic basin.**

1179 Rizzo, V.   IT 020500/53/0001
Osservazioni sistematiche di meteorologia e idrologia agraria nelle regioni Puglia e Basilicata. **Agro–meteorological and hydrological survey in Apulia and Basilicata.** Publications.

1180 Boschi, V.   IT 020500/59/0001
Osservazioni sistematiche di meteorologia ed idrologia agraria del bacino del Panaro (MO). **Agro–meteorological and –hydrological survey of the Panaro river watershed (MO).** Publications.

1181 Boldreghini, P.   IT 022800/77/0004 N
Gestione delle zone umide: il ruolo del Coniglio selvatico nello sviluppo della vegetazione.. **Wetland management: the role of the Wild Rabbit in the development of vegetation..**

1182 Chiusoli, A.   IT 040203/77/0146 R
Studio agrotecnico per interventi conservativi o di ricostituzione del patrimoni vegetale su terreni ad equilibrio alterato. **Agrotechnical study in view of intervening conservatively or reconstituting the vegetable resources of impoverished soils.**

1183 Toni, G.C.   IT 040248/77/0649
Studi sull'erosione di un bacino nel bolognese. **Study on the erosion of a basin in the Bologna region.**

1184 Basile, G.; Palmieri, F.; Violante, P.
IT 040705/78/0001 N
Fluttuazioni annuali del carico inquinante delle acque superficiali. **Annual fluctuations of the organic and inorganic pullutants in flowing waters.** Publications.

1185 Susmel, L.   IT 040809/77/0811
Applicazione di parametri ecologici alla gestione di un territorio alpino. **Ecologic parameters applied to the management of an Alpine area.**

1186 Vaia, F.   IT 042501/77/0623
Ricerche sulla stabilità potenziale dei versanti. **Research on the potential stability of hill sides.**

1187 Stol, Ph.Th.; Gils, J. van   NL 010501/77/7600
Ontwikkeling verwerkingssysteem voor waarnemingsuitkomsten in het cultuurtechnisch onderzoek. **Development of information processing systems in land an water management research.**

D 1310 – Drainage, irrigation and water supply

See also 12, 96, 171, 176, 186, 191, 207, 272, 279, 285, 295, 298, 299, 302, 303, 305, 310, 331, 354, 357, 398, 426, 429, 442, 472, 507, 551, 578, 583, 603, 628, 630, 707, 713, 724, 730, 820, 822, 832, 834, 835, 840, 858, 976, 1100, 1110, 1150, 1746, 1933, 2054, 2055, 2107, 2180, 2374, 3129, 3330, 3846, 4193, 4394, 4398, 4775, 4866, 5107, 5150, 5249, 5315, 5317, 5318, 5379, 5422, 5423, 5442, 5459, 5464, 5469, 5480, 5481, 5514, 5634, 5637, 5667, 5801, 5839, 5864, 5907, 5980, 5994, 10112, 10179, 11428, 15729, 17059, 17076, 19265, 19418, 19681

1188 Sine, J.; Noirfalise, A.; Vermeiren, L.; Calembert, J.; Dendas, J.; Gaspar, S.   BE 010009/58/0004
Etude économique du drainage sur la productivité agricole et du dimentionnement rationnel des réseaux. **Economical study on the repercussion of drainage on agricultural productivity and of the rational size of drainage nets.** Publications.

1189 Sine, L.; Marchal, P.   BE 010009/72/0005 R
Etude de l'altération du milieu de production agricole. **Study of the deterioration of the agricultural environment.** Publications.

1190 Feyen, J.; Deckers, J.; Michels, P.   BE 040403/70/0005 R
Studie van de waterbeheersing in de tuinbouw. **Optimization of crop water use in horticulture.** Publications.

1191 Feyen, J.; Hillel, D.; Belmans, C.; Raes, D.; Ragab, R.
BE 040403/77/0004 R
Dynamische simulatie en beheersing van de bodemvochtbeweging in wortelprofielen. **Dynamic simulation and management of soil moisture movement in rootzones.** Publications.

1192 Maton, A.; Dierickx, W.   BE 070300/70/0031 R
Onderzoek naar de watervoorziening in de witloofforcerie onder dak. **Research on the water supply in the endive chicory culture under a roof.** Publications.

1193 Maton, A.; Dierickx, W.   BE 070300/74/0041
Drainage van sportvelden. **Drainage of sport grounds.** Publications.

1194 Maton, A.; Dierickx, W.; Goossens, F.
BE 070300/78/0048 R
Bepaling van de doorspoeling van voedingselementen als gevolg van drainage. **Determination of the migration of nutrience due to drainage.**

D 1310 – Drainage, irregation and water supply

1195 Maton, A.; Dierickx, W.; Goossens, F.

BE 070300/79/0049 N

Het effekt van moldrainage en diepwoeling op de afwatering door drainage. **The effect of mole drainage and subsoiling on subsurface drainage.**

1196 Van Onsem, J.G.; Gabriels, R. BE 070600/76/0042

Studie van de kwaliteit van artesisch water. **Study of the quality of Artesian water.**

1197 Van Onsem, J.G.; Gabriels, R. BE 070600/76/0043 R

Invloed van de irrigatiewaterkwaliteit op de plantengroei. **Influence of the quality of the irrigation water on plantgrowth.** Publications.

1198 Van Onsem, J.G.; Gabriels, R. BE 070600/76/0044

Inventarisatie en beoordeling van het grondwater gebruikt als irrigatiewater in het Gentse. **Inventarisation and appreciation of the groundwater as irrigation water in the Ghent region.** Publications.

1199 Van Onsem, J.G.; Gabriels, R. BE 070600/76/0045

Beregenings– en grondwateronderzoek. **Irrigation and groundwater analysis.** Publications.

1200 Van Onsem, J.G.; Verdonck, O. BE 070600/76/0071 R

Studie van automatische irrigatie bij sierplanten. **Study of automatic irrigation for ornamental plants.** Publications.

1201 Van Onsem, J.G.; Verdonck, O. BE 070600/76/0072 R

Studie van de waterhuishouding van tuinbouwgronden. **Study of the economy of water of horticultural soils.** Publications.

1202 Biston, R.; Dardenne, P. BE 080900/77/0009 R

L'irrigation en région du Sud–Est. **Irrigation of herbage in the South–Eastern region.** Publications.

1203 Blume, H.–P.; Muljadi, S.; Lacatusu, R.

DE 105050/75/0001

Phosphat–, Nitrat– und Borateutrophierung Berliner Gewässer, Teil Wasser– und Sedimentchemie. **Eutrophism of lakes in Berlin by agricultural and domestic phosphates, nitrates and borates, part water and sediment chemistry.**

1204 Sukopp, H. DE 105060/77/0004

Grundwasserabsenkung bei Baumassnahmen der U–Bahn. **Groundwater lowering during subway building.**

1205 Pahlke, K. DE 105250/78/0002 N

Untersuchungen zur Auswirkung aufgelassener Vorflutsysteme auf die Nutzungsbedingungen der Brackmarschen. **Effects of idle outlet ditches on grassland farming conditions of brackish marsh soils.** Publications.

1206 Wolkewitz, H.; Davier, R.von DE 105250/78/0005 N

Möglichkeiten zur Erhöhung des Wirkungsgrades der Wasserverwendung bei der Feldberegnung mit Hilfe eines kontinuierlich frontal bewegten Reihenregnerverfahrens. **Possibilities for raising the efficiency of water supply in field irrigation using a row sprinkler movable in the front.**

1207 Wesche, J. DE 105255/74/0001

Lysimeteruntersuchungen über Nährstoffauswaschungen und Grundwasserbeeinflussungen bei Verwertung von Abfallkomposten. **Lysimeter analysis of the wash–out of**

nutrient substance and of interferences with ground water caused by the utilization of waste material compost.

1208 Brechtel, M.; Zakosek, H. DE 111050/75/0001

Das forsthydrologische Forschungsgebiet Krofdorf. **The forest–hydrological research district of Krofdorf.**

1209 Thews, J.–D.; Mückenhausen, E.; Zakosek, H.

DE 111050/75/0009

Fossile Böden und fossiler Gesteinszersatz durch aszendentes Wasser. **Fossil soils and fossil decomposition of rocks by ascensive water.**

1210 Kick, H.; Lohse, M. DE 111100/78/0001 N

Beseitigung von Sickerwässern aus Abfalldeponien durch Landbehandlung. **Removal of soakwater out of waste disposal area by land treatment.**

1211 Weiling, F.; Klinke, G. DE 111151/75/0002

Statistisch–biometrische Analyse von physikalischen, chemischen und biologischen Messdaten zur Charakterisierung der Erftwasserqualität unter Berücksichtigung möglicher Trendwirkungen über den Flusslauf und über die Untersuchungsjahre 1963–1974. **Statistic–biometric analysis of physical, chemical and biological measurement data for characterizing the water quality of the Erft with reference to eventual effects of the river over the period under investigation 1963–1974.**

1212 Bick, H.; Brand, G. DE 111202/77/0001

Umweltfaktoren und Populationsdynamik von Protozoen in Süsswassersedimenten. **Environmental factors and population dynamics of protozoa in freshwater sediments.**

1213 Caspers, N. DE 111202/77/0002

Produktivität und Emergenz kleiner Waldbäche. **Productivity and emergence of woodland brooklets.**

1214 Bick, H.; Bauer, M. DE 111202/77/0003

Ökologische Untersuchungen an Süsswasserdiatomeen. **Ecological investigations on freshwater diatoms.**

1215 Bick, H.; Röser, B. DE 111202/77/0004

Limnologische Untersuchung an kleinen Fliessgewässern. **Limnological investigations on brooklets.**

1216 Hundgeburt, H.J. DE 111300/77/0008

Einsatz der Tröpchenberegnung im Gemüsebau. **Drop irrigation for vegetable crops.**

1217 Lenz, F.; Hesse, N. DE 111300/78/0002 N

Wasserhaushalt bei Gemüse in Abhängigkeit von Verfahren und Termin der Bewässerung. **Water regime in vegetables as dependent on technique and timing of irrigation.**

1218 Baitsch, B.; Niemeyer, R. DE 111901/75/0001

Wasserwirtschaftliche Untersuchungen an künstlichen Seen – Baggerseen etc. – im Hinblick auf Nutzungsgrenzen und –prioritäten. **Water management of man–made lakes – gravel lakes, etc. – with regard to the limits and priorities of exploitation.**

1219 Gilles, K.–P. DE 111901/77/0001

Gewässerunterhaltung mit Herbiziden. **Maintenance and chemical weed control.** Publications.

1220 Gilles, K.–P.   DE 111901/77/0002
Modelle zur optimalen Planung und Durchführung der Gewässerunterhaltung. **Models for optimal planning and operation of maintenance.**

1221 Rieser, A.   DE 111901/78/0001 N
Wasserwirtschaftlich–kulturtechnische Planungen in Entwicklungsländern dargestellt an Beispielen im südlichen Sumatra/Indonesien. **Water resources development in developing countries – examples in the region of Southern Sumatra/Indonesia.** Publications.

1222 Flohn, H.   DE 111990/72/0002 R
Untersuchungen zum globalen Wasserhaushalt 1972. **Investigations on global water regime.**

1223 Collins, H.–J.; Christoph, F.; Lhotzky, K.
        DE 114302/75/0001
Der Einfluss der Bodenfeuchte zum Zeitpunkt der Dränausführung auf die hydraulische Leistung der Dräne. **Influence of soil moisture during execution of drain works on hydraulic performance of drains.**

1224 Collins, H.–J.; Spillmann, P.; Regner, J.
        DE 114302/77/0001
Wasser– und Stoffhaushalt in Abfalldeponien und deren Auswirkungen auf Gewässer; Unterthema: Untersuchungen über den Gas– und Wasserhaushalt von Abfalldeponien. **Balances of material in sanitary landfills and the consequences on ground and surface water Subitem: Investigating balance of gas and water in sanitary landfills.**

1225 Collins, H.–J.; Regner, J.   DE 114302/77/0002
Verminderung der Sickerwassermenge aus hochverdichteten Abfalldeponien durch gezielte Ausnutzung der potentiellen Verdunstung. **Reduction of seepage out of garbage disposals by means of utilization of potential evaporation.**

1226 Collins, H.–J.; Lhotzky, K.   DE 114302/78/0001 N
Bodenfeuchtemessungen mit Hilfe eines neu zu entwickelnden ''Hitzdrahtverfahren''. **Determination of soil moisture by means of a heated wire.**

1227 Schröder, W.; Sulser, P.   DE 117301/78/0001 N
Wasserentnahme aus Flüssen in semi–ariden Gebieten. **Intake structures on semi–arid rivers.** Publications.

1228 Schröder, W.   DE 117301/78/0002 N
Biologische Sicherung von Fluss– und Bachufern mit Gehölzen. **Biological protection of river–banks by trees and bushes.** Publications.

1229 Meyer, G.   DE 120200/77/0002
Erschliessung und Entwicklung der ägyptischen Neulandgebiete. **Land reclamation and development of the new lands in Egypt.** Publications.

1230 Moll, W.; Hädrich, F.; Trüby, P. DE 126050/78/0002 N
Der Einfluss intensiver Bewässerung mit Rhein– und Grundwasser auf die Dynamik und die Filtereigenschaften von Zweischicht–Pararendzinen in der Rheinaue bei Hartheim. **The influence of intensive irrigation using Rhine– and ground–water on the dynamics and the filtering properties of a two–layer pararendzina developed on the Rhine flood plain near Hartheim.**

1231 Keller, R.   DE 126675/75/0003
Einfluss des Menschen auf hydrologische Prozesse im südbadischen Oberrheingebiet. **Human influence on hydrological processes in the South–Baden Upper–Rhine region.**

1232 Preusse, H.–U.; Voss, W.   DE 129020/71/0005
Eutrophierung von Oberflächengewässern. **Eutrophication of surface–waters.**

1233 Skirde, W.   DE 129220/77/0001
Be– und Entwässerung von Rasensportflächen. **Irrigation and drainage of sportsturf areas.** Publications.

1234 Kowald, R.; Vogel, C.   DE 129450/75/0001
Regulierung des Bodensalzhaushaltes durch Dränung. **Regulation of the soil salt balance by drainage.**

1235 Wohlrab, B.; Süssmann, W.; Sokollek, V.; Ditter, P.;
Mollenhauer, K.   DE 129450/75/0006
Wirkungen verschiedener Flächennutzung auf das Abflussregime und den Stoffeintrag in Gewässer. **Effects of various types of land use on the runoff and on the transport of nutrients into waters.**

1236 Wohlrab, B.; Mollenhauer, K.   DE 129450/75/0008
Agrarhydrologische und hydropedologische Kriterien zur Festlegung von Empfehlungen und Auflagen für die Bodennutzung in Wasserschutzgebieten. **Agro–hydrologic and hydropedologic criteria for establishing recommendations and conditions regarding land use in water protection areas.**

1237 Meyer, B.; Beese, F.; Wildhagen, H.; Labenski, K.O.
        DE 132032/70/0001
Hydrodynamik und Hydrobilanz der Bodendecke – Wasserforschung IHD – 1970. **Hydrodynamics and water balance in soil surface – research on water content IHD.**

1238 Meyer, B.; Gebhardt, H.; Hugenroth, P.
        DE 132032/70/0002
Wasserhaushalt und Abwässerungsprobleme unter Berücksichtigung der Verhältnisse in Niedersachsen 1968. **Water balance and problems of sewage disposal in Lower Saxony.**

1239 Ulrich, B.; Hetsch, W.   DE 132600/77/0003
Stoffhaushalt in Wassereinzugsgebieten des Oberharzes. **Geochemical balance in water sheds of the upper Harz mountains.**

1240 Benecke, P.   DE 132600/77/0004
Dynamik der Wasserhaushaltskomponenten forstlicher Standorte in Hanglage. **Dynamics of the water budget components of forest sites in sloped areas.**

1241 Fölster, H.; Franco, W.   DE 132602/75/0001
Bodenwasserregime laubwerfender und immergrüner Wälder, Llanos, Venezuela. **Soil water regime of deciduous and evergreen forests, Llanos, Venezuela.**

1242 Ploeg, R.R.van der   DE 132603/77/0002
Entwicklung zweidimensionaler Modelle für den Wasserumsatz im Boden hängiger Fichtenstandorte des Harzes. **Two–dimensional models for the water budget of spruce stands in sloping mountain regions of the Harz.** Publications.

D 1310 – Drainage, irregation and water supply

1243 Ploeg, R.R.van der; Büttner, G.　DE 132603/78/0002 N
Entwicklung von Wasserhaushaltsmodellen für
Staunässeböden. **Development of water balance models for
waterlogged soils.**

1244 Neugebohrn, L.　DE 135053/74/0001
Pflanzensoziologische Untersuchungen an künstlich begrünten
Böschungen – Veränderung der Gesellschaften mit Dauer der
Zeit –. **Plant sociological studies on artificially sodded slopes –
variations in plant associations in the course of time.**

1245 Lillelund, K.; Bemmer, P.; Langer, D.; Parthier, M.;
Schütz, W.　DE 135251/75/0002
Untersuchungen zum Grenzwertproblem bei
Abwassereinleitungen. **Investigations on the tolerance limits of
harmful substances in discharged waste waters.**

1246 Hartge, K.H.; Nissen, J.　DE 138030/77/0002
Darcy–Fluss und elektroosmotische Effekte bei der
ungesättigten Wasserbewegung. **Darcy–flow and
electroosmotic potentials under unsaturated flow conditions.**

1247 Ernst, D.E.W.　DE 138180/77/0001
Vorhersage des Verhaltens von Gewässern. **Prediction of the
behaviour of waters under eutrophication and head charge as
estimated by primary production.** Publications.

1248 Kühn, W.K.G.; Bunnenberg, C.　DE 138181/77/0003
Untersuchungen über die Ausbreitung des Wasserdampfes in
den oberen Bodenschichten und Bestimmung von
Kondensations– und Evaporationsprozessen an der
Bodenoberfläche. **Investigations on the transport of water
vapour in the upper soil layers including evaporation and
transpiration near the soil surface.** Publications.

1249 Bünemann, G.; Rasenack, E.　DE 138240/77/0002
Frostschutzberegnung auf Marschböden des Alten Landes.
Frost protection by irrigation on clay soils of the Altes Land.

1250 Naumann, W.D.　DE 138242/77/0001
Wirkung der Tropfbewässerung bei Erdbeeren bei
unterschiedlichem Termin. **Timing of drip irrigation for
strawberries.**

1251 Hoffmann, B.　DE 138450/72/0004 R
Grundwasserhaushalt im Küstenbereich 1970. **Groundwater
regime in coastal regions.**

1252 Lecher, K.　DE 138450/75/0001
Abflussmodell für Niedrigwasserabflüsse als
Bemessungsgrundlage für die Entwürfe im
landwirtschaftlichen Wasserbau. **Discharge model for low
water discharges as a measuring basis for projects in
agricultural water engineering.**

1253 Lecher, K.; Rickert, K.　DE 138450/77/0001
Auswirkungen von Lebendverbaumassnahmen am Gewässer.
**Influence of plantation on channel side slopes – decrease of
flow discharge –.**

1254 Günther, W.　DE 138450/77/0003
Wasserbilanz im Tidebereich der Ems nach Menge und
Beschaffenheit. **Water balance for water quantity and quality
in tide influenced reaches of the Ems River.**

1255 Ludwig, K.　DE 138450/77/0004
Integralkonzept zur Berechnung der Oberflächenabflüsse in
Flussgebieten. **Integral concept for calculation of surface runoff
in river basins.**

1256 Lecher, K.; Günther, W.; Jansen, M.; Kaller; Koch,
H.W.; Ludwig, K.　DE 138450/78/0001 N
Computergerechte wasserwirtschaftliche Rahmenplanung.
Computer aided water–resources management planning.
Publications.

1257 Lecher, K.　DE 138450/78/0002 N
Die Verteilung des Beregnungswassers im Boden unter
besonderer Berücksichtigung beregnungsbedürftiger und
–würdiger niedersächsischer Böden. **The distribution of the
irrigation–water in the ground with special consideration of
irrigation requirements and worthiness of soils in Lower
Saxony.**

1258 Lecher, K.; Stintzing, W.; Jäschke
DE 138450/78/0003 N
Ermittlung der Häufigkeit von massgebend durch
Schneeschmelze verursachten Hochwasserereignissen.
**Determination of the frequency of flood–events, which are
decisively caused by the snowmelt.**

1259 Lecher, K.; Loof, R.　DE 138450/78/0004 N
Ermittlung von Parametern zur Übertragbarkeit der
Einheitsganglinie und der koaxialen graphischen Darstellung
zur Bemessung von Vorflutern und RHB in kleinen
Einzugsgebieten Niedersachsens. **Determination of parameters
of unit–hydrographs and the antecedent
precipitation–moisture–indices and relations to basin
characteristics for design of river adjustments and storage
basins on small drainage areas in Lower Saxony.**

1260 Lecher, K.; Keser, M.H.　DE 138450/78/0005 N
Die quantitative und qualitative Belastung natürlicher
Vorfluter durch Regen– und Mischwasser aus
Siedlungsgebieten. **Quality and quantity of pollution in natural
streams affected by rainwater and mixed water from urbanized
areas.**

1261 Lecher, K.; Rademacher　DE 138450/78/0006 N
Computergesteuerte Dimensionierung und Bewirtschaftung
von Mehrzweckspeicherverbundsystemen. **Computer
controlled design and control and rationing of multi–purpose
compound storing systems.**

1262 Lecher, K.; Riebe; Habercom; Pau
DE 138450/78/0007 N
Untersuchungen von Tidewässerständen der Deutschen
Nordseeküste im Hinblick auf eventuelle Veränderungen
durch anthropogene Einflüsse. **Research on tide water levels of
the German North Sea coast with regard to possible changes
caused by anthropogenous influences.**

1263 Lecher, K.; Keser, M.H.　DE 138450/78/0008 N
Entwicklung und Installation eines elektronischen
Datenerfassungssystems in Kanalisationen für die Erarbeitung
gesicherter hydrologischer Daten und die Quantifizierung der
hydrologischen Parameter. **Development and installation of an
electronical dataregistration–system in sewage systems for
collection of reliable hydrological data and the quantification of
hydrological parameters.**

D 1310 – Drainage, irregation and water supply

1264 Mull, R.; Battermann   DE 138450/78/0009 N
Wärmeausbreitung bei der Einleitung von abgekühltem
Wasser im Grundwasserleiter. **The heat transfer by the
introduction of cooled water in aquifers.**

1265 Mull, R.; Bröker; Khan   DE 138450/78/0010 N
Berechnung der Veränderung von Hochwasserwellen durch
Retentionseffekte. **Determination of modification of flood
waves by retentioneffects.**

1266 Mull, R.; Barovic; Homagk   DE 138450/78/0011 N
Mathematisch numerisches Modell zur Beschreibung der
Ausbreitung von Inhaltsstoffen im Grundwasser.
**Mathematical–numerical model for the description of the
diffusion of constituents in the groundwater.**

1267 Lecher, K.; Holle; Leonhardt; Träger
    DE 138450/78/0012 N
Hochwasservorhersage in Flussgebieten mit
Regelungssystemen. **The flood–forecast in river basins with
regulation–systems.**

1268 Lecher, K.; Rickert, K.   DE 138450/78/0013 N
Verfahren zur Bestimmung wirtschaftlicher Ausbaugrössen
kleiner Vorfluter in Niedersachsen. **Method for determination
of economical criteria of small draining channels in Lower
Saxony.**

1269 Lecher, K.; Billib, H.; Niemann, E.G.
    DE 138450/78/0014 N
Weltmodell für die Bewirtschaftung des Wassers. **World model
of water management.**

1270 Hoffmann, B.; Meyer, J.   DE 138450/78/0015 N
Nutzung der Wasservorräte auf den Nordseeinseln. **Utilization
of water reserves on the North Sea islands.**

1271 Sieker, F.; Verworn, H.–R.   DE 138450/78/0016 N
Kriterien zur Beurteilung und Vorhersage des Einflusses der
Urbanisierung auf den Hochwasserabfluss. **Criteria for
judgement and prediction of the influence of the urbanization
on flood runoff.**

1272 Hoffmann, B.; Hering   DE 138450/78/0017 N
Untersuchung kurzfristiger Überbelastung eines
GrundwasserReservoirs im Rahmen langfristiger
Wassergewinnungsmassnahmen. **Investigation of a short–run
overload of a groundwater– reservoir during long–term
water–extraction.**

1273 Billib, H.; Verworn, W.   DE 138450/78/0018 N
Digitalgesteuerte, elektro–analoge Simulation des
ungleichförmigen, instationären, reibungsbehafteten
Abflussvorganges in bestehenden, vermaschten und
rückstaubeeinflussten Kanalisationsnetzen. **Digital–controlled
electro–analogous simulation of the discontinuous, intermittent,
friction–influenced discharge procedures in existing,
interconnected sewage networks with backwater effects.**

1274 Lecher, K.; Barleben   DE 138450/78/0019 N
Häufigkeiten bestimmter Wetterlagen und Sturmfluten und
deren Auswirkungen auf bestimmte Abschnitte der
niedersächsischen Küste. **Frequency of certain atmospheric
conditions and storm tides and their effects on specific parts of
the Lower Saxon coast.**

1275 Sieker, F.; Okroy   DE 138450/78/0020 N
Niederschlagsvorhersage. **Rainfall forecast.**

1276 Hoffmann, B.; Hering; Neuss   DE 138450/78/0021 N
Überlastung eines Grundwasser–Reservoirs im Rahmen
langfristiger Wassergewinnungsmassnahmen. **Overloading of a
groundwater–reservoir in the framework of log–term
water–extraction procedures.**

1277 Klenke; Thiem, W.; Radelfahr   DE 138450/78/0022 N
Analyse von Grundwassersystemen unter Ausnutzung
natürlicher Anregungen. **Analysis of ground–water systems
using natural stimulation.**

1278 Mull, R.; Battermann   DE 138450/78/0023 N
Exemplarische Untersuchung des Grundwasser–Dargebotes
im Küstenraum mit mathematisch–numerischen Modellen.
**Exemplary investigations on the available ground–watersupply
in the coastal area with mathematical numerical models.**

1279 Hoffmann, B.   DE 138450/78/0024 N
Vereinfachte numerische Grundwasser–Modelle –
Typ–Modelle – zum Einsatz in der wasserwirtschaftlichen
Rahmenplanung. **Simplified numerical ground–water–models –
type–models – for use in water resources management planning.**

1280 Schlichting, E.; Gaese, D.   DE 144322/71/0002
Untersuchungen über den Wasserhaushalt von Pelosolen.
Investigations into the water regime of pelosols.

1281 Eibach, R.   DE 144450/77/0001
Die Ertrags– und Qualitätsleistung der Rebe bei geringer
Bodenwasserversorgung. **Yield and quality of vitis in low water
supply of soil.**

1282 Alleweldt, G.; Rühl, E.–H.   DE 144450/78/0003 N
Untersuchungen über den Einfluss wechselnder
Bodenwasserversorgung auf die Nettophotosynthese – und
Transpirationsrate – bei Reben. **Investigations on the influence
of changing soil water supply on the net photosynthesis – and
transpiration rate – in grape vines.**

1283 Koch, W.; Philipp, O.   DE 144540/73/0007
Populationsdynamik und Bekämpfung der Wasserhyazinthe
–Eichhornia crassipes– im Sudan. **Population dynamics and
control of water hyacinth –Eichhornia crassipes– in the Sudan.**

1284 Johannes, H.; Hurle, K.   DE 144543/77/0001
Abtrag von Pflanzenschutzmitteln aus Weinbergen in ein
aquatisches System. **Run–off of pesticides from vineyards into
an aquatic system.**

1285 Moser, E.; Sinn, H.   DE 144720/78/0001 N
Strömungstechnische Untersuchungen zur Berechnung von
Tropfbewässerungsanlagen. **Hydrodynamic investigations for
the calculation of trickle irrigation systems.** Publications.

1286 Doppler, W.   DE 144755/75/0002
Die Ökonomie der Allokation von Bewässerungswasser.
Economics of allocation of irrigation water.

1287 Ruthenberg, H.; Walker, H.   DE 144755/77/0002
Die Organisation von Bewässerungsprojekten in Nahost. **The
organisation of irrigation projects in the Near East.**
Publications.

1288 Mosonyi, E.; Koberg, D.   DE 145302/75/0004
Mehrdimensionale Wahrscheinlichkeitsuntersuchungen an Hochwasserwellen. **Multi–variate analysis of flood waves.**

1289 Mosonyi, E.; Hauck, E.   DE 145302/77/0001
Die Anwendung des PMP–Konzepts auf Einzugsgebiete in der BRD. **Application of PMP methods for project basins in the GFR.**

1290 Mosonyi, E.; Eggers, H.   DE 145302/77/0002
Einfluss der Wahl des Schwellenwertes bei der statistischen Analyse von Niedrigwasserdauer und Wassermengendefizit. **Influence of a threshold in analysis of low flow and of deficit in volume of water.**

1291 Plate, E.; Wengefeld, P.   DE 145450/77/0001
Wärmeübergang an der Oberfläche von Flüssen. **Heat loss from the surface of rivers.** Publications.

1292 Plate, E.; Bogardi, J.; Lutz, W.; Binark, M.
    DE 145450/77/0002
Analyse der Ausbaufolge beim Bau von Hochwasserspeichersystemen. **Sequencing and scheduling of flood control reservoir systems.** Publications.

1293 Plate, E.; Beier, M.   DE 145450/77/0003
Sauerstoffübergang an der Oberfläche von Flüssen. **Oxygen transport at the surface of rivers.** Publications.

1294 Plate, E.; Pfaud, A.   DE 145450/78/0001 N
Zusammenwirken qualitativer und quantitativer Parameter bei Analyse und Simulation von Fliessgewässern. **Interdependence of qualitative and quantitative water parameters in analysis and simulation of river systems.** Publications.

1295 Finck, A.; Boysen, P.   DE 148050/73/0002
Nährstoffbelastung von Gewässern aus Boden und Düngung. **Impact on waters by an excess of nutritive substances from soil and fertilization.**

1296 Brümmer, G.; Lichtfuss, R.   DE 148051/74/0001
Gehalte und Bindungsformen toxischer Elemente in fluvialen Sedimenten. **Contents and binding forms of toxic elements in fluvial sediments.**

1297 Börner, H.; Ajang, M.R.   DE 148200/77/0001
Chemische Grabenentkrautung – Paraquat. **Chemical control of aquatic weeds – paraquat.** Publications.

1298 Mann, G.   DE 148550/74/0001
Prüfung der Dränfunktion im Feld und mit Modellen. **Function testing of plastic drain pipes in the field and in the laboratory.**

1299 Schendel, U.   DE 148550/75/0002
Wasser– und Nährstoffbilanz eines kleinen Gewässers im ostholsteinischen Hügelland. **Water and nutrient balance of a small water reservoir in Eastern Holstein upland.**

1300 Schendel, U.; Preuss, E.; Schleich, C.
    DE 148550/75/0005
Einfluss von Meliorationen auf den Wasserhaushalt. **Influence of amelioration on water balance.**

1301 Kretzschmar, R.; Hacker, C.–M.   DE 148550/78/0001 N
Der Einfluss intensiver landwirtschaftlicher Flächennutzung auf die Wasserbilanz und den Nährstoffeintrag in ein kleines Staugewässer Ostholsteins. **The influence of intensive agricultural land utilization on the water regime and nutrient content of a small lake in the east of Schleswig–Holstein.**

1302 Kretzschmar, R.   DE 148550/78/0002 N
Natürliche Entsalzung trockengelegter Watten an der Nordküste Schleswig–Holsteins. **Natural leaching of salts in reclaimed shoal areas by the north coast of Schleswig–Holstein.**

1303 Hüser, R.   DE 160030/75/0002
Einfluss von Klärschlammgaben auf Sickerwasserqualität in Waldbeständen. **Effects of sewage sludge on the chemical composition of percolation water in forest stands.**

1304 Baumgartner, A.; Gietl, G.   DE 160120/77/0001
Niederschlagsuntersuchung im Einzugsgebiet 'Grosse Ohe' im Nationalpark Bayerischer Wald. **Investigation of precipitation in the watershed 'Grosse Ohe' in the national park Bayerischer Wald.**

1305 Baumgartner, A.; Reichel, E.   DE 160121/74/0001
Wasserbilanz der Alpen. **Water balance in the Alps.**

1306 Baumgartner, A.; Kirchner, M.   DE 160121/78/0001 N
Landnutzung und Wasserhaushalt bayerischer Bach– und Flussgebiete. **Land use and water balance of Bavarian watersheds of brooks and rivers.**

1307 Becher, H.H.   DE 161020/73/0007
Messtechnische Eliminierung nichtlinearer Saugspannungsgradienten bei der Messung der ungesättigten Wasserleitfähigkeit nach der Verdunstungsmethode. **Measuring for the elimination of nonlinear suction gradients as arising from measuring unsaturated water conductivity by an evaporation method.**

1308 Kromer, K.–H.; Lechner, E.   DE 161522/77/0001
Einsatz verschiedener Beregnungsverfahren für Sonderkulturen. **Use of different irrigation systems in specialized plantations.**

1309 Marr, G.; Göttle, A.; Bock, E.; Manikarnika, R.
    DE 161860/75/0004
Niederschlagsabfluss und –beschaffenheit in städtischen Gebieten. **Stormwater runoff and stormwater quality in urban areas.**

1310 Reimann, K.; Hajek, P.   DE 161860/77/0001
Wassergütemodell für Fliessgewässer. **Water quality modeling in streams.** Publications.

1311 Dauschek, H.   DE 161860/77/0002
Einfluss von Tausalzen auf Grundwasser und Oberflächengewässer. **Influence of thawing salt on surface– and ground–water in protective ground–water areas.**

1312 Neumann, W.; Rothmeier, F.   DE 161860/78/0002 N
Ermittlung von Niederschlagskenngrössen zur Beschreibung von Modellregen für die Bemessung von Kanalnetzen. **Analysis of rainfall data for storm sewer design for assessment of canal network.** Publications.

1313 Meckelein, W.; May, D.   DE 170251/78/0002 N
Probleme der Desertification am Beispiel einiger Oasen in Südtunesien. **Problems of desertification. Examples of some**

D 1310 – Drainage, irregation and water supply

oases in southern Tunisia.

1314 Tietjen, C.; El–Bassam, N.   DE 201040/71/5007 R
Flächenkompostierung kommunaler Abwässerschlämme:
Nachwirkung hoher Aplikationsraten auf Pflanzeninhaltsstoffe
und Bodenwasser 1971–1979. **Area composting of communal
sewage sludges: effects of considerable application rates on
plant constituents and soil water.**

1315 Dambroth, M.; Czeratzki, W.; Schrödter, H.
                                        DE 201040/73/5001
Ermittlung meteorologischer, bodenphysikalischer und
pflanzenphysiologischer Kennwerte für die technische
Steuerung der Wasserversorgung von Kulturpflanzen.
**Investigation on meteorological, soil physical and plant
physiological parameters for irrigation.**

1316 Bramm, A.; Tietjen, C.   DE 201040/75/0006
Belastbarkeit landwirtschaftlich genutzter Flächen durch
städtische Abwässer und deren Einfluss auf die
Grundwasserbeschaffenheit. **The carrying capacity of
agricultural areas as to municipal waste waters and their
influence on the quality of ground water.**

1317 Sourell, H.   DE 201090/75/0007
Ermittlung technischer Kenndaten von Beregnungs– und
Bewässerungsverfahren. **Determination of characteristic
technical data of irrigation and watering processes.**

1318 Sourell, H.   DE 201090/75/0009
Ermittlung von verfahrensspezifischen Kosten der
Feldberegnung. **Determination of special processing costs for
irrigation.**

1319 Kellermann, M.   DE 208050/74/5001
Anwendung der Indikator–Aktivierungsanalyse zur
Ermittlung der Verteilung radioaktiver Abwässer in Flüssen.
**The use of indicator activation analysis for determining the
distribution pattern of radioactive sewage in rivers.**

1320 Feldt, W.; Lauer, R.; Siebert, W.   DE 208050/74/5002
Entwicklung radiochemischer Trenn– und Messverfahren zur
Bestimmung natürlicher Radioisotope in den Ökosystemen des
Meeres und der Binnen– gewässer. **Development of
radiochemical techniques for separating and measuring natural
radioisotopes in marine and freshwater ecosystems.**

1321 Heidler, G.   DE 215090/77/0005
Entwicklung von Verfahren zur Verringerung der
Wasserbelastung nach Anwendung von Herbiziden an und in
Gewässern. **Development of methods for reduction of impact on
water by use of herbicides by and in waters.**

1322 Kothe, P.; Otto, A.   DE 218000/74/0001
Die Fremdstoffbelastung der Gewässer in der Bundesrepublik
Deutschland durch Land– und Forstwirtschaft. **The pollution
charge on waters in the FRG by foreign substances from
agriculture and forestry.**

1323 Liebscher, H.   DE 218000/77/0002
Untersuchungen über den Einfluss des Waldes auf den
Wasserhaushalt und Abflussvorgang. **Study on the influence of
forests on water balance and runoff process.** Publications.

1324 Kothe, P.; Otto, A.; Wiemers, W.   DE 218000/78/0001 N
Gewässertypologie im Ländlichen Raum. **Typology of flowing**

waters in rural areas.

1325 Braden, H.; Hoyningen–Huene, J.von
                                        DE 301030/77/0001
Thermodynamische Modelle des Wasserhaushalts im System
Boden–Pflanzen–Atmosphäre. **Thermodynamic models of
water budget in the system soil–plant–atmosphere.**
Publications.

1326 Siegert, E.; Schmidt, B.   DE 301030/77/0002
Beregnungsbedürftigkeit landwirtschaftlicher Nutzflächen
bestimmter Standorte. **Irrigation neediness of agricultural
acreage at selected locations.** Publications.

1327 Hoppmann, D.   DE 301050/74/0001
Die orographische Beeinflussung von Starkregen. **The
orographical influence on heavy rains.**

1328 Sioli, H.; Furch, K.   DE 404010/74/0003
Untersuchungen zur Nährstoff– und Spurenelementverteilung
in aquatischen Ökosystemen Zentralamazoniens.
**Investigations on nutrient and trace metal distribution in
aquatic ecosystems of Central Amazonia.**

1329 Scholl, W.; Maier, D.   DE 501010/78/0002 N
Borgehalt in Brunnen– und Flusswasser in Baden. **Content of
boron in the water of wells und rivers in Baden.**

1330 Kannenberg, J.   DE 501106/74/0003
Einfluss von Müllkompost auf die Zusammensetzung von
Sickerwasser, dargestellt anhand von Lysimeterversuchen.
**Lysimetric analysis of the influence of refuse on the
composition of seepage water.**

1331 Evers, F.H.; Bücking   DE 501502/77/0002
Einfluss verschiedener Bestandestypen auf die
Sickerwasserqualität im Naturpark Schönbuch. **Influence of
different soil types on seepage water in Schoenbuch Natural
Reserve.**

1332 Schwarz, O.   DE 501504/75/0003
Wasserabfluss und Bodenabtrag in Waldbeständen. **Running
of water and destruction of soil in forest stands.**

1333 Volk, H.; Schwarz, O.   DE 501504/77/0001
Verbesserung der Grundwasserverhältnisse in der Teninger
Allmend. **Improvement of groundwater conditions in Tening
community.**

1334 Volk, H.; Schwarz, O.   DE 501504/77/0010
Einfluss von Müllklärschlammkompost auf die
Grundwasserqualität. **Influence of sewage sludge compost on
the quality of groundwater.**

1335 Volk, H.; Spahl, H.   DE 501504/78/0002 N
Mitwirkung am ökologischen Gutachten zur Auswirkung
geplanter Speicherbecken im Einzugsbereich der Fils.
**Cooperation in ecological expert opinion on the effect of
storage basins envisaged in the catchment area of the Fils river.**

1336 Volk, H.; Schwarz, O.   DE 501504/78/0008 N
Untersuchungen über die Auswirkungen von
Grundwasserentnahmen auf Waldbestände in den
Forstbezirken Lahr und Ettenheim. **Studies on the effects of
groundwater drawing on forest stands in the forest districts of**

D 1310 – Drainage, irregation and water supply

Lahr and Ettenheim.

1337 Schuch, M.   DE 502050/70/0003
Sackungsbeobachtungen an entwässerten Hoch- und
Niedermooren 1955. **Observation on sagging of drained soils in
high and low marshes.** Publications.

1338 Schuch, M.   DE 502050/70/0010
Bestimmung der Verdunstung in bewachsenem und
unbewachsenem Pseudogley mit Lysimetergefässen 1970.
**Determination of evaporation in overgrown and not
overgrown pseudogley with lysimeter instruments.**

1339 Kern, H.   DE 502050/70/0015
Optimale Wasserführung auf staunassen Böden durch
Systemdränung im Vergleich zur Dränung mit weiten
Dränabständen – 20 – 80 m – in Kombination mit
Untergrundlockerung und Meliorationsdüngung 1967.
**Optimum water volume on moisture–retaining soils by
systematic drainage in comparison with the system of large
drain spacing – 20–80 m – in connection with subsoiling and
improvement fertilization.**

1340 Kern, H.   DE 502050/70/0017
Dränfilterprüfung – Kiesfilter, Styromull, Mutterboden,
Grabenaushub – im Rahmen der kombinierten Dränung,
Tonrohrsammler/Erddränsauger 1969. **Testing of drain filter –
gravel filter, styromull, topsoil, trenching – in view of combined
drainage, earthware pipe collector/earth drain suction pipe.**

1341 Schmid, G.; Borchert, H.   DE 502050/72/0003 N
Einfluss der Tiefendüngung mit NPK auf die Eutrophierung
der Gewässer. **Influence of NPK deep fertilization on the
eutrophication of waters.**

1342 Pfrogner, J.   DE 502050/72/0007 R
Grundwasserdauerlinien des Alchemillo Arrhenatheretum der
Innauen 1972. **Permanent groundwater lines of Alchemillo
Arrhenatheretum in meadows by the Inn river.**

1343 Sommer, G.   DE 502055/73/0001
Die Nährstoffauswaschung aus landwirtschaftlichen
Nutzflächen, am Beispiel der Dränversuche Ottenhofen und
Ellingen. **Nutrient leaching from agricultural areas exemplified
by the field draining trials in Ottenhofen and Ellingen.**

1344 Müller, K.; Peternel, M.   DE 502153/77/0005
Prüfung des Einflusses der anfeuchtenden und
klimatisierenden Beregnung auf das Bestandsklima und damit
auf die Entwicklung und den quantitativen und qualitativen
Ertrag der Reben. **Testing of the influence of moistening and
climatizing irrigation on crop and with that on growth and
quantitative and qualitative yield of vines.**

1345 Karl, J.; Porzelt, M.   DE 502251/71/0002 N
Messung von Abfluss und Abtrag mittels einer transportablen
Regneranlage. **Measurement of runoff and erosion by portable
sprinkler.**

1346 Karl, J.; Schäfer, R.   DE 502251/77/0001
Einfluss der Bodennutzung auf Wasserhaushalt und
Bodenerosion. **Influence of land–use on water balance and soil
erosion.**

1347 Busse, M.; Jung, W.   DE 502551/77/0003
Hydrobakteriologische Untersuchungen im Flusssystem der

Isar zwischen München und Freising, unter besonderer
Berücksichtigung des Speichersees bei Ismaning.
**Hydrobacteriological investigations in the Isar river system
between Munich and Freising with special regard to the
Ismaning reservoir.**

1348 Researcher not indicated   DE 502650/78/0002 N
Wasserwirtschaftliche Rahmenplanung Isar
Chemisch–biologische Untersuchungen am Speichersee und
mittleren Isarkanal. **Water resources political frame planning
on the Isar River. Chemico–biological studies on the reservoir
and on the middle Isar Canal.**

1349 Researcher not indicated   DE 502650/78/0003 N
Wasserwirtschaftliche Rahmenplanung Isar Bakteriologische
Untersuchungen am Speichersee bei Ismaning und am
Isarkanal. **Water resources political frame planning on the Isar
River. Bacteriological studies on the reservoir near Ismaning
and on the Isar Canal.**

1350 Researcher not indicated   DE 502650/78/0004 N
Wasserwirtschaftliche Rahmenplanung Isar Untersuchungen
zur Limnologie des Kochelsees und über die
Nährstoffbelastung durch die Zuflüsse. **Water resources
political frame planning on the Isar River. Studies on the
limnology of Kochel Lake and on the nutrient load by
tributaries.**

1351 Researcher not indicated   DE 502650/78/0005 N
Untersuchungen für ein Wassergütemodell am Tegernsee und
Schliersee. **Analyses for model of water quality of Tegern Lake
and Schlier Lake.**

1352 Siebeck, O.   DE 502650/78/0006 N
Limnologische Erforschung des Königssees. **Limnological
research on Koenigssee.**

1353 Researcher not indicated   DE 502650/78/0007 N
Möglichkeiten und Erfolgsaussichten der Seenrestaurierung.
Possibilities and prospects of restoration of lakes.

1354 Researcher not indicated   DE 502650/78/0008 N
Abwasserchemische und toxikologische Untersuchungen von
Motorölen, Schleifkühlmitteln und Druckwasserzusätzen.
**Sewage–chemical and toxicological analyses of motor oils,
grinding coolants and compressed–water additives.**

1355 Researcher not indicated   DE 502650/78/0009 N
Prüfung der biologischen Abbaubarkeit von aus Bootsmotoren
abgegebenen Rückstandsprodukten. **Testing of biolocical
decomposition of residual products from boat motors.**

1356 Researcher not indicated   DE 502650/78/0013 N
Untersuchungen über den Einfluss von HAB in Kläranlagen
a. Versuche an schwachbelasteten Belebungsanlagen
b. Versuche an einer Oxidationsgrabenanlage c. Versuche an
Tropfkörperanlagen. **Analyses of the influence of HAB in
desilting works a. Experiments on low–stressed activated sludge
plants b. Experiments on an oxidation ditch plant c. Experiments
on trickling filter plants.**

1357 Researcher not indicated   DE 502650/78/0014 N
Untersuchungen über den Einfluss von SASIL auf biologische
Abbauvorgänge. **Studies on the influence of SASIL on
biological decomposition process.**

D 1310 – Drainage, irregation and water supply

1358 Researcher not indicated   DE 502650/78/0015 N
Gewässerökologische Absicherung des Umweltverhaltens von
HAB – SASIL–. **Water–ecological ensuring of environmental
behaviour of HAB – SASIL –.**

1359 Offhaus, K.; Reimann, K.; Wachs, B.; Hoffmann, H.–J.
                                        DE 502651/77/0001
Untersuchungen für ein Wassergütemodell eines
Fliessgewässers am Beispiel des Lech. **Studies on a water
quality model exemplified by Lech river.**

1360 Bohl, M.   DE 502653/75/0001
Reinigung von teichwirtschaftlichen Abwässern aus einer
Fischintensivhaltung mit Kreislaufführung,
Modelluntersuchung zur Reinhaltung von Badeseen. **Cleaning
of pond waste waters from intensive fish keeping with
circulation conduct. Model study on the keeping–clean of
bathing lakes.**

1361 Zakosek, H.; Balzhäuser, H.; Kiefer, W.
                                        DE 506101/77/0002
Untersuchungen über die Tropfbewässerung in Hang– und
Steillagen des Weinbaues. **Investigations on drip irrigation on
viticultural slopes and on steep slopes.**

1362 Rühling, W.; Bäcker, G.   DE 506113/77/0001
Vergleich von Tropfbewässerungssystemen für den Weinbau in
Flach– und Steillagen. **Comparison of drip–irrigation–systems
for vine growing in flats and on slopes.**

1363 Brechtel, H.M.   DE 506452/71/0001
Einfluss von Waldbeständen verschiedener Baumarten– und
Altersklassen auf die Grundwasserneubildung im Stadtwald
Frankfurt/M.. **Influence of forest stands of different tree species
and age classes on the ground water yield in the municipal forest
of Frankfurt/M..**

1364 Brechtel, H.M.; Scheele, G.   DE 506452/71/0003
Hydrologische Bewertung und Klassifizierung der hessischen
Waldgebiete hinsichtlich ihrer Bedeutung als
Hochwasserschutz bei Schneeschmelzen sowie ihrer
Auswirkung auf die Grundwasserneubildung. **Hydrological
evaluation and classification of the Hessian forest regions in
respect of their importance for flood flow reduction and their
effect on the ground water yield.**

1365 Müller, W.; Renger, M.; Voigt, H.   DE 507102/70/0002
Melioration staunasser Böden. **Amelioration of soils with
impeded drainage.**

1366 Müller, W.; Fleige, H.; Fastabend, H.; Renger, M.
                                        DE 507102/73/0001
Untersuchungen über die Auswirkung der Abwasser und
Klärschlammabwasser auf die chemischen und physikalischen
Eigen– schaften wichtiger Böden Niedersachsens und auf das
Grund– wasser. **Investigations on the effects of waste water and
sewage sludge on the chemical and physical characteristics of
important soils and on the ground water in Lower Saxony.**

1367 Renger, M.; Müller, W.; Fleige, H.; Strebel, O.
                                        DE 507102/77/0001
Bestimmung der Beregnungsdürftigkeit wichtiger Standorte
Niedersachsens. **Evaluation of the need for sprinkler irrigation
for the main soil–site types in Lower Saxony.** Publications.

1368 Renger, M.; Müller, W.; Strebel, O.; Voigt, H.

                                        DE 507102/77/0002
Auswirkungen von Absenkungen des Grundwassers auf
Wasserhaushaltskomponenten und Bodennutzung. **Effect of
groundwater table drawdown on the components of water
balance equation and on soil–use.** Publications.

1369 Kuntze, H.; Eggelsmann, R.; Burghardt, W.
                                        DE 507103/70/0001
Grundwasser– und Bodenfeuchtemessungen in einem
Aufforstungsversuch auf Hochmoor mit verschiedenen
Maulwurf–Dränabständen. **Measurements of ground water and
soil moisture in an afforestation test on high bog with different
mole drain spacings.**

1370 Reschke, M.   DE 507250/77/0001
Erprobung eines biologischen Verfahrens zur Beseitigung
unerwünschten Krautwuchses in Gewässern. **Testing of a
biological method to remove undesired weeds in waters.**

1371 Quast, P.   DE 507301/75/0002
Bearbeitung wasserwirtschaftlich–bodenkundlicher Probleme
in den Obstanlagen an der Niederelbe. **Studies on problems of
water utilization and pedology in fruit plantations on the Lower
Elbe.**

1372 Karrenberg, H.   DE 508100/72/0002 R
Deutscher Beitrag zur Hydrogeologischen Karte von Europa
1:1,5 Mill. 1972. **German contribution to the hydrogeological
map of Europe 1:1,5 millions.**

1373 Krämer, F.; Bahr, R.   DE 508302/70/0005 R
Bodennutzung und Pflanzenertrag unter grundwassernahen
und –fernen Verhältnissen in Abhängigkeit von verschiedener
Stickstoffdüngung und bei zusätzlichem Wasserdargebot durch
Beregnung Grundwasserkontamination mit Pestiziden
Hydrologische Untersuchungen zur Ermittlung des
Grundwasserhaushaltes, insbesondere der
Grundwasserneubildung 1968. **Soil utilization and plant yield
under conditions of groundwater being near and far off in
dependence on varying nitrogen fertilization and additional
water supply by irrigation.**

1374 Hansen, L.; Nielsen, C.   DK 010105/58/3406
Afvandingsmetoders betydning for jordstruktur og
dyrkningssikkerhed. **Effects of different drainage methods on
soil structure and crop yield reliability.**

1375 Hansen, L.; Petersen, E.F.   DK 010105/71/3408
Måling af vandmængde og vandkvalitet fra drænede arealer.
**Measurement of water quantity and quality from drained
areas.**

1376 Knudsen, H.   DK 010109/65/3801
Kortlægning af jordbrugets vandingsbehov med henblik på
udarbejdelse af prognoser for det fremtidige vandingsbehov.
**Survey of agricultural water requirements with reference to the
development of prognoses for future water requirements.**

1377 Knudsen, H.   DK 010109/66/3803
Vanding og gødskning af korn, græsmarker og kartofler.
Irrigation and fertilizer use in cereals, grassland and potatoes.

1378 Knudsen, H.; Jørgensen, V.   DK 010109/70/3802
Forskellige plantearters sæsonmæssige vandforbrug med
henblik på en optimal udnyttelse af vandressourcerne.
Seasonal water use of different plant species with reference to

optimal utilization of water resources.

1379 Knudsen, H.; Jensen, F. DK 010109/70/3804
Udnyttelse af spildevand fra levnedsmiddelindustrien –
mejerier og kartoffelmelsfabrikker – til vanding af
landbrugsafgrøder. **Utilization of discharge water from
foodstuff industries – dairies and potato flour factories – for the
irrigation of agricultural crops.**

1380 Knudsen, H.; Jørgensen, V. DK 010109/71/3805
Vanding med saltholdigt vand til landbrugs- og
havebrugsafgrøder. **Irrigation of agricultural and horticultural
crops with saline water.**

1381 Knudsen, H.; Jørgensen, V. DK 010109/73/3806
Udnyttelse af vand som temperaturregulerende faktor. **Use of
water as a temperature–regulating factor.**

1382 Christensen, J. DK 010500/78/0005 N
Vandingsøkonomi og dyrkningssikkerhed. **The economics of
irrigation and the stability of plant production.**

1383 Olesen, S.E.; Andersen, S.Aa. DK 010900/69/0001
Metoder og materialer til dræning af okkerholdig jord.
Methods and materials for drainage of ochreos soils.

1384 Andersen, S.Aa. DK 010900/69/0002
Udvikling og afprøvning af egnede materialer til beskyttelse
mod materiale–indslemning i dræn. **Development and testing of
suitable materials for preventing deposition in drain pipes.**

1385 Olesen, S.E. DK 010900/75/0001
Strukturforbedring og dræning af komprimeret jord. **Structure
improvement and draining of compacted soil.**

1386 Olesen, S.E. DK 010900/76/0003
Behandling af jernholdigt drænvand. **Treatment of ferreous
drainage water.**

1387 Olesen, S.E. DK 010900/76/0008
Drænvandskvantitet og –kvalitet (Suså). **The quantity and
quality of drainage water (Suså).**

1388 Olesen, S.E. DK 010900/77/0001
Næringssaltindhold og grødevækst i vandløb. **Nutrient salt
content and weed growth in streams.**

1389 Olesen, S.E. DK 010900/77/0002
Vandindvinding fra vandløb og indvindingens indflydelse på
vandløbenes kemiske og biologiske forhold. **Water catchment
from streams and the influence of catchment on chemical and
biological conditions in streams.**

1390 Olesen, S.E. DK 010900/77/0004
Dræning af jord med lille hydraulisk ledningsevne. **Draining of
soil with low hydraulic conductivity.**

1391 Olesen, S.E. DK 010900/77/0008
Analyse af danske jordes vandkapacitet. **Analysis of the water
capacity of Danish soils.**

1392 Kristensen, K.J. DK 030143/77/0013
Evapotranspiration (det hydrologiske forskningsprojekt i
Susåens opland). **Evapotranspiration (the hydrological
research project in the Suså area).**

1393 Jensen, H.E. DK 030143/79/0002 N
Potentiel evapotranspiration i relation til globalstråling.
Potential evapotranspiration in relation to global radiation.

1394 Dahl, N.J. DK 030161/75/0002
Ikke–stationære strømninger i åbne ledninger. **Non–stationary
flow currents in open conduits.**

1395 Dahl, N.J. DK 030161/78/0001 N
Basal grundvandsstrømning og vandindvinding til
markvanding. **Base flow and water supply for irrigation.**

1396 Moltesen, P.; Madsen, T.L. DK 030181/77/0020 N
Grundvandstandens indflydelse på rumtæthedsniveauet hos
rødgran. **The influence of the depth of the water–table on the
basic density level of Norway spruce.**

1397 Perrier, A.; Itier, B.; Gosse, G. FR 010101/76/5023 N
Etude du comportement hydrique moyen des couverts
végétaux et conséquences sur les définitions pratiques de
l'évapotranspiration. **Water functionning study of crops and its
consequences on practical definitions for evapotranspiration.**

1398 Archer, Ph. FR 010101/76/5035
Evaporation de l'eau du sol en dehors des périodes de
végétation. **Soil water evaporation between vegetation periods.**

1399 Turc, L. FR 010103/48/6218
Indices climatiques concernant les bilans hydriques et les
potentialités agricoles (généralités sur des opérations de
recherches portant sur ces indices). **Climatic indices concerning
water balance and agricultural potential.**

1400 Turc, L. FR 010103/48/6219
Indices climatiques concernant les bilans hydriques :
complément à fiche précédente. **Climatic indices concerning
water balance.**

1401 Mamy, J. FR 010103/71/6339
Etude de la répartition de l'eau et de ses propriétés
physicochimiques, dans les silicates phylliteux, aux faibles et
aux fortes hydratations. **Study of water distribution and
physicochemical properties in phyllo–silicate for low and strong
hydrations.**

1402 Bétremieux, R. FR 010103/72/6326
Drainage – expérimentation au champ. **Drainage. Field
experimentation.**

1403 Chaussidon, J. FR 010103/73/6332
Etats de l'eau dans les milieux poreux. **Water state in porous
media.**

1404 Le Renard, J. FR 010103/73/6338
Etude directe ou simulée du transfert de l'eau dans les sols.
Direct or simulated study of water flow in soils.

1405 Lemaire, L. FR 010112/65/9107
Utilisation de l'irrigation de complément du maïs–grain dans
l'Ouest de la France. **Study of the economic returns from
complementary irrigation of grain–maize in the West of France.**

1406 Lemaire, L. FR 010112/78/9106
Possibilités de l'utilisation de l'irrigation de complément des
plantes fourragères dans l'Ouest de la France. **Investigation of

the possibilities of improving fodder supply in the West of France through forage irrigation practices.

1407 Lesel, R.; Saboureau, J.L.; Broutin, F.
FR 010204/74/5506
Utilisation de la truite Arc–en–ciel pour l'estimation dëla qualité de l'eau. **Rain bow trout as indicator of water quality.**

1408 Specty, R.
FR 010306/66/3004
Etude de divers niveaux d'irrigation sur différentes rotations de plantes de grande culture. Effets annuels et cumulés.
Experiment on different levels of irrigation on different crop rotations Annual and cumulative effects. Publications.

1409 Gras, R.
FR 010502/76/5006
L'eau et le végétal. Contrôle de l'irrigation des cultures florales sous serre. **Irrigation control of greenhouse bloom.**

1410 Faure, A.
FR 010601/69/0538
Analyse de la portance des sols. **Analysis of the "bearing capacity" of soils.** Publications.

1411 Fies, M. Mme
FR 010601/70/0539
Influence de l'humidité sur la résistance à l'encrasement de pâtons de terre, aux très faibles teneurs en eau. **Effects of moisture on shear strength of soil test–pieces at very low water contents.**

1412 Cabibel, B.
FR 010601/71/0535
Etude expérimentale de la répartition hétérogène de l'eau et des éléments minéraux nutritifs dans le sol sous irrigation goutte à goutte et relations avec l'enracinement. **Experimental study of water and mineral nutrients distribution in soil under trickled irrigation in relation with the development of the roots.**

1413 Fies, M. Mme; Puech, M.
FR 010601/71/0540
Application de la notion de porosité texturale à l'étude de la circulation de l'eau non saturante. **Use of textural porosity for the prediction of unsaturated water flow.**

1414 Cabibel, B.
FR 010601/73/6259
Influence des caracteristiques physiques du sol sur la repartition de l'eau et des sels sous irrigation fertilisante localises sans charge. **Influence of soil physical characteristics on water and salts distribution under null–head fertilizing drip irrigation.**

1415 Cabibel, B.
FR 010601/73/6260
Etude experimentale de la repartition de l'eau dans le sol sous irrigation goutte a goutte et relation avec l'enracinement.
Experimental study of water distribution in soils under trickle irrigation – relation to rooting.

1416 Arnoux, M.; Defrance, H.
FR 010615/73/9038
Etude des potentialités de l'irrigation localisée en arboriculture. **Evaluation of the efficiency of drop system irrigation for fruit tree orchards.**

1417 Durand, J.H.
FR 010704/70/6093
Efficacité des systèmes d'assainissement et de drainage.
Drainage system efficiency.

1418 Juste, C.; Dureau, P.; Tauzin; Menet, M.
FR 010704/72/6090
Lessivage des éléments fertilisants en sols sableux, conséquences sur la pollution de la nappe phréatique et des lacs landais. **Leaching of fertilizers in sandy soils, effects on the pollution of groundwater and lakes in south–western france.**

1419 Durand, J.H.
FR 010704/73/6092
Qualité des eaux d'irrigation. **Irrigation waters quality.**

1420 Robelin, M.
FR 010802/56/6121
Evapotranspiration réelle maximale des differentes cultures.
Maximum real evapotranspiration of differents crops.

1421 Gachon, L. – Robelin, M.; de Montard, F. X.; Triboi, E.
FR 010802/68/6096
Bilans hydriques et chimiques en sols granitiques de différentes profondeurs pour différents systèmes fourragers. **Water and chemical balances in granitic soils for various forrage systems.**

1422 Mingeau, M.; Duthion, C.
FR 010802/73/6117
Influence de l'excès d'eau et de la mauvaise aération sur la plante. **Effect of excess moisture and deficient aeration on plants.**

1423 Mingeau, M.; Robelin, M.
FR 010802/74/6119
Photosynthèse et alimentation hydrique. **Photosynthesis and water supply.**

1424 Specty, R.; Meyer, J.L.; Meyer, J.
FR 010906/66/9062
Etude des effets cumulés et des arrière–effets de plusieurs facteurs agronomiques appliqués dans le cadre de rotations possibles dans la Hardt irriguée. **Study of the cumulated and late effects of several agronomic factors introduced through various rotation systems applicable to the Hardt valley under irrigation.**

1425 Specty, R.; Meyer, J.L.; Mettauer, H.
FR 010906/68/9063
Essais d'irrigation sur diverses cultures en cases lysimétriques : Effet des modifications de l'E.T.R. sur le rendement : recherche de coefficients culturaux E.T.P./E.T.M. **Testing irrigation practices adapted to various crops in lysimetric bins ; measuring the influence of ETR modifications on the yield and evaluating the ITP/ETM cultural coefficients.**

1426 Specty, R.; Meyer, J.L.; Meyer, J.
FR 010906/71/9064
Essais de références en amont et en aval des essais "Irrigation x rotation" d'Algosheim (étudiés par ailleurs). **Testing appropriate cultural and harvesting practices within the framework of the Algosheim Irrigation x Rotation research project.**

1427 Specty, R.; Meyer, J.L.; Meyer, J.
FR 010906/76/9065
Rideau de verdure "protection et loisirs" sous irrigation permanente à "l'eau chaude". **Testing "green curtain" plantations for protection and leisure established through permanent warm water irrigation.**

1428 Concaret, J.; Perrey, C.; Clair, A.; Guyot, J.
FR 011001/65/6285
Etude expérimentale en laboratoire et au champ de la circulation de l'eau en milieu saturé. **Field and laboratory study of water movement in a saturated medium.**

1429 Guyot, J.
FR 011001/73/6288
Influence des cations dans les phénomènes de désorption de l'eau adsorbée par la matière organique. **Influence of cations in desorption phenomenons of water adsorbed by organic matter.**

D 1310 – Drainage, irregation and water supply

1430 Bornand, M.   FR 011211/72/0613
Orientation des aménagements dans le bassin rhodanien.
Programming the reclamation of the Rhone Valley.
Publications.

1431 Favrot, J.C.   FR 011211/72/6317
Etude pédologique en vue du drainage des terres agricoles du
Pays d'Ouche (Eure). **Pedological study for farm land drainage
in Pays d'Ouche (Eure).**

1432 Callot, G.   FR 011211/74/6312
Aménagement hydroagricole de la vallée du Thouet (Deux
Sévres). **Hydro–agricultural management of thouet valley.**

1433 Puech, J.   FR 011401/67/6205
Caractérisation de l'évapotranspiration dans différents
microclimats et répercussions sur les bilans hydriques de
rotations types : application aux avertissements irrigation.
**Evapotranspiration in different microclimats and influence on
hydric balance of crops rotations : application to irrigation
warning.**

1434 Decau, J.; Bouniols, Melle   FR 011401/68/0639
Influence de quelques techniques culturales (irrigation
fertilisation azotée, système de culture) sur la production
protéique des céréales (maïs, sorgho). **Effect of irrigation,
nitrogen fertilization, cropping system on protein production of
maize and sorghum.**

1435 Marty, J.R.; Maertens, C.   FR 011401/69/0641
Etude de l'évolution et de la conservation de la fertilité du sol
dans différentes rotations avec ou sans irrigation; étude des
arrière–effets. **Evolution and conservation of soil fertility under
different rotations, with or without irrigation –Studies on
residual effects.**

1436 Puech, J.; Marty, J.R.   FR 011401/72/0634
Bilan d'évapotranspiration en fonction des systèmes de
cultures et évolution des ressources hydriques à diverses
échelles géographiques. **Evapotranspiration balance in relation
with cropping systems, and evolution of water disponibilities at
various geographic scales.** Publications.

1437 Puech, J.; Marty, J.R.   FR 011401/72/0640
Méthodes et techniques d'irrigation adaptées à divers systèmes
de culture et valorisation de l'eau. **Irrigation practices in
relation with cropping systems and improvment of water use
efficiency.**

1438 Damour, L.; Camus, P.; Lesage, B.   FR 012201/65/9060
Assainissement et drainage des marais côtiers du
Centre–Ouest. **Reclamation and drainage of the coast–line
marsh–lands of central–western France.**

1439 Damour, L.; Camus, P.   FR 012201/70/3020
Drainage, irrigation et systèmes de production applicables
dans les Marais de l'Ouest. **Drainage, irrigation and production
systems in the western Marshland.** Publications.

1440 Damour, L.; Moisan, H.; Garreau, J.; Raichon, C.;
Jeannin, B.   FR 012201/75/3034
Conditions d'adoption de nouvelles techniques
d'assainissement par les agriculteurs. **Conditions of the
adoption of new techniques of drainage by the farmers.**
Publications.

1441 Muller, J.   FR 012210/72/6272
Epandage par aéroaspersion des eaux de féculerie et de
sucrerie. **Spreading by aero sprinklers of waste waters from
starch and sugar works.**

1442 Dutil, P.   FR 012210/74/6273
Transfert de l'eau "sol – nappe". **Water movement between soil
and water–table.**

1443 Ballif, J.L.   FR 012210/74/6275
Ecoulement en milieu poreux crayeux. **Flow in a chalky porous
medium.**

1444 Ballif, J.L.   FR 012210/74/6276
Caractéristiques hydriques des sols de craie. **Water
characteristics of chalky soils.**

1445 Balvay, G.   FR 012218/74/5571
Modalités de mise en oeuvre de l'inventaire du degré de
pollution des lacs et des étangs de France. **Water pollution
criteria of some lakes and ponds in France.**

1446 Feuillade, J.; Feutrie, J.; Feuillade, M.; Blanc, P.
                                       FR 012218/75/5559
Mise au point de tests biologiques pour l'appréciation des
conséquences à long terme de l'action de polluants sur la
qualité des eaux. **Use of algal long time bio–assays with water
pollutants.**

1447 Concaret, J.; De Vaubernier, E.; Jeannin, B.; Teilhard
de Chardin, B.   FR 012224/76/9009
Etude du drainage sur sols lourds de Lorraine. **Evaluation of
the benefits of draining heavy soils in Lorrain.**

1448 Chrétien, J.   FR 012227/72/6258
Composition des eaux de drainage et de captage. **Chemical
contents of drainage and catchment–basin waters.**

1449 Perigaud, S.; Avronsard   FR 012502/75/6143 N
Méthodologie d'étude de projet et de contrôle des épandages
d'eaux usées agroalimentaires.. **Methodology for project
launching and control of food processing wastewaters land
application.**

1450 Aikinson   GB 011008/79/0051 N
Irrigation and water use in orchards.

1451 Garwood   GB 011202/76/0027 R
Irrigation of forage crops : efficiency of water use.

1452 Draycott; Messem   GB 012301/00/0008 R
Irrigation and water usage.

1453 McAllister   GB 040402/00/0009 R
Composition of drainage and rainwater.

1454 McAllister   GB 040402/00/0014 R
**Effect of slurry dressing on soil and drainage water under
varying conditions.**

1455 McAllister; McConaghy   GB 040402/00/0017 R
Effects of large scale drainage on soil conditions.

1456 Smith; Stephens   GB 040801/00/0001 R
**Nitrogen and phosphorus budget of Lough Neagh and its
catchment area.**

D 1310 – Drainage, irregation and water supply

1457 Speirs GB 060114/00/0003 R
Drainage of soils of low permeability.

1458 Farr GB 060219/00/0010 R
Land drainage studies.

1459 Farr; Elliot GB 060219/79/0017 N
Land drainage studies.

1460 McGregor GB 060306/00/0008 R
Development of effective and economic drainage systems for soils of low permeability.

1461 Carey, M.L.; Hendrick, E. IE 050100/60/7204 N
Afforestation techniques including drainage. Publications.

1462 Galvin, L. IE 060300/74/0997
Porous fill in drainage installations. Publications.

1463 Liani, A. IT 011801/76/0004
Prova di irrigazione a goccia al vivaio di pioppo in relazione alla possibilità d'impianto di nuove distanze. **Drip irrigation in poplar nursery related to the use of new planting spacings.** Publications.

1464 Liani, A. IT 011801/79/0006 N
Prova di irrigazione a goccia in impianto specializzato di pioppo in riferimento ai consumi idrici ed alla disposizione dei gocciolatori. **Drip irrigation trial in poplar plantation with relation to water requirements and drop arrangement.**

1465 Cale, M.T.; Romano, E.; Figliolia, A.
 IT 020200/33/0001
Bilanci energetici idrologici in rapporto all'evapotraspirazione. **Energy and water balances in relation to evapotraspiration.** Publications.

1466 Tombesi, L.; Favola, G.; Calè, M.T.; Moretti, R.;
Indiati, R. IT 020200/73/0001
Bilanci meteorologici e idrologici mediante computers per l'ottimizzazione delle riserve idriche. **Meteotological and water balances with computers for the optimum use of water resources.** Publications.

1467 Montorsi, M. IT 020500/73/0006
Qualità delle acque irrigue dell'Emilia–Romagna. **Quality of irrigation water in Emilia–Romagna.**

1468 Lanza, F. IT 020500/74/0574
Studio dei problemi agronomici dell'irrigazione. **Study on agronomic problems in irrigation.**

1469 Venezian Scarascia, M.E. IT 020500/76/0008 N
Confronto di volumi x turni irrigui in girasole. **Sunflowers yield influenced by irrigation volumes and frequencies.**

1470 Colzani, G.; Marsili, A.; Nuccitelli, G.
 IT 020600/77/0004 N
Studio sulle tecnologie per l'irrigazione. Applicazione dell'energia solare al pompaggio dell'acqua.. **About the tecnology of the irrigation, application of solar energy to pump water..**

1471 Marsili, A.; Nuccitelli, G. IT 020600/79/0004 N
Energia solare per gli scopi irrigui. **Solar energy for irrigation purposes.** Publications.

1472 Mariani, G.; Colesanti, F. IT 020800/75/0018 R
Sorgo – Ricerche agronomiche sulla coltura da granella in ambienti semiaridi, con e senza irrigazione, in comparazione con la coltura del mais. **Sorghum – Agronomic researches on cultivation for grain production with and without irrigation in semiarid conditions, in comparison to maize cultivation.** Publications.

1473 Onofrii, M.; Tomasoni, C.; Basta, P.
 IT 020900/79/0004 N
Prove per la determinazione dei periodi di irrigazione del prato permanente polifito in funzione dell'efficienza dell'acqua irrigua. **Trials to determine irrigation periods in permanent polyphytic meadow as effected by irrigation water.**

1474 Onofrii, M. IT 020900/79/0005 N
Ricerche sull'evapotraspirazione effettiva e sui consumi idrici dell'erba medica e del sorgo da foraggio, misurati in cassoni lisimetrici e sue relazioni con quella potenziale. **Comparative study between real evapotranspiration potenzial of lucerne and fodder sorghum measured with lysimeter and potential evapotraspiration determined by other methods.**

1475 Iannelli, P.; D'Andrea, E. IT 020900/79/0008 N
Coltura di medica semirrigua con trasemina di orzo, loglio italico e trifoglio alessandrino. **Cultivation of lucerne partially irrigated with over – seeded barley, italian raygrass and berseem clover.**

1476 Iannelli, P.; Terzano, G.; D'Andrea, E.; Tortone, G.
 IT 020900/79/0009 N
Successioni colturali in regime irriguo e semirriguo. **Fodder cropping systems under irrigation and partial irrigation.**

1477 D'Amore, R.; Petralia, S.; Perella, C.
 IT 021000/79/0007 N
Confronto tra diversi sistemi di irrigazione in coltura pacciamata. **Comparison between different irrigation methods in mulch cultures.**

1478 Cremaschi, D.; D'Amato, A.; Fontana, F.
 IT 021100/79/0007 N
Irrigazione e concimazione azotata in coltura di barbabietola da zucchero. **Irrigation and azotic fertilization in sugar beets cultivation.**

1479 Liuni, C.; Giorgessi, F.; Giovannardi, R.; Colapietra,
M.; Calò, A.; Scalabrelli, G. IT 021300/78/0002 R
Irrigazione della vite. **Vineyard irrigation techniques.**

1480 Fiorino, P. IT 021400/72/1544
Problemi agronomici sulla irrigazione dell'olivo. **Agronomic problems related to the irrigation of olives.** Publications.

1481 Iannotta, N.; Perri, L.; Pantusa, M.
 IT 021400/74/0005 R
Studio dei problemi agronomici dell'irrigazione dell'olivo. **Study of agronomical problems related to olive tree irrigation.**

1482 Colorio, G.; Strabbioni, G.; Bergamini, A.; Manzo, P.;
Cobianchi, D. IT 021500/78/0008 N
Studio dei volumi irrigui, dei vari metodi di distribuzione localizzata dell'acqua, e dei rapporti tra irrigazione e sistemi di lavorazione del terreno nelle piante arboree da frutto. **Study of**

the irrigation volumes, of the localized water distribution systems and of the interaction between irrigation and soil management in fruit tree orchards.

1483 Raciti, G.; Scuderi, A. IT 021600/72/0001 R
Ricerche e sperimentazioni sui nuovi metodi di irrigazione in agrumicoltura. **Research and experiments on new irrigation methods in citriculture.** Publications.

1484 Raciti, G.; Barbagallo, A.; Giuffrida, A.
 IT 021600/79/0003 N
Validità dell'irrigazione localizzata nella pratica della forzatura del limone. **Validity of localized irrigation in the practice of "Verdelli" production of lemon.**

1485 Marzi, V. IT 040101/72/0108
Ricerche di laboratorio e di pieno campo sui problemi dell'irrigezione. **Laboratory and field studies on irrigation problems.** Publications.

1486 Caliandro, A. IT 040101/77/0126 R
Studio sui problemi agronomici dell'irrigazione. **Study of the agronomic problems of irrigation.**

1487 Cavazza, L. IT 040201/72/0088
Ricerche sull'irrigazione. **Research on irrigation.** Publications.

1488 Cavazza, L.; Pesci, C.; Cacchi, D. IT 040201/74/0001
Prova di drenaggio sotterraneo con confronto di distanze tra i dreni. **Drainage trial with comparison between distances between drains.**

1489 Cavazza, L. IT 040201/77/0139 R
Problemi agronomici dell'irrigazione. **Agronomic problems of irrigation.**

1490 Cacchi, D. IT 040201/77/0622 R
Funzione regimante delle scoline. **Effects of outlet channels on the water balance.**

1491 Giari, M. IT 040220/74/0562
Ricerca sperimentale sui metodi per il calcolo delle reti di dranaggio sotterraneo. **Experimental research on the methodology of calculation of subsurface drainage networks.**

1492 Ferrari, C. IT 040226/77/0179 R
Ricerche botaniche sull'irrigazione. **Botanical research on irrigation.**

1493 Amadesi, E. IT 040248/78/1163 N
Erosione e stabilità dei versanti in funzione dell'idrografia e della vegetazione, carta dell'utilizzazione reale del suolo. **Slope erosion and stability depending on hydrographic structure and vegetation ; mapping the effective use of soil.**

1494 Tribulato, E.; Germanà, C.; Alberghina, O.
 IT 040307/79/0001 N
Indagine sperimentale sull'irrigazione localizzata degli agrumi. **Experiments on citrus drip irrigation.** Publications.

1495 Maracchi, G. IT 040502/77/0224 R
Determinazione dell'efficienza idrica di alcune colture. **Determination of water supply efficiency in some cultivations.**

1496 Landi, R. IT 040502/77/0635 R
Dimensionamento della fossa livellare e della maglia

sistematoria. **Size of the catchment basin and of the regulating network.**

1497 Zoli, L. IT 040525/73/0333
Drenaggio sotterraneo del terreno agrario con tubi PVC. **Subsurface drainage of agricultural lands using PVC tubes.**

1498 Grazi, S. IT 040525/77/0631 R
Afflussi meteorici e deflussi, liquidi e solidi e connessi processi erosivi in bacini montani. **Liquid and solid meteorological inflow and outflow and related erosive processes in mountain basins.**

1499 Zoli, L. IT 040525/78/1050 N
Drenaggio sotterraneo del terreno agrario con tubi di Pvc. **The use of Pvc tubes for the underground drainage of agricultural soil.**

1500 Bellia, F. IT 040606/73/0142
Ricerche tecnico–economiche sull'irrigazione a pioggia nell'agrumi–coltura della Sicilia. **Technical and economic research on rainfed irrigation in citrus culture in Sicily.** Publications.

1501 Romita, P.L. IT 040608/72/0548
Ricerca teorico–sperimentale intesa a determinare gli effettivi fabbisogni idrici di alcune coltivazioni irrigue caratteristiche della pianura lombarda. **Theoretical and experimental research in view of determining actual water requirements of some typical irrigated crops of the plain of Lombardy.** Publications.

1502 Barbieri, G.C. IT 040701/78/1030 N
Relazione tra due regimi irrigui e le caratteristiche di un impianto di drenaggio su una coltura di mais da foraggio nella piana del Sele. **Relation between two irrigation levels and the characteristics of a drainage plant in a fodder maize field in the Sele plain.**

1503 Postiglione, L.; Duranti, A. IT 040701/79/0001 N
Migliore utilizzazione dell'acqua a fini irrigui in un comprensorio dell'Italia Meridionale. **Water optimal use in an irrigated land of Southern Italy.**

1504 Carravetta, R. IT 040710/77/0132 R
Studio del moto della fase liquida nei processi di infiltrazione e drenaggio per caratteristiche diverse del sistemaacqua–terreno–pianta. **Study of the liquid phase movement in the processes of infiltration and drainage for different characteristics of the water–soil–plant system.**

1505 Giardini, L. IT 040801/77/0196 R
Comportamento dell'erba medica, del pomodoro e del mais in funzione del regime idrico del terreno, studio dell'evapotraspirazione e della pioggia utile. **Behavior of lucerne, tomato and maize in function of the soil water balance, study of evaporation, transpiration and of rain–fall requirements.**

1506 Giulivo, C. IT 040803/74/0564
Irrigazione della vite. **Irrigation of grapevine.**

1507 Giulivo, C. IT 040803/77/0199 R
Problemi agronomici dell'irrigazione della vite. **Agronomic problems of vine irrigation.**

1508 Benini, G. IT 040807/74/0513

D 1310 – Drainage, irregation and water supply

Indagine probabilistica sulle piogge intense nell'arco alpino. **Prospective investigation on intense rainfalls in the Alps.**

1509 Fattorelli, S.   IT 040807/78/1162 N
Influenza dell'attività antropica sul ciclo idrologico nei bacini di montagna. **The influence of anthropic activity on the hydrologic cycle of mountain basins.**

1510 Della Lucia, D.   IT 040807/78/1182 N
Ricerca sui parametri caratterizzanti l'attività erosiva dei bacini montani. **Mountain basins : research on parameters specific of their erosive action.**

1511 Susmel, L.   IT 040809/77/0648 R
Effetto idrogeologico del bosco. **Hydrogeological influence of woods.**

1512 Ballatore, G.P.   IT 040904/72/0084
Ricerche nel settore dell'irrigazione. Rapporti acqua–suolo–vegetaione. **Research in the field of irrigation. Water–soil –vegetation relations.** Publications.

1513 Lo Cascio, B.   IT 040904/77/0216 R
Ricerche nel settore dell'irrigazione. **Research on irrigation.**

1514 Sarno, R.   IT 040904/78/1109 N
Ricerche nel settore delle colture foraggere asciutte. **Research in the field of dry fodder cultures.**

1515 Fatta Del Bosco, G.   IT 040908/73/0202
Indagini sull'irrigazione a goccia degli agrumi. **Studies on drop irrigation of citrus fruit.**

1516 Fatta Del Bosco, G.   IT 040908/77/0175 R
Indagine sulla irrigazione a goccia degli agrumi. **Research on sprinkling irrigation of citrus plants.**

1517 Santoro, M.   IT 040915/77/0645 R
Aggressività della pioggia nello studio dell'erosione idrica del territorio siciliano. **Rain erosive power in the study of water erosion in the Sicilian territory.**

1518 Stoppini, Z.   IT 041010/77/0647 R
Interrimento di alcuni laghetti collinari umbri. **The filling up of certain small Umbrian hill lakes.**

1519 Natali, S.   IT 041104/72/1546
Ricerche per lo studio dei problemi agronomici sulla irrigazione delle colture legnose del frutto. **Research for the study of agronomic problems related to the irrigation of woody fruit–bearing crops.** Publications.

1520 Natali, S.   IT 041104/77/0320 R
Ricerche per lo studio dei problemi agronomici sull'irrigazione delle colture legnose da frutto. **Research on the agronomic problems related to fruit–tree irrigation.**

1521 Celestre, P.   IT 041107/72/0089
Sistema di irrigazione diuturna ovvero a goccia. **Rain–fed irrigation system.** Publications.

1522 Celestre, P.   IT 041107/77/0141 R
Sistema di irrigazione diurna ovvero a goccia ed affini. **Daytime irrigation sprinkling or similar systems.**

1523 Grossi, P.   IT 041107/77/0632 R

Influenza di alcuni interventi sistematori sul potere regimante del terreno. **Influence of some ground improvements on ground water balance.**

1524 Luppi, G.   IT 041201/72/0093
Problemi fondamentali e applicativi dell'irrigazione. **Fundamental problems and applied problems related to irrigation.** Publications.

1525 Luppi, G.   IT 041201/78/1071 N
Problemi fondamentali e applicativi dell'irrigazione. **Fundamental and practical problems connected with irrigation.**

1526 Tournon, G.   IT 041207/77/0307 R
Ricerche sull'irrigazione sotterranea. **Research on underground irrigation.**

1527 Sasso, G.   IT 041214/72/0127
Ricerche sull'irrigazione delle colture ortive e floreali. **Research on irrigation of horticultural and floral crops.** Publications.

1528 Rivoira, G.   IT 041302/72/0125
Irrigazione. **Irrigation.** Publications.

1529 Rivoira, G.   IT 041302/77/0282 R
Problemi agronomici dell'irrigazione. **Agronomic problems of irrigation.**

1530 Milella, A.   IT 041310/72/0697
Problemi agronomici dell'irrigazione degli agrumi. **Agronomic problems related to the irrigation of citrus fruit.** Publications.

1531 Milella, A.   IT 041310/77/0241 R
Ricerche agronomiche sull'irrigazione. **Agronomical research on irrigation.**

1532 Agabbio, M.   IT 041310/78/1021 N
Studio dei problemi agronomici dell'irrigazione. **Study on the agronomical problems of irrigation.**

1533 Sacchi, C.   IT 041805/77/0806
Zoocenosi delle acque costiere e lacunari. **Coastal waters and lagoon zoocoenosis.**

1534 Barrocu, G.   IT 042401/78/1188 N
Studio sperimentale delle leggi di trasferimento piogge–portate superficiali e sotterranee. **Experimental studies on the transfer laws governing the superficial and underground rain–flow.**

1535 Tedeschi, P.   IT 060700/71/0125
Ricerche sui metodi irrigui. **Studies on irrigation methods.** Publications.

1536 Tedeschi, P.   IT 060700/71/0126
Interazione tra irrigazione e fattori agronomici. **Interaction between irrigation and other agronomic factors.** Publications.

1537 Tedeschi, P.   IT 060700/71/0129
Ricerche sull'irrigazione del carciofo. **Studies on artichoke irrigation.** Publications.

1538 Tedeschi, P.   IT 060700/72/0127
Irrigazione delle colture protette. **Greenhouse crops irrigation.** Publications.

1539 Tedeschi, P.   IT 060700/72/0128
Drenaggio e regimazione idrica dei terreni. **Drainage and water control of soils.** Publications.

1540 Tedeschi, P.   IT 060700/74/0118
Indagine sui sistemi irrigui, i consumi idrici e le produzioni conseguite in zone irrigue del Mezzogiorno. **Survey on irrigation methods, water consumption and yields obtained in irrigation areas of South Italy.**

1541 Tedeschi, P.   IT 060700/74/0120
Ricerche sui potenziali idrici nel sistema terreno–piante–atmosfera. **Research on water–potential of the soil–plant–atmosphere system.** Publications.

1542 Tedeschi, P.   IT 060700/74/0122
Ricerche sulla capacità idrica di campo. **Research on field capacity of water.**

1543 Tedeschi, P.   IT 060700/77/0779
Interazione tra erbicidi e regime idrico. **Herbicide and water balance interaction.**

1544 Tedeschi, P.   IT 060700/77/0987
Distribuzione dell'acqua nel terreno in rapporto ai metodi irrigui. **Distribution of water in the soil in relation to irrigation methods.**

1545 Tedeschi, P.   IT 060700/77/0993
Irrigazione in base ai sintomi di stress idrico su tabacco. **Irrigation based on water stress symptoms in tobacco plants.**

1546 Tedeschi, P.   IT 060700/77/0994
Contenuto idrico del suolo, potenziale idrico della pianta e traspirazione su pesco. **Soil water content, the water potential of plants, transpiration in the peach–tree.**

1547 Tedeschi, P.   IT 060700/77/0995
Studio delle interazioni tra irrigazione e drenaggio in rapporto ai metodi irrigui. **Study on drainage and irrigation interaction in relation to irrigation methods.**

1548 Tedeschi, P.   IT 060700/77/0996
Studio dell'influenza della falda sulla disponibilità idrica delle colture. **Influence of the underground water level on water availability for cultures.**

1549 Blanco, V.V.   IT 062700/77/0940
Ricerche sui fabiisogni idrici e sulla tecnica di irrigazione (cavolo–broccolo, cetriolo, fagiolino, pomodoro). **Research on water requirements and irrigation techniques (broccoli, cucumber, french bean, tomato).**

1550 Foti, S.   IT 063100/77/0898
Riflessi agronomici e biologici dell'irrigazione. **Agronomic and biological consequences of irrigation.**

1551 Friz, C.   IT 063700/77/0784
Effetti idrologici. **Hydrological action.**

1552 Rota, L.; Bompard, F.; Amprimo, G.; Bertolino, D.; Penon, A.; Ansoldo, M.   IT 120100/79/0003 N
Indagine interdisciplinare per il miglioramento del complesso d'irrigazione e di produzione di energia elettrica di Mazzè Canavese (Torino). **Interdisciplinary investigations for the improvement of the irrigation and power plant in Mazze Canavese (Piedmont, Italy).**

1553 Zon, J.C.J. van   NL 010102/72/8927 N
Evaluatie van het gebruik van graskarper bij het beheer van watervegetaties. **Evaluation of the use of grass carp in aquatic vegetation management.** Publications.

1554 Hoogerkamp, M.   NL 010102/77/7993
Onderhoud van de begroeiing van slootkanten. **The management of vegetation on the slopes of waterways.** Publications.

1555 Butijn, G.D.   NL 010102/78/8925 N
De waarde van een aquatisch micro–ecosysteem bij het onderzoek naar enige zoötechnische effecten van het herbicide dichlobenil. **Valuation of an aquatic micro–eco–system for the assessment of some zootoxic effects of the herbicide dichlobenil.**

1556 Delver, P.   NL 010103/74/7230
Druppelbevloeiing in de fruitteelt. **Trickle irrigation in fruit growing.** Publications.

1557 Wallenburg, C. van; Stolp, J.   NL 010119/70/6908 R
Draagkracht van grasland op veengronden en al of niet moerige eerdgronden in West– en Noord–Nederland. **Bearing capacity of grassland on peat and peaty soils in the Western and Northern Netherlands.**

1558 Akker, A.M. van den; Lynden, K.R. van
                    NL 010119/70/7802
Onderzoek grondwaterstanden in herverkavelde en niet–herverkavelde gebieden. **Research into the groundwaterlevels in reallotment and not reallotment areas.**

1559 Westerveld, G.J.W.   NL 010119/73/5993
Onderzoek doorlatendheid van gronden in ruilverkavelingsgebieden. **Permeability of soils in re–allotment areas.**

1560 Sluijs, P. van der; Heesen, H. van   NL 010119/73/5996
Onderzoek vochtleverantie van diverse gronden volgens model Rijtema. **Moisture supply of various soils according to model Rijtema.**

1561 Lynden, K.R. van   NL 010119/74/7801 R
Onderzoek vochtinhoud hoge enkeerdgronden. **Research into the moisture storage capacity of high plaggensoils.**

1562 Bouma, J.; Holst, A.F. van; Heesen, H.C. van; Poelman, J.N.B.   NL 010119/76/7557
Bestudering waterhuishouding Gelderland, t.b.v. de ontwikkeling van een submodel voor de onverzadigde zône tussen bodemoppervlak en grondwater als bijdrage tot de ontwikkeling van een veel omvattende simulatiemodel voor het voorspellen van de gevolgen van grondwateronttrekking voor de hydrologische toestand van het gebied. **The study of the water cycle in the province of Guelre (The Netherlands) on behalf of the development of a simulation model for predicting the consequences of ground–water withdrawal.**

1563 Wallenburg, C. van   NL 010119/76/7806 R
Diepe ontwatering van mosveengronden op een modern weidebedrijf in Durgerdam. **Deep drainage of moss peat soils on a modern dairy farm in Durgerdam.**

D 1310 – Drainage, irregation and water supply

1564 Pape, J.C. NL 010119/77/8305
Onderzoek naar de relatie vegetatie en grondwater. **Research into the relationship between vegetation and hydrology.**

1565 Stolp, J. NL 010119/78/9069 N
Schatting en verificatie van de grondwatertrappen in gronden met slecht doorlatende lagen. **Judging and verification of the groundwaterclasses in soils with slowly permeable soilhorizons.**

1566 Sluijs, P. van der NL 010119/78/9078 N
Inventarisatie en optimalisering landelijk waarnemingsnet voor grondwaterstandsgegevens t.b.v. de Gt-kartering. **Inventory and optimization of a national observation network for groundwaterclasses survey.**

1567 Aendekerk, Th.G.L. NL 010203/70/3759
Onderzoek inzake de kwaliteit van oppervlaktewater voor de boomkwekerij. **Research concerning the quality of open water for the nursery.** Publications.

1568 Valk, G.G.M. van der NL 010205/73/5233 R
De invloed van de vochthuishouding in de grond op ontwikkeling en produktie van bolgewassen. **Soil qualities and water management in bulb crop cultures.** Publications.

1569 Sonneveld, C. NL 010206/64/0939 R
De belasting van oppervlaktewater met anorganische stoffen. **Chemical pollution of open water.** Publications.

1570 Graaf. R. de NL 010206/65/0928
De waterhuishouding van kasgewassen. **Water relations of glass–house crops.** Publications.

1571 Hamaker, P. NL 010206/77/7594
Onderzoek naar de water- en mineralenhuishouding in de glastuinbouw. **Influence of horticulture under glass on the quality of open water.**

1572 Hellings, A.J.; Janssen, J.W.M. NL 010207/70/2628
Toetsing van verdamping uit eenvoudige evaporimeters aan gewasverdamping in het veld. **Testing of evaporisation from simple evaporimeters to crop evaporisation in the field.**

1573 Hellings, A.J. NL 010207/74/5578 R
Invloed van watervoorziening op de opkomst en ontwikkeling van ter plaatse gezaaide groentegewassen. **Effect of irrigation on emergence and development of directly sown vegetable crops.** Publications.

1574 Luten, W. NL 010208/74/8771 N
Invloed van beregening van grasland op grasgroei en benutting van meststoffen. **Effect of sprinkling grassland on grass growth and utilization of fertilizers.** Publications.

1575 Thomas, H. NL 010208/76/8753 N
Graslandbenutting bij ondiepe ontwatering van veengrond. **Grassland utilization on peat soils with a high waterlevel.** Publications.

1576 Geneijgen, J. van NL 010208/78/8724 N
Invloed van beregening in bedrijfsverband op de gebruiksmogelijkheden van grasland op komklei. **Influence of sprinkling of grassland on river basin clay as farm system on the utilization possibilities.** Publications.

1577 Delver, P. NL 010212/74/7230

Druppelbevloeiing in de fruitteelt. **Trickle irrigation in fruit growing.**

1578 Heer, H. de; Leistra, M. NL 010301/74/5906
Fysisch–chemisch gedrag van bestrijdingsmiddelen in oppervlaktewater. **Physio–chemical behaviour of pesticides in water.**

1579 Stakman, W.P. NL 010501/67/0784
Bepaling van vochtkarakteristieken van gronden. **Determination of moisture characteristics of soils.** Publications.

1580 Wesseling, J. NL 010501/67/0786 R
Stroming van water in onverzadigde grond. **Flow of water in unsaturated soil profiles.** Publications.

1581 Rijtema, P.E. NL 010501/67/0787
Methoden ter bepaling van het vochtgehalte in de bodem in het veld. **Methods for the determination of soil moisture content in the field.**

1582 Wesseling, J. NL 010501/67/0790 R
Kleine hydrologische adviezen. **Non–project–bound hydrological counsels.** Publications.

1583 Pomper, A.B. NL 010501/67/0791
Geo–hydrologisch onderzoek ten behoeve van waterbeheersingsplannen. **Geo–hydrological research for watermanagement purposes.** Publications.

1584 Stakman, W.P. NL 010501/67/0792 R
Laboratoriumbepalingen voor de doorlatendheid van diepe bodemprofielen. **Hydraulic conductivity of deep soil profiles.** Publications.

1585 Wesseling, J. NL 010501/67/0794 R
Bepaling van de hydrologische eigenschappen van drainagematerialen. **Determination of the hydrological properties of drainage materials.** Publications.

1586 Wesseling, J. NL 010501/67/0819
De invloed van bouwputbemalingen op de grondwaterstand rond civieltechnische werken en de daaruit voor de landbouw voortvloeiende schade. **Effects of deep drainage construction sites on the ground water level in the neighboorhood and the damage caused to agriculture.** Publications.

1587 Wesseling, J.; Stakman, W.P. NL 010501/67/0827
Ontwikkeling van meetapparatuur voor waterhuishoudkundig onderzoek. **Hydrological measuring implements.** Publications.

1588 Stol, P.T. NL 010501/67/0866
Niet–lineaire vereffening van waarnemingsreeksen van fysische processen. **Non–linear estimation of series of data concerning fysical processes.**

1589 Feddes, R.A. NL 010501/67/9048 N
Afleiden van capillaire eigenschappen van bodemprofielen uit bodemtypen en –eenheden. **Derivation of capillary properties of soil profiles from soil types and soil units.** Publications.

1590 Wesseling, J. NL 010501/67/9051 N
Invloed van bouwputbemalingen en zandwinningen op de grondwaterstand en de daaruit voortvloeiende schade voor landbouw, natuur en landschap en andere objecten. **Influence**

D 1310 – Drainage, irregation and water supply

of pointwells on groundwatertable depth and the effect thereof on agriculture, ecology, landscape and other projects. Publications.

1591 Hellings, A.J.   NL 010501/70/2628
Toetsing van verdamping uit eenvoudige evaporimeters aan gewasverdamping in het veld. **Testing of evaporisation from simple evaporimeters to crop evaporisation in the field.**

1592 Hellings, A.J.   NL 010501/70/2828
Landelijke inventarisatie van de problemen ten aanzien van de waterhuishouding bij de vollegrondsgroenteteelt. **Rural stocktaking of problems concerning water management in horticulture.**

1593 Hellings, A.J.; Janssen, W.J.M.   NL 010501/70/3180
Beregeningsonderzoek bij vollegronds groentegewassen. **Sprinkler irrigation experiments with horticultural crops.**

1594 Wesseling, J.   NL 010501/70/3182 R
Onderhoudskosten van open leidingen. **Maintenance–costs of open water courses.** Publications.

1595 Stakman, W.P.; Veerman, G.J.; Valk, G.A.   NL 010501/70/3184
Laboratoriumbepalingen van vochtgehalten van en vochtspanningen in de grond. **Laboratory determinations of soil moisture characteristics.** Publications.

1596 Pomper, A.B.   NL 010501/72/5202
Kwel en wegzijgingsschade in verband met de verbreding van het Amsterdam–Rijnkanaal. **Seepage and ground water levels in connection with widening the Amsterdam–Rhine Canal.** Publications.

1597 Wesseling, J.   NL 010501/73/5192
Laboratoriumbepalingen van capillair geleidingsvermogen. **Improvement of laboratory determination of capillary conductivity.** Publications.

1598 Ernst, L.F.   NL 010501/73/5193 R
Grondwaterstromingsonderzoek met behulp van modellen. **Research on ground water flow with the aid of models.** Publications.

1599 Ernst, L.F.   NL 010501/73/5194
Theoretische analyse van grondwaterstromingen. **Theoretical analysis of ground water flow.** Publications.

1600 Weerd, B. van der   NL 010501/73/5200
Zoutwaterinfiltratie Calandpolder. **Subsurface irrigation with saline water in the Calandpolder. (prov. Zeeland).**

1601 Hoeks, J.   NL 010501/73/5205
Kolomonderzoek met drainwater van vuilstortplaatsen. **Column experiments with leachate from sanitary landfills.**

1602 Drent, J.   NL 010501/73/5206
Lysimeteronderzoek landbehandeling met afvalwater. **Lysimeter research on land treatment with waste water.** Publications.

1603 Steenvoorden, J.H.A.M.   NL 010501/73/5208 R
Eutrofiëring oppervlaktewater. **Eutrophication of surface waters.** Publications.

1604 Hoeks, J.   NL 010501/73/5214
Kwaliteit van het grondwater en oppervlaktewater in de omgeving van de VAM–vuilstortplaats in Wijster. **Quality of ground water and surface water near the VAM–landfill at Wijster.** Publications.

1605 Hoeks, J.   NL 010501/73/5215
Invloed van vuilstortplaatsen op de kwaliteit van grondwater en oppervlaktewater. **Effect of landfills on quality of ground water and surface water.** Publications.

1606 Steenvoorden, J.H.A.M.   NL 010501/73/5218 R
Zoutbalans van Kennemerland en West–Friesland. **Salt balance of Kennemerland and West–Friesland (Prov. Noord–Holland).**

1607 Hoeks, J.   NL 010501/73/5219
Verontreiniging van bodem en grondwater door olie. **Pollution of soil and ground water by oil components.**

1608 Steenvoorden, J.H.A.M.   NL 010501/73/5220 R
Verontreiniging van bodem, grond– en oppervlaktewater door overmatige bemesting. **Pollution of soil and soil and surface waters by excessive fertilization.** Publications.

1609 Schothorst, C.J.   NL 010501/73/5228
Zakking van veengrond door diepe ontwatering. **Research on subsidence of peat soils.**

1610 Bakker, J.W.   NL 010501/73/5232 R
Reactie van het gewas op luchtvoorziening bij verschillende grondwaterstand en dichtheid. **Reaction of crops on aeration at different soil densities and ground water levels.**

1611 Valk, G.G.M. van der   NL 010501/73/5233
Bodemkwaliteit en waterhuishouding bij bloembollenteelt. **Soil qualities and water management in bulb crop cultures.**

1612 Schothorst, C.J.   NL 010501/73/5234
Grasproduktie van veenweidegronden bij verschillende polderpeilen. **Production of grassland on peat soils depending on ground water level depth.** Publications.

1613 Schothorst, C.J.   NL 010501/73/5240 R
Verbetering van de draagkracht van veenweidegronden door peilverlaging. **Improvement of bearing capacity of peat grassland by increasing drainage depth.**

1614 Wijk, A.L.M. van   NL 010501/73/5242 R
Bodemtechnische ontwerpnormen voor speel– en ligweiden en sportvelden. **Design criteria for soils for sports grounds and playing fields.** Publications.

1615 Steenvoorden, J.H.A.M.   NL 010501/74/5211 R
Processen in bodem– en grondwater na bemesting. **Processes in soil and groundwater after fertilizing and manuring.**

1616 Hellings, A.J.   NL 010501/74/5578
Invloed van watervoorziening op de opkomst en ontwikkeling van ter plaatse gezaaide groentegewassen. **Effect of irrigation on emergence and development of directly sown vegetable crops.** Publications.

1617 Hellings, A.J.   NL 010501/74/5579 R
Berekening van het waterverbruik van groentegewassen in de vollegrond in verschillende teeltcentra over korte perioden met

D 1310 – Drainage, irregation and water supply

behulp van blokevaporimeters. **Estimating of water requirements during short periods of vegetable crops in the open by means of block– evaporimeters.**

1618 Hellings, A.J. NL 010501/74/5580 R
Begeleiding van de uitvoering van beregeningsproeven met vollegrondsgroentegewassen op enkele proeftuinen in het zuiden des lands. **Coaching sprinkler irrigation experiment on vegetable crops in the open in the Southern part of the Netherlands.** Publications.

1619 Pomper, A.B. NL 010501/74/5583
Geohydrologisch onderzoek naar de grondwatervoorraden in samenhang met de geologie van Midden–Brabant. **Geohydrological assessment of ground water reserves and their relationship with geological characteristics in the central part of the province Noord–Brabant.**

1620 Feddes, R.A. NL 010501/74/5584
Geohydrologisch onderzoek naar de grondwatervoorraden in samenhang met de geologie van Oost–Brabant. **Geohydrological assessment of ground water reserves and their relationship with geological charcteristics in the eastern part of the province Noord–Brabant.**

1621 Wesseling, J. NL 010501/74/5585
Grondwateronttrekking in de Gelderse Vallei. **Ground water abstraction in the area of the Gelderse Vallei (central part of the Netherlands).** Publications.

1622 Pomper, A.B. NL 010501/74/6189
Geohydrologisch onderzoek naar de grondwatervoorraden in samenhang met de geologie van Groningen. **Geohydrological assessment of ground water reserves and their relationship with geological characteristics in the province Groningen.**

1623 Schothorst, C.J. NL 010501/74/6192
Draagkracht en detail–ontwateringsproblematiek van kleigrasland. **Problems of bearing capacity and drainage of grass land on clay soils.**

1624 Stol, P.T. NL 010501/75/6645
Correlatiepatroon tussen aangrenzende meetstations voor neerslag–gegevens ten behoeve van een optimale netdichtheid. **Interstation–correlation in precipitation networks.** Publications.

1625 Boheemen, P.J.M. van NL 010501/75/6646 R
Technische en economische mogelijkheden van wateraanvoer. **Technical and economic possibilities of water supply in agriculture.**

1626 Boheemen, P.J.M. van NL 010501/75/6648
Onderzoek in het hydrologisch proefgebied Sleen. **Hydrologic research in the catchment basin Sleen.**

1627 Boheemen, P.J.M. van NL 010501/75/9061 N
Agrohydrologische en technische aspecten van kunstmatige watervoorziening in land– en tuinbouw. **Agrohydrological and technical aspects of water supply in agriculture.**

1628 Schothorst, C.J. NL 010501/76/7205 R
Beregening van veenweidegronden. **Sprinkling irrigation of grassland on peat soils.** Publications.

1629 Boheemen, P.J.M. van NL 010501/76/7206 R

Onderzoek naar de watervoorziening van de landbouw in de zomer van 1976. **Required and given additional agricultural water supply Netherlands 1976.** Publications.

1630 Weerd, B. van der NL 010501/76/7209 R
Eventuele schadelijke gevolgen van inpoldering van De Mosselbanken (provincie Zeeland). **Possible detrimental effects of reclaiming the mud–flats "De Mosselbanken" (province Zeeland).**

1631 Wesseling, J. NL 010501/76/7211
Waterhuishouding en kwaliteit van oppervlakte– en grondwater in Noord–Holland. **Water management and quality of surface and groundwater in the province Noord–Holland.** Publications.

1632 Drent, J. NL 010501/76/7215 R
Verwerking afvalwater van een aardappelschilbedrijf door bevloeiïng. **Irrigation treatment of waste water from a potato–peeling factory.**

1633 Feddes, R.A. NL 010501/77/7207 R
Onderzoek thermisch gedrag van begroeide oppervlakken in relatie tot verdamping en produktie ten behoeve van de "Heat Capacity Mapping Mission" (HCMM) van NASA. **Remote sensing of thermal behaviour of crop surfaces in relation to evaporation and production in connection with the Heat Capacity Mapping Mission (HCMM) of NASA.** Publications.

1634 Bakel, P.J.T. van NL 010501/77/7208 R
Toepassen van numerieke modellen op het grond– en oppervlaktewatersysteem van gebieden van enige omvang. **Use of numerical models on the ground and surface water system of relatively large areas.**

1635 Weerd, B. van der NL 010501/77/7210 R
Vaststelling bestaande zout– en waterhuishouding rondom het toekomstige tracé van het Baalhoekkanaal. **Survey of the present salinity and water situation adjacent to the future Baalhoek canal.**

1636 Hoeks, J. NL 010501/77/7212
Mobiliteit van zware metalen in de bodem en het grondwater. **Mobility of heavy metals in soils and groundwater.** Publications.

1637 Kemmers, R.H. NL 010501/77/7214 R
Bodemfysische en bodemchemische aspekten van halfnatuurlijke vegetaties met betrekking tot bufferzones tussen natuur– en landbouwgebieden. **Soil physical and soil chemical aspects of semi–natural vegetation in connection with buffer zones between nature reserves and agricultural areas.**

1638 Rijtema, P.E. NL 010501/77/7594 R
Onderzoek naar de water– en mineralenhuishouding in de glastuinbouw. **Influence of horticulture under glass on the quality of open water.**

1639 Kemmers, R.H. NL 010501/77/7596 R
Geohydrologisch, landbouw– en vegetatiekundig onderzoek voor het Korenburger Veen en vochtige gebieden in de ruilverkaveling "Winterswijk–West". **Geohydrology, agricultural and vegetation research in connection with the environmental quality of moist and peaty parts of the land consolidation Winterswijk–West.**

D 1310 – Drainage, irregation and water supply

1640 Schothorst, C.J.   NL 010501/77/7598 R
Profielverbetering bij grasland op veengronden met dun
kleidek. **Soil improvement of clay–on–peat soils under grass.**

1641 Bakel, P.J.T. van   NL 010501/77/7890
Waterbeheersing door infiltratie vanuit open leidingen in het
waterschap De Veenmarken. **Water management by
sub–irrigation from open conduits in the drainage district "De
Veenmarken".**

1642 Drent, J.   NL 010501/77/7891
Verregening op kleigronden van afvalwater van een
conservenfabriek. **Sprinkling irrigation on clay soils of waste
water from a canning factory.**

1643 Feddes, R.A.   NL 010501/77/9049 N
Toepassing van door remote sensing verkregen warmte– en
reflectiebeelden in het onderzoek naar de waterhuishouding in
het landelijk gebied. **Application of remote sensing heat and
reflection pictures in agricultural water management.**

1644 Bakel, P.J.T. van   NL 010501/77/9050 N
Ontwikkeling van simulatiemodellen ten behoeve van het
waterbeheer. **Development simulation models for water
management.**

1645 Weerd, B. van der   NL 010501/78/8335
Optimalisering van de polderbemaling. **Optimizing polder
drainage by pumping in saline areas.**

1646 Wesseling, J.   NL 010501/78/8336 R
Optimalisering waterhuishouding in landelijke gebieden.
Optimizing the water management of rural regions.

1647 Steenvoorden, J.A.H.M.   NL 010501/78/8337
Bijdrage van oppervlakte–afvoer aan de belasting van
oppervlaktewater met eutrofiërende verbindingen.
**Contribution of surface and subsurface run–off to the
eutrophication of surface waters.**

1648 Hoeks, J.   NL 010501/78/8338 R
Kwaliteit van grond– en oppervlaktewater bij de vuilstortplaats
Koegorspolder (gem. Terneuzen). **Effect of waste disposal site
Koegorspolder (near the city of Terneuzen) on groundwater and
surface water quality.**

1649 Wind, G.P.   NL 010501/78/8339 R
Simulatie van de vochttoestand van grond bij verschillende
bodemeigenschappen en ontwatering. **Analog simulation of the
unsaturated zone of various soil types and different drainage
conditions.**

1650 Wijk, A.L.M. van   NL 010501/78/8619 N
Effect van ontwatering van verschillende bodemtypen op
opbrengst van landbouwgewassen. **Effect of drainage measures
on various soils in crop yield.**

1651 Weerd, B. van der   NL 010501/78/8620 N
Vooronderzoek van de te verwachten verdroging rondom het
toekomstige Zoommeer'. **Preliminary investigation regarding
the expected dessiccation around the future lake Zoommeer
(Province Zeeland).**

1652 Harmsen, J.   NL 010501/78/8621 N
Identificatie van organische verontreinigingen in water.
Identification of organic polution in water.

1653 Kemmers, R.H.   NL 010501/78/8622 N
Relatie vegetatie–hydrologie in het natuurreservaat 'Groot
Zandbrink'. **Vegetation–hydrology relationship in the nature
conservancy 'Groot Zandbrink'.**

1654 Kemmers, R.H.   NL 010501/78/8623 N
Experimenteel verdampingsonderzoek aan enige wilde
plantesoorten. **Potential evapo–transpiration of wild plant
species.**

1655 Weerd, B. van der   NL 010501/78/8877 N
Onderzoek naar de waterhuishoudkundige gevolgen van de
verbreding en verdieping van het Kanaal door Zuid–Beveland
voor de aangrenzende gronden. **Repercussions of enlarging and
deepening the Canal through Zuid–Beveland on the water
management of the adjacent area.**

1656 Steenvoorden, J.H.A.M.   NL 010501/78/8879 N
Fosfaatbelasting van het oppervlaktewater en fosfaatberging in
het bodemslib. **Phosfate load of surface waters and phosfate
storage in bottom mud.**

1657 Drent, J.   NL 010501/79/8878 N
Onderzoek kanaal Waddinxveen–Voorburg. **Research Canal
Waddinxveen–Voorburg.**

1658 Weerd, B. van de   NL 010501/79/9052 N
Vaststelling van de waterbehoefte bij wateraanvoer in polders
met zoute kwel en de invloed hierop van aanpassingen van het
waterlopenstelsel. **Determination of the water requirement in
polders with saline influent seepage and its effect for the open
water system. Publications.**

1659 Stol, Ph.Th.   NL 010501/79/9056 N
Parameterschattingen in alternatieve formules voor het
verband tussen geregistreerde en gemeten peilen bij
verdronken meetschotten. **Parameter estimation in a discharge
formula for submerged flumes.**

1660 Steenvoorden, J.H.A.M.   NL 010501/80/9053 N
Stikstofconcentraties in bodemvocht en grondwater onder
grasland op zandgrond in afhankelijkheid van
runderdrijfmest– en kunstmeststikstofdosering. **Nitrogen
concentrations in soil solution and groundwater under
grassland on a sandy soil as influenced by the dosage of cattle
slurry and artificial nitrogen fertilizer.**

1661 Burg, J. van den   NL 010601/69/2380
De betekenis van de grondwaterstand voor een aantal
houtsoorten. **The significance of the groundwater level for a
number of tree species. Publications.**

1662 Burg, J. van den; Faber, C.J.   NL 010601/73/3838
De invloed van grondwaterdaling op de groei van bossen. **The
influence of lowering the groundwater table on the growth of
foreststands. Publications.**

1663 Willemsen, J.   NL 010702/75/7032
Onderzoek naar het effect van graskarper, uitgezet ter
bestrijding van overmatige plantengroei, op inheemse
vissoorten. **Research on the effect of grasscarp, stocked to
reduce excessive vegetation, on native fish species. Publications.**

1664 Stroosnijder, L.   NL 020007/70/4970
Bepaling van vochtgehalte en dichtheid aan grondkolommen

m.b.v. gammastraling. **Determination of moisture content and bulk density on laboratory columns with gamma radiation.** Publications.

1665 Nota, D.J.G.   NL 020008/68/4914
Hydrogeologisch onderzoek stroomgebied Gulp. **Watershed project Gulp–river basin.** Publications.

1666 Schaaf, S. van der   NL 020012/68/5082
Elektrisch analogon onderzoek van grondwaterstroming. **Electric analogue investigations of ground water flow.**

1667 Koopmans, R.W.R.   NL 020012/72/4403
Onderzoek naar de toepassing van numerieke berekeningsmethoden voor grondwater stromingsproblemen. **Research on application of numerical methods in groundwater flow.**

1668 Schaaf, S. van der   NL 020012/72/4404
Hydrologie van de zuidelijke Gelderse vallei. **Hydrology of the southern part of the "Gelderse vallei" area.**

1669 Leeuwen, H.P. van   NL 020016/77/8514
Electrometrische bepaling van geringe concentraties zware metalen in aanwezigheid van complexvormers in natuurlijke wateren. **Electrometric determination of low concentrations of heavy metals in the presence of complexing agents in natural waters.** Publications.

1670 Warmerdam, P.M.M.   NL 020022/72/8461
Neerslag–afvoermodel voor het Hupelse Beekgebied als prototype van een stroomgebied in Oost Gelderland. **Rainfall–run–off relationship of the Hupselse Beek catchment area.**

1671 Pitlo, R.H.   NL 020022/73/4973
Onderzoek aan meetinrichtingen en verdeelwerken in open kanalen in de sfeer van de cultuurtechniek. **Research on small hydraulic structures in open channels for measurement and distribution of flow.** Publications.

1672 Pitlo, R.H.   NL 020022/74/6274 R
Regulatie van waterplantvegetaties in sloten. **Regulation of vegetation in waterways (diches).** Publications.

1673 Stricker, J.N.M.   NL 020022/77/8462
Synthese van modellen voor gebiedsverdamping. **Relationship between rainfall, evaporation and discharge in catchment areas.** Publications.

1674 Kraijenhoff van de Leur, D.A.   NL 020022/77/8463
Invloed van verstedelijking op het afvoerproces. **Effects of urbanization on precipitation and catchment run–off.**

1675 Keuning, D.H.   NL 020022/77/8464 N
Wiskundige methoden in de hydrologie en het waterbeheer. **Mathematical methods in hydrology and water management.**

1676 Meys, C.C.A.M.   NL 020033/75/6266
Bloei van kiezelalgen in het Grevelingen–bekken. **Blooming of marine diatoms in the lake Grevelingen.**

1677 Cuppen, J.G.M.   NL 020064/78/8832 N
Biologische beoordeling van de waterkwaliteit, in het bijzonder van wateren in 'de Peel', en onderzoek naar faktoren die de verspreiding van organisman beinvloeden. **Biological assessment of waterquality, in particular of waterbodies in 'the Peel' and investigation of factors which have an influence on the distribution of organisms.**

1678 Vink, N.H.   NL 020065/74/6296
Meten van componenten van de waterbalans van bebouwde en ondebouwde grond. **Measuring components of the waterbalance of cropped and bare soil.**

1679 Vink, N.H.   NL 020065/76/7347
Kwantificeren water opbrengst relaties met toegankelijke meteorologische gegevens. **Quantification water yield relations with standard meteorological data.**

1680 Roscher, K.   NL 020065/78/8835 N
Bedrijfsirrigatie, ontwerp en toepassing. **On–farm irrigation, design and practices.**

1681 Stamhuis, E.   NL 020065/78/8836 N
Sedimenttransport, in het bijzonder in geulen en waterlopen. **Sediment transport, especially in gullies and canals.**

1682 Hendriks, M.   NL 020066/77/8553
Lineaire modellen in verzadigde en onverzadigde grondwaterstroming. **Linear models in saturated and unsaturated groundwater flow.**

1683 Scholten, J.   NL 040007/60/4109
Eigenschappen en bruikbaarheid van drainage materialen. **Properties and utility of drainage materials.**

1684 Zuidema, F.C.   NL 040007/60/4110
Het voorkomen en de sterkte van kwel in de IJsselmeerpolders en de Lauwerszee. **Appearance and rate of seepage in the IJsselmeerpolders and the Lauwerszee.**

1685 Sieben, W.H.   NL 040007/65/4099
De ontwatering van grasland op zware grond. **Drainage of grassland on heavy soils.**

1686 Witteveen, H.   NL 040007/65/4119
Bodemfysisch onderzoek m.b.t. processen bij de waterhuishouding van rijpende gronden in de IJsselmeerpolders. **Soil physical research related to water management processes of ripening soils in the IJsselmeerpolders.**

1687 Berger, C.   NL 040007/66/8432
Eutrofiëringsonderzoek randmeren, Markermeer en IJsselmeer. **Research on eutrophication of the border lakes, Markermeer and IJsselmeer.** Publications.

1688 Berg, J.A. van den   NL 040007/68/4105
Onderzoek naar de waterhuishouding in het stedelijk gebied van Lelystad. **Water management in the urban area of Lelystad.**

1689 Jong, K.G. de   NL 040007/68/4107
De waterhuishouding van sport– en recreatieterreinen op zware gronden. **Water management of sport fields and recreation grounds on heavy (clay) soils.**

1690 Ardon, J.   NL 040007/68/6206
Toepassing van elektrische modellen om de invloed van bepaalde civiel– en cultuurtechnische ingrepen op de waterhuishoudkundige toestand te kunnen voorspellen. **Application of electric model research for predicting the**

influence of civil engineering and land development on water management conditions.

1691 Sieben, W.H.   NL 040007/70/4100
Onderzoek naar de waterhuishouding van beplantingen. **Research on water management of plantations.**

1692 Brinkhorst, W.   NL 040007/70/4111
Onderzoek afvoerverlopen in landelijke gebieden. **Research on discharge graphs in rural areas.**

1693 Berg, J.A. van den   NL 040007/70/4112
Onderzoek t.b.v. waterwinning voor de drink– en industriewatervoorziening in de Zuidelijke IJsselmeerpolders. **Geohydrological research for the water supply for domestic and industrial use in the Southern IJsselmeerpolders.** Publications.

1694 Brinkhorst, W.   NL 040007/70/4115
Waterhuishoudkundig onderzoek t.b.v. inrichtingsplannen van de Markerwaard. **Water management research for land use plans of the Markerwaard.** Publications.

1695 Visser, J.   NL 040007/72/4116
Waterhuishoudkundig onderzoek t.b.v. inrichtingsplannen voor het Grevelingenbekken. **Water management research for land use plans of the Grevelingenbekken.**

1696 Jong, J. de   NL 040007/72/8469
Onderzoek naar de geschiktheid van oppervlaktewater als zwemwater. **Research on the suitability of surface water in Flevoland for swimming purposes.**

1697 Sieben, W.H.   NL 040007/72/8486
De invloed van het waterregime op de aanslag en groei van bomen in het Lauwerszeegebied. **The influence of water management on rooting and growth of trees in the Lauwerszee area.**

1698 Jong, J. de   NL 040007/72/8493
De invloed van een gescheiden rioleringsstelsel op de waterkwaliteit in stedelijke gebieden. **The influence of a separated drainage system on the water quality in urban areas.**

1699 Schultz, E.   NL 040007/73/4108
Onderzoek naar de waterhuishouding van gebieden bestemd voor woonkernen en recreatiegebieden. **Water management of areas intended for house building and recreation.**

1700 Schultz, E.   NL 040007/73/4110
Het voorkomen en de sterkte van kwel in het Almere–gebied. **Appearance and rate of seepage in the Almere area.**

1701 Benning, G.   NL 040007/74/5545
Het waarnemen van de waterbeweging in het natuurgebied Oostvaardersplassen in relatie tot begroeiing en slibtransport teneinde maatregelen te kunnen nemen om de huidige situatie te handhaven. **Observations of water movements in the nature reserve Oostvaardersplassen in relation to vegetations and transport of soil particles in order to take measures to maintain the present situation.**

1702 Sieben, W.H.   NL 040007/75/6200
Waterbeheersingsnormen voor gronden met een verschillende profiel opbouw. **Water management conditions for soils with different profiles.**

1703 Roo, H. de   NL 040007/75/6201
Onderzoek ontwateringstoestand stedelijke gebieden. **Research on drainage conditions in urban areas.**

1704 Schultz, E.   NL 040007/75/6202
Ontwikkeling en verbetering van drainage systemen voor stedelijke gebieden. **Development and improvement of drainage systems for urban areas.**

1705 Jong, J. de   NL 040007/75/8472
De belasting van het polderwater met verontreinigende en/of eutrofiërende stoffen als gevolg van landbouwkundig gebruik van de gronden. **The load on polderwater with polluting and eutrophicating agents as a result of agricultural use of the soil.** Publications.

1706 Jong, J. de   NL 040007/75/8476
Landbehandeling van afvalwater en effluenten. **Land treatment of waste water and effluents.**

1707 Benning, G.   NL 040007/75/8498
Onderzoek ontwateringstoestand Lauwerszee. **Research on drainage conditions of the Lauwerszee–area.**

1708 Ardon, J.   NL 040007/75/8501
Ontwikkeling van meet– en dataverwerkingssystemen t.b.v. het waterhuishoudkundig onderzoek. **Development of measuring and data processing systems for water management research.**

1709 Schultz, E.   NL 040007/76/8477
Onderzoek naar de waterhuishoudkundige aspecten van Markermeer en randmeren. **Research on water management aspects of the Marker Lake and border lakes.**

1710 Schultz, E.   NL 040007/76/8565
Onderzoek wijze van bouwrijpmaken. **Development of methods for fitting of building plots.**

1711 Visser, J.   NL 040007/77/8433
Waterhuishoudkundig onderzoek t.b.v. inrichtingsplannen van het Oosterscheldegebied. **Water management research for land use plans of the Oosterschelde area.**

1712 Rozendaal, H.   NL 040007/77/8468
Onderzoek naar de ontwateringstoestand van landelijke gebieden in Flevoland. **Research on the drainage condition of non–urban areas in Flevoland.**

1713 Rijniersce, K.   NL 040007/77/8483
Bodemfysisch onderzoek m.b.t. waterbewegingen in jonge zware gronden. **Soil physical research on water movements in young heavy soils.**

1714 Benning, G.   NL 040007/77/8495
Waterhuishoudkundig onderzoek t.b.v. de inrichting van het buitengebied van Almere. **Water management research for land use planning of the open spaces outside Almere.**

1715 Jong, J. de   NL 040007/78/8475
De uitwisseling van nutriënten tussen bodem en water i.v.m. eutrofiëring in het IJsselmeergebied. **The exchange of nutrients between soil and water in relation to the eutrophication of the IJsselmeer area.**

D 1310 – Drainage, irregation and water supply

1716 Canton, J.H.; Wegman, R.C.C.   NL 040011/76/7524
Onderzoek naar de toxiciteit van de insecticiden en Omethoaat en de dientengevolge mogelijke verontreiniging van het oppervlaktewater. **Investigation on the toxicity of the insecticides Dimethoate and Omethoate and as a consequence the pollution of surface water.**

1717 Pieterse, A.H.   NL 040012/73/5525
Fysiologie en bestrijding van schadelijke waterplanten. **Biology, physiology and control of tropical aquatic weeds.**

1718 Pieterse, A.H.   NL 040012/75/6597
Geïntegreerde bestrijding van de waterhyacint. **Integrated control of the water hyacinth.**

1719 Ruiter, M.A. de   NL 050401/70/6687
Kwaliteitsverloop oppervlaktewater o.i.v. materie–inbreng. **Influence of various substances in the quality of surface water.** Publications.

1720 Jacobs, A.   NL 060006/71/4383
Studie van de turbulentie van de onderste luchtlagen t.b.v. het bepalen van de verdamping. **Atmospheric boundary layer turbulence.**

1721 Palland, C.L.   NL 060006/71/4384
Ontwikkeling van een verdampingsmeter (verdampingsborstel). **Development of an evaporimeter ("evaporation brush").**

1722 Vossen, G. van de   NL 060006/72/4382
Statistische methoden in de hydrologie. **Statistical methods in hydrology.**

1723 Ouwerkerk, J. van   NL 060006/77/8384
Hydrologie van de onverzadigde zône. **Flow in the unsaturated zône.**

1724 Bouwknegt, J.   NL 060006/77/8386
Ontwikkeling van een computerprogramma voor toepassing in hydraulische netwerken ("Hydra"). **Development of a computer program for hydraulic networks ("Hydra").**

1725 Slijkoord, F.   NL 060006/77/8387
Stadshydrologie. **Urban hydrology.**

D 1320 – Soil improvement

See also 257, 291, 470, 610, 818, 829, 893, 1374, 1438, 1609, 1614, 1710, 1876, 2912, 5152, 5339, 5409, 5968, 15324, 15332, 19786

1726 Maton, A.; Dierickx, W.; Goossens, F.
BE 070300/74/0040
Bodemkonditioneringsmiddelen als hulpmiddel bij drainage. **The influence of soil conditioners on drainage efficiency.** Publications.

1727 Riga, A.; Destain, J.P.   BE 080100/79/0020 N
Etude du rapport isotopique (N) naturel dans les sols traités massivement au lisier. **Natural abundance of nitrogene isotopes in soils haevily traeted with animal slurry.**

1728 Rixhon, L.; Guiot, J.   BE 080800/57/0002
Conservation de la fertilité du sol avec les matières organiques produites à la ferme. **Conservation of soil fertility with farm–produced organic matter.** Publications.

1729 Boeker, P.; Franken, H.   DE 111251/75/0004
Das Verhalten unterschiedlicher Rasentragschichtaufbauten bei Sportplätzen unter Belastung. **The behaviour of different green bearing layers of playing–fields under stress.**

1730 Kern, K.–G.   DE 126600/70/0004 R
Meliorationsversuche auf streugenutzten Kiefernböden der Vorderhaardt – Pfälzer Wald – 1968. **Trials on melioration of pine soils with utilized litter in Vorderhaardt – Pfälzer Wald –.**

1731 Harrach, T.; Werner, G.; Wourtsakis, A.
DE 129020/77/0002
Bodenverbesserung. **Soil melioration.** Publications.

1732 Wohlrab, B.; Ehlers, M.   DE 129450/78/0002 N
Rekultivierung durch Grundwasserabsenkung trockengefallener Niedermoore. **Recultivation of fen soils desiccated because of ground–water drawdown.** Publications.

1733 Jung, L.; Brechtel, R.   DE 129551/70/0001
Erhaltung der durch Überweidung und unsachgemässe ackerbauliche Nutzung in ihrer Existenz gefährdeten türkischen Bö– den. **Conservation of Turkish soils running great risk of erosion caused by overgrazing and inappropriate agriculture.**

1734 Jung, L.; Monadjemi–Darani, R.   DE 129551/74/0001
Verbesserung von Salzalkaliböden durch Schwefel– und Gipsdüngung in der Ghazwin–Ebene. **The improvement of saline–alkali soils by sulphur and gypsum fertizilation in the Ghazwin plain.**

1735 Meyer, B.; Scholz, H.; Fleige, H.; Scheffer, K.
DE 132031/70/0001
Untersuchungen über die Stickstoffmetabolik im Gefolge von Bodenmeliorationsmassnahmen – Tiefpflügen – sowie beim pfluglosen Ackerbau 1969. **Studies on the metabolism of nitrogen in consequence of soil amelioration – deep ploughing – as well as in cultivation without ploughing.**

1736 Hartge, K.H.   DE 138030/71/0001 N
Gefügemelioration. **Textural melioration.**

1737 Hartge, K.H.; Richter, J.   DE 138030/74/0001
Untersuchungen zu Bodenverdichtungen und –verfestigungen infolge Bodenbearbeitung und Viehauftrieb mit Hilfe des Gloor–Penetrographen. **Investigations on soil compaction and consolidation in consequence of soil cultivation and livestock pasturing by means of the Gloor–penetrograph.**

1738 Hartge, K.H.   DE 138030/77/0001
Einfluss der Verwendung von Branntkalk als Stabilisierungsmittel auf neutrale Spannungen in den Aggregaten. **Effect of appliance of burnt lime on neutral stresses in aggregates.**

1739 Volk, H.; Spahl, H.   DE 501504/77/0009
Untersuchungen zum Erfolg von Rekultivierungsmassnahmen. **Studies on successful recultivation measures.**

1740 Diez, T.   DE 502050/70/0013
Kalkmeliorationsversuche zur Verbesserung der Basensättigung, Bodenstruktur, Ertragsleistung und Qualität des Erntegutes auf physiologisch sauren und staunassen Böden

D 1320 – Soil improvement

1961. Trials on lime melioration for improvement of base saturation, soil structure, yield and quality of crop on physiologically acid and moisture–retaining soils.

1741 Diez, T.   DE 502050/70/0014
Primärlössmelioration zur Verbesserung der Basensättigung, Bodenstruktur, Ertragsleistung und Qualität des Erntegutes auf degradierten Bodenbildungen des niederbayerischen Lösslehmgebietes 1963. **Melioration of primary loess for improvement of base saturation, soil structure, yield and quality of crop on degraded loamy loess soils in Lower Bavaria.**

1742 Kern, H.   DE 502050/70/0016
Sanddeckkultur, Klärschlammdeckkultur und Hochflutlehmmischkultur zur Einschränkung der Moorsackung, Ausweitung der Fruchtfolgen und Nematodenbekämpfung im Donaumoos 1967. **Cultivation on sand cover and on sewage sludge cover and mixed cultivation on alluvial loam for restriction of marsh sagging, extension of rotations and control of nematodes in moss by Danube River.**

1743 Kern, H.   DE 502050/70/0018
Regelung der Wasserführung schwerer und staunasser Böden durch Untergrundlockerung und melioratives Pflügen bei Einbringung von Kalk und Mineraldünger in den Untergrund 1967. **Control of water volume of heavy and moisture–retaining soils by subsoiling and meliorative tillage by bringing lime and mineral fertilizers into subsoil.**

1744 Schmid, G.; Weigelt, H.; Diez, T.   DE 502050/70/0019
Anwendung von Müllkompost auf extremen Standorten. Erhöhung der Sorptionskapazität auf leichten und Verbesserung der Bearbeitbarkeit auf schweren Böden 1967. **Application of waste compost to extreme sites. Rise in sorption capacity of light soils and improvement of workability of heavy soils.**

1745 Schuch, M.; Dieter, A.   DE 502050/70/0022
Verringerung bzw. Beseitigung von Erosionsschäden im fränkischen Weinbau durch Aufbringen von Stroh, Müllkompost, Torfen und durch Dauerbegrünung 1970. **Reduction resp. elimination of erosion damage in viticulture in Franconia by application of straw, waste compost, peat moss litter and by permanent green growing.**

1746 Borchert, H.   DE 502055/73/0005
Physiko–chemische Bodenuntersuchung der Lockerungserhaltung nach Tiefenlockerung. **Physico–chemical changes in soil by artificial drainage and deep loosening of clay soils and loess–derived soils.**

1747 Borchert, H.   DE 502055/73/0023
Kalkwanderung nach Meliorationskalkung in Form von Krumenkalkung und Tiefenkalkung. **Distribution of CaCO3 in soil after meliorative liming applicated to top soil and subsoil – in connection with deep loosening –.**

1748 Müller, W.; Renger, M.; Fastabend, H.; Henseler, K.L.
                                     DE 507102/70/0003
Verwendung von Braunkohlepulver in der Landwirtschaft. **Use of powdered lignite in agriculture.**

1749 Krämer, F.   DE 508302/70/0003 R
Bodenentwicklung, Humusanreicherung und Bearbeitbarkeit von Löss–Rohböden bei Zufuhr verschiedener organischer Stoffe 1969. **Soil development, humus accumulation and**

workability of uncultivated loess soils supplying different organic matters.

1750 Schulte–Karring, H.   DE 509101/70/0001
Die Tiefenlockerung von verdichteten Böden und ihre Auswirkung auf den Boden und das Pflanzenwachstum. **Subsoiling of compacted soils and its effects on the soil and on the plant growth.** Publications.

1751 Schulte–Karring, H.   DE 509101/70/0002
Die Erforschung der Strukturerhaltung von tiefgelockerten Böden. **Investigation into the conservation of soil structure after deep subsoiling.** Publications.

1752 Hansen, L.; Kjellerup, C.M.   DK 010105/57/3407
Jordbehandlingens, kalkningens og gødskningens betydning for jordstruktur og dyrkningssikkerhed på marskjorde. **Effects of tillage operations, liming and fertilizer application on soil structure and crop yield reliability on marsh soils.**

1753 Hansen, L.; Stokholm, E.   DK 010105/71/3405
Grøngødningens og halmnedbringningens indflydelse på jordstrukturen. **The influencing af green manuring and straw mulching on soil structure.**

1754 Hansen, L.; Stokholm, E.   DK 010105/72/3402
Undergrundsløsning og –gødskning på lerjord. **Subsoil loosening and fertilizer application in clay soils.**

1755 Olesen, S.E.; Øvig, J.K.   DK 010900/66/0001
Dyb jordbehandling af lagdelte jorde med henblik på forøgelse af planternes rodområde og vandforsyning. **Deep soil cultivation in stratified soils with the purpose of increasing the root depth and water supply of the plants.**

1756 Øvig, J.K.   DK 010900/67/0002
Jordforbedring med organisk materiale for at øge sandjordens kapacitet for tilgængeligt vand og afgrødernes roddybde. **Increase in the available water capacity and crop root depth of sandy soil through soil improvement with organic material.**

1757 Olesen, S.E.   DK 010900/73/0002
Opdyrkning til landbrugsjord eller etablering af plantevækst i tidligere brunkulslejer. **Reclamation of open lignite mines for agricultural production or establishing vegetation.**

1758 Jakobsen, S.T.; Jensen, I.; Nielsen, K.S.
                                     DK 030148/78/0002 N
Markforsøg med forskellig tilførselsmåde af jordbrugskalk. **Effect of surface applied lime compared with the effect of lime ploughed into the soil.**

1759 Faure, A.   FR 010601/68/0537
Mécanisme du compactage des sols. Rôle de la texture. **Soil compaction mechanism – Textural influence.** Publications.

1760 Arnoux, M.; Defrance, H.; Huguet, C.
                                     FR 010615/65/9043
Comparaison de sept modes d'entretien du sol en vergers de pêcher. **Comparing seven soil improvement techniques applied to peach orchards.**

1761 Arnoux, M.; Defrance, H.; Huguet, C.
                                     FR 010615/65/9045
Comparaison de modes d'entretien du sol et de fumure azotée en vergers de Pommier. **Comparison of different soil and**

D 1320 – Soil improvement

fertilizing management practices in apple–tree orchards.

1762 Concaret, J.; Servant, J.　　　　　FR 011001/74/6286
Amélioration des sols salés à alcalis. **Reclamation of alkali soils.**

1763 Langlet, A.　　　　　FR 011401/72/0643
Potentialités et aménagements des terrains de parcours des Causses. **Possibilities of improvement and reclamation of rough grazing in Causses.**

1764 Recamier, A.; Versailles, S.E.I.; Grignon, E.; Grignon, P.　　　　　FR 011801/78/9068
Remise en culture des friches abandonnées. **Agricultural rehabilitation of abandoned wastelands.**

1765 O'Callaghan　　　　　GB 023504/79/0004 N
Modelling soil water movement asociated with land spreading of manures.

1766 Soane; Pidgeon　　　　　GB 030503/00/0005 R
Effect of traffic intensity on production.

1767 Soane; Pidgeon　　　　　GB 030503/00/0006 R
Effect of traffic intensity on production.

1768 Curry, J.P.; Cotton, J.C.F.; Baker, G.; Morgan, M.A.　　　　　IE 120101/78/9078 N
Utilization of earthworms in land reclaimation on cutover peat.

1769 Bellini, P.　　　　　IT 040619/78/1031 N
Miglioramento della struttura del suolo, prove su nuovi prodotti dell'industria chimica atti a stabilizzare i gruppi strutturali. **Improving soil structure; tests on chemical stabilizers of structural groups.**

1770 Sluijs, P. van der; Houben, J.M.M.T.　　　　　NL 010119/73/5972 R
De invloed van cultuurtechnische ingrepen op het poriënvolume en de bewortelbaarheid van diverse gronden. **The rootability of soils that have been subsoiled.**

1771 Smet, L.A.H. de　　　　　NL 010119/78/9072 N
Invloed van het losmaken van zavelige ondergronden op vochtvoorziening, beworteling en produktie van landbouwgewassen. **The influence of the removal of sandy clay subsoil on moisture supply and root development.**

1772 Nicolaï, P.　　　　　NL 010207/71/3273 R
Onderzoek naar de uitwerking van mechanische grondverbeteringsmethoden op de technische mogelijkheden van het akkerbouw- en vollegrondsgroentebedrijf. **Investigations on the effect of mechanical soil improvement methods on farming, concerning arable and outdoor vegetable crops.**

1773 Luten, W.; Roozeboom, L.　　　　　NL 010208/76/6585
Cultuurtechnische ingrepen in grasland. **Improvement of grassland soils.**

1774 Wind, G.P.　　　　　NL 010501/73/5239
Vergroting van de bewortelingsdiepte van gewassen in de veenkoloniën door grondverbeteringsmaatregelen. **Improvement of cut–over high moor peat soils.**

1775 Boels, D.　　　　　NL 010501/74/6194 R

Criteria inzake vochttoestand van de bodem voor het tijdelijk stilleggen van grondverzetswerkzaamheden. **Soil moisture criteria with regards to down time when moving earth for agricultural land layout purposes.**

1776 Boels, D.　　　　　NL 010501/75/6193
Verbetering van veldpodzolgronden. **Amelioration of drought susceptible podzol soils.**

1777 Schothorst, C.J.　　　　　NL 010501/77/7597 R
Profielverbetering bij grasland op zandgrond. **Improvement of sandy soils under grass.**

1778 Tol, G. van　　　　　NL 010601/60/2394
Het verbeteren van door plaggen– en strooiselwinning verarmde grond voor de houtteelt. **Improvement of soils impoverished by sod and litter removal for silviculture.** Publications.

1779 Eppink, L.A.A.J.　　　　　NL 020012/72/5084
Bepaling van de kwantitatieve bodemerosie als gevolg van de hierop van invloed zijnde afzonderlijke factoren, alsmede hun onderlinge wisselwerking. **Determination of the quantitative amount of soil erosion, influenced by its separate causing factors and their interaction.** Publications.

1780 Rijniersce, K.　　　　　NL 040007/64/4091
Onderzoek t.a.v. profielverbetering voor gronden met bestemming landbouw, recreatie of natuur. **Research on improvement of soil profiles on sites for agriculture, recreation or nature reserves.**

1781 Rijniersce, K.　　　　　NL 040007/73/4093
Mogelijke grondverbeteringsmaatregelen op opgespoten terreinen voor groenvoorzieningen in woonkernen. **Soil improvement measures on building sites (with artificial cover of coarse sand) for green belts, etc. in residential centres.**

1782 Rijniersce, K.　　　　　NL 040007/75/8496
Onderzoek naar mogelijkheden tot optimalisering van het grondverzet in stedelijke gebieden. **Research on the possibilities for optimalisation of moving earth in urban areas.**

D 1330 – Surveying

See also 293, 312, 691, 1109, 1189, 20332, 20376

1783 Hanotiaux, G.; Avril, P.; Steffens, R.; Remy, J.　　　　　BE 010016/75/0004 N
Cartographie des sols de Belgique méridionale. **Soil mapping of Southern Belgium.** Publications.

1784 Hanotiaux, G.; Boek, L.; Mathieu, L.; Calembert, J.　　　　　BE 010016/77/0005 N
Etude de la pédogénèse en Belgique et Afrique. **Pedogenesis in Belgium and Africa.** Publications.

1785 Noirfalise, A.; Limbourg, P.　　　　　BE 010024/63/0001 R
Ecologie herbagère en Haute Belgique. **Grassland ecology in High Belgium.** Publications.

1786 Noirfalise, A.; Dethioux, M.; Vanesse, R.　　　　　BE 010024/70/0002 R
Cartographie écologique de la Belgique. **Ecological mapping of Belgium.** Publications.

D 1330 – Surveying

1787 Persoons, E.; Coppens d'Eeckenbrugge, G.; Bazier, G.; De Volder, E.; Danhaive, B.; Verbiese, R.

BE 020400/71/0001 R

Etude d'un système automatique de télémesure et de transmission adapté à l'utilisation du sol et de l'eau en agriculture. **Study of an automatic telemeasuring and transmission system adapted to land and water use in agriculture.** Publications.

1788 Van Onsem, J.G.; Verdonck, O.   BE 070600/67/0068 R
Bodemgeschiktheidsklassificatie voor sierplanten. **Research on soil suitability for ornamentals.** Publications.

1789 Timmermans, J.; Gérard, J.   BE 130000/45/0002 R
Recherches sur l'identification des zones piscicoles des eaux courantes. **Research on the identification of the fish zones in running water.** Publications.

1790 Rogister, J.E.   BE 130000/76/0018 R
Methode voor de ekologische klassering en vergelijking van bosplantengezelschappen. **Method for the ecological classification and comparison of forest communities.** Publications.

1791 Hildebrandt, G.; Al–Homaid; Nasser

DE 126451/74/0004

Informationsgewinnung über Erfassung und Klassifizierung von Brachland durch Luftbildinterpre- tation. **Inventory and classification of fallow land by air–photo interpretation.**

1792 Hildebrandt, G.; Reichert, P.; Lange, G.

DE 126451/77/0002

Aufbau eines digitalen Auswertungssystems für Luftbildund Scanner–Aufzeichnungen. Ermittlung spektraler Signaturen aus digitalen Scanner- Aufzeichnungen. Klassifizierung land– und forstwirtschaftlicher Kulturen aus digitalen Bildinformationen. **Establishment of a digital interpretation system for aerial photographs and scanner records. Determination of spectral signatures from digital scanner records. Classification of cultivations in agriculture and forestry from digital photograph informations.**

1793 Hildebrandt, G.; Reichert, P.   DE 126451/77/0003
Anwendung von multispektralen Scanner–Systemen für quantitative und qualitative Inventuren in Landwirtschaft, Forstwirtschaft und Umwelt. **Application of multispectral scanner systems to quantitative and qualitative inventory in agriculture, forestry and environment.** Publications.

1794 Hildebrandt, G.; Masumy, S.–A.   DE 126451/77/0004
Untersuchungen von Texturparametern zur Erkennung verschiedener Waldtypen. **Studies on texture parameters for determination of different types of forests.** Publications.

1795 Hildebrandt, G.; Sanwald, E.   DE 126451/77/0005
Möglichkeiten des Einsatzes der Fernerkundung zur Erfassung von Pflanzenkrankheiten, dargestellt am Beispiel von Nematodenschäden in Zuckerrübenschlägen. **Possibilities of using long–distance reconnaissance for realization of plant diseases, e.g. of damages to sugar beets by nematodes.** Publications.

1796 Hildebrandt, G.; Carneiro, C.   DE 126451/77/0006
Waldtypenkartierung nach multispektralen SatellitenAufzeichnungen mittels analoger und digitaler Auswerteverfahren. **Cartography of forest types by multispectral satellite records by means of analogue and digital interpretation methods.**

1797 Hildebrandt, G.; Dörfel, H.–J.   DE 126451/77/0007
Phänologische Aspekte der Fernerkundung ausgewählter land– und forstwirtschaftlicher Kulturflächen, Störfaktoren und zusätzliche Differenzierungskriterien. **Phenological aspects of long–distance reconnaissance of selected cultivated areas in agriculture and forestry. Interference factors and additional differentiation criteria.**

1798 Hildebrandt, G.; Lange, G.   DE 126451/77/0008
Dialog–betriebsfähige Datenbank zur Speicherung originaler oder bearbeiteter Multispektral–Bilder. Einschliesslich aller jeweils bekannten Zusatzinformationen mit dem Ziel, auf einzelne oder Gruppen von Luftbildern, auf Grund eines hierarchischen Schlüssels, gezielt zugreifen zu können. **Dialog–operational data bank for storage of original or processed multispectral photographs including all given additional informations aiming at methodical use of single or groups of aerial photographs on basis of hierarchic code.**

1799 Niesslein, E.; Essmann, H.   DE 126500/75/0007
Strukturanalyse für den ländlichen Raum in Österreich. **Structural analysis of rural areas in Austria.**

1800 Schreiber, K.–F.   DE 164401/78/0001 N
Wärmegliederung auf phänologischer Grundlage – Vergleichende Detailkartierung einer Stadt-, Agrar- und Waldlandschaft 1:25000. **Thermal division on phenological basis – comparative detail mapping of urban, agrarian and forest landscapes 1:25000.**

1801 Schreiber, K.–F.   DE 164401/78/0002 N
Phänologische Kartierung und Wärmegliederung von Nordrhein–Westfalen 1:200000. **Phenological mapping and thermal division of Northrhine Westphalia 1:200000.**

1802 Altemüller, H.–J.   DE 201010/71/4013 R
Stabilisierung von Bodengefügen mit organischen und anorganischen Stoffen. Optische und physikalischmechanische Untersuchungen 1971. **Stabilization of soil textures with organic and unorganic matters. Optical and physico–mechanical studies.**

1803 Auweck, F.   DE 502055/78/0006 N
Weiterentwicklung von Kartierungsmethoden zur Erfassung und Beratung von Kleinstrukturen in der Kulturlandschaft und ihre Verarbeitung über EDV. **Continuous improvement of mapping methods for small–sized cultural landscapes using EDP.** Publications.

1804 Bargon, E.; Haupenthal, C.   DE 506051/71/0003
Bodenkarte von Hessen 1:25 000, Bl. 4722 Kassel–Niederzwehren. **Soil map of Hesse scale 1:25 000, sheet no. 4722 Kassel–Niederzwehren.**

1805 Martensen, F.   DK 030149/77/0002 N
Den praktiserende landinspektørs civilretlige ansvar. **The chartered surveyor's responsability with regard to civil law.**

1806 Durand, J.H.   FR 010704/70/6088
Synthèse régionale – Carte des aptitudes des terres de l'Aquitaine. **Synthetic regional study – land capability survey of aquitaine.**

D 1330 – Surveying

1807 Naert, B.   FR 011211/72/0610
Télédétection et méthodologie cartographie appliquée à la pédologie et à l'aménagement du territoire en paysage calcaire. **Remote sensing and mapping applied to pedology and land reclamation in calcareous areas.** Publications.

1808 Draycott; Durrant   GB 012301/00/0009
Soil classification and sugar beet yield.

1809 Researcher not indicated   GB 012401/00/0001
Soil survey of England and Wales.

1810 Grant   GB 030408/00/0001
Soil survey of scotland.

1811 McConaghy   GB 040402/00/0015
Mapping of soil and land capability of pilot areas.

1812 O'Neill   GB 040501/00/0006
An evaluation of the gravel–tunnel drainage system.

1813 Speirs   GB 060114/00/0004
The preparation and evaluation of interpretative and derivative soil maps for the East of Scotland.

1814 Cabot, D.; Sherry, C.; Goodwillie, R.; Webb, R.
     IE 080200/77/7450 R
Ecological mapping of the territory of the European Economic Community. Publications.

1815 Lorenzoni, P.; Rodolfi, G.; Lulli, L.; Magaldi, D.
     IT 020100/79/0003 N
Cartografia di riconoscimento dei suoli della provincia di Rieti. **Reconnaissance soil mapping of the Rieti district.**

1816 Romita, P.L.   IT 040608/77/0643 R
Cartografia automatica di dati climatici idrologici e geopedologici. **Automatic mapping of climatic, hydrological and geopedological data.**

1817 Pietracaprina, A.   IT 041311/77/0642 R
Classificazione e cartografia dei suoli dell'Italia meridionale e insulare. **Soil classification and mapping in Southern Italy and the isles.**

1818 Visser, A.C.   NL 010501/73/5260 R
Toepassing digitaliseringstechnieken op op kaarten vaste te leggen gegevens. **Application of map–digitizing techniques.** Publications.

1819 Vos, W.   NL 010601/76/6684
Karakterisering van reliëf van een aantal Nederlandse landschappen met behulp van de Fourier–analyse. **The characterisation of the relief of some Dutch landscapes by means of a Fourier analysis.**

1820 Gijsen, J.C.O. van   NL 020028/71/5109
Kwantitatieve bepaling van horizontale en verticale refractie. **Quantitative determination of horizontal and vertical refraction.** Publications.

1821 Buiten, H.J.   NL 020028/71/5111
Geautomatiseerde verwerking en opslag van landmeetkundige informatie, verkregen door terreinmetingen door digitalisatie van grafische gegevens of ontleend aan een gegevensbestand. **Electronic dataprocessing and storage of geodetic information,** acquires by terrain surveys by digitizing of graphical representations or by retrieval from a databank. Publications.

1822 Wely, G.A. van   NL 020028/71/5112
Onderzoek naar de mogelijkheden van een grootschalige basiskaart voor Nederland voor technische, beheers– en planologische doeleinden en toepassing van fotokaarten in plaats van lijnenkaarten. **Inquiry into the possibilities of a large–scale basemap of the Netherlands for technical, management en planning purposes and use of photomaps in stead of line–maps.** Publications.

1823 Gijsen, J.C.O. van   NL 020028/77/8459
Systematische invloeden bij 1ste orde waterpasinstrumenten en hun correctie. **Systematic influences on precise levelling instruments and the correction.**

1824 Richardus, P.   NL 020028/77/8460 N
Onderzoek naar de nauwkeurigheid van de hoogtebepaling d.m.v. Doppler satellieten plaatsbepaling. **Investigation into the accuracy of the height of stations on the earth surface, determined by Doppler Satellite positioning.**

D 1500 – Nature conservation

See also 2, 4, 21, 93, 94, 98, 172, 202, 212, 295, 361, 420, 583, 985, 1007, 1112, 1113, 1114, 1117, 1118, 1130, 1132, 1134, 1136, 1140, 1142, 1145, 1181, 1182, 1183, 1184, 1321, 1341, 1353, 1418, 1445, 1446, 1509, 1511, 1517, 1518, 1554, 1663, 1687, 2147, 2365, 2366, 2369, 2378, 2381, 2384, 2390, 2412, 2433, 2551, 2564, 2678, 2687, 2751, 3623, 4739, 4764, 4765, 4806, 4834, 4908, 4909, 5000, 7791, 7826, 9082, 10117, 10147, 10165, 10301, 10311, 10326, 10333, 10338, 10341, 10356, 10389, 11104, 11105, 11117, 11393, 11398, 11440, 11447, 11448, 11449, 11450, 11518, 11521, 11522, 11523, 11524, 11529, 11567, 12727, 12885, 13511, 13695, 13887, 13890, 14802, 14803, 14804, 14805, 14951, 14952, 14953, 14956, 14957, 14961, 15000, 15014, 15015, 15050, 17059, 18217, 18339, 18924, 19596, 19707...

1825 Compère, R.; Thèwis, A.; Pironio, E.
     BE 010003/76/0012 R
Physiologie digestive et comportement alimentaire de Taurotragus Oryx. **Digestive physiology and alimentary behaviour of Taurotragus oryx.**

1826 Breny, R.; Baurant, R.; Magema; Grégoire, M.
     BE 010011/70/0008 R
Etude des événements entomologiques forestiers belges. **Study df entomological events in belgian forest.** Publications.

1827 Breny, R.; Baurant, R.; Fichant, R.; Overal, M.
     BE 010011/71/0009
Etude de l'alimentation des Cervidés en vue des aménagements écologiques et cynégétiques. **Food study of Deer, in relation with ecological and cynegetical management.** Publications.

1828 Leclercq, J.; Gaspar, C.; Verstraeten, C.; Marchal, J.L.
     BE 010020/70/0001 R
Contribution belge à la cartographie des invertébrés européens. **Belgian contribution to the European invertebrate survey.** Publications.

1829 Paul, R.; Impens, R.; Delcarte, E.; Nangniot, P.
     BE 010021/76/0009
Recherches sur la physiologie et l'écologie de plantes

D 1500 – Nature conservation

supérieures, indicatrices de la pollution de l'air. **Research of the physiology and ecology of higherplants, indicators of the air pollution.**

1830 Lambert, J.; Toussaint, B.   BE 020602/72/0006
Observation de jardins phénologiques dans le cadre du programme "Arbeitsgemeinchaft Internationale Phanologische Garten". **Phenological garden observation as a contribution to the "Arbeitsgemeinchaft Internationale Phanologische Garten" program.** Publications.

1831 Van Bladel, M.; Moreale, A.   BE 020700/78/0001 N
Environnement – pesticides dans les argiles et les sols dégradation – adsorption. **Environmental problem of adsorption and degradation of pesticides in clays and soils.**

1832 Gillard, A.; De Maeseneer, J.   BE 030024/68/0007 R
Studie van de watervervuiling en haar weerslag op de dierlijke en plantaardige populaties. **Study of waterpollution and its consequences on plant en animal life.** Publications.

1833 Vanderstappen, R.; Van Hoeyweghen, P.; Guns, M.
          BE 100000/71/0011 R
Recherches sur la pollution par métaux lourds des poissons et des crustacés. **Research upon the pollution by heavy metals of fishes and crustacea.** Publications.

1834 Neirinckx, G.   BE 100000/72/0012 R
Inventaris van de organische polluenten in sedimenten en vissen. **Inventory of the organic polluants in sediments and fish.**

1835 Istas, J.; De Temmerman, L.; Baeten, H.; Raekelboom, E.; Termonia, M.   BE 100000/73/0034 R
Studies van verontreinigingsgevallen. **Studies of pollution cases.** Publications.

1836 De Borger, R.; Van Elsen, Y.   BE 100000/77/0031 R
Studie van de huminezuren in zeewater. **Study about the humic acids of sea–water.**

1837 Vanderstappen, R.; Guns, M.; Hoening, M.
          BE 100000/78/0029 R
Etude des formes chimiques sous lesquelles se présentent les polluants contenant les métaux lourds. **Study of the chemical form under which occur polluants containing heavy metals.** Publications.

1838 Timmermans, J.; Gérard, J.   BE 130000/39/0005 R
Recherches relatives à la pisciculture de repeuplement dans les eaux douces. **Research on fish culture for restocking in inland waters.** Publications.

1839 Timmermans, J.; Gérard, J.   BE 130000/57/0006 R
Contrôle de l'efficacité des repeuplements en poissons des eaux libres. **Efficience of restocking of fish in inland waters.** Publications.

1840 De Crombrugghe, S.   BE 130000/78/0021
Utilisation de données biométriques comme indices de la condition générale des ongulés–gibier. **Use of biometrical datas as indicators of general condition of wild ungulates.**

1841 De Crombrugghe, S.   BE 130000/78/0022
L'affouragement hivernal et l'amélioration des herbages comme facteurs d'équilibre entre les herbivores sauvages et la sylviculture. **Winter feeding and herbage improvement as managing factors of the balance between wild herbivores and sylviculture.**

1842 De Crombrugghe, S.   BE 130000/78/0023
Dynamique des populations des espèces–gibier. **Population dynamics of the game.**

1843 Lebrun, Ph.; De Nayer, J.; Grégoire–Wibo, C.
          BE 140000/72/0051
Influence des pesticides sur la microfaune du sol. **Pesticides influence on soil microfauna.** Publications.

1844 Huygh, A.; Beckers, B.   BE 140000/75/0016 R
Kwaliteit van oppervlaktewateren. **Quality of surface waters.** Publications.

1845 Saar, C.; Gerriets, D.   DE 104500/77/0010
Reproduktion des Wanderfalken – Falco peregrinus – in Gefangenschaft und Wiederansiedlung in verwaisten Biotopen. **Reproduction of the peregrine falcon – Falco peregrinus – in captivity and reintroduction into deserted biotops.**

1846 Sukopp, H.   DE 105060/70/0002
Gefässpflanzenflora von Berlin 1965. **Flora of vascular plants in West Berlin.**

1847 Sukopp, H.; Böcker, R.; Tigges, W.   DE 105060/70/0004
Geobotanische Grundlagenuntersuchungen in Berliner Naturschutz– und Landschaftsschutzgebieten 1970. **Fundamental geobotanical studies on nature preserves and landscape preserves of Berlin.** Publications.

1848 Sukopp, H.; Pitzer, E.   DE 105060/74/0002
Artenschutz in der Bundesrepuplik Deutschland – Farn und Blütenpflanzen –. **Conservation of species in the Federal Republic of Germany – Filicales and Phanerogams –.**

1849 Brande, A.   DE 105060/75/0001
Vegetations– und florengeschichtliche Untersuchungen in Berlin. **Investigations of the vegetational and floral history in Berlin.** Publications.

1850 Sukopp, H.   DE 105060/75/0003
Veränderungen des Röhrichtbestandes der Berliner Havel. **Changes of the reed communities in the Havel river – Berlin –.**

1851 Sukopp, H.; Auhagen, A.   DE 105060/77/0001
Überwachung von Naturschutzgebieten und Vorschläge für den Aufbau eines Regionalsystems von Naturschutzgebieten. **Supervision of nature reserves and proposals for a regional system of nature reserves.**

1852 Sukopp, H.; Trepl, L.   DE 105060/77/0005
Veränderungen der natürlichen Ressourcen bei einem Autobahnbau in Berlin–Tiergarten. **Changes of the natural resources during a highway building in Berlin–Tiergarten.**

1853 Sukopp, H.; Tigges, W.   DE 105060/77/0007
Vegetationskundliche Untersuchungen im Forst Grunewald – Berlin –. **Vegetational analysis in the Grunewald forest – Berlin –.**

1854 Sukopp, H.; Trepl, L.   DE 105060/77/0008
Über Impatiens parviflora D.C. als Agriophyt in Mitteleuropa. **Impatiens parviflora D.C. as an agriophyte in Central Europe.**

1855 Bornkamm, R.; Schuster, J.; Daber, J.
DE 105100/78/0001 N
Analyse der Vegetation ländlicher Gebiete und ihre
Bewertung für Nutzungen. **Vegetation analysis of rural areas
and its evaluation for land use.**

1856 Kiemstedt, H.; Hahn–Herse, G.
DE 105650/75/0002
Auswirkungen von Ferienzentren auf Landschaftshaushalt und
–bild. **Ecological and visual impacts of holiday camps and hotel
complexes.**

1857 Kiemstedt, H.; Sukopp, H.; Hahn–Herse, G.;
Schneider, C.
DE 105650/75/0003
Bewertungsrahmen für Naturschutzplanung. **An approach to
evaluate nature protection areas.**

1858 Kick, H.; Ueberbach; Dörr, R.
DE 111100/75/0004
Vorkommen von polycyclischen Aromaten und anderer
Cancerogene in Klärschlämmen und Komposten aus
Siedlungsabfällen, ihre Aufnahme durch Pflanzen. **Occurrence
polycyclic aromates and of other cancerogens in sewage sludges
and town waste composts, their uptake by plants.**

1859 Drescher, W.; Hasselberg, G.
DE 111201/78/0001 N
Verteilung und Dichte von Wolfsspinnen – Araneae – in
neubesiedeltem, rekultivierten Kohletagebau. **Distribution and
abundance of Lycosid–spiders – Araneae – in a recultivated
coal–mining area.**

1860 Hartfiel, W.; Schnorrenberg, H.J.
DE 111451/77/0001
Vergleichende Verdauungsversuche an Wild– und
Hausschweinen. **Comparative digestion trials of wild boar and
domesticated swine.**

1861 Hartfiel, W.
DE 111451/77/0004
Fütterungsversuche mit Rot– und Rehwild. **Feeding trials with
roe and fallow deer.** Publications.

1862 Hesmer, H.
DE 111950/74/0001
Ermittlung der anthropogenen Veränderungen in der
Bewaldung der Tropen. **Determination of anthropogeneous
changes in the tropic woodlands.**

1863 Müller–Hohenstein, K.; Deil, U.
DE 120200/77/0004
Das Waldkleid des Rifgebirges – Marokko –. **The forests of the
Rif Mountains – Morocco –.**

1864 Marquardt, H.
DE 126100/74/0004
Prüfung eines Spektrums von Mutagenitätstesten als
Vorprüfung auf krebsauslösende Wirkungen von
Umweltchemikalien. **Examination of a spectrum of
mutagenicity tests as a methodical prescreening of carcinogenic
actions of chemicals in the environment.**

1865 Siebert, D.
DE 126100/74/0005
Entwicklung eines Spektrums genetischer Vorteste,
insbesondere mit Hefe, zur Erkennung möglicher
krebsauslösender Umweltchemikalien. **Development of a series
of genetical preli– minary tests using yeast for the characteri–
zation of possibly carcinogenic chemicals in the environment.**

1866 Schmidt–Vogt, H.; Staden, N.von
DE 126300/72/0004
Strukturuntersuchungen und Wertanalysen von
Naturverjüngungsbeständen im südlichen Hochschwarzwald.
Analysis of structure and value referring to natural

regenerations in the high region of southern Black Forest.

1867 Barner, J.; Beschnidt, R.
DE 126400/73/0004
Gravimetrische und konimetrische Untersuchungen von
Luftverunreinigungen im Verdichtungsraum
AalenWasseralfingen. **Gravimetric and conimetric analyses of
air pollution in the agglomeration of Aalen–Wasseralfingen.**

1868 Barner, J.; Sittler, B.
DE 126400/74/0001
Vergleichende Untersuchungen über Bestrebungen und
Organisationen für den Gewässerschutz in der Bundesrepublik
Deutschland. **Comparative studies on trends and organizations
on the conservation of waters in the Federal Republic of
Germany.**

1869 Speidel, G.; Dürr, R.
DE 126450/75/0004
Untersuchung des Beziehungsgefüges
Wald–Schalenwild–Umwelt, dargestellt am Beispiel einiger
Rehwildreviere von BadenWürttemberg im Hinblick auf die
forstliche Planung. **Relationship between forest, deer and
environment with respect to forest planning – presented on the
basis of research work in selected roe deer habitats in the State
of Baden–Württemberg.**

1870 Mantel, K.; Brückner, J.
DE 126500/71/0001
Quantifizierung der Sozialfunktionen des Waldes als Element
der Infrastruktur im Gebiet des Südschwarzwaldes.
**Quantification of the social functions of forestry as a part of
infrastructure.**

1871 Pacher, J.; Hexges, A.
DE 126505/75/0001
Der Kottenforst. Vom Königsforst zum Naturpark. **The
Kottenforst. From Kings–Forest to National Park.**

1872 Mitscherlich, G.; Künstle, E.
DE 126600/74/0002
Die Filterwirkung des Waldes bezüglich Elementgehalt
–Schadstoffe– im Niederschlagswasser und in der Luft. **The
filter effect of forests on the element content –noxious
compounds– in precipitations and in the air.**

1873 Kausch, H.; Gunkel, G.; Heisig, G.   DE 126800/75/0001
Weitergabe, Anreicherung und Wirkung von Schadstoffen in
den Nahrungskettengliedern Primärkonsument –
Sekundärkonsument. **The effects of toxins in primary and
secondary consumers of a food chain, their transmission and
accumulation.**

1874 Müller, H.
DE 126801/75/0002
Einfluss von Umweltfaktoren auf die Bildung algenbürtiger
Schadstoffe sowie die Wirkung dieser Schadstoffe auf andere
Organismen. **The influence of environmental conditions on the
formation of harmful substances of algal origin and their action
on other organisms.**

1875 Schwoerbel, J.; Streit, B.
DE 126802/75/0001
Toxizität und Akkumulation von Pestiziden in der
Nahrungskette: Algen – Ancylus fluviatilis – Glossiphonia
complanata. **Toxicity and accumulation of pesticides in the food
chain: Algae – Ancylus fluviatilis – Glossiphonia complanata.**

1876 Homrighausen, E.; Götze, K.; Budig, M.
DE 129020/70/0010
'Giessener Modell' – gemeinsame Beseitigung fester und
flüssiger Abfallstoffe. **The 'Giessen Model'. Disposal of both
solid and liquid wastes.**

D 1500 – Nature conservation

1877 Simon, U.; Homrighausen, E.   DE 129140/74/0001
Landschaftspflege – Giessener Modell –. **Protection of natural landscape – model Giessen –.**

1878 Simon, U.; Zerwes, K.; Schäfer, K.   DE 129141/74/0001
Der Einfluss der Brachedauer auf Pflanzenbestand und Boden, untersucht auf Dauerflächen. **The influence of the duration of fallowing on plant communities and soils, examined on long–time observation plots.**

1879 Simon, U.; Campino, I.; Schäfer, K.
                                              DE 129141/74/0002
Die Rückentwicklung bisher intensiv genutzter terrestrischer Ökosysteme zu nicht oder extensiv genutzten Ökosystemen in ihrer Wirkung auf Pflanzengemeinschaften und Böden. **The retrogression of intensive cultivation of terrestrial ecosytems to non– or extensive use and consequences to plant communities and the soils.**

1880 Stein, W.; Kelany, J.M.   DE 129200/75/0003
Untersuchungen über Kältetoleranz und Überwinterung synanthroper Fliegen. **Investigations of cold tolerance and hibernation of synanthropic flies.**

1881 Breburda, J.   DE 129382/72/0001
Boden– und Landschaftsschutz in der Sowjetunion 1972. **Conservation of soil and landscape in Soviet Russia.**

1882 Habermehl, K.–H.; Laube, K.   DE 129601/73/0006
Postnatale Entwicklung und Altersbestimmung beim Jagdfasan. **Postnatal development and ageing estimation of game pheasant.**

1883 Hofmann, R.R.; Geiger, G.; Herzog, A.; Blähser, S.
                                              DE 129601/75/0002
Morphologische Untersuchungen über Ursachen des Auftretens von Schalenwildkümmerformen 'Knopfbock, Kurzspiesser'. **Morphological studies on the causes of the occurrence of micro–formations in cloven hoofs 'button–buck, short antler'.**

1884 Hofmann, R.R.; Knab, S.   DE 129602/73/0012
Funktionell–anatomische Anpassung der Nieren von Wildwiederkäuern. **Functional anatomical adaptation of the kidneys of wild ruminants.**

1885 Hofmann, R.R.; Hoffmann, R.   DE 129602/73/0016
Morphologische Untersuchungen am Darm von Reh– und Rotwild. **Morphological studies of roe and red deer intestines.**

1886 Hofmann, R.R.; Geiger, G.   DE 129602/73/0017
Morphologische Untersuchungen am Verdauungsapparat von Wildwiederkäuern – Cerviden, Muffelwild, Dikdiks –. **Morphological studies of the digestive tract of wild ruminants – Cervidae, wild sheep and dik–dik –.**

1887 Herzog, A.; Höhn, H.; Schmidt, I.   DE 129680/74/0001
1.Fortpflanzung beim Reh – Physiologie und Pathologie einschliesslich den Problemen der Eiruhe – 2.Untersuchungen zum sogenannten Knopfbockproblem – Ätiologie und Pathologie – 3.Zytogenetische Untersuchungen an Wildtieren. **1.Reproduction of roe deer – physiology and pathology, including problems of physiological arrest of embryo development 2.Analysis of the problem of the so–called 'button–buck' – etiology and pathology – 3.Cytogenetic studies on game.**

1888 Herzog, A.; Oldenburg, U.   DE 129680/75/0008
Angeborene Anomalien bei Wildtieren in Europa. **Congenital malformations in European game species.**

1889 Herzog, A.; Schulz, R.   DE 129680/78/0008 N
Zytogenetische Untersuchungen bei Wildvögeln. **Cytogenetics in wild birds.**

1890 Welte, E.; Eissa, G.   DE 132064/75/0002
Möglichkeiten zur Bestimmung des Trophiezustandes eines Gewässers durch Algenkulturen. **Possibilities of determining the nutrient content and state of eutrophication in waters by algae cultures.**

1891 Heller, H.   DE 132093/70/0004
Untersuchungen der Biomasse von Bäumen auf den IBP–Probeflächen im Solling. **Investigations of tree biomass on the IBP areas in the Solling.**

1892 Schroeder, F.–G.   DE 132093/70/0005
Waldgesellschaften in Nordwest–Nordcarolina, USA, im Vergleich zu Mitteleuropa. **Forest communities in North Western North Carolina in comparison to Central Europe.**

1893 Schmidt, W.   DE 132093/73/0001
Savannengesellschaften Afrikas und ihre Stickstoff– und Phosphor–Versorgung. **Savanna communities in Africa and their nitrogen and phosphorus supply.** Publications.

1894 Geyger, E.   DE 132093/73/0002
Untersuchungen zum Wasserhaushalt der Puna–Vegetation Nordwest–Argentiniens. **Investigations on water balance of Puna vegetation in the North–West Argentine.**

1895 Schroeder, F.–G.   DE 132093/74/0005
Waldvegetation und Gehölzflora in den Südappalachen –USA–. **Forest vegetation and ligneous flora in the southern Appalachians –USA–.**

1896 Ellenberg, H.; Schad, A.   DE 132093/74/0007
Die Abstufung der Spezies–Diversität vom Freiland zum Grosstadtkern und ihre Ursachen. **The scale of the diversity of species from fields to the centre of cities and ecological causes.**

1897 Dierschke, H.; Riede, U.   DE 132093/74/0009
Lebensbedingungen der Dünenvegetation im Naturschutzgebiet auf der Insel Juist. **Life conditions of dune vegetation in the nature conservancy area on the Isle of Juist.**

1898 Ellenberg, H.; Barth, H.   DE 132093/74/0010
Saugspannungen und osmotische Werte von Halbtrockenrasenpflanzen. **Moisture tensions and osmotic values of Mesobrometum species.**

1899 Freitag, H.   DE 132093/77/0002
Flora und Vegetation im Rahmen eines internationalen und interdisziplinären Forschungsprogrammes über Wüstenökologie im Turan–Schutzgebiet im Iran. **Flora and vegetation as part of an international and interdisciplinary Desert Ecology Prototype Program, Turan Protected Area/Iran.**

1900 Ruthsatz, B.   DE 132093/77/0007
Verteilung der C3– und C4–Pflanzen in den andinen Halbwüsten NW–Argentiniens in Beziehung zu ökologischen Gradienten. **Distribution of C3– and C4–plant species in the**

high Andes of NW–Argentina in relation to ecological gradients. Publications.

1901 Schmidt, W.; Werner, W.   DE 132093/77/0008
Kulturversuche zum Anteil des inneren Stickstoff– und Phosphorkreislaufs bei Wildpflanzen. **The role of internal nitrogen– and phosphorus cycle of native plants tested in cultivation experiments.**

1902 Ellenberg, H.; Zoldan, J.–W.   DE 132093/77/0009
Stickstoffnachlieferung in den Böden von AckerunkrautGesellschaften im südniedersächsischen und nordhessischen Raum. **Nitrogen supply in soils of field weed communities in Southern Lower Saxony and North Hesse.**

1903 Fassbender, H.W.; Steinhard, U.   DE 132600/73/0002
Systemanalytische Untersuchungen in Nebelwald–Ökosystemen Venezuelas. **System analytical studies on forest ecosystems in Venezuela.**

1904 Ulrich, B.; Matzner, E.   DE 132600/78/0001 N
Nährstoffhaushalt eines Heide–Ökosystems. **Nutrient balance of a heath ecosystem.**

1905 Ulrich, B.; Schlichter, T.; Ploeg, R.R.van der   DE 132603/78/0001 N
Wasser– und Energiehaushalt eines Waldökosystems. **Water and energy balance of a forest ecosystem.**

1906 Röhrig, E.   DE 132751/75/0003
Folgen von Waldbrand auf Boden und Vegetation. **Effects of forest fires on soil and vegetation.**

1907 Lamprecht, H.; Jahn, G.   DE 132752/70/0002
Beratung und Mitarbeit bei der waldkundlichen Untersuchung von Naturwaldzellen in Nordrhein–Westfalen. **Collaboration and consultation on forest ecological studies of natural forest areas in Northrhine Westphalia.**

1908 Lamprecht, H.; Jahn, G.   DE 132752/70/0003
Waldkundliche Untersuchungen von Naturwaldparzellen in Rheinland–Pfalz. **Forest ecological studies of natural forest areas in Rhenish Palatinate.**

1909 Lamprecht, H.; Ulrich, B.   DE 132752/73/0002
Standörtliche, waldbauliche und systemanalytische Untersuchungen in Ökosystemen des äquatorialen Nebelwaldes in Venezuela. **Local, silvicultural and system analytical studies on ecosystems in equatorial rain forests in Venezuela.**

1910 Lamprecht, H.; Brun, R.   DE 132752/73/0003
Untersuchungen über Biomassenbestimmung in Naturwäldern. **Accuracy of biomass estimations in natural forests.**

1911 Lamprecht, H.; Jahn, G.   DE 132752/73/0006
Waldkundliche Untersuchungen von Naturwaldreservaten in Niedersachsen. **Forest ecological studies on natural forest areas in Lower Saxony.**

1912 Lamprecht, H.; Bockor   DE 132752/74/0003
Untersuchungen über Waldtypen im andinen Nebelwald Venezuelas. **Studies on tropical forest types in an andine rain forest in Venezuela.**

1913 Jahn, G.   DE 132752/74/0004
Waldgesellschaften im nordwestdeutschen Pleistozän. **Forest communities in the diluvium in Northwestern Germany.**

1914 Berg, C. von   DE 132810/72/0003 R
Radiotelemetrische Untersuchungen zum Raum–Zeit–Verhalten des Rehes. **Radio–telemtric studies on home range and activity of roe.** Publications.

1915 Festetics, A.; Sommerlatte, M.   DE 132810/72/0004
Die Ökologie des Kronenduckers – Sylvicapra grimmia L. – und dessen Einfluss auf junge Kiefernplantagen im Ost–Transvaal/Südafrika. **The ecology of the Grey Duiker – Sylvicapra grimmia, L. – and their effect on young pine plantations in the eastern Transvaal/South Africa.**

1916 Festetics, A.   DE 132810/75/0005
Untersuchungen zum Status jagdbarer Tiere in Niedersachsen – derzeit: Rotmilan –. **Studies on the status of shootable animals in Lower Saxony – at present: red kite –.**

1917 Festetics, A.   DE 132810/75/0006
Untersuchungen zur Lokomotion, zum Jagdverhalten und zur Nahrungswahl einheimischer Raubtiere. **Studies on locomotion, hunting behaviour and food choice of domestic predacious animals.**

1918 Festetics, A.; Wölfel, H.   DE 132810/75/0007
Untersuchungen zur Jugendentwicklung, Mutter–Kind–Bindung und Feindverhalten beim Rothirsch. **Studies on juvenile development, on mother–child–bonds and on enemy behaviour in red deers.**

1919 Festetics, A.; Wölfel, H.   DE 132810/75/0010
Untersuchungen zur Nahrungswahl von Rothirsch und Reh. **Studies on the food choice of red deer and roe deer.**

1920 Festetics, A.; Nöllenheidt, H.   DE 132810/75/0011
Belastbarkeit der Vegetation durch Rothirsch und Reh in einem Revier Nordrhein–Westfalens. **Carrying capacity of vegetation as to red deer and roe deer in a forest district of Northrhine Westphalia.**

1921 Buhse, G.   DE 132810/75/0014
Untersuchungen über die Auswirkungen von Kernkraftwerken auf die Fischfauna der Weser. **Studies on the impact of nuclear power stations on the fish fauna in the Weser.**

1922 Festetics, A.; Sommerlatte, M.   DE 132810/75/0015
Auswirkungen von Feuer und Elefantenherden auf die Vegetation des Chobe National Parks/Botswana. **Effects of fire and of elephant–herds on the vegetation of the Chobe National Park/Botswana.**

1923 Festetics, A.; Riedel, B.   DE 132810/75/0018
Untersuchungen zur Brut– und Zugbiologie des Flussuferläufers in Südniedersachsen. **Studies on the migratory biology of the bank piper in South Lower Saxony.**

1924 Berg, F.–C.von; Ruff, B.   DE 132810/78/0001 N
Ein Simulationsmodell zur Populationsdynamik von Rehwild – Capreolus capreolus –. **Simulation model of roe deer – capreolus capreolus – population dynamics.**

1925 Festetics, A.; Berg, F.–C.von; Sommerlatte, M.
    DE 132810/78/0002 N

D 1500 – Nature conservation

Die Wiedereinbürgerung des Luchses – Lynx lynx – in den österreichischen Ostalpen und ihre Überwachung mittels Funkortung und Ausfährten. **Reintroduction of lynx in the East Austrian Alps and its control by radio– and snow–tracking.**

1926 Kato, F.; Mülder, D.   DE 132840/71/0002
Untersuchung über Anzahl und räumliche Verteilung der Zuwachsträger verschiedener Qualitäten in Buchenbeständen. **Investigation into number and spatial distribution of dominant trees of different quality in beech–stands.**

1927 Mülder, D.   DE 132840/71/0003 R
Welche Entwicklung würden die deutschen Wälder ohne forstliche Bewirtschaftung nehmen 1971. **Which would be the development of German forests without forestry management.**

1928 Zundel, R.; Köpp, H.   DE 132870/72/0003 R
Aufgaben und Ausstattung internationaler staatlicher und nicht–staatlicher Organisationen im Bereich des Umweltschutzes 1971. **Objectives and equipment of international state and organizations in the field of environment protection.**

1929 Lessmann, D.   DE 132870/78/0001 N
Die Betretensregelungen im Bundeswaldgesetz und in den Ländergesetzen. **Regulations on entering the forests in the Federal forest law and in the laws of the Federal states.**

1930 Häberle, S.; Yildirim, M.   DE 132900/75/0001
Einfluss verschiedener Ernteverfahren auf den Nährstoffentzug aus Fichtenreinbeständen 1976–1978. **Influence of different logging practice on nutrient withdrawal in Norway spruce stands.**

1931 Neugebohrn, L.   DE 135053/75/0001
Biotopanalyse für den Verkehrsflughafen Hamburg–Fuhlsbüttel. a.Pflanzensoziologischer Teil. **Biotope analysis for the commercial airport of Hamburg–Fuhlsbüttel a.Plant sociological part.**

1932 Lillelund, K.; Hansen, P.D.   DE 135251/75/0001
Untersuchungen über die Anreicherung von Lindan in Nahrungsketten des Süsswassers. **Investigations on the accumulation of lindane in freshwater food chains.**

1933 Lillelund, K.; Lange, O.   DE 135251/75/0003
Untersuchungen über den Einfluss der Schiffbarmachung der Mosel auf Fische und Fischerei im gestauten Fluss. **Investigations of the influence of the canalization of the Mosel River on fishes and fisheries.**

1934 Braum, E.; Hoffmann, G.   DE 135251/75/0004
Untersuchungen über den Einfluss von Benzalkoniumchlorid auf das Verhalten von Fischlarven und Jungfischen. **Effects of benzalkonium chloride on the behaviour of prelarval and postlarval fishes.**

1935 Braum, E.; Klinger, H.   DE 135251/77/0003
Vergleichende Untersuchungen am Blutbild des Aales Anguilla anguilla L.. **Comparative haematological studies at the European eel – Anguilla anguilla –.**

1936 Lillelund, K.; Brinkmeier, U.   DE 135251/77/0006
Ei, Larval– und Jungfischentwicklung beim Stint und Maifisch in der Elbe im Vergleich zur Situation 1960. **Egg, larval und joungfish development of the smelt in the river Elbe compared**

with investigations from 1960.

1937 Braum, E.; Delventhal, H.; Liemann, F.J.
                    DE 135251/77/0010
Physiologische Untersuchungen über den Hälterungsstress beim Europäischen Aal 'Anguilla anguilla'. **Physiological investigations about environmental stress in European eels 'Anguilla anguilla'.**

1938 Langer, H.; Popp, R.   DE 138600/70/0003
Vergleichende Untersuchungen zur natürlichen Leistungsfähigkeit von Naturräumen 1970. **Comparative investigations on capacity of natural areas.**

1939 Buchwald, K.; Pohl, D.   DE 138600/70/0005
Bewertungssystem zur Ausweisung von Naturschutzgebieten, dargestellt an Beispielen von Niedersachsen 1970. **Evaluation system for identification of nature preserves in Lower Saxony.**

1940 Bierhals, E.; Hard, G.; Nohl, W.; Scharpf, H.
                    DE 138600/72/0002 R
Untersuchung der Auswirkungen landbaulicher Nutzungsformen auf den Landschaftshaushalt und die Benutzbarkeit der Landschaft unter besonderer Berücksichtigung der Brachflächen 1972. **Analysis of effects of agricultural utilization systems on the landscape balance availability of landscape with special regard to fallow land.**

1941 Frank, W.   DE 144130/75/0001
Mykotische Erkrankungen der Haut und der Inneren Organe bei Reptilien. **Mycosis of the skin and of inner organs of reptiles.**

1942 Frank, W.; Keim, A.   DE 144130/75/0007
Über die Wirtsspezifität von Blutegeln. – Untersuchungen mit Hilfe der Disc–Elektrophorese. **Host specificity of leeches 'Hirudo medicinalis' – Disc–electrophoretic investigations.**

1943 Frank, W.; Lucius, R.   DE 144130/75/0009
Untersuchungen zur Biologie und Verbreitung von Dicrocoelium hospes in Westafrika 'Elfenbeinküste'. **Investigations of the biology and distribution of Dicrocoelium hospes in West Africa 'Ivory coast'.**

1944 Frank, W.; Sigmund, U.   DE 144130/75/0011
Vergleichende histopathologische Untersuchungen an Reptilien nach Befall mit Helminthen. **Comparative investigations into histopathology of reptiles after infestation with helminths.**

1945 Frank, W.; Brehm, H.   DE 144130/78/0001 N
Lichtmikroskopische Untersuchungen zum Entwicklungszyklus von Sarcocystis singaporensis in einer Schlange – Python reticulatus – und Ratten – Rattus norwegicus – Stamm Wistar und Sprague Dawley. **Light microscopic studies concerning the life–cycle of Sarcocystis singaporensis in the snake – Python reticulatus – and rat – Rattus norwegicus – of the strains Wistar and Sprague Dawley.**

1946 Frank, W.; Sauter, K.   DE 144130/78/0003 N
Serum–Harnsäurekonzentration bei Reptilien – Frischfängen. **Uric acid concentration in the serum of reptiles in nature.**

1947 Loos–Frank, B.; Schaefer, J.   DE 144130/78/0005 N
Untersuchungen zur Biologie, Pathologie und verzögerten Entwicklung eines Trichostrongyliden der Feldmaus Microtus

arvalis. **Studies on the biology, pathology and delayed development of a trichostrongylid nematode from the vole Microtus arvalis.**

1948 Schubert, G.; Foissner, W.   DE 144151/77/0001
Protopalina–Infektionen in Symphysodon aequifasciata. **Protopalina infections in Symphysodon aequifasciata.**

1949 Schubert, G.; Foissner, W.   DE 144151/77/0002
Peritriche Ciliaten als Parasiten bei Anabantiden. **Peritrichous Ciliates in Anabantidae.**

1950 Schad, F.; Hötzel, H.–J.   DE 144900/73/0002
Rechtsprobleme landschaftspflegerischer Landbewirtschaftung. **Legal problems of landscape–tending land management.**

1951 Kreeb, K.; Müller, J.   DE 144980/75/0006
Experimentell–ökologische Studien zum natürlichen Flechtenvorkommen. – Kausal– und Wechselbeziehungen in den Standortsbedingungen. **Experimental ecological studies on the natural development and growth of lichens – cause and different environmental conditions.**

1952 Kohler, A.; Kutscher, G.; Pfaff, B.   DE 144990/75/0001
Verbreitung und Ökologie submerser Makrophyten auf der Schwäbischen Alb zwischen Bära und Grosser Lauter und im Gebiet zwischen Iller– und Lech–Platten. **Distribution and ecology of submerged vascular macrophytes in the district of the "Schwäbische Alb" –mountains between "Bära" and "Grosse Lauter" and in the region between the "Iller" and "Lech" Plateau.**

1953 Kohler, A.; Labus, B.; Schuster, H.   DE 144990/75/0003
Submerse Makrophyten als Bioindikatoren für Gewässerbelastung mit anionaktiven Tensiden, Schwermetallen und Phenolen. **Submerged macrophytes as biological indicators for water pollution with anionactive surfactants, heavy metals and phenols. Publications.**

1954 Schreiber, K.–F.; Schiefer, J.; Kohler, A.   DE 144990/77/0003
Versuche zur Offenhaltung der Kulturlandschaft. **Experiments with ecological use of uncultivated land.**

1955 Kunick, W.; Konold, W.   DE 144990/77/0004
Die Vegetation der Weinbergsbrachen im Raum Stuttgart und deren Bedeutung für Arten–, Biotopschutz und Erholung. **The vegetation of fallow vineyards in the Stuttgart area and their significance for nature conservation of plant species and biotopes and recreation.**

1956 Kureck, A.   DE 151101/73/0002
Der Einfluss tagesperiodischer Temperaturzyklen auf Entwicklungsgeschwindigkeit, Eiproduktion und Futterverbrauch von Oncopeltus fasciatus –Heteroptera, Lygaeidae–. **The influence of diurnal temperature cycles on development speed, egg production and food consumption of Oncopeltus fasciatus –Heteroptera, Lygaeidae–.**

1957 Thiele, H.U.   DE 151101/77/0001
Laufkäfer – Carabiden – als Bioindikatoren für menschliche Eingriffe in den Wasserhaushalt von Waldgebieten. **Carabids as bioindicators of microclimate changes caused by man in forest habitats. Publications.**

1958 Schwenke, W.; Sievert, M.   DE 160090/74/0003
Zur Verteilung und Populationsdynamik der Nisthöhlen bewohnenden Vögel und Fledermäuse des Ebersberger Forstes. **Distribution and population dynamics of nesting birds and bats in the Ebersberg Forest.**

1959 Schwenke, W.   DE 160090/74/0006
Versuche zur Fledermausansiedlung in Wäldern. Untersuchungen von Populationsdynamik der Fichtengespinstblattwespe, Cephalcia abietis L.. **Experiments on settling bats in forests. Studies on population dynamics of spruce sawfly Cephalcia abietis L..**

1960 Bäumler, W.   DE 160091/72/0006 R
Studien über den Massenwechsel von Kleinsäugern und deren Rolle im Wald–Ökosystem 1972. **Studies of mass–runway of small mammals and of their role in forest ecosystems.**

1961 Haeselbarth, E.   DE 160091/74/0001
Systematische Untersuchungen über bei Coleoptera parasitierende Gruppen der Braconidae –Hymenoptera. **On the taxonomy of Braconidae –Hymenoptera– as parasites of Coleoptera.**

1962 Baumgartner, A.; Mayer, H.   DE 160120/77/0002
Das Mikroklima im Gebirgswald bei verschiedener Bewirtschaftung. **Microclimate in a mountainous forest under different management.**

1963 Burschel, P.; Hohenadl; Veltsistas; Löw, H.   DE 160150/77/0001
Verjüngung des Bergmischwaldes. **Regeneration of mixed stands – spruce, fir, beech, maple – in the Bavarian Alps.**

1964 Franz, F.; Deckelmann, B.   DE 160210/77/0004
Beweissicherung im Bereich des Rhein–Main–Donau–Kanals, Raum Nürnberg/Allersberg. **Investigations on possible effects on forest stands by the Rhine–Main–Danube–Canal in the region of Nuernberg/ Allersberg.**

1965 Schröder, W.   DE 160270/73/0001
Einfluss der grossen Pflanzenfresser auf den Wald im Hochgebirge. **The influence of large herbivores on alpine forests.**

1966 Schröder, W.; Zeimentz, K.   DE 160270/73/0002
Biotopanspruch und Einfluss der Forstwirtschaft auf den Lebensraum des Auerwildes. **Habitat requirements and the influence of forestry on the environmental conditions of capercailzies.**

1967 Schröder, W.; Weyer, U.   DE 160270/73/0003
Nahrungswahl des Gamswildes im Jahresablauf. **Food selection of chamois in a year's cycle.**

1968 Schröder, W.; Georgii, B.   DE 160270/73/0004
Fortpflanzungsleistung bei Reh und Rothirsch. **Reproduction of roe deer and red deer.**

1969 Schröder, W.   DE 160270/74/0001
Experimentelle Reduktion von Rehbeständen – Untersuchungen der Auswirkungen auf die Kondition der Tiere und die Verbissbelastung. **Experimental reduction of roe deer populations – investigations of resulting condition and browsing intensity.**

D 1500 – Nature conservation

1970 Schröder, W.; Georgii, B.　　DE 160270/75/0001
Wohngebiet und Aktivität des Rothirsches. **Home range and activity of red deer.**

1971 Plochmann, R.; Gundermann, E.　　DE 160301/77/0003
Die Beurteilung der Umwelteinwirkungen von Forststrassen im Hochgebirge – eine Delphi Studie. **The evaluation of environmental effects of forest road construction in alpine country – a Delphi study. Publications.**

1972 Brüggemann, J.; Drescher–Kaden, U.
　　DE 160600/73/0004
Analyse des Pheromon–Musters von Wildwiederkäuern mit Hilfe der Gaschromatographie. **Analytical gas chromatographical assay of the pheromone pattern in wild ruminants.**

1973 Drescher–Kaden, U.　　DE 160600/75/0004
Übergang von Organo–Halogen–Verbindungen in Wildtieren. **Contamination of game by organo–halogenides.**

1974 Eisfeld, D.　　DE 160600/77/0001
Die zur Deckung des Energiebedarfes notwendige Nahrungsqualität beim Reh. **Food quality necessary to satisfy the energy requirements of roe deer.**

1975 Ruf, M.; Yediler, A.; Jacobs, J.; Hoffmann
　　DE 160770/77/0001
Die Beeinflussung von Schwermetallanreicherung durch verschiedene Wasserparameter. **The dependence of heavy metal accumulation on different water parameters.**

1976 Ruf, M.; Ollenschläger, B.; Braun, F.
　　DE 160770/77/0002
Die Veränderung der Immunlage in bezug auf die PCBs. **The alteration of the immune situation in view of the PCBs.**

1977 Boch, J.; Erber, M.　　DE 160822/75/0001
Untersuchungen zur Biologie der Sarkosporidien des Wildes. **Studies on the biology of Sarcosporidia of game. Publications.**

1978 Gylstorff, I.; Winteroll, G.　　DE 160855/77/0002
Herpesvirusinfektion bei Psittaciden. **Infections with Herpes virus in psittacine birds.**

1979 Gerlach, H.; Sailstorfer, R.　　DE 160855/77/0008
Die normale Keimflora der Nasenhöhle bei Psittaciden. **On the normal bacterial flora in the nasal cavity of psittacine birds.**

1980 Vollrath, H.; Pattay, P.von　　DE 161255/73/0001
Landschaftsökologische Untersuchungen in der Schwarzachaue/Opf.. **Studies of landscape ecology in the Schwarzach rivermeadows in Upper Palatinate.**

1981 Voigtländer, G.; Bürkle, A.; Spatz, G.
　　DE 161255/77/0001
Erhaltung der Kulturlandschaft mit Schafen im Voralpengebiet. **Landscape conservation by sheep in the prealpine region.**

1982 Spatz, G.; Pletl, L.　　DE 161255/77/0003
Entwicklung von Computerprogrammen zur Auswertung von Vegetationsaufnahmen. **Processing of computer programs in vegetation ecology.**

1983 Haber, W.; Kaule, G.　　DE 161670/74/0001

Kartierung von schutzwürdigen Biotopen in Bayern. **Mapping of biotopes worth protection in Bavaria.**

1984 Neumann, W.; Raff, J.　　DE 161860/78/0001 N
Stickstoffoxidation in Fliessgewässern. **Nitrogen oxidation in rivers.**

1985 Neumann, W.; Hajek, P.　　DE 161860/78/0003 N
Sauerstoffhaushalt der Fliessgewässer. **Oxygen bilance in rivers.**

1986 Acker, L.; Malisch, R.　　DE 164150/78/0001 N
Umweltchemikalien in der Biosphäre. **Environmental chemicals in the biosphere.**

1987 Schreiber, K.–F.; Schiefer, J.　　DE 164401/74/0001
Versuche zur Erhaltung der Kulturlandschaft in Baden–Württemberg. **Efforts in conservation of cultivated landscape in Baden–Wuerttemberg.**

1988 Büttner, K.　　DE 176051/78/0001 N
Untersuchungen zur Verteilung des Rehwildes bei steigendem Jagddruck. **Investigations on the distribution of roe deer under increasing hunting pressure. Publications.**

1989 Maydell, H.–J.von; Panzer, K.F.; Rhody, B.; Schmidt–Lorenz, R.; Wiebecke, C.　　DE 202010/78/0001 N
Agroforstwirtschaft: Beitrag der Forstwirtschaft zur integrierten Landnutzung in von Desertifikation bedrohten Gebieten. **Agroforestry: Contribution of forestry to integrated land use in regions endangered by desertification.**

1990 Erz, W.; Nowak, E.　　DE 205020/75/0001
Ermittlung der gefährdeten Tierarten – "Rote Liste" – und Grundlagen für ein Hilfsprogramm. **Determination of endangered species of animals – "Red List" – and development of fundamentals for an emergency programme.**

1991 Erz, W.; Blab, J.　　DE 205020/77/0003
Modelluntersuchung zur Situation der Amphibienfauna in der Zivilisationslandschaft. **Pilot analysis of the situation of amphibia fauna in civilized landscape.**

1992 Erz, W.; Haarmann, K.　　DE 205020/77/0004
Bestandsaufnahme ökologischer, naturgeschichtlicher und landeskundlicher Daten aus Schutzgebieten und deren Auswertung zur Verbesserung des Gebietsschutzes. **Documentation and processing of ecological, natural historical, and landscape development data on nature reserves.**

1993 Erz, W.; Haarmann, K.　　DE 205020/77/0005 R
Auswahl und Bewertungskriterien für Schutzgebiete zur Sicherung von Lebensräumen und Landschaftsteilen mit wichtigen bio– und geoökologischen Funktionen. **Selection and valuation criteria of protected areas for security of biotopes and of parts of landscape with important bio– and geoecological functions.**

1994 Erz, W.; Bless, R.　　DE 205020/78/0001 N
Situationsanalyse der Fauna in der Zivilisationslandschaft: Faktorenanalyse und Bewertung limnischer Systeme und Entwicklung von Hilfsmassnahmen. **Analysis of the present of fauna in the civilisation landscape. Analysis of factors and evaluation of limnic systems and development of aids.**

1995 Erz, W.; Haarmann, K.   DE 205020/78/0005 N
Fachliche Mitwirkung an internationalen
Naturschutzprojekten in der Bundesrepublik Deutschland.
Collaboration in international projects of nature conservation in the FRG.

1996 Erz, W.; Blab, J.   DE 205020/78/0006 N
Wissenschaftliche Entscheidungshilfen bei Entwicklung,
Durchführung und Verbesserung des Naturschutzrechts.
Scientific aids for decisions in the development, realization and improvement of legal instruments in nature conservation.

1997 Erz, W.; Blab, J.   DE 205020/78/0007 N
Entwicklungskonzept für eine Verbesserung des Artenschutzes
in der Bundesrepublik Deutschland – Konzept für ein
Artenschutzprogramm –. **Developing conception for the improvement of protection of species in the Federal Republic of Germany – conception for programme of protection of species –.**

1998 Trautmann, W.; Henrichfreise, A.; Krause, A.; Wolf,
G.   DE 205030/70/5001
Vegetationskundliche Aufnahme und Beschreibung der
Naturwaldreservate in Nordrhein–Westfalen. **Vegetational survey and description of natural forest reserves in Northrhine Westphalia.**

1999 Trautmann, W.; Bohn, U.; Henrichfreise, A.; Korneck,
D.; Krause, A.; Meisel, K.   DE 205030/70/5008
Herstellung einer Karte der potentiellen natürlichen
Vegetation der Bundesrepublik Deutschland. **Establishment of a map showing the potential natural vegetation of the Federal Republic of Germany.**

2000 Fink, H.G.   DE 205030/71/5002
Datenerhebung und Auswertung für die floristische Kartierung
Mitteleuropas – Regionalstelle Köln/Aachen –. **Data collection and evaluation for floristic cartography of Central Europe – Regional center in Koeln/Aachen.**

2001 Korneck, D.; Trautmann, W.   DE 205030/72/5002
Auswertung und Fortschreibung der 'Roten Liste' der
gefährdeten Farn– und Blütenpflanzen. **Evaluation and recording of the red list of endangered ferns and phanerogams.**

2002 Schröder, L.; Bohn, U.; Krause, A.; Voggenreiter, V.
          DE 205030/73/5002
Ermittlung botanisch wertvoller Flächen und botanische
Bewertung vorhandener Naturschutzgebiete in der
Bundesrepublik Deutschland. **Determination of botanically valuable areas and botanical valuation of existing nature reserves in the Federal Republic of Germany.**

2003 Meisel, K.; Bohn, U.   DE 205030/77/0001
Mitarbeit an der Vegetationskarte der Alpen. **Co–operation in the vegetational map of the Alps.**

2004 Trautmann, W.; Bohn, U.   DE 205030/77/0002
Mitarbeit an einer Vegetationskarte West–Europas 1 : 3 Mill..
Co–operation in a vegetational map of West Europe 1 : 3 million.

2005 Meisel, K.; Wolf, G.   DE 205030/77/0003
Vegetationsuntersuchungen auf Brachflächen. **Studies on vegetation on fallow areas.**

2006 Meisel, K.   DE 205030/77/0004 R
Vegetation als Bioindikator für Auswirkungen anthropgogener
Änderungen des Naturhaushalts landwirtschaftlicher
Nutzflächen. **Vegetation as bioindicator for effects of anthropogenous changes of nature equilibrium of fields.**

2007 Krause, A.   DE 205030/77/0005
Untersuchungen zur Begrünung und biologischen
Ufersicherung von fliessenden und stehenden Gewässern.
Studies on green growing and biological securing of banks of flowing and standing waters.

2008 Trautmann, W.; Fink, H.G.; Gerken, B.; Henrichfreise,
A.; Korneck, D.; Krause, A.   DE 205030/78/0001 N
Zustandserfassung und Bewertung der Vegetation der
badischen Rheinaue unter Naturschutzaspekten.
Determination and valuation of vegetational conditions of the Rhenish lowland in Baden with view to nature protection.

2009 Fink, H.G.; Henrichfreise, A.; Korneck, D.;
Voggenreiter, V.   DE 205030/78/0002 N
Aufbau einer Datei für gefährdete Pflanzenarten –
Artenschutzdatei –. **Collection of data file on endangered plant species – species protection file –.**

2010 Meisel, K.   DE 205030/78/0004 N
Einfluss des alternativen Landbaues auf Acker– und
Grünlandvegetation. **Influence of alternative agriculture on field and grassland vegetation.**

2011 Post, A.; Krefft, G.; Stehmann, M.; Sahrhage, D.
          DE 208010/70/4001
Ichthyologische Bearbeitung verschiedener Fischgruppen.
Ichthyological revision of various fish groups.

2012 Schumacher, A.   DE 208010/77/0001
Fortentwicklung der Methodik für die Abschätzung der
Fischbestände und Untersuchung ihrer Populationsdynamik.
Further development of methods for taxation of fish stock and studies on population dynamics.

2013 Meyer, A.   DE 208010/77/0002
Auswertung der Deutschen Fischereistatistik auf biologischer
Grundlage. **Evaluation of German fishing statistics on biological basis.**

2014 Sahrhage, D.   DE 208010/77/0003
Biologische Bestandsüberwachung der Nutzfischbestände im
Nordatlantik. **Biological control of useful–fish stock in the North Atlantic.**

2015 Sarhage, D.; Bussmann, B.   DE 208010/77/0004
Erforschung und Erschliessung unkonventioneller Meerestiere
für Nahrungszwecke. **Investigations and processing of unconventional sea animals for nutritive purposes.**

2016 Sahrhage, D.   DE 208010/77/0005
Erschliessung neuer Fanggebiete im Nordatlantik für die
Deutsche Seefischerei, Erforschung der Verbreitung, Dichte,
Biologie und des Ertragspotentials bisher nicht genutzter
mariner Lebewesen. **Development of new fishing grounds in the North Atlantic for the German deep–sea fishing, investigations of distribution, density, biology and yield potential of marine animals unused so far.**

2017 Sahrhage, D.   DE 208010/77/0006

D 1500 – Nature conservation

Erschliessung neuer Fangmöglichkeiten für die Deutsche Fernfischerei. Erforschung der Verbreitung, Dichte, Biologie und des Ertragspotentials bisher nicht genutzter mariner Lebewesen. **Development of new fishing possibilities for the German distant fishing. Investigations of distribution, density, biology and yield potential of marine animals unused so far.**

2018 Sahrhage, D.   DE 208010/77/0007
Erforschung und wirtschaftliche Erschliessung der Krill– und Nutzfischbestände in der Antarktis. **Investigations and economic utilization of stocks of krill and useful fish in the Antarctica.**

2019 Schumacher, A.   DE 208010/77/0008
Überwachung der deutschen Fischereiaktivitäten im Rahmen internationaler Abkommen und EG–Fischereiregulierungen. **Control of activities in German fishery within international agreements.**

2020 Stein, M.   DE 208010/77/0010
Hydrographische Untersuchungen auf Fangplätzen und ausgewählten Standardabschnitten. **Hydrographic studies on fishing grounds and selected and standardized zones.**

2021 Thurow, F.   DE 208020/77/0001
Biologisch–statistische Bestandsuntersuchungen der Fangobjekte der deutschen Kutterfischerei in der Ostsee. **Biological–statistical examination of stocks of fishing objects of the German cutter–fishery in the Baltic Sea.**

2022 Tiews, K.   DE 208020/77/0002
Biologisch–statistische Bestandsüberwachung der Fischereiobjekte der deutschen Kutterfischerei in der Nordsee. **Biological–statistical control of stocks of fishing objects of the German cutter–fishery in the North Sea.**

2023 Tiews, K.   DE 208020/77/0003
Erschliessung neuer Fanggebiete und Fangplatzberatung für die deutsche Kutterfischerei. **Development of new fishing grounds and extension services for the German cutter–fishery.**

2024 Halsband, E.   DE 208020/77/0004
Elektrophysiologie und Entwicklung von Fischsperren und Scheuchanlagen im Rahmen des Umweltschutzes. **Electrophysiology and development of fish barriers and scaring plant within environmental protection.**

2025 Dethlefsen, V.   DE 208020/77/0008
Erforschung der Wirkung und Speicherung von Schadstoffen auf Nutztiere des Meeres. **Research on effects and storage of noxious substances on marine useful animals.**

2026 Huschenbeth, E.   DE 208020/77/0009
Überwachung mariner Nutzfische auf chlorierte Kohlenwasserstoffe. **Control of chlorinated hydrocarbons in marine useful fish.**

2027 Steinberg, R.   DE 208030/77/0001
Erschliessung neuer Fanggebiete. **Development of new fishing grounds.**

2028 Feldt, W.   DE 208050/72/4007 N
Die radiologische Belastung der Flüsse und ihrer Estuarien durch die Abfälle der Kernkraftwerke und Anwendung von radioaktiven Isotopen. **The radiological burden of rivers and their estuaries caused by wastes from nuclear energy plants and users of radioactive isotopes.**

2029 Feldt, W.; Vobach, M.   DE 208050/75/0001
Auswirkung der Abwasserwärme des Kernkraftwerks auf die Artenzusammensetzung von Plankton und Fisch in der Unterweser. **Thermal effects of the nuclear power station on the composition of species of plancton and fish in the Lower Weser.**

2030 Schoenhard, G.   DE 215060/78/0001 N
Belastbarkeit von Boden und Pflanze mit den Elementen Mangan, Nickel, Chrom, Kobalt und Vanadin. **Maximum levels of manganese, nickel, chromium, cobalt and vanadium tolerable to soil and plants.**

2031 Leh, H.–O.   DE 215060/78/0004 N
Untersuchungen über Schäden an Strassenbäumen durch Einwirkung von Herbiziden sowie über Baumschäden durch Gasaustritte. **On damages of roadside trees by herbicides and town gas.**

2032 Kapol, F.; Fuss, F.; Puls, A.   DE 301020/75/0001
Einfluss von Nasskühltürmen auf die Niederschläge in der Umgebung von Grosskraftwerken. **Influence of wet cooling towers on precipitations in the surroundings of great power stations.**

2033 Wormuth, H.–J.; Lingk; Schäffer; Franke   DE 305030/75/0056
Tierschutzgerechte Fuchsbaubegasung. **Gassing of fox–earths in accordance with the Act on the Protection of Animals.** Publications.

2034 Sioli, H.; Irmler, U.   DE 404010/74/0001
Ökologie der Bodentiere in zentralamazonischen Überschwemmungswäldern. **The ecology of soil invertebrates in Central Amazonian inundation forests.**

2035 Medina, E.; Klinge, H.   DE 404010/75/0002
Struktur und Funktionen sowie Phytomasse eines Ökosystems im Tropischen Immergrünen Regenwald im Territorium Amazonas, Venezuela. **Structure and functioning as well as phytomass of a tropical evergreen rainforest ecosystem in the Territory Amazonas, Venezuela.**

2036 Mattern, H.   DE 501251/74/0001
Das Landschaftsschutzgebiet mittleres Jagsttal. **The protected landscape of the central Jagst Valley.**

2037 Evers, F.H.; Bücking   DE 501503/75/0001
Grundwasserhaushalt und quantitativer Nährstoff– und Düngemittelaustrag. Im DFG–Schwerpunktprogramm "Quantifizierung der Sozialfunktionen des Waldes als Element der Infrastruktur". **Ground water balance and quantitative loss of nutrients and of fertilizers. – In the framework of the DFG–Programme Quantification of the social functions of the forest as an element of infrastructure'.**

2038 Schwarz, O.; Volk, H.   DE 501504/75/0004
Landschaftsrahmenprogramm – Darstellung der Zielsetzungen und Massnahmen zur Verwirklichung der Grundsätze von Naturschutz, Landschaftspflege und Erholung nach dem neuen Landesnaturschutzgesetz –. **Landscape Programme – Objectives and measures for implementing the principles of nature conservation, landscape management and recreation according to the new Statal law of nature conservation –.**

D 1500 – Nature conservation

2039 Volk, H.; Spahl, H.　　　　DE 501504/78/0001 N
Kartierung der durch den Besucherandrang verursachten
Erosionsschäden im Feldberggebiet – Schwarzwald –
Untersuchungen über die Sanierung der Erosionsflächen.
**Mapping of erosion damages to Feldberg region – Black Forest
– caused by crowds of visitors Investigations on restoration of
erosion areas.**

2040 König, E.; Korsch, J.　　　　DE 501508/78/0004 N
Untersuchungen über die Ursachen des Rückganges der
Waldhühner in Baden–Württemberg. **Studies on the causes of
decrease of capercaillie and hazel hen in Baden–Wuerttemberg.**

2041 Deufel, J.; Berg, R.; Löffler, H.　　DE 501805/78/0002 N
Biologie von Bodenseefischen. **Biology of fishes in the Lake of
Constance.**

2042 Hartmann, J.　　　　DE 501805/78/0003 N
Fischereiliche Veränderungen in kulturbedingt
eutrophierenden Gewässern. **Changes in fishes in waters with
cultural eutrophication.** Publications.

2043 Hartmann, J.　　　　DE 501805/78/0004 N
Populationsdynamik von Bodenseefischen. **Population
dynamics of fishes in the Lake of Constance.** Publications.

2044 Braun, W.　　　　DE 502050/70/0024
Systematik, Ökologie und Verbreitung der
Pflanzengesellschaften der Sümpfe und Moore in Bayern 1968.
**Systematics, ecology and distribution of plant associations in
marshes and bogs in Bavaria.**

2045 Braun, W.　　　　DE 502050/70/0025
Pflanzensoziologie und Ökologie ausgewählter
Naturschutzgebiete Bayerns 1965. **Sociology and ecology of
plants in selected nature preserves in Bavaria.**

2046 Schmid, G.; Weigelt, H.　　　DE 502050/70/0026
Eutrophierung der Seen in Oberbayern durch
landwirtschaftliche Düngungsmassnahmen 1970.
**Eutrophication of lakes in Upper Bavaria by agricultural
fertilizing operations.**

2047 Bauchhenss, J.　　　　DE 502050/72/0012 R
Arbeiten zur Methodik ökologischer Untersuchungen der
Bodenfauna 1971. **On Methodies of ecological investigations on
soil fauna.**

2048 Schmid, G.; Diez, T.; Weigelt, H.　DE 502050/72/0013 R
Einfluss hoher Klärschlamm- und Müllkompostgaben auf den
Nährstoff- und Schwermetallgehalt der Gewässer 1972.
**Influence of high doses of sewage sludge and waste compost on
the nutrient and heavy metal content of waters.**

2049 Dancau, B.; Braun, W.　　　DE 502050/72/0014 R
Untersuchungen über die Zusammenhänge zwischen
Bodentypen und Pflanzengesellschaften 1965. **Investigations on
the correlations between soil types and plant associations.**

2050 Naton, E.　　　　DE 502051/72/0003
Nebenwirkung von Pflanzenschutzmitteln auf Nutzinsekten.
Accessory action of pesticides to useful insects.

2051 Wallnöfer, P.R.　　　　DE 502051/72/0005
Abbau von Pestiziden durch Bodenmikroorganismen.
Degradation of pesticides by soil micro–organisms.

2052 Wallnöfer, P.R.; Poschenrieder, G.; Königer, M.
　　　　DE 502051/74/0002
Interaktionen von Pestizidkombinationen im Boden.
Interactions of combined pesticides in soil.

2053 Kiermeier, F.; Buchberger, J.; Kirchmeier, O.
　　　　DE 502553/75/0004
Zusammenhang zwischen Gencharakteristik der Milchproteine
und quantitativen Aspekten der Milchzusammensetzung bei
Höhenvieh – Fleckvieh –. **Correlations between gene
characteristics of milk proteins and quantitative aspects of milk
composition in brindled mountain cattle.**

2054 Researcher not indicated　　DE 502650/78/0011 N
Bilanzierung von Pestiziden und PCBs in einem Gewässer am
Beispiel des Isarkanals. **Balancing of pesticides and PCBs in a
water, e.g. in the Isar Canal.**

2055 Researcher not indicated　　DE 502650/78/0016 N
Biologische Wasserpflanzenbekämpfung mit Gras– und
Silberfischen Ökologische Untersuchungen über die
interspezifische Konkurrenz von Grasfischen in
Hechtgewässern. **Biological control of hydrophytes with grass
fish – Stenopharyngodon idella – and silver fish –
Hypophthalmichtys molitrix –. Ecological studies on the
interspecific competition of grass fish in waters with pikes.**

2056 Researcher not indicated　　DE 502650/78/0018 N
Erhaltung umweltbedrohter Fische. Aufzucht und künstliche
Besamung des Huchen. **Conservation of fish endangered by
environmental conditions. Rearing and artificial insemination
of huchen.**

2057 Researcher not indicated　　DE 502650/78/0020 N
Aufzucht von Salmoniden und pflanzenfressenden Fischarten
im Hinblick auf genormte steuerbare Umweltbedingungen.
**Rearing of Salmonidae and herbivorous fish species with
regard to standardized control of environmental conditions.**

2058 Offhaus, K.; Braun, W.　　　DE 502651/71/0001
Untersuchungsmethode zur Bestimmung des Quecksilbers und
anderer Schwermetalle im Abwasser, Wasser, Sediment, in
Wasserpflanzen und Fischen. **Methods of investigation for
determination of mercury and other heavy metals in waste
water, water, sediment, water plants and fishes.**

2059 Offhaus, K.; Wachs, B.　　　DE 502651/74/0005
Untersuchungen über die Belastung bayerischer Gewässer mit
Cadmium –Wasser, Wasserpflanzen Flusssedimente.
**Investigations into the impact on waters –water, hydrophytes,
river sediments– in Bavaria bei cadmium.**

2060 Bohl, M.　　　　DE 502653/74/0001
Belastung des Vorfluters durch die teichwirtschaftliche
Fischproduktion. **Impact on outfall drain by pisciculture in
ponds.**

2061 Ruf, M.; Hübel, K.; Herrmann, H.; Lünsmann, W.
　　　　DE 502656/74/0001
Die radiologische Analyse der oberen Donau. **Radiological
analysis of the Upper Danube River.**

2062 Keil, W.; Rossbach, R.　　　DE 506200/72/0001
Untersuchungen zur Abwehr fischereischädlicher Vogelarten.
Investigation into the protection of fisheries against fisheating

D 1500 – Nature conservation

bird species.

2063 Keil, W.; Bauer, W.; Ehlert, H.; Rossbach, R.
DE 506200/72/0003
Erhebungen über Schutzmöglichkeiten bedrohter Vogelarten, insbesondere Graureiher, Rauhfusshühner, Rauhfusskauz, Uhu, Wasser– und Sumpfvogelarten, in bezug auf Biotoperhaltung und Biotopgestaltung. **Investigations into the protection of the exposed position of different bird species, like Ardea cinerea, Tetrao tetrix, Tetrao urogallus, Bonasa bonasia, Aegolius funereus, Athene noctua, Bubo bubo and waterfowl–species in reference to biotop protection and biotop forming.**

2064 Keil, W.; Bauer, W.; Rossbach, R. DE 506200/72/0004
Ausarbeitung von Vorschlägen zur ornithologisch–ökologischen Gestaltung von Kiesbaggerseen und Rückhaltebecken bzw. Teilen davon als "Trittsteine" für wassergebundene Vogelarten. **Elaboration of proposals for the ornithological–ecological formation of gravelpit–lakes and other water–basins for migrating water fowl.**

2065 Rossbach, R.; Ehlert, H. DE 506200/72/0005
Untersuchungen zur Steigerung der Populationsdichte von höhlen– und freibrütenden Vogelarten in unterschiedlichen forstlichen Standorten. **Investigation into general increasing of population–density of granivorous birds in different forestal habitats.**

2066 Keil, W.; Rossbach, R. DE 506200/75/0001
Untersuchungen zur Biologie und Ökologie der Waldschnepfe – Scolopax rusticola – im Hinblick auf die Frühjahrsjagd. **Investigations into the biology and ecology of the woodcock – Scolopax rusticola – with regard to hunting in spring–time.**

2067 Wachendörfer, G.; Manz, D.; Frost, J.; Manteufel, C.
DE 506301/72/0001
Experimentelle und praktische Untersuchungen zur oralen Immunisierung von Füchsen gegen Tollwut unter Labor– und Feldbedingungen. **Experimental and practical studies in oral immunization of foxes against rabies under laboratory and field conditions.**

2068 Brechtel, H.M.; Balazs, A. DE 506452/71/0002
Einfluss der Waldbewirtschaftung – Grundflächenhaltung, Schlagstellung und Baumartenwahl – auf den Wasserertrag von Einzugsgebieten. **Influence of forest management – basal area, cutting system and choice of tree species – on the water yield of watersheds.**

2069 Altenkirch, W.; Huang, P.; Niemeyer, H.
DE 507652/71/0002
Pestizide und Nutzarthropoden: Prüfung der unerwünschten Nebenwirkung von Pflanzenschutzmitteln auf die Kahlrückige Rote Waldameise, Formica polyctena und die Raupenfliege Pales pavida. **Pesticides and beneficial arthropoda: Investigations into undesired side–effects of pesticides on Formica polyctena and Pales pavida.**

2070 Winter, K. DE 507652/75/0002
Untersuchungen über Abundanzdynamik minierender Kleinschmetterlinge in Buchenbeständen des Solling. **Investigation into the population dynamics of mining microlepidoptera in beech stands of the Solling.**

2071 Winter, K. DE 507652/75/0005

Waldbrandfolgen auf die Populationsdynamik der Insektenfauna in Kiefernforsten der Lüneburger Heide. **Consequences of forest fire on population dynamics of the insect fauna in pine forests of the Luneburg Heath.** Publications.

2072 Niemeyer, H. DE 507652/78/0003 N
Nebenwirkungen von Pflanzenschutzmitteln im Walde auf Wirbeltiere. **Side–effects of plant protectives in forests on vertebrates.**

2073 Foerster, E.; Lennartz, H. DE 508301/71/0004 R
Die Pflanzengesellschaften des Grünlandes in NordrheinWestfalen, Systematik und Ökologie 1971. **Plant communities of grassland in Northrhine–Westphalia, systematics and ecology.**

2074 Knabe, W. DE 508303/78/0001 N
Methodische Entwicklung einer immissionsökologischen Waldzustandserfassung Nordrhein–Westfalen. **Development of an ecologic survey of air–pollution effects on forests in Northrhine–Westphalia.**

2075 Berge, H.; Orgis, K. DE 904000/78/0002 N
Die Rauchgasentschwefelung aus biologischer Sicht. **Flue gas desulfurization from biological point of view.** Publications.

2076 Neimann–Sørensen, A.; Thysen, J. DK 010201/74/0015
Produktionssystemer på naturarealer – kvægafgræsningens indflydelse på naturarealers plantesamfund og jordbundsstruktur. **Production systems on natural grazings – the influence of grazing cattle on the plant population and soil structure of natural grazings.**

2077 Holmsgaard, E.; Koch, N.E. DK 020001/75/2315
Skov og folk. Analyse af befolkningens anvendelse af skovene i dens friluftsliv. **Forest and people. Analysis of forest use by the public in its outdoor life.**

2078 Ødum, S. DK 030102/69/0004
Undersøgelser af naturlig skovregeneration på stormfaldsarealer i bøgeskov. **Investigations of natural forest regeneration in storm felled areas of beech wood.**

2079 Christensen, F.G. DK 030102/69/0005
Etablering af systematisk træsamling på forskningscentrets område. **The establishment of systematic tree collecting in the research centre's area.**

2080 Christensen, F.G. DK 030102/74/0001
Opbygning af samling af vedplanter fra Østasien. **The compilation of a collection of woody plants from East Asia.**

2081 Ødum, S. DK 030102/76/0002
Opbygning af samling af vedplanter fra sydlige halvkugle i Torshavn i forbindelse med Nordisk Arboretudvalg. **The compilation in Torshavn of a collection of woody plants from the southern hemisphere in connection with the Scandinavian Arboricultural Committee.**

2082 Vedel, H. DK 030103/57/0009
Successionsundersøgelser i vedplantesamfund. **Investigations of succession in a woody plant community.**

2083 Høst, O. DK 030103/58/0002
Vegetationens afhængighed af lys og edafiske faktorer i den

fredede hegning i Buderupholm. **The dependence of vegetation on light and edaphic factors in the enclosed area at Buderupholm.**

2084 Vedel, H.   DK 030103/65/0008
Forstbotaniske undersøgelser i Suserup Skov. **The first botanical investigations in Suserup forest.**

2085 Vedel, H.   DK 030103/74/0010
Naturarealer og husdyrhold (kreatur– og fåregræsnings indvirkning på vegetation og jordbund i naturarealer). **Nature areas and the keeping of domestic animals (the impact of live stock and sheep grazing on the vegetation and soil).**

2086 Lange, J.   DK 030103/75/0005
Undersøgelse over træers tæppedannende vækst på udsatte steder (stenede og sandede kystnære områder og højfjeldet (i trægrænsen)). **An investigation into tree cover form growth in exposed places (stony and sandy coastal areas and high mountains (at the tree line)).**

2087 Jensen, J.   DK 030103/77/0003
Regeneration af brændt højmosevegetation. **Regeneration of burnt raised bog vegetation.**

2088 Hansen, K.; Petersen, P.M.; Vestergaard, P.
                    DK 030103/78/0001 N
Udvikling af jordbund og vegetation i nyanlagte klitter. **Development of soil and vegetation in new dunes.**

2089 Hansen, K.; Adsersen, H.; Mikkelsen, V.M.
                    DK 030103/79/0001 N
Vegetationsudvikling på forladt agerjord. **Development of vegetation on abandonned agricultural fields.**

2090 Mikkelsen, V.M.   DK 030103/79/0002 N
Vegetationsudvikling på fåregræsset klippehede på Nordbornholm. **Influence of sheep–grassing on heath–vegetation on soil with rock–out crops.**

2091 Friis, P.; Larsen, L.M.   DK 030106/73/0009
Undersøgelser over indholdet af cyanogene forbindelser i planter. **Investigations into the cyanogen content in connection with plants.**

2092 Petersen, L.   DK 030106/79/0005 N
Økologisk undersøgelse i oplandet til Suså–Vendebæk. **Ecological investigation in the "Suså–Vendebæk" basin.**

2093 Jensen, V.   DK 030108/76/0006
Mikrobiologisk nedbrydning af olie og olieprodukter. **Microbiological degradation of oil and oil products.**

2094 Bejer–Petersen, B.   DK 030109/69/0001
Undersøgelse af bestandstæthed og ynglesucces hos hulrugende fugle i Grib Skov. **An investigation of the population density and breeding success of hole brooding birds in Grib Skov.**

2095 Coutin, R.   FR 010106/75/5262
Protection d'insectes menacés de disparition. **Protection of insects which are threatened.**

2096 Timbal, J.   FR 010303/70/4169
Inventaire des séries de végétation et des groupements forestiers de Lorraine. **Inventory of the "séries de végétation"**

and of the forest associations in the Lorraine.

2097 Timbal, J.; Becker, M.; Picard, J.F.   FR 010303/73/4167
Pression humaine et végétation forestière. **Human pressure and forest vegetation.**

2098 Picard, J.F.; Spitz, F.   FR 010303/73/4168
Etude des relations végétation – gibier – sylviculture. **Vegetation – game – forest practices relationships.**

2099 Lavie, P.; Gonnet, M.; Guennelon, G. Mme
                    FR 010608/74/5279
Utilisation des produits de la ruche comme indicateurs de la pollution atmosphérique. **Use of hive products as atmospheric pollution indicators.**

2100 Du Merle, P.; Demolin, G.; Charles, P.J.; Dusaussoy, G.; Cornic, J.F.; Marro, J.   FR 010609/72/4172
Fonctionnement d'écosystèmes forestiers de moyenne altitude en Provence calcaire ; incidences de leur dégradation et de leur reconstitution par l'Homme. **Working of forest ecosystems at middle elevation in "calcareous" Provence. Incidence of their degradation and of their reconstitution by Man.**

2101 Louis, C.; Vago, C.   FR 010612/64/1760
Recherches sur la dissémination de germes pathogènes et sur les maladies des invertébrés en milieu naturel protégé. **Research on spreading of pathogenic germs and diseases of invertebrates in protected natural environments.** Publications.

2102 de Montard, F. X.   FR 010802/69/6099
Amélioration des landes à calluna vulgaris. **Improvment of "Calluna vulgaris" heaths.**

2103 Theriez, P.; Ricoux; Martin–Rosset; Merle, P.L.
                    FR 010802/72/6097
Possibilités d'utilisation et d'amélioration des parcours dégradés de montagne du Massif Central par la pâture. **Possibilities of utilization and improvment of dramaged hill pastures in Massif Central by grazing.**

2104 Mettauer, H.; Delphin; Conesa, A.   FR 010901/73/6124
Caractérisation de l'état du milieu dans la région proche de la centrale nucléaire de Fessenheim. **Caracterisation of the environment state in the area near nuclear power station of fessenheim.**

2105 Marocke, R.; Gros; Conesa, A.   FR 010901/77/6123
Aménagement de zones de cueillette (myrtilles, framboises) dans le massif vosgien. **Planning of picking areas (blueberry, raspberry) in the vosgian mountains.**

2106 Blaisinger, P.; Kienlen, J.C.   FR 010904/73/5023
Action des pesticides sur les arthropodes utiles. Essais en verger. **Effects of pesticides on beneficial arthropods – Orchard tests.**

2107 Bouche, M.; Concaret, J.; Catroux, G.; Duthion, C.; de Guiran, G.   FR 011001/75/6142
Conséquences pour les sols, les plantes et les eaux de l'épandage du lisier de porcherie. **Spreading of semi–liquid pig manure. Consequences for soils, plants and waters.**

2108 Jarrige, R.; Lienard, G.; Gachon, L.; Ricou, G. Mme
                    FR 012220/73/5284
Faune des pâturages d'altitude du Massif Central. **Fauna of the**

altitude Grass–lands in the "Massif Central".

2109 Lefevre, P.; Hulot, J.C.; Gaffet, M.A.; Regnier, P.
FR 012222/64/6039
Recherche des caractéristiques du milieu naturel de la Plaine
Maritime Picarde et étude de leurs influences. **Characteristics
of the environment of the Picardy Coastal Plain.**

2110 Larrere; Poupardin; Decourt, N.   FR 012303/78/8601
Les reboisements "autoritaires" du second Empire (loi de
1860) et leur postérité. **Compulsory afforestation during the
second Empire (1860'sad) and its long term effect.**

2111 Larrere, G.; Petit, F.E.; Decourt, N.; Poupardin, D.
FR 012303/78/8602
L'intervention des pouvoirs publics en matière de reboisement
des régions de montagne, de la libération à nos jours.
**Government actions to control mountains'area –
reafforestation since world war II.**

2112 Lepape, Y.; Pernet, F.   FR 012600/78/8524
Technologie et environnement : analyse des obstacles au
développement des technologies compatibles avec la
sauvegarde de l'environnement. Le cas de l'agriculture
biologique. **Technology and environment : analysis of
contraints of technological development for environment
protection. The case of biological agriculture.**

2113 Hamilton   GB 030903/71/0203 R
Survey of poisoning incidents in wildlife.

2114 Hamilton; Ruthven   GB 030903/72/0202 R
Monitoring the seasonal occurrence of organochlorine and
mercury residues in wild birds.

2115 Symonds   GB 030903/74/0102 R
Availability of insecticide–treated wheat seed to birds after
autumn sowing.

2116 Wright   GB 040301/00/0017 R
Effects of large scale drainage on the flora of catchment areas.

2117 Smith   GB 040801/00/0004 R
Growth kinetics and nutrition of blue–green algae isolated from
Lough Neagh.

2118 Gibson; Stephens   GB 040801/00/0005 R
Factors limiting the growth of blue–green algae in Lough
Neagh.

2119 Kennedy   GB 040901/00/0011 R
The distribution of fish in Lough Neagh in relation to oxygen
levels.

2120 Stronach, B.   IE 050100/68/7219 N
Ecological study of mallard. Publications.

2121 Cross, J.R.   IE 050100/71/7231 N
A study of areas of semi–natural vegetation (woods, heaths,
dunes,etc.) with a view to their conservation and establishment
of nature reserves. Publications.

2122 Wilson, J.   IE 050100/73/9157 N
Distribution and abundance of woodland birds. Publications.

2123 Speight, M.C.D.   IE 050100/74/7226 N
Use of insect indicator groups in assessment of potential faunal
interest of areas proposed for nature conservation.
Publications.

2124 Ryan, J.B.; Heuff, H.   IE 050100/74/7230 N
A study of Irish wetlands to identify areas of conservation
interest.

2125 Wilson, J.   IE 050100/75/7220 N
Study of woodcock winter movements. Publications.

2126 Warner, P.; Collins, T.   IE 050100/76/7222 N
Ecological study of red squirrel. Damage done to commercial
forestry. Publications.

2127 Warner, P.   IE 050100/77/7223 N
Distribution of bats in Ireland. Investigation of roosts.

2128 Warner, P.   IE 050100/78/7224 N
Distribution survey of seals. Publications.

2129 O'Sullivan, A.M.   IE 060200/66/0410 R
Vegetation survey of Co. Carlow. Publications.

2130 O'Sullivan, A.M.; Carton, O.; Moore, J.; Neff, J.
IE 060200/70/0412 N
A botanical and ecological survey of irish heath, hill grassland,
mountain and western bog vegetation. Publications.

2131 O'Sullivan, A.M.   IE 060200/73/0825 N
Vegetation types studies of Ireland. Publications.

2132 O'Sullivan, A.M.   IE 060200/76/1226 N
Survey of the flora and vegetation of south–east Wexford in
relation to the proposed nuclear power project at Carnsore
Point. Publications.

2133 O'Keefe, M.   IE 060500/00/1387 N
Heavy metals as environmentals contaminates. Publications.

2134 Connolly, V.; Crowley, J.; Ribeiro, Do Valle, M.D.V.
IE 060500/79/1429 N
Collection of herbage species from old pastures and other
important ecological habitats in Ireland. Publications.

2135 Cox, P.G.   IE 060600/77/1353 R
EEC directive 268: conservation of countryside and population
maintenance. Publications.

2136 Cabot, D.; Goodwillie, R.; McEvoy, A.
IE 080200/78/9192 N
Money Point generating station : ecological impact assessment.
Publications.

2137 Murphy, J.P.; Moore, D.M.; Heywood, V.H.
IE 110202/76/9152 N
A taxonomic revision of the genus lavundula l. with special
reference to the endemic species of the canary islands.

2138 Prendeville, G.N.; O'Donovan, J.T.; Helleris, A.;
O'Leary, D.   IE 110202/77/9150 N
Effects of oils and detergents on cell membrane permeability in
fucus and laminaria.

2139 Whelan, J.; MacLochlainn C.   IE 120101/79/9175 N
The ecology of red deer (cervus elaphus) in Glenveagh national

park.

2140 Evans, G.O.; Purvis, G. IE 120101/79/9178 N
Terrestrial microarthropods of Carnsore Point, County Wexford, Ireland.

2141 Bracken, J.J. IE 120403/00/9093 N
Baseline studies of Irish lakes and rivers. Publications.

2142 Bracken, J.J. IE 120403/00/9095 N
Cultural eutrophication studies in the Killarney lakes. Publications.

2143 Murray, D.A. IE 120403/00/9096 N
Sedimentology studies in fifty Irish lakes. Publications.

2144 Healy, B.; Bracken, J.J. IE 120403/00/9097 N
Marine study at Carnsore, a nuclear energy plant site. Publications.

2145 Bary, B.McK.; Ryan, T.; Yip, S.Y.; Leahy, P.; Hensey, M. IE 130103/71/9222 N
Interrelationships between sea waters of differing origins a nd the occurrence and abundance of planktonic organisms. Publications.

2146 Reynolds, J.D.; Byrne, R.A.J.; Allott, N.
IE 140202/77/9048 N
Trophic relationships of the biota of turloughs, semi–permanent lakes in karstic limestone. Publications.

2147 Reynolds, J.D.; Bradley, M.D.K. IE 140202/78/9191 N
Ecology and control by macroinvertebrates of microbial slime s (sewage fungus, etc.) in Irish rivers.

2148 Mason, J.; Perasmer, C. IE 140203/77/9063 N
The uptake and fixation of cadmium by the common mussel m. edulis. The transintestinal movement and subsequent organ distribution of m. edulis complexes. Publications.

2149 Bates, R. IE 180111/68/8753 R
Investigation of potential for oyster farming along the Irish coast. Publications.

2150 Lee, T.; Bates, R. IE 180111/68/8754 R
Exploratory fishing for mussels alon the east coast of irel and. Publications.

2151 Persano, L.; Intoppa, F. IT 020400/72/0011
Classificazione della flora spontanea del Lazio. **Wild flora of Lazio.**

2152 Di Giovacchino, Luciano.; Cucurachi, Angelo.; Solinas, Mario.; Mascolo, Antonio. IT 022100/77/0001 N
Valutazione degli inquinamenti dei corsi d'acqua causati dallo smaltimento delle acque di vegetazione delle olive. **Evaluation of the pollution of the rivers caused by discharge of the olive vegetation water from oil mills..**

2153 Spagnesi, M. IT 022800/71/0001 N
Ricerche sperimentali sull'allevamento in cattività della Lepre europea (Lepus capensis). **Experimental researches on the cage rearing of the European Hare (Lepus capensis)..**

2154 Spagnesi, M. IT 022800/74/0001 N
Ricerche sperimentali sulle tecniche di allevamento in stretta

cattività della Pernice sarda (Alectoris barbara). **Experimental researches on the cage rearing of Alectoris barbara.**

2155 Boldreghini, P. IT 022800/74/0002 N
Eto–ecologia riproduttiva e alimentare degli Ardeidi e dei Laridi. **Feeding and breeding etho–ecology of the Herons (Ardeidae) and the Gulls (Laridae)..**

2156 Spagnesi, M. IT 022800/75/0006 N
Censimento della popolazione di Camoscio d'Abruzzo (Rupicapra rupicapra ornata).. **Census and population study of Rupicapra rupicapra ornata (P.N. Abruzzo)..**

2157 Boldreghini, P. IT 022800/75/0007 N
Ecologia alimentare della Volpe nell'Appennino Romagnolo. **Food ecology of the Red Fox (Vulpe vulpes) in the Apenines Mounts of Romagna..**

2158 Spagnesi, M. IT 022800/76/0001 N
Inchiesta faunistica sulla distribuzione di medi e grandi Mammiferi nelle Alpi centro–orientali. **Inquiry on the distribution of the medium and large–size Mammals in the central–east Alps.**

2159 Melotti, P. IT 022800/76/0003 N
Tossicità acuta e a medio termine del Forate e dell'Aldikarb nei confronti di quattro specie ornitiche. **Toxicity of Phorate and Aldikarb to four species of bird..**

2160 Melotti, P. IT 022800/77/0001 N
Analisi delle riprese di Falco pescatore (Pandion haliaetus L.) in Italia. **Analysis of the italian ringing recovery of Osprey (Pandion haliaetus L.)..**

2161 Boldreghini, P. IT 022800/77/0002
Morfologia e distribuzione di alcuni Ofidi mell'Emilia centrale. **Morfology and distribution of some Snakes in the middle Emilia.**

2162 Boldreghini, P. IT 022800/77/0003 N
Ecologia alimentare degli Anatidi (Anatidae) e della Folaga (Fulica atra) nelle valli di Comacchio. **Feeding ecology of the ducks (Anatidae) and the Coot (Fulica atra) in the Comacchio Lagon..**

2163 Melotti, P. IT 022800/79/0001 N
Tecniche di determinazione della DL_{50} di fitofarmaci ed erbicidi su specie ornitiche selvatiche. **Technique for determination of DL_{50} of pesticides and erbicides on wild birds.**

2164 Chiusoli, A. IT 040203/73/0175
Studio agrotecnico per interventi conservativi o di ricostituzione del patrimonio vegetale su terreni ad equilibrio alterato. **Agrotechnical study in view of conserving or reconstituting the vegetation of soils whose balance has been altered.**

2165 Leporati, L.; Trocchi, W.; Melotti, P.; Negrini, F.
IT 040212/78/0002
Effetti dei fitofarmaci sulla selvaggina e su specie ittiche d'acqua dolce. **Pesticides effect on game and fresh water fishes.** Publications.

2166 Pirola, A. IT 040226/73/0279
Progetto per uno studio agrotecnico di interventi conservativi o di ricostituzione del patrimonio vegetale su terreni ad

D 1500 – Nature conservation

equilibrio alterato. **Project for an agrotechnical study in view of conserving or reconstituting the vegetation of soils whose balance has been altered.**

2167 Silva, S.   IT 040406/77/0646 R
Sostanza organica di macroelementi in relazione all'antropizzazione dell'ambiente. **Macro–element constituted organic substance related to man's settlement in the environment.**

2168 Mancini, F.   IT 040510/73/0251
Studio agrotecnico di interventi conservativi o di ricostituzione del patrimonio vegetale in terreni ad. equilibrio alterato. **Agrotechnical study in view of conserving or reconstituting the vegetation of soils whose balance has been altered.** Publications.

2169 Piussi, P.   IT 040516/73/0281
Studio della struttura e della rinnovazione dei boschi subalpini nelle Alpi orientali. **Study on the structure and rennovation of subalpine forests in the eastern Alps.** Publications.

2170 Betti, A.   IT 040519/77/0113 R
Utilizzazione di alcune terre marginali tramite l'incremento del patrimonio faunistico; possibilità economiche dello sfruttamento zootecnico e venatorio. **Utilization of marginal lands through an increment of fauna patrimony, economic possibilities of livestock breeding and hunting exploitation.**

2171 Cappelli, M.   IT 040809/74/0525
Ricerche sul ritmo vegetativo e sulla produttività dell'ecosistema macchia mediterranea. **Research on the vegetative rhythm and the productivity of the mediterranean scrub ecosystem.**

2172 Barbieri, F.   IT 041802/78/1029 N
Indagini ecologiche e parassitologiche sulla riproduzione di uccelli in ambienti ad utilizzazione agricola intensiva. **Ecological and parasitological investigation on bird reproduction in an intensively cultivated agricultural environment.**

2173 Filippello, S.   IT 041805/77/0797
Specie vegetali da proteggere. **Plant species to be protected.**

2174 Crescimanno, F.G.   IT 060900/74/0168
Ricerche sulla biologia fiorale delle specie più rappresentative. **Research on the biological flora of the most representative species.**

2175 Malquori, A.   IT 061200/73/0143
Attacco biologico di alluminosilicati. **Biological weathering of aluminosilicates.** Publications.

2176 Florenzano, G.   IT 061300/77/0967
Ecologia degli autotrofi in rapporto alla selezione dei ceppi ed alle applicazioni della loro attività nella conservazione delle acque ed in agricoltura. **The ecology of autotrophic organisms: selection of sub–species, their activity as applied to water conservation and to agriculture.**

2177 Mondino, G.P.; Dalmasso, G.; Penon, A.
     IT 120100/78/0001 N
Cartografia forestale delle Valli Gesso, Vermenagna, Pesio (provincia di Cuneo) in scala 1:25.000. **Forestry maps of the Gesso, Vermenagna and Pesio Valleys (Cuneo district, Italy)**

drawn to the scale 1:25.000.

2178 Mondino, G.P.; Dalmasso, G.; De Biaggi, E.
     IT 120100/78/0002 N
Carta forestale del Piemonte in scala 1:250.000. **Forestry map of the Piedmont region drawn to the scale 1:250.000.**

2179 Bergh, J.P. van den   NL 010102/71/8933 N
Populatiedynamiek van grassen en kruiden in korte begroeiingen. **Population dynamics of grasses and herbs in short vegetation.** Publications.

2180 Zon, J.C.J. van   NL 010102/72/8928 N
Onderhoud van sloten, droge sloten en taluds, mede in relatie tot natuurwetenschappelijke en landschappelijke waarden. **Maintenance of ditches, temporary dry ditches and slopes, also with respect to their value for the landscape and for natural biota.**

2181 Oomes, M.J.M.; Boer, T.A. de   NL 010104/71/3203
Beheersonderzoek van graslanden met beperkte en niet–landbouwkundige doelstellingen. **Research for development of management systems for semi–agricultural vegetations.** Publications.

2182 Boer, T.A. de; Gooyer, H.H. de   NL 010104/71/3205
Opdracht– en studiekarteringen van korte vegetaties. **Mapping of herbage vegetations.** Publications.

2183 Wolting, H.G.   NL 010108/77/7977
Invloed van de luchtverontreiniging van het wegverkeer op de vegetatie. **Influence of the air pollution from road traffic on the vegetation.**

2184 Pape, J.C.; Bannink, J.C.   NL 010119/73/5974
Onderzoek naar de ontwikkeling van de vegetatie op verlaten akkers. **Development of vegetation on abandoned fields.** Publications.

2185 Haagsma, J.; Laak, E. ter   NL 010401/72/3636
Een onderzoek naar botulismus, speciaal in verband met het optreden bij watervogels in Nederland. **An investigation on botulism, especially in relation to the occurrence in waterfowl in the Netherlands.** Publications.

2186 Smit, T.   NL 010401/75/6523
Opsporen en bestuderen van oorzaken welke aanleiding geven tot ziekte en sterfte bij in het wild levende vogels. **Detecting and studying causes of morbidity and mortality of wild life birds.**

2187 Smit, Th.   NL 010401/76/7742 R
Diagnostiek, epizoötiologie en bestrijding van infectieuze hepatitis bij watervogels. **Diagnostic, epizootology and control of hepatitis in waterfowl.**

2188 Wijngaard, J.K.R. van den   NL 010601/72/3534
Inventarisatie van bos– en opstandstypen op de Veluwe. **Inventory of forest and stand types on the Veluwe.** Publications.

2189 Wijngaard, J.K.R. van den; Burg, J. van den; Harms, W.B.   NL 010601/77/7541
Onderzoek naar de vegetatie–ontwikkeling in een aantal Alno–Padionbossen met van elkaar afwijkende voorgeschiedenis. **A study of vegetation development in major**

D 1500 – Nature conservation

poplar afforestations of different origins.

2190 Leeuwen, C.G. van   NL 010602/57/4565
Onderzoek naar de selectie– en regulatiemechanismen in vegetaties, inclusief de invloed van de mens daarop. **Development of ecological theories on behalf of selection and regulation mechanisms in vegetation, human influence included.**

2191 Rooth, J.; Bergh, L.M.J. v.d.; Timmerman, A.; Smit, J.J.; Beintema, A.J.   NL 010602/57/5625
Bepaling van de ornithologische betekenis van gebieden en landschappen. **Evaluation of areas with ornithological significance.**

2192 Rooth, J.; Timmerman, A.; Renssen, T.A.
    NL 010602/57/5630
Oecologisch–geografisch onderzoek van zwanen. **Distribution and ecology of swans.**

2193 Wijngaarden, A. van   NL 010602/57/5634
Geografisch–oecologisch en autoecologisch onderzoek aan Nederlandse zoogdieren. **Research on the ecology and distribution of terrestial mammal species in the Netherlands.**

2194 Wijngaarden, A. van   NL 010602/57/5635
Geografisch–oecologisch en autoecologisch onderzoek van de Nederlandse amfibieën. **Research on the ecology and distribution of amphibians in the Netherlands.**

2195 Wijngaarden, A. van   NL 010602/57/5917
Inventarisatie van de fauna van diverse terreinen. **Zoological inventory of nature areas.** Publications.

2196 Haaften, J.L. van; Roijnder, P.J.H.   NL 010602/58/4528
Onderzoek naar de oorzaak van populatieschommelingen bij de in onze wateren voorkomende zeehonden. **Causes of fluctuations in the population of the common seal in the Netherlands.** Publications.

2197 Leentvaar, P.   NL 010602/58/4583
Aspecten van de biologische gevolgen van civiel–technische werken op water–planten en –dieren. **Aspects of the influence of technical works on waterplants and –animals.** Publications.

2198 Rooth, J.; Timmerman, A.   NL 010602/60/5626 R
Ornithologische aspecten bij inrichting en beheer van natuurgebieden. **Ornithological aspects of lay–out and management of nature areas.**

2199 Rooth, J.; Timmerman, A.; Beintema, A.J.; Smit, J.J.
    NL 010602/60/5627
Geografisch–oecologisch onderzoek aan ganzen. **Distribution and ecology of geese.**

2200 Niewold, F.J.J.   NL 010602/66/4531
De oecologie van de vos in verband met zijn rol in de verspreiding van de rabies. **Ecology of the fox in relation to his role as a vector of rabies.** Publications.

2201 Broekhuizen, S.   NL 010602/66/4558
Aard en oorzaken van aantal schommelingen in hazenpopulaties. **Nature and causes of fluctuations of numbers in hare–populations.** Publications.

2202 Fuchs, P.   NL 010602/66/4559

Het vóórkomen van residuen van persistente bestrijdingsmiddelen en andere toxische stoffen in een aantal indicator–soorten. **Monitoring of residue levels of organochlorine pesticides and mercury in some indicator–species.**

2203 Londo, G.; Leeuwen, C.G. van; Leijs, H.N.
    NL 010602/66/5603 R
Geobotanisch onderzoek van oecologische proeftuinen. **Geobotanical research on ecological experimental gardens.**

2204 Rooth, J.; Boer, B. de   NL 010602/66/5629
Oecologie en beheer van de flamingo's op Bonaire. **Ecology and management of the flamingos on Bonaire.** Publications.

2205 Rooth, J.   NL 010602/66/5631 R
Het uitzetten van bedreigde vogels (raven, aalscholvers, grauwe gans). **Reintroduction of species which are extinct or seriously threatened in the Netherlands.**

2206 Haaften, J.L. van   NL 010602/67/4530
Beheersproblemen bij edelhert, reewild en moeflon. **Management of red deer, roe deer and mouflon.** Publications.

2207 Fuchs, P.   NL 010602/69/4560 R
De invloed van bestrijdingsmiddelen op de populatie–dynamiek van vogels (eksters). **The effect of pesticides on the population dynamics of birds (c.q. the magpie).**

2208 Fuchs, P.   NL 010602/69/4561
Onderzoek naar de contaminatie van roofvogels en uilen met gechloreerde koolwaterstoffen, PCB's en zware metalen. **Contamination of birds of prey and owls by chlorinated hydro carbons, PCB's and heavy metals.** Publications.

2209 Soet, F. de; Gonggrijp, G.P.   NL 010602/69/4573
Inventarisatie en waardering van Gea–objecten. **Draw up an inventory of geological, geomorfological and soilscience values.** Publications.

2210 Rooth, J.; Jonkers, D.A.   NL 010602/69/4594
Opsporen van invloeden van biociden en andere storingen op vogel– en zoogdier populaties in een aantal proefgebieden. **Impact of pesticides and other disturlant factors on bird and mammal populations.** Publications.

2211 Londo, G.; Reijnders, T.; Leijs, H.N.; Dirkse, G.
    NL 010602/69/5610
De invloed van "niets doen" als interne beheersmaatregel voor landschappen. **Influences of "doing nothing" as internal management measure.** Publications.

2212 Rooth, J.; Timmerman, A.   NL 010602/70/5624
Geografisch–oecologisch onderzoek van bedreigde vogelsoorten. **Ecology of threatened bird species.**

2213 Beintema, A.J.; Smit, J.J.; Bergh, L.J.M. van den
    NL 010602/70/5988
Geografisch–oecologisch onderzoek aan eenden. **Distribution and ecology of ducks.** Publications.

2214 Niewold, F.J.J.   NL 010602/71/4529
Onderzoek naar de rol van kleine roofdieren en van verwilderde katten op de wilde fauna. **Role of small predators and wild domestic cats on natural fauna.**

D 1500 – Nature conservation

2215 Wit, A. de; Laanen, T.J.M. van   NL 010602/71/5602 R
Het samenstellen van verspreidingskaarten van cryptogame
epiphyten in Nederland. **Distribution maps of cryptogamic
epiphytes in the Netherlands.**

2216 Londo, G.   NL 010602/71/5609 R
Het effect van betreding en berijding
(off–the–road–locomotion) op bodem en vegetatie. **The effect
of off–the–road locomotion on soil and vegetation.**

2217 Fuchs, P.; Gussinklo, D.J.; Burgers, J.
    NL 010602/71/5628
Populatieonderzoek van roofvogels en uilen in gebieden met
verschillend gebruik van bestrijdingsmiddelen. **Distribution
and ecology of birds of prey and owls in areas with different use
of pesticides.**

2218 Biezen, J.B. van   NL 010602/71/5915
Biologische natuurwaardering. **Biological evaluation of nature.**

2219 Oosterveld, P.; Londo, G.   NL 010602/71/7222
De effecten van de methode begrazen bij het beheer van
natuurgebieden. **Impact of grazing management on natural
areas in the Netherlands.**

2220 Oosterveld, P.; Londo, G.; Slim, P.A.; Leys, H.N.
    NL 010602/71/7223
De effecten van de beheersmethode begrazing op de vegetatie.
Impact of grazing management on vegetations.

2221 Oosterveld, P.; Slim, P.A.; Hazebroek, E.
    NL 010602/71/7224
De rol van landbouwhuisdieren bij het beheer van
natuurreservaten. **The role of domestic cattle in managing
nature reserves.**

2222 Oosterveld, P.; Broekhuizen, S.; Maaskamp, F.
    NL 010602/71/7225 R
De wisselwerking tussen graasbeheer en de natuurlijke
herbivoren. **Relationships between grazing management and
natural herbivorous species.**

2223 Klomp, H.; Botterweg, P.F.   NL 010602/72/4552
Populatie–dynamica bij dieren, in het bijzonder bij de
dennespanner. **Populations dynamics of animals, the Bordered
White (Bupalus piniarius L.) in particular.** Publications.

2224 Wit, A. de; Laanen, T.J.M. van   NL 010602/72/5614 R
Oriënterend onderzoek naar methodieken en plantensoorten
die kunnen worden gebruikt bij de bestudering van de invloed
van luchtverontreiniging op spontane plantengroei.
**Investigation into methods and plant species to be used in the
study of air pollution influences on spontaneous plant growth.**

2225 Schroevers, P.J.   NL 010602/73/5621 R
Contactsituaties tussen oligotroof en eutroof water in het
Peelgebied. **Contact zones between oligotrophic and eutrophic
waters in the Peel.**

2226 Kersting, K.   NL 010602/73/5623
Neveneffecten van chemische onkruidbetrijdingen in water.
Research on side–effects of chemical weed–control in water.
Publications.

2227 Oosterveld, P.; Dijk, A. van; Leijs, H.N.; Beintema,

A.J.   NL 010602/73/5910 R
De relatie tussen grote herbivoren en de avifauna. **The impact
of grazing management on bird populations.** Publications.

2228 Fuchs, P.; Burgers, J.   NL 010602/73/5916
Onderzoek naar de factoren, die verspreiding,
populatiedichtheid en broedsucces van de steenuil bepalen.
**Analysis of factors which determine dispersion, population
density and reproduction of the little owl.**

2229 Beintema, A.J.; Smit, J.J.; Bergh, L.M.J. van den
    NL 010602/73/5989
Geografisch–oecologisch onderzoek van weidevogels.
Distribution and ecology of meadow birds. Publications.

2230 Beintema, A.J.; Renssen, T.A.   NL 010602/73/5990
De oecologie van het nonnetje. **Ecology of the smew (Mergus
albellus).**

2231 Saaltink, G.J.   NL 010602/73/5992
Het effect van perevuur in meidoorn met betrekking tot het
landschap. **The effect of fire–blight in hawthorn in relation with
the landscape.**

2232 Beintema, A.J.; Bergh, L.M.J. van den
    NL 010602/73/6937
De invloed van ontwatering en landbouwintensivering op de
weide–vogelstand. **Influence of drainage and intensive farming
on meadow birds and their numbers.**

2233 Beintema, A.J.; Bergh, L.M.J. van den; Hazebroek, E.
    NL 010602/73/6938
De invloed van verschillende factoren op het broedseizoen en
het broedsucces van weidevogels. **Influence of some factors on
breeding season and breeding succes of meadow birds.**

2234 Mabelis, A.A.   NL 010602/73/6940 R
De relatie tussen beheersmaatregelen en het voorkomen van
mieresoorten in korte vegetaties. **Relationship between
measures of management and occurrence of ant species in short
vegetations.**

2235 Wijngaarden, A. van; Butot, L.J.M.   NL 010602/74/5636
Geografisch–oecologisch onderzoek van de Nederlandse land–
en zoetwatermollusken. **Distribution and ecology of Dutch land
and fresh water mollucs.**

2236 Wijngaarden, A. van; Butot, L.J.M.   NL 010602/74/6939
De leeftijdsopbouw van een wijngaardslakken–populatie. **The
age structure in a population of the Roman snail (Helix
pomatica linnaeus, 1758).**

2237 Londo, G.   NL 010602/74/7337 N
Geobotanisch onderzoek naar de relatie tussen
plantengemeenschappen en plantesoorten en hun milieu.
**Geobotanical research on the relation between plant
communities and plant species and their habitat.**

2238 Wirdum, G. van; Groot–Veenbaas, G. de; Bodt, J.M.
    NL 010602/75/6933 R
Onderzoek naar de gevolgen van de aanleg van het
Mereveldtracé (RW 27) voor het natuurgebied van
Amelisweerd en de natuurlijke elementen in het agrarisch
gebied. **Research on the impact of a motorway construction
(A27) for the nature area Amelisweerd and for the natural
elements in the agrarian field.**

D 1500 – Nature conservation

2239 Wirdum, G. van   NL 010602/75/7201
Multispectrale scanning en de synecologie van
moerasvegetaties. **Remote sensing and synecology of broad's vegetations.**

2240 Wirdum, G. van; Reijnders, T.   NL 010602/75/7203 R
Waterhuishouding en successie in moerasvegetaties.
Hydrology and succession in mire's vegetations.

2241 Oosterveld, P.   NL 010602/75/7221
Onderzoek naar de ecologische indicatiewaarde van
Taraxacumsoorten voor het beheer van de graslanden.
Ecological research on Taraxacum species as environmental indicators for grassland management.

2242 Wolf, W.J.   NL 010602/75/7395 N
Compilatie en evaluatie van natuurwetenschappelijke kennis
van het Waddengebied. **Compilation of scientific data on the Wadden Sea area.**

2243 Weyland, W.A.   NL 010602/75/8786 N
Onderzoek naar schade (en voorkomen daarvan) aangericht
door het wilde zwijn. **Research on wild boar damage and its prevention.**

2244 Schroevers, P.J.   NL 010602/76/6934
Onderzoek ten behoeve van een watertypologie, gebaseerd op
vrijzwevende mikrofyten. **Investigation on behalf of a typology of waters based on planktonic microphytes.**

2245 Schroevers, P.J.   NL 010602/76/6935
Soortenonderscheiding bij fytosedimentatieplankton,
verzameld ten behoeve van een typologie van wateren.
Differentiation of species of phytosedimentary plankton, collected for a typology of waters.

2246 Wirdum, G. van; Reijnders, T.; Visser, G.J.M.
     NL 010602/76/7202
Vegetatiepatronen en waterhuishouding op Terschelling.
Vegetation and hydrology on the island of Terschelling.

2247 Diemont, W.H.   NL 010602/76/7338 N
Betekenis van afplaggen voor het ontstaan van heidevegetaties
onder verschillende omstandigheden. **The significance of removing top soil for the maintenance or restauration of heath vegetation.**

2248 Diemont, W.H.   NL 010602/76/7339 N
Verkennend onderzoek naar de invloed van uitbaggeren op de
nutriëntenhuishouding van vennen. **Internal regulation of the waterquality by deritus a no n–fertilized and a fertilized fen.**

2249 Kalkhoven, J.T.R.   NL 010602/76/7391 N
Ecologisch onderzoek t.b.v. de ontwikkeling van methoden ter
bepaling en afweging van ruimtelijke aanspraken in een
landschappelijk en natuurwetenschappelijk waardevol gebied
(Midden–Brabant). **Ecological research for the development of methods for determination and evaluation of multiple land–use in areas with high landscape and nature values (Midden Brabant).**

2250 Wirdum, G. van   NL 010602/77/8589 N
Vegetatiepatronen en waterhuishouding in enkele
natuurgebieden in Gelderland. **Vegetation and hydrology of some nature areas in the province of Gelderland.**

2251 Higler, L.W.G.   NL 010602/77/8590 N
De structuur van macrofauna–coenosen in relatie tot de
zuurstofhuishouding in een schone en een vervuilde plas.
Structure of benthic communities in dependence of the oxygen regime in a polluted and non–polluted lake.

2252 Spaans, A.L.   NL 010602/77/8592 N
Verkennend onderzoek naar de ecologische aspecten van
botulisme bij watervogels. **A study on ecological aspects of botulism in aquatic birds.**

2253 Houte de Lange, S.M. ten   NL 010602/77/8596 N
Oriënterend literatuuronderzoek naar de betekenis en
bruikbaarheid van verschillende faunistische gegevens voor
landschapsecologische studies. **Significance and usability of different faunal data in landscape–ecological research.**

2254 Wit, A. de   NL 010602/77/8784 N
Onderzoek naar de oorzaken van de achteruitgangen van
Cantharellus cibarius in Nederland. **Investigation into the decline of Cantharellus cibarius in the Netherlands.**

2255 Wit, A. de   NL 010602/77/8785 N
Onderzoek naar veranderingen in de epifytenrijkdom met
behulp van permanente kwadraten. **Changes in the abundance of epiphytes by means of permanent plots.**

2256 Oosterveld, P.   NL 010602/77/9064 N
De effecten van een extensief graasbeheer op de
vegetatieontwikkeling via de bodem en de bodembewonende
fauna. **Extensive grazing management in relation to vegetation development through effects on soils and on soil living organisms.**

2257 Dankers, N.M.J.A.   NL 010602/78/8787 N
Onderzoek naar de rol van kwelders in de organische
stofhuishouding van de Waddenzee, de Eems en de Dollard.
Research into the role of saltmarshes in the budget of organic matter of the Waddensea, the Ems and the Dollard.

2258 Wolff, W.J.   NL 010602/78/9065 N
Rapport Dollardhaven. **Report on harbour construction in the Dollard estuary.**

2259 Botterweg, P.F.; Eijsackers, H.J.P.
     NL 010602/78/9066 N
Regenwormen als biologische graadmeter van
bodemverontreiniging en de accumulatie van
bodembelastende stoffen. **Earthworms as biological indicator of soil pollution and the bio–accumulation of soil pollutants.**

2260 Minderman, G.   NL 010602/78/9067 N
Bosecosysteem eikebosje Hackfort. **Forestecosystem, oakgrove Hackfort.**

2261 Rooth, J.   NL 010602/78/9068 N
Oriënterend onderzoek naar de overwinterende smienten in
Nederland. **Preliminary ecological investigation of wintering widgeons in The Netherlands.**

2262 Deelder, C.L.   NL 010702/46/7041
Biologisch onderzoek aalvisserij (toestandbeoordeling, intrek
glasaal, metamorphose glasaal/aal/schieraal, aquacultuur,
tegengaan ontsnapping schieraal). **Biological research eelfishing (composition and extent of eelstock,**

D 1500 – Nature conservation

elver–immigration, metamorphoses/elver/eel/silvereel, aquaculture, barring emigrating silvereel). Publications.

2263 Corten, A.A.H.M. NL 010702/47/7075
Onderzoek naar de samenstelling en grootte van haringpopulaties in het noordoost Atlantisch zeegebied en de factoren, die deze samenstelling en grootte beïnvloeden, ten behoeve van een doelmatig beheer van deze populaties. **Research in behalf of efficient management of the composition and extent of herring populations in the north–east Atlantic sea area and the factors influencing this composition and extent.** Publications.

2264 Veen, J.F. de NL 010702/49/7071
Onderzoek naar de samenstelling en grootte van schol–populaties in het noordoost Atlantisch zeegebied en de factoren, die deze samenstelling en grootte beïnvloeden, ten behoeve van een doelmatig beheer van deze populaties. **Research in behalf of efficient management of the composition and extent of plaice populations in the north–east Atlantic sea area and the factors influencing this composition and extent.** Publications.

2265 Veen, J.F. de NL 010702/50/7070
Onderzoek naar de samenstelling en grootte van de tong–populaties in het noordoost Atlantisch zeegebied en de factoren, die deze samenstelling en grootte beïnvloeden ten behoeve van een doelmatig beheer van deze populaties. **Research in behalf of efficient management of the composition and extent of sole populations in the north–east Atlantic sea area and the factors influencing this composition and extent.** Publications.

2266 Daan, N. NL 010702/55/7081
Onderzoek naar de samenstelling en grootte van de kabeljauw–,schelvis–, wijting– en koolvis–populaties in het noordoost Atlantisch zeegebied en de factoren, die deze samenstelling en grootte beïnvloeden, ten behoeve van een doelmatig beheer van deze populaties. **Research in behalf of efficient management of the composition and extent of cod, haddock, whiting and coalfish populations in the north–east Atlantic sea area and the factors influencing this composition and extent.** Publications.

2267 Kerkhoff, M.A.T. NL 010702/55/7127
Hydrografie van de Noordzee, het Nederlandse kust– en binnenwater t.b.v. het biologisch visserijonderzoek, alsmede ter bestudering van transportprocessen in het aquatisch milieu. **Hydrography of the North Sea, the Dutch coastal and inland waters in behalf of the biological fishery investigations and also in connection with the study of transport processes in the aquatic environment.** Publications.

2268 Deelder, C.L. NL 010702/58/7043
Onderzoek naar de bodemfauna in verband met beheersmaatregelen ten behoeve van aalpopulaties. **Research on bottom fauna in relation to the eelstock.**

2269 Corten, A.A.H.M. NL 010702/59/7076
Onderzoek naar de samenstelling en grootte van de makreel–populaties in het noordoost Atlantisch zeegebied en de factoren, die deze samenstelling en grootte beïnvloeden, ten behoeve van een doelmatig beheer van deze populaties. **Research in behalf of efficient management of the composition and extent of mackerel populations in the north east Atlantic sea area and the factors influencing this composition and**

extent. Publications.

2270 Boddeke, R. NL 010702/59/7079
Onderzoek naar de samenstelling en grootte van garnalen–populaties in het noordoost Atlantisch zeegebied en de factoren, die deze samenstelling en grootte beïnvloeden, ten behoeve van een doelmatig beheer van deze populaties. **Research in behalf of efficient management of the composition and extent of shrimp–populations in the north–east Atlantic sea area and the factors influencing this composition and extent.** Publications.

2271 Boddeke, R. NL 010702/68/7086
Onderzoek naar de samenstelling en grootte van populaties van vissoorten in het noordoost Atlantisch zeegebied, welke door de zeevisserij niet of nauwelijks worden geëxploiteerd. **Research on composition and extent of non–exploited fish stocks in the north–east Atlantic sea area.**

2272 Boddeke, R. NL 010702/68/7088
Onderzoek naar de interspecifieke relaties tussen de populaties van de verschillende vissoorten in verband met doelmatige beheersmaatregelen in het noordoost Atlantisch zeegebied. **Research on interspecific relations between populations of marine fishes in behalf of efficient management of fish populations in the north–east Atlantic sea area.** Publications.

2273 Boddeke, R. NL 010702/68/7089
Onderzoek naar de omvang van de vernietiging op zee van niet–marktwaardige en kleine vis door de visserij in verband met het geven van adviezen voor doelmatige beheersmaatregelen in het noordoost Atlantisch zeegebied. **Research on discards resulting from fishing operations in behalf of advising about efficient management of fish populations in the north–east Atlantic sea area.** Publications.

2274 Cazemier, W.G. NL 010702/69/7037
Onderzoek in diverse wateren naar de samenstelling van de visstand ten behoeve van diverse beheersmaatregelen. **Investigations of fishstocks in different waters, in behalf of effective management.**

2275 Kat, M. NL 010702/69/7125
Onderzoek naar de invloed van waterverontreiniging van de Nederlandse binnenwateren op het ontstaan van oscillatoriënbloei en op de verstoring van het gehalte aan opgeloste gassen en naar de samenstelling, verspreiding, de opeenvolging en de hoeveelheid van het fytoplankton in het Nederl. kustwater t.b.v. de biologische waterkwaliteitsbeoordeling. **Investigation into the influence of water pollution in the Dutch inland waters on the development of oscillatoria blooms and on the disturbance of the content of dissolved gasses and investigation into the composition, the distribution, the succession and the amount of phytoplankton in the Dutch coastal waters in relation with the biological water quality assessment.** Publications.

2276 Banning, P. van NL 010702/69/7132
Onderzoek naar het voorkomen en de betekenis van ziekten in vissen, schaal– en schelpdieren. **Investigation into the occurrence and significance of diseases in fish and shellfish.** Publications.

2277 Willemsen, J. NL 010702/71/7025
Onderzoek naar de samenstelling van snoekstanden in de Nederlandse binnenwateren en de factoren die deze

samenstelling beïnvloeden, ten behoeve van doelmatige beheersmaatregelen. **Research in behalf of efficient management on the composition of pike populations in the Netherlands and the factors influencing this composition.** Publications.

2278 Willemsen, J.   NL 010702/71/7026
Onderzoek naar de samenstelling van de stand van snoekbaars, spiering en pos in de Nederlandse binnenwateren en de factoren die deze samenstelling beïnvloeden, ten behoeve van doelmatige beheersmaatregelen, mede gezien de betekenis van spiering en pos als visvoedsel. **Research in behalf of efficient management on the composition of pikeperch–, smelt– and ruffe–populations in the Netherlands and the factors influencing this composition and the value of smelt and ruffe as prey fish.** Publications.

2279 Willemsen, J.   NL 010702/71/7027
Onderzoek naar de samenstelling van baarsstanden in de Nederlandse binnenwateren en de factoren die deze samenstelling beïnvloeden, ten behoeve van doelmatige beheersmaatregelen. **Research in behalf of efficient management, on the composition of perch–populations in the Netherlands and the factors influencing this composition.**

2280 Cazemier, W.G.   NL 010702/71/7031
Onderzoek naar de samenstelling van andere zoetwatervissoortstanden dan van snoek, snoekbaars, baars, spiering en pos in de Nederlandse binnenwateren en de factoren die deze samenstelling beinvloeden. **Investigations of the factors which influence the composition of other freshwater fish populations excluding pike, pikeperch perch, smelt and ruffe in Dutch inland waters, in behalf of management purposes.** Publications.

2281 Willemsen, J.   NL 010702/71/7034
Onderzoek naar de invloed van koelwaterlozingen op de visstand. **Research on the influence of cooling water discharge on the fish population.**

2282 Cazemier, W.G.   NL 010702/71/7036
Onderzoek naar de relatie tussen milieu en visstand (o.a. brasem en blankvoorn) in enkele geselecteerde proefwateren en naar de effecten van doelgerichte beheersmaatregelen. **Investigations of the relationships between environment and fishstocks (a.o. bream and roach) in some selected waterbodies and of the effect of a special management policy.** Publications.

2283 Boddeke, R.   NL 010702/71/7087
Onderzoek naar de biologie van die vissoorten in het noordoost Atlantisch zeegebied, die voor de sportvisserij van belang zijn. **Research on biology of fish populations from the north–east Atlantic sea area which are important for sport fishing.**

2284 Hagel, P.   NL 010702/72/7120
Onderzoek naar de invloed van waterverontreiniging op de gehalten aan zware metalen in vissen, schaal– en schelpdieren en naar de factoren die het zware–metaal gehalte in zeewater bepalen. **Investigation into the influence of water pollution on the content of heavy metals in fish and shellfish and investigation into the factors influencing the heavy metal content of seawater.** Publications.

2285 Berg, R. van den   NL 010702/72/7123
Onderzoek naar de toxiciteit van verontreinigende stoffen op

vissen, schaal– en schelpdieren, mede in verband met het opsporen van oorzaken van vissterften; toepassing van biochemische identificatiemethoden en begeleiding van aquacultuur. **Investigation into the toxicity of pollutants to fish and shellfish, also in connection with the tracing of the origins of fish kills. The application of biochemical identification techniques and the support of aquaculture.**

2286 Cazemier, W.G.   NL 010702/74/7033
Onderzoek naar de voor de visstand in de binnenwateren schadelijke verstoringen van het aquatisch milieu door zand– en grindwinning, de exploratie en exploitatie van olie– en gasvelden en het dumpen van baggerspecie of soortgelijke materialen. **Investigation of the harmfull effects for fish, when the following activities are carried out in freshwater biotopes: sand– and graveldigging, exploration and exploitation of oil– and gasfields and dumping of mud or similar material.** Publications.

2287 Groot, S.J. de   NL 010702/74/7092
Onderzoek naar de voor de visserij schadelijke verstoringen van het aquatisch milieu door zand– en grindwinning en de exploratie en exploitatie van olie– en gasvelden. **Influence of the offshore gas and oil industry as well as the sand and gravel industry on the fish stock at sea.** Publications.

2288 Cazemier, W.G.   NL 010702/75/7038
Onderzoek naar de invloed van stuwen, sluizen en soortgelijke belemmeringen op de visstand en naar de mogelijkheden van vispassages langs deze barrières. **Investigations of the effect of weirs, locks and similar barriers on fish populations and of the application of fishways in such barriers.**

2289 Hagel, P.   NL 010702/76/7124
Onderzoek naar de afbreekbaarheid of anderszins omzetbaarheid van verontreinigende stoffen in het aquatisch milieu. **Investigation into the degradability or otherwise conversibility of pollutants in the aquatic environment.**

2290 Drinkwaard, A.C.   NL 010702/77/7002
Onderzoek naar de samenstelling en bestandsgrootte van kokkels in de Nederlandse kustwateren en de factoren, die deze samenstelling en bestandsgrootte beïnvloeden ten behoeve van een doelmatig beheer ten bate van de visserij. **Biological research concerning cockles and the culture of cockles in the Dutch coastal waters.**

2291 Veen, J.C. van   NL 020013/66/6301
Causaal–analytisch onderzoek naar de numerieke interacties tussen carnivoor levende populaties en hun prooi– resp. gastheerpopulaties. **Investigations into the numerical interactions between carnivorous population and their prey or host populations.** Publications.

2292 Kuchlein, J.H.   NL 020013/69/6302
Experimenteel–taxonomisch onderzoek bij spinnen. **Experimental taxonomic studies on spiders.**

2293 Docters van Leeuwen, W.M.   NL 020013/75/6785
Een vergelijking van samenstelling en structuur van zangvogelgemeenschappen op de Waddeneilanden met overeenkomstige gemeenschappen op het vaste land. **A comparison of songbird communities on the Dutch Northsea islands to those of the Dutch mainland.**

2294 Zwart, K.W.R.   NL 020014/69/4412

D 1500 – Nature conservation

Bewerking van de Ichneumonidae (Hym) van Suriname, en van enkele andere Neotropische gebieden. **Revision of Ichneumonidae (Hym) of Surinam and some other Neotropic countries.**

2295 Cobben, R.H.　　　　　NL 020014/69/5043
Systematische revisies van verschillende Heteropterafamilies. **Revisionary work on various Heteroptera families.** Publications.

2296 Zwart, K.W.R.　　　　　NL 020014/70/4413
Soortomgrenzing en invloed van de gastheer op morfologische kenmerken bij enkele soorten van het geslacht Gelis (Hym.,Ichneumonidae). **Variability of species and induction of morphological characters by the host in some species of the genus Gelis (Hym.,Ichneumonidae).** Publications.

2297 Cobben, R.H.　　　　　NL 020014/72/5042
Variabiliteit en isolatie–mechanismen van graslandcicaden. **Variability and isolation–mechanisms of graminicolous leafhoppers.** Publications.

2298 Zadoks, J.C.　　　　　NL 020018/77/7317
Homoeostasis van waard en pathogenen in een natuurlijke vegetatie van lamsoor (Limonium vulgare Mill.) op de Boschplaat, Terschelling – EPINAT. **Epidemiologic and population–dynamic research on Limonium vulgare and some of its diseases on the Boschplaat, Terschelling.** Publications.

2299 Westra, J.J.　　　　　NL 020021/73/5403
Invloed van mechanisatie op de gesteldheid van de bosbodem, de groei en de gezondheid van de bomen en de welzijnswerkingen van het bos. **Effects of mechanization on forest growth.**

2300 Minderhoud, J.W.; Dirven, J.G.P.　　NL 020025/72/5035
Evolutie en gebruiksmogelijkheden van grasvegetaties bij extensivering van de exploitatie. **Evaluation and utilization possibilities of grass vegetation at decreasing management levels.**

2301 Sloet van Oldruitenborgh, C.J.M.　　NL 020035/68/5052
Oecologische aspecten van verspreiding van ziekten en plagen in natuurlijke vegetaties (o.a. perevuur in meidoorn). **Ecological aspects of distribution and dispersion of epidemic diseases in natural vegetation (a.o. fireblight in hawthorn).** Publications.

2302 Made, J.G. van der　　　　NL 020035/69/5066
Faunistische aspecten van het landschapsoecologische onderzoek, ten dienste van landinrichting, beheer en planologie, in het bijzonder in de Gelderse Vallei, Achterhoek en Zuidwest Nederland. **Zoological aspects of the landscape ecological research as a basis of the Gelderse Vallei, the Achterhoek and the South–west of the Netherlands.**

2303 Kaplan, Y.　　　　　NL 020035/72/4500
Regeneratie in begrazing in het Yahudia Forest (Israel). **Forest–regeneration and grazing in the Yahudia Forest (Israel).**

2304 Mœrzer–Bruyns, M.F.　　　NL 020035/73/5062
Invloeden van recreatie–activiteiten op natuur oecosystemen. **Impact of recreation–activities on ecosystems.** Publications.

2305 Strien, N.J. van　　　　NL 020035/73/5063

Verspreiding en levenswijze van de Sumatraanse neushoorn, de Sumatraanse olifant in Noord–Sumatra. **Distribution and ecology of the Sumatran rhinoceros and the Sumatran Elephant in North Sumatra.** Publications.

2306 Geerling, C.　　　　　NL 020035/73/8661 N
Houtige planten van de West–Afrikaanse savanna: 1) Veldflora met sleutel, beschrijvingen en illustraties 2) als voedselbron van grotere herbivoren. **Woody plants of the West African savanna: 1) flora with key, descriptions and illustrations 2) as a food source of larger herbivores.**

2307 Roos, G.Th. de　　　　NL 020035/74/5059
De invloeden van menselijke activiteiten in het bijzonder van de recreatie op de avifauna van Vlieland. **The impact of human activities, especially of recreation on the avifauna of Vlieland.** Publications.

2308 Tolkamp, H.H.　　　　NL 020035/75/6809
De mikrodistributie van het benthos in laaglandbeken in relatie tot de samenstelling van het substraat. **The microdistribution of the benthos in lowlandstreams in relation to the substrate–composition and substrate–distribution.** Publications.

2309 Mörzer, Bruyns, M.F.　　　NL 020035/75/6810
Natuurbehoud en ruimtelijke ordening in het Waddengebied. **Nature conservation and regional planning in the Waddensea–area.** Publications.

2310 Heringa, A.　　　　　NL 020035/76/6811
Wildbenutting in Mali. **Wildlife utilisation in Mali.** Publications.

2311 Mörzer Bruyns, M.F.　　　NL 020035/76/7561
Onderzoek naar milieubelasting bij bosbouw en houtverwerking. **Environmental impacts of forestry and wood processing.**

2312 Hoek, D. van den　　　　NL 020035/77/8663 N
Beheer natuurterreinen stroomdal Overijsselse Vecht. **Management of nature areas along the river valley of the Overijsselse Vecht.**

2313 Sloet van Oldruitenborgh, C.J.M.　　NL 020035/77/8664 N
Struweel– en bosontwikkeling bij extensief en extensiever wordend grondgebruik. **Scrub– and forest development in relation to extensive forms of land–use.**

2314 Koeman, J.H.　　　　　NL 020055/73/5101
Kwalitatieve en kwantitatieve bepaling van persistente contaminanten in de zeehond. **Qualitative and quantitative determination of persistent contaminations in the seal.**

2315 Koeman, J.H.　　　　　NL 020055/73/5102
De toxicologische betekenis van chroom in het Nederlandse oppervlaktewater. **The toxic potential of chromium in Dutch surface waters.**

2316 Koeman, J.H.　　　　　NL 020055/73/5105
Neveneffecten van helicopterbespuitingen op vissen en vogels in het Noorden van Nigeria. **Side–effects of spraying by helicopter on fishes and birds in the North of Nigeria.** Publications.

2317 Strik, J.　　　　　　NL 020055/74/5104

D 1500 – Nature conservation

Porfyrinogeen vermogen van octachloorstyreen bij vogels. **Porphyrine capacity of octachlorostyrene in birds.**

2318 Koeman, J.H.   NL 020055/74/5106
Onderzoek naar de verspreiding van zink en koper in het aquatisch milieu in Kenya, speciaal met betrekking tot het Nakuru meer. **Monitoring study on zinc and copper in the aquatic environment in Kenya, specially with relations to Lake Nakuru.**

2319 Koeman, J.H.   NL 020055/74/5107
De invloed van organische fosfor–insecticiden op overtebraten in het Nederlandse oppervlaktewater. **The influence of organophosphorous insecticides on crustacea in Dutch surface waters.** Publications.

2320 Putte, I. van de   NL 020055/78/8544
Verband tussen de toxiciteit, opname, weefselverdeling en excretie van chroom en molybdeen bij de regenboogforel. **Relationship between the toxicity, uptake, tissue distribution and excretion of chromium and molybdenum in rainbow trout.**

2321 Richter, C.J.J.   NL 020058/76/7359
Milieufactoren die de aantallen snoek (Esox luceus) in de Wielen en de aantallen pos (gym no cephalus cernua) in het Tjeukemeer beïnvloeden. **Environmental factors influencing the numbers of pike (Esox luceus) in the Wielen and the numbers of the ruff (Gym no cephalus cernua) in lake Tjeukemeer.** Publications.

2322 Doing, H.   NL 020060/56/4945
Landschapskartering op vegetatiekundige grondslag. **Landscape survey with an emphasis on vegetation.** Publications.

2323 Barkman, J.J.; Vries, B.W.L. de   NL 020060/63/5069
Oecologie en sociologie van de paddestoelen van jeneverbesstruwelen in Nederland, N.Duitsland, Denemarken, Z. Zweden en Polen. **Ecology and phytosociology of the fungi of juniper scrub in the Netherlands, N.Germany, Denmark, S.Sweden and Poland.** Publications.

2324 Barkman, J.J.   NL 020060/65/5071
Oecologie van microgezelschappen en van afzonderlijke mossoorten in jeneverbesstruwelen. **Ecology of microcommunities (synusiae) and individual bryophytes species in juniper scrub.** Publications.

2325 Barkman, J.J.   NL 020060/65/6845
Struktuuranalyse van jeneverbesstruweel in relatie tot bodem en mikroklimaat. **Pattern analysis of juniper scrub and its ecological background (microclimate, soil chemistry).** Publications.

2326 Ott, E.C.J.   NL 020060/65/7606
Vegetatiekundig typologie en typen–inventarisatie van hakhoutbossen in Nederland. **History, classification, survey and management of coppice woodland in the Netherlands.**

2327 Kramer, R.N.A.   NL 020060/70/5070
Kwalitatief en kwantitatief vergelijkend onderzoek van de mycoflora van naaldbossen in Drenthe. **Qualitative and quantitative research on the mycoflora of different types of conifer plantations in Drenthe (Neth.).**

2328 Ott, E.C.J.   NL 020060/73/7607

Vegetatiekundige landschapskartering en gedetailleerde vegetatiekartering van beekdalen en oude rivierlopen in Noord–Limburg. **Vegetation survey and mapping, especially regarding the valleys of lowland streams and former beds of the river Meuse (N–Limburg).** Publications.

2329 Arnolds, E.J.M.   NL 020060/75/6844
Vergelijkend mycologisch onderzoek van diverse typen half–natuurlijke en cultuurgraslanden. **Mycological research of semi–natural and permanent grassland.** Publications.

2330 Doing, H.   NL 020060/75/6848
Grondslagen voor een vegetatie–geografisch overzicht van de Nederlandse duinlandschappen. **Ecological basis of a survey of coastal sand dune areas in the Netherlands.**

2331 Jansen, A.E.   NL 020060/76/6846
Kwantitatief en kwalitatief vergelijkend mycologisch onderzoek van loofbossen op min of meer voedselarme bodem, behorend tot het Quercion Robori petraeae. **Quantitative research on the mycoflora of various types of oak wood.** Publications.

2332 Ott, E.C.J.   NL 020060/77/7541
Onderzoek naar de vegetatie–ontwikkeling in een aantal Alno–Padionbossen met van elkaar afwijkende voorgeschiedenis. **A study of vegetation development in major poplar afforestations of different origins.**

2333 Sibbing, F.A.   NL 020071/74/5049
Functionele analyse van het pharyngeale maalapparaat bij de karper (Cyprinus carpio (L.)) en de graskarper (Ctenopharyngodon idellus) en de betekenis van dit apparaat als vormonderdeel in het voedselopnameproces. **Comparative analysis of the structure and function of the pharyngeal teeth apparatus of the carp (Cyprinus carpio(L.)) and the grasscarp (Ctenopharyngodon idellus) to establish the relation between food, chewing and the morphological features of this apparatus.** Publications.

2334 Stroband, H.W.J.   NL 020071/74/5050
Bouw en functie van het spijsverteringsstelsel van de graskarper (Ctenopharyngodon idellus) in relatie tot de aard van zijn voedsel. **Structure and function of the digestive tract of the grascarp (Ctenopharyngodon idellus) in relation with the kind of food.** Publications.

2335 Den Boer, P.J.   NL 020072/59/5409
De betekenis van het verbreidingsvermogen voor het voortbestaan van soorten in een bepaald gebied. **The significance of dispersal power for the persistence of species in a certain area.** Publications.

2336 Boer, P.J. den   NL 020072/59/6849
Theoretisch onderzoek naar de mogelijke betekenis van risicospreiding voor het voortbestaan van populaties en soorten, in vergelijking tot andere gangbare opvattingen. **Theoretical research into the possible significance of spreading of risks for the persistence of populations and species, as compared with other current opinions.** Publications.

2337 Den Boer, P.J.   NL 020072/59/6850
Relaties tussen de belangrijkste "faktoren" die de terreinbinding en die, welke aantalfluktaties bepalen. **Relations between the most important factors responsible for the fitting to habitats and those responsible for the fluctuations**

of numbers in populations. Publications.

2338 Van Dijk, Th.S.   NL 020072/65/6851
Relatief stabiliserende invloed van diverse vormen van
heterogeniteit in de populatie–opbouw op het
fluktualtiepatroon van aantallen. **Stabilization of the
fluctuations of numbers by different forms of diversity in the
composition populations. Publications.**

2339 Eijk, R.H. van   NL 020072/76/6852
De kwantitatieve betekenis van het bestaan van sub–populaties
voor de overlevingskansen van populaties. **The quantitative
significance of the existence of subpopulations for the chances of
survival of populations.**

2340 Mols, P.J.M.   NL 020072/77/8398
De betekenis van polyfage roofkevers voor de
aantalsbeperking van geclusterd levende prooisoorten.
**Significance of polyphageous predatory beetles for the
restriction of numbers in populations of clustering prey species.**

2341 Leeuwangh, P.; Genderen, H. van   NL 030003/73/5887 R
De toxologische betekenis van de chemische verontreiniging
van de Rijn en de door haar gevoede oppervlaktewateren (vis,
kwantitatieve structuur–aktiviteitsrelaties, bioaccumulatie en
eliminatie, toxiciteit van mengsels, pH–invloed). **Toxicological
significance of the chemical pollution of the river Rhine and
connected surface waters in the Netherlands (fish, quantitative
structure–activity relationships, bio–accumulation and
–elimination, toxicity of mixtures, pH–influence).**

2342 Musch, A.   NL 030003/76/5890 R
Onderzoek naar effekten op de fysieke kapaciteiten en
fysiologische parameters van vissen door
milieuverontreinigende stoffen. (milieuverontreiniging,
trifenyltin–chloride, vissen, sublethale effekten, ademhaling,
hartfunktie, zwemvermogen). **Effects of pollutants on physical
capacities and physiological parameters of fishes
(environmental pollution, triphenyltin–chloride, fish, sublethal
effects, respiration, cardiac function, rotatoryflow technique).**

2343 Leeuwangh, P.   NL 030003/77/7757 R
Vóórkomen en toxiciteit van lood bij organismen in het
Nederlandse oppervlaktewater, nabij rijkswegen. **The presence
and toxicity of lead in organisms living in Dutch surface waters,
near highways.**

2344 Musch, A.   NL 030003/78/8951 N
Toxicologisch onderzoek naar gedragsveranderingen bij vissen
door milieuverontreinigende stoffen, afkomstig van de
industrie en de landbouw. (vis, toxicologie,
milieuverontreiniging, gedrag(–s effecten). **Toxicological
research on behavioural effects on fish by environmental
pollutants from industrial and agricultural origin. (fish,
toxicology, environmental pollution, behavioural effects).**

2345 Leeuwangh, P.   NL 030003/78/8953 N
Vergelijkend onderzoek naar de akute toxiciteit van enkele
houtbeschermingsmiddelen, voor jonge ratten.
(houtbeschermingsmiddelen, vleermuizen, ratten).
**Comparative study on the acute toxicity of several wood
preservatives for young rats (wood preservatives, bats, rats).**

2346 Loenen, M.   NL 040007/65/4106
Inventarisatie van de water– en oevervegetatie in de

randmeren van de Flevopolders. **Submerged and emergent
aquatic macrophytes of the lakes bordering the Flevopolders.**

2347 Loenen, M.; Diender, J.   NL 040007/65/4139
Inventarisatie van de vegetatie in de randmeren. **Inventory of
the vegetation in the border lakes.**

2348 Loenen, M.   NL 040007/67/6208
Landschapsecologisch onderzoek Lauwerszee (vegetatie
ontwikkeling, vogelbevolking). **Ecological research in the
Lauwerszee area.**

2349 Loenen, M.   NL 040007/69/6209
Onderzoek naar het effect van beheersmaatregelen in door
natuurbouw verkregen natuurgebieden. **Research on the
impact of management in man–made nature reserves.**

2350 Bij de Vaate, A.   NL 040007/70/4128
Ecologie van algensoorten. **Ecology of algae species.**

2351 Poorter, E.P.R.   NL 040007/70/4142
Oecologisch onderzoek aan ruiende Grauwe Ganzen.
Ecological research on moulting Greylag Geese.

2352 Polman, G.K.R.   NL 040007/70/4299
Ontwikkeling van beheersplannen voor diverse
natuurgebieden. **Development of management plans for various
nature areas.**

2353 Poorter, E.P.R.   NL 040007/71/4148
Voedsel– en populatiebiologie bij de Kleine Zwaan. **Ecology of
the Bewick's Swan.**

2354 Poorter, E.P.R.   NL 040007/71/4158
Oecologie van de N.W. Europese lepelaarpopulatie. **Ecology
of the North–West European Spoonhill population.**

2355 Loenen, M.   NL 040007/71/8428
Beheer van wegbermen en jonge beplantingen. **Management of
road–side verges and young plantations.**

2356 Loenen, M.   NL 040007/72/5563
Vegetatie– en beheersonderzoek Oostvaardersplassen.
**Vegetation and management research in the nature reserve
Oostvaardersplassen.**

2357 Zijlstra, M.   NL 040007/72/5567
Inventarisatie van de loopkever– en zoogdierenbevolking in
aan verandering onderhevige gebieden. **Ground beetle and
mammal population in changing (newly reclaimed) areas.**

2358 Poorter, E.P.R.   NL 040007/73/4158
Oecologie van de N.W. Europese lepelaarpopulatie. **Ecology
of the North–West European Spoonbill population.**

2359 Loenen, M.   NL 040007/74/8434
Landschapsoecologisch onderzoek
Oosterschelde/Grevelingen. **Landscape–ecological research in
the Oosterschelde and Grevelingen area.**

2360 Reitsma, T.; Elburg, H. van; Loenen, M.; Koridon,
A.H.: Diender, J.; Wigbels, V.L.   NL 040007/75/6209
Beheersonderzoek in door natuurbouw verkregen natuur
reservaten. **Reaearch into the functioning of man–made nature
reserves.**

2361 Poorter, E.P.R. NL 040007/75/6212
Oecologie van drie soorten Kiekendieven in Flevoland.
Ecology of circus aeroginosus, circus cyanem and circus pygargus in Flevoland.

2362 Wigbels, V.L. NL 040007/75/6214
Onderzoek naar de successie–stadia in de bosontwikkeling en de ecologie van bostypen t.b.v. de verhoging van de biologische waarde en het natuurlijk aanzien van het bos. **Research on formation and ecology of forest on behalf of improvement of the biological value and a more natural look.**

2363 Zwarts, L. NL 040007/76/8482
Oecologische gevolgen van de landaanwinning op de natuurfunktie van het Waddengebied. **Oecological consequences of land reclamation on the natural function of the Wadden area.**

2364 Vos, R.H. de NL 050301/68/6473
Identificalie en bepaling van residuen van bestrijdingsmiddelen en persistente chemicatien bij de wilde fauna (vogels, zoogdieren). **Identification and determination of residues of pesticides and persistent chemicals in the wild fauna (birds, mammals).**

D 1600 – Planning land use

See also 223, 1124, 1126, 1869, 2038, 2116, 18019, 18069, 18137, 18242, 18299, 18303

2365 Pflug, W.; Leitl, U. DE 101100/71/0001 R
Die Stadt und ihr natürlicher Ausgleichs– und Ergänzungsraum – dargestellt am Beispiel der Stadt Aachen 1971. **Natural urban compensating and complementary zone, exemplified by Aachen city.**

2366 Weckwerth, H. DE 105650/71/0003 R
Erholungssuchender Mensch und Topographie des Freiraums. Eine Untersuchung über die wechselseitige Beeinflussung von Mensch und Kulturlandschaft, unter besonderer Berücksichtigung der Vegetationsstrukturen, dargestellt an ausgewählten Untersuchungsgebieten im Nahbereich der Gross–Stadt Berlin 1971. **Man's demand for recreation and topography of open areas. A study on correlations between man and cultivated landscape with special regard to structures of vegetation exemplified by selected areas in the near surroundings of Berlin.**

2367 Barth, J. DE 105650/71/0004 R
Untersuchungen zur Entwicklung des städtischen Freiraums in Berlin unter besonderer Berücksichtigung der durch Dauervegetation mitbestimmten Flächen 1971. **Studies on the development of urban areas in Berlin with special regard to areas affected by permanent vegetation.**

2368 Mantel, K.; Ott DE 126505/70/0001
Die Forstorganisation in Württemberg mit besonderer Berücksichtigung der Neuordnung im 18. Jahrhundert. **The forest organization in Württemberg under special consideration of the new regulation in the 18th century.**

2369 Pacher, J.; Müller, A. DE 126505/72/0001
Der Stadtwald von Landau, insbesondere seit der Teilung der Oberhaingeraide. **The municipal forest of Landau especially since the division of the Oberhaingeraide.**

2370 Harrach, T.; Schönhals, E.; Werner, G. DE 129020/71/0004
Standortkundliche Grundlagen der Landschaftsentwicklung. **Habitat fundamentals of landscape development.** Publications.

2371 Zundel, R.; Rozsnyay, Z. DE 132870/70/0002 R
Erforschung wissenschaftlicher Grundlagen der forstlichen Beteiligung bei der Regionalplanung 1969. **Research on scientific conditions of forestry participation in regional planning.**

2372 Höschele, K. DE 145050/74/0001 N
Topoklimatologische und lufthygienische Studien zur Landschafts– und Stadtplanung. **Topoclimatological and air quality studies for purpose of landscape and town planning.**

2373 Kaule, G.; Haber, W. DE 161670/78/0002 N
Landschaftsökologische Modelluntersuchung im Raum Ingolstadt. **Model research in landscape ecology in the Ingolstadt area, Bavaria.**

2374 Meckelein, W.; Föhr, C. DE 170251/78/0001 N
Zur Inwertsetzung eines nordsaharischen Oasengebietes – Nefzaoua, Südtunesien –. **Development of an oasis region in the northern Sahara – Nefzaoua, southern Tunisia –.**

2375 Neander, E.; Römert, W. DE 201120/72/5002 N
Vergleich und Beurteilung der Verfahren zur Abschätzung der agrarstrukturellen Entwicklungsmöglichkeiten kleinerer Regionen in ausgewählten agrarstrukturellen Vorplanungen. **Comparison and valuation of methods for estimation of chances of agrarian structural development of small regions in selected agrarian structural pilot planning.**

2376 Eisenhauer, G. DE 202030/75/0001
Erschliessung des Naturwaldes als Voraussetzung für die rationelle Nutzung. **Accessibility of virgin forests as precondition for their rational utilization.** Publications.

2377 Mrass, W.; Winkelbrandt, A. DE 205010/75/0001
Landschaftsplanung und Strassenplanung. **Landscape and road planning.**

2378 Winkelbrandt, A. DE 205010/78/0002 N
Methodik und Verfahrensregelung für die landschaftspflegerische Begleitplanung in der Fachplanung im Sinne des Paragraphen 8 BNatSchG. **Methodology and procedure for landscape planning within sectorial planning with regard to paragraph 8 of the Federal Nature Conservation Act..**

2379 Fritz, G. DE 205010/78/0003 N
Grundlagen für die Sicherung des Erholungspotentials der Landschaft. **Fundamentals for the development and management of landscape recreation potential.**

2380 Mrass, W.; Zvolsky, Z.; Bürger, K. DE 205010/78/0004 N
Erarbeitung von Grundsätzen für Landschaftsprogramme. **Development of principles for landscape programmes.**

2381 Erz, W.; Mader, H.J. DE 205020/78/0002 N
Integration des Arten– und Biotopschutzes in Raumordnungsund Landnutzungsprogramme sowie in die Landschafts– und Eingriffsplanung. **Integration of the protection of species and biotopes into space–relevant and**

D 1600 – Planning land use

2382 Volk, H.; Spahl, H.   DE 501504/78/0003 N
Erarbeitung eines landschaftspflegerischen Begleitplans zur
Gestaltung des Baggersees Nimburg. **Development of an
accompanying landscape management plan on the layout of the
dredging lake of Nimburg.**

2383 Volk, H.; Spahl, H.   DE 501504/78/0006 N
Untersuchung über den Einsatz von
Landschaftsbewertungsmethoden bei forstlichen Gutachten.
**Study on the use of methods of landscape valuation in expert
opinions on forestry.**

2384 Kristensen, V.G.   DK 030181/70/0016
Hugstprognoser for Danmark. **Felling prognoses for Denmark.**

2385 Brugere, D.; Chartier, P.   FR 010112/72/9117
Analyse et classification de la documentation relative à
l'environnement, l'énergie et l'aménagement du territoire
susceptible d'interesser la recherche agronomique. **Analysis
and classification of documents or references pertaining to
environment, energy and regional development problems, as
they may concern agricultural research.**

2386 Brugere, D.; de Casabianca, F.; Arnoux, J.;
Deffontaines, J.P.; Briand, P.   FR 010112/73/9115
Cartographie automatique (système carat) de groupes
homogènes de communes dans les départements des Vosges et
de la Haute Corse. **Automatic mapping (CARAT program) of
homogeneous groups of commons (villages) in the Vosges and
Northern Corsica departments.**

2387 Brun, A.; Houdard, Y.; Deffontaines, J.P.; Brossier, J.;
Osty, P.L.; Bonnemaire, J.   FR 010112/75/9100
Pays, Paysans, Paysages dans les Vosges du Sud. Les pratiques
agricoles et la transformation de l'espace. **Lands, lands–men,
landscapes in the southern Vosges; agricultural practices and
land environment transformation.**

2388 Mainie, P.   FR 010112/77/9103
Enquête sur les attitudes des agriculteurs vis-à-vis de
l'utilisation actuelle et future du marais poitevin. **Survey of the
farmers' outlook on the present and future land use of the Poitou
marshlands.**

2389 Laurent, C.; Brossier, J.   FR 011012/78/8566
Transformation de l'espace et habitat rural dans une zone de
forte concurrence foncière. **Rural space and home in
landownership with high competition.**

2390 Pinet, J.M.   FR 011800/73/5291
Ecologie appliquée aux problèmes de remembrement. **Ecology
applied to the problems of land–regrouping.**

2391 Coulomb, P.   FR 012212/76/8501
Systèmes fonciers et politique foncière. **Land systems and land
policy.**

2392 Petit, F.E.   FR 012212/78/8579 N
La politique française en matière forestière et pastorale dans
les régions de montagne ; ses implications sur l'utilisation de
l'espace. Analyse de quelques cas régionaux. **French policy
about forests and pastures in mountains, results on space use
some regional analysis.**

2393 Lee, J.   IE 060200/78/1402 N
**Agriculture potential of serdo region (counties – Wexford,
Carlow, Kilkenny, Waterford and south Tipperary).**
Publications.

2394 Commins, P.   IE 060600/76/1352 N
Land structure reform. Publications.

2395 Merlo, M.   IT 040804/78/0001 R
La pianificazione rurale in Europa. **Rural Planning in Europe.**
Publications.

D 1610 – Land consolidation and land layout

See also 1187, 1313, 1764, 2089, 2302, 2352, 2573, 2583, 4411,
10554

2396 Pflug, W.; Klett, W.; Lei, H.J.   DE 101100/77/0001
Beziehungen zwischen Baugebieten und einbezogenen bzw.
angrenzenden Waldbeständen, untersucht an Beispielen auf
unterschiedlichen Standorten. **Relationship between building
areas and incorporated resp. adjacent forests, exemplified by
models at different locations.**

2397 Pflug, W.; Jahn, R.; Schramm, A.   DE 101100/77/0002
Beziehungen zwischen Naturhaushalt, Anlage von Strassen
und Strassenverkehr. **Relationship between natural balance,
road building and traffic.**

2398 Sukopp, H.; Böcker, R.; Kunick, W.; Tigges, W.
                    DE 105060/74/0003
Bewertung und Planung von Freiflächen Berlins,
vegetationskundliche Grundlagen. **Valuation and planning of
open spaces in Berlin: vegetation ecological conditions.**
Publications.

2399 Knothe, I.   DE 105650/73/0003
Bestimmung des Erholungswertes von Freiflächen –
exemplarische Untersuchung in Berlin. **Determination of the
recreational value of open spaces in Berlin by way of example.**

2400 Kiemstedt, H.; Heinrich, W.–D.; Keimer, B.; Enis, R.;
Schechter, M.; Tzamir, Y.   DE 105650/75/0001
Planungsmodell Freizeit und Tourismus. **Planning model for
leisure and tourism.**

2401 Wenzel, J.   DE 105650/77/0004
Siedlungsstrukturelle Entwicklung typischer
Fremdenverkehrsgebiete in der Bundesrepublik Deutschland.
**Settlement structural development of typical tourism areas in
the Federal Republic of Germany.**

2402 Kiemstedt, H.; Heinrich, W.-D.; Keimer, B.;
Boetticher, M.; Schwenteit, U.; Tromsdorff, U.
                    DE 105650/77/0005
Erarbeitung eines komplexen Standortbewertungsverfahrens
für Freizeit– und Tourismusplanung. **Development of a complex
location valuation method for off–time and tourism planning.**

2403 Meyer, E.; Schneider, F.; Knothe, I.; Hermann, P.–G.;
Olfe, H.–H.; Jochum, H.   DE 105650/77/0006
Erholung in Wohngebieten. **Recreation in residential districts.**
Publications.

2404 Krensel, H.   DE 108150/71/0001 N

D 1610 – Land consolidation and land layout

Der Wandel der Agrarlandschaft im nördlichen Bergisch–Märkischen Land. **The changing of the rural area in the North of the "Bergisch–Märkisches–Land".**

2405 Zakosek, H.; Warstatt, M. DE 111050/78/0007 N
Untersuchungen über die Standortskartierung im Rahmen der Flurbereinigung. **Study about mapping of the natural environment for land consolidation.**

2406 Kopp, H. DE 120200/75/0001
Agrargeographie der Arabischen Republik Jemen. **Rural geography of the Yemen Arab Republic.** Publications.

2407 Ritter, W.; Beck, D. DE 120250/78/0003 N
Die Flurbereinigung im unterfränkischen Weinbau, unter besonderer Berücksichtigung des Weinortes Sommerhausen. **Land consolidation in viticultural Lower Franconia, especially in Sommerhausen region.**

2408 Speidel, G.; Gerke, R. DE 126450/77/0001
Ermittlung einer volkswirtschaftlich tragbaren Wilddichte im Rahmen einer Grossrauminventur. **Determination of economic game density within large–space inventory.**

2409 Kohler, V. DE 126450/77/0002
Forstliche Planung und Raumordnung. **Planning and area planning in forestry.**

2410 Mitscherlich, G.; Wiedemann, S. DE 126600/77/0002
Verbesserung des Stadtklimas von Freiburg durch Entwaldung des Hirzbergsattels. **Improvement of the urban climate of Freiburg by deforestation of the Hirzberg ridge.**

2411 Mitscherlich, G. DE 126600/77/0003
Lärmmessung in Erholungsgebieten – Wald –. **Noise measuring in recreational areas – forests –.**

2412 Bierhals, E. DE 138600/70/0002
Zur Theorie und Methodik der ökologischen Gliederung 1969. **Estimation of agricultural land–use systems with regard to landscape management – a contribution to theory and methodics of ecological classification.**

2413 Darmer, G. DE 138600/70/0004
Erarbeitung ökologischer Leitbilder für die Rekultivierung industrieller Entnahme– und Schüttflächen durch synoptische Auswertung gesammelter Unterlagen. **Development of ecological models for the recultivation of open cast mining areas applying the method of data comparison and evaluation.**

2414 Buchwald, K. DE 138600/72/0003 R
Vergleichsstudie zur Förderung bundesdeutscher Bergbauerngebiete, durchgeführt in repräsentativen Beispielsgemeinden des Südalpenraumes 1972. **Comparative study on the subvention of Federal upland farmers, exemplified by representative communities in the southern Alps.**

2415 Buchwald, K.; Jakob, H. DE 138600/74/0001
Messung der Erlebnisqualität verschiedener Waldbestandstypen Eine experimental–psychologische Analyse als Beitrag zur Erholungsplanung. **Measuring of the recreational quality of different forest stand types An experimental–psychological analysis as contribution to recreation planning.**

2416 Lendholt, W.; Rautmann, K. DE 138660/72/0002
Untersuchung über Frequentierung städtischer Freiräume unter besonderer Berücksichtigung der dinglichen Ausstattung des Raumes und der sozialen Struktur ihrer Benutzer, dargestellt am Beispiel ausgewählter Freiräume und Benutzergruppen im Untersuchungsraum Hannover. **Analysis of the frequentation of open spaces in towns with special regard to their fittings and the social structure of users, exemplified by selected open spaces and selected user groups in the Hannover area.**

2417 Lendholt, W.; Blecken, F. DE 138660/73/0001
Untersuchung über die Frequentierung und nutzungsbezogene Effizienz von Sportflächen unter besonderer Berücksichtigung der physischen Struktur und der Nutzungsorganisation der Sportflächen, der organisationsbeding– ten Erscheinungsformen der Sportaktivität und der sozialen Struktur der Benutzer. **Analysis of the frequentation rate and efficiency of sports–grounds with special regard to their physical structure and using organization, to organizational types of sporting activities and social structure of users.**

2418 Lendholt, W.; Blecken, F. DE 138660/73/0002
Untersuchung der physischen Struktur von Sportflächen, differenziert nach nutzbaren Spielflächen, Erschliessungsflächen, Flächen für Zuschauer und Nebenflächen; und Er– mittlung der sportfunktionalen Kapazität dieser Flächen. **Analysis of the physical structure of sports–grounds, differentiated into useful playgrounds, developable areas, lookers–on areas and accessory areas, and determination of their action capacity.**

2419 Lendholt, W.; Grundmann, V. DE 138660/73/0003
Untersuchungen über Störungen, die von Kinderspielplätzen ausgehen und auf benachbarte Wohnsiedlungen und deren Bewohner einwirken, unter besonderer Berücksichti– gung des Spiellärms. **Disturbances proceeding from children's playgrounds to neighbouring living quarters and their inhabitants, considering playing noise especially.**

2420 Schad, F.; Krüger, H.–J. DE 144900/72/0001 R
Immissionsschutz und Raumordnung im ländlichen Bereich 1971. **Orotection against immission and area planning in rural areas.**

2421 Schnitzer, U. DE 145250/77/0001
Landwirtschaftliche Neubauten im Verbreitungsgebiet der Schwarzwaldhäuser. **New farm buildings in settlement areas of the Black Forest.**

2422 Knauer, N.; Allers, A. DE 148101/77/0001
Landschaftsnutzung und Umweltbelastung – Versuch einer Bewertung in einer grossräumigen Niederungslandschaft. **Utilization of landscape and strain on environment – approach to valuation of a large–area lowland.**

2423 Löffler, H.; Warkotsch, W.; Pospischil, L.
DE 160311/73/0005 N
Methodische Untersuchungen für Modellplanungen von Holzernte– und Walderschliessungsmassnahmen am Beispiel des bayerischen Landkreises Rosenheim. **Research on developing methods for planning of timber harvest and road–networks in the region of Rosenheim– Bavaria.**

2424 Schwertmann, U.; Schmidt, F.; Bader, S.; Becher, H.H.
DE 161020/78/0002 N
Quantitative Erfassung der Bodenerosion. Abschätzung der

D 1610 – Land consolidation and land layout

Erosionsgefährdung. **Quantitative estimation and prediction of threatening soil loss.** Publications.

2425 Gebhard, H.; Biesterfeld, H.; Brennecke, M.
DE 161560/72/0001
Massnahmen zur Umweltgestaltung in ländlichen Gemeinden unter Berücksichtigung der bürgerlichen Selbsthilfe. **Townplanning in country areas and participation.**

2426 Gebhard, H.; Reichenbach, M.   DE 161560/73/0001
Integrierte Zielsysteme der Gemeindeentwicklung als Planungsinstrumente im Stadt–Umland–Bereich. **Analysis and planning of development in rural areas near to towns as integrated systems of structures and processes.**

2427 Grzimek, G.; Heese, L.   DE 161650/74/0004
Quantitative und qualitative Anforderungen der Freiraumgestaltung an die Bebauung im allgemeinen Wohngebiet. **Quantitative and qualitative demands of open space structure planning on the building–up of general residential areas.**

2428 Haber, W.; Bucerius, M.   DE 161670/78/0001 N
Einführung der Kriteriendatei und der Linearen Planungspraxis als Instrumente der Landschaftsdatenbank. **Introduction of parameter filing and linear programming into planning practice as instruments of landscape data bank.**

2429 Thoss, R.; Burgbacher, W.   DE 164350/73/0001
Ein Flächennutzungsmodell für die Bundesrepublik Deutschland. **A model of land use planning in the Federal Republic of Germany.** Publications.

2430 Schreiber, K.–F.; Durwen, K.–J.; Graf, G.
DE 164401/77/0001
Landschaftsökologische Raumgliederung, Erfassung und Bewertung natürlicher Ressourcen im Hinblick auf aktuelle und potentielle Nutzungen. **Ecological classification of landscape, determination and valuation of resources regarding actual and potential use.**

2431 Müller, P.; Tautz   DE 167200/74/0002
Brachefunktionsplan für den Stadtverband von Saarbrücken. **Functional fallow planning in the municipality of Saarbrücken.**

2432 Rhody, B.   DE 202010/75/0001
Methoden für Grossraum– und Intensivinventuren unter Verwendung von Luftaufnahmen. **Methods for inventorizing large and small areas using aerial photographs.**

2433 Meisel, K.   DE 205030/78/0005 N
Möglichkeiten der Sicherung schutzwürdiger Biotope im Rahmen von Flurbereinigungsverfahren. **Possibilities of ensuring biotopes worth protection within land consolidation procedures.**

2434 Häckel, H.   DE 301100/77/0008
Beurteilungsrahmen über die Eignung zur Intensivnutzung bei Almflächen. **Valuation of alpine acreage for suitability for intensive farming.**

2435 Roether, V.; Zundel, R.   DE 501504/72/0004 R
Quantifizierung der Sozialfunktionen des Waldes als Teil der Infrastruktur – Verhalten und Wünsche von Erholungssuchenden in waldreichen Feriengebieten,

untersucht an den BadenWürttembergischen Gemarkungen Baiersbronn, Todtnauberg und Welzheim 1972. **Quantification of social functions of forests as part of infrastructure – behavior and wishes of persons requiring recreation in well–wooded holiday regions as in Baiersbronn, Todtnauberg and Welzheim in Baden–Wuerttemberg.**

2436 Volk, H.   DE 501504/77/0002
Verhalten und Wünsche von Waldbesuchern im Freiburger Raum und im Südschwarzwald. **Reactions and demands of forest visitors in Freiburg area and Southern Black Forest.**

2437 Volk, H.; Schwarz, O.   DE 501504/77/0011
Ökologisches Gutachten zur Flugplatzerweiterung Stuttgart/Echterdingen. **Ecological expert opinion on the enlargement of the airport Stuttgart/Echterdingen.**

2438 Volk, H.; Spahl, H.   DE 501504/78/0007 N
Untersuchungen über die ökologischen Auswirkungen – Flächenverluste, Lärmauswirkungen, andere Folgewirkungen – grossräumiger Verkehrsplanungen auf den Wald. **Studies on the ecological effects – losses of areas, effects of noise, other consequences – of large–area traffic planning on forests.**

2439 Dietz, P.; Meng, W.   DE 501505/78/0003 N
Feinerschliessung von Beständen – Planung von Maschinenwegnetzen. **Fine method of making stands accessible – planning of machine road network.** Publications.

2440 Dancau, B.; Braun, W.; Schuch, M.; Beck, T.; Bauchhenss, J.   DE 502055/77/0001
Berücksichtigung ökologischer Belange bei der Weinbergsflurbereinigung. **Consideration of ecological moments in consolidation of vineyard plots.**

2441 Dieter, A.   DE 502153/70/0001
Versuche zum Erosionsschutz im Weinbau. **Measures against erosion of steep slopes in grape–vine cultures.**

2442 Henne, A.; Rödig, K.P.   DE 506404/77/0006
Vorränge in der Landnutzung. Die Waldflächen in der Regional– und Forstplanung. **Priorities in land–use. Forest areas in regional and forest planning.** Publications.

2443 Helles, F.   DK 030181/76/0002
Arealanvendelse til land– eller skovbrugsformål. **Area use for agricultural or forestry purposes.**

2444 Lillelund, H.   DK 030181/76/0006
Sylvo–pastoral management.

2445 Folving, S.   DK 030700/79/5135 N
Landskabsøkologisk beskrivelse med henblik på ressourcepotentiale og bæreevne. Evaluering i forbindelse med fysisk planlægning og landskabsproduktion. **Ecological description of the resource potential of the landscape. Evaluation in connexion with physical planning and agricultural production.**

2446 Lee, J.   IE 060200/75/1078 R
The interrelationship between technical productivity of land, market value and griffith valuation of land in Ireland. Publications.

2447 Sorbi, U.   IT 040526/74/0633
Ricerca intorno agli aspetti estimativi della moderna normativa

territoriale anche in relazione al comportamento preferenziale privato nel settore agrario. **Research on land valuation aspects of modern land legislation also in relation with preferential private behaviour in the agricultural sector.**

2448 Sorbi, U.   IT 040526/74/0634
Recente legislazione urbanistico–territoriale come fattore di adattamento o di modifica della teoria estimativa. **Recent legislation on land and urban area management as a factor of adjustment or modification of land valuation theoretics.**

2449 Agostini, D.   IT 040804/74/0501
Indagine comparativa sui problemi dell'utilizzazione e gestione del territorio agricolo in Italia e in Inghilterra. **Comparative investigation on problems of agricultural land use and management in Italy and England.**

2450 Pagella, M.   IT 041205/74/0594
Individuazione e collaudo di una metodologia atta a definire la dimensione economica valida dell'azienda agraria in presenza di vincoli ambientali, economici e sociali. **Establishment and testing of a methodology for estimating the economically viable farm size in the event of environmental, economic and social condition factors.**

2451 Boer, Th.A. de   NL 010102/79/8934 N
Vegetatietypologie gericht op indicatie van groeiomstandigheden, biomassaproductie en visuele aspekten, speciaal van lintvormige begroeiingen. **Vegetation typology to indicate environmental factors, biomass production and visual aspects in particular of linear vegetation.**

2452 Douw, L.   NL 010116/50/0123 N
Onderzoek in ruilverkavelingsgebieden (sociaal–economische schetsen en verkenningen). **Socio–economic research into regions under land consolidation schemes.** Publications.

2453 Bauwens, A.L.G.M.   NL 010116/71/3682 R
Oriënterend onderzoek naar de betekenis van het gemeentelijke planologische beleid voor de ontwikkelingsmogelijkheden in de landbouw. **Trial investigation into the significance of the planning policy of local authorities for the possibilities of development in agriculture.**

2454 Bauwens, A.L.G.M.   NL 010116/73/4465 R
De toekomst van het tuinbouwgebied Vleuten–de Meern in het kader van planologische en tuinbouwstructurele ontwikkelingen. **The future of the horticultural area Vleuten–de Meern in the framework of the developments as regards the town and country planning and horticultural structure.**

2455 Linden, P. van de   NL 010116/73/5937 R
Regionaal gedifferentieerde studie van beperkingen, die de landbouw in het kader van gemeentelijke bestemmingsplannen in het Noorden en Oosten van ons land opgelegd krijgt. **Regional investigation into restrictions imposed on agriculture by municipal planning in the northern and eastern part of the Netherlands.** Publications.

2456 Douw, L.   NL 010116/74/6534 R
Sociaal economisch onderzoek t.b.v. de ontwikkeling van methoden ter bepaling en afweging van ruimtelijke aanspraken in een landschappelijk en natuurwetenschappelijk waardevol gebied (Midden–Brabant). **Socio economic research for the development of methods for determination and evaluation of**

multiple land–use in areas with high landscape and nature values.

2457 Bauwens, A.L.G.M.   NL 010116/75/6882 R
Voorkomen van agrarische gebouwen op de Monumentenlijst en op de lijst van beschermde dorpsgezichten in Zuid–Limburg en de invloed hiervan op de sociale en economische ontwikkeling van bedrijven. **Occurence of agricultural buildings in South Limburg which are scheduled as buildings of historic interest and which are scheduled as protected (village) views and the influence of it on social and economic development of farms.**

2458 Bauwens, A.L.G.M.   NL 010116/75/6930
De bestemming van vrijkomende agrarische gebouwen. **The destination of agricultural buildings by farm liquidation.**

2459 Bauwens, A.L.G.M.   NL 010116/76/7696 R
Agrarisch planologisch onderzoek op regionaal niveau ten behoeve van de voorbereiding van struktuurplannen en bestemmingsplannen buitengebied. **Regional and socio–economic research in connection with physical and structural aspects of town and country planning.**

2460 Bauwens, A.L.G.M.   NL 010116/76/7697 R
De agrarische bedrijfstak in de overgangsgebieden tussen stad en platteland. **Structural developments of agriculture in the peripheric zones between town and country.**

2461 Linden, P. van de   NL 010116/77/8569 N
De betekenis van de bestemmingsplannen voor de ontwikkelingsmogelijkheden van rundvee– en varkensbedrijven in het westelijke kleien veenweidegebied. **The significance of the planning policy of local authorities for the development possibilities of farms with cattle and pigs in the western grassland– on clay and peat areas of the Netherlands.**

2462 Bauwens, A.L.G.M.   NL 010116/77/8572 N
Onderzoek naar de agrarische bedrijfstak ten behoeve van streekplannen. **Structural developments of the agricultural branch in relation to land use planning.**

2463 Oosterveld, H.R.   NL 010116/78/8901 N
Agrarisch planologisch onderzoek in de bufferzones van Zuid Limburg. **Agricultural developments in connection with town and country planning in the buffer area of Zuid–Limburg.**

2464 Oostrom, L.G.J. van   NL 010501/67/0842
Bedrijfsaspecten in verband met de bedrijfsverkaveling in tuinbouwgebieden. **Aspects of farm management in horticulture areas in relation with the layout.** Publications.

2465 Michels, Th.   NL 010501/67/0844 R
Verkeersproductie in plattelandsgebieden in verband met dorpsstructuur en wegpatroon. **Traffic production in rural areas in relation with village structure and road pattern.**

2466 Locht, L.J.   NL 010501/67/0852
Mobiliteit van arbeid en vermogen in verband met cultuurtechnische ingrepen. **Mobility of labour and capital in connection with land consolidation.**

2467 Locht, L.J.   NL 010501/67/0855 R
Economische beoordelingstechnieken in verband met ruilverkavelingen. **Techniques for projectevaluation in**

D 1610 – Land consolidation and land layout

connection with landconsolidation projects.

2468 Visser, A.C.   NL 010501/68/2464
Het optimale plan van landbouwwegen in ruilverkavelingen bij
uiteenlopende investeringsbedragen. **The optimum network of
rural roads in amelioration areas at various investment levels.**

2469 Rheenen, J. van   NL 010501/71/3723 R
Factoren die van invloed zijn op het ontstaan van
suburbanisatie in het algemeen en in het bijzonder op de groei
van woongebieden in de Randstad Holland. **Factors
influencing suburbanization, with special reference to the
growth of built–up areas of the municipalities along the rim of
the province Zuid–Holland ("circle–town Holland").**
Publications.

2470 Locht, L.J.   NL 010501/72/4370 R
Economie van de openluchtrecreatie. **Economics of outdoor
recreation.**

2471 Locht, L.J.   NL 010501/72/4371 R
Economie van milieu en natuur. **Economics of environment.**

2472 Righolt, J.W.   NL 010501/72/4373
Bedrijfstechnische aspecten van de landinrichti ₂g. **Farm
economical aspects of land layout.** Publications.

2473 Righolt, J.W.   NL 010501/72/4374 R
Bedrijfseconomische waardering van de herinrichting van
agrarische gebieden. **farm economical evaluation of planned
changes in land layout.** Publications.

2474 Righolt, J.W.   NL 010501/72/4375
Invloed van beperkingen t.b.v. natuur, milieu en landschap op
landbouwkundige exploitatiemogelijkheden. **Farm economy in
areas with restricted agricultural management in connection
with land lay–out.** Publications.

2475 Michels, Th.   NL 010501/73/5246
Ontwikkeling van een wiskundig model van de relatie tussen
verkeer en grondgebruik in plattelandsgebieden. **Building a
mathematical model of the relationship between traffic and land
use in rural areas.**

2476 Alderwegen, H.A. van   NL 010501/73/5250 R
Inrichting van openluchtrecreatieprojecten. **Layout of outdoor
recreation projects.** Publications.

2477 Alderwegen, H.A. van   NL 010501/73/5252 R
Recreatie in Midden–Delfland. **Recreation in the
"midden–Delfland" area.** Publications.

2478 Verweij, E.J.   NL 010501/73/5256 R
Vraag en aanbodsanalyse van verblijfsrecreatieterreinen.
Supply and demand analysis of recreative dwelling sites.

2479 Bijkerk, C.   NL 010501/73/5257
Cultuurtechnische Inventarisatie Nederland; onderzoek en
ontwikkeling. **Land division survey Netherlands; research and
development.** Publications.

2480 Kik, R.   NL 010501/73/5261 R
Ontwikkeling van methoden voor het maken van het plan van
toedeling voor ruilverkavelingen. **Methodology of making
allotment plans for consolidation projects.** Publications.

2481 Linthorst, Th.J.   NL 010501/73/5262 R
Analyse van de uitvoeringstechnieken bij de uitvoering van
landinrichtingsplannen. **Analysis of execution techniques in
land consolidation projects.** Publications.

2482 Linthorst, Th.J.   NL 010501/73/5263 R
Toepassing van rekentechnieken bij grondverzet. **Application
of calculation techniques to moving earth.**

2483 Oosterom, C.J.G. van   NL 010501/73/5269 R
Ontwikkeling van methoden ter bepaling en afweging van
ruimtelijke aanspraken in een landschappelijk en
natuurwetenschappelijk waardevol gebied. **Development of
methods for determination and evaluation of multiple land–use
in areas with high landscape and nature values.**

2484 Oostrom, C.G.J. van; Vink, L.W.   NL 010501/73/5271
Structuur van tuinbouwgebieden. **Structure of horticultural
areas.**

2485 Michels, Th.   NL 010501/73/5301 R
Rijsnelheden en reistijden van motorvoertuigen op
plattelandswegen in relatie met aard van de weg en
weersomstandigheden en intensiteit en samenstelling van het
verkeer. **Traffic speeds and travel times on rural roads in
relationship with road characteristics and weather conditions
and traffic density and modal split.**

2486 Heester, J.   NL 010501/73/5302 R
Inrichting verblijfsrecreatieterreinen. **Layout of sites for
recreative dwellings.**

2487 Alderwegen, H.A. van   NL 010501/73/8880 N
Bepaling van capaciteitsnormen van recreatievoorzieningen.
**Determination of capacity criteria for outdoor recreation
facilities.** Publications.

2488 Alderwegen, H.A. van   NL 010501/73/8882 N
Regionale analyse van vraag naar en aanbod van
recreatievoorzieningen. **Regional analysis for outdoor
recreation.** Publications.

2489 Michels, Th.   NL 010501/75/6650
Streekdorpen en lintbebouwing in de provincie Groningen.
Ribbon development of villages in the Province Groningen.

2490 Linthorst, Th.J.   NL 010501/77/7599 R
Onderzoek om te komen tot calculatienormen voor het
uitvoeren van egalisaties in het Noordelijk Kleimozaïekgebied.
**Cost norms for levelling in a clay region with a mosaic
parcellation in the northern Netherlands.**

2491 Boels, D.   NL 010501/78/8340
Beslissingsmodel ten behoeve van allocatie zandwinobjecten in
verband met transportkosten. **Decision model to locate
sand–pits in connection with transport costs.**

2492 Alderwegen, H.A. van   NL 010501/78/8881 N
De relatie tussen recreatiepatroon en weersgesteldheid.
**Assessment of the relationship between weather and
recreation pattern.** Publications.

2493 Verweij, E.J.   NL 010501/78/8883 N
Problemen samenhangend met verschijnselen van
verstedelijking in landelijke gebieden. **Problems connected
with processes of urbanisation in rural areas.**

2494 Rheenen, J. van   NL 010501/79/9054 N
Analyse van modellen in de landinrichting door literatuurstudie. **Land use models in rural planning.**

2495 Rheenen, J. van   NL 010501/79/9055 N
Ontwikkeling van methoden voor planvorming op basis van vier sector (deel) adviezen ten behoeve van een structuurschets voor het reconstructiegebied Midden–Delfland. **Methodical framework of a synthesis model for the reconstruction of the Midden–Delfland area.**

2496 Dijkstra, H.   NL 010601/75/6873
Beschrijving en waardering van boerderijen en erven in de Schermer. **Description and evaluation of farms in the "Schermer".**

2497 Boekhorst, J.K.M. te   NL 010601/76/6667
De ruimtelijke invloeden van nederzettingen op hun omgeving. **The spatial influence of human settlements on the environment.**

2498 Boekhorst, J.K.M. te; Harms, W.B.   NL 010601/76/6668
Inventarisatie en evaluatie van bestaande planningsmodellen en ontwerpen op basis van ecologische begrippen. **Inventory and evaluation of existing planning models and plans in ecological terms.**

2499 Goor, C.P. van   NL 010601/77/7878
Bosbouw in relatie tot landinrichting en plattelandsontwikkeling in ontwikkelingslanden. **Forestry in relation to land–use planning and rural development in developing countries.**

2500 Volker, C.M.   NL 010601/77/7879
Voorstudie maatschappelijke achtergronden van landschapsveranderingen. **Pilot study social backgrounds of environmental changes in rural areas.**

2501 Dijkstra, H.   NL 010601/78/7883 R
Onderzoek naar visuele gevolgen van boerderijverplaatsing in de ruilverkaveling Eemland; een bijdrage aan de situering van boerderijen. **–Research on the visual impact of farm buildings in the Eemland consolidation; a contribution to the location of farm uildings.**

2502 Vos, W.   NL 010601/78/8675 N
Voorstudie naar landschapsecologische relaties en de betekenis ervan voor de nationale ruimtelijke planning. **Pilot study on landscape ecological relationships and their significance for the national physical planning.**

2503 Perelman, R.   NL 010601/78/8676 N
Landschapsbouwkundige aspecten van kustgebieden in Europa. **Research on coastal landscapes in a European context.**

2504 Houte de Lange, S.M. ten   NL 010602/77/8595 N
Literatuuronderzoek naar methoden van biologische inbreng in de ruimtelijke ordening in landen buiten Nederland. **Significance and use of biological data in physical planning outside the Netherlands.**

2505 Opdam, P.F.M.   NL 010602/77/8597 N
Onderzoek naar de toepassing van ornithologische gegevens in de planologie. **An investigation of the use of ornithological data in environmental planning.**

2506 Kalkhoven, J.T.R.   NL 010602/77/8601 N
Onderzoek naar de samenhang tussen botanische kenmeken en overige kenmerken van het landschap t.b.v. ruimtelijke planen. **Research on the relationship between the vegetation and other biotic features for the landscape for land–use planning.**

2507 Opdam, P.F.M.   NL 010602/77/8602 N
Ontwikkeling van een methode tot het onderscheiden van vogelgemeenschappen t.b.v. de toepassing in de planologie. **Development of a method for discerning the pattern of bird communities in the landscape for application in land–use planning.**

2508 Jaarsma, C.F.   NL 020012/71/4402
Het bepalen van de verkeersproductie en de verkeerskarakteristieken van plattelands–gebieden en de daarin gelegen bedrijven, rekreatie–objecten, etc. **Determination of production, distribution and nature of traffic in rural areas in connection with agricultural holdings, recreation projects, etc.** Publications.

2509 Lier, H.N. van   NL 020012/77/8645 N
Onderzoekprogramma projectontwikkelingen Landinrichtingsmodel (O.P.O.L.). **Research program to develop land–use–planning method.**

2510 Kleefmann, F.   NL 020040/77/8397
Kennis en methode in de planologie; een onderzoek naar voorwaarden van theorievorming voor de ruimtelijke ordening. **Knowledge and method in planning theory; an investigation into the preconditions of building up theories in physical planning.** Publications.

2511 Kleefmann, F.   NL 020040/78/8867 N
Meenhoven, geen landinrichtingsrecept, maar richting zoeken. **Meenhoven, a goalseeking project instead of a receipt for landuse planning.**

2512 Hidding, M.C.   NL 020040/78/8869 N
Regionale ontwikkeling N.W. Overijssel. **Regional development N.W. Overijssel.**

2513 Groot, J.P. de   NL 020051/70/4262
Snelgroeiende plattelandskernen (plaats, oorzaken van de groei, gevolgen voor de sociale structuur). **Rapid growing rural communities (place, causes of growth, consequences for the social structure).**

2514 Grand, H. le; Elzas, M.S.   NL 020066/73/4898
Onderzoek naar de programmagrondslagen voor de toepassing van interactieve grafische methoden bij het ontwerpen van wooneenheden en landschappen en bij de planologie. **A general programming system for computer aided architectural design and planning.** Publications.

2515 Knoop, H.   NL 020067/77/8872 N
De invloeden van kleur– en textuur op de waardering voor het woonhuis en de omgeving. **The influence of colour and texture on the appreciation of the house and the house–environment.**

2516 Alberts, F.W.   NL 040007/60/4321
Bestemming en inrichting landelijk gebied van Oostelijk Flevoland. **Land use of Oostelijk Flevoland.**

D 1610 – Land consolidation and land layout

2517 Hoeve, H.　　　　　　　NL 040007/60/4329
Onderzoek naar de mogelijke gevolgen van de inpoldering van
Oostelijk en Zuidelijk Flevoland in de randgebieden op het
'oude land'. **Possible influence of the reclamation of Oostelijk
and Zuidelijk Flevoland on the adjacent areas.**

2518 Alberts, F.W.　　　　　　NL 040007/65/4322
Bestemming en inrichting landelijk gebied van Zuidelijk
Flevoland. **Land use of Zuidelijk Flevoland.**

2519 Alberts, F.W.　　　　　　NL 040007/68/4312
Ontwikkelingsmogelijkheden voor Zeewolde als
recreatie–woonkern. **Development possibilities for Zeewolde as
a recreation and living centre.**

2520 Alberts, F.W.　　　　　　NL 040007/68/4323
Onderzoek naar het recreatieverkeer in Flevoland. **Recreation
traffic in Flevoland.**

2521 Polman, G.K.R.　　　　　　NL 040007/69/4297
Technische en economische aspecten van de inrichting van
natuurgebieden in Flevoland en het Grevelingengebied.
**Technical and economical aspects of lay–out and maintenance
of nature reserves in Flevoland and in the Grevelingen area.**

2522 Alberts, F.W.　　　　　　NL 040007/69/4319
Inventarisatie jachthavens langs de randmeren en het
IJsselmeer. **Inventory of yacht havens along the border lakes
and the IJsselmeer.**

2523 Polman, G.K.R.　　　　　　NL 040007/70/4295
De wenselijkheid en de situering van natuurterreinen. **The
desirability and the location of nature reserves.**

2524 Alberts, F.W.　　　　　　NL 040007/70/4316
Inrichtingsplan Markerwaard. **Land use planning of the
Markerwaard.**

2525 Alberts, F.W.　　　　　　NL 040007/72/4313
Het opstellen van een bestemmingsplan voor het landelijk
gebied van Lelystad. **Preparing the physical plan of the rural
area of Lelystad.**

2526 Alberts, F.W.　　　　　　NL 040007/73/8474
Onderzoek recreatieverkeer
Lauwerszeegebied/Grevelingenbekken. **Research on
recreation traffic in the Lauwerszee and Grevelingen area.**

2527 Hoeve, H.　　　　　　　NL 040007/74/8499
Economisch evaluatie van inrichtingsalternatieven voor
bepaalde gebieden. **Economic evaluation of lay–out
alternatives.**

2528 Nagtegaal, P.　　　　　　NL 040007/76/8471
Onderzoek naar de behoefte aan recreatieve voorzieningen
t.b.v. de inrichting van het Grevelingenbekken. **Research on
the need for recreational resources in relation to land use
planning of the Grevelingen area.**

2529 Hoeve, H.　　　　　　　NL 040007/76/8480
Inrichtingsstudie voor het IJmeer. **Land use planning of the
IJmeer area.**

2530 Tigelaar, L.　　　　　　NL 040007/76/8500
Onderzoek systematische planvorming structuurplannen.
Research on systematic structural planning.

2531 Feitsma, K.S.　　　　　　NL 040007/77/8467
Inrichting van de internucleaire ruimte van Almere. **Land use
of open spaces outside Almere.**

2532 Tigelaar, L.　　　　　　NL 040007/77/8470
De recreatieve funktie van het IJsselmeergebied in de regio.
**The recreational function of the IJsselmeer area for the
surrounding areas.**

2533 Tigelaar, L.　　　　　　NL 040007/77/8481
Onderzoek naar de behoefte aan recreatieve voorzieningen
t.b.v. de inrichting van het Oosterscheldegebied. **Research on
the need for recreation resources in relation to land use
planning of the Oosterschelde area.**

2534 Tigelaar, L.　　　　　　NL 040007/77/8492
Struktuurplan IJsselmeergebied. **Land use planning of the total
IJsselmeer area.**

2535 Tigelaar, L.　　　　　　NL 040007/78/8494
Evaluatie plannen Markerwaard. **Evaluation land use plans of
the Markerwaard.**

2536 Alberts, F.W.　　　　　　NL 040007/78/8497
Onderzoek naar de mogelijkheden voor het stichten van een
d.m.v. afvalwarmte te verwarmen kassencomplex in de
buitenruimte van Almere. **Research on the possibilities for
founding of glasshouses in the open spaces of the area outside
Almere, which can be heated by using waste heat.**

2537 Held, J.J. den　　　　　　NL 060006/77/8385
Onderhoudensarm stedelijk groen op natuurlijke basis.
**Introduction of spontaneous vegetation in urban green areas as
a means for reducing costs.**

D 1620 – Landscape gardening

See also 368, 393, 2086, 2300, 2322, 2497, 2498, 2514, 2644,
3624, 4803, 10104, 17345, 18217

2538 Debuisson, J.; Neuray, G.　　　　BE 010004/72/0004
Etude des plantations dans le paysage des régions rurales.
Study of rural landscape planting. Publications.

2539 Huygh, A.; Baeyens, L.; Berben, J.; Berten, R.; De
jamblinne de Meux, A.; Nef, L.　　　BE 140000/75/0017 R
Herstel en heraanpassen van landschappen na
infrastructuurwerken. **Landscape gardening after public works.**
Publications.

2540 Pflug, W.; Beckmann, R.　　　　DE 101100/77/0003
Zur Bedeutung der Hausschutzhecken im Monschauer Land
unter besonderer Berücksichtigung ihrer klimatischen
Auswirkungen. **On the importance of house protective hedges
in the Monschau region with special regard to climatological
effects.**

2541 Sukopp, H.; Markstein, B.　　　　DE 105060/77/0002
Zur Bedeutung ökologischer Kriterien für die
Landschaftsplanung am Beispiel der Gewässerlandschaft der
Havel in Berlin – West –. **On the value of ecological criteria for
landscape planning, e.g. the riverside landscape of the Havel in
Berlin – West –.** Publications.

2542 Kiemstedt, H.; Holtz, E.; Hübler, K.H.

D 1620 – Landscape gardening

DE 105650/77/0001
Ziele, Aufgaben und Durchsetzungsmöglichkeiten der
Landschaftsentwicklung der Bundesrepublik Deutschland und
Frankreichs. **Aims, tasks and carrying–through of landscape
development in the Federal Republic of Germany and in
France.**

2543 Kiemstedt, H.; Hahn–Herse, G. **DE 105650/77/0002**
Ermittlung der Integration von landschaftspflegerischen
Begleitplänen zu Flurbereinigungsverfahren in das
hierarchische System der Landschaftsplanung. **Determination
of integration of landscape management accompanying–plans
for land consolidation measures into the hierarchic system of
landscape planning.**

2544 Kiemstedt, H.; Albrecht, D. **DE 105650/77/0003**
Landschaftsplanung in der Volksrepublik China. **Landscape
planning in the People's Republic of China.**

2545 Richard, W. **DE 105650/78/0001 N**
Entwurfsmethoden und Gestaltungsansätze der
Objektplanung in der Landschaftsentwicklung – Arbeitstitel –.
**Methodes of projects and approaches to object planning in
landscape development.** Publications.

2546 Hanisch, J. **DE 105650/78/0002 N**
Landschaftsplanerischer Beitrag zur ökologischen Planung am
Beispiel des "Ökologischen Gesamtlastplanes
UnterelbeKüstenregion". **Contribution to ecological landscape
planning demonstrated by the "Ecological overall plan on
Lower Elbe coastal region".**

2547 Ermer, K.; Kellermann, B.; Schneider, C.
DE 105650/78/0003 N
Erarbeitung wissenschaftlich–methodischer Grundlagen –
Ziele, Inhalte, Verfahrensweisen – zur Erstellung eines
Landschaftsprogrammes für Berlin. **Elaboration of
scientific–methodical fundamentals – objectives, contents,
methods – for the development of landscape programme on
Berlin.**

2548 Skirde, W. **DE 129220/77/0002**
Golfsportanlage und Landschaftsentwicklung. **Golf
sports–ground and landscape management.** Publications.

2549 Tebrügge, F. **DE 129400/75/0011**
Einsatz von Mulchgeräten in der Landschaftspflege –
Auswirkungen auf Pflanzenbestand, Nährstoffverhältnisse und
Grundwasser. **Utilization of mulching equipment in landscape
management – effects on plants, nutrient balance and ground
water.**

2550 Wohlrab, B.; Ehlers, M. **DE 129450/75/0005**
Modelle für die planmässige Reintegration von Abgrabungen
in die umgebende Kulturlandschaft. **Models for systematic
reintegration of diggings off into the surrounding countryside.**

2551 Zundel, R.; Köpp, H. **DE 132870/72/0002 N**
Verwaltung und Organisation der britischen Nationalparks
und Forstparks als grossräumige Freizeit– und
Erholungsgebiete; eine vergleichende Untersuchung zu
deutschen Naturparks. **Management and organization of
British national parks and forest parks as large–area leisure and
recreational parks; a comparative study on German nature
parks.**

2552 Knauer, N.; Gerth, H. **DE 148101/75/0001**
Produktionssteuerung in Landschaftspflegegebieten durch
Beeinflussung der Vegetationsentwicklung mittels
verschiedener Bewirtschaftungsverfahren. **Production control
in landscape management areas by taking influence on
vegetation development with different systems of cultivation.**

2553 Isensee, E.; Grimm, R. **DE 148150/78/0002 N**
Möglichkeiten von Knickpflegemassnahmen unter besonderer
Berücksichtigung der ökologischen Belange. **Practicabilities to
take charge of quickset hedges in special consideration of the
ecological meaning.**

2554 Runge, M.; Helming, W. **DE 164050/78/0002 N**
Entwicklung bleitoleranter pflanzlicher Populationen an
Strassenrändern. **Evolution of lead–tolerant plant populations
alongside roads.**

2555 Mrass, W.; Koeppel, H.–W.; Arnold, F.
DE 205010/74/5004
Entwicklung und Aufbau eines
Landschaftsinformationssystems auf der Grundlage einer
rasterbezogenen Flächendatenbank. **Development and design
of a landscape information system based on a grid–unit area
data bank.**

2556 Koeppel, H.–W.; Arnold, F. **DE 205010/77/0001**
Erstellung eines benutzerfreundlichen modularen
Programmpaketes für das Landschaftsinformationssystem.
**Preparation of a modular user–adapted packet of programmes
for landscape information system.**

2557 Henke, H.; Krause, C.L. **DE 205010/77/0002**
Entwicklung einer Methode zur ökologischen
Bestandsaufnahme und zur Auswertung planungsrelevanter
ökologischer Unterlagen. **Development of a method for
ecological survey and for evaluation of ecological documents
with relevance to planning.**

2558 Krause, C.L.; Henke, H. **DE 205010/77/0003**
Beurteilung der Eignung von Landschaftsfaktoren und
Auswirkungen von Nutzungsformen auf Landschaftsfaktoren.
**Valuation of suitability of landscape factors as affected by land
use systems.**

2559 Mrass, W.; Handke, M. **DE 205010/77/0006**
Sammlung, Auswertung und Dokumentation von statistischen
Daten, Planungsgrundsätzen und Rechtsfragen zu den
Naturparken. **Collection, evaluation and documentation of
statistical data, planning principles and legal problems on
nature parks.**

2560 Bürger, K. **DE 205010/77/0007**
Möglichkeiten und Grenzen der Ermittlung der Belastbarkeit
des Naturhaushaltes und die Bedeutung für die
Landschaftsplanung. **Possibilities and limits of determining the
carrying capacity of natural balance and importance to
landscape planning.**

2561 Volk, H.; Spahl, H. **DE 501504/77/0003**
Erarbeitung eines landschaftspflegerischen Begleitplanes für
das Flurbereinigungsgebiet Bahlingen a.K.. **Establishment of a
landscape management accompanying plan for land
consolidation region of Bahlingen o.K..**

2562 Henne, A.; Heuser, V. **DE 506404/77/0001**

D 1620 – Landscape gardening

Automatisiertes Landschaftsinformationssystem – Auswertung der Flächenschutzkarte Hessen –. **A computer assisted landscape information system, based on the "Flächenschutzkarte Hessen".**

2563 Damour, L.; Bourgoin, B.; Auge, P.; Chevallier, C.; Allemand, P.; Turpaud, Y.   FR 012201/73/9056
Adaptation végétale des plantes gazonnantes et arbustives sur le littoral Centre–Ouest. **Testing the adaptability to the west central coast–line of turfing and shrubby plants.**

2564 Ricou, G. Mme; Masclet, A.   FR 012220/75/5286
Aménagement des paysages du parc naturel régional de Brotonne. **Landscape remodelling in the regional Park of Brotonne.**

2565 Veer, A.A. de   NL 010119/73/4796
Onderzoek naar methoden van landschapskartering en –waardering. **Investigations on methods for landscape mapping and evaluation.**

2566 Veer, A.A. de   NL 010119/76/7791
Opbouw van een informatiesysteem voor landschapsbeeldgegevens. **Preparation data base landscape physiognomy.**

2567 Schelling, J.   NL 010119/78/8307 R
Landschappelijke aspecten van de opgaande begroeiing in Nederland. **Visual aspects of trees in the Dutch landscape.**

2568 Schelling, J.   NL 010119/78/9071 N
Inventarisatie en classificatie van de Nederlandse beken en laaglandstromen met behulp van geomorfologische, cultuurhistorische en fysiognomische criteria. **Inventory and classification of the Dutch streams by means of geomorphological, cultural–historical and physiognomical parameters.**

2569 Visser, A.C.   NL 010501/73/5258 R
Ontwikkeling van methoden voor de inventarisatie van landschappelijke kenmerken. **Methodology of making inventories of landscape aspects.** Publications.

2570 Dijkstra, H.   NL 010601/73/6661
De voorbereiding en uitvoering van landschapsplannen in ruilverkavelingen. **Landscape planning research in reallotment projects.** Publications.

2571 Berg, A. van den; Toorn, M. van den; Vrijlandt, P.   NL 010601/73/6662
Ontwikkeling van methoden ter bepaling en afweging van ruimtelijke aanspraken in een landschappelijk en natuurwetenschappelijk waardevol gebied (Midden–Brabant) – landschapsonderzoek. **Development of methods for determination and evaluation of multiple land–use in areas with high landscape and nature values – landscape planning research.**

2572 Dijkstra, H.   NL 010601/74/6657
De visuele aspecten van het landschap in een landinrichtingsproject (Midden Brabant). **Visual effects of land reallotment activities on the landscape (Mid Brabant).** Publications.

2573 Harms, W.B.   NL 010601/74/6664
Landschapsecologisch onderzoek m.b.v. vegetatiekundige gegevens in een landinrichtingsgebied. **Landscape ecological research based on vegetation data in a development area.**

2574 Coeterier, J.F.   NL 010601/75/6677
De betekenis van de gebouwde en landschappelijke omgeving voor bewoners van nieuwe stadsuitbreidingen. **Research on the perception of the rural environments by inhabitants of new town settlements.** Publications.

2575 Coeterier, J.F.   NL 010601/75/6678
Midden–Brabant – onderzoek naar de waardering van het landschap door bewoners en gebruikers. **Research on the perception of the landscape by inhabitants and users in Central–Brabant.**

2576 Harms, W.B.; Boekhorst, J.K.M. te   NL 010601/75/7682
Natuurbouw in het kader van landschapsbouw: begripsbepaling en probleemverkenning. **Ecological engineering in the scope of landscape planning.**

2577 Harms, W.B.   NL 010601/75/8668 N
Voorstudie naar de gevolgen van ruilverkavelingen voor het landschap. **Pilot study on the consequences of land consolidation for the landscape.**

2578 Boekhorst, J.K.M. te; Harms, W.B.   NL 010601/76/6666
De mogelijkheden van een ecologische landschapsbeschrijving. **Possibilities for an ecological description of the landscape.**

2579 Coeterier, J.F.   NL 010601/76/6763 N
Onderzoek naar de waarneming en de waardering van het landschap naar visuele kenmerken t.b.v. een nationale kartering van het landschapsbeeld. **Research on landscape perception and evaluation for a national survey of the image of the landscape.** Publications.

2580 Coeterier, J.F.; Boekhorst, J.K.M. te; Schipper–Anderson, E.A.; Teer, M.   NL 010601/77/7543
Ontwikkelen van ontwerpcriteria voor de vormgeving van urbane perifere overgangszones (UPOZ). **Development of criteria for designing urban peripheric zones.**

2581 Vrijlandt, P.   NL 010601/78/7957 N
Landschappelijke criteria voor de ruimtelijke planning van elektriciteitstransport (fase 1). **Guide lines for the planning of transmission lines in the Landscape (phase 1).**

2582 Boekhorst, J.K.M. te   NL 010601/78/7974
Ontwikkelen van criteria voor de planning van bossen ten behoeve van het woonmilieu. **Development of guidelines for the design of forests for the benefit of the living environment.**

2583 Tideman, P.   NL 010601/78/8666 N
Een poging tot het fenomenologisch beschrijven van landschap t.b.v. de ruimtelijke planning. **A phenomenological way of description of the landscape on behalf of planning.**

2584 Harms, W.B.   NL 010601/78/8669 N
Een onderzoek naar de relatie tussen classificaties van bossen vanuit ecologische en omgevingspsychologische invalshoek. **Ecology and perception of the environment in classification of forests.**

2585 Teer, M.   NL 010601/78/8764 N
Landschappelijke criteria voor de ruimtelijke planning van de

weg in het landschap. **Guidelines for the planning of the road in the landscape.**

2586 Kerkstra, K.   NL 020029/77/8813 N
De ontwikkeling van het natuurbegrip, en de betekenis
daarvan voor inrichting en vormgeving. **The development of the
concept of nature, and its influence upon landscape
development and design.**

2587 Struik, J.B.   NL 020029/77/8814 N
De vormpotenties van een aantal bodemgebruiksprocessen.
The landscape design opportunities of some landuse processes.

2588 Kerkstra, K.   NL 020029/78/8816 N
Landschapbeschrijving en analyse ten behoeve van ruimtelijke
planning en vormgeving. **Landscape description and analysis.**

2589 Warnau, H.   NL 020029/78/8817 N
Instandhouding en reconstructie van het Vondelpark.
Conservation and reconstruction of the "Vondel"park.

2590 Warnau, H.   NL 020029/78/8818 N
Recente geschiedenis van de tuin– en landschapsarchitectuur,
in Nederland. **Recent history of the Dutch garden and
landscape design.**

2591 Struik, J.B.   NL 020029/78/8819 N
Landschapsplanning in kustgebieden. **Landscape planning in
coastal areas.**

2592 Boerwinkel, H.W.J.   NL 020048/74/7567
Ontwikkeling en psychologische bijdragen aan planning,
vormgeving en beheer van de groene omgeving. **Development
of psychological contributions to planning, design and
management of the green environment.**

2593 Loenen, M.   NL 040007/76/8478
Oecologische landschapswaardering van gebieden in de
IJsselmeerpolders. **Ecological classification and evaluation of
landscapes in the IJsselmeer area.**

2594 Sieben, W.H.   NL 040007/77/8485
Onderzoek landschapsbeeld in de IJsselmeerpolders in relatie
tot bodem en waterhuishouding (landschapsbeeldkartering).
**Research on landscape physiognomy of the IJsselmeerpolders
in relation to soil and water management (landscape mapping).**

D 2100 – Plant production general and crop husbandry

See also 106, 441, 1075, 1125, 11481, 16641

2595 Ferauge, M.; Siaens, M.　　BE 010002/60/0004
Etude de la couverture du sol en culture fruitière. **Study of the soil coverture in fruitculture.**

2596 Nyns, E.J.; Naveau, H.; Binot, R.　　BE 020105/78/0005
Production de méthane par digestion anaérobie d'algues cultivées dans les eaux de refroidissement des centrales électriques. **Methane production through anaerobic digestion of algae grown in cooling waters from powerplants.**

2597 Van Parijs, R.; Callebout, A.　　BE 030006/76/0001
Werkingsmechanismen van natuurlijke groeistoffen in hogere planten, het optreden van somatische polyploidie. **Biochemical mode of action of growth substances in higher plants; the occurrence of somatic polyploidy in plants.** Publications.

2598 Vlassak, K.; Reynders, L.　　BE 040201/75/0007 N
Associatieve stikstoffixatie door Azospilillum sp. **Associative nitrogen fixation by Azospirrillum sp.** Publications.

2599 Van Onsem, J.; Heursel, J.　　BE 070600/66/0003
Onderzoek van de invloed van groeiregulatoren in de azaleateelt met het oog op kwaliteitsverbetering, verkorte levensduur of arbeidsbesparing. **Research on the influence of growth regulators in the azalea culture to become a quality improvement and a shorter cultivation time or laboursaving.** Publications.

2600 Detroux, L.; Haquenne, W.　　BE 080700/76/0022
Etude des régulateurs de croissance en cultures maraîchères, fruitières, céréalières et betteravières. **Study of growthregulators in vegetable and fruit crops, cereals and beets.**

2601 Rixhon, L.; Droeven, G.; Couvreur, L.; Crohain, A.; Raimond, Y.; Guiot, J.　　BE 080800/75/0011
Influence de la valorisation agronomique du lisier sur les rendements, sur la qualité des récoltes, sur les principales propriétés du sol et sur la protection du milieu. **Influence of the agricultural use of the slurry on crop yield and quality and on the principal soil properties while minimizing environmental pollution hazards.**

2602 Dermine, E.; Dubuisson, J.; Gruselle, P.; Tréfois
　　BE 080900/00/0014 N
Bouturage en horticulture. **Propagation of ornementals by cutting.** Publications.

2603 Dardenne, P.　　BE 080900/79/0012 N
Utilisation en horticulture des eaux chaudes de rejet industriel. **Utilization of waste heat water from industry in crop production.** Publications.

2604 Stadelbauer, J.　　DE 126700/74/0001
Agrargeographische Studien in Transkaukasien zu Problemen der subtropischen Landwirtschaft in der UdSSR. **Studies on agricultural geographical conditions in Transcaucasia in relation to problems of subtropic agriculture in the USSR.**

2605 Dreyling, G.　　DE 135053/74/0003
Untersuchungen über die Eignung zur Massenkultur von mikroskopischen Meeresalgen. **The suitability of microscopic algae for mass cultivation.**

2606 Matzner, F.　　DE 161266/75/0001
Querterrassierung im Weinbau. **Transverse terraces in viticulture.**

2607 Horstmann, K.　　DE 176051/75/0001
Regulation des Nahrungseintrags der Waldameise. **Regulation of food consumption of wood ants.** Publications.

2608 Hodapp, W.; Siebenbürger　　DE 501506/78/0001 N
Ertragslage der Hauptbaumarten im Staatsforstbetrieb von Baden–Württemberg. **Crop situation of main tree species in the state forest enterprise of Baden–Wuerttemberg.**

2609 Kern, H.　　DE 502055/73/0028
Welche Wirkungen zeigen unterschiedliche Methoden mechanischer, chemischer und biologischer Art bei der Strohverwer– tung auf dem Acker a)auf die physikalischen und chemischen Bodeneigenschaften und b)auf die Ertragsleistung getreidestarker Fruchtfolgen. **On the effects of different methods of mechanical, chemical, and biological straw processing in the field, a.**

2610 Olesen, J.　　DK 010701/75/0010
Kollektive læplantningers virkning og betydning som produktionsfaktor i landbruget. **Effect of collective shelterbelts and their importance as a production factor in agriculture.**

2611 Aslyng, H.C.; Kristensen, K.J.　　DK 030143/77/0002
Strålingsklima og planteproduktion. **Radation conditions and plant production.**

2612 Jensen, N.J.B.　　DK 030181/78/0002 N
Fremstilling og anvendelse af containerplanter (dækrodsplanter). **Production and use of containerplants in forestry.**

2613 Dauple, P.; Vergniaud, P.　　FR 010610/70/3000
Expérimentation d'innovations techniques en cultures maraîchères sous abris plastiques et en plein champ dans la basse Vallée du Rhône. **Experiments on technical innovations in vegetable production, in the field or under plastic shelters, in the lower part of the Rhone Valley.** Publications.

2614 Perry　　GB 010601/76/0170 R
Develop biological techniques to improve meat production quality in female animals.

2615 Sharples; Johnson　　GB 011004/00/0002
Determine the best orchard practices and postharvest treatments for storage.

2616 Landsberg; Huxley　　GB 011503/00/0002 R
Topographical effects on microclimate. Relate to standard weather data. Windbreaks and airflow.

2617 Tompsett　　GB 011510/00/0021 R
Cone production in Sitka spruce.

2618 Dewar; Fletcher　　GB 012004/78/0039 R
Population dynamics and diversity of pests and beneficial invertebrates in cereal farming systems.

D 2100 – Plant production general and crop husbandry

2619 Taylor; Woiwod GB 012004/78/0053 R
Aggregation migration and diversity of plant and animal populations.

2620 Lofty; Edwards GB 012004/78/0057 R
Biology ecology and taxonomy of earthworms.

2621 Vincent GB 023908/79/0002 N
Factors affecting the stabilisation of insect cuticce.

2622 Calvert GB 030204/77/0027 R
The autonomic nervous control of substrate supply.

2623 Garrett GB 040403/00/0005
Photosynthetic reclamation of slurry nutrients.

2624 McElroy GB 041206/00/0004
Production of cellulose on disadvantaged land.

2625 Researcher not indicated GB 050130/00/0003 R
Hay: techniques of haymaking.

2626 Researcher not indicated GB 050130/00/0004 R
Silage: techniques of silage making.

2627 Researcher not indicated GB 050170/00/0003
Hay: techniques of haymaking.

2628 Researcher not indicated GB 050170/00/0004
Silage: techniques of silage making.

2629 Edwards GB 060103/00/0005
Dry matter and other losses in farm silos.

2630 Holmes GB 060106/00/0027
Effects of farmyard manure on crop yield and response of crops to fertiliser application.

2631 Holmes GB 060106/00/0028
Development of new tillage methods into whole farm systems.

2632 Langley GB 060109/00/0005
Handling and storage of big bales.

2633 Young GB 060214/00/0001
Effect of silage making techniques on silage quality and on subsequent performance of beef cattle.

2634 Galbraith GB 060219/78/0014 N
The mode of action of anabolic compounds in improving meat production from sheep.

2635 O'Brien, D. IE 050100/76/9242 N
Effect of silviculture on wood density.

2636 O'Brien, D. IE 050100/76/9252 N
Effect of silviculture on wood density.

2637 Calo', A.; Costacurta, A. IT 021300/75/0006
Studio dei rapporti fra Monosaccaridi presenti nelle Bacche della vite. **Relation among the monosaccharides of the vine berries.**

Banks, shores, dikes and their vegetation (B 1530)

See also 648, 3616, 3626

Man–made recreational resources (B 1600)

2638 Aarts, H.F.M. NL 010207/79/9010 N
De invloed van bedrijfssystemen in de landbouw op natuur en landschap. **The influence of farming systems on nature and landscape.**

Parks, gardens, urban greenspaces, plantations (B 1610)

See also 2179, 3616, 3618, 3623, 3624, 4608, 4685, 4687, 4713, 4759, 4764, 4848, 4850, 4983, 5000, 5001, 10283

2639 De Kerchove d'Exaerde, H.; Semal, J.; Cottenie, A.;
Delcarte, E.; Impens, R.; Nangniot, P. BE 140000/74/0043
Conditions de vie de la végétation urbaine et remèdes à y apporter. **Life conditions of urban vegetation and protection methods.** Publications.

2640 Volk, H. DE 501504/77/0013
Waldbauliche Gundsätze für die Bewirtschaftung im Erholungswald. **Silvicultural principles for management of recreational forests.**

2641 Hansen, R.; Müssel, H.; Sieber, J. DE 502107/72/0004 R
Die Ansiedlung von Frühlingsgeophyten im öffentlichen Grün 1969. **Planting of spring geophytes in public greens.**

2642 Kiermeier, P. DE 502107/78/0001 N
Zur Problematik stadtfester Gehölze. **On problems of trees in urban areas.** Publications.

2643 Ødum, S. DK 030102/76/0003
Etablering af forsøgsplantninger i sydvestlige Grønland, hovedsagelig med materiale indsamlet på ekspedition til Rocky Mountains 1971. **The establishment of experimental planting in south west Greenland, particularly with material collected on an expedition to the Rocky Mountains in 1971.**

2644 Hentgen, A.; Damour, L.; Chevalier, C.; Guimbard, C.;
Billot, C. FR 010112/75/3029
Techniques d'établissement et d'entretien de gazons (ornement – sports). **Establishment and management of lawns and turfs.** Publications.

2645 Dissen, H.D. NL 010601/63/2383
Bosaanleg opgespoten terreinen, vuilstortplaatsen, veengronden en andere voor bosbouw onbekende gronden. **Afforestation of "artificial" soils and other in forestry unknown kinds of sites. (raised areas, refuse heaps and on wet peat–bogs).** Publications.

2646 Tol, G. van NL 010601/65/2385
Aanleg en verzorging van loofhoutbeplantingen. **Formation and tending of stands of broadleaved species.** Publications.

2647 Tol, G. van NL 010601/65/2387
Onderzoek naar de houtsoortenkeuze in verband met de groeiplaats. **Research on the choice of tree species in relation to the site.** Publications.

2648 Dissen, H.D. NL 010601/67/2389 R
Onderzoek naar de samenstelling en doelmatigheid van windsingels in bosbeplantingen. **Research on the composition of**

D 2100 – Plant production general and crop husbandry

windbreaks and their effect on forest plantations. Publications.

2649 Kopinga, J. NL 010601/78/8670 N
Onderzoek naar de vitaliteitskenmerken bij stadsbomen.
Research on the characteristics of the vitality of amenity trees.

2650 Tol, G. van NL 010601/78/8671 N
Onderzoek naar de toepasbaarheid van verschillende vormen
van onkruidbeheer in stedelijke beplantingen. **A survey into
the possibilities of weed vegetation management in urban
plantations.**

Arboreta and botanical gardens (B 1620)

See also 4823

Sportfields, play and camping grounds (B 1630)

See also 2644, 3389, 3402, 3616, 3623, 3624

Plants and parts of plants in general (B 2100)

See also 10, 1415, 2596, 2678, 2876

2651 Rünger, W. DE 105110/71/0002
Einfluss von Temperatur und Licht auf das Wachstum von
Jungpflanzen. **Influence of temperature und light on growth of
young plants.**

2652 Döring, H.–W. DE 105202/74/0001 N
Wirkung externer Faktoren – hohe Substrat–Salzgehalte – auf
Teile des Mineralstoffwechsels höherer Pflanzen. **Effect of
external conditions – high substrate salinity – on parts of the
mineral metabolism in higher plants.**

2653 Hoyningen–Huene, J.von; Schrödter, H.; Braden, H.
DE 301030/73/0001
Mikrometeorologische Untersuchungen zum Problem des
Wasserhaushalts im System Boden–Pflanze–Atmosphäre.
**Microclimatic investigations on problems of the water balance
in the soil–plant–atmosphere system.** Publications.

2654 Nissen, T.V. DK 010117/74/4606
Mikroorganismernes antal, arter, aktivitet og forekomstens
fasthed på blade og rødder samt mulige praktiske betydning.
**Number, species, activity and stability of occurrence of
microorganisms on leaves and roots, and possible practical
applications.**

2655 Zubr, J. DK 030143/78/0001 N
Planters fotomorfogenetiske tilpasning til stråling med
varierende spektralsammensætning. **Photomorphogenic
adaption of plants to radiation of various spectral composition.**

2656 Farestveit, B. DK 030171/71/0005 N
Undersøgelser over celledelingsforløb og vævsdifferentiering
hos højere planter. **Investigations of the course of cell division
and tissue differentiation in higher plants.**

2657 Cornillon, P. FR 010601/70/6051
Effets de la température des racines sur la croissance et le
développement de la plante. **Effects of root temperature on the
growth and development of the plant.**

2658 Cornillon, P. FR 010601/76/6052
Désaturation des acides gras et adaptation du végétal aux

basses températures. **Desaturation of fatty acids and adaptation
of the plant to low temperature.**

2659 Obaton, M. FR 011002/73/6303
Ecologie des rhizobiums. **Ecology of rhizobium.**

2660 Nicotra, A.; Moser, L.; Damiano, C.; Bergamini, A.;
Albertini, A.; Cobianchi, D. IT 021500/78/0010 N
Impiego di sostanze rizogene per ottenere piante da talee
legnose ed erbacee e studio sui meccanismi della rizogenesi
avventizia. **Use of rooting substances to obtain trees by soft and
hardwood cuttings and study of the adventitious rooting.**

2661 Damiano, C. IT 021500/79/0005 N
Propagazione "in vitro" delle piante. **Tree fruit propagation "in
vitro".**

2662 Cocucci, S. IT 040603/78/1178 N
Fattori di controllo della dormienza nei vegetali superiori.
Control factors of seed dormancy in superior plants.

2663 Massantini, F. IT 041101/77/0755
Proteine da foglie. Ricerche di carattere agronomico. **Leaf
proteins. Agronomic research.**

2664 Gregorini, G. IT 041115/77/0664 R
Propagazione per coltura in vitro di apici vegetativi.
Propagation by "in vitro" culture of germinal apices.

2665 Braber, J.M. NL 010104/71/3202
Veranderingen in de proteïnen van bladeren bij het ouder
worden. **Changes in leaf proteins upon ageing of leaves.**

Plant communities as ecological systems (B 2200)

See also 2611, 4788

2666 Noirfalise, A.; Thill, A. BE 010024/66/0003 R
Ecologie et la productivité des essences forestières en
Belgique. **Ecology and productivity of forest trees in Belgium.**
Publications.

2667 Veri, G. IT 060200/77/0782
Caratterizzazione del microclima e fisiologia delle piante
adatte. **Characterisation of micro–climate. The physiology of
plants suited to it.**

Animal and plant communities as ecological systems
(B 2500)

See also 3514

Animal – plant interrelationships (B 2600)

See also 2598

Crops in general (B 3000)

See also 10, 159, 174, 189, 203, 449, 492, 532, 878, 887, 897,
1064, 1382, 1424, 1426, 1432, 1536, 1542, 1740, 1741, 1750,
2030, 2393, 2597, 2601, 2602, 2604, 2993, 6055, 7813, 7821,
11071, 15633, 15636, 15777, 15935, 15981, 17377

2668 Van Hee, L. BE 070400/69/0018
Beproeven van nieuwe introduktiegewassen. **Plant
introduction trials.**

D 2100 – Plant production general and crop husbandry

2669 Limberg, P.; Perwitz, S.   DE 105201/70/0004
Untersuchungen zur Ermittlung des Produktivitätstypes einiger Zwischenfrüchte. **Investigations to find out the type of productivity of some catch–crops.**

2670 Krzysch, G.   DE 105201/70/0010
Untersuchungen des Strahlungshaushaltes – kurz– und langwelliger Strahlungsströme – im Bereich landwirtschaftlicher Nutzpflanzenbestände. **Investigations of the radiation–balance in crops.**

2671 Krzysch, G.   DE 105201/70/0011
Untersuchungen des CO2–Umsatzes und der Stoffproduktion landwirtschaftlicher Nutzpflanzenbestände. **Investigations of the CO2–balance and the dry matter production of agricultural crops.**

2672 Krzysch, G.   DE 105201/70/0012
Untersuchungen der Bestandsklimaausbildung und deren Abhängigkeit von Bestandseinflüssen und Makroklima. **Investigations of the development of microclimate in crops and their relation to plant development and macroclimate.**

2673 Krzysch, G.   DE 105201/77/0004
Der Bodenwasserhaushalt unter Nutzpflanzenbeständen und methodische Untersuchungen zur Bestimmung der Evaporation und Evapotranspiration. **Soil water balance under crop and methodical determination of evaporation and evapotranspiration.**

2674 Limberg, P.; Hegewald, B.   DE 105201/77/0005
Die gegenseitige Beeinflussung von Pflanzenarten im Mischfruchtanbau. **The mutual influence of plant species in mixed cultivation.**

2675 Limberg, P.; Perwitz, S.   DE 105201/77/0014
Einfluss langjähriger Bewirtschaftungsmassnahmen auf Eigenschaften und Ertragsleistung eines Sandbodens in halbkontinentaler Klimalage. **The effect of long standing cultural and plant–production measures upon sandy soil and the productivity of this location in semi–continental climate.** Publications.

2676 Daunicht, H.–J.   DE 105203/70/0003
Die Wirkung von Klimafaktoren und Speicherungsorganen auf die primäre Stoffproduktion von Nutzpflanzen im Laufe ihrer Entwicklung - kompensative Photosynthesemessung –. **The effect of climatic factors and storage organs on the primary organic production of crop plants in course of their development – compensational photosynthesis measurement –.**

2677 Daunicht, H.–J.; Fritz, U.   DE 105203/78/0003 N
Physikalische Wirkungen von transparenten Folien zur Pflanzenkultur. **Physical effects of transparent plastic films for crop growing.**

2678 Kick, H.; Sonnleitner, B.   DE 111100/75/0002
Die Aufnahme und Wirkung von Tensiden – oberflächenaktiven Stoffen –, Weichmachern, Textilhilfsstoffen bei Pflanzen. **The uptake and efficiency of detergents – surfactants –, builders, emulgators in plants.**

2679 Boeker, P.; Schulte, D.   DE 111251/74/0003
Wurzelentwicklung verschiedener Zwischenfrüchte in Abhängigkeit von Bodentyp und Saattermin. **Growth of the roots of different intermediate crops in dependence on soil type and sowing time.**

2680 Boeker, P.; Opitz–von–Boberfeld, W.   DE 111251/75/0003
Die Ausbildung der Wurzelmasse verschiedener Arten und Sorten in unterschiedlichen Medien im Hinblick auf die Auswahl geeigneter Methoden zur Vereinfachung von Wurzelgewichtsfeststellungen – Wurzelentwicklung in verschiedenen Medien –. **Formation of root matter of different species and varieties in different nutrient solutions with respect to the selection of appropriate methods for simpler determination of root weight – Formation of root matter in different nutrient solutions –.**

2681 Judel, G.–K.   DE 129046/75/0002
Einfluss steigender Blei–, Cadmium– und Nickelgaben auf das Pflanzenwachstum. **Influence of increasing amounts of lead, cadmium and nickel on plant growth.**

2682 Stählin, A.; Stählin, L.   DE 129140/72/0003
Samen und Früchte der Nutzpflanzen unter besonderer Berücksichtigung der Tropen und Subtropen. **Seeds and fruits of cultivated plants especially of the tropics and subtropics.**

2683 Przemeck, E.; Lüdtke, M.   DE 132061/73/0001
Proteine in verschiedenen Stärken. **Proteins in different starches.**

2684 Baeumer, K.; Böhm, W.   DE 132181/75/0003
Wurzelwachstum von Feldfrüchten in bearbeiteten und unbearbeiteten Ackerböden. **Root growth of field crops as affected by soil tillage intensity.**

2685 Baeumer, K.; Köpke, U.; Böhm, W.   DE 132181/75/0006
Methodenvergleich bei Untersuchungen des Wurzelwachstums von Feldfrüchten in bearbeiteten und unbearbeiteten Löss–Parabraunerden. **Comparison of methods for root investigations of field crops in tilled and untilled loess soils.**

2686 Rehm, S.; Hassan, M.   DE 132240/77/0001
Endogene Wuchsstoffe in schwer zu bewurzelnden Stecklingen tropischer Kulturpflanzen. **Endogenous growth substances in hard–to–root cuttings of tropical crops.**

2687 Götz, V.; Heins, H.H.   DE 132450/78/0001 N
Umweltschutz und pflanzliche Produktion. **Environment protection and plant production.**

2688 Fendrik, I.; Bors, J.; Kiselnic, L.   DE 138180/73/0001
Wirkung kurzzeitiger und langzeitiger Strahlenbelastungen auf die Ertragsleistung der heimischen Nutzpflanzen. **Effects of acute and chronic irradiation impact on the yield production of indigenous useful plants.** Publications.

2689 Stösser, R.; Anvari, S.   DE 144445/77/0021
Befruchtungsbiologie, Pollenschlauchwachstum und Alterung der Samenanlagen unter dem Einfluss von Wuchsstoffen. **Studies on fertility, pollen tube growth and senescence of ovules as influenced by growth substances.** Publications.

2690 Kahnt, G.; Kübler, E.   DE 144485/78/0001 N
Wirkung verschiedener Fruchtarten auf den Wasser– und N–Haushalt des Bodens in verschiedenen Fruchtfolgen. **Effect of different crops in crop–rotations on water and N–content of soils.**

D 2100 - Plant production general and crop husbandry

2691 Kahnt, G.　　　　　DE 144485/78/0007 N
Vergleich konventioneller und biologischer Landbau.
Comparison of crop production in "conventional" and "biological" farming.

2692 Moser, E.; Sinn, H.　　　　　DE 144720/70/0001 R
Biotechnische Eigenschaften von Produkten aus Intensivkulturen 1970. **Biotechnical properties of products from intensive cultivation.**

2693 Reiner, L.; Mangstl, A.　　　　　DE 161241/74/0002
Pflanzenbauliche Erhebungen in Form einer Schlagkartei als Informationssystem. **Collection of field data for decisions in plant production.**

2694 Fischbeck, G.; Geiger, H.　　　　　DE 161250/77/0005
Untersuchungen zur Variabilität des Einsatzes von zugekauftem Saatgut, Dünge- und Pflanzenschutzmitteln in der landwirtschaftlichen Praxis. **On the variability using additional seeds, fertilizers and protectives in farming.**

2695 Fritz, D.; Franz, C.　　　　　DE 161260/70/0003
Anpassung von Wildpflanzen an Kulturbedingungen und Entwicklung geeigneter Anbautechniken. **Investigations to cultivate various wildlings and development of appropriate cultivation methods.**

2696 Harms, H.　　　　　DE 201010/74/0005
Metabolisierung polycyclischer, aromatischer Kohlenwasserstoffe in pflanzlichen Zellsuspensionskulturen und in steril kultivierten Kulturpflanzen. **Metabolization of polycyclic aromatic hydrocarbons in vegetal cell suspension cultures and in sterile-cultivated plants.**

2697 Grahl, A.; Hondelmann, W.; Lange, N.; Seidewitz, L.
　　　　　DE 201040/73/5006
Informationen zur Beschreibung des Genmaterials, zur Langzeitlagerung und Vermehrung der eingelagerten Muster, zur Vermittlung und zum Austausch des Genmaterials einschliesslich Kartoffeln. **Information concerning the description of gene material, long–term storage and multiplication of stored patterns, communication and exchange of gene material including potatoes.**

2698 Hondelmann, W.　　　　　DE 201040/73/5009
Förderung und Beteiligung von bzw. an international bzw. regional organisierten Sammelexpeditionen. **Promotion and participation of/in international or regional collecting expeditions.**

2699 Czeratzki, W.　　　　　DE 201040/75/0001
Einfluss der Beregnung auf die Ertragsbildung landwirtschaftlich genutzter Kulturpflanzen. **The influence of spray irrigation on the yield of agricultural cultivated plants.**

2700 Standke, K.–H.; Radatz, W.　　　　　DE 201040/75/0015
Ermittlung der Bedingungen zur Langzeitlagerung und Restitution von Zellkulturen, insbesondere vegetativ vermehrter Kulturpflanzen. **Research on conditions in vitro for long–term storage and restitution in particular of vegetatively propagated crop plants.**

2701 Zach, M.　　　　　DE 201040/75/0017
Entwicklung von Messverfahren zur Quantifizierung der Wechselwirkungen zwischen Pflanzenentwicklung und bodenphysikalischen Parametern. **Development of measurement methods for quantifying the interactions between plant growth and soil physical parameters.**

2702 Schoedder, F.　　　　　DE 201060/78/0003 N
Verbessern von Trocknungs– und Konservierungsprozessen im Hinblick auf ein Herabsetzen des Energiebedarfs. **Improvement of drying and preservation processes with respect to reduction of energy consumption.**

2703 Steiner, K.G.　　　　　DE 215090/77/0003
Erarbeitung wissenschaftlicher Grundlagen über die Wirkungsweise und Wirksamkeit von Wachstumsreglern sowie Entwicklung von Prüfungsverfahren für diese Stoffe als Voraussetzung für die Zulassung. **Establishment of scientific principles on mode of action and on efficiency of growth regulators and development of test methods for these substances as condition of admittance.**

2704 Hoyningen–Huene, J.von　　　　　DE 301030/72/0006
Untersuchungen über den Einfluss des Energiehaushalts auf die Produktivität von Kulturpflanzen unter Berücksichtigung des Standraumanspruchs 1972. **Investigations into the influence of energy balance on the productivity of cultivated plants with regard to growing–space requirements.** Publications.

2705 Hoyningen–Huene, J.von　　　　　DE 301030/75/0001
Wasserdargebot und unproduktive Verdunstung. **Water supply and unproductive evaporation.**

2706 Hoyningen–Huene, J.von; Ohlmeyer, P.; Garbrecht, G.
　　　　　DE 301030/77/0005
Biometrische Modelle des Wasserbedarfs von Kulturpflanzen unter ariden Bedingungen. **Biometrical models of consumptive water use of plants under arid conditions.** Publications.

2707 Zunker, E.–J.　　　　　DE 301050/75/0001
Untersuchungen der Auswirkungen hoher Wärmeumsätze und Sättigungsdefizite auf die Entwicklung der Pflanzen. **Investigations of the reaction of plants to high heat balances and high saturation deficits.**

2708 Brandtner, E.　　　　　DE 301050/75/0005
Phänologisch–meteorologische Untersuchungen zur Vorhersage des Eintritts von Pflanzenentwicklungsstufen. **Phenologico–meteorological studies for forecasting the appearance of phytophenological phases.**

2709 Dancau, B.; Braun, W.　　　　　DE 502050/72/0011 R
Untersuchungen zur natürlichen Sukzession auf landwirtschaftlichen Problemflächen 1970. **Investigations on natural succession on agricultural problematic areas.**

2710 Kraus, A.; Diercks, R.; König, K.; Diez, T.; Klein, W.
　　　　　DE 502050/78/0001 N
Erarbeitung ökologischer Grundlagen für die Entwicklung von integrierten Produktionssystemen im Ackerbau – Prognose und wirtschaftliche Schadensschwelle –. **Ecological fundamentals for the development of integrated systems of plant production – prognosis and economic detrimentous threshold –.**

2711 Pommer, G.　　　　　DE 502053/74/0003
Erarbeitung von Produktionssystemen für verschiedene landwirtschaftliche Kulturpflanzenarten zur Koordinierung von Acker– und Pflanzenbau– sowie Pflanzenschutzmassnahmen im Sinne eines ökonomisch

D 2100 – Plant production general and crop husbandry

sinnvollen und ökologisch vertret– baren Landbaus. **Development of production systems in diverse agricultural crop species for the coordination of crop farming and plant cultivation and measures of plant protection in accordance with reasonable economy and balanced ecology.**

2712 Rosopulo, A.; Hahn, M.   DE 502055/73/0021
Die Bestimmung der Schadstoffe Blei, Cadmium, Chrom und Nickel in verschiedenen Kulturpflanzen mittels Atomabsorption. **Determination of the pollutants lead, cadmium, chromium and nickel in different cultivated plants by means of atomic absorption.**

2713 Schuch, M.   DE 502055/73/0022
Entwicklung einer Registriervorrichtung zur digitalen Erfassung der potentiellen Verdunstung. **A device for digital registration of potential evaporation.**

2714 Schmid, G.; Weigelt, H.   DE 502055/74/0005
Einfluss unterschiedlicher Bewirtschaftungssysteme auf die Gesundheit und Qualität der erzeugten Erntegüter in einem Grossfeldversuch. **Influence of different cultivation systems on quality and quantity of produced crops tested in field trial.**

2715 Hansen, L.; Rasmussen, K.J.   DK 010105/68/3404
Jordbearbejdningssystemer ved reduceret jordbehandling. **Tillage systems for minimal cultivation.**

2716 Hansen, L.; Rasmussen, K.J.   DK 010105/72/3403
Forårsbearbejdningens betydning for jordstruktur og plantevækst. **Effects of spring tillage operations on soil structure and plant growth.**

2717 Rasmussen, P.; Svendsen, O.   DK 010106/50/0001 N
Indsamling af data om planter med henblik på at vurdere planternes værdi som trækplanter for honningbier og andre bestøvende insekter. **The nectar– and pollen value of various plants for honey bees and other pollinating insects estimated on basis of collected plant data.**

2718 Sørensen, C.   DK 010117/63/4616
Udvikling og afprøvning af planteanalyseringsmetoder. **Development and testing of methods for the chemical analysis of plants.**

2719 Olesen, J.   DK 010701/73/0001
Jordbehandlingsforsøg bl.a. til belysning af mulighederne for at indskrænke den mekaniske jordbehandling. **Tillage experiments to elucidate the possibilities of limiting mechanical soil treatment.**

2720 Olesen, J.   DK 010701/74/0002
Afgrødernes udbytte og kvalitet belyst ved udbyttemålinger og omfattende afgrødeanalyser. **Crop yield and quality as determined by yield measurements and comprehensive yield analyses.**

2721 Andersen, A.S.; Rajagopal, V.; Veirskov, B.   DK 030104/77/0023 N
Roddannelse i stiklinger. I. Vækststofbestemmelse i stiklinger med forskellig roddannelsesevne. **Root initiation in cuttings. I. Determination of hormones in cuttings with different rooting capabilities.**

2722 Fischer, P.; Veierskov, B.; Karlsen, P.; Eriksen, E.N.; Andersen, A.S.   DK 030104/79/0006 N

Stiklingeformering i vandkultur. **Propagation by cuttings in hydroponics.**

2723 Andersen, A.S.; Veierskov, B.   DK 030104/79/0007 N
Roddannelse i stiklinger. V. Kulhydratfordeling i stiklinger. **Root initiation in cuttings. V. Carbohydrate distribution in cuttings.**

2724 Andersen, A.S.   DK 030104/79/0008 N
Ethylen som stress hormon eller symptom. **Ethylene as a stress hormone or symptom.**

2725 Hansen, G.K.   DK 030143/77/0014 N
Afgrøders udnyttelse af de primære fotosynteseprodukter til vækst og respiration i relation til miljø og udvikling. **Utilization of primary photosyntetic products for growth and respiration in relation to environment and crop production.**

2726 Hovmand, M.F.   DK 030202/79/9157 N
Afgrødernes fastlæggelse og optagelse af atmosfærisk metalnedfald. **Fixation and absorption of atmospheric metallic fall–out in crops.**

2727 Gosse, G.; Jaussely; Perrier, A.; Itier, B.   FR 010101/76/5006
Mesure directe de la photosynthèse à l'échelle parcellaire. Analyse de l'efficience en eau des cultures. **Direct measure of photosynthésis in the fields. Analysis of water efficiency of crop.** Publications.

2728 Turc, L.   FR 010103/62/6220
Indices climatiques concernant les potentialités agricoles. **Climatic indices concerning agricultural potential.**

2729 Mennessier, P.; Lemaire, L.; Teilhard de Chardin, B.; Houdard, Y.   FR 010112/65/3027
Expérimentation de nouveautés techniques dans le domaine des grandes cultures, étude de leur insertion dans des systèmes de production. **Experiments on technical innovations in crop plants and study of their introduction in production systems.** Publications.

2730 Houdard, Y.; Lemaire, L.; Mennessier, P.; Teilhard de Chardin, B.   FR 010112/70/9098
Etudes sur plantes de grande culture. **Developing improved farming of major field crops.**

2731 Levy, G.   FR 010303/66/4150
Amélioration de la production de plants par les méthodes traditionnelles. **Improvement of traditionnal methods for producing conifers plants.**

2732 Bachacou, J.   FR 010312/75/8295
Compétition en plantation régulière et en peuplements naturels (3). **Competition patterns in regular plantations and natural stands.**

2733 Maîa, E.; Beck, D.   FR 010501/68/1014
Régénération d'espèces florales par multiplication végétative. **Floral species regeneration by vegetative multiplication.** Publications.

2734 Galichet, P.F.; Marie, R.; Guennelon, G. Mme; Jourdheuil, P.   FR 010602/73/5091
Introduction d'insectes auxiliaires dans l'agrosystème du Delta du Rhône. **Introduction of auxiliary insects in the Rhône Delta**

agrosystem.

2735 Brun, R.; Peyriere, J.; Rico, F.   FR 010616/70/9087
Introduction de cultures de remplacement dans un assolement
maraîcher traditionnel régional. **Introducing alternative crops
within a traditional regional market–garden rotation.**

2736 Robelin, M.   FR 010802/56/0580
Evapotranspiration maximale des différents couverts vegetaux.
Maximum evapotranspiration of various plant canopies.
Publications.

2737 Robelin, M.; Mingeau, M.   FR 010802/62/0579
Croissance et développement des plantes cultivées en
condition de sécheresse. **Growth and development of crops
under dry conditions.** Publications.

2738 Robelin, M.   FR 010802/70/0577
Influence de la structure et de l'âge du couvert sur les échanges
gazeux. **Effect of structure and age of canopy on gaseous
exchanges.** Publications.

2739 Renard, H.A.   FR 011103/66/0296
Etudes méthodologiques concernant le pouvoir germinatif.
Methodological studies on the germination.

2740 Renard, H.A.   FR 011103/70/0294
Etudes méthodologiques concernant la vigueur de
germination. **Methodologic studies on the germination vigor.**

2741 Blanchard, M.; Arnoux, M.   FR 011103/71/0191
Exploitation des résultats d'analyses concernant le contrôle des
semences. **Utilization of analysis results concerning seed
control.**

2742 Marty, J.R.; Hutter, W.   FR 011401/69/6204
Comportement de différentes espèces en rotations culturales
adaptées à différents systèmes de culture avec ou sans
irrigation. **Comportment of various species in crop rotations
suited to different irrigated and non irrigated cultural systems.**

2743 Hutter, W.; Marty, J.R.   FR 011401/72/0642
Préhension et ordonnancement des informations relatives à
l'essai "systèmes de cultures" de la station. **Collecting and
processing data in an experiment on cropping systems.**

2744 Hutter, W.; Marty, J.R.   FR 011401/75/6206
Bilan d'énergie dans différentes rotations culturales. **Energy
balance in various crop rotations.**

2745 Bosc, M.; Maertens, C.   FR 011401/76/6193
Cinétique de l'enracinement des plantes cultivées et relation
avec leur alimentation minérale et hydrique. **Kinetic of root
system developement of cultivated plants and relation to their
mineral and hydric alimentation.**

2746 Charpenteau, J.L.   FR 011409/75/8271
Systèmes de culture (1). **Agronomic systems.**

2747 Fleury, A.; Roncin, F.   FR 011701/68/6173
Etude des relations sociales chez les plantes annuelles cultivées
selon différentes structures de peuplement. **Studies on effects of
spacing design in annuel crops.**

2748 Capillon, A.; Fleury, A.; Sebillotte, M.
   FR 011701/75/6172

Recherche de systèmes de techniques culturales cohérents avec
différents types d'objectifs socio–économiques pour une
rotation de cultures donnée. **Research for systems of cultural
techniques coherent with socio economical aims, for a given
crop rotation.**

2749 Bruckler; Sebillotte, M.; Boiffin, J.; Manichon, H.
   FR 011701/76/6169
Etude des comportements des lits de semence – Elaboration
d'un modèle prédictif. **Study on seed beds behaviour :
elaboration of a predictive model.**

2750 Damour, L.; Jeannin, B.; Garreau, J.; De Ch. Teilhard,
B.; Guy, P.   FR 012201/70/9061
Techniques et successions culturales sur parcelles de marais
drainé. **Testing various cultural practices and crop rotations on
plots of drained marsh lands.**

2751 Gachon, L.; Lienard, G.; Decourt, N.; Jarrige, R.;
Ricou, G.   FR 012220/76/5287
Equilibres agro–sylvo–pastoraux en Margeride. **Study of the
agro–sylvo–pastoral equilibriums in Margeride.**

2752 Lefevre, G.; Arnoux; Regnier, P.; Hulot, J. C.
   FR 012222/73/6041
Etude des systèmes de culture dans les conditions de la
pratique agricole. **Study of agricultural systems.**

2753 Guimbard, C.; Attonaty, J.M.; Corre, J.; Spartel, G.
   FR 012228/74/9052
Etude technico–économique comparée des rotations en
serre–verre et sous grands abris plastique. **Technico–economic
comparative study of crop rotations under glasshouses and
under plastic shelters.**

2754 Jackson   GB 010502/00/0007 R
**Responses of crop plants to waterlogged soil conditions,
anti–transpirants and hormones.**

2755 Barber   GB 010502/75/0013 R
The release of organic material by plant roots into the soil.

2756 Shone; Wood   GB 010502/77/0016 R
**Absorption and release of water and solutes by different parts of
the root system.**

2757 Barlow   GB 010502/78/0017 R
Controls of division and development of cells in root apices.

2758 Cannell   GB 010503/00/0001
**Development and evaluation of methods for assessing plant
performance in relation to soil conditions.**

2759 Cannell; Belford   GB 010503/00/0007
**Effects of variation in water–table on soil conditions and crop
growth.**

2760 Ellis   GB 010503/79/0019 N
**Maximisation of the benefits of direct drilling in farming
systems.**

2761 Smith   GB 010802/00/0008 R
**Influence of repeated applications of MCPA, tri–allate,
simazine and linuron on "fertility" of soil.**

2762 Peters   GB 010804/00/0004 R

Interaction of factors affecting competition between crops and weeds.

2763 Chancellor; Froud   GB 010804/00/0005 R
Weed ecology.

2764 Lutman; May   GB 010811/00/0006 R
Effect of high organic matter soils on use of herbicides .

2765 Skene; Slater   GB 011008/00/0024 R
Studies on chloroplast structure by electron microscope.

2766 Acock   GB 011102/00/0001 R
Relation of plant form and function to tissue water potentials.

2767 Grange; Fitter   GB 011102/00/0003 R
Light interception by crop canopies.

2768 Acock; Hand   GB 011102/00/0008 R
Carbon dioxide assimilation by crop canopies.

2769 Ludwig   GB 011102/00/0013 R
Photosynthesis and respiration of leaves.

2770 Acock; Ludwig   GB 011102/00/0045 R
Photosynthetic characteristics and water potentials of hard and soft leaves.

2771 Ludwig   GB 011102/00/0048 R
Glasshouse measurement of leaf photosynthesis.

2772 Cockshull   GB 011102/00/0049 R
Effect of light quality on plant growth and development.

2773 Grange   GB 011102/76/0065 R
Morphogenetic control of vascular differentiation.

2774 Garwood; Tyson   GB 011202/76/0029
Interactions between forage crops and cash crops : effects of crop sequence on soil condition and yields.

2775 Hills   GB 011403/00/0003 R
Cell wall biogenesis.

2776 Holloway; Baker   GB 011504/00/0005 R
Analysis and chemistry of waxes, cutins and suberins from plant surfaces.

2777 Webb; Purves   GB 011507/00/0009 R
Improve crop productivity by critical path analysis of biological processes.

2778 Abbott; Whiteley   GB 011509/00/0006 R
Growth and morphogenesis in tissue and cell cultures – microvegetative propagation of fruits and ornamentals.

2779 Hoad; Challice   GB 011510/00/0012 R
Mechanisms by which growth and fruiting is controlled by endogenous hormones and growth regulators.

2780 Smith, B.D.; Kendall   GB 011514/00/0002 R
Assess crop losses due to inadequate pollination; develop methods to improve pollen transfer by insects.

2781 Pain   GB 011704/00/0006
Problems associated with slurry disposal on land.

2782 Pain; Leaver   GB 011704/00/0007
Slurry utilisation in crop production.

2783 Dyer; Bowman   GB 011905/00/0014 R
The biochemical study of cytoplasmic organelles.

2784 Bunting   GB 011910/00/0006
Agronomic studies to assess possibility of commercial cultivation of minor crops in UK.

2785 Holden   GB 012001/00/0003 R
Breakdown of chlorophyll.

2786 Miflin; Bright   GB 012001/00/0016 R
The selection of mutants of higher plants in tissue culture and their subsequent regeneration.

2787 Hill   GB 012001/00/0018 R
The active site of diamine oxidase.

2788 Holden   GB 012001/00/0020 R
Studies on the presence and metabolism of analogues of protein amino acids in plants.

2789 Thomas; Petzing   GB 012001/76/0025 R
Regulation of crop plants from protoplasts.

2790 Mattingly   GB 012003/00/0009
Compare chemical forms of phosphate fertilisers.

2791 Stevenson   GB 012005/00/0034 R
Minimising the adverse effects of agricultural chemicals (particularly insecticides) on beneficial insects.

2792 Lord; Wheeler   GB 012005/00/0046 R
Investigation of movement and metabolism of growth regulators in cereals.

2793 Legg   GB 012008/00/0001
Response of farm crops to irrigation.

2794 French   GB 012008/00/0002 R
Water use by irrigated and unirrigated crops.

2795 Long   GB 012008/00/0003 R
Compare measured water use with meteorological estimates.

2796 Legg   GB 012008/00/0006 R
Field study of assimilation and respiration of crops.

2797 Parkinson; Leach   GB 012008/00/0007 R
Laboratory studies of assimilation and respiration.

2798 Legg; McCartney   GB 012008/78/0021 N
Air flow and turbulence amongst crop foliage; the movement and deposition of fungal spores and liquid crops.

2799 Mosse; Hayman   GB 012010/00/0003 R
Mycorrhizal studies.

2800 Brown   GB 012010/00/0007 R
Rhizosphere studies.

2801 Davis   GB 012010/00/0008 R
Rhizobium culture collection.

2802 Johnston GB 012101/00/0021
Potentiality trials to assess yield and adaptation of new crops to western conditions.

2803 Evans GB 012701/77/0019 R
Biogenic amines in the insect nervous system.

2804 Yeoman GB 021001/79/0009 N
Scanning electron microscopy of cellular interactions during graft formation.

2805 Monteith GB 023706/79/0012 N
Assessment of crop photosynthetic efficiency from the spectral composition of reflected radiation.

2806 Whatley GB 023802/79/0003 N
Studies on the effects of sulphur dioxide on plant metabolism.

2807 Sagar GB 024603/79/0003 N
Earthworms grazing on root systems.

2808 Robertson, R. GB 030401/00/0014 R
The use of peat and peat products in agriculture and horticulture.

2809 Burridge GB 030402/00/0002 R
Trace element uptake by plants: distribution in different species and plant parts.

2810 Dyson GB 030406/00/0007
Growth, development, nutrient accumulation and yield of field crops: effects of environment and management.

2811 Soane; Pidgeon GB 030503/00/0011 R
Effect of tractor wheels and tracks on soil conditions and crop growth.

2812 Lawson; Wiseman GB 030701/00/0029
Weed control in crop rotations.

2813 Tompson; Taylor GB 030701/00/0050
Control of growth, yield and quality of protein and other seed crops used for feed manufacture.

2814 Cooper; Warwick GB 030901/72/0301 R
Investigation into methods of distinguishing morphologically similar seeds.

2815 Rennie; Don GB 030901/74/0203 R
Aspects of germination physiology in relation to seed testing.

2816 Warwick; Cooper GB 030901/76/0101 R
Identification and prescription of best methods of laboratory sampling of seed submitted for testing.

2817 Hodges GB 040302/79/0018 N
Evaluation in Northern Ireland of available cultivars of unusual or exotic crop species.

2818 Harper GB 060105/00/0004
Rapid multiplication of horticultural stock by in–vitro techniques.

2819 Harper GB 060105/00/0005 R
Physiology of seedlings and young plants.

2820 Addison GB 060213/00/0013
Maintenance of woodlands field, Craibstone.

2821 Paterson; Tilley GB 060314/00/0018
The assessment of direct drilling for crop and grass land establishment in West Scotland.

2822 O'Brien, D.; Fitzsimons, B. IE 050100/75/7234 N
Analytical projections of forest production. Publications.

2823 Brogan, J.C. IE 060200/73/0834 R
Liming of tillage crops. Publications.

2824 MacNaeidhe, F.; Prendergast, A.G. IE 060201/76/1155 R
The production of arable crops on peat soil. Publications.

2825 Lamb, J.G.D.; Kelly, J. IE 060300/69/0108 R
The use of plastic in plant propagation. Publications.

2826 O'Flaherty, T. IE 060300/74/1038 R
Review of the energy situation in regard to glasshouse crops production. Publications.

2827 Maher, M.J.; O'Flaherty, T. IE 060300/76/0557 N
Crop production under plastic. Publications.

2828 Jeffares, M. IE 060300/77/1292 R
Plant proteins. Quantity, quality, and texturisation of proteins extractable from a range of home–grown plants. Publications.

2829 O'Riordan, F. IE 060300/77/1309 R
Micropropagation of plants under sterile conditions. Publications.

2830 Neenan, M.; Connolly, J.F. IE 060500/62/0186 R
Investigations into new crops for Ireland. Publications.

2831 Rice, B. IE 060500/76/1235 N
Effect of row width and plant density on yield and harvest losses. Publications.

2832 D'Ambrosio, M.; Ricciardelli, G.; Intoppa, F. IT 020400/79/0003 N
Potenziale nettarifero della flora mellifera spontanea e coltivata. Honey potential of wild and cultivated bee plants.

2833 Veri, G. IT 020600/70/0086
Evapotraspirazione, fotosintesi e bilancio di energia in relazione al microclima ed allo stato idrico nel suolo e nelle piante. Evapotranspiration, photosynthesis and energy balance in relation to microclimate and water status in plant and soil. Publications.

2834 Terranova, G.; Caruso, A. IT 021600/77/0005
Prove di semina e coltivazione di semenzali in paperpotos. Seeding trials of seedlings in paperpots.

2835 Giametta, G. IT 040103/78/1062 N
Meccanizzazione delle colture ortive e industriali. Mechanization of market garden and industrial cultivations.

2836 Scaramuzzi, F. IT 040119/73/0306
Morfogenesi ed organogenesi in vitro, con particolare riferimento all'influenza degli ormoni in piante di interesse

agrario. **Morphogenesis and organogenesis in vitro with specific reference to the influence of hormones in plants of interest to agriculture.**

2837 Toderi, G.; Giordani, G.　　　IT 040201/66/0002
Confronto tra diverse rotazioni colturali in differenti condizioni di concimazione organica e minerale. **Comparison between different crop rotations under different organic and mineral fertilizer regimes.** Publications.

2838 Lovato, A.; Montanari, M.　　　IT 040201/78/0005
Studi sulla germinabilità e conservazione della vitalità delle sementi. **Research on the seed germination and vitality.**

2839 Cainelli, G.F.　　　IT 040243/77/0480 R
Sintesi di nuove sostanze a potenziale carattere giovanilizzante e a potenziale carattere abscissico–mimetico. **Synthesis of new chemicals with a potential rejuvenating action and a potential abscission–like action.**

2840 Foschi, S.　　　IT 040247/77/0823
Interazione fitofarmaci e fitoregolatori. **Pesticides and phyto–regulators interaction.**

2841 Orsi, S.　　　IT 040502/78/1093 N
Ricerche sulla produzione sementiera. **Research on nursery seed production.**

2842 Tognoni, F.　　　IT 040701/77/0670 R
Nuove tecniche di propagazione in vitro e coltura di apici vegetativi. **New "in vitro" techniques and culture of vegetative apices.**

2843 Duranti, A.　　　IT 040701/78/1049 N
Ordinamenti colturali ed evoluzione della fertilità. **Cultural policies and the evolution of fertility.**

2844 Giannattasio, M.　　　IT 040704/73/0222
Ruolo dei fitoregolatori naturali e sintetici nei processi fisiologici della crescita, fruttificazione e rizogenesi delle piante di interesse agrario. **Role of natural and synthetic plant regulators in the physiological processes of growth, fructification and rooting of agriculturally important plants.**

2845 Marano, B.; Palmieri, G.; Imperato, F.　　　IT 040705/73/0001
Efficienza della fotosintesi e potenziale agronomico nell'ambiente del basso volturno. **Photosintetic efficiency and agronomic potential of "Basso Volturno".** Publications.

2846 Antonelli, M.　　　IT 041003/78/1024 N
Relazione fra produttività e caratteristiche citologiche, istoanatomiche e fisiologiche in piante erbacee di interesse agrario. **Relationship between productivity and the cytological, and histological–anatomical, physiological characteristics of grasses used in agriculture.**

2847 Basoccu, L.　　　IT 041214/73/0139
Studio di alcuni fitoregolatori. Effetti sul radicamento e la fioritura, azioni di sinergismo e di antagonismo, metabolismo e residui dei vegetali. **Study on some plant regulators. Effects on rooting and flowering, synergetic and antagonistic activity, metabolism and plant residues.**

2848 Bosticco, A.　　　IT 041216/77/0001 N
Agricoltura ed allevamento animale in un comprensorio collinare della Provincia di Asti. **Agriculture and animal breeding in a hill – zone of Asti province..**

2849 Ballio, A.　　　IT 041609/77/0471
Nuovi fitoregolatori prodotti da microorganismi. **New phytoregulators produced by micro–organisms.**

2850 Di Marco, G.　　　IT 060200/70/0081
Regolazione della Omoserina deidrogenasi in diverse sorgenti vegetali. **Aspects of Homoserine dehydrogenase regulation in higher plants.** Publications.

2851 Sarti, A.　　　IT 060200/70/0089
Studio della relazione tra relazione luminosa e sviluppo delle piante d'interesse agrario durante le varie stagioni. **Growth of some crop plants as influenced by the light intensity and time of the season.** Publications.

2852 Tomati, U.　　　IT 060200/72/0079
Ruolo fisiologico della rodanese sulle piante. **Physiological role of rhodanese in higher plants.** Publications.

2853 Toponi, M.A.　　　IT 060200/74/0084
Alcuni aspetti biochimici e fisiologici delle diverse fasi dell'organogenesi. Determinazione di antigeni e equilibri ormonali in culture "in vitro" e in talee. **Some biochemical and physiological aspects of various organogenesis phases. Antigens determinations and hormonal bioassay "in vitro" culture and cuttings.**

2854 Migliaccio, F.　　　IT 060200/74/0085
Studio dell'azione dei fitormoni (IAA,ABA e Kinetina) su lo influsso,l'efflusso, l'accumulo e la traslocazione ionica in piante di interesse agrario sottoposte a soluzioni saline di variabile composizione e concentrazione. **Study of the effect of phythormons (IAA,ABA and Kinetin) on ion influx, efflux, accumulation and translocation in plants interesting to agriculture submitted to saline solution variable in composition and concentration.**

2855 Porceddu, E.　　　IT 060500/71/0174
Moltiplicazione e valutazione delle collezioni acquisite. **Multiplication and valuation of collected materials.** Publications.

2856 De Leo, P.　　　IT 060500/72/0176
Studio su alcune attività enzimatiche che accompagnano la perdita di vitalità dei semi. **Enzymatic activities in ageing seeds.** Publications.

2857 Porceddu, E.　　　IT 060500/74/0177
Studio del controllo ormonale dei processi che determinano la maturazione e la senescenza nei semi. **Phytohormons in ripening and ageing of seeds.**

2858 Porceddu, E.　　　IT 060500/77/1005
Studi sui meccanismi che portano alla perdita della vitalità nei semi. **Studies on the mechanisms inducing the loss of vitality in seeds.**

2859 Tedeschi, P.　　　IT 060700/74/0119
Studio degli apparati radicali di colture erbacee. **Studies on root systems of herbaceous crops.**

2860 Tedeschi, P.　　　IT 060700/74/0130
Epoca di semina ed economia idrica in relazione a colture

diverse. **Seeding time and water use economy with relation to different crops.**

2861 Bottalico, A.   IT 061700/77/0824 R
Produzione di nuovi fitoregolatori. **Production of new phyto–regulators.**

2862 Bini, G.   IT 062800/73/0181
Ricerche sulla propagazione per seme e biologia della riproduzione. Indagini sull'influenza della cultivar impollinatrice sulla facoltà germinativa dei semi. **Researches on seed propagation and on reproduction biology. The influence of pollinator cvs. on germinative ability of seeds.** Publications.

2863 Bartolini, G.   IT 062800/73/0185
Ricerche sulla propagazione per radicazione diretta.Propagazione per talea. **Researches on vegetative propagation.Propagation by cutting.** Publications.

2864 Bellini, E.   IT 062800/73/0186
Ricerche sulla propagazione per radicazione diretta. Propagazione per margotta. **Researches on vegetative propagation. Propagation by layer.** Publications.

2865 Bini, G.   IT 062800/73/0187
Ricerche sulla propagazione per radicazione diretta. Moltiplicazione per meristemi apicali. **Researches on vegetative propagation. Propagation by apical meristhems.** Publications.

2866 Scaramuzzi, F.   IT 062800/73/0189
Ricerche sulla propagazione per innesto. Ricerche sull'innesto. **Researches on graft propagation. Researches on grafting.** Publications.

2867 Scaramuzzi, F.   IT 062800/73/0190
Ricerche sulla propagazione per innesto. Ricerche sui portinnesti. **Researches on graft propagation. Researches on rootstocks.** Publications.

2868 Casini, E.   IT 062800/74/0192
Problemi di tecnica vivaistica. Pacciamatura delle piante in vivaio. **Nursery technical problems. Mulching in nursery.**

2869 Vredenberg, W.J.; Schapendonk, A. NL 010102/67/1099
Biofysisch onderzoek naar de relatie tussen actief ionentransport en fotosynthetisch electronentransport in groene planten. **Biophysical examination of relations between active ion transport and photosynthetic electron transport in green plants.** Publications.

2870 Heringa, J.W.   NL 010102/70/7304
Scheikundig en fysiologisch onderzoek naar stoffen uit oogstresten, die een nadelige invloed uitoefenen op de ontwikkeling van een volggewas. **Chemical and physiological investigations of products derived from crop residues causing retardation of the development of the following crop.**

2871 Breman, H.   NL 010102/75/6292
Plantaardige produktie in de Sahel; Oecologische en fysiologische aspecten van de primaire produktie in de Sahel. **Primary productivity in the Sahel zone; Ecological and physiological aspects of the primary production in the Sahel.** Publications.

2872 Sibma, L.   NL 010102/76/6736 R
Onderzoek naar de invloed van de duur van de gewasgroei op

de opbrengst. **Research on theinfluence of the duration of crop growth on yield.** Publications.

2873 Versteeg, M.N.   NL 010102/76/7303 R
Factoren die de produktie van voedsel– en voedergewassen op geïrrigeerde woestijngronden in Peru beïnvloeden. **Factors influencing the production of forage and arable crops on irrigated desert soils in Peru.**

2874 Heemst, H.D.J. van   NL 010102/77/7995
Onderzoek naar de plantenteeltkundige aspecten van de wereldvoedselproduktie. **Research on crop growth aspects of world food production.** Publications.

2875 Louwerse, W.   NL 010102/78/7986
Beschrijving en verklaring van fotosynthese–, respiratie– en transpiratiesnelheid in hun onderlinge relatie, mede in samenhang met de invloed van de voorbehandeling. **Interrelations between rate of photosynthesis, respiration and transpiration, also as influenced by pretreatment.**

2876 Wit, C.T. de   NL 010104/69/2504
Simuleren van de groei van planten en gewassen. **Simulation of the growth of plants and crops.** Publications.

2877 Bakermans, W.A.P.   NL 010104/73/3968
De invloed van planteresten en bodembedekkende gewassen op bodem en gewas bij vastegrondsteelt en de interactie met het gebruik van chemische middelen. **Mulching and cropping systems with zero–tillage.** Publications.

2878 Veen, B.W.   NL 010104/75/6227
Onderzoek naar de temperatuurbehoefte van verschillende delen van de plant in verband met de toepassing van energiebesparende teelttechnieken. **Influence of temperature on the physiological activity of different parts of the plant in connection with the optimal use of energy in greenhouses.** Publications.

2879 Veen, B.W.   NL 010104/75/6228
De invloed van milieu–omstandigheden en inwendige factoren op de groei, morfologie en fysiologische activiteit van het wortelstelsel. **The influence of external and internal factors on growth, morphology and physiological activity of the root system.**

2880 Hoekstra, O.   NL 010207/71/3267 R
Onderzoek naar de onderlinge relatie en beïnvloeding van gewassen. **Investigations on interactions between crops.** Publications.

2881 Velde, H.A. te   NL 010207/71/3270 R
Onderzoek naar de toepassingsmogelijkheden van groenbemestingsgewassen. **Investigations on the cultivation of green manure crops in crop rotations.** Publications.

2882 Maenhout, C.A.A.A.   NL 010207/72/3495 R
Onderzoek naar de gevolgen van grondontsmetting voor het bouwplan. **Evaluation of soil disinfection in crop rotation.**

2883 Lumkes, L.M.   NL 010207/76/6622
De invloed van werktuigen en bewerkingstechnieken op grond en gewas. **Research to thesuitability of farm implements and tillage technics and the influence of soil and crop.** Publications.

2884 Maenhout, C.A.A.A.   NL 010207/77/7631 R

D 2100 – Plant production general and crop husbandry

Onderzoek naar de grondsoort– en gebiedsgebonden aspecten van vruchtwisseling en bouwplan. **Influence of regional aspects and soil type on the production capacity of crops in intensive rotations.**

2885 Kruistum, G. van   NL 010207/78/7667 N
Verbetering van de veldopkomst. **Improvement of the field emergence.**

2886 Nicolaï   NL 010207/79/9009 N
Onderzoek naar de mogelijkheden en gevolgen van het gebruik van afvalwarmten in de land– en tuinbouw. **Research on the possibilities and consequences of the utilization of reject heat in agriculture and horticulture.**

2887 Hoekstra, F.A.   NL 020041/77/8448
Pollen als eencellig eukaryoot modelsysteem voor mitochondriënontwikkeling en voor de activering en vorming van eiwitten. **Pollen as an unicellular eukaryotic model system for mitochondrial development and for the activation and synthesis of proteins.**

2888 Pieters, G.A.   NL 020042/72/5028
Onderzoek van de warmtebalans, de lichttransmissie en de transportweerstanden van het blad van o.a. populier, opgekweekt onder verschillende condities van licht, temperatuur en wortelmilieu. **The heat balance, light transmission, and diffusion resistances of leaves of poplar and other plants cultivated under different conditions of light, temperature and root environment.** Publications.

2889 Pieters, G.A.   NL 020042/72/5029
Onderzoek van de energiebalans van o.a. populier gemeten aan de fotosynthese en ademhaling op de groeiplaats en aan transport van assimilaten naar verschillende groeizones van de plant, in samenhang met groei op orgaan– en celniveau. **The energy balance of poplar and other plants, as measured in situ on the rates of photosynthesis and respiration, the transport of assimilates to the growing zônes, in connection to growth on the level of the cell and organ.** Publications.

2890 Penning de Vries, F.W.T.   NL 020054/75/6292
Plantaardige produktie in de Sahel; Oecologische en fysiologische aspecten van de primaire produktie in de Sahel. **Primary productivity in the Sahel zone; Ecological and physiological aspects of the primary production in the Sahel.** Publications.

2891 Goudriaan, J.   NL 020054/78/7986
Beschrijving en verklaring van fotosynthese–, respiratie– en transpiratiesnelheid in hun onderlinge relatie, mede in samenhang met de invloed van de voorbehandeling. **Interrelations between rate of photosynthesis, respiration and transpiration, also as influenced by pretreatment.**

2892 Laar, H.H. van   NL 020054/78/8543
Simulatie van waterverbruik, droge stof produktie en morfogenetische aspecten van gewassen. **Simulation of water use, dry matter production and morphogenetic aspects of crops.**

2893 Wessel, M.   NL 020056/76/7610 R
De invloed van groei omstandigheden tijdens jeugd op de latere ontwikkeling bij tropische gewassen. **Influence of early growth conditions on later development of tropical crops.**

2894 Staritsky, G.   NL 020056/78/8861 N
Vegetatieve vermeerdering van tropische gewassen door middel van weefselkweektechnieken. **Vegetative propagation of tropical crops in vitro.**

2895 Jong, G.J. de   NL 040007/62/4341
Vruchtopvolging in de IJsselmeerpolders. **Crop rotation in the IJsselmeerpolders.**

Cereals in general (B 3100)

See also 491, 493, 1439, 2600, 2823, 3282, 3390, 3926, 4362, 5386, 5545, 6100, 6110, 15855

2896 Laloux, R.; Dohet, J.; Falisse, A.; Nijst, P.; Poelaert, J.
                                            BE 010008/64/0003
Phytotechnie et fumure des céréales et adaptation aux différentes conditions régionales (Belgique, France, Allemagne, Angleterre). **Phytotechnical problems, fertilization and the adaptation on different environmental conditions of cereals (Belgium, France, Germany and United Kingdom).** Publications.

2897 Moes, A.; Seilleur, P.   BE 010014/51/0001 R
Etude de la mutagenèse chez les céréales et le glaïeul et l'analyse physico–chimique des mutants obtenus. **Mutation breeding study in cereals and gladiolus and the physico–chemical analyses of induced mutants.** Publications.

2898 Nijs, L.; Crohain, A.; Detroux, L.; Biston, R.; Parmentier, G.; Salembier   BE 080900/70/0007 R
Phytotechnie générale et spéciale des fumures et des traitements phytosanitaires de la pomme de terre et des céréales (dont l'épeautre) en Haute Belgique. **General and special phytotechniques for fertilizing and chemically treating by potatoes and cereals (spelt, etc. . .).** Publications.

2899 Heyland, K.–U.; Fewson, S.; Jacob, D.; Braun, H.
                                            DE 111252/77/0002
Bestimmung des Anbauwertes verschiedener Saatgutqualitäten bei Getreide. **Determination for the use of cereals seed with different quality.**

2900 Heege, H.J.; Mülle, G.   DE 111852/75/0001
Einzelkornsaat von Getreide. **Single–grain sowing of cereals.**

2901 Höfner, W.; Brückner, U.   DE 129044/78/0003 N
Einfluss von Wachstumsregulatoren auf Ertrag und Assimilationsleistung verschiedener Getreidearten. **Influence of growth retardants on yield and assimilation productivity of several cereals.**

2902 Mengel, K.; Judel, G.–K.   DE 129047/78/0003 N
Einfluss der Belichtungsintensität auf den Energiestatus und auf die am Stärkeaufbau beteiligten Enzyme in sich entwickelnden Weizenkörnern. **Influence of light intensity on the energy charge and the starch forming enzymes in developing wheat grains.**

2903 Pospelowa, G.; Fliess, H.   DE 129384/72/0001 R
Anbau und Produktivität der Getreidekulturen in Polen – biologisch–technischer Fortschritt – 1972. **Cultivation and productivity of cereals in Poland – biological–technical progress.**

2904 Graebe, J.E.   DE 132070/72/0001 R

D 2100 – Plant production general and crop husbandry

Gaschromatographische Auswertung von Veränderungen der Pflanzenhormone in Getreidekaryopsen während ihrer Entwicklung 1972. **Gas chromatographic evaluation of changes in plant hormones in cereal caryopses during their development.**

2905 Graebe, J.E.; Rademacher, W.; Ropers, H.–J.
DE 132070/75/0001
Die Rolle von Hormonen bei der Speicherung im Getreideendosperm und die Biosynthese von Gibberellinen. **The role of plant hormones in the formation of storage endosperm in cereals and gibberellin biosynthesis.**

2906 Michael, G.; Goldbach, H.   DE 144400/78/0006 N
Phytohormone im Getreidekorn und deren Rolle bei der Regulation der Speicherungsvorgänge, der Korngrösse und des Kornertrages. **Hormonal regulation of ripening processes, size and yield of cereal grains.**

2907 Steiner, A.M.   DE 144411/74/0001
Untersuchungen zu physiologischen Grundlagen der Triebkraft von Saatgut. **Studies on the physiological basis of vigour in seed.**

2908 Kahnt, G.; Bausch, R.   DE 144485/75/0001
Auswirkung von Bodenbearbeitungs- und Bestellverfahren auf den Feldaufgang von Getreide. **Effect of tillage and cultivation techniques on field sprouting of cereals.**

2909 Kahnt, G.; Kübler, E.   DE 144485/78/0003 N
Sommergetreide- und Wintergetreideanteil in der Fruchtfolge. **Effect of different percentages of spring- and winter–cereals in crop rotations.**

2910 Fischbeck, G.; Knittel, H.   DE 161250/77/0003
Vergleich anbautechnischer Differenzierungen in Getreidefruchtfolgen – Dauerversuche –. **Comparative differentiations of cultivation of cereal crop rotations – permanent experiments –.**

2911 Zach, M.   DE 201040/75/0018
Möglichkeiten zur Verbesserung der mechanischen Bodensetzung zur Optimierung der Pflanzenentwicklung in getreidestarken Fruchtfolgen. **Possibilities of improving mechanical land subsidence for optimizing the plant growth in crop rotations where cereals are prevailing.**

2912 Müller, S.; Bramm, A.; El–Bassam, N.
DE 201040/78/0006 N
Wechselwirkungen zwischen bodenphysikalischen Parametern und der Wurzelausbreitung sowie der Bestockungsintensität. **Interaction between soil–physical parameters, root development and tillering intensity of cereals.**

2913 Gerstenkorn, P.; Bolling, H.   DE 209020/77/0001
Einfluss von Düngung, Sorte und Umwelt auf die Qualität von Getreide. **Quality of cereals as affected by fertilization variety and environmental conditions.**

2914 Sanchez, W.   DE 301100/77/0009
Mikroklima in Getreidebeständen im Hinblick auf Wachstum und Ertrag. **Microclimate in cereal crops regarding growth and yield.**

2915 Pommer, G.   DE 502050/70/0028
Acker- und pflanzenbauliche Untersuchungen in engen

Körnerfruchtfolgen und in Getreidemonokulturen 1967. **Investigations on crop growing in close rotations of cereals and in grain one–crop system.**

2916 Pommer, G.   DE 502053/74/0001
Langjährige Ertragsentwicklung der Getreidearten in einer "Alten Dreifelder–Fruchtfolge" bei unterschiedlicher Anbauintensität. **Many years' yield development in cereal species in an "old three–field crop rotation" with varying intensity of cultivation.**

2917 Ohms, J.–P.   DE 507050/77/0001
Bestimmung der Sortenzugehörigkeit und Sortenechtheit von Getreide mittels Gelelektrophorese. **Grain variety diagnosis by use of proteins and isoenzymes electrophoretic patterns.**

2918 Haeder, H.–E.   DE 902003/77/0003
Ertragsbildung und Hormonhaushalt bei Getreide. **Yield formation and phytohormone metabolism in cereals.**

2919 Jepsen, H.M.   DK 010107/59/3602
Forskellige sædskifters og mellemafgrøders indflydelse på kornudbytte og kvalitet samt på jordens dyrkningsegenskaber. **Influence of different rotations and cash crops on cereal yields and quality and on soil conditions.**

2920 Jepsen, H.M.; Hansen, P.F.   DK 010107/70/3601
Forskellige så- og høsttider i korn og disses indflydelse på udbytte og kvalitet. **Different sowing and harvesting dates in cereals and their influence on yield and quality.**

2921 Olesen, J.   DK 010701/74/0010
Belysning af sædskiftemæssige foranstaltninger med henblik på at opnå fortsat stigende udbytter af de enkelte kornarter. **Evaluation of crop rotation measures as a means of increasing the yields of cereal species.**

2922 Chesneaux, M.T.   FR 011103/76/0426
Etude de l'influence de la dimension de la graine sur les caractéres morphologiques des plantules. Application aux céréales (orge–blé–maïs) et aux plantes potagères (pois–haricots–radis). **Seeds dimension influence on morphological characters of plantlet application for cereales (barley – wheat – corn) and legumes (peas – beans – radish).**

2923 Recamier, A.; Huet; Fourbet, J.F.; Rapilly
FR 011801/58/9070
Rotations céréalières. **Cereal rotations.**

2924 Guibert, P.; Jeannin, B.; De Vaubernier, E.
FR 012224/67/3014
Etude expérimentale de divers types de rotations applicables en Lorraine. **Experimental study of possible crop rotations in Lorraine.** Publications.

2925 Ellis; Cannell   GB 010503/00/0002
Reduced cultivation. Joint project with WRO.

2926 Cannell   GB 010503/00/0008
Reduced cultivation. Long term experiments on direct drilling.

2927 Gales   GB 010503/77/0014 R
Effects of soil conditions on plant water relationships and yield of cereals.

2928 Ellis; Bragg   GB 010503/79/0018 N

Root and shoot growth of cereals under field conditions.

2929 Ellis; Graham   GB 010503/79/0022 N
Sowing time and the growth of cereals.

2930 Cussans; Moss   GB 010811/00/0002
Effect of changes in tillage systems on the growth and control of
unwanted plant material in cereals.

2931 Elliott; Pollard   GB 010811/00/0004
Growth of cereals in reduced tillage systems.

2932 Tottman; Cussans   GB 010811/00/0008 R
Tolerance of cereals to herbicides.

2933 Bennett; Smith J   GB 011905/00/0005 R
Ultra–structural studies of plant cells.

2934 Kirby; Appleyard   GB 011907/00/0001 R
Factors affecting the growth and development of the ear as part
of physiology of yield.

2935 Austin   GB 011907/00/0002 R
Simulation models of growth and development to assist in the
establishment of selection criteria.

2936 Austin; Ford   GB 011907/00/0004 R
Effect of temperature and photoperiod on growth and
development.

2937 Austin; Morgan   GB 011907/00/0005 R
Interrelationships between photosynthesis, respiration and
growth of cereals.

2938 Quarrie   GB 011907/00/0006 R
Abscisic acid concentration in cereals in relation to genotype
water status and other environment factors.

2939 Austin   GB 011907/00/0007 R
The role of abscisic acid in mediating the responses of tropical
grain crops to water stress.

2940 Murphy   GB 011907/77/0011 R
Gibberellins in cereals in relation to development.

2941 Parker   GB 011907/77/0012 R
Histology and ultrastructure of cereals in relation to their
development, physiology and biochemistry.

2942 Innes; Blackwell   GB 011907/79/0013 N
Study the effects of morphological and physiological characters
on the water economy of cereals.

2943 Miflin   GB 012001/00/0013 R
The biosynthesis of amino acids, particularly lysine, in cereals.

2944 Welbank; Taylor   GB 012002/00/0005 R
Root growth and its relationship to uptake of plant nutrients
and depletion of soil water in cereals.

2945 Whittngham; Parry   GB 012002/00/0006 R
The effect of aerial pollutants, mainly sulphur dioxide and
fluoride on crop growth.

2946 Thorne; Taylor   GB 012002/79/0019 N
The physiological basis of yield determination in winter wheat

and barley.

2947 Thorne; Thomas   GB 012002/79/0020 N
Control of cereal grain yield by physiological factors affecting
production and distribution of carbohydrate.

2948 Widdowson; Welbank   GB 012003/00/0015
Compare rates of growth and yields of wheat and barley on
contrasting sites and in different seasons.

2949 Edwards; Lofty   GB 012004/78/0036 R
Effects of agricultural practice especially minimum tillage on
soil fauna.

2950 Prew; Hornby   GB 012009/00/0033
Disease risks in reduced cultivation systems.

2951 Soane   GB 030503/00/0007 R
Reduced cultivation techniques.

2952 Soane   GB 030503/00/0014 R
Incidence of clods in cereal seedbeds.

2953 Hayter   GB 030804/78/0002 R
Survey physiological characters related to crop performance in
barley and oats and construct breeding models.

2954 Allison   GB 030804/78/0009 R
Study biochemical components of barley and oat grains related
to malting, feeding and processing quality.

2955 Allison   GB 030804/78/0011 R
Study enzymic hormonal and other biochemical factors
affecting cereal development performance and yield.

2956 Researcher not indicated   GB 030901/00/0407
Cereal seed certification – methods of varietal purity assessment
and the classification of impurities.

2957 Researcher not indicated   GB 030901/00/0413 R
Development of an indoor vernalisation test for barley and
wheat varieties.

2958 Stewart; White   GB 040303/00/0012
Investigation of factors affecting the yield of cereal cultivars.

2959 White   GB 040303/79/0016 N
An investigation of provenance and its effect on development
and yield of cereals.

2960 Easson   GB 041308/77/0006
Factors effecting the performance of precission drilled cereals.

2961 Easson   GB 041308/77/0007
Husbandry factors affecting the development of yield in cereals.

2962 Researcher not indicated   GB 050121/00/0001 R
Cereals: rotations and continuous cropping.

2963 Researcher not indicated   GB 050121/00/0002 R
Cereals: cultivations and seeding techniques.

2964 Researcher not indicated   GB 050121/00/0003 R
Cereals: harvesting losses.

2965 Researcher not indicated   GB 050121/00/0007 R

D 2100 – Plant production general and crop husbandry

Cereals: manuring.

2966 Researcher not indicated   GB 050121/00/0008 R
Cereals: seed quality.

2967 Researcher not indicated   GB 050121/00/0012 R
Cereals: straw disposal.

2968 Researcher not indicated   GB 050121/00/0013 R
Cereals: varieties (other than ADAS/NIAB collaborative trials).

2969 Researcher not indicated   GB 050161/00/0001
Cereals: rotations and continuous cropping.

2970 Researcher not indicated   GB 050161/00/0002
Cereals: cultivation and seeding techniques.

2971 Researcher not indicated   GB 050161/00/0003
Cereals: harvesting losses.

2972 Researcher not indicated   GB 050161/00/0007
Cereals: manuring.

2973 Researcher not indicated   GB 050161/00/0008
Cereals: seed quality.

2974 Researcher not indicated   GB 050161/00/0012
Cereals: storage.

2975 Researcher not indicated   GB 050161/00/0013
Cereals: varieties (other than adas/niab collaborative trials).

2976 Researcher not indicated   GB 050161/00/0015
Straw disposal.

2977 Duffus   GB 060101/00/0006 R
Plant biochemistry including cereal grain development and germination.

2978 Gill   GB 060106/00/0010
Cereal variety trials(except NLT).

2979 Gill   GB 060106/00/0011
Continuous cereal production.

2980 Holmes   GB 060106/00/0012
Cultivations and soil conditions in relation to cereal production.

2981 Holmes   GB 060106/00/0013
Maximising yield of cereals.

2982 Lockhart   GB 060106/00/0015
Grain quality in cereals.

2983 Blackett   GB 060213/00/0003
Cereal husbandry.

2984 Morrice   GB 060213/00/0005 R
Monitoring of grain quality.

2985 Vallance   GB 060226/00/0006
Seedbed preparation systems and their influence on cereal yields.

2986 Paterson   GB 060314/00/0009
Evaluation of cereal varieties for various husbandry systems and environments.

2987 Paterson   GB 060314/00/0015
Surveys of farming practices and problems (cropping).

2988 MacNaeidhe, F.   IE 060201/00/1599 N
An investigation of cereal production on cutaway peat. Publications.

2989 Conry, M.J.; Thomas, T.M.   IE 060500/00/1420 N
Winter and spring cereal comparison. Publications.

2990 Leonard, T.F.   IE 060500/65/0189 R
The influence of agronomic practices on grain production in continous barley growing systems. Publications.

2991 Rice, B.   IE 060500/71/1221 R
Precision sowing of cereals. Publications.

2992 Leonard, T.F.   IE 060500/72/0191 R
Optimum grain production and economic land utilisation in continuous barley growing. Publications.

2993 Fortune, A.   IE 060500/73/0855 R
Comparison of various cultivation and seeding systems for cereals and other crops. Publications.

2994 Fortune, A.   IE 060500/76/1194 R
An investigation of the effect of various cultivation and sowing techniques on the growth and yield of cereals. Publications.

2995 Dunne, W.   IE 060600/75/1228 R
Spatial distribution of feed grain production, processing and utilization in Ireland. Publications.

2996 Gallagher, E.J.; Bird, G.; Doyle, A.; Cahill, P.; Green, T.; Wright, T.J.   IE 120105/59/9043 R
Cereal improvement programme which includes investigation of optimum input level, varietal evaluation, fertilization levels, growth regulation, disease and quality control. Publications.

2997 Lanza, F.; Boschi, V.; Onofrii, M.; Spallacci, P.
      IT 020500/72/0001
Confronto di ordinamenti cerealicolo–industriali in avvicendamento e in monocoltura e relativo bilancio della fertilità. **Rotation and monoculture of cereal–industrial farming systems: fertility balance.**

2998 Bianchi, G.   IT 041803/77/0117
Caratterizzazione della struttura e delle funzioni delle cere epicuticolari nei cereali. **Characterization of the structure and functions of epicuticular waxes in cereals.**

2999 Dantuma, G.; Klein Hulze, J.A.   NL 010102/67/6730
Vergelijkend onderzoek naar fysiologische produktiviteitskenmerken van enkele zaadgewassen. **Comparative research on physiological productivity characteristics of some seed crops.** Publications.

3000 Dantuma, G.   NL 010104/72/3582
Vergelijkend onderzoek naar fysiologische en ecologische produktiviteitskenmerken van granen en enkele andere zaadgewassen. **Comparative research on physiological and ecological productivity characteristics of cereals and some other seed crops.**

D 2100 – Plant production general and crop husbandry

3001 Vos, N.M. de   NL 010104/75/6229
Eco–fysiologische onderzoek bij graangewassen betreffende
ontwikkeling en gewasopbouw. **Eco–physiologic research with
grain crops in relation to development and crop structure.**
Publications.

3002 Vos, N.M. de; Sinke, J.   NL 010104/75/6230
Opbrengstvermogen van graangewassen en schade door enkele
biotische en abiotische factoren. **Yield capacity of cereal crops
and yield losses due to some biotic and abiotic factors.**
Publications.

3003 Hag, B.A. ten; Kuizenga, J.; Darwinkel, A.
      NL 010207/75/7639
Betekenis van multilines en mengteelt van rassen voor de
oogstzekerheid van granen. **Importance of multilines and
mixtures of varieties for the stability of cereal yields.**

3004 Darwinkel, A.   NL 010207/78/8400 R
Mogelijkheden voor verbetering van de brouwgerstteelt.
Possibilities for improvement of malting barley growing.

3005 Kupers, L.J.P.; Scholte, K.   NL 020025/65/5040
Onderzoek naar de oorzaken van het ontstaan van schade als
gevolg van nauwe vruchtopvolgingen bij granen. **Possibilities of
continuous cropping of cereals.**

3006 Jong, G.J. de   NL 040007/67/4347
De invloed van het uitstel van het oogsttijdstip op de opbrengst
en kwaliteit van koolzaad en granen. **The influence of delaying
the harvest on the yield and quality of rape–seed and cereals.**

3007 Jong, G.J. de   NL 040007/73/4342
De invloed van plantenregulatoren en fungiciden op de
opbrengst van granen. **The influence of plant regulators and
fungicides on the yield of cereals.**

Barley (B 3110)

See also 3056, 3079, 3096, 3666, 6159, 6428, 17446

3008 Bockstaele, L.; Maddens, K.; Ampe, S.; Derolez, J.
      BE 140000/71/0022 R
Rassenstudie, fytopathologie en fytotechnie van
brouwersgerst. **Study of races, phytopathological and
phytotechnological aspects of brewery barley.** Publications.

3009 Limberg, P.   DE 105201/71/0004
Wirkung von Reihenentfernung und Standraum auf die
ertragsbildenden Eigenschaften von Sommergerste. **Effect of
distance of rows and spacing upon the properties of spring
barley influencing yield.**

3010 Limberg, P.; Schildbach, R.   DE 105201/77/0001
Die klima– und standortbedingte Modifikabilität der
Brauqualität und der Ertragsbildung von Sommergerste. **The
modifiability of brewing quality and the yield formation of
spring barley as influenced by climate and location.**

3011 Schildbach, R.; Kimottho, S.; Limberg, P.
      DE 105201/77/0002
Beziehungen zwischen Ertrag und Qualitätseigenschaften von
Braugerste aus geographisch unterschiedlichen Gebieten
Europas. **Correlations between yield and quality of malting
barley from different geographical regions of Europe.**

3012 Schildbach, R.   DE 105201/77/0003
Prüfung von Wintergersten auf Anbauwürdigkeit und
Brauqualität. **Testing of winter barley for cultivation suitability
and brewing quality.**

3013 Limberg, P.; Schildbach, R.; Mokhtare, F.; Köhn, W.
      DE 105201/77/0012
Erforschung des Produktivitätstyps bei Gerste. **Studies on the
productivity type of barley.** Publications.

3014 Aufhammer, W.; Kochs, H.–J.; Schulte–Geers, C.;
Solansky, S.   DE 111252/73/0012
Untersuchungen zur Ertragsbildung und ihrer
Beeinflussbarkeit durch physiologisch wirksame Substanzen
bei der Sommergerste. **Investigations on the yield of spring
barley and its response to growth substances.**

3015 Heyland, K.–U.; Block, K.; Braun, H.
      DE 111252/73/0018
Auswirkung variierter Saatmengen in den Randreihen von
Fahrgassen auf den Reihenertrag bei Wintergerste und
Winterweizen. **Effect of varying seed rates in marginal rows of
driving lanes on the row yield of winter barley and winter
wheat.**

3016 Schön, W.J.; Krausse, H.–J.   DE 132182/75/0001
Untersuchungen zur Dynamik der Speicherungsprozesse in
reifenden Gerstenkaryopsen mit Hilfe von 14CO2.
**Investigations of the dynamics in developing barley grains by
the application of 14CO2.**

3017 Rajogopal, R.; Madsen, A.   DK 030104/79/0001 N
Faktorer, der påvirker udfoldning af unge bygblade. **Factors
affecting the unfolding of barley leaves.**

3018 Jensen, S.E.   DK 030143/79/0003 N
Evapotranspiration i byg i relation til bladareal og rodvækst.
**Evapotranspiration in barley in relation to leaf area and root
development.**

3019 Doll, H.; Køie, B.   DK 030146/73/0005
Produktiviteten af byg med højt indhold af lysin i proteinet.
**The productivity of barley with high lysine content in the
protein.**

3020 Kreis, M.   DK 030146/76/0006
Syntesen af stivelse i byg med højt indhold af lysin. **The
synthesis of starch in barley with a high lysine content.**

3021 Jensen, E.S.; Jakobsen, S.T.   DK 030148/79/0005 N
Karforsøg med stråforkortningsmidlet Terpal anvendt til
vårbyg. **Effect of "Terpal" on growth of barley in pot
experiments.**

3022 Recamier, A.; Fourbet, J.F.   FR 011801/58/3009
Etude expérimentale de rotations céréalières simplifiées et de
monoculture de céréales. **Experimental study of simplified
cereal rotations and monocultures of cereals.** Publications.

3023 Gallagher; Thorne   GB 012002/78/0018 R
Control of potential grain size in barley.

3024 Soane; Pidgeon   GB 030503/00/0004 R
Effect of primary cultivations on continuous barley growing.

3025 Researcher not indicated   GB 030901/00/0406 R

D 2100 – Plant production general and crop husbandry

Investigation of a possible source of variation in grow winter barley varieties.

3026 Researcher not indicated   GB 040402/76/0020 R
The effect of ploughing date on barley production.

3027 Easson   GB 041308/78/0004
The long term effects of direct drilling.

3028 Researcher not indicated   GB 050121/00/0010 R
Barley: spring and winter; ADAS/NIAB variety trials.

3029 Researcher not indicated   GB 050161/00/0010
Barley: spring and winter; adas/niab variety trials.

3030 Holmes   GB 060106/00/0014
Site and seasonal variation in yield and quality of barley.

3031 Thomas, T.M.   IE 060500/77/1422 N
Winter barley evaluation. Publications.

3032 Conry, M.J.   IE 060500/78/1421 N
Winter barley husbandry. Publications.

3033 Tafuri, F.   IT 061600/74/0027
Influenza di alcuni erbicidi sull'attività dell'IAA–ossidasi nell'orzo. Herbicidal influence on IAA–oxidase in barley.

3034 Kuizenga, J.   NL 010207/75/7642
Mogelijkheid van latere zaai bij wintergerst. Late sowing of winter barley.

3035 Wilten, W.   NL 050304/78/8990 N
Invloed van abscissinezuur (ABA) en gibberellinezuur (GA) op de kiemrijpheid van brouwgerst. The influence of abscisic acid (ABA) and gibberilic acid (GA) on germination dormancy and malt ripeness of brewing barley.

Maize (B 3120)

See also 494, 522, 1434, 3022, 3177, 6269, 16611

3036 Lambert, J.; Legros, P.; Toussaint, B.
     BE 020602/77/0015
Possibilités d'implantation du maïs dans la province de Luxembourg. Possibilities of maize inplantation in the province of Luxemburg.

3037 Livens, J.; Vlassak, K.; Van Holm, L.
     BE 040201/75/0006 R
Studie van de mais–soja mengkultuur. Study of intercropping maize soybeans. Publications.

3038 Limberg, P.   DE 105201/73/0002
Untersuchungen über die Tageslängeninduzierbarkeit in der Embryonalphase bei Mais. Day length induction into the embryonal state of maize.

3039 Caesar, K.; Rudat, H.   DE 105201/74/0002
Bildung und Verteilung der Trockenmasse in der Maispflanze am tropischen Standort. Dry matter formation and distribution in maize in the tropics.

3040 Heyland, K.–U.; Eversheim, F.; Kochs, H.–J.
     DE 111252/77/0001
Massnahmen zur Reifebeschleunigung bei Körnermais.

Treatment in advancement of the ripening of grain–maize.

3041 Jahn, W.   DE 129123/75/0001 N
Die technologische Qualität des Maiskornes in Abhängigkeit von Genotyp und Umwelteinfluss. The technological quality of the maize–grain as a function of genotype and environmental influence.

3042 Michael, G.; Zink, F.; Wilberg, E.   DE 144400/78/0010 N
Physiologische Untersuchungen an verschiedenen neuen Maisformen. Physiological studies on various new corn varieties.

3043 Gliemeroth, G.; Kübler, E.   DE 144485/73/0001 R
Mais als Reinigungsfrucht in getreidestarken Fruchtfolgen 1972–1978. Maize as cleansing crop in rotations rich in cereals.

3044 Stegemann, H.   DE 215140/77/0006
Makromolekulare Komponenten in Mais. Beziehungen zu phytopathologischen und genetischen Eigenschaften. Macromolecular components in maize. Relations to phytopathological and genetic properties.

3045 Olesen, J.   DK 010701/75/0012
Dyrkningsteknik og sortsvalg i silomajs. Growing methods and choice of cultivar in maize for silage.

3046 Bonhomme; Robelin, M.   FR 010802/76/6115
Influence de l'orientation des feuilles par rapport au rang et de l'épe mâle sur la photosynthèse d'un couvert de maïs. Photosynthesis of a maize canopy : influence of leaf orientation with respect to the row and of male spike.

3047 Manichon, H.; Sebillotte, M.   FR 011701/70/6166
Etude de la monoculture du maïs. Study of continuous corn.

3048 Roncin, F.   FR 011701/74/6174
Comportement du maïs en conditions difficiles (basses températures du sol) en début de végétation. Reaction of young maize to bad conditions (low temperature of soil).

3049 Researcher not indicated   GB 050103/00/0016 R
Sweet corn, production techniques and variety trials.

3050 Researcher not indicated   GB 050161/00/0014
Maize for grain; production including varieties (other than adas/niab collaborative trials).

3051 Venezian Scarascia, M.E.; Losavio, N.; Leo, S.
     IT 020500/75/0002
Morfologia di una coltura di mais e intercettazione della radiazione. Morphology of a maize–crop and radiation interception.

3052 Mariani, G.; Desiderio, E.   IT 020800/73/0003
Mais – Effetti ambientali e genetici sul contenuto di proteina della granella. Maize – Environmental and genetic effects on grain protein content.

3053 Desiderio, E.; Scarino, D.   IT 020800/77/0004 R
Mais – Revisione e nuova definizione delle classi di maturità FAO. Maize – Revision and new definition of FAO maize maturity classes.

3054 Maggiore, T.; Bertolini, M.   IT 020800/79/0004 N
Mais – Agrotecnica: investimenti e concimazione azotata.

Maize – Agronomic researches on nitrogen fertilization and sowing densities.

3055 Ghidini, G.; Bezzi, A.　　　IT 021800/78/0005
Prove di coltivazioni comparative di ibridi di mais a ciclo breve e di graminacee foraggere in colture monofite a 1.000 m s.l.m. **Cultivation comparative tests between short cycle maize hybrids and Gramineae foragers to monophyto cultivation at 1000 meters from sea level.** Publications.

3056 Righetti, P.G.　　　IT 040604/77/0281 R
Studi sulle proteine di riserva dei cereali mais e orzo. **Research on the reserve proteins of maize and barley.**

3057 Postiglione, L.; Cuocolo, L.　　　IT 040701/79/0005 N
Epoche di semina su diversi ibridi di mais e sorgo in secondo raccolto. **Effect of sowing dates on several hybrids of the maize and sorghum as second crops.**

3058 Baltadori, A.　　　IT 041006/73/0138
Effetto della temperatura sull'accrescimento del mais, e di altri cereali durante il sottoperiodo germinazione–fogliazione e il periodo fogliazione–maturazione. **Effect of temperature on the growth of maize and other cereals during the sub period of germination–foliation and the period of foliation–maturation.** Publications.

3059 Soave, C.　　　IT 060300/74/0029
Biosintesi delle varie frazioni proteiche in endospermi di mais normale e ricco in lisina. **Biosynthesis of different protein fractions in endosperms of normal and high lysine maize.**

3060 Torti, G.　　　IT 060300/77/0869
Biosintesi delle varie frazioni proteiche in endosperma di mais normale e ricco di lisina. **Biosynthesis of the various protein fractions in the endosperm of normal and lysine rich maize.**

3061 Tedeschi, P.　　　IT 060700/77/0989
Epoca di semina ed economia idrica in mais e girasole. **Sowing time and water saving in maize and sunflower plants.**

3062 Deinum, B.　　　NL 020025/75/6796
Onderzoek naar de groeiomstandigheden die in Nederland in het voorjaar bij optimale bemesting de produktie en voederwaarde van snijmaïs beperken. **Productionpattern of maize.**

Oats (B 3130)

See also 3079, 6273

3063 Antoine, A.; Compère, R.　　　BE 010003/76/0011 R
Production et utilisation de l'avoine fourragère. **Production and utilization of oat as forage.**

3064 Grimme, K.　　　DE 132093/77/0004
Der Gang des Gesamtwasserpotentials und des potentiellen osmotischen Druckes in Haferbeständen auf bearbeiteter und unbearbeiteter Löss–Parabraunerde. **Pressure potential and osmotic potential of expressed cell sap in oat–plants growing on tilled and untilled loess soil.**

3065 Baeumer, K.; Capelle, A.; Heuer, K.　DE 132181/77/0003
Ertragsbildung von Hafer in Abhängigkeit vom Wasser–, Wärme– und Stickstoffhaushalt bearbeiteter und unbearbeiteter AckerParabraunerden. **Growth and yield of oats as influenced by water regime, soil temperature and nitrogen transformation in tilled and untilled loessial soil.**

3066 Steiner, A.M.　　　DE 144411/77/0001
Untersuchungen zur Echtheitsbestimmung bei Saatgut von Agrostis. **Studies on the verification of species and/or cultivar in Agrostis seeds.**

3067 Strass, F.　　　DE 502058/78/0005 N
Verbesserung der Züchtungsgrundlagen bei Hafer. **Improvement of oats breeding.**

3068 Rajagopal, R.; Olsen, C.E.　　　DK 030104/79/0002 N
Isolering og identificering af auxinbeskyttende forbindelser fra havre koleoptiler. **Isolation and identification of auxin protectors from oat coleoptiles.**

3069 Researcher not indicated　　　GB 050121/00/0011 R
Oats: spring and winter; ADAS/NIAB variety trials.

3070 Researcher not indicated　　　GB 050161/00/0011
Oats: spring and winter; adas/niab variety trials.

Rice (B 3140)

See also 5432, 6305

3071 Caesar, K.; Daubitz–Kühnel, H.　　　DE 105201/77/0011
Die ökologischen und ökonomischen Bedingungen des Reisanbaus im nördlichen Mittelmeerraum. **The ecological and economic conditions of rice cropping in the northern area of the Mediterranean Sea.**

3072 Atanasiu, N.; Thiagalingam, K.　　　DE 129552/75/0002
Physiologische Untersuchungen über die Salztoleranz bei Reis. **Physiological studies on salt tolerance of rice plants.**

3073 Marie, R.; Grillard, M.　　　FR 011201/75/0287
Etude du développement de la plante de riz en milieu pollué contrôlé. **Study of the rice plant development in controlled conditions of polluted environment.**

3074 Fossati, G.; Fantone, G.C.; Mazzini, F.; Pela, E.; Facelli, L.　　　IT 011400/74/0004
Studio delle relazioni fra produzione di sostanza secca, produzione di granella, assorbimento ed accumulo degli elementi nutritivi in coltivazioni di riso. **Studies on the interrelationships among dry matter production, grain yield, uptake and storage of nutrients in rice fields.**

3075 Russo, S.; Licata, V.　　　IT 020800/74/0001
Riso – Effetti dell'interramento della paglia sulla crescita delle piante e la produzione della granella. **Rice – Effects on growth and paddy yield of straw incorporation into the ground.** Publications.

3076 Tano, F.　　　IT 040619/74/0639
Prove di tecnica culturale sul riso. **Trials on rice cultivation practices.** Publications.

3077 Tano, F.　　　IT 040619/77/0304 R
Prova di tecnica colturale del riso. **Test on rice growing.**

3078 Finassi, A.　　　IT 063300/74/0067
Semina del riso sul terreno asciutto. **Rice seeding on dry soil.**

D 2100 – Plant production general and crop husbandry

Rye (B 3150)

See also 3093

3079 Linser, H.; Kühn, H.   DE 129040/75/0003
Halmverkürzung bzw. Lagerungshemmung bei Roggen, Hafer und Gerste durch Wachstumsregulatoren. **Inhibition of lodging of cereals – rye, oats, barley – by growth regulators.**

3080 Wricke, G.; Schmidt–Stohn, G.   DE 138300/77/0001
Untersuchungen zur Proteinqualität beim Roggen. **Investigations on protein quality in rye.**

3081 Strass, F.   DE 502058/78/0002 N
Verbesserung der Anbautechnik im Roggenanbau. **Improvement of rye production.** Publications.

Sorghum (B 3160)

See also 1434, 3057

3082 Langlet, A.   FR 011401/69/0645
Etude de la résistance à la sécheresse du Sorgho grain. **Studies on drougth resistance of grain–sorghum.** Publications.

3083 Parker   GB 010809/00/0003 R
Study of the resistance of sorghum and millet varieties to a range of Striga species and strains.

3084 Martillotti, F.; Francia, U.; Verna, M.
IT 020700/79/0004 N
Studio sulle caretteristiche chimiche di 2 varietà di sorgo, destinate all'insilamento come pianta intera a diversi stadi di maturazione nonchè sul valore nutritivo degli insilati. **Chemical characteristics of two sorghum varieties ensiled as whole plant at different stages, and nutritive value of silage ripeness.**

3085 Stefanelli, G.   IT 062400/72/0107
Ibridi, investimento e distanza tra le file nella coltura del sorgo da granella. **Hybrids, plant population and row distances in grain sorghum crops.** Publications.

Wheat (B 3170)

See also 494, 526, 3015, 3022, 3177, 3873, 5449, 6342, 6391, 6428, 6429, 9411, 10204

3086 Heyland, K.–U.; Block, K.; Jacob, D.; Eversheim; Solansky, S.   DE 111252/73/0001
Einflüsse von pflanzenbaulichen Massnahmen, insbesondere Wirkstoffapplikationen auf die Assimilationsintensität und die Assimilateverteilung bei Sommerweizen mit Hilfe von 14C. **Influence of cultivation measures, especially applied substances, on photosynthesis and translocation using 14C.**

3087 Heyland, K.–U.; Gehlen, W.; Goldhammer, T.; Volger, B.; Kochs, H.–J.   DE 111252/73/0003
Einfluss der Zwischenfrucht–Anbautechnik auf die Ertragsleistung von Weizen und Zuckerrüben. **Influence of intercropping technique on yield performance of wheat and sugar beets.**

3088 Heyland, K.–U.; Block, C.; Schulte–Geers, C.
DE 111252/77/0003
Anbauverfahren bei Triticale. **Cultivation–technique of Triticale.**

3089 Heyland, K.–U.; Socansky, S.   DE 111252/78/0001 N
Einfluss zunehmender Weizenmonokultur auf die Kornertragsbildung. **The influence of increasing monoculture of wheat on the grain yield formation.** Publications.

3090 Müller, K.; Ngakoutou, P.   DE 132062/75/0006
Die Wirkung einer Blattdüngung mit Magnesium und Mangan auf Eiweissgehalt und Qualität von Weizen und Futtergerste bei hohen und geteilten Stickstoffgaben. **The effect of magnesium and manganese applications to leaves on the protein content and quality of wheat and fodder barley as a function of high and divided nitrogen supply.**

3091 Knoppik, D.   DE 161120/77/0001
Netto–Assimilation einzelner Weizenblätter während einer Wachstumsperiode. **Net assimilation of individual wheat leaves during one growth period.**

3092 Fischbeck, G.; Knittel, H.   DE 161250/78/0001 N
Vergleichsanbau von zertifiziertem Saatgut und selbsterzeugtem Saatgut mehrerer Winterweizensorten. **Comparative cultivation of certified seeds and self–produced seeds of several winter wheat varieties.**

3093 Weipert, D.; El–Baya, A.W.   DE 209020/77/0015
Veränderung der Inhaltsstoffe und Qualität von Weizen, Roggen und Triticale während der Reife. **Changes in constituents and quality of wheat, rye and Triticale during ripening.**

3094 Seemann, J.   DE 301010/73/0001
Internationale Weizenanbauversuche mit verschiedenen Aussaatzeiten und Düngergaben zur Ermittlung von Klima–Einflüssen. **International wheat cultivation experiment with different periods of sowing and fertilization for estimation of the climatic influences.**

3095 Schrödter, H.; Grahl, A.   DE 301030/72/0003
Untersuchungen über den Einfluss der Witterung auf die Keimruhedauer des Weizens als Grundlage für einen Auswuchs– warndienst. **Influence of weather factors on seed dormancy of wheat as basis for a sprouting warning service.** Publications.

3096 Andersen, A.S.   DK 030104/78/0007 N
Vækstretarderende stoffers indflydelse på prolinindhold i hvede og byg. **Growth retardants and proline content of wheat and barley.**

3097 Renard, H.A.   FR 011103/70/0292
Etudes sur la dormance psychrolabile des graminées, spécialement du blé. **Study of the water sensitive dormancy of the graminaceae, especially on wheat.**

3098 Simon, M.; Melle Chesneaux, M.T.; Cavenel, B.
FR 011103/74/0112
Etude des réactions variétales de céréales, principalement blé tendre, au stade plantule en présence de nouveaux herbicides et fongicides de contact ou systémiques. **Varietal reactions in cereals, especially soft wheat, at the "seedling" stage with new contact or systemic herbicides and fungicides.**

3099 Felix, L.; Rauzy, G.   FR 011104/75/0227
Origines et consequences agro–techniques des amplitudes de variation de la grosseur du grain de blé. **Agrotechnics origins**

and consequences of variation amplitudes of the size of grain of wheat. Publications.

3100 Huet, Ph.; Recamier, A.　　　FR 011801/71/9073
Problèmes posés par la monoculture du blé tendre. **Problems in continuous wheat cropping.**

3101 Fourbet, J.F.　　　FR 011801/74/9071
Introduction de cultures fourragères de courte durée dans une succession de cultures de type céréalier, implantées en technique de travail minimum totale ou partielle. **Introduction of a short cycle forage crop into a cereal type rotation carried out totally or pastially without tillage.**

3102 Houdard, Y.; Barlier, J.　　　FR 012224/71/9005
Amélioration de la technique de production du blé d'hiver dans l'Est de la France. **Improving winter wheat production techniques in the East of France.**

3103 Huges; Oliver　　　GB 011901/01/0030 N
Winter wheat, investigate response to specific environments and develop screening procedure.

3104 Miller; Worland　　　GB 011905/00/0025 R
Maintenance of precisely defined genetic stocks for experimental purposes.

3105 Payne; Corfield　　　GB 011907/77/0008 R
Biochemistry of the storage proteins of wheat grains in relation to bread–making quality.

3106 Murphy　　　GB 011907/77/0010 R
Metabolism of abscisic acid in wheat.

3107 Miflin　　　GB 012001/00/0015 R
Factors affecting the quality and quantity of protein synthesised in wheat grains.

3108 Whittingham; Radley　　　GB 012002/00/0002 R
Estimation of naturally occuring growth substances in developing wheat grains.

3109 Whittingham　　　GB 012002/77/0017 R
Physiological studies on maximum yield of cereals.

3110 Researcher not indicated　　　GB 050121/00/0009 R
Wheat: spring and winter; ADAS/NIA–b variety trials.

3111 Researcher not indicated　　　GB 050161/00/0009
Wheat: spring and winter; adas/niab variety trials.

3112 Leonard, T.F.; Cunningham, P.C.　IE 060500/76/1272 R
Continuous winter wheat and feed barley (spring). Publications.

3113 Thomas, T.M.; Dunne, B.　　　IE 060500/77/1295 R
Effect of seeding rate, timing of nitrogen application and fungicide application on the development and yield of winter wheat. Publications.

3114 Rizzo, V.; Lanza, F.　　　IT 020500/76/0003 N
Controllo agronomico di grani duri migliorati per la resistenza al freddo. **Agronomical trials of durum wheat lines "resistant to cold".**

3115 Perniola, M.; Convertini, G.; Ferri, D.

IT 020500/79/0002 N
Influenza dell'ambiente sulla composizione chimica del frumento con particolare riferimento alla frazione lipidica e proteica. **Environmental influence on wheat chemical composition particularly on lipid and protein contents.**

3116 Wittmer, G.; Baldelli, G.P.; De Stefanis, E.; Brando, A.
IT 020800/76/0004 R
Frumento duro – Analisi comparata della crescita in differenti grani duri (ed altre specie di cereali) con caratterizzazione del microclima. **Durum wheat – Comparative analysis of growth in several varieties (and other cereal species) with microclimate caracterization.**

3117 Wittmer, G.; Li Destri Nicosia, D.; Baldelli, G.P.; Brando, A.　　　IT 020800/78/0002 R
Frumento duro. Analisi della crescita: studio dell'apparato radicale in relazione al diverso comportamento varietale. **Durum wheat. Growth analysis: root system in relation to different behaviour of the varieties.**

3118 Wittmer, G.; De Stefanis, E.; Li Destri Nicosia, O.
IT 020800/78/0004
Frumento duro. Comparazione dei livelli di clorofilla presenti a differenti stadi di crescita nelle diverse varietà con caratterizzazione del microclima. **Durum wheat. Comparative analysis of chlorophyll in several varieties during different stages of growth with microclimate characterization.**

3119 Fortini, S.; D'Egidio, M.G.; Galterio, G.; Nardi, S.; Sgrulletta, D.　　　IT 020800/79/0001 N
Grano duro – Nuovi micrometodi per valutare la qualità tecnologica delle semole. **Durum wheat – New micromethods to evaluate semolina technological properties.** Publications.

3120 Maracchi, G.P.　　　IT 040502/74/0584
Importanza dello studio dei microclimi in agricoltura. Microclima di una coltura di grano. **The importance of microclimate studies in agriculture. Microclimate of a wheat crop.**

3121 Manenti, G.　　　IT 040618/77/0223 R
Ricerche sui meccanismi della resistenza all'allettamento del grano duro–tri–ticum Durum desf–. **Research on the mechanisms of resistance to the laying of durum–tri–ticum Durum desf– wheat.**

3122 Di Marco, G.　　　IT 060200/73/0082
Fisiologia di culture di campo di grano tenero e duro con particolare riguardo ad enzimi carbossilanti, nitrato riduttasi, proteine ed amminoacidi. **Carboxylating enzymes, nitrate reductase, protein and aminoacids in hard and soft wheat cultivated in open field.** Publications.

3123 Grappelli, A.　　　IT 060200/74/0078
Effetto di estratti ormonali di Arthrobacter su alcuni enzimi della germinazione e radicazione del grano. **Effect of growth regulators from Arthrobacter on the germination and root formation of wheat.**

3124 Torti, G.　　　IT 060300/77/0870
Regolazione della sintesi di proteine in semi di frumento. **Regulation of protein synthesis in wheat seeds.**

3125 Porceddu, E.　　　IT 060500/77/1006
Indagine qualitativa delle proteine in specie diverse di

frumento. **Qualitative evaluation of proteins in different species of wheat.**

3126 Samoggia, L.   IT 120500/77/0396
Produttività del frumento duro. **Hard wheat yields.**

3127 Darwinkel, A.; Hag, B.A. ten; Kuizenga, J.
NL 010207/73/7640
Invloed gewasstructuur op ziekteaantasting en opbrengst van tarwe. **Influence of crop structure on disease attack and grain yield of wheat.** Publications.

3128 Darwinkel, A.; Kuizenga, J.; Maenhout, C.A.A.A.; Cevaal, P.K.   NL 010207/76/7641
De teelt van wintertarwe bij vroege inzaai. **Growing of winter wheat after early sowing.**

3129 Hag, B.A. ten; Kuizenga, J.; Titulaer, H.H.H.
NL 010207/77/7637
Opbrengstbeperkende factoren bij de tarweteelt op zandgrond. **Limiting factors for high wheat yields on sandy soils.**

3130 Spiertz, J.H.J.; Ellen, J.   NL 020025/74/4468
Gewasstructuur en produktievermogen van tarwe in afhankelijkheid van morfologische en fysiologische rasverschillen. **Crop structure and potential production of a wheat crop as related to varietal differences.**

3131 Vos, J.   NL 020025/74/7346
De invloed van de temperatuur op produktie en verbruik van assimilanten alsmede op verouderingsverschijnselen bij zomertarwe na de bloei. **The influence of temperature on production and utilization of assimilates and also on phenomena of senescence in spring wheat after anthesis.**

Other cereals (B 3190)

See also 3083, 10210

Fibre plants and oil crops in general (B 3200)

See also 2999, 3324

3132 Laloux, R.; Falisse, A.; Cors, F.   BE 010008/78/0005
Phytotechniques sur les protéagineux et oléagineux. **Research on proteinaceous and oleaginous.**

3133 Specty, R.   FR 010906/75/3003
Etude sur les possibilités d'utilisation d'eaux chaudes produites par des centrales nucléaires pour la production de plantes à cellulose. **Studies on the possibility of using hot water rejected by nuclear electric power plants for growing fibre crops.**

3134 Veen, B.W.   NL 010104/70/2841
Fysiologische en anatomische aspecten van de vorming van vezelige bestanddelen in planten. **Physiological and anatomical aspects of the formation of fibrous elements in plants.**

3135 Vreeke, S.   NL 010207/59/3256
Teeltonderzoek aan oliehoudende gewassen (zaaimethoden, bemesting, ziektebestrijding). **Research on agronomic aspects of oilseed crops (sowing methods, fertilization, disease control).** Publications.

Flax (B 3210)

3136 Bockstaele, L.; Derolez, J.; Maddens, K.
BE 140000/71/0021 R
Rassenstudie en fytotechnie van vlas. **Race culture and phytotechnological aspects of flax.** Publications.

3137 Bockstaele, L.; Derolez, J.   BE 140000/79/0064 N
Verbetering en fytotechnie van het vlas en mechanisatie van de vlasteelt. **Breeding and phytotechnie of flax and mecanisation of flax culture.** Publications.

3138 Menoux Boyer, Y.; Bethenod; Rode, J.C.; Chartier, Ph.
FR 010101/75/5018
Activité photosynthétique en relation avec l'accélération de la croissance des plantes après une sécheresse temporaire modérée. **Growth and CO2 exchange as influenced by a moderate and time limited water stress in flax.**

Olive (B 3220)

See also 4105, 4106, 4108, 4109, 4113, 4114, 4115, 4116, 4264, 4266, 4273, 4474, 4839, 6455, 6459, 6460, 6461, 6467, 6468, 6470, 15267

3139 Lombardo, N.; Sposati, I.; Ciliberti, A.
IT 021400/74/0003 R
Prove di raccolta meccanica delle olive con l'impiego di sostanze facilitanti l'abscissione. **Trials of mechanical harvesting of olives with chemical aids to induce olive abscission.**

3140 Lombardo, N.; Ciliberti, A.; Parlati, M.; Turco, D.; Sposati, I.   IT 021400/77/0002 R
Influenza del periodo di raccolta sulla differenziazione a fiore delle gemme di olivo. **Influence of harvested period on the flower bud differentiation of olive tree.**

3141 Pannelli, G.; Filippucci, B.   IT 021400/77/0003 R
Fattori che influenzano la radicazione dell'olivo: ambiente, sostanze rizogene, apporti nutrizionali, tipo di talea. **Factors that influence the olive rooting: medium, root substance, nutitional brought, cutting type.**

3142 Pannelli, G.; Filippucci, B.   IT 021400/77/0004 R
Correlazioni tra lo stato vegetativo della pianta madre e la radicazione nell'olivo. **Correlation between the vegetative state of mother plant and the radication in olive tree.**

3143 Petruccioli, G.   IT 021400/77/0521 R
Individuazione dei limiti di efficacia sull'azione degli etilen–promotori nell'abscissione delle olive, influenze dei trattamenti sull'Habitus vegetativo e produttivo, residui nell'olivo. **Assessment of the efficiency range of ethylen–promotors in olive abscission; effects of dry treatment on the vegetative and productive behaviour, residues in olive trees.**

3144 Di Martino, E.   IT 021400/77/0690 R
Meccanizzazione della raccolta delle olive, aspetti agronomici, raccolta meccanica delle olive da mensa. **Mechanization of olive harvesting, agronomic aspects, mechanised harvesting of table olives.**

3145 Lombardo, N.; Diana, G.; Briccoli–Bati, C.
IT 021400/79/0003 N
Propagazione in vitro dell'olivo. **In vitro propagation of the**

D 2100 – Plant production general and crop husbandry

olive tree.

3146 Pannelli, G.; Filippucci, B.   IT 021400/79/0004 N
Propagazione dell'olivo e fattori che influenzano la radicazione
e la crescita. **Propagation of the olive and factors which have
influence on radication and growth.**

3147 Pannelli, G.; Silvestri, B.; Filippucci, B.
IT 021400/79/0007 N
Costituzione di oliveti idonei alla raccolta meccanica.
Constitution of olive trees fit for the mechanical harvesting.

3148 Corti, R.   IT 040505/72/0435
Propagazione dell'olivo e della vite. **Propagation of the olive
and the vine.** Publications.

3149 Antognozzi, E.   IT 041005/77/0468 R
Controllo della maturazione e dell'abscissione delle drupe di
olivo. **Control of maturation and abscission in olive drupes.**

3150 Tombesi, A.   IT 041005/77/0669 R
Effetti della potatura e dell'ombreggiamento sulla
fruttificazione dell'olivo e del nocciuolo, ricerca di
impollinatori efficienti per la CV. di nocciuolo–tonda
romana–. **Olive–tree and filbert : the effects of pruning and
shading on fruit–bearing ; research for efficient pollinators of
the filbert CV. – tonda romana –.**

3151 Jacoboni, N.   IT 041005/77/0696 R
Meccanizzazione della raccolta delle olive; aspetti agronomici
raccolta meccanica. **Mechanization of olive harvesting;
agronomic aspects of mechanical harvesting.**

3152 Guerriero, R.   IT 041104/77/0203 R
Differenziazione fiorale nell'ulivo. Influenza dei fattori
accrescimento dei rami e disponibilità idrica. Nel quadro
dell'accordo di collaborazione Italia–Francia. **Floral
differentiation in olive trees. Influence of branch growth and
water supply. Research above within the frame–work of the
Italian–French collaboration agreement.**

3153 Vitagliano, C.   IT 041104/77/0720 R
Meccanizzazione della raccolta delle olive, aspetti agronomici
raccolta meccanica. **Mechanization of olive harvesting; the
economics of mechanical harvesting.**

3154 Trentadue, A.   IT 041306/73/0323
Trattamenti procascola e raccolta meccanica delle olive
mediante l'impiego della macchine scuotitrici. **Procascola
treatment and mechanized olive harvesting using shaking
machines.**

3155 Vodret, A.   IT 041309/73/0329
La composizione dell'olio vergine di oliva in relazione all'uso
di sostanze atte a facilitare la raccolta meccanica delle drupe.
**The composition of virgin olive oil in relation to the use of
substances which would facilitate the mechanical harvesting of
the drupes.**

3156 Cresti, M.   IT 041402/77/0656 R
Biologia dell'impollinazione e fecondazione, ricerche
ultrastrutturali e biochimiche sull'autoincompatibilità
dell'olivo e altre piante arboree autoincompatibili. **Pollination
and fertilization biology, ultrastructural and biochemical
research on the self–incompability of the olive–tree and other
self–incompatible trees.**

3157 Fontanazza, G.   IT 061500/77/0818
Ricerche sulla propagazione per talea dell'olivo. **Research on
olive tree propagation by stem rooting.**

3158 Jacoboni, N.   IT 061500/77/0965
Indagini sulla biologia fiorale e tecnica colturale dell'olivo.
**Research on the floral biology and the cultural technique of the
olive tree.**

3159 Jacoboni, N.   IT 061500/77/0966
Indagini sulla propagazione e selezione dei portainnesti di
olivo. **Research on the propagation and selection of olive
root–stocks.**

3160 Casini, E.   IT 062800/74/0183
Ricerche sulla propagazione per seme e biologia della
riproduzione. Contributo allo studio della scarsa percentuale di
allegagione nell'olivo:eventuali interventi per aumentarla.
**Researches on seed propagation and on reproduction biology.
Contribution to the knowledge of the low fruiting in olive
trees: possibilities to increase it.**

Rape (B 3230)

See also 494, 3006

3161 Brinkmann, W.; Gehlen, W.   DE 111850/72/0008 R
Probleme des vereinzelungslosen Rübenbaus 1968. **Problems
of beet–growing without singling.**

3162 Kahnt, G.; Kübler, E.   DE 144485/78/0002 N
Raps als Haupt– und Zwischenfrucht in der Fruchtfolge. **Effect
of winter rape as main– and intercrop in the crop–rotation.**

3163 Beinhauer, R.; Stoltenberg   DE 301060/77/0002
Rapserträge in Abhängigkeit von Witterungsfaktoren in
Schleswig–Holstein. **Rape seed yield depending on weather
factors in Schleswig–Holstein.**

3164 Rasmussen, P.; Nordestgaard, A.   DK 010106/73/3511
Udbringningstid, kvælstofmængder, såmængde og
rækkeafstande, høsttid m.m. til gul sennep og rasp. **Time and
rate of nitrogen application, sowing rate and distance, time of
harvest, etc., in mustard and rape.**

3165 Mortensen, G.; Holm, S.N.; Madsen, E.
DK 030145/77/0022
Bestøvning, frøsætning og frøkvalitet i vårraps. **Pollination,
seed–setting and seed quality in spring rape.**

3166 Mesquida, J.   FR 011306/76/5132
Pollinisation des Crucifères par les abeilles. **Pollination of
crucifers by bees.**

3167 Researcher not indicated   GB 050127/00/0003 R
Oil seed rape: winter and spring; production techniques.

3168 Researcher not indicated   GB 050127/00/0006 R
Oil seed rape: spring and winter; ADAS/NIAB variety trials.

3169 Researcher not indicated   GB 050167/00/0003 R
Oil seed rape: winter and spring; production techniques.

3170 Researcher not indicated   GB 050167/00/0004 R
Oil seed rape: spring and winter; ADAS/NIAB variety trials.

D 2100 – Plant production general and crop husbandry

Soyabean (B 3240)

See also 3037, 3633, 3664, 3900, 3901, 5463, 6484, 6485

3171 Laeremans, R.   BE 040204/77/0002 R
Toepassingsmogelijkheden van sojabonen in België.
Adaptation possibilities of soyabeans in Belgium.

3172 Kahnt, G.   DE 144485/78/0005 N
Realisierung des genetischen Ertragspotentials von Soja.
Possibilities of realizing the genetic potential of soybeans.

3173 Rasmussen, P.; Flengmark, P.   DK 010106/67/3505
Prøvedyrkning af nye arter og sorter af bælgsæd – soyabønner
m.v.. **Preliminary trials of new species and varieties of pulses –
soyabeans, etc..**

3174 Billot, C.   FR 010615/76/9028
Amélioration des techniques de production du soja dans la
région du Sud–Est. **Improving agricultural practices for
soybean production in the South–East of France.**

3175 Blanchet, R.; Bosc, M.; Marty, J.R.; Clavier; Puech, J.;
Obaton, M.   FR 011401/73/6196
Adaptation de divers types variétaux de soja aux conditions
écologiques et agronomiques du sud de la France
(température, sécheresse, systèmes de culture). **Adaptation of
varietal types of soybean to ecological and agronomical
conditions of south of France (température, drought, cropping
systems).**

3176 Puech, J.   FR 011401/74/6198
Floraison et fructification du soja – Répercussions sur les
composantes du rendement. **Flowering and fructification of
soybean. Répercussion on the yield components.**

3177 Roncin, F.   FR 011701/75/6168
Insertion du soja dans les systèmes de culture – Etude de l'effet
précédent "soja" sur les cultures du blé et du maïs. **Study on
the back effects of soybeans on wheat and corn.**

3178 Pirani, V.   IT 021100/79/0028 N
Epoche di semina e densità in coltura di soia. **Times of seeding
and density in the cultivation of soya–bean.**

3179 Favilli, R.   IT 041101/74/0550
Ricerche sulla coltivazione della soia da granella. **Research on
soybean for use as grain fodder.**

3180 Izzo, R.   IT 041129/78/1065 N
Influenza delle radiazioni luminose sull'evoluzione degli steroli
in Soja hispida. **Influence of light radiations on the evolution of
Soja hispida sterols.**

Sunflower (B 3250)

See also 3061, 5468

3181 Renard, H.A.   FR 011103/75/0293
Etudes physiologiques sur le vieillissement et la conservation
des semences de tournesol. **Physiological studies on the
vieillissement and the conservation of sunflower seeds.**

3182 Laureti, D.   IT 021100/79/0024 N
Epoche di semina e densità nella coltura del girasole. **Times of

seeding and density in the cultivation of the sunflower.**

3183 Bruinsma, J.   NL 020041/73/4878
De hormonale regulatie van de fototropie bij zonnebloemen.
**Hormonal regulation of phototropism with Helianthus annuus
seedlings.** Publications.

Other fibre plants and oil crops (B 3290)

See also 5014

3184 Demol, J.; De Langhe, E.; Raes, G.; Waterkeyn, L.;
Verschaege, L.   BE 010019/76/0005 R
Etude du développement des "fibres" chez le cotonnier et des
facteurs écologiques et physiologiques susceptibles
d'influencer les propriétés de celles–ci. **Research on the
development of cotton "fibres" as influenced by physiological
and ecological factors.** Publications.

3185 De Langhe, E.; Vincke, H.   BE 040900/78/0001 R
Studie van de ontwikkeling van de katoenvezel en van de
ecologische en fysiologische faktoren die de
vezeleigenschappen kunnen beïnvloeden. **Study on cotton
development and on the ecological and physiological factors
influencing the fibre properties.** Publications.

3186 Bünsow, R.; Khafaga, E.   DE 132070/78/0001 N
Entwicklungsphysiologische Grundlagen des Anbaus der
Baumwolle – Gossypium –. **Principles of developmental
physiology in the cultivation of cotton – Gossypium –.**

3187 Rehm, S.; Hamza, A.   DE 132240/77/0003
Einfluss atmosphärischer Faktoren auf die Salzresistenz von
Baumwolle mit und ohne Mykorrhiza. **Influence of
atmospheric factors on salt resistance of cotton with and without
mycorrhiza.**

3188 Laureti, D.   IT 021100/79/0030 N
Tecnica colturale del ricino. **Technique of the castor bean
cultivation.**

3189 Ferweda, J.D.; Samson, J.A.   NL 020056/68/4611
Onderzoek inzake de teelt van Carthamus tinctorius L.
Research on growing of Carthamus tinctorius L. Publications.

3190 Mutsaers, H.J.W.   NL 020056/78/8859 N
Groeisimulatie van katoen. **Simulation of growth of cotton.**

**Sugarbeets and starch producing plants in general
(B 3300)**

See also 2924

3191 Researcher not indicated   GB 050127/00/0002 R
Sugar beet: production techniques.

3192 Researcher not indicated   GB 050127/00/0005 R
Sugar beet: ADAS/NIAB variety trials.

3193 Researcher not indicated   GB 050167/00/0002 R
Sugar beet: production techniques.

Potatoes (B 3310)

See also 2898, 3303, 3708, 3873, 3902, 15390

D 2100 – Plant production general and crop husbandry

3194 Limberg, P. DE 105201/71/0002 N
Physiologische Veränderungen der Kartoffel beim Anbau unter verschiedenen ökologischen Bedingungen. **Physiological changes of potato by planting under different ecological conditions.**

3195 Delhey, R.; Carls, J. DE 105201/77/0008
Untersuchungen über die Anpassungsfähigkeit von Kartoffelsorten an tropische Tieflandbedingungen insbesondere Temperatur und Tageslänge betreffend. **Studies on the adaptability of potato cultivars to lowland tropics conditions – temperature, day length –.**

3196 Delhey, R.; Carls, J. DE 105201/77/0009
Interaktionen von Virusinfektionen und Temperatur auf die Leistungsfähigkeit von Kartoffelsorten. **Interactions of virus infections and temperature on the yielding ability of potato cultivars.**

3197 Caesar, K.; Randeni, G. DE 105201/78/0001 N
Einfluss der Boden– und Lufttemperatur auf die Knollenbildung bei Kartoffelpflanzen. **Influence of soil– and air–temperature on tuberisation of potatoes.**

3198 Marschner, H.; Krauss, A. DE 144400/78/0007 N
Physiologische Studien über Knollenbildung bei der Kartoffel. **Physiological studies on potato tuber initiation and growth.**

3199 Marschner, H.; Mares, D. DE 144400/78/0008 N
Knolleninduktion und Stärkespeicherung in wachsenden Kartoffelknollen. **Tuber initiation and the accumulation of starch in growing potato tubers.**

3200 Dambroth, M. DE 201040/73/5011
Assimilatspeicherung der Kartoffelpflanze unter dem Einfluss exogener Wachstumsregulatoren bei verschiedenen edaphischen und atmosphärischen Bedingungen. **The influence of growth regulators on the translocation and accumulation of assimilates in the potato plant under different edaphic and atmospheric conditions.**

3201 Heilinger, F.; Dambroth, M. DE 201040/75/0027
Definition der Zeiträume für Keimruhe und Keimbereitschaft von Kartoffelknollen. **Definition of the periods of germ dormancy and of germination capacity in potato tubers.**

3202 Geike, F. DE 215060/78/0006 N
Untersuchungen zur Ursache des Schwarzkochens bzw. der Verfärbung von Kartoffeln und Knollengemüse und Möglichkeiten zur Verminderung. **On the causes of blackening and discoloration of potatoes and tuberous vegetables and procedures to reduction.**

3203 Langerfeld, E. DE 215120/77/0002
Untersuchungen über Wechselwirkungen verschiedener Faktoren und Krankheitserreger auf das Resistenzverhalten von Kartoffelsorten im Hinblick auf Lager– und Pflanzguteigenschaften. **Investigation on reciprocal effects of different factors and disease agents on the resistance of potato varieties with regard to storing and plant properties.**

3204 Loeschcke, V. DE 215140/77/0001
Untersuchungen über den Erbgang von Isoenzym–Mustern und ihre mögliche Eignung als Marker für Resistenzgene der Kartoffel. **Investigations on the heredity of iso–enzyme patterns and their potential suitability as markers for resistance genes in potato.**

3205 Hunnius, W.; Munzert, M. DE 502050/72/0009 N
Untersuchungen über Wechselwirkungen verschiedener Faktoren und Krankheitserreger auf das Resistenzverhalten von Kartoffelsorten im Hinblick auf Lager– und Pflanzguteigenschaften. **Investigations on interactions of different factors and agents of diseases on resistance varieties regarding storing and planting characteristics.**

3206 Hunnius, W.; Munzert, M. DE 502058/78/0003 N
Verbesserung der Produktionstechnik im Kartoffelbau. **Improvement of potato production technics.** Publications.

3207 Hansen, S.E.; Østergaard, P.S. DK 010111/73/4001
Forskellige dyrknings– og håndteringsmetoders indflydelse på udbytte og kvalitet af kartofler. **Influence of different growing and handling methods on the yield and quality of potatoes.**

3208 Torregrossa, J.P.; Pointel, J. FR 011605/77/5273
Pollinisation des solanées cultivées par Exomalopsis ogilviei. **Pollination of cultivated solonceae by Exomalopsis ogilvei.**

3209 Davin, A.; Davin, S. FR 012203/70/2257
Recherche des conditions optimales de production des tubercules en vue de leur transformation industrielle. **Studies on the optimal conditions for porato tuber production destined for industrial processing.**

3210 Lutman; Cussans GB 010811/00/0007 R
Control of potato groundkeepers.

3211 Challice; Jarrett GB 011510/78/0024 R
Physiology of earliness in potatoes.

3212 Bleasdale; Wurr GB 011804/00/0011
Production methods and physiology of seed potatoes in relation to subsequent crop yield.

3213 MacDonald GB 030902/67/0105 R
Assessment of varietal characteristics as an aid to variety identification.

3214 Hall GB 030902/77/0101 R
Characterisation of microscopic features of potato sprouts.

3215 Hall GB 030902/77/0201
Improvement of nodal shoot production from mother plants.

3216 Hutchinson GB 041101/00/0011
Plant population studies on potatoes grown for seed.

3217 Hutchinson GB 041101/00/0012
Sprouting potatoes for seed production.

3218 Hutchinson GB 041101/00/0015
The effect of date of haulm destruction and harvest on physiological ageing of potatoes.

3219 Hutchinson GB 041101/00/0023
ADAS blueprint for maximum ware potato production.

3220 Hutchinson GB 041101/77/0024
Plant population for blueprint seed crops.

3221 Researcher not indicated GB 050127/00/0007 R

Potatoes: maincrop; production techniques, varieties (excluding ADAS/NIAB collaborative trials).

3222 Researcher not indicated   GB 050127/00/0008 R
Potatoes: early ware; production techniques, varieties (excluding ADAS/NIAB collaborative trials).

3223 Researcher not indicated   GB 050127/00/0009 R
Potatoes: for seed; production techniques.

3224 Researcher not indicated   GB 050127/00/0010 R
Potatoes: maincrop, second early and early; ADAS/NIAB variety trials.

3225 Researcher not indicated   GB 050167/00/0005
Potatoes: maincrop; production techniques, varieties (excluding ADAS/NIAB collaborative trials).

3226 Researcher not indicated   GB 050167/00/0006
Potatoes: early ware; production techniques, varieties (excluding ADAS/NIAB collaborative trials).

3227 Researcher not indicated   GB 050167/00/0007
Potatoes: for seed; production techniques.

3228 Researcher not indicated   GB 050167/00/0008
Potatoes: maincrop, second early and early; ADAS/NIAB variety trials.

3229 Researcher not indicated   GB 050167/00/0009
Minor cash crops: production techniques.

3230 Lang   GB 060106/00/0017
Potato variety trials.

3231 Lang   GB 060106/00/0018
Tuber numbers and size distribution and yield of potatoes.

3232 Addison   GB 060213/00/0009
Potato husbandry.

3233 Duggan, J.J.   IE 060300/70/0026 R
Control of potato cyst nematode, using chemicals, resistant varieties and cultural practices. Publications.

3234 Dowley, L.J.   IE 060500/64/0239 R
Propagation and maintainance of virus free stocks of new potato seedlings. Publications.

3235 Rice, B.   IE 060500/69/0177 R
Investigation of potato harvesting and storage systems. Publications.

3236 Kehoe, H.W.   IE 060500/76/1163 N
The effect of seed size, treatment, spacing and row width on the optimum yield of new varieties of potatoes. Publications.

3237 Cremaschi, D.; Laureti, D.; Affatato, E.; D'Amato, A.; Cerato, C.   IT 021100/75/0005 R
Messa a punto dei criteri colturali atti alla produzione di patata da seme. **Set–up of cultural criteria for production of seed potatoes.** Publications.

3238 Cremaschi, D.; Affatato, E.   IT 021100/76/0012 N
Produzione di patata da seme con raccolta anticipata.. **Seed potato yield with advanced harvest.**

3239 Cremaschi, D.; Bartocci, S.; Affatato, E.
IT 021100/76/0013 N
Fallanze e produzione in coltura di patata.. **Missing plants and yield in a potato growing..**

3240 Cremaschi, D.; Vender, C.   IT 021100/79/0018 N
Densità di semina e fertilizzazione della patata da fecola. **Density of seeding and fertilization of the plant potato.**

3241 Pratella, G.C.; Biondi, G.; Menniti, A.M.
IT 040216/76/0002 N
Taglio e cicatrizzazione delle patate da seme. **Cutting and healing of potato seeds.**

3242 La Malfa, G.   IT 063100/77/0820
Anticipo della raccolta nella patata precoce. Aumento dell'allegagione ed anticipo della raccolta nel pomodoro da mensa. **Anticipated harvesting of the early potato. Increased fruit setting and anticipated harvesting of table tomatoes.**

3243 Foti, S.   IT 063100/77/0900
Ricerche sulla patata in ciclo autunno–primaverile. **Research on the potato grown in the autumn–spring cycle.**

3244 Bodlaender, K.B.A.   NL 010102/68/7990
Groei, ontwikkeling en kwaliteit van aardappelen bij optimale en stress–omstandigheden. **Growth, development and quality of potatoes under optimal and stress conditions.** Publications.

3245 Bodlaender, K.B.A.   NL 010102/68/7991
Beginontwikkeling, knolzetting en knolsortering van pootaardappelen; snelle vermeerdering van plantmateriaal. **First development, tuber initiation and tuber size distribution of seed potatoes; rapid multiplication of planting material.** Publications.

3246 Lamers, J.G.   NL 010207/73/7630 R
Onderzoek naar de produktiecapaciteit van aardappelen en suikerbieten in nauwe rotaties en continuteelt. **Investigations on the production capacity of potatoes and sugar beet in intensive crop rotations and monoculture.**

3247 Schepers, A.   NL 010207/74/7815 R
Toepassing van groeiregulerende stoffen bij poot– en consumptieaardappelen. **Application of growth regulators on seed and ware potatoes.** Publications.

3248 Schepers, A.   NL 010207/74/7816 R
Toepassing van groeiregulatoren bij fabrieksaardappelen. **Application of growth regulators on potatoes for starch industry.** Publications.

3249 Loon, C.D. van   NL 010207/74/7817 R
De invloed van de fysiologische leeftijd van pootgoed op de groei en de ontwikkeling van aardappelen. **Influence of physiological age of seed potatoes on growth and development of the crop.**

3250 Loon, C.D. van; Houwing, J.F.   NL 010207/77/7635
Onderzoek naar mogelijkheden tot verbetering van de kwaliteit van consumptieaardappelen. **Research on possibilities of improvement of the quality of ware potatoes.**

3251 Bus, C.B.   NL 010207/78/9016 N
Onderzoek naar verschillen in opbregst en kwaliteit van

aardappelen in het zuidwesten van Nederland. **Investigations on differences in yield and quality of potatoes in the South–Western part of the Netherlands.**

3252 kupers, L.J.P.; scholte, K. NL 020025/66/4450
Produktiepatronen van aardappelen en suikerbieten. **Production patterns in potatoes and sugar beets.** Publications.

Sugarbeets and other sugar crops (B 3320)

See also 684, 2600, 2997, 3087, 3246, 3252, 6595

3253 Martens, M.; Van Steyvoort, L.; Roussel, N.; Vigoureux, A.; Vanstallen, R. BE 140000/74/0026 R
Voeding, fysiologie en produktiviteit van suikerbieten in één, twee en driejarige vruchtwisseling. **Nutrition physiology and the productivity of sugarbeets in different numbers of years between sugarbeets in rotation.** Publications.

3254 Schildbach, R.; Lurz, E. DE 105830/78/0001 N
Der Einfluss einiger qualitätsbestimmender Faktoren – Jahre, Standorte, N–Düngung, Sorten – auf die Ertragsleistung und die technologische Verarbeitbarkeit der Zuckerrübe. **Influence of factors that determine the quality of sugar beets – years, locations, N–fertilization, varieties – on the yield and the technological processability.**

3255 Schulze, E. DE 111250/75/0001
Ertrags– und Qualitätsverbesserungen im Zuckerrübenbau über Verlängerung der Wachstumszeit durch Pflanzenanzucht im Freiland, Bodenabdeckung und Umpflanzen der Rüben. **Improvements of yield and quality in the cultivation of sugar beets by prolonging the growth period by means of outdoor cultivation, soil covering and transplanting of the beets.**

3256 Schulze, E. DE 111253/75/0006
Vergleichende Untersuchungen zwischen Zuckerrohr– und Winterzuckerrübenbau in Khuzistan/Iran. **Comparative studies on the cultivation of sugar cane and winter sugar beets in Khuzistan/Iran.**

3257 Schulze, E. DE 111253/75/0007
Saat–, Bewässerungs– und Beregnungsversuche im Winter– und Frühjahrsanbau von Zuckerrüben im Iran. **Sowing, irrigating and sprinkling experiments in winter and spring cultivation of sugar beets in Iran.**

3258 Brinkmann, W.; Gehlen, W. DE 111850/72/0007 R
Entwicklungstendenzen der Zuckerrübenernte 1971. **Trends of sugar–beet harvest.**

3259 Baeumer, K.; Märländer, B. DE 132181/77/0002
Jugendentwicklung und Ertragsbildung von Zuckerrüben in Abhängigkeit von Bodentemperatur und Stickstofftransformation unterschiedlich bewirtschafteter Ackerböden. **Seedling growth and yield of sugar beet as influenced by soil temperature and nitrogen transformation in tilled and untilled soil.**

3260 Zach, M.; Dambroth, M. DE 201040/75/0020
Feldaufgang und Ertragsleistung von Zuckerrüben bei variierter Bodenbearbeitungsintensität. **Shooting and yield performance of sugar beets as a function of varied tillage intensity.**

3261 Heger, K. DE 301030/73/0002 N

Meteorologisch–biologisches Modell für das Wachstum der Zuckerrübe. **Meteorological–biological model for the growth of sugar beets.**

3262 Bürcky, K.; Winner, C. DE 907010/78/0001 N
Vegetative Entwicklung der Zuckerrübe in Abhängigkeit von Witterung und Nährstoffangebot. **Vegetative development of the sugar beet depending on weather conditions and nutrient supply.**

3263 Winner, C.; Schäufele, W.R. DE 907030/74/0001
Wurzelerkrankungen, Ertrag und Qualität der Zuckerrübe in unterschiedlicher Fruchtfolge. **Root diseases, yield and quality of sugar beet in different crop rotations.**

3264 Rasmussen, P.; Augustinussen, E. DK 010106/72/3512
Forskellige gødsknings–, dyrknings– og opbevaringsmetoders indflydelse på sukkerroesaftens kvalitet. **The influence of fertilizers, growing and storage conditions on the sugar purity of sugar beet.**

3265 Christiansen, K. DK 030104/77/0022 N
Undersøgelse af sukkerroemitochondriers stofskifteprocesser. **Examination of mitochondrial metabolism in sugar beet.**

3266 Lenton; Pocock GB 012002/00/0003 R
Identification and physiological effects of plant growth regulators in sugar beet.

3267 Milford; Pocock GB 012002/00/0009 R
Environmental factors and internal mechanisms controlling growth development and yield of sugar beet.

3268 Draycott; Messem GB 012301/00/0006 R
Crop rotation and forms of lime.

3269 Jaggard GB 012301/00/0012 R
Soil physical conditions and root development.

3270 Jaggard GB 012301/00/0013 R
Cultivations and growing practices (including NIAB variety trials).

3271 Longden; Johnson GB 012301/00/0027 R
Agronomic factors affecting seed yield and quality.

3272 Longden; Johnson GB 012301/00/0029 R
Seed grading, advancement, washing and storage.

3273 Longden; Johnson GB 012301/00/0030 R
Seedling germination, seedling emergence and crop growth.

3274 Coulter, B.S. IE 060200/78/1425 N
Survey of sugar beet yields and related variables. Publications.

3275 McEntee, M.; Burke, W. IE 060300/74/1220 R
Relationship between sugar beet yields and climatic conditions. Publications.

3276 Gibbons, J.; Thomas, T.M.; Barry, P. IE 060500/00/1424 N
Effect of spring transplanting of sugar beet on yield, sugar content and sugar extractability. Publications.

3277 Fortune, A.; McEntee, M.; Jelley, R.M. IE 060500/73/1240 R

Cultivation treatments of sugar beet in a variety of soil and climatic conditions. Publications.

3278 O'Connor, L.; Fitzgerald, P.   IE 060500/74/1335 N
Assessment of sugar beet varietal differences in field emergence on different soil types. Publications.

3279 O'Connor, L.; Fitzgerald, P.   IE 060500/74/1337 R
Assessment of sugar beet varieties for yield and quality characteristics under planting–to–a–stand regime. Publications.

3280 Rice, B.; Power, R.; Burke, G.   IE 060500/75/1222 R
Sugar beet harvesting and storage. Publications.

3281 Gibbons, J.; Thomas, T.M.; Sherrington, J.
   IE 060500/77/1340 R
An in–depth localised survey on factors affecting sugar beet yield and quality on farms. Publications.

3282 Rice, B.; Cunney, M.B.; Power, R.; Williams
   IE 060500/78/1389 N
Mechanization costs in sugar beet and cereal harvesting. Publications.

3283 Boschi, V.; Spallacci, P.   IT 020500/75/0010
Prove di coltivazione della bietola da zucchero nell'ambiente collinare. Yield and fitting of sugar–beet in hilly areas.

3284 Casarini, B.; Fontana, F.; Di Candilo, M.
   IT 021100/75/0011 N
Effetti della coltivazione della barbabietola da zucchero sullo stesso terreno.. Effects of the sugarbeet growing recurring it self in same soil.

3285 Casarini, B.   IT 021100/77/0676 R
Meccanizzazione della raccolta della barbabietola da zucchero del pomodoro e del pisello, produttività ed epoche raccolta barbabietole, nuove cultivar adatte raccolta meccanica, pomodoro–pisello. Mechanization of sugar beet, tomato and pea harvesting, sugar beet yields and harvesting times, new cultivars suited to mechanised harvesting, tomatoes–peas.

3286 Pirani, V.   IT 021100/79/0009 N
Impiego di antitraspiranti miscelati o non con prodotti anticercosporici su barbabietola da zucchero. Use of anti–transpiration products mixed or not with products against the sugar beet leaf spot.

3287 Pirani, V.   IT 021100/79/0010 N
Effetto dell'Ethrel e di composti dipiridilici su barbabietola da zucchero. Effect of ethrel and dipiridilic compounds on sugar beet.

3288 Lovato, A.; Montanari, M.   IT 040201/78/0004
Tecniche di produzione di seme di barbabietola, medica e di piante ortive. Research on the seed production techniques for sugar beet, lucerne and vegetable crops.

3289 Mantovani, G.   IT 042001/77/0699 R
Meccanizzazione della raccolta della barbabietola da zucchero, raccolta, trasformazione, aspetti chimici. Mechanization of sugar beet harvesting, harvest, processing, chemical aspects.

3290 Macrì, F.   IT 062200/74/0134
Ricerche sul metabolismo del saccarosio in barbabietole

rizomani. Sucrose metabolism in sugar beet roots affected by Rizomania.

3291 Kromwijk, P.A.M.   NL 010207/75/6262 R
Ontwikkeling van methodieken ter bepaling van de vitaliteit van suikerbietenzaad. Development of methods for determination of the vitality of sugar beet seeds.

3292 Boer, J.   NL 010207/76/6620
Bestudering van de invloed van groeifactoren (incl. teeltmaatregelen) op de opbrengst van suikerbieten op een aantal praktijkpercelen. Yield of sugarbeets as influenced by growth characters (including cultural methods) in practice.

3293 Kromwijk, P.A.M.   NL 010207/77/7636
Verliezen bij de produktie en de verwerking van suikerbieten. Production, harvesting and processing losses of sugar beet. Publications.

3294 Boer, J.   NL 010207/78/8414
Verbetering van de opkomst en groei van de suikerbieten d.m.v. grondbindende en grondbedekkende middelen en methoden. Improvement of the emergence and development of sugar beet by means of soil conditions and stabilizers.

3295 Smit, A.L.   NL 020025/75/6795
De kwantitatieve betekenis van teelkundige en genetische factoren, die tot het "schieten" van bieten in hun eerste groeijaar leiden. The quantitative importance of cultivation methods and genetic factors influencing premature bolting in beets.

3296 Jorritsma, J.   NL 060003/70/7904
Toepassing van groeiregulatoren bij de suikerbietenteelt. Application of growth regulating substances in sugarbeet growing. Publications.

3297 Jorritsma, J.   NL 060003/74/7717
Ontwikkeling en verbetering van methoden ten behoeve van de oogstverwachting voor suikerbieten. Development and improvement of methods for the yield forecast of sugar beet. Publications.

3298 Jorritsma, J.   NL 060003/78/8989 N
Plantafstand en plantontwikkeling bij suikerbieten. Plant distance and plant development in sugarbeet. Publications.

Other starch producing plants (B 3390)

See also 3202, 3205, 3206

3299 Rixhon, L.; Delhaye, R.   BE 080800/78/0014
Etude des techniques culturales pour la fèverole d'hiver. Study of crop husbandry for winter horse beans.

3300 Carls, J.; Delhey, R.   DE 105201/77/0007
Ertragsphysiologische Untersuchungen an tropischen und subtropischen Knollenfrüchten. Physiological studies on tropical and subtropical root crops.

3301 Delhey, R.; Carls, J.   DE 105201/77/0010
Ertragsstruktur des Topinamburs – Helianthus tuberosus –. Yield components of Jerusalem artichoke – Helianthus tuberosus –.

3302 Lange, N.   DE 201040/75/0013

Erstellung einer Sammlung von Kartoffel–Wildarten, Primitivformen und Kultursorten in Form von Samen. **Establishment of a collection of wild potato species, of primitive types and of cultivated varieties in the form of seeds.**

3303 McNulty, P.B.; Al–jubouri, K.   IE 120301/75/9058 N
Vibratory potato digging.

3304 Bruijn, G.H. de   NL 020056/66/4612
Vergelijkend onderzoek naar de teelt van waterrijke zetmeelleverende gewassen. **Comparative research into the cultivation of some waterstarch crops.** Publications.

3305 Flach, M.   NL 020056/78/8860 N
Relatie tussen groei van bovengrondse en ondergrondse delen van cassave. **Relationship between top and rootgrowth of cassava.**

Grasses and forage crops in general (B 3400)

See also 1451, 1476, 1514, 2924, 3101, 3132, 5513, 10860, 10877, 11745, 12312, 13790, 15371, 15374, 15382, 15860, 15862

3306 Laloux, R.; Dohet, J.; Nijst, P.; Poelaert, J.; Falisse, A.
BE 01008/63/0001
Etude des plantes fourragères orientée vers le choix des cultivars. **Study of fodder crops, specially the choice of cultivated varieties.** Publications.

3307 Behaeghe, T.; De Baets, A.   BE 030017/76/0004 R
Netto–productiviteit van grasland en groenvoeders in praktijkomstandigheden. **Net productivity of grassland and fodder crops on farms.**

3308 Andries, A.; Carlier, L.; Van Hee, L.
BE 070400/68/0019 R
Bruto– en nettoproduktie van grasland en groenvoedergewassen onder praktijkomstandigheden. **Gross– and netto production of grassland and green fodder crops on a practical scale.** Publications.

3309 Boeker, P.; Opitz–von–Boberfeld, W.
DE 111251/74/0006
Einfluss des Mähertyps – Kreisel– und Balkenmäher – auf die Zusammensetzung und den Trockensubstanzertrag des Pflanzenbestandes in Abhängigkeit von der N–Düngung. **The influence of the type of mower – circular mower and bar cutter – on composition and dry–matter yield of plant stock in dependence on nitrogen fertilization.**

3310 Boeker, P.; Tennigkeit, E.   DE 111251/75/0002
Die Entwicklung der Blattfläche verschiedener Arten und Sorten im Hinblick auf die Produktionsrate und eine mögliche Sortenfrühidentifizierung. **Development of the leaf surface of various species and varieties with respect to production rate and a possible early variety identification.**

3311 Stählin, A.; Schäfer, K.   DE 129140/70/0001
CO2–Gaswechsel des Sprosses und die Wurzelatmung ganzer, intakter Graspflanzen. **Net assimilation and respiration in shoots and respiration in roots of whole and intact grass plants.**

3312 Simon, U.; Daniel, P.   DE 129140/72/0005
Einfluss der Bewirtschaftungsmassnahmen auf Ertrag und wertbestimmende Inhaltsstoffe von Futterpflanzenarten und –sorten. **Effect of cultivation practices on yield and quality constituents of forage species and varieties.**

3313 Kühbauch, W.   DE 161255/75/0001
Die Kohlenhydratqualität von Futterpflanzen aus Höhenlagen. **Carbohydrate quality of forage plants grown in highland areas.**

3314 Voigtländer, G.; Kloskowski, J.; Kühbauch, W.
DE 161255/77/0004
In vivo–Verdaulichkeit und Verwertung von frischem und konserviertem Frühjahrs– und Herbstfutter sowie von Bergheu mit besonderer Bearbeitung der Nichtstrukturkohlenhydrate. **In vivo digestibility and utilization of fresh and conserved forages grown in spring and autumn and of wilted fodder grown in highland areas, both with special working on non–structural carbohydrates.**

3315 Voigtländer, G.; Kühbauch, W.   DE 161255/78/0001 N
Gesamtanalyse von Futterpflanzen; Wachstum – Inhaltsstoffe – Verdaulichkeit. **Total analysis of forage crops; growth – chemical composition – digestibility.**

3316 Wermke, M.   DE 201030/77/0005
Einfluss unterschiedlicher Wirtschaftsweisen auf die Werteigenschaften des Grundfutters. **Influence of different cultivation systems on the quality of basic ration.**

3317 Wermke, M.   DE 201030/77/0007
Anpassungsfähigkeit von Futterpflanzenarten des gemässigten Klimas unter verschiedenen Standortbedingungen Koreas; Beitrag zur Kausalanalyse der Adaptionsfähigkeit. **Adaptability of forage plants in temperate climate under different site conditions in Korea; contribution to causal analysis of adaptability.**

3318 Sonnenberg, H.   DE 201070/78/0003 N
Technische Verfahren der Applikation von Konservierungsstoffen für Futterpflanzen. **Technical procedures for application of preservatives for fodder plants.**

3319 Schöllhorn, J.; Szokolai, P.   DE 501200/70/0007
Der Einfluss des Reifestadiums auf die Gärfutterqualität verschiedener Futterpflanzen. **The stage of growth and its influence on the quality of silage of different kinds of fodder plants.**

3320 Scheller, H.; Rieder, J.B.   DE 502050/70/0032
Untersuchungen über die Eignung verschiedener Feldfuttermischungen bei Anbau unter unterschiedlichen Standort– bedingungen für die Heisslufttrocknung 1970. **Investigations on the suitability of different green forage mixtures grown under different ecological conditions for hot–air drying.**

3321 Lennartz, H.   DE 508301/73/0005
Konkurrenzverhalten von in Deutschland zugelassenen Sorten von Deutschem Weidelgras, Wiesenschwingel, Lieschgras und Wiesenrispe. **Competition between gramineae as admitted in Germany of common ryegrass, meadow fescue, Timothy grass, and rough–stalked meadow grass.**

3322 Lütke–Entrup, N.   DE 508301/75/0004
Ertragsvergleich verschiedener Futterpflanzen nach dem 1. und 2. Welschgrasschnitt. **Comparison of yield of various forage plants after the first and the second Italian ryegrass cutting.**

D 2100 – Plant production general and crop husbandry

3323 Lütke–Entrup, N.   DE 508301/77/0004
Einfluss unterschiedlicher Bodentemperaturen auf Ertrag und Inhaltsstoffe von ein– und mehrjährigen Futterpflanzen. **Effects of different temperatures in soils on crop yield and constituents of annual and perennial forage plants.**

3324 Olesen, J.   DK 010701/74/0004
Forsøg med dyrkning af frøafgrøder og industriafgrøder. **Growing trials with seed and industrial crops.**

3325 Dennis, B.; Nielsen, H.M.; Johansen, B.R.; Lysgaard, C.P.; Petersen, H.L.   DK 030145/73/0008
Produktionsfaktorer i fodergræsser. **Production factors in forage grasses.**

3326 Hentgen, A.; Jeannin, B.   FR 010112/70/3028
Nouveautés techniques dans le secteur de la production fourragère et leur insertion dans des systèmes de production. **Technical innovations in forage production: their possibilities of introduction in production systems.** Publications.

3327 Billot, C.; Allerit   FR 010615/72/9029
Etude des potentialités de diverses espèces fourragères. **Comparison of the growth and yield of various forage crops.**

3328 Niqueux, M.; Arnaud, R.   FR 010801/66/0132
Détermation de l'adaptation des plantes fourragères semées en altitude. **Assessment of the adaptability of fodder plants sawn in altitude.** Publications.

3329 Mettauer, H.   FR 010901/70/6122
Etude des potentialités de nouvelles espèces en systèmes fourragers dans une vallée vosgienne. **Forage potentiality of new species in vosgian valley.**

3330 Mennessier, P.; Jeannin, B.; Damour, L.; Garreau, J.   FR 012201/65/9058
Potentialité de cultures annuelles, fourragères ou destinées à la vente, sur marais drainés. **Testing the potentiality of annual crops for fodder or for sale on drained marsh–lands.**

3331 Jeannin, B.; Lafon, E.   FR 012201/74/9057
Potentialité des plantes fourragères pérennes implantées sur marais drainé, en sec et à l'irrigation. **Testing the potentiality of perennial forage plants on drained marsh–lands with and without irrigation.**

3332 Corrall; Terry   GB 011202/76/0025 R
Collation of data on the yield and quality of forage crops in relation to time of harvest and management.

3333 Sheldrick   GB 011202/76/0028 R
Sequences of forage crops to maximise yields for silage, zero grazing or drying enterprises.

3334 Corrall   GB 011202/76/0044 R
Bio–economic assessments of forage species and varieties and management.

3335 Wilkinson   GB 011202/76/0045 R
Bioeconomic assessment of forage conservation methods.

3336 Garwood   GB 011202/77/0043 R
Effects of edaphic factors particularly water status, on forage crop growth and nutrient loss.

3337 Thomas   GB 012104/76/0025 R
Temperature, light and water limitations to growth of contrasting herbage populations in the field.

3338 Bartholomew   GB 041308/76/0003 R
Examination of the potential total seasonal dry matter output of different crop sequences.

3339 Briant   GB 060213/00/0007 R
Forage crop husbandry.

3340 Paterson; Heppel   GB 060314/00/0012 R
Evaluation of root and forage crop varieties for various husbandry systems and environments.

3341 Edgar; Harkess   GB 060314/79/0019 N
Effect of application of liquid digested sewage sludge on grassland production and quality.

3342 Maguire, M.F.; Dempster, J.F.; Griffiths, T.W.; Wilson, R.K.; Gibney, M.J.   IE 060100/75/1199 R
The extraction and utilisation of protein from forage crops. Publications.

3343 Thomas, T.M.   IE 060500/71/0198 R
Growing arable fodder crops for animal feed. Publications.

3344 Onofrii, M.; Lanza, F.; Boschi, V.; Spallacci, P.   IT 020500/66/0001
Prati pascolo:specie,miscugli,resistenza al pascolamento. **Pasture fields:types,mixes,resistance to grazing.** Publications.

3345 Venezian Scarascia, M.E.; Losavio, N.   IT 020500/76/0001 N
Ritmo di accrescimento di colture a differente efficienza fotosintetica. **Growth rate of cultivations at different photosynthetic efficiency..**

3346 Lanza, F.; Ferri, D.; Perniola, M.; Lopez, G.   IT 020500/79/0004 N
Influenza di differenti modalità di interventi agronomici sul valore nutritivo delle produzioni foraggere in diversi ambienti pedo–climatici. **Influence of different agronomical practices on forage plants nutritive value in various pedo–climatic environments.**

3347 Iannelli, P.   IT 020900/70/0006
Tecnica delle colture da seme delle principali foraggere prative. **Technique of the main grassland crops grown for seeds.**

3348 Corleto, A.   IT 040101/73/0182
Ricerche sulla produzione foraggera. **Research on forage production.**

3349 Corleto, A.   IT 040101/77/0152 R
Ricerche sulla produzione foraggera. **Research on fodder production.**

3350 Bianco, P.   IT 040119/73/0149
Ricerche sulla flora pabulare dei pascoli naturali della Murgia pugliese, ai fini del miglioramento della produzione. **Research on the pabulary flora of natural pastures in the Murge of Apulia in view of incrementing the production.**

3351 Giardini, A.   IT 040201/72/0470

Ricerche sulla produzione foraggera. **Research on forage production.** Publications.

3352 Giardini, A.   IT 040242/77/0195
Ricerche sulla produzione foraggere. **Research on fodder production.**

3353 Foti, S.   IT 040305/73/0212
Ricerche sulla produzione foraggera. **Research on forage production.**

3354 Foti, S.   IT 040305/77/0186 R
Ricerche sulla produzione foraggera. **Research on fodder production.**

3355 Paris, P.   IT 040401/73/1198
Limiti e conseguenze agronomiche dell'impiego di alte dosi di concimi minerali nella foraggicoltura emiliana. **Limits and agronomic consequences of the use of large doses of mineral fertilizers in forage cultivation in Emilia.** Publications.

3356 Talamucci, P.   IT 040501/73/0315
Competizione fra le piante foraggere. **Competition between forage plants.**

3357 Talamucci, P.   IT 040501/77/0303 R
Competizione fra le piante foraggere. **Fodder crops competition.**

3358 Talamucci, P.   IT 040501/77/0812
Catene foraggere, allungamento del periodo di disponibilità di foraggere. **Fodder chains, lengthening the fodder availability period.**

3359 Orsi, S.   IT 040502/73/0268
Ricerche sulla produzione foraggera e determinazioni qualitative in laboratorio. **Research on forage production and qualitative analysis in the laboratory.**

3360 Orsi, S.   IT 040502/77/0251 R
Ricerche sulla produzione foraggera. **Research on fodder crop production.**

3361 Toniolo, L.   IT 040801/77/0717 R
Meccanizzazione della raccolta dei foraggi per gli allevamenti vacca–vitello, utilizzazione stocchi, recupero zone marginali, aspetti agronomici. **Mechanization of fodder crops harvesting for brood–cow farms, uses of stalks, reclaiming marginal lands, agronomical aspects.**

3362 Bianchi, A.A.   IT 041001/77/0116 R
Ricerche sulle foraggere.. **Research on fodder crops..**

3363 Favilli, R.   IT 041101/73/0203
Ricerche su colture foraggere. Valutazione bio–agronomica di ecotipi italiani e della festuca arundinacea. Valutazione fertilità residua del prato da vicenda. Moltiplicazione del seme di specie graminacee. **Research on forage crops. Bio–agronomic evaluation of Italian ecotypes and of Festuca arundinacea. Evaluation of the residual fertility of alternate grasslands. Seed multiplication of the Gramineae species.**

3364 Pardini, G.   IT 041101/74/0596
Ricerca sulla produzione foraggera. Miglioramento genetico della sulla. Reperimento e valutazione bio–agronomica di ecotipi di specie foraggere. Concimazione azotata phalaris tuberosa. Ricerche sul più conveniente ritmo di utilizzazione della festuca arindinacea. **Research on forage production. Genetic improvement of Hedysarum coronarium. Identification and bio–agronomic evaluation of forage species ecotypes. Nitrogen fertilizing of phalaris tuberosa. Research on the economically optimal rate of exploitation of festuca arindinacea.**

3365 Cavallero, A.   IT 041201/74/0532
Ricerche sulla produzione foraggera. **Research on forage production.**

3366 Cavallero, A.   IT 041201/77/0138 R
Produzione foraggera. **Fodder production.**

3367 Bosticco, A.   IT 041216/77/0674 R
Meccanizzazione della raccolta dei foraggi per gli allevamenti vacca da latte, fienagione e insilamento, aspetti zootecnici. **Mechanization of fodder crops harvesting for milch–cow farms, hay–making and ensilage, aspects related to live–stock rearing.**

3368 Rivoira, G.; Bullitta, P.; Caredda, S.   IT 041302/75/0006
Confronto fra 10 tipi di erbai con diverse modalità di utilizzazione invernale. **Comparison between 10 types of forage crops with different systems of winter use.** Publications.

3369 Lucifero, M.   IT 041313/73/0246
Ricerche sulla produzione foraggera. **Research on forage production.** Publications.

3370 Dattilo, M.   IT 041313/77/0683 R
Meccanizzazione della raccolta dei foraggi per gli allevamenti ovini e caprini, aspetti zootecnici, raccolta prodotti. **Mechanization of fodder crops harvesting for sheep and goat farms, aspects related to livestock rearing, the collecting of products.**

3371 Montanari, M.   IT 062400/70/0105
Confronto tra diverse soluzioni foraggicole nella pianura irrigua e nella collina asciutta del bolognese. **Comparison of various fodder crop solutions on the irrigated plains and dry hills around Bologna.** Publications.

3372 Baldoni, R.   IT 062400/77/0948
Confronto tra diverse impostazioni foraggicole in ambiente siccitoso della pedecollina bolognese. **A comparison of different fodder cultivation systems in the droughty environment of the Bolognese foothills.**

3373 Broekhoven, L.W. van   NL 010102/77/7994
Vorming van nitrosaminen in de bodem en tijdens nitraatreductie in plant en dier. **Formation of nitrosamines in soil and during nitrate reduction in plants and animals.** Publications.

3374 Spiertz, J.H.J.   NL 010102/79/0936 N
Onderzoek naar de benutting van stikstof en zonne–energie door ruwvoedergewassen in de teeltsystemen met en zonder vlinderbloemigen. **Research on the utilization of nitrogen and solar energy by forage crops in cropping systems with legumes.**

3375 Velde, H.A. te   NL 010207/71/3271
Onderzoek naar de inpassing van groenvoedergewassen in het bouwplan. **Investigations on the use of forage crops in crop rotations.** Publications.

D 2100 – Plant production general and crop husbandry

3376 Boeker, P.; Opitz–von–Boberfeld, W.
DE 111251/74/0005
Untersuchungen zum Konkurrenzverhalten verschiedener Sorten von Lolium perenne, Festuca pratensis und Phleum pratense in Mischungen. **Studies on the competitive behaviour of divers varieties of Lolium perenne, Festuca pratensis and Phleum pratense in mixed culture.**

3377 Simon, U.; Park, B.H.
DE 129140/78/0002 N
Untersuchung zur Ontogenese von Futtergräsern. **Investigation on the ontogenesis of forage grasses.**

3378 Simon, U.; Daniel, P.; Behrend, M.C.
DE 129140/78/0003 N
Optimaler Nutzungszeitraum von Futtergräsern. **Optimum period of utilization of forage grasses.** Publications.

3379 Simon, U.; Förster, E.
DE 129142/75/0001
Pflanzenbauliche Untersuchungen an Goldhafer – Trisetum flavescens L. P. B. –. **Cultivation research on Trisetum flavescens L. P. B..**

3380 Ellenberg, H.; Didden–Zopfy, B.   DE 132093/78/0001 N
Tagesgänge von Grössen des Wasserhaushaltes bei Lolium perenne L. und Arrhenatherum elatius L. M. unter verschiedenen Bedingungen. **Daily courses of parameters of water balance in Lolium perenne L. and Arrhenatherum elatius L. M. under different conditions.**

3381 Grimme, K.
DE 132093/78/0002 N
Transpiration, Gesamtwasserpotential und potentieller osmotischer Druck von Alopecurus pratensis und Brachypodium pinnatum unter abgestuften Bodenwasserverhältnissen. **Transpiration, total potential and potential osmotic pressure of expressed cell sap of Alopecurus pratensis and Brachypodium pinnatum growing under different soil water supply.**

3382 Kobabe, G.; Bugge, G.
DE 132183/71/0001
Verbesserung von Qualitätseigenschaften – Verdaulichkeit, Gehalt an Kohlehydraten – bei Futtergräsern. **Improvement of quality characters – digestibility, carbohydrate content – in fodder grasses.**

3383 Kobabe, G.; Bugge, G.
DE 132183/75/0001
Verbesserung des Eiweissgehaltes und der Eiweissqualität bei Futtergräsern. **Improvement of the protein content and the quality of protein in fodder grasses.**

3384 Weihe, K.von
DE 135053/71/0001 R
Systematik, Ökologie und Physiologie der für den Küstenschutz wichtigen Arten und Ökotypen der Gattungen Puccinellia, Festuca und Agrostis 1968. **Systematics, ecology and physiology of ecotypes of Puccinellia, Festuca and Agrostis of importance to coastal protection.**

3385 Neugebohrn, L.
DE 135053/74/0002
Keimung und Keimlingswachstum von Puccinellia maritima –Huds.– Parl. bei verminderter Sauerstoffversorgung. **Germination and growth of seedling in Puccinellia maritima**
–Huds.– Parl. **under reduced oxygenization.**

3386 Pirson, H.; Schering, J.
DE 135054/77/0001
Sortenfrüherkennung bei Gräsern in Rasenmischungen. **Quick diagnosis of grass cultivars in turf mixtures.**

3387 Schulz, H.
DE 144500/78/0002 N
Verhalten früher und später Sorten von Knaulgras und Deutschem Weidelgras in Mischungen. **The effect of early and late varieties of Dactylis glomerata and Lolium perenne in mixtures.**

3388 Knauer, N.; Häger, H.
DE 148101/77/0002
Ertragsbildung verschiedener Futtergräserarten und –sorten. **Crop formation of diverse species and varieties of feed grasses.**

3389 Voigtländer, G.; Mehnert, C.   DE 161255/73/0003 N
Pflanzensoziologische Veränderungen auf Fussballsport– rasen unter besonderer Berücksichtigung von Bodenaufbau, Bespielungsintensität und Pflegemassnahmen. **Plant sociological changes in football field green with special regard to soil structure, playing intensity, and tending operations.**

3390 Wermke, M.
DE 201030/77/0003
Jahreszeit und Nettoenergiegehalt von Gramineen. **Season and net energy content of gramineae.**

3391 Honig, H.; Schild, G.
DE 201030/77/0011 R
Vergleichende Untersuchungen zur Leistung von Konservierungsverfahren bei höchster Nutzungsintensität von Gräsern und Luzerne. **Comparative studies on the efficiency of preserving methods by greatest intensity of utilization of gramineae and Lucerne.**

3392 Geiger, K.; Schottdorf, W.
DE 502153/78/0002 N
Einfluss verschiedener Gräser auf den Boden und die vegetative Entwicklung sowie den Ertrag der Rebe. **Studies on the influence of different gramineae on soil and vegetative growth and yield of vines.**

3393 Ziegenbein, G.
DE 506155/75/0001
Untersuchungen mit Wachstumsregulatoren auf extensiven Rasenflächen, im Zierrasen und im Grassamenbau. **Studies with growth regulators on extensive lawns, in ornamental lawn and in grass seed production.**

3394 Lütke–Entrup, N.
DE 508301/75/0007
Einjähriges– und Welsches Weidelgras im Zwischenfruchtbau. **Annual– and Italian ryegrass in catch–crop–growing.**

3395 Lütke–Entrup, N.
DE 508301/77/0007
Untersuchungen über den Einfluss der Saatzeit und zeitlich differenzierter Herbstvornutzungen auf die Ertragsleistung des Welschen Weidelgrases im Hauptnutzungsjahr. **Crop yield of perennial ryegrass in the main year of farming as affected by seed time and differentiated preceding cropping.**

3396 Lütke–Entrup, N.
DE 508301/78/0001 N
Einfluss der Saatstärke auf den Ertrag von diploiden und tetraploiden Sorten des Einjährigen und Welschen Weidelgrases im Haupt– und Zwischenfruchtbau. **Influence of seed intensity on the yield of diploid and tetraploid varieties of annual and Italian ryegrass in main and catch–crop growing.**

3397 Lütke–Entrup, N.
DE 508301/78/0002 N
Ertragsleistung und Nährstoffgehalt von Sudangrassorten im

D 2100 – Plant production general and crop husbandry

Vergleich zu Silomais – im Hauptfruchtbau – bzw. Grünmais – im Zwischenfruchtbau –. **Yield performance and nutrient content of Sudan grasses – Sorghum sudanense – compared to silage maize – in main cropping resp. green maize – in catch cropping –.**

3398 Lütke–Entrup, N.   DE 508301/78/0003 N
Ertragsleistung, Ertragsverteilung, Nährstoffgehalt sowie Verdaulichkeit von Mischungen und unterschiedlichen Sortentypen des Deutschen Weidelgrases. **Performance and distribution of yield and nutrient content and digestibility of mixtures and different varietal types of perennial ryegrass.**

3399 Mølle, K.G.; Winther, P.   DK 010114/74/4301
Forskellige græsarter og græssorter i renbestand og i blandinger til slæthøstning. **Different grass species and varieties in monoculture and in mixtures for cutting.**

3400 Nielsen, G.   DK 030146/78/0002 N
Isoenzymundersøgelser i græsser. **Investigation of isoenzymes in grasses.**

3401 Larsen, I.   DK 030148/77/0005 N
Lysforholdenes betydning i karforsøg. **The influence of uneven light conditions in pot experiments on yield and content of some plant nutrients in italian ryegrass.**

3402 Hentgen, A.; Chevallier, C.   FR 010112/68/9097
Etude des techniques d'installation et d'entretien de gazons (ornement, exploitation, gibier, terrains de sport) et de diverses plantes ornementales. **Testing various techniques for the establishment and upkeep of swards (ornemental, playground, livestock and game feeding) and various ornamental plants.**

3403 Robelin, M.   FR 010802/71/0576
Influence des facteurs climatiques sur la photosynthèse des graminées fourragères. **Effect of climatic factors on photosynthesis in forage grasses.**

3404 Mingeau, M.   FR 010802/72/0578
Résistance des graminées fourragères à l'excès d'eau. **Resistance of forage grasses to water excess.**

3405 Coppenet, M.   FR 012208/75/6175
Composition minérale (éléments majeurs) des plantes fourragères. Influence des facteurs culturaux et climatiques sur la composition minérale du Ray–Grass italien au stade du pâturage. **Effects of cultural and climatic factors on mineral composition of lolium multiflorum at pasture stage.**

3406 Coppenet, M.; More, E.   FR 012208/75/6176
Composition minérale (oligoéléments) des plantes fourragères – Cu, Co, Zn, Mn, influence des facteurs culturaux et climatiques sur la composition minérale du Ray–Grass italien au stade du pâturage. **Effects of cultural and climatic factors on mineral composition of Lolium multiflorum at pasture stage (Cu, Co, Zn, Mn).**

3407 Jewiss   GB 011202/00/0010 R
Mechanisms which control tillering in the Gramineae.

3408 Smith   GB 011202/00/0012
Longevity of sown herbage crops and transition towards permanent grassland.

3409 Sheldrick; Corrall   GB 011202/76/0026
Sown grasses : yield and quality of species and varieties in relation to cutting, management and environment.

3410 Sinclair   GB 011202/76/0041
Prediction of grass yield from environmental measurements.

3411 Woledge   GB 011203/00/0001 R
Factors affecting the photosynthesis and respiration of individual leaves of grasses and legumes.

3412 Leafe   GB 011203/00/0002 R
Physiology of growth of grass and legume crops in the field.

3413 Ryle   GB 011203/00/0003 R
Carbon utilization and growth in the Graminae and Leguminosae.

3414 Robson; Parsons   GB 011203/00/0004 R
Growth and carbon metabolism of grass and legume crops: use of simulated swards in controlled environments.

3415 Harris   GB 011203/00/0006 R
Factors affecting water loss from leaves and tillers after cutting.

3416 Peacock   GB 011203/00/0007
Effects of temperature on the growth of grasses in the field.

3417 Sheehy   GB 011203/00/0008 R
Role of physical properties of forage grass and legume canopies in the utilization of solar energy for growth.

3418 Jones   GB 011203/75/0017 R
Physiological basis of the response of forage crops to management in the field.

3419 Ryle; Arnott   GB 011203/75/0018 R
Developmental morphology of legumes.

3420 Woledge   GB 011207/79/0016 N
Factors affecting the photosynthesis and respiration of individual leaves of grasses and legumes.

3421 Green   GB 011209/00/0001 R
Grassland intelligence and surveys.

3422 Forbes; Dibb   GB 011210/00/0001 R
Identification of factors (environmental, biological and sociological) which limit use productivity of permanent grassland.

3423 Davies   GB 012104/76/0020 R
Variations in tillering and regrowth after defoliation of forage grasses in relation to persistency.

3424 Troughton   GB 012104/76/0021 R
Variation in root systems of grasses and clovers in relation to productivity of herbage and to persistency.

3425 Davies   GB 012105/00/0009 R
Differences in the pattern of tiller and stem development in relation to conservation and grazing systems.

3426 Hughes   GB 012105/76/0023
Agronomic evaluation and management of potential varieties of ryegrasses and their hybrids.

3427 Hughes   GB 012105/76/0024
Agronomic evaluation and management of potential varieties of
cocksfoot, timothy and fescue species.

3428 Walters; Evans   GB 012105/76/0026 R
Evaluate feed quality characteristics of potential varieties of
ryegrass and their hybrids.

3429 Walters; Evans   GB 012105/76/0027 R
Evaluate feed qulaity characteristics of potential varieties of
cocksfoot, timothy and fescue species.

3430 Hayward; Breese   GB 012106/00/0002 R
Genetic control of variation of ecotypes and long term natural
selection. Survey breeding material.

3431 Breese; Hill   GB 012106/00/0004 R
Significance of genotype–environment interactions in forage
crops.

3432 Charles   GB 012106/00/0007 R
General and specific adaptation of herbage cultivars.

3433 Griffiths; Roberts   GB 012109/00/0006
Assess seed yield potential of varieties under various
management and fertiliser practices.

3434 Roberts; Lewis J   GB 012109/00/0007
Investigate harvesting, drying and storage of seeds.

3435 Bean   GB 012109/00/0013 R
Assess environmental effects on inflorescence development,
floret fertility and seed quality in grasses.

3436 Bean   GB 012109/00/0014 R
Assess effect of selection and breeding for forage yield and
quality on reproductive growth.

3437 Stoddart; Jones TWA   GB 012110/00/0001 R
Biochemistry of floral induction and initiation in forage plants.

3438 Stoddart   GB 012110/00/0003 R
Effect of environment and hormones on rates and siting of
enzyme formation in plants.

3439 Stoddart   GB 012110/75/0005 R
The mechanisms of plant hormone action with particular
reference to gibberellins and ethylene.

3440 Stoddart; Thomas   GB 012110/77/0009 R
The biochemistry of low temperature growth and winter
hardiness in forage plants.

3441 Pearson   GB 030901/00/0401 R
Evaluation of characteristics for identifying and classifying
varieties of red fescue.

3442 Sutton; Pearson   GB 030901/00/0402 R
Evaluation of characteristics and techniques for identifying
varieties in some herbage species.

3443 Laidlaw   GB 040301/00/0013
Suitable varieties for in–bye and marginal land improvement.

3444 Laidlaw; Faulkner   GB 040301/00/0020
Paraquat resistant cultivar of perennial ryegrass in a grazing
system.

3445 Hayes   GB 040301/00/0023 R
Physiological response of plants to environmental change.

3446 Hayes   GB 040301/00/0024 R
Effects of various managements on plant and pasture
persistency.

3447 Laidlaw   GB 040301/76/0018 R
Study of productivity and biology of legumes uncommon in
Northern Ireland.

3448 Camlin   GB 040303/00/0009 R
Varietal susceptibility to frost damage in ryegrass.

3449 Camlin   GB 040303/00/0010 R
Relative competitive ability of grass species and cultivars.

3450 Camlin   GB 040303/00/0011 R
Seedling establishment and tiller production in ryegrass
cultivars.

3451 Adams   GB 040402/00/0024
Blueprint for grass production.

3452 Harvey   GB 040403/76/0012 R
Mechanisms of herbicide tolerance in resistant grass species.

3453 Bartholomew   GB 041309/00/0004
The influence of method of sowing on the performance of
ryegrass.

3454 Researcher not indicated   GB 050130/00/0001 R
Grassland: production from leys and permanent grass.

3455 Researcher not indicated   GB 050130/00/0002 R
Forage: crop dehydration.

3456 Researcher not indicated   GB 050130/00/0005 R
Grasses: ADAS/NIAB grass/legume variety trials.

3457 Researcher not indicated   GB 050130/00/0006 R
Herbage seeds: production and herbage variety studies
excluding ADAS/NIAB collaborative trials.

3458 Researcher not indicated   GB 050130/00/0007 R
Grassland: establishment of swards: weeds :pests.

3459 Researcher not indicated   GB 050130/00/0008 R
Hill grazing: production and management.

3460 Researcher not indicated   GB 050130/00/0009 R
Maize for silage: production, varieties excluding ADAS/NIAB
collaborative trials.

3461 Researcher not indicated   GB 050130/00/0010 R
Maize: ADAS/NIAB variety trials.

3462 Researcher not indicated   GB 050130/00/0011 R
Fodder brassicas, other feed roots and arable forage crops:
production and techniques for utilisation.

3463 Researcher not indicated   GB 050130/00/0012 R
Fodder brassicas, other feed roots and arable forage crops:

ADAS/NIAB variety trials.

3464 Researcher not indicated　　　GB 050170/00/0001
Grassland: production from leys and permanent grass.

3465 Researcher not indicated　　　GB 050170/00/0002
Forage: crop dehydration.

3466 Researcher not indicated　　　GB 050170/00/0005
Grasses: ADAS/NIAB grass/legume variety trials.

3467 Researcher not indicated　　　GB 050170/00/0006
Herbage seeds: production and herbage variety studies.

3468 Researcher not indicated　　　GB 050170/00/0007
Grassland: establishment of swards; weeds; pests.

3469 Researcher not indicated　　　GB 050170/00/0009
Maize: for silage; production, varieties excluding ADAS/NIAB collaborative trials.

3470 Researcher not indicated　　　GB 050170/00/0010
Maize: ADAS/NIAB variety trials.

3471 Researcher not indicated　　　GB 050170/00/0011
Fodder brassicas, other feed roots and arable forage crops: production and techniques for utilisation.

3472 Researcher not indicated　　　GB 050170/00/0012
Fodder brassicas, other feed roots and arable forage crops: ADAS/NIAB variety trials.

3473 Habeshaw　　　GB 060105/00/0001 R
Frost resistance in grasses.

3474 Holmes　　　GB 060106/00/0002
Management studies on Italian ryegrass.

3475 Swift, G.　　　GB 060106/00/0003 R
Ear emergence,yield and quality of grass varieties.

3476 Morrison　　　GB 060106/00/0008
NLT grasses and clovers.

3477 Briant　　　GB 060213/00/0012 R
Turnip and swede survey.

3478 Naylor　　　GB 060216/00/0002 R
Establishment of grass swards.

3479 Murray　　　GB 060307/00/0003
A study of the effect of Ethrel on selected agricultural, horticultural and soil–stabilising grasses.

3480 Leclerc, M.H.　　　IE 060200/76/1139 R
The effect of early grazing,nitrogen and cutting dates on the growth and physiology of a pure grass sward in the first half of the year. Publications.

3481 O'Sullivan, M.　　　IE 060200/76/1229 R
Study of sward factors that affect animal intake. Publications.

3482 Brereton, A.J.　　　IE 060200/77/1267 R
Seasonal analysis of leaf and root efficiency in grass – under grazing and cutting conditions. Publications.

3483 Burke, W.; Jelley, R.M.　　　IE 060300/77/1380 R
Investigation of grass yields on reclaimed bog at Clonsast.
Publications.

3484 Ribeiro, Do Valle, M.D.V.; Leonard, T.F.
　　　IE 060500/75/1000 N
Study of relationships between stage of growth and silage quality. (For different ryegrasses). Publications.

3485 Crowley, J.; Fortune, A.; O'Rourke, C.; Daly, P.J.;
Murphy, W.E.　　　IE 060500/77/1338 R
Sward establishment and renovation by direct drilling.
Publications.

3486 Keane, G.P.　　　IE 120105/76/9042
The effect of harvesting schedule, nitrogen fertilisation and sward composition on the production of grass/clover swards .

3487 Lanza, F.; Venezian, M.E.; Losavio, N.
　　　IT 020500/79/0007 N
Prove di consociazione tra foraggere graminacee e leguminose nell'ambiente meridionale. **Mixture trials between grasses and forage legumes in southern Italy.**

3488 Iannelli, P.　　　IT 020900/69/0001
Composizione di erbai stagionali. **Components suitable for fodder catch crops.** Publications.

3489 Sardara, M.　　　IT 020900/70/0001
Tecnica dei prati poliennali semplici e misti in asciutto.
Techniques of long duration meadows, mono – or poliphytic, in dry land. Publications.

3490 Sardara, M.　　　IT 020900/70/0002
Composizione di erbai vernini e primaverili in asciutto.
Components of winter and spring fodder catch crops in dry land.

3491 Iannelli, P.　　　IT 020900/72/0001
Tecnica delle colture da seme di specie da erbaio. **Techniques of fodder catch crops grown for seeds.** Publications.

3492 Onofrii, M.; Paoletti, R.; Locatelli, C.
　　　IT 020900/79/0003 N
Rigenerazione e miglioramento delle vecchie cotiche erbose nell'Appennino settentrionale. **Renewal and improvement of old grassland swards in the northern Appennine.**

3493 Piano, E.　　　IT 020900/79/0006 N
Studio della stabilità fenotipica in popolazioni locali di festuca arundinacea. **Study on phenotipic stability in local populations of tall fescue.**

3494 Postiglione, L.; Basso, F.　　　IT 040701/79/0002 N
Effetto della quantità di seme in erbai semplici e misti per foraggio fresco. **Effect of the rate of seeding on yield of single and mixed grasslands for green fodder.**

3495 Postiglione, L.; Basso, F.　　　IT 040701/79/0004 N
Influenza dell'epoca di semina su erbai misti autunno primaverili per foraggio fresco. **Effect of sowing date on mixed grasslands for green fodder in winterspring.**

3496 Caporali, F.　　　IT 041101/77/0793 R
Indagini sulle possibilità produttive di cotici erbosi naturali ed artificiali nei terreni marginali. **Investigation on the possible**

D 2100 – Plant production general and crop husbandry

production of natural or artificial turf on marginal lands.

3497 Ciotti, A.   IT 063300/74/0061
Epoca di sfalcio primaverile del prato in relazione alle modalità di raccolta del foraggio. **Spring cutting time of grass in relation to different harvesting techniques.**

3498 Alberda, T.; Sibma, L.; Louwerse, W.
NL 010104/75/6226
De invloed van de structuur van een grasgewas op het gedrag van de afzonderlijke spruiten, speciaal in relatie tot veranderingen in groeisnelheid. **The influence of the structure of a grass crop on the performance of its composing tillers, especially in relation to changes in crop growth rate.** Publications.

3499 Meÿer, W.J.M.   NL 010207/74/7651
Bepaling van het optimale oogsttijdstip met behulp van het vochtgehalte bij Engels raaigras. **Determination of the optimum harvest time using the moisture content in perennial ryegrass.**

3500 Meijer, W.J.M.; Vreeke, S.   NL 010207/74/7652
Bestudering van groei en ontwikkeling van roodzwenk, Festuca rubra L., in relatie tot de zaadopbrengst. **Studies on growth and development of red fescue (Festuca rubra L.) in relation to seed production.**

3501 Meÿer, W.J.M.   NL 010207/74/7653
Bepaling optimale oogsttijdstip bij roodzwenk. **Determination of the optimum harvest time in chewings fescue.**

3502 Meijer, W.J.M.; Vreeke, S.   NL 010207/74/7654
Bestudering van groei en ontwikkeling van veldbeemd (Poa pratensis L.) in relatie tot de zaadopbrengst. **Studies on growth and development of Kentucky bluegrass (Poa pratensis L.) in relation to seed production.**

3503 Vreeke, S.   NL 010207/76/7649
Verjongen van graszaadgewassen. **Stimulating regrowth of grass for seed production.** Publications.

3504 Luten, W.; Woldring, J.J.; Roozeboom, L.
NL 010208/71/3885
Vergelijking van grasmengsels en –soorten bij verschillende gebruikswijzen en milieuomstandigheden. **Comparison of mixtures and species of grasses under different utilization and environment conditions.** Publications.

3505 Deinum, B.; Dirven, J.G.P.   NL 020025/73/5041
Onderzoek naar de relatie tussen voederwaarde en morfologische opbouw van grassen en andere voedergewassen uit gematigde en tropische gebieden in afhankelijkheid van uitwendige omstandigheden. **Research into the relationship between nutritive value and morphological development of grasses and other forage crops from temperate and tropical regions as influenced by environmental factors.** Publications.

Pastures, grassland (B 3420)

See also 558, 1124, 1439, 2179, 3307, 3308, 4824, 5617, 5633, 6731, 6735

3506 Lambert, J.; Toussaint, B.   BE 020602/74/0007 R
Influence de la fumure azotée croissante sur la pérennité de variétés herbagères soumises à différents modes d'exploitation. **Influence of increasing nitrogen manure on the perennity of**

grass varieties submitted to different methods of utilization. Publications.

3507 Behaeghe, T.; Traets, J.; Van Bockstaele, E.
BE 030017/63/0001
Seizoenvariatie in de grasgroei met studie van de ecologische fenologische en fysiologische invloeden. **Seasonal pattern of grassgrowth; viewing the study of the influence of ecological, phaenological and physiological factors.** Publications.

3508 Boeker, P.; Bürger, E.   DE 111251/74/0011
Der Einfluss des Grundwassers auf die Pflanzengesellschaft und ihren Trockensubstanzertrag in Abhängigkeit von Jahreswitterung, Düngungsintensität und Nutzungsart. **The influence of groundwater on plant communities and their dry–matter yield in dependence on seasonal climatic conditions, fertilizing intensity, and land use system.**

3509 Boeker, P.; Opitz–von–Boberfeld, W.
DE 111251/74/0013
Wie wirken sich verschiedene Wirkstoffe bei unterschiedlichem Aufwand auf den Ertragsanteil von Agropyron repens und den Weiderest aus und welche Zeiten sind zwischen der Herbizidapplikation und der folgenden Neuansaat einzuhalten. **What about reactions of different active substances in different levels of application on the yield share of Agropyron repens and other pasture plants, and what about the intermediate terms to keep between herbicide application and subsequent resowing.**

3510 Boeker, P.; Opitz–von–Boberfeld, W.
DE 111251/77/0001
Auswirkung unterschiedlicher Nutzungsfrequenzen in Abhängigkeit von der Höhe der Stickstoffgabe auf Ertrag und Bestandszusammensetzung unter dem Aspekt der intensiven Standweidennutzung. **Effects of different grazing frequencies depending on degree of nitrogen fertilization on crop yield and stand composition in view of intensive continuous grazing.**

3511 Simon, U.; Daniel, P.   DE 129140/72/0004
Ertragsvergleich zwischen Dauerwiese und Feldfutterbau. **Comparative yield evaluation of natural meadow and ley.**

3512 Simon, U.; Campino, I.; Schäfer, K.
DE 129140/78/0001 N
Untersuchung des Detritusanteiles der Nettoprimärproduktion in Grünlandökosystemen verschiedener Nutzungs– bzw. Pflegeintensität. **Investigation on the significance of the detritus with respect to the net primary production of grassland ecosystems at different levels of cultivation intensity.**

3513 Dierschke, H.; Vogel, A.   DE 132093/77/0001
Klimabedingungen und Nährstoffversorgung von Wiesengesellschaften verschiedener Höhenstufen des Westharzes. **Climatic conditions and nutrient supply of meadow communities in different attitudes of the Western Harz mountains.**

3514 Schulz, H.   DE 144500/72/0001 N
Einfluss von Mahd und Beweidung auf Pflanzenbestand, Ertrag und Qualität des Grünlandaufwuchses. **Influence of mowing and grazing on plants, crop and quality of growing grassland.**

3515 Jacob, H.   DE 144500/74/0003
Untersuchungen zur Zusammensetzung von

Wiesenansaatmischungen. **Investigations on the composition of mixed meadow seeds.**

3516 Jacob, H.   DE 144500/74/0004
Prüfung von verschiedenen Weidemischungen. **Testing of different pasture seeds.**

3517 Jacob, H.; Schulz, H.   DE 144500/78/0004 N
Untersuchungen zur Erhaltung von Beständen kurzlebiger Gräser durch fortgesetzte Nachsaat. **Investigations on preservation of swards of short–lived grasses by permanent oversowing.**

3518 Jacob, H.; Schulz, H.   DE 144500/78/0005 N
Untersuchungen zur Dauergrünlanderneuerung durch Ansaat ohne Bodenbearbeitung. **Investigations for grassland–renovation by seeding without tillage.** Publications.

3519 Jacob, H.; Schulz, H.   DE 144500/78/0006 N
Untersuchungen zur Einsaat kurzlebiger Gräser in Knaulgrasbestände zur Verbesserung des Futterwertes. **Investigations for sowing short–lived grasses into permanent grassland for improvement of feeding value.**

3520 Spatz, G.; Kau, M.; Voigtländer, G.   DE 161255/73/0002
Untersuchungen zur Schafälpung im Hochgebirge. **Investigations of sheep grazing on alpine pastures.**

3521 Spatz, G.   DE 161255/74/0001
Untersuchungen zur Nutzungsintensität von Almflächen unter besonderer Berücksichtigung ihrer Auswirkung auf den Pflanzenbestand, die Bodenfaktoren, den Erosionsschutz und die Wirtschaftlichkeit. **Investigations on alpine pasture management intensity in special consideration of influences on plant communities, soil conditions, erosion control and economy.**

3522 Bogner, H.; Mayer, J.; Hofmann, P.; Voigtländer, G.; Kühbauch, W.   DE 161255/78/0003 N
Qualität, Futteraufnahme und Leistungen von Mähweidegras in frischer und konservierter Form. **Quality, forage intake and milk yields from pasture grass as fresh and preserved forage.**

3523 Voigtländer, G.; Krischke, H.   DE 161255/78/0004 N
Produktionstechnische Massnahmen zur Verbesserung der Futterproduktion in Monsungebieten. **Improvement of forage production measures in monsoon regions.**

3524 Häckel, H.   DE 301100/73/0002
Mikroklimatische Untersuchungen zur Nutzungsintensität von Almweiden. **Microclimatological researches on the alpine pasture use intensity.**

3525 Wagner, C.   DE 301100/78/0003 N
Beeinflussung des Wachstumsbeginns auf Grünland durch verschiedene meteorologische Parameter. **Influence of different meteorological parameters on the beginning of growing of grassland.**

3526 Schöllhorn, J.   DE 501200/70/0005
Welchen Einfluss hat die Trocknungsgeschwindigkeit auf die Qualität und den Nährstoffgehalt des Endproduktes. **The influence of the drying speed on the quality and the amount of nutrients of hay.**

3527 Schöllhorn, J.; Szokolai, P.   DE 501200/70/0009

Untersuchungen über die Trocknungsgeschwindigkeit von Futterpflanzen unter besonderer Berücksichtigung blattreicher Bestände der Mähweiden. **Experiments on drying speed of fodder–plants with special regard to fodder from hay pastures with high parts of foliose plants.**

3528 Rieder, J.B.   DE 502058/78/0010 N
Die Verbesserung von ertragsarmem und entartetem Grünland durch Nachsaat und umbruchloser Neuansaat. **Improvement of poor and deteriorated grassland by complementary seed and resowing without ploughing.**

3529 Beckhoff, J.   DE 508301/73/0004 N
Einfluss der Schnitthöhe auf Ertrag und Narbenzusammensetzung. **Relations between cutting height and yield and sward composition.**

3530 Ernst, P.   DE 508301/75/0010
Leistung der Umtriebs– und Standweide bei hoher Stickstoffdüngung und Besatzstärke; Beweidungsversuche mit Jungrindern und Milchkühen. **Production of rotational and continuous grazing both with high nitrogen fertilizing and high stocking rate; grazing experiments with young cattle and dairy cows.**

3531 Ernst, P.   DE 508301/77/0001
Zusammenhang zwischen Temperaturentwicklung ab 1. Januar und Wachstumsbeginn auf Dauergrünland. **Relation between temperature development from the first of January and the beginning of grass growth in permanent pasture.**

3532 Ernst, P.   DE 508301/77/0005
Einfluss von Nachsaat auf den Pflanzenbestand von Dauergrünland. **Effect of overseeding on sward composition of permanent pasture.**

3533 Ernst, P.   DE 508301/77/0008
Einfluss unterschiedlicher Bodentemperaturen auf Ertrag und Inhaltsstoffe einer Dauerweide. **Effect of different soil temperature on yield and contents in permanent pasture.**

3534 Jeannin, B.   FR 010112/75/3031
Intensification du pâturage dans les exploitations d'élevage des zones à prairie permanente dominante. **Greazing intensification on farms located in districts with a high proportion of natural grassland.** Publications.

3535 Laissus, R.; Marty, J.   FR 010115/71/0075
Etude des possibilités de régulation estivale de la production de fourrage en prairie permanente. **Studies on the possibility improving summer production in permanent pasture.**

3536 Millier, C.   FR 010312/71/8288
Modélisation du système fourrager (3). **Simulation of a herd–pasture agrosystem.**

3537 Niqueux, M.; Arnaud, R.   FR 010801/69/0316
Fertilisation et exploitation de prairies permanentes en zone de montagne. **Fertilizing and management of mountain perennial pastures.**

3538 Montard de, F.X..; Gachon, L.   FR 010802/66/0569
Potentialités des pâturages de montagne. **Potentialities of mountain grasslands.**

3539 Gachon, L.; Montard de, F.X.   FR 010802/70/0575

Aménagement des pâturages de montagnes de l'Auvergne. **Determination of the valuable grazing areas in the Auvergne mountains.** Publications.

3540 de Montard, F. X.; Dejou, J. FR 010802/70/6104
Action des facteurs climatiques sur la croissance de l'herbe des prairies de montagne. **Herbaceous growth in hill herbages in relation between climatic factors.**

3541 Niqueux, M.; de Montard, F.X.; Demarquilly
  FR 010802/71/6105
Calendrier de production de la prairie permanente de fauche en montagne. **Calendar of production in hill meadows.**

3542 Renault, P.; Jeannin, B.; Louyot, J.M.
  FR 010804/70/3013
Intensification de la production fourragère en montagne (Cantal 1000 m). **Intensive grass production in mountainous district (Cantal 1000m).** Publications.

3543 Louyot, J.M.; Jeannin, B.; de Ch. Teilhard, B.
  FR 010804/78/9024 N
Essai de rénovation de prairie dégradée.. **Rehabilitation of highland pasture..**

3544 Guibert, P.; Jeannin, B.; Faivre, P. FR 012224/74/3015
Etude expérimentale des possibilités d'intensification fourragère en Lorraine. **Experimental study of the possibilities of intensive grassland production in Lorraine.**

3545 Jeannin, B.; Parrassin, P.; Guinot, J.P.; De Vaubernier, E. FR 012224/74/9001
Recherches sur l'utilisation rationnelle d'une surface de prairie permanente du Plateau Lorrain par le pâturage tournant d'un troupeau de génisses gestantes et la fauche. **Evaluation of the optimum use of a permanent meadow of the "Plateau Lorrain" through rotational grazing and or mowing.**

3546 Jeannin, B.; Parrassin, P.; Guinot, J.P.; De Vaubernier, E. FR 012224/74/9006
Recherches sur les systèmes de conduite au pâturage de génisses en croissance en zones difficilement mécanisables du Plateau Lorrain. **Research on pasture management systems for developing heifers in areas of the "Plateau Lorrain" which are difficult to improve mechanically.**

3547 Oswald GB 010812/00/0001 R
The agroecology and control of important broadleaved weeds including bracken in grass legume swards.

3548 Haggar; Kirkham GB 010812/00/0002 R
The role of herbicides in manipulating sward composition with particular reference to clover encouragement.

3549 Squires; Boatman GB 010812/00/0003 R
Minimum cultivation/herbicide systems for establishing grasses, legumes fodder crops in existing swards.

3550 Morrison; Denehy GB 011202/76/0022 R
The management and cultural requirements of white clover – grass associations.

3551 Sheldrick; Lavender GB 011202/76/0023 R
Legumes (excl. white clover) : yield and quality of species and varieties ; optimum management for cutting.

3552 Large; Tallowin GB 011202/76/0030 R
The effects of plant type and animal management on the growth and harvested yield of grazed swards.

3553 Tetlow; Wilkinson GB 011202/76/0037 R
Field studies in the production of hay of high nutritive value.

3554 Morrison; Russell GB 011202/76/0042 R
The production of white clover/grass swards in relation to environment.

3555 Jewiss GB 011202/79/0047 N
The manipulation of yield of grasses in mid–season.

3556 Clements GB 011202/79/0712 N
Significance of invertebrates in swards established by minimum cultivation techniques (with Wro and Res).

3557 Munro; Davies DA GB 012105/00/0019 R
Determine grass and clover performance in relation to soil and climatic factors in hills and uplands.

3558 Walters GB 012105/76/0029 R
Assess agronomic performance of new herbage varieties when grown in mixtures.

3559 Hughes GB 012105/76/0030 R
Assess performance of new grass and clover varieties in different farming systems.

3560 Hill GB 012106/00/0005 R
Competition and co–operation within and between grass and clover swards.

3561 Charles GB 012106/00/0006 R
Effects of management on sward composition in experiments and farm systems.

3562 Davies, W.E.; Mytton GB 012106/00/0020 R
Effect of management on legume/grass associations.

3563 King GB 030304/00/0006 R
Effect of patterns and intensity of use on growth and regrowth of native and improved hill swards.

3564 Grant GB 030304/77/0010 R
Effect of utilisation by grazing hill sheep and beef cattle on growth and production of hill pastures.

3565 McAllister GB 040402/00/0010 R
Long term effects of slurry dressings on grassland.

3566 Gracey GB 041101/00/0002 R
Long term effects of applying pig slurry to grassland.

3567 Stewart GB 041101/78/0026 R
Direct drilling as a method of crop establishment and pasture improvement.

3568 Gracey GB 041101/79/0029 N
The phosphorus and potassium requirements of intensively managed silage swards.

3569 Gracey GB 041101/79/0031 N
The effect of time and interval between slurry application and harvesting on grass production.

3570 Chestnutt   GB 041309/78/0001 R
Effects of clover and nitrogen on the contributions of grass species to yield.

3571 Swift, G.   GB 060106/00/0004 R
Yield and quality potential of grasses and clovers in mixtures.

3572 Herriot   GB 060106/00/0005 R
Hill sward improvement by surface cultivations, lime and slag oversowing, sod–sowing and bracken eradication.

3573 Herriot   GB 060106/00/0006 R
Grass and clover species and varieties for hill pastures.

3574 Holmes   GB 060106/00/0009 R
A survey of grassland in the East of Scotland.

3575 Weddell; Copeman   GB 060214/00/0002 R
Screening and testing of herbage species varieties and mixtures.

3576 Young   GB 060214/00/0003 R
Preliminary screening of miscellaneous materials for use on grass or grass products.

3577 Younie   GB 060214/00/0005 R
Survey of surface seeded swards in crofting areas.

3578 Mackie; Weddell   GB 060214/00/0007 R
Relationships between herbage yield and quality.

3579 Weddell; Mackie   GB 060214/00/0008 R
Salt spray damage to grassland.

3580 Younie   GB 060214/00/0010 R
Methods of improving efficiency of use on rotational grass, long term swards and surface seeded areas.

3581 Young   GB 060214/00/0012 R
Physical effects of slurry on grassland.

3582 Mackie; Younie   GB 060214/00/0013 R
Sward deterioration and renovation.

3583 Younie; Dunn   GB 060214/00/0014 R
Establishment of grass and assessment of potential of grass production on the machair.

3584 Grant   GB 060226/00/0005 R
Study of grassland husbandry on machair soils.

3585 McCreath; MacLeod   GB 060311/00/0006 R
Hill land improvement in Argyll.

3586 Boyd   GB 060314/00/0001 R
The comparative potentiality of herbage plant varieties.

3587 Harkess; Frame   GB 060314/00/0002 R
Effects of grazing and cutting management on sward productivity.

3588 Frame   GB 060314/00/0005 R
Effects of different grazing management systems on sward characteristics and productivity.

3589 Frame; Tiley   GB 060314/00/0008 R

Improvement of hill land.

3590 Collins, D.P.   IE 060103/76/1211 R
Land spreading of animal manures on grazed pasture. Publications.

3591 O'Sullivan, A.M.   IE 060200/65/0411 R
A botanical and ecological survey of the Irish lowland grasslands. Publications.

3592 Murphy, W.E.   IE 060200/71/0408 R
Production of early grass. Publications.

3593 Leclerc, M.H.   IE 060200/72/0427 R
The effect of two intensities of grazing on the production of a grass clover sward. Publications.

3594 Coulter, B.S.; Brogan, J.C.   IE 060200/75/1124 R
Mapping and contouring of soil nutrient levels of pasture from the results of soil analysis. Publications.

3595 Murphy, W.E.   IE 060200/78/2090 N
Efficiency of n used for grazing for cattle. Publications.

3596 McCarthy, D.D.; Kiely, J.   IE 060400/76/1270 R
Comparative output of resown and permanent pasture under an intensive system of dairying. – Mullinahone. Publications.

3597 McCarthy, D.D.; Arkins, S.   IE 060400/77/1408 N
Techniques for the measurement of pasture output. Publications.

3598 Daly, P.J.   IE 060700/74/1028 R
The role of the sod seeding technique in improving pasture production for sheep and cattle and shallow limestone soils. Publications.

3599 Daly, P.J.; Thomas, T.M.; Flanagan, S.P.; Hanrahan, S.P.   IE 060702/77/1367 R
Sheep–tillage development project. Publications.

3600 Pegazzano, F.; Nannelli, R.   IT 020400/77/0003 N
Il ruolo degli acari nella trasformazione della materia organica nel prato–pascolo.. **The role of mites in the transformation of the organic matter of pastures.**

3601 Iannelli, P.   IT 020900/69/0002
Miscugli semplici per colture prative in asciutto. **Simple mixtures for non irrigated meadows.** Publications.

3602 Haussmann, G.   IT 020900/70/0007
Miscugli polifiti seminati con modalità differenti. **Different methods of sowing poliphytic mixtures.** Publications. –

3603 Sardara, M.   IT 020900/73/0001
Miglioramento dei pascoli e il loro impianto. **Improvement of pastures and their establishment.** Publications.

3604 Zannone, L.   IT 020900/78/0003
Studio della competizione interspecifica in foraggere graminacee per la costituzione di prati polifiti. **Study of interspecific competition for polyphytic meadow constitution with fodder grasses.**

3605 Ropelato, A.   IT 021800/75/0001
Ricerche sui pascoli di Pampeago. **Research on the pastures of**

D 2100 – Plant production general and crop husbandry

Pampeago.

3606 Bezzi, A.; Ghidini, G.   IT 021800/75/0002
Ricerche sui pascoli della Val Rendena. **Research on the pastures of the Rendena Valley.**

3607 Bezzi, A.; Ghetti, S.   IT 021800/76/0006 N
Studio dei pascoli di Pescocostanze in Provincia dell'Aquila. **Study on the Pescocostanze's pastures in District of Aquila..**

3608 Bezzi, A.; Orlandi, D.; Ghidini, G.   IT 021800/78/0004
Ricerche sui pascoli della Val di Non: tipologia, produttività e miglioramenti. **Research on the pastures of Non Valley: vegetation type, productivity and improvements.**

3609 Rivoira, G.; Bullitta, P.; Caredda, S.   IT 041302/75/0007
Infittimento pascoli con graminacee poliennali consociate con medica e Trifolium subterraneum. **The thickening of pasture land with pluriannual fodder grasses combined with alfalfa and Trifolium subterraneum.** Publications.

3610 Rivoira, G.; Bullitta, P.; Caredda, S.   IT 041302/75/0009
Prova di decespugliamento dei pascoli: modalita' e epoca. **Test on scrub clearing in pasture land: ways and period.**

3611 Rivoira, G.; Bullitta, P.; Caredda, S.   IT 041302/75/0010
Confronto tra diverse tecniche di infittimento dei pascoli. **Comparison between different tecniques of pasture land thickening.** Publications.

3612 Vanzetti, C.   IT 042301/78/1116 N
Censimento ed indagine conoscitiva sui pascoli montani del Veneto. **Census–taling and explorative research on the Venetian mountain pastures.**

3613 Nassimbeni, P.; Parente, G.; Menegon, S.
     IT 090701/79/0004 N
Etude comparée de l'influence des bovins et des ovins sur la composition botanique, le sol et le rendement végétal et animal d'un pâturage de montagne. **Comparative test on the influence of cattle and sheep on the botanical composition, soil, herbage and livestock production of an upland pasture.**

3614 Parente, G.; Menegonx, S.   IT 090701/79/0005 N
Régéneration des herbages au moyen de semis avec et sans destruction de l'ancien gazon. **Regeneration of meadows, seeding, with or without old turfs destruction.**

3615 Ennik, G.C.   NL 010102/69/7302
Onderzoek naar de invloed van de voorvrucht op de zodesluiting bij Engels raaigras en naar factoren die de persistentie beïnvloeden. **Research on the influence of the preceding crop on sward density of perennial ryegrass and on factors determining persistency.** Publications.

3616 Bergh, J.P. van den; Elberse, W.Th. NL 010102/76/6735
Vergelijkend onderzoek naar het groeipatroon en de onderlinge beïnvloeding van enkele graslandplanten bij beperkende omstandigheden. **Comparative studies on growthpatern and interference of some grassland species at limiting conditions.**

3617 Bergh, J.P. van den   NL 010102/77/7997 R
Oecofysiologisch onderzoek aan plantesoorten in semi–aride weide gebieden. **Eco–physiological research on plant species of semi–arid grassland areas.**

3618 Oomes, M.J.M.   NL 010102/79/8935 N
Extensief graslandgebruik in het kader van natuurbeheer door landbouwbedrijven. **Extensive use of grassland in relation to nature conservancy by agriculture.**

3619 Boer, T.A. de   NL 010104/71/3204
Eigenschappenonderzoek van graslanden met landbouwkundig gebruik. **Research on characteristics for evaluation of grasslands.** Publications.

3620 Hoogerkamp, M.   NL 010104/74/5387
De invloed van herinzaai en van een tijdelijke akkerbouwperiode op de produktiviteit van grasland. **The influence of resowing and ley/arable farming on the productivity of grassland.** Publications.

3621 Schukking, S.   NL 010208/77/8769 N
Diverse aspecten van het bloten van grasland. **Several aspects of pasture topping.**

3622 Korevaar, H.   NL 010208/79/8767 N
Invloed van beheersbeperkingen op grasland op de bruto–produktie van grasland. **Influence of restrictions of grassland use on gross grassland production.**

3623 Dirven, J.G.P.; Neuteboom, J.H.   NL 020025/74/5037
De oecologie van graslandplanten, in het bijzonder de grassen. **Ecology of grassland plants, particularly the grasses.** Publications.

3624 Minderhoud, J.W.   NL 020025/74/5338
Onderzoek betreffende grasland– en grasveldaanleg. **Research on turfgrass and pasture establishment.** Publications.

3625 Deinum, B.   NL 020025/78/8651 N
De productiviteit van gras bij beweiding en andere vormen van beschadiging en factoren die daarop van invloed zijn. **The productivity of grass during grazing and other damages and the factors that affect it.**

3626 Jong, G.J. de   NL 040007/65/4345
Ontwikkeling en instàndhouding van een grasmat op dijken zonder beweiding. **Development and maintenance of grass turf on dams without grazing.**

Mangolds (B 3430)

See also 3295, 3818

3627 Lütke–Entrup, N.   DE 508301/75/0005
Vergleichsanbau von Futterrüben, Silomais und Markstammkohl. **Comparative crop–growing of fodder beets, maize for silage and marrowstem kale.**

3628 Rasmussen, P.; Lyngby, S.P.   DK 010106/74/3506
Såmetoder og –afstande for bederoer. **Sowing methods and distances for sugar beet.**

3629 Lysgaard, C.P.   DK 030145/79/0002 N
Frøalderens betydning for planteproduktionskapaciteten hos bederoer. **The influence of seed–age on the development and yielding capacity of fodder beets.**

Legumes in general (B 3440)

D 2100 – Plant production general and crop husbandry

See also 1439, 3487, 3899, 3900, 3901, 3911, 3920, 3926, 3931, 3933

3630 Biston, R.; Casimir, J.; Dardenne, G.; Fabry, J.
BE 080900/78/0008 R
Cultures fourragères protéagineuses en région du Sud–Est.
Proteaginous cultivations in the South–Eastern region.
Publications.

3631 Kahnt, G.; Gutzmann, H. DE 144485/78/0008 N
N–Bindung durch Leguminosen und Wirkung als
Zwischenfrucht auf Hauptfrüchte. **N–fixation by
intercrop–legumes and the effect of intercrops on main crops.**

3632 Wermke, M. DE 201030/77/0006
Ertragsaufbau und Nutzungseigenschaften von Gramineen und
Leguminosen in der Bundesrepublik–Deutschland. **Yield
increase and utilization properties of gramineae and
leguminosae in the Federal Republic of Germany.**

3633 Decau, J.; Puech, J.; Blanchet, R.; Marty, J.R.
FR 011401/74/6209
Insertion des protéagineux dans les systèmes de culture.
Economie d'azote et production protéique au niveau de la
rotation. **Legumes introduction in agricultural systems.
Nitrogen economy and protein production.**

3634 Bate–smith GB 010301/00/0033 R
**The distribution and determination of tannins in plants:
specifically in legumes.**

3635 Clements GB 011202/79/0711 N
**The abundance of invertebrates in grass and legume crops and
their effect on herbage yield (with Res).**

3636 Cooper GB 040105/75/0005 R
**The role of biotypes of rhizobia in the establishment of legumes
on soil.**

3637 Murphy, P.; Masterson, C.L. IE 060200/00/1594 N
**Maximising symbiotically produced forage legume protein in
the presence of mineral nitrogen.** Publications.

3638 Lanza, F.; Venezian, M.E.; Losavio, N.
IT 020500/79/0006 N
Valutazione delle principali caratteristiche agronomiche di
essenze foraggere graminacee e leguminose. **Evaluation of
agronomical characteristics in forage legumes and grasses.**

Grassland legumes (B 3441)

See also 1475, 3288, 3420, 3486, 3609, 6773

3639 Olesen, J. DK 010701/71/0002
Forskellige dyrkningsmetoder og gødningstilførsel til græs,
kløvergræs og grønmajs. **Different growing methods and
fertilizer rates for grass, clover–grass and forage maize.**

3640 Tasei, J.N.; Delaude, A. FR 010114/72/1701
Essai d' acclimatation et de multiplication d'une population de
Megachile Rotundata pollinisateur de la luzerne.
**Acclimatization and multiplication of Megachile Rotundata as
pollinators of alfalfa.** Publications.

3641 Obaton, M. FR 011002/70/6302
Amélioration de la fixation symbiotique chez la luzerne.

Improvment of symbiotic fixation of alfalfa.

3642 Woodward; Sheehy GB 011203/76/0019 R
**The influence of temperature on the growth of legumes in the
field and in the controlled environment.**

3643 Hughes GB 012105/76/0025 R
**Agronomic evaluation and management of potential varieties of
red and white clovers.**

3644 Roberts; Griffiths GB 012109/76/0020 R
**Assess seed yield potential of legume varieties under various
management and fertiliser practices.**

3645 Laidlaw; Mc Bratney GB 040301/00/0005 R
Establishment, productivity and utilisation of red clover.

3646 Laidlaw GB 040301/76/0019 R
**The influence of the components of grazing upon the white
clover content of mixed swards.**

3647 Swift, G. GB 060106/00/0007 R
Establishment, growth and yield of lucerne.

3648 Holding GB 060113/00/0003 R
Inoculation of white clover seed with rhizobia for hill pastures.

3649 Younie GB 060214/00/0011 R
**The agronomy of white clover with special reference to its
unpredictability of yield and persistence.**

3650 Martin GB 060307/00/0001 R
Lotus species in Scottish agriculture.

3651 Frame; Simpson GB 060314/00/0007 R
Growth and productivity of red clover.

3652 Venezian, M.E.; Losavio, N. IT 020500/77/0001
Aspetti fisiologici della produzione di erba medica.
Physiological aspects of the production of alfalfa.

3653 Zannone, L. IT 020900/78/0004
Studio delle relazioni tra contenuto in saponina nella pianta e
nel terreno, quantità di tubercoli radicali, grado di eterozigosi
e vigore nell'erba medica. **Study of relationship between
saponin content in plant and soil, quantity of root nodules,
heterozigosity level and vigour in lucerne.**

3654 Monotti, M. IT 041001/77/0374 R
Messa a punto della tecnica colturale per la produzione del
seme di sulla. Costituzione di nuove varietà di sulla. **Perfecting
the cultural technique of Hedysarum coronarium L. Seed
production. Breeding new varieties of Hedysarum coronarium
L..**

Other legumes (B 3449)

See also 3179, 3391, 3627, 3897, 4233

3655 Heyland, K.–U.; Vogel, S.; Sayampol, N.;
Schulte–Geers, C.; Kamnalrut, A. DE 111255/75/0002
Serologische Testverfahren zur Prüfung der spezifischen
Kombinationseignung verschiedener Ackerbohnenlinien –
Vicia faba minor – im Hinblick auf eine Ertrags– und
Qualitätssteigerung und –sicherung. **Investigations into the
specific combining ability of different lines of Vicia faba minor**

D 2100 – Plant production general and crop husbandry

L. by serological methods with regard to increasing and securing the yield and quality.

3656 Röbbelen, G.; Frauen, M.; Paul, C. DE 132184/75/0005
Genetische Variabilität der Ertragsstruktur und –stabilität bei Ackerbohnen. **Genetic variability of yield structure and stability in horse beans.**

3657 Steinhauser, H.; Kraxner, H. DE 161440/75/0002
Zur Ökonomik der Lieschkolbenschrotsilage. **Economic aspects of ground corn–cob ear silage.**

3658 Kittlitz, E.von DE 501700/78/0001 N
Einfluss des Befruchtungsverhaltens von Inzuchtlinien auf die Homogenität und die Ertragsfähigkeit von synthetischen Sorten bei Ackerbohnen – Vicia faba –. **Influence of fertilization behaviour of inbreed lines on homogeneity and productivity of synthetic varieties of Vicia faba.**

3659 Hübner, R.; Wagner, F. DE 506153/70/0002
Silomais–Monokultur im Vergleich zu vier verschiedenen Fruchtfolgen mit Silomais in ihren Auswirkungen auf Pflanze und Boden bei chemischer und mechanischer Unkrautbekämpfung 1965. **Maize–for–ensilage one–crop farming in comparison with four different maize silage rotations and effects on plants and soil with chemical and mechanical weed control.**

3660 Ziegenbein, G.; Morgner, F. DE 506155/75/0005
Prüfung verschiedener Leguminosenarten auf ihre Eignung für den Zwischenfruchtbau. **Testing of different leguminosae species for their suitability for catch crop–growing.**

3661 Mennessier, P.; Houdard, Y.; Teilhard de Chardin, B. FR 010112/75/9118
Etude régionale sur le pois d'hiver et de printemps : techniques culturales. Utilisation fermière du pois récolté. **Regional study of winter and spring forage pea : cultural practices and farm use of the harvested crop.**

3662 Billot, C.; Lenoble, M. FR 010615/77/9033
Adaptation de la culture du lupin à la région du Sud–Est. **Testing the adaptability of winter lupine to the South–East of France.**

3663 Storey, T.S.; Barry, P. IE 120105/73/9040 R
An investigation into the production of field beans (vicia faba l.) under Irish conditions. Publications.

3664 Favilli, R. IT 041101/77/0177 R
Ricerca sulla coltivazione della soia da granella. **Research on the cultivation of soya beans.**

Cereals used for forage (B 3450)

See also 493, 1439, 1475, 3045, 3055, 3090, 3112, 3390, 3397, 3542, 3627, 3926, 4233, 5545, 6110, 6812, 6814, 6815

3665 Labouesse, F.; Gachet, J.P. FR 011205/72/8581
Etude des conditions d'intensification de systèmes de production céréales–élevage en sec en Tunisie du Nord. **How to intensify the production systems of grain–cattle dry farming in the North–Tunisia.**

3666 Spillane, P.A. IE 060300/75/1197 R
Screening survey. Feeding barley crop. Publications.

3667 Toniolo, L. IT 040801/72/0573
Ricerche sull'adattabilità e sulle migliori tecniche di sfruttamento delle piante foraggere. **Research on the adaptability and techniques for a better utilization of forage plants.** Publications.

3668 Rivoira, G.; Bullitta, P.; Caredda, S. IT 041302/74/0003
Graminacee foraggere poliennali: comportamento biologico e produttivo. **Pluriannual fodder grasses: biological and productive behaviour.** Publications.

3669 Baldoni, R. IT 062400/77/0949
Ricerche sull'epoca di raccolta dell'erbaio di sorgo. **Research on the harvesting time for sorghum lawn.**

3670 Schepers, J.H. NL 010103/75/6251
De teelt van mais met een voorgewas (o.a. het effect op zaaibedvoorbereiding, groei en opbrengst). **The effect of a preceding forage crop as to seed–bed preparation on growth and yield of maize.**

3671 Velde, H.A. te NL 010207/75/6251
De teelt van mais met een voorgewas (o.a. het effect op zaaibed voorbereiding, groei en opbrengst). **The effect of a preceding forage crop as to seed–bed preparation on growth and yield of maize.**

3672 Hag, B.A. ten; Velde, H.A. te; Haan, G.H. de NL 010207/76/7645
Effect van een rogge/gras–voorteelt op de oogstzekerheid van snijmais. **Effects of precropping with an early fodder crop on yield and quality of silage maize.**

3673 Hag, B.A. ten; Haan, G.H. de; Titulaer, H.H.H. NL 010207/76/7646
Consequenties.van continuteelt voor groei en opbrengst van mais. **Continuous maize growing and its effects on growth and yield of silage maize.**

3674 Hag, B.A. ten; Haan, G.H. de NL 010207/76/7647
Invloed van oogsttijdstip op opbrengst en kwaliteit van snijmais. **Effects of harvest time on yield and quality of silage.**

3675 Deinum, B. NL 020025/76/7345
De invloed van groeiomstandigheden in de nazomer op de produktiviteit en voederwaarde van snijmais. **The effect of growthfactors in autumn on productivity and nutritive value of maize for silage.**

Turnips (B 3460)

See also 3166

3676 Lütke–Entrup, N. DE 508301/77/0006
Vergleich verschiedener Rapssorten für Grünnutzung im Sommerzwischenfruchtanbau – Erntezeitstaffelung, Nährstoffgehalt –. **Comparison of different rape varieties for forage farming in spring intercropping – graduation of harvest time, nutrient content –.**

3677 Tasei, J.N. FR 012223/75/5085
Etude des rapports entre la floraison du colza et les Apoïdes. **Relations between bees and rape flowers.**

Other forage crops (B 3490)

D 2100 – Plant production general and crop husbandry

See also 3639

3678 Rasmussen, P.; Pedersen, K.E. DK 010106/72/3509
Prøvedyrkning af nye arter og sorter af grønfoderplanter –
solsikke, fodermarvkål, foderraps m.v.. **Preliminary trials of
new species and varieties of forage plants – sunflower, kale,
forage rape, etc..**

3679 Rasmussen, P.; Nordestgaard, A. DK 010106/74/3508
Plantetætheds– og høsttidsforsøg i majs til grønfoder og til
modenhed. **Plant density and date of harvest in maize grown for
forage and grain.**

3680 Sheldrick GB 011202/76/0024 R
**Annual forages (eg. maize, brassicas, roots and catch–crops):
yield, quality and cultural management.**

3681 Grant GB 030304/00/0005 R
**The effect of seasonal patterns of dfferent intensities of
utilisation on the growth of heather.**

3682 Easson GB 041308/77/0005 R
**Husbandry factors in Brassica crops direct drilled or sown with
conventional cultivation.**

3683 Mclauchlan GB 041401/79/0007 N
The growing harvesting and feeding of main crop swedes.

3684 Thow GB 060106/00/0020 R
Factors affecting the growth and development of swedes.

3685 Harper GB 060106/00/0024 R
The agronomy of forage Brassica crops.

3686 Paoletti, R.; Locatelli, C. IT 020900/79/0002 N
Studio sull'introduzione e relative tecniche colturali del cavolo
da foraggio nella pianura padana. **Study on introduction and
culture techniques of fodder kale in the Po Valley.**

3687 Piano, E. IT 020900/79/0007 N
Tecnica di produzione di seme di Phalaris tuberosa a diverse
modalità di impianto e a diverse dosi di concimazione azotata.
**Research on seed production techniques for Phalaris tuberosa
with different planting systems and levels of nitrogen.**

3688 Bonciarelli, F. IT 041001/77/0792
Impianti arbusteti. **Setting up shrubberies.**

Vegetables in general (B 3500)

See also 932, 1477, 2600, 2603, 3208, 3288, 3373, 4043, 4091,
4362, 4363, 4518, 6866, 15350, 15352, 17564

3689 Dermine, E. BE 080200/76/0037 R
Etude de production légumière en Gaume. **Studies on
vegetable production in the Gaume region.**

3690 Dermine, E.; Tréfois, R. BE 080200/78/0044 R
Mise au point des conditions et modalités des cultures en
containers. **Growing plants in containers.**

3691 Bernier, G.; Bodson, M.; Kinet, J.M.; Parmentier, A.
BE 140000/72/0046
Physiologie de la floraison et de la mise à fruits chez certaines
plantes horticoles. **Physiology of flower and fruit production by**

some vegetables and crops. Publications.

3692 Van der Linden, L.; Benoit, F.; Ceustermans, N.
BE 140000/74/0034
Fytotechnische aspekten van belangrijke groenten onder glas
en in open lucht. **Phytotechnical aspects of important vegetables
under glass and in the open air.** Publications.

3693 Daunicht, H.–J.; Adaros, G. DE 105203/73/0001
Entwicklung eines Wüsten–Gewächshaussystems mit
geringstmöglichem Süsswasserverbrauch. **Development of a
desert greenhouse system with minimum sweet water need.**
Publications.

3694 Daunicht, H.–J.; Adaros, G. DE 105203/77/0001
CO_2–Versorgung von Pflanzenkulturräumen durch Lüftung
bzw. künstliche CO_2–Zufuhr. **CO_2–supply of plant cultivation
rooms by ventilation or artificial CO_2–injection.**

3695 Daunicht, H.–J. DE 105203/78/0001 N
Die Wirkung von Witterungsverlauf und Kulturmassnahmen
auf Ertrag und Qualität von Freilandgemüsen. **The effect of
weather and phytotechnic measures on yield and quality of field
vegetables.**

3696 Daunicht, H.–J.; Pour–Esmailiyeh, K.
DE 105203/78/0004 N
Wirkungen reduzierter Sauerstoffkonzentration auf die
Produktivität von Gemüsepflanzen und deren
Enzymstoffwechsel. **Effects of reduced O2–concentration on
productivity of vegetable plants and their enzyme metabolism.**

3697 Franke, W.; Lawrenz, M.; Kraus, D.
DE 111152/78/0001 N
Über die Aminosäurezusammensetzung des Eiweisses von
Wildgemüsepflanzen. **On the composition of amino acids in the
protein of edible wild vegetable plants.**

3698 Schmidt, H.H.; Schering, J. DE 135054/70/0003
Feststellung der Anbauwürdigkeit verschiedener ausländischer
Gemüsesorten und landwirtschaftlich genutzter Sorten für die
Bundesrepublik 1960. **Determination of cropping of various
foreign vegetables and of cultivated varieties in the Federal
Replie of Germany.**

3699 Wehrmann, J.; Reiter, M. DE 138060/75/0001
Einfluss des Feldgemüsebaus auf den Mineralstickstoffgehalt
der Böden. **Relation between cultivation of vegetables and the
mineral nitrogen content of soils.**

3700 Krug, H.; Wiebe, H.–J.; Fölster, E.; Hurka, W.; Liebig,
H.–P.; Lorenz, H.–P. DE 138210/74/0001
Ermittlung von Temperaturreaktionskurven für den
geschützten Anbau wichtiger Gemüsearten unter praxisnahen
Anbaubedingungen. **Determination of temperature reactions
curves for the protected cultivation of important vegetables
under commercial conditions.**

3701 Wiebe, H.–J. DE 138210/78/0001 N
Saatgutvorbehandlungen zur Verfrühung des Auflaufens
verschiedener Gemüsearten. **Pre–seed treatments for earlier
emergence of several kinds of vegetables.**

3702 Fritz, D.; Venter, F. DE 161260/77/0004
Wirkung von Sorte, Erntetermin und N–Düngung auf
Rohfaser in verschiedenen Gemüsearten. **Crude fiber contents**

of vegetables as influenced by cultivars, harvesting dates, and nitrogen fertilization.

3703 Kromer, K.–H.; Lechner, E. DE 161522/75/0001
Mechanisierung der Feldgemüseproduktion durch Anbau unter Folien und vollmechanische Ernte. **Mechanization of field vegetables production by cultivation under plastic film covers and fully mechanized harvesting.**

3704 Duden, R. DE 211020/78/0008 N
Einfluss verschiedener Anbauverfahren auf die Peroxydaseaktivität in Gemüsen. **Influence of various production methods on the peroxydaseactivity in vegetables.**

3705 Kretschmer, M. DE 506107/75/0005
Saatgutqualität einiger gärtnerischer Kulturpflanzen in Abhängigkeit vom Positionseffekt des Samenträgers. **Seed quality of some horticultural cultivated plants in dependence on the effect of position of the seed on the plant.**

3706 Rasmussen, P.; Jørgensen, I. DK 010106/67/3510
Gødningsanvendelse og dyrkningsteknik til grønsager på Lammefjorden. **Fertilizer use and growing methods for vegetables in the Lamme–fjord area.**

3707 Andersen, A.; Eriksen, E.N. DK 030171/78/0001 N
Undersøgelse af lyskilders virkning ved tiltrækning af frøplanter og på roddannelse hos stiklinger. **Comparison of various fluorescent lamps as light sources in growth rooms for production of rooted cuttings and seedlings for commercial purpose.**

3708 Ginoux, G. FR 010610/71/9082
Technique du greffage en tête des solanacées. **Head–grafting technique for solanum plants.**

3709 Peyriere, J.; Brun, R.; Rico, F. FR 010616/69/9089
Amélioration des principales productions légumières régionales. **Improvement of the main regional vegetable corp productions.**

3710 Clairon, M.; Dumas, Y. FR 011602/73/6033
Appréhension de quelques problèmes d'agronomie des principales cultures maraîchères en milieu tropical. **Study of some agronomical problems for the main horticultural crops in tropical area.**

3711 Zinsou, C. FR 011603/77/5060
Adaptation des plantes maraîchères au climat tropical. **Adaptation of vegetables in tropical climate.**

3712 Boa GB 011603/00/0014 R
The establishment of crops by transplanting.

3713 Rowse; Stone GB 011802/76/0024 R
Influence of deep incorporation of nutrients and loosening of the sub–soil on growth of vegetable crops.

3714 Greenwood GB 011802/77/0025
Site to site variation on the yields of vegetable crops.

3715 Currah GB 011804/00/0003
Effect of planting pattern, time of emergence and time of harvest on variability of size and yield.

3716 Gray GB 011804/00/0007

Seed treatment and sources as affecting the establishment of vegetable seed.

3717 Thomas; Biddingtn GB 011804/00/0012 R
Role of hormones in controlling growth and development of crop.

3718 Brocklhurst; Currah GB 011804/00/0023 R
Cultural techniques affecting the establishment of seed.

3719 Salter; Gray GB 011804/00/0024
Preliminary appraisal of novel production systems or crops.

3720 Hardwick; Andrews GB 011804/00/0025 R
Yield potential and energy conversion efficiency of major vegetable crops.

3721 Cox; McKee GB 011804/00/0027
Factors affecting the establishment of transplants.

3722 Currah; Brocklehurst GB 011804/78/0028 R
Factors affecting the establishment of plants from seeds and tissue culture.

3723 Wurr; Darby GB 011804/78/0029 R
Determine the importance of environmental factors during early stages of growth on variation in yield.

3724 Hegarty GB 030701/00/0004
Germination and establishment of vegetable seeds in relation to moisture and temperature.

3725 Thompson; Taylor GB 030701/00/0030
Control of growth, yield and quality of vegetable crops by cultural methods and choice of genotype.

3726 Mackenon; Gill GB 030701/00/0049 R
Effects of weather conditions on growth, yield and quality of vegetable crops.

3727 Hegarty; Royle GB 030701/00/0051
Effect of soil structure on germination and emergence of vegetable seeds.

3728 Hegarty GB 030701/00/0052
Effects of seed production conditions on germination and establishment of vegetables.

3729 Richardson GB 041201/00/0005
The use of flexible plastic films for tunnel houses.

3730 Richardson GB 041201/76/0007
Fuel conservation in protected crops.

3731 Dawson GB 041206/00/0003
Production techniques for continuity of supply of vegetables.

3732 Dawson GB 041206/00/0005
Crop responses to irrigation.

3733 Researcher not indicated GB 050102/00/0005 R
Newer vegetable crops, production techniques and variety trials (including crop rotations).

3734 Researcher not indicated GB 050103/00/0001 R
Asparagus, production techniques and variety trials.

3735 Researcher not indicated   GB 050103/00/0017 R
Minor vegetable crops (marrows, swedes, chicory and radish) production techniques and variety trials.

3736 Researcher not indicated   GB 050103/00/0018 R
Vegetable storage.

3737 Researcher not indicated   GB 050103/00/0021 R
Vegetables, plant establishment from seed.

3738 Researcher not indicated   GB 050142/00/0004 R
Newer vegetable crops, production techniques and variety trials (including crop rotations).

3739 Researcher not indicated   GB 050143/00/0015
Vegetable storage.

3740 Researcher not indicated   GB 050143/75/0017
Vegetables, plant establishment from seed.

3741 Gill   GB 060106/00/0021
Production of various roots, brassicas and pulses.

3742 Thow   GB 060106/78/0020 N
Factors affecting the growth and development of brassicas.

3743 Turner   GB 060112/00/0004
Vegetable crop production and storage.

3744 Berridge   GB 060215/00/0002
Vegetables–methods of husbandry.

3745 Potts   GB 060314/00/0016 R
Potential value of new crops including field scale vegetables.

3746 MacNaeidhe, F.; Cassidy, J.C.; Prendiville, M.D.; Murphy, R.F.   IE 060201/67/0424 R
Vegetable crop production on cutover peat. Publications.

3747 Jeffares, M.; Gormley, T.R.   IE 060300/71/1244 N
Development and modification of fruit and vegetables and related products and process development. Publications.

3748 Cassidy, J.C.   IE 060300/77/1318 R
Use of temporary natural shelter in vegetable cropping. Publications.

3749 Prendiville, M.D.; Murphy, R.F.; Cassidy, J.C.   IE 060300/77/1325 R
Investigations into the use of plastic in the production of early and out–of–season vegetables. Publications.

3750 Prendiville, M.D.; Kenny, T.A.   IE 060300/77/1326
Investigation into growth and uses of some unusual vegetables and herbs.

3751 Uncini, L.; Ferrari, V.   IT 021000/79/0009 N
Ricerca di tecniche agronomiche per il miglioramento della produzione delle colture da seme. Agronomic techniques researches for development of seed culture production.

3752 D'Amore, R.; Uncini, L.; Ferrari, V.
        IT 021000/79/0011 N
Studio delle relazioni tra diverse epoche e densità di semina trapianto e precocità, qualità e quantità della produzione di varietà ed ibridi di specie coltivate. Study on relations between different sowing, pricking out, earliness time and density, production quantity and quality of hybrid varieties of cultivated species.

3753 Porcelli, S.; Perella, C.; Petralia, S.; D'Amore, R.
        IT 021000/79/0013 N
Utilizzazione di fonti energetiche alternative in orticoltura protetta. Utilization of alternate energy sources in protected cultures.

3754 Porcelli, S.; Petralia, S.; Perella, C.   IT 021000/79/0016 N
Colture su substrati artificiali in ambiente protetto. Protected cultures on artificial substratum.

3755 D'Amore, R.; Uncini, L.; Restaino, F.; Petralia, S.; Perella, C.   IT 021000/79/0017 N
Studio di tecniche agronomiche atte a favorire la meccanizzazione delle operazioni colturali (raccolta) riguardanti varietà e ibridi di specie ortive. Study of agronomic techniques to improve harvesting mechanization, with regard to varieties and hybrids of horticultural plants.

3756 Lombardi, D.; Scaramucci, S.   IT 021000/79/0021 N
Studio delle relazioni gametofito–sporofito in specie ortive. Study on gametophyte–sporophyte relations in horticultural plants.

3757 Giordani, G.   IT 040212/77/0663 R
Ricerche sull'impollinazione controllata di piante orticole e frutticole entomofile in colture aperte e sotto serra, con api mellifiche. Research on controlled pollination of entomophile horticultural crops in open fields and greenhouses by honeybees.

3758 La Malfa, G.   IT 040313/73/0240
Ricerche su problemi relativi alla biologia, ed alla tecnica di coltivazione di colture ortive e floreali di pien aria e di serra. Research on problems related to the biology and cultivation techniques of horticultural and floral crops in the open air and in glasshouses. Publications.

3759 Borrelli, A.   IT 040701/72/0402
Ricerche sulle colture ortofloricole coperte. Researches on horticulture in glasshouses. Publications.

3760 Pisani, P.L.   IT 040803/73/0944
Ricerche sull'applicazione dei fitoregolatori in ortoflorofrutti–coltura. Research on the use of plant growth regulators in floriculture and fruit and vegetable production.

3761 Caruso, P.   IT 040914/72/0417
Ricerche sull'orticoltura in serra etc. Research on horticulture in glasshouses, etc. Publications.

3762 Caruso, P.   IT 040914/77/0133 R
Tecnica di produzione delle specie da orto e da fiore. Growing techniques in flower and vegetable garden species.

3763 Loreti, F.   IT 041104/73/1818
Ricerche sull'applicazione dei fitoregolatori in ortoflorofrutticoltura. Research on the use of plant growth regulators in floriculture and fruit and vegetable production.

3764 Tognoni, F.   IT 041115/73/1827
Ricerca sull'applicazione dei fitoregolatori in ortofloricoltura.

D 2100 – Plant production general and crop husbandry

Research on the use of plant growth regulators in horticulture and floriculture.

3765 Moschini, E.   IT 041115/74/0591
Ricerche per lo studio delle tecniche di produzione delle specie da orto e da fiore. **Research relating to the study of horticultural species and flower production techniques.**

3766 Quagliotti, L.   IT 041235/77/0275
Tecniche di produzione di specie da orto e da fiore. **Vegetable and flower species production techniques.**

3767 Foti, S.   IT 063100/77/0899
Ricerche sulla validità del mezzo serra ai fini di unaulteriore evoluzione dell'orticoltura del litorale tirrenico della Sicilia;. **Research on the usefulness of greenhouses to further the evolution of horticulture on the tyrrhenian coast of Sicily.**

3768 Challa, H.   NL 010102/72/7296
Fysiologisch onderzoek ten behoeve van het optimaliseren van de produktie en het energiegebruik in de kasteelt. **Physiological research to improve glasshouse crop production and energy consumption.** Publications.

3769 Bokhorst, D.   NL 010106/73/5361
Onderzoek naar de invloeden van geschermde kasdekken op het gewas. **Research into the influences of screened greenhouse bays on the climate and the crop.** Publications.

3770 Bokhorst, D.   NL 010106/73/5362
Onderzoek naar signalen voor automatische klimaatregelingen. **Signals for automatic control of the greenhouse climate.**

3771 Meer, Q.P. van der   NL 010114/73/3825
Oriënterend onderzoek aan nieuwe groentegewassen. **Tentative research on new vegetables.** Publications.

3772 Klapwijk, D.   NL 010206/49/1149
De invloed van licht en andere milieu–factoren op de groei en de bloei van glasgewassen. **The influence of light and other environment factors on growth and flowering of glasshouse crops.** Publications.

3773 Mol, C.   NL 010206/55/1144 R
Teeltonderzoek van op kleine schaal geteelde groenten onder glas. **Research on growing of various crops grown on a small scale under glass.** Publications.

3774 Berkel, N. van   NL 010206/60/1150
Het effect van koolzuurgasvoorziening op stofproduktie en stofdistributie van het gewas bij kasgewassen. **The effect of CO_2 application and dry matter production and distribution in glasshouse crops.**

3775 Spithost, L.S.   NL 010206/72/4665 R
De invloed van de omstandigheden in het wortelmilieu op groei van glasgewassen. **The influence of root environment conditions on the growth of glasshouse crops.**

3776 Ravestijn, W. van   NL 010206/72/4666
De werking en het gebruik van groeiregulatoren onder praktijkomstandigheden op kasgewassen. **Functions and applications of growth regulators in practical vegetable growing under glass.**

3777 Ravestijn, W. van   NL 010206/72/4667
Vruchtzetting en vruchtontwikkeling bij kasgroenten. **Fruitsetting and development in glasshouse vegetables.**

3778 Berkel, N. van   NL 010206/72/4668 R
Fysiogene afwijkingen bij glasgewassen. **Physiological disorders in glasshouse crops.**

3779 Heij, G.   NL 010206/72/4669 R
Toepassingsmogelijkheden van kunstlicht bij glasgewassen. **Possibilities for artificial light on glasshouse crops.** Publications.

3780 Vooren, J. van de   NL 010206/72/5429 R
Regeling van het kasklimaat. **Climate control in glasshouses.** Publications.

3781 Holsteijn, G.P.A. van   NL 010206/73/5361
Onderzoek naar de invloeden van geschermde kasdekken op het gewas. **Research into the influences of screened greenhouse bays on the climat and the crop.** Publications.

3782 Holsteijn, G.P.A. van   NL 010206/73/5362
Onderzoek naar signalen voor automatische klimaatregelingen. **Signals for automatic control of the greenhouse climate.**

3783 Esch, H.G.A. van   NL 010206/78/8980 N
Kwaliteitsonderzoek bij groenten onder glas (Invloed van produktie, oogst–, sorteer– en verpakkingshandelingen en afzet op de kwaliteit). **Quality research with vegetables grown in glasshouses (Influence of production, harvesting, sorting, packaging and marketing on product quality).**

3784 Kruistum, G. van   NL 010207/76/7660
Verbetering van de toepassingsmethodiek van plastic folie ter vervroeging van vollegrondsgroentegewassen. **Improvement of the application of plastic film on outdoor vegetables.**

3785 Pierik, R.L.M.   NL 020057/69/7370 R
Ontwikkeling van nieuwe methoden voor de vegetatieve vermeerdering van tuinbouwgewassen in vivo en in vitro met bijzondere aandacht voor de rol die groeiregulatoren kunnen spelen. **Development of new methods for vegetative propagation of horticultural crops in vivo and in vitro with special attention to the role of growth regulators.** Publications.

3786 Bierhuizen, J.F.   NL 020057/69/7378 R
De invloed van milieufactoren (o.a. temperatuur, licht, zaaidiepte, vochtgehalte van de grond) op kieming en opkomst van tuinbouwgewassen. **The influence of environmental factors (a.o. temperature, light, sowing depth, soil moisture content) on germination and early growth of horticultural crops.** Publications.

3787 Bierhuizen, J.F.   NL 020057/70/7371 R
De invloed van fysische groeifactoren (o.a. temperatuur, licht, vochtvoorziening) op groei, ontwikkeling fotosynthesse, transpiratie en waterhuishouding van vollegronds– en glasgroenten. **The influence of physical growth factors (a.o. temperature, light, waterrelations) on growth, development, photosynthesis, transpiration and water management of glasshouse and outdoor vegetables.** Publications.

3788 Hopmans, P.A.M.   NL 020057/70/7374
Ontwikkeling en verbetering van apparatuur en methoden als

hulpmiddel voor het onderzoek naar de invloed van milieufactoren op groei, fotosynthese en waterhuishouding van tuinbouwgewassen. **Development and improvement of apparature and methods for research on the effects of environmental factors on growth photosynthesis and waterrelationships of horticultural crops.** Publications.

3789 Kronenberg, H.G.　　　　　NL 020057/75/7373 R
De geschiktheid van het Nederlands klimaat in het bijzonder wat betreft straling en temperatuur, voor de tuinbouwproduktie in vergelijking met die van de ons omringende landen. **Suitability of the Dutch climate, specially the influences of received radiation and temperature, for horticultural production as compared with those in nearby countries.** Publications.

Root, tuber and bulb vegetables (B 3510)

See also 2823, 3166, 3202, 3300, 3822, 3835, 3838, 3839, 3878, 3966, 3977

3790 Bockstaele, L.; Vulsteke, G.　　　　BE 140000/72/0033
Rassenstudie en onderzoek van de fytotechnie van vollegrondsgroenten en van de teelt van champignons. **Study of races and phytotechnological aspects of open air vegetables and mushrooms.** Publications.

3791 Neumann, K.–H.　　　　　DE 129041/70/0001
Untersuchungen über den Einfluss von Wechselwirkungen zwischen Hormonspritzungen und photoperiodischen Bedingungen auf die Entwicklung von Tabak und Karotten 1967. **Effects of correlations between hormone injections and photoperiodic conditions on the growth of tobacco and carrots.**

3792 Wonneberger　　　　　DE 507701/78/0007 N
Einfluss von 6 Kohlrabisorten unter verschiedenen Folien auf Ertrag und Qualität. **Influence of six varieties of kohlrabi and different plastics on yield and quality.**

3793 Vergniaud, P.　　　　　FR 010610/77/9080
Influence du poids moyen des semences et des distances de plantation sur le rendement et la taille des bulbes d'ail et d'échalote. **Influences of the average seed weight and plantation spacing on the yield and size of garlic and shallot bulbs.**

3794 Acock　　　　　GB 011102/76/0067 R
Lettuce physiology and response to environment.

3795 Brewster　　　　　GB 011804/00/0010 R
Major determinants of growth rate, bulb yield and flowering.

3796 Biddingtn; Thomas　　　　GB 011804/00/0013 R
Inhibitors to seed germination and ways of overcoming their effects.

3797 Kelso　　　　　GB 041404/79/0014 N
Yield and quality of red beet juice concentrates and powders.

3798 Researcher not indicated　　　GB 050103/00/0002 R
Beetroot, production techniques and variety trials.

3799 Researcher not indicated　　　GB 050103/00/0006 R
Carrots, production techniques and variety trials.

3800 Researcher not indicated　　　GB 050103/00/0008 R
Celery, production techniques and variety trials.

3801 Researcher not indicated　　　GB 050103/00/0009 R
Leeks, production techniques and variety trials.

3802 Researcher not indicated　　　GB 050103/00/0012 R
Onions (bulb), production techniques and variety trials.

3803 Researcher not indicated　　　GB 050103/00/0013 R
Onions (salad), production techniques and variety trials.

3804 Researcher not indicated　　　GB 050143/00/0001
Beetroot, production techniques and variety trials.

3805 Researcher not indicated　　　GB 050143/00/0005
Carrots, production techniques and variety trials.

3806 Researcher not indicated　　　GB 050143/00/0007
Celery, production techniques and variety trials.

3807 Researcher not indicated　　　GB 050143/00/0008
Leeks, production techniques and variety trials.

3808 Researcher not indicated　　　GB 050143/00/0011
Onions (bulb), production techniques and variety trials.

3809 Researcher not indicated　　　GB 050143/00/0012
Onions (salad), production techniques and variety trials.

3810 Johnston; Cheffins　　　　GB 060318/00/0002
Glasshouse lettuce: winter production.

3811 Prendiville, M.D.; Cassidy, J.C.　IE 060300/67/0014 R
Cultivar screening trials in onions and french beans. Publications.

3812 Prendiville, M.D.　　　　IE 060300/68/0011 R
Carrot production trials on peat and mineral soil for season long fresh market and processing requirements. Publications.

3813 Cassidy, J.C.　　　　　IE 060300/75/1320 R
Early onion production from autumn sowing. Publications.

3814 Cassidy, J.C.　　　　　IE 060300/77/1319 R
Production of leeks for fresh market and processing. Publications.

3815 Cassidy, J.C.　　　　　IE 060300/77/1321 R
Effect of time and method of harvesting on onion quality. Publications.

3816 Cassidy, J.C.　　　　　IE 060300/77/1322 R
Propagation and transplanting systems for certain vegetable crops e.g. onions and leeks. Publications.

3817 Tiernan, P.I.　　　　　IE 120108/77/9101 N
Use of some seed treatments to improve germination, growth and development of celery (apium graveolens).

3818 Tiernan, P.I.　　　　　IE 120108/78/9100 N
Some effects of sowing date and seeding rate on the size of roots of beta vulgaris.

3819 Buishand, T.　　　　　NL 010207/77/7655
Invloed van zaai– en opkweekmethoden op groei, opbrengst en kwaliteit van knolselderij. **The influence of growing conditions on growth, yield and quality of celeriac.**

D 2100 – Plant production general and crop husbandry

3820 Vlug, J. NL 010207/78/8401
Teeltmethodenonderzoek bij koolrabi. **Research on the establishment of kohlrabi.**

3821 Buishand, T. NL 010207/78/8402
Bestudering van de mogelijkheid van de teelt van wortelpeterselie. **Research on growing root–parsley.**

Greens and leafy vegetables (B 3520)

See also 3790, 3963, 3972, 3977, 7140, 17233

3822 Germain, R.; Louant, B.P.; Bochkoltz–Maufroid, C.
BE 020303/78/0005 R
Mécanismes de la reproduction et transmission des caractères génétiques chez la chicorée de Bruxelles (Cichorium intybus L.). **Reproductions problems and heritability of some caracters in the cultivated Cichorium intybus L. (Brussels endive).** Publications.

3823 Scheys, G.; Van Nerum, K. BE 040202/70/0007
Onderzoek van witloofforcerie zonder dekgrond. **Research on soilless chicon culture.** Publications.

3824 Lamberts, D.; Deckers, T.; Lettani, L.
BE 040202/77/0017 R
Forcerie in containers van asperges. **Container –culture of asparagus.**

3825 Maton, A.; Lips, J. BE 070300/67/0033
Studie van de precisiezaai in de witloofteelt. **Research on the precision drilling in Belgian endive culture.** Publications.

3826 Maton, A.; Lips, J. BE 070300/69/0037 R
Studie van het forceren van witloof in een witloofschuur. **Forcing of endive chicory in a barn.** Publications.

3827 Maton, A.; Lips, J. BE 070300/70/0034
Studie van de witloofteelt op ruggen. **Research on the cultivation of Belgian endive roots on ridges.** Publications.

3828 Maton, A.; Lips, J. BE 070300/72/0038 R
Studie van het forceren van witloof in containers en in waterkultuur. **Forcing of endive chicory in containers and in hydroculture.** Publications.

3829 Plumier, W.; Valette, R. BE 080200/60/0020 R
Forçage et densité de peuplement de la chicorée de Bruxelles. **Forcing and density of chicory plants in the field.** Publications.

3830 Dermine, E.; Ferauge, M.; Plumier, W.; Valette, R.
BE 080200/76/0032
Recherche des relations entre l'époque du semis et la montaison chez le witloof. **Research about relation between sowing time and bolting by chicory.**

3831 Krug, H.; Liebig, H.–P. DE 138210/72/0001
Komplexwirkung von Wachstumsfaktoren auf Wachstum und Ertragsbildung am Beispiel von Raphanus sativus radicula. **Complex effect of growth factors on growth and yield, with the example of Raphanus sativus radicula.**

3832 Krug, H.; Wiebe, H.–J. DE 138210/73/0001
Untersuchungen zur physiologischen Reaktion von Blumenkohlsorten als Grundlage für die Anbauplanung. **Investigations into physiological reactions of cauliflower varieties as a basis for the production programming.**

3833 Krug, H.; Wiebe, H.–J. DE 138210/73/0002
Erstellung eines Simulationsmodells für den Einfluss der Temperatur auf die Kulturdauer von Blumenkohl im Hinblick auf Terminkultur und Ertragsvorausschätzung. **Working–out of a simulation model for the influence of temperature on the growing season of cauliflower with regard to continuous production and yield prognosis.**

3834 Wiebe, H.–J.; Westhoff, B. DE 138210/78/0002 N
Wirkung von Vernalisation und Photoperiode auf die Blütenanlage von Chinakohl. **Effect of vernalization and photoperiodism on flower differentiation in Chinese cabbage.**

3835 Buchloh, G.; Kerber, E. DE 144440/78/0001 N
Glucosinolatbestimmung in Samen von Raphanus– und BrassicaArten. **Determination of glucosinolate in seeds of Raphanus and Brassica species.**

3836 Buchloh, G.; Liegel, W. DE 144445/77/0008
Qualitätsuntersuchungen, insbesondere die Oxalsäurebildung, bei Rhabarber. **Investigations of quality in rhubarb, with special reference to oxalic acid.**

3837 Fritz, D.; Weichmann, J. DE 161260/78/0004 N
Einfluss der Düngung, Standweite und des Erntetermines auf das Lagerverhalten und die Qualitätserhaltung von Chinakohl. **Influence of fertilization, plant density and harvesting date on storage ability and quality of Chinese cabbage.**

3838 Frenz, F.–W.; Andresen, F. DE 502104/70/0002
Fruchtwechsel und Spezialisierung bei Blumenkohl und Porree 1963. **Crop rotation and specialization of cauliflower and leek.**

3839 Frenz, F.–W.; Andresen, F. DE 502104/70/0003
Fruchtwechsel und Spezialisierung bei Kopfsalat und Knollensellerie 1968. **Crop rotation and specialization of lettuce and celeriac.**

3840 Hartmann, H.D.; Wuchner, A. DE 506107/72/0003 R
Forschungsschwerpunkt Spargelanbau 1967. **Focus of research: asparagus growing.**

3841 Reuther, G. DE 506111/77/0004
Verklonung von Spargelzuchtstämmen durch Gewebekultur. **Cloning of Asparagus breeding strains by tissue culture.** Publications.

3842 Wonneberger DE 507701/78/0005 N
Einfluss von 6 Blumenkohlsorten unter verschiedenen Folien auf Ertrag und Qualität. **Influence of six varieties of cauliflower and different plastics on yield and quality.**

3843 Willumsen, J. DK 010113/66/4211
Styring af vand– og næringstilførslen til væksthuskulturerne potteplanter, salat, tomat og A. Plumosus i vandkultur og på dyrkningssubstrat. **Control of water and nutrient supplies to glasshouse cultures of pot plants, lettuce, tomato and A. Plumosus in water culture and on growth media.**

3844 Bacher, E.; Bredmose, N.; Adriansen, E.; Christensen, O.V. DK 010113/71/4208
Vækst– og blomstringsregulering af tomater, agurk, salat, potteplanter, snit–chrysanthemum og freesia i væksthus v.hj.a.

D 2100 – Plant production general and crop husbandry

lys, daglængde og temperatur. **Regulation of growth and flowering in tomato, cucumber, lettuce, pot plants, chrysanthemums for cutting and freesia under glasshouse conditions with the aid of light, daylength and temperature.**

3845 Bruno, J.F.   FR 010112/70/3024
Etude de nouveautés techniques en production maraîchère et de leur insertion dans des systèmes de production. **Experiments on technical innovations in vegetable production and study of their introduction in production systems.** Publications.

3846 Brun, R.; Peyriere, J.   FR 010610/71/3002
Recherche des systèmes de production applicables sous abris plastiques dans le Roussillon (production de légumes). **Studies on the possible production systems under plastic shelters in the Roussillon (vegetable production).** Publications.

3847 Lubet; Juste, C.; Dureau, P. Mme   FR 010704/66/6082
Etude de l'influence de la densité de plantation et des fumures organiques et minérales sur le rendement et la qualité des asperges cultivées en sols sableux. **Effect of the density of planting and of the type of fertilization organic and inorganic— on the yield and quality of asparagus cropped on sandy soils.**

3848 Soyer, J.P.; Routchenko, W.   FR 010704/70/0556
Contribution relative de plusieurs méthodes de diagnostic à l'étude de la nutrition minérale des plantes. Application à une rotation annuelle de cultures sous serre (laitue et tomate). **Contribution of several diagnosis methods to the study of plant mineral nutrition. Application on an annual rotation under greenhouse (lettuce and tomato).**

3849 Richardson   GB 041201/00/0002 R
Lettuce growing under plastic film and glass.

3850 Researcher not indicated   GB 050102/00/0002 R
Lettuce, production techniques.

3851 Researcher not indicated   GB 050103/00/0003 R
Brussels sprouts, production techniques and variety trials.

3852 Researcher not indicated   GB 050103/00/0004 R
Cabbage, production techniques and variety trials.

3853 Researcher not indicated   GB 050103/00/0005 R
Calabrese, production techniques and variety trials.

3854 Researcher not indicated   GB 050103/00/0007 R
Cauliflower, production techniques and variety trials.

3855 Researcher not indicated   GB 050103/00/0011 R
Lettuce (outdoor), production techniques and variety trials.

3856 Researcher not indicated   GB 050103/00/0015 R
Rhubarb, production techniques and variety trials, natural and forced crops.

3857 Researcher not indicated   GB 050142/00/0002 R
Lettuce, production techniques.

3858 Researcher not indicated   GB 050143/00/0002
Brussels sprouts, production techniques and variety trials.

3859 Researcher not indicated   GB 050143/00/0003
Cabbage, production techniques and variety trials.

3860 Researcher not indicated   GB 050143/00/0004
Calabrese, production techniques and variety trials.

3861 Researcher not indicated   GB 050143/00/0006
Cauliflower, production techniques and variety trials.

3862 Researcher not indicated   GB 050143/00/0010 R
Lettuce (outdoor), production techniques and variety trials.

3863 Murphy, R.F.; Prendiville, M.D.   IE 060300/68/0012 R
Green broccoli production for processing and fresh market. Publications.

3864 Prendiville, M.D.   IE 060300/69/0013 R
Cultivar screening of minor crops, celeriac, asparagus, celery etc. Publications.

3865 Feely, L.   IE 060300/70/0057 R
Lettuce culture and variety testing. Publications.

3866 Murphy, R.F.; Ryan, E.W.   IE 060300/71/0004 R
Cauliflower production the year round. Publications.

3867 Murphy, R.F.   IE 060300/74/1200
Cabbage for dehydration and freezing. Publications.

3868 Murphy, R.F.; Ryan, E.W.   IE 060300/75/1201 R
The use of polythene tunnels for brassica plant propogation. Publications.

3869 Murphy, R.F.   IE 060300/77/1310 R
Production of winter cabbage outdoors from january – beginning april. Publications.

3870 Petralia, S.; Perella, C.; D'Amore, R.
                IT 021000/79/0012 N
Nuove tecniche di forzatura della cicoria in ambiente protetto. **Chicory new forcing techniques in protected cultures.**

3871 Sarti, A.   IT 060200/72/0087
Studio della efficienza fotosintetica su Lactuca c.v. romana in rapporto a differenti livelli in luce durante l'accrescimento ed in rapporto alle variazioni del contenuto in clorofilla e della disposizione dei cloroplasti. **Growth and photosynthetic activity of lactuca sativa c.v. romana cultivated in three daylight intensities and its relationships with chlorophyll content.** Publications.

3872 Vanadia, S.   IT 062700/77/0821 R
Anticipo della produzione ed ottenimento piante sane nel carciofo. Aumento della produzione ed ottenimento della contemporaneità di maturazione nel pomodoro da industria. Controllo della fioritura nelle bulbose da fiore. **Early production of healthy artichoke plants. Yield increase and simultaneous maturation of tomatoes for processing. Flowering regulation in flower bulbs.**

3873 Blanco, V.V.   IT 062700/77/0938
Ricerche sulla successione di colture in relazione ad alcune tecniche colturali (spinacio–fagiolino–cetriolocavolo–broccolo; patata–frumento–pisello–frumento). **Research on crop rotation in relation to certain cultural techniques (spinach, french bean, cucumber, broccoli; potato, wheat, green pea, wheat).**

3874 Foti, S.   IT 063100/77/0905
Studio di aspetti della tecnica di coltivazione del carciofo.

Study of certain technical aspects of artichoke cultivation.

3875 Foti, S. IT 063100/77/0907
Studio dei fattori che condizionano l'epoca di differenziazione dei capolini del carciofo. **A study of the factors determining the differenciation time in artichoke flowerheads.**

3876 Maaswinkel, R.H.M. NL 010206/54/1132 R
Teeltonderzoek bij kassla. **Research on growing of lettuce in glasshouses.** Publications.

3877 Snoek, N.J. NL 010207/71/8406
Verkorting van de oogstperiode en verbetering van de planningsmogelijkheden bij de teelt van bloemkool. **Shortening of the harvest period and improvement of the planning of cauliflower.**

3878 Snoek, N.J. NL 010207/78/7661
Bestudering van de mogelijkheden van winterteelten bij enkele sluitkoolrassen en prei. **The culture of varieties of winter headed cabbage and winter leek.**

3879 Dekker, P.H.M. NL 010207/78/8403
Bestudering van de mogelijkheid van de teelt van winterspinazie bestemd voor de conservenindustrie. **Research on growing winter spinach for processing.**

3880 Neuvel, J.J. NL 010207/78/8404
Bestudering van de mogelijkheid van de teelt van knolvenkel. **Research on growing fennel.**

3881 Neuvel, J.J. NL 010207/78/8407
Bestudering van de mogelijkheid van de teelt van chinese kool. **Research on growing Chinese cabbage.**

3882 Snoek, N.J. NL 010207/78/8408 R
Toetsing van verschillende planningsmogelijkheden bij eenmalige oogst van spruitkool. **Testing of a planning scheme of single harvested Brussels sprouts.**

3883 Kruistum, G. van NL 010207/79/9018 N
Onderzoek naar de kwantitatieve invloed van factoren die van belang zijn tijdens het forceren van witlof op water. **Research on the quantitative influence of factors playing an important role during the hydroculture of witloof chicory.**

3884 Went, J.L. van NL 020046/76/8452
Ultrastructureel en histochemisch onderzoek van de gametofyt ontwikkeling en bevruchtingsprocessen bij Impatiens en Spinacia. **Ultrastructural and histochemical study of the gametophyte development and fertilization in Impatiens and Spinacia.** Publications.

Leguminous vegetables (B 3531)

See also 2922, 3285, 3633, 3658, 3790, 3811, 3845, 3873, 3964, 3973, 3977, 7053, 13512, 13513

3885 Weber, W.E. DE 138300/72/0001
Untersuchung von quantitativen Eigenschaften der Erbse. **Investigations of quantitative characters in Pisum.**

3886 Weber, W.E. DE 138300/78/0003 N
Untersuchungen zum Proteingehalt bei Erbsen – Pisum sativum –. **Protein content in peas – Pisum sativum –.**

3887 Buchloh, G.; Rodmanis, J. DE 144445/77/0007
Bildung und Stoffwechsel von Oligosacchariden in Leguminosen in Abhängigkeit von Reifestoffwechsel und Kulturmassnahmen. **Formation and metabolism of oligosaccharide in leguminoses depending on metabolism during ripening and on cultivation.**

3888 Wende, E.; Weidemann, S. DE 206000/71/4029 R
Untersuchungen von Faktoren, die das Platzen der Samenschale frischer Körner von Pisum sativum verursachen können 1971–1980. **Investigations on factors that may cause the bursting of seed husk of fresh grains of Pisum sativum.**

3889 Wende, E.; Hansen, H.; Weidemann, S.
DE 206000/75/0002
Untersuchungen über das Auftreten von "blonds" – gelben Körnern – bei frischen und reifen Markerbsen in grünsamigen Genotypen von Pisum sativum. **Studies on the appearance of "blonds" in vining and mature wrinkled peas in green seeded genotypes of Pisum sativum.**

3890 Junge, H.; Pressler, I. DE 206000/75/0005
Chemische und physiologische Untersuchungen zur Differenzierung von Pal– und Markerbsen, insbesondere bezüglich des unterschiedlichen Keimungsverhaltens. **Chemical and physiological investigations with respect to the differentiation of smooth and wrinkled peas especially referring to different germinating behaviour.**

3891 Wende, E. DE 206000/75/0008
Untersuchungen über den Einfluss der Trockenkornfarbe auf den Aufgang und die Entwicklung von grünkörnigen Markerbsen. **Studies on the influence auf grain colour on germination and development of green grained wrinkled peas.**

3892 Henningsen, K.W.; Jacobsen, S.–E.
DK 030101/78/0002 N
Genetiske og biokemiske undersøgelser af bakterieknoldenes udvikling hos ært. **Genetic control of root nodula function in pea.**

3893 Fischer, P. DK 030104/77/0021 N
Anatomiske undersøgelser over roddannelse på ærtestiklinger. **Anatomical investigations of root initiation in pea cuttings.**

3894 Mennessier, P.; Houdard, Y.; Teilhard de Chardin, B.
FR 010112/74/3025
Etude sur la culture et l'utilisation de variétés de pois d'hiver et de printemps en tant que protéagineux. **Experiments on the production and the utilization of winter–and spring peas for protein production.** Publications.

3895 Tasei, J.N.; Delaude, A. FR 010114/71/1702
Problème de la pollinisation de la Fèverole. **Broad bean pollination problems.** Publications.

3896 Vergniaud, P.; Chavagnat FR 010610/77/9081
Etude des techniques propres à développer l'installation des cultures légumières par semis direct. **Development of appropriate practices for successful direct sowing of vegetable crops.**

3897 Billot, C.; Mennessier, P.; Cousin FR 010615/71/9046
Mise au point des techniques de production du Pois source de protéines. **Developing proper agricultural practices for winter pea production as a source of protein.**

3898 Lefebvre, J.M.; Lavielle, G.; Jolivet, E.
FR 011001/72/6138
Influence des traitements agronomiques sur la qualité nutritionnelle des légumes de serres et de pleine terre. **Influence of agronomical treatments on nutritional quality of glasshouse and open field vegetables.**

3899 Amarger, N.; Lagacherie
FR 011002/72/6293
Utilisation des propriétés symbiotiques pour améliorer le rendement des légumineuses à graines, soja exclu. **Use of symbiotic properties to improve the yield of seed–legumes, soyabean excluded.**

3900 Bouniols, A.; Decau, J.
FR 011401/74/6200
Fructification de quelques légumineuses à graines. Effet de régulateurs de croissance sur le développement et la production. **Fructification of some légumes. Effect of growth regulators on development and production.**

3901 Bouniols, A.; Decau, J.; Marty, J.R. FR 011401/74/6201
Fructification de quelques légumineuses à graines. Influence de la nutrition azotée en conditions hydriques différentes sur le rendement et la qualité protéïque du grain. **Fructification of some legumes. The influence of nitrogen nutrition in different hydric conditions upon yield and seed protein quality.**

3902 Clairon, M.; Dumas, Y.
FR 011602/74/6034
Agronomie des plantes à tubercules en zone tropicale humide. **Tuber crops agronomy in wet tropical area.**

3903 Lencrerot, P.
FR 011602/75/6032
Agronomie des plantes riches en protéine : Date de semis et densité de peuplement Réponse de certaines cultures (Vigna notamment) à l'alimentation minérale (N.P.K. Ca). **Agronomy of rich proteins plants – Spacing and time of planting – Performance of some crops specially Vigna to N.P.K. Ca application.**

3904 Lefevre, G.; Hiroux, G.
FR 012222/65/0502
Qualité. Influence de facteurs agronomiques et génétiques sur la qualité du pois. **Quality. Effect of agronomic and genetic factors on quality of peas. Publications.**

3905 Guimbard, C.; Debil, J.P.; Le Jeune, B.; Pennors, J.
FR 012228/70/9051
Amélioration des techniques de production des cultures classiques de la zone légumière du Nord de la Bretagne. **Improving the agricultural practices of typical vegetable production in the traditional area of North Brittany.**

3906 Guimbard, C.; Coppenet, M.; Le Jeune, B.; Guyot; Penors, J.; Lesaint, C.
FR 012228/74/9050
Etude de rotations légumières intensives de plein champ derrière des écrans brise–vent de porosités différentes. **Testing various truck–crops vegetable rotations for intensive cultivation adapted to different types of windbredks.**

3907 Guimbard, C.
FR 012228/78/9053
Production hors sol de légumes et fleurs sous serres et abris recouverts de plastique. **Hydroponic cultivation of vegetables and flowers under glasshouses and under plastic shelters.**

3908 Browning
GB 011401/76/0022 R
Peas, hormones and seed development.

3909 Hardwick; Hardaker
GB 011804/00/0008 R
The physiology of cold tolerance in species and cultivars of Phaseolus.

3910 Hole; Hardwick
GB 011804/00/0009 R
Environmental factors which affect yield, their importance and stability under field conditions.

3911 Miflin
GB 012001/00/0014 R
The biosynthesis of amino acids, particularly methionine, in beans and peas.

3912 Researcher not indicated
GB 050103/00/0010 R
Legumes, production techniques and variety trials.

3913 Researcher not indicated
GB 050127/00/0001 R
Field beans: spring and winter; production techniques and variety trials other than ADAS/NIAB trials.

3914 Researcher not indicated
GB 050127/00/0004 R
Field beans: winter and spring; ADAS/NIAB variety trials.

3915 Researcher not indicated
GB 050127/00/0011 R
Minor cash crops: production techniques.

3916 Researcher not indicated
GB 050143/00/0009
Legumes, production techniques and variety trials.

3917 Researcher not indicated
GB 050167/00/0001
Field beans: spring and winter; production techniques and variety trials other than ADAS/NIAB trials.

3918 Lockhart
GB 060106/00/0023
NLT beans.

3919 Lockhart
GB 060106/00/0025
NLT peas.

3920 Masterson, C.L.; Murphy, P.
IE 060200/75/1176 R
Legume production (yields, management effects, nitrogen availability). Publications.

3921 Murphy, R.F.; Ryan, E.W.
IE 060300/74/0886 R
Factors affecting yield and maturity of late sown green peas in Ireland. Publications.

3922 Curran, P.L.
IE 120101/78/9072 N
The selective effect of pea cultivars on indigenous soil–borne rhizobium.

3923 Uncini, L.
IT 021000/79/0018 N
Possibilità di impiego della vernalizzazione ai fini del conseguimento di primizie di fava (Vicia faba minor). **Methods to improve broad been winter crops (Vicia faba minor).**

3924 Lombardi, D.
IT 021000/79/0026 N
Studio delle basi fisiologiche ed agronomiche della produzione del fagiolo (Ph. vulgaris). **Study on agronomic and physiological bases of bean production (Ph. vulgaris).**

3925 Casarini, B.; Ranalli, P.; Di Candilo, M.
IT 021100/79/0021 N
Correlazioni tra produzione di granella verde e caratteristiche fenotipiche delle piante nel pisello da industria. **Correlations between the green seed yield and the plant phenotypical characteristics of pea for processing.**

D 2100 – Plant production general and crop husbandry

3926 Landi, R.   IT 040502/73/0241
Ricerca sulla competizione nutrizionale delle associazioni
leguminose–graminacee. **Research on nutritional competition
of associations of Leguminosae–Gramineae.** Publications.

3927 D'Amato, F.   IT 041114/77/0657 R
Ricerche fisiologiche e biochimiche sullo sviluppo dello
embrione e del seme nel fagiolo. **Physiological and biochemical
research on the development of the bean embryo and seed.**

3928 Alpi, A.   IT 041115/78/1023 N
Controllo dello sviluppo in piante da orto e da fiore,
metabolismo ormonale durante lo sviluppo del seme in
Phaseolus sp. Aspetti fisiologici della termoinduzione a fiore in
Fresia hybrida. **Development regulation in market garden and
flower plants; hormonal metabolism during the growth of the
seed in Phaseolus sp. . Physiological aspects of flower
thermo–induction in Fresia Hybrida.**

3929 Quagliotti, L.   IT 041201/77/0385 R
Tecniche di produzione del seme di fagiolo e di fava.
Techniques of bean seed production.

3930 Di Leo, P.   IT 060500/77/0658
Ricerche fisiologiche e biochimiche sullo sviluppo
dell'embrione e del seme nel fagiolo. **Physiological and
biochemical research on the development of the bean embryo
and seed.**

3931 Nuti, M.P.   IT 061900/73/0013
Fisiologia della simbiosi Rizobio–Leguminosa: Presenza di
DNA batterico entro la radice dell'ospite. **Physiology of the
Rhizobium–legume association: Presence of bacterial DNA
within plant root.** Publications.

3932 Blanco, V.V.   IT 062700/77/0939
Ricerche su alcuni problemi di tecnica colturale (cimatura,
modalità di raccolta, orientamento delle file, densità disemina
su fava, pomodoro, mais dolce). **Research on certain problems
of cultural techniques (clipping, harvesting systems, bed
orientation, density of sowing) concerning beans, tomatoes,
sweet corn.**

3933 Dantuma, G.   NL 010102/79/8931 N
Productiviteit en oogstzekerheid van eiwithoudende
zaadgewassen. **Productivity and yield stability of protein
producing seed crops.**

3934 Sparenberg, H.   NL 010105/73/8356 R
De invloed van rassenkeuze en teeltomstandigheden op de
eiwitopbrengst, de kwaliteit en verwerkbaarheid van veld– en
tuinbonen en andere peulvruchten. **The influence of variety
choice and growing conditions on the protein yield, the quality
and the processing suitability of Vicia Faba beans and other
Leguminoseae.** Publications.

3935 Kruistum, G. van   NL 010207/76/7663 R
Verhoging van de opbrengst bij stamslaboon door meer
peulzetting met behulp van regulatoren. **Increasing the yield of
snap bean by increased pod set caused by growth regulators.**

3936 Neuvet, J.J.   NL 010207/76/8411 N
Bestudering van het opbrengstverloop bij verschillende typen
tuinbonen. **Research on yield capacity of different types of
broad beans.**

3937 Dekker, P.H.M.   NL 010207/78/8410
Bestudering van de mogelijkheid van de teelt van
winterdoperwten. **Research on growing winter peas.**

Tomatoes (B 3532)

See also 3242, 3285, 3843, 3844, 3845, 3846, 3848, 3872, 3932,
5756, 7124, 7140

3938 Lamberts, D.; Lettani, L.; Deckers, T.
                BE 040202/77/0015 R
Hoge plantdichtheid bij tomaat in kassen. **High density
planting of tomatoes in glass houses.**

3939 Bangerth, F.; Sjut, N.; Bünger   DE 144445/78/0006 N
Endogene Hormone in samenhaltigen und parthenokarpen
Tomatenfrüchten und ihre mögliche regulatorische
Bedeutung. **Endogenous hormones in seeded and
parthenocarpic tomato fruits and their possible regulatory
functions.** Publications.

3940 Hartmann, H.D.; Zengerle, K.–H.   DE 506107/75/0001
Tomatenkultur in Folienkontainern. **Cultivation of tomatoes in
plastic film containers.**

3941 Bacher, E.; Moes, E.   DK 010113/73/4204
Produktionsplanlægning på grundlag af standardiserede
dyrkningsprogrammer for tomater, agurker og potteplanter i
væksthus. **Planned production on the basis of standardized
growing programmes for tomato, cucumber and pot plants in
glasshouses.**

3942 Bredmose, N.; Bacher, E.   DK 010113/74/4210
Udvikling af temperaturprogrammer og kulturmetoder til
energibesparelse for potteplanter, snit–chrysanthemum, A.
plumosus og tomater i væksthus. **Development of temperature
programmes and growth methods for energy–saving in the
production of pot plants, chrysanthemums for cutting, A.
plumosus and tomato in glasshouses.**

3943 Ginoux, G.   FR 010610/77/9083
Influence du peuplement sur la productivité et la qualité des
fruits de tomates à croissance indéterminée greffés sur
l'hybride hollandais K.N.V.F. **Influence of the plantation
density upon the fruit productivity and quality of tomato
varieties grafted on the Dutch variety K.N.V.F., irrespedive of
their respective precocities.**

3944 Brun, R.; Rico, F.; Gauvrit, D.   FR 010616/78/9091
Comportement de la tomate sous serre ombrée en fin de
culture de printemps sous abri–serre plastique. **Behaviour of
spring tomato crops grown under plastic shelter after late
shading arrangement.**

3945 Brun, R.; Rico, F.; Gauvrit, D.   FR 010616/78/9092
Comportement de la tomate en culture hors–sol. **Behaviour of
the tomato plant grown hydroponically.**

3946 Brun, R.; Rico, F.; Gauvrit, D.   FR 010616/78/9093
Comportement de la tomate en fonction de différentes
conduites d'ébourgeonnage basées soit sur une cadence de
passage, soit sur la grosseur de l'axillaire. **Behaviour of the
tomato plant according to different disbudding techniques
involving either various lapses of time between operations or
various thicknesses of the axillaries concerned.**

3947 Hand   GB 011102/00/0010
Environmental control for glasshouse crops.

3948 Slack; Hand   GB 011102/00/0057
Effects of selected truss and leaf removal on fruiting of
tomatoes.

3949 Russell; Rees   GB 011102/00/0060 R
Physiological control of inflorescence development.

3950 Tucker   GB 011103/00/0009 R
The effects of synthetic and naturally occurring growth
regulators on glasshouse edible crops.

3951 Hobson   GB 011103/00/0011 R
Chemical and genetic manipulation of the maturation and
ripening of glasshouse fruits.

3952 Cooper   GB 011106/00/0015 R
Root environment control in glasshouse tomato production. .

3953 Richardson   GB 041201/00/0003
Tomato growing under plastic film and glass.

3954 Researcher not indicated   GB 050102/00/0001 R
Tomatoes, production techniques.

3955 Researcher not indicated   GB 050142/00/0001
Tomatoes, production techniques.

3956 Johnston   GB 060318/00/0001
The effect of different temperatures on the growth of tomatoes.

3957 Johnston   GB 060318/00/0008
Effects of culture practices on the production and quality of
tomatoes.

3958 Wilson   GB 060322/00/0001
Factors affecting the quality and production of tomatoes on
commercial holdings.

3959 Morgan, J.V.; O'Haire, R.; Roche, N.
       IE 120108/68/9113 N
The development and use of an archway concept (a–frame) for
the production of short–term crops of tomato, cucumber and
pepper at high plant densities. Publications.

3960 Morgan, J.V.; Scanlon, F.; Moustafa, A.T.; Doolan, D.
       IE 120108/77/9111 N
Factors affecting the propagation of tomato, chrysanthemum
and strawberry plants for hydroponic culture. 0000000.
Publications.

3961 Soressi, G.P.; Badino, M.   IT 021000/79/0020 N
Fitoregolatori nello sviluppo della pianta e della bacca di
pomodoro. **Phytoregulators in development of tomato plant and
fruit.**

3962 Palmieri, S.; Soressi, G.P.   IT 021000/79/0030 N
Studio del meccanismo di controllo della germinabilità in semi
di pomodoro. **Study on control mechanism of germination
capacity of tomato seeds.**

3963 Dellacecca, V.   IT 040101/77/0686 R
Meccanizzazione della raccolta di pomodoro e carciofo, semina

e prove, aspetti agronomici, pomodoro, raccolta meccanica,
aspetti agronomici, carciofo. **Mechanization of tomato and
artichoke harvesting, sowing trials, agronomic aspects,
tomatoes; mechanised harvesting, agronomic aspects,
artichokes.**

3964 Postiglione, L.   IT 040701/77/0708 R
Meccanizzazione della raccolta di pomodoro e fagiolino.
Epoche, impianto e prove raccolta, aspetti agronomici.
**Mechanization of tomato and green bean harvesting. Planting
calendar and harvesting trials. Agronomic aspects.**

3965 Caruso, P.   IT 040914/74/0527
Tecniche di produzione delle specie da orto e da fiore
pomodoro da mensa – fragola – rosa – sterlizia. **Techniques of
vegetable and flower species production – table tomato –
strawberry – rose – sterlization.**

3966 Alpi, A.   IT 041115/77/0467 R
Impiego fitoregolatori in pomodoro, cipolla e bulbose da fiore.
**The use of phyto–regulators applied to tomatoes, onions, flower
bulbs.**

3967 Moschini, E.   IT 041115/77/0702 R
Meccanizzazione della raccolta del pomodoro e peperone,
prove agronomiche raccolta meccanica peperone.
**Mechanization of tomato and Capsicum Grossum
harvesting. Agronomic trials of Capsicum Grossum mechanical
harvesting .**

3968 Soave, C.   IT 060300/74/0032
Fitoregolatori nell'allegagione e nello sviluppo della bacca del
pomodoro partenocarpico. **Plant growth substances in fruit
development in parthenocarpic tomatoes.**

3969 Torti, G.   IT 060300/77/0872
Fitoregolatori nella allegaggione e nello sviluppo della bacca
del pomodoro partenocarpico. **Phyto–regulators in fruit setting
and in the development of the parthenocarpic tomato pericarp.**

3970 Torti, G.   IT 060300/78/1148 N
Fitoregolatori nello sviluppo della pianta e delle bacche di
pomodoro. **Tomatoes: The action of phyto–regulators on plant
and fruit growth.**

3971 Blanco, V.V.   IT 062700/77/0936
Ricerche sui fitoregolatori (Pomodoro). **Research on
phyto–regulators (tomato).**

3972 Blanco, V.V.   IT 062700/77/0937
Ricerche sulla forzatura in serra (pomodoro, cetriolo, lattuga).
Research on greenhouse forcing (tomato, cucumber, lettuce).

3973 Porretta, A.   IT 070500/77/0707 R
Meccanizzazione della raccolta di pomodoro, fagiolino,
pisello. Valutazione prodotto raccolto. **Mechanization of
tomato, green bean and pea harvesting; crop assessment.**

3974 Buitelaar, K.   NL 010206/59/1126
Teeltonderzoek bij tomaat. **Research on growing of tomatoes.**
Publications.

3975 Varga, A.   NL 020041/70/4513
Voedings– en hormonale factoren in de vruchtgroei en –rijping
van tomaat in vivo en in vitro. **Nutritional and hormonal factors
in the in vivo and in vitro development and ripening of tomato**

fruit. Publications.

Cucumbers (B 3533)

See also 3844, 3941, 3959

3976 Krug, H.; Lorenz, H.–P.　　　DE 138210/75/0001
Wirkung fluktuierender Klimafaktoren auf das Wachstum von
Gewächshausgurken. **Influence of fluctuating climatic factors
on the growth of Cucumis sativus.**

3977 Jørgensen, M.B.; Jensen, J.; Henriksen, K.
　　　　　　　　　　　DK 010115/72/4503
Udvikling af dyrkningsmetoder til asieagurk, bønner, ærter,
blomkål, kepaløg, porre og salat på friland. **Development of
growing methods for pickling cucumbers, beans, peas,
cauliflower, onion, leek and lettuce as outdoor crops.**

3978 Slack; Hand　　　　　　　　GB 011102/00/0043
Environmental effects on cucumber growth and development.

3979 Researcher not indicated　　　GB 050102/00/0003 R
Cucumbers,production techniques.

3980 Researcher not indicated　　　GB 050142/75/0011
Cucumbers, production techniques.

3981 Smith　　　　　　　　　　GB 060318/00/0009
Varieties and culture of cucumbers.

3982 Uffelen, J.A.M. van　　　　NL 010206/55/1130
Teeltonderzoek bij komkommers. **Research on growing of
cucumbers.** Publications.

Other vegetable fruits (B 3539)

See also 3873, 3932, 3959, 3967, 3972, 15388

3983 Daunicht, H.–J.; Awad, G.　　DE 105203/78/0002 N
Die Wachstumsdynamik von Paprika. **Growth dynamics of
papricas.**

3984 Krug, H.; Fölster, E.　　　　DE 138210/77/0001
Modellversuche zur Temperaturwirkung auf Wachstum,
Blüten– und Fruchtbildung sowie Ertragsleistung von
Gemüsepaprika. **Influence of temperature on growth,
flowering, fruit set and yield of paprica.**

3985 Krug, H.; Thiel, F.　　　　　DE 138210/77/0003
Zur Wirkung der Bodentemperatur auf Wachstum und
Ertragsleistung von Gewächshausgurken in Abhängigkeit von
der Lufttemperatur und Strahlung. **The effect of soil
temperature on growth and yield of cucumis sativus depending
on air temperature and radiation.**

3986 Ginoux, G.　　　　　　　　FR 010610/71/9084
Greffage de l'aubergine: étude des effets liés à la mise en
oeuvre de cette technique. **Eggplant grafting: study of the
effects connected with its implementation.**

3987 Hand　　　　　　　　　　GB 011102/00/0038 R
**Environmental effects on growth development and cropping of
sweet peppers.**

3988 Hitchon; Johnston　　　　　GB 060318/00/0003 R
Glasshouse production of sweet peppers.

3989 Basoccu, L.　　　　　　　IT 041201/77/0473 R
Stimolo della fioritura e della allegagione nel peperone,
stimolo dello sviluppo della lamina fogliare in calathea
makoiana, potatura chimica nel garofano miniature. **Inducing
flowering and fruit setting in Capsicum grossum, stimulating
the development of the leaf–blade in Calathea makoiana,
chemical pruning of the miniature pink.**

3990 De Donato, M.　　　　　　IT 041201/77/0684 R
Meccanizzazione della raccolta del peperone, aspetti
agronomici raccolta meccanica. **Mechanization of Capsicum
Grossum harvesting, agronomic aspects of mechanized
harvesting.**

3991 Foti, S.　　　　　　　　　IT 063100/77/0901
Studio dei possibili mezzi per modificare i rapporti tra attività
vegetativa e riproduttiva nella melanzana in funzione dei tipi e
degli indirizzi produttivi. **Feasibility studies with a view to
modifying the relationship between vegetative and reproductive
activity in egg–plants according to types and to productivity
aims.**

3992 Uffelen, J.A.M. van　　　　NL 010206/65/1135
Teeltonderzoek bij paprika. **Growing conditions of
sweet–peppers.** Publications.

3993 Janssen, G.A.J.　　　　　　NL 010206/67/1148
Teeltonderzoek bij augurken in kassen. **Growing conditions of
gherkins in glasshouses.** Publications.

3994 Vries, K.J. de　　　　　　NL 010207/69/2323 R
Onderzoek naar de aanpassing van de teelttechniek van de
augurk voor de éénmalige oogst. **Research on cultural
requirements for once–over harvested pickling cucumbers.**
Publications.

3995 Vries, K.J. de　　　　　　NL 010207/72/4005 R
Beheersing van de stengelgroei en het bloeigedrag van augurk
door regulatoren t.b.v. de machinale oogst. **Managing of
vine–growth and flowering habit of pickling cucumber by
growth regulators on behalf of mechanical harvesting.**
Publications.

3996 Vlug, J.　　　　　　　　　NL 010207/78/8412
Vergelijking van enkele plukpatronen voor de handoogst bij
augurk. **A comparison of various harvest intervals of
hand–harvested pickling cucumbers.**

3997 Ferwerda, J.D.　　　　　　NL 020056/76/7611 R
Jeugdgroei in het gewas okra (Hibiscus esculentus L.) (invloed
groeiomstandigheden tijdens jeugdop latere groei en
ontwikkeling). **Early growth in okra (Hibiscus esculentus
L.)(Influence of early growth conditions on succesive growth
and development in okra).**

Mushrooms and other edible fungi (B 3540)

See also 3790, 4477, 5769

3998 Welvaert, W.; Poppe, J.　　BE 030011/64/0005 R
Domestikatie, ziekte resistentieonderzoek en ziektebestrijding
bij eetbare paddestoelen. **Domestication, disease resistance
research and control on edible mushrooms.** Publications.

3999 Esser, K.; Stahl, U.; Meinhardt, F.　DE 108051/77/0002

Züchtung von holzabbauenden Speisepilzen. **Breeding of wood–disintegrating edible mushrooms.** Publications.

4000 Laborde, J.   FR 010703/69/0558
Fructification des champignons saprophytes comestibles: qualité physique du milieu (eau, gobetage) et de l'environnement (CO₂, substances volatiles). **Fructification of edible saprophytic mushrooms: physical quality of substratum (water, casing soil) and environment (CO₂, volatile substances).** Publications.

4001 Delmas, J.; Poitou, N.   FR 010703/70/0563
Conditions écologiques de croissance et de fructification de la truffe Tuber melanosporum en relation avec l'arbre–hôte. **Ecological conditions of growth and fructification in Tuber melanosporum in relation with the host–tree.** Publications.

4002 Delmas, J.; Poitou, N.   FR 010703/70/6067
Conditions écologiques de croissance et de fructification de la truffe Tuber melanosporum en relation avec l'arbre–hôte. **Ecological conditions for growth and fruiting of truffle (Tuber melanosporum) in relation with the host tree.**

4003 Delmas, J.; Poitou, N.   FR 010703/72/0562
Facteurs de croissance et de fructification de la morille. **Growth and fructification factors of Morel (Morchella).**

4004 Poitou, N.; Delmas, J.; Delpech, P.   FR 010703/72/6058
Facteurs de croissance et de fructification de la Morille (Morchella). **Growth factors and fruiting of Morchella sp.**

4005 Delmas, J.; Poitou, N.   FR 010703/73/0564
Conditions écologiques de croissance et de fructification des champignons mycorrhiziens et notamment du cèpe de Bordeaux. **Ecological conditions of growth and fructification in mycorrhizal fungi especially Boletus Edulis.**

4006 Laborde, J.; Leplae, M.; Pradet, M.   FR 010703/73/6056
Evaluation de la biomasse mycélienne de champignons saprophytes croissant sur milieux solides. **Determination of the mycelium biomass of saprophytic fungi growing on solid media.**

4007 Imbernon, M.; Brian, C.   FR 010703/75/0238
Etude de la germination des spores de différents champignons. **Spores germination of different mushrooms.**

4008 Delmas, J.; Laborde, J.   FR 010703/75/0391
Etude des caractéristiques et du comportement des matériaux employés comme terre de gobetage en culture du champignon de couche. **Caracteristics and behaviour of casing soils used for mushroom growing.**

4009 Imbernon, M.; Guinberteau, J.; Brian, C.; Pirobe, L.
                    FR 010703/75/6054
Etude de la germination des spores de différents champignons comestibles. **Spore dormancy germination studies of some edible fungi.**

4010 Laborde, J.; Imbernon, M.; Delmas, J.; Brian, C.
                    FR 010703/75/6065
Technologie de la culture des champignons saprophytes en vue de l'obtention d'un cycle court de production. **Technology of the cultivation of edible saprophytic fungi leading to a short production cycle.**

4011 Brian, C.; Guimberteau, J.   FR 010703/76/0360

Mise au point d'une technique de culture et amélioration génetique de la pholiote du peuplier. **Cultural practice and breeding for agrocybe aegerita.**

4012 Brian, C.; Fleury, J.; Guimberteau, J.   FR 010703/76/0361
Mise au point d'une technique de culture et amélioration génetique de l'agaricus arvensis. **Cultural practice and breeding for agaricus arvensis.**

4013 Goulas, J.P.   FR 010703/76/0383
Recherche d'un milieu optimal pour la culture du mycelium d'agaricus bisporus au laboratoire. **Determination of an optimal medium for growing the mycelium of agaricus bisporus under laboratory conditions.**

4014 Goulas, J.P.   FR 010703/76/6057
Recherche d'un milieu optimal pour la culture du mycélium d'Agaricus bisporus au laboratoire. **Determination of an optimal medium for growing the mycelium of Agaricus bisporus under laboratory conditions.**

4015 Poitou, N.; Delmas, J.   FR 010703/76/6066
Nutrition des associations arbres/champignons mycorhiziens. **Nutrition of mycorhizal fungi/host trees associations.**

4016 Manning   GB 011103/00/0013 R
Cellulolytic activity of mushroom and mushroom composts.

4017 Flegg; Smith   GB 011106/00/0001 R
Controlled cropping of mushrooms.

4018 Flegg   GB 011106/00/0002
Water relations of the mushroom crop.

4019 Flegg; Smith   GB 011106/00/0003 R
The development of new cropping systems for mushrooms.

4020 Flegg   GB 011106/00/0012 R
Assessment and development of new methods, materials and equipment for mushroom growing.

4021 Randle; Flegg   GB 011106/00/0017 R
Microbiology and chemistry of mushroom compost.

4022 Gandy   GB 011109/00/0022
Investigations into methods of pasteurisation and post–harvest disinfection of composts.

4023 Wood   GB 011109/00/0024 R
Cellulolytic activity of mushroom and mushroom composts.

4024 Dawson   GB 041202/00/0001
Use of slurries with cereal straws and rough herbage as mushroom compost materials.

4025 Dawson   GB 041202/77/0002 R
Initiation of sporophores in Agaricus bisporus.

4026 Dawson   GB 041202/78/0003 N
Bulk pasteurisation spawn–running and cropping mushroom compost.

4027 Researcher not indicated   GB 050102/00/0006 R
Mushrooms, production techniques, pests and diseases.

4028 Researcher not indicated   GB 050142/00/0005

D 2100 – Plant production general and crop husbandry

Mushrooms, production techniques, pest and disease control.

4029 O'Riordan, F.; MacCanna, C.   IE 060300/74/1184 R
Cultivation of additional edible fungi to agaricus bisporus
(mushroom). Publications.

4030 Grappelli, A.   IT 060200/74/0077
Effetto di alcuni metaboliti prodotti da un ceppo di
Arthrobacter sullo sviluppo di funghi eduli. **Effect of some
metabolites from an Arthrobacter strain on mushrooms growth.**

4031 Ceruti, A.   IT 061400/70/0161
Coltivazione in vitro dei funghi superiori e specialmente di
tartufi. **Cultures "in vitro" of higher fungi and above all of
truffles.** Publications.

4032 Ceruti, A.   IT 061400/77/0915
Coltivazione in vitro di funghi superiori e specialmente di
tartufi. **"In vitro" cultivation of higher species of fungi with
special regard to truffles.**

4033 Palenzona, M.   IT 120100/75/0001
Programma di studio per la conservazione e la diffusione del
tartufo bianco del Piemonte, Tuber magnatum Pico. **Research
programme for the preservation and diffusion of the Piedmont
white truffle Tuber magnatum Pico.**

4034 Gerrits, J.P.G.   NL 010204/66/6686 R
Het ontwikkelen van de teelt van compost– en houtbewonende
paddestoelen, die niet tot het geslacht Agaricus behoren.
**Development of the cultivation of compost– and
wood–inhabiting mushrooms, not belonging to the genus
Agaricus.** Publications.

4035 Visscher, H.R.   NL 010204/70/7270
De functie die dekaarde vervult bij de teelt van eetbare
paddestoelen, voornamelijk van het geslacht Agaricus. **The
function of the casing layer in the culture of edible mushrooms,
mainly of the genus Agaricus.** Publications.

4036 Visscher, H.R.   NL 010204/70/7271
Het beïnvloeden van de knopvorming en de ontwikkeling van
vruchtlichamen van eetbare paddestoelen, voornamelijk van
het geslacht Agaricus. **Studying influences on fructification and
development of edible mushrooms, mainly of the genus
Agaricus.** Publications.

4037 Pompen, T.G.M.   NL 010204/78/7889
De invloed van teelt– en oogstomstandigheden op kwaliteit en
slinkverlies van geconserveerde champignons. **The influence of
growing and harvesting conditions on quality and shrinkage of
preserved mushrooms.**

Fruits in general (B 3600)

See also 932, 1482, 3156, 3373, 3705, 3707, 3747, 3757, 3758,
3759, 3760, 3761, 3763, 3765, 3766, 3769, 3770, 3780, 3781,
3782, 3785, 3786, 3788, 3789, 4839

4038 Karnatz, A.   DE 105204/72/0001
Einfluss der Wurzeltemperatur auf die Entwicklung
vegetativer und generativer Organe bei Obstgehölzen.
**Influence of root–temperature on vegetative and generative
development of young fruit woods.**

4039 Karnatz, A.; Yang, D.   DE 105204/74/0004
Vergleich der Pflanzenentwicklung nach Aussaat und nach
vegetativer Vermehrung von Obstgehölzen. **Comparative
studies on woody fruit plants as to their development after
sowing and after vegetative propagation.**

4040 Engel, G.   DE 111300/77/0002
Anbau– und Erntemassnahmen bei Obst. **Crop management
and harvesting methods for fruit crops.**

4041 Lenz, F.; Pietsch, M.   DE 111300/77/0005
Untersuchungen über Nachbauprobleme bei Obst.
Investigation on replant problems in fruit culture.

4042 Lenz, F.; Herborn, A.   DE 111300/77/0006
Einfluss des Fruchtbehanges auf den Gasdiffusionswiderstand
bei Blättern. **Effect of fruit load on the gaseous diffusive
resistance in leaves.**

4043 Lenz, F.   DE 111300/77/0007
Photosynthese, Dunkelatmung und Photorespiration bei Obst
und Gemüse. **Photosynthesis, dark respiration and
photorespiration in fruit and vegetable crops.**

4044 Zachariae, A.   DE 111300/77/0009
Farbstoffsynthese bei Früchten. **Pigment synthesis in fruits.**

4045 Henze, J.; Werner, H.; Lenz, F.   DE 111300/78/0001 N
Abwehr von Schäden durch Blütenfrost im Obstbau. **Measures
for frost protection of flowers in fruit crops.**

4046 Bünemann, G.; Struklec, A.   DE 138240/72/0004 R
Beeinflussung der Fruchtqualität an 4 Sorten in Ertragsanlagen
auf starken Unterlagen durch Eingriffe im Winter und im
Sommer. **Influence of potassium on the quality of fruits.**

4047 Roemer, K.   DE 138240/77/0003
Einfluss der Samen auf bestimmte Qualitätsmerkmale der
Frucht. **Certain characteristics of fruit quality as influenced by
seeds.** Publications.

4048 Schwerdtfeger, G.   DE 144445/77/0004
Exine Struktur der Pollenkörner unserer Obstarten –
REM–Aufnahmen –. **Structure of the exine of pollen grains of
cultivated fruit species.**

4049 Buchloh, G.; Dietz, F.; Neubeller, J. DE 144445/77/0006
Untersuchung physiologischer Effekte der Generativ– und
Vegetationsknospen auf den Gesamtstoffhaushalt an
Obstgehölzen, bei gezielter Veränderung einzelner
Öko–Faktoren Methoden: URAS; Gaschromatographie.
**Investigations on physiological effects of generative and
vegetative buds on the metabolism of fruit trees, with special
reference to ecological factors.**

4050 Neubeller, J.; Buchloh, G.   DE 144445/77/0017
Veränderungen des Aromas im Verlauf der Reife. **Changes of
aroma during ripening.** Publications.

4051 Neubeller, J.; Stösser, R.   DE 144445/77/0018
Mobilisierung und Akkumulierung von Reservesubstanzen wie
Zuckern und Fett. **Mobilisation and accumulation of reserve
substances as sugars and fats.** Publications.

4052 Stösser, R.   DE 144445/77/0022
Untersuchungen zur Reduktion von Haltekräften bei Früchten

auf chemischem Wege. **Investigations on the reduction of fruit removal forces in fruits by chemicals.** Publications.

4053 Bangerth, F.; Bufler, G.　　DE 144445/78/0001 N
Hormonale und enzymatische Regulation der Fruchtreife und –aromabildung. **Hormonal and enzymatic regulation of fruit ripening and aroma production.** Publications.

4054 Bangerth, F.　　DE 144445/78/0002 N
Beeinflussung der Transpiration und der Photosynthese durch Phytohormone. **Effect of phytohormones on transpiration and photosynthesis.**

4055 Bangerth, F.　　DE 144445/78/0004 N
Untersuchungen zur Wirkung von Äthylen und Abscisinsäure auf die Fruchtentwicklung an der Pflanze. **Effects of ethylene and abscisic acid on the development of fruits.** Publications.

4056 Bangerth, F.; Ebert, A.　　DE 144445/78/0005 N
Auswirkungen von Wachstumsregulatoren auf Fruchtbehang und –entwicklung und ihre Rückwirkungen auf endogene Hormongehalte. **Action of growth regulators on fruit set and development and their effect on endogenous hormones.**

4057 Liebster, G.　　DE 161265/70/0001 R
Nomenklatur der Fachausdrücke des Obstbaumschnittes 1963. **Nomenclature of terms of fruit–tree pruning.**

4058 Liebster, G.; Schmid, P.; Häckel, H.; Wagner, C.
　　DE 161265/70/0003 R
Versuche zur künstlichen Frostung von Obstgehölzen 1969. **Experiments on artificial freezing of fruit–trees.**

4059 Borcherdt, C.; Müller, I.　　DE 170255/78/0001 N
Agrarstrukturelle Wandlungen des Obstbaugebietes im östlichen Bodenseeraum. **Changes in agricultural structures of the fruiticultural region of the eastern Lake of Constance area.**

4060 Quast, P.　　DE 507301/70/0001 R
Schaffung der ernährungsphysiologischen Grundlagen für die Erzeugung von Qualitätsobst an der Niederelbe. **Determination of nutrition–physiological standards for production of quality fruit on the Lower Elbe.**

4061 Tiemann, K.–H.; Bockstedte, W.　　DE 507302/75/0001
Technik im Obstbau. **Technique in fruit production.**

4062 Graf, H.　　DE 507303/73/0001
Wachstumsregulatoren im Obstbau. **Growth regulators in fruit growing.**

4063 Schwarz, K.G.; Rimmele, H.　　DE 509156/78/0001 N
Betriebswirtschaftliche Erhebungen in Sonderkulturbetrieben in Rheinland–Pfalz zur Ermittlung der Wirtschaftlichkeit – Arbeitsaufwand, Kosten, Erträge und Erlöse im Obst– und Tabakanbau. **Investigation on economy of special crop farms in Rheinland–Pfalz for the determination of profitability – work input, costs, yields and returns in fruiticulture and tobacco growing.** Publications.

4064 Poulsen, E.; Hansen, P.; Grauslund, J.
　　DK 010102/68/3105
Vækststof–fysiologiske undersøgelser til at klarlægge skudvækst, blomstring, frugtsætning og frugtkvalitet i frugttræer. **Physiological investigations of growth substance activity to elucidate shoot growth, flovering, fruit setting and fruit quality in fruit trees.**

4065 Poulsen, E.; Christensen, J.V.　　DK 010102/71/3102
Udvikling af rationelle metoder til sanerings– og nyplantningssystemer i frugtplantager. **Development of rational methods for clearing and replanting orchards.**

4066 Billot, C.; Arnoux, M.　　FR 010610/70/3006
Etude de l'entretien des sols en vergers irrigués. **Study of soil cultivation in irrigated orchards.** Publications.

4067 Arnoux, M.; Billot, C.; Defrance, H.; Huguet, C.
　　FR 010615/70/9040
Etude de trois systèmes de culture en Moyenne Vallée du Rhône. **Testing three combinations of fruit tree cultivation systems for the Central Rhône Valley.**

4068 Arnoux, M.; Marboutie, G.　　FR 010615/77/9048
Comparaison de traitements chimiques en verger et de traitements thermiques après cueillette sur les accidents de conservation. **Evaluating the effects on fruit conservation mishaps due to chemical sprays in the orchards and to thermal treatments after cropping.**

4069 Huguet, J.G.　　FR 010704/68/0550
Etude de la sève de printemps des arbres fruitiers. **Study on spring sap of fruit–trees.** Publications.

4070 Huguet, J.G.; Giraudon　　FR 010704/73/6075
Influence du milieu et des techniques culturales sur l'enracinement des végétaux pérennes ligneux. **Effects of environment and management on the fruit–trees root system.**

4071 Marocke, R.　　FR 010901/75/6129
Relations porte–greffe/greffon. **Relations between stock and scion.**

4072 Clay; Davison　　GB 010806/00/0004
Tolerance of fruit crops to soil and foliage applied herbicides.

4073 Davison; Bailey　　GB 010806/00/0006
Response of newly planted fruit crops and nursery stock (ornamental) to weed competition and herbicides.

4074 Fulford; Mousdale　　GB 011006/00/0003 R
Regulation of shoot growth.

4075 Priestley　　GB 011006/00/0005 R
Growth and cropping potential in relation to turnover of carbohydrates and other organic resources.

4076 Fulford　　GB 011006/00/0006 R
Regulation of flowering.

4077 Mousdale　　GB 011006/00/0007 R
Regulation of interactions between fruit and shoots.

4078 Fulford; Jewer　　GB 011006/00/0012 R
Control of cambial activity.

4079 Atkinson　　GB 011008/00/0006 R
Root growth and function with especial reference to distribution and function of perennial roots.

4080 Corke; Hunter　　GB 011508/00/0005 R
Effects of benomyl on pollination, fruit set, fruit drop and

D 2100 – Plant production general and crop husbandry

microflora of the flower. Assay uptake.

4081 Williams; Challice, A.H.   GB 011510/00/0014 R
Characterise phenolics of pome fruits and study distribution, metabolism and function.

4082 Williams, R.; Wood   GB 011510/77/0023 R
Growing apples and pears for cider and juice manufacture.

4083 Researcher not indicated   GB 050101/00/0003 R
Fruit (excluding apples) variety collections.

4084 Researcher not indicated   GB 050141/75/0011
Minor fruit crops.

4085 O'Kennedy, N.D.   IE 060302/60/0495 R
Shelter species for fruit plantations. Publications.

4086 Branzanti, E.   IT 021500/72/0407
Portainnesti delle piante da frutto. **Grafting of fruit–bearing plants.** Publications.

4087 Manzo, P.; Monastra, F.; Limongelli, F.; Temperini, O.; Bergamini, A.; Cobianchi, D.   IT 021500/79/0004 N
Studio delle forme di allevamento atte all'intensificazione degli impianti frutticoli. **Study of training systems for high density planting of fruit trees.**

4088 Donno, G.   IT 040112/74/0546
Ricerca sui problemi agronomici della frutticoltura del Mezzogiorno. **Research on agronomic problems in fruit crop production in the Mezzogiorno.**

4089 Marangoni, B.   IT 040203/77/0225 R
Ricerche sui portinnesti degli alberi da frutto. **Research on the root–stocks of fruit trees.**

4090 Damigella, P.   IT 040307/72/0443
Portinnesti delle piante arboree da frutto. **Grafting of fruit–bearing trees.** Publications.

4091 La Malfa, G.   IT 040313/77/0210 R
Problemi biologici e tecnici di colture ortive e floreali in piena aria ed in serra. **Technical and biological problems of floral and vegetable cultivation in the open air and in glass–houses.**

4092 Casini, E.   IT 040507/74/0529
Ricerche su alcuni problemi dell'arboricoltura collinare. **Research on some of the problems related to arboriculture in hilly areas.**

4093 Poma Treccani, C.   IT 040605/73/0286
Influenza dell'acido 2–cloroetilfosfonico, ethephon, irrorato, preraccolta sul metabolismo respiratorio, sull'abscissione, sull'accelerazione della maturazione etc.. **Influence of sprinkled ethephon 2–chloroethyl phosphonic acid. Data on the respiratory metabolism, abscission, on the acceleration of maturation, etc..** Publications.

4094 Eccher, T.   IT 040605/77/0002 N
Studio delle condizioni ottimali per la radicazione di talee di specie frutticole, ornamentali e forestali.. **Studies on rooting of various fruiting ornamental and forestry cuttings.**

4095 Poma Treccani, C.   IT 040605/78/0003
Influenza della GA$_3$ irrorata a diverse dosi e diverse epoche

sulla induzione fiorale e sulla durata del sacco embrionale. **Effects of different concentrations and times of GA$_3$ sprays on flower induction and effective pollination period.**

4096 Sottile, I.   IT 040908/74/0635
Ricerche sui problemi della frutticoltura nel Mezzogiorno. **Research on fruit crop production problems in the Mezzogiorno.**

4097 Tombesi, A.   IT 041005/74/0641
Biologia applicata al miglioramento delle produzioni frutticole. **Biology as applied to the improvement of fruit crops.**

4098 Vitagliano, C.   IT 041104/72/0134
Porta innesti dei fruttiferi. **Grafting of fruit–bearing plants.** Publications.

4099 Romisondo, P.   IT 041218/74/0615
Biologia applicata al miglioramento delle produzioni frutticole. **Biology as applied to the improvement of fruit crops.**

4100 Agabbio, M.   IT 041310/77/0101 R
Problemi agronomici della frutticoltura nel Mezzogiorno.. **Agronomic problems of fruit–growing in Southern Italy.**

4101 Sanna, M.   IT 041310/78/1108 N
Problemi agronomici della frutticoltura nel Mezzogiorno. **Fruit growing in Southern Italy: agronomic problems.**

4102 Casini, E.   IT 062800/73/0182
Ricerche sulla propagazione per seme e biologia della riproduzione. Impiego dei fitoregolatori per la germinazione dei semi di specie fruttifere ed ornamentali. **Researches on seed propagation and on reproduction biology. The use of growth regulators in seed germination of fruit and ornamental species.** Publications.

4103 Bartolini, G.   IT 062800/73/0188
Ricerche sulla propagazione per radicazione diretta. Studi sulle relazioni tra rizogenesi e livelli ormonici in talea di specie da frutto. **Researches on vegetative propagation. Studies on rhyzogenesis and hormonical level in cutting of fruit trees.** Publications.

4104 Fiorino, P.   IT 062800/77/0817 R
Prove di radicazione di talee di piante fruttifere e ornamentali sia con il metodo della nebulizzazione che con quello del riscaldamento basale, integrate da prove di moltiplicazione in vitro. **Experimental rooting of ornamental and fruit plant cuttings boosted by nebulisation or basal heating and integrated "in vitro" multiplication trials.**

4105 Scaramuzzi, F.   IT 062800/77/0931
Selezione materiale di propagazione. **Selection of material for propagation.**

4106 Scaramuzzi, F.   IT 062800/77/0932
Problemi di tecnica vivaistica. **Problems of nursery techniques.**

4107 Scaramuzzi, F.   IT 062800/77/0933
Propagazione per innesto. **Propagation by grafting.**

4108 Scaramuzzi, F.   IT 062800/77/0934
Propagazione per radicazione diretta (Rigenerazione). **Propagation by direct rooting (Regeneration).**

4109 Scaramuzzi, F. IT 062800/77/0935
Propagazione per seme (Biologia della riproduzione).
Propagation by seeds (Biology of reproduction).

4110 Sansavini, S. IT 063000/70/0014
Potatura meccanica degli alberi da frutto. **Fruit trees mechanical pruning.** Publications.

4111 Cristoferi, G. IT 063000/70/0017
Fisiologia dei fitoregolatori. **Growth regulators physiology.**
Publications.

4112 Baldini, E. IT 063000/70/0018
Applicazione dei fitoregolatori agli alberi da frutto.
Applications of growth regulators to fruit trees. Publications.

4113 Baldini, E. IT 063000/77/0926
Ricerche di base sui fitoregolatori. **Basic research on phyto–regulators.**

4114 Baldini, E. IT 063000/77/0927
Ricerche sulle micropropagazioni. **Research on micro–propagations.**

4115 Baldini, E. IT 063000/77/0928
Ricerche applicative sui fitoregolatori. **Applied research on phyto–regulators.**

4116 Baldini, E. IT 063000/77/0930
Ricerche sui sistemi di allevamento e sulla potatura. **Research on breeding systems and on pruning.**

4117 Bargioni, G. IT 102801/74/0510
Ricerche sulla frutta da industria. **Research on industrial fruit crops.**

4118 Wertheim, S.J. NL 010212/76/8316 R
Toetsing bijzondere fruitgewassen. **Screening unusual fruit crops.**

4119 Wertheim, S.J. NL 010212/76/8317 R
Teeltonderzoek in de vruchtboomkwekerij. **Research on culture in fruit tree nurseries.**

4120 Jonkers, H. NL 020057/66/7377 R
De invloed van verschillende factoren (temperatuur, licht, vochtvoorziening, groeistoffen) op groei en ontwikkeling van boomkwekerijgewassen en op waterhuishouding en vruchtzetting van fruitteeltgewassen. **The influence of various factors (temperature, light, moisture supply, growth substances) on growth and development of nursery plants and on watermanagement and fruitsetting of fruitcrops.**
Publications.

Top fruit in general (B 3610)

See also 2595, 2600, 4474, 9759, 10283

4121 Crabbe, J.; Lakhoua, H.; Arias, O. BE 010002/66/0005
Intervention de corrélations internes dans la morphogénèse des végétaux ligneux, en particulier des arbres fruitiers.
Intervening of internal correlations in the morphogenesis of woodyplants, particularly fruit trees. Publications.

4122 Soenen, A.; Marcelle, R.; Simon, P.; Pittevils, J.; Gilles,
G.; Porreye, W. BE 140000/72/0003 R

Studie van fytotechnische problemen in de fruitteelt. **Study of phytotechnical problems in orcharding.** Publications.

4123 Soenen, A.; Marcelle, R.; Simon, P.; Pittevils, J.; Gilles,
G.; Porreye, W. BE 140000/72/0004 R
Fysiologische studie van fruit en fruitbomen. **Physiological study on fruit and fruittrees.** Publications.

4124 Lüdders, P.; El–Sayed, M. DE 105204/78/0003 N
Einfluss von Unterlage, K–Ernährung und Fruchtbehang auf Wasserverbrauch, Triebwachstum und Fruchtertrag bei bei Obstgewächsen. **Effect of rootstock, K nutrition and fruit load on water consumption, shoot growth and yield of fruit trees.**

4125 Lenz, F. DE 111300/78/0004 N
Beziehung zwischen dem vegetativen und generativen Wachstum bei Kernobst. **Relationship between the vegetative and reproductive growth of pome fruit.**

4126 Küster, E.; Korn, S. DE 129080/77/0004 R
Artenzusammensetzung der Rhizosphärenflora von Obstbäumen verschiedenen Alters. **Species composition of the microflora of the rhizosphere of fruit trees of different age.**

4127 Bünemann, G. DE 138240/72/0003
Wachstum und Fruchtbarkeit von Kernobstbäumen und Auswirkung auf Lagerfähigkeit der Früchte. **Growth and fertility of pome fruit trees and their effects upon storage quality of the fruit.**

4128 Stösser, R. DE 144440/78/0002 N
Befruchtungsbiologische Untersuchungen beim Steinobst.
Pollination studies on stone fruit.

4129 Tiemann, K.–H.; Dammann, H.–J. DE 507306/77/0002
Unterlagenfragen bei Kernobst. **Problems of rootstocks of pome fruits.**

4130 Tiemann, K.–H.; Zahn, F.–G. DE 507307/75/0002
Artgerechte Kronenbehandlung des Steinobstes. **Specific treatment of the tops of stone fruit trees.**

4131 Tiemann, K.–H.; Blank, H.–G. DE 507308/72/0002
Untersuchung der Einflüsse von Standort, Unterlage, Düngung, Herbizide und Pflanzenschutzmassnahmen auf die Lagerfähigkeit von Kernobst. **Study on the influences of site conditions, substratum, fertilization, herbicides and plant protective measures on the storage ability of pome–fruit.**

4132 Jones; Higgs GB 011003/00/0001
Water relations.

4133 Jones; Hopgood GB 011006/00/0008 R
Mechanisms involved in rootstock–scion interactions.

4134 Parry; Preston GB 011008/00/0004
Agronomic studies of new EMRS varieties and improved source material.

4135 Jackson; Oehl GB 011008/00/0005
Soil fumigation treatments and other methods for the control of replant disorders.

4136 Hamer GB 011008/00/0007
Control and mitigation of frost damage by physical methods.

4137 Webster; Chapman — GB 011008/00/0018
Causes and control of variability in crop yield and quality of plums and cherries.

4138 Skene; Mackenzie — GB 011008/00/0022 R
Histological studies with especial reference to graft unions and rootstock effects.

4139 Quinlan; Preston — GB 011008/00/0047 R
Relationship between apical and lateral shoots; chemical control.

4140 Quinlan; Preston — GB 011008/00/0048 R
Relationship between apical and lateral shoots; chemical control.

4141 Howard; Shepherd — GB 011010/00/0001
Propagation of fruit plants.

4142 Williams; R.R. — GB 011510/00/0006
Develop improved methods of pollination of pome fruits.

4143 Luckwill; Child — GB 011510/00/0007
Use growth regulators to improve quality, quantity and regularity of pome fruits.

4144 Luckwill; Child — GB 011510/00/0009
Develop techniques of production based on mechanical harvesting.

4145 Stott — GB 011510/00/0019
Control of ground clover and effect of herbicides on quality and yield of crop.

4146 Researcher not indicated — GB 050101/00/0004 R
Top fruit variety testing (including clones).

4147 Researcher not indicated — GB 050141/75/0009
Top fruit, variety testing (including clones).

4148 O'Kennedy, N.D.; McDonnell, P.F.
IE 060302/62/0500 R
High density apple plantings. Publications.

4149 O'Kennedy, N.D. — IE 060302/71/0494 R
Fruit set on apples and pears. Publications.

4150 O'Kennedy, N.D. — IE 060302/72/0502
Pollination of apples–movable pollinators. Publications.

4151 Cappellini, P. — IT 021500/74/0526
Rapporto tra i bionti nelle piante da frutto e loro influenza sulla produzione in riferimento al pesco, albicocco, pero. Biont interrelationships in fruit trees and biont influence on production with special reference to peach, apricot, and pear trees.

4152 Poma Treccani, C. — IT 040605/77/0266 R
Fisiologia del calcio nella insorgenza della butteratura amara e fattori pre e post raccolta influenzanti altre alterazioni fisiologiche da conservazione. Calcium physiology in the genesis of apple scald and pre and post harvesting factors producing other physiological alterations due to storage.

4153 Tromp, J. — NL 010212/62/1801
Onderzoek naar de achtergrond van de relatie groei –

vruchtbaarheid bij vruchtbomen. Investigations on the backgrounds of the relation between the vegetative and generative development of fruit trees. Publications.

4154 Tromp, J. — NL 010212/70/2721
De fysiologie van de vrucht voor de oogst in verband met de vruchtkwaliteit. Pre–harvest physiology of fruits as related to fruit quality. Publications.

4155 Oosten, H.J. van — NL 010212/73/4453
Onderstammen– en tussenstammenonderzoek bij pit– en steenvruchten. Rootstock and interstock experiments with pit– and stone fruits. Publications.

4156 Wertheim, S.J.; Lemmens, J.J. — NL 010212/73/4454
Teeltonderzoek groot fruit. Cultural experiments in top fruit. Publications.

4157 Wertheim, S.J. — NL 010212/73/4455
Regulatie van vruchtzetting, –groei, -val en vegetatieve groei. Regulation of fruitset, fruit development, fruit drop and vegetative growth. Publications.

Apple (B 3611)

See also 4213, 4221, 4255, 4332, 7252

4158 Monin, A. — BE 080200/65/0010
Sous–types de Cox's. Clonal types of Cox's.

4159 Monin, A. — BE 080200/67/0008
Comparaison de spurs et de variétés de pommiers greffés sur différents sujets porte–greffe. Comparison of spur types and apple varieties grafted on different stocks.

4160 Monin, A. — BE 080200/70/0002
La greffe intermédiaire chez le pommier et le poirier. Intermediate grafting on apple and peartrees.

4161 Monin, A. — BE 080200/76/0041
Essai de taille de pommiers et de poiriers. Training apple and peartrees in high density systems.

4162 Lüdders, P.; Firuzen, P. — DE 105204/75/0001
Einfluss von Düngung und Bodenpflege auf Apfelbäume im Iran. Effect of fertilization and soil management on apple trees in Iran.

4163 Naumann, W.D.; Al–Rawi, A. — DE 138242/78/0002 N
Abscisinsäuregehalt in Blättern von Apfelsämlingen bei Wasserstress. Abscisic acid contents in apple leaves as influenced by water stress.

4164 Liegel, W. — DE 144445/77/0009
Untersuchungen über Beziehungen des pH–Verlaufs und Entwicklungszustandes einer Apfelfrucht. Investigations on the relation of pH–changes and stage of development of an apple fruit. Publications.

4165 Liegel, W. — DE 144445/77/0010
Wirkung von Cd– und Pb–verbindungen bei zwei Apfelsorten. Effect of Cd and Pb compounds on two apple cultivars. Publications.

4166 Buchloh, G.; Liegel, W. — DE 144445/77/0015
Zur Remobilisierung von Ca im Parenchym von

Apfelfrüchten. **On the remobilisation of Ca in parenchyma of apple fruits.** Publications.

4167 Reimann–Philipp, R.; Hofmann, K.; Laskawy, W.; Schmidt, H.   DE 206000/75/0007
Abschliessende Wertermittlung von Nachkommenschaften aus Apfel– und Kirsch–Kreuzungen in bezug auf die Auslese neuen Zuchtmaterials oder für eine Sorteneignungsprüfung. **Evaluation of offsprings from apple and cherry crosses with respect to selection of new starting material or to propagation for variety trials.**

4168 Tiemann, K.–H.; Dammann, H.–J.   DE 507306/77/0001
Anbausysteme, Pflanzabstände, Kronenerziehung und Schnittbehandlung bei neuen Apfelsorten. **Cropping systems, plant spacing, training and pruning of new apple tree varieties.**

4169 Poulsen, E.; Hansen, P.   DK 010102/64/3104
Fysiologiske undersøgelser af vekselbæringens årsager i æbletræer. **Physiological investigations of the causes of alternate fruit bearing in apple trees.**

4170 Villemur, P.   FR 011201/66/0004
Contribution à l'étude de la croissance et du développement chez le pommier. **Contribution to the study of growth and development in apple–tree.**

4171 Lefevre, P.   FR 012222/63/0505
Amélioration de la qualité du fruit du pommier. Variété Cox's Orange. **Improvement of the fruit quality of Cox's Orange apple.** Publications.

4172 Goode; Hyrycz   GB 011003/00/0016
Nutrition in relation to irrigation.

4173 Jackson; Knight   GB 011008/00/0001
Control of fruit initiation, development and quality by chemical means.

4174 Jackson; Palmer   GB 011008/00/0010
Radiation reception and utilisation in orchards.

4175 Atkinson   GB 011008/00/0013
Spacing effects on root distribution, nutrient and water uptake.

4176 Parry; Jackson   GB 011008/00/0015
Orchard systems.

4177 Jackson   GB 011008/00/0017
Causes and control of variability in crop yield and apple quality.

4178 Jackson; Preston   GB 011008/00/0030
Rootstock evaluation and selection – effects on growth, crop yield and quality and on propagation.

4179 Knight; Jarvis   GB 011008/00/0031
Chemical fruit thinning and the chemical induction of abscission of fruits for mechanical harvesting.

4180 Quinlan   GB 011008/00/0032
Control of growth by pruning, plant hormones and other chemical growth regulators.

4181 Atkinson; White   GB 011008/00/0039 R
Herbicide and nutritional effects in modern apple orchards.

4182 White; Atkinson   GB 011008/00/0050
Nutrition in relation to irrigation.

4183 Powell; Thorpe   GB 011503/00/0003 R
Experimental and theoretical studies on the water relations of woody plants and their effects on growth.

4184 Landsberg; Thorpe   GB 011503/00/0004 R
Model of the growth and production cycle of apple trees, based on physiological and environmental studies.

4185 Abbott   GB 011510/00/0008 R
Factors influencing flower initiation, fruit set and cropping. Develop method of fruit forecasting.

4186 Ross   GB 041205/00/0003
Growth regulators in the production of Bramley seedlings.

4187 Ross   GB 041205/00/0004
Pollination of Bramley seedlings.

4188 Researcher not indicated   GB 050101/00/0001 R
Apples, production techniques, crop handling and storage.

4189 Researcher not indicated   GB 050101/00/0002 R
Apples, dessert and culinary variety collection.

4190 McDonnell, P.F.; O'Kennedy, N.D.
                                          IE 060302/65/0507 N
Growth regulators (apple trees). Publications.

4191 McDonnell, P.F.   IE 060302/66/0506 N
Chemical thinning of apple fruitlets. Publications.

4192 Hennerty, M.J.   IE 120108/71/9104 N
High density production systems for apples. Publications.

4193 Cappellini, P.; Colorio, G.; Simeone, A.; Strabbioni, G.
                                          IT 021500/77/0008 R
Studio sull'epoca di fioritura del melo e dell'albicocco in relazione al contenuto in ABA nelle gemme dormienti, e alla possibilità di applicazione dell'irrigazione termoregolata. **Study on apple and apricot blooming time in relation to ABA dormant bud content and to overhead sprinkler climatizing irrigation.**

4194 Monastra, F.   IT 021500/77/0515 R
Trattamenti con fitoregolatori sulla cultivar annurca. **Phyto–regulators used on "annurca" cultivar.**

4195 Nicotra, A.   IT 021500/77/0517 R
Diradamento dei frutti di melo su CV. Spur, in ambiente montano e della CV. annurca in Campania, diradamento dei frutti di pesco e loro metabolismo in rapporto ai prodotti applicati. **Thinning CV. Spur apples on mountainous ground and CV. Annurca apples in Campania, thinning peaches, the metabolism of this fruit in relation to chemical applications.**

4196 Nicotra, A.; Bergamini, A.; Albertini, A.; Cobianchi, D.; Cappellini, P.; Monastra, F.   IT 021500/78/0006 N
Impiego dei fitoregolatori su melo, pero e ciliegio, per contenere lo sviluppo vegetativo delle piante. **Use of the growth regulators to control the tree size of apple, pear and sweet cherry trees.**

4197 Limongelli, F.; Monastra, F.; Cappellini, P.; Temperini, O. IT 021500/78/0009 N
Studio del tipo di innesto più idoneo per il melo e per il noce. **Comparison of different graftings for apple and walnut.**

4198 Monastra, F.; Cobianchi, D.; Manzo, P.; Nicotra, A.; Bergamini, A.; Rivalta, L. IT 021500/79/0002 N
Studio dei portinnesti del melo, susino, ciliegio dolce, albicocco, pesco, nettarine e mandorlo e pero. **Trials on apple, plum, almond, apricot, sweet scherry, nectarine, peach, and pear rootstocks.**

4199 Gorini, F.; Sozzi, A.; Eccher Zerbini, P.
IT 021900/79/0003 N
Ricerca nella differenziazione della raccolta per una migliore classificazione commerciale delle mele Golden Delicious. **Research on grading for scalar harvesting of Golden Delicious apples.**

4200 Sansavini, S. IT 040203/74/0625
Miglioramento qualitativo delle pomacee e del pesco. **Quality improvement of pome and peach trees.**

4201 Sansavini, S. IT 040203/77/0668 R
Incremento fertilità nel melo nel pesco e nel susino, ricerche sull'autofecondazione e sulla sterilità fattoriale nel melo e nelle nettarine, relazione tra germinabilità e contenuto ormonale nel polline. **Improving the fertility of the apple–tree, peach–tree and plum–tree, research on self–fertilization and factorial sterility of the apple and nectarines, relation between germinability and pollen hormonal content.**

4202 Lalatta, F. IT 040605/74/0572
Ricerche sui fattori influenzanti la vitalità del sacco embrionale nel melo e loro riflessi sull'allegazione. **Research on the factors influencing the viability of apple tree embryonal bag and their side effects on fruit setting.**

4203 Lalatta, F. IT 040605/74/0573
Biologia della fruttificazione del melo. Fattori influenzanti la vitalità del sacco embrionale e del polline del melo. **Biology of fruit formation in apple trees. Factors influencing the viability of apple tree embryonal bag and pollen.**

4204 Roversi, A. IT 040605/74/0620
Influenza dell'ambiente di pianura, di collina, di montagna sulle caratteristiche qualitative delle mele spur. **Influence of plain, hill and mountain environment on spur apple quality characteristics.**

4205 Eccher, T. IT 040605/77/0491 R
Influenza dei fitoregolatori sulla qualità delle mele. **Effects of growth–regulators on apple quality.**

4206 Lalatta, F. IT 040605/77/0665 R
Biologia fiorale del melo, fattori condizionanti la durata del periodo effettivo di impollinazione e la allegazione. **Floral biology of the apple–tree, factors determining the length of the actual period of pollination and fruit–setting.**

4207 Poma Treccani, C. IT 040605/78/1141 N
Effetti di fitoregolatori esogeni sull'insorgenza di alterazioni fisiologiche da conservazione e sull'evoluzione della maturazione delle pomacee. **The action of phyto–regulators on physiological alterations due to storage and on the evolution of maturation in species of Pomaceae.**

4208 Eccher, T. IT 040605/79/0001 N
Studio sulla rugginosità e sulla forma del frutto nelle Golden Delicious. **Studies on russeting and on fruit form in Golden Delicious apple.**

4209 Avanzi, S. IT 041131/78/1180 N
Studio a livello citogenetico, citofotometrico ed autodiografico della sterilità maschile nel melo. **Cyto–genetic, cyto–photometric and auto–diographic study of male sterility in the apple–tree.**

4210 Jonkers, H.J. NL 020057/72/7375 R
Onderzoek inzake het optreden van bladvlekken bij Golden Delicious en bodemmoeheid bij Cox's Orange Pippin en James Grieve. **Research on leafspot of Golden Delicious and soil exhaustion with Cox's Orange Pippin and James Grieve.** Publications.

Pear (B 3612)

See also 4160, 4161, 4190, 4196, 4198, 4207, 4257, 7252

4211 Deckers, J. BE 040202/68/0003
Snoeiproeven op Doyenné du Comice peer. **Prunning trials on doyenné du comice pear.** Publications.

4212 Drescher, W.; Engel, G. DE 111201/75/0001
Der Einfluss der Honigbiene auf den Fruchtansatz selbststeriler Birnensorten. **The influence of the honeybee on the yield of self–sterile pear–varieties.**

4213 Barbier, E.; Tasei, J.N. FR 010608/75/5281
Pollinisation des poiriers et pommiers dans les vergers du Sud–Est de la France. **Pollination of pear – and apple–trees in French southeastern orchards.**

4214 Parry; Knight GB 011008/00/0002
Control of fruiting by chemical sprays, pollination etc.

4215 Parry GB 011008/00/0033
Rootstock evaluation and selection – effects on growth, crop yield and quality and on propagation.

4216 Parry; Jackson GB 011008/00/0034
Control of growth by pruning and chemical means.

4217 Atkinson GB 011008/78/0042 R
Root growth water and nutrient use and requirements in intensive pear orchards.

4218 Parry GB 011008/78/0045
Intensive cropping systems for pears.

4219 Researcher not indicated GB 050101/00/0006 R
Pears, production techniques.

4220 Researcher not indicated GB 050141/00/0003
Pears, production techniques.

4221 Cristoferi, G. IT 063000/77/0822
Stimolo allegagione e fruttificazione in pero, melo, pesco e ciliegio. **Inducing fruit setting and fructification in the pear tree, the apple tree, the peach tree and the cherry tree.**

4222 Tosi, T. IT 102801/77/0529 R

Controllo allegagione nel pero. **The regulation of fruit development in pear–trees.**

Other top fruit (B 3619)

See also 850, 4167, 4193, 4195, 4196, 4198, 4200, 4201, 4221, 4332, 7270, 7275

4223 Scheys, I.; Deckers, J.; Keulemans, J.; Joukers, G.
BE 040202/77/0013 R
Onderzoek over cultivars, onderstammen en teelttechniek bij steenfruit. **Research on varieties, rootstocks and culture technics of stone fruit.** Publications.

4224 Scheys, I.; Deckers, J.; Keulemans, J.
BE 040202/77/0018 N
Fytotechnisch onderzoek om de oogstzekerheid bij de teelt van zure en zoete kers, pruim en bessen te verbeteren.
Fytotechnical research to eliminate the variations of the yield of sour and sweet cherry, plums and currants. Publications.

4225 Monin, A.; Tréfois, R. BE 080200/78/0045 R
Etude du comportement de vergers de cerises industrielles dans le sud du pays et traitement de la récolte. **Study of cherry orchards for industry in the south part of Belgium.**

4226 Heinze, W. DE 105110/71/0001
Einfluss von Umweltbedingungen auf Blütenbildung und Ablauf der Ruhe bei der Prunus glandulosa 'Albuplena'.
Influence of environmental conditions on flower formation and rest of Prunus glandulosa 'Albuplena'.

4227 Bünemann, G.; Lee, C.–L. DE 138240/78/0001 N
Pollenschlauchwachstum bei Pflaumensorten – Prunus domestica –. **Pollen tube growth in plums – Prunus domestica –.**

4228 Hartmann, W. DE 144445/77/0005
Vermehrungsmethoden zur Anzucht geklonter, wurzelechter Hauszwetschentypen. **Methods for propagation of cloned 'Hauszwetsche' grown on their own roots.**

4229 Feucht, W.; Nachit, M. DE 161265/75/0003
Methoden zur Frühselektion bei Prunus–Unterlagen. **Methods for pre–selection of Prunus–rootstocks.** Publications.

4230 Feucht, W. DE 161265/75/0005
Phenolische Inhaltsstoffe von in vitro kultiviertem Prunus–Kallus. **Phenolic compounds of Prunus callus tissue.**

4231 Feucht, W.; Schmid, P. DE 161265/78/0001 N
Polyphenole in Sprossen von Kirscharten und –sorten und deren Beeinflussung durch Veredlung. **Polyphenols in sprouts of species and varieties of cherries as affected by grafting.**

4232 Kohstall, H.; Wirth, H. DE 507701/77/0001
Der Einfluss verschiedener Unterlagen und Zwischenveredlungen auf das Wachstum und den Ertrag von Süsskirschen. **The influence of different rootstocks and interstocks on growth and yield of different sweet cherry varieties.**

4233 Billot, C.; Arnoux, M. FR 010610/70/3007
Etude de systèmes de productions mixtes: arboriculture fruitière, Cultures intercalaires, applicables dans la Vallée du Rhône. **Study of production systems including orchards and**

other crops in the Rhone Valley. Publications.

4234 Arnoux, M.; Defrance, H.; Huguet, C.
FR 010615/65/9042
Effets combinés de la taille, des doses d'azote et de l'éclaircissage des fruits en vergers de pêcher. **Combined effects of pruning, nitrogen fertilization and thinning in peach orchards.**

4235 Arnoux, M.; Morvan; Defrance, H.; Vigouroux; Castellain; Leclan FR 010615/71/9041
Essais de synthèse en vergers d'Abricotier. **Optimizing all aspects of apricot orchard management.**

4236 Bernhard, R.; Grassely, Ch.; Olivier, G.
FR 010701/68/0099
Recherche des possibilités de production de plants grâce à la production de semences hybrides F1 interspécifiques.
Possibilities of seedling production thanks to interspecific hybrid F1 seeds (peach X almond tree rootstocks).

4237 Mesnier, Y.; Renaud, R. FR 010701/68/0233
Sélection de porte–greffes à croissance rapide compatibles pour quelques variétés importantes du groupe des reine–claude. **Selection of quickly growth rootstocks compatible with important plums "reine–claude" varieties.** Publications.

4238 Huguet, J.G. FR 010704/72/0549
Etude de l'enracinement du prunier d'ente selon la nature du sol et le porte–greffe. **Study on the root development of prune in relation with soil and rootstock.** Publications.

4239 Clanet, H.; Salles, J.C. FR 011207/69/0441
Recherche d'une technique d'éclaircissage chimique des fleurs ou des jeunes fruits du pêcher. **Flowers or small fruits thinning on peach: research for an effective and practical method.**

4240 Clanet, H.; Salles, J.C. FR 011207/70/0442
Recherche sur l'action de l'acide gibberellique sur l'évolution des bourgeons à fleur de pêcher. **Investigation of gibberellic acid effects on evolution of flower buds of the peach tree.**

4241 Hugard, J.; Clanet, H. FR 012207/71/0011
Recherches sur les facteurs d'adaptation au milieu des cultivars d'abricotier. **Research on the adaptability to environment of apricot cultivars.**

4242 Webster GB 011008/00/0003
Control of growth and cropping by pruning and chemical growth regulators.

4243 Webster; Chapman GB 011008/00/0035
Rootstock evaluation selection – effects on growth, disease resistance, crop yield quality, propagation.

4244 Webster GB 011008/00/0036
Control of growth and cropping by pruning and chemical growth regulators.

4245 Webster; Chapman GB 011008/00/0037
Rootstock evaluation and selection – effects on growth, disease resistance, crop yield and quality.

4246 Atkinson GB 011008/78/0041 R
Root growth water and nutrient use and requirements in intensive cherry orchards.

4247 Webster GB 011008/78/0043
Intensive cropping systems for plums.

4248 Webster GB 011008/78/0044
Intensive cropping systems for cherries.

4249 Researcher not indicated GB 050101/00/0007 R
Plums, production techniques.

4250 Researcher not indicated GB 050101/00/0008 R
Cherries, production techniques.

4251 Researcher not indicated GB 050141/00/0004
Plums, production techniques.

4252 Researcher not indicated GB 050141/00/0005
Cherries, production techniques.

4253 Monastra, F.; Fideghelli, C. IT 021500/71/0006
Prova portainnesti della nettarina "nectarose". **Trial of different rootstocks of "Nectarose" nectarine.**

4254 Nicolli, C.; Albertini, A. IT 021500/71/0007
Confronto portainnesti ed intermedi per il ciliegio. **Interstock trial for sweet cherry.**

4255 Grassi, G.; Fideghelli, C.; Bergamini, A.; Cappellini, P.; Monastra, F.; Strabbioli, G. IT 021500/72/0015 R
Correlazione tra fattori climatici e produttività di pesco, albicocco, susino e melo. **Correlation between climatic factors and productivity of peach, apricot, plum and apple cultivars.**

4256 Colorio, G.; Manzo, P.; Bergamini, A.; Monastra, F. IT 021500/77/0025 R
Prove di potatura meccanica, su albicocco e pesco. **Mechanical pruning trials on apricot and peach trees.**

4257 Cappellini, P. IT 021500/77/0654 R
Biologia fiorale dell'albicocco e del pero. **Floral biology of the apricot–tree and the pear–tree.**

4258 Monastra, F.; Recupero, S.; Cappellini, P.; Fideghelli, C. IT 021500/78/0004 N
Studio sulle cause patologiche, agronomiche e fisiologiche, determinanti l'insorgenza di una grave forma di deperimento del pesco. **Study of the phitopathologic, agronomic and phisiologic causes of the peach dieback.**

4259 Bagni, N. IT 040226/74/0505
Studio sulla dormienza delle gemme del pesco. Contenuto e stato di aggregazione dei ribosomi. **Study on the dormancy of peach tree buds. Ribosome content and state of aggregation.**

4260 Bagni, N. IT 040226/77/0652 R
Studio dei parametri ambientali e fisiologici che influenzano e caratterizzano la vitalità del polline di diverse cultivars di pesco del gruppo nettarine. **Study of the environmental and physiological parameters influencing and characterizing the vitality of the pollen in different peach cultivars belonging to the nectarine group.**

4261 Serafini Fracassini, D. IT 040226/78/1112 N
Controllo della maturazione del frutto di pesco mediante lo studio della variazione dei ribosomi e dell'attività di sintesi proteica. **The regulation of peach maturation : a study of rhibosome variation and protein synthesis.**

4262 Roversi, A.; Valli, R. IT 040403/70/0006 R
Ricerche sui fruttiferi da industria (percoche, nocciole, cotogne, ciliegie). **Research on fruit–trees for industry (cling peach, hazelnuts, quince, sweet cherries).** Publications.

4263 Roversi, A. IT 040403/78/1102 N
Raccolta meccanizzata e trasformazioni industriali dei frutti di ciliegio dolce. **The mechanical harvesting and industrial processing of sweet cherries.**

4264 Ponchia, G. IT 040507/78/1142 N
Controllo della maturazione di frutti e trattamenti cascolanti al ciliegio, olivo, vite, potatura chimica vite. **The regulation of fruit maturation, treatments against premature fruit setting applied to the cherry–tree, the olive–tree and vines; chemical pruning of vines.**

4265 Pisani, P.L. IT 040803/77/0523 R
Diradamento chimico del pesco con particolare riferimento allo studio delle basi fisiologiche e biochimiche dell'abscissione dei frutticini. **Chemical thinning of the peach–tree with particular regard to a study of the physiological and bio–chemical bases of the abscission of newly formed fruits.**

4266 Zucconi, F. IT 041104/77/0533 R
Diradamento chimico del pesco, controllo maturazione e trattamenti cascolanti olivo. **Chemical pruning of the peach tree, maturation control and treatments preventing floral abscission.**

4267 Bellini, E. IT 062800/74/0184
Ricerche sulla propagazione per seme e biologia della riproduzione. Indagini sulla biologia fiorale della cv. di susino "Burmosa" e di alcune sue cultivar impollinatrici. **Researches on seed propagation and on reproduction biology. The flower biology of the plum cv. "Burmosa" and of some of its pollinators.**

4268 Sansavini, S. IT 063000/73/0016
Allevamento del pesco in coltura protetta. **Peach growing under plastic.** Publications.

4269 Filiti, N. IT 063000/77/0819
Anticipo messa a frutto del ciliegio; diradamento chimico frutti pesco e melo. **Early fruit bearing of the cherry tree; chemical thinning of peaches and apples.**

4270 Bargioni, G. IT 102801/77/0472 R
Anticipo messa a frutto ciliegio, controllo maturazione e trattamenti cascolanti ciliegio. **Early fruit setting in the cherry tree, maturation control and thinning treatment of the cherry tree.**

4271 Bargioni, G. IT 102801/77/0653 R
Biologia fiorale del ciliegio. **Floral biology of the cherry tree.**

4272 Bargioni, G.; Tosi, T. IT 102801/79/0002 N
Comportamento del ciliegio dolce su nuovi portinnesti. **Behaviour of sweet cherry on new rootstocks.**

4273 Schiaparelli, A. IT 121400/77/0525 R
Diradamento chimico del pesco, abscissione frutti olivo. **Chemical thinning of the peach–tree; abscission of olive fruits.**

D 2100 – Plant production general and crop husbandry

Soft fruit (berries and cane fruits) (B 3620)

See also 2105, 3777, 3960, 3965, 4224, 4474, 5832, 5833, 7297, 7916

4274 Scheys, I.; Deckers, J.; Keulemans, J.; Jonkers, G.
BE 040202/77/0012 R
Onderzoek over cultivars, onderstammen en teelttechniek bij kleinfruit. **Research on varieties, rootstocks and culture technics of soft fruit.** Publications.

4275 Lamberts, D.; Lettani, L.; Deckers, T.
BE 040202/77/0016 R
Hoge plantdichtheid bij aardbeien in kassen. **High density planting of strawberries in glass houses.**

4276 Linden, R.
BE 080200/70/0030
Etude de la dormance du fraisier. **Dormancy of strawberry plants.**

4277 Linden, R.
BE 080200/76/0031
Détection des stades et de l'époque de l'induction des ébauches florales chez les nouvelles variétés de fraisiers. **Research on stage and time of bud initiation of new strawberry varieties.**

4278 Karnatz, A.; Heesch, W.
DE 105204/78/0001 N
Stomatawiderstand bei Schwarzer Johannisbeere und Kaffee in Abhängigkeit von Bodentemperatur und Herbizideinwirkung. **Stomata resistance in black currant and coffee as influenced by soil temperature and herbicide.**

4279 Naumann, W.D.; Synowski, B.
DE 138242/77/0002
Die Wirkung der Mikrovermehrung auf Wuchs und Ertrag bei Erdbeeren und Gehölzen. **Growth and crop yield of strawberries and woody perennials as influenced by micropropagation.**

4280 Hartmann, W.
DE 144445/77/0003
Verhinderung des Verrieselns durch Wuchsstoffbehandlungen bei der schwarzen Johannisbeere. **The effect of some growth substances on fruit drop in black currants.**

4281 Stösser, R.; Schmidt, G.
DE 144445/78/0003 N
Untersuchungen über einige im Zusammenhang mit der maschinellen Ernte von Erdbeeren wichtige Kriterien. **Investigations on some parameters which are important in mechanical harvesting of strawberries.**

4282 Feucht, W.; Schimmelpfeng, H.
DE 161265/71/0001 R
Erzielung von Schwachwüchsigkeit bei Edelsorten aus den Kreisen Prunus avium, Prunus cerasus und Prunus domestica durch Unterlagen und Zwischenveredlungen 1969. **Attainment of slender grothw of grafted varieties of Prunus avium, Prunus cerasus and Prunus domestica by stocks and intermediate grafting.**

4283 Schmid, P.; Christ, E.
DE 161265/72/0002
Untersuchungen der Möglichkeiten des Anbaues der Preiselbeere - Vaccinium vitis idaea L. - auf Ackerboden 1972. **english title not indicated.**

4284 Christ, E.
DE 161265/72/0005 R
Züchterische Bearbeitung der Preiselbeere – Vaccinium vitis idaea L. – und der Cranberry – Vaccinium macrocarpon Ait. – 1972. **Breeding of cranberry Vaccinium vitis idaea L. and Vaccinium macrocarpon Ait..**

4285 Schimmelpfeng, H.
DE 161265/73/0001
Verwendung von Frigo–Material im Erdbeeranbau in klimatischen Grenzlagen. **Use of Frigo plants in strawberry growing under unfavourable climatic conditions.** Publications.

4286 Schimmelpfeng, H.
DE 161265/74/0001
Spezialfragen des grossflächigen Erdbeeranbaues. **Special problems of strawberry cultivation on a large scale.**

4287 Feucht, W.; Schimmelpfeng, H.
DE 161265/74/0002
Fragen eines neuzeitlichen Brombeeranbaues. **Problems of a modern blackberry culture.**

4288 Bauckmann, M.
DE 506106/77/0002
Einfluss von Pflanzgut aus üblicher Pflanzenvermehrung und Meristemkultur auf den Ertrag von Erdbeeren. **Influence of usual plant propagation and meristem cultivation on the yield of strawberries.**

4289 Bauckmann, M.
DE 506106/77/0003
Einfluss von Schnittmethode und Standweite auf die Ertragsleistung von Brombeeren. **Influence of cut method and different distances on the yield of blackberries.**

4290 Zahn, F.–G.
DE 507307/72/0001
"Stufenweises Absetzen" von Hauptästen zur Prophylaxe des Kirschbaumsterbens und zur Erhaltung einer Niedrigen Baumform. **"Removing by steps" of main branches for prevention of dieback of cherry trees and for keeping down the tree top.**

4291 Vang–Petersen, O.
DK 010102/78/0001 N
Sorts–, kultur– og ernæringsforsøg med buskfrugt. **Variety trials, nutrition experiments and cultural practice in soft fruits.**

4292 Jørgensen, M.B.; Thuesen, A.
DK 010115/72/4502
Kulturforanstaltninger til styring af bær– og planteproduktion i jordbær. **Growing methods for steering berry and plant production in strawberries.**

4293 Clay; Davison
GB 010806/00/0007
New herbicides for the control of annual and perennial weeds in strawberries.

4294 Jones; James
GB 011006/00/0017 R
Regulation of root initiation and establishment of vegetative propagules.

4295 Jones; James
GB 011006/00/0018 R
Role of endogenous factors in root initiation and establishment of vegetative propagules.

4296 James
GB 011006/78/0020 R
Regulation of shoot and root initiation in Rubus Ribes and Fragaria species.

4297 Jackson; Palmer
GB 011008/00/0012
Radiation reception and utilisation particularily with regard to new systems of growing.

4298 Atkinson
GB 011008/00/0014
Root distribution, water and nutrient uptake.

4299 Holloway; Knight
GB 011008/00/0027
Rubus agronomy.

4300 Oehl GB 011008/00/0028
Strawberry agronomy.

4301 Holloway; Jarvis GB 011008/00/0038 R
Agronomic studies of new EMRS varieties and improved source material.

4302 Guttridge GB 011510/00/0002 R
Improve cropping by resolving major factors that limit productivity.

4303 Guttridge; Anderson GB 011510/00/0003
Develop cultural methods to improve yield of cultivars.

4304 Mackerron; Waister GB 030701/00/0001 R
Effects of weather conditions on growth, yield and quality of soft fruit crops.

4305 Cormack; Waister GB 030701/00/0012
Ecology of new fruit crops for Scotland.

4306 Cormack; Waister GB 030701/00/0014
Physiological and cultural factors affecting the mechanical harvesting of soft fruits.

4307 Cormack; Waister GB 030701/00/0018
Control of growth, yield and quality of raspberries by cultural methods and choice of genotype.

4308 Cormack; Waister GB 030701/00/0019
Control of growth, yield and quality of strawberries by cultural methods and choice of genotype.

4309 Ross GB 041204/00/0002
The evaluation of soft fruit crop management techniques.

4310 Ross GB 041204/00/0003
The evaluation of soft fruit varieties.

4311 Researcher not indicated GB 050101/00/0005 R
Soft fruit variety testing (including clones).

4312 Researcher not indicated GB 050101/00/0009 R
Strawberries, production techniques.

4313 Researcher not indicated GB 050101/00/0010 R
Bush fruits, production techniques.

4314 Bd GB 050101/75/0012
Cane fruit, production techniques.

4315 Researcher not indicated GB 050141/00/0002
Soft fruit, variety testing (including clones).

4316 Researcher not indicated GB 050141/00/0006
Strawberries, production techniques.

4317 Researcher not indicated GB 050141/00/0007
Cane fruit, production techniques.

4318 Researcher not indicated GB 050141/75/0010
Bush fruits, production techniques.

4319 Turner GB 060112/00/0001
Raspberry production and harvesting.

4320 Turner GB 060112/00/0002
Strawberry production.

4321 Turner GB 060112/00/0003
Production and harvesting of miscellaneous soft fruit crops.

4322 Sutherland GB 060215/00/0003
Soft fruit variety testing.

4323 Sutherland GB 060215/00/0004
Soft fruit–methods of husbandry.

4324 Lamb, J.G.D. IE 060300/58/0097 R
Blueberry trials (yield). Publications.

4325 Rath, N.; O'Callaghan, T.F.; Kavanagh, T.
IE 060300/73/0879 R
Raspberry cultural methods. Harvesting. Publications.

4326 O'Callaghan, T.F.; Rath, N.; O'Riordan, F.
IE 060301/65/0088 R
Blackcurrants–cultivar testing and cultural methods. Publications.

4327 O'Callaghan, T.F.; Rath, N.; O'Riordan, F.
IE 060301/65/0089 R
Gooseberries–cultivar testing and cultural methods. Publications.

4328 Rath, N.; O'Callaghan, T.F. IE 060301/68/0090 R
Cultural trials on strawberries. Publications.

4329 Rath, N.; O'Callaghan, T.F.; Kavanagh, T.
IE 060301/70/0091 R
Production of early strawberries (plant spacing). Publications.

4330 Colorio, G.; Manzo, P.; Bergamini, A.; Monastra, F.;
Pennone, F.; Recupero, S. IT 021500/77/0027 R
Studio di nuove forme di allevamento del lampone e del melo, atte alla raccolta meccanica. **Study of new training systems of raspberry bush and apple tree suitable for mechanical harvesting.**

4331 Eynard, I. IT 041203/73/0199
Introduzione di nuove cultivar di vaccinium e rubus, e studi relativi agli ambienti più idonei di coltura, alla meccanizzazione della raccolta ed alla prima elaborazione dei prodotti. **Introduction of new cultivars of Vaccinium and Rubus and related studies on the most adapted environment for their cultivation, mechanization of harvesting and preliminary processing of products.**

4332 Rosati, P. IT 063000/77/0667
Androgenesi, colture aploidi, micropropagazione nella fragola ed in alcune specie da frutto, albicocco, pesco, melo. **Androgenesis, haploid cultures, micro-propagation of strawberries and other fruit plants (apricot tree, peach tree, apple tree).**

4333 Ravestijn, W. van NL 010206/74/6865
Opbrengst vervroeging en verbetering bij glasaardbeien. **Increasing production and earliness of glasshouse–strawberries.**

4334 Dijkstra, J. NL 010212/58/1792

Teeltonderzoek aardbei. **Cultural experiments strawberry.**
Publications.

4335 Dijkstra, J.   NL 010212/58/1794
Teeltonderzoek bij frambozen, bessen en bramen. **Cultural experiments with raspberries, currants and blackberries.**
Publications.

Citrus fruit (B 3630)

See also 1484

4336 Lenz, F.; Noga, G.   DE 111300/78/0005 N
Ursachen der Rauhschaligkeit bei Satsumas – Citrus unshiu, Marc. –. **Reasons for skin roughness in Satsuma – Citrus unshiu, Marc. –.**

4337 Jullian, P.; Attonaty, J.M.   FR 011808/77/8529
Etude technique et économique sur la rénovation du verger cidricole. **Technical and economical study about cider orchard restoration.**

4338 Blondel, L.; Vittori, F.; Jacques, C.; Mond
FR 012202/65/0215
Régulation de la production du clémentinier par l'utilisation de substances de croissance– Cueillette mecanique. **Utilization of growth substances for regulating Clementine production – Mechanical harvest.**

4339 Blondel, L.; Jacquemond, C.   FR 012202/71/0015
Etude sur les porte–greffes des agrumes appelés à remplacer le bigaradier. **Study on citrus–rootstocks other than bitter orange.**

4340 Russo, F.; Reforgiato, G.   IT 021600/58/0001
Prova portainnesti per agrumi. **Citrus rootstock tests.**
Publications.

4341 Di Martino, E.; Lanza, G.   IT 021600/69/0005 N
Studi sulle fisiopatie dei frutti. **Studies on physiological troubles of citrus fruits..**

4342 Di Pinna, S.   IT 021600/71/0001
Ricerche fenologische su agrumi. **Phenological researchs on citrus.**

4343 Russo, F.; Amato, R.; Reforgiato, G.   IT 021600/73/0012
Studio della sporogenesi in relazione all'apirenia degli agrumi. **Development of male and female gametes and relation to citrus seedlessness.**

4344 Russo, F.; Reforgiato, R.; Starrantino, A.
IT 021600/73/0014 N
Prove di portinnesti e di coltivazione protetta del cedro. **Rootstocks trials and protected cultivation of citron.**

4345 Terranova, G.; Caruso, A.   IT 021600/74/0007 N
Prove di portinnesti del bergamotto e del cedro. **Roostock trials of bergamot and citron.**

4346 Caruso, A.   IT 021600/76/0001 N
Gli oli essenziali del bergamotto in relazione all'ambiente di coltivazione ed allo stato nutritivo delle piante.. **Essential oils of bergamot in relation to environment and nutritive status of trees.**

4347 Lo Giudice, V.   IT 021600/76/0003 N
Influenza dei trattamenti erbicidi al terreno sulla biologia delle piante di agrume. **Influence of herbicides on citrus trees biology.**

4348 Lanza, G.; Lo Giudice, V.   IT 021600/76/0004 N
Studio sulle micorrize degli agrumi. **Research on citrus mycorrihyzes.**

4349 Starrantino, A.; Russo, F.   IT 021600/77/0001 N
Studio sulla incompatibilità d'innesto negli agrumi mediante la tecnologia dei "microinnesti in vitro". **Investigation on graft incompatibility in citrus by "in vitro micrografting" technology.**

4350 Raciti, G.B.   IT 021600/77/0709 R
Meccanizzazione della raccolta degli agrumi, raccolta integrata, aspetti tecnico–colturali. **Mechanization of citrus fruit harvesting, integrated harvesting, technical and cultural aspects.**

4351 Di Martino, E.; Lo Giudice, V.; Lanza, G.; Benfatto, D.; Di Martino Aleppo, E.   IT 021600/79/0004 N
Indagine sull'attività dei fitoregolatori. **Research on citrus growth regulators activity.**

4352 Damigella, P.   IT 040307/74/0541
Ricerche sul pompelmo, sull'avocado e sulla papaia. **Research on grapefruit, avocado and papaya.**

4353 Continella, G.; Tribulato, E.   IT 040307/79/0002 N
Influenza del clima sull'evoluzione della maturazione dei frutti di agrumi. **Influence of the climate on maturity of citrus fruits.**

4354 Crescimanno, F.G.; Sottile, I.; De Michele, A.; Barone, F.   IT 040908/74/0001
Ricerche sulla biologia fiorale e delle fruttificazione negli agrumi. **Researches on citrus flower and fruit biology.**
Publications.

4355 Crescimanno, F.G.; Di Marco, L.; Sottile, I.
IT 040908/76/0001
Prove di copertura con plastica nei vivai di agrumi. **Trials on plastic covering citrus nursery.** Publications.

4356 Deidda, P.   IT 041310/77/0685 R
Meccanizzazione della raccolta degli agrumi, raccolta integrata, aspetti tecnico–colturali. **Mechanization of citrus fruit harvesting, integrated harvesting, technical aspects of cultivation.**

4357 Crescimano, F.G.   IT 060900/77/0891
Prove agronomiche di portainnesti di nuova introduzione in combinazione con limone, arancio W. Navel e mandarino tardivo di Ciaculli. **Agricultural tests on newly introduced root–stocks for the lemon, the Washington Navel orange and the late Ciaculli tangerine.**

4358 Crescimano, F.G.   IT 060900/77/0893
Ricerca sul microinnesto ed allevamento in vitro di tessuti di giovani embrioni di agrumi e studio sulla trasmissione dei virus in piante così ottenute. **Research on micro–grafting and "in vitro" growth of young citrus embryo tissues; a study on virus transmission in plants obtained thereof.**

4359 Crescimano, F.G.   IT 060900/77/0895

D 2100 – Plant production general and crop husbandry

Osservazioni di pieno campo su nuovi portainnesti, in combinazione con diverse specie e cultivar di agrumi. **Field observation on new root–stocks used with different citrus species and cultivars.**

4360 Crescimano, F.G.   IT 060900/77/0896
Prosecuzione delle indagini sul patrimonio agrumicolo italiano. **Follow–up on the Italian citrus patrimony.**

Tropical and sub–tropical fruits (B 3640)

See also 2613, 3773, 3846, 4352

4361 De Langhe, E.; Vincke, H.; Naku–Mbumba.
      BE 040900/78/0002
Fysiologische studie van het kiemingsgedrag van de meelbanaan. **Physiological study of the shooting behaviour of the Plantain Banana.**

4362 Atanasiu, N.; Alkämper, J.; Westphal, A.
      DE 129552/75/0005
Einfluss tropischer Standorte auf Ertrag und Qualität verschiedener tropischer Kulturpflanzen. **Influence of tropical locations on yield and quality of various tropical crops.**

4363 Moawad, M.; Yantasath, K.; Abo–Shoba, L.
      DE 132240/75/0001
Der Einfluss atmosphärischer Faktoren auf Wachstum und Nährstoffaufnahme tropischer Kulturpflanzen. I.Lichtintensität und Wind. **The influence of atmospheric factors on growth and nutrient uptake of tropical crops. I.Light intensity and wind.**

4364 Vogel, R.   FR 012202/59/0101
Amélioration de la culture des arbres fruitiers exotiques. **Improvement of exotic fruit–trees cultivation.** Publications.

4365 Pratella, G.C.; Biondi, G.   IT 040216/76/0003 N
Maturazione controllata dei Kaki. **Controlled ripening of persimmons.**

4366 Samson, J.A.   NL 020056/73/6356
Onderzoek aan tropische vruchten (verbetering van oculatiemethoden). **Tropical fruit research (improvement of grafting methods).** Publications.

Grapes (B 3650)

See also 2441, 2606, 2637, 3148, 3392, 4264, 7388, 7392, 15817, 17006

4367 Kausch, W.; Gotthold, J.; Schlöder, F.–R.
      DE 111150/72/0001
Öko–physiologische Grundlagen von Wachstum und Ertrag bei der Rebe. **Ecophysiological basis of growth and yield of vine.**

4368 Henze, J.; Welches, H.G.; Becker, H.
      DE 111301/77/0003
Wuchs und Qualität von Rebenzüchtungen an verschiedenen Standorten. **Growth and quality of new grape varieties at different locations.**

4369 Alleweldt, G.; Lupold, H.   DE 144450/78/0001 N
Mykorrhiza der Rebe. **Mykorrhiza of grape vine.** Publications.

4370 Alleweldt, G.; Zierock, R.   DE 144450/78/0002 N
Physiologische Nachwirkungen der Thermotherapie bei Reben. **Physiological after–effects of thermotherapy in grape vine.**

4371 Eichhorn, K.W.; Müller, F.   DE 144541/74/0003
Untersuchungen über den Einfluss von physiologischen Störungen auf das Absterben von Freilandreben. **Investigations on the influence of physiological disturbances on the die–back of Vitis vinifera in the field.**

4372 Matzner, F.; Schubert, C.von   DE 161265/75/0001
Steilhangbewirtschaftung im Weinbau. **Cultivation of steep slopes in viticulture.**

4373 Matzner, F.; Geiger, K.   DE 161266/73/0001
Untersuchungen über den Austriebszeitpunkt von Vitis vinifera ssp. sativa als Kriterium für die Ampelographie. **Investigations on the point of sprouting in Vitis vinifera ssp. sativa as criterion for ampelography.**

4374 Matzner, F.   DE 161266/73/0005
Möglichkeiten und Grenzen der Dauerbegrünung im Weinbau. **Possibilities and limits of covering with permanent grass in viticulture.**

4375 Alleweldt, G.; Düring, H.; Gebbing, H.
      DE 204000/75/0008
Untersuchungen zur Funktion der Abscisinsäure bei der Beerenreife. **Analysis of the action of abscisic acid in ripening grapes.**

4376 Alleweldt, G.; Herwig, K.   DE 204000/77/0005
Untersuchungen über die Affinität zwischen Unterlage und Reis. **Studies on affinity between rootstock and shoot.**

4377 Düring, H.   DE 204000/77/0007
Untersuchungen zum Transpirationsverhalten einzelner Sorten und Arten der Rebe. **Studies on transpiration of individual varieties and species of vine.**

4378 Alleweldt, G.   DE 204000/78/0001 N
Einfluss der Klimafaktoren auf die Ertrags– und Qualitätsleistung von Rebsorten. **Influence of climatic factors on yield and must quality of grape vine varieties.**

4379 Alleweldt, G.; Zierock, R.   DE 204000/78/0003 N
Die Auswirkung der Wärmebehandlung auf Wachstum und Ertragsleistung der Rebe. **After effect of thermotherapy on growth and yield of vines.**

4380 Bachmann, O.; Blaich, R.   DE 204000/78/0010 N
Bestimmung von Inhaltsstoffen pilzresistenter Rebsorten. **Determination of substances contained in fungus resistant vine varieties.**

4381 Bachmann, O.; Blaich, R.   DE 204000/78/0011 N
Vergleichende Untersuchungen zur Phytoalexinbildung bei pilzresistenten und pilzanfälligen Rebsorten. **Comparative investigations on the formation of phytoalexins in vines.**

4382 Rapp, A.; Knipser, W.   DE 204000/78/0012 N
Untersuchungen über die Aromastoffe in Blättern verschiedener Rebsorten. **Investigations on flavour compounds in leaves of different vine varieties.**

4383 Herwig, K. DE 204000/78/0015 N
Der Einfluss von Pfropfkombinationen auf das Wachstum und die Nährstoffaufnahme der Reben. **The influence of grafting combinations on growth and nutrient uptake in grape plants.**

4384 Hoppmann, D.; Zunker, E.–J. DE 301050/72/0001
Bestandsklimatische Untersuchungen in Reben–Normalund Weitraumanlagen. **Climatic investigations in vineyards with narrow and wide spaced rows.** Publications.

4385 Hoppmann, D. DE 301050/75/0002
Untersuchung der Beziehung zwischen Wärmehaushalt und Qualitätsleistung von Weinbaustandorten. **Relation between heat balance and quality production of vine–growing sites.** Publications.

4386 Hoppmann, D. DE 301050/75/0003
Kaltluftverhalten in weinbaulich genutzten Hang– und Tallagen. **Cold air behaviour at vine–growing sites on slopes and in valleys.**

4387 Hoppmann, D. DE 301050/75/0004
Wärme– und Wasserhaushalt im Weinberg. **Heat balance and water balance in vineyards.**

4388 Hoppmann, D. DE 301050/77/0001
Die klimatischen Verhältnisse im Foliengewächshaus und die Auswirkungen auf die Entwicklung von Pfropfreben. **The climatic conditions in polyethylene foil house and the effect on the growth of grafted vines.** Publications.

4389 Staudt, G. DE 501100/74/0003
Cytologische Untersuchungen an reifenden Weinbeeren. **Cytological investigations on ripening grape berries.**

4390 Staudt, G. DE 501100/74/0004
Untersuchungen über den Zeitpunkt der Bestäubung und Befruchtung bei verschiedenen Rebensorten. **Investigations on the time of pollination and fertilization in different grape vine varieties.**

4391 Staudt, G.; Kassemeyer, H. DE 501100/77/0001
Untersuchungen über den Einfluss von Befruchtung und Samenentwicklung auf den Beerenansatz der Weinrebe. **Investigations on the influence of fertilization and seed development on berry set in vines.**

4392 Becker, N.J. DE 501102/78/0001 N
Einfluss ökologischer Faktoren auf Rebenentwicklung, Inhaltsstoffe der Trauben und Weinqualität. **Influence of ecological factors on development of vines, constituents of grapes and wine quality.** Publications.

4393 Fischbeck, G.; Wahl, K. DE 502151/77/0001
Untersuchungen über den Einfluss von Böden verschiedener Ausgangsgesteine auf physiologische Leistungsmerkmale und auf den Weincharakter ausgewählter Ertragsrebsorten. **Investigations on the influence of various geological strata on physiological characteristics and on the wine from selected grape varieties.**

4394 Müller, K.; Peternel, M. DE 502153/77/0003
Erörterung des Einflusses der Beregnung auf die Bodenwasserund Bodenwärmegehalt und damit auf die Entwicklung der Rebe und auf die Menge und Güte des Traubenertrages bei verschiedenen Bodenbedeckungen.
Discussion on the influence of irrigation on poisture and heat in soil and with that on growth of vine and on quantity and quality of grape yield using different mulching. Publications.

4395 Geiger, K.; Schottdorf, W. DE 502153/78/0003 N
Untersuchungen über den Einfluss unterschiedlicher Zeilenund Stockabstände auf die Leistung der Rebsorte Müller–Thurgau bei konstantem Anschnitt von 10 Augen 2 pro Quadratmeter. **Investigations on the influence of different row and vine distances on growth and yield of the vine variety Mueller– Thurgau by pruning 10 buds/ 2 square metre of area.**

4396 Schmitt, A.; Rothbächer, H. DE 502157/78/0001 N
Untersuchungen über den Einfluss verschiedener Inhaltsstoffe auf sensorische Merkmale des Weines unter besonderer Berücksichtigung unterschiedlicher Mengenerträge. **Research on the influence of different constituents on sensory wine quality in special regard of quantitative grape yield.**

4397 Kiefer, W.; Weber, M. DE 506101/70/0002
Der Einfluss der Rebenerziehung auf Menge und Güte des Ertrages 1965. **Influence of vine training on quantity and quality of crop.**

4398 Kiefer, W.; Steinberg, B.; Bettner, W.
DE 506101/70/0003
Untersuchungen über die Wasseransprüche der Rebe und über die Beeinflussung des Wasserhaushaltes im Boden durch verschiedene Kulturmassnahmen 1970. **Investigations on the demands of vine for water supply and on the water volume in soil as affected by different measures of cultivation.**

4399 Steubing, L.; Arneth, A.; Kiefer, W. DE 506101/77/0001
Untersuchungen über den Einfluss unterschiedlicher Bodenbedeckungsarten auf den Boden und die Rebe. **Investigations on the influence of different mulching on soil and vine.**

4400 Kiefer, W.; Slamka, P.; Gruppe, W. DE 506101/77/0003
Die Entwicklung der Rebenerziehung in osteuropäischen Ländern. **Development of vine shape in Easteuropaen countries.**

4401 Reuther, G.; Schneider, F. DE 506111/71/0001
Die Kinetik des Kohlenhydratmetabolismus und des Aminosäurehaushalts bei frostresistenten Rebsorten. **The kinetics of carbohydrate and amino acid metabolism in frost–resistant grape varieties.**

4402 Wienhaus, H. DE 506111/71/0003
Beeinflussung der die Traubenreife bestimmenden Stoffwechselvorgänge durch Phytohormone und Stoffwechselinhibitoren – Phytopathologische und entwicklungsphysiologische Folgen von Wuchsstoff– und Atmungshemmstoffbehandlung der Rebe während der Reifephase. **Possibilities of taking influence on metabolic processes of the maturation of grapes by phytohormones and metabolic inhibitors – phytopathological and growth effects by treat– ment of vine with hormones and metabolic inhibitors on the maturation of berries.**

4403 Reuther, G.; Schneider, F. DE 506111/72/0004
Histologische und histochemische Untersuchungen der Holzreife und der Beziehungen zum physiologischen Verhalten frostresistenter Rebsorten. **Histological and histochemical investigations of the matu– rity of shoots and the**

correlation to physiological be– haviours of frost–resistant grape varieties.

4404 Reuther, G.; Schneider, F.   DE 506111/77/0002
Die spezifischen Temperaturoptima der Photosynthese von Rebsorten. **The specific optimum temperatures of photosynthesis of grape varieties.**

4405 Wienhaus, H.   DE 506111/77/0006
Die Böschungsbegrünung an Weinbergsterrassen des Kaiserstuhles nach der Flurbereinigung: Ergebnisse eines Aussaatverfahrens und Regeneration der natürlichen Flora dieser Standorte. **Green covering of vineyard slopes after land consolidation in the region "Kaiserstuhl": Results of a sowing technique and the regeneration of the natural vegetation of these habitats.**

4406 Rühling, W.; Bäcker, G.; Steinberg, B.
    DE 506113/75/0001
Untersuchungen zur Verbesserung und Rationalisierung der Bodenbearbeitung am Steilhang. **Studies on the possibilities of improving the tillage – mechanization in vine growing on slopes.**

4407 Kalinke, H.; Brendel, G.; Stumm, G. DE 506114/77/0001
Ökologische und pflanzensoziologische Untersuchungen weinbaulich genutzter Standorte am Kaiserstuhl hinsichtlich standortgerechter und für die Rebkulturen verträglicher Schutzund Ausgleichspflanzungen verbunden mit einer Kosten–Nutzen– Analyse. **Ecological and plant sociological studies on viticultural locations in the Kaiserstuhl region with regard to protective and compensation plantations depending on location and accommodating vine cultivation in connection with a cost–utility–analysis.**

4408 Schumann, F.   DE 509154/75/0001
Untersuchungen zur Verwendung von Mulchfolien in Rebschulen. **Mulching sheets in vine nurseries.**

4409 Porsch, M.   DE 509154/75/0003
Der Einfluss der lang– und kurzwelligen Strahlung des Energieangebotes am Standort auf die Zusammensetzung des Traubenmostes – Ein Beitrag zur Produktionsökologie im Weinbau. **The influence of long–wave and short–wave radiation of energy–budget in vineyards on the composition of grape must – A contribution to production ecology in viticulture.**

4410 Beran, N.   DE 509154/77/0001
Untersuchungen über den Einfluss der Bodentemperatur auf den Gasstoffwechsel der Rebe – Vitis vinifera –. **Studies of the influence of soil temperature of the gas exchange in Vitis vinifera.**

4411 Adams, K.; Fader, W.   DE 509155/75/0002
Untersuchungen über die Mechanisierungsmöglichkeiten und Arbeitsverfahren für die Umtriebsarbeiten im Weinbau unter besonderer Berücksichtigung von Grossflächenverfahren und deren Einsatz im Rahmen von Flurbereinigungen. **Studies on possibilities of mechanization and on working methods for rotational works in viticulture with special reference to large area processes and to their employment in the framework of land consolidation.**

4412 Fader, W.   DE 509155/78/0002 N
Vegetative und generative Leistung von Reben bei verschiedenen Anlage– und Erziehungsformen. **Vegetative and generative performance of vines in different systems of cropping and training.**

4413 Remoue, M.; Lemaitre, Cl.   FR 010405/71/0232
Etude des conséquences physiologiques des variations des éléments qui constituent le mode de conduite de la vigne. **Physiological consequences of variations of elements in vine cultivation.**

4414 Remoue, M.; Lemaitre, Cl.   FR 010405/75/0358
Ecologie viticole. **Ecology of vine.**

4415 Carbonneau, A.; Casteran, P.; Leclair, Ph.
    FR 010702/74/0312
Etude de l'influence des facteurs du milieu sur le développement de la vigne et la maturation des raisins: application à la mise au point de système de conduite. **Study of environmental influence on grape vine development and grape–ripening : determining behaviour systems.**

4416 Lefort, P. L.; Leglise, N.; Boussion, C.
    FR 010702/75/0362
Etude de la vigueur propre et de la vigueur conférée par le porte–greffe au greffon, dans le cadre de la création de variétés nouvelles de porte–greffes de vigne. **Rootstock–scion relationships in vine. Studies in the determination of scion's vigor by the rootstock.**

4417 Delas, J.; Molot, P.; Pouget (Viticulture Bordeaux)
    FR 010704/72/6073
Relations entre porte–greffe et greffon chez la vigne. **Relations between scion and rootstock in vine.**

4418 Marocke, R.   FR 010901/68/0582
Vigne. Alsace. Expérimentation au champ. Insuffisances. Fertilisation corrective. Production végétale. Qualité technologigue. **Vine. Alsace. Field experimentation. Deficiencies. Fertilization needs. Yield. Technological quality.**

4419 Marocke, R.; Huglin, P.   FR 010901/75/6128
Relations entre la phytotechnie et la qualité chez la vigne. **Relations between cultural technics and quality for the vine.**

4420 Balthazard, J.   FR 010902/63/0374
Recherches sur la germination des graines de vigne. **Research for grape–vine germination. Publications.**

4421 Ancel, J.; Fuchs, V.   FR 010902/64/0274
Amélioration des techniques de production des plants de vigne. **Improvment of production technics of vine plantations. Publications.**

4422 Huglin, P.; Balthazard, J.   FR 010902/72/0273
Recherches de modes de conduites "modernes" de la vigne, favorables à la qualité de produits. **Research about vine "modern" cultivation method, favorable to products quality.**

4423 Huglin, P.; Wagner, R.; Balthazard, J.
    FR 010902/73/0272
Essai international d'écologie viticole. **International trial of vine ecology.**

4424 Balthazard, J.   FR 010902/74/0271
Amélioration de la maturation des graines de vigne. **Ripening improvment of vine seeds.**

4425 Lefebvre, J.M. FR 011001/74/6140
Prémultiplication de la vigne et contrôle sanitaire en cultures hydroponiques. **Vine premultiplication and sanitary control in hydroponic culture.**

4426 Truel, P.; Vergnes, A. FR 011204/69/0078
Amélioration de la production du vin rouge dans les situations les plus élevées du vignoble méridional. **Improving red wine production in meridional vineyards established at a high elevation.**

4427 Thomas; Farrar GB 012201/00/0001
Improved cultural techniques.

4428 Thomas GB 012201/00/0013 R
Effect of environment on growth and development.

4429 Thomas GB 012201/00/0015
Effect of hormone applications on yield and quality.

4430 Farrar GB 012201/00/0023
Vineyards. Establishment, management and variety selection leading to improvements in cultural methods.

4431 O'Kennedy, N.D. IE 060302/72/0501 N
Vine culture. Publications.

4432 Iannini, B.; Liuni, C.; Poppi, M.; Ridomi, A.; Moretti, G. IT 021300/71/0001 R
Studio degli effetti di sostanze ad azione fitoregolatrice su particolari combinazioni d'innesto nella vite. **Investigation of growth–regulators effects on some vine graft–combinations.** Publications.

4433 Liuni, C.; Catalano, V. IT 021300/72/0002
Osservazioni sulla dormienza della Vitis vinifera in Puglia. **Observations on the Vitis vinifera dormancy in Puglia.** Publications.

4434 Calò, A.; Costacurta, A.; Cancellier, S. IT 021300/72/0003
Studio sulle fasi fenologiche della vite; correlazione tra caratteri fenologici e fisiologici. **Research on phenological vine stages; correlations between phenological and physiological characters.** Publications.

4435 Iannini, I.; Ridomi, A.; Poppi, M.; Liuni, C.; Pol, R.; Pezza, L. IT 021300/74/0001 R
Studio del rendimento metabolico dei portinnesti e di alcune combinazioni d'innesto nella vite; studio sui caratteri morfo–fisiologici di portinnesti e combinazioni d'innesto in diversi ambienti pedoclimatici. **Investigation on metabolic output of some vine rootstocks and graft–combinations; investigation on morpho–physiological characters of vine rootstocks and of some graft–combinations in several pedoclimatic environments.**

4436 Iannini, B.; Costacurta, A.; Liuni, C.; Poppi, M. IT 021300/74/0002
Ricerca sui portinnesti capaci d'imprimere caratteri brachizzanti alle viti. **Research on the rootstocks able to impress dwarfing characters to vines.**

4437 Calo', A.; Costacurta, A.; Cancellier, S. IT 021300/74/0003
Prove di ecologia viticola e rilevazioni dei fattori climatici ambientali. **Trials of viticultural ecology and remarks of the environmental climatic factors.**

4438 Iannini, B.; Ridomi, A.; Poppi, M.; Moretti, G.; Liuni, C.; Pezza, L. IT 021300/74/0004 R
Studio delle caratteristiche fisiologiche di alcuni portinnesti e delle interazioni che si manifestano in seguito all'innesto con le principali varietà ad uva da vino e da tavola. **Investigation on physiological characters of some rootstocks and in interaction subsequent to their grafting with the main wine and table–grape varietes.** Publications.

4439 Iannini, B.; Ridomi, A.; Poppi, M.; Pol, R.; Moretti, G.; Pezza, L. IT 021300/74/0005 R
Rilievi sull'attività traspiratoria e sulla produzione di elaborati nei portinnesti. **Remarks on the transpiration activity and metabolites production in rootstocks.** Publications.

4440 Iannini, B.; Liuni, C.; Poppi, M. IT 021300/74/0006
Studio delle variazioni di alcuni parametri durante la rizogenesi di talee di viti e ricerca di metodi idonei alla individuazione di fitoregolatori. **Study of modifications of some parameters during scions rhizogenesis and search of suitable methods for growth–regulators singling out.**

4441 Liuni, C.; Stramaglia, L. IT 021300/74/0009
Prova di confronto fra portinnesti più diffusi in Puglia. **Comparison among the main Puglia Rootstocks.** Publications.

4442 Liuni, C.; Iannini, B.; Stramaglia, L.; Ridomi, A.; Moretti, G. IT 021300/74/0011 R
Studio dell'interazione portainnesto – sesto d'impianto e forme di allevamento della vite in diversi ambienti. **Investigation on the interaction: vine rootstock – spacing and trainings in several sites.**

4443 Iannini, B.; Poppi, M.; Egger, E. IT 021300/75/0010
Studio di tecniche idonee alla moltiplicazione, in ambiente controllato di talee erbacee. **Research of suitable techniques for multiplication, under controlled environment, of herbaceous scions.**

4444 Calò, A.; Costacurta, A.; Cancellier, S. IT 021300/77/0003
Studi sulla germinabilità dei vinaccioli. **Study of grape–stones germinating–capacity.** Publications.

4445 Liuni, C.; Corino, L.; Serra, G.; Iona, IT 021300/77/0004 R
Ricerche sul momento dell'induzione fiorale nel Barbera. **Investigation on vine floral induction stage.** Publications.

4446 Lavezzi, A.; Cargnello, G. IT 021300/77/0005 R
Fitoregolatori spollonanti, defoglianti della vite. **Growth regulator, sucker–controlling and defoliating substances on vine.**

4447 Jannini, B. IT 021300/77/0697 R
Meccanizzazione della raccolta dell'uva, sistemi di allevamento per raccolta meccanica. **Mechanization of grape–gathering, growing methods permitting mechanical harvesting.**

4448 Egger, E.; Borgo, M.; Roncador, I. IT 021300/79/0001 N
Indagine sulla comparsa del "Disseccamento del rachide" nella vite. **Investigation on the appearance of vine "stiellhäme".** Publications.

D 2100 - Plant production general and crop husbandry

4449 Egger, E.; Borgo, M.; Tocchetti, G.
IT 021300/79/0002 N
Indagini sul "legno riccio" nella vite. **Investigations on vine "stem pitting" and "stem growing".** Publications.

4450 Lovino, R.
IT 022000/79/0005 N
Impiego di un prototipo di macchina vendemmiatrice per uve per vini bianchi in Puglia. **Prototype mechanical harvester for grapes for white wines in Puglia.**

4451 Interesse, F.
IT 040107/77/0208 R
Sull'attività della polifenolossidasi nelle uve meridionali. **On polyphenoloxydasis activity in grapes in southern Italy.**

4452 Intrieri, C.
IT 040203/77/0003
Tecnologia della meccanizzazione integrale della potatura della vite: aspetti fisiologici e meccanici. **Physiological and mechanical aspects of pruning in grapevines (mechanical pruning done by cutting machines).** Publications.

4453 Fregoni, M.; Scienza, A.; Miravalle, R.; Zamboni, M.; Boselli, M.; Dorotea, G.
IT 040403/70/0003 R
Sistemi di allevamento della vite adatti alla vendemmia meccanica. **Vine growing methods suitable for mechanical harvesting.** Publications.

4454 Fregoni, M.; Scienza, A.; Miravalle, R.; Zamboni, M.; Boselli, M.; Dorotea, G.
IT 040403/70/0005 R
Disseccamento del rachide della vite. Cause e terapia. **Vine cluster desiccation: causes and cure.** Publications.

4455 Scienza, A.
IT 040403/74/0628
Relazione fra ecosistemi viticoli – clima, terreno e vitigno – e caratteristiche qualitative dell'uva e del vino. **Relationship between grapevine ecosystems – climate, soil and plant – and quality characteristics of grapes and wine.**

4456 Scienza, A.
IT 040403/74/0629
Ricerche sulle cause del disseccamento del rachide della vite. **Research on the causes of grapevine rachis desication.**

4457 Fregoni, M.; Scienza, A.; Miravalle, R.; Zamboni, M.; Boselli, M.; Volpe, B.
IT 040403/76/0001 R
Ricerche sui portinnesti della vite (riduzione della vigoria, resistenza alla siccità ed al calcare). **Vine rootstock (reduction in vigour and resistance to drought and calcareous).**

4458 Visai, C.
IT 040605/74/0644
Indagini sulle cause del disseccamento del rachide del grappolo di vite. **Investigations on the causes of desiccation of the rachis in grape bunches.**

4459 Scienza, A.
IT 040605/78/0002
Prove sulla tecnica colturale della vite nell'Oltrepò Pavese. **Management trials on wine–grapes in Oltrepò Pavese.**

4460 Gerola, F.M.
IT 040618/77/0661 R
Biologia fiorale della vite, fattori condizionanti la scarsa allegazione di alcune cultivar –picolit moscato rosa ecc.–. **Floral biology of the grapevine, factors determining poor fruit–setting in certain cultivars –picolit, moscato rosa–.**

4461 Sarcinelli, S.
IT 040913/77/0714 R
Meccanizzazione della raccolta dell'uva, trasformazione vigneti e prove vendemmiatrici. **Mechanization of grape–gathering, transformation of vineyards and harvesting trials.**

4462 Jacoboni, N.
IT 041005/72/0488
Propagazione della vite. **Propagation of vines.** Publications.

4463 Jacoboni, N.
IT 041005/77/0209 R
Problemi inerenti alla propagazione della vite. **Problems inherent to the propagation of vines.**

4464 Puppi, G.; Riess, S.; Giovannetti, M.
IT 041607/79/0001 N
Endomicorrize della vite (Vitis vinifera) nell'Italia centrale: identificazione tassonomica dei simbionti fungini e descrizione delle strutture micorriziche. **Endomycorrhizas of vine (Vitis vinifera) in central Italy: taxonomic identification of the fungal symbionts and description of mycorrhizal structures.**

4465 Di Marco, G.
IT 060200/74/0083
Andamento stagionale di enzimi carbossilanti, zuccheri ed acidi organici in tre tipi di allevamento della vite. **Seasonal patterns of carboxylating enzymes, sugars and organic acids in Vitis vinifera in three different types of cultivation.**

4466 Eynard, I.
IT 062100/71/0152
Ricerche sul meccanismo di azione nella vite dei fitormoni e dei fitoregolatori. **Studies on behaviour of phytormones and growth regulators in vitis.** Publications.

4467 Eynard, I.
IT 062100/74/0153
Ricerche sulle tecniche di moltiplicazione della vite. **Vitis multiplication techniques.** Publications.

4468 Carlone, R.
IT 062100/77/0956
Moltiplicazioni delle migliori cultivar di vite ottenute e realizzazione di vigneti sperimentali in varie zone a vocazione viticola. **Multiplication of the best vine cultivars obtained, development of experimental vineyards in different areas dedicated to vine growing.**

4469 Carlone, R.
IT 062100/77/0959
Ricerche sulla biologia fiorale della vite con tecniche istochimiche ed istoautoradiografiche. **Research on grapevine floral biology using histologicalchemical and histological–autoradiographical techniques.**

4470 Carlone, R.
IT 062100/77/0960
Ricerche sul meccanismo di azione nella vite dei fitormoni e fitoregolatori. **Research on the mechanisms of action of phyto–hormones and of growth regulators.**

4471 Carlone, R.
IT 062100/77/0961
Ricerche sulle tecniche di moltiplicazione della vite. **Research on multiplication techniques applied to the grapevine.**

4472 Carlone, R.
IT 062100/77/0962
Identificazione ed analisi degli acidi nucleici di vitigni a diversa affinità d'innesto con studio parallelo delle reazioni in campo. **Identification and analysis of nucleic acids in vine stocks with different grafting compatibility, parallel study of reactions in the field.**

4473 Baldini, E.
IT 063000/77/0724
Sistemi allevamento vite e nuove vendemmiatrici. Raccolta meccanica della frutta. **Vine growing systems and new grape harvesters. The mechanical harvesting of fruit.**

D 2100 – Plant production general and crop husbandry

4474 Baldini, E.   IT 063000/77/0929
Ricerche sulla raccolta meccanica dell'uva, della frutta; delle olive e delle fragole. **Research on the mechanical harvesting of grapes, fruits, olives and strawberries.**

4475 Lisa, L.   IT 063300/73/0063
Raccolta dell'uva in collina a scuotimento verticale su filo. **Grape harvesting by vertical shaking of the trellis wire, in hillside vineyards.** Publications.

Edible nut fruits (B 3660)

See also 3150, 4197, 4198, 4236, 4262, 5012, 5013, 7275, 7462, 7465, 7471, 7481, 10294

4476 Arnoux, M.; Defrance, H.; Huguet, C.
                    FR 010615/74/9037
Culture du Noyer en sol de colluvionnement. **Testing the introduction of walnut cultivation in colluvial tracts of land.**

4477 Arnoux, M.; Defrance, H.; Chevallier; Tabardel, J.P.
                    FR 010615/75/9047
Essais combinés Noisetier x Truffe. **Testing combined hazel–nut and truffle production.**

4478 Germain, E.; Leglise, P.; Delort, F.   FR 010701/72/0100
Etude des processus de pollinisation (autostérilité, intercompatibilité, interincompatibilité pollinique) et de fécondation chez le noisetier (corylus avellana) dans le but d'augmenter et de régulariser en verger la production de cette espèce. **Study of pollination (self–sterility, intercompatibility, pollen interincompatibility) and fertilization in hazel–tree (Corylus Avellana) for improving and regularizing the production in orchards.** Publications.

4479 Mauget, J. C.; Robelin, M.   FR 010802/74/6114
Photosynthèse et structure de l'arbre (noyer). **Photosynthesis and tree structure (walnut tree).**

4480 Fulford; Justin   GB 011006/00/0009 R
Vegetative propagation of coconut palm trees.

4481 Ghidini, G.; Ghetti, S.   IT 021800/76/0004 N
Studio sulla morfometria delle foglie di Juglans nigra, Juglans regia e Juglans nigra x regia.. **Morfometric study on the leaves of Juglans nigra, Juglans regia and Juglans nigra x regia..**

4482 Bagnaresi, U.   IT 040238/74/0504
Riordinamento produttivo e sanitario dei castagneti nell'Appennino Emiliano–romagnolo. **The management of chestnut tree orchards for improved production and plant health in the Appennini in Emilia–Romagna.**

4483 Sottile, I.   IT 040908/74/0636
Biologia. La propagazione e la coltivazione del nocciolo. **Biology. Hazel propagation and cultivation practices.**

4484 Tombesi, A.   IT 041005/72/0571
Biologia, propagazione e tecnica colturale del nocciolo. **Biology, propagation and cultivation techniques of hazelnut.** Publications.

4485 Romisondo, P.   IT 041218/73/0297
Biologia, propagazione e tecnica colturale del nocciolo. **Biology, propagation and cultivation techniques of the hazelnut.** Publications.

4486 Romisondo, P.   IT 041218/77/0666 R
Biologia fiorale del nocciuolo. **Floral biology of the filbert.**

Ornamentals and ornamental products in general (B 3700)

See also 932, 3402, 3690, 3705, 3707, 3758, 3759, 3760, 3761, 3762, 3763, 3764, 3765, 3766, 3769, 3770, 3780, 3781, 3782, 3785, 3786, 3788, 3789, 4091, 4094, 4102, 4104

4487 Boesman, G.; Flamee, M.   BE 030015/66/0003 N
Bloeibeïnvloeding bij sierplanten. **Flowering regulation of ornamental plants.** Publications.

4488 Van Onsem, J.G.; Verdonck, O.   BE 070600/70/0070 R
Kultuur van sierplanten in inerte substraten. **Culture of ornamental plants in inert substrata.** Publications.

4489 Van Onsem, J.G.; Mekers, O.   BE 070600/76/0066 R
Karakterisering van milieu – en kultuuromstandigheden bij nieuwe teelten. **Determination of the environmental characteristics and the cultural requirements of new crops.**

4490 Van Onsem, J.G.; Verdonck, O.   BE 070600/79/0081 N
Studie van het optimale vochtgehalte van substraten voor sierplanten. **Study of the optimal moisture content of substrates used in the production of ornamentals.**

4491 Dermine, E.; Dubuisson, J.; Gruselin, P.; Tréfois,
                    BE 080900/79/0013 N
Culture de plantes ornementales en container. **Culture of ornamentals in containers.** Publications.

4492 Stautemas, E.; Beel, E.; Blomme, R.
                    BE 140000/74/0032
Fytotechnische aspecten van warme kasplanten, coniferen, sierheesters snijbloemen en azalea's. **Phytotechnical aspects of hot house plants, conifers, ornamental shrubs, cut flowers and azaleas.** Publications.

4493 Franke, W.; Bömeke, H.   DE 111152/75/0003
Beitrag zur Frage der Haltbarkeit von Schnittblumen insbesondere beim Verbraucher. **Contribution to the problem of the keeping time of cutflowers, especially with the consumer.**

4494 Lange, P.   DE 161270/70/0002 R
Erfassung der Wachstumsfaktoren und des Wachstumsverlaufes bei Zierpflanzen in Abhängigkeit von wechselnden Umweltbedingungen. **Influence of different environmental conditions on growth development of ornamental plants.**

4495 Hansen, R.; Müssel, H.   DE 502107/77/0002
Xerophile Stauden für oligotrophe Standorte zur Dachbegrünung. **Xerophile perennials for oligotrophic sites in roof greens.** Publications.

4496 Papenhagen, A.; Strotmann, G.   DE 508201/75/0002
Untersuchungen zur geregelten Kühlung der Blätter von Zierpflanzen durch direkte Messung der Temperaturdifferenz zwischen Blatt und umgebender Luft. **Studies on controlled cooling of leaves of ornamental plants by direct measurement of the temperature difference between leaf and ambient air.**

4497 Papenhagen, A.; Strotmann, G.   DE 508201/75/0003
Ermittlung der Temperaturansprüche von Zierpflanzen in

Gewächshäusern bei suboptimalen natürlichen Lichtbedingungen mit Hilfe von Gaswechselmessungen. **Determination of temperature requirements of ornamental plants in greenhouses at suboptimal natural light conditions by gas exchange measurements.**

4498 Hurd   GB 011102/00/0002 R
Control of leaf growth by environmental factors in glasshouse crops.

4499 Hurd; Gay   GB 011102/00/0011 R
Stomatal and cuticular resistance to carbon dioxide diffusion.

4500 Cockshull   GB 011102/00/0016 R
Photoperiodism and the control of plant processes.

4501 Nichols; Hammond   GB 011102/00/0018 R
Effect of environment on post–harvest development of cut flowers.

4502 Nichols   GB 011102/00/0019 R
Translocation in cut flowers with reference to preservatives.

4503 Nichols   GB 011102/00/0020 R
Effect of growth substances and regulators on flower senescence.

4504 Acock; Nichols   GB 011102/00/0044 R
Pre and post harvest water potentials of flowers and leaves.

4505 Menhennet   GB 011103/00/0010 R
The effects of synthetic and naturally occurring growth regulators on glasshouse ornamentals.

4506 Whalley; Loach   GB 011106/78/0022 N
Factors affecting growth and establishment of ornamental nursery stock.

4507 Richardson   GB 041201/77/0008
Nutrient film technique.

4508 Researcher not indicated   GB 050102/00/0010 R
Minor cut flowers, production techniques and variety trials.

4509 Researcher not indicated   GB 050102/00/0012 R
Bedding plants, production techniques.

4510 Researcher not indicated   GB 050102/00/0015 R
Protected crops, post–harvest handling and storage.

4511 Researcher not indicated   GB 050142/00/0008
Bedding plants, production techniques.

4512 Mccoll   GB 060215/00/0009
Protected crop production.

4513 Dixon   GB 060215/78/0010 N
Rose rootstock evaluation.

4514 McKelvie   GB 060216/00/0005 R
Vegetative propagation of ornamental crops of commercial importance in Scotland.

4515 Smith   GB 060318/00/0005
Flower and vegetable production in plastic tunnels.

4516 Cheffins; Johnston   GB 060318/00/0006
Evaluation of substrates other than soil for glasshouse production.

4517 Feely, L.   IE 060300/75/1312 R
Cultivar and husbandry trials on ornamental crops. Publications.

4518 Borrelli, A.   IT 040701/78/1035 N
Ricerche sulle colture ortofloricole protette. **Research on protected cultures of flowers and market garden produce.**

4519 Pimpini, F.   IT 040801/72/0530
Studio sull'influenza dell'illuminazione artificiale sul radicamento e sullo sviluppo delle piante ornamentali e da fiore. **Study on the influence of artificial lighting on rooting and on the development of ornamental and flowering plants.** Publications.

4520 Pimpini, F.   IT 040801/77/0261 R
Studi sull'influenza dell'illuminazione artificiale e di sostanze rizogene sul radicamento e sullo sviluppo di piante ornamentali e da fiore. **Studies on the effect of artificial lighting and rhizogene compounds on the rooting and development of ornamental and flower plants.**

4521 Pol, P.A. van de   NL 020057/69/7372 R
De invloed van fysische groeifactoren (o.a. temperatuur, licht, vochtvoorziening) en groeiregulatoren op groei, ontwikkeling, fotosynthese, transpiratie en waterhuishouding van bloemisterijgewassen en bloembollen. **The influence of phisical growth factors (o.a. temperature, light, moisture supply) and growthregulators on growth, development, photosynthesis, transpiration and watermanagement of flower crops transpiration management of flower crops and flower bulbs.** Publications.

Bulbs (B 3710)

See also 2897, 3872, 3928, 3966, 4608, 4660, 9919, 15390

4522 Van Onsem, J.; Haegeman, J.   BE 070600/60/0015 R
Onderzoek naar de toepassingsmogelijkheden van groeiregulatoren bij knolbegonia's. **Application of growth regulators on tuberous begonias.** Publications.

4523 Van Onsem, J.; Haegeman, J.   BE 070600/60/0016
Studie van dormancyverschijnselen en bewaring van begoniaknollen. **Study of dormancy and storage of tuberous Begonias.**

4524 Van Onsem, J.G.; Haegeman, J.   BE 070600/78/0076
Energiebesparing in de knolbegoniateelt. **Energysaving in tuberous Begonia culture.**

4525 Van Onsem, J.G.; Haegeman, J.   BE 070600/78/0078
Fotoperiodische reakties van knolbegonia – zaailingen. **Fotoperiodic reaction of tuberous Begonia seedlings.**

4526 Carow, B.; Pieper, B.; Zimmer, K.   DE 138270/77/0010
Untersuchungen zur klonalen Vermehrung von Gloriosa in vitro. **Studies on clonal propagation of Gloriosa in vitro.**

4527 Carow, B.   DE 138270/77/0011
Keimuntersuchungen an Gloriosa rothschildiana. **Studies on germination of Gloriosa rothschildiana.**

D 2100 – Plant production general and crop husbandry

4528 Horn, W.; Wallbruch, D.　　　　DE 138300/70/0005
Züchtungsverfahren für vegetativ vermehrte Pflanzen mit mehrjähriger Generationsdauer – Modellobjekt Tulipa –. **Breeding methods for vegetatively propagated plants with a long generation interval. Model – Tulipa –.**

4529 Preil, W.; Hoffmann, M.; Engelhardt, M.; Engelhardt, K.; Häfen, K.von; Kunze, H.　　DE 206000/78/0007 N
Untersuchungen zur Regenerationsfähigkeit von Meristemen und Blütengewebe von Convallaria majalis. **Regeneration ability of meristems and flowerbud tissue of Convallaria majalis.**

4530 Jørgensen, M.B.; Rasmussen, E.　　DK 010115/58/4507
Kulturmetoder til narcis og tulipan. **Growing methods for narcissus and tulips.**

4531 Jørgensen, M.B.; Rasmussen, E.　　DK 010115/64/4508
Udvikling af metoder til drivning og post–harvest til narcis og tulipan. **Development of forcing methods and post–harvest treatment in narcissus and tulips.**

4532 Cohat, J.　　　　　　　　　　FR 012207/72/0229
Etude de la croissance et du développement du glaîeul. **Study of the growth and the development of gladiolus.**

4533 Rees; Ludwig　　　　　　　　GB 011102/00/0023
Growth of bulb plants in field.

4534 Davies　　　　　　　　　　GB 011103/00/0005 R
Changes in bulb constituents in relation to onset and breaking of dormancy.

4535 Rees; Hanks　　　　　　　　GB 011106/78/0019 N
Effects of growth regulators on bulbs.

4536 Rees　　　　　　　　　　　GB 011106/78/0020 N
Growth of bulb plants in the field.

4537 Rees　　　　　　　　　　　GB 011106/78/0021 N
Physiology of bulb forcing.

4538 Lawson; Wiseman　　　　　　GB 030701/00/0023
Weed ecology and control in flower bulbs.

4539 Thompson; Taylor　　　　　　GB 030701/00/0037
Control of growth, yield and quality of flower bulb crops by cultural methods.

4540 Richardson　　　　　　　　GB 041201/00/0006
The production of early outdoor flowers from prepared daffodil bulbs.

4541 Researcher not indicated　　　GB 050104/00/0001 R
Narcissus, production techniques and variety trials.

4542 Researcher not indicated　　　GB 050104/00/0002 R
Tulips, production techniques and variety trials.

4543 Researcher not indicated　　　GB 050104/00/0003 R
Anemones, production techniques and variety trials.

4544 Researcher not indicated　　　GB 050104/00/0004 R
Other bulbs, production techniques and variety trials.

4545 Researcher not indicated　　　GB 050144/75/0002
Narcissus, production techniques and variety trials.

4546 Turner　　　　　　　　　　GB 060112/00/0005
Production and storage of bulb crops.

4547 Turner　　　　　　　　　　GB 060112/00/0006
Propagation of virus tested stocks of narcissus.

4548 Shiel　　　　　　　　　　　GB 060215/00/0006
Evaluation of methods of bulb husbandry.

4549 Duncan; Shiel　　　　　　　GB 060215/00/0008
Twin scale propagation of virus–free narcissus.

4550 Puccini, G.; Pergola, G.　　　IT 021200/74/0001
Esperienze di trattamenti a freddo dei bulbi di Nerine bowdenii per ritardare la fioritura. **Cool treatments to Nerine bowdenii bulbs to retard the flowering.**

4551 De Ranieri, M.; Grassotti, A.　　IT 021200/77/0010 N
Attitudine alla moltiplicazione agamica di alcune varietà di Lilium in ambienti diversi. **Disposition to agamic multiplication in different environments of some Lilium varieties..**

4552 Cirrito, M.; Provenzale, M.G.　　IT 021200/77/0016 N
Prove di vernalizzazione dell'Iris. **Trials on the Iris vernalisation.**

4553 Cirrito, M.; De Vita, M.　　　IT 021200/77/0017 N
Prove di densità colturale sull'ingrossamento dei bulbetti del gladiolo.. **Trials on the influence of the cultivation density in gladiolus bulblets enlargement..**

4554 Cirrito, M.; Zizzo, G.; De Vita, M.　IT 021200/78/0002 R
Volumi idrici e pacciamatura nella coltura di tuberosa per l'ingrossamento dei bulbi. **Different water amounts and mulching trials, on the cultivation for tuberosa bulb enlargement.**

4555 Cirrito, M.; Provenzale, M.G.　　IT 021200/79/0005 N
Studio sulla biologia della Tuberosa. **Research on the biology of the Tuberosa.**

4556 Sciortino, A.　　　　　　　IT 040914/73/0307
Ricerche sulla fisiologia e sulla tecnica dell'infrossamento dei bulbi di specie floricole. **Research on the physiology and technique of swelling flower bulbs.** Publications.

4557 Sciortino, A.　　　　　　　IT 040914/77/0297 R
Ricerche sulla fisiologia e sulle tecniche di ingrossamento dei bulbi di specie floricole. **Research on the physiology of flower bulbs and on the methods of increasing their size.**

4558 benschop, M.　　　　　　　NL 010102/74/5908
Simulatie van de droge–stofproduktie bij bolgewassen. **Simulation of dry–matter production of bulbous crops.**

4559 Sytsema, W.　　　　　　　　NL 010201/62/0948
De invloed van temperatuur tijdens de bolbewaring, planttijd en teeltmethode op de bloei van Nerine. **Influence of temperature during bulb storage, time of planting and cultural methods on flowering of Nerine.** Publications.

4560 Hoogeterp, P.　　　　　　　NL 010205/65/1487 R
De invloed van bolbehandeling, teeltmethoden en

D 2100 – Plant production general and crop husbandry

ziektenbestrijding op het bloeiresultaat bij de bloementeelt van tulp, hyacint en narcis. **The influence of bulb handling, growing techniques and disease control on yield and flowering results of tulip, hyacinth and narcissus flowers.** Publications.

4561 Beijersbergen, J.C.M. NL 010205/65/1491
Koolhydraatmetabolisme in verband met de bloeirealisatie van de tulp. **Carbohydrate metabolism in relation to flowering in tulips.** Publications.

4562 Meeteren, U. van NL 010205/65/1494 R
Ontwikkeling en bloei van Hollandse irissen. **Development and flowering of Dutch irises.** Publications.

4563 Meeteren, U. van NL 010205/67/1506 R
Ontwikkeling en bloei van lelies. **Development and flowering of lilies.** Publications.

4564 Alkema, H.Y. NL 010205/70/3445
Vermeerdering van bloembollen door middel van vorming van adventief–knoppen. **Propagation by adventitious bud formation in bulbs.** Publications.

4565 Groen, N.P.A. NL 010205/71/3572
De invloed van knolbehandeling, teeltmethoden en gewasbeschermingsmaatregelen op opbrengst en kwaliteit bij de knollen– en bloementeelt van gladiolen. **The influence of bulb handling, growing techniques and crop protection on yield and quality of gladiolus.** Publications.

4566 Boontjes, J. NL 010205/71/3573
De invloed van bolbehandeling, teeltmethoden en gewasbeschermingsmaatregelen op opbrengst en kwaliteit bij de bollen– en bloementeelt van lelies. **The influence of bulb handling, growing techniques and crop protection on yield and quality of lilies.** Publications.

4567 Koster, J.; Knoppien, P. NL 010205/72/3575
De invloed van plantgoedbehandeling, teeltmethoden en gewasbescherming op opbrengst en kwaliteit bij de bollenteelt van tulpen. **The influence of bulb handling, growing techniques and crop protection on yield and quality of tulip bulbs.** Publications.

4568 Winter, J.A.Th. de NL 010205/73/3939 R
De invloed van bolbehandeling, teeltmethoden en gewasbeschermingsmaatregelen op opbrengst en kwaliteit bij de bollen– en bloementeelt van bijgoed. **The influence of bulb handling, growing techniques and crop protection on yield and quality of miscellaneous bulbs.** Publications.

4569 Benschop, M. NL 010205/74/5908
Simulatie van de droge–stofproduktie bij bolgewassen. **Simulation of dry–matter production of bulbous crops.** Publications.

4570 Schipper, J.A.; Schipper, J.A. NL 010205/75/6990 R
De invloed van bolbehandeling, teeltmethoden en gewasbeschermingsmaatregelen op opbrengst en kwaliteit bij de bollen– en bloementeelt van irissen. **The influence of bulb handling, growing techniques and crop protection on yield and quality of iris bulbs and flowers.** Publications.

4571 Alkema, H.Y. NL 010205/75/6993
Plantgoedkeuze en produktieanalyse bij bolgewassen. **Choice of plant material and production analysis of bulbous crops.**

Publications.

4572 Vreeburg, P.J.M. NL 010205/76/6991 R
De invloed van plantgoedbehandeling, teeltmethoden en gewasbeschermingsmaatregelen op opbrengst en kwaliteit bij de bollenteelt van narcissen. **The influence of bulb handling, growing techniques and crop protection on yield and quality of narcissus bulbs.** Publications.

4573 Vreeburg, P.J.M. NL 010205/76/6992 R
De invloed van plantgoedbehandeling, teeltmethoden en gewasbeschermingsmaatregelen op opbrengst en kwaliteit bij de bollenteelt van hyacinten. **The influence of bulb handling, growing techniques and crop protection on yield and quality of hyacinth bulbs.**

4574 Meeteren, U. van NL 010205/77/7554 R
Ontwikkeling en bloei van bijgoedgewassen. **Development and flowering of miscellaneus crops.** Publications.

4575 Dijkhuizen, T. NL 010206/72/4661
Teeltonderzoek bij onder glas geteelde bol– en knolgewassen, met name freesia en amaryllis. **Research on growing of ornamental bulb and tuber crops under glass, especially freesia and amaryllis.** Publications.

Flowers and pot plants (B 3720)

See also 2599, 3772, 3774, 3775, 3778, 3779, 3843, 3844, 3907, 3941, 3942, 3960, 3965, 3989, 4575, 4686, 5888, 5889, 5890, 5905, 5907, 5910, 7575, 15352

4576 Lamberts, D.; Lettani, L.; Deckers, T.
BE 040202/77/0014 R
Hoge plantdichtheid bij gerbera in kassen. **High density–planting of gerbera in glass houses.**

4577 Van Onsem, J.G.; Mekers, O.; Thomas, F.
BE 070600/68/0030
Bloeibeheersing bij Bromeliaceae. **Flowering control on Bromeliads.** Publications.

4578 Van Onsem, J.; Heursel, J. BE 070600/72/0004 R
Onderstammenproef met uit Japan (Hirado–eiland) ingevoerde Rhododendron scabrum hybriden. **Rootstock trial with Rhododendron scabrum hybrids from Japan (Hirado–Island).**

4579 Van Onsem, J.G.; Meneve, I. BE 070600/76/0055 R
Stekproeven net klim– en heesterrozen. **Rooting experiments with shrubroses and climbers.**

4580 Van Onsem, J.G.; Mekers, O. BE 070600/76/0058
Bloeibeïnvloeding bij Stromanthe sanguinea. **Flower induction on Stromanthe sanguinea.**

4581 Van Onsem, J.G.; Mekers, O. BE 070600/76/0060 R
Bloeibeheersing bij Bromeliaceae. **Flowering control of Bromeliads.**

4582 Van Onsem, J.G.; Mekers, O.; Thomas, F.
BE 070600/77/0073
Snijbloemen–produktie bij Bromeliaceae. **Production of cut flowers on Bromeliacea.**

4583 Van Onsem, J.G.; Haegeman, J. BE 070600/78/0077

D 2100 – Plant production general and crop husbandry

Teelt van Gloxinia's (Sinningia) in turf. **Gloxinia (Sinningia) culture in peat moss.**

4584 Rünger, W.; Führer, H.   DE 105110/70/0001
Einfluss von Umweltbedingungen auf die Blütenbildung von Zierpflanzen. **Influence of environmental conditions on the flower formation of ornamental plants.**

4585 Franke, W.; Langhans, D.   DE 111152/75/0001
Zur Frage der Biosynthesewege der Ascorbinsäure – Vitamin C – in Wundgeweben von Sprossknollen am Beispiel von Solanum tuberosum und Helianthus tuberosus. **The pathways of biosynthesis of ascorbic acid – vitamin C – in wounded tissues of stem tubers as of Solanum tuberosum and Helianthus tuberosus.**

4586 Zimmer, K.   DE 138270/70/0003
Untersuchungen zum Keimverhalten von Kakteen. **Germination of cactus seeds.**

4587 Zimmer, K.   DE 138270/72/0003
Untersuchungen zur Entwicklung und Blütenbildung verschiedener Campanula–Arten 1972. **Studies of development and flower formation of different species of Campanula.**

4588 Zimmer, K.; Pieper, W.   DE 138270/73/0003
Untersuchungen zur meristematischen Vermehrung von Bromeliaceae und anderer Pflanzen. **Research on tissue culture of Bromeliaceae and other plants.**

4589 Zimmer, K.; Krebs, O.   DE 138270/74/0003
Untersuchungen zur spektralen Abhängigkeit der Störlichtwirkung bei einigen Begonien. **Studies on some Begonia spp. as to their spectral dependency on night breaks.**

4590 Zimmer, K.; Krebs, O.   DE 138270/77/0001
Untersuchungen zur Blütenbildung von Begonia boweri. **Flower formation in Begonia boweri.**

4591 Zimmer, K.   DE 138270/77/0002
Zur photoperiodischen Abhängigkeit der Blütenbildung von Salvien–Sorten. **Photoperiodic response of Salvia varieties.**

4592 Zimmer, K.   DE 138270/77/0004
Entwicklung von Phytolacca dodecandra in Abhängigkeit von Licht und Temperatur. **Growth of Phytolacca dodecandra as influenced by light and temperature.**

4593 Zimmer, K.   DE 138270/77/0006
Untersuchungen zum Kältebedürfnis von Campanula pyramidalis. **Chilling requirement of Campanula pyramidalis.**

4594 Zimmer, K.   DE 138270/77/0007
Untersuchungen zur Blütenbildung von Crassula rubicunda. **Flower formation of Crassula rubicunda.**

4595 Zimmer, K.; Carow, B.   DE 138270/77/0008
Untersuchungen an Nopalxochia phyllanthoides. **Experiments on Nopalxochia phyllanthoides.**

4596 Carow, B.   DE 138270/77/0009
Einfluss der Temperatur auf das Blühen von Chrysanthemen. **Effect of temperature on flowering of Chrysanthemum.** Publications.

4597 Carow, B.   DE 138270/77/0013

Untersuchungen zur Entwicklung und Blütenbildung von Leonotis leonurus. **Studies on growth and flowering of Leonotis leonurus.**

4598 Bachthaler, E.   DE 138270/77/0014
Untersuchungen zur Keimung und Lagerung von Anthurium scherzerianum–Samen. **Germination and storage of Anthurium scherzerianum seeds.** Publications.

4599 Bachthaler, E.   DE 138270/77/0015
Einfluss von Wachstumsregulatoren auf die vegetative und generative Entwicklung verschiedener Pelargonium–ZonaleF1–Hybriden. **Effects of growth–retardants on vegetative and generative growth of different Pelargonium–Zonale–F1–hybrids.**

4600 Zimmer, K.   DE 138270/78/0001 N
Zur photoperiodischen Beeinflusssung der Ausläuferbildung bei Chlorophytum. **Runner formation in Chlorophytum.**

4601 Fritz, D.; Schultze, J.   DE 161260/77/0007
Inhaltsstoffe von Gentiana lutea in Abhängigkeit von Ökotyp, Anbaustandort, Pflanzenalter und Jahreszeit; ausserdem Vergleich zwischen Wild– und Feldernte. **Active compounds of Gentiana lutea as dependent on ecotype, growing location, plant age and season; in addition, comparison between wild and cultivated plants.**

4602 Preil, W.   DE 206000/73/4006
Versuche zur Erkennung des Merkmals "frühe Blüte" im Sämlingsstadium bei Chrysanthemen. **Experiments of identifying the character "early flowering" in Chrysanthemum seedlings.**

4603 Preil, W.; Engelhardt, M.; Hoffmann, M.;
Reimann–Philipp, R.   DE 206000/75/0006
Untersuchungen zur Entmischung von Chimärenstrukturen bei Azaleen und Poinsettien durch Meristem– und Gewebeschüttelkulturen. **Studies on the separation of chimerical tissues in Azalea and Poinsettia by meristem and suspension cultures.**

4604 Junge, H.; Mattiesch, L.   DE 206000/75/0014 R
Biochemische Untersuchungen in schwachlichtsensitiven Pflanzen: Freie Aminosäuren und Enzymaktivitäten in Chrysanthemen. **Studies on basic metabolism in poor light sensitive plants: Free amino acids and activities of enzymes in chrysanthemum.**

4605 Hoffmann, M.; Engelhardt, M.; Engelhardt, K.; Häfen,
K.von; Kunze, H.; Lübbers, H.   DE 206000/78/0006 N
Regenerierung haploider Gerbera aus in vitro Kulturen von Blütenköpfchen. **Regeneration of haploid Gerbera from in vitro cultured capitulum explants.**

4606 Penningsfeld, F.; Fast, G.   DE 502101/70/0009
Kulturmethodik und Nährböden bzw. –lösungen zur Meristemvermehrung wichtiger Orchideengattungen. **Methods of cultivation and nutritional substrata respectively nutrient solutions for meristem propagation of important orchids.**

4607 Penningsfeld, F.; Forchthammer, L.   DE 502101/71/0001
Substrat– und Nährstoffansprüche von Gerbera jamesonii. **Substratum and nutrient requirements of Gerbera jamesonii.**

4608 Penningsfeld, F.; Kurzmann, P.; Müller, L.; Reis, A.

D 2100 – Plant production general and crop husbandry

DE 502101/73/0001

Versuchsanstellung über unterschiedliche Anzuchtsysteme, Kultursubstrate sowie Düngungs– und Bewässerungsmethoden bei der Begrünung von Dachgärten und der Pflanzenkultur in Grosscontainern, die in den Fussgängerzonen von Städten Verwendung finden. Prüfung von Pflanzenkombinationen, die sich im Stadtklima bewähren. **Experiments on different cultivation systems, culture substrata as well as on fertilizing and irrigation methods for growing green in roof–gardens and for plant cultures in large containers as in use in pedestrian zones in towns. Tests of plant combinations standing the climatic conditions in town.**

4609 Penningsfeld, F.; Forchthammer, L.; Kalthoff, F.
DE 502101/78/0003 N

Wuchsstoffbehandlung von Gerbera zur Verbesserung der Stecklings– und Blütenproduktion sowie Prüfung der Nachwirkung auf Ertrag und Blütenqualität. **Application of plant growth promoters to Gerbera to improve the production of cuttings and flowers, and examination of after–effect on yield and flower quality.**

4610 Sieber, J. DE 502107/72/0006 R
Das Verhalten von Rosensorten in verschiedenen Klimagebieten 1952. **Behavior of rose species in different climatic zones.**

4611 Reimherr, P.; Müller–Haslach, W. DE 502156/77/0001
Variabilitäts– und Ertragsuntersuchungen bei Schnittstauden. **Studies on variability and crop of cut perennials.**

4612 Hentig, W.–U.von; Köhler, K. DE 506108/72/0002
Einfluss von Tageslänge und Temperatur auf Wachstum und Entwicklung der Mutterpflanzen, die Stecklingsproduktion, –bewurzelung und Adventivtriebbildung bei Begonia–ElatiorHybriden. **Influence of day length and temperature on growth and development of mother plants, the production and rooting of cuttings, the formation of adventitious shoots of elatior begonias.**

4613 Hentig, W.–U.von; Fischer, M. DE 506108/77/0003
Einfluss von Alter und Kulturmethode bei Mutterpflanzen von Hibiscus rosa–sinensis auf die Stecklingsproduktion und –lagerung. **Influence of age and cultivation method of Hibiscus rosa–sinensis motherplants on production and storage of cuttings.**

4614 Hentig, W.–U.von; Köhler, K. DE 506108/77/0004
Einfluss von Blattalter, –grösse und –stiellänge auf den Vermehrungserfolg bei Saintpaulia ionantha. **Influence of age, size and length of petiole of saintpaulia leaf cuttings on the propagation success.** Publications.

4615 Hentig, W.–U.von; Fischer, M.; Röber, R.
DE 506108/77/0005
Einfluss der Substrattemperatur auf die Bewurzelung und weitere Entwicklung von Chrysanthemum und anderen Zierpflanzen. **Influence of substratum temperature on rooting and further development of chrysanthemum and other ornamental plants.**

4616 Hentig, W.–U.von; Köhler, K. DE 506108/77/0006
Einfluss verschiedener Kulturmethoden auf den Stecklingsertrag von Mutterpflanzen einiger Zierpflanzenarten. **Influence of different cultivation methods on the cutting yield of motherplants of some ornamental plants.**

4617 Hentig, W.–U.von; Fischer, M.; Röber, R.
DE 506108/78/0002 N
Einfluss von Temperatur und Druck auf die Lagerfähigkeit von Stecklingen bei Zierpflanzen. **Influence of temperature and pressure on the storage ability of cuttings of ornamental plants.**

4618 Röber, R.; Fischer, M. DE 506108/78/0003 N
Einfluss verschiedener N–Formen auf die Entwicklung und das Wachstum von Azaleen – Rhododendron simsii Planch. –. **Influence of different N–forms on the development and growth of azaleas – Rhododendron simsii Planch –.**

4619 Reuther, G.; Hentig, W.–U.von; Röber, R.; Reuther, G.; Hentig, W.–U.von; Röber, R. DE 506111/77/0001
Einfluss der Belichtung und Ernährung auf die Photosynthese, Produktivität und Stecklingsqualität bei Pelargonienmutterpflanzen. **Influence of radiation and nutrition on photosynthesis, productivity and quality of cuttings of Pelargonium stock plants.**

4620 Reuther, G. DE 506111/77/0003
Sortenspezifische Regenerationspotenz von Sprossspitzenexplantaten verschiedener Pelargoniumvarietäten in Gewebekultur. **Variety specific regeneration potency of shoot tip explants of various Pelargonium cultivars in tissue culture.** Publications.

4621 Geier, T. DE 506111/77/0005 N
Regulation der Morphogenese und Regeneration von Pflanzen aus in vitro kultivierten Organfragmenten von Cyclamen persicum. **Regulation of morphogenesis and plant regeneration from organ fragments of Cyclamen persicum cultured in vitro.**

4622 Geier, T. DE 506111/77/0007
In vitro Kulturen von Antheren und isolierten Pollenkörnern bei einigen Vertretern der Gesneriaceen. **In vitro cultures of anthers and isolated pollen in some members of the Gesneriaceae.**

4623 Escher, F.; Strech, H. DE 507701/74/0001
Einsatz chemischer Präparate zur Blütenbildung bei verschiedenen Bromeliaceen. **The use of chemicals for flower induction in different Bromeliaceae.**

4624 Adriansen, E.; Jensen, H.–E.K. DK 010113/72/4207
Vækststoffer til regulering af vækst og blomstring hos potteplanter i væksthus. **Substances for regulating plant growth and flowering in pot plants in glasshouses.**

4625 Christensen, O.V.; Bredmose, N. DK 010113/73/4205
Produktionsplanlægning på grundlag af produktionstiden for potteplanter og A. plumosus i væksthus. **Planned production on the basis of production time for pot plants and A. plumosus in glasshouses.**

4626 Adriansen, E. DK 010113/78/9001
Fysiologiske årsager til for tidlig knop–, blomster– og bladfald hos potteplanter. **Physiological causes of premature bud, flower and leaf loss in pot plants.**

4627 Klougart, A. DK 030171/78/0002 N
Holdbarhed af afskårne blomster i skummateriale. **Keepability of cutflower placed in foam material.**

4628 Anstett, A. FR 010115/71/6222

Recherche de substrats simples et stables pour la culture d'orchidées. **Study of elementary and stable substrates for orchidea's cultures.**

4629 Goujon, C.   FR 010505/73/0202
Obtention de porte–greffes multipliés par semis destinés à l'exploitation hivernale. **Breeding rose rootstocks propagated from seeds for winter production.**

4630 Chavagnat, A.   FR 011103/75/0430
Prétraitements et germination des semences de rosa. **Preteatments and germination of rosa seeds.**

4631 Cockshull; Hand   GB 011102/00/0036
Environmental requirements for glasshouse rose flower production.

4632 Hand   GB 011102/00/0058
Capillary watering of pot plants.

4633 Davies   GB 011103/18/0015 N
Quality in the pot chrysanthemum in relation to the environment.

4634 Bunt; Powell   GB 011106/00/0005 R
Effects of environment and growth regulators on quality and yield of carnation flowers.

4635 B'.nt   GB 011106/00/0006
Effect of temperature and daylength on spray varieties of carnation.

4636 Powell; Bunt   GB 011106/00/0007
New subjects and cultural methods for pot plants.

4637 Bunt   GB 011106/78/0006 N
Effect of temperature and daylength on spray varieties of carnation.

4638 Ward   GB 041203/00/0001
Production and management of rose root stocks.

4639 Researcher not indicated   GB 050102/00/0007 R
Chrysanthemums, production techniques and variety trials.

4640 Researcher not indicated   GB 050102/00/0008 R
Carnations, production techniques and variety trials.

4641 Researcher not indicated   GB 050102/00/0009 R
Roses, production techniques and variety trials.

4642 Researcher not indicated   GB 050102/00/0011 R
Pot plants (excluding chrysanthemums), production techniques and variety trials.

4643 Researcher not indicated   GB 050105/00/0001 R
Roses, propagation and production techniques.

4644 Researcher not indicated   GB 050142/00/0006
Chrysanthemums, production techniques and variety trials.

4645 Researcher not indicated   GB 050142/75/0012
Carnations, production techniques and variety trials.

4646 Researcher not indicated   GB 050142/75/0013
Pot plants (excluding chrysanthemums), production techniques and variety trials.

4647 Researcher not indicated   GB 050145/00/0001
Roses, propagation and production techniques.

4648 Researcher not indicated   GB 050145/00/0005
Cut flowers, production in the open, herbaceous plants.

4649 Turner   GB 060112/00/0007
The production of chrysanthemums.

4650 Duncan   GB 060215/00/0005
Commercial production of rose rootstocks.

4651 Hitchon   GB 060318/00/0004
The production of pot plants.

4652 Lamb, J.G.D.; Kelly, J.   IE 060300/67/0099 R
Ground cover plants (cultivar comparison). Publications.

4653 Seager, J.C.R.; Gallagher, P.A.   IE 060300/76/1314 R
Flowering pot plant production – longevity studies. Publications.

4654 Seager, J.C.R.; Lamb, J.G.D.; Kelly, J.
    IE 060300/77/1313 R
Extending the season of azalea production. Publications.

4655 Morgan, J.V.; Moustafa, A.; McGarrigle, M.
    IE 120108/78/9110 N
Factors affecting the production of spray chrysanthemums at high densities in nutrient solution culture on raised benches.

4656 De Ranieri, M.; Grassotti, A.   IT 021200/77/0012 N
Prove di coltivazione di cinque cv di garofani "miniature". Rispondenza della pacciamatura. **Cultivation trial of 5 miniature carnation varieties and plastic mulch effects..**

4657 De Ranieri, M.; Grassotti, A.   IT 021200/77/0013 N
Effetto di alcuni trattamenti geosterilizzanti sulla coltivazione di differenti cv di garofano.. **Effects of different soil disinfectant treatments on the cultivation of carnation different varieties.**

4658 Cirrito, M.; Zizzo, G.   IT 021200/77/0015 N
Esperienze sulle densità colturali della tuberosa coltivata per l'ingrossamento dei bulbetti. **Trials on cultivation densities in tuberosa bulblets enlargement..**

4659 Pergola, G.   IT 021200/77/0520 R
Controllo dello sviluppo di crisantemo ed Euphorbia fulgens a mezzo di fitoregolatori. **Growth regulation of chrysanthemums and Euphorbia fulgens with phyto–regulators.**

4660 Cocozza, M.   IT 040101/77/0150 R
Ricerche applicate alla vernalizzazione e al controllo chimi–co ormonale della senescenza dei fiori recisi e delle bulbo se. **Research applied to vernalization and to the hormonal control of senescence in cut flowers and bulb forming plants.**

4661 De Donato, M.   IT 041201/77/0487 R
Micropropagazione del crisantemo, studio delle variazioni del potere rizogeno in talee di azalea durante il ciclo di sviluppo della pianta. **Chrysanthemum micro–propagation, study on the rhizogenic power of azalea cuttings during the growing cycle of the plant.**

4662 Veen, H. NL 010102/71/7979 R
Veroudering bij hogere planten (snijbloemen). **Senescence of higher plants (cut flowers).** Publications.

4663 Sytsema, W. NL 010201/62/0947
De invloed van groeiregulatoren op groei en bloei van bloemisterijgewassen. **The influence of growth regulators on growth and flowering of floricultural crops.** Publications.

4664 Vonk Noordegraaf, C. NL 010201/69/2864
Bloeibeïnvloeding van potplanten. **Flower regulation of potplants.**

4665 Vonk Noordegraaf, C. NL 010201/70/3148
Ontwikkeling, groei en bloei van de Orchid Flowering Alstroemeria hybride "Walter Fleming". **Development, growth and flowering of the orchid flowering Alstroemeria hybrid "Walter Fleming".** Publications.

4666 Leffring, L. NL 010201/71/3884
Teelt- en selectie–onderzoek bij Anthurium andreanum. **Cultivation, breeding of Anthurium andreanum.**

4667 Berg, G.A. van der NL 010201/72/3802
Klimaatonderzoek bij kasrozen. **Research on the glass–house climate for roses.**

4668 Vonk Noordegraaf, C. NL 010201/72/3883
Teeltkundige aspecten van snijbloemen. **Growing aspects of cutflowers.**

4669 Leffring, L. NL 010201/75/6931 N
Weefselkweekonderzoek bij bloemisterijgewassen. **Research on tissue culture of floricultural crops.**

4670 Hoeven, A.P. van der NL 010206/70/3430 R
Teeltonderzoek bij snijgroen. **Growing conditions of Asparagus plumosus.**

4671 Hoeven, A.P. van der NL 010206/72/4660
Teeltonderzoek bij chrysant. **Research on growing of chrysanthemum.** Publications.

4672 Spithost, L.S. NL 010206/73/5596
Groei en ontwikkeling van trosanjers onder glas. **Growth and development of spray–carnations (Dianthus caryophyllus L).** Publications.

4673 Winden, C.M.M. van NL 010206/75/6863
Teeltonderzoek bij diverse bloemisterijgewassen in kassen. **Research on growing conditions of different flower crops in glasshouses.** Publications.

4674 Hoekstra, F.A. NL 020041/77/8447
De kweek van bloemen van Petunia in vitro. **The in vitro culture of Petunia flower buds.**

4675 Doorenbos, J. NL 020057/67/7376 R
De invloed van interne (genetische) en externe (klimaat, waterhuishouding, chemische stoffen) factoren op kwaliteit en houdbaarheid van bloemisterijgewassen (gerbera, begonia). **The influence of internal (genetical) and external (climate, water balance, growth regulators) factors on quality and post–harvest life of floricultural crops (gerbera, begonia).** Publications.

Ornamental shrubs (B 3730)

See also 2641, 3393, 4105, 4106, 4108, 4109, 4120, 4279, 4616, 4777, 4778, 4781, 4782, 4783, 4784, 4785, 4839, 4869, 4870, 4871, 4965

4676 Van Onsem, J.; Meneve, I.; Istas, W.
BE 070600/69/0019 R
Vegetatieve vermeerdering van coniferen, sierbomen en sierheesters. **Vegetative propagation of conifers, ornamental trees and shrubs.** Publications.

4677 Van Onsem, J.; Meneve, I.; Istas, W.
BE 070600/69/0022 R
Onderzoek inzake de teelt van sierheesters en coniferen in container. **Research about container growing of conifers and ornamental schrubs.** Publications.

4678 Van Onsem, J.G.; Meneve, I.; Istas, W.
BE 070600/76/0075 R
Introduktie van nieuwe sierheesters. **Introduction of new ornamental shrubs.** Publications.

4679 Rehm, S.; El–Afry, M.M.F. DE 132240/73/0001
Bildung wertbestimmender Inhaltsstoffe in Karkadeh – Hibiscus sabdariffa var. sabdariffa –. **Synthesis of essential constituents in karkadeh – Hibiscus sabdariffa var. sabdariffa –.**

4680 Schmidt, H.H. DE 135054/70/0002
Die Untersuchung von Gras– und Klee–Grasmischungen 1969. **Investigations on grass and clover mixtures.**

4681 Bünemann, G.; McCarthy, D.; Wennemuth, G.
DE 138241/78/0001 N
Beobachtungen zum Wachstum von Zwerggehölzen. **Obsvervations on growth patterns of dwarf shrubs and trees.**

4682 Bachthaler, E. DE 138270/70/0005
Untersuchungen zur Keimung und Lagerung von Rhododendron simsii–Samen. **Germination and storage of Rhododendron simsii–seeds.**

4683 Preissel, H.G.; Krebs, O. DE 138270/75/0001
Untersuchungen zum Einfluss der Umweltbedingungen auf die Blattfärbung von Codiaeum. **Studies on the influence of environmental factors on leaf colour of Codiaeum.**

4684 Liegel, W. DE 144445/77/0012
Eiweissspektren in der Gattung Cotoneaster. **Protein patterns in genus Cotoneaster.**

4685 Hansen, R.; Müssel, H. DE 502107/72/0005 R
Die Verwendung von Stauden auf ökologischer Grundlage in öffentlichen Grünanlagen 1971. **Ecological aspects of using shrubs in public greens.**

4686 Knapp; Franke, E.; Hansen, R. DE 502107/73/0001
Die Festlegung wichtiger Wuchsgebiete für eingeführte Park– und Gartenpflanzen in Westdeutschland. **Marking out of important growth areas for park and garden plants imported into West Germany.**

4687 Kolb, W. DE 502156/77/0002
Untersuchung des Pflegeaufwandes von bodendeckenden Gehölzen im innerörtlichen Grün in Abhängigkeit von

D 2100 – Plant production general and crop husbandry

Pflanzenart und Pflanzabstand sowie Beobachtung der Alterungsbeständigkeit. **Studies on maintenance input in soil covering shrubs in central greens depending on species and spacing of plants and observations on the steadiness of ageing.** Publications.

4688 Maurer, M.   DE 507701/78/0003 N
Staudenanzucht aus Samen. **Propagation of perennials by seeding.**

4689 Groven, I.; Bøvre, O.   DK 010104/76/0020 N
Formerings– og dyrkningsforsøg med stauder. **Propagation and production of perennials.**

4690 Gilly, G.   FR 010502/72/0525
Etude de l'enracinement du rosier. **Studies on the root system of rosetree rootstocks.**

4691 Preston; Quinlan   GB 011008/00/0026
Control of growth by pruning and chemical regulators.

4692 Howard; Oehl   GB 011008/00/0029
Replant disorders in soil re–used for hardy ornamental nursery crops.

4693 Howard; Shepherd   GB 011010/00/0002
Propagation of ornamental plants.

4694 Cockshull   GB 011102/00/0047 R
Photoperiodism, photosynthesis and translocation in hardy ornamental nursery stock.

4695 Winsor; Adatia   GB 011105/00/0007
Growth of calcifuge plants under calcareous conditions.

4696 Loach; Whalley   GB 011106/00/0009 R
Environmental factors affecting propagation of ornamental nursery stock.

4697 Loach   GB 011106/00/0013
Container production of nursery stock.

4698 Ward   GB 041203/75/0003
Shrub propagation techniques.

4699 Ward   GB 041203/75/0004
Post propagation techniques for shrubs.

4700 Ward   GB 041203/75/0005
Tree production methods.

4701 Researcher not indicated   GB 050105/00/0002 R
Hardy ornamental stock, propagation (excluding roses).

4702 Researcher not indicated   GB 050105/00/0003 R
Nursery stock, container growing.

4703 Researcher not indicated   GB 050105/00/0004 R
Hardy ornamental trees and shrubs (excluding roses).

4704 Researcher not indicated   GB 050105/00/0005 R
Cut flowers, production in the open, herbaceous plants.

4705 Researcher not indicated   GB 050145/00/0002
Hardy ornamental nursery stock, propagation (excluding roses).

4706 Researcher not indicated   GB 050145/00/0003
Nursery stock, container growing.

4707 Researcher not indicated   GB 050145/00/0004
Hardy ornamental trees and shrubs (excluding roses).

4708 Mccoll   GB 060215/00/0007
Nursery stock production.

4709 Johnston; Percy   GB 060318/00/0007
Shrub and tree growing in exposed areas and in high pH soils.

4710 Lamb, J.G.D.; Kelly, J.   IE 060300/67/0102 R
Mist propagation of hardy trees and shrubs. Publications.

4711 Kelly, J.; Lamb, J.G.D.   IE 060300/67/0106 R
Rhododendrons and azaleas – production techniques under Irish conditions. Publications.

4712 Kelly, J.; Lamb, J.G.D.   IE 060300/68/0109 N
Nursery stock production on peatland. Publications.

4713 Lamb, J.G.D.   IE 060300/69/0100 R
Propagation and cultivar trials of street trees. Publications.

4714 Lamb, J.G.D.   IE 060300/69/0104
Production of trees and shrubs from seed. Publications.

4715 Lamb, J.G.D.; Kelly, J.   IE 060300/69/0107 R
Tree and shrub propagation without artificial heat. Publications.

4716 Lamb, J.G.D.; Kelly, J.   IE 060300/70/0105 R
Production of container grown trees and shrubs. Publications.

4717 Seager, J.C.R.; Leclerc, M.H.; Lamb, J.G.D.
     IE 060300/76/1316 R
Assessment of outdoor shrubs as low temperature ornamental crops. Publications.

4718 Chiusoli, A.   IT 040203/77/0482 R
Controllo dello sviluppo di arbusti ornamentali impiegati nelle siepi. **Controling the development of ornamental shrubs used in hedges.**

4719 Lorenzi, R.   IT 041115/77/0508 R
Studi sulla propagazione delle piante ornamentali sempreverdi ed arbustive. **Studies on the propagation of ornamental evergreen shrubs.**

4720 Elk, B.C.M. van   NL 010203/64/3756
Groei en bloeibeïnvloeding met behulp van chemische middelen. **Flowering of nursery plants with chemical products.** Publications.

4721 Elk, B.C.M. van   NL 010203/66/3746
Algemeen oriënterend onderzoek omtrent het stekken van boomkwekerijgewassen. **General research concerning the propagation of ornamental plants by grafting.** Publications.

4722 Elk, B.C.M. van   NL 010203/66/3747 R
De invloed van groeistoffen bij het stekken van boomkwekerijgewassen. **The influence of growth regulators on cuttings of ornamental trees and shrubs.** Publications.

D 2100 – Plant production general and crop husbandry

4723 Elk, B.C.M. van NL 010203/66/3748 R
Het stekken van Rhododendron. **Propagation of Rhododendron by cuttings.** Publications.

4724 Elk, B.C.M. van NL 010203/66/3750
Algemeen oriënterend onderzoek omtrent het enten van boomkwekerijgewassen. **General research concerning propagation of ornamental shrubs from cuttings.** Publications.

4725 Elk, B.C.M. van NL 010203/66/3752
Teeltonderzoek bij boomkwekerijgewassen in plastic potten en zakken. **Growing of nursery stock in small containers.** Publications.

4726 Elk, B.C.M. van NL 010203/70/3755
De teelt van boomkwekerijgewassen onder staand glas. **Growth of nursery stock in a glasshouse.** Publications.

4727 Wijnands, D.O. NL 020044/78/8627 N
Een taxonomische studie van de in cultuur zijnde Rosa–soorten van de gematigde luchtstreken en hun cultuurvormen. **Taxonomic study of the cultivated hardy Rosa–species and their cultivars.** Publications.

Other ornamentals and ornamental products (B 3790)

See also 3942, 4625

4728 Rünger, W.; Führer, H. DE 105110/75/0001
Einfluss der Temperatur auf die Blütenentwicklung von Schlumbergera – Zygocactus –. **Influence of the temperature on the flower development of Schlumbergera – Zygocactus –.**

4729 Kluge, M. DE 117030/74/0001 N
Biochemie und Ökologie des Säurestoffwechsels der Sukkulenten – Crassulaceen Acid Metabolism = CAM – ; Regulation des CAM; CAM bei einheimischen Sukkulenten. **Biochemistry and ecology of Succulents, Crassulaceae Acid Metabolism, CAM; regulation of CAM; CAM in indigenous succulents.**

Forests in general (B 3800)

See also 750, 930, 1461, 1862, 1907, 1908, 1910, 1911, 1927, 2068, 2098, 2100, 2110, 2111, 2299, 2608, 2612, 2635, 2636, 2666, 2916, 3133, 4015, 4094, 4105, 4106, 4108, 4109, 7651, 10309, 10990, 15756, 15757

4730 Nanson, A. BE 130000/50/0016
Tests de descendance en sylviculture. **Progeny tests in sylviculture.** Publications.

4731 Nanson, A. BE 130000/58/0017
Conception, mise au point des techniques opérationnelles et réalisation de vergers à graines d'arbres forestiers. **Conception, fitting of operational technics and realization of forest tree seed orchards.** Publications.

4732 Delvaux, J. BE 130000/78/0019
Les premières coupes d'éclaircie dans les jeunes plantations. **Early thinnings in young forest plantations.** Publications.

4733 Huygh, A.; Baeyens, L.; Beckers, B.; Berben, J.; de Jamblinne de Meux, A.; Nef, L. BE 140000/77/0014 R
Fytotechnische problemen bij de bosexploitatie.

Phytotechnical problems of forest exploitation. Publications.

4734 Schmidt–Vogt, H.; Mall, B. DE 126300/73/0001
Zur Theorie der natürlichen Verjüngung im Wirtschaftswald. **A contribution to the theory of natural regeneration in productive forests.**

4735 Schmidt–Vogt, H.; Unger, H. DE 126300/78/0003 N
Elektrische Eigenschaften von Forstpflanzen bei unterschiedlicher Wasserversorgung und bei verschiedenen Temperaturen. **Electrical properties of forest plants at various degrees of water supply and at various temperatures.**

4736 Schmidt–Vogt, H.; Carneiro, J.G.A.
 DE 126300/78/0004 N
Morphologische und physiologische Untersuchungen über die Qualität von Forstpflanzen. **Morphological and physiological studies on plant stock quality in forests.**

4737 Barner, J.; Hoernstein, P. DE 126400/73/0003
Untersuchungen über den Stammabfluss an Bäumen. **Investigations on stem flow of trees.**

4738 Mantel, K.; Schoch, O. DE 126505/72/0002
Die Herleitung waldgeschichtlicher Typen und ihre Anwendung in der forstlichen Standortsanalyse. **The deduction of forest types in historical view and their application in forest site analysis.**

4739 Mitscherlich, G.; Künstle, E.; Ullrich, C.H.
 DE 126600/70/0001 R
Untersuchungen über den Gaswechsel in Waldbeständen 1968. **Investigations on gas interchange in forest stands.**

4740 Müller, G. DE 132690/75/0001
Untersuchungen zur Einschätzung von Selbst– und Fremdbefruchtungswahrscheinlichkeiten in Waldbaumpopulationen und deren Bedeutung für die genetische Struktur. **Estimation of probabilities of self– and cross–fertilization and its significance for the genetic structure of forest tree populations.** Publications.

4741 Gregorius, H.R.; Ziehe, M. DE 132690/77/0002
Paarungssystem von Waldbäumen und sein Einfluss auf genetische Eigenschaften des Saatgutes. **Impacts of the mating system in forest trees on genetic properties of the seed.**

4742 Hattemer, H.; Glock, H. DE 132690/78/0001 N
Versuche zur in–vitro–Kultur von Einzelzellen und Protoplasten bei Waldbäumen und ihre Regeneration zu ganzen Pflanzen. **In vitro–culture of cells and protoplasts of forest trees and the production of regenerated plants.**

4743 Kramer, H.; Athari, S. DE 132720/77/0001
Die Regenerationsfähigkeit rauchgeschädigter Bäume aus ertragskundlicher Sicht. **The regenerative capability of smoke–damaged trees in respect of forestry yield science.**

4744 Röhrig, E.; Jahn, G. DE 132752/77/0001
Sukzession nach Waldbrand. **Succession after forest fire.**

4745 Lamprecht, H.; Marmillod, D. DE 132752/78/0002 N
Strukturelle und waldbauliche Untersuchungen im peruanischen Amazonaswald. **Structural and silvicultural research in Amazonian forest of Peru.**

4746 Sachsse, H.　　　　DE 132780/77/0001
Mikroskopische Holzartenbestimmungen an mittelalterlichen
Skulpturen im westfälischen Raum. **Microscopic determination
of wood species of medieval sculptures from Westphalia.**

4747 Mülder, D.; Häberle, S.; Kratsch, H.–D.
　　　　　　　　　　DE 132840/71/0005 R
Mechanisierung als Mittel forstbetrieblicher Rationalisierung
1971. **Mechanization as means of forestry rationalization.**

4748 Mülder, D.　　　　DE 132840/74/0001
Die Netzplantechnik als Hilfsmittel zur Lösung komplexer
Aufgaben in Forstbetrieben. **Grid systems technique as an
expedient for solving complex problems in forestry enterprises.**

4749 Häberle, S.　　　　DE 132900/77/0001 N
Stückmassegesetz. **Piece mass law in forestry.**

4750 Koch, W.; Kerner, H.　　　　DE 160062/70/0001
Ökologie der forstlichen Produktion. **Ecology of forest
production.**

4751 Baumgartner, A.; Gietl, G.　　　　DE 160120/73/0002
Quantifizierung klimatischer Funktionen von Wäldern.
Quantification of climatic functions of forests.

4752 Huss, J.　　　　DE 160150/73/0001
Erprobung von Containerpflanzen unter Praxisbedingungen.
Testing of container plants under practical conditions.

4753 Franz, F.; Preuhsler, T.　　　　DE 160210/74/0003
Wachstumsgang, Ertragsleistung und Verjüngung in
oberbayerischen Bergmischwaldbeständen unter dem Einfluss
verschiedener Behandlung. **Process of growth, yield and
regeneration of mixed mountain forest stands in Upper Bavaria
under the influence of various treatment methods.**

4754 Franz, F.; Preuhsler, T.　　　　DE 160210/74/0004
Einfluss von Grundwasserabsenkungen auf Struktur und
Wachstum der Waldbestände im Donau–Lechgebiet. **The
influence of artificial lowerings of groundwater level on
structure and growth of forest stands in the region between the
Danube and the Lech.**

4755 Zöhrer, F.; Kennel, E.　　　　DE 160210/77/0001
Beweissicherung im Kernkraftwerk – Bereich Grafenrheinfeld
bei Schweinfurt. **Investigations on possible effects on forest
stands caused by the construction of the atomic power plant in
the area of Grafenrheinfeld near Schweinfurt.**

4756 Franz, F.　　　　DE 160210/77/0005 N
Entwicklung von Struktur– und Leistungstafeln auf der Basis
von Bestandes– und Wachstumssimulatoren für die wichtigsten
Baumarten in Bayern. **Development of advanced yield tables
and growth models based on growth simulation for the main
tree species in Bavaria.**

4757 Franz, F.; Meyer, F.　　　　DE 160210/78/0002 N
Entwicklung der Dimensionsgliederung der Hauptbaumarten
in Bayern. **Development of the distribution of dimensions of the
main tree species in Bavaria.**

4758 Franz, F.; Flurl, H.　　　　DE 160210/78/0005 N
Schaftkurvensysteme für die Hauptbaumarten Bayerns. **Stem
curve systems for the main tree species in Bavaria.**

4759 Agerer, R.; Kottke, J.; Sautter, C.　DE 173051/78/0004 N
1.Versuche zur Mykorrhiza–Bildung forstwirtschaftlich
wichtiger Bäume im Naturpark Schönbuch.
2.Bestandsaufnahme von parasitischen und Mykorrhiza–Pilzen
in ausgewählten Probeflächen des Naturparks Schönbuch. **1.
Trials on Mykorrhiza generation of trees of forestry importance
in the nature park Schoenbuch 2.Inventory of parasitic and
Mykorrhiza fungi in selected sample areas of the nature park
Schoenbuch.**

4760 Brünig, E.F.; Schneider, T.W.　　　　DE 202010/77/0003
Umweltgerechte optimale Bestandesaufbauformen und
Betriebszieltypen. **Environment–protective optimum
composition of stands and management goal types.**

4761 Brünig, E.F.　　　　DE 202010/77/0004
Entwicklung, Struktur und Funktionen von Waldökosystemen.
Development, structure and functions of forest ecosystems.

4762 Brünig, E.F.; Heuveldop, J.　　　　DE 202010/77/0005
Produktionsökologische Untersuchungen von Waldbeständen,
insbesondere im Hinblick auf den Strahlungs– und
Wasserhaushalt. **Production ecological studies on forest stands
with special regard to radiation and water economy.**

4763 Brünig, E.F.; Müllerstael, H.　　　　DE 202010/77/0006
Physio–ökologische Kriterien von Baumarten zur Optimierung
von Waldökosystemen und Frühdiagnose von Umweltschäden
und Belastbarkeit. **Physio–ecological criteria of tree species for
optimization of forest ecosystems and early diagnosis of
environmental injuries and carrying capacity.**

4764 Moosmayer, H.U.　　　　DE 501500/77/0008 N
Waldwachstumskundliche Beiträge zum Forschungsprojekt
'Naturpark Schoenbuch'. **Contributions of forest mensuration
to the research project 'Naturpark Schoenbuch'.**

4765 Kenk, G.; Müller, S.　　　　DE 501502/74/0003
Quantitative Erfassung der Bodenschutzfunktionen des
Waldes. **Quantification of soil protecting functions of forests.**

4766 Dieterich, H.; Löffler, H.　　　　DE 501502/78/0001 N
Ausarbeitung von Herkunftsempfehlungen für forstliches
Vermehrungsgut auf der Grundlage der regionalen Gliederung
in Baden–Württemberg. **Development of recommendations for
provenance of forestry propagation goods based on regional
structures in Baden–Wuerttemberg.**

4767 Dietz, P.; Tritschler, A.　　　　DE 501505/74/0004
Erprobung von Pflanzmaschinen. Verfahren für
Erstdurchforstung.. **Testing of tree–planting machines. Method
for initial thinning.**

4768 Kenk, G.; Altherr, E.　　　　DE 501509/78/0001 N
Ausarbeitung eines EDV–Programms zur Berechnung von
Jahrringanalysen in Düngungsversuchen. **Development of
EDP–programme for calculation of year–ring analyses in
fertilization trials.**

4769 Kenk, G.　　　　DE 501509/78/0002 N
Zuwachsgang im Altbestand und Entwicklung der
Naturverjüngung bei verschiedenen Bestockungsdichten.
**Trend of increment in old forest crop and development of
natural regeneration in different stock densities.**

4770 Henne, A.; Riebeling, R.　　　　DE 506403/77/0003 N

Auswirkungen von Umweltveränderungen – insbesondere Grundwasserabsenkungen – auf Waldbestände. **Effects of environmental changes – especially of groundwater use – on forest stands.** Publications.

4771 Henne, A.; Riebeling, R.   DE 506403/77/0004 N
Wuchsmodelle für verschiedene Baumarten auf hessischen Standorten. **Growth models of different tree species on forest sites in Hesse.**

4772 Weisgerber, H.   DE 506451/75/0001
Untersuchungen zur Blühinduktion von Waldbäumen. **Stimulation of flowering of forest tree species.**

4773 Schober, R.; Seibt, G.   DE 507651/70/0002 R
Forstwissenschaftliche Untersuchungen über die Primärproduktion im Buchen– und Fichtenwald im Rahmen des 'Solling–Projekts' der DFG 1968. **Forest scientific investigations on primary production in beech and spruce forests according to the Solling project of the DFG.**

4774 Kleinschmit, J.; Muhle, O.   DE 507653/78/0002 N
Untersuchungen über die Saattechniken. **Investigations on seeding techniques.**

4775 Günther, K.–H.   DE 508300/70/0001 N
Auswirkungen einer grossräumigen Grundwasserabsenkung auf den Wald im Erfttal – westl. Köln –. **Effects of sinking of the groundwater–table to the forests in Erft–valley – near Köln –.**

4776 Groven, I.; Brander, P.E.   DK 010104/57/0021 N
Afprøvning af planter til læ. **Tests of plants for windbreaks and shelters.**

4777 Groven, I.; Larsen, O.N.; Knoblauch, F.
          DK 010104/69/0011 N
Specifikke plantegrupper for træer og buske. **Propagation of specific groups of trees and shrubs for landscaping.**

4778 Groven, I.; Brander, P.E.   DK 010104/70/0007 N
Introduktion af nye landskabsplanter, træer og buske. **Introduction of new trees and shrubs for landscaping.**

4779 Groven, I.; Sønderhousen, E.   DK 010104/70/0022 N
Etablering, dyrkning og vedligeholdelse af læhegn. **Establishment and maintenance of shelterbelts.**

4780 Groven, I.; Larsen, O.N.; Knoblauch, F.
          DK 010104/72/0010 N
Roddannelse og knopbrydning hos træer og buske. **Root formation and bud break in cuttings of trees and shrubs for landscaping.**

4781 Groven, I.; Bøvre, O.   DK 010104/74/0016 N
Produktionsmetoder og udarbejdelse af produktionsprogrammer for træer og buske. **Production methods and production programmes for landscape plants.**

4782 Groven, I.; Bøvre, O.   DK 010104/75/0017 N
Overvintring af containerplanter; træer og buske. **Wintering trees and shrubs for landscaping in containers.**

4783 Groven, I.; Brander, P.E.   DK 010104/76/0005 N
Sortssamling af landskabsplanter, træer og buske. **Collection of cultivars of trees and shrubs and herbaceous plants for**

landscaping.

4784 Groven, I.; Brander, P.E.   DK 010104/76/0006 N
Kåring af landskabsplanter, træer og buske. **Election of trees and shrubs for landscaping on basis of phenotype.**

4785 Groven, I.; Brander, P.E.; Bøvre, O.
          DK 010104/78/0023 N
Etablering, dyrkning og funktion af landskabsplanter på blivestedet. **Establishment, growing and function of landscape plants on their growing place.**

4786 Henriksen, H.A.   DK 030181/66/0008
Undersøgelse over randvirkning i skov. **Investigation of border effects in forests.**

4787 Ranfelt, L.W.   DK 030181/76/0007
Systemdynamik indenfor dansk skovbrug. **System dynamics within Danish forestry.**

4788 Jacob, C.; Rodolphe, F.   FR 010214/76/8259
Dynamique population dans parcelles bocagères (3). **Population dynamics within bocage.**

4789 Aussenac, G.; Granier, A.   FR 010301/65/4055
Etude du bilan hydrique et de l'évapotranspiration réelle des peuplements forestiers. **Water balance and evapotranspiration of forest stands.**

4790 Oswald, H.; Garbaye, J.; Bouchon, J.; Ottorini, J.M.;
Le Tacon, F.   FR 010301/65/4065
Effets des différentes modalités d'éclaircie. **Effects of different types of thinning.**

4791 Aussenac, G.; Ducrey, M.; Granier, A.
          FR 010301/67/4056
Influence des traitements sylvicoles sur les microclimats forestiers dans les jeunes plantations et régénérations naturelles. **Influence of the sylvicultural treatments on the microclimate of young forest plantations and natural regeneration.**

4792 Aussenac, G.   FR 010301/68/4059
Etude de l'influence des accidents climatiques et des facteurs du microclimat sur la croissance des jeunes arbres forestiers. **Influence of climatic accidents or microclimatic factors on the growth of young forest trees.**

4793 Ducrey, M.   FR 010301/69/4057
Etude de la distribution du rayonnement solaire dans différents peuplements. **Solar radiation distribution of various forest stands.**

4794 Le Goff, N.; Decourt, N.; Ottorini, J.M.; Aussenac, G.;
Bouchon, J.; Millier, C.   FR 010301/70/4061
Accroissement et croissance des arbres et des peuplements forestiers. **Increment of trees and stands.**

4795 Aussenac, G.; Granier, A.   FR 010301/72/4050
Etude de la transpiration chez les jeunes plants en relation avec l'eau et la lumière. **Transpiration of young seedling in relation to water and light.**

4796 Ducrey, M.   FR 010301/72/4052
Etude de l'écophysiologie des échanges gazeux de la photosynthèse chez les jeunes plants forestiers. **Ecophysiology**

D 2100 – Plant production general and crop husbandry

of the photosynthetic gaz exchange of forest seedlings.

4797 Ducrey, M. FR 010301/73/4058
Influence des traitements sylvicoles sur la photosynthèse nette des peuplements forestiers. **Influence of the silvicultural treatments on net photosynthesis of forest stands.**

4798 Granier, A.; Aussenac, G. FR 010301/74/4051
Flux de sève brute dans les troncs. **Sap flow in the tree stems.**

4799 Levy, G.; Riedacker, A.; Le Tacon, F.; Garbaye, J.
FR 010301/74/4070
Production de plants forestiers sur tourbe et en conteneurs. **Production of forest seedlings on peat moss and in pots.**

4800 Auclair, D. FR 010301/74/4071
Influence de l'empoussièrement sur la photosynthèse. **Effects of dust deposit on photosynthesis.**

4801 Auclair, D. FR 010301/75/4053
Mesure de la photosynthèse in situ. **In situ evaluation of photosynthesis.**

4802 Michel M.France. FR 010301/75/4054
Phytoncides, antibiotiques produits et diffusés par les arbres forestiers. **Phytoncides, antibiotics produced and diffused by forest trees.**

4803 Ottorini, J.M.; Millier, C.; Bouchon, J.; Le Goff, N.
FR 010301/75/4062
Indices pour la modélisation des peuplements forestiers et de leur évolution. **Index for stand models and forest dynamics.**

4804 Parde, J.; Bouchon, J.; Le Goff, N.; Le Tacon, F.;
Ottorini, J.M.; Oswald, H. FR 010301/76/4066
Potentialités forestières des taillis et taillis sous futaie du Nord de la France. **Stand production of coppice and coppice with standards in the North of France.**

4805 Becker, M.; Picard, J.F.; Timbal, J. FR 010303/68/4165
Caractérisation mésologique des stations forestières et autécologie des espèces indicatrices. **Mesologic characterization of forest sites and autecology of indicative species.**

4806 Picard, J.F.; Le Tacon, F.; Becker, M.; Levy, G.;
Garbaye FR 010303/73/4166
Interventions sylvicoles et dynamisme de la végétation spontanée. **Forest practices and dynamism of spontaneous vegetation.**

4807 Nepveu, G.; Teissier du Cros, E. FR 010304/75/4205 N
Validité de tests précoces d'appréciation de certains critères de qualité du bois. **Accuracy of early testings to estimate some wood quality criteria.**

4808 Bachacou, J.; Monestiez, P. FR 010312/75/8294
Echantillonnage dans les peuplements forestiers (3). **Sampling in forest stands.**

4809 Poitou, N.; Delmas, J.; Mousain, D. (Clermont Fd.)
FR 010703/75/6068
Influence des champignons mycorhiziens sur la croissance des espèces forestières. **Effect of mycorhizal fungi on the growth of forest tree species.**

4810 Bonnet–Masimbert FR 012301/68/4006
Induction florale précoce chez les végétaux ligneux. **Precocious flowering induction in woody plants.**

4811 Cornu; Aubert; Birot; Lacaze FR 012301/70/4007
Multiplication végétative des arbres forestiers. **Vegetative propagation of forest trees.**

4812 Selby GB 040301/79/0038 N
The micropropagation of coniferous forest trees through tissue culture.

4813 Milner; Dickson GB 041001/00/0001
Effect of fertilisers and herbicides on growth and nutrient uptake of forest species.

4814 Milner GB 041001/00/0002
Effects of ground preparation and drainage treatments on tree growth and stability.

4815 Milner GB 041001/00/0003
Effects of spacing and thinning on forest stands.

4816 Milner GB 041001/00/0004
Best age and size of planting stock to use on various sites.

4817 Milner GB 041001/00/0006 R
Tatter flags to measure the effect of exposure.

4818 Milner GB 041001/76/0008
Effects of mixing hardwoods and conifers on flora, fauna, soil, timber production and visual amenity.

4819 Currò, P.; Eccher, A.; Valenziano, S.; Duranti, G.
IT 011801/79/0001 N
Prove di utilizzazione dei boschi cedui: aspetti di raccolta, selvicolturali e tecnologici (in collaborazione con altri organismi). **Natural coppice trials: silvicultural, exploitation and technological aspects (in collaboration with other organizations).**

4820 Patrone, G. IT 012002/74/0598
Ricerca sulle leggi generali che regolano l'accrescimento delle fustaie coetanee a lento accrescimento. **Research on general rules governing the growth of slow growing even–aged coppices.**

4821 Tocci, A. IT 021700/68/0004
Tecnica vivaistica. **Nursery techniques.**

4822 Gambi, G.; Tocci, A. IT 021700/69/0001
Tecniche di rimboschimento. **Reafforestation techniques.**

4823 Tocci, A. IT 021700/78/0003
Controllo sistematico e gestione arboreti dendrologici. **Systematic control and management of dendrological arboreta.**

4824 Gambi, G.; Sulli, M.; Amorini, E.; Buresti, E.; Fabbio,
G. IT 021700/79/0001 N
Prove di conversione di cedui in altofusti pascolabili. **Experiments on the conversion of coppices to High forest stands for grazing purposes.** Publications.

4825 Castellani, C. IT 021800/77/0003
Indagine pluriennale sull'accrescimento diametrico stagionale

D 2100 – Plant production general and crop husbandry

delle più importanti specie forestali italiane. **Survey on the seasonal diameter increment of more important Italian forestal species.**

4826 Todeschini, E.   IT 021800/79/0002 N
Influenza sulla produttività di alcuni popolamenti forestali, indotta da caratteristiche pedologiche e morfologiche del territorio. **Effect of soil and topographic characteristics on productive potential of forest land.**

4827 Giannini, R.   IT 040116/77/0001
Ricerche sull'interazione genotipo/ambiente in specie forestali. **Researches on genotype–environment interaction in forest trees.** Publications.

4828 Giordano, E.; Scarascia Mugnozza, G.; Santostasi, M.; Manzari, R.   IT 040116/79/0001 N
Possibilità di estensione della coltura specializzata da legno nell'Italia Meridionale. **Possibility of expanding specialized wood cultivation in Southern Italy.**

4829 Giordano, E.; Scarascia Mugnozza, G.; Santostasi, M.; Manzari, R.   IT 040116/79/0002 N
Ricerche sulla possibilità di utilizzazione mediante il pascolo ovino delle formazioni forestali e dei rimboschimenti in Puglia ed in Basilicata. **Researches on potential utilization of the sheep pasture in natural forests and reforested areas in Apulia and Basilicata.**

4830 Bernetti, G.   IT 040504/78/1032 N
Analisi dendrometrica del bosco ceduo. **Dendrometric analysis of coppices.** •

4831 Piussi, P.   IT 040504/78/1087 N
Indagine su alcuni cedui dell'Italia settentrionale e centrale, loro trattamento passato ed odierno, produttività, possibilità di miglioramento. **Research on some coppices of Northern and Central Italy; their past and present management; productivity, possibilities of improvement.**

4832 Bernetti, G.; la Marca, O.; Paganucci, L.; Piussi, P.; Hermanin, L.; Zanzi–Sulli, A.   IT 040516/76/0001
Ricerche sulle fustaie e sui cedui avviati all'alto fusto nell'Italia centrale. **Researches on the high forests and the conversion of coppice forests of central Italy.**

4833 Loreti, F.   IT 041104/78/1139 N
Comparazione gassosa del substrato e radicazione, rapporti tra nutrizione minerale e radicazione, studio sulla variazione stagionale della radicazione in alcune specie arboree. **Gas comparison of substratum in relation to radication, relation between mineral nutrition and radication; a study of the seasonal variations of radication in certain tree species.**

4834 Rosa, G.   IT 041308/77/0286 R
Ricerca sperimentale sugli apparati radicali delle piante forestali nei comprensori rimboschiti della Sardegna per la valutazione delle specie più adatte ai fini della sistemazione idraulico–forestale. **Experimental research on the root system of forest plants in the reafforested areas of Sardinia to determine the most suitable species for hydrological–forestal settlement.**

4835 Ceruti, A.   IT 061400/77/0917
Nutrizione dei funghi micorrizogeni. **Nutrition of mycorrhizogenous fungi.**

4836 Ceruti, A.   IT 061400/77/0918
Individuazione dei funghi micorrizogeni. **Identification of mycorrhizogenous fungi.**

4837 Ceruti, A.   IT 061400/77/0919
Strutture submicroscopiche delle micorrize. **Sub–microscopical stuctures of mycorrhiza.**

4838 Ceruti, A.   IT 061400/77/0920
Sintesi di micorrize. **Mycorrhiza synthesis.**

4839 Serra, G.   IT 092002/77/0526
Controllo della rizogenesi nella propagazione per talea. **Rhizogenesis control in propagation by cuttings.**

4840 Faber, P.J.   NL 010601/50/2415
De invloed van de dunning op de houtproduktie. **The influence of thinning on wood production.** Publications.

4841 Faber, P.J.   NL 010601/60/2414
De invloed van de zuivering op de ontwikkeling van de houb opstand. **The influence of precommercial thinnings on the development of forest stands.** Publications.

4842 Faber, P.J.   NL 010601/63/2412
De invloed van de plantafstand op de ontwikkeling van de hout opstand. **The influence of spacing on the development of forest stands.** Publications.

4843 Heybroek, H.M.   NL 010601/65/1865
Maatregelen en middelen voor bloeibevordering bij verschillende houtsoorten. **Flowering control of different tree species.** Publications.

4844 Tol, G. van   NL 010601/73/4456
Oriëntering naar mogelijkheden voor groeiverbetering voor zaailingen van bosplantsoen. **Orientation after possibilities for improvement of growth of forest tree seedlings.**

4845 Leek, N.A.   NL 010601/77/7544
Kwantitatieve inventarisatie van herbebossingssystemen. **Quantitative inquiry of reforestation systems.**

4846 Vries, P.G. de   NL 020009/47/4399
De invloed van houtsoort, herkomst, plantverband en dunning op de grootte en aard van de houtproduktie, zowel in Nederland als in Suriname. **Spacing/thinning experiment with Caribbean pine in Surinam.** Publications.

4847 Vries, P.G.de   NL 020009/47/4824
Productie van verschillende Nederlanse houtsoorten. **Yield research of various tree species in the Netherlands.**

4848 Broekhuizen, J.T.M. van   NL 020021/60/5402
Geschiktheid van exotische bosbomen voor aanplant in het Nederlandse bos. **The value of exotic trees for Dutch purposes.**

4849 Boerboom, J.H.A.   NL 020021/61/8646 N
Anthropogene ingrepen in het oesosysteem tropisch regenwoud. **Human interference in the tropical rainforest ecosystem.**

4850 Boeijink, D.E.   NL 020021/68/4442
Vegetatieve vermeerdering d.m.v. stekken bij bosbomen. **Vegetative propagation by means of cuttings of forest trees.**

D 2100 – Plant production general and crop husbandry

4851 Boerboom, J.H.A. NL 020021/72/4826
Ecologische en houtteeltkundige studies met betrekking tot gedegradeerde bergterreinen op Java en geëxploiteerd bos in Kalimanten (Indonesië). **Ecological and silvicultural studies concerning deteriorated mountain areas on Java and exploited forest of Kalimanten (Indonesia). Publications.**

4852 Wiersma, J.H. NL 020021/77/7369
Onderzoek naar de in Nederlandse en West–Europese arboreta en pineta aanwezige exoten, die van belang zouden kunnen zijn voor de Nederlandse bosbouw. **Investigation of the exotics, present in Dutch and West–European arboreta and pineta, which could be of interest for the Dutch forestry.**

Pine forests in general (B 3810)

4853 Delvaux, J. BE 130000/25/0011 R
Etudes sur la productivité des essences résineuses cultivées en Belgique. **Productivity study of the coniferous trees grown in Belgium. Publications.**

4854 Delvaux, J. BE 130000/60/0012 R
Etude de la compétition intraspécifique considérée comme fondement de la technique de l'éclaircie des peuplements (résineux). **Intraspecific competition seen as a foundation of thinnings (of coniferous stands). Publications.**

4855 Mitscherlich, G.; Oberfeld, B. DE 126600/77/0001
Saugspannungsmessungen an Kiefern–Nadeln – Pinus silvestris und Pinus nigra – im Tagesgang. **Moisture tension measurements on needles of Pinus silvestris and Pinus nigra during course of day.**

4856 Mitscherlich, G.; Lohner, P. DE 126600/77/0004
Saugspannungsmessungen an Kiefern–Nadeln – Pinus silvestris und Pinus nigra – im Jahresgang. **Moisture tension measurements on needles of Pinus silvestris and Pinus nigra during course of year.**

4857 Kramer, H.; Tuyll van Serooskerken, C.N.van
DE 132720/75/0002
Der Einfluss des Pflanzverbandes auf die Massen– und Wertleistung, die Produktionssicherheit und die spätere Bestandesbehandlung von Nadelholzbeständen. **The influence of the planting arrangement on volume production and financial yield, on regularity of production and later management of coniferous stands.**

4858 Knigge, W.; Hapla, F. DE 132780/77/0004
Auswirkung unterschiedlicher Pflanzverbandsweiten auf die Holzeigenschaften verschiedener Nadelbaumarten. **Effects of spacing on wood quality of different softwoods.**

4859 Häberle, S.; Eisele, F.B. DE 132900/72/0003
Entwicklung von Produktionssystemen für Nadelschwachholz in Hanglagen. **Development of production systems for conifers in montainous regions.**

4860 Schuck, H.J.; Lo, H.C. DE 160061/74/0002
Anatomische Untersuchungen des Markbereichs verschiedener Koniferen. **Anatomy of the pith of different conifers.**

4861 Huss, J. DE 160150/73/0002
Erprobung neuer Verfahren bei der chemischen Läuterung, besonders von Nadelbäumen. **Testing of new methods in chemical cleaning especially of coniferous trees.**

4862 Huss, J.; Schmidt; Burschel, P. DE 160151/74/0001
Begründung von Kiefernbeständen mit verschiedenartigen Pflanzensortimenten und Bodenbearbeitungen. **Production of pine woods with diverse plant assortments and different methods of soil cultivation.**

4863 Aufsess, H.von DE 160333/74/0002
Untersuchungen über die Auswirkung von Rückeschäden bei der Durchforstung junger Nadelholz– bestände. **Investigations on the effect of logging damages in thinning young coniferous stands. Publications.**

4864 Abetz, P.; Prange, H. DE 501509/72/0002 N
Anwuchsverhalten verschiedener Kiefernpflanzen–Sortimente. **Growing behaviour of different assortments of pine.**

4865 Jestaedt, M. DE 506451/71/0005
Untersuchungen über die auto– und heterovegetative Vermehrung von Laub– und Nadelbäumen. **Research on auto- and heterovegetative propagation of coniferous and broadleaved tree species.**

4866 Moltesen, P.; Madsen, T.L. DK 030181/77/0022 N
Kunstig vandings indflydelse på Pinus contorta's rumtæthedsniveau. **The influence of irrigation on the basic density level of Pinus contorta.**

4867 Becker, M.; Timbal, J. FR 010303/72/4164
Etude écologique des sapinières vosgiennes. **Ecological studies in the fir forests of the Vosges.**

4868 Le Tacon, F.; Mousain; Clement, A.; Chevalier; Garbaye, J.; Birot, Y. FR 010303/75/4153
Production, comportement et nutrition des résineux sur les sols calcaires. **Production, behaviour and nutrition of conifers on calcareous soils.**

4869 O'Brien, D.; Lynch, T,; Fitzsimons, B.
IE 050100/58/7518 N
Spacing in conifers. Publications.

4870 O'Brien, D.; Lynch, T.; Gallagher, G.
IE 050100/62/7217 N
Thinning in conifers. Publications.

4871 O'Brien, D.; Lynch, T.; Lynch, T. IE 050100/63/7216 N
Pruning in conifers. Publications.

4872 O'Driscoll, J.; Pfeiffer, A. IE 050100/64/7209 N
Vegetative reproduction of conifers. Publications.

4873 Eccher, A. IT 011801/72/0002
Adattabilità, limiti ecologici e capacità produttiva di specie e provenienze di conifere sotto sperimentazione preliminare. **Adaptability, ecological limitations and productivity of conifer species and provenances under preliminary investigation.**

4874 Eccher, A.; Currò, P. IT 011801/79/0002 N
Confronto di diversi metodi di diradamento in piantagioni di conifere a finalità produttive: aspetti selvicolturali, bioecologici e di utilizzazione. **Comparison of different thinning methods in conifer production stands: silvicultural,**

bioecological and exploitation aspects.

4875 Rambelli, A.   IT 041602/72/0542
Studio dei rapporti microbiologici che intercorrono tra
micorrizia e rizosfera in relazione al fenomeno di ingiallimento
del pinus radiata. **Study of the microbiological relations
between Mycorrhiza and the Rhizosphere with respect to the
yellowing phenomena of Pinus radiata.** Publications.

4876 Tol, G. van   NL 010601/63/2391
Zaaien, verspenen en afpennen in kwekerijen van
bosplantsoen. **Seeding, transplanting and undercutting in forest
nurseries.** Publications.

4877 Tol, G. van   NL 010601/63/2397
Onderzoek naar de betekenis van kwaliteitsnormen van
bosplantsoen in relatie tot de bosaanlegmethodiek. **Research
on the significance of grading criteria of material in relation to
afforestation techniques.** Publications.

4878 Wiersma, J.H.   NL 020021/52/4807
Photoperiodiciteits onderzoek bij verschillende herkomsten
van grove den en andere boomsoorten. **Research on
photoperiodicity in different provenances of Scots pine (Pinus)
silvestris L.) and other trees.** Publications.

Fir forests (B 3811)

See also 4916, 4973, 4975

4879 Schmidt–Vogt, H.; Farrokhpur, B. DE 126300/78/0005 N
Waldbaulich–ökologische Untersuchungen bei der
Femelschlagverjüngung in
Fichten–Tannen–BuchenMischbeständen.
**Silvicultural–ecological investigations on mixed stands of
spruce, white fir and beech in a group selection regeneration.**

4880 Kramer, H.; Bierg, N.   DE 132720/77/0003
Anlage von Versuchsflächen für die Behandlung von
Kiefern–Weitverbänden. **Establishment of sample plots for the
treatment of widely spaced Scotch pine.**

4881 Lamprecht, H.; Vergos, S.   DE 132752/78/0001 N
Waldbauliche und waldkundliche Untersuchungen in
natürlichen Schwarzkiefernwäldern Nordwest–Griechenlands.
**Silvicultural and forest ecological studies on natural forests of
Black Pine – Pinus nigra – in the northwest of Greece.**

4882 Burschel, P.; Schmidt, H.; Huss, J. DE 160150/75/0001
Erprobung neuer Verfahren bei der Kulturbegründung der
Kiefer. **Testing of new methods for establishing Scotch pine –
Pinus sylvestris – plantations.** Publications.

4883 Burschel, P.; Seitz, R.   DE 160151/78/0001 N
Die natürliche Verjüngung der Kiefer in Süddeutschland.
Natural regeneration of Scots pine in Southern Germany.
Publications.

4884 Schönborn, A.von; Bleymüller, H. DE 160180/73/0001
Untersuchungen über die Förderung der Blühinduktion durch
Wirkstoffbehandlung, Propfung und Umweltveränderung bei
Fichte und Kiefer – Picea abies L. Karst und Pinus silvestris L..
**Investigations on the promotion of flower induction by
treatment with active substances, by grafting and changes in
environment of Norway spruce – Picea abies L. Karst – and
Scotch fir – Pinus silvestris L. –.**

4885 Franz, F.; Forster, H.   DE 160210/78/0001 N
Ertragstafeln für die Aleppo–Kiefer des Aures–Gebietes–
Algerien –. **Yield tables for Pinus halepensis in the Aures area –
Algeria –.**

4886 Franz, F.; Oliveira, A.   DE 160210/78/0003 N
Wachstumselemente junger Kiefern–Bestände. **Growth
elements of young pine stands.**

4887 Franz, F.; Deckelmann, B.; Hirschfelder; Blank
                                        DE 160210/78/0004 N
Schätzrahmen für den Durchforstungsanfall in Fichten– und
Fichten–Kiefern–Mischbeständen in Süd–Bayern. **Tables for
estimation of thinning volume in spruce– and mixed
spruce–pine–stands in Southern Bavaria.**

4888 Moosmayer, H.U.   DE 501500/77/0007 N
Untersuchungen über die Ertragsleistung der Douglasie.
Fichte. **Investigations about growth and volume production of
Douglas fir.**

4889 Lemoine, B.   FR 010301/66/4064
Estimation et explication de la croissance des peuplements de
Pin maritime. **Estimation and explanation of the growth of the
Pinus pinaster stands.**

4890 Toth, J.; Bouchon, J.; Ottorini, J.M. FR 010301/68/4068
Production et régénération naturelle du Pin Noir d'Autriche
(Pinus nigra austriaca). **Yield, reproduction and natural
regeneration of Pinus nigra, var. nigricans.**

4891 Warner, P.   IE 050100/72/7507 N
Ecological study and distribution survey of pine marten.
Publications.

4892 Eccher, A.   IT 011801/65/0002
Influenza della densità d'impianto e dei diradamenti selettivi
sull'accrescimento e sulla forma di Pinus radiata e di Pinus
pinaster. **Influence of stand density and selective thinning on
Pinus radiata and Pinus pinaster growth and form.**
Publications.

4893 Ferrari, G.; Scaramuzzi, G.; Curro, P.
                                        IT 011801/70/0005
Qualificazione del legno del Pinus radiata e di altre conifere di
interesse cartario. **Wood quality assessment of Pinus radiata
and other conifers suitable for pulping.**

4894 Eccher, A.   IT 011801/70/0006
Confronto di provenienze di Pseudotsuga menziesii e di Pinus
ponderosa in diverse stazioni dell'Appennino. **Comparison of
Pseudotsuga menziesii and Pinus ponderosa provenances in
different Apennines sites.**

4895 Pirazzi, R.; Cavalcaselle, B. IT 011801/74/0006
Individualizzazione di nuovi simbionti attivi in micorrize di
Pinus radiata. **Identification of new active symbionts in Pinus
radiata mycorrhizae.** Publications.

4896 Pirazzi, R.; Cavalcaselle, B. IT 011801/74/0008
Indagine morfologica sulle micorrize di Pinus radiata.
Morphological study of Pinus radiata mycorrhizae.
Publications.

4897 Ferrari, G.; Scaramuzzi, G.   IT 011801/75/0003
Indagine preliminare sulla variabilità individuale e di stazione

delle caratteristiche del legno in Pinus radiata. **Preliminary investigation on individual and site variations of wood characteristics in Pinus radiata.**

4898 Ferrari, G.; Scaramuzzi, G.　　IT 011801/75/0010
Indagine sulle possibilità di valutazione delle caratteristiche del legno di piante in piedi e di una loro valutazione precoce in Pinus radiata. **Wood qualities assessment of standing trees and possibility of their early assessment in Pinus radiata.**

4899 Ferrari, G.; Scaramuzzi, G.　　IT 011801/75/0012
Esame delle correlazioni esistenti tra i principali caratteri di qualità del legno nel Pinus radiata. **Investigation of the correlations existing between main wood characteristics in Pinus radiata.**

4900 Lubrano, L.　　IT 011801/76/0016
Ricerche sulla propagazione per talea di Pinus radiata e P.halepensis. **Investigations on Pinus radiata and P.halepensis propagation by cutting.**

4901 Ciancio, O.; Avolio, S.; Menguzzato
　　IT 021700/68/0003 R
Sfollamento e diradamento nei rimboschimenti di pini mediterranei. **First thinning trials on reafforestation stands of Mediterranean pines.** Publications.

4902 Ciancio, O.; Avolio, S.　　IT 021700/69/0003
Diradamento di Pinus nigra Laricio. **Thinning of Pinus nigra Laricio stands.**

4903 Gambi, G.; Amorini, E.; Guidi, G.; Preto, G.
　　IT 021700/71/0002
Trattamento delle pinete di pino nero. **Management of Pinus nigra artificial stands.**

4904 Avolio, S.　　IT 021700/78/0001
Studi ecologici sul pino loricato. **Ecological studies on Pinus leucodermis.**

4905 Guidi, G.; Fusaro, E.　　IT 021700/78/0002
Ricerche su Pinus radiata. **Researches on Monterey pine.**

4906 Castellani, C.; Tosi, V.　　IT 021800/79/0004 N
Costruzione di tavole di cubatura e di produzione per il Pinus halepensis. **Volume and yield table construction for Pinus halepensis.**

4907 Giordano, E.; Scarascia, G.　　IT 040116/74/0001
Ricerche sulla germinazione del seme di Pino d'Aleppo. **Researches on seed germination of Aleppo Pine.**

4908 Giordano, A.　　IT 040116/77/0197 R
Indagini sulla rinnovazione naturale delle pinete di pino d'aleppo. **Research on the natural renewal of Aleppo pine woods.**

4909 Giordano, E.; Scarascia, G.　　IT 040116/78/0001
Indagine sulla rinnovazione spontanea delle pinete di Pino d'Aleppo. **Natural regeneration of Aleppo pine forests.**

4910 Burg, J. van den　　NL 010601/73/4461
De invloed van verschillende wintertemperaturen op fijnspar en groveden. **The influence of different winter temperatures on Norway spruce and Scots pine.** Publications.

4911 Willemse, M.T.M.　　NL 020046/75/8449
Zaadknop ontwikkeling bij Pinus. **Ovule development of Pine.**

Spruce and fir forests (B 3812)

See also 1396, 4879, 4884, 4887, 4894, 4910, 7677

4912 André, P.; Blérot, P.　　BE 020301/69/0007
Intensité d'éclaircies en jeunes pessières. **Intensity of thinning out young epicea stands.**

4913 Braun, H.J.; Vanselow, G.　　DE 126250/77/0001
Wurzelwachstum von Fichtenstecklingen. **Root growth of Norway spruce cuttings.**

4914 Schmidt–Vogt, H.; Rautanen, J.　　DE 126300/72/0001
Wuchsverhalten verschiedener Fichtenklone im südlichen Finnland. **Growth of different spruce clones in southern Finland.**

4915 Schmidt–Vogt, H.; Trauth, K.　　DE 126300/78/0002 N
Untersuchungen über die Widerstandsfähigkeit der Fichte gegen Schneebelastung in Beständen verschiedener Behandlung und Struktur. **Investigations on the resistance of spruce to snow–load in woods of different silvical management and of different structure.**

4916 Schmidt–Vogt, H.; Farrokhpur, B. DE 126300/78/0006 N
Untersuchungen zum Fruchtwechsel bei der Naturverjüngung in Fichten–Tannen–Buchen–Mischbeständen. **Investigations on the rotation of crops in mixed stands of spruce, white fir and beech during natural regeneration.**

4917 Schmidt–Vogt, H.; Pham–Nguyen, T.
　　DE 126300/78/0007 N
Ökophysiologische Untersuchungen bei verschiedenen Koniteren – Picea abies, Pseudotsuga menziesii, Abies alba – bei kontrollierter Wasserversorgung unter Freilandbedingungen. **Ecophysiological investigations on coniferous trees – Picea abies, Pseudotsuga menziesii, Abies alba – at controlled degrees of water supply under field conditions.**

4918 Barner, J.; Al–Kawaz, S.　　DE 126400/75/0001
Untersuchungen über die Wurzelausbildung der Fichte und Erle in Fichten–Erlenmischbeständen. **Studies on the root formation of spruce and alder in mixed stands of spruce and alder.** Publications.

4919 Abetz, P.; Krezdorn, R.　　DE 126600/78/0001 N
Untersuchungen über die Bildung von Reaktionsholz bei stärkeren Eingriffen in Fichten–Jungbeständen in Baden–Württemberg. **Investigations on the formation of reaction wood after intensive thinning of young spruce stands in Baden–Wuerttemberg.**

4920 Bergmann, F.　　DE 132690/71/0001 N
Untersuchungen zur Bestimmung des Ursprungs von Fichtensamen mit Hilfe biochemischer Polymorphismen. **Studies on the determination of the provenience of spruce seed by means of biochemical polymorphisms.**

4921 Kramer, H.　　DE 132720/75/0005
Der Einfluss der Stammzahlhaltung auf den Wachstumsablauf von Fichtenbeständen. **The influence of stem counting management on the progressive development of spruce stands.**

D 2100 – Plant production general and crop husbandry

4922 Huss, J.　　　　　　　　DE 132751/71/0001
Die Wirkung der Graskonkurrenz auf das Wachstum wurzelnackter und ballierter Fichten und Douglasien 1971. **english title not indicated.**

4923 Larsen, J.B.　　　　　　DE 132751/77/0001
Untersuchungen über Trockenresistenz und Transpirationsverhalten bei Koniferen, insbesondere bei Pseudotsuga menziesii und Abies grandis. **Studies on drying resistance and water–relations of conifers, especially of Pseudotsuga menziesii and Abies grandis.**

4924 Koltzenburg, C.　　　　　DE 132780/77/0002
Aufkommen und Verwendung von in Westdeutschland erzeugtem Douglasienholz. **Production and utilization of Douglas fir grown in Western Germany.**

4925 Lewark, S.　　　　　　　DE 132780/77/0005
Auswirkung von Vererbung, Standorteinflüssen und Alter auf Ästigkeit, Rohdichte und Jahrringbreiten der Fichte. **Influence of heritability, site and age on branching wood, density and width of growth ring of Norway spruce.**

4926 Sachsse, H.　　　　　　　DE 132780/77/0008
Histologische Untersuchungen an maschinell sommergeästeten Douglasien. **Histological investigations of Douglas fir pruned in summer by machine.**

4927 Foerst, K.　　　　　　　DE 160030/78/0001 N
Standortsansprüche, Ernährungszustand und Wuchsleistung älterer bayerischer Bestände der Grünen Douglasie – Pseudotsuga menziesii 'Mirb.' Franco var. menziesii –. **Site requirements, nutrition and growth production of older Bavarian Douglas fir stands – pseudotsuga menziesii 'Mirb.' Franco var. menziesii –.**

4928 Huss, J.; Schmidt, H.; Reinhardt; Burschel, P.
　　　　　　　　　　　　　　DE 160150/74/0001
Anlage und Auswertung von Durchforstungsversuchsflächen in Fichten–Stangenhölzern. **Layout and utilization of experimental thinning areas in spruce pole timber.**

4929 Huss, J.　　　　　　　　DE 160150/75/0002
Untersuchungen über die Begründung von Douglasien–Kulturen. **Investigations of the establishment of Douglas fir plantatitons.**

4930 Burschel, P.; Huss, J.　　　DE 160150/75/0005
Entwicklung von zweckmässigen Durchforstungseingriffen in dichten Fichtenjungbeständen. **Development of suitable measures for thinning dense stands of young spruces.**

4931 Schönborn, A.von; Werner, H.　DE 160180/75/0002
Kombinierte Hydrokulturversuche und Gaswechselmessungen an Fichtenkeimlingen und –jungpflanzen. **Investigations of hydroculture and gas exchange of seedlings and young plants of Picea abies.**

4932 Schulz, H.; Aydin, I.　　　DE 160330/75/0002
Beziehungen zwischen der Rohdichte und Härte des Fichtenholzes unter besonderer Berücksichtigung des Standorteinflusses. **On the relationship between the density and hardness of spruce wood with special reference to environmental influences.**

4933 Gärtner, E.　　　　　　　DE 506451/78/0001 N
Auswahl potentieller Fichten–Saatguterntebestände in Hessen und deren Überprüfung durch Nachkommenschaftsprüfung als Vorbereitung zur Zulassung für die Gewinnung von forstlichem Vermehrungsgut. **Selection of potential seed stands of Norway spruce in Hesse and their examination through a progeny test as preparation for the approval as source of forest reproductive material.**

4934 Siebert, H.; Jungbluth, H.J.; Dimitri, L.
　　　　　　　　　　　　　　DE 506453/72/0001
Untersuchungen zur Ermittlung von bestgeeigneten Pflanzverfahren bei der Douglasie für die hessischen Standorte. **Research on planting methods for Douglas fir specifically suited for site conditions prevailing in Hesse.**

4935 Dimitri, L.; Vaupel, O.　　DE 506453/73/0001 N
Untersuchungen zur Erforschung optimaler Verfahren für die Bestandesbegründung bei Fichte, insbesondere auf sturmgefährdeten Standorten. **Research on the optimum proceeding in establishing Norway spruce stands, in particular on sites exposed to storm damages.**

4936 Gussone, H.A.　　　　　　DE 507651/78/0002 N
Untersuchungen über die Anbauwürdigkeit der Abies grandis auf weitverbreiteten Standorten Niedersachsens. **Studies on growing suitability of Abies grandis on widespread sites in Lower Saxony.**

4937 Kleinschmit, J.　　　　　DE 507653/73/0002
Klärung der Grundlagen und Weiterentwicklung von Verfahren der vegetativen Vermehrung der Douglasie und der Lärche durch Stecklinge bis zur Praxisreife. **Developing principles and improvement of methodical vegetative propagation of Douglas fir and larch cuttings.**

4938 Kleinschmit, J.　　　　　DE 507653/75/0001
Wurzeluntersuchungen bei Fichten–, Lärchen– und Douglasien–Stecklingen. **Investigations into the root development of cuttings of spruce, larch and Douglas fir.**

4939 Jensen, N.J.B.; Jakobsen, S.T.　DK 030148/79/0004 N
Karforsøg med dækrodsplanter af Abies grandis. **Pot experiments with factors influencing growth of Abies grandis after transplantation from nursery.**

4940 Sanojca–Abrahamer, K.　　DK 030181/68/0009
Naturlig foryngelse af ædelgran. **Natural regeneration of Abies nobilis.**

4941 Moltesen, P.; Madsen, T.L.; Olesen, P.O.
　　　　　　　　　　　　　　DK 030181/77/0021 N
Klimaets indflydelse på rødgranens rumtæthedsniveau. **The influence of the climate on the basic density level of Norway spruce.**

4942 Olesen, P.O.　　　　　　　DK 030181/78/0003 N
Trakeidebreddens og rumtæthedsniveauets variation i frø– og vegetativt formerede rødgraner. **The variation of the tracheid–width and the basic density level in vegetatively and sexually propagated Norway spruce.**

4943 Moltesen, P.　　　　　　　DK 030181/78/9091
Hugststyrkens og vækstbonitetens betydning for dansk rødgrans styrkeegenskaber. **The importance of grade of thinning and growth quality for strength properties of Danish**

D 2100 – Plant production general and crop husbandry

Norway spruce.

4944 Joyce, P.M.; Clear, T.　　　　IE 120107/68/9088 N
Spacing trial with Norway spruce.

4945 Stuart, M.R.; Meany, M.　　　　IE 120401/78/9164 N
An investigation of rhizoplane fungi associated with sitka s
pruce mycorrhizae.

4946 Ciancio, O.; Cappelli, F.; Nocentini, S.
　　　　　　　　　　　　　　　IT 021700/77/0001
Trattamento di Pseudotsuga douglasii. **Management of douglas
fir stands.**

4947 Del Favero, R.; Scrinzi, G.; Tabacchi, G.
　　　　　　　　　　　　　　　IT 021800/79/0001 N
Studio sulle variazioni di massa in giovani popolamenti di Picea
abies. **Study on volume variations in the young stands of Picea
abies.**

4948 Piussi, P.　　　　　　　　IT 040504/78/1088 N
Studio della struttura e della rinnovazione dei boschi di Picea
nelle Alpi orientali. **Study on the structural aspect and the
renewal of Picea woods in the eastern Alps region.**

4949 Hellrigh, B.　　　　　　　IT 040809/74/0569
Ricerca su alcune relazioni di biomassa per l'abete e la picea
nelle Alpi orientali. **Research on some biomass relations for fir
and spruce in the Eastern Alps.**

4950 Kriek, W.; Evers, P.W.　　　　NL 010601/77/7535
Vegetatieve vermeerdering van douglas in vitro. **Vegetative
propagation of Douglas fir in vitro.**

4951 Vries, P.G.de　　　　　　　NL 020009/23/4823
Productie van verschillende herkomsten van de Douglasspar.
**Yield research of various Douglas fir provenances in the
Netherlands. Publications.**

4952 Pierik, R.L.M.; Evers, P.W.　　　　NL 020057/77/7535
Vegetatieve vermeerdering van douglas in vitro. **Vegetative
propagation of Douglas fir in vitro.**

Larch forests (B 3813)

See also 4902, 4937, 4938, 7677

4953 Knigge, W.; Broese, V.; Hamring　　DE 132780/77/0003
Untersuchung der Auswirkung der Säbelwüchsigkeit des
unteren Schaftteils der Lärche unterschiedlichen Alters auf
deren Holzeigenschaften. **Investigation of the effects of
abnormal form of the lower part of the trunk of European larch
on wood quality and lumber properties.**

4954 Sachsse, H.　　　　　　　DE 132780/77/0007
Untersuchung über die anatomischen und physiologischen
Auswirkungen maschineller Wertästung bei Lärche.
**Investigation of the anatomical and physiological consequences
of pruning of larch by machine.**

4955 Mülder, D.; Ruppertshofen, H.　　DE 132840/71/0001
Die Krummwüchsigkeit der europäischen Lärche im
Jugendstadium als erworbene Eigenschaft und Möglichkeiten
der Abhilfe. **Curcked growth of European larch in the juvenile
stage as an invironmentally conditioned trait and ways of
remedy.**

4956 Westra, J.J.　　　　　　　NL 020021/62/5404
Stamanalyse van lariks ten behoeve van vaststelling externe
invloeden op groei. **Research on external effects on growth of
larch by means of trunk analysis.** Publications.

Other pine forests (B 3819)

See also 4975

4957 Toth, J.　　　　　　　FR 010301/69/4069
Rénégération et traitement sylvicole des peuplements de
Cèdre. **Regeneration and sylviculture of Cedrus atlantics.**

Leafwoods in general (B 3820)

See also 4002, 4865

4958 Roisin, R.; Liard　　　　　BE 140000/79/0071 N
Reproduction végétative d'essences ligneuses. **Reproduction
from wood species by stecking.**

4959 Lamprecht, H.; Rastin, N.; Jahn, G.
　　　　　　　　　　　　　　　DE 132752/78/0003 N
Die ökologischen Bedingungen der Laubwälder in der
Kaspischen Ebene und Vorschläge zur Verbesserung ihrer
waldbaulichen und wirtschaftlichen Leistung. **Ecological
conditions of hardwoods in the Caspian Plain and proposals for
their silvicultural and economic improvement.**

4960 Huss, J.　　　　　　　DE 160150/75/0003
Untersuchungen über die Begründung von Laubholzkulturen –
in Containern und wurzelnackt, mit und ohne Düngetabletten
–. **Investigations of the establishment of hardwood plantations –
containerized and bare root, with and without nutrient tablets
–.**

4961 Brandtner, E.　　　　　　DE 301050/75/0007
Anwendung der Phänologie von Laubgehölzen für
lokalklimatische Untersuchungen. **Application of the
phenology of broad–leaved trees to investigations of local
climates.**

4962 Keller, R.; Nepveu, G.　　　　FR 010304/67/4210 N
Elagage de branches vivantes. **Green pruning.**

4963 Polge, H.; Keller, R.　　　　FR 010304/71/4209 N
Formation de la tige et qualité du bois d'essences feuillues
cultivées en absence de concurrence. **Stem formation and wood
quality of hardwoods grown out of competition.**

4964 Gambi, G.; Guidi, G.; Ciancio, O.　IT 021700/69/0002
Conversione e miglioramento dei cedui. **Management of
coppice forests. Publications.**

4965 Heybroek, H.M.　　　　　NL 010601/58/1854
Vegetatieve vermeerdering van diverse houtsoorten.
Vegetative propagation of different tree species. Publications.

4966 Wigbels, V.L.　　　　　　NL 040007/66/4151
Onderzoek m.b.t. aanleg en beheer van loofhoutbossen op
zware kleigronden. **Research on formation and management of
broad leaved stands on heavy clay soils.**

Oak tree stands (B 3821)

D 2100 – Plant production general and crop husbandry

See also 1926, 4879, 4916, 4985

4967 Researcher not indicated DE 126000/77/0001 R
Vergleichende Untersuchungen über die Wuchsleistung,
Frostund Trockenresistenz SO–europäischer Buchenherkünfte
als Grundlage für den Anbau in der Bundesrepublik
Deutschland 1977–1978. **Comparative investigations on groth
production, frost and dry resistance of south–east European
beech provenances as basis for forest planting in the Federal
Republic of Germany.**

4968 Röhrig, E.; Schaper, C. DE 132751/75/0006
Ansprache des Aufbaus und der morphologischen
Eigenschaften von Eichendickungen und deren Pflege.
**Structure and morphological properties of young oak growths
and their tending.**

4969 Röhrig, E.; Lüpke, B.von DE 132751/77/0002
Wachstum von Forstpflanzen und Bodenvegetation unter dem
Schirm eines Buchen–Altbestandes unterschiedlicher Dichte.
**Growth of forest plants and soil vegetation under the cover of
old beech crop of varying density.**

4970 Franz, F.; Flurl, H.; Lutz DE 160210/78/0006 N
Kronendimensionen und Zuwachsleistung von
Eichen–Beständen in der Rheinpfalz. **Crown dimensions and
current increment in oak stands in the Rheinpfalz area.**

4971 Schober, R.; Warth DE 507651/75/0003
Untersuchungen über Wachstum und Ertrag der Roteiche.
Research on growth and yield of red oak – Quercus borcalis –.

4972 Holm, E.; Petersen, H. DK 030108/69/0002 N
Biologisk nedbrydning af bøgeløv. **Biological decomposition of
beech litter.**

4973 Bouchon, J.; Le Goff, N.; Oswald, H.; Keller, R.;
Ottorini, J.M.; Le Tacon, F. FR 010301/69/4063
Normes de sylviculture et compétition pour le hêtre, le chêne
et le pin sylvestre. **Growing stock in beech, oak and pine stands.**

4974 Oswald, H.; Aussenac, G.; Le Tacon, F.; Ducrey, M.;
Garbaye, J.; Bouchon, J. FR 010301/69/4067
Régénération naturelle et/ou artificielle du hêtre et du chêne
(rouvre et pédonculé). **Treatment of natural or/and artificial
regenerations of beech and oak.**

4975 Riedacker, A. FR 010301/73/4080
Morphogénèse et croissance des systèmes racinaires du Chêne,
du hêtre, du cédre et du Pin laricio. **Root morphogenesis and
growth of Oak, Beech, Corsican Pine and Cedrus atlantica.**

4976 Garbaye, J.; Oswald, H.; Levy, G.; Bouchon, J.; Picard,
J.F. FR 010303/67/4155
Le Chêne – Différents types de chênaies, caractérisation par le
sol et la flore – Relation entre
sol–nutrition–production–qualité du bois – Régénération.
**Oakstands types, characterized by soil and vegetation –
Relationships between soil, mineral nutrition, production,
woodquality – Natural regeneration.**

4977 Le Tacon, F.; Keller, R.; Timbal, J.; Becker, M.;
Oswald, H.; Malphettes, C.B. FR 010303/72/4156
Le Hêtre – Différents types de hêtraies; caractérisation par le
sol et la flore – Relation entre sol; nutrition, production,

qualité du bois – Régénération naturelle. **The Beech –
Beechstands types, characterized by soil and vegetation –
Relationships between soil, mineral nutrition, woodquality –
Natural regeneration.**

4978 Huber, F. FR 010304/75/4201 N
Déterminisme physiologique des variations du plan ligneux
dans les bois de chêne et de hêtre. **Physiological determinism
of anatomical variation in oak and beech wood structure.**

4979 Guidi, G.; Fusaro, E.; Ranalli, A. IT 021700/71/0001 R
Rinnovazione delle cerrete. **Regeneration of Quercus cerris
forests.**

4980 Gambi, G.; Guidi, G.; Amorini, E. IT 021700/77/0002
Trattamento delle faggete di produzione e di protezione.
Management of productive and protective beech forests.

4981 Corti, R. IT 040505/77/0153 R
Ricerca sulla determinazione di alcune esigenze ecologiche e
fisiologiche di semenzali di faggio di diverse provenienze
italiane. **Reserch aiming to establish some of the ecological and
physiological needs of beech seedlings of various Italian origins.**

4982 De Philippis, A. IT 040516/77/0163 R
Indagine sulla biomassa e mineralomassa dei cedui di quercia.
Research on the biomass and mineralmass of oak coppices.

4983 Tol, G. van NL 010601/77/8673 N
De invloed van gras– en kruidenvegeties op slaging van groei
van eikenbezaaliïngen. **The influence of grass and herbaceous
vegetations on survival and growth of seedings.**

4984 Faber, P.J. NL 010601/78/7881
Selectieve verzorging in opstanden van eik en es. **Selective
tending in stands of oak and ash.**

Ash tree stands (B 3822)

See also 4984, 5992

4985 Bierg, N. DE 132720/77/0002
Wachstum von Ahorn und Buche in jungen
BuchenEdellaubholz–Pflanzbeständen. **Growth of common
maple and beech in young beech–maple plantations.**

Poplar tree stands (B 3823)

See also 382, 4733, 4918, 5993, 7697, 7700, 10032

4986 Sachsse, H. DE 132780/78/0001 N
Untersuchung technisch bedeutsamer Holzeigenschaften von
Balsampappeln und Balsam–Schwarzpappelbastarden.
**Investigation of wood properties of technological importance
of balsam poplars and of balsam black–poplar–hybrids.**

4987 Bonnemann, A.; Vaupel, O.; Schumann, G.
DE 506453/77/0001
Anbauversuche mit verschiedenen Pappel–Arten bzw. –Klone
als Zeitmischung in Laub– und Nadelholzbeständen und in
Kurzumtrieben. **Plantation trials with various species and
clones of poplars for temporary mixture in stands of conifers
and latifolious trees and for mini–rotation.**

4988 Weisgerber, H.; Jestaedt, M. DE 912000/72/0002 R
Anlage und Auswertung von Pappelversuchsflächen in der

D 2100 – Plant production general and crop husbandry

BRD 1965. **Layout and evaluation of experimental poplar plots in the Federal Republic of Germany.**

4989 Siebert, H.   DE 912000/72/0003 R
Verbesserung der Äsungsverhältnisse im Walde mit Hilfe von Weiden und sonstigen Weichlaubhölzern 1969. **Improvement of browsing conditions by planting willows and other weed–trees.**

4990 Baumeister, G.   DE 912000/73/0001
Untersuchungen zur Stimulierung von Polyploidie bei Pappeln. **Studies on polyploid stimulation in poplar.**

4991 Weisgerber, H.; Bonnemann, A.   DE 912000/75/0004
Kurzumtrieb von Pappeln zur Steigerung der Holzproduktion. **Mini–rotation of poplars for increasing timber production.** Publications.

4992 Jestaedt, M.   DE 912000/77/0001
Untersuchungen über Qualitätsnormen von Pappeln der Sektionen Tacamahaca und Leuce. **Studies on quality standards of poplars of the sections Tacamahaca and Leuce.**

4993 Holmsgaard, E.   DK 020001/79/5041 N
Udredning af mulighederne for dyrkning af poppel og andre hurtigtvoksende løvtræarter i Danmark. **Evaluation of the possibilities of growing poplar and other fast growing hardwood trees in Denmark.**

4994 Ferrari, G. Curro, P.; Scaramuzzi, G.   IT 011801/60/0002
Qualificazione del legno dei pioppi di interesse colturale. **Wood quality assessment of poplars of cultivation interest.** Publications.

4995 Avanzo, E.   IT 011801/68/0001
Relazione tra distanze d'impianto, turni ed assortimenti in Populus alba e Populus nigra. **Spacing/rotation/sorting relationship in Populus alba and P.nigra.**

4996 Avanzo, E.   IT 011801/70/0001
Prove di governo a ceduo del pioppo. **Coppice trials with poplars.** Publications.

4997 Ferrari, G.; Scaramuzzi, G.   IT 011801/72/0003
Indagini sulle possibilità di valutazione delle caratteristiche del legno di piante di pioppi in piedi. **Wood qualities assessment of standing poplars.** Publications.

4998 Ferrari, G.; Scaramuzzi, G.   IT 011801/73/0001
Indagine sull'influenza delle tecniche colturali sulle caratteristiche del legno nei pioppi. **Investigations on the effects of cultural techniques on wood characteristics in poplars.**

4999 Avanzo, E.   IT 011801/73/0003
Prove d'introduzione di alcuni pioppi esotici (P.trichocarpa, P.maximowiczii) nell'Italia centro–meridionale. **Introduction trials of certain exotic poplars (P.trichocarpa, P.maximowiczii) in central and southern Italy.**

5000 Rambelli, A.; Riess, S.   IT 041607/78/0001
Ricerche sulle micorrize delle betulle nel Parco Nazionale d'Abruzzo. **Researches on birch mycorrhizas in Abruzzo National Park.**

5001 Dissen, H.D.   NL 010601/50/2382
Onderzoek naar de bruikbaarheid van Tacamahaca en

Aigeiros klonen van populier op verschillende groeiplaatsen. **Research on the suitability of Tacamahaca and Aigeiros clones of poplar for different sites.** Publications.

5002 Burg, J. van den   NL 010601/66/2373
De relatie tussen groeiplaats en groei van Aigeiros populieren. **Relationship between site and growth of Aegeiros poplars.**

5003 Faber, P.J.   NL 010601/66/2413
Groeiruimte–onderzoek van populier. **Research on the optimum growing space of poplar.**

5004 Dissen, H.D.   NL 010601/71/3531
Houtteeltkundig onderzoek van nieuwe populierenklonen. **Silvicultural research of new poplar clones.**

5005 Burg, J. van den   NL 010601/76/7787
Onderzoek naar de relatie tussen groeiplaats en groei van de "Robusta" populier. **Research on growth–site relationships of "Robusta" poplar.**

5006 Broekhuizen, J.T.M. van   NL 020021/56/4806
Groei en morfologische aspecten van populiererasson in verband met bodem en klimast. **Growth and morphological aspects of poplar clones in relation to soil and climatological factors.**

5007 Glastra, T.F.   NL 040007/66/2413
Groeiruimte–onderzoek van populier. **Research on the optimum growing space of poplar.** Publications.

Other leafwoods (B 3829)

See also 4989, 5992

5008 Weisgerber, H.   DE 912000/75/0001
Provenienzversuch von europäischen und nordamerikanischen Aspen. **Provenance experiments with European and North American aspen.** Publications.

5009 Stott; Jefferies   GB 011510/00/0018 R
Improve the culture of basket, cricket bat, tree and ornamental willows.

5010 Pirazzi, R.   IT 011801/74/0009
Studio morfologico dei noduli radicali e del simbionte azotofissatore in Alnus cordata. **Morphological study of root nodules and nitrogen–fixing symbiont in Alnus cordata.** Publications.

5011 Gemignani, G.   IT 011801/75/0004
Ricerche sull'influenza della distanza d'impianto sulla produzione di Alnus cordata e Platanus orientalis. **Trials on the influence of spacing on Alnus cordata and Platanus orientalis productivity.**

5012 Magini, E.   IT 040516/78/1072 N
Studi sulle possibilità di riconversione di cedui di castagno in castagneti da frutto. **Feasibility studies on the conversion of chestnut tree coppices back to fruit producing woods.**

5013 Palenzona, M.   IT 120100/77/0003
Programma di studio ed interventi per l'utilizzazione, la rigenerazione e la trasformazione delle foreste castagni Piemontesi. **Research and operative programme for the utilization, regeneration and transformation of chestnut forests**

D 2100 – Plant production general and crop husbandry

of Piedmont. (Italy).

Other forests (B 3890)

See also 1909

5014 Atanasiu, N.; Thiagalingam, K.　　DE 129552/75/0003
Untersuchungen über Blattanalyse – Grenzwerte bei Ölpalmen
im Jugendstadium. **Studies on leaf analysis – Critical values of
oil palm during the early growth period.**

5015 Curro, P.; Scaramuzzi, G.; Ferrari, G.
　　　　　　　　　　　　　　　IT 011801/60/0001
Qualificazione del legno degli eucalitti di interesse colturale.
Wood quality assessment of eucalyptus of cultivation interest.
Publications.

5016 Curro, P.　　　　　　　　IT 011801/61/0001
Indagini sulle possibilità di riduzione delle deformazioni
longitudinali dovute alle tensioni di accrescimento nei segati di
eucalitti. **Investigations on the possibility of reducing
longitudinal deformations due to growth–stresses in eucalyptus
sawn timber.** Publications.

5017 Gemignani, G.　　　　　　IT 011801/69/0001
Ricerche sull'influenza della distanza d'impianto sulla
produzione degli eucalitti. **Trials on the influence of spacing on
eucalyptus productivity.**

5018 Gemignani, G.　　　　　　IT 011801/69/0002
Ricerche sulla adattabilità degli eucalitti resistenti al freddo.
**Investigations on the adaptability of cold – resistant
eucalyptus.** Publications.

5019 Gemignani, G.　　　　　　IT 011801/70/0002
Ricerche sull'influenza delle lavorazioni prima dell'impianto,
sulla resistenza al vento e sullo sviluppo degli eucalitti.
**Investigations on soil preparation techniques for planting, on
windresistance and growth of eucalyptus.**

5020 Lubrano, L.　　　　　　　IT 011801/76/0015
Ricerche sulla propagazione per talea di Eucalyptus trabutii ed
E.dalrympleana. **Investigations on Eucalyptus trabutii and
E.dalrympleana propagation by cutting.**

5021 Ferrari, G.; Scaramuzzi, G.　　IT 011801/77/0001
Indagini sulle possibilità di valutazione delle caratteristiche del
legno di Eucalyptus in piedi. **Wood qualities assessment of
standing trees in Eucalyptus.**

5022 Ferrari, G.; Scaramuzzi, G.　　IT 011801/77/0002
Indagine sull'influenza delle tecniche colturali sulle
caratteristiche del legno in Eucalyptus globulus ed E. Trabutii.
**Investigations on the effects of cultural techniques on wood
characteristics in Eucalyptus globulus and E. Trabutii.**

5023 Ferrari, G.; Scaramuzzi, G.　　IT 011801/77/0003
Indagine preliminare sulla variabilità individuale e di stazione
delle caratteristiche del legno in Eucalyptus globulus ed E.
Trabutii. **Preliminary investigation on individual and site
variations of wood characteristics in Eucalyptus globulus and E.
Trabutii.**

Stimulant crops (B 3910)

See also 4278, 5044

5024 Thomas; Farrar　　　　　GB 012201/00/0002 R
**Effect of plant spacing, height of wirework and time of picking
on yield and alpha acid production.**

5025 Neve　　　　　　　　　　GB 012201/00/0022 R
**Organisation and recording of farm trials testing new
developments.**

5026 Researcher not indicated　　GB 050106/00/0001 R
Hops, production techniques and variety trials.

Spice and seasoning plants of warm climates (B 3920)

See also 5035

5027 Hohmann, B.　　　　　　　DE 135055/77/0003
Zur mikroskopischen Diagnostik von Blättern verschiedener
Gewürze liefernder Pflanzen. **Microscopic diagnosis of leaves of
different spices supplying plants.**

5028 Fritz, D.; Freytag, W.; Franz, C.　　DE 161260/77/0006
Ertrag und Alkaloidgehalt des Schöllkrauts in Abhängigkeit
von Ökotyp, Düngung, Anbau– und Erntetermin. **Yield and
alkaloid contents of greater celandine as dependent on ecotype,
fertilization, and seeding and harvesting date.**

5029 Jansen, P.C.M.　　　　　　NL 020044/74/8625 N
Kruiden, specerijen en medicinale planten van Ethiopië, hun
taxonomie en landbouwkundige betekenis. **Spices, condiments
and medicinal plants in Ethiopia, their taxonomy and
agricultural significance.**

5030 Waard, P.W.F. de; Dijck, J.I. van　　NL 040012/72/5522
Oculatie en verenting van peper (Piper nigrum). **Grafting and
budgrafting of pepper (Piper nigrum).**

Spice and seasoning plants of temperate climates (B 3930)

See also 5027

5031 Maton, A.; Pieters, M.　　　BE 070300/64/0021 R
Studie van niet bewerkte velden, de oogst, windschermen en
nieuwe veldstrukturen, van de hop. **Research on
non–cultivated field, the hopharvest wind screens, and new
hopfieldstructures.** Publications.

5032 Bockstaele, L.; Maddens, K.; Ampe, G.; Derolez, J.
　　　　　　　　　　　　　　　BE 140000/72/0023 R
Rassenstudie, fytopathologie van hop. **Study of races and
phytopathological aspects of hop.** Publications.

5033 Fritz, D.; Franz, C.; Schröder, F.–J.　DE 161260/75/0013
Untersuchung über den Einfluss genetischer und ökologischer
Faktoren auf den Sekundärstoffwechsel und Ertrag von
Matricaria chamomilla und Silybum marianum. **Investigations
of the influence of genetic and ecological factors on the
secondary metabolism and on the yield of Matricaria
chamomilla and Silybum marianum.**

5034 Fritz, D.; Schröder, F.–J.　　　DE 161260/75/0014
Schwankungen des Gehaltes und der Zusammensetzung des
ätherischen Öles in Einzelpflanzen von drei
Kamillenherkünften – Matricaria chamomilla L. –. **Variations
in content and composition of the essential oils in single plants of**

three cultivars of camomile – Matricaria chamomilla L. –.

5035 Fritz, D.; Franz, C.   DE 161260/75/0015
Abhängigkeit der Qualität von Gewürzkräutern von Standort und Aufbereitung. **Quality of spice herbs depending on location and processing.**

5036 Kohlmann, J.; Christl, J.   DE 502059/70/0001
Der Einfluss verschiedener Anbaumethoden des Hopfens auf Bodenstruktur, Ertrag und Qualität 1968. **Influence of different methods of hops growing on soil structure, crop yield and quality.**

5037 Kohlmann, J.   DE 502059/71/0001 R
Untersuchung über den Standraum beim Hopfen in Hinblick auf Ertrag, Qualität und Arbeitsaufwand 1971. **Studies on the growing space of hops with regard to crop yield, quality and working input.**

5038 Maier, J.   DE 502059/74/0003
Untersuchungen zur Sortenbestimmung bei Hopfen nach chemisch erfassbaren Merkmalen, insbesondere in Sortenmischungen. **Studies on determining hop varieties esp. in variety mixtures, by chemical characteristics.**

Perfume plants (B 3940)

See also 4555

5039 Blanc, D. Mme; Bellenand– Mayeur, P.
     FR 010502/72/0515
Etude du lavandin abrial. **Study on abrial lavander.**

5040 Gras, R.; Bellenand–mayeur, P.; Moulinier, H.
     FR 010502/73/6002
Etude du fonctionnement d'un peuplement effets du milieu et des techniques culturales sur le déperissement du lavandin. **Effects of medium and growing techniques on lavender decay.**

5041 Gilly, G.   FR 010502/73/6004
Amélioration des techniques de production des plantes à parfum. **Improvment of fragans and condiment plants cultivation.**

5042 Marocke, R.   FR 010901/70/0583
Houblon. Alsace. Bourgogne. Variétés nouvelles. Expérimentation champ. Croissance et développement. Alimentation. Production végétale. Qualité technologique. **Hop. Alsace. Burgundy. New varieties. Field experimentation. Growth and development. Nutrition. Yield. Technological quality.**

5043 Chavagnat, A.; Chesneaux, M.T.   FR 011103/75/0429
Germination des semences de lavandula. **The germination of lavandula seeds.**

Rubber, gum, wax and resin plants (B 3950)

5044 Giesberger, G.   NL 040012/73/5520
Weefselcultuur van rubber (Hevea brasiliensis), cacao en koffie. **Tissue culture of rubber (Hevea brasiliensis), cacao and coffee.**

Drugs and medicine plants (B 3970)

See also 3791, 4063, 5029, 6002, 7728

5045 Bezzi, A.; Orlandi, D.   IT 021800/77/0002
Ricerche sulla coltivazione di piante officinali in ambiente alpino. **Research project on officinal plant cultivation at Alpine place.**

5046 Cammilli, A.   IT 022300/76/0005 N
Cimatura meccanica del Tabacco Bright. **Mechanical topping on Bright tobacco.**

5047 Cammilli, A.   IT 022300/77/0675 R
Meccanizzazione della raccolta del tabacco, cantieri raccolta aspetti agronomici. **Mechanization of tobacco harvesting, harvesting yards, agronomic aspects.**

5048 Liguori, O.; Blago, A.   IT 022300/79/0003 N
Influenza della densità di investimento e della concimazione azotata su tabacco Erzegovina. **Spacing and nitrogen fertilization influence on the Erzegovina tobacco.**

5049 Vardabasso, A.; Carotenuto, R.; Tremola, M.G.
     IT 022300/79/0008 N
Prove propedeutiche di raccolta meccanizzata influenza della concimazione azotata, delle distanze di trapianto e della cimatura sulle caratteristiche quali–quantitative del tabacco Burley. **Mechanical harvesting propaedeutic trials: effects of nitrogen fertilization spacing and topping on quali–quantitative characteristics of Burley tobacco.**

5050 Postiglione, L.; Cuocolo, L.; Mucci, F.
     IT 040701/79/0006 N
Prove di investimento su tabacchi orientali. **Effect of plant population on oriental tobacco.**

Other crops (B 3990)

5051 Rehm, S.; Sieverding, E.G.   DE 132240/75/0002
Die Rolle der vesikulär–arbuskulären Mykorrhiza für den Wasserhaushalt tropischer Kulturpflanzen. **The role vesicular–arbuscular mycorrhiza for the water balance of tropical crops.**

5052 Sioli, H.; Esteves, F.   DE 404010/75/0001
Primärproduktion aquatischer Makrophyten. **Primary production of aquatic macrophytes.**

5053 Harper, F.   GB 060106/00/0026 R
New crops–production and conservation techniques.

5054 Blaak, G.   NL 040012/78/8941 N
Vegetatie vermeerdering en stekelstudies in Bactris gasipaes H.B.K.. **Vegetative propagation and spine studies on Bactris gasipaes H.B.K..**

Plants for experimental purposes (B 9130)

See also 2700

D 2200 – Plant nutrition and fertilization

See also 227, 239, 248, 285, 292, 618, 662, 675, 780, 2634

5055 Liard, O.   BE 010002/58/0006
Etude technologique des problèmes relatifs à l'éclaircissage des fruits. **Technological study of the problems relative to fruit thinning.** Publications.

5056 Ferauge, M.; Trzcinski, M. BE 010002/60/0003
Etude de la nutrition des arbres fruitiers. **Study of fruit tree nutrition.** Publications.

5057 Laloux, R.; Dohet, J.; Falisse, A.; Nijst, P.; Poelaert, J.
BE 010008/76/0004
Essais de fertilisation de longue durée concernant les éléments N, P et K. **Long–term trial with fertilizers concerning N, P and K.** Publications.

5058 Hanotiaux, G.; Heck, J.P.; Marlier–Geets, O.
BE 010016/69/0002 R
Dynamique des éléments nutritifs dans les sols, en particulier l'étude des phénomènes d'absorption et de désorption des ions et l'étude du comportement des végétaux vis-à-vis des formes chimiques des éléments nutritifs. **Dynamics of soils nutrients particularly the study of ions adsorption and desorption phenomena and the study of plant behaviour towards chemical forms of soil nutrients.** Publications.

5059 André, P.; Lheureux, C.; Giot–Wirgot, P.; Bailleux, P.; Khairia el Sayed; Abu el Khair BE 020301/69/0005
Influence de la fertilisation sur la croissance de l'épicéa (Picea abies Karst). **Influence of fertilization on the growth of Picea (Picea abis Karst).** Publications.

5060 Lambert, J.; Toussaint, B. BE 020602/77/0016 R
Utilisation de certains produits fatals de l'industrie comme engrais ou amendements applicables en prairie. **The use of some by–products of industry as fertilizers or improvements of grassland.** Publications.

5061 Livens, J.; Vlassak, K.; Verstraeten, L.; Delaere, L.
BE 040201/73/0005 R
Omzettingen van traagwerkende N–meststoffen in de bodem en hun effect op gewassen (tarwe, grassen, rijst...). **Transformations of slow release N–fertilizers in soil and their effects on crops (wheat, grasses, rice).** Publications.

5062 Van Assche, C. BE 040203/77/0021 R
Fytotechnische mogelijkheden van het gebruik van huisvuilcompost in land- en tuinbouw. **Phytotechnical possibilities of the input of domestic waste compost in agriculture en horticulture.** Publications.

5063 Droeven, G.; Darcheville, M. BE 080100/73/0018 R
Etude de la fumure phosphopotassique bloquée pour plusieurs années. **Study of the periodic application of the fertilizer requirements.**

5064 Darcheville, M.; Crohain, A. BE 080100/75/0012
Etude de la fumure phosphopotassique optimale dans différentes régions agricoles du pays. **Study of the optimal P–K manuring in different agricultural regions of the country.**

5065 Rixhon, L.; Crohain, A. BE 080800/70/0001
Etude de la fumure minérale des plantes de grande culture. **Study of field crops fertilizing.** Publications.

5066 Süss, A.; Beck, T. DE 502055/73/0015
Beeinflussung der Mikroflora und ihrer Umsetzungsaktivität bei verschiedenen Böden nach Ausbringung von unbehandeltem, hitzebehandeltem und bestrahltem Klärschlamm. **Effects on microflora and its mineralization activity in different soils by the application of untreated, heat treated and irradiated sewage sludge.**

5067 Sørensen, C.; Mortensen, J.V. DK 010117/76/0008 R
Kvælstofsammensætningen i ensilage– og pressesaft fra planter (del af EF–projekt). **Nitrogen composition in silage effluent and extracted plant sap (part of EEC project).**

5068 Olesen, S.E. DK 010900/73/9014 N
Slamdeponering i nåletræsplantager med sigte på belysning af gødnings– og giftvirkning samt risiko for grundvandsforurening. **Sludge deposition in conifer plantations with reference to the eludicidation of fertilizer and toxic effects and the risk of groundwater pollution.**

5069 Mogensen, V.O. DK 030143/75/0011
Studier over atmosfærens CO_2 koncentration. **Studies of the atmospheric CO_2 concentration.**

5070 Mogensen, V.O. DK 030143/76/0012
Optimale vandingstidspunkter for byg ved tilførsel af begrænsede vandmængder. **Optimal irrigation times for barley with the application of limited water quantities.**

5071 Sørensen, H. DK 030146/75/0018
Mineraliseringen af organisk bundet kvælstof. **Mineralisation of organic nitrogen.**

5072 Pinstrup–Andersen, P. DK 030149/78/0001 N
Plantenæringsstoftilførslens betydning for fødevareforsyningen. **The importance of fertilizer in food production.**

5073 Moltesen, P.; Olesen, P.O. DK 030181/74/0019
Gødskningens indvirkning på vedkvaliteten. **Influence of fertilizers on wood quality.**

5074 Hughes GB 012105/00/0012 R
Grazing studies: animal/pasture relationships of contrasting species and varieties.

5075 Researcher not indicated GB 040402/00/0011
Evaluation of local dolomite as a source of magnesium and as a liming material.

5076 Dickson GB 040402/00/0016
Studies on the availability of nitrogen for tree growth on different soil types.

5077 Gracey GB 041101/79/0030 N
Comparison of insoluble phosphate fertilisers.

5078 McLaren GB 060115/00/0002
Soil and plant cobalt and copper (partly in relation to soil series).

5079 Downey, N.E.; Moore, J.F. IE 060100/75/1174 R
Investigation into the possibility that spreading slurry may increase with helminth parasites of livestock. Publications.

5080 Ryan, M.J. IE 060200/67/0413 R
Soil productivity experiment: response to lime NPK on grass on different soils. Publications.

5081 Murphy, W.E. IE 060200/71/0402 R
Fertilizer use survey. Publications.

5082 Tunney, H.; Murphy, W.E. IE 060200/72/0390
Animal manures. Production, storage and disposal.
Publications.

5083 O'Nuallain, T.; Brophy, P.O.; Caffrey, P.J.
IE 120207/70/9226 N
Beef production from spring born calves using an intensive g
razing system. **Field studies on the epidemiology of ostertag iasis
and dictyocaulosis using a leader follower system.**

Banks, shores, dikes and their vegetation (B 1530)

5084 Dubois, J. P. FR 012218/77/5566
Fixation du phosphore par le roseau. **Phosphorus uptake by
caires.**

Man–made recreational resources (B 1600)

5085 Boeker, P.; Hiller, H. DE 105250/78/0003 N
III.Rasendüngungsversuch der Deutschen Rasengesellschaft
e.V., Bonn. **III.Lawn fertilization experiment of Deutsche
Rasengesellschaft e.V. in Bonn.** Publications.

Parks, gardens, urban greenspaces, plantations (B 1610)

See also 4608, 5925, 5991

Sportfields, play and camping grounds (B 1630)

5086 Skirde, W. DE 129220/77/0003
Verwertung von Siedlungsabfällen im Grünflächen– und
Sportplatzbau. **Utilization of domestic refuse in constriction of
green areas and sports–grounds.** Publications.

5087 Riem Vis, F. NL 010103/74/5591
Stikstofbemesting van grassportvelden; hoeveelheid, verdeling
over het seizoen en meststofvorm. **Nitrogen fertilization of
sports fields; amounts, distribution over the season, and
fertilizer materials.**

5088 Riem Vis, F. NL 010103/74/5592
Onderzoek naar de optimale pH en de optimale fosfaat– en
kalivoorziening bij aanleg en onderhoud van grassportvelden.
**Search for the optimum pH and optimum phosphorus and
potassium supply desired for establishment and maintenance of
sports fields.** Publications.

Plants and parts of plants in general (B 2100)

See also 1087, 1422, 2659, 5064, 5231, 5951, 6032

5089 Massart, D.; François, M. BE 140000/76/0037 R
Opname– en translocatiemechanismen van sporenelementen
bij planten. **Uptake and translocationmechanism of trace
elements by plants.** Publications.

5090 Clijsters, H.; Van Assche, F. BE 140000/76/0038
Fotosynthese en –respiratie. **Photosynthesis and respiration.**
Publications.

5091 Döring, H.–W. DE 105202/71/0009
Einfluss der Mineralstoffversorgung auf pflanzliche
Inhaltsstoffe. **Influence of mineral nutrition on organic
compounds in plants.**

5092 Döring, H.–W.; Rathert, G.; Witt, J.
DE 105202/74/0002
Einfluss hoher NaCl–Konzentrationen auf
Kohlenhydratbildung und –verlagerung in unterschiedlich
salztoleranten Pflanzenarten. **Effect of high NaCl
concentrations on carbohydrate metabolism in plant species
with different salt tolerance.**

5093 Hentschel, G. DE 144400/70/0002
Einfluss unterschiedlicher Faktoren auf Aufnahme und
Transport von Harnstoff. **The influence of different nutritional
factors on the uptake and translocation of urea.**

5094 Willumsen, I. DK 010113/72/7003 N
Formulering og styring af næringsopløsningers kemiske
sammensætning ved dyrkning af planter i vandkultur.
**Formulation and control of the chemical composition in water
culture.**

5095 Andersen, A.S. DK 030104/73/0010
Roddannelse i stiklinger. II. Samspillet mellem miljø og
hormoner. **Root formation in cuttings. 2. Interaction between
the environment and hormones.**

5096 Rajagopal, V.; Andersen, A.S.; Rasmussen, S.;
Josefsen, A. DK 030104/77/0014 R
Roddannelse i stiklinger. III. Indflydelse af induceret
vandstress på roddannelsen. **Root–formation in cuttings. 3. The
influence of induced water stress on root–formation.**

5097 Poulsen, A.; Andersen, A.S.; Eriksen, E.N.
DK 030104/77/0015 R
Roddannelse i stiklinger. IV. Indstrålingens indflydelse på
roddannelse i Hedera helix. **Root formation in cuttings. 4. The
influence of direct sun light on root formation in Hedera helix.**

5098 Sørensen, H. DK 030106/76/0016
Katabolisme af glucosinolater i planter. **Catabolism of
glucosinolates in plants.**

5099 Nielsen, N.E. DK 030148/75/0012
Formulering og styring af næringsstofopløsningers kemiske
sammensætning ved dyrkning af planter i vandkultur.
**Formulation and control of the chemical composition of
nutrient solutions for plant production in water culture.**

5100 Karlsen, P. DK 030171/74/0003
Formulering og styring af næringsstofopløsningers kemiske
sammensætning ved dyrkning af planter i vandkultur.
**Formulation and control of the chemical composition of
nutrient solutions for plant production in water culture.**

5101 Blanc, D. Mme; Otto, C.Mme FR 010502/73/6016
Influence de la nutrition azotée sur la composition en
oligoéléments du végétal. **Effects of nitrogen supplies on
composition of plants in trace elements.**

5102 Blanc, D. Mme; Otto, C. Mme FR 010502/76/6014
Influence de l'absorption de l'ion ammonium sur la
composition en oligo–éléments des végétaux. **The effect of
ammonium absorption on the mimor elements content of plant.**

5103 Adamowicz, S. FR 010502/76/6015
Relations de la nutrition azotée avec les facteurs climatiques et
nutritionnels. **Nitrogen nutrition connections with climate and
mineral nutrition.**

5104 Morizet, J.   FR 010802/70/0572
Absorption de l'eau par la plante. **Water uptake by plant.**
Publications.

5105 Morizet, J.   FR 010802/70/6112
Transferts et bilans de l'eau dans la plante en fonction des
conditions de nutrition minérale. **Water transport and balance
in the plant. Effects of mineral nutrition.**

5106 Morizet, J.   FR 010802/76/6113
Résistance racinaire au transfert de l'eau. **Root resistance to
water transport.**

5107 Duthion, C.   FR 011001/71/6141
Réactions des plantes aux excès d'eau. **Effect of excess moisture
on plant behaviour.**

5108 Maertens, C.   FR 011401/70/0629
Etude du rôle des racines dans l'alimentation minérale et
hydrique des plantes, par l'influence des conditions du milieu
sur les possibilités physiologiques d'absorption. **Roots in
mineral and water plant nutrition. Effects of environmental
factors on physiological capacities for absorption of the
elements.** Publications.

5109 Blanchet, R.; Bosc, M.   FR 011401/70/0635
Etude des processus de l'alimentation des plantes dans le sol à
partir des systèmes racinaires et de la rhizosphère. **Studies on
the process of plant nutrition in the soil by roots and
rhizosphere.** Publications.

5110 Blanchet, R.; Bosc, M.   FR 011401/72/0636
Nutrition minérale et valorisation de l'eau à travers les
relations sol–plante–atmosphère. **Mineral nutrition and
improvment of water use efficiency through
soil–plant–atmosphere relationships.**

5111 Maertens, C.   FR 011401/76/6191
Détermination des profils de fonctionnement racinaire.
Determination of the profile of the roots system work.

5112 Ferrari, G.; Cocco, G.; Passera, C.; Renosto, F.;
Maggioni, A.   IT 040802/74/0001
Fattori genetici e ambientali nella regolazione dell'
assorbimento nutrizionale dei vegetali. **Genetic and
environmental factors affecting nutrient uptake efficiency by
plant roots.** Publications.

5113 Veri, G.   IT 060200/77/0985 R
Assorbimento ionico. **Ionic absorption.**

5114 Wiersum, L.K.   NL 010103/72/3561
Bepaling van het zuurstofverbruik van een wortelstelsel
gedurende zijn ontwikkeling. **Measurement of the oxygen
consumption of a developing root system.**

5115 Wiersum, L.K.   NL 010103/74/5588
Meting van doorvoersnelheden van mineralen (ion flux) en
water aan het worteloppervlak. **Measurement of the flux of ions
and water into the root at its surface.**

Plant communities as ecological systems (B 2200)

See also 5577, 5950

5116 Kristensen, K.J.   DK 030143/69/0008

Planteproduktion og aktuel fordampning i relation til
vegetationstæthed og til jordens udtørringsgrad. **Plant
production and actual evaporation in relation to vegetation
density and soil moisture content.**

5117 Jensen, H.E.   DK 030143/73/0006
Planteproduktion og miljø. Potentiel planteproduktion,
energi–, vand– og kvælstofbalance. **Plant production and
environment. Potential plant production and energy, water and
nitrogen balances.**

5118 Willumsen, J.; Hansen, M.; Karlsen, P.; Knoblauch, F.;
Nielsen, N.E.; Frits–Nielsen, B.   DK 030143/75/0003 R
Formulering og styring af næringsstofopløsningers kemiske
sammensætning ved dyrkning af planter i vandkultur.
**Formulation and control of the chemical composition of
nutrient solutions for plant production in water culture.**

5119 Jensen, C.R.   DK 030143/77/0005
Planteproduktion i relation til jordvandets osmotiske potential
og trykpotential. **Plant production in relation to the osmotic
potential and pressure potential of soil water.**

Crops in general (B 3000)

See also 17, 121, 122, 123, 153, 176, 183, 224, 237, 311, 319,
333, 337, 338, 342, 444, 448, 450, 468, 472, 502, 506, 523, 524,
528, 539, 605, 609, 706, 716, 753, 778, 801, 1339, 1420, 1433,
1540, 2694, 2727, 2737, 2742, 2745, 2796, 2806, 2837, 2860,
2871, 2873, 2877, 2881, 5058, 5062, 5065, 5067, 5072, 5077,
5115, 10349, 16351, 17061, 17080, 17086, 17088, 19710

5120 Darcheville, M.; Destain, J.P.   BE 080100/67/0003
Etude de la relation entre les fumures minérales généralement
utilisées et les rendements des cultures en région limoneuse.
**Study of the mineral fertilizers level generally used in the
loamy region as related to the crop yields.** Publications.

5121 Van Ruymbeke, M.; Baert, L.; Piot, R.; Boon, R.;
Deventer, J.; Biermans, V.   BE 140000/77/0007 R
Onderzoek naar de verbetering van de bemestingsadviezen in
welbepaalde streken. **Research to the improvement of
fertilization advices in well defined regions.** Publications.

5122 Krzysch, G.   DE 105201/77/0013
Pflanzennährstoffe in Niederschlägen. **Plant nutrients in
precipitations.**

5123 Döring, H.–W.; Alexander, A.; Plarre, W.
                    DE 105202/77/0003
Auswirkungen inverser K/Na–Verhältnisse bei konstanter
K/NaGesamtkonzentration auf Wachstum, Chlorophyllgehalt
und auf Teilbereiche des Kohlenhydratstoffwechsels sowie auf
Enzymaktivitäten. **Effects of inverse K/Na ratio and constant
K/Na total concentration on growth, chlorophyll content and on
parts of the carbohydrate metabolism as well as on enzyme
activities.**

5124 Döring, H.–W.; Rathert, G.; Witt, J. DE 105202/77/0004
Einfluss des K/Na–Verhältnisses sowie des Begleitanions – Cl–
und SO4– – bei hohen Substrat-Salzgehalten auf den
Kohlenhydratstoffwechsel bei unterschiedlich salztoleranten
Pflanzenarten. **Effect of K/Na–ratio and the counter anion Cl-
and SO4 2– at high substrats salinity on carbohydrate
metabolism in plant species of different salt tolerance.**
Publications.

5125 Hecht–Buchholz, C. DE 105202/77/0006
Einfluss der Mineralstoffernährung auf die Feinstruktur von Pflanzen. **Influence of mineral nutrition on fine structure of plants.**

5126 Bussler, W.; Anversa, M. DE 105202/78/0001 N
Diagnose mehrfacher Ernährungsstörungen durch Ca– und B–Mangel bzw. –Überschuss. **Diagnosis of multiplied deficiency resp. toxicity diseases according to B and Ca.**

5127 Bussler, W.; Fischer, G. DE 105202/78/0002 N
Diagnose mehrfacher Ernährungsstörungen durch N– und P–Mangel bzw. –Überschuss. **Diagnosis of multiplied deficiency resp. toxicity diseases according to N and P.**

5128 Rahimi, A. DE 105202/78/0003 N
Der Einfluss unterschiedlicher Zn–Gaben auf die Entwicklung von höheren Pflanzen sowie die Aktivität der Carboanhydrase. **Influence of different Zn amounts on the growth of higher plants as well as on the acitivity of carboanhydrase.** Publications.

5129 Bussler, W.; Leseberg, S. DE 105202/78/0006 N
Diagnose von Stickstoffmangel mit Hilfe von Mikrosymptomen und Schnelltests. **Diagnosis of N–deficiency by microsymptoms and quick tests.**

5130 Hampe, T. DE 105202/78/0007 N
Einfluss von Salz– und Trockenstress auf unterschiedlich salzresistente Pflanzenarten. **Effect of salt– and drought–stress on plant species differing in salt resistance.**

5131 Sommer, K. DE 111100/72/0004
Pflanzenernährung auf sauren, humiden Tropenböden. **Nutrition of plants on acid tropic soils.**

5132 Sommer, K.; Ndoreyaho, V. DE 111100/73/0003
Untersuchungen zur Versorgung von Pflanzen mit notwendigen Nährstoffen auf humid–tropischen Böden der Bong Mining Company, Liberia, Westafrika. **Investigations into the supply of plants with necessary nutrients in humid–tropical soils of the "Bong Mining Company" in Liberia, West Africa.**

5133 Kick, H.; Bagdadhi, N. DE 111100/75/0006
Wirkung der Phosphatdüngung bei sehr hohen Kalkgehalten des Bodens. **The effect of phosphate fertilization on soil with very high calcium carbonate contents.**

5134 Sauerbeck, D.; Allard, J.L. DE 111101/78/0002 N
Stoffproduktion und Stoffumsatz von Pflanzenwurzeln unter dem Einfluss verschiedener Umweltfaktoren. **Production and turnover of organic substances by plant roots under the influence of different environmental factors.**

5135 Sauerbeck, D.; Nonnen, S. DE 111101/78/0003 N
Assimilatebedarf und –umsatz von Pflanzenwurzeln in Nährlösung. **Photosynthate requirement and turnover of plant roots in nutrient solution.**

5136 Sauerbeck, D. DE 111101/78/0004 N
Einfluss von Pflanzenwurzeln auf dem Umsatz bodeneigener organischer Substanzen und Phosphorverbindungen. **Influence of plant roots on the turnover of soil organic matter and phosphorus compounds.**

5137 Boeker, P.; Opitz–von–Boberfeld, W. DE 111251/74/0009
Wie wirken sich verschiedene Phosphatformen in Abhängigkeit von der verabreichten Menge auf den P–Gehalt des Aufwuchses, den Ertrag und die Bestandsentwicklung aus unter Berücksichtigung von NP–Lösungen. **What about the effects of different forms of phosphate depending on the applied quantity on the P content of growing–on plants, on yield and growth of crop in consideration of NP–solutions.**

5138 Boeker, P.; Opitz–von–Boberfeld, W. DE 111251/75/0006
Vergleich von flüssigen – ANH–Lösung – und festen – Kalkammonsalpeter – N–Düngern in Abhängigkeit von Aufwandmenge und Applikationszeitpunkt. **Comparison of liquid – ANH solution – and solid – calcium ammonium nitrate – N–fertilizers in relation to expenditure and moment of application.**

5139 Heumann, W.; Kamberger, W.; Burkardt, H.J. DE 120101/77/0003
Die Wechselwirkung zwischen Rhizobium und Wirt. **The recognition between Rhizobia and host plant.**

5140 Homrighausen, E.; Eschraghi, I. DE 129020/73/0001
Untersuchungen auf dem Gebiet der Anwendung von Komposten aus Siedlungsabfällen. **Investigations on the application of urban waste compost.**

5141 Mengel, K.; Arnecke, W.–W. DE 129040/75/0005
Interaktion zwischen Wasserversorgung und Kaliumernährung und deren Einfluss auf Wachstumsrate und Ertragsbildung. **Interaction between water supply and potassium supply and its effect on growth rate and yield formation.**

5142 Mengel, K.; Drews, J.–U. DE 129040/75/0006
Die Aneignung von Kalium aus Zwischenschichten der Tonminerale in Abhängigkeit vom Energiestoffwechsel der Pflanze. **The absorption of potassium from interlayers of clay minerals in relation to the energy metabolism of the plant.**

5143 Mengel, K.; Behring, J. DE 129040/75/0007
Aufnahmemechanismus von Ammonium–Stickstoff. **Uptake mechanism of ammonium nitrogen.**

5144 Höfner, W.; Peters, U. DE 129044/72/0001
Speicherungsprozesse in Kulturpflanzen und deren Regulation 1972. **Storing processes and regulation in cultivated plants.**

5145 Alkämper, J.; Westphal, A.; Do–Van–Long; Henrichs, J.; Girefe, S. DE 129552/74/0002
Einfluss der Verunkrautung auf Nährstoffaufnahme und Ertragsbildung von Kulturpflanzen. **Effect of weed population on the nutrient uptake and yield production of cultivated plants.**

5146 Przemeck, E. DE 132061/73/0003
Veränderungen im Aktivitätsbestand einiger Enzyme des Aminosäurestoffwechsels im Wachstumsverlauf von Kulturpflanzen in Abhängigkeit von der Mn–Ernährung. **Alterations in the activities of some enzymes of the amino acid metabolism during the growth of plants and dependent on Mn–nutrition.**

5147 Timmermann, F.; Stille, M. DE 132063/78/0001 N

Ermittlung der Nährstofffrachten in einem landwirtschaftlich genutzten Wassereinzugsgebiet der südniedersächsischen Mittelgebirgslandschaft in Abhängigkeit von Düngung, Bodenart und Fruchtfolge. **Investigations to quantify the leaching of nutrients due to fertilization, soil type and crop rotation in the catchment area of a rural landscape of southern Lower Saxony.**

5148 Timmermann, F. DE 132065/78/0002 N
Untersuchungen zur Verbesserung der P–Düngewirkung durch Wurzel– und Blattapplikation von Nährstoffen. **Studies on improvement of P–fertilizing efficiency by root– and foliar application of nutrients.**

5149 Baeumer, K.; Capelle, A. DE 132181/75/0002
Verfügbarkeit und Aufnahme von Stickstoff durch Feldfrüchte im Ackerbausystem mit reduzierter Bodenbearbeitung. **Availability of nitrogen in soil and fertilizer and N–uptake by crops in a zero–tillage system.**

5150 Ehlers, W.; Ploeg, R.R.van–der DE 132181/77/0001
Computersimulation der Wasserbewegung in Ackerböden und der Wasseraufnahme durch Kulturpflanzen. **Computer simulation of water flow in agricultural soils and of water uptake by crops.**

5151 Moawad, M.; Blume, E.; Krone, W.; Mikhail, N.B.
DE 132240/73/0003
Rolle der Mykorrhiza für die Nährstoffversorgung tropischer Pflanzen. **The role of mycorrhiza for the nutrient supply of tropical crops.**

5152 Schürmann, B. DE 135051/71/0001
Untersuchungen zu Fragen der Bodenverbesserung und der Humusdüngung. **Investigations into problems of soil improvement and humus fertilizing.**

5153 Jungk, A. DE 138060/73/0001
Quantitative Charakteristika der Mineralstoffaufnahme von Pflanzenwurzeln. **Quantitative characteristics of the mineral uptake by plant roots.**

5154 Wehrmann, J.; Scharpf, C. DE 138060/74/0001
Kriterien für die Beurteilung der N–Versorgung von Kulturpflanzen. **Criteria for the evaluation of N supply in cultivated crops.**

5155 Bors, J.; Fendrik, I.; Pahlow, R. DE 138181/77/0004
Untersuchung der Wirkungsmechanismen der Fixierung von Luftstickstoff, insbesondere an Nichtleguminosen, und Optimierung dieser Prozesse. **Investigations on the mechanism of N–fixation in non–leguminous plants and possible optimization of processes involved.**

5156 Martin, P. DE 144400/70/0003
Verlagerung von Stickstoff in der Pflanze Einfluss von Mineralstoffen auf die Reduktion von NO3. **Distribution of nitrogen within the plant Influence of mineral nutrients on nitrate reduction.**

5157 Martin, P.; Schröder, M. DE 144400/75/0004
Stickstoff–Umlagerung in Pflanzen. **Retranslocation of nitrogen in plants.**

5158 Marschner, H.; Römheld, V. DE 144400/78/0002 N
Mechanismus der Eisenaufnahme bei verschiedenen Pflanzenarten. **Mechanism of iron uptake by different plant species.**

5159 Marschner, H.; Hentschel, G.; Martin, P.
DE 144400/78/0003 N
Möglichkeiten und Grenzen der Anwendung von Klärschlammkompost und Klärschlamm. **Possibilities and limits for the application of sludge and sludge–compost.**

5160 Kahnt, G.; Gutzmann, H. DE 144485/78/0004 N
PK–Optimierung für verschiedene Pflanzenarten, Fruchtfolgen und Ertragsniveaus. **Potassium– and phosphorus–optimum for different plants, crop rotations and yields.**

5161 Kahnt, G.; Gutzmann, H. DE 144485/78/0006 N
Flüssigmistaufbereitung und –anwendung in Fruchtfolgen und Mineralstoffausgleichsmassnahmen. **Preparation and application of liquid dung in different crop rotations and mineral fertilizer compensation.**

5162 Zeller, E.; Strauch, D. DE 144610/78/0004 N
Entstehung und Ausbreitung von Bakterienaerosolen bei der Verregnung von Gülle. **Formation and spreading of bacterial aerosols by sprinkler irrigation of semi–liquid manure.**

5163 Moser, E.; Ganzelmeier, H.; Göhlich, H.; Grossmann, F.; Koch, W. DE 144720/78/0003 N
Optimierung verschiedener Pflanzenschutz– und Düngeverfahren hinsichtlich der biologischen Wirkung, der Wirtschaftlichkeit und der Umweltbelastung. **Optimization of different methods of plant protection and fertilization regarding biological effect, profitability and impact on environment.** Publications.

5164 Finck, A.; Franck, E.von DE 148050/77/0001
Zinkversorgung landwirtschaftlicher Kulturpflanzen. **Zinc supply of crops.**

5165 Finck, A.; Bunje, G. DE 148050/77/0002
Kälteresistenz in Abhängigkeit von der SpurenelementErnährung der Pflanzen. **Cold tolerance in dependence on supply with trace elements.**

5166 Finck, A.; Pissarek, H.–P. DE 148052/70/0002
Möglichkeiten einer mikroskopischen Diagnose von Nährelementmangelerscheinungen. **Possibilities of microscopic diagnosis of nutrient deficiencies.**

5167 Schendel, U.; Schleich, C. DE 148550/74/0005 R
Kennzeichnende Grössen des Wasserhaushaltes für den Bewässerungsbedarf. **Ratio of water balance and irrigation requirement.** Publications.

5168 Amberger, A.; Gutser, R.; Schweiger, P.; Wünsch, A.
DE 161040/74/0001
Einfluss verschiedener mineralischer N–Düngung bzw. Müllklärschlammkompostgaben auf den vertikalen Mineralstoff– transport, gemessen an Saugkerzen – nach Czeratzki – in verschiedenen Tiefen. **The influence of different mineral nitrogen fertilization resp. using sewage sludge compost on the vertical mineral transport as measured with suction candles in different depths of soil according to the method of Czeratzki.**

5169 Amberger, A.; Amann, C. DE 161040/77/0001

D 2200 – Plant nutrition and fertilization

Phosphatmobilität unter dem Einfluss einer organischen Düngung. **Mobilization of phosphate by organic chelating agents.**

5170 Amberger, A.; Renneisen, C.   DE 161040/77/0002
Zn–Ernährung und N–Stoffwechsel in höheren Pflanzen. **Zn–nutrition and nitrogen metabolism in higher plants.**

5171 Fischbeck, G.; Knittel, H.   DE 161250/78/0002 N
Auswirkungen differenzierter Stickstoffdüngung – Menge und Zeitpunkt – bei unterschiedlicher Bodenbearbeitung. **Effects of differentiated nitrogen fertilization – quantity and time – after different soil cultivation.**

5172 Söchtig, H.   DE 201010/71/4023
Wirkung organischer Dünger – N–Lignin, aufbereiteter Tierkot – auf Wachstum und Ertrag von Pflanzen und auf die organische Substanz des Bodens. **Effects of organic fertilizers – N–lignin, treated animal faeces – on growth and crop yield of plants and on organic matter in soil.**

5173 Stühmeier, K.   DE 201010/77/0014
Ertragsgleichung zur Beschreibung von bei intensiver Nährstoffanwendung zu beobachtenden Ertragsdepressionen. **Yield equation for description of yield depressions in intensive application of nutrients.**

5174 Heinemeyer, O.; Draeger, S.; Jagnow, G.
          DE 201020/75/0005
Beeinflussung der mikrobiellen Stickstoffbindung durch Pestizide. **The influence of pesticides on the microbial fixation of molecular nitrogen.**

5175 Tietjen, C.   DE 201040/71/5001
Einfluss von Müllkompost und Stalldünger bei hoher Phosphatdüngung auf Pflanzenertrag und Humusdynamik. **Plant crop and humus dynamics as affected by rich phosphatic fertilizer in waste compost and manure.**

5176 Czeratzki, W.; Bramm, A.   DE 201040/74/0001
Neue Anbautechnologie zur Verminderung der Nitratauswaschung bei der pflanzlichen Produktion. **New technologies for the reduction of nitrate leaching in plant production systems.**

5177 Dambroth, M.   DE 201040/75/0024
Optimierung der Nährstoffzufuhr bei hohem Ertragsniveau von Kulturpflanzen unter besonderer Berücksichtigung der Ertragsstabilität. **Optimizing the nutrient supply of cultivated plants with a high level of output with special reference to regularity of yield.**

5178 Bramm, A.   DE 201040/78/0001 N
Ermittlung pflanzenphysiologischer Kennwerte für die Steuerung der Beregnung von Kulturpflanzen. **Determination of physiological criteria of plants for irrigation control of field crops.**

5179 Sommer, C.   DE 201040/78/0002 N
Untersuchungen über den Einfluss der Bodenwasserspannung in dem durchwurzelten Bodenraum auf Wachstum und Wasserverbrauch der Pflanzen. **Investigations on the influence of soil–water–potential with regard to growth and water consumption of cultural plants.**

5180 Bramm, A.   DE 201040/78/0003 N

Ertragsverhalten von Genotypen auf unterschiedliche Wasserversorgung und Bestimmung der Reaktionsnormen. **Yield and yield structure of genotypes with regard to water supply and determination of their reactions.**

5181 Bramm, A.   DE 201040/78/0004 N
Einfluss unterschiedlicher Bewässerungsverfahren auf die Ertragsbildung und Ertragsstruktur von Kulturpflanzen. **Influence of different irrigation systems on the yield and yield structure of field crops.**

5182 Bramm, A.   DE 201040/78/0005 N
Einsparung von Wasser durch Einsatz der Tropfbewässerung als Ersatz herkömmlicher Beregnungsverfahren. **Saving of water by use of drip irrigation instead of common irrigations systems.**

5183 Frank, H.K.; Bohling, H.; Hansen, H.; Overbeck, G.; Schwerdtfeger, E.; Wedler, A.   DE 211010/78/0001 N
Einfluss der Düngung auf verschiedene Qualitätskriterien bei Nahrungspflanzen. **Influence of fertilization on different quality criteria in food crops.**

5184 Schwerdtfeger, E.   DE 211010/78/0002 N
Die Aktivität der Nitratreduktase in Nahrungspflanzen; Verteilung in der Pflanze und thermische Stabilität. **The activity of nitrate reductase in food plants; distribution within the plant and thermic stability.**

5185 Kloke, A.   DE 215060/77/0004
Untersuchungen über die Aufnahme von Schadstoffen aus Industrie– und Siedlungsabwässern bzw. –schlämmen durch Nutzpflanzen. **Analyses on the uptake of noxious substances by useful plants from industrial and domestic waste waters resp. sludges.**

5186 Schiff, H.; Siegert, E.   DE 301030/75/0009 N
Agrarmeteorologische Probleme der künstlichen Feldberegnung unter Berücksichtigung technischer und wirtschaftlicher Aspekte. **Agrometeological problems of sprinkler irrigation of fields with reference to technical and economical aspects.**

5187 Diez, T.   DE 502053/73/0002
Langjähriger Einfluss der P–Düngerformen auf Boden und Pflanze auf Ackerland 1970. **The longstanding influence of P–fertilizers on soil and plants on arable lands .**

5188 Beck, T.   DE 502055/72/0008
Die Wirkung von Nitrifikationshemmern auf die N–Ernährung der Pflanzen. **The effect of nitrificides on N–uptake by plants.** Publications.

5189 Süss, A.; Schurmann, G.   DE 502055/73/0012
Erprobung einer Versuchsbestrahlungsanlage zur Entkeimung von Klärschlamm und Wirksamkeit verschieden behandelter Klärschlämme von Pflanze und Boden. **Testing of an experimental radiation facility for sterilizing sewage sludge and the effectivity of differently treated sewage sludges to plants and soil.**

5190 Süss, A.; Rosopulo, A.   DE 502055/73/0013
Aufnahme von Pflanzennährstoffen, Spurenelementen und Schadstoffen durch verschiedene Kulturpflanzen und Bestimmung der jeweiligen Ausnützungsgrade. **Uptake of nutrients, minor elements and pollutants by diverse cultivated**

D 2200 – Plant nutrition and fertilization

plants and determination of their specific utilization ability.

5191 Diez, T.   DE 502055/74/0010
Einsatz von flüssigen Mehrnährstoff- und
Spurenelementdüngern – Blattdünger – zur Ertragssteigerung.
Use of liquid fertilizers containing the main nutrients and trace-elements for increasing yield.

5192 Sommer, G.   DE 502055/78/0004 N
Schadstoffe in Siedlungsabfällen und ihre Wirkung auf das
Wachstum und die Gehalte in Kulturpflanzen. **Injurious substances in domestic waste and their influence on growth and content in cultivated plants.** Publications.

5193 Diez, T.; Rosopulo, A.; Wurzinger
                DE 502055/78/0005 N
Schwermetallaufnahme landwirtschaftlicher Kulturpflanzen.
Absorption of heavy metals by cultivated plants.

5194 Mackroth, K.   DE 506113/78/0004 N
Die Wirkung verschiedener Giesswasserqualitäten auf
verstopfungsempfindliche Tropfsysteme. **The influence of various water qualities to drip systems.**

5195 Kuntze, H.; Feige, W.; Pluquet   DE 507103/77/0001
Tolerierbare Schwermetallgehalte von Abwasserfaulschlamm
in Abhängigkeit variabler Bodeneigenschaften und
Pflanzenverträglichkeit. **Tolerable heavy metal contents of sewage sludge depending on changeable soil conditions and plant compatibility.**

5196 Vetter, H.; Steffens, G.   DE 507400/77/0001
Der Einfluss grosser Wirtschafts–Düngermengen auf die
Qualität des Grundwassers, des Oberflächenwassers und des
Drainagewassers sowie auf den Ertrag und die Ertragsqualität.
The influence of high amounts of manure on the quality of groundwater, surface water and drainage water and on yield and quality of crops.

5197 Vetter, H.; Schulte–im–Walde, W.   DE 507400/77/0002
Möglichkeiten zur Herabsetzung der Schwermetallgehalte von
Pflanzen in Belastungsgebieten und Wege zur Minderung ihrer
Schadwirkung. **Possibilities of reducing the content of heavy metals in plants in polluted areas and methods of diminishing their injurious effects.**

5198 Seibt, G.; Reemtsma, J.B.   DE 507651/78/0001 N
Auswirkungen von N– und NK–Düngung nach früherer Ca–
und CaP–Düngung in Neuenheerse und Oerrel. **Effects of fertilization with N and NK after previous fertilization with Ca and CaP in Neuenheerse and Oerrel.**

5199 Gudehus, H.–C.   DE 507701/78/0004 N
Entwicklung von Einheiten zur Regelung der
Düngerkonzentration bei Fliesskulturen über pH–Wert und
Leitfähigkeit. **Design of units for control of nutrient film systems, especially pH and conductivity level.**

5200 Krämer, F.   DE 508302/70/0004 R
Bodennährstoffgehalt und –verfügbarkeit von Löss–Rohböden
und ihre Wirkung auf die Entwicklung und den Ertrag
landwirtschaftlicher Kulturpflanzen 1959. **Nutrient content and availability in uncultivated loess soils and effects on the development and crop yield of cultivated plants.**

5201 Bahr, R.; Sunkel, R.; Krämer, F.; Wittkötter, U.

                DE 508302/70/0006 R
Lysimeteruntersuchungen: Auswirkungen verschiedener
Grundwasserverhältnisse nach Mittelwert und
Schwankungsamplitude auf Pflanzenentwicklung und
Pflanzenertrag Gesamtwasserhaushalt, insbesondere
Verdunstung und Grundwassererneuerung
Grundwasserkontamination mit Düngemitteln und
Pflanzenschutzmitteln 1968. **Examinations with lysimeter: effects of different groundwater conditions by medium value and varying amplitude on growth and yield of plants. Total water regime, especially evaporation and supply of groundwater. Contamination of groundwater with fertilizers and plant protectives.**

5202 Barmann, C.; Munk, H.   DE 901000/73/0001
Phosphatausnutzung von Kot und Schwemmist. **Utilization of phosphate in dung and liquid manure.**

5203 Beringer, H.; Koch, K.   DE 902000/70/0007
K–Ernährung und Assimilation von anorganischem N.
Potassium nutrition and assimilation of inorganic N.

5204 Behringer, H.; Nemeth, K.; Forster, H.
                DE 902000/77/0001
Verfügbarkeit von K nach Stroh– und Rübenblattdüngung. **K availability after manuring with straw and sugar beet leaves.**

5205 Beringer, H.; Grimme, H.; Nemeth, K.
                DE 902001/73/0001
Beziehungen zwischen Düngung, Nährionen der Bodenlösung
und Ertragsbildung unter Freilandbedingungen. **The relationship between fertilization, nutrient ions in the soil solution, and crop yield under field conditions.**

5206 Nemeth, K.; Rex, M.   DE 902001/77/0001
Die Bedeutung der Tiefe des durchwurzelbaren Raumes für
die Höhe des Nährstoffbedarfs im Boden. **Importance of rooting depth in relation of the critical nutrient level in soils.**

5207 Nemeth, K.   DE 902001/77/0002
Einfluss der Kalkung auf die K–, Mg– und P–Dynamik im
Boden. **Liming effects on the K–, Mg– and P–dynamics in soil.**

5208 Grimme, H.   DE 902001/77/0003
Magnesiumaufnahme der Pflanzen als Funktion von
Bodeneigenschaften und Beitrag des Magnesiums zur
Ertragsbildung. **Magnesium uptake in relation to soil properties and as a factor of yield formation.**

5209 Grimme, H.; Koch, K.   DE 902001/78/0001 N
Wasser– und Nährstofflieferung an die Pflanzenwurzel und
Nährstoffdynamik im durchwurzelten Raum. **Water and nutrient supply to the plant root and nutrient dynamics in the rhizosphere.**

5210 Forster, H.   DE 902003/78/0001 N
Auswirkungen verschiedener konstanter und unterbrochener
Mg–Ernährung auf Ertragsbildung und Kationenaufnahme.
Effect of different constant and interrupted Mg nutrition on yield formation and uptake of cations.

5211 Trolldenier, G.; Kraffczyk, I.   DE 902004/77/0002
Einfluss der Ernährung der Pflanze auf Wurzelausscheidungen
und Mikroorganismen der Rhizosphäre. **Influence of plant nutrition on root excretions and microorganisms of the rhizosphere.**

D 2200 – Plant nutrition and fertilization

5212 Trolldenier, G.; Rheinbaben, W.von
DE 902004/78/0001 N
Denitrifikation in der Rhizosphäre in Abhängigkeit vom
Ernährungszustand der Pflanze und Bodenfaktoren.
**Denitrification in the rhizosphere as related to the nutritional
status of the plant and to soil factors.**

5213 Werner, W.; Solle, A.
DE 903000/78/0006 N
Löslichkeit und Pflanzenverfügbarkeit von Phosphat aus
synthetischen Al–P–Verbindungen mit amorpher Struktur.
**Rate of solubility and plant–availability of phosphate from
amorphous synthetic Al–P–compounds.**

5214 Kofoed, A.D.
DK 010101/01/9401 N
Sammenligning af virkninger af staldgødning og kunstgødning i
fastliggende forsøg samt sædskifte og gødningstilførslens
indflydelse på jordens humusindhold. **Fertilizing value of
farmyard manure in comparison to commercial fertilizers in
permanent experiments and humus content influenced by
rotations of crops and fertilization.**

5215 Kofoed, A.D.; Fogh, H.T.
DK 010101/59/3006
Faste og flydende kvælstofgødningers indflydelse på
landbrugsplanternes udbytte og kvalitet samt drænvandets
næringsstofindhold og jordens kalktilstand. **Influence of solid
and fluid nitrogenous fertilizers on the yield and quality of
agricultural crops, nutrient contents in drainage water, and soil
lime status.**

5216 Kofoed, A.D.; Klausen, P.S.
DK 010101/65/3004
Fosfor-, kalium–, og magnesiumgødninger samt
mikronæringsstoffer og kalkningsmidlers indflydelse på
landbrugsafgrøders udbytte og kvalitet. **Influence of
phosphorus, calcium and magnesium fertilizers,
micronutrients, and limes on the yield and quality of
agricultural crops.**

5217 Kofoed, A.D.; Højmark, J.V.
DK 010101/71/3005
PK– og NPK–gødningers indflydelse på landbrugsafgrøders
udbytte og kvalitet. **Influence of PK and NPK fertilizers on the
yield and quality of agricultural crops.**

5218 Knudsen, H.; Gregersen, A.K.
DK 010109/78/0001 N
Sædskifte og akkumuleret vandingseffekt. **Crop rotation and
accumulated effect of irrigation.**

5219 Sørensen, C.
DK 010117/64/4610
Næringsstoftilførslens indflydelse på planternes
aminosyrekomposition og kemiske sammensætning iøvrigt.
**Influence of nutrient supply on the amino acid composition and
general chemical composition of plants.**

5220 Nielsen, J.D.
DK 010117/70/4601
Jordens indhold af K– og P–forbindelser og deres
plantetilgængelighed. **Soil contents of K and P compounds and
their availability to plants.**

5221 Lind, A.M.
DK 010117/71/4602
Nitratreduktion i de øvre jordlag. En undersøgelse af faktorer,
der medvirker til at skabe de anaerobe betingelser for såvel
kemisk som biologisk reduktion. **Nitrate reduction in the upper
soil layers. An investigation of factors contributing to the
anaerobic conditions for chemical and biological reduction.**

5222 Sørensen, C.
DK 010117/71/4611

Urinstoffets optagelse, transport og indflydelse på planternes
kvælstofsammensætning. **Absorption and transport of urea and
influence on the nitrogen composition of plants.**

5223 Sørensen, C.; Kyllingsbæk, A.
DK 010117/73/4613
Udvikling af metoder til bestemmelse af organiske syrer i
planterne og disses indhold heraf ved varierede
vækstbetingelser. **Development of methods for qualitative and
quantitative estimation of organic acids in plants under various
growth conditions.**

5224 Kyllingsbæk, A.
DK 010117/74/4614
Næringsstofoptagelse gennem blade ved forskellige
klimaforhold og næringsstofindholdet til forskellig tid efter
tilførsel af næringsstofopløsninger. **Nutrient absorption by
leaves under different climatic conditions, and the nutrient
content at different times after the application of nutrient
solutions.**

5225 Olesen, J.
DK 010701/64/0002
Gødningsforsøg til belysning af de nye gødningstypers virkning
ved fortsat anvendelse gennem en længere årrække. **Fertilizer
experiments to elucidate the effect of new fertilizer types when
used continuously over a number of years.**

5226 Andersen, A.J.
DK 030146/75/0016
Plantearter og –sorters optagelse af fosfor. **Phosphorus uptake
by plant species and cultivars.**

5227 Jakobsen, S.T.; Bille, S.W.
DK 030148/78/0001 N
Virkning af flyveaske på udbytte af og næringsstofoptagelse i
forskellige plantearter. **Effects of fly ashes on yield and nutrient
uptake by different plant species.**

5228 Jakobsen, S.T.; Salvo, R.
DK 030148/78/0003 N
Vekselvirkning mellem calcium og fosfat ved optagelse i
planter. **Interaction between uptake of phosphate and calcium
in plants.**

5229 Jensen, I.
DK 030148/78/0004 N
Virkning af fosfor i staldgødning til forskellige afgrøder. **The
nutritional value of phosphorus in farmyard manure to
different crops.**

5230 Roos, S.A.; Nielsen, N.E.
DK 030148/78/0006 N
Virkningen af placeringen af primært calciumfosfat i
varierende andele af jordvolumet på planters optagelse af
næringsstoffer. **The effect of phosphorus fertilizer placement in
soil on phosphorus uptake by plants.**

5231 Kaufholz, H.; Nielsen, N.E.
DK 030148/79/0006 N
Planterødders optagelse af jod med særligt henblik på I–129 i
miljøet. **Iodine uptake by plant roots with special reference to
I–129 in the environment.**

5232 Nielsen, N.E.; Andersen, A.
DK 030148/79/0010 N
Planteegenskaber, der påvirker effektiviteten af planters
udnyttelse af jord som næringsstofkilde. **Plant factors
influencing the efficiency by which plants utilize soil as a source
of nutrients.**

5233 Jakobsen, I.; Jensen, A.
DK 030602/79/5048 N
Vesikulær–arbuskulær mykorrhiza's betydning for
plantevæksten. **The importance of vesicular–arbuscular
mycorrhiza for plant growth.**

5234 Trocme, S.; Boniface, Mme. FR 010103/46/0664
Arriére action de P, K, oligo–èlèments. **Late actions of P,K, and trace–elements.**

5235 Lemaire, F. FR 010103/67/0655
Effet spécifique de la matière organique sur l'alimentation minérale des plantes. **Specific effects of organic matter on plant mineral nutrition.**

5236 Boniface, R. mme; Trocme, S. FR 010103/69/0647
Arrière –action des engrais azotés minéraux. **Residual effects of mineral nitrogen fertilization.** Publications.

5237 Simon, G. Mme FR 010103/69/0661
Incidence d'une fertilisation soufrée répétée sur les cycles biologiques du sol (soufre et azote). **Effect of replicated sulfur fertilization on biological cycles in the soil (Sulfur and nitrogen).** Publications.

5238 Trocme, S.; Boniface, Mme FR 010103/70/0663
Carences et toxicité en zinc. **Zinc deficiencies and toxicities.** Publications.

5239 Trocme, S.; Boniface, R. Mme FR 010103/76/6216
Risques de phytotoxicité dus à des teneurs élevées en bore des ordures ménagères, des composts urbains et des boues d'épuration. **Risks of phytotoxicity due to content of boron of municipal composts and sewage sludges.**

5240 Blanc, D.Mme; Bellenand–Mayeur, P.
FR 010502/70/0518
Etude des substrats. **Studies on substrata.**

5241 Blanc, D. Mme; Blondel, L. FR 010502/70/0519
Etude de la nutrition ammoniaco–nitrique. **Studies on ammonia and nitrate uptake.**

5242 Guennelon, R. FR 010601/71/0531
Propriétés adsorbantes de milieux artificiels (Pollution et de contamination). **Adsorption in artificial substrata (pollution and cleaning).**

5243 Monnier, G.; Fies, M. Mme FR 010601/73/0526
Recherches des caractéristiques physiques et mécani–ques optimales pour les substrats artificiels de culture. **Research on optimum Physical and Mecanical characteristics of Artificial Substrata.**

5244 Juste, chr. FR 010704/69/0544
Technologie de la fertilisation potassique en sol sableux. **Technology of potassium fertilization in sandy soils.** Publications.

5245 Soyer, J.P.; Chignon FR 010704/69/6077
Diagnostic de la nutrition minérale des végétaux basé sur l'analyse des extraits frais des tissus conducteurs. **Mineral crop nutrition diagnosis by analysis of the sap freshly extracted from the vascular system of plants.**

5246 Juste, C.; Delas, J. FR 010704/70/0551
Utilisation en agriculture de divers sous– produits organiques de la consommation ou de l'industrie. **Agricultural use of urban and industrial organic waste products.** Publications.

5247 Juste, C.; Dureau, Mme FR 010704/72/0545
Lessivage des éléments fertilisants en sol sableux. Pollution de la nappe phréatique des landes de Gascogne. **Leaching of fertilizing elements in sandy soils. Pollution of ground–water table in Gascony sandy moor.** Publications.

5248 Gachon, L. FR 010802/60/0565
Rotation et fumures azotées. **Crop rotation and nitrogen fertilization.** Publications.

5249 Mingeau, M.; Robelin, M. FR 010802/66/6118
Croissance et développement des plantes cultivées en condition de sécheresse. **Effect of water stress on growth and development of plants.**

5250 Morizet, J. FR 010802/67/0568
Action de la nutrition minérale sur la dynamique de l'eau dans la plante. **Effect of mineral nutrition on water dynamics in plant.** Publications.

5251 Hebert FR 010901/65/0585
Etude sur les engrais : composition, analyses, échantillonnage. **Studies on fertilizers – Composition, analysis, sampling.**

5252 Maertens, C. FR 011401/69/0630
Influence des propriétés du sol sur la croissance et le développement des racines; répercussion sur l'alimentation hydrique et minérale des plantes cultivées. **Effect of soil properties on root growth and development. Effect on mineral and water nutrition in crops.** Publications.

5253 Maertens, C.; Puech, J. FR 011401/71/0631
Modalités d'utilisation de l'eau du sol en relation avec l'enracinement de plantes cultivées avec ou sans irrigation. **Use of soil water in relation with the root development of crops grown under or without irrigation.** Publications.

5254 Maertens, C. FR 011401/75/6189
Etude des facteurs endogènes et exogènes de l'absorption des éléments minéraux par les racines. Conséquences sur l'alimentation des cultures. **Study of interior and exterior factors of the mineral uptake by the roots and the consequences on nutrition of the crops.**

5255 Maertens, C. FR 011401/75/6190
Détermination des vitèsses et des pouvoirs d'extraction de l'eau du sol par différentes espèces et variétés cultivées. **Determination of the water extraction rate and power by different cultivated species and varieties.**

5256 Morel, R.; Chabouis, Mme FR 011801/00/0620
Influence de la fertilisation sur le rendement et la qualité des récoltes (Dispositif Deherain) Essai de longue durée. **Effect of fertilization on yield and quality of harvested crops (Deherain long term experiment).** Publications.

5257 Fourbet, J.P. FR 011801/75/9072
Etude approfondie de l'alimentation minérale de quelques végétaux en conditions de travail du sol réduit. **Research about mineral feeding of some plants cultivated under reduced soil tillage conditions.**

5258 Bussieres, Ph.; Coppenet, M. FR 012208/74/6178
Etude de nouveaux amendements calcaires marins. **Study of new sea calcareous amendments.**

5259 Duval, Y.; Masclet, A. FR 012219/72/6186
Fertilisation magnésienne. **Magnesium and Potassium**

Fertilization in Lehm of Haute–Normandie.

5260 Lefevre, G.; Lefevre, P.   FR 012222/71/0500
Azote et matières organiques – Système cultural et économie de la fertilisation azotée. **Nitrogen and organic matter. Cropping system and economy of nitrogen fertilization.**

5261 Hebert, J.   FR 012226/63/0607
Définition et méthode d'évaluation des fertilisants. **Type of fertilizers : their description and evaluation.**

5262 Remy, J.C.   FR 012226/70/0602
Sort des engrais contenant de l'urée. **Evolution of urea fertilizers, particularly in liquid manures.** Publications.

5263 Herbert, J.; Remy, J.C.   FR 012503/56/6151
Effet de différents engrais phosphatés sur les rendements de plantes de grande culture et sur la teneur en phosphore assimilable des sols a ph élevé. **Effect of various phosphate fertilizers on the yields of crops and on the rate of available phosphorus of alcaline soils.**

5264 Hebert, J.   FR 012503/60/6153
Données sur l'analyse et la valeur agronomique des matières fertilisantes en vue de servir à la réglementation. **Data on the analysis and agronomic value of fertilizing materials for their policy.**

5265 Marin–Lafleche, A.; Remy, J.C.   FR 012503/76/6152
Appréciation au niveau de l'exploitation agricole des régles générales de la fertilisation. **Use of general rules in fertilization for a farm.**

5266 Galliard; Wardale   GB 010203/00/0049 R
Sub–cellular localisation of lipid–degrading enzymes in plant tissue.

5267 Clarkson; Sanderson   GB 010502/00/0001
Ion uptake by root systems related to fine structure, nutrient status, temperature and mineral toxicity.

5268 Lynch   GB 010502/00/0005 R
Products of soil micro–organisms in relation to plant growth.

5269 Drew   GB 010502/00/0008
Effects of nutrient supply on root development and nutrient uptake.

5270 Lee   GB 010502/75/0014
Response of crop plants under waterlogged conditions to applications of nitrogen fertilizers.

5271 Lee   GB 010502/77/0015 R
Metabolic responses of roots to mineral nutrient deficiency with special reference to absorption of ions.

5272 Researcher not indicated   GB 011102/00/0012 R
Mechanisms of assimilate translocation in fruiting crops.

5273 Barratt   GB 011507/00/0008 R
Effect of nitrogen fertilizers and growth regulators on plant composition, growth and yield.

5274 Hewitt   GB 011509/00/0001 R
Inorganic nitrogen metabolism.

5275 Johnston; Mattingly   GB 012003/00/0008 R
Fertilises and farm–yard manure in long–term experiments and the valve of their residues for arable crops.

5276 Widdowson; Penny   GB 012003/00/0019 R
Plant nutrient balance sheets in relation to soil reserves.

5277 Nowakowsk   GB 012003/00/0020
Effect of N, K, Na, Mg, P and S fertilisers on nitrogenous and carbohydrate fractions of grasses.

5278 Page; Talibudeen   GB 012003/00/0021 R
Activity ratios of ions in soil solution related to uptake by crops.

5279 Bolton; Mattingly   GB 012003/75/0039 R
Effects of fertilizers, organic manures and soil pH on uptake of micronutrients by crops.

5280 Bolton; Williams   GB 012003/75/0040 R
Plant nutrients from the atmosphere.

5281 Tinker; Gildon   GB 012003/77/0042 R
The effect of micro–organisms on the uptake of phosphate and trace elements by crops.

5282 Barraclough; Tinker   GB 012003/79/0078 N
Processes in nutrient uptake from soil.

5283 Walker; MacDonald   GB 012010/00/0004 R
Nitrification.

5284 Dart   GB 012010/00/0005 R
Free living nitrogen fixing microorganisms.

5285 Nutman; Dart   GB 012010/00/0006 R
Symbiotic nitrogen fixing associations.

5286 Wingfield   GB 012011/77/0010 R
Effect of ligands for potassium and sodium on seedlings (beans and wheat).

5287 Chatt   GB 013001/00/0001 R
Chemistry and biology of nitrogen fixation.

5288 Dekock   GB 030404/00/0001
Iron and copper metabolism of plants.

5289 Dekock   GB 030404/00/0002
Uptake and physiological effects of chelated trace elements on plants.

5290 Dekock   GB 030404/00/0008
Nitrate reductase and molybdenum–copper interactions in plants.

5291 Williams   GB 030406/00/0003 R
Available nitrogen in soils.

5292 Knight; Booth   GB 030406/00/0006
Inorganic and organic constiuents in crops: forms, patterns and balance in relation to age and yield.

5293 Reith; Crooke   GB 030406/00/0010
Assess lime and nutrient status of soil.

5294 Shepherd   GB 030406/00/0013

Development and application of radioactive techniques.

5295 McAllister GB 040402/00/0012 R
Evaluation of the availability of the potassium and phosphorus in slurry.

5296 Steele GB 041406/00/0006 R
The fate of nitrogen applied as fertiliser or slurry to the soil.

5297 Smith, K. GB 060114/78/0005 N
Efficiency of nitrogen fertiliser use.

5298 Herlihy, M. IE 060200/73/0824 R
The effect of soil texture, moisture and organic matter on the mineralisation and availability of soil and fertiliser nitrogen. Publications.

5299 Tunney, H. IE 060200/75/1121
Fertiliser value of dung and seepage. Publications.

5300 Kiely, P.V.; Tunney, H. IE 060200/75/1180 R
Land spreading of animal manures. Effect of time and rate of applicationof pig and cattle slurry. Publications.

5301 Tunney, H. IE 060200/76/1182 R
Slurry nitrogen factorial experiment. Publications.

5302 Fleming, G.A.; Duff, C.; MacNaeidhe, F.
IE 060200/78/1419 N
Crop nutrition on cutover peatland. Publications.

5303 Gallagher, P.A.; Maher, M.J. IE 060300/70/0053 R
Developmental and nutritional studies on the use of peat in protected cultivation. Publications.

5304 Hennerty, M.J.; Titus, J.S. IE 120108/78/9106 N
Protein degradation and reutilisation in higher plants.

5305 Cuddy, M.; Boylan, T.; O'Muircheartaigh, I.
IE 130201/79/9221 N
Functional form of fertiliser demand functions. Publications.

5306 Romanin, M. IT 020200/77/0004 N
Utilizzazione dei liquami suini mediante lo spargimento sul terreno. **Utilisation of pig slurry by land spreading.** Publications.

5307 Raciti, G.; Di Martino Aleppo, E.; Intrigliolo, F.
IT 021600/77/0008 R
Prove di nuovi prodotti per la cura della clorosi ferrica dell'arancio. **Trials on new products for the control of ferric clorosis of orange.**

5308 Rossi, N. IT 040202/73/0299
Ricerche sullo stato nutrizionale delle colture agrarie della regione emiliana. **Research on the nutritional status of agricultural crops in the region of Emilia.** Publications.

5309 Silva, S. IT 040406/74/0631
Studio degli effetti di microelementi contenuti in fertilizzanti NPK sul metabolismo vegetale. **Study of the effects on plant metabolism of microelements contained in NPK fertilizers.**

5310 Giovannozzi–Sermanni, G. IT 040705/77/0198 R
Regolazione dell'assorbimento dei nitrati in piante a fotosintesi C3 e C4 in pieno campo. **The regulation of nitrate absorption in plants with C3 and C4 photosynthesis in tilled land.**

5311 Ferrari, G. IT 040802/73/2556
Fattori genetici e ambientali nella regolazione dell'assorbimento nutrizionale dei vegetali. **Genetic and environmental factors in regulating the nutrient absorption of plants.**

5312 Lucci, G.C. IT 042001/78/1069 N
Inibizione e stimolazione dell'assorbimento e della traslocazione di elementi fertilizzanti in piante coltivate. **Inhibition and stimulation of the absorption of fertilizing elements and of their translocation in cultivated plants.**

5313 Cacciari, I. IT 060200/70/0074
Studio dei fattori ambientali che possono influenzare la azotofissazione di batteri appartenenti alla specie Arthrobacter. **Study of environmental factors affecting the nitrogen fixation by some Arthrobacter strains.** Publications.

5314 Grappelli, A. IT 060200/73/0076
Studio dei fattori ambientali che possono influenzare la produzione di fitormoni da parte dei batteri appartenenti alla specie Arthrobacter. **Study of environmental factors affecting the production of growth regulators by Arthrobacter sp.** Publications.

5315 Tedeschi, P. IT 060700/70/0124
Determinazione dei volumi di adacquamento e dei fabbisogni idrici stagionali per colture diverse. **Optimal watering volumes and seasonal irrigation needs for different crops.** Publications.

5316 Tedeschi, P. IT 060700/71/0123
Esigenze idriche delle colture in funzione delle fasi del ciclo biologico. **Water needs of crops with relation to biological cycle.** Publications.

5317 Tedeschi, P. IT 060700/77/0986
Studio degli apparati radicali in relazione ai metodi irrigui. **A study of root systems in relation to irrigation methods.**

5318 Tedeschi, P. IT 060700/77/0992
Variabilità del flusso traspirativo di colture diverse in funzione delle stato idrico dell'atmosfera e del terreno. **Variability of the transpiration flow in different cultures in relation to the atmospheric and soil water content.**

5319 Bakermans, W.A.P. NL 010102/66/6739
De teelt van groenbemesters en dekvrucht in verband met de stikstof huishouding van de grond, de chemische onkruidbestrijding en de mogelijkheden voor geïntegreerde bestrijding van insektenplagen in de kasteelt. **The husbandry of green manuring plants with and without a cover crop in relation to soil nitrogen, chemical weed control and the possibilities of integrated control of insects in glasshouse farming.** Publications.

5320 Steiner, A.A. NL 010102/68/7978
Aanvullend onderzoek en adviezen voor methoden van plantenteelt zonder aarde. **Complementary research and advice about methods of soilless culture.** Publications.

5321 Dijkshoorn, W. NL 010102/76/6731
Werking van opgenomen meststofbestanddelen op de groei van planten. **Effect of absorbed fertilizer elements on the growth of plants.**

D 2200 – Plant nutrition and fertilization

5322 Steiner, A.A.   NL 010102/77/7306
Toelaatbare grenzen van ionenhoeveelheden in voedingsoplossingen. **Admissible limits for quantities of ions in nutrient solutions.**

5323 Lubbers, J.   NL 010103/49/0473
Waardering van veeljarige organische bemesting op zand– en dalgrond. **Evaluation of continued organic manuring on sandy soils and reclaimed peat soils.** Publications.

5324 Grootenhuis, J.A.   NL 010103/49/0474
Waardering van veeljarige organische bemesting op klei– en zavelgrond en toepassing in bedrijfsverband. **Evaluation of many years organic manuring on clay and silty clay soils and the application in farming practice.** Publications.

5325 Prummel, J.   NL 010103/49/0506 R
Wijze en tijd van toediening van kunstmeststoffen. **Method and time of application of fertilizers.** Publications.

5326 Driel, W.van   NL 010103/58/0462
De rol van synthetische en natuurlijke metaalchelaten bij de voeding van de plant. **The role in plant nutrition of synthetic and natural metal chelates.** Publications.

5327 Wiersum, L.K.   NL 010103/60/0458
Verband tussen bodemstructuur en wortelontwikkeling. **Relation between soil structure and root development.** Publications.

5328 Lande Cremer, L.C.N. de la   NL 010103/60/0512
Bepaling van normen en gebruikswaarde van meststofbalansen. **Establishment of criteria for and evaluation of nutrient balance sheets.** Publications.

5329 Loman, H.   NL 010103/61/0533
Onderzoek naar de betekenis van de kalktoestand van de ondergrond. **Investigations on the importance of the lime status of the subsoil.** Publications.

5330 Smilde, K.W.   NL 010103/62/0567
Diagnose en vastlegging van verschijnselen van gebrek en overmaat in land– en tuinbouwgewassen. **Diagnosis and recording of deficiency and toxicity symptoms in agricultural and horticultural crops.** Publications.

5331 Prummel, J.   NL 010103/64/0562 R
Verloop van de kalitoestand op bouwland. **Course of potash status on arable land.** Publications.

5332 Haan, S.de   NL 010103/65/0468 R
Betekenis van organische meststoffen voor de stikstofvoeding van de gewassen en de humushuishouding van de grond. **Significance of organic fertilizers for the nitrogen nutrition of crops and the organic matter content of the soil.** Publications.

5333 Dilz, K.   NL 010103/69/2782
Toetsing van de landbouwkundige waarde van in ontwikkeling zijnde meststoffen. **Testing of the agricultural value of fertilizers in state of development.**

5334 Werkhoven, C.H.E.   NL 010103/69/2793
Waardering van langzaamwerkende stikstofmeststoffen. **Evaluation of slow–release nitrogen fertilizers.**

5335 Driel, W. van   NL 010103/70/2791
Betekenis van de zware–metaalcontaminatie voor de plantevoeding. **Significance of environmental heavy–metal contamination for plant nutrition.** Publications.

5336 Werkhoven, C.H.E.   NL 010103/70/2914
Nadere karakterisering van de beschikbaarheid van K en Ca voor het gewas, rekening houdend met hun voorraad, nalevering en beweging naar het worteloppervlak. **Further characterization of K and Ca availability to crops, taking into consideration total available quantity in the soil, rate of release, and movement to the root surface.**

5337 Herwerden, C.H. van   NL 010103/70/3000
Oriënterend onderzoek ten behoeve van een optimale uitvoering van pot– en vakproeven. **Exploratory investigations to achieve maximum efficiency in the execution of pot and plot experiments.**

5338 Loman, H.   NL 010103/71/3002
Beheersing van het kalkniveau in de grond. **Maintenance of a favourable soil calcium level.** Publications.

5339 Haan, S. de   NL 010103/71/3003 R
Toepassingsmogelijkheden van stedelijke en industriële organische afvalstoffen als meststof en grondverbeteringsmiddel. **Possibilities for utilization of organic household and industrial wastes as fertilizers and amendments for soil improvement.** Publications.

5340 Ris, J.   NL 010103/71/7973
Opstelling van een adviesschema voor de stikstofbemesting van bouwland op basis van resultaten van grondonderzoek. **Establishment of an advisory scheme for nitrogen fertilization of arable land on the basis of soil testing.** Publications.

5341 Dilz, K.   NL 010103/73/4357 R
Bemestingswaarde en toepassingsmogelijkheden van vloeibare meststoffen, in het bijzonder van meststofoplossingen. **Agricultural evaluation and possible utility of liquid fertilizers, especially of fertilizer solutions.** Publications.

5342 Driel, W. van   NL 010103/74/6013
Betekenis van zware metalen in havenslib, gestort op loswallen, voor de plantegroei. **Plant growth on harbour sludge dumping grounds as affected by heavy metals.** Publications.

5343 Goor, B.J. van   NL 010103/75/6246
Transport van zware metalen in de plant. **Translocation of heavy metals in the plant.** Publications.

5344 Goor, B.J. van   NL 010103/75/6247
Zware metalen in de landbouw in relatie tot de voor de menselijke voeding toelaatbare gehalten. **Heavy metals in agriculture in relation to human nutrition.** Publications.

5345 Smilde, K.W.   NL 010103/76/6588
Maatregelen ter voorkoming van boriumgebrek bij verschillende gewassen. **Control of boron deficiency in various crops.** Publications.

5346 Noordwijk, M. van   NL 010103/76/6864
Minimaal benodigd wortelvolume bij de teelt van gewassen op kunstmatige substraten. **Minimum volume of roots required in crop culture on artificial substrates.** Publications.

D 2200 – Plant nutrition and fertilization

5347 Noordwijk, M. van   NL 010103/77/7231
Ontsluiting van de grond door wortelstelsels bij verschillen in verdelingspatroon. **Accessibility of soil reserves to root systems with different patterns of distribution.** Publications.

5348 Ris, J.   NL 010103/77/7692 R
Onderzoek naar de efficiëntie van de voorjaarsgift kunstmeststikstof op bouwland. **Investigation into the efficiency of spring applied fertilizer nitrogen to arable land.**

5349 Sissingh, H.A.; nl04001778618   NL 010103/77/8618 N
Methoden van grondonderzoek ten behoeve van het fosfaatbemestingsadvies voor éénjarige gewassen in tropische ontwikkelingslanden. **Soil testing methods to establish phosphorus fertilizer recommendations for annual crops in developing countries in the tropics.**

5350 Lande Cremer, L.C.N. de la   NL 010103/78/7970
De meest gunstige afstand ten opzichte van het gewas voor het injecteren van dunne mest op bouwland. **The optimum distance with respect to the crop for injecting slurry in arable land.**

5351 Desmet, G.   NL 010110/67/7287
Root and foliar uptake of radioactive contaminants by crops; their transport and redistribution towards different plant organs (e.g. edible parts, reproductive organs). Publications.

5352 Dorp, F. van   NL 010110/72/7939 R
De minerale voeding van landbouwgewassen met betrekking tot de samenstelling van de voedingsmatrix met speciale aandacht voor de nitraat en fosfaat absorptie, Invloed van klimatologische omstandigheden. **The mineral nutrition of agricultural crops in relation to the composition of the nutritional matrix with emphasis on the nitrate and phosphate absorption. Influences of climatic parameters. Related model studies including nitrogen and energy balance studies.**

5353 Breteler, H.   NL 010110/76/7943
Geïntegreerde studie van de opname, verwerking en verdeling van stikstof in planten. **An integrated study of uptake, assimilation and distribution of nitrogen in plants.** Publications.

5354 Roorda van Eysinga, J.P.N.L.   NL 010206/62/0567
Diagnose en vastlegging van verschijnselen van gebrek en overmaat in land– en tuinbouwgewassen. **Diagnosis and recording of deficiency and toxicity symtoms in agricultural and horticultural crops.** Publications.

5355 Titulaer, H.H.H.   NL 010207/71/7626
Bemestingsonderzoek in bedrijfsverband. **Fertilization research within crop production systems and crop rotations.**

5356 Titulaer, H.H.H.   NL 010207/71/7629
De invloed van de toediening van organische afvalstoffen op de fysische en chemische bodemvruchtbaarheid binnen vruchtwisselingssystemen. **The influence of organic waste matter on the physical and chemical soil fertility within crop rotations.**

5357 Titulaer, H.H.H.   NL 010207/75/7627
De invloed van verschillende hoeveelheden en soorten drijfmest op de kwalitatieve en kwantitatieve opbrengsten van gewassen (o.a. granen, aardappelen, suikerbieten, groenten). **The influence of different amounts and sorts of animal slurry on the qualitative and quantitative yields of arable crops (a.o. cereals, potatoes, sugar beet and outdoor vegetables).**

Publications.

5358 Titulaer, H.H.H.   NL 010207/75/7628
De invloed van de toediening van afvalwaterzuiveringsslib op de opbrengst en de kwaliteit van landbouwgewassen. **The influence of the use of sewage on the yield and quality of arable crops and outdoor vegetables.**

5359 Rijtema, P.E.   NL 010501/67/0808
Invloed van zout water op de plantengroei by kasteelten. **Influence of saline water on plant growth of glasshouse crops.** Publications.

5360 Ploegman, C.   NL 010501/77/7213
Opname van zware metalen door het gewas. **Uptake of heavy metals by plants.**

5361 Houba, V.J.G.   NL 020007/74/4968 R
Tolerantie van planten voor "zware" metalen die als verontreiniging in de bodem geraken. **Tolerance of plants for heavy metals accumulated in the soil as pollutants.**

5362 Keltjens, W.G.   NL 020007/78/8844 N
Opname en transport van water en zouten in de plant. **Absorption and translocation of water and salts in the plant.**

5363 Jong, G.J. de   NL 040007/62/4343
Stikstofbemesting van diverse akkerbouwgewassen. **Nitrogen–fertilizing on various arable crops.**

5364 Jong, G.J. de   NL 040007/62/4344
De invloed van fosfaatbemesting op groei, opbrengst en kwaliteit van akkerbouwgewassen. **Influence of phosphate–fertilizing on the growth, yield and quality of arable crops.**

5365 Muller, A.   NL 040012/77/8618 N
Methoden van grondonderzoek ten behoeve van het fosfaatbemestingsadvies voor éénjarige gewassen in tropische ontwikkelingslanden. **Soil testing methods to establish phosphorus fertilizer recommandations for annual crops in developing countries in the tropics.**

5366 Jorritsma, J.   NL 060003/71/7973
Opstelling van een adviesschema voor de stikstofbemesting van bouwland op basis van resultaten van grondonderzoek. **Establishment of an advisory scheme for nitrogen fertilization of arable land on the basis of soil testing.** Publications.

5367 Jorritsma, J.   NL 060003/75/7627
De invloed van verschillende hoeveelheden en soorten drijfmest op de kwalitatieve en kwantitatieve opbrengsten van gewassen (o.a. granen, aardappelen, suikerbieten, groenten). **The influence of different amounts and sorts of animal slurry on the qualitative and quantitative yields of arable crops (a.o. cereals, potatoes, sugar beet and outdoor vegetables).** Publications.

5368 Dilz, K.   NL 060008/69/2782
Toetsing van de landbouwkundige waarde van in ontwikkeling zijnde meststoffen. **Testing of the agricultural value of fertilizers in state of development.**

5369 Dilz, K.   NL 060008/72/3567
De werking van ureum als stikstof meststof. **Urea as a nitrogen fertilizer.**

5370 Dilz, K.   NL 060008/73/4357
Bemestingswaarde en toepassingsmogelijkheden van vloeibare
meststoffen, in het bijzonder van meststofoplossingen.
**Agricultural evaluation and possible utility of liquid fertilizers,
especially of fertilizer solutions.** Publications.

Cereals in general (B 3100)

See also 486, 2896, 2898, 2913, 5061, 5488, 5545, 6661, 9302

5371 Solansky, S.   DE 111252/78/0002 N
Untersuchungen über die 14C – Assimilateeinspeicherung in
die Körner in Abhängigkeit von deren Insertion innerhalb der
Getreideähre. **Investigations on the translocation of 14C
assimilates into the grains of different insertion within the ear.**

5372 Wehrmann, J.; Molitor, H.–D.   DE 138060/77/0001
Ermittlung des N–Düngerbedarfs bei Getreide im Löss– und
Geestgebiet von Niedersachsen mit Hilfe der Nmin–Methode.
**Fertilizer nitrogen requirements of small grains as affected by
the mineral nitrogen content of soils in spring.**

5373 Beringer, H.; Hess, G.   DE 144400/75/0003
Ausarbeitung eines Schnelldienstes zur Diagnose des
N–Ernährungszustandes von Getreide. **Development of a rapid
test method for the diagnosis of the N–nutrition status of
cereals.**

5374 Martin, P.; Glatzle, A.   DE 144400/78/0005 N
Stickstoff–Fixierung in der Rhizosphäre von Gramineen.
Nitrogen fixation in the rhizosphere of Gramineae.

5375 Michael, B.; Zink, F.   DE 144400/78/0009 N
Phytingehalt von Getreidekörnern unter dem Einfluss einer
variierten Phosphatdüngung. **Effect of varying phosphate
fertilization on the phytin content of cereal grains.**

5376 Fischbeck, G.; Estler, M.   DE 161250/78/0003 N
Untersuchungen über den Einfluss ungleichmässiger
Verteilung von N–Düngemitteln in Getreidebeständen.
**Studies of the influence of disproportionate distribution of
N–fertilizers in cereal crops.**

5377 Beringer, H.; Haeder, H.–E.   DE 902000/78/0001 N
Wechselwirkungen zwischen K–Ernährung, Wasserhaushalt
und Abscisinsäurebildung bei Getreide. **Interactions between
K nutrition, water status and abscisic acid synthesis in cereals.**

5378 Beringer, H.; Schacherer, A.   DE 902002/78/0001 N
Einfluss der Düngung auf die Anlage von Endospermzellen im
Getreidekorn. **Influence of mineral nutrition on the formation
of endosperm cells in cereal grain.**

5379 Hansen, L.; Bennetzen, F.   DK 010105/74/3409
Vandbalance og kvælstofbalance ved dyrkning af korn og græs
på sand– og lerjord. **Water and nitrogen balances in cereal and
grass crops grown on sand and clay soils.**

5380 Sørensen, C.   DK 010117/71/4612
Kornsorternes variation i kvælstofindhold,
farvebindingskapacitet, amidindhold og
aminosyresammensætning i relation til årstid og
kvælstoftilførsel. **Variation between cereal varieties in nitrogen
content, dyebinding capacity, amide content and amino acid
composition in relation to season and nitrogen supply.**

5381 Andersen, A.J.; Haahr, V.; Sandfær, J.
                                           DK 030146/76/0002
Vækstanalyse af vår– og vintersæd. **Growth analysis in spring
and winter cereals.**

5382 Engvild, K.   DK 030146/77/0003
Kernefyldningens fysiologi. **The physiology of grain filling.**

5383 Nielsen, J.M.; Frederiksen, J.; Skriver, K.; Gosvig, V.
                                           DK 030148/79/0009 N
Gødskning af vårsæd ud fra planteanalyser i landbrugspraksis.
**Applying a newly developed fertilization system for
spring–sown cereals to agricultural practice.**

5384 Studer, R.; Morlat   FR 012209/76/6132
Recherche des causes des accidents nutritionnels observés sur
céréales en terres d'aubues (sols bruns calcaires sur craie
argileuse). **Causes of nutritional desorders of cereals in
calcareous soils issued from clayed chalk.**

5385 Miflin; Bahramian   GB 012001/79/0026 N
**Studies on the isolation of the genes specifying the storage
proteins of cereals.**

5386 Lawlor   GB 012002/76/0015 R
Water relations of cereal crops.

5387 Johnston; Widdowson   GB 012003/00/0016 R
**Compare rates, times and methods of application of solid and
liquid fertilisers for wheat and barley.**

5388 Benzian   GB 012003/00/0030
**Effects of nitrogen fertilisers on concentration of nitrogen in
cereal grain.**

5389 Johnston; Widdowson   GB 012003/75/0037 R
**Fertiliser requirements of cereals related to crop rotation and
interaction with agricultural chemicals.**

5390 Blackett   GB 060213/00/0002
Response of cereals to fertiliser.

5391 Paterson   GB 060314/00/0010 R
Manurial requirements of tillage crops (cereals).

5392 Gately, T.F.   IE 060200/70/0380 R
**Fertiliser nitrogen trial on wheat, oats and barley grown
alongside each other for several years (1 site).** Publications.

5393 Lanzani, G.A.   IT 040603/74/0575
Correlazione tra biosintesi proteica durante la germinazione di
semi di cereali, nutrizione azotata, effetti ormonali.
**Correlations between protein biosynthesis during cereal seed
germination, nitrogen nutrition, hormonal effects.**

5394 Spiertz, J.H.J.   NL 010102/76/8938 N
De invloed van stikstofaanbod en fungicidenbehandelingen op
het productiepatroon, de assimilatiehuishouding en de
korrelopbrengst van graangewassen. **The influence of nitrogen
and fungicides on production pattern, assimilate utilization and
grain yields of cereals.**

5395 Dilz, K.   NL 010103/73/4358
De invloed van groeiregulatoren en
gewasbeschermingsmiddelen op het met stikstofbemesting te

bereiken opbrengstniveau van granen. **The effect of growth regulators and pesticides on the yield level of cereals attainable through nitrogen fertilization.** Publications.

5396 Darwinkel, A.; Kuizenga, J.; Hag, B.A. ten
NL 010207/75/7638
Invloed van fractionering van de stikstofbemesting op de ziekteaantasting, opbrengst en kwaliteit bij granen. **Effect of fractionating N–fertilization on disease attack, grain yield and grain quality of cereals.**

5397 Schouls, J. NL 020025/68/4473
De drogestof productie in de graanplant, speciaal i.v.m.de stikstofvoeding. **Dry matter production in the cereal plant, especially in relation to nitrogen nutrition.**

5398 Ellen, J. NL 020025/76/8938 N
De invloed van stikstofaanbod en fungicidenbehandelingen op het productiepatroon, de assimilatiehuishouding en de korrelopbrengst van graangewassen. **The influence of nitrogen and fungicides on production pattern, assimilate utilization and grain yields of cereals.**

5399 Dilz, K. NL 060008/73/4358
De invloed van groeiregulatoren en gewasbeschermingsmiddelen op het met stikstofbemesting te bereiken opbrengstniveau van granen. **The effect of growth regulators and pesticides on the yield level of cereals attainable through nitrogen fertilization.**

Barley (B 3110)

See also 5070, 5226, 5437, 5454

5400 Timmermann, F. DE 132065/78/0001 N
Feldversuche zur Stickstoffdüngerbedarfsprognose für Winterweizen und Wintergerste auf Löss–Parabraunerden des Göttinger Raumes. **Field experiments on forecasting nitrogen requirements of winter wheat and winter barley cultivated on grey brown podsolic soils in the Goettingen area.**

5401 Beringer, H.; Koch, K. DE 902003/77/0002
Einfluss der K–Ernährung auf Mehltaubefall und N–Stoffwechsel bei Gerste. **Influence of K–nutrition on infection of barley by mildew and its interaction with N–metabolism.**

5402 Andersen, S.; Stølen, O.; Lysgaard, C.P.
DK 030145/74/0011
Byggens kvælstofindhold og proteinkvalitet. **Nitrogen content and protein quality in barley.**

5403 Andersen, A.J.; Haahr, V. DK 030146/74/0015
Optagelse og udnyttelse af kvælstof i byg. **The uptake and utilisation of nitrogen in barley.**

5404 Køie, B.; Nielsen, G. DK 030146/76/0008
Hordein i bygkernen. **Hordein in the barley grain.**

5405 Doll, H.; Kreis, M.; Køie, B. DK 030146/77/0007
Relationen mellem indlejringen af hordein og stivelse i bygkernen. **Relationship between deposition of hordein and starch in the barley grain.**

5406 Jensen, A.; Jacobsen, I.; Andersen, A.J.
DK 030146/79/0004 N

Visikulær–arbuskulær mykorrhiza's betydning for plantevæksten, (byg). **Effect of vesicular–arbuscular mycorrhiza on plant growth (barley).**

5407 Nielsen, N.E. DK 030148/78/0005 N
Udvikling og fordeling af rodsystemet hos byg under markforhold og forløbet af næringsstofoptagelsen. **Growth and distribution of, and nutrient uptake by barley roots in the field.**

5408 Jakobsen, S.T.; Jensen, M.T. DK 030148/79/0001 N
Vækst af byg og næringsstofoptagelse efter ændring af henholdsvis pH og tilført calcium. **Growth and nutrient uptake of barley as influenced by changes in pH and/or applied calcium–salts.**

5409 Emanuelsson, J.; Jakobsen, S.T. DK 030148/79/0002 N
Rodudvikling af byg og kemisk miljø i pløjelaget efter forskellig kalktilførsel. **Effect of lime application on root development of barley and soil chemical condition in the upper 30 cm soil layer.**

5410 Murali, N.S.; Lemmich, B.; Møller, T.B.; Nielsen, J.M.
DK 030148/79/0007 N
Gødskning af vinterhvede ud fra planteanalyser i landbrugspraksis. **Applying a newly developed fertilization system for winter–wheat to agricultural practice.**

5411 Shewry; Mifliin GB 012001/00/0022 R
Factors affecting the quality and quantity of protein synthesised in barley grains.

5412 Adams; Hodges GB 040402/00/0023
The use of plant analysis to determine the nitrogen status of cereals.

5413 Dilz, K. NL 010103/70/5325
Vaststelling van de behoefte aan kunstmeststikstof van brouwgerst. **Establishment of the nitrogen requirement of malting barley.** Publications.

5414 Kuizenga, J. NL 010207/75/6263 R
Invloed van stikstof–bemesting en ziektenbestrijding op opbrengst en kwaliteit van wintergerst. **Influence of nitrogen fertilization and disease–control on yield and quality of winter barley.** Publications.

5415 Wilten, W. NL 050304/70/5325
Vaststelling van de behoefte aan kunstmeststikstof van brouwgerst. **Establishment of the nitrogen requirement of malting barley.** Publications.

5416 Dilz, K. NL 060008/75/5325
Vaststelling van de behoefte aan kunstmeststikstof van brouwgerst. **Establishment of the nitrogen requirement of malting barley.** Publications.

Maize (B 3120)

See also 522, 682, 788, 1434, 3061, 5434, 5448, 5454, 6246, 7844

5417 Höfner, W.; Scherer, H.–W. DE 129044/75/0003
Eisen–Mangan–Wechselwirkungen als Ursachen von Chlorosen und ihre Beziehung zum Carbonsäuregehalt von Mais und Sonnenblumen. **Iron–manganese–antagonism in relation to chlorosis and carbon acid content in maize and sunflowers.**

5418 Höfner, W.; Kovanci, I.; Hakerlerler, H.

DE 129044/78/0001 N

Einfluss von Müllkompost und Eisenverbindungen auf die Eisen–, Zink– und Manganaufnahme von Sonnenblumen und Mais im Gefässversuch. **Influence of refuse compost and iron compounds on the uptake of iron, zinc and manganese by sunflower and maize in pot experiments.**

5419 Pollmer, W.G.; Klein, D.; Schwab, A.

DE 144430/77/0002

Stickstoffaufnahme und Stickstoffumlagerung bei Maispflanzen mit hohem Proteingehalt. **Nitrogen uptake and translocation in high–protein maize plants.**

5420 Trolldenier, G.

DE 902000/75/0005

Interaktionen zwischen Kalium– und Eisenernährung bei Mais. **Interactions between potassium and iron nutrition of maize.**

5421 Venezian Scarascia, M.E.; Losavio, N.; Leo, S.

IT 020500/75/0003

Stress idrico e manifestazioni morfo–fisiologiche in piante di mais di primo raccolto. **Water stress and morpho–physiological consequences on maize plants (as first crop).**

5422 Iannelli, P.

IT 020900/76/0001 N

Colture di mais ibridi ed altri erbai estivi irrigui.. **Hybrid mais cultivation and other summer fodder catch–crops under irrigation..**

5423 Postiglione, L.; Barbieri, G.

IT 040701/79/0003 N

Consumi idrici del mais in secondo raccolto. **Water consumption of the maize as second crop.**

5424 Tedeschi, P.

IT 060700/77/0991

Rese produttive di mais, pomodoro,tabacco e barbabietola in relazione a periodi di deficit idrico di diversa intensità e durata. **Yield of maize, tomato, tobacco and beetroot in relation to water shortage of differing severity and duration.**

5425 Beusichem, M.L. van

NL 020007/76/6776

Relaties tussen opname en verwerking van nitraat in mais (wortels). **Relations between uptake, reduction and assimilation of nitrate in maize (roots).**

Oats (B 3130)

5426 Baeumer, K.; Ehlers, W.; Stülpnagel, R.; Böhm, W.

DE 132181/75/0005

Einfluss von Bodenstruktur, Wassergehalt und Wurzeldichte von Hafer auf dessen Wasseraufnahme in bearbeiteten und unbearbeiteten Löss–Parabraunerden. **Water uptake of oats as influenced by soil structure, moisture regime, root density in tilled and untilled loess soil.**

5427 Mortensen, J.V.

DK 010117/74/4615

Næringsstoffers (Mo, Mn og Mg) betydning for nitratreduktion og nitratindhold i spinat og havre. **Significance of nutrients (Mo, Mn and Mg) for nitrate reduction and nitrate content in spinach and wheat.**

Rice (B 3140)

See also 3075

5428 Atanasiu, N.; Alkämper, J.; Westphal, A.; Silva–Montenegro, M.

DE 129552/74/0001

Untersuchungen über die Wirkung verschiedener N–Düngerformen auf die Nitrifikation, Denitrifikation, Stickstoffaufnahme und Ertragsbildung bei Bewässerungsreis. **Investigations on the effects of N–fertilizer forms on nitrification, denitrification, nutrient uptake and yield production of paddy.**

5429 Fendrik, I.; Othman, M.bin; Bors, J.   DE 138180/77/0003

Untersuchungen zur biologischen N–Fixierung bei Reis. **Investigations on the utilisation of biological nitrogen fixation in rice.**

5430 Beringer, H.; Trolldenier, G.

DE 902004/74/0001

Einfluss der Pflanzenernährung auf Redox–Verhältnisse und mikrobielle Stickstoffbindung in der Rhizosphäre von Sumpfreis. **The influence of plant nutrition on redox conditions and microbial nitrogen fixation in the rhizosphere of paddy.**

5431 Nielsen, J.M.; Paiboon, P.; Murali, N.S.; Sumitra, P.; Alva, A.K.

DK 030148/76/0010 N

Risgødskning på grundlag af kemiske planteanalyser. **Paddy fertilization based on chemical plant analyses.**

5432 Russo, S.; Licata, V.

IT 020800/74/0003 R

Riso – Influenza delle dosi di azoto e dell'epoca di distribuzione sulla produzione e le caratteristiche qualitative della granella. **Rice – Influence of nitrogen rate and time of application on the yield and chemical composition of the kernel.** Publications.

5433 Finassi, A.

IT 063300/69/0066

Trinciatura della paglia di riso, modalità di interramento in funzione della concimazione azotata. **Chopping of straw from rice and different systems of its under–plowing in relation to nitrogenous fertilizing.** Publications.

5434 Finassi, A.

IT 063300/72/0065

Tecniche di impiego dei fertilizzanti liquidi nel riso e nel mais. **Different techniques for applying liquid fertilizers in rice and maize cultivation.** Publications.

5435 Becking, J.H.

NL 010110/72/8977 N

Symbiontische stikstofbinding van Azolla t.b.v. de toepassing in de rijstcultuur en selectie van Rhizobiumstammen t.b.v. de veredeling op stikstofbinding bij veldbonen (Vicia faba L.). **Symbiontic nitrogen fixation, including selection of Rhizobium strains (symbiontic nitrogen fixation of Azolla for use in tropical rice production and selection of Rhizobium strains for use in breeding Vicia faba on nitrogen fixation capacity).** Publications.

Sorghum (B 3160)

See also 1434

Wheat (B 3170)

See also 788, 3090, 3129, 7844, 9411

5436 Riga, A.; François, E.

BE 080100/79/0019 N

Influence du fractionnement de la fumure minérale azotée sur le développement radiculaire du froment d'hiver. **Influence of N. fertilization spitting on the root development of winter whaet.**

5437 Schildbach, R.   DE 105201/70/0008
Untersuchungen zur Optimierung der Ertragsbildung durch intensive Düngung bei Winterweizen und Sommergerste und deren Einfluss auf die Qualitätseigenschaften. **Investigations for the optimum yield development by intensive fertilization to winter–wheat and spring barley and its effect upon quality criterions.**

5438 Przemeck, E.; Eissa, S.   DE 132061/75/0004
Über den N–Stoffwechsel in der Karyopse von Sommerweizen bei ausschliesslicher Ammonium– oder Nitraternährung. **On the N metabolism within the caryopse of summer wheat exclusively fed with ammonia or nitrate.**

5439 Müller, K.; Beyer, P.   DE 132062/75/0002
Einfluss stark differenzierter Mineralstoffgaben auf das Spektrum qualitätsbestimmender Inhaltsstoffe und die Bildung biogener Amine in Weizen, Gemüse und Kartoffeln. **Influence of highly differing mineral fertilizing on the spectrum of quality determining components and on the formation of biogenous amines in wheat, vegetables and potatoes.**

5440 Harms, H.   DE 201010/71/4019
Aufnahme und Umwandlung von in unterschiedlichen Stellungen 14C–markierten Ligninabbauprodukten durch Weizenkeimpflanzen und deren Abbau in pflanzliche Zellsuspensionskulturen. **Intake and transformation of differently 14C–labelled lignin metabolites by wheat germ plants and their degradation into vegetal cell suspension cultures.**

5441 Trolldenier, G.   DE 902004/77/0001
Einfluss der Mineralstoffernährung und von Bodeneigenschaften auf die Schwarzbeinigkeit bei Weizen. **Influence of mineral nutrition and soil conditions on take–all of wheat.**

5442 Hoque, R.   DK 030143/79/0004 N
Vandingsplanlægning ved hvedeproduktion. **Irrigation scheduling in wheat production.**

5443 Nielsen, J.M.; Skriver, K.; Gosvig, V.
                    DK 030148/74/0013 N
Vinterhvedegødskning udfra planternes kemiske sammensætning. **Fertilization of winter–wheat based on chemical plant analyses.**

5444 Gachon, L.; Triboi, E.   FR 010802/72/6109
Fertilisation azotée du blé d'automne en Limagne. **Nitrogen fertilization of wheat in Limagne.**

5445 Libois, A.   FR 011001/75/6136
Contribution à l'étude de la fertilisation azotée du blé tendre d'hiver. **Contribution to the study of nitrogen fertilization of winter wheat.**

5446 Muller, J.   FR 012210/70/6270
Prévision de la fumure azotée du blé d'hiver. **Forecasting the nitrogen fertilizer needs of winter wheat.**

5447 Remy, J.C.; Hebert, J.   FR 012503/69/6161
Prévision de la fumure azotée sur blé d'hiver. **Forecasting nitrogen fertilization on winter wheat.**

5448 Remy, J.C.; Dutil, P.; Gachon, L.; Trocme, S.; Lefevre,

G.; Studer, R.   FR 012503/75/6149
Optimisation de la fertilisation en rotation simple. **Optimisation of the fertilization in rotation including wheat, sugar beet and maize.**

5449 Thorne; Thomas   GB 012002/00/0013 R
The effect of nitrogen fertilisation on the physiology and biochemistry of grain growth in wheat.

5450 Reilly, M.L.; Cullen, F.; Flynn, V.   IE 120101/72/9074 N
Enzymes, nitrogen metabolism and yield relationships in wheat. Publications.

5451 Venezian Scarascia, M.E.; Leo, S.   IT 020500/76/0002 N
Influenza dei fattori ambientali ed agronomici sulla produzione quali–quantitativa di alcune varietà di frumento duro nel Meridione. **Influence of environmental and agronomic factors on the quali–quantitative production of some varieties of durum wheat in Southern Italy..**

5452 Mariani, G.; Colesanti, F.; D'Egidio, M.G.
                    IT 020800/74/0011 R
Frumento duro – Effetti della concimazione sulla produzione e la qualità della granella di varietà ad alto rendimento. **Durum wheat – Effects of fertilization on grain yield and quality of high yielding varieties.**

5453 Manzocchi, L.A.   IT 060300/74/0030
Regolazione della sintesi di proteine in semi di frumento anche in relazione alla concimazione azotata tardiva. **Regulation of protein synthesis in wheat grains as affected by late spring nitrogen fertilization.**

5454 Del Zan, F.; Snidaro, M.   IT 090701/76/0002
Prove di concimazione N,P,K, su frumento, orzo, mais, patata, peperone in ambienti pedoclimatici diversi della Regione Friuli–Venezia Giulia. **Manuring tests of N.P.K. on wheat, barley, maize, potato, pepper in different planes and climates of the Friuli–Venezia Giulia district.**

5455 Arnold, G.H.   NL 010103/73/4359
Invloed van de stikstofbemesting op de opbrengst en de bakkwaliteit van tarwe. **Influence of nitrogen fertilization on yield and baking quality of wheat.** Publications.

Fibre plants and oil crops in general (B 3200)

See also 3135

Olive (B 3220)

5456 Parlati, M.; Turco, D.; Petruccioli, G.; Manca, M.G.
                    IT 021400/78/0003 N
Influenza della concimazione fogliare sulla produttività dell'olivo. **Leaf fertilizing influence on the olive tree productivity.**

5457 Parlati, M.; Petruccioli, G.; Manca, M.G.
                    IT 021400/78/0004 N
La concimazione minerale dell'olivo: effetti semplici e combinati dei principali elementi fertilizzanti sulla produttività. **The mineral fertilizing of the olive tree: simple and combined effects on the productivity of principal fertilizer.**

5458 Lombardo, N.   IT 021400/79/0006 N
Emendamenti al terreno per aumentare la produttività

D 2200 – Plant nutrition and fertilization

dell'olivo. **Corrections to soil to increase the olive tree productivity.**

Rape (B 3230)

See also 3164

5459 Lemaire, L.; Morice, J.　　　　FR 010112/76/9109
Etude des besoins en eau du colza oléagineux. **Investigation of the water requisements of oleaginoux rape–seed.**

5460 Nye; White　　　　GB 023801/79/0005 N
Isolation and identification of phosphate solubilising compounds in the rhizosphere of rape (brassica napus).

Soyabean (B 3240)

See also 3633, 3901

5461 Juste, C.; Lubet; Delas, J.; Menet, M.
　　　　FR 010704/73/6081
Adaptation du soja aux sols acides du sud–ouest atlantique. **Adaptation of soyabean to french south–west acid soils.**

5462 Lagacherie, B.; Obaton, M.　　FR 011002/67/0599
Implantation du soja en France. Amélioration de la symbiose Rhizobium – Soja. **Growing soyabean in France. Improvment of soyabean – Rhizobium symbiosis.** Publications.

5463 Decau, J.; Lencrerot, I.　　　　FR 011401/72/0638
Production oléo–protéique du soja selon les techniques de culture (alimentation hydrique, inoculation, fertilisation). **Effect of water supplies, fertilization, inoculation on oil and protein production of soyabean.** Publications.

5464 Puech, J.; Blanchet, R.; Maertens, C. FR 011401/72/6197
Alimentation hydrique du soja et efficience de l'eau consommée. **Water supply of soybean and water efficiency.**

5465 Pirani, V.　　　　IT 021100/76/0005 N
Fertilizzazione della soia.. **Fertilizer application on soybean.**

5466 Ciafardini, G.　　　　IT 021100/76/0008 N
Efficacia di un ceppo di "rhizobium japonicum" su cultivars di soia. **Effectiveness of a "rhizobium japonicum" strain on soybean cultivars..**

Sunflower (B 3250)

See also 3061, 5417, 5418

5467 Decau, J.; Lencrerot, I.　　　　FR 011401/71/0637
Protéogénèse, rendement et équilibre oléo–protéique du tournesol selon les techniques culturales appliquées. (Irrigation, alimentation minérale). **Effect of irrigation and mineral nutrition on proteogenese yield and oil–protein balance in sunflower.** Publications.

5468 Marano, G.; Tarallo, V.; Palmieri, G. IT 040705/77/0001
Azione della concimazione e della competizione sulla composizione gliceridica e sterolica degli olii di girasole. **Action of fertilization and competition on gliceridic and sterolic composition of sunflower oils.** Publications.

5469 Tedeschi, P.　　　　IT 060700/77/0988
Studio delle curve rese/volume su girasole. **A study of the**

yield/volume curve in sunflower plants.

Other fibre plants and oil crops (B 3290)

5470 Bussler, W.; Ghoneim, M.F.　　DE 105202/78/0004 N
Zinklöslichkeit in verschiedenen Lösungsmitteln bei unterschiedlich ernährter Baumwolle. **Solubility of Zn in different extractants from cotton under different nutritional conditions.**

5471 Muller, A.　　　　NL 040012/78/8939 N
Invloed van NPK meststoffen op bodemeigenschappen en samenstelling van voedingsstoffen in bladeren en bladstengels van katoen. **Effect of NPK fertilizers on yield of cotton, soil properties and nutrient composition of cotton leaves and petioles from a cotton field experiment of UNDP (Cotton Development Project Nepal).**

Potatoes (B 3310)

See also 552, 976, 2898, 3240, 5439, 5454, 5733

5472 Amberger, A.; Etman, A.W.　　DE 161040/77/0003
N–Stoffwechsel in Kartoffeln unter dem Einfluss verschiedener N–Ernährung. **Influence of different nitrogen–nutrition on the nitrogen metabolism in potatoes.**

5473 Behringer, H.; Forster, H.　　DE 902003/77/0001
K–Aufnahmerate verschiedener Pflanzen in Abhängigkeit von Pflanzenart und Sorte. **K–uptake rate of different plants and varieties.**

5474 Josefsen, A.B.　　　　DK 030143/78/0002 N
Tørkeperioders indflydelse på udbytte og kvalitet af kartofler. **Production and quality of potatoes in relation to water supply at different growth stages.**

5475 Widdowson; Penny　　　　GB 012003/00/0035 R
Compare rates, times and methods of application of solid and liquid fertilisers for potatoes.

5476 Hutchinson　　　　GB 041101/77/0025 N
Fertiliser rates for blueprint crops of seed and ware potatoes.

5477 Addison; Briant　　　　GB 060213/00/0010
Response of potatoes and root crops to fertiliser.

5478 Schepers, J.H.　　　　NL 010103/72/3566 R
Gedeelde stikstofbemesting bij aardappelen. **Split application of nitrogen to potatoes.** Publications.

5479 Smilde, K.W.　　　　NL 010103/77/7688 R
Onderzoek naar de mogelijkheden van bladanalyse t.b.v. de bepaling van de meststofbehoefte voor de aardappelteelt in ontwikkelingslanden. **Possibilities of foliar analysis for determination of the fertilizer requirements for potato growing in developing countries.**

5480 Loon, C.D. van　　　　NL 010207/71/3973
Invloed van bodemkundige factoren (o.a. vochtvoorziening) en beworteling op opbrengst en kwaliteit van consumptie– aardappelen. **Influence of the properties of soils (a.o. water supply) and root development upon yield and quality of ware potatoes.**

5481 Loon, C.D. van　　　　NL 010207/74/9014 N

Aanpassing van het aardappelgewas aan droogte en hoge tenperaturen. **Adaption of the potato crop to drought and high temperatures.**

5482 Loon, C.D. van   NL 010207/79/9015 N
Optimalisering van de stikstofvoeding van aardappelen. **Optimizing nitrogen nutrition of potatoes.**

Sugarbeets and other sugar crops (B 3320)

See also 1478, 3253, 3254, 3264, 3283, 5424, 5448, 10223

5483 Willenbrink, J.   DE 111120/72/0003 N
Zur Physiologie des Transports von Assimilaten in die wachsende Zuckerrübe. **On the physiology of transport of assimilates into growing sugar–beets.**

5484 Heyland; K.–U.; Braun, B.; Volger, B.
DE 111254/75/0001
Einfluss agrotechnischer Massnahmen auf die Stickstoffaufnahme sowie Ertrags– und Qualitätsbildung der Zuckerrübe unter besonderer Berücksichtigung der Nitratverfügbarkeit im Boden. **Influence of cultivation measures on nitrogen uptake as well as formation of yield and quality of sugar beets with special reference to availability of nitrate in the soil.**

5485 Fischbeck, G.; Büechl, A.   DE 161250/78/0006 N
Der Einfluss unterschiedlicher Stickstoffversorgung auf Mineralstoffaufnahme, Rüben– und Blattertrag sowie Qualität der Zuckerrüben auf verschiedenen Standorten. **Influence of different nitrogen supply on mineral uptake, beet and leaf yield and on quality of sugar–beets on different sites.**

5486 Fischbeck, G.; Berger, F.   DE 161250/78/0007 N
Wirkung unterschiedlicher Stickstoffversorgung bei hoher Kaliumversorgung auf die Ertrags– und Qualitätsbildung der Zuckerrübe auf verschiedenen Standorten Südbayerns. **Effect of Na–fertilization after high K–supply on crop yield and quality of sugar–beets on different sites in South Bavaria.**

5487 Dambroth, M.   DE 201040/75/0025
Reduzierung des Blattapparates von Zuckerrüben durch eine modifizierte Applikationsweise der Stickstoffdüngung als Basis für verbesserte Ertragsleistungen. **Reduction of the leaf apparatus of sugar beets by a modified method of applying N–fertilizers as a basis for improved yield production.**

5488 Meinhold, K.; Kleinhanss, W.   DE 201100/78/0004 N
Zur Frage der optimalen Düngungsintensität im Zuckerrübenund Getreidebau. **On optimal intensities of fertilizer application to sugar beets and cereals.**

5489 Merkel, D.   DE 507051/75/0001
Bodenuntersuchung auf Mineralstickstoff als Hilfsmittel für die optimale Bemessung der Düngerhöhe bei Zuckerrüben. **Evaluation of the optimum nitrogen nutrition of sugar beets by means of an estimation of the mineral nitrogen content of the soil.**

5490 Winner, C.; Müller, A.von; Feyerabend, I.
DE 907000/75/0002
Stickstoffaufnahme der Zuckerrübe aus verschiedenen Bodenschichten im Verlauf der Vegetationszeit bei unterschiedlicher N–Düngung. **N–uptake of sugar beets from different soil depths during the vegetation period influenced by nitrogen fertilization.**

5491 Winner, C.; Bürcky, K.   DE 907010/77/0001
Bedeutung des Nährstoffangebotes – N, K, Na – für die Qualität der Zuckerrübe – Modellversuche –. **Importance of nutrient supply – N, K, Na – for the quality of sugarbeet – model trials –.**

5492 Mettauer, H.; Conesa, A.; Vuittenez FR 010901/76/6130
Etude globale des rendements en sucre de betterave en relation avec les conditions physiques et les disponibilités en azote minéral des sols. **Study of sugar beet yield relationship with physical conditions and mineral nitrogen availability in soils by mean of surwey method.**

5493 Arnoux, M.   FR 011201/74/0220
Etude de l'amélioration hydrique. **Study of hydrous weeds in breeding Provence–Cane for higher productivity.**

5494 Arnoux, M.   FR 011201/74/0221
Physiologie de la nutrition azoté de la canne ob Provence. **Study of nitrogen nutrition physiology in breeding Provence – Cane for high productivity.** Publications.

5495 Hebert, J.; Machet, J.M.   FR 012503/76/6160
Prévision de la fumure azotée de la betterave à sucre. **Predicting nitrogen fertilizer needs for sugar beet.**

5496 Draycott   GB 012301/00/0001 R
Nitrogen fertilization of sugar beet.

5497 Draycott; Durrant   GB 012301/00/0002 R
Potassium and sodium fertilization of sugar beet.

5498 Draycott   GB 012301/00/0004 R
Interactions and long–term experiments with N,P,K, and sodium on rotations.

5499 Draycott   GB 012301/00/0005 R
Magnesium fertilozation of sugar beet.

5500 Draycott; Farley   GB 012301/00/0007 R
Trace element nutrition (manganese, copper, boron).

5501 Brogan, J.C.; Herlihy, M.; Blagden, P.; McEntee, M.; Jelly, R.M.; Lee, J.   IE 060200/78/1423 N
Soil fertility and sugar beet yield and quality. 1.Crop studies. 2.Soil and nitrogen studies. 3.Soil physical factors. 4.The role of weather. Publications.

5502 Curry, J.P.; Coleman, D.   IE 120101/79/9179 N
The influence of cattle manure on the arthropod fauna of sugar beet.

5503 Cremaschi, D.; Casarini, B.; Giordano, I.; Fontana, F.; Alessandrini, L.; Ciafardini, G.   IT 021100/75/0008 R
Criteri di fertilizzazione della barbabietola da zucchero. **Fertilization criteria for sugar beet.** Publications.

5504 Biancardi, E.; Leoni, O.; Dalla Pace; Casalicchio; De Boni   IT 021100/79/0015 N
Prove di concimazione potassica, fosfatica e con microelementi su barbabietola da zucchero. **Essays of potassic and phosphatic fertilization and with microelements on the sugar beet.**

5505 Ciafardini, G.; Cremaschi, D.; Buzzoni, M.T.; Fontana,

D 2200 – Plant nutrition and fertilization

F.   IT 021100/79/0016 N
Azotofissazione libera con inoculo di SPIRILLUM lypoferum
su barbabietola da zucchero. **Free fixation of nitrogen with
inoculation of sugar beet with spirillum lypoferum.**

5506 Ferrari, Th.J.   NL 010103/60/0540
Oorzaken van de verschillen in opbrengst van suikerbieten in
de Noordoostpolder. **Factors of yield differences of sugar beet
in the North–East Polder.**

5507 Boer, J.   NL 010207/78/8415
Verbetering van de opbrengst bij suikerbieten door laagsgewijs
pH verhoging onder de bouwvoor door middel van injecteren
van kalkmeststof in vloeibare vorm. **Improvement of yield of
sugar beet by means of increasing of pH of the subsoil by means
of injection of liquid chalk fertilizer.**

Other starch producing plants (B 3390)

See also 5474

5508 Müller, K.; Achmed, S.S.   DE 132062/75/0007
Vergleichende Untersuchungen über den Alkaloidgehalt in
Solanum tuberosum und Solanum laciniatum nach Abfolge
differenzierter Mineralstoffgaben. **Comparative studies on the
alkaloid content in Solanum tuberosum and Solanum
laciniatum as a function of differentiated mineral supply.**

Grasses and forage crops in general (B 3400)

See also 3309, 3331, 3344, 3364, 11973

5509 Schöllhorn, J.   DE 501200/71/0001
Der Einfluss der Teilung der Nährstoffgaben auf die Menge,
den zeitlichen Anfall und die Qualität der Futtererträge. **The
distribution of fertilizers, its consequences on the yield of
forage, its levelling during the summer and its content of
nutrients.**

5510 Bohle, H.; Puffe, D.   DE 506153/74/0002
Reaktion von Futterpflanzen – Entwicklung, Ertrag, Gehalte –
und Standort – Nährstoff–, Wasserhaushalt – auf
Gülledüngung. **Reaction of forage crop – growth, yield,
contents – and ecological nutrient and water balance to
fertilization with semi–liquid manure.**

5511 Nielsen, G.G.   DK 030146/72/0020
Foderplanters selenindhold. **Selenium content of fodder plants.**

5512 Laissus, R.   FR 010115/66/0263
Analyse des variations de la production et de la flore de deux
prairies permanentes sous l'effet de doses croissantes d'azote.
**Analysis of the production variations of the flora of two
permanent meadows induced by increasing nitrogen supply.**

5513 Laissus, R.; Leau, G.   FR 010115/72/0269
Etude du comportement de diverses espèces fourragères de la
prairie permanente en zone océanique après intensification.
**Study of the behaviour of various fodder plant species of the
permanent meadow in the oceanic area after intensification.**

5514 Dumas, Y. – Clairon, M.; Sobesky, O. Mme
                                           FR 011602/73/6030
Alimentation hydrique et minérale des fourrages tropicaux.
Water and mineral alimentation of tropical forages.

5515 Duval, L.   FR 012208/68/0623
Variation de la teneur en molybdène des plantes fourragères
sous l'influence de divers facteurs. **Variation in the
molybdenum content of forage plants under the influence of
various factors.** Publications.

5516 Duval, L.   FR 012208/68/6180
Facteurs influençant la teneur en molybdène des plantes.
Carence végétale, toxicité animale. **Molybdenum content of
forage plants ; deficiency, toxicity.**

5517 More, E.; Coppenet, M.   FR 012208/73/6179
Teneurs en se des plantes fourragères en Bretagne. Effets des
apports de sélénite. **Se contents of forage plants in Brittany
Effects of applied selenite.**

5518 Duval, Y.   FR 012219/76/6185
Influence de la date et de la dose d'apport de P sur la
production et la composition de l'herbe. **Influence of rate and
frequency of applying phosphorus fertilizer on yield and
composition of herbage plants.**

5519 Jones   GB 011201/00/0002 R
**Uptake and utilisation of N.K.S. and trace elements by forage
plants.**

5520 Harvey; Wright   GB 040301/79/0037 N
Salt tolerance in grasses.

5521 Fleming, G.A.; Ryan, M.J.   IE 060200/75/1108 R
Mineral composition of herbage. Publications.

5522 Herlihy, M.; Murphy, W.E.   IE 060200/77/1265 N
**Efficiency of nitrogen use and herbage production in relation to
fertiliser source, times of application and climatic variables.**
Publications.

5523 Murphy, W.E.; Flynn, A.V.   IE 060200/78/1426 N
Evaluation of urea as a fertiliser for silage. Publications.

5524 Montorsi, M.; Spallacci, P.   IT 020500/74/0007
Studio della qualità dei foraggi ottenuti dalle prove di
fertilizzazione con liquami suini. **Quality of forages manured
with pig wastes.** Publications.

5525 Landi, R.   IT 040502/78/1066 N
Ricerca sulla competizione nutrizionale delle associazioni
leguminose–graminacee. **Research on nutritional competition
in pulse – grass combinations.**

5526 Dijkshoorn, W.   NL 010102/77/7298
Beoordeling van de fosfaatvoorziening van tropische gewassen
door bladanalyse. **Evaluation of the phosphorus status of
tropical herbage by leaf analysis.**

Grasses (B 3410)

See also 3489, 3490, 3530, 3542, 5061, 5374, 5379, 5571, 5578,
5613, 5787, 5789, 6661

5527 Cordiez, E.; Lambert, J.; Nogarede, P.; Forceille, M.J.;
Bienfait, J.   BE 020602/74/0011
Etude comparative des méthodes de récolte et de conservation
permettant la valorisation maximale des protéines en
provenance des associations graminées + légumineuses
soumises à différents niveaux de fertilisation. **Comparative**

study of the harvesting methods permitting a maximal valorization of the proteins originated in companion crops of graminaceae and leguminous plants and this under different levels of fertilizers. Publications.

5528 Lambert, J.; Forceille, M.J. BE 020602/75/0009 R
Etude de la qualité des productions végétales (graminées, légumineuses, productions horticoles) en fonction du niveau d'intensification (fumure azotée et emploi de pesticides). **Study of the quality of vegetal productions (graminaceae, legumes and horticulture) in regard with intensification on level (N manure and use of pesticides).**

5529 Ellenberg, H.; Sharifi, M.R. DE 132093/75/0001
Ökologisches und physiologisches Verhalten von Bromus erectus, Alopecurus pratensis und Arrhenatherum elatius bei unterschiedlicher Stickstoff- und Wasserversorgung. **Ecological and physiological behaviour of Bromus erectus, Alopecurus pratensis and Arrhenatherum elatius at various nitrogen and water supply levels.**

5530 Gütte, J.O.; Helfferich, B. DE 132305/74/0001
Einfluss der Düngung auf den Mineralstoffgehalt und Futterwert des Weidegrases. **The influence of fertilization on mineral content and feeding value of pasture grass.**

5531 Jacob, H. DE 144500/74/0002
Eignung einzelner Dauergrünlandarten und -sorten zur Vielschnittnutzung bei verschiedener Düngungsintensität. **The qualification of single species and varieties of permanent grasses for multi-cut utilization in dependence on different intensity of fertilization.**

5532 Jacob, H. DE 144500/78/0003 N
N–Verwertungsvermögen verschiedener Gräser. **The effects of different rates of nitrogen application on different species of grasses.**

5533 Boeker, R.; Müssel, H. DE 502107/74/0001
Erprobung zahlreicher, im Handel zur Rasendüngung empfohlener Düngemittel auf ihre spezielle Eignung bei Rasen, getrennt nach 3 Gruppen: herbizidhaltige Produkte – orga – nische Dünger – synthetisch organische und leichtlösliche Dünger. **Testing of numerous turf fertilizers usual in trade for their special suitability i.e. of fertilizers classified into herbicide–containing products, organic manures, and synthetic organic and easily soluble fertilizers.**

5534 Wasshausen, W. DE 507350/78/0004 N
Wirkung der herbstlichen N–Düngung auf das Frühjahrswachstum des Grünlandes. **The influence of autumnal nitrogen–fertilization on grass growth in spring.**

5535 Rasmussen, P.; Nordestgaard, A. DK 010106/67/3513
Forskellige kvælstofmængder, såmængder og rækkeafstande, såtider, dækafgrøder, forårsslæt, udnyttelse af genvækst efter frøhøst m.m. ved frøavl af græsser, roer, spinat, radis m.v.. **Rate of nitrogen application, sowing rate and distance, date of sowing, cover crops, effect of spring cutting, utilization of regrowth after seed harvest, ect., in seed crops of grasses, sugar beet, spinach, radish, etc..**

5536 Salette, J.; Lemaire, G. FR 010403/75/6020
Dynamique de l'absorption de l'azote par un peuplement de graminées pur ou une prairie naturelle sans légumineuses. **Dynamics of nitrogen uptake by a sown grass–sward or a**

permanent pasture without legumes.

5537 Lemaire, G. FR 010403/76/6023
Mise au point de dispositifs expérimentaux pour étudier l'évolution de la consommation hydrique des graminées fourragères tout au long de leur croissance. **Study of an experimental device for grass and water field trials.**

5538 Salette, J.; Lemaire, G. FR 010403/77/6021
Interactions azote x génotype sur les graminées fourragères : étude du premier cycle. **Nitrogen and genotype interactions in forage grasses : study of the primary growth of italian rye–grass.**

5539 Champeroux, A. Mme FR 010502/66/0522
Etude du potentiel d'utilisation de l'azote par le ray–grass. Recherche de tests de sélection basés sur la physiologie de la plante. **Study on the possibilities of nitrogen utilization by rye–grass. Breeding tests based upon plant physiology.** Publications.

5540 Niqueux, M.; Arnaud, R. FR 010801/69/0365
Fertilisation et exploitation de prairies permanentes en zone de montagne. **Perennial mountain grassland : fertilization and exploitation.**

5541 Fourbet, J.F. FR 011801/75/9075
Incidence de la fertilisation phosphatée en terrain argilo–limoneux sur les qualités nutritionnelles d'une fétuque élevée. **Incidence of phosphorus fertilizing in a clay–loam soil, on feed values of Festuca arundinacea.**

5542 Coppenet, M.; Moré, E. FR 012208/65/0625
Influence de divers facteurs sur la teneur des plantes fourragères (maïs, ray–grass) en oligo–éléments: Mn, Cu, Co, Zn. **Effect of various factors on trace element content (Mn, Cu, Co, Zn) in forage plants (maize, rye–grass). Publications.**

5543 Coppenet, M.; Moré, E. FR 012208/68/0626
Enrichissement des plantes fourragères (maïs, ray–grass) en minéraux majeurs et mineurs (P, Mg, Cu, Co, Zn) par apport de sels au sol et sur les feuilles. **Increasing forage plants content (maize, rye–grass) in major and minor elements (P, Mg, Cu, Co, Zn) by salts supplies to soil and leaves. Publications.**

5544 Moré, E.; Coppenet, M. FR 012208/73/0628
Enrichissement des plantes fourragères par apport de séléniates au sol. Teneurs des maïs et ray–grass. Cause de variations. **Increasing selenium content of forage plants by seleniates supplies to the soil. Variations in maize and Rye–grass content.**

5545 Gillet, M.; Sauvion, A.; Jadas–Hecart, J.
  FR 012223/73/0416
Etude du déterminisme nutritionnel de la production précoce des graminées au printemps. **Research of the nutritional determinism of early spring growth of grasses.**

5546 Robson; Parsons GB 011203/00/0013 R
Entry, partition and utilization of carbon in grass crop production; C14 as label and tracer.

5547 Ryle; Gordon GB 011203/75/0015 R
The biochemistry of carbon utilisation during the growth of graminaceous plants.

5548 Pollock; Lloyd GB 012110/75/0006 R
Synthesis and export of sucrose through the developmental sequence of individual leaves of lolium spp.

5549 Pollock GB 012110/76/0007 R
The synthesis and metabolism of polyfructans in grasses.

5550 Newbould GB 030304/00/0003
Nutrient requirements of white clover and sown grasses in hill soils.

5551 Laidlaw GB 040301/00/0015 R
High nitrogen tolerance in white clover cultivars.

5552 Laidlaw; Hayes GB 040301/00/0016 R
Measurement of nitrogen fixation of white clover using acetylene reduction test.

5553 Harvey GB 040301/75/0025 R
Genetic variation of magnesium uptake by grasses.

5554 Garrett GB 040403/75/0009 R
Photorespiration in grass.

5555 Forbes GB 060216/00/0001
Mineral content of herbage in relation to botanical composition.

5556 Voss GB 060306/00/0003
Effect of season and of fertiliser nitrogen level and type on trace element content of S24 perennial ryegrass.

5557 Voss GB 060306/00/0004
Effect of soil pH on trace element uptake of herbage.

5558 Murphy, P. IE 060200/74/1048 R
Characteristics of the organic constituents (especially proteins and carbohydrates) of the newer grass and clover cultivars with particular reference to effects of nitrogen fertiliser. Publications.

5559 Herlihy, M. IE 060200/74/1098 N
Efficiency of use of commercial nitrogen fertilisers in relation to method of application, soil texture, ph and free carbonates. Publications.

5560 Tunney, H. IE 060200/76/1183 R
Comparative effects of pig slurry, cattle slurry and fertiliser on yield and quality of grass silage. Publications.

5561 Meijer, W.J.M.; Bor, N.A. NL 010207/76/7650
De invloed van stikstofbemesting op de groei en de zaadopbrengst van raaigrassen. The effect of nitrogen applications on growth and seed yield of ryegrass.

5562 Meijer, W. NL 010207/78/9011 N
De invloed van stikstofbemesting op groei en zaadproduktie van roodzwenkgras (Festuca rubra). The effect of nitrogen applications on growth and seed production of red fescue (Festuca rubra).

5563 Meijer, W. NL 010207/79/9012 N
De invloed van stikstofbemesting op groei en zaadproduktie van veldbeemdgras (Poa pratensis). The effect of nitrogen applications on growth and seed production of smooth stalked meadow grass (Poa pratensis).

Pastures, grassland (B 3420)

See also 11, 1574, 3506, 3508, 3513, 3530, 3537, 3542, 3543, 3544, 3590, 3592, 3601, 5060, 5074, 5080, 5536, 10247, 12002

5564 Lambert, J.; Denudt, G.; Marot, J. BE 020602/68/0003 R
Etude phytosociologique des prairies permanentes améliorées Lolieto Cynosuretum et de la relation entre la fumure et la composition botanique. Phytosociological study of old permanent grasslands Lolieto Cynosuretum and the relationship between applied fertilizers and botanical composition. Publications.

5565 Lambert, J.; Marot, J. BE 020602/69/0001 R
Etude de la compétition sur les macronutrients et quelques oligoéléments en prairie semi–naturelle. Studies on competition for macronutrients and some trace–elements in old permanent grasslands. Publications.

5566 Lambert, J.; Toussaint, B.; Marot, J.; Genot, P. BE 020602/71/0004 R
Recherche des différentes fumures amenant la valorisation maximale des prairies temporaires de fauche. Effect of different fertilization levels viewing a maximal valorisation of temporary grassland. Publications.

5567 Lambert, J.; Toussaint, B. BE 020602/71/0005
Analyse du lisier et premières observations sur son effet fertilisant en prairie. Chemical analysis of slurry and first observation on its effect on grasslands as fertilizer. Publications.

5568 Bussler, W.; Jung, Y.K. DE 105202/78/0005 N
Verbesserung von Grünland durch Spurennährstoffdüngung. Improvement of grassland by fertilization with trace elements.

5569 Boeker, P.; Opitz–von–Boberfeld, W. DE 111251/74/0007
Wie verhalten sich Narben aus Ökotypen im Vergleich zu Narben aus Zuchtsortengemischen bei unterschiedlichem N–Aufwand im Hinblick auf Narbendichte, Ertrag und Bestandesentwicklung. What about the reaction of ecotypological swards in comparison with swards of mixed selected varieties to different levels of nitrogen fertilization with regard to sward density, crop yield, and plant growth.

5570 Boeker, P.; Opitz–von–Boberfeld, W. DE 111251/74/0008
Einfluss verschiedener Kalidüngemittel auf die Na– und Mg–Gehalt des Weideaufwuchses in Abhängigkeit von der verabreichten Kalimenge und –form und der Höhe der N–Düngung. The effects of different potash fertilizers on the Na and Mg content in pasture plants in dependence on the applied amount and form of potash and on the level of N–fertilization.

5571 Boeker, P.; Opitz–von–Boberfeld, W. DE 111251/74/0010
Wie verhalten sich in Mähweidemischungen di– und tetraploide Lolium perenne–Sorten unterschiedlicher Reifegruppen in Abhängigkeit von differenziertem N–Aufwand im Hinblick auf Konkurrenzkraft, Ausdauer und Ertrag. What about reactions of diploid and tetraploid varieties of Lolium perenne of different stage of maturity in pasture plant mixtures in dependence on differentiated nitrogen fertilization with regard to competitive power, tenacity and yield.

5572 Boeker, P.; Opitz–von–Boberfeld, W.
DE 111251/75/0001
Ertragsleistung und Weiderest verschiedener Festuca arundinacea–Sorten eines Artbastardes und von Lolium perenne – NFG – in Weidemischungen bei unterschiedlicher Stickstoffdüngung. **Yield production and pasture remainder of different varieties of Festuca arundinacea of an interspecific hybrid and of Lolium perenne – NFG – in pasture plant mixtures in relation to diverse nitrogenous fertilization.**

5573 Gehring, M.; Thalmann, H.; Voigtländer, G.; Kühbauch, W.
DE 161255/78/0002 N
Wirkungen belüfteter Rindergülle auf Pflanzenbestand und Ertrag auf Grünlandflächen. **Effects of aerated cattle slurry on sward composition and yield of permanent grassland.**

5574 Schöllhorn, J.; Müller, A.
DE 501200/70/0002
Die Wirkung von Drei–Nährstoffdüngern im Vergleich zu Einzeldüngern bei Schnittwiesen auf Altmoräneböden. **Compound fertilizers compared with straight fertilizers in their results on hay meadows.**

5575 Schöllhorn, J.; Müller, A.
DE 501200/70/0003
Der Einfluss von Wasserzusätzen zu Schwemmist auf vierschürigen Wiesen. **Liquid farmyard manure with different parts of water 1969 (Forts..**

5576 Diez, T.
DE 502055/74/0009
Verwertung von Klärschlamm aus 2– und 3–stufigen Kläranlagen auf Grünland im Hinblick auf Bodenfruchtbarkeit, Ertrag und Futterqualität. **Disposal of sludge from sewage plants on pasture land affecting soil fertility, yield and quality of feed.**

5577 Rieder, J.B.
DE 502058/78/0008 N
Einfluss steigender Schnittfrequenz und Düngung auf unterschiedliche Pflanzengesellschaften des Dauergrünlandes. **Influence of increasing cutting utilization and fertilization on different plant communities of permanent grassland.**

5578 Speidel, B.
DE 506151/75/0001
Bioelement–Umsätze einer verschieden gedüngten Goldhaferwiese und eines optimal gedüngten Feldgrasbestandes. **Residual bio–elements of a differently fertilized meadow of golden oat–grass and of an optimally fertilized ley–farming stand.**

5579 Wasshausen, W.
DE 507350/78/0005 N
Erhöhung des Natriumgehaltes im Weideaufwuchs durch natriumhaltige Düngemittel. **The increase of sodium content in pasture growth by sodium containing fertilizers.** Publications.

5580 Wasshausen, W.
DE 507350/78/0006 N
Untersuchungen zum N– und P–Bedarf von Mähweiden. **Investigations on the demand for nitrogen and phosphorus in hay pastures.**

5581 Neuhaus, H.
DE 507351/78/0001 N
Nährstoffaustrag bei gedränten Brackmarschböden in Grünlandnutzung. **Nutrient loss in drained brackish marsh soils in grassland farming.** Publications.

5582 Neuhaus, H.
DE 507351/78/0002 N
Wirkung extrem hoher Güllegaben auf den Ertrag und die Narbenzusammensetzung von Mähweiden im nordwestdeutschen Marschgebiet. **Effect of extremely high distribution of liquid manure on crop and sward composition of hay pastures in the marshes in Northwest Germany.** Publications.

5583 Neuhaus, H.
DE 507351/78/0003 N
Wirkung der Gülledüngung auf Mähweiden während der Hauptvegetationszeit im Vergleich zu mineralischer Düngung. **Effect of fertilization with liquid manure on hay pastures during main vegetation period in comparison to mineral fertilization.** Publications.

5584 Mott, N.
DE 508301/70/0006 N
Einflub der P–K–Düngung auf Mähweiden mit guter und überhöhter P–K–Versorgung auf Ertrag und Mineralstoffgehalt. **Influence of P–K fertilization on mowing pastures with convenient and overdosed P–K supply on yield and mineral content.**

5585 Ernst, P.
DE 508301/75/0009
Vergleich von flüssigen und festen Mineraldüngern; Einfluss auf Ertrag und Inhaltsstoffe einer Dauerweide. **Comparison of liquid and solid mineral fertilizers; influence on yield and contents in permanent pasture.**

5586 Ernst, P.
DE 508301/75/0011
Einfluss des pH–Wertes auf Ertrag und Inhaltsstoffe einer Dauerweide; Vergleich von Hüttenkalk und Kalkmergel. **The influence of the pH value on yield and contents of a permanent pasture; comparison of metallurgical calcium and of marly lime.**

5587 Ernst, P.
DE 508301/77/0002
Einfluss des Zeitpunktes der ersten Stickstoffgabe auf den Weideertrag Anfang Mai. **Effect of date of the first nitrogen fertilization on the grass yield at the beginning of May.**

5588 Ernst, P.
DE 508301/77/0003
Einfluss der Stickstoffdüngung – Termin und Menge – auf den Grünlandertrag im Spätsommer und Herbst. **Effect of nitrogen fertilizing – date and amount – on the grass yield in late summer and autumn.**

5589 Niqueux, M.
FR 010801/65/0315
Production des pâturages d'estive du Massif Central. **Yielding summer pastures in Massif Central.**

5590 Niqueux, M.; de Montard, F.X.; Jeannin; Petit
FR 010802/65/6101
Action de la fertilisation azotée sur la croissance de l'herbe des pâturages de montagne, conséquences sur les performances des bovins à l'estive. **Nitrogen fertilizer in hills pasture and its consequence on growth of live stock.**

5591 Périgaud, S.; de Montard, F. X.
FR 010802/72/6102
Oligo–éléments des fourrages de prairies permanentes. **Micro elements in Forages of permanent meadows.**

5592 Dejou, J.; de Montard, F.X.; Arnaud
FR 010802/74/6100
Action des fertilisants minéraux sur les prairies de fauche en montagne volcanique. **Fertilizing trials on Meadows in Massif Central.**

5593 Louyot, J.M.; Jeannin, B.; De Montard, F.
FR 010804/72/9012 N
Etude des formes et doses de fumure organique à appliquer en

zones pastorales d'altitude. **Study of types and amount of manure to be used on highlands.**

5594 Louyot, J.M.; Jeannin, B.; Niqueux, A.
FR 010804/72/9014 N
Rythme d'exploitation de prairies naturelles et prairies temporaires en relation avec la fertilisation azotée.. **Grassland management (ley pastures and perennial pastures) in connection with nitrogen fertilization..**

5595 Louyot, J.M.; Jeannin, B.; Renault, P.
FR 010804/75/9013 N
Etude des niveaux et interaction de P et K pour une prairie naturelle (pâturée et fauchée). **Study of P and K dressings 'interactions on a natural pasture.**

5596 Jeannin, S.E.I.B.; Lavalette, B.   FR 012224/74/9007
Etude de la fumure azotée et des modes de conduite de la prairie permanente dans les conditions du Plateau Lorrain. **Evaluation of optimum nitrogen fertilisation of permanent pastures for best pasture management in "Plateau Lorrain" conditions.**

5597 Guibert, S.E.I.P.; Lavalette, B.   FR 012224/76/9008
Etude de la fumure potassique et de l'intéraction azote–potasse sur la prairie permanente pâturée. **Evaluation of the optimum potash fertilization and of the potash nitrogen interaction on pastured permanent meadows.**

5598 Cowling   GB 011201/00/0003 R
Circulation of substances (compounds of N. S. etc) in environment of grasses and legumes.

5599 Morrison   GB 011202/00/0003 R
Utilisation of fertilizer nitrogen by perennial ryegrass.

5600 Jonstone; Mattingly   GB 012003/00/0034 R
Fertiliser and farmyard manure residues in long–term experiments and their value to grass.

5601 Widdowson; Ashworth   GB 012003/00/0036 R
Compare rates, times and methods of application of solid and liquid fertilisers for grass.

5602 Reid   GB 030201/00/0012 R
Effects of date and rate of application of nitrogenous fertilizer on grassland production.

5603 Floate   GB 030304/00/0007 R
Cycling of nutrients in grazed hill pastures and its influence on requirements for lime and fertilisers.

5604 Linehan   GB 030403/00/0009 R
The effect of organic constituents of soil on growth and nutrition with special reference to root processes.

5605 Reith   GB 030406/00/0008 R
Field responses to nutrients: soil type effects and prediction of fertilizer requirements.

5606 Reith   GB 030406/00/0009 R
Trace element status of soils and crops: effects of soil type: diagnosis of deficiencies and excesses.

5607 Adams; Adams   GB 040402/78/0026 R
Liming grassland.

5608 McAllister; Laidlaw   GB 040402/78/0027 R
Nitrogen manuring experiment on ryegrass/white clover.

5609 Gracey   GB 041101/75/0018 R
Value of slurry applied to grassland during the non–growing season.

5610 Leclerc, M.H.   IE 060200/64/0426 R
Clover–nitrogen balance trial for summer milk production. Publications.

5611 Blagden, P.   IE 060200/72/0405 R
Cycling of nutrient – potassium (grassland, soil). Publications.

5612 Brogan, J.C.   IE 060200/75/1113 R
Liming of grazed pasture. Publications.

5613 Gately, T.F.   IE 060200/75/2054 R
Evaluation of the role of new legumes and grass cultivars in pastures at increasing rates of fertiliser nitrogen for milk production. Publications.

5614 Murphy, W.E.   IE 060200/76/1428 N
Evaluation of urea as fertilizer for grazing. Publications.

5615 Hammond, R.F.; Murphy, W.E.   IE 060201/78/1379 R
Nitrogen for grassland on midland organic soils. Publications.

5616 Curry, J.P.; Bolger, T.; O'Brien V.   IE 120101/76/9077 N
The effects of animal manures applied as slurry on the invertebrate fauna of grassland. Publications.

5617 Rivoira, G.; Bullitta, P.; Caredda, S.   IT 041302/73/0001
Esperienze miglioramento pascoli con concimazione, infittimento e turnazioni. **Tests on improving pasture lands through fertilizing, thickening and allowing a rationed grazing.**

5618 Rivoira, G.; Bullitta, P.; Caredda, S.   IT 041302/75/0008
Esperienza concimazione pascoli con diverse dosi di azoto e fosforo. **Tests on fertilizing pasture lands with different quantities of nitrogen and phosphorus.** Publications.

5619 Meer, H.G. van der   NL 010102/78/8923 N
Kwantitatieve analyse van de stikstofhuishouding van grasland. **Quantitative analysis of the nitrogen economy in grassland.**

5620 Meer, H.G. van der   NL 010102/79/8937 N
Het verloop van stikstofopname door een grasgewas en de stikstof accumulatie in de zode en de bodem. **The course of nitrogen uptake by the crop and nitrogen accumulation in the sward and the soil.**

5621 Lande Cremer, L.C.N. de la   NL 010103/64/0577
De invloed van dunne mest op de kalihuishouding en op de botanische samenstelling van grasland. **Influence of liquid manure (slurry) on the potassium economy and on the botanical composition of grassland.** Publications.

5622 Postmus, J.   NL 010103/69/2780 R
Invloed van stikstof op de grasgroei in het vroege voorjaar. **Influence of nitrogen on the grass growth in early spring.** Publications.

5623 Prins, W.H.   NL 010103/72/4361

Het groeiverloop van grasland en de stikstofreactie gedurende het seizoen. **The growth rate of grassland herbage and the nitrogen response of grassland in the course of the growing season**, Publications.

5624 Prins, W.H.   NL 010103/73/4354
Invloed van de veebezetting op de stikstofbehoefte en hergroeisnelheid van gras. **Effect of stocking rate on nitrogen fertilizer requirement and regrowth rate of grassland herbage.**

5625 Prins, W.H.   NL 010103/73/4355
De hergroeisnelheid van grasland in afhankelijkheid van de zwaarte van de voorafgaande snede en van de hoogte van de voorafgaande stikstofgift. **The regrowth rate of grassland herbage in relation to the yield of the preceding cut and the rate of the preceding nitrogen application.** Publications.

5626 Arnold, G.H.   NL 010103/73/4360
De werking van kalkammonsalpeter (magnesiumhoudend) in verband met de minerale samenstelling van gras onder omstandigheden van hoge K–aanvoer. **The effect of ammonium nitrate lime (containing magnesium) on the mineral composition, and especially the magnesium content, of grass under conditions of heavy potassium supply.** Publications.

5627 Prins, W.H.   NL 010103/73/4362 R
De invloed van het gebruik van zeer grote hoeveelheden dierlijke organische mest op de behoefte aan kunstmeststikstof van grasland, de kwaliteit van de zode en van het gras, en op de indringing van mineralen in het bodemprofiel. **The effect of application of very large quantities of animal organic manure on the nitrogen fertilizer requirement of grassland, on sward quality and the quality of the herbage, and on the movement of mineral elements into the profile.** Publications.

5628 Rauw, G.J.G.   NL 010103/73/4363
De invloed van de pH van de grond op het effect van stijgende stikstofgiften op de opbrengst en de kwaliteit van het gras. **The influence of soil pH on the effect of rate of nitrogen application on yield and quality of grassland herbage.** Publications.

5629 Arnold, G.H.   NL 010103/75/6249 R
Onderzoek naar de fosfaatwerking van dierlijke organische mest op grasland. **Investigation into the effectiveness of animal organic manure as a source of phosphorus for grassland.**

5630 Lande Cremer, L.C.N. de la   NL 010103/78/7969
De invloed van rundvee– en kippedrijfmest ondergebracht bij het scheuren van oud grasland op het nitraat gehalte van heringezaaid gras bemest met verschillende N–hoeveelheden. **The effect of cattle and poultry slurry applied before ploughing up old grassland on the nitrate content of reseeded grass fertilized with different amounts of nitrogen.**

5631 Lande Cremer, L.C.N. de la   NL 010103/78/7971
De behoefte aan kunstmeststikstof van grasland onder invloed van het jaarlijks gebruik van dunne mest van rundvee en van kippen. **The fertilizer nitrogen requirement of grassland as affected by annual application of cattle and poultry slurry.**

5632 Kemp, A.; Dijkshoorn, W.   NL 010104/73/3963
Accumulatie van zware metalen in graslandplanten en de eventuele gevolgen daarvan voor het vee. **Accumulation of heavy metals in grass and its influence on cattle.** Publications.

5633 Alberda, T.; Sibma, L.   NL 010104/73/3965

De invloed van de stikstofbemesting en de duur van de groeiperiode op de uitstoeling van een grasgewas. **The influence of nitrogen fertilization and growth period on tiller formation in a grass sward.** Publications.

5634 Geneijgen, J. van   NL 010208/73/8760 N
Invloed van verregenen van rundermest op grasland op de benutting van de N en op het milieu. **Influence of sprinkling irrigation of slurry on grasland, on N–utilization and on environment.** Publications.

5635 Luten, W.   NL 010208/74/8770 N
Injecteren van drijfmest op grasland ter voorkoming van stankoverlast. **Injection slurry on grassland to prevent stench nuisance.**

5636 Boer, D.J. den   NL 010208/76/6944 R
Bepaling van de invloed van een verschillende bemesting op de pH van de bodem, de graslandopbrengst en –samenstelling en de kwaliteit van de zode. **The influence of various dressings on pH of the soil, grass yield, grass composition and sward quality.**

5637 Steenvoorden, J.H.A.M.   NL 010501/77/7595 R
Stikstofuitspoeling op grasland. **Leaching of nitrogen on grassland.**

5638 Postmus, J.   NL 060008/69/2780
Invloed van stikstof op de grasgroei in het vroege voorjaar. **Influence of nitrogen on the grass growth in early spring.** Publications.

5639 Prins, W.H.   NL 060008/72/4361
Het groeiverloop van grasland en de stikstofreactie gedurende het seizoen. **The growth rate of grassland herbage and the nitrogen response of grassland in the course of the growing season.** Publications.

5640 Prins, W.H.   NL 060008/73/4354
Invloed van de veebezetting op de stikstofbehoefte en hergroeisnelheid van gras. **Effect of stocking rate on nitrogen fertilizer requirement and regrowth rate of grassland herbage.**

5641 Prins, W.H.   NL 060008/73/4355
De hergroeisnelheid van grasland in afhankelijkheid van de zwaarte van de voorafgaande snede en van de hoogte van de voorafgaande stikstofgift. **The regrowth rate of grassland herbage in relation to the yield of the preceding cut and the rate of the preceding nitrogen application.** Publications.

5642 Prins, W.H.   NL 060008/73/4362
De invloed van het gebruik van zeer grote hoeveelheden dierlijke organische mest op de behoefte aan kunstmeststikstof van grasland, de kwaliteit van de zode en van het gras, en op de indringing van mineralen in het bodemprofiel. **The effect of application of very large quantities of animal organic manure on the nitrogen fertilizer requirement of grassland, on sward quality and the quality of the herbage, and on the movement of mineral elements into the profile.** Publications.

5643 Prins, W.H.   NL 060008/75/6249
Onderzoek naar de fosfaatwerking van dierlijke organische mest op grasland. **Investigation into the effectiveness of animal organic manure as a source of phosphorus for grassland.**

5644 Boer, D.J. den   NL 060008/76/6944 R
Bepaling van de invloed van een verschillende bemesting op de

pH van de bodem, de graslandopbrengst en –samenstelling, en de kwaliteit van de zode. **The influence of various dressings on pH of the soil, grassyield, grass composition and shard quality.**

Mangolds (B 3430)

See also 5535

5645 Nielsen, J.M.; Smed, N.N.; Skriver, K.
　　　　　　　　　　　　　　　DK 030148/74/0014 N
Gødskning af fabriksroer udfra planternes kemiske sammensætning. **Fertilization of sugar beets based on chemical plant analyses.**

5646 Maurice, J.　　　　　　　　FR 012208/75/6183
La nutrition en bore de la betterave fourragère. **Boron in fodder beets.**

Legumes in general (B 3440)

See also 3631, 3633, 3901, 5527, 5528, 5613, 5746, 5748, 5750

5647 Obaton, M.; Lagacherie, B.　　　FR 011002/67/0597
Technologie des inoculum et de l'inoculation des légumineuses avec des Rhizobium. **Producing Rhizobium and inoculation of legumes.**

5648 Amarger, N.　　　　　　　　FR 011002/68/0594
Fixation symbiotique : Relations entre Rhizobium et légumineuses dans les processus d'infection. **Symbiotic N. fixation : legume – Rhizobium relationships during infection.**

5649 Dixon　　　　　　　　GB 021001/79/0011 N
Efficiency of nitrogen fixation in root nodules.

5650 Murphy, P.; Masterson, C.L.; Sherwood, M.
　　　　　　　　　　　　　　　IE 060200/76/1181 R
Plant physiological factors controlling nitrogen fixation in legumes. Publications.

5651 Masterson, C.L.　　　　　IE 060200/77/1290
The fertiliser requirements of forage legumes grown for conservation.

5652 Masterson, C.L.　　　　　IE 060200/78/1407 N
The effect of lime and nitrogen on legumes and their nodule bacteria in field soils. Publications.

5653 Lie, T.A.　　　　　　　　NL 020033/58/4710
Symbiontische stikstofbinding bij leguminosen: fysiologie van de wortelknolvorming. **Symbiotic nitrogen fixation with legumes: physiology of root–nodule formation.** Publications.

5654 Mulder, E.G.　　　　　　NL 020033/65/4709
Invloed van bemesting en bodemfactoren op de stikstofbinding van vlinderbloemige planten. **Influence of fertilization and soil factors on nitrogen fixation of leguminous plants.** Publications.

5655 Egeraat, A.W.S.M. van　　　NL 020033/65/4712
De invloed van plante–uitscheidingen (van vnl. leguminosen) op de micropopulatie in rhizosfeer. **The influence of root exudates (especially from leguminous plants) on the rhizosphere micro–population.** Publications.

5656 Lie, T.A.　　　　　　　　NL 020033/68/5001
Symbiontische stikstofbinding onder sub–optimale (stress) condities. **Symbiotic nitrogen fixation under stress conditions.** Publications.

Grassland legumes (B 3441)

See also 3639, 3641, 5558, 5687

5657 Scheller, H.; Rieder, J.B.　　DE 502050/70/0030
Untersuchungen über den Einfluss hoher Düngergaben bei Vielschnittnutzung auf den Pflanzenbestand des Grünlandes und die Qualität von Grünmehl 1968. **Investigations on grassland plants and quality of forage meal as affected by high doses of fertilizer for multi–cut use.**

5658 Laissus, R.; Lecomte, D.; Leau, G.　FR 010115/75/0265
Utilisation rationnelle du trèfle blanc en tant que fournisseur d'azote dans les prairies paturées. **Rational use of white clover as nitrogen supplier in pastures.**

5659 Menet, M.; Delas, J.; Tauzin; Juste, C.
　　　　　　　　　　　　　　　FR 010704/66/6080
Carence en soufre et en molybdéne de la luzerne en sols argilo–calcaires des Charentes. **Sulphur and molybdenum deficiency of alfalfa in calcareous soils of the charente area.**

5660 Guy, P.　　　　　　　　FR 012223/70/0300
Etude des relations luzerne–rhizobium au niveau du rendement. **Relationships between lucerne and rhizobium for the yield.**

5661 Mytton　　　　　　　GB 012106/00/0021 R
Variation in the effectiveness of isolates of rhizobium associated with white clover.

5662 Haystead　　　　　　GB 030304/74/0008 R
Factors affecting the fixation and transfer of nitrogen by white clover in hill pasture.

5663 Newbould　　　　　　GB 030304/77/0009 R
Microbiological requirements of white clover growing in hill soils.

5664 Gracey　　　　　　　GB 041101/75/0019 R
Value of cattle slurry for the production of red clover.

5665 Murphy, W.E.; Brogan, J.C.　IE 060200/74/0850 R
Liquid fertilisers for grassland. Publications.

5666 Masterson, C.L.　　　　IE 060200/75/1097
The effect of sulphur on legume nitrogen fixation. Publications.

Other legumes (B 3449)

See also 5435, 5742, 6794

5667 Lenoble, M.; Felix, L.; Lemaire, L.; Durey, P.;
Meriaux, S.　　　　　　　　　FR 010112/78/9111
Etude des besoins en eau du Lupin blanc. **Investigation of the water requirements of white lupin.**

5668 Austin; Morgan　　　　GB 011907/79/0014 N
Study carbon economy of field beans to improve definition of ideal model plant for high stable yield.

Cereals used for forage (B 3450)

D 2200 – Plant nutrition and fertilization

See also 3090, 3542, 5420, 5422, 5542, 5543, 5544, 5545, 6661, 6814

5669 Sommer, K.; Six, R.   DE 111100/74/0007
Futtergerstenanbau mit Ammonium oder Nitrat als
Stickstoffquellen. **Cultivation of fodder barley with ammonium
or nitrate as nitrogen sources.**

5670 Dilz, K.   NL 010103/73/4356
Invloed van stikstofhoeveelheid op ontwikkeling, opbrengst en
kwaliteit van maïs. **Effect of rate of nitrogen application on
growth, yield, and quality of fodder maize.**

5671 Dilz, K.   NL 010103/73/4364
Invloed van het gebruik van grote hoeveelheden dierlijke
organische mest bij de verbouw van maïs op de
stikstofbehoefte van het gewas en op de indringing van
voedingsstoffen in het bodemprofiel. **Effect of application of
large quantities of animal manure on the nitrogen requirement
of maize and on the leaching of nutrients into the soil profile.**
Publications.

5672 Arnold, G.H.   NL 010103/75/6250
De invloed van de hoeveelheid en de frequentie van toepassing
van dierlijke mest op het effect van een rijenbemesting met
fosfaat bij snijmais. **Influence of quantity and frequency of
application of amimal manure on the effect of band placement
of phosphorus on the growth of silage maize.**

5673 Haan, G.H. de   NL 010207/72/3736
Invloed van grote giften drijfmest op de groei, opbrengst en
samenstelling van snijmais en op de bodem–en
waterverontreiniging. **Effect of large quantities of dung on
growth, yield and composition of fodder maize and on soil and
water polution.** Publications.

5674 Hag, B.A. ten; Haan, G.H. de   NL 010207/76/7643
Bemesting van snijmais. **Fertilization of silage maize.**

5675 Luten, W.; Krist, G.   NL 010208/72/3736
Invloed van grote giften drijfmest op de groei, opbrengst en
samenstelling van snijmais en op de bodem–en
waterverontreiniging. **Effect of large quantities of dung on
growth, yield and composition of fodder maize and on soil and
water pollution.** Publications.

5676 Dilz, K.   NL 060008/73/4356
Invloed van stikstofhoeveelheid op ontwikkeling, opbrengst en
kwaliteit van mais. **Effect of rate of nitrogen application on
growth, yield, and quality of fodder maize.**

5677 Dilz, K.   NL 060008/73/4364
Invloed van het gebruik van grote hoeveelheden dierlijke
organische mest bij de verbouw van mais op de
stikstofbehoefte van het gewas en op de indringing van
voedingsstoffen in het bodemprofiel. **Effect of application of
large quantities of animal manure on the nitrogen requirement
of maize and on the leaching of nutrients into the soil profile.**
Publications.

5678 Arnold, G.H.   NL 060008/75/6250
De invloed van de hoeveelheid en de frequentie van toepassing
van dierlijke mest op het effect van een rijenbemesting met
fosfaat bij snijmais. **Influence of quantity and frequency of
application of animal manure on the effect of band placement of
phosphorus on the growth of silage maize.**

Other forage crops (B 3490)

See also 3639, 3687

5679 Lemaire, L.; Du Crehu, G.   FR 010112/71/9108
Etude des besoins en eau du chou fourrager. **Investigation of
the water requirements of fodder kale.**

Vegetables in general (B 3500)

See also 1, 3702, 3706, 3775, 3778, 5439, 5528, 5761, 5911,
16090

5680 De Boodt, M.; De Vleeschouwer, D.; Verdonck, O.
     BE 030013/70/0002 R
Fysische karakterisatie en optimalisatie van de organische en
de inerte substraten gebruikt in de tuinbouw. **Physical
caracterisation and optimalisation of organic and inert
substrates for use in horticulture.** Publications.

5681 Wendt, T.   DE 111300/77/0004
Einfluss von Bodentemperatur auf Wachstum,
Wasserverbrauch, Nährstoffaufnahme und Qualität bei
Gemüse. **Effect of root temperature on growth, water
consumption, nutrient uptake and quality of vegetables.**

5682 Fritz, D.; Venter, F.   DE 161260/71/0001
Einfluss der Düngung auf Nitratgehalt in verschiedenen
Gemüsearten. **Influence of fertilization on the nitrate content in
different vegetables.**

5683 Fritz, D.; Venter, F.   DE 161260/74/0004
Einfluss der Klärschlammdüngung auf den Schwermetallgehalt
von Gemüse. **The influence of sewage sludge fertilization on the
heavy metal content in vegetables.**

5684 Fritz, D.; Maync, A.   DE 161260/77/0005
Anbau von Gemüse mit unterschiedlicher organischer
Düngung – Dauerversuch für spätere Qualitätsuntersuchungen
–. **Different amounts of organic fertilizers applied to vegetables
and quality of vegetables.**

5685 Fritz, D.; Venter, F.   DE 161260/78/0001 N
Einfluss der Düngung mit Siedlungsabfällen auf die Qualität
von Gemüse. **Quality of vegetables as affected by fertilization
with garbage.**

5686 Leh, H.–O.   DE 215060/77/0002
Untersuchungen über Standort– und Sortenabhängigkeit von
Calciummangel bei Gemüse. **Studies on calcium deficiency of
vegetables in dependence on location and variety.**

5687 Leh, H.–O.   DE 215060/78/0003 N
Untersuchungen über die Aufnahme von Blei aus dem Boden
durch Kulturpflanzen, speziell Gemüsepflanzen und Luzerne,
unter besonderer Berücksichtigung der
Aufnahmebeeinflussung durch Hauptnährstoffe. **On the
uptake of lead from contaminated soil by crop plants, especially
vegetables and alfalfa as influenced by main nutrient supply.**

5688 Penningsfeld, F.; Kurzmann, P.; Müller, L.
     DE 502101/70/0003
Wirkung gestaffelter Gaben von Stallmist, Torf und
Mineraldünger, allein und in Kombination, auf den Ertrag von
Freilandgemüse. **Effect of increased dressings of manure, peat**

and mineral fertilizer, alone and in combination, on the yield of outdoor cultivated vegetables.

5689 Frenz, F.-W.; Andresen, F.   DE 502104/70/0001
Die Wirkung verschiedener organischer Dünger auf den Ertrag einiger Gemüsearten 1958. **Effects of different manures on crop yield of some vegetables.**

5690 Frenz, F.-W.; Lechl, P.   DE 502104/73/0001
Optimale Wasserversorgung bei mehreren Gemüsearten unter Verwendung einer Bewässerungs- und Düngungsautomatik. **Optimum water supply of divers vegetables by automatic irrigation and fertilization.**

5691 Hege, H.; Rannertshauser, J.   DE 502110/78/0001 N
Tropfbewässerung für Gemüsekulturen im Freiland. **Trickle irrigation for vegetable crops in fields.**

5692 Klein, E.   DE 502156/77/0003
Einfluss von vier verschiedenen Intensitätsstufen der Düngung und des Pflanzenschutzes bei Gemüse auf den Gehalt an Inhaltsstoffen und Schwermetallen in der Pflanze, sowie auf Pflanzenrückstände, Unkrautflora, biologische Aktivität des Bodens und den Ertrag. **Effects of four different intensity levels of fertilization and plant protection in vegetables on constituents and heavy metals in plant and on residues of protectives, weed flora, biological activity in soil and crop.**

5693 Niemann, J.; Alt, D.   DE 507701/71/0001
Der Einfluss gestaffelter P–K– und Mg–Düngung auf den Ertrag von Gemüsekulturen und den Verlauf der Nährstoffgehalte des Bodens. **Influence of gradually applicated P–, K–, and Mg nutrients on mass production of vegetable plants and on mineral nutrient content of soil.**

5694 Maurer, M.; Alt, D.   DE 507701/78/0002 N
Düngung von gärtnerischen Samenträgern. **Fertilizing of horticultural seed crops.**

5695 Anstett, A.; Bats, J.   FR 010115/70/6221
Utilisation des écorces de pin et celles de feuillus comme amendements organiques en culture sous serre. **Using the bark of pine and of deciduous trees as organic soil conditionning material under glass cropping.**

5696 Blanc, D.; Otto, C.; Adamowicz, S.   FR 010502/76/6017
Distribution et réduction des nitrates au niveau des différents organes des espèces maraîchères en fonction de la nutrition azotée. **Localization and reductions of nitrates in the different parts of the plant in relation to nitrogen nutrition.**

5697 Blanc, D. Mme; Mars, S.; Gilly, G.   FR 010502/76/6018
Accumulation des nitrates dans les productions maraîchères. **Nitrate accumulation in vegetables.**

5698 Juste, C.; Gomez, A.; Dureau, P. Mme; Solda; Lubet; Lasserre   FR 010704/76/6087
Etude des possibilités d'utilisation des composts d'ordures ménagères en cultures maraîchères sous serre et en plein champ. **Utilisation of town–compost for vegetable crops under glass or in open field.**

5699 Greenwood; Scaife   GB 011802/00/0001
Develop and test models for the influence of soil and weather conditions on crop growth.

5700 Burns   GB 011802/00/0002
Movement of nutrients in soil and their uptake by plant roots.

5701 Turner   GB 011802/00/0007
Characterise crop responsiveness to added fertilizer. Fit models to results.

5702 Niendorf   GB 011802/00/0008
Compile data bank of results of UK fertilizer experiments for ready access and collation of information.

5703 Page; Scaife   GB 011802/00/0009
Improvement of the effectiveness of applied fertilizer: type of fertilizer and method of application.

5704 Page; Stone   GB 011802/00/0011
Soil environment and seedling emergence.

5705 Greenwood; Cleaver   GB 011802/00/0012
Predict fertilizer response of each vegetable crop on different soils: compare with experiment.

5706 Collier   GB 011802/00/0022
Calcium, magnesium and micro–nutrient disorders of vegetable crops.

5707 Scaife   GB 011802/00/0023
Effects of plant composition and growth rate on nutrient uptake.

5708 Maher, M.J.   IE 060300/76/1179 R
Nutrient solution culture(nsc) of glasshouse crops. Publications.

5709 Uncini, L.; D'Amore, R.; Petralia, S.
    IT 021000/79/0008 N
Studio degli effetti della concimazione organica, minerale e oligominerale, su qualità e quantità della produzione. **Study of organic, mineral and oligomineral manuring, on production quality and quantity.**

5710 Driel, W.van   NL 010103/59/0505 R
Oorzaken van de schadelijke invloed van stadsvuilcompost op tuinbouwgewassen. **Causes of the detrimental effect on horticultural crops of town refuse compost.**

5711 Prummel, J.; Barnau Sijthoff, P.A. van
    NL 010103/62/0534
Bemesting van grove tuinbouwgewassen op landbouwgronden. **Manuring of vegetables in agriculture.** Publications.

5712 Roorda van Eysinga, J.P.N.L.   NL 010103/62/0548
Bepaling van de optimale bemesting van de minder verbreide groenten onder glas. **Assessment of the optimum fertilization of less common vegetables cultivated under glass.** Publications.

5713 Boon, J. van der   NL 010103/64/0574
Kweekmogelijkheden van tuinbouwgewassen in potten in verband met bemestingsproeven. **Research on optimal development of horticultural crops in pots with a view to fertilizer trials.**

5714 Pieters, J.H.   NL 010103/71/3563
Kritische bewerking van resultaten van oude bemestingsproeven in de groenteteelt in de volle grond. **Critical evaluation of the results of old fertilizer experiments with outdoor vegetables.**

5715 Roorda van Eysinga, J.P.N.L. NL 010103/75/6253
Optimale bemesting van bloemisterijgewassen geteeld onder glas in rotatie met groentegewassen. **Establishment of the optimum fertilization of ornamental crops grown under glass in rotation with vegetable crops.** Publications.

5716 Roorda van Eysinga, J.P.N.L. NL 010103/75/6590 R
De chemische samenstelling van onder glas geteelde groenten in verband met hun consumptiekwaliteit. **Relation between chemical composition and quality of glasshouse vegetables.** Publications.

5717 Sonneveld, C. NL 010206/60/0935
Verhouding en concentratie van voedingselementen in het bevloeiingswater bij kasteelten. **Ratio and concentration among nutrients in irrigation water for glasshouse crops.** Publications.

5718 Boertje, G.A. NL 010206/61/2580 R
De chemische en fysische gesteldheid van de potgrond bij de opkweek van plantmateriaal voor kasteelten. **Chemical and physical conditions of the potting media in the raising of planting material for glasshouse crops.** Publications.

5719 Roorda van Eysinga, J.P.N.L.; Nederpel, W.A.C.
NL 010206/62/0548
Bepaling van de optimale bemesting van de minder verbreide groenten onder glas. **Assessment of the optimum fertilization of less common vegetables cultivated under glass.** Publications.

5720 Sonneveld, C. NL 010206/66/0933
De nadelige invloed van zouten bij kasteelten. **The detrimental effect of salts on glasshouse crops.** Publications.

5721 Meijs, M.Q. van der NL 010206/66/0938
Invloed van de grond en teeltmethoden op de wortel–ontwikkeling van kasgewassen. **Influence of the soil and growing methods on the root development of glasshouse crops.** Publications.

5722 Sonneveld, C. NL 010206/70/3429
Het gebruik van substraten bij kasteelten. **Application of substrates for glasshouse crops.** Publications.

5723 Sonneveld, C. NL 010206/74/6861 R
Opname van mangaan, ijzer en zink bij kasteelten. **The uptake of manganese, iron and zinc by greenhouse crops.** Publications.

5724 Roorda van Eysinga, J.P.N.L. NL 010206/75/6253
Optimale bemesting van bloemisterijgewassen geteeld onder glas in rotatie met groentegewassen. **Establishment of the optimum fertilization of ornamental crops grown under glass in rotation with vegetable crops.** Publications.

5725 Roorda van Eysinga, J.P.N.L. NL 010206/75/6590
De chemische samenstelling van onder glas geteelde groenten in verband met hun consumptie kwaliteit. **Relation between chemical composition and quality of glasshouse vegetables.** Publications.

Root, tuber and bulb vegetables (B 3510)

See also 3790, 5535, 5739

5726 Neumann, K.-H.; Bender, L. DE 129041/72/0003

Untersuchungen über den Hormonstoffwechsel von Karottengewebekulturen unter besonderer Berücksichtigung des Äthylens und der IES 1972. **Studies of hormone metabolism in carrot tissue cultures with special regard to ethylene.**

Greens and leafy vegetables (B 3520)

See also 1549, 3790, 3847, 3848, 5427, 5535

5727 Buchloh, G.; Onggo, T.; Liegel, W. DE 144445/77/0014
Einfluss von Bleisalzen auf die Ertragsleistung und auf die Qualität von Spinat. **Effect of lead salts on yield and quality of spinach.**

5728 Fritz, D.; Weichmann, J. DE 161260/78/0003 N
Wirkung unterschiedlicher Luftzusammensetzungen auf das Lagerverhalten und Qualitätsveränderung verschiedener Rosenkohlsorten. **Influence of different CA–conditions on storage ability and quality changes of different cultivars of Brussels sprouts.**

5729 Hartmann, H.D.; Born, H.U. DE 506107/77/0001
Der Magnesiumhaushalt in Liliaceen, dargestellt am Spargel, Asparagus officinalis L.. **Magnesium supply of liliaceae presented by Asparagus officinalis L..**

5730 Jakobsen, S.T. DK 030148/79/0003 N
Orienterende undersøgelser over forekomst af calciummangel i iceberg salat og kinakål. **Pot experiments with factors influencing deficiency of calcium in vegetables.**

5731 Karlsen, P.; Nielsen, N.E. DK 030171/77/0011 N
Døgnvariation af ionoptagelse hos salat. **Diurnal fluctuation in ion uptake in lettuce.**

5732 Morgan, J.V.; O'Haire, R.; Roche, N.
IE 120108/76/9112 N
The use of nutrient solution culture for the production of lettuces at high densities on an arch framework. Publications.

5733 Blanco, V.V. IT 062700/77/0941
Ricerche sulla concimazione (cavolo–broccolo, cetriolo, fagiolino, patata, pisello). **Research on manure application (broccoli, cucumber, french bean, potato).**

5734 Pieters, J.H. NL 010103/67/0582 R
Bepaling van het tijdstip en de hoeveelheid van overbemesting met stikstof bij spruitkool. **Assessment of time and quantity of topdressing with nitrogen for Brussels sprouts.** Publications.

5735 Kruistum, G. van NL 010207/74/6516 N
Bemesting van witlof tijdens de trek. **Fertilization of chicory during the forcing.** Publications.

5736 Diest, A. van NL 020007/77/7325
Onderzoek naar de oorzaken en naar de mogelijkheden tot vermijden van hoge nitraat gehalten in spinazie. **Investigation into the causes and the possibilities of circumventing high nitrate contents of spinach.** Publications.

Leguminous vegetables (B 3531)

See also 187, 516, 1549, 3633, 3790, 3898, 3901, 3903, 3907, 5435, 5653, 5654, 5655, 5656, 5733

5737 Döring, H.-W.; Rusitzka, G.; Plarre, W.

DE 105202/77/0002

Auswirkungen inverser K/Na–Verhältnisse bei konstanter K/NaGesamtkonzentration auf Mineralstoffgehalt sowie auf Teilbereiche des Kohlenhydrat– und Säurestoffwechsels bei den Phaseolussorten RED KIDNEY und CARLOS FAVORITENKONA S 69. **Effects of inverse K/Na ratio and constant K/Na total concentration on mineral content as well as on parts of carbohydrate and organic acid metabolism in the bean varieties Red Kidney and Favorite Black.**

5738 Marschner, H.; Berger–Geiger, B.   DE 144400/78/0001 N
Genotypische Unterschiede in der Aluminium–Empfindlichkeit bei Cowpea. **Genotypic differences in susceptibility of cowpea to aluminium–toxicity.**

5739 Jørgensen, M.B.; Steen, T.N.   DK 010115/74/4504
Gødskning af bønner, rødbeder og porre. **Fertilizers for beans, beetroot and leeks.**

5740 Eriksen, E.N.   DK 030171/71/0007 N
Undersøgelser over roddannelse på ærtestiklinger. **Investigations of root formation on pea propagules.**

5741 Felix, M.; Simon, G. Mme   FR 010103/72/0660
Fertilisation soufrée et qualités nutritionnelles de la féverole. **Sulfur fertilization and nutritional value of field–bean.**

5742 Lemaire, L.; Mériaux, S.   FR 010112/70/9110
Etude des besoins en eau de la féverole de printemps. **Investigation of the water requirements of spring horse–bean.**

5743 Amarger, N.   FR 011002/72/0595
Utilisation des propriétés symbiotiques pour améliorer le rendement de la féverole. **Improving the yield of field bean by Rhizobium: Survey of Rhizobium populations breeding of efficient strains.**

5744 Beringer; Johnson   GB 011402/75/0008
Genetics of rhizobium.

5745 Lloyd–Jones; Hill–Cottingham   GB 011505/00/0004
Nitrogen–15 isotope investigations on the uptake, distribution and metabolism of nitrogen in plants.

5746 O'Gara, F.; Hynes, C.; Robeson, D.; Meade, J.; Casey, C.; Buckley, A.   IE 110102/78/9142 N
Enhancing legume nitrogen fixation by genetically engineering more efficient rhizobium strains. Publications.

5747 Prendeville, G.N.   IE 110202/78/9181 N
Interactions between herbicides and na++, ca++, mg++, k+ ions in terms of membrane integrity in phaseolus and lemna minor.

5748 Picci, G.; Lepidi, A.; Nuti, M.P.   IT 041119/74/0001
Ricerche sui primi stadi della nodulazione nelle leguminose e miglioramento della fissazione simbiontica dell'azoto. **Studies on the first stages of legume nodule formation and improvement of symbiotic nitrogen fixation.** Publications.

5749 Dekhuijzen, H.M.; Verkerke, D.R.   NL 010102/77/7297
Opname en reductie van stikstof door veldbonen. **Uptake and reduction of nitrate by Vicia faba.** Publications.

5750 Diest, A. van   NL 020007/78/8843 N
Verband tussen luchtstikstofverbinding, kationen–

anionenbalans en benutting van ruwe fosfaten bij leguminosen. **Relationship between atmosphere N fixation, cation–anion balance and utilization of rock phosphates by leguminous crops.**

5751 Frings, J.F.J.   NL 020033/71/4711
Remming van symbiose tussen Rhizobium en erwt door hoge temperatuur. **Effect of temperature on Rhizobium – legume (pea) symbiosis.**

5752 Kammen, A. van   NL 020068/78/8629 N
De rol van een groot plasmide in Rhizobium bacterieën bij het tot stand komen van de symbiotische stikstofbinding. **The role of a large plasmid occurring in Rhizobium bacteria for the symbiotic nitrogen fixation.**

5753 Kammen, A. van   NL 020068/78/8630 N
De regulering van de synthese van het enzym nitrogenase in Rhizobium bacteroïden in wortelknollen van de erwt. **The regulation of the synthesis of nitrogenase in Rhizobium bacteroids in pea root nodules.** Publications.

Tomatoes (B 3532)

See also 1549, 3848, 5424

5754 Krug, H.; Fölster, E.   DE 138210/77/0002
Eignung verschiedener Substrate – Steinwolle, Torf, Stroh, Einheitserde, Grundbeet – für die Tomatenkultur. **Suitability of different substrates – rock wool, peat, straw, uniform earth, basal bed – on the production of tomatoes.**

5755 Ginoux, G.   FR 010610/75/9085
Lutte contre la maladie des racines liégeuses de la tomate "Corky–root". Etude comparative de deux procédés de lutte. **Corky–root control on tomatoes: comparative study of two control methods.**

5756 Hurd   GB 011102/75/0061 R
Nutrient film culture : investigation of late spring tomato fruiting failure.

5757 Besford   GB 011103/00/0003
Role of potassium in tomato growth and development.

5758 Adams   GB 011105/00/0001
Effects of nutrition on yield, quality and composition of glasshouse tomatoes in peat beds.

5759 Massey; Windsor   GB 011105/00/0012
Nutrition of tomatoes and other glasshouse crops in re–circulating solution culture.

5760 Researcher not indicated   GB 050142/00/0009
Crop nutrition, relationship with substrate and leaf analysis.

5761 Gallagher, P.A.   IE 060300/72/0055 R
Nutrition of glasshouse food crops, particularly tomato. Publications.

5762 Maher, M.J.   IE 060300/77/1300 R
Calcium nutrition of tomatoes in peat. Publications.

5763 Cremaschi, D.; Casarini, B.; Giordano, T.; D'Amato, A.   IT 021100/79/0019 N
Concimazione organica del pomodoro. **Organic fertilization of**

D 2200 – Plant nutrition and fertilization

the tomato.

5764 Goor, B.J. van. NL 010103/69/0543
Fysiologische ziekten van tuinbouwgewassen: Storingen in het calciummetabolisme van appels (stip) en tomaten (neusrot). **Physiological diseases of horticultural crops: Disturbances in the calcium metabolism of apples (bitter pit) and tomatoes (blossom–end rot).** Publications.

Cucumbers (B 3533)

5765 Adams GB 011105/00/0009
Effects of nutrition on yield, quality and composition of glasshouse cucumbers in peat beds.

5766 Researcher not indicated GB 050102/00/0013 R
Crop nutrition, relationship with substrate and leaf analysis.

Other vegetable fruits (B 3539)

See also 1549, 5454, 5733

Mushrooms and other edible fungi (B 3540)

See also 3790, 4000, 4008, 4013

5767 Laborde, J.; Imbernon, M. FR 010703/67/0559
Nutrition des champignons saprophytes comestibles à partir de déchets végétaux compostés. **Nutrition of edible saprophytic mushrooms with composted plant residues.** Publications.

5768 Laborde, J.; Delmas, J. FR 010703/67/6061
Etude de la fermentation des substrats solides destinés à la culture du champignon de couche. **Analysis of the fermentation process of solid substrates for the cultivated mushroom.**

5769 Imbernon, M.; Laborde, J.; Goulas, J.P.
FR 010703/68/6060
Microbiologie de substrats organiques compostés et évolution de la microflore au cours de l'envahissement de ces composts par le mycélium de champignons. **Microbiology of composted organic substrates and microflora evolution during the growing phase mushroom mycelium inside the compost.**

5770 Delmas, J.; Imbernon, M.; Poitou, N. FR 010703/70/6064
Utilisation des déchets organiques de l'industrie du papier et des composts urbains pour la culture des champignons. **Use of organic wastes of Paper Industry and composted town refuse for mushroom culture.**

5771 Laborde, J.; Imbernon, M. FR 010703/72/0560
Nutrition des champignons saprophytes comestibles à partir de déchets végétaux non compostés. **Nutrition of edible saprophytic mushrooms with uncomposted plant residues.** Publications.

5772 Laborde, J.; Delmas, J. FR 010703/72/6063
Préparation de substrats non fermentés pour la culture de quelques champignons saprophytes comestibles. **Substrate preparation without fermentation for the cultivation of several edible saprophytic fungi.**

5773 Delmas, J.; Laborde, J. FR 010703/75/6062
Etude des caractéristiques et du comportement des matériaux employés comme terre de gobetage en culture du champignon de couche. **Caracteristics and behaviour of casing soils used for**

mushroom growing.

5774 Goulas, J.P. FR 010703/76/6059
Disponibilité comparée de l'azote protéique microbien pour le mycélium d'Agaricus bisporus et pour la microflore des composts. **Comparative availability of microbial proteic nitrogen for Agaricus bisporus mycelium and for compost microflora.**

5775 MacCanna, C. IE 060300/69/0065 R
Development of synthetic composts for mushrooms. Publications.

5776 Gerrits, J.P.G. NL 010204/73/3859
Het verder ontwikkelen van synthetische compost voor de champignon–cultuur op basis van stro (en andere materialen). **Further development of synthetic compost for the cultivated mushroom on a basis of straw (and other materials).** Publications.

5777 Gerrits, J.P.G. NL 010204/73/3860
Het bestuderen van het composteringsproces en de voedingsbehoeften van de champignon om te komen tot een optimaal substraat, uitgaande van paardemest en andere natuurlijke mestsoorten. **Study of the composting process and the nutritional needs of the cultivated mushroom to achieve the optimum substrate, based on horse manure and other natural manures.** Publications.

Fruits in general (B 3600)

See also 1, 4060, 4066, 4100, 5680, 5708, 5710, 5713

5778 Roemer, K. DE 138240/72/0006
Einfluss der Kaliumversorgung auf die Qualität von Früchten 1972. **english title not indicated.**

5779 Huguet, C. Mme FR 010601/68/0533
Mécanismes de l'absorption des éléments minéraux chez les arbres fruitiers. Effets combinés de la nutrition magnésienne et calcique sur la physiologie des arbres et la production de fruits. **Uptake of mineral elements by fruit–trees. Joint effects of magnesium and calcium nutrition on tree physiology and fruit yield.** Publications.

5780 Pouget, R.; Ottenwaelter, M.; Mme Gazeau
FR 010702/72/0111
Etude de l'influence du porte–greffe sur l'alimentation minérale du greffon: application à la sélection des porte–greffes. **Rootstock influence on the scion mineral supply: application to rootstock selection.**

5781 Huguet, J.G.; Giraudon FR 010704/69/6074
Absorption, mise en réserve et mobilité des éléments minéraux dans les arbres fruitiers. **Absorption, storage and mobility of mineral ions in fruit–trees.**

5782 Huguet, J.G.; Arbo, F.; Giraudon FR 010704/75/6076
Interactions fertilisation/état sanitaire des arbres fruitiers. **Interactions between fruit–trees nutrition and non animal pest susceptibility.**

5783 Avery; Hadlow GB 011006/00/0001 R
Mechanisms controlling partition of assimilates between crop and other plant components.

5784 Avery; Treharne GB 011006/00/0004 R
Environmental and internal factors affecting carbon assimilation.

5785 Bould; Parfitt GB 011507/00/0001
Leaf analysis as a guide to mineral nutrient status of black currants, apples, pears and plums.

5786 Colorio, G.; Strabbioli, G.; Manzo, P.; Cobianchi, D.; Monastra, F.; Bergamini, A. IT 021500/78/0001 N
Rapporti tra metodologia di distribuzione, quantità, qualità dei fertilizzanti, e qualità della frutta. Study on the relationship among distribution, quantity and quality of the fertilizers and fruit quality.

5787 Delver, P. NL 010103/55/0491 R
Landelijke bodembehandelingsproeven in de fruitteelt. Soil management experiments in orchards. Publications.

5788 Delver, P. NL 010103/59/0511
Betekenis van stikstof voor opbrengst en kwaliteit van fruitgewassen. Importance of nitrogen for yield and quality of fruit crops. Publications.

5789 Delver, P. NL 010212/55/0491
Landelijke bodembehandelingsproeven in de fruitteelt. Soil management experiments in orchards. Publications.

5790 Delver, P. NL 010212/59/0511
Betekenis van stikstof voor opbrengst en kwaliteit van fruitgewassen. Importance of nitrogen for yield and quality of fruit crops. Publications.

Top fruit in general (B 3610)

See also 2595, 4124, 5055, 5056

5791 Lüdders, P.; Helm, H.–U.; Ramarmurthy, S.
DE 105204/78/0002 N
Wirkung der N–Ernährung auf Obstbäume mit unterschiedlichem Virusbefall. Effect of N nutrition on fruit trees with different virus infection.

5792 Lüdders, P.; Ohme, J. DE 105204/78/0004 N
Einfluss von Unterlage, N–Ernährung und Fruchtbehang auf die Mineralstoffaufnahme und –verteilung bei Obstgewächsen. Effect of rootstock, N nutrition and fruit load on mineral uptake and distribution in fruit trees.

5793 Poulsen, E.; Vang–Petersen, O. DK 010102/67/3103
Gødskning og bladanalyser til æble–, pære–, blomme– og kirsebærtræer. Fertilizer application and foliar analyses in apple, pear, plum and cherry trees.

5794 Greenham GB 011003/00/0003
Nutrient uptake and utilisation.

5795 Allen GB 011003/00/0007
Soil composition effects on uptake of mineral ions.

5796 Ford GB 011003/00/0008
Regulation of calcium nutrition of the tree in relation to fruit storage quality.

5797 Greenham; Ford GB 011003/00/0015
Nutrient supply in relation to fruit development, yield and quality.

5798 Samuelson GB 011003/00/0017
Diagnosis and control of mineral nutrient imbalance.

5799 Campbell; Bould GB 011510/00/0017
Effect of latent viruses on growth of apples and pears under different nutritional regimes.

5800 McDonnell, P.F. IE 060302/62/0514 R
Apple nutrition. Publications.

5801 Visser, J. NL 040007/63/4101
Onderzoek naar de in economisch opzicht optimale ontwateringstoestand en bemesting bij appelbomen. Research on the optimum drainage conditions and fertilization of apple trees. Publications.

Apple (B 3611)

See also 396, 1761, 4162, 4171, 5764, 8733

5802 Deckers, J. BE 040202/67/0002 R
Ca, K en Mg bemestingsproeven op appel. Ca, K and Mg fertilising trials on apple.

5803 Lüdders, P.; Ramarmurty, S. DE 105204/73/0004
Die Wirkung gleichmässiger und jahreszeitlich unterschiedlicher NH4plus– und NO3minus–Versorgung auf Apfelbäume. The effect of uniform and seasonally varied NH4plus and NO3minus supply on apple trees.

5804 Lüdders, P. DE 105204/74/0006
Einfluss von N– und K–Düngung auf Ertrag und Fruchtqualität bei Äpfeln und Birnen. Effect of N– and K–fertilization on yield and fruit quality of apples and pears.

5805 Kick, H.; Gross, K.–J.; Lenz, F. DE 111100/77/0002
Unmittelbare Anwendung von Calciumverbindungen durch Einbringung in den Saftstrom von Apfelbäumen. Immediate application of calcium compounds by injection in the sap stream of apple trees.

5806 Schmidle, A.; Dickler, E.; Seemüller, E.; Krczal, H.
DE 215210/77/0006
Untersuchungen über den Einfluss organischer und mineralischer Düngung auf Pilzkrankheiten sowie Schad– und Nutzarthropoden beim Apfel. Investigations on the influence of organic and mineral fertilization on fungal diseases as well as on injurious and useful arthropods in apple.

5807 Huguet, C. Mme FR 010601/67/0532
Relations entre les techniques culturales, le rendement et la qualité des fruits. Modèle mathématico–statistique pour l'étude des vergers de pommiers. Manuring, yield and fruit quality relationships. A mathematical model for apple orchards.

5808 Huguet, C. FR 010601/71/6048
La nutrition minérale du pommier: effets des cations sur la production et la qualité des fruits. Apple mineral nutrition: cationic role in the production of apples and incidence on fruit disorders.

5809 Huguet, C. FR 010601/74/6049
Aptitude des racines de différents porte–greffes du pommier à

assurer l'alimentation cationique de la plante. **Cationic accumulation in different apple outting roots.**

5810 Arnoux, M.; Defrance, H.; Huguet, C.
FR 010615/65/9044
Comparaison de modes d'apport de la fumure potassique en vergers de pommier. **Comparing the effects of various procedures of potash fertilization of apple trees.**

5811 Redmond   GB 011507/00/0007
Uptake and movement of nutrients, with particular reference to calcium.

5812 O'Kennedy, N.D.   IE 060302/62/0499 N
Production systems for apples including irrigation.
Publications.

5813 Hennerty, M.J.   IE 120108/71/9103 N
Post harvest urea applications to apple trees. Publications.

5814 Hennerty, M.J.   IE 120108/71/9105 N
Trickle irrigation of apples.

5815 Boon, J.van der   NL 010103/61/0538
Bestrijding van stip in appels door verbetering van de watervoorziening en van de calcium- en boriumvoeding. **Control of bitterpit by improvement of water supply, calcium and boron nutrition.** Publications.

5816 Delver, P.   NL 010103/69/2592
Relatie tussen kaliumgehalte van appelblad en grond. **Relationship between potassium contents of apple leaves and soil.** Publications.

5817 Delver, P.   NL 010103/72/3558 R
Invloed van de stikstofvoeding van de waardplant op de ontwikkeling van mijtenpopulaties (Panonychus ulmi (Koch)) en op de produktie en de kwaliteit van appels. **Influence of nitrogen nutrition of apple trees on the development of spider mite populations (Panonychus ulmi (Koch)) and on the yield and quality of the crop.** Publications.

5818 Delver, P.   NL 010103/74/6014
Stip in appel (invloed van bodem en bemesting). **Bitter pit in apples (effects of soil and fertilization).** Publications.

5819 Delver, P.   NL 010212/69/2592
Relatie tussen kaliumgehalte van appelblad en grond. **Relationship between potassium contents of apple leaves and soil.** Publications.

Pear (B 3612)

See also 5802, 5804, 8733

Other top fruit (B 3619)

See also 1546, 4234, 5869, 8733

5820 Lüdders, P.   DE 105204/77/0001
Wirkung von Düngung und Bodenpflegemassnahmen auf Sauerkirschen. **Effect of fertilization and soil cultivation on sour cherries.**

5821 Werner, H.   DE 111300/77/0003
Ernährungs– und Qualitätsprobleme bei Sauerkirschen.

Nutritional and qualitative problems in sour cherries.

5822 Matzner, F.   DE 161265/72/0007 R
Einfluss steigender Stickstoffgaben auf die Entwicklung von Baum, Frucht und Fruchtqualität bei Äpfeln 1972. **Influence of increasing nitrogen fertilization on growth of tree, fruit and on fruit quality of apples.**

5823 Matzner, F.; Maurer, E.   DE 161266/75/0002
Einfluss steigender Stickstoffgaben auf die Entwicklung von Baum, Frucht und Fruchtqualität bei Sauerkirschen. Korrelation zwischen vegetativen Merkmalen, dem Fruchtertrag und den Gütemerkmalen und Inhaltsstoffen der Früchte der 'Schattenmorelle' in Abhängigkeit von steigenden N–Gaben. **Influence of increasing nitrogen supply on the development of tree, fruit and fruit quality of sour cherries. Correlation between vegetative characteristics, fruit yield, quality characteristics and constituents of the fruits of the 'morello cherry' as a function of increasing nitrogen supply.** Publications.

5824 Lalatta, F.   IT 040605/73/0001
Ricerche sullo stato di nutrizione del pesco con il controllo della diagnostica fogliare. **Peach trees nutritional status as determined by leaf analysis.**

5825 Crescimanno, F.G.; Barone, F.   IT 040908/76/0002
Ricerche sulla nutrizione minerale del Nespolo del Giappone (Eriobotrya japonica). **Researches on mineral nutrition of loquat trees (Eriobotrya japonica).**

Soft fruit (berries and cane fruits) (B 3620)

See also 4291, 5712, 5719, 7296

5826 Naumann, W.D.; Krüger, E.   DE 138242/78/0001 N
Entwicklung von Methoden zur Ermittlung des Ernährungszustandes von Vaccinium corymbosum und Vaccinium vitis idaea. **Methods to control the mineral nutrient supply of Vaccinium corymbosum and Vaccinium vitis idaea.**

5827 Bauckmann, M.   DE 506106/78/0001 N
Erprobung verschiedener Bewässerungsmethoden bei Beerenobst. **Testing of various irrigation methods by soft fruit.**

5828 Wirth, H.; Alt, D.   DE 507701/75/0001
Verteilung und Höhe der Stickstoffgaben bei Erdbeeren Sorte: Red Gauntlet. **Seasonal distribution and amount of nitrogen with strawberry 'Red Gauntlet'.**

5829 Callesen, O.   DK 030171/76/0004 N
Ernæringsforsøg med haveblåbær (Vaccineum corymbosum). **Nutrition trials with garden bilberry (Vaccineum corymbosum).**

5830 Greenham   GB 011003/78/0019
Nutrient uptake and utilisation.

5831 Priestley; Avery   GB 011006/00/0013 N
Productivity of bush and cane fruits in relation to assimilation and utilization of carbon resources.

5832 Webb; Purves   GB 011507/00/0002 R
Nutritional, cultural and environmental conditions on performance and yield of strawberry and blackcurrant.

5833 Webb; Purves   GB 011507/00/0003 R
Cultural and environmental conditions on nutrient uptake, growth and fruiting potential of strawberry runners.

5834 Morgan, J. V.; Doolan, D.   IE 120108/74/9107 N
Production of strawberries on a year round scale on raised arches using nutrient solution culture. Publications.

Citrus fruit (B 3630)

See also 231, 4346

5835 Meriaux, S.; Bourzeix; Wagner   FR 010601/73/6053
Effets de la sécheresse sur la vigne. **Effects of draught on the wine.**

5836 Cassin, J.   FR 012202/66/0108
Contrôle de la fertilisation du clémentinier. **Fertilization controc of the clementine.** Publications.

5837 Raciti, G.; Scuderi, A.   IT 021600/74/0006 N
Studio sull'impiego dei concimi liquidi ed idrosolubili in agricoltura.. **Research on liquid and hidrosoluble fertilizers in citruculture.**

5838 Raciti, G.; Scuderi, A.; Intrigliolo, F.
IT 021600/79/0001 N
Indagine sui livelli nutrizionali dell'arancio e sull'efficacia delle varie forme azotate. **Research on nutritional levels on orange and efficacy of various nitrogen formulations.**

5839 Intrigliolo, F.; Raciti, G.; Di Martino Aleppo, E.
IT 021600/79/0002 N
Riflessi sulla nutrizione minerale dell'arancio, in relazione al sistema di distribuzione dell'acqua ed ai volumi di adacquamento. **Mineral nutrition of orange as influenced by water distribution system and watering volume.**

5840 Tropea, M.   IT 040304/78/1115 N
Prove comparative sull'assorbimento radicale di alcuni chelati di ferro in piante di agrumi affette da clorosi ferrica da calcare. **Comparative tests on the root absorption of some iron organic compounds in citrus plants affected by calcium iron chlorosis.**

Tropical and sub–tropical fruits (B 3640)

See also 5721

5841 Höfner, W.; Hakerlerler, H.; Kovanci, I.
DE 129044/78/0002 N
Ursache der Chlorose an Mandarinen der ägäischen Region. **Cause of chlorosis of mandarines in the Aegean region.**

Grapes (B 3650)

See also 286, 1479, 4390, 4418, 4454, 8785, 8789, 9869

5842 Mengel, K.; Malissiovas, N.   DE 129047/78/0001 N
Eisenchlorose bei Reben – Vitis vinifera – in Abhängigkeit von der Nährstoffkonzentration in der Rhizosphäre. **Iron chlorosis of vine – Vitis vinifera – in relation to the nutrient concentration in the rhizosphere.**

5843 Matzner, F.   DE 161265/78/0004 N
Einfluss steigender Stickstoffgaben auf die vegetative sowie generative Entwicklung von Reben und die Traubenqualität.

Influence of increasing nitrogen supply on vegetative and generative development of vines and on grape quality.

5844 Alleweldt, G.; Rapp, A.; Herwig, K.
DE 204000/78/0002 N
Einfluss der N–Düngung auf Ertrag und Mostqualität von Rebsorten. **Influence of nitrogen fertilization on yield and must quality of grape vine varieties.**

5845 Gärtel, W.   DE 215220/77/0001
Untersuchungen über die Verfrachtung der mit Düngemitteln in Weinbergsböden eingebrachten Anionen in Grundwasser und Flüsse. **Studies on the transfer of anions brought into vineyard soils via fertilizers into groundwater and rivers.**

5846 Gärtel, W.   DE 215220/77/0002
Einfluss von Metallen aus Siedlungsabfällen auf Wachstum und Ertrag der Reben – Einlagerungen toxischer Metalle in Trauben und ihr Übergang in Most und Wein. **Influence of metals from domestic refuse on growth and crop of vines. – Storage of toxic metals in grapes and their transfer into must and wine.**

5847 Gärtel, W.   DE 215220/77/0003
Untersuchungen über die Bedingungen, unter welchen Phosphatüberschuss Eisen– und Zinkmangel auslöst– Verfahren zur Behebung dieser Mangelerscheinungen. **Investigations on the causes of phosphate excess, Fe and Zn deficiency – methods for elimination of these deficiencies.**

5848 Gärtel, W.   DE 215220/77/0004
Untersuchungen über Ernährungsbedingungen, die Symptome der infektiösen Vergilbung der Rebe auslösen – kausale Zusammenhänge der analogen Symptomausprägung. **Investigations on nutritive causes of symptoms of infectious yellow virosis of vines – causal correlations between analogous symptoms.**

5849 Gärtel, W.   DE 215220/77/0005
Untersuchungen über anatomische und histologische Veränderungen in Sprossen und Blütenständen bei beginnendem Bormangel – ihr Einfluss auf Ertrag und Holzreife. **Investigations on anatomical and histological changes in sprouts and flowers in initial boron deficiency – influence on crop and mellowing.**

5850 Enkelmann, R.   DE 501104/75/0001
Aufnahme von Schwermetallen – Zn, Cd, Cu, Pb, – durch Reben, die mit Müllklärschlammkompost gedüngt wurden. **Contamination of vines with heavy metals – Zn, Cd, Cu, Pb, – by manuring with dust–mud–remainders.**

5851 Enkelmann, R.   DE 501104/78/0001 N
Aufnahme der Schwermetalle Mn und Hg durch Reben, die mit Müllklärschlammkompost gedüngt wurden. **Contamination of grape vines with heavy metals Mn an Hg by manuring with clear sewage sludge compost.**

5852 Kannenberg, J.   DE 501106/74/0002
Auswirkungen von Mineralstickstofformen, insbesondere langsam wirkender N–Dünger auf Rebenwachstum und Ertrag. **Studies on the effects especially of slow–release nitrogenous fertilizers on growth and yield of vines.**

5853 Kannenberg, J.   DE 501106/78/0001 N
Untersuchungen zur Verwendung von Trestern im Weinbau.

D 2200 - Plant nutrition and fertilization

Investigations on the influence of grape-skins in vineyard soils.

5854 Müller, K.; Peternel, M.   DE 502153/77/0004
Erörterung der Möglichkeiten und der optimalen Zeitpunkte der jährlichen N–P–K–Düngung als Flüssigkeitsdüngung über die Beregnungsanlage. **Discussion on possibilities and optimum date of N–P–K fertilization every year as liquid fertilization via irrigation system.**

5855 Geiger, K.; Schottdorf, W.   DE 502153/78/0001 N
Einfluss verschiedener Formen der Strohanwendung auf den Boden und die vegetative Entwicklung sowie den Ertrag der Rebe. **Studies of the influence of different straw–mulching on soil and vegetative growth and yield of vines.**

5856 Schumann, F.   DE 509154/73/0002
Untersuchungen zur Düngung in Rebschulen. **Studies on manuring in vine nurseries.**

5857 Schaefer, H.   DE 509154/74/0001
Untersuchungen über den Stoffwechsel von Rebenwurzeln und –holz. **Studies on the metabolism in roots and wood of grape vine.**

5858 Morlat, R.; Salette, J.   FR 010403/76/6029
Etude régionale de la chlorose de la vigne. **Vine chlorosis conditions in the medium loire valley.**

5859 Delas, J.; Molot, M.   FR 010704/59/0546
Entretien de la fertilité des sols viticoles du Bordelais. **Vineyard soil fertility maintenance in the Bordeaux district (Bordelais).** Publications.

5860 Delas, J.; Molot, P.; Viticulture Bordeaux   FR 010704/59/6069
Fertilisation des vignobles de cru bordelais. **Fertilization of high quality vineyards in the bordeaux–area.**

5861 Juste, C.; Delas, J.   FR 010704/63/0552
Etude de la chlorose calcaire des plantes pérennes et notamment de la vigne. **Study on calcium induced chlorosis in perennial plants especially in vine.** Publications.

5862 Marocke, R.; Huglin, P.   FR 010901/73/6127
Besoins de la vigne et caractérisation des troubles de la nutrition. **Requirements of the vineyard and caracterisation of deficiencies.**

5863 Iannini, B.; Giorgessi, F.; Cappelleri, G.; Liuni, C.; Calò, A.; Zazzi, A.   IT 021300/60/0001 R
Prove di nutrizione delle piante. **Vine nutrition trials.** Publications.

5864 Liuni, C.; Antonacci, D.   IT 021300/74/0008 R
Prove di concimazione in regime irriguo. **Fertilization trials with irrigation regime.**

5865 Liuni, C.; Stramaglia, L.   IT 021300/74/0010
Prova di esami degli accumuli su diverse forme d'allevamento e diverse varietà in clima caldo–arido. **Analysis of accumulations on different varietes and trainings in warm–dry climate.**

5866 Fregoni, M.; Scienza, A.; Miravalle, R.; Zamboni, M.; Boselli, M.; Dorotea, G.   IT 040403/70/0004 R
Nutrizione minerale ed idrica della vite in relazione alla qualità. **Mineral and water nutrition of vines in relation to quality.** Publications.

5867 Bozzini, S.   IT 040403/77/0122
Acido abscissico e nutrizione idrica nella vite. **Abscissic acid and water nutrition in vines.**

5868 Fregoni, M.   IT 040403/77/0660 R
Nutrizione e riproduzione della vite. **Grapevine nutrition and reproduction.**

Edible nut fruits (B 3660)

See also 7473, 7478

5869 Arnoux, M.; Grosclaude, C.; Huguet, C. Mme
  FR 010601/76/6050
Relations entre la nutrition des arbres à noyau, la qualité des fruits et les maladies physiologiques ou parasitaires. **Interrelations between stone–fruit nutrition and fruit quality or physiological, fungal and bacterial deseases.**

Ornamentals and ornamental products in general (B 3700)

See also 1, 5680, 5694, 5708, 5710, 5713

5870 Besford; Manning   GB 011103/00/0006 R
Enzymes as indicators of plant nutritional status.

5871 Massey; Winsor   GB 011105/00/0005
Nitrogen requirement of glasshouse crops as influenced by light.

5872 Moorby   GB 011105/75/0013
Interactions between nutrition and root temperature.

5873 McGregor   GB 060306/00/0005
The effect of nutritional factors on the production and quality of glasshouse crops.

Bulbs (B 3710)

See also 4608

5874 Klougart, A.   DK 030171/78/0004 N
Tulipandrivning i vandkultur, særlig med henblik på calciumoptagelse. **Tulip forcing in hydroponic with special reference to calcium uptake.**

5875 Klougart, A.; Johansen, H.C.   DK 030171/79/0003 N
Calciumoptagelse i seks tulipansorter. **Calcium uptake of six tulipa cultivars.**

5876 Cirrito, M.; De Vita, M.   IT 021200/78/0001
Esperienze sull'influenza di diversi tipi di concimazione nell'ingrossamento di bulbi di tuberosa. **Trials on different manuring qualities effects in the enlargement of tuberosa bulbs.**

5877 Boon, J. van der   NL 010103/69/2597
Onderzoek naar de periodiciteit van de stikstofopname van bloembollen. **Investigation into the periodicity of nitrogen uptake by bulbs.** Publications.

5878 Boon, J. van der   NL 010103/70/2789
De schadelijkheid van magnesiumgebrek in bloembollen op

duinzandgrond. **The harmful effect of magnesium deficiency in flower–bulbs on dune sand.**

5879 Ploegman, C.   NL 010501/67/0811 R
Invloed van zout beregeningswater op de produktie en kwaliteit van bloembollen. **Effect of saline sprinkling water on yield and quality of flower bulbs.** Publications.

Flowers and pot plants (B 3720)

See also 3775, 3778, 3907, 4606, 4607, 4608, 4619, 4655, 4658, 5712, 5715, 5717, 5718, 5719, 5720, 5721, 5722, 5723, 5724

5880 Van Onsem, J.; Gabriels, R.   BE 070600/68/0028
Onderzoek naar de minerale voedingsbehoeften van sierplanten. **Research into the mineral nutritional requirements of ornamental plants.** Publications.

5881 Heinze, W.   DE 105110/70/0003
Wachstum und Nährstoffaufnahme von Zierpflanzen bei verschiedenen Luft– und Substrattemperaturen. **Growth and nutrient uptake of ornamental plants at different air and substrate temperatures.**

5882 Zimmer, K.; Preissel, H.G.   DE 138270/72/0001
Untersuchungen zur Wirkung des Stickstoffes während induktiver Bedingungen auf das Blühen einiger Zierpflanzen 1972. **Studies of nitrogen effects during inductive conditions on blooming ornamental plants.**

5883 Horn, W.; Kämpf, A.   DE 161270/78/0004 N
Untersuchungen zur Nährstoffaufnahme bei Bromeliaceen. **Investigations on nutrient absorption in bromeliads.**

5884 Sauthoff, W.   DE 215070/77/0001
Einfluss der Düngung auf den Verlauf der Verticillium Welkekrankheit der Chrysanthemen. **Influence of fertilization on the process of Verticillium wilt in Chrysanthemum.**

5885 Penningsfeld, F.; Fast, G.   DE 502101/70/0001
Substrat– und Nährstoffansprüche marktwichtiger Orchideen bei Aussaat, im Jungpflanzenstadium und zur Zeit der Blüte. **Substrate and nutrient requirements of commercially important orchids in the seed stage, at the time of development and when flowering.**

5886 Penningsfeld, F.; Kurzmann, P.   DE 502101/70/0006
Prüfung neuer langsamwirkender Dünger zu Topfpflanzen und Schnittblumen. **Examination of the effect of new slow release fertilizers on pot plants and cut flowers.**

5887 Penningsfeld, F.; Forchthammer, L.   DE 502101/70/0008
Bedeutung von Klonauslese, Klimagestaltung sowie Bewässerungsund Düngungstechnik für Gesundheitszustand, Ertrag und Qualität von Gerbera jamesonii. **Importance of clone selection, climate regulation as well as irrigation and fertilization techniques for health, yield and quality of Gerbera jamesonii.**

5888 Penningsfeld, F.; Kurzmann, P.; Kalthoff, F.
     DE 502101/78/0001 N
Hydrokulturversuche mit verschiedenen Depotdüngern, Nährlösungen und Kulturgefässen. Versuchspflanzen: Verschiedene Orchideen–Gattungen und Blattpflanzen. **Trials with various slow–release fertilizers, nutrient solutions and special pots and containers, used for hydroculture. Test plants: various orchids and foliage plants.**

5889 Penningsfeld, F.; Kurzmann, P.   DE 502101/78/0002 N
Prüfung neuer Fertigsubstrate auf Torfbasis zu Schnittgrün– Arachniodes adiantiformis – und Topfpflanzen – Anthurien, Begonien, Cyclamen, Pelargonien, Gloxinien, Saintpaulien –. **Examination of new–developed commercial substrates basing on peat – Arachnodes adiantiformis, Anthurium, Begonia, Cyclamen, Pelargonium, Gloxinia, Saintpaulia –.**

5890 Penningsfeld, F.; Fischer, P.; Kalthoff, F.
     DE 502101/78/0004 N
Einfluss unterschiedlicher Kalkherkünfte, –körnungen und Giesswässer auf den pH–Wert von Torfkultursubstraten bei Begonia–Elatior–Hybriden, Hibiscus, Pelargonium und Saintpaulia. **The pH of peat substrates as influenced by different lime proveniences and particle sizes and the quality of irrigation water. Test plants: Begonia–Elatior–hybrids, Hibiscus, Pelargonium and Saintpaulia.**

5891 Penningsfeld, F.   DE 502101/78/0005 N
Untersuchung der Ansprüche an Bodenreaktion und Eisenversorgung von Ixora coccinea in Torfkultur). **Investigations on pH and iron requirements of Ixora coccinea in peat as growing medium.**

5892 Röber, R.; Fischer, M.   DE 506108/77/0001
Einfluss der Stickstoff– und Kaliumernährung von Chrysanthemen–Mutterpflanzen auf die Stecklingsquantität und –qualität. **Influence of nitrogen and potassium nutrition of chrysanthemum motherplants upon quantity and quality of cuttings.** Publications.

5893 Röber, R.; Eismann, I.; Köhler, K.   DE 506108/77/0002
Stickstoffernährung von Mutterpflanzen und Lagerfähigkeit der Stecklinge von Hydrangea macrophylla und Pelargonium–zonale–Hybriden. **Nitrogen nutrition of motherplants and storage capability of cuttings of Hydrangea macrophylla and Pelargoniumzonale–hybrids.**

5894 Röber, R.; Fischer, M.   DE 506108/78/0001 N
Wirkung von N–, P– und K–Steigerung auf das Wachstum und Blühen von Zierpflanzen in Hydrokultur. **Influence of increasing concentrations of N, P, and K on the growth and the flowering of ornamental plants in hydroponics.** Publications.

5895 Mackroth, K.; Bambach, G.   DE 506113/78/0003 N
Die kapillare Wasseraufnahme von Topfpflanzen bei verschiedenen Topfböden. **The capillary water uptake in potted plants bei various pot bottoms.**

5896 Andersen, A.   DK 030171/73/0005
Undersøgelser i vækstkammer af klimafaktorernes indflydelse på nettofotosyntesen hos forskellige potteplanter. **Growth chamber investigations of the influence of climatic factors on net photosynthesis in different pot plants.**

5897 Dartigues, A.   FR 010403/75/6027
Problèmes posés par la culture de l'hortensia en vert : substrats et fertilisation. **Problems of hydrangea culture before flowering : mineral nutrition and substrats.**

5898 Dartigues, A.   FR 010403/75/6028
Nutrition minérale des plantes en pots : bruyères, pélargonium, primevères. **Mineral nutrition of pot ornementals**

: heather – pelargonium – primrose.

5899 Moulinier, H.   FR 010502/71/0514
Etude de la fertilisation du rosier sous serre. **Study on the fertilization of greenhouse rosetree.**

5900 Blanc, D. Mme; Mars, S.   FR 010502/71/0524
Etude des réserves glucidiques chez le rosier de serre. **Studies on carbohydrates reserves in greenhouse rose–tree.**

5901 Moulinier, H.; Montarone, M.   FR 010502/74/6001
Besoins du gerbera en éléments minéraux. **Mineral requirements of gerbera.**

5902 Winsor   GB 011105/00/0002
Chrysanthemum nutrition with reference to effect of micronutrients on flowering.

5903 Adams; Winsor   GB 011105/00/0003
Nutrition of carnations in peat beds.

5904 Gallagher, P.A.; Seager, J.C.R.   IE 060300/72/0054 R
Nutrition of glasshouse ornamentals. Publications.

5905 Volpi, L.; Paterniani, T.   IT 021200/77/0006 N
Ricerche sulla concimazione e tecnica colturale dell' Euphorbia fulgens.. **Study on the fertilization and culturae technique of the Euphorbia fulgens.**

5906 Volpi, L.   IT 021200/77/0007 N
Nutrizione piante madri per la produzione di piante ornamentali in vaso. **Trials on the fertilization of mother plants for the production of ornamental pot plants cuttings.**

5907 De Ranieri, M.; Grassotti, A.   IT 021200/77/0011 N
Confronto fra investimenti unitari ed altre tecniche colturali sul garofano in coltura estiva. **Comparison between cultivation densities and other cultural techniques in summer production carnation.**

5908 De Ranieri, M.; Grassotti, A.   IT 021200/78/0003
Influenza di diversi apporti idrici nell'irrigazione del garofano in produzione estiva. **Influence of different water amounts, on the irrigation of the carnation for summer harvest.** Publications.

5909 Farina, E.; Lupi, R.   IT 021200/79/0003 N
Effetti di prodotti influenzanti l'assimilazione dell'azoto sulle caratteristiche qualitative e quantitative della produzione del garofano mediterraneo. **Results of products affecting nitrogen uptake and metabolism on qualitative and quantitative characteristics of mediterranean carnation production.**

5910 Volpi, L.; Sulis, S.   IT 021200/79/0004 N
Ricerche sulla concimazione e tecnica colturale dell' Antirrhinum majus. **Research on fertilisation and cultural techniques of Antirrhinum majus.**

5911 Smilde, K.W.   NL 010103/67/0575
De spoorelementenvoorziening van planten op veensubstraten. **Trace element supply of plants growing on Sphagnum peat substrates.** Publications.

5912 Arnold Bik, R.   NL 010103/67/0580 R
Bemesting van kasrozen. **Fertilization of glasshouse roses.** Publications.

5913 Arnold Bik, R.   NL 010103/67/0945
Bemestingsproblemen bij bloemisterijgewassen. **Fertilization problems in ornamentals.** Publications.

5914 Arnold Bik, R.   NL 010103/69/2594
De invloed van substraat, vochttoestand van de grond en bemesting op de houdbaarheid van cyclamen. **The influence of substrate, soil moisture status and fertilization on the durability of cyclamen.** Publications.

5915 Arnold Bik, R.   NL 010103/71/3009
Onderzoek van fysische en chemische kwaliteit van handelspotgronden. **Research on the physical and chemical quality of commercial potting mixtures.** Publications.

5916 Arnold Bik, R.   NL 010103/72/3565
Bemesting van gerbera. **Fertilization of gerberas.** Publications.

5917 Arnold Bik, R.   NL 010201/67/0580
Bemesting van kasrozen. **Fertilization of glasshouse roses.** Publications.

5918 Arnold Bik, R.   NL 010201/67/0945
Bemestingsproblemen bij diverse bloemisterijgewassen. **Fertilization problems in various ornamentals.** Publications.

5919 Arnold Bik, R.   NL 010201/69/2594
De invloed van substraat, vochttoestand van de grond en bemesting op de houdbaarheid van cyclamen. **The influence of substrate, soilmoisture status and fertilization on the durability of cyclamen.** Publications.

5920 Arnold Bik, R.   NL 010201/71/3009
Onderzoek van fysische en chemische kwaliteit van handelspotgronden. **Research on the physical and chemical quality of commercial potting mixtures.** Publications.

5921 Arnold bik, R.   NL 010201/72/3565
Bemesting van gerbera. **Fertilization of gerberas.** Publications.

5922 Berg, G.A. van der; Valentin, J.C.M. NL 010201/74/5901
Watergeefsystemen bij potplanten. **Watersupply systems for potplants.**

Ornamental shrubs (B 3730)

See also 5508, 5953, 5956, 5957, 5958, 5960

5923 Van Onsem, J.G.; Gabriels, R.   BE 070600/78/0079 R
Bemestingsproeven met boomkwekerijgewassen in pot. **Fertilizing trials with woody nursery stock cultivated in containers.**

5924 Schmidt, H.H.; Schering, J.   DE 135054/70/0001
Rasendüngungsversuche 1969. **Trials on fertilization of lawns.**

5925 Wennemuth, G.   DE 138241/77/0001
Der Einfluss einer Spätstickstoffdüngung auf die Überwinterungsqualität und Zuwachsleistung der Koniferen und Immergrünen Laubgehölzen. **The wintering quality and growth of conifers and evergreen deciduous trees in dependence on late nitrogen fertilization.**

5926 Penningsfeld, F.; Kalthoff, F.   DE 502101/70/0002
Boden– und Nährstoffansprüche von immergrünen

Ziergehölzen, insbesondere Koniferen und Rhododendron im Baumschulstadium. **Soil and nutrient requirements of evergreen shrubs especially conifers and rhododendron in the nursery stage.**

5927 Penningsfeld, F.; Kurzmann, P.; Kalthoff, F.; Reis, A.
DE 502101/75/0001 N
Wirkung von NPK–Mangel und –steigerung auf Wuchs, Gesundheitszustand, Ertrag, Blühdauer und Haltbarkeit der geschnittenen Blüte von 14 Schnittstaudenarten bzw. Sorten. **The influence of deficiency and increase of N, P and K on growth, health, yield, flowering period and keeping–time of cut blossoms of 14 species resp. varieties of cut perennials.**

5928 Alt, D.
DE 507701/71/0002
Stickstoff– und Kaliumernährung von Schnittstauden. **N– and K–nutrition of perennial herbs.**

5929 Nielsen, J.M.; Villumsen, J.; Selvaratnam, V.
DK 030148/79/0008 N
Gødskning af Hedera canariensis udfra planternes kemiske sammensætning. **Fertilization of Hedera canariensis based on chemical plant analyses.**

5930 Klougart, A.; Johansen, H.C.
DK 030171/78/0003 N
Indflydelse af pH og saltkoncentration på vækst og produktion af Adiantum. **Growth and production of Adiantum in media with variated pH and salinity.**

5931 Lemaire, F.; Salette, J.; Dartigues, A.
FR 010403/76/6024
Exportations minérales des plantes de pépinières d'ornement. **Mineral uptake of different ornemental trees and conifers.**

5932 Winsor; Adatia
GB 011105/00/0011
Nutrition of ornamental nursery stock.

5933 Whalley
GB 011106/00/0011
Nutrition of ornamental nursery stock.

5934 Boon, J. van der
NL 010103/70/2792 R
Bemesting en vochtvoorziening van boomteeltgewassen in potten. **Fertilization and irrigation of ornamental container–grown plants.** Publications.

5935 Das, A.
NL 010103/75/6520
Onderzoek naar de oorzaak van een groeistoornis bij de opkweek van Laburnum watereri cv. "vossii" (gouden regen). **Search for the cause of a growth disorder in the culture of Laburnum watereri cv. "vossii".**

5936 Boon, J. van der
NL 010103/76/6587
Optimale groei van coniferen en ericaceeën op zand door toevoeging van veenprodukten en verlaging van pH. **Optimum growth of conifers and Erica species on sandy soils by adding peat and lowering pH.**

5937 Boon, J. van der
NL 010103/76/6589
Opkweek van laanbomen in steenwol. **Cultivation of avenue trees in rock wool.**

5938 Boon, J. van der
NL 010103/77/7687 R
Bemesting van boomteeltgewassen op zandgrond. **Fertilization of tree nursery stock on sandy soil.**

5939 Elk, B.C.M. van
NL 010203/64/3754

Onderzoek naar de voedingsstoffenbehoefte van boomkwekerijgewassen in plastic potten, boven de grond. **Research on the fertiliser requirement of nursery stock cultivated in plastic pots above the soil.** Publications.

5940 Elk, B.C.M. van
NL 010203/65/3753
De invloed van potgrondmengsels op de ontwikkeling van boomkwekerijgewassen in stenen potten in de grond. **The influence of potting composts on the growth of nursery–crop in stone pots in the soil.** Publications.

5941 Elk, B.C.M. van
NL 010203/69/3749 R
Het gebruik van natuurlijke en kunstmatige stekmedia in de boomkwekerij. **The use of natural compounds, such as peat and sand, and artificial cuttingmedia in the nursery.** Publications.

5942 Aendekerk, Th.G.L.
NL 010203/73/5914
Gebreksverschijnselen bij boomkwekerijgewassen. **Deficiency diseases in ornamental shrubs.** Publications.

Other ornamentals and ornamental products (B 3790)

5943 Penningsfeld, F.; Kurzmann, P.; Reis, A.
DE 502101/70/0005
Boden– und Nährstoffansprüche beliebter Alpen– und Steingartenpflanzen. **Soil and nutrient requirements of favoured alpine and rock garden plants.**

Forests in general (B 3800)

See also 559, 560, 4833

5944 Manil, G.; Delecour, F.; Van Praag, H.; Weisse, F.
BE 010016/57/0003
Recherches sur l'amélioration, en particulier, par la fertilisation chimique et la conversation de la fertilité des sols forestiers de la Belgique méridionale. **Study of improvement in particulary by chemical fertilization and conservation of forest soils fertility in Southern Belgium.** Publications.

5945 Huygh, A.; Baeyens, L.; Beckers, B.; Berben, J.
BE 140000/77/0018 R
Recuperatie voor de bossen en de groenvoorzieningen van afvalverwerkingsresten. **Recuperation of waste products in forests and greenspaces.** Publications.

5946 Moll, W.; Pietrowicz, P.; Stahr, K. DE 126050/78/0005 N
Untersuchungen der Auswirkungen von Müllkompostgaben zu sorptionsschwachen Sandböden im Forstamt Schwetzingen. **The effects of refuse compost application to sandy soils of low cation exchange capacity in the forest district of Schwetzingen.** Publications.

5947 Moll, W.; Schwarz, O.
DE 126050/78/0012 N
Untersuchungen zur Verwendung von Müllkompost auf forstlichen Rekultivierungsflächen im Forstamt Rastatt. **Investigations on the use of refuse compost on recultivated forest sites in the forest district of Rastatt.**

5948 Braun, H.J.
DE 126250/70/0001
Physiologie und Ökologie des Wasserhaushaltes von Bäumen. **Physiology and ecology of water consumption of trees.**

5949 Kreutzer, K.
DE 160030/72/0003
Düngungsversuche im Fichtelgebirge. **Fertilization**

D 2200 – Plant nutrition and fertilization

experiments in the Fichtelgebirge.

5950 Kreutzer, K.   DE 160030/75/0003
Erfassung des Stoffeintrages mit den Niederschlägen und der
Stoffauswaschung mit dem Sickerwasser in typischen
Waldökosystemen. **Investigations of input by precipitation and
output by percolation water in typical forest ecosystems.**

5951 Runge, M.   DE 164050/78/0001 N
Pflanzenökologische Bedeutung des Aluminiums in
Waldböden. **Plant ecological significance of aluminum in
woodland soils.**

5952 Evers, F.H.; Bücking   DE 501502/77/0001
Klärschlamm–Versuche in Waldbeständen. **Experiments with
sewage sludge in forest stands.**

5953 Schaller, C.; Röber, R.; Rohde, J.   DE 506115/78/0001 N
Einfluss unterschiedlicher Aufwandmengen von
Rindenkompost auf das Wachstum verschiedener
Gehölzarten. **Influence of different quantities of
hardwood–bark–compost on the growth of various species of
trees and shrubs.**

5954 Siebert, H.; Dimitri, L.   DE 506453/77/0002
Untersuchungen über die Auswirkungen von Klärschlamm auf
das Wachstum verschiedener Baumarten. **Research on the
influence of sewage sludge on the growth of various tree species.**

5955 Reemtsma, J.B.   DE 507651/77/0002
Nachwirkungen der P–Düngung bei wiederholter N–
K–Düngung. **Effects of P–fertilization after repeated N–
K–fertilization.**

5956 Kohstall, H.; Alt, D.   DE 507701/78/0001 N
Einfluss von Müllkompostbeimischungen zu Torfsubstraten
auf das Wachstum von Gehölzen in Containern. **Effect of
mixtures of peat substrates and municipal compost on the
growth of container–grown trees and shrubs.**

5957 Groven, I.; Knoblauch, F.   DK 010104/67/0015 N
Ernæringsfysiologi vedrørende træer og buske. **Nutritional
physiology of trees and shrubs for landscaping.**

5958 Groven, I.; Bøvre, O.   DK 010104/69/0018 N
Dyrkningssubstrater til containerplanter. **Growing medium for
trees and shrubs in containers.**

5959 Groven, I.; Knoblauch, F.   DK 010104/72/0013 N
Ernæring og vanding af træer og buske. **Nutrition and
irrigation of trees and shrubs for landscaping.**

5960 Groven, I.; Knoblauch, F.   DK 010104/75/0014 N
Ernæring og rodvækst hos træer og buske. **Nutrition and root
growth of trees and shrubs for landscaping.**

5961 Bonneau, M.; Garbaye, J.; Le Tacon, F.
                                           FR 010303/61/4152
Fertilisation de plantations résineuses et feuillues. **Fertilization
of conifers and hardwood plantations.**

5962 Le Tacon, F.; Bonneau, M.; Garbaye, J.
                                           FR 010303/75/4160
Utilisation en forêt ou en pépinière des déchets urbains
organiques. **Use of organic town wastes in forests or in
nurseries.**

5963 Carey, M.L.; Hendrick, E.   IE 050100/62/7203 N
Nutrition of forest crops and their fertiliser requirements.
Publications.

5964 Smilde, K.W.   NL 010103/69/2596 R
Spoorelementproblemen in de bosbouw. **Trace–element
problems in forestry.** Publications.

5965 Burg, J. van den   NL 010601/76/6874
Interpretatie van grond– en gewasanalyses t.b.v.
bosbouwkundige problemen. **Interpretation of soil and foliar
analytical data in relation with problems in forestry.**
Publications.

Pine forests in general (B 3810)

See also 4868, 5068, 5695

5966 Zöttl, H.W.; Reissmann, B.   DE 126050/78/0008 N
Nährelementversorgung von Koniferen–Aufforstungen in
Südbrasilien. **Nutrient supply of young coniferous stands in
Southern Brazil.**

5967 Moll, W.; Genser, H.   DE 126050/78/0009 N
Nadelanalytische Untersuchungen zur Kennzeichnung der
Ernährungssituation von Pinus–Beständen in der Provinz
Parana/Brasilien. **Investigations for the determination of
nutritional status of Pinus–stands using needle analysis, in the
Parana state, Brazil.**

5968 Zöttl, H.W.   DE 126050/78/0011 N
Untersuchung von Rindensubstraten zur Koniferen–Anzucht
und als Bodenverbesserungsmittel. **Investigations on the use of
bark residues for the nursing of conifers and as a substrate for
soil amelioration.**

5969 Rehfuess, K.E.; Baum, U.; Rodenkirchen, H.
                                           DE 160030/78/0002 N
Wirkung von Meliorationsmassnahmen auf den
Nährelementhaushalt eines streugenutzten
Kiefernwaldökosystems in der Oberpfalz. **Effect of meliorative
operations on nutrient balance of litter–raked pine forest
ecosystem in the Upper Palatinate.** Publications.

5970 Reemtsma, J.B.   DE 507651/75/0001
Untersuchungen über die Möglichkeit langfristiger
Ertragssteigerung von Nadelholzbeständen auf nährstoffarmen
Böden durch Düngung. **Possibilities of long–term increase in
yield of softwood stands on oligotrophic soils by means of
fertilizing.**

5971 Nys, C.; Le Tacon, F.   FR 010303/71/4154
Peuplements adultes de résineux dans le Massif Central :
relations entre le sol, la nutrition, la production ligneuse –
Essais de fertilisation. **Adult coniferous stands in Massif
Central : soil–nutrition–production relationships – Fertilization
experiments.**

5972 Miller, H.   GB 030401/00/0017 R
Nutrient deficiencies in conifers: diagnosis and amelioration.

5973 Miller, H.   GB 030401/74/0015 R
**Conifer nutrition: nutrient cycling, tree growth and influence of
fertilizers.**

D 2200 - Plant nutrition and fertilization

5974 Burg, J. van den; Tol, G. van   NL 010601/57/2395
Onderzoek naar de oorzaak en de mogelijkheid tot verbetering van groeistoornissen bij houtsoorten veroorzaakt door het gehalte aan sporenelementen in de grond. **Research on the cause and the possibilities for improvements of growth disturbances of tree species related to the content of trace elements.** Publications.

Fir forests (B 3811)

5975 Zöttl, H. W.; Moll, W.   DE 126050/78/0010 N
Nadelanalytische Kontrolle von Düngungsversuchen in Kiefernbeständen des Pfälzer Waldes. **Needle analysis for the evaluation of fertilizer experiments on pine plantations in the Palatinate Forest - Pfälzer Wald -.**

5976 Matzner, E.; Reemtsma, J.B.; Ulrich, B.
DE 132600/77/0006
Nährstoffaustrag unter Beständen von Eiche, Kiefer und Fichte im nordwestdeutschen Flachland. **Nutrient loss under cover of oak, pine and spruce in Northwest German flat land.**

5977 Hüser, R.   DE 160030/73/0003
Einfluss von Klärschlamm auf die chemische Zusammensetzung von Fichten- und Kiefernnadeln. **The influence of sewage sludge on chemical compounds in spruce and pine needles.**

5978 König, E.; Olberg-Kallfass, R.   DE 501508/72/0002
Untersuchungen über Wachstumsförderung von Fichtenkulturen durch Einsatz von Herbiziden und Düngemitteln 1970. **english title not indicated.**

5979 Ulrich, B.; Reemtsma, J.B.; Hetsch, W.
DE 507651/77/0001
Nährstoffaustrag unter Beständen von Eiche, Kiefer und Fichte im nordwestdeutschen Flachland. **Nutrient loss under stands of oak, pine and spruce in the Northwest German lowlands.**

5980 Liani, A.   IT 011801/76/0009
Calcolo dei fabbisogni idrici di semenzali di Pinus radiata allevati in fitocella. **Assessment of water requirements of Pinus radiata seedlings grown in plastic bags.** Publications.

Spruce and fir forests (B 3812)

See also 4927, 5976, 5977, 5979, 7948

5981 André, P.; Lheureux, C.; Giot-Wirgot, P.; Bailleux, P.; Abu el Kaiir, K.   BE 020301/69/0010 N
Influence de la fertilisation sur la croissance de l'épicea (Picea abies Karst). **Influence of fertilization on the growth of Picea (Picea abies Karst).** Publications.

5982 Zöttl, H.W.; Ferraz, J.   DE 126050/75/0006 R
Nährelementversorgung und Spurenelementverteilung in Fichtenbeständen des Kristallinschwarzwaldes. **Nutrient supply and trace element distribution in spruce stands in the crystalline Black Forest.**

5983 Zöttl, H.W.; Glotzbach, N.; Stahr, K.
DE 126050/78/0004 N
Auswirkungen kleinräumiger Standortsunterschiede auf den Spuren- und Nährelementgehalt der Nadeln junger Fichtenbestände des Bärhaldegranit-Gebietes -

Südschwarzwald -. **The effects of small-scale spatial site differences on trace element and macronutrient contents in the needles of young spruce stands in the Bärhalde-granite area in southern Black Forest.**

5984 Henne, A.; Riebeling, R.   DE 506403/70/0003
Die Düngung von Fichtenbeständen auf verschiedenen hessischen Standorten in ertragskundlicher und betriebswirtschaftlicher Sicht. **The influence of fertilisation on the yield of older spruce stands on different sites of Hesse.**

5985 Siebert, H.; Jungbluth, H.J.; Dimitri, L.
DE 506453/72/0002
Untersuchungen über die Wirkung verschiedener Düngungsverfahren auf den Erfolg und Leistung der Douglasie im Kultur- stadium. **Research on techniques of fertilizer application and their influence on the success and growth of Douglas fir plantations.**

5986 Dickson   GB 040402/00/0001 R
Nutritional requirements of Sitka spruce on oligotrophic blanket peat.

5987 Adams   GB 040402/00/0002 R
Nutritional requirements of Sitka spruce on mesotrophic blanket peat.

5988 Dickson   GB 040402/00/0003 R
Nutritional requirements of Sitka spruce and other species planted on mineral soils.

5989 Dickson   GB 040402/00/0004 R
Nutrient cycling in sitka spruce forests.

5990 Gardiner, J.J.; Coen, R.C.; Geoghegan, M.J.
IE 120107/77/9089 N
The influence of fertilizers upon microbial activity in sitka spruce litter.

Larch forests (B 3813)

See also 4955

Leafwoods in general (B 3820)

See also 4960, 4966, 5695

5991 Smilde, K.W.   NL 010103/74/5587 R
Bruikbaarheid van rioolslib als substraat voor loofhoutsoorten. **Utilization of sewage sludge as a substrate for broad-leaved tree species.**

Oak tree stands (B 3821)

See also 4976, 4977, 5976, 5979

Ash tree stands (B 3822)

5992 Levy, G.; Picard, J.F.; Aussenac, G.   FR 010303/73/4157
Frêne, Merisier, Noyer, Erable: relation entre la nature du sol, la nutrition et la production ligneuse - Typologie phytoécologique des stations favorables à leur culture. **Ash, Cherry tree, walnut, maple: relationships between soil, mineral nutrition and yield, Phytoecological description of suitables sites.**

D 2200 – Plant nutrition and fertilization

Poplar tree stands (B 3823)

5993 Garbaye, J.; Picard, J.F.   FR 010303/72/4158
Le Peuplier – Différents types de peupleraies – Caractérisation
par le sol et la flore – Relations entre sol, nutrition, production
– Possibilités d'extension de la culture du peuplier. **Poplars –
Characterization of poplar sites by soil and vegetation –
Relationships between soil, nutrition, production. Suitability of
different sites for increase of cultivation area.**

5994 Liani, A.   IT 011801/76/0002
Prova di irrigazione a goccia in vivaio di pioppo. **Drip irrigation
trial in poplar nursery.** Publications.

5995 Akkermans, A.D.L.   NL 020033/71/4714
Symbiotische stikstofverbinding door niet–leguminosen (els).
Symbiotic nitrogen fixation by not–leguminous plants (alder).
Publications.

Other leafwoods (B 3829)

See also 5992

Other forests (B 3890)

5996 Hase, H.; Fölster, H.   DE 132600/77/0005
Bioelementverluste nach Kahlschlag und
Bioelement–Bevorratung in jungen Teakplantagen, Llanos –
Venezuela. **Bioelement loss after forest clearing, and
bioelement fixation in young teak plantations, Llanos –
Venezuela.**

Stimulant crops (B 3910)

5997 Talibudeen; Wickramasinghe   GB 012003/76/0041 R
Transformation of nitrogen in Ceylon soils under tea.

Spice and seasoning plants of warm climates (B 3920)

See also 5028

Spice and seasoning plants of temperate climates (B 3930)

See also 7724

5998 Kohlmann, J.; Zwack, F.   DE 502059/70/0002
Untersuchungen über den Einfluss organischer Düngung auf
Ertrag und Qualität beim Hopfen 1970. **Influence of manuring
on crop and quality of hops.**

5999 Niemann, J.; Escher, F.   DE 507701/72/0002
Die Wirkung von Branntkalk und Magnesiumbranntkalk auf
die Reaktion und den Nährstoffhaushalt des Bodens sowie auf
das Wachstum verschiedener Schnittstauden. **Influence of CaO
and CaO + MgO on pH, mineral nutrient content of soil and on
growth rate of different perennial herbs.**

6000 Marocke, R.; Trouvelot, A.   FR 010901/74/6126
Nutrition minérale du houblon – Rendement et qualité de la
production. **Mineral nutrition of the hop. Yield and quality for
the production.**

Perfume plants (B 3940)

See also 5042

Drugs and medicine plants (B 3970)

See also 1545, 5048, 5049, 5424

6001 Neumann, K.–H.; Stoltenberg, G.   DE 129041/72/0004
Vergleichende Untersuchungen über die Alkaloidsynthese
ganzer Pflanzen und Gewebekulturen von Papaver
somniferum L. 1972. **Comparative studies of alkaloid synthesis
of complete plants and tissue of Papaver somniferum L.**

6002 Alkämper, J.; Manos, G.   DE 129552/78/0001 N
Einfluss der Düngung und der Erntetechnik auf Ertrag und
Qualität von Tabak. **Influence of fertilizer use and harvest
technique on yield and quality of tobacco.**

6003 Vardabasso, A.; Aversano, B.   IT 022300/78/0011 R
Impiego di alcuni diserbanti, in miscela ai concimi liquidi, nella
coltivazione del tabacco. **Employ of some herbicides, mixed to
liquid fertilizers on tobacco.** Publications.

Plants for experimental purposes (B 9130)

See also 5189

D 2300 – Plant breeding

See also 2617

6004 Deltour, J.; Nisen, A.; Coutisse, S.; Nijskens, J.
          BE 010018/78/0003 N
Economie d'énergie en cultures sous abri verre ou plastique.
Energy in cultures under shelter or in glasshouses.
Publications.

6005 Tilkin, V.E.; Philippot, R.   BE 080200/68/0022 R
Etude des propriétés des semences potagères. **Studies on the
characteristics of vegetable seeds.** Publications.

6006 Rixhon, L.; Frankinet, M.; Guiot, J.; Crohain, A.
          BE 080800/64/0007 R
Etude de la succession des cultures en relation avec la
production et les maladies. **Influence of crop rotation on crop
production and plant diseases.** Publications.

6007 Fouarge, G.   BE 080900/60/0003 R
Promotion de la production de plants de pomme de terre de
base en Belgique. **Promotion of the basic potato seed
production in Belgium.** Publications.

6008 Tahon, J.   BE 081100/72/0007 N
Recherches bio–écologiques sur les oiseaux présumés nuisibles
à l'agriculture. **Bio ecological research on bird species liable to
cause damage to agriculture.**

6009 Christensen, F.G.   DK 030102/72/0007
Udvikling af tjørneresistens mod ildsot. **The development of
Hawthorn resistance against fire blight.**

6010 Jensen, H.P.; Jørgensen, J.H.   DK 030146/76/0011
Spontane meldugresistensgener i byg. **Spontaneous genes for
mildew resistance in barley.**

6011 Klougart, A.   DK 030171/73/0004
Indsamling af slægter og arter af potteplanter egnede for

handelskulturer. **The collection of genera and species of pot plants suitable as commercial crops.**

6012 O'Kennedy, N.D.   IE 060302/64/0497
Production of virus free apple planting material. Publications.

6013 Crimella, C.   IT 040638/78/1154 N
Polimorfismi immunologici e biochimici nei bovini ed equini.
Immunological and bio–chemical polymorphism in bovine races and horses.

6014 Finzi, A.   IT 040907/78/1155 N
Possibilità di miglioramento genetico della attitudine lattifera
della razza modicana. **Possible genetic improvement of lactation in the "modicana" race.**

6015 Brandano, P.   IT 041313/78/1153 N
Indagine sulle popolazioni rustiche bovine, ovine, equine e
caprine della Sardegna. **Sardegna : investigation on the local bovine, sheep and horse population.**

Parks, gardens, urban greenspaces, plantations (B 1610)

See also 4713, 6009, 6020, 6707, 6723, 6728, 7655, 7689, 7693,
7708, 7709, 7710, 7712, 7713, 7714

6016 Beuster, K.–H.; Hiller, H.   DE 105250/78/0004 N
Ergänzende Rasengräsersorten–Prüfung auf
Landschaftsraseneignung der besonderen Anbauprüfungen auf
Rasennutzung. **Supplementary testing of turf grass varieties for suitability for landscape greens in special cultivation tests for turf utilization.**

6017 Heybroek, H.M.   NL 010601/28/1849
Veredeling van iepen (o.a. op resistentie voor de
iepziekteschimmel). **Selection and breeding of elms (e.g. for resistance to the elm disease fungi).** Publications.

6018 Dissen, H.D.   NL 010601/71/3530
Bijzondere boom– en struiksoorten in recreatieve
beplantingen. **Trees and shrubs in recreation areas.**
Publications.

6019 Heybroek, H.M.; Gremmen, J.   NL 010601/73/3839
Veredeling van meidoorn i.v.m. resistentie tegen perevuur.
Breeding haw thorn (Crataegus) for resistance to fire blight (Erwinia amylovora). Publications.

Sportfields, play and camping grounds (B 1630)

See also 6707, 6723, 6728, 7627

6020 Dijkstra, J.   NL 010120/70/7425 R
Veredelingsonderzoek aan Engels raaigras–roodzwenkgras
bastaarden voor recreatieve doeleinden. **Breeding research in perennial ryegrass–red fescue hybrids for recreational purpose.**
Publications.

Plants and parts of plants in general (B 2100)

See also 6620, 7534

6021 Semal, J.; Sommereyns, G.; Dutrecq, A.
                    BE 010022/74/0004
Sélection in vitro de plantes, résistantes ou tolérantes à l'égard

de champignons pathogènes. **In vitro selection of plants resistant or tolerant towards pathogenic fungi.**

6022 Seilleur, P.   BE 010022/78/0007 N
Régénération des plantes supérieures par culture in vitro
Analyse génétique des régénérats. **Regeneration of plants by tissue cultures. Genetical analysis of the regenerated plants.**
Publications.

6023 Schell, J.; Van Montagu, M.; De Beuckeleer, M.;
Dewilde, M.   BE 140000/76/0036
D.N.A. – tranfert systeem. **D.N.A. tranfert system.**
Publications.

6024 Opdedrynck, A.; Weenen, A.; Ceulemans, E.; Janvier,
A.; Delannay, J.   BE 140000/77/0001 R
Mitochondrien complementatie. **Complementation of mitochondria.**

6025 Weiling, F.   DE 111151/70/0004
Untersuchungen zur Vorgeschichte der Genetik und der
Wechselbeziehung zwischen Statistik und Auffindung der
Vererbungs– regeln. **Researches on the pre–history of genetics and the interaction between statistics and the rules of heredity.**
Publications.

6026 Weiling, F.; Unger, C.   DE 111151/72/0003
Die Bestimmung und Analyse der Leistungsstabilität von
Pflanzentypen. **The determination and analysis of the stability of genotypequalities.**

6027 Larsen, K.   DK 030101/71/0005
Genetiske undersøgelser af selv–inkompatibilitets mekanismer
hos angiospermerne: Selvinkompatibilitet hos bederoen, Beta
vulgaris L. **Genetic investigations of self–incompatibility mechanisms in the angiosperms: Self–incompatibility in the beet, Beta vulgaris L.**

6028 Østerbye, U.   DK 030101/71/0006
Genetiske undersøgelser af selv–inkompatibilitets mekanismer
hos angiospermerne: Selv–inkompatibilitet hos ranunculus
acris L. **Genetic investigations of the self–incompatability mechanisms in the angiosperms: Ranunculus acris L.**

6029 Henningsen, K.W.; Kaufmann, U.; Stummann, B.
                    DK 030101/75/0001 R
Molekylær genetiske undersøgelser af de nukleære og
extranukleære geners betydning ved kloroplast biogenesen.
Molecular–genetical investigations of the nuclear and extra nuclear genes and their importance in chloroplast biogenesis.

6030 Larsen, K.   DK 030101/77/9062
Biokemiske– og fysiologiske undersøgelser af
selvinkompatibilitetsmekanismen hos bederoen, Beta vulgaris
L. **Biochemical and physiological investigations of the self–incompatibility mechanism in beet, Beta vulgaris L.**

6031 Østerbye, U.   DK 030101/78/9052
Biokemiske og cellebiologiske studier af mekanismen i
komplekse selvinkompatibilitetssystemer hos angiospermerne
undersøgt hos Ranunculus acris L. **Biochemical and cell biological studies of the mechanism in complex self–incompatibility systems in angiosperms investigated in Ranunculus acris L.**

6032 Tedeschi, P.   IT 060700/77/0998

D 2300 – Plant breeding

Studio della variabilità inter e intraspecifica riguardo al valore critico del potenziale idrico fogliare della pianta per lallungamento fogliare e l'accrescimento in genere. **A study of inter– and intra–species variability as to the critical value of leaf water content in relation to the lengthening of a plant leaves and its general growth.**

Plant communities as ecological systems (B 2200)

See also 6736

Animal and plant communities as ecological systems (B 2500)

See also 6008

Crops in general (B 3000)

See also 2697, 2786, 2814, 5311, 6004, 6022, 6160, 8143, 9128, 9133, 20011

6033 Semal, J.; Dutrecq, A.; Sommereyns, G.
BE 010022/77/0005
Utilisation de toxines fongiques en vue de la sélection de plantes résistantes à l'infection fongique. **Use of fungal toxins in the selection of plants resistant to the infection by fungi.** Publications.

6034 Van Sumere, C.; Hanselaer, R.   BE 140000/76/0039 R
Onderzoek over het belang van natuurlijke voorkomende fenolen met betrekking tot de eiwitbiosynthese. **Research of the importance of natural phenols in relation to proteinsynthesis.** Publications.

6035 Kobabe, G.; Koehler, U.   DE 132183/75/0002
Entwicklung neuer Zuchtverfahren mit Hilfe der männlichen Sterilität. **Development of new breeding methods by using male sterility.**

6036 Kühbauch, W.   DE 161255/77/0005
Physiologische Kenndaten der Zellwand von wachsendem Pflanzengewebe als Auswahlkriterien für die Züchtung. **Physiological characteristics of the cell wall of growing plant tissue as screening tools in plant breeding.**

6037 Hondelmann, W.; Grahl, A.   DE 201040/73/5007
Koordinierung der systematischen Testung – Evaluierung – von Genmaterial auf Resistenz und Werteigenschaften. **Coordination of systematic testing – evaluation – of gene material for resistance and value properties.**

6038 Grahl, A.; Seidewitz, L.   DE 201040/73/5008
Vermehrung und Vermittlung von Genmaterial. **Multiplication of gene material in the genetic resources centre of the FRG.**

6039 Kristensen, K.; Rudemo, M.   DK 030107/77/0001 N
Statistisk analyse af sortsforsøg. **Statistical analysis of variety trials.**

6040 Andersen, K.; Andersen, S.   DK 030145/20/0001
Sortssamling af kulturplanter og undersøgelser i denne. **Collection of cultivars of crop plants and related investigations.**

6041 Kruse, A.   DK 030145/64/0003
Arts– og slægskrydsninger. Metoder og teknik til fremstilling af arts– og slægshybrider inden for vore kulturplanter, i særlig grad hybrider mellem kornarterne. **Interspecific and intergeneric hybrids. Methods and techniques for the production of interspecific and intergeneric hybrids within crop plants, chiefly cereal species.**

6042 Vear, F.; Leclercq, P.; Guillaumin, J.J.
FR 010801/73/0160
Méthode de sélection précoce pour la résistance au Botrytis et au Sclerotinia. **Methods of early selection for resistance to Botrytis and Sclerotinia.**

6043 Wagner, R.; Bronner, A.; Charrier; Balthazard, J.
FR 010902/75/0318
Etude de la coulure accidentelle et du millerandage d'origine climatique en vue de mettre au point un test de sélection permettant de mieux préciser les différences variétales. **Study of occasionnal abortion and climatical grape–failure : application to a selection test allowed to precise better varietal differences.**

6044 Champion, R.   FR 011103/76/0427
Etude de la résistance des variétés aux maladies. Mise au point de méthodes pour le contrôle de la résistance. **Studies related to resistance of cultivars to diseases, methods for the testing of cultivars for resistance.**

6045 Grimbly   GB 011101/00/0010
Use of periclinal chimaeras and interspecific hybridisation.

6046 Cullis   GB 011401/00/0010 R
Genetics and environmentally induced variation in DNA of higher plants.

6047 Sunderland   GB 011401/00/0012
The production of haploid plants by pollen culture.

6048 Watts   GB 011403/00/0001
Study of plant cell ultrastructure; protoplast biology.

6049 Lamont–Fisher   GB 011403/76/0010
Cell protoplasts : cell fusion and surface interaction.

6050 Dunn; Trigg   GB 011404/00/0001 R
Constituents of plant nucleic acids.

6051 Snape; Simpson   GB 011905/78/0029
Use of dihaploid lines for genetic analysis.

6052 Law; Snape Worland   GB 011905/78/0031 R
Cytoplasmically inherited variation.

6053 Starr; Morgan   GB 011908/00/0005 R
To devise or modify biochemical methods suitable for analysing samples generated in plant breeding programme.

6054 Smith; Thorburn   GB 011908/78/0007
Quality components of potential interest to plant breeders.

6055 Bright; Miflin   GB 012001/00/0017 R
The selection of biochemical mutants and crop plants.

6056 Trewaves   GB 021001/79/0010 N
Transcriptional regulation of the cellulase gene by auxins.

6057 England   GB 030806/78/0001

D 2300 – Plant breeding

Study of trial designs and field management for plant breeding.

6058 Weatherup   GB 040201/75/0002
Discrimination between biological populations using multivariate techniques.

6059 Scarascia Veneziani, M.E.   IT 020500/77/0399 R
Prove di adattabilità ambientale, valutazione qualitativa di linee in selezione di frumento duro. **Tests of adaptability to environment, qualitative evaluation of selected lines of hard wheat.**

6060 Sardara, M.   IT 020900/70/0003
Adattamento ambientale di specie e varietà di nuova introduzione. **New introductions of species and varieties and their adaptation to the environment.**

6061 Russo, F.; Starrantino, A.; Donini, B.
       IT 021600/75/0003 N
Induzione artificiale di mutazioni mediante l'impiego cobalto 60.. **Artificial induction of mutations with Co 60.**

6062 Scossiroli, R.E.   IT 040214/77/0808
Produzione ed adattabilità ecotipi, selezione, moltiplicazione varietà sintetiche. **Ecotypes production and adaptability, selection, multiplication, synthetic varieties.**

6063 Ghidoni, A.   IT 040320/77/0193
Produzione di poliploidi e di aneuploidi in piante coltivate. **Production of polyploids and aneuploids in cultivated plants.**

6064 Rivoira, G.   IT 041302/77/0390 R
Attuazione di incroci fra cultivar e linee di interesse agronomico, selezione su materiale già disponibile. **Hybridisation of cultivars and lines of agronomic value, selection on material already available.**

6065 Lucci, G.   IT 042001/72/0496
Effetti del poliploidismo sulle attività enzimatiche con particolare riguardo all'assorbimento e al trasporto degli elementi fertilizzanti. **Effects of polyploidism on enzymatic activity with specific reference to the absorption and transport of fertilizing elements.** Publications.

6066 Porceddu, E.   IT 060500/74/0179
Studio della variabilità genetica in popolazioni di piante autogame. **Genetic variability in self pollinated species.**

6067 Porceddu, E.   IT 060500/77/0981
Reperimento e raccolta di germoplasma in Nord Africa e nelle penisole balcanica e iberica. **Finding and collection of germ plasm in North Africa the Balkan and the Iberian peninsulas.**

6068 Lamberti, F.   IT 060600/77/0455
Resistenza genetica ai nematodi o a malattie da essi trasmesse. **Genetic resistance to nematodes and nematode–borne diseases.**

6069 Lamberti, F.   IT 060600/77/0910
Analisi delle reazioni varietale e clonale agli attacchi dei nematodi. **Analysis of variety and clonal reactions to infesting nematodes.**

6070 Crescimanno, F.G.   IT 060900/72/0166
Prosecuzione delle osservazioni su nuove specie e cultivar in collezione. **Continuing of observations on new species and c.v.**

Publications.

6071 Eynard, I.   IT 062100/71/0148
Introduzione degli incroci dell'Istituto Nacional de Tecnologia Agropecuaria di S. Rafael (Mendoza) Argentina. **Introduction of cross breedings from Istituto Nacional de Tecnologia Agropecuaria di S. Rafael (Mendoza) Argentina.**

6072 Bellini, E.   IT 062800/74/0194
Selezione del materiale di propagazione. Valutazione delle caratteristiche delle cv. in rapporto alla costituzione di piante madri. **Selection of propagation materials. Evaluation of the cultivar characters in order to obtain mother–plants.**

6073 Andreotti, R.   IT 070500/77/0327 R
Valutazione delle selezioni ai fini della trasformazione industriale. **Evaluation of selected produce for industrial processing.**

6074 Leoni, C.   IT 070500/77/0366 R
Controllo del materiale in selezione ai fini dell'utilizzazione industriale, controllo analitico, valutazione della materia prima, trasformazione e valutazione dei prodotti finiti del materiale in fase di selezione avanzata. **Screening of produce selected for industrial processing, analytical check, evaluation of the raw material, processing and evaluation of the products at an advanced stage of selection.**

6075 Broertjes, C.   NL 010110/59/3850
Coördinatie van mutatieveredeling bij vegetatief vermeerderde gewassen. **Coordination of mutation breeding in vegetatively propagated crops.** Publications.

6076 Broertjes, C.   NL 010110/74/7946
Methodologische aspecten van mutatieveredeling in vegetatieve en zaadvermeerderde gewassen. **Methodological aspects of mutation breeding in vegetatively and seed propagated crops.** Publications.

6077 Dellaert, L.   NL 010110/76/7876
Het effect van snelle neutronen, vergeleken met dat van X–stralen op mutatie spectrum en frekwentie in Arabidopsis thaliana L. en Hordeum vulgare L. in verband met de evaluatie van de BARN–reactor. **The effect of fast neutrons, as compared with X–rays, upon mutation spectrum and frequency in Arabidopsis thaliana L. and Hordeum vulgare L. in relation to evaluation of the BARN–reactor.**

6078 Sree Ramulu, K.   NL 010110/78/8975 N
Het vergelijken van snelle neutronen en X–stralen m.b.t. genetische effecten welke zich voordoen bij de totstandkoming van chromosoom afwijkingen i.v.m. de evaluatie van de BARN–reactor. **Comparison of fast neutrons and X–rays in respect to genetic effects accompanying induced chromosome aberrations in relation to evaluation of the BARN–reactor.**

6079 Werry, P.A.Th.J.   NL 010110/78/8976 N
Chromosoomtransplantatie in planten. **Somatic cell genetics in plants.**

6080 Duyvendak, R.; Vos, H.; Schneider, F.
       NL 010113/43/2456
Algemene methodiek en regelingen betreffende het rassenonderzoek. **General methodics and regulations concerning research on varieties.** Publications.

D 2300 – Plant breeding

6081 Garretsen, F. NL 010114/55/2006
Bestudering wiskundige methoden ten behoeve van het veredelingsonderzoek. **Study of mathematical methods to be used in breeding research.** Publications.

6082 Kaulen, H.A. van NL 010114/67/2005
Het uitwerken van snelle chemische bepalingsmethodes ten behoeve van de veredeling. **Development of rapid chemical determinations for breeding purposes.**

6083 Varekamp, H.Q. NL 010114/74/5169
Taxonomie van cultuurgewassen. **Taxonomy of crops.** Publications.

6084 Gelder, W.M.J. van NL 010120/67/1345
Chemisch onderzoek ten dienste van de veredeling op kwaliteitskenmerken. **Chemical research in the field of breeding for quality characteristics.** Publications.

6085 Post, J. NL 010120/67/1346
Toepassing van wiskundig–statistische kennis en computergebruik in de plantenveredeling. **Application of mathematical–statistical science and the use of computers in plant breeding.**

6086 Wal, A.F. van der NL 010120/77/8889 N
Ecofysiologisch onderzoek bij landbouwgewassen gericht op toepassingen in de plantenveredeling. **Ecophysiological research on arable crops for plant breeding purposes.**

6087 Miedema, P. NL 010120/79/8888 N
Toepassing van weefselkweektechnieken in de veredeling van landbouwgewassen. **Application of tissue culture techniques in breeding of agricultural crops.** Publications.

6088 Bos, I. NL 020045/72/4517
De populatie–genetica van de S–allelen in een sporofytisch systeem. **Population genetica of S–alleles in a sporophytic incompatibility system.** Publications.

6089 Parlevliet, J.E. NL 020045/78/8853 N
Selectiemethodieken voor horizontale resistentie. **Selection methodology for horizontal resistance.**

6090 Harten, A.M. van NL 020045/78/8856 N
Inductie, selectie en gebruik van mutaties betreffende cytoplasmatische mannelijke steriliteit. **Induction, selection and utilization of mutations concerning cytoplasmic male sterility.**

6091 Keuls, M. NL 020066/75/6316
Enkele schatters van het overkruisingspercentage. **Some estimators of linkage.**

6092 Jong, G.J. de NL 040007/73/4340
Rassenonderzoek van akkerbouwgewassen onder in de Ijsselmeerpolders heersende omstandigheden. **Research of varieties of arable crops under the prevailing conditions in the Ijsselmeerpolders.**

Cereals in general (B 3100)

See also 2897, 2913, 2935, 2957, 3001, 3002, 3003, 6041, 6570, 6661, 8454, 9313, 20069

6093 Noulard, L.; Derenne, P. BE 080300/73/0007

Etude d'hybrides interspécifiques et intergénériques et de matériel aneuploïde chez les céréales. **Study of interspecific and intergeneric hybrids and aneuploid material in cereals.**

6094 Van Looveren, K.; Niclaes, J.; Kempeneers, L. BE 180000/24/0006
Kweken van betere kultivars van graangewassen (tarwe, rogge, gerst haver). **Breeding research for better varieties of cereals (wheat, rye, barley, oats).** Publications.

6095 Przemeck, E. DE 132061/74/0002
Vergleichende Untersuchungen über den Aminosäuren–Pool in der Entwicklung der Karyopse einiger Getreidearten. **Comparative investigations on the amino acid pool during the development of the caryopsis of some cereal species.**

6096 Niemann, E.–G.; Sharma, T.R. DE 138181/73/0003
Schnelle Untersuchungsmethoden zur Bestimmung der Proteinqualität in Pflanzenzüchtungsprogrammen für Getreide und Leguminosen. **Rapid screening techniques for the evaluation of protein quality in plant–breeding samples of cereals and legumes.**

6097 Steiner, A.M.; Werth, H. DE 144411/74/0002
Untersuchungen zur Methodik des Topographischen Tetrazoliumtests auf Keimfähigkeit. **Studies on the method of topographical tetrazolium testing of germination capacity in seed.**

6098 Meinhold, K.; Bühner, T. DE 201100/78/0005 N
Ökonomische Analyse der Qualitätszüchtung bei Getreide aus einzelbetrieblicher Sicht. **Micro–economic analysis of quality improvements of cereals by genetics.**

6099 Bartels, G. DE 215120/75/0003
Verfahren zur Selektion und Züchtung von Getreidesorten mit einer von Erregerrassen unabhängigen relativen Krankheitsresistenz. **Investigations of methods for selecting and breeding cereal cultivars with non–specific – horizontal – disease–resistance.**

6100 Pommer, G. DE 502053/73/0001 R
Acker– und pflanzenbauliche sowie phytosanitäre Untersuchungen in getreidereichen Fruchtfolgen und im Getreidedaueranbau 1959–1980. **Research on crop rotation with mainly cereals and permanent cereals, with relation to arable farming and botany, as well as phytosanitation .**

6101 Rasmussen, F.; Rasmussen, J. DK 010112/74/4101
Sortsforsøg i rug. **Variety trials with rye.**

6102 Rasmussen, F.; Rasmussen, J. DK 010112/74/4105 R
Sortsforsøg i vinter– og vårbyg. **Variety trials with spring and winter barley.**

6103 Rasmussen, F.; Rasmussen, J. DK 010112/74/4107
Sortsforsøg i havre. **Variety trials with oats.**

6104 Rasmussen, F.; Rasmussen, J. DK 010112/74/4406
Sortsforsøg i vinter– og vårhvede. **Variety trials with spring and winter wheat.**

6105 Stølen, O.; Hermansen, J.E. DK 030145/76/0018
Tiltrækning af og undersøgelser over multiline–sorter. **Development and studies of multiline varieties.**

6106 Cauderon, Y. Mme; Dauge, M.; Gay, G.
FR 010104/69/0433
Etude de l'influence du cytoplasme sur la fertilité des triticale. **Comparative study of triticale with different cytoplasma.**

6107 Bernard, M.; Rousset, M.; Dommergues, M.
FR 010801/76/0373
Mise au point d'une méthode d'exploitation de la variabilité interspécifique chez les triticinae. **Management method of interspecific variability for triticinae.**

6108 Bernard, M.; Trottet, M.; Doussinault, G.; Rousset, M.
FR 010801/76/0398
Utilisation d'un schéma de sélection commun à plusieurs laboratoires pour l'obtention de lignées améliorées de blé et de triticale. **The use, in a few laboratories, of a same selection method for wheat and triticale breeded lines.**

6109 Dommergues, P.; Bernard, M.; Bodergat, R.
FR 011003/74/0439
Augmentation de la variabilité chez les triticales traitements mutagènes sur lignées. **Increase in varability in the triticale mutagenic treatments of lines.**

6110 Felix, L.; Rauzy, G.; Gosselin, M. FR 011104/74/0192
Jugement des aptitudes culturales pour des gèniteurs et cultivars INRA en puissance. **Assessment of cultural abilities of potential INRA cereal parents and cultivars.**

6111 Autran, J.C.; Bourdet, A. FR 012213/69/2265
Variabilité génotypique des protéines du gluten: gliadines. **Genotypic variability of gluten proteins in Gliadines.** Publications.

6112 Jackson GB 011901/00/0029
Local external cereal trials.

6113 Bennet; Smith J GB 011905/00/0004 R
Studies of reproductive cells and grain development in cereals.

6114 Bennett; Smith J GB 011905/00/0007 R
The influence of DNA mass and associated characters on developmental processes in plants.

6115 Simpson GB 011905/00/0010 R
Use of dihaploids in barley breeding in cereals.

6116 Flavell GB 011905/00/0013 R
The organisation and function of the DNA and protein in plant chromosome.

6117 Finch GB 011905/78/0028
Investigation of wide crosses in cereals.

6118 Wolfe; Bennett GB 011906/00/0025
Introduction of sources of disease resistance into breeding material and tedting for resistance.

6119 Lawes GB 012101/00/0005
Genetic systems controlling yield and its components to develop improved breeding methods for cereals.

6120 Clifford GB 012101/00/0011
The relationship between host and systemic fungicides in expression of disease resistance in cereals.

6121 Griffiths GB 012101/00/0020
Collaborative trials to assess disease susceptibility and yield potential of advanced breeding material.

6122 Habgood GB 012101/76/0022
Assessment of selection criteria for cereals under western conditions.

6123 Valentine GB 012101/76/0023
Investigations of early generation testing in cereal breeding.

6124 Welch; Thomas GB 012102/75/0011
Evaluation and development of analytical techniques for cereal breeding programmes.

6125 Catherall GB 012107/00/0009
Genetics of cereal host virus interactions.

6126 Simmonds GB 012107/00/0010
Screening cereals for resistance to virus diseases.

6127 Hayter GB 030804/78/0001
Collect, assess and maintain oat and barley genotypes of use to breeders. Use computer–based data systems.

6128 Hayter GB 030804/78/0003
Study inheritance of cereal performance characters. Design procedures to maximise and exploit variability.

6129 Hayter GB 030804/78/0004
Evaluate techniques for choosing parents and selecting offspring. Design data handling system for breeders.

6130 Hayter GB 030804/78/0005
Test cereals locally from this and other institutes. Explore potential of unfamiliar crops.

6131 Hayter GB 030804/78/0006
Survey virulence genes in pathogens of oat and barley. Design strategies for disease resistance breeding.

6132 Hayter GB 030804/78/0007
Study mechanisms of partial resistance of oats and barley to Erysiphe and their use in resistance breeding.

6133 Hayter GB 030804/78/0008
Improve methods to establish oat and barley disease nurseries. Assemble virulence genes of main pathogens.

6134 Allison GB 030804/78/0012
Investigate inheritance of biochemical components of significance in breeding oats and barley.

6135 Hayter GB 030804/78/0015
Produce pure seed stocks of new cultivars. Investigate diagnostic features of oats and barleys.

6136 Researcher not indicated GB 030901/00/0409 R
The use of progeny rows in uniformity testing of cereal varieties.

6137 Researcher not indicated GB 030901/00/0410 R
Evaluation of characteristics and techniques for distinguishing closely similar cereal varieties.

6138 Lloyd; Hall GB 030901/00/0414 R

The agronomic evaluation of common mutant types in barley varieties.

6139 Chadwick; Barlow GB 030901/00/0415 R
Description and classification of cereal varieties.

6140 Lockhart GB 060106/00/0016 R
NLT cereal varieties.

6141 Blackett GB 060213/00/0001 R
Cereal variety testing.

6142 Morrice GB 060213/00/0004 R
National list trials–cereals.

6143 Spillane, P.A.; Dwyer, E. IE 060300/59/0157 N
Quality evaluation of varieties and selections from dept. of agriculture wheat breeding programme.

6144 Duggan, J.J. IE 060300/69/0028 R
Studies on the distribution of cereal cyst eelworm pathotypes in this country and on the control of the pest using resistant barley varieties. Publications.

6145 Ahloowalia, B.S. IE 060500/70/0205 R
Genetic and physiological studies on the semi–dwarf character in bread wheat, triticum aestivum. Publications.

6146 Zitelli, G.; Pasquini, M.; Forni, C.; Cecchi, V.
IT 020800/65/0001 R
Cereali – Raccolta, studio e utilizzazione di fonti di resistenza all'oidio, alle ruggini nera e bruna e alla septoria. Cereals–Collection, study and use of resistance sources to mildew, to stem and leaf rusts and to septoria. Publications.

6147 Mariani, B.M.; Manmana Novaro, P.; Stefanini, R.
IT 020800/75/0002
Cereali – Applicazione di metodi statistici al miglioramento genetico mediante l'uso di un calcolatore elettronico. Cereals–Application of statistical and computer systems in plant breeding. Publications.

6148 Bianchi, A. IT 020800/77/0115 R
Miglioramento genetico qualitativo di cereali. Genetic improvement of the quality of cereals.

6149 Porceddu, E. IT 060500/77/1000
Studi sulle relazioni tra le specie del genere Triticum. A study on Triticum species interrelations.

6150 Wassenaar, R.; Duyvendak, R. NL 010113/43/0615
Rassenonderzoek granen. Research on varieties of cereals. Publications.

6151 Speckmann, G.J. NL 010120/51/7431
Kortlopende cytologische vraagstukken en diverse onderzoekingen voor de produktie van polyploid en haploid materiaal (o.a. autotetraploidie vastzadig karwij, produktie amphidiploide geslachtsbastaarden bij grassen, produktie van haploiden in granen en aardappelen.). Short–term cytological investigations and various studies on the production of polyploid and haploid plantmaterial for breeding (a.o. autotetraploidy caraway, production of amphidiploid intergeneric hybrids in grasses, production of haploids in cereals and potatoes). Publications.

6152 Lange, W. NL 010120/62/1339
Cytogenetisch onderzoek in granen. Cytogenetical research in cereals. Publications.

6153 Sneep, J. NL 020045/12/6287 R
Selectie in vroege generaties bij zelfbevruchtende granen. Selection in early generation of self–fertilizing cereals. Publications.

6154 Zeven, A.C. NL 020045/76/7343 R
Evolutie van het ras: aard, snelheid en oorzaken van verandering van graanrassen. Evolution of cultivars: Direction and extent of changes of cereal varieties.

6155 Parlevliet, J.E. NL 020045/78/8854 N
Onderzoek naar de graan–roest relaties, andere dan gerst–dwergroest. Investigations into the cereals–rust relationship other than barley–leafrust relationship.

6156 Wilten, W. NL 050304/72/5324 N
Rassenonderzoek van wintergerst op landbouwkundige en brouweigenschappen. Variety trials on agricultural and brewing properties of winter barley. Publications.

Barley (B 3110)

See also 3008, 3019, 3020, 3034, 6010, 6094, 6323, 6339, 6395, 6409, 6428, 9353, 10198

6157 Noulard, L.; Froidmont, F. BE 080300/67/0006
Création d'orge d'hiver à deux rangs de qualité brassicole. Selection of two rows winter barley viewing brewery quality.

6158 Noulard, L.; Froidmont, F. BE 080300/68/0004
Amélioration de la teneur en protéines des orges de printemps. Breeding to increase the protein content in spring barley.

6159 Rixhon, L.; Crohain, A.; Guiot, J.; Couvreur, L.
BE 080800/76/0015 N
Techniques modernes de production d'orge d'hiver. Modern crop husbandry for winter barley.

6160 Glansdorff, N.; Jacobs, M.; Reynaerts–Cattoir, A.;
Degrijse, E. BE 140000/75/0019
Onderzoek omtrent de selectie op celniveau van planten met een verhoogde voedingswaarde (Arabidopsis, wortel en gerst). Study of the selection on cellevel of plants with an increased nutritional value. (Arabidopsis, roots en barley). Publications.

6161 Van Looveren, K.; Niclaes, J. BE 180000/70/0005
Veredeling van gerst op tolerantie op resistentie tegen meeldauw. (Erisyphe graminis J.sp. Hordei Marchal). Breeding research for tolerance or resistance of barley against mildew (Erisyphe graminis f.sp. Hordei Marchal). Publications.

6162 Mai, E. DE 104600/71/0002 R
Erzeugung neuen Züchtungsmaterials für die Resistenzzüchtung gegen den Mehltau der Gerste unter gleich–zeitiger Berücksichtigung der Gelbrostresistenz 1969–1978. english title not indicated.

6163 Schildbach, R. DE 105201/70/0001
Sommergersten – Sortenprüfungen unter ökologisch unterschiedlichen Bedingungen. Testing of spring barley varieties under different ecological conditions.

D 2300 – Plant breeding

6164 Heyland, K.–U.; Kochs, H.–J.; Schulte–Geers, C.

DE 111255/75/0001

Statistische Masszahlen als Selektionskritierien in jungen Generationen, dargestellt am Beispiel von Sommergerste. **Statistical numbers as selection criteria in young generations, demonstrated by spring barley.**

6165 Röbbelen, G.; Reinhold, M.; Hafez, A.

DE 132184/75/0006

Auslösung und Charakterisierung von Resistenzmutanten gegen Mehltau in horizontal resistenten Gerstenformen. **Induction and characterization of mutations for resistance to powdery mildew in horizontally resistant lines of barley.**

6166 Fischbeck, G.; Häuser, H. DE 161250/77/0004

Lokalisierung von Translokationen und mutierter Gene bei Gerste. **Localization of translocations and mutated genes in barley.**

6167 Schnell, F.W.; Kling, C.; Schmütz, W.

DE 501700/73/0002

Erblichkeit und korrelierte Erblichkeit von Qualitätseigenschaften bei Sommergerste. **Heritability and correlated heritability of quality characters in spring barley.**

6168 Ulonska, E.; Baumer, M. DE 502058/73/0015

Verbesserung der Brauqualität bei Sommer– und Wintergerste 1973. **Improvement of the brewing quality of spring barley and winter barley.**

6169 Ulonska, E.; Baumer, M. DE 502058/73/0016

Züchterische Bekämpfung von Mehltau, Rost und Blattfleckenkrankheiten bei Gerste 1973. **Breeding control of mildew, rust and net blotch in barley.**

6170 Ulonska, E.; Baumer, M. DE 502058/73/0017

Züchterische Entwicklung von Nacktgersten mit Futter– und Brauqualität 1973. **Breeding development of gymnospermous barley with feeding and brewing quality.**

6171 Ulonska, E.; Baumer, M. DE 502058/73/0018

Züchterische Weiterentwicklung von tetraploiden Gersten mit Hilfe mutagener Behandlungen – Röntgen, ÄMS – 1973. **Breeding development of tetraploid barley by means of mutagenic treatment – X–rays, Aems –.**

6172 Ulonska, E.; Baumer, M. DE 502058/73/0019

Züchterische Nutzung von Kleinmutationen aus mutagenen Behandlungen mit Röntgenstrahlen und ÄMS bei Gerste 1973. **Breeding use of small mutations in barley caused by mutagenic treatment with X–rays and Aems.**

6173 Stølen, O. DK 030145/70/0006

Blomsterbiologi og krydsbefrugtning i byg. **Flower biology and cross–pollination in barley.**

6174 Andersen, S.; Andersen, K. DK 030145/74/0012 N

Kobling af gener i byg. **Linkage of marker genes in barley.**

6175 Mortensen, G.; Dennis, B. DK 030145/77/0026 N

Bestandsgeometriske undersøgelser hos byg. **Seed rate and competitive ability in barley.**

6176 Haahr, V.; Wettstein, D. von DK 030146/68/0001

Stråegenskaber hos byg. **Straw characteristics in barley.**

6177 Jensen, H.P.; Jørgensen, J.H. DK 030146/69/0010

Meldugresistente bygmutanters egenskaber. **Characteristics of mildew resistant barley mutants.**

6178 Linde–Laursen, I. DK 030146/73/0014

Båndfarvning af bygkromosomer. **Band staining of barley chromosomes.**

6179 Doll,.H.; Jensen, J. DK 030146/75/0012

Genetiske undersøgelser af højlysin mutanter i byg. **Genetic investigations of high lysine mutants in barley.**

6180 Jensen, C.J. DK 030146/75/0013

Monoploider i byg. **Monoploids in barley.**

6181 Pedersen, J.T.; Haahr, V.; Doll, H.

DK 030146/77/0018 N

Sortsvariation i bygkernens kemiske sammensætning. **Variation in the chemical composition of barley grain.**

6182 Smedegaard–Petersen, V. DK 030147/77/0012 N

Undersøgelse over resistensprocessernes indflydelse på planternes energiforbrug og udbytte. **The energy consumption and yield of susceptible and highly mildew – resistant barley varieties.**

6183 Berbigier, A. FR 010801/70/0026

Diversification des orges demi–naines. **Breeding new types semi–dwarf barley.**

6184 Berbigier, A.; Jestin, L.; Doussinault, G.; Derieux, J.

FR 010801/75/0372

Etude des interactions génotype x milieu chez l'orge. **European spring barley adaptability nursery.**

6185 Larambergue, R. de FR 011003/70/0025

Obtention de variétés d'orge de printemps résistantes à l'oidium. **Breeding of spring barley varieties resistant to downy mildew.**

6186 Chery, J. FR 011201/70/0282

Création de variétés d'orges riches en protéines et en lysine. **Breeding of barley varieties with a high protein and lysine content.**

6187 Chéry, M. FR 011201/71/0024

Recherche sur la variabilité génétique de la teneur en protéines et en acides aminés indispensables chez l'orge cultivée. Obtention de variétés productives et à haute valeur nutritive. **Research on the genetic variability of protein and amino–acid content in cultivated barley. Breeding productive varieties with a high nutritive value.**

6188 Jacques, C.; Gerard, R. FR 011201/74/0213

Amélioration de l'orge à grain nu. **Breeding bailey for bare grain.**

6189 Chery, J.; Puech, M. FR 011201/75/0283

Création de variétés d'orge bien adaptées aux conditions méridionales. **Breeding of barley varieties adapted to meridional conditions in France.**

6190 Riggs; Hanson GB 011901/00/0005 R

Breeding spring barley for improved yield.

D 2300 – Plant breeding

6191 Rhodes   GB 011901/00/0009 R
Breeding barley for improved protein quality.

6192 Riggs; Hanson   GB 011901/00/0010 R
Breeding spring barley for improved malting quality.

6193 Researcher not indicated   GB 011901/79/0031 N
Breeding winter barley.

6194 Finch   GB 011905/78/0027 R
The study of chromosome behaviour and stability in barley and related species.

6195 Gothard; Smith   GB 011908/00/0002
Biochemistry of barley grain relevant to malting quality and the assessment of genotypic variation.

6196 Habgood   GB 012101/00/0003
Breed high yielding, stiff strawed, disease resistant barleys for livestock use.

6197 Foster   GB 012101/00/0006
Breeding F1 hybrid barley.

6198 Pickering   GB 012101/00/0019
Breeding new varieties of barley using interspecific hybridization and embryo culture techniques.

6199 Thomas; H.U.   GB 012103/00/0005 R
Chromosome elimination in interspecific hybrids of Hordeum.

6200 Hayter   GB 030804/78/0013
Breed malting and feed barley cultivars.

6201 Hodges   GB 040302/00/0011 R
Barley breeding.

6202 Stanca, A.M.; Odoardi, M.; Delogu, G.
     IT 020800/74/0008 R
Orzo – Miglioramento genetico per la resistenza all'allettamento di varietà per uso zootecnico. **Barley – Genetic improvement for lodging resistance of animal feeding varieties.**

6203 Odoardi, M.   IT 020800/76/0002 R
Orzo – Miglioramento, mediante mutagenesi artificiale, per incrementare la qualità delle proteine endospermiche. **Barley – Breeding for improvement of endosperm protein quality by artificial mutagenesis.**

6204 Delogu, G.; Stanca, A.M.   IT 020800/79/0006 N
Orzo – Studio dell'eredità del carattere precocità. **Barley – Inheritance of earliness.**

6205 Stanca, A.M.; Sage, G.; Roffey, A.P.
     IT 020800/79/0007 N
Orzo – In semina autunnale: selezione parallela a Fiorenzuola d'Arda (PC) e a Cambridge (Inghilterra). **Winter Barley – Concurrent selection in Fiorenzuola d'Arda (PC) and Cambridge (England).**

6206 Stanca, A.M.; Delogu, G.   IT 020800/79/0008 N
Orzo – Prove di confronto varietale per produzione di granella e di trinciato integrale. **Barley – National trials for grain and silage production.**

6207 Ceccarelli, S.   IT 040102/78/1043 N
Miglioramento genetico dell'orzo da granella. **Genetic improvement of grain barley.**

6208 Panella, A.; Ceccarelli, S.; Lorenzetti, F.
     IT 041002/73/0001
Miglioramento genetico dell'orzo da granella. **Barley breeding.** Publications.

6209 Balkema–Boomstra, A.G. van   NL 010120/50/7407
Beheer en bestudering van rassencollecties van zomergerst, wintergerst en haver. **Management and evaluation of germ plasm collections of spring barley, winter barley and oats.** Publications.

6210 Balkema, A.G.   NL 010120/50/7408 R
Organisatie van internationale gerst "disease nurseries" en veredelingsonderzoek naar de mogelijkheden van integratie van genetische resistentie en fungicidenbehandeling van meeldauw bij gerst in internationaal verband. **The organisation of international barley disease nurseries and breeding research on possibilities of integration of genetic resistance and fungicide treatment of mildew in barley.** Publications.

6211 Balkema–Boomstra, A.G. van   NL 010120/50/7412
Veredelingsonderzoek naar enkele kwaliteitsbepalende factoren in brouw– en voergerst. **Breeding research about several quality aspects of barley for malt and feed.** Publications.

6212 Balkema, A.G.   NL 010120/50/7413 R
Verzamelproject veredeling zomergerst, met name onderzoek naar de combineerbaarheid en de synthese van resistentie tegen ziekten (meeldauw, gele roest, dwergroest, Rhynchosporium secalis, graancystenaaltje, stuifbrand) en andere eigenschappen (kort stro, vroegheid, pH–tolerantie). **Various projects on breeding in spring barley, i.e. research about the combining ability and synthesis of disease resistances (Erysiphe graminis, Puccinia striiformis, Puccinia hordei, Rhynchosporium secales, Cereal Cyst Nematode, and smuts) and other characteristics (short straw, earliness and pH tolerance).** Publications.

6213 Balkema, A.G.   NL 010120/50/7414 R
Veredelingsonderzoek naar winterhardheid, ziekteresistentie en brouwkwaliteit bij wintergerst. **Breeding research with winterbarley about winterhardiness, disease resistance and malting quality.** Publications.

6214 Balkema, A.G.   NL 010120/66/7409 R
Veredelingsonderzoek naar niet–specifieke resistentie tegen meeldauw bij gerst. **Breeding research about non–specific resistance to powdery mildew in barley.** Publications.

6215 Balkema, A.G.   NL 010120/69/7410 R
Onderzoek naar een voor de opbrengst optimaal ontwikkelingsritme en naar de toepassing hiervan bij de veredeling van zomergerst. **Research about a for yield, optimal development rhythm, and about the application in breeding spring barley.** Publications.

6216 Balkema, A.G.   NL 010120/76/7411 R
Onderzoek naar de mogelijkheid om op stikstofefficientie te veredelen in gerst. **Research about the possibility to breed for nitrogenefficiency in barley.**

6217 Balkema–Boomstra, A.G.   NL 010120/79/8887 N
Veredelingsonderzoek naar dwergroest resistentie bij gerst.

D 2300 – Plant breeding

Breeding research for resistance to leaf rust (Puccinia Hordei) in barley.

6218 Dieleman, F.L. NL 020014/74/4851
Veredeling op bladluizen resistentie in gerst. **Breeding for aphid resistance in barley.**

6219 Spitters, C.J.T. NL 020045/74/6289
Concurrentie bij gerst en bij suikerbieten en de consequentie voor selektie. **Competition in barley and in sugar beet, the consequences for selection.**

6220 Kramer, Th. NL 020045/76/7341 R
Fysiologische aspecten van dwergroest tolerantie bij gerst. **Physiological aspects of barley leaf rust tolerance.**

6221 Niks, R.E. NL 020045/78/8852 N
Histologische aspecten van de gerst–dwergroest relatie. **Histological aspects of the barley–leafrust relationship.**

6222 Parlevliet, J.E. NL 020045/78/8855 N
Onderzoek naar de gerst–dwergroest relatie. **Investigations into the barley–leafrust relationship.**

6223 Wilten, W. NL 050304/34/5323
Beoordeling van zomergerstrassen op landbouwkundige en brouweigenschappen. **Variety trials on agricultural and malting properties of spring barley.** Publications.

6224 Wilten, W. NL 050304/34/5328
Bepaling van de identiteit van gerstrassen. **Identification of barley varieties.** Publications.

6225 Wilten, W. NL 050304/56/5327 R
Beoordeling van de brouwwaarde van nieuw aangeboden brouwgerstrassen t.b.v. de bevordering van het kweken van brouwgerst. **Evaluation of the malting quality of new offered malting barley varieties for promoting the malting barley growing.** Publications.

Maize (B 3120)

See also 3052, 6347, 6348

6226 Goffeau, A.; Briquet, M.; Boutry, M.; Faber, A.M. BE 020105/79/0006 N
Etude des bases moléculaires de la stérilité mâle chez les plantes cultivées. **Basic molecular study of male sterility in plants.**

6227 Schnell, F.W.; Janssen, D. DE 144420/74/0004
Interaktionen von Maispopulationen verschiedenen Inzuchtgrades mit pflanzenbaulichen Faktoren. **Interactions between maize populations with different degree of inbreeding and methods of cultivation.**

6228 Schnell, F.W.; Schmidt, W. DE 144420/74/0005
Untersuchungen zur Zuchtwertschätzung und Testerwahl bei Mais. **Studies on the estimation of breeding value and choice of tester in maize.**

6229 Schnell, F.W.; Schmidt, C. DE 144420/74/0006
Die Korrelation von Frühreife mit Ertrag während der generativen Entwicklung von Maishybriden. **The correlation of early maturity with yield during the generative development of maize hybrids.**

6230 Pollmer, W.G.; Klein, D.; Eberhard, D.; Dhillon, B.S. DE 144430/71/0001
Entwicklung von Maisformen mit verbesserter Kornproteinqualität. **Development of maize genotyps with improved grain protein quality.** Publications.

6231 Pollmer, W.G.; Klein, D.; Frölich, W.; Mann, C. DE 144430/77/0001
Prüfung von Maisformen mit hohem Proteingehalt auf ihre Eignung für den westeuropäischen Maisanbau. **Adaptation of high–protein maize genotypes to West European growing conditions.**

6232 Pollmer, W.G.; Klein, D.; Eberhard, D. DE 144430/77/0003
Entwicklung von Maisformen mit verbessertem Proteingehalt im Korn und in der Gesamtpflanze. **Development of maize genotypes with improved protein content of the grain and of the whole plant.**

6233 Pollmer, W.G.; Goertz, P.; Frölich, W. DE 144430/77/0004
Untersuchungen über Möglichkeiten der Verbesserung der Proteinqualität und –quantität in floury–1–Populationen von Mais in den Hochlandregionen der Anden in Kolumbien, Ecuador, Peru und Bolivien. **Investigations on the possibilities of improvement of the protein quality and quantity in floury–1 maize populations in the Andean regions of Colombia, Ecuador, Peru, and Bolivia.**

6234 Hepting, L. DE 502058/73/0020 R
Entwicklung früher bis mittelfrüher Inzuchtlinien zur Silo– und Körnermaisnutzung in Grenzlagen des Maisanbaues unter Betonung von Kältetoleranz und Jugendentwicklung 1973. **Breeding of early to medium–early inbreeding lines for silage–maize and grain–maize utilizatiin in marginal sites of maize growing in special consideration of cold tolerance and growth of early stages.**

6235 Derieux, M. FR 010104/70/0020
Comparaison des méthodes de sélection réciproque et récurrente et de sélection généalogique réciproque classique pour la création de lignées et hybrides précoces de maïs. **Comparison between method of reciprocal and recurrent selection and classic reciprocal genealogic selection for breeding early lines and hybrids in maize.**

6236 Berville, A.; Cassini, R.; Greneche, M. FR 010105/73/0327
Les mécanismes moléculaires réglant l'expression des cytoplasmes dits mâle–stérile. Cas du maïs. Aspects pathologiques. **Molecular control of male sterility. Approach of mechanism by corn.**

6237 Anglade, P.; Vible, J.C. FR 010707/61/1787
Participation à la sélection du maïs résistant à la pyrale. **Corn breeding for resistance to european corn borer.** Publications.

6238 Anglade, P.; Vible, J.C. FR 010707/70/5212
Participation entomologique à la sélection de variétés de maïs résistantes ou tolérantes à la pyrale du maïs. **Entomological participation to breeding program for resistance or tolerance of Maize to the European Corn Borer.**

6239 Pollacsek, M.; Caenen, M. FR 010801/72/0023

D 2300 – Plant breeding

Diversification du maïs pour la teneur en protéines. **Research of new types of maize with a high protein content.**

6240 Deshayes, A.; Vuillaume, E.; Cornu, A.
FR 011003/75/0379
Recherche par mutagenèse, appliquée in vitro, d'une résistance ou tolérance à Helminthosporium Maydis, race T. parmi des maïs à cytoplasme mâle stérile texas. **Induction by mutagenesis applied in vitro of texas maize, resistant to Helminthosporium Maydis, race t.**

6241 Rautou, S.; Feillet, P.
FR 011201/70/0022
Amélioration de la qualité des protéines du grain de maïs. **Breeding for protein quality in maize grain.**

6242 Rautou, S.; Panouille, A.
FR 011201/70/0124
Obtention de variétés précoces tolérantes aux attaques de pyrale et adaptées à la zone nord de la culture du maïs. **Selection of early varieties tolerant to the european corn borer attacks and adequate for the northern area of maize cultivation.** Publications.

6243 Rautou, S.; Panouille, A.
FR 011201/73/0199
Recherches et utilisation de nouveaux systèmes d'androstérilité génétique pour la production de semences hybrides de maïs. **Research and utilization of new genetic systems for maize hybrid seed production.**

6244 Gunn
GB 011910/00/0005
Definition of breeding objectives, development of selection criteria and study of breeding and testing methods.

6245 Gunn
GB 011910/00/0010
Breeding forage maize varieties.

6246 Lanza, F.; Onofrii, M.; Boschi, V.; Trenti, M.P.; Spallacci, P.
IT 020500/67/0003
Monocoltura di mais da granella: ibridi, concimazioni, produzioni, fertilità del terreno e infestanti. **Continuous grain–corn: hybrids, fertilization, yields, soil fertility and weeds.** Publications.

6247 Mariani, G.; Desiderio, E.
IT 020800/70/0005 R
Mais–Miglioramento genetico per la produzione di granella, in semina estiva dopo frumento. **Maize –Breeding for grain production in summer sowing conditions after wheat.** Publications.

6248 Mariani, G.; Colesanti, F.
IT 020800/72/0005 R
Mais – Ricerche agronomiche sulla coltivazione a semina estiva dopo frumento per la produzione di granella. **Maize – Agronomic researches on cultivation for grain production in summer sowing conditions after wheat.** Publications.

6249 Mariani, G.
IT 020800/72/0505
Miglioramento genetico del mais, specifico per la coltura dopo frumento nel mezzogiorno. **Genetic improvement of maize specifically for its cultivation after wheat in southern Italy.** Publications.

6250 Bertolini, M.; Maggiore, T.; Motto, M.
IT 020800/75/0001
Mais – Ottenimento di linee pure normali, opaco–2 e fl2 e determinazione della loro attitudine combinatoria. **Maize–Obtention of normal, opaque–2 and fl2 inbreds and evaluation of their general combining ability.**

6251 Maggiore, T.; Gentinetta, E.; Motto, M.; Manmana Novaro, P.
IT 020800/75/0003
Mais – Miglioramento di sintetiche ad elevata qualità e quantità di proteine (o_2, high protein). **Maize – Breeding of synthetics for high protein quality and quantity (o_2, high protein).** Publications.

6252 Maggiore, T.; Stefanini, R.; Bertolini, M.
IT 020800/75/0005
Mais – Prove comparative di ibridi commerciali. **Maize – Comparative trials of commercial hybrids.**

6253 Salamini, F.; Bianchi, G.
IT 020800/75/0008
Mais – Studi sulla chimica delle cere nelle mutazioni glossy. **Maize – Studies on the wax chemistry of the glossy mutants.** Publications.

6254 Gentinetta, E.; Salamini, F.
IT 020800/75/0010
Mais – Metabolismo dei carboidrati in mutanti con endosperma difettoso. **Maize – Carbohydrate methabolism in defective endosperm mutants.** Publications.

6255 Fortini, S.; Mariani, B.M.; Galterio, G.; Sgrulletta, D.; Manmana Novaro, P.; Nardi, S.
IT 020800/75/0013 R
Mais–Studio delle relazioni tra attività enzimatiche ed eterosi. **Maize – Relationships between enzyme activities and heterosis.** Publications.

6256 Bertolini, M.; Stefanini, R.
IT 020800/75/0014
Mais – Miglioramento delle sintetiche GD e BS5 per precocità. **Maize – Improvement of the synthetics GD e BS5 for earliness.**

6257 Motto, M.; Mariani, B.M.; Maggiore, T.; Bertolini, M.
IT 020800/75/0015
Mais – Miglioramento di 3 popolazioni sintetiche per numero di ranghi, spiga lunga, seme grande e di 7 popolazioni sintetiche a base genetica stretta. **Maize – Improvement of 3 synthetics for kernel rows, ear length, kernel weight, and of 7 genetically narrow–based synthetics.**

6258 Salamini, F.
IT 020800/75/0017
Mais – Instabilità genetica al locus opaco–2. **Maize – Genetic instability of the opaque–2 locus.** Publications.

6259 Coppolino, F.
IT 020800/79/0002 N
Mais–Studio e utilizzazione di resistenze genetiche per il controllo dell'Ostrinia nubilalis. **Maize–Study and utilization of genetical resistance to control Ostrinia nubilalis.**

6260 Salamini, F.; Maggiore, T.; Motto, M.; Bertolini, M.
IT 020800/79/0005 N
Mais – Sviluppo di tipi adatti al secondo raccolto. **Maize– Breeding for crop production in summer sowing conditions.**

6261 Mariani, G.; Desiderio, E.
IT 020800/79/0011 N
Mais – Prove di adattamento di ibridi commerciali da granella in semina estiva dopo cereale autunno–primaverile. **Maize – Trials of hybrids for grain production in summer sowing after small grains.**

6262 Baldoni, R.
IT 040201/74/0507
Ricerche per il miglioramento genetico del mais e del girasole. **Research on genetic improvement of maize and sunflower.**

6263 Baldoni, R.
IT 040201/78/1027 N

D 2300 – Plant breeding

Miglioramento genetico del mais. **Genetic improvement of maize.**

6264 Lorenzoni, C.; Fogher, C.   IT 040405/79/0001 N
Intervento genetico e fisiologico per il miglioramento della produttività del mais. **Improvement of the productivity in maize through genetic and physiological means.**

6265 Lorenzoni, C.   IT 040415/72/0106
Conversione di linee pure di mais normali in linee epaco–2 e ricerca delle combinazioni tra le stresse che eliminino gli effetti della mutazione 02 sulla produttività. **Conversion of pure lines of normal maize into Epaco – 2 lines and research on combinations between those which eliminate the effects of 02 mutation on productivity.** Publications.

6266 Lorenzoni, C.   IT 040415/77/0217 R
Impiego di linee opaco–2 per la costituzione di ibridi di mais, individuazione delle combinazioni migliori e sviluppo di sintetiche 02 ad alta frequenza di modificatori favorevoli. **Utilization of opaque–2 lines for the constitution of maize hybrids, identification of the best combinations and development of 02 synthetics with high frequency of favorable modifiers.**

6267 Barigozzi, C.   IT 040615/72/0387
Miglioramento genetico del mais da foraggio e da granella. **Genetic improvement of fodder–maize and grain maize.** Publications.

6268 Gavazzi, G.   IT 040615/78/1059 N
Studi di mutagenesi ai fini del miglioramento del mais. **A mutagenesis study aiming at improving maize.**

6269 Toniolo, L.; Mosca, G.; Sattin, M.   IT 040801/79/0001 N
Intervento genetico, fisiologico e agrotecnico per il miglioramento della produttività del mais in Italia. **Breeding–physiology and Agronomic factors on the improvement of corn in Italy.**

6270 Giardini, A.   IT 062400/71/0106
Confronto tra ibridi di mais e sorgo in coltura irrigua e asciutta in diversi ambienti dell'Italia Centro–settentrionale. **Comparison of corn and sorghum hybrids under irrigated and dry conditions in different environments of Northerncentral Italy.** Publications.

6271 Miedema, P.   NL 010120/74/6579
Fysiologisch onderzoek over de invloed van lage temperaturen op mais t.b.v. veredeling van mais. **Physiological research about the influence of low temperatures on maize for breeding purposes.** Publications.

Oats (B 3130)

See also 3067, 6094, 6209

6272 Noulard, L.; Clamot, G.   BE 080300/71/0001
Amélioration de l'avoine par mutations induites. **Breeding of oats by induced mutations.**

6273 Rixhon, L.; Delhaye, R.   BE 080800/78/0013 R
Etude des techniques culturales pour l'avoine d'hiver. **Study of crop husbandry for winteroats.**

6274 Schmidt, B.; Leist, N.   DE 501014/78/0002 N

Versuche zur Klärung der genetischen Variabilität bei Flughafer– Avena fatua L. – und Sorten von Saathafer– Avena sativa L. –. **Experiments to clarify the genetic variability of wild oat – Avena fatua L. – and of cultivars of oat –Avena sativa L. –.**

6275 Grignac, P.   FR 011201/74/0123
Amélioration de l'avoine d'hiver à partir de croisements avec Avena Sterilis. **Improvement of winter oats through breeding with Avena Sterilis.**

6276 Grignac, G.   FR 011201/74/0205
Amélioration de l'avoine d'hiver à partir de croisements avec Avena Sterilis. **Breeding winter oats from crossing with Avena Sterilis.**

6277 Grignac, P.   FR 011201/74/0284
Amélioration de l'avoine d'hiver à partir de croisements avec Avena Sterilis. **Improvment of winter oat from crossings with Avena Sterilis.**

6278 Saur, L.; Doussinault, G.   FR 011301/73/0118
Obtention de variétés d'avoine à haute valeur énergétique. **Selection of highly nutritive dat varieties.**

6279 Hanson; Williams   GB 011901/00/0003 R
Breeding oats for improved yield.

6280 Valentine   GB 012101/00/0001
Breed winter oats for yield, stiff straw, hardness; resistance to mildew, rust, nematodes and mosaic virus.

6281 Lawes   GB 012101/00/0002
Breed spring oats for improved yield, quality, stiff straw; resistance to mildew, rust and nematodes.

6282 Thornton   GB 012101/77/0026
Development of high yielding huskless oats.

6283 Thomas; Leggett, H.U.   GB 012103/00/0004 R
Cytogenetic relationships in Avena. Transfer of desirable characters from wild species.

6284 Hayter   GB 030804/78/0014
Breed spring oat cultivars.

6285 Researcher not indicated   GB 030901/00/0411 R
Identification of Avena spp. by means of caryopsis characters.

6286 Hodges   GB 040302/00/0010 R
Oat breeding.

Rice (B 3140)

See also 5435, 10202, 10203

6287 Marie, R.; Grillard, M.   FR 011201/70/0028
Sélection de variétés de riz plus productives grâce à un meilleur départ en végétation. **Breeding more productive rice varieties, with better start in growth.**

6288 Marie, R.; Grillard, M.   FR 011201/75/0285
Etude du comportement d'une collection de variétés de riz vis–à–vis de Sclerotium sp. à des fins de sélection. **Behaviour study of a collection of rice varieties against Sclerotium sp. for the selection.**

6289 Marie, R.; Grillard, M. FR 011201/75/0286
Amélioration de la qualité culinaire du riz et son aptitude à l'appertisation. **Improvement of cooking and canning quality of rice.**

6290 Baldi, G.; Giovannini, G.; Sandoli, M.
IT 011400/65/0001 R
Programma di miglioramento per l'ottenimento di varietà migliorate di riso. I principali caratteri sotto selezione sono: 1) precocità; 2) taglia; 3) resistenza alle malattie; 4) capacità produttiva. **Breeding programme for obtaining improved rice varieties. Main characters under selection are: 1) earliness; 2) height; 3) diseases resistance; 4) yielding ability.** Publications.

6291 Baldi, G.; Giovannini, G.; Sandoli, M.
IT 011400/66/0001 R
Programma di miglioramento per l'ottenimento di varietà di riso precocissime (110–120 gg.) da utilizzare in semina tardiva (25/5–5/6) dopo il taglio di una coltura foraggera. **Breeding programme for obtaining very early rice varieties (110–120 days) for late sowing (25 may–5 june) after a forage crop harvest.** Publications.

6292 Baldi, G.; Giovannini, G.; Mazzini, F.
IT 011400/67/0001 R
Programma di miglioramento per l'ottenimento di varietà di riso adatte alla coltivazione con irrigazione turnata. **Breeding programme for obtaining rice varieties adapted to the conditions of periodical irrigation.** Publications.

6293 Baldi, G.; Giovannini, G.; Fossati, G.; Fantone, G.C.
IT 011400/69/0001 R
Programma di miglioramento per l'ottenimento di varietà di riso con caratteristiche merceologiche particolari, principalmente granello lungo e traslucido. **Breeding programme for obtaining improved rice varieties with particular quality characteristics, mainly long and translucent grain.**

6294 Baldi, G.; Fossati, G.; Fantone, G.C.; Giovannini, G.
IT 011400/69/0002 R
Programma di miglioramento per l'ottenimento di varietà di riso ad alto contenuto proteico. **Breeding programme for obtaining high–protein rice varieties.** Publications.

6295 Baldi, G.; Moletti, M.; Villa, B. IT 011400/72/0001 R
Programma di reincrocio per il trasferimento dei caratteri di resistenza alla Pyricularia oryzae dalla varietà di riso Roncarolo a 8 linee pure selezionate al Centro per alta produttività. **Back–cross programme to transfer the resistance to Pyricularia oryzae from the resistant Roncarolo rice variety to eight new lines selected at the Centre for high–yielding capacity.**

6296 Moletti, M.; Villa, B.; Lucchelli, N. IT 011400/74/0005
Studi sulla resistenza alla malattia virale del riso denominata "giallume" utilizzando l'infezione artificiale con l'afide vettore Rhopalosiphum padi. **Studies on resistance to a rice virus–disease called "giallume" by artificial infection with the vector aphid Rhopalosiphum padi.** Publications.

6297 Mazzini, F.; Bonandin, E.; Baldi, G.; Feccia, S.
IT 011400/75/0002 R
Studi su meccanismi fisiologici caratterizzanti genotipi di riso adatti alle condizioni di irrigazione turnata. Ricerche sui

possibili cambiamenti a livello di alcune attività enzimatiche. **Studies on some physiological aspects of rice genotypes adapted to conditions of periodical irrigation. Researches on the possible changes of some enzymatic activities.**

6298 Moletti, M.; Villa, B. IT 011400/79/0001 N
Valutazione della resistenza al fungo patogeno del riso Fusarium moniliforme di varietà e di linee pure selezionate al Centro mediante infezione artificiale. **Evaluation of resistance to the rice pathogenic fungus Fusarium moniliforme by artificial infection on varieties and pure lines selected at this Centre.**

6299 Moletti, M.; Villa, B. IT 011400/79/0002 N
Valutazione della resistenza di varietà di riso al fungo patogeno Sclerotium oryzae mediante infezione artificiale. **Evaluation of resistance to the pathogenic fungus Sclerotium oryzae by artificial infection in rice varieties.**

6300 Moletti, M.; Villa, B. IT 011400/79/0003 N
Valutazione della resistenza al fungo patogeno del riso Helminthosporium oryzae di varietà e di linee pure selezionate al Centro mediante infezione artificiale. **Evaluation of resistance to the rice pathogenic fungus Helminthosporium oryzae by artificial infection on varieties and pure lines selected at this Centre.**

6301 Baldi, G.; Moletti, M.; Giovannini, G.
IT 011400/79/0004 N
Trasferimento della resistenza al giallume da 5 varietà di riso resistenti a varietà e linee pure selezionate al Centro suscettibili, ma agronomicamente interessanti. **Program to transfer the resistance to a rice virus disease called "yellowing" from 5 resistant genotypes to agronomically valid varieties and pure lines selected at this Centre.**

6302 Baldi, G.; Moletti, M.; Giovannini, G.; Villa, B.
IT 011400/79/0005 N
Studio della ereditarietà della resistenza al virus del giallume nel riso. **Genetic studies on the resistance to a rice virus disease called "yellowing".**

6303 Russo, S.; Nutolo, C.; Maglio, D. IT 020800/74/0004 R
Riso – Miglioramento genetico per la elevata produttività e la qualità del granello di tipo "indica". **Rice–Genetic improvement for productivity and grain quality of "indica" type.**

6304 Russo, S.; Lupotto, E.; Giani, P. IT 020800/74/0006 R
Riso – Induzione di piante aploidi da colture "in vitro" di antere e loro utilizzazione nel miglioramento genetico. **Rice–Induction of haploid plants from "in vitro" cultures of anthers and their utilization in genetic improvement.** Publications.

6305 Russo, S.; Nutolo, C. IT 020800/78/0001 R
Riso – Valutazione agronomica di nuove linee per il miglioramento della produttività e della qualità. **Rice–Agronomical evaluation of new lines for productivity and quality improvement.**

6306 Pacucci, G. IT 040115/72/0113
Ricerche sul miglioramento genetico del frumento duro, del girasole e del carciofo. **Research on the genetic improvement of hard wheat, sunflowers and artichokes.** Publications.

6307 Torti, G. IT 060300/74/0031
Meccanismi fisiologici in genotipi di riso adattati a crescere in

coltura non sommersa. **Studies on physiological mechanisms in rice genotypes adapted to grow in non paddy soils.**

6308 Torti, G. IT 060300/77/0871
Meccanismi fisiologici in genotipi di riso adatti a crescere in coltura non sommersa. **Physiological mechanisms in rice genotypes adapted to growth in non flooded fields.**

Rye (B 3150)

See also 6094, 6321, 6330, 6338

6309 Wricke, G.; Peters, R.; Meyer, H. DE 138300/70/0001
Untersuchungen zur Selbstfertilität und Zuchtmethodik beim Roggen. **Investigations of self–fertility and breeding methods in rye.**

6310 Kuckuck, H. DE 138300/72/0002 N
Genetische und züchterische Untersuchungen an der Roggenwildart Secale vavilovii und ihren dipoloiden und tetraploiden Nachkommen aus den Kreuzungen mit Kulturroggen Secale cereale. **Genetic and breeding studies on the wild rye species Secale vavilovii and diploid and tetraploid progeny from crossbreeding with cultivated Secale cereale.**

6311 Zeller, J. DE 161256/78/0001 N
Herstellung einer Chromosomenkarte des Roggens mit Hilfe von Trisomen und Genmutationen. **Mapping of rye chromosomes by means of trisomes and gene mutations.**

6312 Geiger, H.H.; Morgenstern, K. DE 501700/73/0001
Beiträge alleler und nichtalleler Genwechselwirkungen zur Heterosis des Roggens. **Contribution of allelic and nonallelic gene interactions to heterosis in rye.**

6313 Strass, F. DE 502058/73/0022
Anwendung des Hybridzuchtverfahrens bei Roggen. **Application of the hybrid breeding method to rye.**

6314 Berbigier, A.; Lestrange, G. de FR 010801/71/0027
Mise au point d'une nouvelle méthode d'amélioration du seigle. **Testing a new method of breeding rye.**

6315 Sybenga, J. NL 020015/61/4418
Kwantitatieve analyse van meiotische mechanismen (rogge). **Quantitative analysis of meiotic mechanisms (rye).** Publications.

6316 Bos, I. NL 020045/74/7342
Massa–selectie methoden bij rogge. **Mass–selection methods for rye (Secale cereale).**

Sorghum (B 3160)

See also 3083, 6270

6317 Rautou, S. FR 011201/60/0200
Obtention d'hybrides précoces de sorgho–grain. **Breeding early grain sorghom varieties.**

6318 Lenoble, M.; Porcheron, P. FR 012223/71/0073
Sélection pour la tolérance au froid de printemps des sorghos. **Breeding spring frost–hardiness in sorghum.**

6319 Lenoble, M. FR 012223/74/0298
Utilisation des croisements interspécifiques dans le genre sorghum. **Utilization of interspecific crossings in the sorghum genus.**

6320 Mariani, G.; Pezzoli, M.; Desiderio, E.
IT 020800/79/0010 N
Sorgo – Prove di adattamento di ibridi da granella in differenti ambienti. **Sorghum – Trials with grain hybrids in different environments.**

Wheat (B 3170)

See also 3103, 3104, 6094, 6247, 6248, 9413

6321 Casier, J.; De Paepe, G. BE 040501/70/0001
Selectieonderzoek op tarwe en rogge cultivars met verbeterde bakwaarde met respectievelijk een verhoogd en een verlaagd pentosangehalte. **Selection research on wheat and rye cultivars to improve baking value based respectively on a increased and a reduced pentosancontent.**

6322 Noulard, L.; Vandam, J. BE 080300/61/0003
Etude de la résistance du froment vis–à–vis du Cercosporella Herpotrichoïdes Fron. **Study of the wheat resistance against Cercosporella Herpotrichoïdes Fron.** Publications.

6323 Dubois, J. BE 080300/76/0008
Culture de tissus végétaux (blé, orge, lin). **Plant tissue culture (wheat, barley, flax).**

6324 Van Looveren, J.; Niclaes, J. BE 180000/61/0002
Veredeling van tarwe op resistentie tegen meeldauw (Erisyphe graminis f.sp. Tritici D.C.). **Breeding Research for resistance against mildew Erisyphe graminis f.sp. Tritici D.C.).** Publications.

6325 Van Looveren, K.; Niclaes, J. BE 180000/61/0003
Veredeling van tarwe op tolerantie op resistentie tegen gele roest (Puccinia strüformis Westend). **Breeding for tolerance or resistance of wheat against yellow rust (Puccinia strüformis Westend).** Publications.

6326 Van Looveren, K.; Niclaes, J. BE 180000/61/0004
Veredeling van tarwe op resistentie tegen bruine roest (Puccinia reconditta Rob.). **Breeding to introduce resistance against brown rust (Puccinia recondita Rob.) in wheat.** Publications.

6327 Van Looveren, K.; Niclaes, J. BE 180000/70/0001
Veredeling van tarwe op resistentie tegen de oogvlekkenziekte (Cercosporella herpotrichoides Fron). **Breeding research to introduce resistance against eyespot disease (Cercosporella herpotrichoïdes Fron) in wheat.** Publications.

6328 Niclaes, J.; Kempeneers, L. BE 180000/77/0009 N
Verbetering bakwaarde niveau winter– en zomertarwe. **Improvement of bread making quality of winter and spring wheat.**

6329 Röbbelen, G.; Elbedawy, R. DE 132182/70/0001
Herstellung von Monosomen–Sortimenten und ihre Nutzung für cytogenetische Analysen beim hexaploiden Weizen. **Establishment of monosomic series and their use for cytogenetic analyses in hexaploid wheat.**

6330 Röbbelen, G.; Rimpau, J.; Scheidl, G.
DE 132182/70/0002

Cytogenetische Analyse an Additions– und Substitutionslinien vom hexaploiden Weizen mit Roggenchromosomen. **Cytogenetic analysis of addition and substitution lines of hexaploid wheat with rye chromosomes.**

6331 Röbbelen, G.; Lelley, T.; Rimpau, J.; Feldman, M.
DE 132182/73/0003
Verteilung, Paarung und Austausch homologer Chromosomen in der Meiose des Weizen. **The distribution, pairing, and crossing–over of homologous chromosomes in meiosis of common wheat.**

6332 Röbbelen, G.; Rimpau, J.   DE 132182/77/0001
Bestimmung und genetische Analyse negativer Teigeigenschaften beim Weizen. **Determination and genetic analysis of sticky dough character in wheat.**

6333 Utz, H.F.; Snoy, M.–L.   DE 144420/74/0001
Untersuchungen zur Variabilität und Selektionsmethodik bei Weizen. **Studies on wheat variability and selection methods.**

6334 Fischbeck, G.; Chae, Y.A.   DE 161250/77/0002
Versuche zur Lokalisierung von Erbfaktoren für den quantitativ verminderten Mehltaubefall der Blätter der Weizensorte Diplomat. **Localization of hereditary factors in quantitatively reduced mildew on the leaves of wheat variety 'diplomate'.**

6335 Zeller, F.J.   DE 161250/77/0009
Herstellung von Substitutionen einzelner Weizenchromosomen durch Aegilops longissima–Chromosomen und die Evolution des Weizen–B–Genoms. **Production of substitutions of single wheat chromosomes by Aegilops longissima chromosomes and evolution of wheat genome B.**

6336 Günzel, G.   DE 161250/77/0010
Untersuchungen über Umfang und Ursachen der Verbesserung des Aufmischwertes deutscher A–Weizensorten durch Sortenmischung. **Investigations on extent and causes of improving the quality of German wheat A varieties for blending.**

6337 Fischbeck, G.; Zehatschek, W.; Günzel, G.; Zeller, F.J.
DE 161250/78/0004 N
Genetische Analyse der Kleberproteine des Weizens. **Genetic analysis of gluten proteins in wheat.**

6338 Grahl, A.; Schrödter, H.   DE 201040/75/0023
Erarbeitung physiologischer Grundlagen der Auswuchsresistenz bei Weizen und Roggen. **Elaboration of physiological fundamentals of the outgrowth resistance in wheat and rye.**

6339 Keydel, F.; Münzer, W.   DE 502058/73/0009
Grundlagenarbeiten auf dem Gebiet der Hybridzüchtung bei Selbstbefruchtern – Weizen, Gerste – 1967. **Fundamental studies on breeding of autogamous hybrids – wheat, barley –.**

6340 Hoeser, K.; Oppitz, K.   DE 502058/73/0011
Züchterische Bekämpfung von Mehltau und Rostkrankheiten bei Weizen 1967. **Breeding control of mildew and rust in wheat.** Publications.

6341 Hoeser, K.; Oppitz, K.   DE 502058/73/0013
Erarbeitung von Grundlagen zur Züchtung auf Toleranz gegen

Cercosporella herp., Ophiobolus und Septoria bei Weizen 1973. **Development of fundamental conditions for breeding of wheat for tolerance of Cercosporella Herp., Ophiobolus and Septoria.**

6342 Keydel, F.   DE 502058/78/0004 N
Untersuchungen zur Hybridsaatguterzeugung bei Weizen. **Hybrid seed production of wheat.**

6343 Auriau, Ph.; Pluchard, P.   FR 010104/64/0016
Utilisation de la vigueur hybride chez le blé. **Using hybrid vigor in wheat Breeding.**

6344 Auriau, Ph.; Pluchard, P.   FR 010104/70/0017
Création de variétés productrices de blé de force de printemps. **Breeding productive varieties of spring wheat with a high baking strength.**

6345 Cauderon, Y. Mme; Tempe, J.   FR 010104/70/0434
Essai de transfert des nènes de résistance aux rouilles issus d'Agropyron Intermedium sur les chromosomes du blé tendre. **Induction of Agropyron Intermedium/wheat transfers by homeologous pairing.**

6346 Auriau, Ph.   FR 010104/73/0122
Création de variétés demi–naines de blé d'hiver tolérantes aux maladies. **Breeding winter wheats for lodging and disease resistance.**

6347 Essad, S.   FR 010104/75/0317
Activité mitocyclique et hétérosis. **Mitocyclic activity and heterosis.**

6348 Berville, A.   FR 010105/72/0189
Hérédité cytoplasmique chez le blé et le maïs. **Cytoplasmic heredity in wheat and maize.**

6349 Bernard, M.   FR 010801/74/0129
Recherche de gènes de restauration à action complémentaire qui, à l'état hétérozygote, confèrent aux hybrides F1 de blé tendre une bonne fertilité pollinique: localisation chromosomique. **A search for complementary restorer genes which, in heterozygous state, give good pollen fertility in wheat f1 hybrids: chromosomal localization.**

6350 Rousset, M.   FR 010801/74/0130
Recherches de gènes de restauration à action complémentaire qui, à l'état hétérozygote, confèrent aux hybrides F1 de blé tendre une bonne fertilite pollinique. **A search for complementary restorer genes which, in heterozygous state, give good pollen fertility in wheat F1 hybrids.**

6351 Bernard, M.   FR 010801/74/0138
Accroissement de la variabilité génétique et chromosomique chez le triticale. **En largement of genetical and chromosomal variability in triticale.**

6352 Bernard, M.   FR 010801/74/0139
Etude physiologique et génétique de l'échaudage du grain chez le triticale. **Physiological and genetical studies of seed shrivelling in triticale.**

6353 Bernard, M.   FR 010801/74/0140
Orientation de lignées adaptées aux conditions climatiques et culturales françaises. **Breeding of triticale lines adapted to french climatic and cultural conditions.**

6354 Rousset, M.; Trottet, M.; Doussinault, G.; Saur, L.
FR 010801/76/0399
Comparaison de differents schémas de sélection applicablés à l'amélioration du blé tendre. **Comparaison in different selection schemes applied to wheat breeding.**

6355 Koller, J. FR 011003/68/0115
Obtention de variétés de blé demi–naines. **Selection of walf–dwarf wheat varieties.**

6356 Koller, J. FR 011003/70/0018
Obtention de variétés de blé résistantes aux herbicides. **Breeding herbicides resistant wheat.**

6357 Touvin, H.; Leblanc, D. FR 011003/71/0119
Mutants de résistance à l'atrazine. **Atrazine resistant wheat mutants.**

6358 Touvin, H.; Leblanc, D. FR 011003/71/0120
Production par traitements mutagènes de mutants courts. **Selection of short–strawed mutants through mutagene treatments.**

6359 Koller, J. FR 011003/72/0114
Obtention de lignées de blé résistantes au froid. **Selection of frost resistant wheat lines.**

6360 Touvin, H.; Leblanc, D. FR 011003/72/0121
Recherche de géniteurs résistants au Septoria Nodorum. **Researches for the selection of parents resistant to Septoria Nodorum in wheat.**

6361 Autran, J.C.; Simon, M. FR 011103/74/0113
Identification des variétés de blé tendre par électrophorèse des gliadines. **Gliadine electrophoresis to identify soft wheat varieties.**

6362 Grignac, P.; Tomas, A.; Poux, G. FR 011201/74/0209
Création de variétés de blé tendre à productivité élevée et bien adaptées aux conditions du sud de la France grâce à une résistance élevée aux adversités climatiques et aux principales maladies. **Breeding soft wheat varieties for yield and adaptation to the southern france conditions through resistance to climatic adversities and the main diseases.**

6363 Grignac, P.; Tomas, A.; Poux, G. FR 011201/74/0210
Création de variétés de blé tendre à grains riches en protéines et à valeur technologique élevée. **Breeding soft wheat varieties for high protein content and good technological value.**

6364 Grignac, P.; Tomas, A.; Poux, G. FR 011201/74/0211
Création de variétés très productives et adaptées aux différentes conditions agronomiques de la France. **Breeding high yielding soft wheat varieties for France.**

6365 Grignac, P.; Poux, G.; Tomas, A. FR 011201/74/0212
Création de variétés à grains de très bonne qualité semoulière et pastière de blé dur. **Breeding high quality hard wheat varieties.**

6366 Doussinault, G.; Dosba, F. FR 011301/71/0019
Obtention de variétés de blé tendre tolérantes au pietin–verse. **Breeding eyespot tolerant soft– wheat varieties.**

6367 Dosba, F.; Doussinault, G.; Lemaire, L.
FR 011301/73/0116
Obtention, étude et utilisation des lignées d'addition disomiques d'AE. Ventricosa sur génotype blé dur ou blé tendre. **Selection, study and utilization of AE. Ventricosa disomic alien addition lines on hard wheat or soft wheat genotype.** Publications.

6368 Dosba, F.; Doussinault, G.; Lemaire, L.
FR 011301/73/0117
Obtention de variétés de blé tendre tolérentas à la septoriose provoquée par septoria nodorum. **Selection of soft wheat varieties tolerant to the glume blotch induced by septoria nodorum.**

6369 Dosba, F.; Trottet, M.; Bourgeois, F.; Doussinault, G.
FR 011301/75/0314
Mise en évidence et identification des translocations réciproques chez des blés ou des aegilops utilisés pour la création de géniteurs résistants aux principaux parasites du blé. **Determining and identifying reciprocal translocations in wheat and aegilops used to new genitors resistant to the principal wheat parasites.**

6370 Trottet, M.; Rapilly, F.; Doussinault, G.; Dosba, F.
FR 011301/76/0460
Diversification des sources de tolérance durable à la rouille jaune (puccinia glumarum) chez le blé. **Diversification of durable tolerance source against wheat puccinia glumarum.**

6371 Doussinault, G.; Pauvert, P.; Dosba, F.; Trottet, M.
FR 011301/76/0461
Diversification des sources de tolérance durables à l'oïdium (erisiphe graminis FSP tritici) chez le blé. **Diversification of durable tolerance sources against oidium wheat erisiphe graminis FSP tritici.**

6372 Doussinault, G.; Trottet, M.; Dosba, F.; Trottet, M.; Jahier, J.; Lemaire, L. FR 011301/76/0462
Localisation chromosomique des gènes responsables de quelques caractéristiques agronomiques du géniteur v.p.m. 1. **Chromosomic implantation of genes responsible for some agronomic character of genitor vpm1.**

6373 Bourdet, A. FR 012213/70/2264
Recherche d'une méthode aisément applicable facilitant l'identification des variétés. **Search into easily applied method for the identification of wheat varieties.** Publications.

6374 Autran, J. C. FR 012213/70/2266
Les désoxyribonucléoprotéines et les histones du grain de blé. **Desoxyribonucleoproteins and histones of wheat grain.** Publications.

6375 Lupton; Bingham GB 011901/00/0001
Breeding winter wheat for improved yield.

6376 Angus GB 011901/00/0002 R
Breeding spring wheat for improved yield.

6377 Blackman GB 011901/00/0007
Breeding for improved grain quality in wheat.

6378 Huges GB 011901/00/0011 R
Breeding hybrid wheat.

6379 Gregory GB 011901/00/0013

D 2300 - Plant breeding

Breeding tetraploid wheats.

6380 Chapman; Miller GB 011905/00/0001 R
Genetic control of meiotic chromosome pairing in wheat and related natural and synthetic species.

6381 Chapman; Miller GB 011905/00/0002 R
Exploit the control of chromosome pairing in the transfer of useful genes from related species into wheat.

6382 Chapman; Miller GB 011905/00/0003 R
Chromosome and genetic relationships within wheat and related species.

6383 Flavell GB 011905/00/0015 R
The study of ribosomal RNA genes, nucleolar organiser activities and protein synthesis.

6384 Gale; Marshall GB 011905/00/0017 R
The genetics of height control and associated characters in wheat and barley.

6385 Gale; Marshall GB 011905/00/0019 R
Developmental genetics of hormone control in wheat.

6386 Law; Worland GB 011905/00/0020 R
Development of aneuploid lines and intervarietal chromosome substitutions.

6387 Law GB 011905/00/0021 R
The study of aneuploids and whole chromosome substitution lines.

6388 Snape; Worland GB 011905/00/0023 R
Biometrical investigations of quantitative inheritance.

6389 Dyer; Bowman GB 011905/78/0030 R
The genetics and biochemistry of isoenzyme and endosperm protein variation in wheat.

6390 Walsh, E.J. IE 120105/73/9212 N
Development of improved varieties and breeding techniques in spring wheat (tritacum aestivum).

6391 Onofrii, M.; Lanza, F.; Trenti, M.P.; Boschi, V.
  IT 020500/67/0005
Frumenti duri a Nord: confronto di cultivar anche in relazione al frumento tenero in pianura e in collina. **Durum wheat yield and fitting in Nothern Italy. Variety trials and comparison with bread wheat in both upland and lowland areas.** Publications.

6392 Boggini, G.; Corbellini, M. IT 020800/60/0002
Frumento duro – Miglioramento genetico per produttività e qualità per gli ambienti dell'Italia settentrionale. **Durum wheat – Genetic improvement for the conditions of the Po Valley.** Publications.

6393 Zitelli, G.; Biancolatte, E.; Alessandroni, A.; D'Egidio, M.G. IT 020800/61/0001 R
Frumento duro – Miglioramento genetico per produttività, altri caratteri agronomici, qualità della granella e resistenza alle più importanti malattie crittogamiche. **Durum wheat – Genetic improvement for yield, other agronomic traits, grain quality and resistance to important diseases.** Publications.

6394 Zitelli, G.; Pasquini, M.; Ceoloni, C.; Gras, M.; Forni,

C. IT 020800/65/0003 R
Frumento e specie affini– Ricerche sui fattori genetici che condizionano i diversi tipi di resistenza nelle specie ospiti rispetto a Erisyphe graminis, P. graminis, P. recondita. **Wheat and related species–Genetics of host resistance to Er. gr. tritici, to Puccinia graminis tritici and to Puccinia recondita.** Publications.

6395 Zitelli, G.; Pasquini, M.; Ceoloni, C.; Cecchi, V.; Biancolatte, E.; Gras, M. IT 020800/68/0003 R
Frumento ed orzo – Effettività dei fattori di resistenza in relazione alla variabilità dei fattori di virulenza nelle popolazioni di Erysiphe graminis f. sp. graminis e f. sp. hordei, P. graminis e P. recondita. **Wheat and Barley – Effective resistance factors with relation to the variability of populations of Erisyphe graminis f. sp. graminis and f. sp. hordei, P. graminis and P. recondita.** Publications.

6396 Zitelli, G.; Alessandroni, A.; Mariani, B.M.; Biancolatte, E.; Manmana Novaro, P. IT 020800/70/0006 R
Frumento duro – Analisi genetiche di caratteri quantitativi. **Durum wheat – Genetic analysis of quantitative characters.** Publications.

6397 Zitelli, G.; Pasquini, M.; Ceoloni, C.; Forni, C.
  IT 020800/72/0004 R
Frumento duro – Studi sulla resistenza specifica, in confronto o in addizione a quella di tipo generale, rispetto alla P. uccinia recondita. **Durum wheat – Study on specific resistance to P. recondita in comparison or in addition to the general type.** Publications.

6398 Borghi, B.; Boggini, G.; Cattaneo, M. IT 020800/72/0006
Frumento tenero – Miglioramento genetico per produttività e qualità. **Common wheat – Genetic improvement for yield and quality.** Publications.

6399 Borghi, B.; Boggini, G.; Corbellini, M.; Cattaneo, M.; Pogna, N. IT 020800/72/0007 R
Frumento tenero – Studio della base genetica della produttività e della qualità mediante analisi biometriche, uso degli aneuploidi e linee di sostituzione. **Common wheat – Genetical investigations by means of biometrical analysis, aneuploids and substitution lines.** Publications.

6400 Zitelli, G.; Manenti, S.; Wittmer, G.; Cecchi, V.; D'Egidio, M.G.; Paradisi, F. IT 020800/73/0004
Frumento duro – Prove comparative effettuate in diverse località dell'Italia centro–meridionale per studiare il comportamento e l'adattabilità di varietà diffuse e di recente e nuova costituzione. **Durum wheat – Comparative trials designed in several locations of central and southern Italy to measure the performance and adaptation of current and new varieties.** Publications.

6401 Borghi, B.; Corbellini, M.; Boggini, G.; Cattaneo, M.
  IT 020800/73/0005
Frumento – Valutazione agronomica e tecnologica delle recenti costituzioni di frumento tenero e duro nell'Italia centro–settentrionale. **Wheat – Performance trials and technological evaluation of common and durum wheats in North Italy.** Publications.

6402 Borghi, B. IT 020800/73/1165
Induzione di sterilità maschile nel frumento tenero mediante l'impiego di un gametocida, l'ethrel acido 2–cloroetilfosfonico,

per la produzione di seme ibrido F1. **Induction of male sterility in common wheat. Use of a gametocide, ethrel 2–chloroethyl phosphonic acid, for the production a hybrid seeds, F1.** Publications.

6403 Zitelli, G.; Biancolatte, E.; D'Egidio, M.G.; Cecchi, V.; Stefanini, R. ' IT 020800/75/0007 R
Frumento duro – Criteri di selezione per l'alto contenuto in proteine, fin dalle prime generazioni. **Durum wheat – Criteria on improvement of high protein content, since early generations.**

6404 Wittmer, G.; Li Destri Nicosia, O.; De Stefanis, E. IT 020800/76/0001
Frumento duro – Costituzione di nuove varietà per le Puglie: precocità, resistenza combinata a fitopatie, attitudine pastificatoria. **Durum wheat – New varieties adapted to Puglia conditions: Earliness, resistance to diseases, ability to give good quality spaghetti.**

6405 Zitelli, G. IT 020800/77/0409 R
Miglioramento genetico del frumento duro particolarmente nei confronti della resistenza alle malattie. **Genetic improvement of hard wheat especially as regards disease resistance.**

6406 Wittmer, G.; Brando, A.; De Stefanis, E.; Ciocca, L. IT 020800/78/0003 R
Frumento duro. Studio del profilo proteico e della composizione aminoacidica sul materiale proveniente dai piani di miglioramento genetico. **Durum wheat. Analysis of proteins and amino acid composition in experimental material from genetic improvemnet program.**

6407 Scarascia Mugnozza, G.T. IT 040115/73/1179
Proposte per accrescere il polimorfismo genetico dei frumenti tetraploidi. **Proposals for incrementing the genetic polymorphism of tetraploid wheats.** Publications.

6408 Blanco, A. IT 040115/74/0519
Proposte per accrescere il polimorfismo genetico dei frumenti tetraploidi duri. **Proposals for the increase of genetic polymorphism of tetraploid coarse grains.**

6409 Blanco, A. IT 040115/77/0118 R
Ricerche esplorative di genetica e miglioramento genetico in due cereali di importanza riconosciuta o potenziale per il Mezzogiorno d'Italia – Frumento duro e orzo. **Preliminary research on the genetics and genetical improvement of two cereals of recognized or potential importance in Southern Italy – Durum wheat and barley.**

6410 Blanco, A. IT 040115/77/0332 R
Valutazione quali–quantitativa di linee e varietà di frumento duro di nuova costituzione in differenti ambienti. **Qualitative and quantitative evaluation of newly developed durum wheat lines and varieties in different environments.**

6411 Cassaniti, S. IT 040305/78/1187 N
Selezione per la produttività del frumento duro. **Durum wheat : selection for improved yield.**

6412 Barbieri, R. IT 040701/74/0509
Miglioramento genetico del grano duro. **Genetic improvement of durum wheat.**

6413 Mucci, F. IT 040701/77/0377 R

Costituzione di nuove cultivar di frumento duro adatte per grandi zone di coltura, di collina e di pianura. **Development of new cultivars of hard wheat suitable for extensive growing on hills and level ground.**

6414 Di Prima, G. IT 040904/77/0348 R
Costituzione di nuove varietà di frumento duro dotate di elevate produttività qualitativamente pregiate e resistenti alle avversità ambientali. **Breeding new varieties of high–yielding, higher quality hard wheat resistant to environmental hazards.**

6415 Briganti Giannoni, G.M.; Piano, E.; Lorenzetti, F.; Arcioni, S.; Ceccarelli, S. IT 041002/70/0001
Miglioramento genetico del grano tenero per la qualità. **Breeding of soft wheat for quality.** Publications.

6416 D'Amato, F. IT 041114/77/0345 R
Impiego delle colture in vitro a fini di miglioramento genetico – fru mento duro–. **The use of in vitro cultivation for genetic improvement – hard wheat – .**

6417 Marras, G.F. IT 041302/73/1170
Miglioramento genetico del grano duro, pomodoro e carciofo. **Genetic improvement of hard wheat, tomatoes and artichokes.**

6418 Rivoira, G.; Milia, M.; Marras, G.F. IT 041302/75/0005
Grano duro – prova varietale. **Durum wheat – Varietal test.** Publications.

6419 Perrino, P. IT 060500/71/0173
Studi sui rapporti esistenti tra le specie del genere Triticum. **Studies on relationships among Triticum species.** Publications.

6420 Porceddu, E. IT 060500/71/0180
Lunghezza del celeoptile in popolazioni di frumento duro. **Coleoptile length in durum wheat.** Publications.

6421 Porceddu, E. IT 060500/72/0175
Analisi della collezione del frumento duro per la resistenza alle ruggini ed all'oidio e per il contenuto proteico. **Analysis of a durum wheat collection for rust and mildew resistance and protein content.** Publications.

6422 Perrino, P. IT 060500/73/0170
Esplorazione e raccolta di vecchie popolazioni di frumento nel Nord Africa. **Exploration and collection of wheat old land races in North Africa.** Publications.

6423 Porceddu, E. IT 060500/77/0449
Costituzione di nuove linee di frumento duro caratterizzate da produttività elevata, resistenza a malattie, buona qualità tecnologiche ai fini della pastificazione; elevato valore nutritivo; studio per un eventuale sfruttamento dell'eterosi. **Breeding of high yielding, disease resistant Durum wheat with good technological qualities for Italian paste making and a high nutritional value; study on the possible use of heterosis.**

6424 Porceddu, E. IT 060500/77/0450
Valutazione quantitativa e qualitativa di linee e varietà di frumento duro di nuova costituzione in differenti condizioni ambientali e di tecniche colturali. **Quantitative and qualitative evaluation of Durum wheat new lines and varieties under different environmental conditions; evaluation of different cultural techniques.**

D 2300 – Plant breeding

6425 Porceddu, E. IT 060500/77/0451
Studio della struttura genetica di vecchie popolazioni di
frumento duro. **Study on the genetical structure of old durum
wheat cultivations.**

6426 Porceddu, E. IT 060500/77/0452
Identificazione di linee di frumento duro con elevate
caratteristiche qualitative e studio del loro determinismo
genetico. **Identification of high quality lines of durum wheat, a
study of their genetical determinism.**

6427 Porceddu, E. IT 060500/77/0999
Studi sulla diversificazione dei frumenti mediterranei. **A study
on mediterranean wheat diversification.**

6428 Porceddu, E. IT 060500/77/1001
Moltiplicazione e valutazione di accessioni di frumenti e
d'orzo. **Multiplication and evaluation of accessions in wheat and
barley.**

6429 Porceddu, E. IT 060500/77/1002
Seconda moltiplicazioe e valutazione di accessioni di frumenti
raccolte in Etiopia. **Second multiplication and evaluation of
accessions in wheat harvested in Ethiopia.**

6430 Porceddu, E. IT 060500/77/1003
Valutazione della collezione del frumento duro per la
resistenza alle rugini e all'oidio. **Evaluation of durum wheat
collection with special regard to its resistance to rusts and to
oidium.**

6431 Giorgi, B. IT 070101/78/1193 N
Ottenimento di varietà di frumento duro resistenti al freddo, e
di altre resistenti alla siccità. **Development of hard wheat
varieties resistant to cold and of other varieties resistant to
drought.**

6432 Deidda, M. IT 092002/78/1186 N
Costituzione di nuove cultivar di frumento duro dotate di
elevate capacità produttivie e di caratteristiche qualitative più
rispondenti ai fini della pastificazione. **Development of new
high yielding durum wheat cultivars best suited for pasta
making.**

6433 Maliani, C. IT 120300/77/0370
Miglioramento genetico del grano duro. **Genetic improvement
of hard wheat.**

6434 Mesdag, J. NL 010120/50/7399 R
Het instandhouden, uitbreiden en bestuderen van een collectie
tarwerassen ten behoeve van het veredelingsonderzoek.
**Maintenance, expansion and evaluation of a collection of
wheat varieties to support the breeding research.**

6435 Mesdag, J. NL 010120/50/7401 R
Veredelingsonderzoek naar resistentie tegen bladziekten bij
tarwe (gele– en bruine roest, meeldauw, Septoria tritici,
Septoria (Leptospheria) nodorum). **Breeding research on
resistance to leaf diseases of wheat (stripe rust (Puccinia
striiformis), leaf rust (Puccinia recondita), mildew, Septoria
tritici, Septoria (Leptospheria) nodorum).** Publications.

6436 Mesdag, J. NL 010120/50/7405 R
Veredelingsonderzoek naar resistenties tegen abiotische
factoren bij tarwe (schot, koude, legering, hoge zuurgraad van
de grond) en naar worteleigenschappen (wortelontwikkeling

en wortelactiviteit). **Breeding research for resistance to abiotic
factors in wheat (sprouting in the ear; freezing; lodging; soil
acidity) and to root characteristics (development of the roots
and root activity).** Publications.

6437 Mesdag, J. NL 010120/60/7404
Veredelingsonderzoek naar resistenties tegen dierlijke
parasieten van tarwe (tarwestengelgalmug, graancystenaaltje,
luizen). **Breeding research on insect resistance in wheat (Saddle
gall midge (Haplodiplosis equestris); cereal cyst nematode;
aphids).**

6438 Mesdag, J. NL 010120/60/7406 R
Veredelingsonderzoek naar kwaliteitseigenschappen bij tarwe:
1. Technologische kwaliteit (bakkwaliteit); 2. Voedings– en
voederkwaliteit (gehalte aan essentiële aminozuren en aan
eiwit van de korrel). **Breeding research on quality
characteristics in wheat: 1. Technological quality (baking
quality); 2. Quality of food and feed (content of essential amino
acids and of protein in the kernel).** Publications.

6439 Mesdag, J. NL 010120/65/7402
Veredelingsonderzoek naar resistentie tegen voetziekten bij
tarwe (oogvlekkenziekte en tarwehalmdoder). **Breeding
research on resistance to eye spot (Cercosporella graminis) in
wheat.** Publications.

6440 Mesdag, J. NL 010120/68/7403 R
Veredelingsonderzoek naar resistentie tegen afrijpingsziekte
van de aar bij tarwe (Fusarium culmorum). **Breeding research
on resistance to Fusarium head blight of wheat (caused by F.
Culmorum).** Publications.

6441 Mesdag, J. NL 010120/72/7400 R
Onderzoek naar de invloed van het ontwikkelingsritme op de
opbrengst bij tarwe. **Research into the relation between the
rythm of development and the yielding capacity of wheat.**
Publications.

6442 Balkema–Boomstra, A.G. van; Mesdag, J.
NL 010120/73/7415
Veredelingsonderzoek naar niet–specifieke resistentie tegen
bruine roest in tarwe. **Breeding research for race–non specific
resistance to leaf rust of wheat (Puccinia recondita).**
Publications.

6443 Kramer, Th. NL 020045/76/6768 R
Efficientie van eiwitproductie bij wintertarwe. **Efficiency of
protein production of winter wheat.**

6444 Kramer, T. NL 020045/76/8850 N
Fysiologische selectiecriteria voor opbrengst in tarwe.
Physiological selection criteria for grain yield in wheat.

6445 Zeven, A.C. NL 020045/77/8653 N
Evaluatie en benutting van Aegilops squarossa en landrassen
(WTC) voor de tarweveredeling. **Evaluation and use of
Aegilops squarossa and landraces (WTC) for wheat breeding.**

Other cereals (B 3190)

See also 3083, 6482

6446 Gregory GB 011901/00/0012 R
Breeding Triticale.

D 2300 – Plant breeding

6447 Plonka, F.   FR 010104/70/0161
Obtention variétés F1 de lin. **Breeding flax F1 varieties.**

6448 Plonka, F.; Travers, P.   FR 010104/76/0436
Production de graines F1 de lin. **F1 seeds production in flax.**

6449 Billot, C.; Mennessier, P.   FR 010615/74/9030
Etude du comportement du lin oléagineux. **Testing linseed flax varieties.**

6450 Chesneaux, M.T.; Monties; Simon, M.
FR 011103/76/0425
Etude quantitative et qualitative des pigments de colorations contenues dans les lignées. **Quantitative and qualitative study on colorations pigments enclosed in hybrid varieties lines.**

6451 Wassenaar, R.; Koster, H.   NL 010113/43/0617
Rassenonderzoek vlas. **Research on flax varieties.**

6452 Jongmans, M.A.   NL 010120/50/6691
Veredelingsonderzoek naar resistentie tegen schimmelziekten bij vlas. **Research on breeding for resistance to fungus diseases of fibre flax.** Publications.

6453 Jongmans, M.A.   NL 010120/50/6693
Selectie op en synthese van landbouwkundige eigenschappen bij vlas. **Selection and synthesis of agricultural characters of fibre flax.** Publications.

Olive (B 3220)

6454 Villemur, P.; Gonzalez, A.; Delmas, J.M.
FR 011207/76/0444
Sélection clonale de l'olivier. **Clonal selection of olive tree.**

6455 Silvestri, B.; Filippucci, B.   IT 021400/73/0010 R
Selezione e allevamento di olivi devianti e nanizzanti. **Selection and breding of olive trees which are deviated and dwarf.**

6456 Parlati, M.; Scarponi, E.; Petruccioli, G.; Turco, D.
IT 021400/77/0001 R
Selezione di cloni di olivo ad alta resa, in prodotto e in olio. **Selection of olive clones to high yield, in product and in oil.**

6457 Lombardo, N.   IT 021400/77/0367 R
Miglioramento genetico dell'olivo. **Genetic improvement of the olive-tree.**

6458 Silvestri, B.; Filippucci, B.   IT 021400/78/0001 N
Selezione di cloni di olivo resistenti alle avversità climatiche, patologiche ed entomologiche. **Selection of olive clones resisting to the climatic, pathological and entomological adversities.**

6459 Silvestri, B.; Filippucci, B.   IT 021400/78/0002 N
Studio dei processi fecondativi dell'olivo. **Study of olive fecundated processes.**

6460 Lombardo, N.; Briccoli-Bati, C.; Diana, G.
IT 021400/79/0001 N

Selezione di portinnesti nanizzanti e di cloni di cultivar di olivo di limitata vigoria. **Selection of dwarfing rootstooks and clones of olive tree of limited vigour.**

6461 Parlati, M.; Manca, M.G.; Scarponi, E.
IT 021400/79/0002 N
Influenza della filloptosi sull'induzione autogena e differenziazione a fiore nell'olivo. **Filloptosi's influence on the inside induction and differentiation flower on the olive tree.**

6462 Diana, G.; Lombardo, N.; Garofalo, M.G.
IT 021400/79/0005 N
Prove di germinazione di semi di olivo di alcune cultivar calabresi. **Germination tests of olive seeds of some cultivar in Calabria.**

6463 Cucurachi, A.   IT 022100/77/0344 R
Individuazione di CV suscettibili di valorizzazione come olive da mensa e studio delle tecniche di lavorazione. **Identification of olive cultivars susceptible to be improved for table use and study of processing techniques.**

6464 Lamparelli, F.   IT 040107/77/0365 R
Valutazione merceologica di olive da mensa prodotte da CV. migliorate. **Commercial value of table olives grown on improved CV.**

6465 Ferrara, E.   IT 040112/77/0352 R
Selezione genetica per il miglioramento dell'olivo per frutti da mensa. **Genetic selection to improve olive trees for the production of table olives.**

6466 Alberghina, O.   IT 040307/77/0326 R
Olivo per frutti da mensa, individuazione e selezione clonale osservazioni comparative. **Olive trees for growing table olives, identification and clonal selection, comparative observations.**

6467 Guerriero, R.   IT 041104/77/0361 R
Miglioramento della coltura dell'olivo da mensa. **The improvement of table olive growing.**

6468 Fontanazza, G.   IT 061500/77/0463
Individuazione e selezione nell'ambito delle popolazioni esistenti di cultivar di olive da mensa di pregio. Individuazioni di portainnesti nanificanti e di aree di coltivazioni. **Identification and selection of high quality table olive cultivars in existing populations. Identification of dwarfing rootstocks and of cultivation areas.**

6469 Jacoboni, N.   IT 061500/77/0964
Selezione delle cultivar di olivo e miglioramento varietale. **Selection of olive cultivars and varietal improvement.**

6470 Roselli, G.C.   IT 062800/77/0464
Selezione portainnesti, incrocio e mutagenesi nell'olivo da mensa. **Rootstocks selection; hybridization and mutagenesis in table olive trees.**

Rape (B 3230)

6471 Van Hee, L.   BE 070400/72/0010
Veredeling van koolzaad. **Breeding of rape.**

6472 Odenbach, W.   DE 104600/75/0002

D 2300 – Plant breeding

Merkmalskorrelationen und Selektionsgewinn bei der Züchtung auf einen neuen Sortentyp beim Winterraps, Brassica napus L.. **Correlations of characteristics and produce of selection in the breeding of winter rape, Brassica napus L., for a new type of variety.**

6473 Röbbelen, G.; Thies, W.   DE 132182/70/0004
Züchterische Selektion auf Fettsäurequalität beim Raps. **Breeding selection for fatty acids in rape.**

6474 Röbbelen, G.; Thies, W.; Jürges, K.; Gland, A.
            DE 132184/75/0002
Biogenese und Selektion auf Glukosinolate in Samen und Blättern bei Raps. **Biogenesis and selection for glucosinolates in seeds and leaves of rape.**

6475 Morice, J.   FR 010104/70/0029
Sélection pour l'absence de substances goîtrigènes dans le tourteau de colza. **Breeding rape for low content in thioglucosides.**

6476 Morice, J.; Renard, M.   FR 011301/76/0459
Etude de la stérilité mâle cytoplasmique et de l'hétérosis chez le colza. **Studies on cytoplasmic male sterility and heterosis in rape.**

6477 Morice, J.; Renard, M.   FR 011301/76/0463
Sélection pour une diminution de la teneur en acide linolénique de l'huile de colza. **Breeding for decreasing linolenic acid content in rape oil.**

6478 Mesquida, J.; Tasei, J.N.   FR 011306/75/5131
Pollinisation du Colza en production de semence hybride. **Pollination of rape for hybrid seed production.**

6479 Thompson; Capitain   GB 011910/00/0019 R
Breed oilseed rape as an oil and protein source.

6480 Thompson; Capitain   GB 011910/79/0023 N
Reproduction of commercial hybrids of oilseed rape.

6481 Benvenuti, A.   IT 041101/77/0112 R
Miglioramento genetico delle specie oleaginose erbacee secondarie, colza, ricino, arachide, sesamo, cartamo. **Genetic improvement in secondary herbaceous oleaginous species, cole–seed, castor–oil plant, peanut, sesame, safflower..**

6482 Wassenaar, R.; Baltjes, H.J.   NL 010113/43/0618
Rassenonderzoek handelsgewassen (winter koolzaad, kanariezaad, karwij, blauwmaanzaad). **Research on varieties of commercial crops (rape seed, canary seed, caraway, poppy seed). Publications.**

Soyabean (B 3240)

See also 6481

6483 Clavier, C.; Lacombe, P.; Vidal, A.; Angevain, M.
            FR 011201/74/0466
Recherche de géniteurs et création de nouvelles variétés de soja. **Investigation of genetic types and breeding for new varieties in soybean.**

6484 Clavier, C.; Lacombe, P.; Vidal, A.; Angevain, M.
            FR 011201/74/0469
Etude du potentiel variétal et des techniques culturales du soja. **Studies on the varieties potentialy and soybean production.**

6485 Decau, J.   FR 011401/75/6199
Comparaison intervariétale des qualités oléo–protéïques des graines de soja soumis à différentes conditions culturales. **Oil and protein content and quality in soybean grain. Comparative study of varieties grown in different conditions.**

6486 Laureti, D.   IT 021100/79/0026 N
Individuazione di mutazioni indotte in seme irradiato di soia. **Singling out of induced mutations in irradiated soybean seed.**

6487 Laureti, D.   IT 021100/79/0027 N
Confronto di varietà di soia. **Comparison of different kinds of soya–bean.**

6488 Toderi, G.; Amaducci, M.T.; Quaquarelli, S.
            IT 040201/78/0002
Studio comparativo di varietà di soia. **Comparative study of soybean varieties.**

6489 Toderi, G.; Amaducci, M.T.; Quaquarelli, S.
            IT 040201/79/0002 N
Studio sulla efficacia di diversi ceppi di Rhizobium japonicum su alcune varietà di soia. **Comparative analysis of several strains of Rhizobium japonicum on different varieties of soyabean.**

Sunflower (B 3250)

See also 6262, 6306, 7146

6490 Leclercq, P.; Vear, F.   FR 010801/71/0032
Etude génétique de la résistance du tournesol au mildiou (Plasmopara Helianthi). **Genetic study on sunflower resistance to mildew (Plasmopara Helianthi). Publications.**

6491 Leclercq, P.   FR 010801/71/0033
Etude de la stérilité mâle cytoplasmique du tournesol. **Study on cytoplasmic male sterility in sunflower. Publications.**

6492 Vear, F.; Tersac, M.; Leclercq, P.; Philippon, J.
            FR 010801/75/0367
Etude de la résistance du tournesol au Macrophomina Phaseoli. **Macrophomina Phaseoli resistance in sunflowers.**

6493 Vear, F.; Leclercq, P.; Philippon, J.   FR 010801/75/0368
Etudes de la résistance du tournesol au mildiou. **Downy mildew resistance in sunflowers.**

6494 Bouchet, M.; Mme Moreau,; Saulas, M.; Champion, R.
            FR 011103/73/0204
Monographie des variétés de tournesol cultivées en France. **Description of sunflower varieties grown in France.**

6495 Piquemal, G.; Cleomene, J.   FR 011201/66/0159
Création de lignées à entrenoeuds courts. **Selection of lines with short internodia in sunflowers.**

6496 Piquemal, G.; Cleomene, J.   FR 011201/70/0155
Développement de lignées à stérilité-mâle cytoplasmiques ou restauratrices. **Development of male sterile cytoplasmic or restorer lines in sunflowers.**

6497 Piquemal, G.; Cleomene, J.   FR 011201/70/0157

Création de lignées possédant plusieurs gènes de résistance au mildiou. **Selection of lines with several genes for resistance to downy mildew.**

6498 Piquemal, G.; Cleomene, J.   FR 011201/71/0031
Obtention par mutagenèse de nouveaux gènes de résistance au mildiou du tournesol. **Mutation breeding in sunflower for mildew resistance.**

6499 Piquemal, G.; Cleomene, J.   FR 011201/72/0030
Etude de la tolérance du tournesol à Botrytis Cinerea. **Study on sunflower tolerance to Botrytis Cinerea.**

6500 Piquemal, G.   FR 011201/74/0156
Recherche de facteurs de résistance aux principales maladies dans les diverses espèces du genre Helianthus. **Research of resistance factors foward the main diseases in the various species of the Helianthus genus.**

6501 Piquemal, G.; Cleomene, J.   FR 011201/74/0158
Création d'un tournesol "prolifique". **Selection of a "productive" sunflower.**

6502 Piquemal, G.; Cleomene, J.   FR 011201/75/0289
Sélection pour la tolérance à Botrytis Cinerea. **Selection for Botrytis Cinerea tolerance.**

6503 Piquemal, G.; Cleomene, J.   FR 011201/76/0464
Amélioration génetique du tournesol: obtention d'hybrides de tournesol autogames. **Genetical improvement of sunflower: breeding of autogamous sunflower hybrids.**

6504 Pirani, V.; Mengarelli, A.   IT 021100/75/0006
Miglioramento genetico del girasole. **Sunflower genetic improvement.** Publications.

6505 Pirani, V.   IT 021100/76/0006 N
Confronto di varieta' di girasole. **Varietal comparison with sunflower..**

6506 Pirani, V.; Zazzerini, M.   IT 021100/79/0023 N
Valutazione della resistenza alla peronospora de linee parentali di girasole attraverso inoculazione del patogeno al seme. **Evaluation of the resistance to the mildew of sunflower parental lines by means of inoculation of seed with the pathogenous agent.**

6507 Alba, E.   IT 040115/77/0325 R
Costituzione di varietà migliorate, varietà sintetiche ed ibridi di girasole da utilizzare in primo e secondo raccolto. **Breeding of new improved varieties, synthetic varieties and hybrids of sunflower to be used for the first and second harvest.**

6508 Benvenuti, A.   IT 041101/77/0331 R
Miglioramento delle varietà di girasole di diretta costituzione, costituzione di ibridi semplici. **Improvement of direct constitution sunflower varieties. Simple hybrid constitution.**

6509 Benvenuti, A.   IT 090901/74/0514
Miglioramento genetico delle piante erbacee oleifere con particolare riguardo al girasole e al colza. **Genetic improvement of herbaceous oil plants with special reference to sunflower and Brassica napus (rape).**

Other fibre plants and oil crops (B 3290)

See also 6151, 6481, 6482

6510 Maréchal, R.   BE 010019/63/0002 R
Recherches sur les possibilités d'amélioration du cotonnier par diverses méthodes d'introgression de caractères à partir d'espèces diploïdes sauvages. **Research on the possibilities of improving cotton by different breeding methods using introgression of characters from wild diploid species.** Publications.

6511 Demol, J.; Maréchal, R.; Verschraeghe, L.
BE 010019/76/0006
Etude systématique des caractéristiques de cotonniers allohexaploides obtenus par croisements entre G. Hirsutum et diverses espèces sauvages. **Systematic studies on the characteristic of cotton allohexaploids issued from crosses between G. hirsutum L. and different wild species.** Publications.

6512 Rasmussen, P.; Flengmark, P.   DK 010106/72/3504
Sortsafprøvning med olie– og spindplanter. **Variety trials of oil and fibre crops.**

6513 Allavena, D.   IT 021100/75/0013 N
Miglioramento genetico della canapa (cannabis sativa). **Hemp genetic improvement (cannabis sativa)..**

6514 Laureti, D.   IT 021100/79/0029 N
Miglioramento varietale del ricino. **Improvement of variety of the castor bean.**

6515 Venturi, G.; Allavena, D.   IT 021100/79/0031 N
Miglioramento genetico del Kenaf (HIBISCUS cannabinus). **Genetic improvement of the Kenaf (HIBISCUS cannabinus).**

Potatoes (B 3310)

See also 3233, 6151, 7191, 8523, 8524, 9470, 9491, 9492

6516 Schnell, F.W.; Kameke, K.von   DE 144420/74/0003
Untersuchungen zur quantitativen Variabilität in drei Kreuzungsnachkommenschaften der Kartoffel. **Studies on the quantitative variability of three potatoe crosses.**

6517 Stegemann, H.; Loeschcke, V.   DE 215140/77/0003
Selektion und Verringerung der genetisch interessanten Kartoffelklone durch molekularbiologische Verfahren. **Selection and reduction of potato clones of genetic interest by molecular biological methods.**

6518 Stegemann, H.; Loeschcke, V.   DE 215140/77/0004
Erweiterung der Basis für genetische Studien und Resistenzzüchtung bei Kulturkartoffeln. Virusverseuchung bei Wildkartoffeln. **Enlargement of basis for genetic studies and breeding for resistance in cultivation of potatoes. Virus infestation of wild potatoes.**

6519 Hunnius, W.   DE 502058/73/0003
Züchterische Verbesserung der Speisequalität bei Kartoffeln 1973. **Breeding improvement of the quality of cooking potato and special studies on correlations between protein content and criteria of cooking quality.** Publications.

6520 Hunnius, W.   DE 502058/73/0004
Untersuchungen zur genetischen Streubreite des Eiweissgehaltes der Kartoffel 1973. **Investigations on the**

genetic scattering range of protein content in potato . Publications.

6521 Hunnius, W.; Munzert, M.   DE 502058/73/0005
Untersuchungen zur Erfassung der Vollernteverträglichkeit der Kartoffel und deren züchterische Verbesserung 1973. **Investigations on determination and breeding development of combining tolerance in potato.** Publications.

6522 Hunnius, W.; Munzert, M.; Scheidt, M.
    DE 502058/78/0006 N
Haploidiezüchtung bei Kartoffeln. **Haploid–plant breeding with potatoes.**

6523 Hansen, S.E.; Bach, Aa.   DK 010111/74/4003
Sortsforsøg i kartofler. **Variety trials with potatoes.**

6524 Foldø, N.E.   DK 020700/79/5078 N
Forædlingsprogram for fremstilling af proteinrige kartoffelsorter. Undersøgelse og udvikling af metoder til tidligst mulig udvælgelse i forædlingsprogrammet af stivelses– og proteinrige kartoffelsorter. **The development of methods for selection of potato varieties rich in starch and protein.**

6525 Perennec, P.   FR 012207/67/0040
Obtention et utilisation de dihaploïdes dans l'amélioration de la pomme de terre. **Breeding potato using dihaploids.**

6526 Perennec, P.   FR 012207/67/0291
Création de variétés de pommes de terre pour la transformation industrielle. **Breeding varieties of potatoes suitable for processing.**

6527 Madec, P.   FR 012207/70/0038
Recherche d'une méthode de sélection au stade semis pour la précocité chez la pomme de terre. **Testing a breeding method for earliness of tuber formation applied on potato seedlings.**

6528 Perennec, P.   FR 012207/70/0039
Obtention de variétés de pommes de terre résistantes aux virus X, Y, A. **Breeding potato–varieties resistant to X, Y, A. virus.**

6529 Perennec, P.; Madec, P.   FR 012207/70/0228
Selection de varietes de pommes de terre pour le bassin mediterraneen. **Breeding of potato varieties for mediterranean areas).**

6530 Cole; Thomson   GB 011910/00/0011
Breed maincrop potatoes.

6531 Cole   GB 011910/00/0012
Breed first early potatoes.

6532 Cole   GB 011910/00/0013
Breed potatoes for processing quality.

6533 Howard; Fuller   GB 011910/00/0014
Breeding for resistance to potato cyst nematode species, both Globodera rostochiensis and G. pallida.

6534 Thomson   GB 011910/00/0016 R
Breeding methods (including use of dihaploids) and selection criteria in potatoes.

6535 Davidson   GB 030805/78/0001
Breed maincrop potato cultivars for quality disease resistance

and yield for fresh use and for processing.

6536 Davidson   GB 030805/78/0002
Breed early potato cultivars for early yield and quality in relation to fresh use , crisping and canning.

6537 Davidson   GB 030805/78/0003
Maintain and multiply healthy breeding and experimental stocks. Develop and apply improved health control procedure.

6538 Davidson   GB 030805/78/0005
Evaluate advanced potato selections in field trials in Scotland, England and Wales.

6539 Davidson   GB 030805/78/0006
Study biometrical genetics of potato characters and devise improved breeding schemes.

6540 Davidson   GB 030805/78/0007
Research into design and predictive efficiency of potato field trials and into G x E intreractions.

6541 Davidson   GB 030805/78/0008
Evaluate potato selection procedures and devise improvements for application in breeding programme.

6542 Holden   GB 030805/78/0009
Establish and manage computerised data bank on clones under selection in potato breeding programme.

6543 Glendinning   GB 030805/78/0016
Manage the commonwealth potato collection of Latin American origin. Liase with the Dutch German gene bank.

6544 Glendinning   GB 030805/78/0017
Breed Neotuberosum potatoes from andigena origin for use in breeding cultivars.

6545 Glendinning   GB 030805/78/0018
Evaluate Neotuberosum potatoes as parental material for use in breeding cultivars.

6546 Glendinning   GB 030805/78/0019
Breed diploid potatoes and evaluate as potential parents for diploid and tetraploid cultivars.

6547 Glendinning   GB 030805/78/0020
Produce breed and maintain collection of dihaploid potatoes. Use dihaploids to enhance disease resistance.

6548 Costelloe   GB 040302/00/0014
Breeding and testing of commercial potato varieties.

6549 Addison   GB 060213/00/0008 R
Potato variety testing.

6550 Paterson   GB 060314/00/0014 R
Potato variety trials.

6551 Kehoe, H.W.   IE 060500/62/0216 R
Potato breeding and evaluation. The breeding and evaluation of varieties for all aspects of the potato trade. Publications.

6552 Kehoe, H.W.   IE 060500/71/0217
Potato variety evaluation. Publications.

6553 Lombardi, D.   IT 021000/79/0004 N
Costituzione di cloni di patata adatti a basse temperature e ad
opportuna luminosità. **Constitution of potato clones suitable for
low temperatures and right luminosity.**

6554 Cremaschi, D.; Affatato, E.   IT 021100/76/0004 N
Confronto di varieta' di patata. **Varietal comparison with
potato.**

6555 Cremaschi, D.; Cerato, C.; Vender, C.; Affatato, E.
                                              IT 021100/79/0017 N
Verifica agronomica e merceologica di linee di patata
ambientate ed in via di risanamento. **Agronomic and
technological check of potato lines fitted for our environment
conditions or with restoration to health in progress.**

6556 Wassenaar, R.; Baltjes, H.J.   NL 010113/43/0614
Rassenonderzoek aardappelen. **Research on varieties of
potatoes.** Publications.

6557 Tazelaar, M.F.   NL 010120/48/1314
Veredelingsonderzoek naar resistentie tegen Phytopthora
infestans bij aardappelen. **Breeding for resistance to
Phytophthora infestans in potatoes.** Publications.

6558 Wiersema, H.T.   NL 010120/50/1310 R
Veredelingsonderzoek naar resistentie tegen virusziekten bij
aardappelen. **Breeding research on resistance to virus diseases
of potatoes.** Publications.

6559 Bouma, W.F.   NL 010120/50/1319
De synthese en produktie van uitgangsmateriaal van
aardappelen voor de kweekbedrijven. **Synthesis and
production of initial potato material for plant breeding stations.**

6560 Huisman, C.A.   NL 010120/51/1312 R
Veredelingsonderzoek naar resistentie tegen
aardappelmoeheid. **Research on breeding for resistance to the
potatocyst nematode in potatoes.** Publications.

6561 Hermsen, J.G.T.   NL 010120/55/1320
Beheer en onderzoek van de Wageningse aardappelcollectie.
**Management and research of the Wageningen potato
collection.** Publications.

6562 Wiersema, H.T.   NL 010120/57/1311 R
Veredelingsonderzoek naar resistentie tegen schurft bij
aardappelen. **Breeding research on resistance to common scab
of potato.** Publications.

6563 Wiersma, H.T.   NL 010120/60/1313 R
Veredelingsonderzoek naar kwaliteitsbepalende faktoren bij
aardappelen. **Breeding research on factors governing the
qualities of the potato.** Publications.

6564 Suchtelen, N. van   NL 010120/62/1315
Onderzoek naar de mogelijkheden van het gebruik van
haploiden voor de aardappelveredeling. **Research on the
possibilities of haploids in potato breeding.** Publications.

6565 Hermsen, J.G.T.   NL 010120/65/1321
Inductie en toepassing haploïden bij aardappelen. **Induction
and application of haploids in potatoes.** Publications.

6566 Wagenvoort, M.   NL 010120/66/1340
Cytogenetisch onderzoek in aardappelen. **Cytogenetical
research in potatoes.** Publications.

6567 Maris, B.   NL 010120/67/1316
Aanpassingsmogelijkheid van Solanumsoorten aan het
Nederlandse cultuurmilieu. **Studies concerning the possibilities
of adaptation of Solanum–species to environmental conditions
in the Netherlands.**

6568 Maris, B.   NL 010120/67/1318
Onderzoek met haploïden naar de overerving van de
resistentie tegen de aardappelwratziekte. **Studies with potato
dihaploids on the inheritance of resistance to wart disease.**

6569 Wiersema, H.T.   NL 010120/69/2794 R
Veredelingsonderzoek op tolerantie voor mechanische
beschadigingen en resistentie tegen wondparasieten
(knolziekten) bij aardappelknollen. **Tolerance to mechanical
damage and resistance to tuber diseases in potato tubers.**
Publications.

6570 Wal, A.F. van der   NL 010120/77/8891 N
Toepassing voor droogteresistentie van landbouwgewassen, in
het bijzonder van de aardappel. **Screening methods for drought
resistance of arable crops particularly the potato.**

6571 Hermsen, J.G.T.   NL 020045/55/1320
Beheer en onderzoek van de Wageningse aardappelcollectie.
**Management and research of the Wageningen Potato
Collection.** Publications.

6572 Hermsen, J.G.T.   NL 020045/65/1321
Inductie en toepassing haploïden bij aardappelen. **Induction
and application of haploids in potatoes.**

6573 Harten, A.M. van   NL 020045/69/4518
Stralingsgevoeligheid, mutabiliteit en verwante onderwerpen
(inclusief histogenetische aspecten) bij aardappel (Solanum
tuberosum L.). **Radiation sensitivity, mutability and related
subjects (including histogenetic aspects) in potato (Solanum
tuberosum L.).** Publications.

6574 Hermsen, J.G.Th.   NL 020045/70/4854 R
Aneuploidie en genlocalisatie bij Solanum species. **Aneuploidy
and gen localisation in Solanum species.** Publications.

6575 Hermsen, J.G.T.   NL 020045/71/4850 R
Incompatibiliteit bij Solanum. **Incompatibility of Solanum.**
Publications.

6576 Hermsen, J.G.Th.   NL 020045/75/8848 N
Incongruentie en genoomverwantshap bij Solanum.
Incongruency and genome relationship in Solanum.

6577 Ramanna, M.S.   NL 020045/75/8849 N
Sexuele polyoidisatie bij Solanum. **Sexual polyoidisation in
Solanum.**

Sugarbeets and other sugar crops (B 3320)

See also 3254, 3270, 3279, 3285, 3295, 5494, 6219, 6570, 9531

6578 Kruse, A.   DK 030145/58/0002 N
Rene linier i bederoer. **Pure lines in beet.**

6579 Denizot, J.P.; Vincent, A.   FR 011003/73/0219
Production de formules hybrides balancées, triploïdes et

tétraploïdes par croisements industriels de lignées stables. **Using industrial crossings of stable lines for new triploid and tetraploid hybrids.** Publications.

6580 Denizot, J.P.; Vincent, A.   FR 011003/73/0328
Production de formules hybrides balancées, triploïdes et tétraploïdes par croisements industriels de lignées stables. **Balanced hybrids production, triploïdes and tetraploid, by stable lines industrial crossings.**

6581 Arnoux, M.   FR 011201/74/0214
Etude du comportement en milieu salé. **Study of "Provence– Cane" behaviour in salt euvironment.**

6582 Arnoux, M.   FR 011201/74/0222
Amélioration clonale et variétale. **Clonal selections among Provence – Cane varieties.** Publications.

6583 Arnold   GB 011904/00/0001 R
Breeding for yield quality and disease resistance.

6584 Arnold   GB 011904/00/0002 R
Systems and techniques for breeding.

6585 Brown; Cook   GB 011904/00/0011 R
Studies on resistance and tolerance to sugar beet pests and diseases.

6586 Fitzgerald, P.; O'Connor, L.   IE 060500/58/0220 R
Sugar beet cytological investigations. Publications.

6587 O'Connor, L.; Fitzgerald, P.   IE 060500/58/0222 R
Breeding high purity sugar beet varieties. Publications.

6588 O'Connor, L.; Fitzgerald, P.   IE 060500/58/0225 R
Breeding bolting resistant sugar beet varieties. Publications.

6589 O'Connor, L.; Fitzgerald, P.   IE 060500/58/0226 R
Field testing of commercial and experimental sugar beet varieties. Publications.

6590 Fitzgerald, P.; O'Connor, L.   IE 060500/59/0224 R
Breeding for resistance to peronospora farinosa. Publications.

6591 O'Connor, L.; Fitzgerald, P.   IE 060500/64/0227 N
Breeding polyploid (anisoploid) sugar beet varieties. Publications.

6592 Fitzgerald, P.; O'Connor, L.   IE 060500/65/0229 R
Breeding genetic monogerm sugar beet varieties (a) conventional breeding (b) hybrid breeding. Publications.

6593 Fitzgerald, P.; O'Connor, L.   IE 060500/69/0228 R
Breeding program for synthesis of pollinators for hybrid monogerm varieties. Publications.

6594 O'Connor, L.; Fitzgerald, P.   IE 060500/71/0223 R
Breeding winter sugar beet varieties. Publications.

6595 O'Connor, L.; Fitzgerald, P.   IE 060500/75/1336 R
International co–operative joint studies on 'sugar beet variety – environment interactions. Publications.

6596 Fitzgerald, P.; O'Connor, L.   IE 060500/77/1339 R
Recurrent selection in sugar beets. Publications.

6597 Rusconi Camerini, G.; De Biaggi, M.; Biancardi, E.; Ranalli, P.; Cerato, C.; Giordano, I.   IT 021100/75/0002 R
Miglioramento genetico della barbabietola da zucchero. **Sugar beet genetic improvement.** Publications.

6598 Rusconi Camerini, G.; Ranalli, P.; De Biaggi, M.
          IT 021100/79/0001 N
Studio citogenetico del cariotipo in BETA vulgaris. **Cytogenetic study of the karyotype in BETA vulgaris.**

6599 Rusconi Camerini, G.; Giordano, I.; Ranalli, P.; De Biaggi, M.   IT 021100/79/0002 N
Saggio agronomico di famiglie diploidi plurigermi di barbabietola da zucchero idonee alla semina autunnale. **Agronomic essay of diploid plurigerm families of sugar beet fit for the autumnal seeding.**

6600 Rusconi Camerini, G.; Ranalli, P.; De Biaggi, M.
          IT 021100/79/0003 N
Introduzione dei caratteri monogermia, maschiosterilità e reperimento del tipo "O" in altre linee di barbabietola da zucchero. **Introduction of the features: monogerm, male sterility and test of the type "O" in other lines of the sugar beets.**

6601 Rusconi Camerini, G.; Ranalli, P.; De Biaggi, M.
          IT 021100/79/0004 N
Conservazione in stato di purezza di stock maschiosterili monogermi con i relativi "O–Type". **Preservation in state of purity of male sterile monogerm stocks with the "O" types connected.**

6602 Rusconi Camerini, G.; De Biaggi, M.; Pirani, V.; Laureti, D.; Ranalli, P.   IT 021100/79/0005 N
Impiego delle linee maschiosterili monogermi per l'ottenimento di seme ibrido da saggiare in prove sperimentali di campo. **Use of the malesterile monogerm lines for the obtainment of ibrid seed to try by experiments of field.**

6603 Cremaschi, D.; Casarini, B.; Fontana, F.; Pirani, V.; Giordano, I.; Cerato, C.   IT 021100/79/0006 N
Confronto varietale di barbabietole da zucchero a semina primaverile, autunnale ed invernale. **Comparison of variety of sugar beet with spring, autumnal and winter seeding.**

6604 D'Ambra, V.   IT 062200/77/0860
Caratterizzazione di isolati di Cercospora bieticola Sacc. tolleranti il Benomyl. **Characterisation of Cercospora bieticola Sacc. colonies tolerant to Benomyl.**

6605 Bakker, J.J.; Baltjes, H.J.   NL 010113/43/0620
Rassenonderzoek suikerbieten. **Research on varieties of sugar–beets.** Publications.

6606 Cleij, G.   NL 010120/54/1328
Veredelingsonderzoek naar resistentie tegen de vergelingsziekte bij bieten. **Breeding research on virus yellows in beets.** Publications.

6607 Cleij, G.; Lekkerkerker, B.   NL 010120/56/1329
Verbreding van de genetische basis van de cultuurbiet d.m.v. soortkruisingen in het geslacht Beta. **Broadening of the genetic basis of the cultivated beet by means of species crosses in the genus Beta.** Publications.

6608 Cleij, G.   NL 010120/67/1330

D 2300 – Plant breeding

Onderzoek naar de factoren die de overgang van het vegetatieve naar het generatieve stadium in het geslacht Beta beheersen. **The study of factors governing the transition from the vegetative to the generative stage in the genus Beta.**

6609 Mesken, M. NL 010120/67/7417 R
Veredelingsonderzoek aan enkele kwaliteitseigenschappen van suikerbieten (suikergehalte, sapzuiverheid, wortelvorm, monogermie). **Breeding research on some quality aspects of sugar beet (sugar content, price purity, root shape, monogermity).** Publications.

6610 Mesken, M. NL 010120/69/7432 R
Onderzoek naar mogelijkheden van selektie op kophoogte in suikerbieten. **Research about possibilities of selection for crown height in sugar beets.** Publications.

6611 Cleij, G. NL 010120/70/3177
Onderzoek naar het voorkomen en de overerving van apomixie in het geslacht Beta. **Study on the occurrence and heredity in the genus Beta.** Publications.

6612 Speckman, G.J. NL 010120/74/6577 R
Cytogenetisch onderzoek aan soortbastaarden van bieten. **Cytogenetical studies on interspecific hybrids in beet.** Publications.

6613 Mesken, M.; Jongh de Leeuw, M.J. de; Dieleman, J. NL 010120/74/7418
Instandhouden van inteeltmateriaal en oude rassen van suiker- en voederbieten. **Maintenance of inbred material and old varieties of sugar and fodder beets.**

6614 Cleij, G. NL 010120/75/6694
Veredelingsonderzoek naar resistentie tegen het bietencystenaaltje (Heterodera schachtii) bij het geslacht Beta. **Research into breeding for resistance to the beeteelworm disease (Heterodera schachtii) in the genus Beta.** Publications.

6615 Mesken, M.; Jongh de Leeuw, M.J. de; Dieleman, J. NL 010120/75/7433
Onderzoek naar genotype–milieu–interaktie bij suikerbieten met betrekking tot opbrengstfaktoren. **Research on genotype–environment–interaction of yield factors in sugar beets.**

6616 Mesken, M. NL 010120/77/7877
Veredelingsonderzoek over droogtetolerantie bij bieten. **Breeding research on drought tolerance in beets.**

6617 Jorritsma, J. NL 060003/24/7711
Onderzoek naar de gebruikswaarde van suikerbietenrassen. **Research on cultural value of sugar beetvarieties.** Publications.

6618 Heijbroek, W. NL 060003/52/6694
Veredelingsonderzoek op resistentie tegen het bietecystenaaltje bij suikerbieten. **Breeding research for resistance to the beet cystnematode in sugar beet.** Publications.

Other starch producing plants (B 3390)

See also 6522, 6524, 17123

6619 Dambroth, M.; Heilinger, F. DE 201040/75/0011
Prüfung von Neuzuchtstämmen und Sorten auf Keimverhalten, Rohbreiverfärbung und Neigung zur Blauverfärbung des Knollenfleisches im Rahmen des Kartoffelschwerpunktes. **Testing of new populations and varieties of potatoes for germination behaviour, for changing of colour of the raw puree and for blueing of the tuber flesh.**

6620 Lange, N. DE 201040/75/0014
Durchführung von Kreuzungen schwierig kreuzbarer Wildarten bei Kartoffeln. **Implementation of crossbreedings of wild potato species difficult to cross.**

6621 Arnolin, R.; Degras, L.; Poitout FR 011601/74/0303
Sélection de l'igname cousse–couche (dioscorea trifida). **Yam selection (dioscorea trifida).** Publications.

Grasses and forage crops in general (B 3400)

See also 3327, 3344, 3364, 6291

6622 Simon, U.; Daniel, P. DE 129140/72/0002
Erarbeitung von Selektionsgrundlagen für die züchterische Verbesserung des Futterwertes. **Selection criteria for the genetic improvement of forage quality.**

6623 Scheller, H. DE 502058/77/0001 N
Untersuchungen über die Durchführung von Erhaltungszüchtung von synthetischen Sorten und Hybridsorten bei Futterpflanzen. **Studies on the realization of research on maintenance of synthetic varieties and hybrids of forage plants.**

6624 Niqueux, M. FR 010801/74/0366
Plantes fourragères pérennes pour ensilage. **Perennial forage plants for silage.**

6625 Dumas, Y.; Clairon, M.; Sobesky, O. Mme FR 011602/74/6031
Critères de choix rapide de fourrages tropicaux adaptés aux conditions locales. **Tests for quick choice of tropical forages adapted to local conditions.**

6626 Vincourt, P.; Gallais, A. FR 012223/74/0152
Vigueur des variétés synthétiques selon la génération, le nombre et l'origine des constituants. Interprétation et synthèse de nombreux résultats expérimentaux. **Vigour of the synthetic fodder plant varieties according to the generation, the number and origin of the components. Interpretation and synthesis of numerous experimental results.**

6627 Dulphy, J.P.; Huguet, L.; Demarquilly, C. FR 012223/76/0420
Etude préliminaire comparative sur l'appréciation des résultats de la sélection des fourrages pour la qualité par différents herbivores (ovins, bovins, caprins). **Preliminary comparative study of the evaluation of quality selection results in forage with differents herbivores (sheep, cattle, goats).**

6628 Huguet, L. FR 012223/76/0421
Utilisation de la chèvre et production pour mesurer l'efficacité de la sélection des fourrages pour la valeur alimentaire et définir des critères de sélection. **The use of milking goats to measure effectiveness of selection in forages for feed value and to define selection criteria.**

6629 Jones, D.I.H. GB 012102/00/0002 R
Research into analytical techniques for breeding and evaluation in forage grasses.

D 2300 – Plant breeding

6630 Eagles; Wilson   GB 012104/76/0022 R
Genetic variation in , and the effects of selecting for, photosynthetic activity in forage grasses and legumes.

6631 Wilson; Eagles   GB 012104/76/0023 R
Genetic variation in and the effects of selecting for respiratory activity in forage grasses and legumes.

6632 Eagles   GB 012104/76/0026 R
Genetic variation and selection for cold hardiness and growth at low temperature in grasses and clovers.

6633 Wilson   GB 012104/76/0027 R
Genetic variation in and effects of selecting for water use efficiency in grasses and clover.

6634 Briant   GB 060213/00/0006 R
Forage crop variety testing.

6635 Delforno, G.   IT 020900/73/0003
Caratteristiche qualitative di foraggere in selezione.
Qualitative characteristics of fodder plants under breeding.

6636 Parrini, P.   IT 040801/77/0379 R
Miglioramento della produzione foraggera mediante interventi genetici. **Improvement of fodder crop yield by genetical procedures.**

6637 Perrino, P.   IT 060500/77/0459
Raccolta moltiplicazione e valutazione di accessioni di piante foraggere. Messa a punto di metodi di moltiplicazione del seme. **Forage harvesting, multiplication and growth evaluation. Development of seed multiplication practices.**

6638 Panella, A.   IT 062000/70/0104
Miglioramento genetico delle piante foraggere. **Breeding of forage plants.** Publications.

Grasses (B 3410)

See also 3376, 3428, 3429, 3430, 3431, 3436, 3448, 3492, 3493, 5531, 5538, 5539, 5553, 6151, 6783, 6796, 6808, 6818, 6819, 7626, 9555, 20149

6639 Van Bogaert, G.   BE 070400/34/0001 R
Veredeling van nieuwe diploïde varieteiten van grassen en vlinderbloemigen voor het Belgisch cultuur milieu. **Breeding of new diploid varieties of grasses and papilionaceous flowers for the environmental conditions in Belgium.** Publications.

6640 De Roo, R.   BE 070400/55/0002 R
Veredeling van nieuwe polyploide varieteiten van grassen en klavers, die voor België van belang zijn. **Breeding of new polyploid grasses and clovers for the environmental conditions in Belgium.** Publications.

6641 Vyncke, A.; Vandepitte, H.   BE 070400/71/0015 R
Onderzoek naar de invloed op raszuiverheid van een stijgend aantal generaties bij de zaadvermeerdering van enkele R.v.P. cultivars. **Research on the influence of an increasing number of seed generations in a seed multiplication program on the varietal purity of R.v.P.**

6642 Vyncke, A.   BE 070400/76/0027 R
Zaadteeltproeven met Engels raaigras. **Seed yield experiments for Perennial ryegrass.**

6643 Vyncke, A.   BE 070400/78/0029 R
Zaadteeltproeven met Westerwolds raaigras. **Seed yield experiments for westerwolds ryegrass.**

6644 Boeker, P.; Opitz–von–Boberfeld, W.
DE 111251/75/0005
Nachsaat zur Narbenverbesserung mit verschiedenen Lolium perenne–Sorten. **Reseeding with different Lolium perenne varieties for sward improvement.**

6645 Simon, U.; Daniel, P.; Nowruzian, H.
DE 129140/72/0001
Vergleichende Untersuchungen der Verdaulichkeit von Gras– und Kleearten und –sorten in Abhängigkeit vom Entwicklungsstadium. **Comparative studies on the digestibility of species and varieties of grass and clover depending on the stage of growth.**

6646 Jacob, H.; Schulz, H.   DE 144500/78/0001 N
Prüfung verschiedener Sorten von Dactylis glomerata, Festuca pratensis, Festuca rubra, Lolium perenne, Phleum pratense, Poa pratensis auf Persistenz und Standorteignung. **Testing of several varieties of Dactylis glomerata, Festuca pratensis, Festuca rubra, Lolium perenne, Phleum pratense, Poa pratensis for persistence and suitability of site.**

6647 Schöllhorn, J.; Szokolai, P.; Müller, A.
DE 501200/70/0010
Vergleichende Untersuchungen an tetraploiden Sorten der Weidelgräser auf Ertragsleistung, Gehalt an Inhaltsstoffen und ihre Eignung zur Konservierung. **Comparing test on tetraploid species of Italien rye–grass with regard to the yield, the content of nutrients and the suitability for ensilaging.** Publications.

6648 Scheller, H.   DE 502058/73/0025 R
Prüfung von Sortimenten in– und ausländischer Sorten und Ökotypen bei Gräsern 1973. **Testing of assortements of domestic and foreign species and ecotypes of grasses.**

6649 Ziegenbein, G.   DE 506155/75/0003
Züchtung auf Virustoleranz bei Lolium perenne. **Breeding of Lolium perenne for virus tolerance.**

6650 Rasmussen, F.; Rasmussen, J.   DK 010112/74/4104
Sortsafprøvning med græsarter. **Variety trials of grasses.**

6651 Petersen, H.L.   DK 030145/75/0013
Konkurrenceevne og persistens mod alm. rajgræs. **Competition and persistency in perennial ryegrass.**

6652 Dennis, B.; Nielsen, H.M.; Johansen, B.R.; Petersen, H.L.; Lysgaard, C.P.   DK 030145/76/0015
Genotype–milieu vekselvirkning i alm. rajgræs ved forskellige plantebestande. **Genotype–environment interaction in perennial ryegrass at different plant densities.**

6653 Frandsen, K.J.   DK 030145/76/0016
Variation og nedarvning af fordøjelighed hos fodergræsser. **Variation and inheritance of digestibility in forage grasses.**

6654 Frandsen, K.J.   DK 030145/76/0017
Undersøgelser over top–cross metoden ved forædling af græs. **Studies of the use of the top–cross method in grass breeding.**

6655 Petersen, H.L.   DK 030145/77/0023
Nedarvning af persistens hos alm. rajgræs. **Inheritance of persistency in perennial ryegrass.**

6656 Andersen, S.   DK 030145/79/0001 N
Frøgivende egenskaber hos græsser. **Factors affecting seed yield in grasses.**

6657 Billot, C.; Chevallier, C.; Cousin   FR 010615/69/9032
Etude du comportement de variétés de graminées à gazon et des techniques culturales à leur appliquer dans le cadre d'implantation de gazons d'ornement et utilitaires. **Testing the behaviour of lawn grass varieties and the agricultural practices required for their use in decorative lawns and play grounds.**

6658 Billot, C.; Chevallier, C.; Bourgoin, B.
FR 010615/76/9031
Etude de la résistance au piétinement de variétés de graminées á gazon. **Evolution of the resistance to tramping of various lawn grass varieties.**

6659 Gallais, A.   FR 012223/68/0070
Prévision et utilisation de la vigueur hybride dans différentes structures variétales chez le dactyle. **Forecasting and using hybrid vigor in different types of cultivar in Cocksfoot.**

6660 Mansat, P.; Betin, M.   FR 012223/68/0071
Devenir des relations génétiques après tétraploïdisation artificielle de la fétuque des prés. **Genetic relations in artificial tetraploids in meadow fescue.**

6661 Gillet, M.; Robelin, M.; Sauvion, A.   FR 012223/72/0415
Etude d'un critère de sélection des graminées pour l'utilisation de l'azote: recherche des causes de leur mortalité par l'excès d'azote. **Study of a selection criterion of grasses for nitrogen utilization: research of the causes of their death by excess of nitrogen.**

6662 Gillet, M.; Jadas–Hecart, J.   FR 012223/73/0150
Sélection de fétuques élevées hybrides européennes multipliées par méditerranéennes. **Selection of hybrid weed fescues european mediterranean.** Publications.

6663 Bourgoin, B.; Mansat, P.   FR 012223/73/0175
Utilisation des croisements entre sous–espèces botaniques au sein de la fétuque rouge. **Use of crossings between subspecies of red fescue.**

6664 Bourgoin, B.; Mansat, P.   FR 012223/74/0407
Sélection de géniteurs tolérants au pietinement chez différentes espèces à gazons (fétuque rouge et ray–grass anglais). **Breeding for wear tolerances in differents turfgrass species (red fescue–perennial ray–grass).**

6665 Gillet, M.; Betin, M.   FR 012223/75/0409
Intérêt d'une sélection du ray–grass d'Italie pour une culture de très courte durée en dérobée. **The value of breeding italian rye–grass for very short term leys.**

6666 Bourgoin, B.; Mansat, P.   FR 012223/76/0408
Tolérance d'espèces et cultivars de gazons au piétinement réel de l'homme en comparaison avec le piétinement artificiel. **Wear tolerance of turfgrass species and cultivars against real trading and artificial one.**

6667 Wright; Webb   GB 011910/00/0001 R

Grass breeding variety maintenance.

6668 Lewis   GB 012103/00/0001 R
Cytogenetic studies of species relationships in Lolium–Festuca as basis for genetic exchange.

6669 Lewis   GB 012103/00/0002 R
The stabilisation of chromosome behaviour in novel synthetic grass species.

6670 Clarke   GB 012103/77/0007
The monitoring of chromosome behaviour in polyploid breeding material.

6671 Lewis; E J   GB 012103/77/0008
The synthesis of breeding material combining interspecific characters in the Lolium–Festuca complex.

6672 Rhodes   GB 012104/76/0018
Genetic variation in and effects of selecting for canopy characteristics in forage grasses.

6673 Rhodes; Davies   GB 012104/76/0019
Effects of genetic variation in morphology of grasses and clovers on growth of grass/clover associations.

6674 Eagles   GB 012104/76/0024 R
Genetic variation in growth and development of grasses and clovers in response to light and temperature.

6675 Goodman   GB 012104/76/0028
Genetic variation in and effects of selecting for efficient use of mineral nutrients in forage grasses.

6676 Goodman   GB 012104/76/0029
Genetic variation in, and effects of selecting for, efficient use of N, P, K, and Mg in forage legumes.

6677 Wilson   GB 012104/76/0030
Effects of selecting for anatomical, physical and chemical components of nutritive value in forage grasses.

6678 Wilson   GB 012104/76/0031
Development of experimental populations of forage grasses and legumes using physiological indices.

6679 Munro; Davies   GB 012105/76/0031 R
Evaluate grass varieties and potential breeders material for hill and upland requirements.

6680 Hayward   GB 012106/00/0001
Efficiency of selection techniques for outbreeding grasses and genetic assessment of breeding material.

6681 Breese; Hayward   GB 012106/00/0003 R
Genetic control and consequences of sexual and asexual reproduction in ryegrass populations.

6682 Tyler   GB 012106/00/0008
Collect material to extend genetic resources in herbage plants. **Initial screening.**

6683 Borrill   GB 012106/00/0010
Race and species relationships in Lolium and Festuca; stabilise hybrid combination by al opolyploidy selection.

6684 Lewis; EJ Borrill   GB 012106/00/0011
Race and species relationships in dactylis; scope for genetic
exchange within and between ploidy levels.

6685 Borrill   GB 012106/00/0012
Improve digestibility and growth by wide crossing in Dactylis.

6686 Lewis; EJ Tyler   GB 012106/00/0013
Develop intergeneric hybrids of ryegrass and fescue.

6687 Hides   GB 012106/00/0014 R
Genetic and physiologic control of seed quality in Lolium
multiflorum.

6688 Hides   GB 012106/00/0015
Breed Lolium multiflorum for productivity, quality and winter
hardiness.

6689 Charles   GB 012106/00/0016
Breed Lolium perenne for winter hardiness. Effects of
management and physiological control.

6690 Charles   GB 012106/00/0017
Select for persistence of Lolium perenne under high stock
density and high N application.

6691 Breese; Thomas AC   GB 012106/00/0018
Select for factors affecting nutritive value in herbage plants.

6692 Stephens; Breese   GB 012106/00/0019
Produce and develop hybrid ryegrass.

6693 Davies; WE Hill   GB 012106/00/0024
Breed red clover for yield, persistance and resistance to disease,
especially Sclerotinia trifoliorum.

6694 Dale; Breese Jones   GB 012106/76/0028
Application of tissue culture techniques to herbage plant
breeding.

6695 Dale   GB 012106/77/0029
Research and development of tissue culture techniques for plant
breeding especially of herbage species.

6696 Humphreys   GB 012106/78/0030
Breeding grasses for amenity purposes.

6697 Lewis; Bean, J.   GB 012109/00/0009
Select for improved seed production in grasses.

6698 Griffiths; Pegler   GB 012109/00/0010
Select for seed setting in polyploids and hybrids of Lolium and
Festuca.

6699 Griffiths; Pegler   GB 012109/00/0011 R
Determine extent of crossing in hybrids of Lolium, Festuca etc.

6700 Jones TWA   GB 012110/77/0010
Operation and development of isoenzyme screening facilities.

6701 Faulkner   GB 040302/00/0015
Perennial ryegrass breeding.

6702 Faulkner   GB 040302/00/0016
Breeding for herbicidal tolerance in grasses.

6703 Faulkner   GB 040302/00/0017
Breeding new cultivars of tall fescue.

6704 Spoor   GB 060105/00/0002
Breeding systems in grasses.

6705 Morrison, M.   GB 060106/00/0001 R
Grass variety trials (except NLT).

6706 Ryan, M.J.; Diamond, J.J.; Sugrue, J.J.
                IE 060200/76/1293 R
Evaluation of ryegrass cultivars. Publications.

6707 Kavanagh, T.; Burke, W.   IE 060300/73/0870 R
Turfgrass; cultural, management and disease problems in
Ireland. Publications.

6708 Crowley, J.   IE 060500/66/0213 R
Italian ryegrass breeding. Publications.

6709 Ribeiro, Do Valle, M.D.V.   IE 060500/68/0199 N
Breeding of tall fescue and ryegrass x tall fescue hybrids.
Publications.

6710 Crowley, J.   IE 060500/68/0212 R
Perennial ryegrass breeding. Hybridization programme
between north African ecotypes and European varieties.
Publications.

6711 Ahloowalia, B.S.   IE 060500/70/0202 R
Genetic studies on interspecific hybrids of lolium. Selection
within tetraploid perennial x italian ryegrass populations.
Publications.

6712 Ahloowalia, B.S.   IE 060500/70/0204 N
Cytogenetic studies on aneuploids of perennial ryegrass.
Publications.

6713 Ahloowalia, B.S.   IE 060500/71/0203 R
Tetraploid perennial ryegrass: selection in the fourth generation
tetraploid progenies and multiplication of new strains.
Publications.

6714 Crowley, J.; Connolly, V.; Ribeiro, Do Valle, M.D.V.
                IE 060500/71/0214 R
Evaluation of herbage varieties, species and mixtures under
cutting systems using small plots. Publications.

6715 Iannelli, P.   IT 020900/68/0001
Miglioramento genetico di specie da erbaio. Breeding of fodder
catch crops. Publications.

6716 Rotili, P.   IT 020900/78/0002
Costituzione di varietà di Dactylis glomerata altamente
produttive nella stagione estiva. Constitution of Dactylis
glomerata varieties highly yielding in summer.

6717 Longo, G.   IT 040305/72/0492
Miglioramento genetico delle piante erbacee coltivate. Genetic
improvement of cultivated grass plants. Publications.

6718 Longo, G.   IT 040305/77/0368 R
Miglioramento genetico foraggere, dactylis glomerata. Genetic
improvement of the fodder crop: dactylis glomerata.

6719 Ghisleni, P.L.   IT 040619/77/0356 R

Miglioramento genetico di lolium multiflorum L. **Genetic improvement of lolium multiflorum L.**

6720 Parrini, P.   IT 040801/72/0528
Ricerche sul miglioramento genetico delle piante erbacee. **Research on the genetic improvement of grass plants.** Publications.

6721 Panella, A.; Ceccarelli, S.; Lorenzetti, F.; Falcinelli, M.
                    IT 041002/73/0003
Valutazione di graminacee foraggere. **Evaluation of grass varieties.** Publications.

6722 Moschini, E.   IT 041115/72/0521
Miglioramento genetico delle piante erbacee. **Genetic improvement of grass plants.** Publications.

6723 Bakker, J.J.; Duyvendak, R.   NL 010113/43/0621
Rassenonderzoek van grassen in verband met voeder–en recreatieve doeleinden. **Research on varieties of grasses in relation to forage and recreational use.** Publications.

6724 Dijkstra, J.   NL 010120/55/7424 R
Veredelingsonderzoek aan Engels raaigrasbeemdlangbloem bastaarden voor voederdoeleinden. **Breeding research in ryegrass–meadow fescue hybrids for forage.** Publications.

6725 Dijkstra, J.   NL 010120/59/7423 R
Veredelingsonderzoek aan raaigras–rietzwenkgras–bastaarden voor voederdoeleinden. **Breeding research in ryegrass–tallfescue hybrids for forage.** Publications.

6726 Dijk, G.E. van   NL 010120/65/7420 R
Bestudering van soortkruising als middel tot de veredeling van de apomictische soort veldbeemgras. **Interspecific crosses as a tool in breeding Poa pratensis.** Publications.

6727 Dijk, G.E. van   NL 010120/66/7419 R
Bestudering van methoden voor het veredelen op opbrengst en oogstzekerheid bij Engels raaigras. **A study of the methods of breeding Lolium perenne for productivity and reliability of yield.** Publications.

6728 Dijk, G.E. van   NL 010120/66/7422 R
Veredelingsonderzoek naar resistentie tegen ziekten bij grassen. **Breeding research for disease resistance in grasses.**

6729 Speckmann, G.J.   NL 010120/68/1341 R
Cytogenetisch onderzoek aan geslachtsbastaarden in grassen. **Cytogenetical research on intergenetic hybrids in grasses.**

6730 Dijk, G.E. van   NL 010120/72/7421
Veredelingsonderzoek ten behoeve van de zaadteelt van veldbeemdgras. **Breeding research on the seed growing of smooth stalked meadowgrass.** Publications.

Pastures, grassland (B 3420)

See also 3603, 3608, 6660, 6723

6731 Gillet, M.; Huguet, L.   FR 012223/72/0417
Etude des problèms posés par le choix et l'exploitation d'une chaine simplifiee de production d'herbe verte au printemps. Conséquences pour les critères de sélection. **Study of problems posed by the choice and management of a simplified series of spring pastures. Implications for selection criteria for**

grasses.

6732 Weddell; Copeman   GB 060214/00/0004 R
National list trial variety testing–herbage.

6733 Weddell; Copeman   GB 060214/00/0006 R
Turf cultivar testing.

6734 Connolly, V.; Murphy, W.E.; Crowley, J.; Leonard,
T.F.   IE 060500/72/0211
Grass variety evaluation under grazing in terms of animal productivity. Publications.

6735 Haussmann, G.   IT 020900/69/0003
Miglioramento e utilizzazione dei pascoli montani. **Improvement and exploitation of mountain pastures.** Publications.

6736 Bezzi, A.; Castellani, C.   IT 021800/79/0003 N
Ricerca di cultivar e consociazioni adatte alla ricostruzione di cotiche da destinare al pascolo. **Study on cultivars and plant communities suitable to range reconstitution.**

Mangolds (B 3430)

See also 3295, 6606, 6608, 6611, 6612, 6613, 6614

6737 Rousseau, M.   BE 070400/60/0006 R
Kweken van mechanisch oogstbare monogerme voederbieten. **Breeding of monogerm fodder beets and their suitability of mechanical harvesting.**

6738 Rasmussen, P.; Petersen, K.E.   DK 010106/62/0017
Sortsforsøg med rodfrugter. **Variety trials with root crops.**

6739 Le Cochec, F.   FR 011301/67/0036
Obtention d'hybrides F_1 monogermes de betteraves fourragères. **Breeding monogerm F_1 hybrid fodder beet.**

6740 Bakker, J.J.; Baltjes, H.J.   NL 010113/43/0619
Rassenonderzoek voedergewassen en groenbemestingsgewassen. **Research on varieties of forage crops.** Publications.

6741 Vries, A.P. de   NL 020045/74/6288 R
Directe versus indirecte selectie bij veldboon. **Direct in respect of indirect selection in field bean.**

Legumes in general (B 3440)

See also 6639, 6640, 6645, 6740, 6817

6742 Rasmussen, F.; Rasmussen, J.   DK 010112/74/4102
Sortsafprøvning med kløver, lucerne mv.. **Variety trials of herbage legumes.**

6743 Rosenstand, A.   DK 030101/72/0007
Forsøg på udvikling af metoder til hybridisering af bælg–sædarter inden for tribus Fabeae, Fabaceae.
Experimental development of methods for the hybridisation of Tribus fabeae, Fabaceae.

6744 Guldager, P.   DK 030101/75/0002
Genetiske undersøgelser af proteinindhold hos bælgplanter. **Genetic investigation of the protein content of leguminous plants.**

D 2300 – Plant breeding

6745 Clavier, C.; Angevain, M.; Lacombe, P.
FR 011201/74/0467
Recherche de nouvelles sources de protéines chez les légumineuses. **Investigation of high protein level plants in legumes.**

6746 Lawes GB 012101/00/0013 R
Breeding high yielding, self fertile lines of field beans.

6747 Connolly, V. IE 060500/67/0209 R
Variety evaluation, legumes. Publications.

6748 Ropelate, A. IT 021800/76/0005 N
Selezione di ecotipi di piante foraggere.. **Selection of ecological types of forager plants.**

Grassland legumes (B 3441)

See also 3654, 6096, 6818, 7869, 9573

6749 Dennis, B.; Nielsen, H.M.; Johansen, B.R.; Petersen, H.L.; Lysgaard, C.P. DK 030145/76/0014
Udbyttepotentiel og konkurrenceevne hos tetraploid rødkløver. **Yield potential and competitive ability in tetraploid red clover.**

6750 Holm, S.N.; Mortensen, G. DK 030145/77/0025 N
Selektion for nektarproduktion i rødkløver. **Selection for nectar production in red clover.**

6751 Niqueux, M.; Arnaud, R. FR 010801/74/0133
Sélection d'une variété de fléole très tardive. **Breeding a very late common timothy variety.**

6752 Picard, J. FR 011003/70/0074
Obtention de variétés synthétiques tétraploides de trèfle violet. **Breeding synthetic tetraploid varieties of red clover.**

6753 Clavier, C.; Angevain, M.; Lacombe, P.
FR 011201/72/0208
Recherche de géniteurs résistants à Colletotrichum Trifolii chez la luzerne. **Search for resistance to Colletotrichum Trifolii in alfalfa.**

6754 Clavier, C.; Angevain, M.; Lacombe, P.
FR 011201/73/0207
Création de variétés améliorées de luzerne pour la zone méditerranéenne. **Breeding alfafa varieties for the mediterrannean.**

6755 Clavier, C.; Angevain, M.; Lacombe, P.
FR 011201/75/0468
Maîtrise de la culture du fenu–grec (Trigonella Poenum–Graecum l.) et recherche des meilleurs cultivars. **Studies on the fenu greck (Trigonella Poenum–Graecum l.) production and investigation of the best cultivars.**

6756 Guy, P.; Genier, G. FR 012223/68/0072
Sélection d'une variété de luzerne résistante au verticillium, recherche de gènes de résistance. **Breeding verticillium resistant varieties of lucerne.**

6757 Gallais, A.; Bertholleau, J–C. FR 012223/69/0069
Facteurs de la variabilité de la vigueur hybride chez la luzerne. **Studies on hybrid vigor in lucerne.**

6758 Guy, P.; Genier, G. FR 012223/70/0153
Recherche et utilisation de la stérilité mâle cytoplasmique pour l'amelioration de la luzerne. **Research and utilization of the cytoplasmic male sterility for improving lucerne.**

6759 Guy, P.; Genier, G. FR 012223/73/0145
Sélection d'une variété de luzerne résistante à la fois au verticillium (recherches de gènes de résistance 71.093) et au colletotrichum (complément 1973). **Selection of a lucerne variety resistant to verticillium (researches of 71.093 resistance genes) as well as to colletotrichum (complement 1973).**

6760 Guy, P.; Genier, G.; Gondran, J. FR 012223/73/0400
Déterminisme génétique de la résistance au colletotrichum de la luzerne. Relations phénotype – génotype. **Inheritance for tolerance to colletotrichum trifolii in medicago sativa correspondance between genotype and phenotype.**

6761 Guy, P.; Ahmim, M.; Genier, G.; Vuillaume, E.
FR 012223/75/0403
Dynamisme de l'évolution et du maintien de la variabilité dans des populations naturelles et artificielles du genre médicago. ATP CNRS 1892. **Dynamics of the evolution and maintenance of variability in natural and artificial populations of the genus medicago.**

6762 Genier, G.; Caubel, G.; Guy, P.; Gondran, J.
FR 012223/76/0401
Sélection d'une variété de luzerne résistante à de multiples parasites et accidents: verticillium, colletotrichum, mématode, et recherche de géniteurs résistants au nematode, ditylenchus dipsaci. **Selection of a lucerne variety simultaneously resistant to lodging and several pathogens; verticillium, colletotrichum, nematodes.**

6763 Rogers; Aubury GB 011910/00/0004 R
Breed for yield and quality and for resistance to Verticillium and bacterial wilt in lucerne.

6764 Toynbee–Clarke GB 011910/00/0009 R
Breed red clover for forage yield, eelworm stem rot resistance, quality components; tetraploid seed yield.

6765 Wright GB 011910/78/0020 R
Population improvement in red clover.

6766 Munro; Young GB 012105/76/0032 R
Evaluate legume varieties and potential breeders material for hill and upland requirements.

6767 Mytton GB 012106/00/0022 R
Select for improved symbiosis between rhizobium and white clover.

6768 Davis, W.E.; Evans, D. GB 012106/00/0023 R
Breed white clover for yield, length of season and tolerance to high nitrogen regimes.

6769 Davis, W.E.; Hill GB 012106/00/0026 R
Breed lucerne for yield, quality, creeping root and resistance to Verticillium, eelworm and other diseases.

6770 Mackie; Weddell GB 060214/00/0009 R
Development of tetraploid red clover.

D 2300 – Plant breeding

6771 Connolly, V.   IE 060500/75/1198 R
White clover breeding. Publications.

6772 Rotili, P.   IT 020900/72/0551
Miglioramento genetico dell'erba medica. Studio delle possibilità produttive degli ibridi doppi. **Genetic improvement of lucerne. Establishment of synthetics with self–fertilized material. Verification of the validity of the polycross–test in the genetic improvement of lucerne.** Publications.

6773 Zannone, L.   IT 020900/76/0003 N
Comportamento di varietà ed ecotipi di medica sottoposti a diversi regimi di taglio. Studio di caratteri quantitativi e qualitativi. **Performance of lucerne varieties and ecotypes under different cutting systems. Analysis of quantitative and qualitative characters.**

6774 Rotili, P.   IT 020900/77/0001 N
Costituzione di sintetiche di medica a più alto contenuto proteico e con più basso contenuto in saponine.. **Constitution of synthetic varieties of lucerne having a higher protein and lower saponin content..**

6775 Iannelli, P.   IT 020900/77/0362 R
Miglioramento genetico della veccia, selezione genealogiche e massali. **Genetic improvement of the vetch, genealogical and mass selection.**

6776 Rotili, P.   IT 020900/78/0001
Costituzione di varietà di medica da usare nell'industria di disidratazione e nella produzione di leaf protein. **Constitution of lucerne varieties to be used for dehydration and leaf protein production.** Publications.

6777 Pacucci, G.   IT 040101/77/0378 R
Costituzione di nuove varietà di veccia. **Development of new varieties of vetch.**

6778 Sarno, R.   IT 040904/77/0397 R
Costituzione di nuove varietà di sulla. **Breeding new varieties of Hedysarum coronarium L..**

6779 Ribaldi, M.   IT 041026/77/0389
Miglioramento dell'erba medica per la resistenza alla fusariosi. **Improvement of lucerne resistance to Fusarium wilt.**

6780 Pardini, G.   IT 041101/77/0255 R
Miglioramento genetico della sulla. **Genetic improvement of Hedysarum coronarium L.**

6781 Lorenzetti, F.   IT 062000/77/0456
Costituzione di varietà di Sulla (Hedysarium coronarium L.) e messa a punto dei metodi per la loro moltiplicazione. **Development of new Hedysarium coronarium L. varieties and of methods for their multiplication.**

6782 Lorenzetti, F.   IT 062000/77/0457
Costituzione di sintetiche di erba medica resistenti alla "fusariosi". **Development of Fusarium resistant synthetic varieties of lucern.**

6783 Lorenzetti, F.   IT 062000/77/0458
Costituzione e moltiplicazione di erba medica, Lolium perenne, Dactylis glomerata, Festuca arundinacea. **Development and multiplication of lucern, Lolium perenne, Dactylis glomerata, Festuca arundinacea.**

6784 Panella, A.   IT 062000/77/0851
Miglioramento genetico dell'erba medica. **Genetic improvement of lucerne.**

Other legumes (B 3449)

See also 5435, 7082, 7089

6785 Röbbelen, G.; Frauen, M.; Paul, C.; Thies, W.   DE 132182/73/0002
Auslese methioninreicher Mutanten in Ackerbohnen, Vicia faba. **Selection of mutants rich in methionine in horse bean – Vicia faba –.**

6786 Steiner, A.M.; Käser, H.   DE 144411/75/0001
Untersuchungen zur Echtheitsbestimmung bei Saatgut von Vicia faba L.. **Studies on the verification of species and/or cultivar in Vicia faba L..**

6787 Lysgaard, C.P.   DK 030145/78/0002 N
Virkningen af selektion i lucerne for tolerance over for lavt pH i jorden. **The effect of selection in lucerne on tolerance to low soil–pH.**

6788 Laissus, R.; Leconte, D.   FR 010115/74/0134
Etude de la variation de la composition de diverses variétés de pois fourragers entre le stade début floraison et la grainaison en vue de l'utilisation du pois sous forme d'ensilage plante entière. **Composition variations in various field pea varieties between the "beginning of flowering" stage and the seminiferous stage, seen in the light of using the plast as a whole as a form of tillage.**

6789 Picard, J.   FR 011003/73/0136
Obtention de féveroles sans tannins. **Selection of tannin–free horse beans.**

6790 Berthaut, J.; Sixdenier, G.; Sigwalt, C.   FR 011003/73/0137
Recherche et création de variétés protéagineuses de lupin. **Research and selection of protein–yielding lupine varieties.**

6791 Berthelem, P.; Le Guen, J.   FR 011301/73/0154
Etude fondamentale de la stérilité mâle chez la féverole. **Fundamental study of male sterility in horse beans.**

6792 Lenoble, M.; Papineau, J.   FR 012223/73/0144
Amélioration de la productivité en graines et de la teneur en protéines des graines de lupins annuels: lupinus albus, lupinus luteus, lupinus angustifolius. **Improvement of grain productivity and protein content of grains in annual lupines: lupinus albus, l.lutens, l.angustifolius.**

6793 Lenoble, M.; Papineau, J.   FR 012223/74/0424
Sélection de L. mutabilis pour l'absence d'alcaloïdes. **Breeding sweet lines of L. mutabilis.**

6794 Bond; Lockwood   GB 011910/00/0008 R
Study of plant models; inheritance of nutritional quality; inter and intraspecific crosses in field beans.

6795 Walsh, E.J.   IE 120105/75/9044 R
Breeding field bean (vicia faba) cultivars with improved yield and other agronomic characters for Irish conditions.

6796 Rotili, P. IT 020900/77/0394 R
Miglioramento genetico della veccia e della Festuca arundinacea. **Genetic improvement of Vicia and of Festuca arundinacea.**

6797 Stringi, L. IT 040904/77/0406 R
Miglioramento della fava da granella per uso zootecnico. **Improvement of field beans for fodder crops.**

6798 Porceddu, E. IT 060500/77/0982
Studi di diversificazione in Vicia faba. **Studies on Vicia faba diversification.**

6799 Heringa, R.J. NL 010120/75/6695
Veredelingsonderzoek in veldbonen (Vicia faba) ter verhoging van de kwantiteit en de kwaliteit van het zaad en van het eiwit in het zaad. **Breeding research on fieldbeans (Vicia faba) to improve the quantity and quality of seed and the protein of seed.** Publications.

6800 Heringa, R.J. NL 010120/75/6697
Veredelingsonderzoek in lupinen (lupinus luteus en lupinus albus) ter verhoging van de kwantiteit en de kwaliteit van het zaad en van het eiwit in het zaad. **Breeding research on improvement of quantity and quality of seedproduction and of protein quality of the seed in lupins (lupinus luteus and lupinus albus).**

6801 Heringa, R.J. NL 010120/76/6696
Veredelingsonderzoek naar de mogelijkheden tot verbetering van de invloed van stikstofbindende bacteriën (Rhizobium) op de zaadopbrengst bij veldbonen (Vicia faba). **Breeding research on the possibilities for improvement of the influence of fixing bacteria (Rhizobium) on the seed production of field beans (Vicia faba).** Publications.

6802 Heringa, R.J. NL 010120/76/6698
Veredelingsonderzoek naar mogelijkheden tot verbetering van de invloed van stikstofbindende bacteriën (Rhizobium) op de zaadopbrengst bij lupine (Lupinus luteus en Lupinus albus). **Breeding research on the possibilities for improvement of the influence of fixing bacteria (Rhizobium) on the seed production of lupins (Lupinus luteus and Lupinus albus).**

Cereals used for forage (B 3450)

See also 6110, 6148, 6149, 6206, 6211, 6267, 6271, 6351, 6352, 6353, 6661, 6740, 6748

6803 Schön, W.J.; Röbbelen, G. DE 132182/70/0006
Biochemische Untersuchungen und züchterische Selektion auf Qualität von Getreideproteinen und deren physiologische Bedeutung. **Biochemical investigations of the quality of barley proteins and their physiological significance.**

6804 Ulonska, E.; Baumer, M. DE 502058/73/0014
Verbesserung der Proteinmenge und Proteinqualität bei Futtergerste 1973. **Increase in content and quality of protein in fodder barley.**

6805 Laissus, R.; Leau, G. FR 010115/74/0266
Etude de la valeur fourragère sous climat océanique de différents bromes d'origine étrangère. **Study of the forage value of various foreign bromines under the oceanic climate.**

6806 Rautou, S.; Feillet, P. FR 011201/70/0021

Amélioration de la teneur en protéines du grain de maïs destiné à l'alimentation des ruminants. **Breeding for protein content in maize grain for feeding of ruminants.**

6807 Gallais, A.; Picard, J.; Pollacsek; Demarquilly, C.; Barloy, J. FR 012223/72/0151
Création et étude de la valeur alimentaire d'hybrides de maïs pauvre en lignine. **Selection of ligning–lacking maize hybrids and study of their nutritive value.**

6808 Salette, J.; Gillet, M. FR 012223/73/0143
Mise au point de critères de sélection de graminées pour l'utilisation de l'azote. **Development of criteria for grasses breeding for nitrogen consumption.**

6809 Gallais, A.; Pollacsek, M.; Bertholeau, J.C.; Derieux, M.; Berthet, H.; Panouille, A. FR 012223/73/0404
Recherche de variabilité génétique pour des critères de valeur alimentaire du maïs fourrage, plante entière. **Study of genetic variability for some traits influencing the feeding value of forage maize.**

6810 Lenoble, M. FR 012223/74/0299
Utilisation de la polyploïdie pour la sélection des sorghos fourragers. **Use of the polyploidy for the foddersorghums selection.**

6811 Gallais, A.; Mousset, C. FR 012223/74/0402
Efficacité de quelques méthodes de sélection récurente pour l'amélioration du rendement en matière séche de variétés synthétiques chez le dactyle. **Efficiency of some recurrent selection schemes for the yield of synthetic varieties of cocksfoot.**

6812 Lanza, F.; Venezian, M.E.; Losavio, N. IT 020500/79/0005 N
Adattamento e potenzialità produttiva di ibridi di sorgo nell'ambiente meridionale in funzione dell'epoca e dell'investimento di semina. **Influence of seedingtime and seed density on adaptation and yielding of sorghum hybrids in southern Italy.**

6813 Salamini, F.; Maggiore, T.; Gentinetta, E.; Motto, M. IT 020800/79/0003 N
Mais – Sviluppo di tipi adatti alla coltura per trinciati. **Maize– Breeding for silage.**

6814 Onofrii, M.; Tomasoni, C.; Basta, P. IT 020900/79/0001 N
Prove di confronto varietale, epoca, densità di semina, e concimazione in avena ed orzo da foraggio. **Trials on fodder barley and oat; comparison of varieties, dates and density of sowing, fertilizations.**

6815 Iannelli, P.; Rossi, O.; Franco, V. IT 020900/79/0010 N
Cereali minori: Orzo. Prova agronomica, accertamento produttività di cv. sottoposte a diverse modalità di utilizzazione. **Secondary fodder cereals: Barley. Agronomic trial, evaluation of productivity in cultivars differently utilized.**

6816 Ottaviano, E. IT 040615/77/0252 R
Miglioramento genetico del mais da foraggio e da granella. **Genetic improvement of fodder maize and grain maize.**

6817 Perrino, P. IT 060500/71/0171
Esplorazione e raccolta di graminacee foraggere, leguminose

coltivate e spontanee, e varietà obsolete di piante ortensi e di specie selvatiche del genere Brassica in Italia. **Exploration and collection of grasses, legums and obsolete garden varieties and wild species of genius–brassica in Italy.** Publications.

6818 Panella, A.   IT 062000/77/0850
Moltiplicazione dei tipi migliorati di graminacee foraggere e erba medica. **Multiplication in improved types of cereal fodder and lucerne.**

6819 Panella, A.   IT 062000/77/0852
Miglioramento genetico delle graminacee foraggere. **Genetic improvement of fodder cereals.**

6820 Dolstra, O.   NL 010120/72/4378 R
Veredelingsonderzoek naar koudetolerantie, voederwaarde en ziekteresistentie bij snijmais. **Breeding resistance for cold tolerance, fodder value and disease resistance in forage mais.** Publications.

6821 Balkema–Boomstra, A.G. van   NL 010120/75/7416
Veredelingsonderzoek naar eiwitkwaliteit– en–kwantiteit, kortstro en vroegheid bij haver. **Oat breeding research about protein– quality and –quantity short straw and earliness.**

Turnips (B 3460)

See also 6578, 6740

6822 Rousseau, M.   BE 070400/71/0004 R
Selektie van rapen op resistentie tegen verschillende fysio's van plasmodiophora Brassicae. **Breeding of turnips to introduce resistance against some physios of Plasmodiophora Brassicae.**

6823 Van Hee, L.   BE 070400/72/0005 R
Selektie van groene bladkool. **Breeding of fodder rape.**

6824 Petersen, H.L.   DK 030145/73/0010
Indavl af kålroer. **Inbreeding in swedes.**

6825 McNaughton   GB 030803/78/0012 R
Breed turnip cultivars especially for Scottish uplands.

6826 Dekhuijzen, H.M.   NL 010102/70/7987
Kwantitatieve analyse van endogene groeiregulatoren in wortels en callusweefsel van gewassen in verband met ziekte–aantastingen. **Quantitative analysis of endogenous growth regulators in roots and callus tissue of crops in relation to plant diseases.** Publications.

6827 Toxopeus, H.   NL 010120/52/7428 R
Veredelingsonderzoek naar resistentie tegen knolvoet (Plasmodiophora brassicae Wor.) bij Cruciferen. **Research into breeding for resistance to clubroot disease (Plasmodiophora brassicae Wor.) in Cruciferae.** Publications.

6828 Toxopeus, H.   NL 010120/72/7427 R
Analyse en de studie van de vererving van factoren die de drogestofopbrengst bepalen van kruisbloemige stoppelgewassen. **Analysis and study of the inheritance of factors determining total dry matter yield of Cruciforous stubble crops.** Publications.

6829 Toxopeus, H.   NL 010120/73/7429
Instandhouding en uitbreiding van de werkcollectie kruisbloemigen en de genebank van stoppelknollen en consumptieraapjes. **Maintenance and expansion of the Collection of Cruciferous crops and the turnip gene–bank.** Publications.

6830 Dolstra, O.   NL 010120/74/6578
Cytogenetisch onderzoek aan soort– en geslachtsbastaarden van landbouwcruciferen. **Cytogenetical research on interspecific and intergeneric hybrids in cruciferous crops.** Publications.

6831 Toxopeus, H.   NL 010120/74/7426 R
Genetische maximalisatie van snelle jeugdontwikkeling bij late zaai van kruisbloemige groenbemesters. **Genetic maximalisation of rapid juvenile development after late sowing of cruciferous green manure crops.** Publications.

6832 Toxopeus, H.   NL 010120/74/7430
Veredelingsonderzoek op resistentie tegen het bietencystenaaltje (Heterodera schachtii) binnen de familie Cruciferae. **Research into breeding for resistance to the beeteelworm disease (Heterodera schachtii) in the family Cruciferae.** Publications.

6833 Heijbroek, W.   NL 060003/74/7430
Veredelingsonderzoek op resistentie tegen het bietecystenaaltje (Heterodera schachtii) binnen de familie Cruciferae. **Research into breeding for resistance to the beet eelworm disease (Heterodera schachtii) in the family Cruciferae.** Publications.

Other forage crops (B 3490)

See also 6740, 6826, 6827, 6828, 6829, 6830, 6831, 6832, 6833

6834 Rasmussen, P.; Pedersen, K.E.   DK 010106/72/3503
Sortsafprøvning med majs og andre grønfoderarter. **Variety trials of maize and other forage plants.**

6835 Rasmussen, P.; Flengmark, P.   DK 010106/73/0008
Sortsforsøg i ærter, hestebønner og vikker. **Variety trials with peas, field beans and wetch.**

6836 Du Crehu, G.   FR 011301/61/0037
Création de variétés hybrides doubles de choux fourragers. **Breeding double hybrid varieties of fodderkale.**

6837 Taylor   GB 011910/00/0018 R
Breeding marrow–stem kale.

6838 Johnston   GB 012101/00/0014 R
Breed leafy kales, rape, swede and radish for livestock use in situ.

6839 Johnston   GB 012101/00/0015 R
Improve methods of breeding forage brassicae by study of genetics and growth; yield and quality.

6840 McNaughton   GB 030803/78/0001 R
Exploit interspecific and intergeneric crosses as sources of variation for Brassica and radicole breeding.

6841 Allison; McNaughton   GB 030803/78/0002 R
Develop and apply screening tests for useful and harmful biochemical components in brassicas and related spp.

6842 McNaughton   GB 030803/78/0003 R

Collect, assess and maintain genetic material of use to brassica breeders.

6843 McNaughton GB 030803/78/0004 R
Agronomic,physiological, biochemical and genetic investigations to formulate brassica breeding objectives.

6844 McNaughton GB 030803/78/0005 R
Identify and maintain s–alleles in Brassicas. study their strength and dominance relations.

6845 McNaughton GB 030803/78/0006 R
Survey virulence genes in pathogens of brassicas. Design and initiate strategies for resistance breeding.

6846 McNaughton GB 030803/78/0008 R
Assemble and test genetic sources of resistance to diseases of Brassicas. Produce improved parents.

6847 McNaughton GB 030803/78/0009 R
Breed F1 hybrid and inbred swede cultivars.

6848 McNaughton GB 030803/78/0010 R
Breed rape cultivars from natural and artificial genotypes of Brassica napus and related species.

6849 McNaughton GB 030803/78/0011 R
Breed kale and fodder cabbage cultivars.

6850 McNaughton GB 030803/78/0013 R
Breed Brassica and radish cash crops for late sowing and autumn grazing. Breed fodder radish cultivars.

6851 McNaughton GB 030803/78/0014 R
Breed radicole cultivars as substitutes for rape from hybrids of Raphanus and Brassica.

6852 McNaughton GB 030803/78/0015 R
Test and multiply brassica radish and radicole cultivars.

6853 Rodger GB 060106/00/0019 R
Swede variety trials (except NLT).

6854 Lockhart GB 060106/00/0022 R
N.L.T. Swede variety trials.

6855 Morrice GB 060213/00/0011 R
National list miscellaneous crops.

Vegetables in general (B 3500)

See also 3752, 3785, 6005, 7169, 7495

6856 Boesman, G.; Debergh, P.; Maene, L.
BE 030015/77/0001 N
Studie van de voortplanting bij tuinbouwgewassen. **Research on the propagation of horticultural plants.**

6857 Boxus, Ph.; Druart, Ph. BE 080200/76/0029 R
Cultures de tissus d'espèces horticoles. **Horticultural plants tissue culture.**

6858 Huhnke, W.; Engelhardt, M.; Hoffmann, M.; Preil, W.
DE 206000/72/5001 N
Versuche zur Anwendung moderner Methoden auf dem Gebiete der Meristem–, Kallus– und Embryokultur zur

Unterstützung der Zuchtarbeit bei gärtnerischen Kulturpflanzen, insbesondere der massenhaften vegetativen Vermehrung 1972. **Experiments on the application of modern methods to meristemic, callus and embryonic cultivation for support of breeding horticultural plants, especially for support of mass–vegetative propagation.**

6859 Hallig, V.A. DK 010113/77/0014
Vegetativ formering af væksthusplanter. **Vegetative propagation of glasshouse plants.**

6860 Blankholm, E. DK 010115/75/4512
Nyhedsafprøvning af enårige grønsager. Fællesudvalget for prøvedyrkning af køkkenurters sortsafprøvninger og efterkontrol. **Assessment of the value of annual vegetable varieties. The Joint Committe for Test–growing of Vegetables.**

6861 Faulkner GB 011801/00/0015
Evaluation of new NVRS cultivars.

6862 Crisp GB 011801/76/0025
Genetic investigations into vegetable crops of potential commercial interest.

6863 Researcher not indicated GB 050102/00/0004 R
Variety trials of tomatoes, lettuce and cucumber.

6864 Researcher not indicated GB 050142/00/0003 R
Variety trials of tomatoes, lettuce, cucumbers.

6865 Berridge GB 060215/00/0001 R
Vegetable variety testing.

6866 Uncini, L.; Restaino, F.; D'Amore, R.; Allavena, A.
IT 021000/79/0010 N
Indagine sull'adattabilità bio–agronomica di nuove costituzioni orticole a ciclo primaverile–estivo e autunno–vernino allevate in pien'aria. **Investigations on bio–agronomic adaptability of new horticultural constitutions, with spring–summer and autumn–winter cycle, grown in open air.**

6867 Uncini, L.; Soressi, G.P.; Allavena, A.; Restaino, F.
IT 021000/79/0019 N
Riproduzione conservativa di costituzioni dell'Istituto. **Conservative reproduction of the Institute constitutions.**

6868 Uncini, L.; Restaino, F.; Soressi, G.P.; Allavena, A.; Falavigna, A.; Badino, M. IT 021000/79/0029 N
Riproduzione e conservazione di popolazioni, varietà e linee di varie specie orticole. **Reproduction and storage of populations, varieties and lines of different horticultural species.**

6869 La Malfa, G.; Ruggeri, A.; Lipari, V.; Noto, G.
IT 040313/77/0001
Individuazione, nell'ambito di specie ortive diverse, di tipi a ridotte esigenze termiche. **Identification of the types of different vegetable species, with low thermic requirements.**

6870 Schneider, F.; Stolk, J.H. NL 010113/68/2480
Rassenonderzoek groenten onder glas (excl. tomaat, sla, komkommer en augurk). **Variety testing of glasshouse vegetables (excluding tomato, lettuce, cucumber and gherkin).** Publications.

6871 Roggen, H.P.J.R. NL 010114/69/2817
Biochemische methoden voor het bepalen van de

verwantschap en combinatiegeschiktheid van inteeltlijnen van hybriderassen van groentegewassen. **Biochemical methods to determinate relationships and combining ability of inbred lines of hybrid varieties of vegetable crops.**

6872 Custers, J.B.M. NL 010114/72/3830
Onderzoek in vitro cultuur ten behoeve van de veredeling van tuinbouwgewassen. **Research on in vitro culture for breeding of horticultural crops.** Publications.

6873 Roelofsen, H.J. NL 010114/73/7936
Genenbank en collecties van tuinbouwgewassen. **Genetic resources of horticultural crops.** Publications.

6874 Riepma, P.; Snoek, N.J.; Kraker, J. de; Kanters, F.M.L.
NL 010114/74/6511
Gebruikswaarde onderzoek van diverse nieuwe groentegewassen. **Research on cultural value of new vegetables.**

6875 Keulen, H.A. van NL 010114/75/6196
Chemisch onderzoek ten behoeve van de veredeling van tuinbouwgewassen. **Chemical research for the breeding of horticultural crops.**

6876 Meer, Q.P. van der NL 010114/79/8789 N
Onderzoek over mannelijke steriliteit in kool in verband met het kweken en de zaadteelt van hybriden. **Research on male sterility in cole crops in connection with the breeding and seed production of hybrids.**

6877 Giessen, A.C. van der NL 010114/79/8791 N
Onderzoek ten behoeve van de veredeling op ziekteresistentie van tuinbouwgewassen. **Research for disease resistance of horticultural crops.**

6878 Stolk, J.H. NL 010206/68/2480
Rassenonderzoek groenten onder glas (excl. tomaat, sla, komkommer en augurk). **Variety testing of glasshouse vegetables (excluding tomato, lettuce, cucumber and gherkin).** Publications.

Root, tuber and bulb vegetables (B 3510)

See also 3792, 6829, 6946, 7191, 7302

6879 Van Hee, L. BE 070400/68/0009 R
Veredeling van cichorei. **Breeding of Chicory.**

6880 Van Hee, L. BE 070400/72/0012
Veredeling van schorseneer. **Breeding of salsify.**

6881 Van Hee, L. BE 070400/73/0017 R
Onderzoek naar de mogelijkheden van selektie op schieterresistentie bij cichorei en schorseneer in het jeugdstadium. **Research concerning the possibilities for selection in the youth phase on bolting resistance in chicory and salsify.**

6882 Wricke, G.; Spickernagel, K. DE 138300/75/0001
Untersuchungen zum Ertrag und Karotingehalt bei Möhrenhybriden. **Investigations of yield and carotene content in carrot hybrids.**

6883 Wricke, G.; Freese, L. DE 138300/78/0001 N
Untersuchungen zum Resistenzverhalten und zur Auslese resistenter Formen gegen Meloidogyne hapla bei Möhren. **Investigations on resistance behaviour of carrots and selection of resistant forms against Meloidogyne hapla.**

6884 Wricke, G.; Frees, L. DE 138300/78/0004 N
Probleme der Züchtung von Hybridsorten mit Hilfe pollensteriler Formen bei Möhren. **Problems in breeding hybrid varieties of carrots by the aid of pollen-sterile forms.**

6885 Fischbeck, G.; Konvicka, D. DE 161250/77/0006
Untersuchungen über die Mechanismen der cytoplasmatisch bedingten Pollensterilität bei Allium-Arten. **On the mechanism of pollen sterility in allium depending on cytoplasma.**

6886 Handke, S.; Orlovius, I. DE 206000/71/4034 R
Züchtung für Treibanbau unter Schwachlichtbedingungen unter Berücksichtigung der für diese Selektion vorrangigen Merkmale bei Radies 1970-1982. **Breeding for forcing cropping under Low-Light conditions of characteristics of little radish of priority for that selection.**

6887 Schubert, G.; Jüngling, H. DE 502200/70/0001
Sortenprüfungen bei Radies, Tomaten, Kopfsalat, Gurken, Sellerie, Porree, Stangenbohnen, Blumenkohl, Kohlrabi, Rosenkohl, Meerrettich 1970. **Testing of varieties of radish, tomato, lettuce, cucumber, celeriac, leek, pole bean, cauliflower, kohlrabi, Brussels sprouts, and wild horse radish.**

6888 Philipsen, H. DK 030109/77/0010
Resistens i gulerødder over for gulerodsfluer. **The resistance of carrots to carrot flies.**

6889 Schweisguth, B. FR 010104/62/0042
Création de variétés hybrides d'oignon. **Breeding hybrid varieties of onion.**

6890 Schweisguth, B. FR 010104/68/0043
Création de variétés hybrides de poireau. **Breeding hybrid varieties of leek.**

6891 Bonnet, A. FR 010603/65/0049
Utilisation de la stérilité mâle cytoplasmique chez la carotte. **Using cytoplasmic male sterility in carrot breeding.**

6892 Bonnet, A. FR 010603/69/0050
Mise au point de diverses méthodes de sélection appliquées au radis. **Perfecting various breeding methods for radish.**

6893 Bonnet, A. FR 010603/69/0051
Introduction d'une stérilité mâle cytoplasmique dans des variétés françaises de radis. **Transfering cytoplasmic male sterility to french varieties of radish.**

6894 Vergniaud, P. FR 010610/75/9078
Etude variétale d'oignon pour l'industrie et le marché en région méditerranéenne. **Onion variety studies for both the canning industry and the fresh market in the Mediterranean region.**

6895 Smith GB 011101/00/0003
Genetic control of lettuce reproduction.

6896 Barber GB 011401/00/0007
Selection of haploid plants. Produce uniform F1 hybrids.

6897 Dowker; Jackson GB 011801/00/0008
Development of improved breeding methods.

6898 Dowker; Jackson GB 011801/00/0009
Investigate the genetics of resistance to carrot fly.

6899 Dowker; Fennell GB 011801/00/0010
Breeding methods in onions.

6900 Ockendon GB 011801/00/0013 R
Factors affecting self- and cross-fertilization in onions and carrots.

6901 Dowker; Fennel GB 011801/00/0019
Genetic variability in autumn sown onions.

6902 Cowker; Jackson GB 011801/78/0022 N
Investigate genetical basis of improved establishment and root size uniformity in carrots.

6903 Prendiville, M.D.; Jeffares, M. IE 060300/69/0015 R
Onion breeding (yield, keeping quality). Publications.

6904 Uncini, L.; Restaino, F.; Interlandi, G.
IT 021000/68/0001 R
Miglioramento genetico del finocchio. **Fennel breeding.** Publications.

6905 Porcelli, S.; Uncini, L. IT 021000/71/0001
Possibilità di utilizzare la poliploidia indotta nella bietola da costa. **Exploiting artificially induced poliploidy in beet (Beta cicla).**

6906 Badino, M. IT 021000/76/0019 R
Miglioramento della cipolla "Dorata di Parma" per la resistenza al Fusarium oxysporum f. sp. cepae. **Breeding of "Dorata di Parma" onion for resistance to Fusarium oxysporum f. sp. cepae.**

6907 Badino, M.; Soressi, G.P. IT 021000/76/0020 R
Ricostituzione della varietà di cipolla "Dorata di Parma". **Reconstitution of the "Dorata di Parma" onion variety.**

6908 Greco, N. IT 060600/73/0046
Analisi delle reazioni varietali agli attacchi dei nematodi. Varietà di cipolla a Ditylenchus dipsaci. **Analysis of the varietal reactions to nematode attacks. Onion varieties to Ditylenchus dipsaci.**

6909 Greco, N. IT 060600/73/0047
Analisi delle reazioni varietali agli attacchi dei nematodi. Varietà di carota a Heterodera carotae. **Analysis of the varietal reactions to nematode attacks. Carrot varieties to Heterodera carotae.**

6910 Ouden, H. den NL 010108/73/3822 R
Veredeling van de wortel op resistentie tegen de wortelvlieg (Psila rosae). **Breeding for resistance to the carrot fly (Psila rosae).** Publications.

6911 Schneider, F.; Riepma, P. NL 010113/71/3033
Gebruikswaarde onderzoek bij knolselderij. **Research on cultural value of celeriac.** Publications.

6912 Schneider, F.; Riepma, P. NL 010113/75/6723
Gebruikswaarde onderzoek prei. **Research on cultural value of**

leeks. Publications.

6913 Schneider, F.; Riepma, P. NL 010113/77/7668
Gebruikswaarde–onderzoek schorseneer. **Variety testing of scorzonera.**

6914 Schneider, F.; Riepma, P. NL 010113/78/8419
Gebruikswaarde–onderzoek kroot. **Variety testing of beetroot.**

6915 Schneider, F.; Riepma, P. NL 010113/78/8420
Gebruikswaarde–onderzoek radijs. **Variety testing of radish.**

6916 Schneider, F.; Riepma, P. NL 010113/78/8421
Gebruikswaarde–onderzoek koolrabi. **Variety testing of kohlrabi.**

6917 Nieuwhof, M. NL 010114/58/0720
Veredelingsmethodiek voor het kweken van hybriderassen bij wortel. **Breeding methods for the breeding of hybrids of carrots.**

6918 Meer, Q.P. van der NL 010114/64/0726
Veredeling op ziekteresistentie bij de ui. **Breeding for disease resistance in onions.** Publications.

6919 Meer, Q.P. van der NL 010114/67/0719
Onderzoek over mannelijke steriliteit in verband met het kweken en de zaadteelt van hybride–rassen bij ui. **Research on male sterility in connection with the breeding and seed growing of hybrid varieties of onion.** Publications.

6920 Ponti, O.B.M. NL 010114/73/3822 R
Veredeling van de wortel op resistentie tegen de wortelvlieg (Psila rosae). **Breeding for resistance to the carrot fly (Psila rosae).** Publications.

6921 Meer, Q.P. van der NL 010114/73/3827
Soortkruisingen bij Allium. **Species crosses of Allium.**

6922 Ponti, O.M.B. de NL 010114/74/5165 R
Veredeling van de ui op resistentie tegen de uievlieg (Hylemya antiqua Mg). **Breeding onion for resistance to the onion fly (Hylemya antiqua Mg).**

6923 Nieuwhof, M. NL 010114/76/6643
Veredelingsonderzoek aan radijs gericht op snellere groei in de winter en grotere uniformiteit. **Breeding of radish for better growth in winter and higher uniformity.** Publications.

6924 Meer, Q.P. van der NL 010114/78/7931 R
Veredeling op resistentie tegen ziekten bij prei. **Breeding for disease resistance in leek.**

6925 Meer, Q.P. van der NL 010114/78/7932 R
Veredelingsmethodiek voor het kweken van hybriderassen bij prei. **Methodology of breeding hybrid varieties in leek.**

6926 Nieuwhof, M. NL 010114/79/8788 N
Veredeling op resistentie tegen valse meeldauw (Peronospora parasitica (Pers.ex.Fr.) Fr.) bij radijs. **Breeding for resistance to downy mildew (Peronospora parasitica (Pers. ex Fr.) Fr.) in radish.**

6927 Riepma, P. NL 010207/71/3033
Gebruikswaarde onderzoek bij knolselderij. **Research on cultural value of celeriac.** Publications.

D 2300 – Plant breeding

6928 Riepma, P.   NL 010207/75/6723
Gebruikswaarde onderzoek prei. **Research on cultural value of leeks.** Publications.

6929 Riepma, P.   NL 010207/77/7668
Gebruikswaarde–onderzoek schorseneer. **Variety testing of scorzonera.**

6930 Riepma, P.   NL 010207/78/8419
Gebruikswaarde–onderzoek kroot. **Variety testing of beetroot.**

6931 Riepma, P.   NL 010207/78/8420
Gebruikswaarde–onderzoek radijs. **Variety testing of radish.**

6932 Riepma, P.   NL 010207/78/8421
Gebruikswaarde–onderzoek koolrabi. **Variety testing of kohlrabi.**

Greens and leafy vegetables (B 3520)

See also 6306, 6417, 6826, 6827, 6829, 6887, 7140, 7191, 7302, 9633

6933 Laeremans, R.; Muyldermans, L.   BE 040204/77/0001
Onderzoek van de fertiliteit van witloofzaaddragers. **Fertility of seed bearing plants of Belgian endive.**

6934 Laeremans, R.   BE 040204/78/0003 N
Produktie van asperge–kultivars (Asparagus officinalis) met een hoge graad van eenvormigheid. **Breeding of varieties of asparagus (Asperagus officinalis) with a high level of homogenety.**

6935 Plumier, W.; Férauge, M.; Valette, R.
                    BE 080200/54/0019 R
Amélioration de la chicorée de Bruxelles en tant que légume à consommer frais. **Breeding of Brussels chicory as vegetable for fresh consumption.** Publications.

6936 Van Looveren, K.; Niclaes, J.; Kempeneers, L.
                    BE 180000/24/0007
Kweken van betere kultivars van sla. **Breeding research for better varieties of lettuce.** Publications.

6937 Van Looveren, K.; Niclaes, J.; Kempeneers, L.
                    BE 180000/24/0008
Kweken van betere kultivars van witloof. **Breeding research for better varieties of Brussels chicory.** Publications.

6938 Wricke, G.; Löptien, H.   DE 138300/74/0004
Genetische und cytologische Untersuchungen zur Frage der Geschlechtsbestimmung bei Spargel und Spinat. **Genetical and cytological studies on sex determination of asparagus and spinach.**

6939 Handke, S.; Bandze, E.   DE 206000/71/4038 R
Züchtung eines Sommersalates mit Resistenz gegen Salatmosaikvirus und falschen Mehltau in Verbindung mit dem Versuch der Auslese einer Resistenz R1 bis R5 1970–1982. **Breeding of spring lettuce resistant to mosaic virus and downy mildew in connection with the attempt to select a resistance from R1 to R5.**

6940 Handke, S.; Orlovius, I.   DE 206000/71/4040 R
Züchtung von monözischen F1–Sorten mit Resistenz gegen falschen Mehltau und Gurkenmosaikvirus für verschiedene Anbau–zeiten – Frühjahr, Sommer, Herbst, Winter und Treibanbau der Spinatsorten – 1966. **Breeding of monoecious F1–varieties resistant to downy mildew and cucumber mosaic virus for different periods of cultivation – spring, summer, autumm, winter and forcing cropping of spinach varieties.**

6941 Handke, S.; Orlovius, I.   DE 206000/75/0015
Auslese von Genotypen des Freilandspinates mit sehr langer Vegetationsdauer als monözische und diözische Formen. **Selection of genotypes in spinach for the open field with very long vegetative growth period.**

6942 Hartmann, H. D.; Kretschmer, M.   DE 506107/75/0003
Untersuchungen zu Apikaldominanz bei Spargel. **Studies on apical dominance in asparagus.** Publications.

6943 Thevenin, L.; Teisseire, G.   FR 010104/63/0044
Obtention d'hybrides F_1 entre lignées homozygotes d'Asperge. **Breeding F_1 hybrids between homozygotes lines in Asparagus.**

6944 Thevenin, L.   FR 010104/63/0169
Création d'hybrides "doubles" d'asperges à partir de 4 plantes hétérozygotes (2 mâles et 2 femelles). **Breeding "double" asparagus hybrids from four heterozygous plants (two male and two female plants).**

6945 Bannerot, H.   FR 010104/69/0171
Obtention de variétés hybrides F1 d'endives pour le forçge en salle climatisée. **Breeding witloof F, varieties for forcing in air. conditioned room.**

6946 Bannerot, H.; Cauderon, Y. Mme   FR 010104/69/0172
Recherche d'une stérilité mâle utilisable pour la fabrication d'hybrides F1 chez Brassica Oleracea. **Research of a male sterility for F, production in Brassica Oleracea.**

6947 Thevenin, L.; Dore, C.   FR 010104/70/0168
Obtention et multiplication de géniteurs par culture in vitro de tissus d'asperge. **Selection and multiplication of parents through in vitro culture of asparagus tissues.**

6948 Bannerot, H.; Boulidard, L.   FR 010104/72/0170
Création de variétés de laitues pour culture d'hiver sous abri. **Breeding lettuce varieties for winter cultivation under shelter.**

6949 Foury, C.   FR 010603/68/0048
Etude des possibilités de création et d'utilisation de variétés d'artichaut issues de semences. **Studies on the possibility of breeding and growing artichoke varieties using seeds.**

6950 Herve, Y.   FR 011301/71/0045
Création de variétés de chou–fleur d'automne. **Breeding autumn cauliflower varieties.**

6951 Herve, Y.   FR 011301/71/0046
Création de variétés de chou–fleur d'hiver. **Breeding winter cauliflower varieties.**

6952 Herve, Y.   FR 011301/71/0047
Etude de la physiologie du développement du chou–fleur. **Study on the developmental physiology of cauliflower.**

6953 Smith   GB 011101/00/0002 R
Breeding of various types of lettuce.

6954 Johnson; Smith B   GB 011801/00/0001
Breeding of F1 hybrids.

6955 Crisp; Gray   GB 011801/00/0002
Genetic improvement of autumn cauliflower.

6956 Crisp; Gray   GB 011801/00/0003
Breed for uniformity, quality and maturity periods suitable for south–west counties (UK).

6957 Johnson; Crisp   GB 011801/00/0005
Genetics of resistance to cabbage root–fly among Brassicae.

6958 Johnson; Crisp   GB 011801/00/0006
Conservation of genetic stocks of Brassica oleracea.

6959 Crisp   GB 011801/00/0007 R
Tissue culture techniques for the clonal propagation of brassicae and asparagus.

6960 Johnson; Norwood   GB 011801/00/0012 R
Develop lettuce breeding material that is resistant to pests, diseases and physiological disorders.

6961 Faulkner   GB 011801/00/0014
Problems associated with hybrid seed production.

6962 Ockenden; Gates   GB 011801/00/0018
Genetics and physiology of self–incompatibility in Brassicae.

6963 Crisp   GB 011801/00/0024
Creation of genetic stocks for the development of improved cultivars of calabrese.

6964 Buczacki; Crisp   GB 011807/78/0027
Investigations into the mechanisms and exploitation of resistance in Brassica oleracea to clubroot.

6965 Hodgkin; Wiseman   GB 030703/00/0010 R
Genetics of s–allele incompatibility system in Brassica oleracea.

6966 Hodgkin   GB 030703/00/0011
Breeding hybrid varieties.

6967 Redfern   GB 030703/00/0012
Breed hybrid varieties.

6968 Wills; Wiseman   GB 030703/00/0013
Isoenzyme analysis in Brassica oleraceae.

6969 Wills; Smith   GB 030703/00/0015 R
Genetics and cytology of Brassica oleracea in relation to linkage groups.

6970 Wills   GB 030703/00/0019
Breed calabrese varieties adapted to N. European conditions.

6971 Murphy, R.F.; Ryan, E.W.   IE 060300/73/0883 R
Brussels sprouts for freezing and fresh market. Publications.

6972 Uncini, L.; Restaino, F.   IT 021000/67/0001 R
Miglioramento genetico della scarola per produzioni autunnali e invernali. **Endive breeding for autumn and winter crops.**
Publications.

6973 Uncini, L.; D'Amore, R.   IT 021000/71/0007 R

Miglioramento del cavolfiore per produzioni precoci autunnali.
Cauliflower breeding for early autumn crops.

6974 Falavigna, A.; Soressi, G.P.   IT 021000/76/0008 R
Sviluppo di una varietà sintetica di Asparago. **Development of a synthetic variety of Asparagus.**

6975 Falavigna, A.   IT 021000/76/0018 R
Applicazione di tecniche di coltura in vitro per l'ottenimento e la riproduzione di linee pure di asparago. **Application of in vitro culture techniques for obtainment and reproduction of asparagus pure lines.**

6976 Falavigna, A.; Soressi, G.P.   IT 021000/79/0001 N
Combinazioni d'incrocio tra fenotipi di asparago "Precoce d'Argenteuil". **Cross combination between phenotype of "Precoce d'Argenteuil" asparagus.**

6977 Uncini, L.   IT 021000/79/0002 N
Miglioramento delle popolazioni di cavolfiore "Fanese" medio–tardivo e tardivo. **Populations breeding of middle late and late "Fanese" cauliflower.**

6978 Uncini, L.   IT 021000/79/0003 N
Miglioramento del cavolfiore per resistenza all'Alternaria brassicicola e brassicae. **Cauliflower breeding for resistance to Alternaria brassicicola and brassicae.**

6979 Uncini, L.; Ferrani, V.   IT 021000/79/0022 N
Indagine sui meccanismi di auto e allo–incompatibilità presente in linee selettive e varietà commerciali di cavolfiore. **Investigations on mechanisms of self and allo–incompatibility in selective lines and cauliflower commercial varieties.**

6980 Conti, S.; Paradisi, U.; Bigelli, G.   IT 040214/71/0001
Miglioramento genetico del cavolfiore. **Cauliflower breeding.**

6981 Foti, S.   IT 063100/77/0903
Studio della biologia della riproduzione ai fini della conservazione e della utilizzazione della variabilità genetica di popolazioni locali di cavolo–broccolo. **A study of the biology of reproduction in local populations of broccoli with a view to preservation and to exploiting their genetic variability.**

6982 Foti, S.   IT 063100/77/0906
Miglioramento genetico e trasmissibilità dei caratteri del carciofo. **Genetic improvement and characters transmissibility in the artichoke.**

6983 Schneider, F.; Stolk, J.H.   NL 010113/48/2144
Rassenonderzoek glassla. **Variety testing of indoor lettuce.**
Publications.

6984 Schneider, F.; Riepma, P.   NL 010113/68/1547
Gebruikswaarde onderzoek spinazie. **Variety testing of spinach.** Publications.

6985 Schneider, F.; Riepma, P.   NL 010113/68/1548
Gebruikswaarde onderzoek spruitkool. **Variety testing of Brussels sprouts.**

6986 Schneider, F.; Riepma, P.   NL 010113/68/2097
Gebruikswaarde onderzoek bij kropsla. **Variety testing of lettuce.** Publications.

6987 Schneider, F.; Riepma, P.   NL 010113/68/2349

Gebruikswaarde onderzoek groene savooiekool. **Variety testing of green savoy cabbage.**

6988 Schneider, F.; Riepma, P.   NL 010113/68/2353
Gebruikswaarde – onderzoek witte kool (zuurkool). **Variety testing white cabbage (sauerkraut).** Publications.

6989 Schneider, F.; Riepma, P.   NL 010113/69/1544
Gebruikswaarde onderzoek bloemkool. **Variety testing of cauliflower.** Publications.

6990 Schneider, F.; Riepma, P.   NL 010113/74/6509 N
Gebruikswaarde onderzoek rode kool. **Research on cultural value of red cabbage.** Publications.

6991 Schneider, F.; Riepma, P.   NL 010113/75/6724
Gebruikswaarde onderzoek witlof. **Research on cultural value of chocory.** Publications.

6992 Schneider, F.; Riepma, P.   NL 010113/75/7669
Gebruikswaarde–onderzoek boerenkool. **Variety testing of kale.** Publications.

6993 Schneider, F.; Riepma, P.   NL 010113/76/7670
Gebruikswaarde–onderzoek spitskool. **Variety testing of spring cabbage.** Publications.

6994 Schneider, F.; Riepma, P.   NL 010113/78/7671
Gebruikswaarde–onderzoek witte kool (verse markt). **Variety testing of white cabbage (fresh market).**

6995 Schneider, F.; Riepma, P.   NL 010113/78/8422
Gebruikswaarde–onderzoek rabarber. **Variety testing of rhubarb.**

6996 Schneider, F.; Riepma, P.   NL 010113/78/8423
Gebruikswaarde–onderzoek winterbloemkool. **Variety testing winter cauliflower.**

6997 Schneider, F.; Riepma, P.   NL 010113/78/8424
Gebruikswaarde–onderzoek savooiekool (gele). **Variety testing of savoy cabbage (yellow).**

6998 Eenink, A.H.   NL 010114/55/0700
Veredeling op ziekteresistentie bij spinazie. **Breeding for disease resistance of spinach.** Publications.

6999 Meer, Q.P. van der   NL 010114/58/7928 R
Veredeling op ziekteresistentie bij kool. **Breeding for resistance in cabbage crops.**

7000 Eenink, A.H.   NL 010114/59/0699
Veredelingsmethodiek witlof. **Breeding methods of chicory.** Publications.

7001 Eenink, A.H.   NL 010114/64/0686
Onderzoek naar de mogelijkheid trager doorschietende andijvierassen te kweken. **Research into the possibility of breeding slower bolting endive varieties.** Publications.

7002 Visser, D.L.   NL 010114/67/0717
Incompatibiliteit bij koolgewassen in verband met het kweken van hybride rassen. **Incompatibiliy in cabbage crops in connection with the breeding of hybride varieties.** Publications.

7003 Eenink, A.H.   NL 010114/68/2152

Veredeling op resistentie tegen bladluizen in sla. **Breeding methods for aphid resistance in lettuce.** Publications.

7004 Roggen, H.P.J.R.   NL 010114/68/2483
Biochemie van de incompatibiliteit bij koolgewassen. **Biochemistry of incompatibility in cole crops.** Publications.

7005 Eenink, A.H.   NL 010114/68/2484
Onderzoek naar de mogelijkheden andere slatypen dan botersla te winnen voor de teelt onder glas. **Research into the possibilities of raising lettuce types other than butterhead lettuce for growing under glass.** Publications.

7006 Eenink, A.H.   NL 010114/69/2814
Onderzoek naar de mogelijkheid van het winnen van voor Nederland geschikte rassen van "Pain de Sucre"(Zuckerhut). **Research into the possibilities of raising varieties of "Pain de Sucre" (Zuckerhut) suitable for the Netherlands.** Publications.

7007 Smeets, L.   NL 010114/70/2824
Vaststellen van selectiecriteria bij sla door groeifactoren–analyse. **Determination of selection criteria with lettuce by an analysis of the growth factors.** Publications.

7008 Riepma, P.; Kanters, F.M.L.   NL 010114/70/6510
Gebruikswaarde onderzoek rabarber. **Research on cultural value of rhubarb.** Publications.

7009 Eenink, A.H.   NL 010114/71/3417
Veredeling op ziekteresistentie (mozaîek, wit en rand) bij kropsla. **Breeding for disease resistance (mosaic, Bremia lactucae and tip burn) in head lettuce.** Publications.

7010 Roggen, H.P.J.R.   NL 010114/71/3538
Inulase–activiteit als parameter voor de trekrijpheid van witlofwortels. **inulase–activity as measure for optimal shoot formation of chicory roots.**

7011 Meer, Q.P. van der   NL 010114/72/3543
Veredeling van koolrabi voor de kasteelt. **Breeding kohlrabi for glasshouse crops.** Publications.

7012 Eenink, A.H.   NL 010114/73/3820
Veredelingsmethodiek bij sla. **Breeding methods of lettuce.** Publications.

7013 Roelofsen, H.J.   NL 010114/73/7927
Genenbank voor sla. **Gene bank for lettuce.**

7014 Roelofsen, H.J.   NL 010114/75/6197
Genenbank voor koolgewassen. **Genebank for cole crops.**

7015 Eenink, A.H.   NL 010114/77/7272
Veredeling van sla op geringere energiebehoefte. **Breeding lettuce with lower energy requirements.** Publications.

7016 Meer, Q.P. van der   NL 010114/78/7984 R
Veredeling van Chinese kool. **Breeding of Chinese cabbage.**

7017 Stolk, J.H.   NL 010206/48/2144
Rassenonderzoek glassla. **Variety testing of indoor lettuce.** Publications.

7018 Jansen, G.A.J.   NL 010206/72/3543
Veredeling van koolrabi voor de kasteelt. **Breeding kohlrabi for glass–house crops.**

D 2300 – Plant breeding

7019 Riepma, P.   NL 010207/68/1547
Gebruikswaarde onderzoek spinazie. **Variety testing of
spinach.** Publications.

7020 Riepma, P.   NL 010207/68/1548
Gebruikswaarde onderzoek spruitkool. **Variety testing of
Brussels sprouts.** Publications.

7021 Riepma, P.   NL 010207/68/2097
Gebruikswaarde onderzoek bij kropsla. **Variety testing of
lettuce.** Publications.

7022 Riepma, P.   NL 010207/68/2349
Gebruikswaarde onderzoek groene savooiekool. **Variety
testing of green savoy cabbage.** Publications.

7023 Riepma, P.   NL 010207/68/2353
Gebruikswaarde–onderzoek witte kool (zuurkool). **Variety
testing of white cabbage (sauerkraut).** Publications.

7024 Riepma, P.   NL 010207/69/1544
Gebruikswaarde onderzoek bloemkool. **Variety testing of
cauliflower.** Publications.

7025 Riepma, P.   NL 010207/74/6509 N
Gebruikswaarde onderzoek rode kool. **Research on cultural
value of red cabbage.** Publications.

7026 Riepma, P.   NL 010207/75/6724
Gebruikswaarde onderzoek witlof. **Research on cultural value
of chicory.** Publications.

7027 Riepma, P.   NL 010207/75/7669
Gebruikswaarde–onderzoek boerenkool. **Variety testing of
kale.** Publications.

7028 Riepma, P.   NL 010207/76/7670
Gebruikswaarde–onderzoek spitskool. **Variety testing of spring
cabbage.** Publications.

7029 Riepma, P.   NL 010207/78/7671
Gebruikswaarde–onderzoek witte kool (verse markt). **Variety
testing of white cabbage (fresh market).**

7030 Riepma, P.   NL 010207/78/8422
Gebruikswaarde–onderzoek rabarber. **Variety testing of
rhubarb.**

7031 Riepma, P.   NL 010207/78/8423
Gebruikswaarde–onderzoek winterbloemkool. **Variety testing
of winter cauliflower.**

7032 Riepma, P.   NL 010207/78/8424
Gebruikswaarde–onderzoek savooiekool (gele). **Variety
testing of savoy cabbage (yellow).**

7033 Dieleman, F.L.   NL 020014/68/2152
Veredelingsmethodiek bladluisresistentie in sla. **Breeding
methods for aphid–resistance in lettuce.**

7034 Sneep, J.   NL 020045/75/6767 R
Productie triploide hybriede bij spinazie. **Production of triploid
hybrid in spinach.**

Leguminous vegetables (B 3531)

See also 3285, 3894, 3904, 3923, 3927, 5435, 6096, 6570, 6741,
6745, 6791, 6798, 6817, 6887, 7191, 9652, 9661, 9666

7035 Le Marchand, G.; Maréchal, R.; Otoul, E.; Homble;
Baudoin, J.P.   BE 010019/75/0008 R
Recherches sur les possibilités d'amélioration de Phaseolus
lunatus L. par croisements interspécifiques. **Research on the
possibilities of improving luniabean (Phaseolus lunatus L.) by
interspecific crosses.** Publications.

7036 Le Marchand, C.   BE 010019/76/0003 R
Recherches sur les possibilités d'accroître, par unité de surface
cultivée, le potentiel alimentaire protidique (valeur biologique
et rendement protidique) des graines des espèces comestibles
de la sous–tribu des Phaseolinae (haricots et espèces
apparentées). **Investigation in the possibilities of increasing, per
cultivated unite area, the nutritional protidic potential
(biological value and protidic yield) of the edible grain legumes
from the Phaseolinae subtribe (beans and related species).**
Publications.

7037 Le Marchand, G.; Maréchal, R.; Otoul, E.; Homble;
Baudoin, J.P.; Gepts, P.   BE 010019/77/0009 R
Recherches sur les possibilités d'amélioration de Phaseolus
vulgaris L. par croisements interspécifiques. **Research on the
possibilities of improving Kidney bean (Phaseolus vulgaris L.)
by interspecific crosses.** Publications.

7038 Dermine, E.; Tilkin, V.   BE 080200/46/0017
Amélioration du pois et du haricot pour l'industrie des
conserves. **Breeding of pea and bean for industrial use.**
Publications.

7039 Fischbeck, G.; Abdalla, M.M.F.   DE 161250/77/0007
Erbliche Differenzierung der Selbstauslösung und ihre
Bedeutung für den Kornansatz bei Vicia faba. **Hereditary
differentiation of self–release and importance to graining of
Vicia faba.**

7040 Handke, S.; Bandze, E.   DE 206000/71/4039 R
Züchtung von tagneutralen Markerbsen mit Eignung für das
Tiefgefrieren und die Dosenkonservierung für den
Spätsommeranbau mit Resistenz gegen echten Mehltau,
Adern– mosaikvirus und Fusarium 1969. **Breeding of
day–neutral wrinkled peas suitable for deepfreezing and
canning for cultivation in late spring with resistance to powdery
mildew, streak mosaic virus and Fusarium.**

7041 Wende, E.; Hansen, H.   DE 206000/71/5028 R
Sensorische Analysen zur Unterstützung der Auslese von
Genotypen mit hoher Qualität bei Erbsen, Buschbohnen und
Obst, insbesondere Äpfel, sowie zur Selektion von
Zuchtstämmen bei der Kombinationszüchtung 1969. **Sensorial
analyses for support of selection of genotypes with high quality
in peas, bush beans and fruit, especially apples, as well as for
selection of breeding stocks in combination breeding.**

7042 Wende, E.; Martens, H.; Hansen, H.
                                 DE 206000/71/5030 R
Züchtung von Buschbohnensorten mit spezieller Eignung für
das Tiefgefrieren und andere Konservierungsarten 1969–1978.
**Breeding of bush–bean varieties with special suitability for
deep–freezing and other preserving methods.**

7043 Wende, E.; Weidemann, S.; Hansen, H.

DE 206000/71/5031 R
Züchtung von Markerbsen mit spezieller Eignung für die Gefriertrocknung auf dem Wege der Kombinationszüchtung 1969–1979. **Breeding of wrinkled peas with special suitability for deep–freezing by combination breeding.**

7044 Reimann–Philipp, R.; Martens, H.   DE 206000/73/5004
Herstellung von Artbastarden zwischen Phaseolus vulgaris und Phaseolus coccineus zur Schaffung von Zuchtmaterial mit spezieller Eignung für verschiedene Konservierungsverfahren und verbesserte Resistenzeigenschaften. **Species hybridization between Phaseolus vulgaris and Phaseolus coccineus for producing a breeding material especially suitable for different preserving methods –canning, freezing, freezedrying– and for improved resistance.**

7045 Wende, E.; Hansen, H.; Weidemann, S.
DE 206000/75/0001
Untersuchungen über die Vererbung qualitativer Merkmale bei Markerbsen, Pisum sativum, insbesondere bezüglich der Konservierungseignung. **Studies on the genetics of qualitative characteristics in wrinkled peas, particularly with respect to processing requirements.**

7046 Wende, E.; Hansen, H.   DE 206000/75/0003 R
Kreuzungen von Phaseolus vulgaris vom Typ Prinzessbohnen mit Phaseolus coccineus und Rückkreuzungen mit Artbastarden zur Verbesserung der Hülsenqualität. **Crossing of Phaseolus vulgaris of stringless bean type with Phaseolus coccineus and back–crossing with interspecific hybrids for improvement of quality of legumes.**

7047 Wende, E.; Hansen, H.; Weidemann, S.; Engelhardt, K.; Wirth, G.   DE 206000/78/0005 N
Versuche zur Verbesserung der Samenqualität von Buschbohnen, insbesondere bezüglich ihrer "Widerstandsfähigkeit gegen niedrige Keimtemperaturen" und der "Unempfindlichkeit gegen mechanische Beschädigung". **Studies on the improvement of the seed quality of dwarf beans – Phaseolus vulgaris – with respect to tolerance to low germination temperature and stability to mechanical damage of the seeds.**

7048 Jørgensen, M.B.; Jensen, J.; Thuesen, A.
DK 010115/75/4511
Nyhedsafprøvning af ært og jordbær på friland. **Assessment of the merits of pea and strawberry varieties for outdoor crops.**

7049 Bannerot, H.   FR 010104/63/0173
Création de variétés à rames de haricots verts pour la culture en serre. **Breeding pole beans varieties for greenhouse cultivation.**

7050 Cousin, R.   FR 010104/65/0041
Obtention de variétés de pois d'hiver. **Breeding new varieties of winter peas.**

7051 Cousin, R.   FR 010104/73/0142
Obtention de variétés de pois protéagineux. **Breeding of protein – yielding pea varieties.**

7052 Fouilloux, G.; Le Tan, T.   FR 010104/74/0267
Obtention de variétés de haricot filet récoltables en cueille unique et si possible mécaniquement. **Breeding green beans varieties for a unique and, if possible, mechanical harvest.**

7053 Fouilloux, G.; Le tan, T.   FR 010104/74/0305
Obtention de variétés de haricot filet récoltables en cueille unique et si possible mécaniquement. **Green beans varieties breeding, with a unique gathering, and if it is possible mechanicaly.**

7054 Fouilloux, G.; Le Tan, T.; Eteve, G.   FR 010104/75/0268
Obtention de variétés de haricot grain à fort rendement et adaptées aux conditions françaises de culture et d'utilisation. **Obtention of grain–bean varieties with high yield and adapted to french conditions of cultivation and utilization.**

7055 Fouilloux, G.; Le Tan, T.; Eteve, G.   FR 010104/75/0269
Obtention de variétés de haricot résistantes aux quatre principales maladies et ayant de bonnes qualités agronomiques et technologiques. **Obtention of bean varieties resistant to four main diseases and with good agronomic and technologic qualities.**

7056 Fouilloux, G.; Le Tan, T.; Eteve, G.   FR 010104/75/0306
Obtention de variétés de haricot grain à fort rendement et adaptées aux conditions francaises de culture et d'utilisation. **Grain–bean varieties breeding with a high yield and adapted to french cultiuation conditions and french utilization.**

7057 Fouilloux, G.; Le Tan; Eteve, G.   FR 010104/75/0307
Obtention de variétés de haricot résistantes aux quatre principales maladies et ayant de bonnes qualités agronomiques et technologiques. **Breeding of bean varieties resistant to the main four diseases and with good agronomfic and technologic qualities.**

7058 Berville, A.; Thiellement, H.   FR 010105/76/0325
Les mécanismes moléculaires réglant l'expression des cytoplasmes dits mâle–stérile. Etude sur la féverole. **Molecular control of cytoplasmic male sterility. An approach of the mecanism with field bean.**

7059 Berthaut, J.; Sixdenier, G.   FR 011003/73/0141
Choix et obtention de variétés protéagineuses de pois. **Selection and breeding of protein–yielding pea varieties.**

7060 Berthelem, P.   FR 011301/64/0034
Obtention de variétés hybrides de Féveroles d'hiver. **Breeding hybrid varieties of winter field–bean.**

7061 Berthelem, P.; Picard, J.   FR 011301/66/0035
Obtention de variétés hybrides de Féveroles de printemps. **Breeding hybrid varieties of spring field bean.**

7062 Snoad   GB 011401/00/0005
Breeding and components of yield.

7063 Snoad   GB 011401/00/0006
Manipulate genetic characters. Maintain gene bank.

7064 Harvey   GB 011401/76/0016
Peas: breeding of leafless forms.

7065 Casey   GB 011401/76/0018
Peas: breeding for defined seed proteins.

7066 Davies   GB 011401/76/0019 R
Peas– breeding and control of seed size.

7067 Matthews   GB 011401/76/0023

D 2300 – Plant breeding

Peas breeding for disease resistance.

7068 Innes   GB 011801/00/0020
Genetic improvement of dry Phaseolus beans.

7069 Bond; Toynbee–Clarke   GB 011910/00/0007
Breeding and evaluation of inbred lines and synthetic varieties for yield stability, pest and disease resistance.

7070 Lawes   GB 012101/76/0025
Breeding field beans for improved symbiotic efficiency.

7071 Soressi, G. P.; Salamini, F.; Allavena, A.
       IT 021000/70/0006 R
Introduzione di resistenza genetica al virus del mosaico comune (BCMV) in fagiolo (Phaseolus vulgaris). **Introduction of genetic resistance to common mosaic virus (BCMV) of beans (Phaseolus vulgaris).** Publications.

7072 Allavena, A.; Soressi, G.P.   IT 021000/71/0011 R
Introduzione di resistenza genetica a Pseudomonas phaseolicola in varietà di fagiolo resistenti a BCMV. **Introduction of genetic resistance to Pseudomonas phaseolicola in bean varieties resistent to BCMV.**

7073 Salamini, F.; Allavena, A.; Soressi, G.P.
       IT 021000/71/0012 R
Conversione a taglia nana del fagiolo di Spagna rampicante (Ph. coccineus). **Runner bean (Ph. coccineus) conversion to bush type.**

7074 Salamini, F.; Soressi, G.P.; Allavena, A.
       IT 021000/72/0019 R
Riconversione a taglia nana della varietà di fagiolo mangiatutto Meraviglia di Venezia (M. V.). **Reconversion of french bean variety "Meraviglia di Venezia" (M.V.) to bush type.**

7075 Allavena, A.; Salamini, F.   IT 021000/76/0010 R
Utilizzazione del mutante glossy nel miglioramento genetico del fagiolo. **Utilization of glossy mutant in bean breeding.**

7076 Uncini, L.   IT 021000/76/0011 R
Costituzione di varietà di fava da mensa per il consumo di semi allo stato fresco e di varietà da industria adatte alla surgelazione. **Constitution of table broad bean varieties for fresh seeds consumption respondent to domestic and foreign market needs.**

7077 Badino, M.; Allavena, A.; Soressi, G.P.; Falavigna, A.
       IT 021000/79/0024 N
Miglioramento delle proteine del fagiolo con particolare riguardo agli aminoacidi solforati. **Breeding of bean proteins with regard to sulphurate aminoacids.**

7078 Casarini, B.; Ranalli, P.; Di Candilo, M.
       IT 021100/75/0003
Miglioramento genetico del pisello da industria. **Genetic improvement of pea for processing.** Publications.

7079 Ranalli, P.   IT 021100/77/0386 R
Miglioramento genetico del pisello. **Genetic improvement of peas.**

7080 Casarini, B.; Giordano, I.   IT 021100/79/0022 N
Studio della risposta varietale all'epoca di semina del pisello nel Meridione. **Study of the answer of the variety at the seeding time of the pea in the South.**

7081 Ciccarese, F.   IT 040106/78/1189 N
Miglioramento genetico del pisello per la resistenza verso le malattie. **Genetic improvement of disease resistant peas.**

7082 Scarascia Mugnozza, G.T.; Greco, S.; Filippetti, A.; Dellagatta, C.; Depace, C.   IT 040115/75/0001
Miglioramento genetico di vicia per l'alimentazione umana ed animale. **Breeding studies for improving viciae species of relevant importance for animal and human nutrition.**

7083 Filippetti, A.   IT 040115/78/1184 N
Costituzione di cultivar di favino e pisello dotate di elevata produttività e precocità. **French beans and peas : development of high yielding, early cultivars.**

7084 Pritoni, G.; Amaducci, M.T.   IT 040201/79/0001 N
Confronto varietale fra cultivar di favino (Vicia faba minor) in diverse località ed epoche di semina. **Comparison among horse bean varieties in different localities and sowing date.**

7085 Tonini, Giuseppe; Maccaferri, Massimo
       IT 040216/77/0005 N
Adattabilità varietale dei fagiolini alla surgelazione. **Quality of green beans for freezing.**

7086 Foti, S.   IT 040305/77/0353 R
Miglioramento genetico della fava. **Genetic improvement of beans.**

7087 La Malfa, G.   IT 040313/77/0363 R
Ottenimento di varietà di fave da mensa per consumo fresco. **Establishment of new bean varieties for table use as fresh vegetables.**

7088 Cervato, A.   IT 040401/77/0338
Selezione di linee di pisello adatte alla conservazione industriale della granella verde. **Selection of pea lines suitable for the industrial processing of green peas.**

7089 Monti, L.   IT 040701/77/0376 R
Miglioramento genetico del pisello per uso industriale e della fava per uso zootecnico. **Genetic improvement of peas for processing and of field beans for fodder crops.**

7090 Caruso, P.   IT 040914/77/0336 R
Miglioramento genetico della fava da mensa e del fagiolo da granella. **Genetic improvement of broad beans and haricot beans.**

7091 Polignano, G.B.   IT 060500/78/1195 N
Costituzione di nuove cultivar di pisello e fava caratterizzate da elevata produttività e da buona precocità e resistenza ad avversità ambientali (siccità, freddo). **Development of new pea and bean cultivars with high productivity, early crops and resistance to environmental hazards (drought, cold).**

7092 Di Vito, M.   IT 060600/73/0048
Analisi delle reazioni varietali agli attacchi dei nematodi. Varietà di pisello a Heterodera gottingiana. **Analysis of the varietal reactions to nematode attacks. Pea varieties to Heterodera gottingiana.**

7093 Bianco, V.V.   IT 062700/77/0453
Selezione e coltivazione di cultivar di pisello e fava

D 2300 – Plant breeding

caratterizzate da uniformità di maturazione, elevata produttività e ottime caratteristiche qualitative ai fini della preparazione di surgelati. **Selection and cultivation of high yielding, uniformly maturing pea and bean cultivars with excellent qualities for processing as frozen foods.**

7094 Bianco, V.V.   IT 062700/77/0454
Costituzione di nuove cultivar di pisello e fava caratterizzate da elevata produttività e da buona precocità e che abbiano buone capacità tecnologiche ai fini della utilizzazione industriale e resistenza ad avversità ambientali. **Development of new pea and bean cultivars characterized by their high yielding capacity, early maturation, good processing qualities and their resistance to adverse environmental factors.**

7095 Foti, S.   IT 063100/77/0904
Analisi dei caratteri che regolano il ritmo e la contemporaneità di maturazione dei fagioli da legume mangiatutto. **Analysis of the characters regulating the rythm and simultaneous maturation of string beans.**

7096 Saccardo, F.   IT 070101/78/1192 N
Mutagenesi e miglioramento genetico del pisello da industria. **Mutagenesis and the genetic improvement of peas for processing.**

7097 Del Zan, F.; Murgut, G.; Tonetti, I.   IT 090701/79/0002 N
Miglioramento genetico del fagiolo rampicante. **Genetic improvement of pole bean.**

7098 Wassenaar, R.   NL 010113/43/0616
Rassenonderzoek peulvruchten. **Research on varieties of pulses.** Publications.

7099 Schneider, F.; Riepma, P.   NL 010113/69/1545
Gebruikswaarde onderzoek stamslabonen. **Variety testing of bush beans.** Publications.

7100 Schneider, F.; Riepma, P.   NL 010113/75/6727
Gebruikswaarde–onderzoek doperwten. **Research on cultural value of green peas.** Publications.

7101 Schneider, F.; Riepma, P.   NL 010113/78/8425
Gebruikswaarde–onderzoek tuinboon. **Variety testing broad bean.**

7102 Schneider, F.; Riepma, P.   NL 010113/78/8427
Gebruikswaarde–onderzoek stokboon. **Variety testing of runner bean.**

7103 Drijfhout, E.   NL 010114/55/0676
Veredelingsonderzoek ziekte–resistentie bij bonen (bonevirus 1 en 2). **Breeding research on disease resistance in bean (beanvirus 1 and 2).** Publications.

7104 Riepma, P.   NL 010114/69/2351
Gebruikswaarde–onderzoek tuinboon. **Variety testing of broad beans.** Publications.

7105 Riepma, P.; Kraker, J. de   NL 010114/74/6508
Gebruikswaarde–onderzoek snijboon. **Research on cultural value of dwarf slicing bean.**

7106 Drijfhout, E.   NL 010114/77/7274
Veredeling op ziekteresistentie van droge bonen voor Latijns Amerika. **Breeding for disease resistance in dry beans for Latin America.**

7107 Riepma, P.   NL 010207/69/1545
Gebruikswaarde onderzoek stamslabonen. **Variety testing of bush beans.** Publications.

7108 Riepma, P.   NL 010207/75/6727
Gebruikswaarde–onderzoek doperwten. **Research on cultural value of green peas.** Publications.

7109 Riepma, P.   NL 010207/78/8425
Gebruikswaarde–onderzoek tuinboon. **Variety testing of broad bean.**

7110 Riepma, P.   NL 010207/78/8427
Gebruikswaarde–onderzoek stokboon. **Variety testing of runner bean.**

Tomatoes (B 3532)

See also 3285, 6417, 6887, 7041, 7191

7111 Plumier, W.   BE 080200/63/0021 R
Amélioration de la tomate pour la consommation à l'état frais. **Breeding of tomato as vegetable for fresh consumption.** Publications.

7112 Persiel, F.; Rockstroh, K.   DE 206000/71/4008 R
Die genetischen Beziehungen zwischen den spontan auftretenden Nekrosen und der Resistenz gegen Cladosporium fulvum bei Tomaten 1966–1979. **Genetic correlations between spontaneous necroses and resistance of tomatoes to Cladosporium fulvum.**

7113 Sibi, M.   FR 010105/76/0471
Recherche de phénovariants par culture de tissus in vitro sur lycopersicum. **In vitro tissue culture of lycopersicum esculentum. Phenovariants formation, heredity in selfings and crossings.**

7114 Laterrot, H.   FR 010603/71/0058
Etude des possibilités d'utilisation de facteurs de résistance au virus de la mosaïque du tabac chez la tomate. **Studies on the possibility of using resistance factors to tobacco mosaic virus in tomato breeding.** Publications.

7115 Philouze, J.   FR 010603/71/0059
Etude de la stérilité mâle pour la production d'hybrides F_1 chez la tomate. **Studies using male sterility in breeding F_1 hybrids of tomato.**

7116 Philouze, J.   FR 010603/71/0060
Recherche de variétés de tomate adaptées à la culture sous serre. **Research on tomato varieties for glasshouse–cultivation.**

7117 Philouze, J.   FR 010603/71/0061
Recherche de variétés à croissance déterminée pour la culture non tuteurée, pour l'expédition. **Research on tomato varieties for the market with a defined growth growing without support.**

7118 Laterrot, H.   FR 010603/72/0003
Etude des possibilités d'utilisation de différentes origines de résistance à la maladie des racines liégeuses chez la tomate. **Study on the possibility of using different sources of resistance to corky–root in tomato – breeding.**

7119 Laterrot, H.   FR 010603/72/0092
Etude des possibilités d'utilisation de différentes origines de résistance à la maladie des racines liégeuses chez la tomate. **Investigation on the utilization of different sources of resistance to corky roots in tomatoes.**

7120 Laterrot, H.   FR 010603/74/0223
Etude de possibilités de sélection de la tomate pour un haut niveau de résistance au Corynebacterium Michiganense. **Breeding tomatoes for good resistance to Corynebacterium Michiganense.**

7121 Vergniaud, P.   FR 010610/78/9079
Identification de nouvelles variétés de référence de tomates de conserve en fonction de leur emploi industriel. **Searching appropriate canning tomato reference (control) varieties suited to the different industrial products aimed at.**

7122 Darby   GB 011101/00/0004
Breeding tomato varieties for contrasting crop production systems.

7123 Taylor   GB 011101/00/0005
Genetic elimination of tomato sideshoots.

7124 Grimbly   GB 011101/00/0006 R
Genetic and environmental causes of silvering of tomato plants.

7125 Smith   GB 011101/00/0007
Genetic control of winter tomato fruiting.

7126 Hall   GB 011101/00/0008
Virus strain competition in tomato.

7127 Darby; Ritchie   GB 011101/00/0009
Develop isogenic tomato lines.

7128 Hesling   GB 011108/00/0016
Breeding for resistance to potato cyst–eelworm.

7129 Feely, L.; Gallagher, P.A.   IE 060300/72/0056 N
Tomato cultivar testing. Publications.

7130 Rizzo, V.   IT 020500/77/0391 R
Valutazione agronomica di nuove linee e varietà di pomodoro. **Agronomical value of new lines and varieties of tomatoes.**

7131 Soressi, G.P.; Falavigna, A.; Uncini, L.; Ferrari, V.; D'Amore, R.   IT 021000/70/0001 R
Utilizzazione del gene Nor² per la sintesi di genotipi ibridi con particolari caratteristiche di conservabilità e qualità. **Utilization of Nor² gene for hybrid genotype synthesis with quality and storage traits.** Publications.

7132 Porcelli, S.; Soressi, G.P.; Uncini, L.   IT 021000/70/0007
Miglioramento genetico del pomodoro per concentrato e pelato. **Tomato breeding for paste and peeling.** Publications.

7133 Uncini, L.; Ferrari, V.; Restaino, F.; Soressi, G.P.
                    IT 021000/70/0008 R
Costituzione di varietà ed ibridi di pomodoro da mensa per la coltura di piena aria e protetta. **Constitution of table tomato varieties and hybrids for open air and protected cultures.** Publications.

7134 Uncini, L.; Restaino, F.; Fiume, F.; Interlandi, G.;

Ferrari, V.   IT 021000/71/0008 R
Introduzione di resistenze genetiche alle principali malattie da funghi, virus ed ai nematodi in linee di pomodoro. **Introduction of genetic resistance (TMV) to main fungi deseases, virus, nematodes in tomato lines.** Publications.

7135 Soressi, G.P.; Falavigna, A.; Badino, M.
                    IT 021000/72/0015 R
Utilizzazione del carattere di partenocarpia "pat" in pomodoro. **Utilization of the partenocarpic character "pat" in tomato.** Publications.

7136 Soressi, G.P.; Falavigna, A.   IT 021000/75/0004
Tecnica di produzione seme ibrido di pomodoro. **Production technique of tomato hybrid seed.**

7137 Soressi, G.P.; Uncini, L.; Falavigna, A.; D'Amore, R.
                    IT 021000/75/0005 R
Costituzione di varietà ed ibridi di pomodoro da industria idonei alla meccanizzazione delle operazioni colturali con riferimento alla raccolta. **Constitution of canning tomato varieties and hybrids suitable for cultural operations mechanization with reference to mechanical harvesting.**

7138 Soressi, G.P.; Uncini, L.; Falavigna, A.; D'Amore, R.
                    IT 021000/76/0012 R
Costituzione di varietà di pomodoro da industria con architettura modificata della pianta idonei all'alto investimento e alla raccolta meccanica. **Constitution of canning tomato varieties with plant modified shape suitable for high investment and mechanical harvesting.**

7139 Falavigna, A.   IT 021000/77/0350 R
Costituzione di ibridi F1 da pelato e concentrato, pomodoro. **Breeding of new F1 hybrids for processing into peeled tomatoes and tomato paste.**

7140 Porcelli, S.   IT 021000/77/0706 R
Meccanizzazione della raccolta di pomodoro e cavolfiore, nuove cultivar adatte a raccolta meccanica, pomodoro, aspetti agronomici, raccolta meccanica, cavolfiore. **Mechanization of tomato and cauliflower harvesting, new cultivars suited to mechanical harvesting, tomatoes, agronomic aspects, mechanical harvesting, cauliflowers.**

7141 Uncini, L.; Porcelli, S.; D'Amore, R.; Ferrari, V.
                    IT 021000/79/0006 N
Costituzione di varietà di pomodoro idonee alla preparazione di concentrato e triturato. **Constitution of tomato varieties suitable for concentrate and triturate.**

7142 Casarini, B.; Ranalli, P.; Di Candilo, M.
                    IT 021100/75/0004
Miglioramento genetico del pomodoro da industria. **Genetic improvement of tomato for processing.** Publications.

7143 Giordano, I.   IT 021100/75/0012 N
Confronto di varieta' di pomodoro. **Varietal comparison with tomato.**

7144 Di Candilo, M.   IT 021100/78/1185 N
Costituzione di CV. di pomodoro per concentrato e pelato resistenti alle malattie e corrispondenti alle nuove esigenze colturali italiane. **Tomato : development of disease resistant cultivars for processing as peeled tomatoes or tomato paste and satisfying the new requirements of Italian cultures.**

7145 Cirulli, M.　　　　　IT 040106/77/0340 R
Miglioramento genetico del pomodoro da industria verso le malattie. **Genetical improvement of disease resistant tomatoes for processing.**

7146 Conti, S.　　　　　IT 040201/77/0342 R
Costituzione di varietà di pomodoro omozigoti a taglia ridotta maturazione sincrona con frutti a colorazione intensa ed uniforme e prive di articolazione sul penducolo, ottenimento di varietà e di ibridi F1 di girasole adatti alle condizioni pedoclimatiche della bassa collina emiliana e marchigiana a ciclo medio. **New varieties of small sized tomato homozygotes maturing synchronically, intensely and uniformly coloured, jointless; establishment of new sunflower varieties and F1 hybrids with an intermediate growth cycle, suited to the local climate of the lower hills of Emily and the Marches.**

7147 Restuccia, G.　　　　　IT 040305/77/0388 R
Costituzione di varietà di pomodoro in grado di assicurare in Sicilia rese soddisfacenti e dotate delle indispensabili caratteristiche per le diverse utilizzazioni industriali del prodotto. **Breeding new varieties of tomatoes which will guarantee a satisfactory yield in Sicily and possess the necessary requirements for various processing procedures.**

7148 Marchesi, G.　　　　　IT 040415/72/0504
Miglioramento genetico del pomodoro, ricerche per costituire cultivar per la raccolta meccanica. **Genetic improvement of tomatoes; reserach in view of creating a cultivar which can be mechanically harvested.** Publications.

7149 Marchesi, G.　　　　　IT 040415/77/0226 R
Costituzione di cultivar di pomodoro da raccogliersi a macchina con la introduzione del mutante jointless, studio sulla eredità della resistenza al cracking con l'impiego del mutante fleshy calyx. **Establisment of tomato cultivars for machine harvesting with the introduction of the jointless mutant; shedy on hereditary resistance to cracking using the fleshy calyx mutant.**

7150 Postiglione, L.　　　　　IT 040701/74/0607
Miglioramento genetico del pomodoro da industria. **Genetic improvement of industrial tomato crops.**

7151 Monti, L.　　　　　IT 040701/77/0375 R
Ottenimento di varietà di pomodoro da industria più adatte agli ambienti meridionali. **Development of tomato varieties for processing better suited to a southern environment.**

7152 Ceccarelli, S.　　　　　IT 041002/72/0425
Miglioramento genetico del pomodoro da mensa e del peperone. **Genetic improvement of table tomatoes for export. Research on sterile male strains.** Publications.

7153 Tesi, R.　　　　　IT 041115/77/0407 R
Costituzione di nuove cultivar di pomodoro da industria adatte alla raccolta meccanica. **Development of new tomato cultivars for processing suitable for mechanical harvesting.**

7154 Marras, G.F.　　　　　IT 041302/77/0371 R
Attuazione di incroci di cultivar di pomodoro e linee di interesse agronomico. **Forming hybrids of tomato cultivars and lines of agronomical interest.**

7155 Foti, S.　　　　　IT 063100/77/0902

Ulteriore caratterizzazione dei pomodori da mensa costituiti dal centro sotto il profilo della adattabilità a condizioni diverse e della resistenza ai parassiti. **Table tomato cultivars produced by the centre: further characterisation regarding their adaptability to varying conditions and their resistance to parasites.**

7156 Saccardo, F.　　　　　IT 070101/78/1184 N
Ottenimento di varietà di pomodoro da industria a maturazione contemporanea e di varietà di pomodoro resistenti a malattie. **Tomatoes : breeding varieties for processing characterized by simultaneous maturation and varieties resistant to diseases.**

7157 Schneider, F.; Stolk, J.H.　　　　　NL 010113/51/2147
Rassenonderzoek tomaat. **Variety testing of tomato.** Publications.

7158 Smeets, L.　　　　　NL 010114/63/0723
Produktie factoren–analyse vroege stooktomaat. **Analysis of production factors in early forcing tomato.**

7159 Boukema, I.W.　　　　　NL 010114/70/2815
Veredeling van tomaat op resistentie tegen Cladosporium fulvum, Didymella lycopersici en Phytopthora nicotianae. **Breeding of tomatoes resistant to Cladosporium fulvum, Didymella lycopersici and Phytopthora nicotianae.** Publications.

7160 Hogenboom, N.G.　　　　　NL 010114/70/2996
Veredeling van tomaat op resistentie tegen tabaksmozaïekvirus en kurkwortel. **Breeding resistance in tomato to TMV and corky root.** Publications.

7161 Ponti, O.B.M. de; Hogenboom, N.G.
　　　　　NL 010114/72/3821
Veredeling op resistentie tegen de witte vlieg (Trialeurodes vaporariorum Westw.) in tomaat. **Breeding for resistance against greenhouse white fly (Trialeurodes vaporiorum) in tomato.** Publications.

7162 Hogenboom, N.G.　　　　　NL 010114/74/5166
Veredeling van tomaat op geringere warmtebehoefte. **Tomato breeding for lower heat requirements.** Publications.

7163 Hogenboom, N.G.; Pet, G.　　　　　NL 010114/74/5167
Ontwikkeling van een tomaat voor een minder arbeidsintensieve teelt. **Tomato breeding for lower requirement.**

7164 Roelofsen, H.J.　　　　　NL 010114/78/7930
Genenbank voor tomaat. **Gene bank for tomato.**

7165 Stolk, J.H.　　　　　NL 010206/51/2147
Rassenonderzoek tomaat. **Variety testing tomato.** Publications.

Cucumbers (B 3533)

See also 6887

7166 Schulte, H.-K.; Kunkel, U.　　　　DE 138300/70/0007 R
Die genetischen, physiologischen und züchterischen Grundlagen zur Verbesserung der Gurkensorten für den Freilandanbau. **Genetic, physiological and breeding conditions for the improvement of cucumber varieties for field cultivation.**

D 2300 – Plant breeding

7167 Wricke, G.; Franken, S. DE 138300/75/0002
Untersuchungen zur Vererbung von Internodienlänge, Hermaphroditismus und Ertragsfähigkeit bei Freilandgurken mit determiniertem Wuchstyp. **Investigations of the inheritance of length of internodes, hermaphroditism and yield potential in pickling cucumbers of determinate habit.**

7168 Grimbly GB 011101/00/0011
Breeding cucumber varieties for contrasting crop production systems.

7169 Feely, L.; Maher, M.J. IE 060300/70/0058 N
Cucumber, pepper and other glasshouse crops cultivar testing. Publications.

7170 Schneider, F.; Stolk, J.H. NL 010113/52/2139
Rassenonderzoek komkommer. **Variety testing of cucumber.** Publications.

7171 Nijs, A.P.M. den NL 010114/63/3548
Veredeling op ziekteresistentie bij komkommer. **Breeding for disease resistance in cucumber.** Publications.

7172 Ponti, O.B.M. de NL 010114/72/3540
Veredeling van komkommer op resistentie tegen bone spintmijt en kas witte vlieg. **Breeding for resistance to the two spotted spider mite and glass house white fly in cucumber.** Publications.

7173 Nijs, A.P.M. den NL 010114/74/5168
Veredeling van komkommer op geringere warmtebehoefte. **Breeding of cucumbers for lower heat requirement.** Publications.

7174 Nijs, A.P.M. den NL 010114/77/7273
Soortkruisingen bij Cucurbitaceae. **Species crosses of Cucurbitaceae.**

7175 Roelofsen, H.J. NL 010114/78/7933
Genenbank voor komkommer. **Gene bank for cucumber.**

7176 Stolk, J.H. NL 010206/52/2139
Rassenonderzoek komkommer. **Variety testing of cucumber.** Publications.

Other vegetable fruits (B 3539)

See also 7169, 7174, 16651

7177 Reimann–Philipp, R.; Laskawy, C.; Timmann, E.–M.
DE 206000/71/5003
Züchtung von Freilandgurken zur Kombination folgender Merkmale: vorwiegend weiblich, bitterfrei, virustolerant, parthenokarp, mehltauresistent, gestauchtwüchsig. **Breeding of pickling cucumber for combination of the following characters: gynomonoecious – predominantly female – bitterfree, virustolerant, parthenocarpic fruit, mildew–resistant, determinate growth.**

7178 Pochard, E.; Serieys, A. FR 010603/75/0244
Sélection de l'aubergine pour la culture en serre: étude de l'influence de l'état hybride des plantes sur les capacités de production. **Aubergine breeding for greenhouse cultivation: influence of hybrid state of plants on the production ability.**

7179 Pochard, E.; Serieys, A. FR 010603/75/0323

Sélection de l'aubergine pour la culture en serre: étude de l'influence de l'état hydrique des plantes sur les capacités de production. **Breeding aubergine for cultivation under glass: study of plants water states influences on yielding capacities.**

7180 Pochard, E.; Florent, A.; Marchoux, G.; Migliori, A.
FR 010603/76/0359
Etude de la stratégie d'emploi des gènes de résistance au virus Y (PVY) chez le piment (Capsicum Annuum). **Strategy for the use of different resistance genes to PVY (virus Y) in the pepper (Capsicum Annuum).**

7181 Uncini, L. IT 021000/67/0003 R
Costituzione di varietà e ibridi di zucchino idonei alla coltura di pien'aria e protetta, con frutti rispondenti alle esigenze dei mercati nazionali ed esteri. **Constitution of marrow varieties and hybrids suitable for protected and open air cultures, with fruits respondent to domestic and forcing markets.** Publications.

7182 Restaino, F. IT 021000/70/0002
Miglioramento della melanzana per precocità a mezzo di mutageni chimici e fisici. **Egg–plant breeding for earliness by induced mutations.** Publications.

7183 Restaino, F.; Uncini, L. IT 021000/70/0003 R
Miglioramento genetico della melanzana. **Egg–plant breeding.** Publications.

7184 Uncini, L.; Ferrari, V.; Restaino, F. IT 021000/72/0010 R
Studio dell'idoneità alla surgelazione e alle diverse tecniche di conservazione di specie ortive. **Study on deep freezing suitability and storage tecniques of horticultural plants.**

7185 Restaino, F.; Fiume, F.; Interlandi, G.; Uncini, L.
IT 021000/72/0014 R
Introduzione di resistenze genetiche ai virus più dannosi e a Verticillium in linee e varietà di peperone e melanzana. **Introduction of genetic resistance to most noxious virus and Verticillium in egg–plant and pepper lines and varieties.**

7186 Uncini, L.; Restaino, F.; D'Amore, R.; Ferrari, V.
IT 021000/75/0002 R
Costituzione di varietà ed ibridi di peperone dolce e piccante adatti alla coltura protetta e di pien'aria, per consumo fresco, conservazione e trasformazione. **Constitution of varieties and hybrids of sweet and hot pepper suitable for open air and protected culture, for fresh consumption, storage and processing.**

7187 Uncini, L. IT 021000/79/0005 N
Sviluppo di nuove forme di architettura delle piante di peperone. **New shape development of pepper plants.**

7188 Lombardi, D.; Interlandi, G. IT 021000/79/0023 N
Applicazione di tecniche di colture in vitro per il reperimento di nuove fonti di resistenza alla tracheomicosi del peperone. **Applications of in vitro techniques for the obtainment of new resistance sources to pepper mycosis wilt.**

7189 Conti, S.; Concilio, L. IT 040201/79/0003 N
Miglioramento genetico della melanzana. **Genetic improvement of eggplant.**

7190 Lepori, G.; Franceschetti, U.; Quagliotti, L.; Nassi, M.O. IT 041212/73/0001

D 2300 – Plant breeding

Miglioramento genetico del peperone per la resistenza alle virosi. **Pepper breeding for virus resistance.**

7191 Blanco, V.V.　　　　　　　IT 062700/77/0942
Ricerche sul comportamento agronomico delle cultivar (cetriolo, cipolla, fava, fagiolo, lattuga, mais dolce, patata, pomodoro, porro, sedano–rapa). **Research on the cultural behaviour of cultivars (cucumber, onion, beans, lettuce, sweet corn, potato, tomato, leek, celery rape).**

7192 Del Zan, F.; Tonetti, I.; Murgut, G.　IT 090701/79/0001 N
Prova di confronto tra cultivar di melanzana in serra fredda. **Tests between egg–plant varieties under plastic greenhouse.**

7193 Schneider, F.; Riepma, P.　　　　NL 010113/68/4000
Gebruikswaarde onderzoek augurk. **Variety testing of pickling cucumber.**

7194 Schneider, F.; Stolk, J.H.　　　　NL 010113/72/4659
Rassenonderzoek kasaugurken. **Variety testing of glasshouse gherkins.** Publications.

7195 Nijs, A.P.M. den　　　　　　NL 010114/64/3546
Veredeling van kasaugurken. **Breeding indoor pickling cucumber.** Publications.

7196 Hogenboom, N.G.; Pet, G.; Boukema, I.W.
　　　　　　　　　　　　　　　NL 010114/72/3823
Veredelingsonderzoek aan paprika(vroege rassen en arbeidsarme teelt). **Breeding research on peppers (early varieties, low labour requirements).** Publications.

7197 Hogenboom, N.G.　　　　　NL 010114/73/3824 R
Bloembiologisch onderzoek aan paprika. **Floral biological research on peppers.** Publications.

7198 Stolk, J.H.　　　　　　　　NL 010206/72/4659
Rassenonderzoek kasaugurken. **Variety testing of glasshouse gherkins.** Publications.

7199 Riepma, P.　　　　　　　　NL 010207/68/4000
Gebruikswaarde onderzoek augurk. **Variety testing of pickling cucumber.** Publications.

Mushrooms and other edible fungi (B 3540)

See also 3998, 4011, 4012

7200 Brian, C.; Imbernon, M.　　　　FR 010703/75/0237
Obtention de nouvelles variétés d'agaricus bisporus par association de mycelium haploïdes. **Breeding of new varieties of agaricus bisporus in association of haploid myceliums.**

7201 Brian, C.; Imbernon, M.　　　　FR 010703/75/0239
Isolement d'auxotrophes à partir des souches d'Agaricus Bisporus et sélection de génotypes dépourvus d'altérations biochimiques. **Auxotrophs isolation from Agaricus Bisporus strains and selection of genotypes without biochemical alterations.**

7202 Brian, C.; Imbernon, M.　　　　FR 010703/75/0240
Obtention de souches de Pleurotus Ostreatus ne produisant pas de spores. **Breeding of Pleurotus Ostreatus strains without spores production.**

7203 Brian, C.; Imbernon, M.　　　　FR 010703/75/0241

Evaluation des potentialités d'un certain nombre d'espèces de champignons sauvages comestibles. **Potentialities evaluation of wild edible mushrooms strains.**

7204 Brian, C.; Imbernon, M.　　　　FR 010703/75/0242
Obtention de souches améliorées de Pleurotus Ostreatus par association de mycelium haploïdes. **Breeding of improved strains of Pleurotus Ostreatus in association of haploid myceliums.**

7205 Imbernon, M.; Brian, C.; Pirobe, L.　FR 010703/75/6055
Obtention de souches améliorées de Pleurotus ostreatus par association de mycéliums haploïdes. **Pleurotus ostreatus breeding.**

7206 Elliott　　　　　　　　　GB 011101/00/0001 R
Genetics of cultivated mushroom.

7207 Schneider, F.　　　　　　　NL 010113/77/8314
Rassenonderzoek champignons. **Variety testing of mushrooms.** Publications.

7208 Fritsche, G.　　　　　　　NL 010204/73/3861
Veredeling van champignons en de daarmee in verband staande vragen zoals broedbereiding en instandhouden van rassen. **Breeding of mushrooms (Agaricus) and research in spawnmaking and maintenance of strains, related with the breeding work.** Publications.

Fruits in general (B 3600)

See also 3156, 3785, 6857, 6858, 6872, 6873, 6875, 6877, 7495, 7654

7209 Boxus, Ph.; Druart, Ph.　　　　BE 080200/64/0015 R
Assainissement viral des espèces arborescentes et production de matériel assaini. **Fruit tree virus control and production of virus free material.** Publications.

7210 Monin, A.; Seynaeve, R.　　　　BE 080200/76/0042
Pollinisation artificielle. **Artificial pollination.**

7211 Weiling, F.; Unger, C.　　　　DE 111151/70/0002
Kriterien der Klonenauslese 1969. **Criteria of clone selection.** Publications.

7212 Groven, I.; Larsen, O.N.; Bøvre, O.　DK 010104/57/3302
Formering og produktion af frugttræer og frugtbuske. **Propagation and production of fruit trees and fruit bushes.**

7213 Salesses, G.; Mouras, A.　　　　FR 010701/75/0320
Essai de mise au point d'une technique d'identification des chromosomes chez les arbres fruitiers. **Test about chromosomes identification practice in fruit–trees.**

7214 Hunter; Murray　　　　　　GB 011007/75/0004 R
Chemical and biochemical composition of plants, pests, predators and pathogens with reference to interaction on fruits and hops.

7215 Campbell; Lacey　　　　　　GB 011510/00/0016
Induction and selection of mutant forms of fruit trees using irradiation.

7216 Intrieri, C.　　　　　　　IT 040203/74/0570
Portinnesti degli alberi da frutto, albicocco, ciliegio e pesco e

D 2300 – Plant breeding

sovrinnesto del pero. **Rootstocks of fruit trees, apricot, cherry and peach trees and overgrafting of pear tree.**

7217 Sansavini, S.; Costa, G. IT 040203/77/0002
Miglioramento genetico e studio varietà di fruttiferi. **Fruit breeding and varieties testing program.**

7218 Poma Treccani, C. IT 040605/74/0606
Influenza del portinnesto, del tipo spur e del sadh sulle variazioni stagionali dell'equilibrio endogeno delle gibberelline e dell'acido abscissico. **Influence of rootstock, spur and sadh type, on the seasonal variations in the endogenic balance of giberellins and abscissic acid.**

7219 Paglietta, R. IT 041203/74/0595
Porta innesti dei fruttiferi. **Rootstocks of fruit trees.**

7220 Eynard, L. IT 041203/77/0171 R
Introduzione di nuove cultivar di vaccinium, rubus e ribes e studi relativi agli ambienti piu idonei di coltura, alla meccanizzazione della raccolta ed alla prima lavorazione dei prodotti. **Introduction of new cultivars of vaccinium, rubus and ribes. Studies on the most suitable places for cultivation, on the mecchanization of harvesting and on initial processing.**

7221 Romisondo, P. IT 062100/69/0151
Induzione di mutazioni con agenti mutageni e successiva selezione. **Mutations induced by mutagen agents and selection of treated material.** Publications.

7222 Bargioni, G. IT 102801/74/0511
Miglioramento genetico della frutta da industria. **Genetic improvement of industrial fruit crops.**

7223 Kiès, B. NL 010113/77/8315
Rassenonderzoek fruit. **Variety testing of fruit.** Publications.

Top fruit in general (B 3610)

See also 8700, 8702

7224 Tiemann, K.–H.; Dammann, H.–J.; Palm, G.
DE 507306/75/0002
Sortenprüfung Kernobst. **Variety testing of pome fruit.**

7225 Poulsen, E.; Rasmussen, P.M. DK 010102/17/3101
Sortsafprøvning af æble, pære, blomme og kirsebær. **Variety trials with apples, pears, plums and cherries.**

7226 Watkins GB 011002/00/0005
Breed new varieties of top fruit rootstocks.

7227 O'Kennedy, N.D. IE 060302/62/0496
Evaluation on new apple cultivars. Publications.

7228 Nicotra, A.; Damiano, C.; Moser, L.; Ricciardi, P.; Monastra, F.; Grassi, G. IT 021500/78/0007 N
Selezione e costituzione di nuovi portinnesti per: drupacee, pero e castagno. **Breeding new rootstocks for stone fruits, pear and chestnut trees.**

7229 Goddrie, P.D. NL 010212/68/3545
Gebruikswaarde–onderzoek groot fruit. **Variety testing of top fruit.** Publications.

7230 Oosten, H.J. van NL 010212/73/4452

Virus– en klonenonderzoek bij groot fruit. **Investigations on variability in top fruit induced by viruses and clonal differences.** Publications.

Apple (B 3611)

See also 4167, 4171, 7287

7231 Monin, A. BE 080200/65/0009
Recherches sur hybrides Cox's xPresident Roulin. **Study of hybrides Cox's x Pr. Roulin.**

7232 Monin, A. BE 080200/76/0043
Essai sur Reinette de France basse–tige. **Dwarfed trees of Reinette de France.**

7233 Monin, A.; Boxus, Ph. BE 080200/76/0046 R
Etudes de divers clônes du type M 9. **Clonal types rootstocks Malling 9.**

7234 Karnatz, A. DE 105204/70/0001 N
Genetische Analyse von Selbstungsnachkommenschaften bei Äpfeln. **Genetical analysis of selfing progenies in apples.**

7235 Schmidt, H.; Hofmann, K.; Laskawy, W.; Engelhardt, K.; Wirth, G.; Münzel, K. DE 206000/78/0002 N
Apfelunterlagenzüchtung unter Ausnutzung der in polyploiden Wildformen vorkommenden Apomixis. **Apple rootstock breeding using apomixis from polyploid wild Malus species.**

7236 Schmidt, H.; Hofmann, K.; Laskawy, W.; Engelhardt, K.; Wirth, G. DE 206000/78/0003 N
Züchtung leistungsfähiger, krankheitsresistenter Apfelsorten mit hoher Fruchtqualität. **Breeding of productive apple cultivars with resistance to diseases and high fruit quality.**

7237 Schmidt, H.; Hofmann, K.; Laskawy, W.; Engelhardt, K.; Wirth, G.; Münzel, K. DE 206000/78/0004 N
Abschliessende Wertermittlung von Apfelsämlingsnachkommenschaften aus der, Obstbauversuchanstalt Jork. **Final evaluation of apple seedling progenies from the Obstbauversuchsanstalt Jork.**

7238 Kunze, L.; Krczal, H. DE 215210/70/4017
Untersuchungen über die Triebsucht des Apfels. **Investigations into the proliferation of apple.**

7239 Tiemann, K.–H.; Zahn, F.–G. DE 507307/75/0001
Sortenprüfung Steinobst. **Variety testing of stone fruit.**

7240 Wirth, H.; Ladebusch, H. DE 507701/78/0006 N
Sorten– und Baumformversuch zu Äpfeln auf Unterlage M 9. **Trials of apple varieties and tree forms on rootstock M 9.**

7241 Christensen, J.V. DK 030171/65/0001 N
Mutationsforædling med æble og kirsebær. **Mutation breeding with apple and cherry.**

7242 Decourtye, L.; Lantin, B. FR 010401/67/0001
Obtention par hybridation de nouvelles variétés de pommier résistantes à la tavelure et peu sensibles a l' oïdium. **Breeding new hybrid varieties of apple scab–resistant and not too susceptible to powdery mildew.** Publications.

7243 Lespinasse, Y.; Decourtye, L.; Renoux, A.
FR 010401/72/0096

Obtention de variétés de pommier résistantes à la tavelure associant un contrôle monogénique et un contrôle polygénique. **Apple breeding for scab resistance combining monogenic and polygenic control.** Publications.

7244 Lefevre, P.; Gaffet, M.A. FR 012222/63/6047
Recherches sur la qualité de deux variétés de pommes: a)Cox's Orange Pippin, b) Golden Delicious. **Researchs about the quality of the apples: a) Cox's Orange Pippin, b) Golden Delicious.**

7245 Watkins; Alston GB 011002/00/0001
Breed new varieties.

7246 Alston GB 011002/00/0002
Breed new varieties.

7247 Fideghelli, C.; Monastra, F.; Donini, B.
IT 021500/71/0017
Miglioramento genetico del melo. **Apple breeding.** Publications.

7248 Rosati, P.; Faedi, W.; Tagliani, F. IT 021500/72/0027
Osservazioni su piante "spur", piante normali e piante "spur" regredite di melo. **Observations on spur and standard apple tree and on spur regressed to standard.**

7249 Cobianchi, D.; Faedi, W.; Turci, E.; Limongelli, F.
IT 021500/77/0012 R
Studio sulla differenziazione delle gemme e sulla longe–vità del sacco embrionale delle cultivar Granny Smith, Granny Smith spur e Annurca, innestate su alcuni portinnesti clonali. **Study of the flower bud formation and of the embryo–sac life of the Granny Smith and Annurca apple varieties, grafted on several clonal rootstocks.** Publications.

7250 Limongelli, F.; Monastra, F.; Temperini, O.
IT 021500/78/0005 N
Autofertilità e interfertilità della cv. "Annurca". **Self and interfertility of the Annurca apple variety.**

7251 Fidechelli, C.; Nicotra, A.; Manzo, P.; Monastra, F.; Cobianchi, D.; Bergamini, A. IT 021500/79/0001 N
Valutazione agronomica delle migliori e recenti cultivar, italiane ed estere, delle seguenti specie: melo, pero, pesco, albicocco, nettarine, mandorlo, ciliegio dolce, ciliegio acido, susino, fragola, lampone, mirtillo gigante, rovo, ribes, noce, nocciolo, pecan, sambuco, fico, castagno, pistacchio, nespolo, giapponese, kaki, actinidia. **Variety testing trial of the best and new italian and foreign cultivar of the following species: apple, pear, peach, apricot, nectarine, almond, sweet and sour cherry, plum, strawberry, raspberry, blueberry, blackberry, current, walnut, filbert, pecan, sambucus, fig, chestnut, pistachio, loquat, persimmon, kiwi.**

7252 Sansavini, S. IT 040203/74/0624
Ricerche sul miglioramento genetico e qualitativo del melo e del pero. **Research on genetic and quality improvement of pear and apple trees.**

7253 Visser, T. NL 010114/69/2808 R
Het kweken van nieuwe appel– en pererassen. **Breeding new apple and pear varieties by crossing.** Publications.

7254 Visser, T. NL 010114/69/2809
Overerving– en voorselectiestudies bij appel en peer.

Inheritance and pre–selection studies with apple and pear. Publications.

7255 Visser, T. NL 010114/69/8792 N
Veredeling op schurft– en meeldauwresistentie bij appel. **Breeding for scab and mildew resistance in apple.**

Pear (B 3612)

See also 7233, 7239, 7251, 7252, 7253, 7254

7256 Thibault, B.; Hermann, L. FR 010401/63/0005
Obtention par hybridation de nouvelles variétés de poirier à maturité tardive. **Breeding new late Ripening pear varieties.**

7257 Bidabe, B.; Le Lezec, M. FR 010401/65/0002
Héritabilité des exigences thermiques des bourgeons du pommier. **Heritability of thermic requirements of apple buds.** Publications.

7258 Brossier, J.; Michelesi, J.C. FR 010401/65/0006
Sélection de porte–greffes clonaux dans l'espèce Pyrus Communis. **Selection of clonal rootstocks in Pyrus communis.**

7259 Thibault, B.; Hermann, L.; Belouin, A.
FR 010401/74/0095
Obtention par hybridation de variétés de poirier à floraison tardive. **Pear breeeding late flowering varieties.**

Other top fruit (B 3619)

See also 4167, 4236, 4237, 7239, 7241, 7251, 10379

7260 Monin, A. BE 080200/76/0039 R
Sélection clonale de la prune d'industrie Quetsche commune (Altesse imple). **Clonal selection of common quetsche.**

7261 Gruppe, W.; Schmidt, H. DE 129161/75/0001
Züchtung schwachwüchsiger Prunusunterlagen. **Breeding of dwarfing Prunus rootstocks.**

7262 Hartmann, W. DE 144445/77/0001
Selektion von wurzelechten Hauszwetschen in Baden–Württemberg. **Selection of 'Hauszwetsche' grown on their own roots in Baden–Württemberg.** Publications.

7263 Schmid, P.; Feucht, W. DE 161265/78/0002 N
Isoelektrische Trennung von Proteinen und Enzymen von verschiedenen Kirscharten und –sorten und deren Beeinflussung durch Veredlung. **Isoelectric separation of proteins and enzymes of different species and varieties of cherries as affected by grafting.**

7264 Schmidt, H.; Laskawy, W.; Hofmann, K.; Engelhardt, K.; Wirth, G.; Münzel, K. DE 206000/78/0001 N
Züchtung schwachen Wuchs induzierender Unterlagen für Süsskirschen. **Breeding of dwarfing rootstocks for sweet cherries.**

7265 Reimann–Philipp, R.; Brand; Göttmann
DE 402000/78/0003 N
Bonitur der Züchtungsbastarde von Sauerkirschen. **Taxation of hybrids of sour cherries.**

7266 Christensen, J.V. DK 030171/63/0002 N
Forædling af hyld. **Breeding of elderberry.**

7267 Sanfourche, G. FR 010701/60/0253
Recherche sur l'interpollinisation des variétés de cerisiers. **Research on interpollinization of cherry–tree.**

7268 Monet, R. FR 010701/64/0007
Création de variétés de pêcher ayant un comportement satisfaisant en France et dont les qualités organoleptiques sont améliorées. **Breeding new peach varieties well adapted to french conditions and improved for flavour.**

7269 Renaud, R.; Persais, J.P. FR 010701/65/0256
Sélection variétale de la prune de table. **Varietal selection of plums.**

7270 Mesnier, Y.; Bernhard, R. FR 010701/65/0304
Sélection de porte–greffe de prunier d'ente conférant des développements réduits. **Plum–tree "d'ente" rootstock selection giving short developments.** Publications.

7271 Salesses, G. FR 010701/66/0008
Recherches sur les possibilités de l'hybridation interspécifique pour améliorer les pruniers domestiques et créer des porte–greffes nouveaux. **Research about possibilities of interspecific hybridization for improvement of prunus domestica and breeding of new rootstocks.** Publications.

7272 Renaud, R.; Persais, J.P. FR 010701/66/0255
Sélection de prunes de sechage de maturité décalée. **Selection of prunes with shifted maturity.**

7273 Sanfourche, G.; Sarger, J. FR 010701/67/0254
Orientation de variétés nouvelles de cerisier. **Orientation of new varieties of cherry – tree.**

7274 Salesses, G.; Bonnet, A. FR 010701/68/0009
Recherches sur l'asphyxie radiculaire et la fatigue des sols. Création de porte–greffes de pêcher tolérants à l'asphyxie. **Research on root asphyxia and soil fatigue. Breeding peach rootstocks tolerant to asphyxia.** Publications.

7275 Grassely, C.; Olivier, G. FR 010701/68/0098
Obtention de nouveaux clones de pêcher x amandier présentant une bonne aptitude au bouturage ligneux. **Breeding of new peach x almond tree clones with a good ability for woody cutting propagation.**

7276 Couranjou, J. FR 010701/69/0010
Création de variétés d'abricotier en vue de régulariser et d'accrôitre la production et de l'étaler dans la saison. **Breeding apricot varieties of different earliness, for regular and increased yields.**

7277 Marenaud, Cl. FR 010701/71/0279
Etude de la variabilité génétique du caractère de sensibilite aux virus chez les prunus. **Study of the genetic variability of the character of sensibility to virus in prunus.**

7278 Saunier, R. FR 010701/71/0380
Sélection de cultivars de pêcher pour la production fruitière française. **Selection of peach cultivars for french fruit production.**

7279 Mesnier, Y. FR 010701/73/0234
Suite des travaux de sélection clonale du mirabellier en ce qui concerne plus particuliérement les objectifs de la conserve et de la distillation. **Clonal selection of mirabelle–plum–tree chiefly for preserved food aims and the distillation.**

7280 Salesses, G.; Mouras, A. FR 010701/75/0321
Culture d'anthères chez les arbres fruitiers à noyau. **Cultivating anthers on stone – fruit trees.**

7281 Matthews; Dow GB 011401/00/0002
Breed for resistance to bacterial canker and for self fertility.

7282 Jordan GB 011508/76/0012
Breeding / screening plums for disease resistance.

7283 Wilson; Stott GB 011510/00/0004
Breed cultivars and develop cultural methods to make plums an economic crop.

7284 Fideghelli, C. IT 021500/71/0022
Miglioramento genetico dell'albicocco. **Apricot breeding.**

7285 Fideghelli, C. IT 021500/73/0032
Miglioramento genetico del susino. **Breeding of prunes.**

7286 Fideghelli, C. IT 021500/75/0001
Miglioramento genetico del ciliegio dolce. **Sweet cherry breeding.**

7287 Monastra, F.; Bergamini, A.; Cooianchi, D.; Liverani, L.; Della Strada, G.; Albertini, A. IT 021500/78/0002 N
Idoneità delle principali cultivar di albicocco, ciliegio, pesco, susino e melo, a sottostare a processi di conservazione e trasformazione industriale. **Study on the suitability of the most important apple, apricot, cherry, peach and plum varieties to processing.**

7288 Fideghelli, C.; Quarta, R.; Della Strada, G. IT 021500/78/0003 N
Costituzione di cultivar di pesco, nettarine resistenti alla bolla e all'oidio. a maturazione precoce e tardiva, a vegetazione compatta, ad elevata attitudine rizogena. **Nectarine and peach breeding for resistance to Taphrina deformans and Sphaerotheca pannosa, early and late ripening, "compact" growth habit, self rooting ability.**

7289 Pratella, Giancarlo; Giuseppe, Tonini IT 040216/77/0007 N
Adattabilità varietale delle pesche percoche alla surgelazione. **Quality of cling peaches for freezing.**

7290 Eccher, T. IT 040605/78/0005
Confronti varietali di ciliegi spur. **Comparison among various spur cherry varieties.**

7291 Marro, M. IT 040605/79/0002 N
Osservazioni sul comportamento di diverse cultivar di pesco di recente introduzione. **Research on the behaviour of recently introduced peach varieties.**

7292 Bargioni, G. IT 102801/77/0001 N
Selezione di cultivar di ciliegio dolce per la raccolta meccanica. **Selection of sweet cherry cultivars for mechanical harvesting.**

Soft fruit (berries and cane fruits) (B 3620)

See also 4291, 4326, 4327, 7048, 7212, 7251, 8759

7293 Linden, R.   BE 080200/52/0011
Création de nouvelles variétés de fraisiers. **Breeding new strawberry varieties.** Publications.

7294 Monin, A.; Tréfois, R.   BE 080200/62/0001
Nanification du cerisier. **Dwarfing cherry trees.** Publications.

7295 Boxus, P.   BE 080200/67/0016 R
Assainissement viral du fraisier et production de plants sains. **Strawberry virus control and production of virus–free plants.** Publications.

7296 Dubuisson, J.; Lemaitre, R.   BE 140000/72/0058 N
Problèmes spécifiques posés par la culture des fraisiers dans la Vallée de la Meuse. **Breeding of strawberries in the Meuse Valley.** Publications.

7297 Van der Linden, L.; Aerts, J.; Benoit, F.; Van Looy, J.
                              BE 140000/73/0030 R
Selectie en studie van de teelttechnieken voor aardbeien. **Selection and study of the cultivation technics of strawberries.** Publications.

7298 Dubuisson, J.; Lemaitre, R.   BE 140000/79/0067 N
Introduction de la culture de ronces sans épines dans la Vallée de la Meuse. **Introduction of blackberry–bush culture without thorns in de Meuse Valley.**

7299 Gruppe, W.   DE 129161/75/0006
Züchtung von Strauchbeerenobst. **Breeding of bush–fruit.**

7300 Schubert, G.; Paluschka, S.   DE 502200/71/0001 R
Sortenprüfung von Erdbeeren 1971. **Varietal testing of strawberry.**

7301 Wirth, H.   DE 507701/72/0001
Prüfung von 12 Kulturheidelbeersorten unter verschiedenen Standortverhältnissen mehrerer europäischer Länder. **Twelve varieties of highbush blueberry tested under different ecological conditions of several European countries.**

7302 Jørgensen, M.B.; Thuesen, A.   DK 010115/70/4501
Forædling af hindbær, stikkelsbær, asparges, jordbær, peberod, Asparagus plumosus. **Breeding of raspberry, gooseberry, asparagus, strawberry, horse radish and Asparagus plumosus.**

7303 Jørgensen, M.B.; Thuesen, A.   DK 010115/73/4509
Sortsafprøvning af jordbær og asparges på friland. **Variety trials of strawberry and asparagus for outdoor crops.**

7304 Lantin, B.   FR 010401/72/0195
Obtention par hybridation de variétés de cassis résistantes a l'oïdium et ayant de faibles exigences en froid hivernal. **Breeding black currant varieties resistant to oïdium with low winter cold requirement through hybridization.**

7305 Risser, G.   FR 010603/71/0052
Sélection de variétés de fraisier résistantes à Phytophthora Cactorum et adaptées aux besoins français. **Breeding Phytophthora resistant strawberry varieties adapted to the french market.**

7306 Keep   GB 011002/00/0003
Breed new varieties.

7307 Keep   GB 011002/00/0004
Breed new varieties.

7308 Jordan; Tarr   GB 011508/00/0004
Screen seedlings for resistance to verticillium wilt and Sphaerotheca macularis.

7309 Wilson; Jones   GB 011510/00/0001
Breed processing and dessert strawberry cultivars for extended season and processing.

7310 Wilson; Jones   GB 011510/00/0005
Breed heavy yielding, disease resistant cultivars.

7311 Gooding; McNicol   GB 030703/00/0001
Breeding and associated genetic studies.

7312 Gooding; McNicol   GB 030703/00/0003
Breeding systems at different ploidy levels.

7313 Jennings   GB 030703/00/0006
Breeding and associated studies.

7314 Jennings   GB 030703/00/0008
Breed early, erect blackberries and other hybrid Rubus berries.

7315 Anderson   GB 030703/00/0009
Breed black currants for northern regions of the U.K.

7316 MacLachlan, J.B.; O'Callaghan, T.F.; Kavanagh, T.; Jeffares, M.   IE 060301/68/0723 R
Evaluation of raspberry cultivars. Publications.

7317 MacLachlan, J.B.; O'Callaghan, T.F.; Kavanagh, T.; Jeffares, M.   IE 060301/69/0093 R
Strawberry cultivar testing and evaluation. Publications.

7318 MacLachlan, J.B.; O'Callaghan, T.F.; Kavanagh, T.; Jeffares, M.   IE 060301/69/0094 R
Strawberry breeding (improved marketing and harvesting characteristics). Publications.

7319 MacLachlan, J.B.; O'Callaghan, T.F.; Lamb, J.G.D.; O'Riordan, F.   IE 060301/73/0771 R
The evaluation and breeding of blackberry cultivars. Publications.

7320 MacLachlan, J.B.; O'Callaghan, T.F.; Kavanagh, T.; Gormley, T.R.   IE 060301/75/1083 N
Breeding, selection and evaluation of everbearing strawberries and methods for production of late fruit. Publications.

7321 Tonini, Giuseppe; Maccaferri, Massimo
                              IT 040216/77/0006 N
Adattabilità varietale delle fragole alla surgelazione. **Quality of strawberries for freezing..**

7322 Lalatta, F.   IT 040605/77/0001 N
Indagini sul comportamento di nuove varieta'di mirtillo gigante in lombardia.. **Researches on behaviour of new bilberry varieties in lombardia..**

7323 Eccher, T.   IT 040605/78/0004
Studio del comportamento di nuove cultivar di lampone in Lombardia. **Studies on behaviour of new raspberry varieties in**

D 2300 – Plant breeding

Lombardia.

7324 Wassenaar, L.M.   NL 010114/67/2153
Het kweken van rassen van éénmaal dragende aardbeien in de open grond. **Breeding of June–bearing strawberries in the open.** Publications.

7325 Wassenaar, L.M.   NL 010114/67/2154
Het kweken van rassen van éémaal dragende aardbeien onder glas. **Breeding of June–bearing strawberries under glass.** Publications.

7326 Wassenaar, L.M.   NL 010114/67/2155
Het kweken van remonterende aardbeirassen. **Breeding of remontant strawberries.**

7327 Wassenaar, L.M.   NL 010114/71/3220
Veredeling bessen. **Breeding of bush fruits.**

7328 Dijkstra, J.   NL 010212/58/1793
Gebruikswaarde–onderzoek aardbeirassen. **Variety testing of strawberries.** Publications.

7329 Dijkstra, J.   NL 010212/58/1795
Gebruikswaarde–onderzoek framboze–, besse– en braamsrassen. **Variety testing of raspberries, currants and blackberries.** Publications.

Citrus fruit (B 3630)

7330 Vogel, R.   FR 012202/59/0102
Sélection sanitaire des agrumes. **Citrus selection for resistance to virus and mycoplasma diseases.**

7331 Blondel, L.; Vittori, F.; Jacques, C.; Mond   FR 012202/63/0216
Sélection clonale du clémentinier, du mandarinier 'satsuma' et du kumquat. **Clonal selections of Clementina–, Mandatine– and Kumquat–trees.** Publications.

7332 Cassin, J.; Ciccoli, H.   FR 012202/66/0014
Régénération des variétés d'Agrume par sélection de lignées nucellaires et création de nouvelles variétés par hybridation. **Regeneration of citrus varieties by selection of nucellar progenies and breeding new varieties by hybridization.**

7333 Russo, F.; Starrantino, A.   IT 021600/49/0001
Miglioramento genetico dell'arancio, mandarino, pompelmo, cedro e bergamotto. **Genetic improvement of orange, mandarin, grapefruit, citron and bergamot orange.** Publications.

7334 Russo, R.; Puglisi, A.; Starrantino, A.; Reforgiato, R.   IT 021600/68/0003
Prove comparative di diversi cloni delle cultivar di arancio "Tarocco" "Washington Navel" e "Biondo" a maturazione mediotardiva di clementino di mandarino "Avana" e "Sastuma". **Comparative tests among different clones of "tarocco" "Washington navel" and medium –late "Biondo" varietes of orange mandarins, and of Clementine "Avana and Satsuma".**

7335 Russo, F.; Starrantino, A.; Puglisi, A.   IT 021600/68/0004
Prove comparative tra vecchi e nuovi cloni (nucellari) di agrumi. **Comparative trials of old and young lines (nucellar) of citrus.** Publications.

7336 Russo, F.; Reforgiato, G.   IT 021600/70/0002
Miglioramento genetico e selezione portainnesti agrumi. **Citrus rootstock breeding and selection.**

7337 Russo, F.; Starrantino, A.; Reforgiato, G.   IT 021600/74/0005 N
Ricerche sulle scelte varietali in agrumicoltura in relazione alle esigenze dei mercati esteri.. **Research on cultivar choice in citrus industry according to foreign markets.**

7338 Terranova, G.; Caruso, A.   IT 021600/77/0003
Indagine varietale sul bergamotto. **Varietal survey on bergamot.**

7339 Russo, F.   IT 021600/77/0404 R
Reincrocio e mutagenesi del limone. **Lemon recrossing and mutagenesis.**

7340 Salerno, m.   IT 040106/77/0395 R
Miglioramento genetico del limone con particolare riguardo alla resistenza al mal secco. **Genetic improvement of the lemon tree with special regard to its resistance to dry rot.**

7341 Catara, A.   IT 040302/72/0423
Miglioramento genetico degli agrumi. **Genetic improvement of citrus fruit.** Publications.

7342 Catara, A.   IT 040302/77/0137 R
Miglioramento genetico degli agrumi. **Genetic improvement of citrus species.**

7343 Perotta, G.   IT 040302/77/0381 R
Selezione cultivar di limone resistenti al mal secco. Aspetti biochimici della resistenza al mal secco. **Selection of lemon cultivars resistant to dry rot. Biochemical aspects of resistance to dry rot.**

7344 Tribulato, E.   IT 040307/74/0642
Rapporti tra i bionti nelle piante innestate. Ricerche sui portinnesti degli agrumi e del mandorlo. **Relationships between bionts in grafted plants. Research on citrus fruit and almond tree rootstocks.**

7345 Tribulato, E.   IT 040307/77/0408 R
Ricerca di individui di limone resistenti al Phoma–tracheiphila, produzioni di frutti tendenzialmente apireni, miglioramento attuale standard produttivo. **Research on developing lemon specimens resistant to Phoma–tracheiphila, growing fruits tending to "piplessness", improvement of the present production standard.**

7346 Somma, V.   IT 040905/77/0400 R
Miglioramento genetico del limone con particolare riguardo alla resistenza al mal secco. **Genetic improvement of the lemon tree with special regard to its resistance to dry rot.**

7347 Calabrese, Francesco; Di Marco, Luigi; Crescimanno, F.G.; De Michele, Andrea   IT 040908/78/0001
Confronti tra cloni di arancio Navel. **Comparison between clones of Navel orange.**

7348 Marras, F.   IT 041307/72/0562
Miglioramento genetico degli agrumi. **Genetic improvement of citrus fruit.** Publications.

D 2300 – Plant breeding

7349 Milella, A.   IT 041310/72/0514
Miglioramento genetico degli agrumi. **Genetic improvement of citrus fruit.** Publications.

7350 Deidda, P.   IT 041310/74/0545
Ricerca collegiale coordinata per il miglioramento genetico degli agrumi. **Coordinated cooperative research on genetic improvement of citrus fruit.**

7351 Deidda, P.   IT 041310/77/0164 R
Miglioramento genetico degli agrumi. **Genetic improvement in citrus fruits.**

7352 Agabbio, M.   IT 041310/77/0324 R
Selezione clonale e sanitaria del limone in Sardegna. **Clonal and health selection of the lemon tree in Sardinia.**

7353 Crescimanno, F.G.   IT 060900/72/0165
Osservazioni di pieno campo su nuovi portinnesti, in combinazione con diverse specie e cultivar di agrumi. **Grafting and observations on new citrus rootstocks, in combination with different species and c.v. of citrus fruit.** Publications.

7354 Crescimanno, F.G.   IT 060900/77/0465
Ricerca di individui resistenti al mal secco, in grado di fornire buone produzioni. **Production trials of dry rot resistant and high yielding specimens.**

7355 Crescimano, F.G.   IT 060900/77/0892
Creazioni di triploidi di mandarini e limoni, da genitori diploidi e loro studio dal punto di vista agronomico, commerciale e fitosanitario. **New triploids of tangerine and lemon created from diploid parents. Study from an agricultural, commercial and health selection point of view.**

7356 Crescimano, F.G.   IT 060900/77/0894
Osservazioni su specie e cultivar di agrumi in collezione e loro diffusione. **Observations on species and cultivars of citrus plants grown in a botanical garden; diffusion of these plants.**

7357 Fici, P.   IT 091902/76/0003
Selezione massale e clonale della varietà di uva: "Moscato Binaco" di Siracusa. **Clonal and mass selection of "Moscato Bianco di Siracusa" grape variety.**

Tropical and sub–tropical fruits (B 3640)

See also 6870, 6878, 7251

7358 Dermine, E.; Tilkin, V.   BE 080200/76/0033
Amélioration et sélection d'une variété de melon. **Breeding and selection of a melon variety.**

7359 Bauckmann, M.   DE 506106/77/0001
Sortenprüfung bei Kiwis – Actinidia chinensis –. **Testing of different varieties of Kiwis – Actinidia chinensis –.**

7360 Risser, G.   FR 010603/60/0053
Etude de la résistance au Fusarium Oxysporum F. Melonis chez le melon (cucumis melo). Sélection de variétés résistantes. **Studies in cantaloup (cucumis melo) resistance to fusarium oxysporum melonis. Breeding resistant varieties.** Publications.

7361 Risser, G.   FR 010603/70/0054
Etude des possibilités d'utilisation des lignées gynoïques de melon. **Studies on the possibility of using gynoic lines in cantaloup breeding and growing.**

7362 Dumas de Vaulx, R.   FR 010603/71/0055
Etude des possibilités d'utilisation de la polyploïdie chez le melon. **Studies on the possibility of using polyploidy in cantaloup breeding.** Publications.

7363 Risser, G.; Pitrat, M.; Rode, J.C.   FR 010603/73/0174
Etude des méthodes de sélection pour la résistance au virus 1 du concombre chez le melon (Cucumis Melo). **Methods of selection for cucumber virus 1 resistance in Cucumis Melo.**

7364 Uncini, L.; Ferrari, V.   IT 021000/73/0003 R
Costituzione di varietà ed ibridi di melone per la coltura di pien'aria, dotati di elevata serbevolezza, pregevoli proprietà organolettiche e di resistenze alle tracheomicosi e all'oidio. **Constitution of melon varieties and hybrids for open air culture, storable, with precious taste properties, and oidium and mycosis wilt resistance.**

7365 De Michele, A.   IT 040908/77/0162 R
Prove di confronto di cultivar di avocado e altri fruttiferi tropicali e subtropicali. **Comparison test on avocado cultivar, and other tropical and subtropical fruit trees.**

Grapes (B 3650)

See also 4382, 4416, 4418, 4423, 4443, 4448, 4449, 5868

7366 Weiling, F.; Unger, C.   DE 111151/72/0004
Statistische Analyse von Testweinproben zum Zwecke der Rebenselektion. **Statistical analysis of wine–tests for selection of vine–clones resp. vine–varieties.** Publications.

7367 Matzner, F.   DE 161265/74/0003
Untersuchungen zur Schüttelfähigkeit der Trauben verschiedener Rebensorten – Vitis vinifera –. **Investigations into the shaking suitability of grapes of different vine varieties– Vitis vinifera –.**

7368 Rapp, A.; Hastrich, H.   DE 204000/71/4001
Einfluss der Beerenreife und des Standortes auf die sortenspezifischen Inhaltsstoffe der Weinbeeren. **Variety–specific constituents of grapes as affected by ripeness and location.**

7369 Koepchen, W.   DE 204000/71/4015
Untersuchungen über die Güte–Menge–Relation bei Rebenneuzuchten. **Investigations concerning the relation between quality and yield of new grape varieties.**

7370 Koepchen, W.; Hessberg, W. von; Freytag, G.   DE 204000/71/4016
Untersuchungen zur Ökovarianz von Rebneuzüchtungen. **Investigations on eco–variance of new grape varieties.**

7371 Hahn, H.   DE 204000/71/4018
Resistenzprüfung gegen den roten Brenner – Pseudopeziza tracheiphila M.Th. –. **The testing of vines for resistance against Pseudopeziza tracheiphila M.Th..**

7372 Hahn, H.   DE 204000/71/4019
Die Entwicklung einer Selektionsmethode für die Resistenzzüchtung gegen Rebvirosen, besonders die Reisigkrankheit. **The development of selection methods of**

breeding for resistance against grape viruses, especially fan leaf disease.

7373 Alleweldt, G.; Grossmann, I.; Meyer, S.; Koepchen, W.
DE 204000/71/4027
Züchtung winterfrostresistenter Reben. **Breeding of frost–resistant vines.**

7374 Alleweldt, G.; Koepchen, W. DE 204000/71/4029
Resistenzzüchtung bei Reben gegen Plasmopara und Reblaus bei hohen Qualitäts– und Ertragsleistungen. **Breeding of vines for resistance to Plasmopara viticola and phylloxera with high quality and yield.**

7375 Alleweldt, G.; Koepchen, W. DE 204000/71/4037
Die Züchtung von Qualitätssorten auf Vitis vinifera–Basis. **The breeding of quality varieties on Vitis vinifera base.**

7376 Hahn, H.; Alleweldt, G. DE 204000/72/4006
Der Aufbau einer Genbank für Reben. **The collection of gene resources of Vitis species.**

7377 Rapp, A.; Ziegler, A.; Bachmann, O.
DE 204000/75/0005
Untersuchung zur sortenspezifischen Einlagerung von Saccharose in die Weinbeeren. **Studies on variety–specific storage of saccharose in grapes.**

7378 Blaich, R.; Wind, R. DE 204000/75/0007
Züchtung haploider Reben. **Breeding of haploid vines.**

7379 Alleweldt, G.; Grossmann, I. DE 204000/77/0003
Variabilität der Rebenneuzuchten. **Variability of new vine varieties.**

7380 Koepchen, W. DE 204000/77/0004
Die Erhaltungszüchtung neuer Rebsorten. **Maintenance breeding of new vine varieties.**

7381 Rapp, A.; Gutzler, A.; Ziegle, A.; Kupfer, G.
DE 204000/77/0009
Prüfung neuer Sorten zur Herstellung von Mostkonzentrat. **Testing of new varieties for production of must concentrate.**

7382 Rapp, A.; Klenert, M. DE 204000/78/0013 N
Einfluss der Temperatur auf das Äpfelsäure–Weinsäure Verhältnis bei Weinbeeren verschiedener Rebsorten. **Effect of temperature on the ratio of malic and tartaric acid in grape berries of different varieties.**

7383 Rapp, A.; Steffan, H.; Alleweldt, G.
DE 204000/78/0014 N
Untersuchungen zum sortenspezifischen Äpfelsäure Weinsäure–Verhältnis verschiedener Rebsorten. **Investigations on the ratio of malic and tartaric acid in different grape–vine varieties.**

7384 Staudt, G.; Schneider, W. DE 501100/74/0005
Untersuchungen über die Heritabilität bei Reben. **Studies on the heritability in vines.**

7385 Staudt, G. DE 501100/78/0002 N
Untersuchungen über die Kreuzbarkeit von Vitis armata und die Bedeutung dieser Art für die Rebenzüchtung. **Investigations about the crossability of Vitis armata and the significance of the species for grapevine–breeding.**

7386 Benda, I.; Wahl, K. DE 502151/70/0005
Untersuchungen über die Kompatibilitätsleistungen alter wie neuer Unterlagsrebsorten. **Investigations into the grafting compatibility of old and new stock varieties.**

7387 Beeskow, H. DE 502151/73/0001
Übertragung von Frostresistenzgenen auf fränkische Qualitätssorten. **Increase in frost hardiness of qualified Franconian vine varieties by crossing with pertinent types.**

7388 Geiger, K.; Schottdorf, W. DE 502153/78/0004 N
Prüfung unterschiedlicher Augenzahlen pro Quadratmeter bei verschiedenen Rebsorten auf den Ertrag und die Weinqualität. **Testing of different numbers of buds per square metre of area of different vine varieties for crop yield and wine quality.**

7389 Kiefer, W.; Bäder, G.; Gruppe, W. DE 506101/75/0001
Untersuchungen über die Regulierung von Menge und Güte des Ertrages bei den wichtigsten Rebsorten im deutschen Weinbau. **Studies on the regulation of quantity and quality of yield of the most important vine varieties in German viticulture.**

7390 Schumann, F. DE 509154/72/0002
Untersuchungen zur Verwendung von Antitranspirantien in der Rebenveredlung. **Studies on antitranspirants by grafting.**

7391 Fader, W. DE 509155/74/0001
Vergleichende Untersuchungen der Leistungseigenschaften von Standardrebsorten und Neuzüchtungen. **Comparative studies on the performance properties of standard and of new vine varieties.**

7392 Schöffling, H.; Servaty, E. DE 509203/75/0001
Umstellung von Rebflächen des Obermoselgebietes auf qualitativ bessere Sorten. **Substitution of Upper Mosel vine aereas by new varieties of better quality.**

7393 Puissant, A.; Asselin, C. FR 010404/70/2234
Evolution des colorants au cours de la maturation dans les raisins de divers cépages. **Evolution of pigments during ripening of various grape varieties.**

7394 Remoue, M.; Lemaitre, Cl. FR 010405/70/0231
Sélection et étude génétique de clones de chenin de cabernet franc et de chardonnay. **Selection and genetic study of chenin of cabernet franc and chardonnay clones.**

7395 Bouquet, A. FR 010701/74/0310
Hybridation interspécifique Vitis vinifera et Vitis Rotundifolia. **Interspecific hybridization between Vitis Vinifera and Vitis Rotundifolia.**

7396 Leclair, Ph. FR 010702/50/0076
Sélection de clones de vigne à partir de variétés cultivées. **Clonal selection in vine–cultivars.** Publications.

7397 Pouget, R.; Ottenwaelter, M. FR 010702/60/0084
Création de variétés de porte–greffes de vigne. **Breeding vine rootstocks.**

7398 Doazan, J.P.; Ottenwaelter, M. FR 010702/66/0082
Création de variétés de raisin de table apyréne. **Breeding seedless table grape varieties.**

7399 Doazan, J.P. FR 010702/70/0079
Création de variétés de vigne à raisin de cuve rouge pour la zone viticole européenne. **Breeding red vine varieties for EEC.**

7400 Doazan, J.P.; Bouquet, A. FR 010702/74/0311
Sélection de variétés de vigne résistantes à l'excoriose et à la pourriture grise. **Breeding vine varieties for resistance to excoriose and "grey mold".**

7401 Huglin, P.; Balthazard, J.; Bisson, J. FR 010902/46/0375
Sélection clonale génétique de la vigne. **Genetic clonal selection of grape–vine.**

7402 Wagner, R. FR 010902/63/0085
Etude des semis de vigne ; application à l'amélioration variétale en région septentrionale. **Studies on vine seedlings : application to vine breeding in northern regions.** Publications.

7403 Wagner, R.; Balthazard, J. FR 010902/63/0090 R
Etude des semis de vigne : application à l'amélioration variétale en régions septentrionales. **Investigations on grapevine seedling frogenics: application to varietal improvement in Northern regions.**

7404 Wagner, R. FR 010902/65/0081
Création, pour les vignobles septentrionaux, de variétés de raisin de cuve fruitées ou aromatiques, précoces et rustiques. **Breeding flavoured and aromatic, early and hardy vine–varieties, for northern vineyards.** Publications.

7405 Balthazard, J.; Bronner, A.; Meyer, J.P. FR 010902/71/0376
Expérimentation de nouvelles variétés de raisins de cuve allemandes. **New varieties experiments in german vat grapes.**

7406 Huglin, P.; Balthazard, J. FR 010902/72/0270
Etude de l'hétérogénéité de vieilles variétés de vigne, création de réserves de gènes. **Study of the heterogeneity of old vine varieties – creation of gene reserves.**

7407 Bayonove, C.; Ratier, R. FR 011202/70/2231
Déterminisme génétique de la transmission du caractère musqué. **Genetic determinism of muscat character transmission in grapes.** Publications.

7408 Truel, P.; Vergnes, A.; Domergue FR 011204/65/0109
Sélection de nouvelles variétés de vigne améliorant la qualité des vins dans le Midi. **Breeding of new vine varieties for improving the wine quality in the south of France.**

7409 Branas, J.; Truel, P. FR 011204/70/0077
Reconversion des vignobles où la culture des hybrides producteurs était étendue. **Reconversion of vineyard: using vitis vinifera cultivars instead of hybrids.**

7410 Wagner, R.; Domergue; Truel, P.; Rennes, C. FR 011204/75/0452
Transmission héréditaire des principaux caractères pris en compte lors des travaux d'amélioration variétale de la vigne. **Hereditery transmission of characters important in grape selection.**

7411 Boubals, D.; Guiraud, E.; Bernard, A.; Mur, G. FR 011204/75/0456
Etude des causes de la résistance des baies de raisins à la pourriture grise (Botrytis Cinerea). **Berry resistance to Botrytis Cinerea in grape.**

7412 Samson, C.; Pistre, R.; Wagner, R.; Vergnes, A. FR 011204/76/0450
Influence des facteurs du milieu sur les variétés de vigne: essai international d'écologie viticole. **Study of environmental influence on the grape varieties: international experimentation of viticultural ecology.**

7413 Wagner, R.; Samson, C. FR 011204/76/0451
Appréciation objective de la qualite technologique du raisin et du vin dans le cadre des travaux d'amélioration variétale chez la vigne. **Objective estimation of the organoleptic quality microvinification of grapes and wines for breeding for breeding purposes.**

7414 Boubals, D.; Pistre, R. FR 011213/68/0080
Etude du mode de transmission héréditaire du goût du Cabernet: Obtention de variétés de vigne productives et à vin de qualité. **Heredity of Cabernet flavour. Breeding high productive vine cultivars for quality wine.**

7415 Boubals, D.; Pistre, R. FR 011213/71/0083
Résistance des Vitacées au nématode Xiphinema index et à l'infection par le virus du court–noué. **Vitis resistance to nematode Xiphinema index and to viral contamination of grape fanleaf virus.**

7416 Bourzeix, M.; Heredia, N. FR 011403/70/2214
Identification et teneur individuelle des composés phénoliques et des acides organiques dans les principaux cépages. **Identification and content of the phenolic compounds and organic acids in principle grape varieties.** Publications.

7417 De Sanctis, F.; Barba, M. IT 020300/76/0002 N
Selezione clonale e sanitaria della vite e risanamento con termoterapia. **Clonal and sanitary selection of the grapevine and production of virus–free propagating material..**

7418 Cosmo, I.; Calo', A.; Liuni, C.; Costacurta, A.; Egger, E. IT 021300/50/0001
Descrizioni ampelografiche vitigni ad uva da vino e da tavola e portinnesti. **Ampelografic descriptions of Wine, Table and Rootstocks Vine Varietes.** Publications.

7419 Calò, A.; Costacurta, A.; Cancellier, S. IT 021300/66/0001
Miglioramento genetico della vite per via gamica: studio del carattere precocità. **Genetical vine improvement by gamic reproduction: study of earliness character.**

7420 Calò, A.; Costacurta, A.; Cancellier, S. IT 021300/71/0002
Miglioramento genetico della vite per via gamica: studio morfo–fisiologico e sanitario sulle F_1 ed F_2 allevati in collezione. **Genetical vine improvement by gamic reproduction: morpho–physiological and sanitary remarks on F_1 and F_2.**

7421 Calo' A.; Costacurta, A.; Egger, E. IT 021300/72/0001
Collezione clonale Portinnesti, introdotti dai Paesi CEE. **Clonal collection of rootstocks introduced from C.E.E. countries.**

7422 Cosmo, I.; Calo', A. IT 021300/75/0011
Descrizione ampelografica di nuovi vitigni ad uva da tavola

ottenuti dall'Istituto. **Amphelografic descriptions of table Institute new table–grape varieties.**

7423 Calò, A.   IT 021300/77/0334 R
Miglioramento dell'uva da vino mediante selezione clonale genetica sanitaria tecnologica. **Improvement of wine grapes through genetic clonal selection, health and technological selection.**

7424 Corino, L.; Marosco, A.; Pogna, A.   IT 021300/79/0003 N
Analisi cromosomica di specie diverse del genere Vitis. **Chromosomal analysis of different species of the Vitis genus.**

7425 Martelli, G.   IT 040106/77/0373 R
Selezione clonale e sanitaria vitigni uva da vino. **Clonal and sanitary selection of vine–plants for wine grapes.**

7426 La Notte, E.   IT 040107/77/0364 R
Ricerche sulla vite, selezione clonale. **Research on vines, clonal selection.**

7427 Donno, G.   IT 040112/72/0449
Sul miglioramento del patrimonio varietale di uve da tavola in Puglia ivi comprese le varietà apirene. **On the improvement of table grape varieties of Apulia, including the apirene varieties.** Publications.

7428 Godini, A.   IT 040112/77/0358 R
Miglioramento della vite ad uva da vino mediante selezione clonale. **Improvement of wine producing grape–vines through clonal selection.**

7429 Faccioli, F.   IT 040202/77/0349 R
Miglioramento genetico vitigni emiliano–romagnoli, selezione genetica. **Genetic improvement of vine–stocks growing in Emily and Romagna, genetic selection.**

7430 Giunchedi, L.   IT 040211/77/0357 R
Miglioramento sanitario della vite mediante selezione e termoterapia. **Health improvement of grapevines through selection and thermotherapy.**

7431 Refatti, E.   IT 040302/77/0387 R
Miglioramento genetico della vite ad uva da vino mediante selezione clonale, selezione sanitaria. **Genetic improvement of the wine producing grape–vine by clonal and sanitary selection.**

7432 Fregoni, M.; Zamboni, M.; Scienza, A.; Miravalle, R.; Boselli, M.; Dorotea, G.   IT 040403/70/0001 R
Selezione clonale della varietà di vite della provincia di Piacenza e Parma. **Clone selection of the varieties of vine in the province of Piacenza e Parma.** Publications.

7433 Fregoni, M.   IT 040403/77/0354 R
Selezione clonale dei vitigni piacentini, alessandrini e valtellinesi. **Clonal selection of vine–stocks growing in the areas of Piacenza, Alessandria and Valtellina.**

7434 Scienza, A.   IT 040403/78/1110 N
Incremento del valore alimentare dell'uva. Ricerche sul controllo genetico ed ormonale dell'accumolo dello zucchero nelle bacche. **Enhancement of the nutritive value of grapes. Research on the genetic and hormonal control of the sugar concentration in the grape.**

7435 Stella, C.   IT 040511/77/0405 R

Attitudine enologica di uva da vino. **Oenological suitability of wine grapes.**

7436 Marro, M.   IT 040605/77/0372 R
Selezione clonale della vite interessante la regione lombarda. **Clonal selection of vines in Lombardy.**

7437 Fortusini, A.   IT 040612/77/0330 R
Miglioramento della vite da vino mediante selezione clonale sanitaria. **Improvement of wine producing vines through clonal health selection.**

7438 Rosciglone, B.   IT 040905/77/0393 R
Miglioramento della vite ad uva da vino mediante selezione clonale e fitosanitaria. **Improvement of wine grape vines by clonal and health check selection.**

7439 Sottile, I.   IT 040908/78/1183 N
Selezione clonale dei vitigni ad uva da vino. **Clonal selection of vitis vinifera vine–stocks.**

7440 Cartechini, A.   IT 041005/77/0335 R
Selezione clonale vitigni umbri e laziali. **Clonal selection of vine–stocks growing in Umbria and Latium.**

7441 Basso, M.   IT 041104/77/0329 R
Selezione clonale dei vitigni da vino coltivati in Toscana litoranea. **Clonal selection of vine–stocks grown in the coastal areas of Tuscany.**

7442 Scaramuzzi, G.   IT 041110/77/0398 R
Miglioramento uva da vino mediante selezione clonale, selezione sanitaria. **Improvement of wine grapes by clonal and health check selection.**

7443 Triolo, E.   IT 041110/78/1191 N
Selezione sanitaria dei vitigni ad uva da vino in Toscana, Umbria e Lazio. **Health selection of vitis vinifera vine–stocks in Tuscany, Umbria and Latium.**

7444 Gandini, A.   IT 041211/77/0355 R
Miglioramento della vite ad uva da vino mediante selezione clonale, valutazioni delle attitudini enologiche. **Improvement of wine producing grape–vines through clonal selection, evaluation of their oenological qualities.**

7445 Prota, U.   IT 041307/77/0384 R
Miglioramento della vite ad uva da vino mediante selezione clonale, selezione sanitaria vitigni sardi. **Improvement of the wine producing grape–vine by clonal selection. Sanitary selection of Sardinian vine plants.**

7446 Deidda, P.   IT 041310/77/0347 R
Miglioramento genetico della vite ad uva da vino mediante selezione clonale in Sardegna. **Genetic improvement of wine producing grape–vines through clonal selection in Sardinia.**

7447 Conti, M.   IT 060100/77/0461
Miglioramento della vite ad uva da vino mediante selezione clonale. Selezione fitosanitaria. **Improvement of wine grapevines by clonal selection. Health selection.**

7448 Eynard, I.   IT 062100/68/0149
Esecuzione di nuovi incroci fra i migliori cloni disponibili. **Breeding work among the best clones.** Publications.

7449 Eynard, I.   IT 062100/71/0147
Selezione degli incroci del professor Giovanni Dalmasso.
Selection of professor Dalmasso's breedings.

7450 Eynard, I.   IT 062100/77/0460
Miglioramento della vite ad uva da vino mediante selezione
clonale con selezione genetica. **Improvement of wine
grapevines by clonal selection with genetic selection.**

7451 Carlone, R.   IT 062100/77/0953
Selezione degli incroci di vite Dalmasso. **Selection of crosses of
Dalmasso vines.**

7452 Carlone, R.   IT 062100/77/0954
Introduzione degli incroci di vite dell'Instituto Nacional de
Tecnologia Agropecuaria di S. Rafael (Mendoza) Argentina e
di altri vitigni. **Introduction of crosses obtained from vines
developed by the Istituto Nacional de Tecnologia Agropecuaria
di S. Rafael (Mendoza) Argentina and other vine–stocks.**

7453 Carlone, R.   IT 062100/77/0955
Esecuzione di nuovi incroci tra i migliori cloni di vite
disponibili. **Development of new crosses between the best vine
clones available.**

7454 Carlone, R.   IT 062100/77/0957
Caratterizzazione cariologica ed istologica del patrimonio
ereditario delle diverse cultivar di vite. **Caryological and
histological characterisation of the hereditary endowment of
different vine cultivars.**

7455 Carlone, R.   IT 062100/77/0958
Induzione di mutazioni nella vite con agenti mutageni e
successiva selezione. **Induction of mutations in vines through
mutagenetic agents and subsequent selection.**

7456 Carlone, R.   IT 062100/77/0963
Selezione clonale delle principali cultivar di vite da vino
piemontesi. **Clonal selection of the main Piedmontese wine
producing grapevine cultivars.**

7457 Casini, E.   IT 062800/77/0462
Selezione clonale dei principali vitigni da vino diffusi in
Toscana. **Clonal selection of the main wine grape stocks present
in Tuscany.**

7458 Bambara, G.; Picciolo, F.; Refatti, E.   IT 091901/75/0001
Selezione clonale e fitosanitaria cultivar "Malvasia di Lipari".
Selection of cultivars of "Malvasia di Lipari".

7459 Refatti, E.; Picciolo, F.; Bambara, G.; Granata, F.
                                   IT 091901/77/0002
Selezione clonale e sanitaria di vitigni siciliani. **Clonal and
sanitary selection of grapevine cultivars in Sicily.**

Edible nut fruits (B 3660)

See also 4236, 7228, 7251, 7275, 7280, 7344

7460 Grassely, Ch.; Olivier, G.   FR 010701/60/0012
Obtention de variétés d'amandier résistantes aux maladies
cryptogamiques, de floraison tardive et de bonne résistance
aux gelées de printemps. **Breeding late flowering varieties of
almond, resistant to spring frost and fungus diseases.**
Publications.

7461 Solignat, G.; Chapa, J.   FR 010701/60/0013
Obtention et sélection de châtaigniers producteurs directs ou
porte – greffes résistants à l'encre et au chancre à l'Endothia.
**Breeding self–rooted or rootstocks of chesnut resistant to
canker and chesnut bark disease. (Endothia).**

7462 Germain, E.; Jalinat, J.L.; Marchou, M.
                                   FR 010701/70/0097
Sélection par voie sexuée, dans l'espèce Juglans Nigra, de
portegreffe permettant d'obtenir une homogénéité
satisfaisante des arbres dans le cadre de noyeraies intensives.
**Rootstock sexual breeding in Juglans Nigra species for an
improved tree homogeneity in intensive walnut orchards.**

7463 Germain, E.; Leglise, P.; Delord, F. FR 010701/70/0257
Création par hybridation de variétés de noisetier mieux
adaptées aux conditions climatiques et aux besoins du marché
français. **Hybridization of hazel–tree varieties for an improved
adaptation to the french climate and market.**

7464 Germain, E.; Leglise, P.; Delord, F. FR 010701/70/0258
Sélection parmi les principales variétés étrangères de noisetier
de cultivars permettant de compléter et d'améliorer la gamme
variétale existante. **Selection of cultivars among the main
foreign hazel–tree varieties allowing to complete and improve
the present varietal spectrum.**

7465 Germain, E.; Jalinat, J.L.; Marchou, M.
                                   FR 010701/70/0262
Sélection par voie sexuée, dans l'espèce Juglans Regia, de
porte–greffes bien adaptées aux zones de culture traditionelle
du Périgord. **Sexual breeding of rootstocks in the species
Juglans Regia for their adaptability to the "Perigord" area.**

7466 Grassely, C.   FR 010701/71/0236
Obtention de variétés d'amandiers autofertiles. **Breeding of
self–fertile varieties of almond–trees.**

7467 Germain, E.; Jalinat, J.L.; Marchou, M.
                                   FR 010701/71/0260
Sélection de variétés de noyer, d'origine étrangère, très
productives et bien adaptées aux exigences climatiques
françaises. **Selection of walnut varieties of foreign origin for
their productivity and adaptability to the french climate.**

7468 Germain, E.; Jalinat, J.L.; Marchou, M.
                                   FR 010701/71/0261
Sélection de pollinisateurs pour les trois principales variétés
françaises de noyer à floraison tardive: franquette, corne, et
parisienne. **Selection of pollinators for the three main french
walnut varieties with late flowering: franquette, corne and
parisienne.**

7469 Germain, E.; Leglise, P.; Delort, F. FR 010701/72/0259
Induction florale femelle chez le noisetier: étude de l'influence
de la luminosité, de la vigueur et de l'origine des rameaux.
**Female flower induction in hazel–trees: influence of
luminosity, vigour and branch beginning.**

7470 Salesses, G.; Bonnet, A.   FR 010701/72/0322
Etude de quelques variétés cultivées à fertilité réduite. **Study of
some cultivated varieties with low fertility.**

7471 Solignat, G.; Chapa, J.   FR 010701/74/0252
Obtention et sélection de porte–greffes de châtaignier
d'origine c. sativa (indigénes). **Obtention and selection of**

D 2300 – Plant breeding

chesnut. tree rootstocks from c. sativa.

7472 Arnold; Dickinson GB 012005/00/0038 R
Development of methods for extracting, storing and applying coconut pollen for hybridization.

7473 Rizzo, V.; Convertini, G.; Perniola, M.; Ferri, D.; De Giorgio, D. IT 020500/70/0004 R
Germoplasma del mandorlo e confronto varietale per rendimento e qualità del frutto. **Almond–tree germoplasm conservation and investigation and varieties comparison trials for yield and quality of fruits.** Publications.

7474 Fideghelli, C.; Monastra, F. IT 021500/73/0027
Miglioramento varietale Mandorlo. **Varietal improvement of almond.**

7475 Godini, A. IT 040112/77/0359 R
Miglioramento del mandorlo mediante selezione clonale. **Improvement of almond–trees through clonal selection.**

7476 Catara, A. IT 040302/77/0337 R
Selezioni cloni esenti da virus, miglioramento genetico del mandorlo. **Selection of virus free clones, genetical improvement of the almond–tree.**

7477 Damigella, P. IT 040307/77/0346 R
Selezione clonale del mandorlo e nocciolo nell'ambito delle popolazioni esistenti nella Sicilia centro–orientale. **Clonal selection of almond trees and hazel–nut trees among the existing strains of east–central Sicily.**

7478 Crescimanno, F.G.; Barone, F.; Di Marco, L. IT 040908/76/0003
Ricerche sull'adattamento varietale e la nutrizione minerale di alcune cultivar di mandorlo in Sicilia. **Behaviour and mineral nutrition of some almond varieties in Sicily.** Publications.

7479 Fatta Del Bosco, G. IT 040908/77/0351 R
Mandorlo, raccolta germoplasma, selezione clonale, moltiplicazione e conservazione cloni nocciolo, selezione clonale, introduzione cultivar straniere. **Almond–tree, harvesting of germplasm, clonal selection, multiplication and conservation of hazel clones, clonal selection, introduction of foreign cultivars.**

7480 Preziosi, P. IT 041005/77/0383 R
Selezione clonale CV. di nocciolo tonda romana. **Clonal CV. selection of "tonda romana" hazel–nut trees.**

7481 Loreti, F. IT 041104/77/0369 R
Selezione delle cultivar locali e straniere e influenza dei portinnesti sul loro comportamento vegetativo e produttivo. Miglioramento genetico del mandorlo. **Selection of indigenous and foreign cultivars; influence of root–stocks on their growth and yield. Genetic improvement of the almond–tree.**

7482 Romisondo, P.; Radicati, L.; Me, G.; Iacurti, G. IT 041218/73/0001 R
Miglioramento e potenziamento della coltura del noce da frutto in Piemonte mediante la selezione di varietà locali e l'introduzione di specie e varietà esotiche. **Improvement and development of Walnut culture in Piedmont (Italy) through the selection of local varieties and the introduction of foreign varieties and species.**

7483 Frau, A.M. IT 041310/77/0323 R
Selezione cultivar locali di mandorlo, introduzione di nuove cultivar prove su portainnesti adatti alla mandorlicoltura irrigua. **Selection of local cultivars of almond trees, introduction of new cultivars, tests on root stocks suitable for the irrigated cultivation of almond orchards.**

Other fruits (B 3690)

See also 7251

7484 Duquesne, J.; Delmas, J.M. FR 011207/74/0290
Etude de l'héritabilité du caractère compatibilité à l'union de l'abricotier greffé sur pêcher et sur prunier. **Inheritability of the compatible character for the union of a grafted apricot–tree on peach–tree and plum–tree.**

Ornamentals and ornamental products in general (B 3700)

See also 3785, 4517, 6817, 6856, 6858, 6859, 6872, 6873, 6875, 6877

7485 Boesmans, G.; Flamee, M. BE 030015/66/0002 N
Vegetatieve en generatieve vermenigvuldiging van sierplanten. **Vegetative and generative multiplication of ornamental plants.** Publications.

7486 Horn, W.; Bundies, H. DE 138300/70/0006
Züchtungsforschung und Entwicklung von Züchtungsverfahren bei Zierpflanzen. **Research on ornamental plant breeding.**

7487 Horn, W.; Potthoff, H. DE 138300/75/0003
Selektion und Isolation induzierter Mutanten bei vegetativ vermehrten Zierpflanzen. **Selection and isolation of induced mutants in vegetatively propagated ornamentals.**

7488 Horn, W. DE 161270/78/0002 N
Züchtung neuer Zierpflanzen. **Breeding of new ornamental flower crops.**

7489 Horn, W. DE 161270/78/0003 N
Cytologische und genetische Untersuchungen bei Zierpflanzen. **Cytological and genetical investigations on ornamental flower crops.**

7490 Horn, W.; Schlegel, G. DE 161270/78/0005 N
Induktion von Mutationen bei vegetativ vermehrten Zierpflanzen in vitro. **Induction of mutations in vegetatively propagated ornamental flower crops in vitro.**

7491 Farestveit, B. DK 030171/77/0006 N
Induktion af mutanter i Pachystachys lutea. **Induction of mutants in Pachystachys lutea.**

7492 Schmidt, J.P. DK 030173/63/0002
Undersøgelse af egenskaber som knytter sig til kloner af lignoser som er i handel i Danmark. **Investigation of characters associated with clones of lignoses commercially available in Denmark.**

7493 Schmidt, J.P. DK 030173/75/0001
Kåring af isolerede forekomster af lignoser og disses afprøvning. **Selection and testing of isolated occurrences of lignoses.**

7494 Stickland; Harrison   GB 011402/75/0007 R
Genetics and biochemistry of flower pigments.

7495 Quagliotti, L.   IT 041212/72/0540
Miglioramento genetico delle piante ortensi e floreali. **Genetic improvement of horticultural and floral crops.** Publications.

Bulbs (B 3710)

See also 2897, 4528, 4560, 4565, 4566, 4570, 7602

7496 Van Onsem, J.G.; Haegeman, J.   BE 070600/60/0037
Veredeling van grootbloemig – dubbel knolbegonia cultivars. **Breeding of large flowering double tuberous begonia cultivars.** Publications.

7497 Van Onsem, J.G.; Haegeman, J.   BE 070600/60/0038
Veredeling van kleinbloemige knolbegoniacultivars. **Breeding of small flowering tuberous begonia cultivars.** Publications.

7498 Van Onsem, J.; Haegeman, J.   BE 070600/67/0018
Rassenvergelijkingsproeven met knolbegonia's. **Comparative variety trials with tuberous begonias.** Publications.

7499 Van Onsem, J.G.; Haegeman, J.   BE 070600/76/0039 R
Kollektie knolbegoniaspecies en – cultivars. **Collection of tuberous begonia–species and hybrids.** Publications.

7500 Reuther, G.   DE 506111/71/0002 R
a. Züchterische Bearbeitung von Art– und Sektionsbastarden der Gattung Iris. b. Züchtung diploider Cyclamen, kleinblumige Wildtypen und Rokokocyclamen. **a. Breeding of species and section bastards of iris. b. Breeding of diploid cyclamen, wildtypes and rococyclamen.**

7501 Reuther, G.   DE 506111/72/0001
Züchtung diploider Cyclamen und neuer polyploider Irissorten. **Breeding of diploid Cyclamen and new polyploid varieties of Iris.**

7502 Clausen, G.   DK 010113/65/4202
Sortsafprøvning af liljer, stauder, chrysanthemum, bunddækkeplanter og prydgræsser på friland. **Variety trails of lilies, herbaceous perennials, chrysanthemums, creeping plants and ornamental grasses for outdoor use.**

7503 Poisson, C.   FR 010505/73/0164
Tentative de création de glaîeuls adaptés à la culture hivernale sous serre par exploitation de la variabilité interspécifique. **Attempt to breeding gliadioli adapted to winter cultivation under glass using the inter specific variability.**

7504 Berninger, E.   FR 010505/73/0309
Création d'une variété de semis en strelitzia. **Strelitzia varieties breeding.**

7505 Cohat, J.   FR 012207/67/0065
Obtention de variétés de glaïeul adaptées à la floraison à contre–saison. **Breeding Gladiolus varieties flowering out of season.**

7506 Le Nard, M.   FR 012207/67/0167
Obtention de variétés de tulipe possédant une bonne aptitude au forçage (production de fleurs en serre). **Breeding tulip varieties with a good ability to forcing (production of**

greenhouse flowers).

7507 Le Nard, M.; Jolivet.   FR 012207/71/0064
Etude des possibilités de forçage précoce des diverses variétés d'Iris bulbeux. **Studies on the adaptability to forcing of different varieties of bulbous Iris.**

7508 Hussey; Dunwell   GB 011401/00/0009
Bulb propagation. Vegetative propagation techniques.

7509 North; Tulloch   GB 030703/00/0018
Breed disease resistant lily varieties.

7510 North; Tulloch   GB 030703/00/0020
Breeding basic material of narcissus for further selection.

7511 Schiva, T.   IT 021200/79/0002 N
Analisi genetica di caratteri quantitativi su Calla, (Zantedescia Aethiopica (L.)Spreng). **Genetic analysis of quantitative characters on Calla (Zantedeschia Aethiopica (L.) Spreng).**

7512 Barendrecht, C.J.   NL 010113/77/8311
Rassenonderzoek bloembolgewassen. **Variety testing of bulbs.** Publications.

7513 Tuyl, J.M. van   NL 010114/64/0604
Veredelingsmethodiek bij de hyacinth, gericht op forceerbaarheid en kleur. **Methodology of breeding hyacinth for forcing ability, colour and disease resistance.** Publications.

7514 Tuyl, J.M. van   NL 010114/64/0605
Veredelingsmethodiek bij de boliris gericht op forceerbaarheid en kleur. **Breeding methods of Dutch iris directed towards forcing capacity and colour.** Publications.

7515 Eijk, J.P. van   NL 010114/64/0613
Selectie van tulpen op correlatieve kenmerken. **Selection of tulips for correlative characters.** Publications.

7516 Eijk, J.P. van   NL 010114/65/2157
Opwekken van mutaties bij tulp door bestraling. **Induction of mutations of tulip by irradiation.** Publications.

7517 Eijk, J.P. van   NL 010114/65/6198
Soortkruisingen bij tulp. **Interspecific crosses in tulip.** Publications.

7518 Tuyl, J.M. van   NL 010114/71/7934
Onderzoek naar de toepassingsmogelijkheden van mutatieveredeling bij lelie en hyacinth. **Research into the possibilities of mutation breeding with lily and hyacinth.**

7519 Eijk, J.P. van   NL 010114/72/3541
Veredeling op kwaliteitskenmerken (ziekte resistentie, bloemkleur, houdbaarheid, forceerbaarheid) en produktievermogen bij de tulp. **Breeding of tulip for quality characters (disease resistance, flower colour, keeping quality, forcing ability) and bulb production.** Publications.

7520 Varekamp, H.Q.   NL 010114/72/3542
Genenbank voor tulp. **Genebank for tulip.**

7521 Tuyl, J.M. van   NL 010114/72/3544
Introductie van botanische soorten en rassen van het geslacht Lillium en het verrichten van soortkruisingen. **Introduction of botanic species and varieties of the genus Lillium and the**

D 2300 – Plant breeding

production of interspecific crosses. Publications.

7522 Kroon, G.H.   NL 010114/73/3832
Cytogenetische grondslagen van de nerineveredeling.
Cytogenetic basis of nerine breeding. Publications.

7523 Tuyl, J.M. van   NL 010114/78/7935
Veredelingsmethodiek bij de lelie. **Methodology of breeding of
lily.**

Flowers and pot plants (B 3720)

See also 4578, 4666, 4675, 5887, 6011, 7502, 7601, 7622

7524 Debuisson, J.; Neuray, G.; Henrard, G.
                                        BE 010004/60/0001
Etude de la croissance et du développement de Stephanotis
Floribunda, Gloriosa Rothschildiana Euphorbia Fulgens.
**Study on the growth and developpement of cyclamen
persicum, Stephanotis Floribunda, Gloriosa Rothschildiana,
Euphorbia Fulgens.** Publications.

7525 Debuisson, J.; Neuray, G.   BE 010004/68/0003
Recherche relative à la croissance, au développement et à
l'obtention de lignées pures de Cyclamen. **Research on the
growth, development of pure lines of Cyclamen.**

7526 Van Onsem, J.G.; Meneve, I.   BE 070600/55/0031
Veredeling van rozen met het oog op het bekomen van nieuwe
cultivars. **Breeding of new rose–cultivars.**

7527 Van Onsem, J.G.; Meneve, I.; Istas, W.
                                        BE 070600/55/0032
Veredeling van de enkelbloemige hibiscus syriacus. **Breeding
of single flowered hibiscus syriacus.**

7528 Van Onsem, J.G.; Heursel, J.   BE 070600/62/0034
Het winnen van cultivars van azalea met een zeer hoog rode
kleurstof gehalte in de bloem. **Breeding of cultivars of azalea
with a very high content of red colorants in the flowers.**
Publications.

7529 De Loose, R.   BE 070600/65/0029
Mutatie– veredeling bij de hybriden van Rhododendron simsii
Planch (Azalea) indica. **Mutation breeding on the hybrids of
Rhododendron simsii Planch (Azalea indica).** Publications.

7530 Van Onsem, J.; Mekers, O.   BE 070600/68/0007 R
Introduktie van nieuwe potplanten. **Introduction of new
potplants.** Publications.

7531 Van Onsem, J.; Heursel, J.   BE 070600/72/0002 R
Rassenvergelijkingsproeven met Azalea indica L.
(Rhododendron simsii Planch). **Variety trials with new Azalea
indica L. (Rhododendron simsii Planch) cultivars.** Publications.

7532 Van Onsem, J.G.; Meneve, I.   BE 070600/75/0074
Rassenproef met 2 nieuwe rozencultivars. **Comparison of two
new rose cultivars.**

7533 Van Onsem, J.G.; Heursel, J.   BE 070600/76/0040 R
Verdere opbouw en uitwerking van de Rhododendron
kollektie. **Further development of the Rhododendron
collection.** Publications.

7534 Van Onsem, J.G.; Gabriels, R. BE 070600/76/0041 R

Onderzoek naar de optimale minerale samenstelling van
voedingsbodems voor weefselkulturen van bromeliaceae.
**Study of the optimal mineral composition of substrates for
tissue culture of bromeliads.**

7535 Van Onsem, J.G.; Meneve, I.; Istas, W.
                                        BE 070600/76/0050
Het winnen van nieuwe Callicarpa cultivars. **Breeding of new
Callicarpa cultivars.**

7536 Van Onsem, J.G.; Meneve, I.   BE 070600/76/0051
Verwekken van tetraploidie bij diploide rozenspecies.
Inducing tetraploid rose–species.

7537 Van Onsem, J.G.; Meneve, I.   BE 070600/76/0052
Studie over de overerving van bepaalde kenmerken na
kruising. **Study on the inheritance of certain rose
characteristics.**

7538 Van Onsem, J.G.; Mekers, O.   BE 070600/76/0059
Onderzoek naar snellere vermeerderingsmethodes voor
Bromeliacea. **Research on rapid propagation methods for
Bromeliaceae.** Publications.

7539 Van Onsem, J.G.; Mekers, O.   BE 070600/76/0063
Chromosomenonderzoek bij Bromeliaceae. **Chromosome
countings on Bromeliads.**

7540 Van Onsem, J.G.; Mekers, O.   BE 070600/76/0067
Veredeling van Bromeliaceae. **Breeding of Bromeliads.**

7541 Wricke, G.   DE 138300/70/0004
Genetisch–züchterische Untersuchungen bei Gerbera
jamesonii. **Genetic and breeding studies in Gerbera jamesonii.**

7542 Wricke, G.; Trang, Q.S.   DE 138300/74/0005
Quantitativ–genetische Untersuchungen an Tetraploiden –
Modellobjekt Viola. **Investigations on quantitative genetic
characters in tetraploids – model test of Viola.**

7543 Weber, W.E.; Stümper, M.   DE 138300/78/0002 N
Schätzung von quantitativ–genetischen Parametern
einschliesslich der Epistasie bei Levkojen – Matthiola incana –
über Generationsmittelwerte. **Estimation of
quantitative–genetic parameters including epistasis for
Matthiola incana by analysis of generation means.**

7544 Horn, W.   DE 161270/78/0001 N
Züchtungsforschung bei Pelargonien. **Breeding research on
pelargonium.**

7545 Reimann–Philipp, R.; Ebbinghaus, R.; Jordan, C.
                                        DE 206000/71/4004 R
Züchtung von F1–Hybriden – auch triploid – bei Ageratum
houstonianum mit Hilfe einer von uns gefundenen Form der
Pollensterilität. Spezielle Bemühungen zur Auffindung eines
"restorers", der im Falle einer cytoplasmatisch–genischen
Bedingtheit der Pollensterilität ihre Aufhebung je nach Bedarf
ermöglichen und das Verfahren perfektionieren könnte.
Gleichlaufende Arbeiten auch bei Tetraploiden 1970–1979.
**Breeding of F1 hybrids – triploid also – of Ageratum
housonianum by means of a form of pollen sterility detected by
us. Special efforts to find a 'restorer' which could render the
removal of pollen sterility possible if required under
cytoplasmaticgenic conditions and could perfect the process.
Parallel tests with tetraploids also.**

D 2300 – Plant breeding

7546 Preil, W.; Lorenz, A.   DE 206000/71/4022 R
Möglichkeiten der Ausnutzung von freiliegenden
Samenanlagen bei Begonia semperflorens zur Ausschaltung
von Inkompatibilitätsstörungen bei der Herstellung von Artbastarden 1970–1978. **Possibilities of utilization of open seeds in
Begonia semperflorens for elimination of disturbances in
incompatibility in production of hybrids.**

7547 Reimann–Philipp, R.; Lorenz, A.   DE 206000/71/4035 R
Züchtungsarbeiten an Euphorbia pulcherrima zur Schaffung
von Sorten mit frühem Blühbeginn unter natürlichen
Tageslängenverhältnissen 1970–1982. **Breeding experiments
with Euphorbia pulcherrima for production of varieties with
early start of flowering period under natural day-Length
conditions.**

7548 Reimann–Philipp, R.; Jordan, C.; Satory, M.
                                     DE 206000/71/5001 N
Züchtung von Chrysanthemensorten mit spezieller Eignung für
Kultur in den lichtarmen Wintermonaten auf dem Weg der
Kombinationszüchtung. **Breeding of chrysanthemum varieties
with special suitability for cultivation in winter months wanting
in light by coombination breeding.**

7549 Reimann–Philipp, R.; Hofmann, K.
                                     DE 206000/71/5002 R
Versuche zur Addition oder Substitution derjenigen
Komponenten des Genoms von Begonia schmidtiana zum
Genom von Begonia semperflorens, die den
"gracilis"-Charakter bedingen 1971–1979. **Experiments on
addition or substitution of those components of the genom of
Begonia schmidtiana to the genom of Begonia semperflorens
effecting the character gracilis.**

7550 Reimann–Philipp, R.; Ebbinghaus, R.; Jordan, C.
                                     DE 206000/71/5023 R
Versuche zur Züchtung von Cinerarien–Genotypen mit gelber
Blütenfarbe bei gleichzeitiger normaler Fertilität zur Erhöhung
des Farbeffektes in Farbmischungssorten, wie z.B. "Hansa"
1964–1978. **Experiments on breeding Cineraria genotypes with
yellow flowers along with normal fertility for increasing the
effect of colour in varietal mixtures of colour, e. g. 'Hansa'
variety.**

7551 Reimann–Philipp, R.; Lorenz, A.; Jordan, C.
                                     DE 206000/71/5026 R
Einsparung der Kastrationsarbeiten bei der Herstellung von
triploiden F1–Hybriden durch Verwendung eines männlich
sterilen, mütterlichen Kreuzungspartners – Begonien –
1965–1979. **Saving the castration process in producing triploid
F1–hybrids using a male-sterile, maternal crossing partner –
begonia–.**

7552 Reimann–Philipp, R.; Martens, H.   DE 206000/75/0010
Cytogenetische Analysen an Rosenartbastarden –
insbesondere mit Rosa multiflora – im Hinblick auf
Resistenzzüchtung. **Cytogenetical investigations of interspecific
hybrids in roses particularly including R. multiflora with
respect to breeding for resistance.**

7553 Jordan, C.; Reimann–Philipp, R.   DE 206000/75/0011
Züchtung von generativ vermehrten Sorten von Gerbera
jamesonii mit einheitlichen Blütenfarben. **Breeding of seed
propagated varieties of Gerbera jamesonii with uniform flower
colours.**

7554 Reimann–Philipp, R.; Ebbinghaus, R.
                                     DE 206000/75/0012
Untersuchungen über die Vererbung des "Rokoko"-Merkmals
bei diploiden Alpenveilchen – Cyclamen persicum splendens–.
**Studies on the genetics of the "Rokoko" character in diploid
Cyclamen persicum splendens.**

7555 Schubert, G.; Rost, J.   DE 502200/78/0001 N
Sortenversuche mit Gerbera, Viola, Schnittstauden,
Sommerblumen. **Varietal experiments on Gerbera, Viola, cut
perennials, summer–flowers.** Publications.

7556 Groven, I.; Larsen, O.N.   DK 010104/69/0002 N
Sortsforsøg med roser. **Variety trials with roses.**

7557 Christensen, O.V.; Bredmose, N.   DK 010113/69/4206
Udvikling af ensartet plantemateriale af potteplanter, A.
plumosus og Gerbera til brug ved produktionsplanlægning i
væksthus. **Development of uniform plant material of pot plants,
A. plumosus and Gerbera for use in planned production in
glasshouses.**

7558 Christensen, O.V.; Bredmose, N.   DK 010113/72/4201
Sortsafprøvning af nelliker, freesia og poinsettia i væksthus.
**Variety trials of carnations, freesia and poinsettia in
glasshouses.**

7559 Clausen, G.   DK 010113/74/4203
Nyhedsafprøvning af Euphorbia fulgens, Euphorbia
pulcherrima, Allamanda, Impatiens, Freesia og Pelargonium i
væksthus. **Assessment of the merits of Euphorbia fulgens,
Euphorbia pulcherrima, Allamanda, Impatiens, Freesia and
Pelargonium in glasshouses.**

7560 Goujon, C.   FR 010505/69/0201
Sélection de clones de porte–greffes de rosiers favorisant
l'exploitation hivernale. **Breeding clonal rose rootstocks for
winter production.**

7561 Meynet, J.   FR 010505/70/0063
Obtention d'hybrides de Gerbera pour la culture sous serre.
Breeding Gerbera hybrids for glasshouse–cultivation.

7562 Meynet, J.   FR 010505/71/0062
Création d'hybrides de Renoncule. **Breeding Ranunculus
hybrids.**

7563 Goujon, C.   FR 010505/72/0203
Mise au point de méthodes précoces de sélection permettant
d'améliorer la valeur culturale des variétés de rosiers de serre.
**Development of early tests for breeding greenhouse rose
varieties.**

7564 Maizonnier, D.; Cornu, A.; Moessner, A.
                                     FR 011003/70/0180
Utilisation des remaniements chromosomiques pour
l'établissement de la carte génique du pétunia. **Use of
chromosomal rearrangements for establishing the genetic map
of petunias.**

7565 Dommergues, P.; Bodergat, R.; Cornu, A.; Giraud, P.;
Maizonnier, D.   FR 011003/70/0182
Etude génétique du pétunia hybrida. **Genetic study of petunia
hybrida.**

D 2300 – Plant breeding

7566 Touvin, H.; Leblanc, D. FR 011003/71/0067
Production d'haploides androgénétiques chez le Pétunia. **Breeding androgenetic haploids in Petunia.**

7567 Touvin, H.; Leblanc, D. FR 011003/71/0068
Obtention de Pétunia fleurissant en jours courts. **Breeding short–days flowering Petunia.**

7568 Maizonnier, D.; Cornu, A.; Moessner, A. FR 011003/71/0181
Recherche et utilisation de la série trisomique primaire chez le pétunia. **Research and utilization of the primary trisomic series in petunia.**

7569 Dommergues, P.; Pelletier, A.; Cornu, A.; Bodergat, R. FR 011003/71/0183
Etude des événements génétiques induits par les traitements mutagènes. **Study of the genetic phenomenons induced by the mutagenetreatments in higher plants.**

7570 Cornu, A.; Pelletier, A.; Dulieu, H.; Paynot, M. FR 011003/71/0185
Etude des phénomènes d'instabilité aux loci A et R et pétunia. **Study of instability phenomenons on the petunia a and r loci.**

7571 Touvin, H.; Leblanc, D. FR 011003/72/0066
Recherche de Pétunia contenant du Pelargonidol. **Research on Petunia containing pelargonidol.**

7572 Dommergues, P.; Cornu, A.; Bodergat, R. FR 011003/72/0184
Efficacité des agents mutagènes chimiques. **Efficiency of the chemical mutagene agents in higher plants.**

7573 Cornu, A.; Pelletier, A.; Dulieu, H.; Vallade, J. FR 011003/73/0186
Utilisation des néoformations en mutagenèse. **Utilization of neo–formations through mutagenesis in higher plants.**

7574 Piquemal, G.; Cleomene, J. FR 011201/69/0166
Création d'hybrides de tournesols ornementaux. **Ornemental sunflower hybrids.**

7575 Langton GB 011101/00/0012 R
The genetics of physiological characters in ornamental plants.

7576 Arthur GB 011401/76/0021
Carnations : select for fusarium resistance and improved productivity.

7577 Hedley; Arthur GB 011401/78/0020 N
Selection techniques and physiological genetics of commercial flowering crops.

7578 Schiva, T. IT 021200/73/0008
Miglioramento genetico della gerbera. **Breeding of gerbera.** Publications.

7579 Cirrito, M.; Schiva, T.; Volpi, L.; De Ranieri, M. IT 021200/75/0005
Ricerca sull'origine di fiori semplici in infiorescenze di Polyanthes tuberosa a fiore doppio. **Research on the origin of single flowers on the inflorescence of double Polianthes tuberosa.**

7580 Schiva, T.; Dalla Guda, C. IT 021200/79/0001 N

7581 Quagliotti, L.; Lepori, G.; Franceschetti, U. IT 041212/74/0001
Miglioramento genetico della gerbera per l'uniformità del colore dei fiori. **Gerbera breeding for uniformity of flower colour.**

7582 Barendrecht, C.J. NL 010113/77/8312
Rassenonderzoek bloemisterijgewassen. **Variety testing flowers and potplants.** Publications.

7583 Sparnaaij, L.D. NL 010114/55/0600
Onderzoek naar de produktiemogelijkheden en de gebruikswaarde van legitiem en vrijbestoven zaad van tetraploïde freesia's. **Production and cultural value of legitimate and open–pollinated seed of tetraploid fresias.** Publications.

7584 Sparnaaij, L.D. NL 010114/60/0599
Het kweken van kasanjers resistent tegen Phialophora en Fusarium. **Breeding of Phyalophora and Fusarium–resistant carnations in greenhouses.** Publications.

7585 Jong, J. de NL 010114/63/0601
Veredeling op bloemproduktie en bloemkwaliteit bij gerbera. **Breeding for flowerproduction and flowerquality of gerbera.** Publications.

7586 Jong, J. de NL 010114/65/0611
Ontwikkeling van nieuwe potplanten. **Development of new potplants.**

7587 Vries, D.P. de NL 010114/66/0594
Het kweken van specifieke kasrozen. **Breeding of specific glasshouse roses.** Publications.

7588 Sparnaaij, L.D. NL 010114/67/2158
Toepassingsmogelijkheden van mutatieveredeling bij anjers. **Possibilities of mutation breeding for carnations.** Publications.

7589 Jong, J. de NL 010114/74/8790 N
Kweken van rassen van gloriosa met een betere bloemkleur en planttype. **Improving flower colour and plant type of gloriosa by breeding.**

7590 Sparnaaij, L.D.; Demmink, J.F. NL 010114/75/6195
Verbetering van het kasanjersortiment. **Improvement of glasshouse carnations.** Publications.

7591 Jong, J. de NL 010114/76/6644
ne–Veredelingsonderzoek aan de chrysant gericht op produktie en kwaliteit in de winter. **Breeding chrysanthemums for productivity and quality of the winter crop.** Publications.

7592 Belgraver, W.; Kleinhesselink, C.; Marsbergen, W. van; Bonnyai, J. NL 010201/60/0973
Sortimentsonderzoek bij bloemisterijgewassen. **Investigations on the assortment of floricultural crops.** Publications.

7593 Leffring, L. NL 010201/68/0959 N
Veredeling van gerbera voornamelijk gericht op kwaliteitsverbetering en produktieverhoging in de winter. **Breeding of gerbera especially for production in winter and**

flower quality. Publications.

7594 Hoekstra, F.A. NL 020041/77/7368
Genetische en hormonale regulatie van pigmentvorming in kroonbladeren van Petunia. **Genetic and hormonal regulation of pigment synthesis in Petunia petals.**

7595 Hensen, K.J.W. NL 020043/50/4655
Taxonomisch onderzoek van sortimenten Delphinium, Dianthus en siergrassen (Hosta, Vinca). **Taxonomical research of assortments of Delphinium, Dianthus and ornamental grasses (Hosta, Vinca).**

7596 Legro, R.A.H. NL 020043/57/4652
Ontwikkeling van Delphinium cultivars op basis van cytotaxonomische informatie. **Development of Delphinium cultivars based on cytotaxonomical information.**

7597 Legro, R.A.H. NL 020043/66/4651
Cytotaxonomie van Begonia (wilde soorten en cultivars.). **Cytotaxonomy of Begonia (cultivated plants and related wild species).**

7598 Zeven, A.C. NL 020045/70/8847 N
Onderzoek naar de soortaffiniteit bij Streptocarpus. **Investigation into the species affinity in Streptocarpus.**

7599 Sneep, J. NL 020045/78/8851 N
Mannelijke steriliteit in Impatiens Walleriana. **Male sterility in Impatiens Walleriana.**

7600 Marrewijk, G.A.M. van NL 020045/78/8857 N
Integrale karakterisering van mannelijke fertiel, cytoplasmatische mannelijk steriele en hersteld fertiele genotypen van de petunia. **Integrated characterization of male fertile, cytoplasmatic male sterile and restored fertile genotypes of the garden petunia.**

Ornamental shrubs (B 3730)

See also 4678, 4713, 6019, 7535, 7555, 7587, 7640, 7641, 7642, 7643, 7644, 7658

7601 De Loose, R. BE 070600/76/0049
Inbrengen van het genetisch materiaal (DNA) van een geelbloemige Rhododendron of Mollis azalea, in het genetisch materiaal van rhododendron simsii. **Introduction of the genetic material (DNA) of a yellow flowering Rhododendron or Mollis azalea, in the gentic material of Rhododendron simsii.**

7602 De Loose, R. BE 070600/77/0080 R
Biochemische karakterisatie van cultivars van Rhododendron simsii Planch en begonia hybriden. **Biochemical "Fingerprinting" of cultivars of Rhododendron simsii Planch and tuberous begonia hybrids.** Publications.

7603 Schmidt, H.; Laskawy, W. DE 206000/71/5015
Artkreuzungen bei Cotoneaster zur Züchtung verbesserter bodenbedeckender Rasenersatzpflanzen. **Interspecific hybridization in Cotoneaster with respect to the improvement of soil covering varieties for replacing lawns.**

7604 Persiel, F.; Laskawy, W. DE 206000/73/5001
Resistenzzüchtung gegen Feuerbrand – Erwinia amylovora – bei Cotoneaster durch Auslese aus nicht apomiktisch entstandenen Sämlingsnachkommenschaften auf einem Versuchsfeld im Befallsgebiet. **Breeding for resistance to fireblight – Erwinia amylovora – in Cotoneaster by selection from non-apimictical seedling offsprings in a naturally infected experimental field.**

7605 Preil, W. DE 206000/74/0001
Analysen der Erbgänge von physiologischen Merkmalen bei Rhododendron simsii. **Investigations on the genetics of physiological characters in Rhododendron simsii.**

7606 Groven, I.; Brander, P.E. DK 010104/76/0019 N
Sortsforsøg med landskabsplanter – urter. **Variety trials with herbaceous plants for landscaping.**

7607 Rasmussen, F. DK 010112/74/4103
Sortsafprøvning med plænegræsser. **Variety trials of turf grasses.**

7608 Jørgensen, M.B.; Thuesen, A. DK 010115/72/4510
Værdiafprøvning af plænegræsser. **Assessment of the value of turf–grass varieties.**

7609 Cadic, A.; Decourtye, L.; Martin, D. FR 010401/72/0165
Obtention de variétés de berberis à feuillage rouge et persistant. **Breeding berberis varieties for red and everlasting foliage.**

7610 Decourtye, L.; Renoux, A. FR 010401/72/0226
Recherche de formes nouvelles dans le genre weigela par mutagénèse provoquée. **Mutagenesis in weigela shrubs.** Publications.

7611 Decourtye, L.; Renoux, A. FR 010401/74/0225
Recherche de nouvelles formes de plantes de haies dans le genre thuya. **New plants for hedges in the genrus thuya.**

7612 Lemoine, J.; Duron, M.; Morand, J.C.
FR 010401/76/0354
Sélection sanitaire des plantes ornementales ligneuses. **Improvement of woody ornementals for virus free stocks.**

7613 Watkins; Keep GB 011002/75/0006
Selection of ornamental plants occuring in breeding lines of bush and tree fruits.

7614 Harrison; Carpenter GB 011402/00/0005 R
Unstable genes in Antirrhinum.

7615 Lamb, J.G.D. IE 060300/68/0098 R
Eucalyptus cultivar trials. Publications.

7616 Schneider, F. NL 010113/77/8313
Rassenonderzoek boomkwekerijgewassen. **Variety testing of arborous nursery crops.** Publications.

7617 Heyting, J. NL 010114/61/2569
Het winnen van nieuwe cultuurvariëteiten van sierheesters door middel van kruisingen. **The raising of new cultivated varieties of ornamental shrubs.** Publications.

7618 Heyting, J. NL 010114/61/2570
Opwekken van mutaties in sierheester–cultivars door bestraling en colchicinebehandeling. **Inducing mutations in cultivars of ornamental shrubs by irradiation and colchicine treatment.**

D 2300 – Plant breeding

7619 Heyting, J.   NL 010203/61/2569
Het winnen van nieuwe cultuurvariëteiten van sierheesters door middel van kruisingen. **The raising of new cultivated varieties of ornamental shrubs.** Publications.

7620 Heyting, J.   NL 010203/61/2570
Opwekken van mutaties in sierheester–cultivars door bestraling en colchicinebehandeling. **Inducing mutations in cultivars of ornamental shrubs by irradiation and colchicine treatment.**

7621 Laar, H.J. van de   NL 010203/66/2572
De identificatie en het gebruikswaarde–onderzoek van het boomkwekerijsortiment. **Research concerning the registration and cultural value of ornamental shrubs.** Publications.

7622 Legro, R.A.H.   NL 020043/70/4656
Cytotaxonomisch en experimenteel–taxonomisch onderzoek van houtige siergewassen en vasteplanten. **Cytotaxonomic and biosystematic studies in woody ornamentals and hardy plants.**

7623 Brandenburg, W.A.   NL 020043/76/7565 R
Cytotaxonomisch en experimenteel–taxonomisch onderzoek van Clematis. **A cytotaxonomic and biosystematic approach of the genus Clematis.** Publications.

Other ornamentals and ornamental products (B 3790)

See also 7502, 7557, 7595

7624 Bourgoin, B.; Mansat, P.   FR 012223/70/0176
Etude de la variété nouvelle dans les différentes espèces de plantes à gazon. **Study of the new variety in the various species of sward plants.**

7625 Bourgoin, B.; Mansat, P.   FR 012223/72/0177
Comparaison de techniques d'appréciation de plantes à gazon, utilisables en sélection. **Comparison of sward plant evaluation techniques suitable for selection.**

7626 Bourgoin, B.; Mansat, P.   FR 012223/73/0178
Effets de différents niveaux de consanguinité sur les caractéristiques importantes pour la valeur en gazon chez la fétuque rouge. **Consequences of different inbreeding levels on the characteristics essential for the sward value of red fescue.**

7627 Bourgoin, B.; Mansat, P.   FR 012223/75/0301
Etude du comportement d'espèces et cultivars de gazons au piétinement du cheval. **Study of turfgrass species and cultivars behaviour : horse treading experiments.**

Forests in general (B 3800)

See also 4731, 4741, 4831

7628 Nanson, A.   BE 130000/65/0014
Génétique quantitative appliquée à la sélection des arbres forestiers. **Applied quantitative genetics in forest tree breeding.** Publications.

7629 Nanson, A.   BE 130000/65/0015 R
Recherches sur les provenances les plus appropriées à la sylviculture de la Belgique. **Provenances research to determine the best adapted forms in Belgian sylviculture.** Publications.

7630 Muhs, H.–J.; Krusche, D.; Illies, Z.M.; Reck, S.   DE 202020/70/5006
Klärung der genetischen Grundlagen für die Züchtung, inklusive Konkurrenzprobleme sowie Schätzung genetischer Varianzen in Populationen. **Clearing of genetic principles for breeding including problems of competition and estimation of variances in populations.**

7631 Melchior, G.H.; Herrmann, S.   DE 202020/71/5005 N
Genresourcen für Züchtungsvorhaben bei Waldbaumarten. **Forest tree gene resources for breeding programs.**

7632 Muhs, H.–J.; Reck, S.; Scholz, F.   DE 202020/72/5001
Identifizierung von Populationen, Sorten und Genotypen. **Identification of populations, cultivars and genotypes.**

7633 Reck, S.; Krusche, D.; Muhs, H.–J.; König, A.; Stephan, B.R.; Mohrdiek, O.   DE 202020/73/5001
Variationsmuster und Prüfung der Anbaueignung exotischer und einheimischer Waldbaumarten. **Variation patterns and control of adaptation of indigenous and exotic forest tree species on given sites.**

7634 Reck, S.; Stephan, B.R.; Scholz, F.; Krusche, D.; Behrens, V.; Mohrdiek, O.   DE 202020/73/5004
Methoden zur Frühdiagnose und Frühbeurteilung bei Waldbäumen. **Methods for early diagnosis and valuation of forest trees.**

7635 Melchior, G.H.; Herrmann, S.; Illies, Z.M.; König, A.; Krusche, D.; Scholz, F.   DE 202020/74/5001
Rationalisierung und Entwicklung von Verfahren im forstlichen Züchtungswesen. **Rationalization and development of methods in forest breeding.**

7636 Reck, S.; Muhs, H.–J.; Melchior, G.H.; Stephan, B.R.; Krusche, D.; Scholz, F.   DE 202020/77/0001
Züchtungsstrategien und Züchtung bei einheimischen und exotischen Waldbaumarten. **Breeding strategies and breeding of indigenous and exotic forest tree species.**

7637 Stephan, B.R.; Scholz, F.; Reck, S.   DE 202020/77/0002
Resistenzeigenschaften bei Waldbaumarten. **Resistance properties of forest tree species.**

7638 Melchior, G.H.   DE 202020/78/0001 N
Forstliche Saatgutversorgung und Züchtungsprogramme in den Tropen und Subtropen. **Seed supply and breeding programmes in tropical and subtropical forestry.**

7639 Melchior, G.H.; Mohrdiek, O.; Krusche, D.; König, A.; Reck, S.   DE 202020/78/0002 N
Sorten– und Klonprüfung bei Baumarten. **Variety and clone testing of tree species.**

7640 Groven, I.; Brander, P.E.   DK 010104/69/0008 N
Selektion i dyrket plantemateriale af træer og buske. **Selection in cultivated material of trees and shrubs for landscaping.**

7641 Groven, I.; Brander, P.E.   DK 010104/70/0001 N
Sortsforsøg med landskabsplanter – træer og buske. **Variety trials with trees and shrubs for landscaping.**

7642 Groven, I.; Brander, P.E.; Larsen, O.N.   DK 010104/75/0004 N
Nyhedsafprøvning af landskabsplanter – træer og buske.

D 2300 – Plant breeding

Registration test of trees and shrubs for landscaping.

7643 Groven, I.; Brander, P.E. DK 010104/76/0003 N
Proveniensforsøg med landskabsplanter – træer og buske.
Trials with seed sources of trees and shrubs for landscaping.

7644 Groven, I.; Larsen, O.N.; Knoblauch, F.
DK 010104/76/0009 N
Moderplanter og formeringsmateriale af træer og buske.
Motherplants and propagation material in trees and shrubs for landscaping.

7645 Feilberg, L. DK 030102/69/0011
Blomstring og bestøvning. **Flowering and pollination.**

7646 Nepveu, G.; Arbez, M.; Birot, Y.; Baradat, Ph.
FR 010304/73/4203 N
Application des méthodes de génétique quantitative aux
critères de qualité du bois. **Quantitative genetic studies of wood quality criteria.**

7647 Nepveu, G.; Birot, Y. FR 010304/74/4204 N
Variabilité intraspécifique des critères de qualité du bois.
Intraspecific variability of wood quality criteria.

7648 Milner GB 041001/00/0005
Species and provenance trials for forest trees for Northern Ireland.

7649 O'Driscoll, J.; Pfeiffer, A. IE 050100/60/7210 N
Conifer and broadleaf species trials. Publications.

7650 Tocci, A.; Pelizzo, A. IT 021700/72/0002 R
Trattamenti selettivi in boschi da seme. **Selection in tree seed stands.**

7651 Giannini, R. IT 040116/77/0194 R
Studi sull'interazione genotipo–ambiente in specie forestali.
Study of genotype–habitat interaction in forest species.

7652 Magini, E. IT 040516/74/0579
Studi sulla variazione intraspecifica e sull'ereditabilità di alcuni
caratteri in specie forestali. **Studies on the intraspecific
variability and on the heritability of some genetic characteristics
of forest species.**

7653 Magini, E. IT 040516/77/0222 R
Studi sulla variazione intraspecifica e sull'ereditabilità di alcuni
caratteri in specie forestali. **Studies on the intraspecific
variation and on the heredity of some characteristics in forest
species.**

7654 Cartechini, A. IT 041005/73/0165
Selezione clonale delle specie arboree. **Clonal selection of tree
species.** Publications.

7655 Koster, R. NL 010601/68/2734
Samenstelling van een lijst van teelt– en uitgangsmateriaal
voor bosen landschapsbouw. **Editing a list of reproductive and
basic material for forestry and landscaping.** Publications.

Pine forests in general (B 3810)

7656 Baumeister, G. DE 506451/71/0001
Prüfung des Erbwerts bei verschiedenen Laub– und
Nadelbaumarten. **Tests of their hereditary characteristics of**

coniferous and broadleaved tree species.

7657 Baumeister, G. DE 506451/71/0003
Züchtung neuer Sorten durch intra– und interspezifische
Kreuzung. **Breeding of new cultivars of coniferous by interand
intra–species hybridization.**

7658 O'Driscoll, J.; Pfeiffer, A. IE 050100/62/7212 N
Selection and breeding of conifers. Publications.

7659 O'Driscoll, J.; Pfeiffer, A. IE 050100/62/7213 N
Provinence studies of conifers. Publications.

7660 Rota, L.; De Vecchi, E.; Palenzona, M.; Zeppegno,
G.F.; Ferraris, P. IT 120100/78/0003 N
Verifica dell'adattamento e produzione legnosa delle parcelle
di conifere a rapida crescita costituite dall'Istituto Nazionale
Piante da Legno. **Checking the adaptability and productivity of
fast growing conifer plots established by the "Istituto Nazionale
Piante da Legno" in Italy.**

7661 Tol, G. van NL 010601/63/2392
Praktische en theoretische aspecten van de plantsoensortering.
Practical and theoretical aspects of stock grading. Publications.

Fir forests (B 3811)

See also 10010

7662 Schütt, P.; Schuck, H.J. DE 160060/72/0003 R
Terpenzusammensetzung bei Pinus nigra verschiedener
Herkunft 1971. **Composition of terpene in Pinus nigra of
different provenience.**

7663 Lang, K.J. DE 160060/72/0007
Terpenzusammensetzung bei verschiedenen
Lärchenherkünften 1972. **english title not indicated.**

7664 Feilberg, L. DK 030102/68/0008
Forædling af pinus contorta. **The improvement of Pinus
contorta.**

7665 Eccher, A. IT 011801/72/0005
Selezione massale e individuale del Pinus radiata. **Pinus radiata
stand and individual selection.**

7666 Eccher, A. IT 011801/75/0008
Analisi delle discendenze di piante plus di Pinus radiata e
P.halepensis per la costituzione di arboreti da seme. **Tests of
Pinus radiata and P.halepensis plus trees progenies for the
establishment of seed orchards, arboreta.**

7667 Eccher, A. IT 011801/79/0003 N
Indagine sulle possibilità di introduzione di Pinus muricata in
condizioni marginali per il P.radiata. **Investigation on the
possibility of Pinus muricata introduction under marginal site
conditions for P.radiata.**

7668 Eccher, A. IT 011801/79/0004 N
Indagine sulle possibilità di introduzione di Pinus eldarica.
Investigation on the possibility of Pinus eldarica introduction.

7669 Eccher, A. IT 011801/79/0005 N
Indagine sulla variabilità genetica di Pinus radiata e
P.muricata. **Investigation on Pinus radiata and P.muricata
genetic variation.**

D 2300 - Plant breeding

7670 Mittempergher, L.   IT 061100/73/0208
Miglioramento genetico di alcune specie di pino a due aghi al
Cronartium flaccidum. **Genetic improvement of some two
needle pine species for resistance to Cronartium flaccidum.**
Publications.

7671 Kriek, W.   NL 010601/52/2405
Veredeling van grove den. **Breeding and tree improvement of
Scots pine.** Publications.

7672 Kriek, W.   NL 010601/63/2409
Veredeling van Pinus nigra ssp., Pinus contorta, Pinus strobus
en Pinus divazicata. **Breeding and tree improvement of Pinus
nigra, Pinus contorta, Pinus strobus and pinus divazicata.**
Publications.

Spruce and fir forests (B 3812)

See also 7611, 10018, 10023

7673 Steinhauer, A.; Hattemer, H.   DE 132690/71/0002 R
Herstellung haploider Pflanzen bei Fichte und Birke als Ersatz
für Inzucht 1970. **Studies on the determination of the
provenience of spruce seed by means of biochemical
polymorphisms.**

7674 Bergmann, F.   DE 132690/72/0001
Genetische Untersuchungen bei Douglasie mit Hilfe von
Isoenzym–Polymorphismen zur Bestimmung geographischer
Variationsmuster und zur Identifizierung von
Samen–Herkünften. **Genetic investigations on Douglas fir by
means of isoenzyme polymorphisms for the analysis of
geographic variation patterns and for the identification of seed
provenances.**

7675 Bergmann, F.   DE 132690/77/0001
Beschreibung des geographisch strukturierten genetischen
Variationsmusters der Fichte mit Hilfe von
IsoenzymGenhäufigkeiten. **Analysis of the
geographically–dependent genetic variation pattern of Norway
spruce by means of isozyme gene frequencies.** Publications.

7676 Schönborn, A.von; Braun, G.   DE 160180/72/0001
Über die Ursachen und Kriterien der Immissionsresistenz bei
Fichte – Picea abies L. Karst. **Causes and criteria of immission
resistance in Picea abies L. Karst.**

7677 Kleinschmit, J.   DE 507653/77/0001
Kontrollierte Kreuzung von Fichten und Lärchenarten,
Prüfung der Nachkommen und vegetative Vermehrung der
besten Hybriden. **Controlled pollination of spruce and larch
species for hybrid–production, progeny testing and selection of
superior hybrids for vegetative propagation.**

7678 Kaufmann, U.; Wellendorf, H.   DK 030101/75/0004 R
Undersøgelser af genetiske mekanismer i rødgran, der har
hæmmende virkning på vækst af rodfordærversvampen.
**Investigations of genetical mechanisms in Norway sprue which
have a restrictive influence on the growth of heart–rot fungus.**

7679 Wellendorf, H.   DK 030102/50/0009
Forædling af rødgran. **The improvement of Norway spruce.**

7680 Søegaard, B.   DK 030102/51/0015
Forædling af thuja. **The improvement of Thuja.**

7681 Roulund, H.   DK 030102/58/0013
Forædling af douglasgran. **The improvement of Douglas spruce.**

7682 Roulund, H.   DK 030102/68/0012
Forædling af sitkagran. **The improvement of Sitka spruce.**

7683 Raddi, P.   IT 061100/73/0211
Miglioramento genetico del Cupressus sempervirens per la
resistenza al Coryneum cardinale. **Breeding Cupressus
sempervirens for resistance to Coryneum cardinale disease.**
Publications.

7684 Kriek, W.   NL 010601/52/2406
Veredeling van douglas. **Breeding and tree improvement of
douglas fir.** Publications.

7685 Kriek, W.   NL 010601/66/2407
Veredeling van fijnspar en Sitkaspar. **Breeding and tree
improvement of Norway spruce and Sitka spruce.** Publications.

7686 Kriek, W.   NL 010601/76/6675
Veredeling van Abies grandis. **Breeding and tree improvement
of Grand fir.**

Larch forests (B 3813)

See also 7677

7687 Keiding, H.   DK 030102/50/0014
Forædling af lærk. **The improvement of larch.**

7688 Kriek, W.   NL 010601/66/2408
Veredeling van lariks. **Breeding and tree improvement of
larchs.** Publications.

Leafwoods in general (B 3820)

See also 7656

7689 Koster, R.   NL 010601/52/2401 R
Veredeling van ander loofhout dan populier, wilg, els, berk en
eik. **Selection of other broadleaved species than poplar, willow,
alder, birch and oak.**

7690 Heybroek, H.M.   NL 010601/56/1850
Mutatie–onderzoek en polyploidieteelt bij loofbomen.
Research on mutation and polyploidy in broadleaved species.

Oak tree stands (B 3821)

7691 Kleinschmit, J.; Muhle, O.   DE 507653/78/0001 N
Aufnahme und Auswertung der von Prof. Krahl–Urban
begründeten Buchen–Herkunftsversuche. **Survey and
evaluation of beech provenance trials established by Prof.
Krahl–Urban.**

7692 Tocci, A.; Pelizzo, A.   IT 021700/62/0001 R
Miglioramento genetico di specie forestali arboree (Quercus
borealis, Populus alba e Platanus orientalis, Juglans regia).
**Forest tree improvement (Quercus borealis, Populus alba e,
Platanus orientalis, Juglans regia).**

7693 Kriek, W.; Verwey, J.A.   NL 010601/52/7538
Selectie en veredeling van zomer– en wintereik (Quercus robur

en Q. petraea). **Breeding of oaks (Quercus robus and Q. petraea).**

Poplar tree stands (B 3823)

See also 4995, 7673, 7692

7694 Steenackers, V. BE 071000/76/0001
Veredeling en teelttechnisch onderzoek van populus sp.
Breeding and research on growing techniques of populus sp.

7695 Impens, I. BE 140000/79/0059 N
Uitwerken van ecofysische en ecofysiologische methoden
dienstig voor de identificatie van planten met een hoog
groeivermogen. **Elaborating of ecofysical and ecofysiological
methods for the identification of plants with an high growth
capacity.**

7696 Dieterich, H.; Schlenker, G. DE 501502/73/0002
Prüfung von Nachkommenschaften der Populus trichocarpa –
Originalsaatgut aus JUFR–Kollektion 1973 – und aus eigenen
kontrollierten Kreuzungen – P.szechuanica x trichocarpa –.
**Testing of the progenies of Populus trichocarpa from the
original seeds of the JUFR collection 1973 and resulting from
own controlled breeds as Populus szechuanica x trichocarpa.**

7697 Kleinschmit, J.; Spethmann, W. DE 507653/77/0002
Auslese von Birken–Zuchtbäumen, Prüfung durch
Einzelstammsaaten, Hybridzüchtung und Begründung von
Samenplantagen. **Selection of birch–plus–trees,
open–pollinated single–tree progeny test, hybrid–production
and seed orchard establishment.**

7698 Weisgerber, H.; Hoffmann, E. DE 912000/74/0002
IUFRO–Provenienzversuch Populus trichocarpa. **IUFRO
provenance experiment on Populus trichocarpa.**

7699 Ridé, M.; Poutier, F. FR 010402/58/1006
La résistance du chancre bactérien dans les programmes de
sélection du peuplier. **Poplar resistance to bacterial canker its
use in plant breeding programs.** Publications.

7700 Clear, T.; Joyce, P.M.; Gardiner, J.J.
IE 120107/64/9087 N
Poplar clonal trials.

7701 Avanzo, E. IT 011801/58/0001
Miglioramento genetico del Populus deltoides e selezione di
cloni adatti all'Italia centro–meridionale. **Genetic improvement
of Populus deltoides and selection of clones suitable for central
and southern Italy.** Publications.

7702 Avanzo, E. IT 011801/61/0002
Miglioramento genetico di altri pioppi della Sezione Leuce per
l'ambiente di montagna dell'Italia centro–meridionale. **Genetic
improvement of other Section Leuce poplars suitable for the
mountainous environment of central and southern Italy.**
Publications.

7703 Avanzo, E. IT 011801/63/0001
Miglioramento genetico del Populus x euramericana e
selezione di cloni adatti all'Italia centro–meridionale. **Genetic
improvement of Populus x euramericana and selection of clones
suitable for central and southern Italy.** Publications.

7704 Avanzo, E. IT 011801/64/0001

Miglioramento genetico del Populus nigra e selezione di cloni
adatti all'Italia centro meridionale. **Genetic improvement of
Populus nigra and selection of clones suitable for central and
southern Italy.** Publications.

7705 Avanzo, E. IT 011801/65/0003
Miglioramento genetico del Populus alba e selezione di cloni
adatti all'Italia centro–meridionale. **Genetic improvement of
Populus alba and selection of clones suitable for central and
southern Italy.**

7706 Sekawin, M. IT 011802/72/0050
Confronto fra mescolanze di cloni e cloni puri di pioppo.
Comparative test of clone mixtures an pure clones of poplar.

7707 Magini, E.; Giannini, R. IT 040516/71/0001
Studio sulla variabilità intraspecifica dell'ontano napoletano
(Alnus cordata, Lois.). **Research on the intraspecific variability
of Italian Alder (Alnus cordata, Lois.).**

7708 Koster, R. NL 010601/48/2399
Veredeling van zwarte en balsempopulier. **Selection and
breeding of black and balsam poplar.**

7709 Kriek, W.; Verwey, J.A. NL 010601/52/7536
Selectie en veredeling van els (Alnus glutinosa, A. incana en
A. cordata). **Breeding of alder (Alnus glutinosa, A. incana and
A. cordata).**

7710 Kriek, W.; Verwey, J.A. NL 010601/52/7537
Selectie en veredeling van berk (Betula pendula en B.
pubescens). **Breeding of birch (Betula pendula and B.
pubescens).** Publications.

7711 Broekhuizen, J.T.M. van NL 020021/55/4805
Morfologische beschrijving en identificatie van populieren en
nilgen. **Morphological discription and identification of poplars
and willows.** Publications.

7712 Glastra, T.F. NL 040007/48/2399
Veredeling van zwarte en balsempopulieren. **Selection and
breeding of black and balsem poplars.** Publications.

Other leafwoods (B 3829)

See also 7711

7713 Koster, R. NL 010601/52/2400
Veredeling van wilg en grijze abeel. **Selection and breeding of
willow and Populas canescens.**

7714 Kriek, W.; Verwey, J.A. NL 010601/77/7539
Herkomstonderzoek van Nothofagus spp. **Provenance research
of Nothofagus spp.**

Other forests (B 3890)

— **7715** Keiding, H. DK 030102/65/0017
Forædling af tropiske og subtropiske træarter. **The
improvement of tropical and subtropical tree species.**

7716 Barner, H.; Keiding, H. DK 030102/69/0016
Forædling af tropiske og subtropiske træarter. **The
improvement of tropical and subtropical tree species.**

7717 Gemignani, G. IT 011801/70/0004

D 2300 - Plant breeding

Confronto di 24 provenienze di Eucalyptus camaldulensis (Progetto FAO/SCM/CRFM/6). **Comparison of twenty-four provenances of Eucalyptus camaldulensis (FAO/SCM/CRFM/6 project).**

7718 Gemignani, G.   IT 011801/75/0007
Analisi delle discendenze di piante plus di E. bicostata, E. camaldulensis, E. globulus, E. maideinii, E. x trabutii, E. viminalis per la costituzione di arboreti da seme. **Tests of Eucalyptus bicostata, E. camaldulensis, E. globulus, E. maidenii, E. x trabutii and E. viminalis plus trees progenies for the establishment of seed orchards, arboreta.** Publications.

Stimulant crops (B 3910)

See also 6872

7719 Neve; Royle   GB 012201/00/0004 R
Breed hops for resistance to wilt and mildews, combined with high yields and brewing value.

7720 Neve; Thomas   GB 012201/00/0005 R
Induction of seedlessness by elimination of males and breeding for sterility.

Spice and seasoning plants of warm climates (B 3920)

7721 Pochard, E.; Chambonnet, D.   FR 010603/71/0056
Etude de méthode de création de variétés de piment résistantes au Phytophthora Capsici et au virus 1 du concombre. (C.V. 1). **Studies on method of breeding red pepper varieties resistant to Phytophthora Capsici and cucumber virus 1. (C.V. 1).** Publications.

7722 Pochard, E.; Dumas de Vaulx, R.   FR 010603/71/0057
Mise au point de méthodes nouvelles d'obtention de plantes haploïdes chez Capsicum Annuum. **Perfecting new methods of obtaining haploid plants in red pepper (Capsicum annuum).** Publications.

7723 Lorenzetti, F.   IT 041002/74/0577
Miglioramento genetico del peperone. **Genetic improvement of pepper.**

Spice and seasoning plants of temporate climates (B 3930)

See also 5032

7724 Dubuisson, J.; Lemaitre, R.   BE 140000/79/0066 N
Essai d'introduction de la culture d'ail dans la Vallée de la Meuse. **Introduction of garlic (allium) in the Meuse Valley.**

7725 Kohlmann, J.; Zwack, F.   DE 502059/73/0001 R
Quantitative Erfassung und Merkmale des Habitus, Krankheitsanfälligkeit, Ertrag und Qualität verschiedener Hopfensorten 1971. **Quantification and characteristics of habitus, susceptibility to diseases, crop yield and quality of different species of hops.**

7726 Maier, J.; Kremheller, H.T.; Pichelmaier, K.
                                       DE 502059/73/0003
Stoffwechselphysiologische Untersuchungen zur Toleranz verschiedener Hopfensorten gegenüber Verticillium alboatrum und Verticillium dahliae als Erreger der Hopfenwelke.

Nutritive metabolism studies on the tolerance of different hops varieties of Verticillium alboatrum and of Verticillium dahliae as causal agents of hops wilt.

Perfume plants (B 3940)

See also 5042

7727 McDonnell, P.F.   IE 060302/65/0513 R
Evaluation of hop varieties. Publications.

Drugs and medicine plants (B 3970)

See also 7572

7728 Stryckers, J.; De Baets, A.   BE 030017/50/0003 R
Verbetering van de teelttechniek van tabak met inbegrip van selektie op ziekteresistentie en betere gebruikseigenschappen op landbouwkundig en industrieel vlak. **Study on the improvement of cultural practices for the tobacco culture and breeding for desease resistance and improved agricultural and industrial properties.** Publications.

7729 Mainil, J.; Fraiture, A.; Vermeulen, H.
                                       BE 140000/78/0073 N
Etude des variétés et clônes de Vinca minor et techniques culturales. **Study of Vinca minor varieties and culture technics.**

7730 Schmidt, J.A.; Stier, F.   DE 501122/77/0001
Überprüfung und Selektion von neuen Tabaksorten auf den Gehalt an gesundheitsschädlichen Stoffen – Nikotin, Teer, CO u.a. –). **Examination and selection of new sorts of tobacco by content of unhealthy substances – nicotine, tar, CO a.o. –.** Publications.

7731 Deshayes, A.; Delbut, J.; Dulieu, H.; Bruneau, R.
                                       FR 011003/72/0187
Recherche sur la stabilité génétique des cellules somatiques des plantes supérieures. **Research on the genetic stability of the somatic cells in higher plants.**

7732 Morice, J.; Gourvest, C.   FR 011301/75/0313
Sélection de variétés de papaver bracteatum donnant un rendement élevé en thebaine. **Breeding papaver bracteatum varieties for thebaine high yielding.**

7733 Avigliano, M.   IT 022300/75/0017
Indagine sulla resistenza a virosi e Peronospora tabacina A. di linee di tabacco italiane e straniere. **Study on various tobacco cvs. resistance to blue–mould and viroses.** Publications.

7734 Liguori, O.   IT 022300/76/0003 N
Selezione e confronto agronomico di alcune linee di tabacchi orientali autofertili e mediamente resistenti alla Peronospora tabacina A.. **Study on the selection for the agronomical characteristics of some sun–cured tobacco resistant to Peronospora tabacina A..**

7735 Di Muro, A.; Sorrentino, C.; Liguori, O.; Insero, O.; Piro, F.; Blago, A.   IT 022300/78/0001 R
Campi confronto preliminari di ibridi commerciali di prima generazione di vari tipi di tabacco. **Preliminary field test of tobacco lines (F1 hybrids and other cvs. from abroad).** Publications.

7736 Di Muro, A.; Donini, B.; Sorrentino, C.; Saccardo, A.;

Avigliano, M. IT 022300/78/0002
Costituzione di linee isogeniche di tabacco. **Tobacco isogenic
lines costitution.**

7737 Di Muro, A.; Vardabasso, A.; Aversano, B.; Tonini, A.;
Tumminello, M.; Avigliano, M. IT 022300/78/0005 R
Determinazione degli standard varietali dei principali tipi di
tabacco italiano. **Varietal standards determination of main
tobacco italian types.** Publications.

7738 Di Muro, A. IT 022300/79/0001 N
Costituzione di varietà di tabacco che diano un fumo più
povero di sostanze dannose. **Breeding of tobacco for reduced
health hazard.**

7739 Di Muro, A.; Piro, F.; Sorrentino, C.
IT 022300/79/0006 N
Ibridazione e selezione per la costituzione di nuove varietà di
tabacco. **Breeding of tobacco.**

7740 Vardabasso, A.; Carotenuto, R.; Tremola, M.G.
IT 022300/79/0007 N
Campi di confronto di linee Burley e Bright. **Test fields of
Burley and Bright tobacco lines.**

7741 Lamberti, F. IT 060600/73/0049
Analisi delle reazioni varietali agli attacchi dei nematodi.
Varietà di tabacco a Meloidogyne incognita. **Analysis of the
varietal reactions to nematode attacks. Tobacco varieties to
Meloidogyne incognita.**

Other crops (B 3990)

7742 England GB 030806/78/0002
**Studies the contribution which plant breeding may make to the
development of crops new to Scotland.**

7743 Trevisan, M. IT 020400/68/0001
Indagine sull'acclimamento e il comportamento di nuove
varieta' di Morus. **The acclimation and the behaviour of new
varieties of Morus.**

D 2400 - Plant protection

7744 Laloux, R.; Dohet, J.; Nijst, P.; Poelaert, J.; Falisse, A.
BE 010008/64/0002 R
Etude de la lutte chimique contre les adventices, les maladies
et les arthropodes en céréales des régions tempérées. **Study on
the chemical control of weeds, diseases and arthropodes in
cereals cultivated in temperate areas.** Publications.

7745 Breny, R.; Biernaux, J. BE 010011/62/0002
Etude sur la biologie et les dégâts causés par les Myriapodes–
Diplopodes. **Study of the biology and domages caused by the
Myriapodes– Diplopodes.** Publications.

7746 Breny, R.; Biernaux, J.; Quoilin, BE 010011/72/0003
Etude de la biologie et des moyens de lutte contre la
Cécidomyie du chou (Contorinia na sturtii KIEFF). **Study of
the biology and the disase prevention against the Swede midge
(Contorinia na sturtii KIEFF).** Publications.

7747 Van den Bruel, W.; Masson, M. BE 010017/71/0001 R
Recherches sur l'utilisation des biopréparations pour détruire
les insectes nuisibles. **Researches about the use of
biopreparations for the control of noxious insects.**

7748 Martens, P.; Caussin, R.; Copin, A.; Zenon–Roland, L.;
Fraselle, J. BE 010021/58/0002
Evolution des dépôts de pesticides agricoles en résidus.
Evolution in agricultural pesticide deposits to residus.
Publications.

7749 Semal, J.; Kummert, J. BE 010022/78/0006
Identification des virus de plantes et étude de la relation
virus–hôte. **Identification of plant viruses and study of
virus–host interaction.** Publications.

7750 Verhoeyem, M. BE 020302/79/0004 N
Recherches sur les propriétés antigéniques et études des
phages des bactéries phytopathogènes. **Research on the
antigene properties and study on the phages of
phytopathogenics bacteria.**

7751 Maroquin, C. BE 080600/72/0005
Virus des arbres fruitiers. **Viruses of fruit trees.** Publications.

7752 Maroquin, C. BE 080600/78/0011
Virus des céréales. **Viruses of cereals.**

7753 Henriet, J.; Galoux, M. BE 080700/78/0023
Recherche sur l'évolution des dépôts de pesticides dans les
fruits et légumes. **Study of pesticides residues behaviour in fruit
and vegetables.** Publications.

7754 Latteur, G. BE 081100/70/0005
Etude sur les pucerons des céréales et leurs ennemis naturels.
Study of cereal aphids and their natural ennemies.
Publications.

7755 Nøddegård, E. DK 010116/77/0020
Pesticiders fytotoksiske virkning på diverse væksthuskulturer.
**The phytotoxic effect of pesticides on diverse glasshouse
cultures.**

7756 Olesen, J. DK 010701/72/0010
Forsøg med bekæmpelse af ukrudt samt plantesygdomme og
skadedyr i praksis med henblik på brug i vejledningsarbejdet.
**Weed, disease and pest control trials in practice for use in
advisory work.**

7757 Rivière, J.L. FR 010106/69/1782
Histologie et histochimie des lésions provoquées par les
insecticides chez les insectes. **Histology and histochemistry of
insect lesions induced by insecticides.** Publications.

7758 Chater; Wright GB 011402/00/0001 R
Genetics and differentiation of actinomycetes.

7759 Sawicki GB 012005/00/0029
**Development of pesticide management programmes for
preventing or delaying development of resistance.**

7760 Graham–Bryce; Briggs GB 012005/00/0031
**Principles determining movement and persistence of pesticides
in soil.**

7761 Briggs; Evans GB 012005/00/0042
**Effects of using combinations of crop protection chemicals
repeatedly on their efficacy and on soil fertility.**

7762 Blackett GB 060213/79/0014 N

D 2400 – Plant protection

Screening of crop protection chemicals.

7763 McGrath, D.　　　　　IE 060200/71/0398 R
The influence of soil factors on the phytoxicity of soil–applied herbicides. Publications.

7764 McGrath, D.　　　　　IE 060200/71/0399 R
The persistence of herbicides in soil. Publications.

7765 Duggan, J.J.　　　　　IE 060300/64/0031 R
Stem nematodes – ditylenchus dipsaci – survey and certification. Publications.

7766 Duggan, J.J.　　　　　IE 060300/64/0032 R
Leaf and bud nematodes – aphelenchoides spp. – survey and certification. Publications.

7767 Duggan, J.J.　　　　　IE 060300/66/0035 R
Nematodes – xiphinema and longidorus spp. – Survey and soft–fruit certification. Publications.

7768 Dunne, R.M.　　　　　IE 060300/72/0022 R
Biological control of glasshouse whitefly with encarsia formosa and red spider mite with phytoseiulus persimilis. Publications.

7769 Eades, J.F.K.　　　　　IE 060500/64/0259 R
The effects of pesticides and other pollutants on the environment. Publications.

7770 Cunningham, P.C.　　　　　IE 060500/66/0233 R
Investigation of variability, virulence and host–pathogen environment relationships of soil–borne pathogens of cereals. Publications.

7771 Cunningham, P.C.　　　　　IE 060500/67/0234 R
Assessment of yield losses caused by soil borne pathogens of cereals. Publications.

Parks, gardens, urban greenspaces, plantations (B 1610)

See also 2639, 7791

Plants and parts of plants in general (B 2100)

See also 104

7772 Welvaert, W.　　　　　BE 030011/74/0008 N
Beschadiging van planten veroorzaakt door luchtverontreiniging, i.h.b. fluoriden, S02, Ozon. **Damage to plants caused by air pollution esp. fluorides, SO_2, Ozon.**

7773 Bertossi, F.　　　　　IT 040226/74/0649
Studio biochimico del differenziamento cellulare nei vegetali. **Biochemical study of cellular differentiation in plants.**

7774 Raggi, V.　　　　　IT 041015/74/0609
Variazione dell'attività fotorespiratoria, del punto di compensazione e loro ripercussione sul metabolismo della pianta malata, composti fenolici nelle piante e resistenza alle malattie. **Variations of photorespiratory activity, of compensation point and their effect on the metabolism of sick plants, phenolic compounds in plants and disease resistance.**

7775 Sempio, C.　　　　　IT 041015/74/0630
Studio delle alterazioni metaboliche nelle piante malate.

Ricerche citologiche ed enzimologiche con particolare riguardo ai meccanismi della resistenza ai parassiti. **Study of metabolic alterations in sick plants. Cytological and enzymological research with special reference to the mechanisms of resistance to parasites.**

Plant communities as ecological systems (B 2200)

See also 7745, 7746

7776 Thorup, S.; Permin, O.　　　　　DK 010119/70/4813
Udvikling af sprøjteteknik til forebyggelse af vinddrift. **Development of spraying techniques to prevent wind drift.**

Other subjects related to plants and animals in general (B 2900)

7777 Bernard, J.; Hoyoux, J.M.　　　　　BE 081100/78/0008 N
Etude de méthodes physiques de protection des récoltes contre les dégâts d'oiseaux. **Study of physical methods of plant protection against birds.** Publications.

Crops in general (B 3000)

See also 108, 196, 397, 398, 1069, 2694, 2840, 5163, 7747, 7748, 7749, 7750, 7755, 7762, 7772, 9093

7778 Van Assche, C.; Van Wambeke, E.; Vanachter, A.
　　　　　BE 040203/72/0009 R
Studie over de nawerking en de residu's van preculturale en culturale chemische grondontsmettingsmiddelen. **Study on effects and residues of precultural and cultural chemical soildesinfectants.** Publications.

7779 Koch, W.; Nau, N.; Welker, O.　　　　　DE 144540/73/0005
Vergleichende Untersuchungen über die biologische Wirksamkeit verschiedener Pflanzenschutzverfahren mit Grossgeräten. **Comparative studies of the biological effects of different methods in crop protection with field applicators.**

7780 Börner, H.; Vaagt, G.; Westphal, R.
　　　　　DE 148200/78/0001 N
Beeinflussung des mikrobiellen Herbizidabbaues im Boden durch pestizide Zweitkomponenten. **Influence of pesticides on the microbial herbicide degradation in soil.**

7781 Ebing, W.　　　　　DE 215010/77/0001
Entwicklung automatisch arbeitender Apparaturen und Einheitsverfahren zur Aufbereitung pflanzlicher Rohextrakte für die Analytik multipler Pestizidrückstände. **Development of automatic apparatus and of standard methods for processing vegetal raw extracts for the analysis of multiple pesticidal residues.**

7782 Ebing, W.　　　　　DE 215010/77/0004
Langzeitschicksal konjugierter Endmetaboliten. **Long–run residue of conjugated end metabolites.**

7783 Ebing, W.　　　　　DE 215010/77/0005
Entwicklung von Spurenanalysenmethoden für spezielle Rückstandsuntersuchungen von Pflanzenschutzmitteln und deren Metaboliten. **Development of methods of trace analysis for special examinations of residues of plant protectives and their metabolites.**

7784 Kloke, A.; Schoenhard, G.　　　　　DE 215060/75/0002

Untersuchungen über den Einfluss von Schadelementen im Boden auf den Ertrag und deren Gehalt in Pflanzen. **Investigations on the influence of contamination elements in the soil on crop yield and their content in plants.**

7785 Geike, F.   DE 215060/77/0007
Untersuchungen über den Einfluss von Umweltchemikalien auf Inhaltsstoffe und Enzymaktivitäten in Pflanzen. **Studies on the influence of plant protectives on content and activities of enzymes in plant.**

7786 Schütte, F.   DE 215180/77/0010
Erprobung und Weiterentwicklung verbesserter Methoden für den Meldedienst. **Acquisition and further development of improved methods for warning service.**

7787 Scholz, M.   DE 215240/77/0003
Untersuchen über wirtschaftliche Aspekte von Pflanzenschutzmassnahmen und von Schaderregerauftreten in der Pflanzenproduktion. **Studies on economic aspects of plant protective measures and of incidence of pests in plant production.**

7788 Hille, M.   DE 215240/77/0004
Erhebungen über Art und Menge der in verschiedenen Kulturen ausgebrachten Pflanzenschutzmittel. **Survey of type and quantity of plant protectives applied to different cultivations.**

7789 Wallnöfer, P.R.; Engelhardt, G.   DE 502051/77/0003
Abbau von aromatischen Pestizidmetaboliten durch Bodenorganismen. **Degradation of aromatic pesticide metabolites by soil microorganisms.** Publications.

7790 Astier, S.; Cornuet, P.   FR 010107/70/1026
Etude des relations enzyme–modèle chez les RNA polymérases, RNA dépendantes. **Studies on model. Enzyme relationships in RNA Polymerases RNA dependent.** Publications.

7791 Allemand, P.; Augé, P.   FR 010501/69/1011
Etude de la résistance des plantes aux embruns sur le littoral méditerranéen français. **Studies on plant resistance to sea spindrift along French mediterranean coast.**

7792 Taylor   GB 010801/00/0008 R
Improvement of methods for the application of herbicides.

7793 Baker; Holloway   GB 011504/00/0006
Cuticular, environmental and developmental factors in relation to penetration of foliar applied chemicals.

7794 Thomas   GB 011505/00/0001
Physical characteristics of surfactants in relation to spraying crops with pesticides.

7795 Skerrett; Batt   GB 011505/00/0008
Pesticide analytical studies in collaboration with M.A.F.F.

7796 Skerrett   GB 011505/00/0009
Phytotoxicity of pesticide spray materials.

7797 Morgan; Herrington   GB 011511/00/0001
Requirements for effective spray applications.

7798 Cooke; Herrington, B.K.   GB 011511/00/0003

Development of spray application assessment techniques.

7799 Pickard; Stringer   GB 011511/00/0004
Pesticides in the environment.

7800 Cooke; Herrington   GB 011511/00/0005
Crop protection overseas.

7801 Smith; Singer, B.D.   GB 011514/00/0003
Factors, including growth regulators, and cultural changes influencing the resistance of plants to pests.

7802 Dunning   GB 012301/00/0026 R
Pests and disease surveys.

7803 Webb   GB 012301/00/0032
Fungicides on barley.

7804 Jones   GB 030405/00/0008 R
Soil–borne fungal parasites.

7805 Gordon; Woodford   GB 030705/00/0001
Ecology and control of horticultural and agricultural pests particularly raspberry cane midge and mite.

7806 Henderson   GB 030903/75/0408 R
Chemical and other methods of rabbit control.

7807 Cuthbert   GB 030903/75/0412 R
Crop protection from damage by wild geese.

7808 Courtney; Johnston   GB 040301/79/0036 N
The evaluation of soil sterilants for weed control in Northern Ireland.

7809 Lovett   GB 050201/77/0022 N
Studies of new methods of formulation analysis prior to collaborative work.

7810 Shaw   GB 060218/00/0005
Selective assessment of control measures.

7811 Paton   GB 060220/00/0005
Agricultural implications associated with the use of a selective bactericide.

7812 Ponti, I.   IT 023500/79/0003 N
Disinfezione delle sementi. **Seeds disinfection.**

7813 Flori, P.   IT 040211/78/1135 N
Studio dell'interazione tra fitofarmaci e fitoregolatori. **A study of the interactions of pesticides and phyto–regulators.**

7814 De Carolis, C.   IT 040612/77/0161 R
Studio delle variazioni della potenza radiante in piante colpite da malattie diverse. **Study of the variations of radiation power in plants stricken by various diseases.**

7815 Galbiati, C.   IT 040612/78/1056 N
Ricerche sulla relazione fra attività fotosintetica, sviluppo e sporificazione di alcuni patogeni obbligati. **Research on the relation between photo–synthetic activity, growth and sporification of certain strict pathogenes.**

7816 De Carolis, C.   IT 040612/78/1133 N
Analisi dei residui e dei metaboliti, messa a punto metodi di

analisi per la valutazione quali–quantitativa dei residui di fitofarmaci, dosaggio residui nel terreno e in organi vegetali a seguito di trattamenti con fitofarmaci. **Analysis of residues and metabolites; elaboration of analytical methods for the qualitative and quantitative evaluation of pesticide residues; dosage of the residues present in the soil and in plant organs after treatment with pesticides.**

7817 Gambogi, G.   IT 041110/77/0191 R
Patologia del seme. **Seed pathology.**

7818 Lamberti, F.   IT 060600/77/0909
Rapporti tra nematodi ed altri agenti fitopatogeni. **Relations between nematodes and other plant parasites.**

7819 Foschi, S.   IT 061800/77/1007
Trattamenti a "turno fisso" ed a "turno fisso allungato" nella lotta contro la Venturia inaequalis e la Podosphaera leucotricha. **"Fixed schedule" and "prolonged fixed schedule" treatments in Venturia inaequalis and Podosphaera leucotricha control.**

7820 Foschi, S.   IT 061800/77/1008
Ricerche sull'attività dei nuovi geodisinfestanti. **Research on the action of new soil pesticides.**

7821 Marrè, E.   IT 064200/78/1147 N
Test fisiologici e biochimici per la classificazione di fitofarmaci e fitoregolatori. **Physiological and biochemical tests for the classification of pesticides and phyto–regulators.**

7822 Noordink, J.Ph.W.   NL 010108/60/7533
Biologisch onderzoek met behulp van isotopen en ioniserende bestraling ten behoeve van gewasbescherming. **Biological research using isotopes and ionising radiation on behalf of crop protection.**

7823 Maas, P.W.Th.   NL 010300/51/9092 N
Verzamelen van informatie t.a.v. vóórkomen, verspreiding, bestrijding en economische betekenis van belangrijke aaltjesaantastingen in en buiten Nederland. **Information on occurrence, distribution, control and economic importance of major nematode diseases.**

7824 Leistra, M.; Smelt, J.H.; Houx, W.N.   NL 010301/74/6101
Fysisch–chemisch gedrag van bestrijdingsmiddelen in de bodem en opname in gewassen. **Physic–chemical behaviour of pesticides in soil and uptake by crops.** Publications.

7825 Duym, J.   NL 040007/70/8502
Fenologische waarnemingen m.b.t. de werking van bestrijdingsmiddelen m.i.v. waarnemingen m.b.t. de werking van spuitapparatuur (bedekkingsgraad, drift e.d.). **Phenological observations in relation to the action of pesticides including observations in relation to the functioning of spraying equipment (coverage, drift a.o.).**

7826 Duym, J.   NL 040007/74/5551
Onderzoek naar na– en nevenwerkingen van bestrijdingsmiddelen t.a.v. volggewassen en milieu. **Research on side–effects of pesticides for following crops and environment.**

7827 Duym, J.   NL 040007/74/5552
Onderzoek naar niet–chemische (alternatieve) gewasbeschermingsmiddelen en –methoden. **Research on crop protection without using pesticides.**

7828 Vonk, J.W.   NL 050104/76/7570
Omzetting van pesticiden in planten en in het milieu. **Conversion of pesticides in plants and in the environment.**

Cereals in general (B 3100)

See also 6144, 7744, 7752, 7754, 7770, 7771

7829 Parmentier, G.; Cavelier, M.   BE 080600/58/0001
Lutte intégrée contre les principales maladies des céréales. **Integrated control against cereal diseases.** Publications.

7830 Defosse, L.   BE 080600/58/0010
Lutte biologique contre les parasites du pied des céréales. **Biological control against soil borne diseases of cereals.** Publications.

7831 Rixhon, L.; Crohain, A.; Guiot, J.; Frankinet, M.
        BE 080800/70/0008 R
Lutte intégrée contre les parasites des céréales. **Integrated control of cereal parasites.**

7832 Heyland, K.–U.; Kochs, H.–J.; Fewson, S.
        DE 111252/73/0006
Einfluss phytosanitärer Massnahmen bei Getreidearten auf Krankheitsbefall und Ertragsbildung. **Influence of phyto–sanitary treatments on disease attack and yield performance of cereal crops.**

7833 Abrook   GB 012107/00/0012
Insect resistance in cereals and fodder crops.

7834 Carr; Potter   GB 012107/00/0013
Interaction between virus and fungus infections in Graminae.

7835 Richardson   GB 030902/69/0501 R
Survey of cereal diseases to assess potential yields total loss and to partition loss between various causes.

7836 Richardson; Whittle   GB 030902/73/0502 R
Yield loss relationships in cereals.

7837 Courtney   GB 040301/79/0035 N
Herbicide tolerance and yield response to weed control in the cereal crop.

7838 Cunningham, P.C.   IE 060500/67/0236
Screening of barley varieties for resistance to disease. Publications.

7839 Leonard, T.F.   IE 060500/68/0190
Control of perennial weeds in continuous barley (malting) growing. Publications.

7840 Dunne, B.   IE 060500/71/0245 R
Assessment of yield losses caused by foliar diseases of cereals. Publications.

7841 Zadoks, J.C.; Rijsdijk, F.H.   NL 020018/70/5075
Het samenstellen van een atlas van ziekten en plagen in de graanteelt in Europa. **Composing an atlas of cereal diseases and pests in Europe.** Publications.

Barley (B 3110)

D 2400 – Plant protection

7842 Sly; Greaves GB 050314/00/0005
Control of aphids and mildew on spring barley.

Maize (B 3120)

See also 7844

Wheat (B 3170)

7843 Müller, F.; Frahm, J. DE 144541/77/0004
Einfluss von Herbiziden auf Winterweizen im Zusammenhang mit dessen Anfälligkeit für Pflanzenkrankheiten und tierische Schädlinge. **Influence of herbicides on winter wheat in connection with susceptibility to plant diseases and pests.**

7844 Covarelli, G. IT 041001/77/0485 R
Diversi prodotti contro le malattie del frumento applicati in diverse epoche e in diverse CV. con differenti processi colturali, abbinati o no alla concimazione azotata liquida o al diserbo chimico, diserbo chimico selettivo del frumento e mais, effetti residui diserbanti mais e frumento. **Various pesticides used against wheat diseases at different periods, on different CV. grown subsequently to various cultures, also used in combination or not with liquid nitrogen fertilisation or chemical weeding; selective chemical weeding of wheat and maize; action of weedkiller residues on maize and wheat.**

7845 Tafuri, F. IT 061600/77/0875
Determinazione gas–cromatografica dei residui di Vitavax (carbossina) e di Plantvax (ossicarbossina) nel grano. **Gas chromatography determination of Vitavax (carboxine) and Plantvax (oxycarboxine) residues in wheat.**

Fibre plants and oil crops in general (B 3200)

See also 7871

Potatoes (B 3310)

See also 3233

7846 Bockstaele, L.; Derolez, J.; Maddens, K.
BE 140000/79/0063 N
Bestrijding van ziekten en parasieten bij pootaardappelen. **Control of diseases and parasites of seed–potatoes.** Publications.

7847 Cayley GB 012009/00/0035
Plant protective chemistry of potato crops.

7848 Mabbott; Howell GB 030902/75/0304
Importance of potato ground keepers in persistence of potato cyst nematode and other pests and diseases.

7849 Lelliott; Knight GB 050322/73/0021 R
Bioclimatology of Solanaceous wilt.

7850 Duggan, J.J. IE 060300/58/0025 R
Potato cyst nematode (heterodera rostochiensis) survey of occurrence and certification. Publications.

7851 Duggan, J.J. IE 060300/60/0034 R
Potato tuber nematode (ditylenchus destructor) – survey and certification. Publications.

7852 Duggan, J.J. IE 060300/67/0024 R
Studies on the occurrence and distribution of pathotypes of potato cyst nematode in Ireland. Publications.

7853 Frost, M.C. IE 060500/61/0188 R
Evaluation of fungicides for the protection of potato crops against potato blight phytophthora infestans. Publications.

7854 Bruin, Th. de NL 010300/78/9111 N
Invloed van mengsels van insecticiden, fungiciden en minerale oliën op de werkzaamheid van iedere component bij de bestrijding van ziekten en plagen in aardappels. **The effect of mixtures of insecticides, fungicides and mineral oils on the mode of action of each component in pest and disease control of potatoes.**

Sugarbeets and other sugar crops (B 3320)

See also 6585

7855 Martens, M.; Van Steyvoort, L.; Roussel, N.; Vigoureux, A.; Vanstallen, R. BE 140000/74/0025
Onkruidbestrijding en strijd tegen vijanden en ziekten van de suikerbiet. **Study of the control of weeds, diseases and enemies of sugarbeets.** Publications.

7856 Bonnemaison, L. FR 010106/68/1780
Traitement des cultures betteravières avec de nouveaux pesticides. **New pesticides in sugar beet pest control.** Publications.

7857 Byass; Sharp GB 011609/79/0009 N
Air carried spraying sugar beet.

7858 Feeney, A. IE 060500/65/0247 N
The occurrence and control of seedling pests in sugar beet with particular reference to atomaria linearis pigmy mangel beetle. Publications.

7859 Heijbroek, W. NL 060003/30/7726
Onderzoek t.b.v. acuut optredende ziekten en plagen in suikerbieten. **Research concerning acute occurring pests and diseases in sugar beet.**

7860 Heijbroek, W. NL 060003/75/7725
Bestrijding van vergelingsziekte in suikerbieten door cultuurmaatregelen. **Control of virus yellows in sugar beet by cultural measures.**

7861 Jorritsma, J. NL 060003/78/7905
De praktische toepassing van gewasbeschermingsprodukten en hun invloed op de groei van suikerbieten. **The practical application of plant protection products and their effects on sugar beet growing. Soil fumigation.** Publications.

Grasses and forage crops in general (B 3400)

7862 Van Breuseghem, R.; Pelsener, J.; Villers, S.
BE 140000/76/0044 N
Lutte microbiologique par les Entomoptorales pathogènes pour les pucerons. **Microbiological control against aphis.**

Grasses (B 3410)

7863 Plumb; Gibson GB 012009/00/0005
Epidemiology of virus diseases of grasses.

D 2400 – Plant protection

7864 Carr; Wilkins GB 012107/00/0003
Assess yield loss due to fungal pathogens.

7865 Harvey GB 040301/76/0028 R
Mechanisms of herbicide tolerance in resistant grass species.

7866 Connolly, V. IE 060500/69/0210 R
Hybrid varieties – development of technique to facilitate the
production of hybrid grass and white clover. Publications.

Pastures, grassland (B 3420)

7867 Mitchell, B. IE 060500/66/0192
Weed control in grassland. Publications.

Legumes in general (B 3440)

See also 7744

7868 Maroquin, C. BE 080600/78/0012 R
Virus des légumineuses. **Viruses of leguminous.**

Grassland legumes (B 3441)

7869 Zannone, L. IT 020900/76/0002 N
Prova di resistenza ai parassiti su erba medica. EUCARPIA,
Gruppo Erba Medica.. **Trial on lucernes resistant to parasites.**
EUCARPIA, Medicago Sativa Group..

7870 Ribaldi, M. IT 041026/77/0280 N
Ricerche sulle fitopatie dell'erba medica in Umbria ed in altre
regioni dell'Italia centrale. **Research on fungus diseases of
lucerne in Umbria and other regions of central Italy.**

Turnips (B 3460)

7871 Prendergast, A.G. IE 060500/67/0721 R
Plasmodiophora brassicae, worl, disease of crucifers (clubroot)
in Ireland. Publications.

Other forage crops (B 3490)

7872 McNaughton GB 030803/78/0007 R
Improve methods for assessing brassica diseases and for
estimating yield losses caused by them.

Vegetables in general (B 3500)

See also 5692, 7753, 7768, 19130

7873 Gillard, A.; Pelerents, C.; Hertveldt, L.
BE 030024/72/0004 R
Biologische bestrijdingsmogelijkheden van insekten schadelijk
aan groenten. **Biological control possibilities against insects
parasiting vegetables.** Publications.

7874 Riethus, H.; Bau, H. DE 105203/73/0002
Applikationsbedingte Einflüsse von Pestiziden auf
Verträglichkeit und Inhaltsstoffe bei Gemüsepflanzen.
**Influences of pesticide application on the susceptibility and
composition of vegetable plants.**

7875 MacNaeidhe, F. IE 060201/66/0421
The control of annual and perennial weeds in vegetable crops on
peat soil. Publications.

7876 Cassidy, J.C. IE 060300/64/0010
Annual weed control in vegetable crops on mineral soils.
Publications.

7877 Cirulli, M. IT 040106/72/0090
Ricerche sulla patologia delle piante ortensi. **Research on the
pathology of horticultural plants.** Publications.

7878 Goidanich, G. IT 040211/73/0231
Patologia delle piante ortensi. **Pathology of horticultural
plants.**

7879 Noviello, C. IT 040707/77/0248 R
Patologia piante ortensi. **Pathology of market–garden plants.**

7880 Matta, A. IT 041215/73/0936
Patologia delle piante ortensi. **Pathology of horticultural
plants.**

7881 Matta, A. IT 041215/77/0234 R
Patologia delle piante ortensi. **Pathology of market–garden
plants.**

7882 Bakel, J.M.M. van NL 010108/77/7634
Onderzoek naar de mogelijkheden van gewasbescherming in
de zogenaamde kleine groentegewassen. **Possibilities of crop
protection in the little vegetable crops.**

7883 Theune, D. NL 010206/50/4675
Toepassingswijze van bestrijdingsmiddelen incl.
grondontsmetting in kassen. **Application techniques of
pesticides, including soil disinfection in glasshouse crops.**

7884 Theune, D. NL 010206/72/4676
Residu–onderzoek van bestrijdingsmiddelen in kasgewassen.
Residu analysis of pesticides in glasshouse crops.

7885 Bakel, J.M.M. van NL 010207/77/7634 R
Onderzoek naar de mogelijkheden van gewasbescherming in
de zogenaamde kleine groentegewassen. **Possibilities of crop
protection in the little vegetable crops.**

7886 Bakel, J.M.M. van NL 010207/78/9044 N
Onderzoek naar de invloed van grondontsmetting op het
voorkomen van ziekten, plagen en onkruiden en op de
kwaliteit en het opbrengstniveau van groentegewassen.
**Research on the influence of soil disinfection on the occurence
of diseases, pests and weeds and on quality and yield of
vegetable crops.**

Root, tuber and bulb vegetables (B 3510)

See also 3790

7887 Bosmans, P.; Kamoen, O. BE 070500/74/0010 R
Studie van schimmelziekten en hun bestrijding in de
groenteteelt. (prei– schorseneer– sla). **Study and control of the
most important diseases of vegetables caused by fungi (spec.
Leek, Scorzonera and Lettuce).** Publications.

7888 Duggan, J.J. IE 060300/68/0030 N
Control of stem nematode (ditylenchus dipsaci) in onions.
Publications.

7889 Dunne, R.M. IE 060300/70/0021 N

D 2400 – Plant protection

Biology and control of carrot fly (psila rosae). Publications.

Greens and leafy vegetables (B 3520)

See also 3790, 7746, 7871, 7887

7890 Cornuet, P.; Astier, S.　　　　FR 010107/70/1028
Rôle physioloqique d'une RNA polymérase RNA dépendante isoleé de choux. **Physiological role of a RNA polymerase, RNA dependent isolated from cabbage.** Publications.

7891 Walkey; Dance　　　　GB 011807/75/0024
Storage necrosis of Dutch white cabbage.

7892 Johnston　　　　GB 012101/00/0016 R
Resistance to club root, mildew and insect pests in Brassicae.

7893 Staunton, W.P.; Ryan, E.W.　　　　IE 060300/71/0051 N
Control of diseases of lettuce. Publications.

Leguminous vegetables (B 3531)

See also 3790

7894 Tilkin, V.E.; Férauge, M.　　　　BE 080200/76/0035 R
Etude des manifestations du gel sur les fleurs de pois. **Study of frost effect on pea flowers.**

Tomatoes (B 3532)

7895 Van Assche, C.; Vanachter, A.　　　　BE 040203/76/0013 R
Effect van preinoculatie en premunitie op Fusariumverwerking bij tomaat. **Effect on preinoculation and premunition Fusarium wilt expression on tomatoes.** Publications.

7896 Staunton, W.P.　　　　IE 060300/69/0050 R
Diseases of tomatoes (resistant varieties). Publications.

Mushrooms and other edible fungi (B 3540)

See also 3790

Fruits in general (B 3600)

See also 5782, 7209, 7751, 7753, 7878, 7879, 7880, 7881, 7941, 19130

7897 Austin; Murray　　　　GB 011007/00/0001
Chemistry of pesticides, herbicides and other chemical aids used on fruits and hops.

7898 Murray; Allen　　　　GB 011007/00/0002
Analytical methodology at trace level in fruit and hops.

7899 Hunter; Allen　　　　GB 011007/00/0003
Chemical deposition and persistence on fruit and hops.

7900 Hunter; Warman　　　　GB 011007/76/0005
Phytotoxicity and other undesirable effects of agrochemicals.

7901 Jones; Morgan, K.G.　　　　GB 011511/00/0002
Development of spraying techniques for plantation crops.

7902 Smith　　　　GB 011514/76/0007
Bird damage on fruit trees.

7903 Researcher not indicated　　　　GB 050101/00/0011 R
Fruit crops, pest and disease control.

7904 Cp Ld Bd　　　　GB 050101/75/0013
Wind breaks and shelter for fruit plantations.

7905 Researcher not indicated　　　　GB 050141/00/0008
Fruit crops, pest and disease control.

7906 Govi, G.　　　　IT 040211/74/0566
Miglioramento sanitario dei portinnesti delle varie specie da frutto. **Sanitary improvement of grafts of various fruit tree species.**

Top fruit in general (B 3610)

See also 7920

7907 Tiemann, K.–H.　　　　DE 507300/75/0001
Untersuchungen über den Einfluss von Fluor–Immissionen auf Obstpflanzen an der Niederelbe. **Research on the effects of fluorine immissions on the fruit trees on the Lower Elbe.** Publications.

7908 Blaisinger, P.; Simonis, M.T.　　　　FR 010904/70/1747
Lutte intégrée dans les vergers de pruniers du Nord–Est. **Integrated control in plum orchards in North–eastern France.** Publications.

7909 Muir; Flegg　　　　GB 011001/00/0016
Birds – evaluation of repellants.

7910 Thresh; Adams　　　　GB 011005/00/0008
Produce healthy clones.

7911 Kavanagh, T.; O'Kennedy, N.D.　　　　IE 060302/60/0048 R
Control of apple diseases. Publications.

Apple (B 3611)

7912 Populer, C.　　　　BE 080600/76/0008 R
Résistance du pommier et du poirier aux maladies cryptogamiques. **Resistance of apple and pear to cryptogamic diseases.**

7913 Bernard, J.　　　　BE 081100/73/0006 R
Recherches sur la possibilité d'appliquer la lutte intégrée en verger de pommiers. **Research on the integrated control in apple orchards.** Publications.

Pear (B 3612)

See also 7912

Soft fruit (berries and cane fruits) (B 3620)

See also 7295, 7296, 7767

7914 Kamoen, O.; Jamart, G.　　　　BE 070500/74/0011 R
Bestrijding van Botrytis, vruchtrot, bij aardbeien. **Control of Botrytis, fruit rot, on strawberries.** Publications.

7915 Mussillon, P.; Martin, C.; Carre, M.　　　　FR 011007/74/5013
Obtention de plants sains de framboisiers. **Production of pathogen free saplings of rasberry bush.**

D 2400 – Plant protection

7916 Mason   GB 030701/00/0008 R
Physiological disorders of soft fruit.

7917 Lyons   GB 030702/00/0022
Epidemiology and etiology: harvest disorders of soft fruit.

7918 Kavanagh, T.; O'Callaghan, T.F.; Rath, N.; Staunton,
W.P.; Jeffares, M.   IE 060300/63/0044 R
Control of diseases of strawberries. Publications.

7919 Duggan, J.J.   IE 060300/70/0027 R
Control of foliar nematodes and stem nematode in strawberries.
Publications.

7920 O'Riordan, F.   IE 060300/71/0041 R
Bush fruit diseases (leaf spot, botrytis, powdery mildew).
Publications.

7921 Rath, N.; O'Callaghan, T.F.; Robinson, D.W.
     IE 060301/68/0092
Weed control in soft fruit crops. Publications.

Citrus fruit (B 3630)

7922 Terranova, G.; Cutuli, G.; Starrantino, A.
     IT 021600/73/0005 R
Indagine fitosanitaria sul bergamotto e cedroanche ai fini del
risanamento delle virosi. Phytosanitary survey of bergamot and
citron also with the aim of recovery from viruses. Publications.

7923 Nucifora, A.; Viggiani, G.; Barbagallo, S.; Mineo, G.;
Di Martino, E.; Inserra, S.   IT 040301/79/0001 N
Area pilota di lotta integrata nei limoneti della Sicilia. Pilot
area of integrated control in lemon Orchards in Sicily.
Publications.

Grapes (B 3650)

See also 4425, 7430

7924 Müller, K.; Peternel, M.   DE 502153/77/0002
Ergründung der Möglichkeiten und der Technik eines
erfolgreichen Rebschutzes über die Beregnungsanlage;
Vergleich von herkömmlichen Verfahren zu verschiedenen
Methoden der Pflanzenschutzmittelverregnung und zum
Hubschraubereinsatz. Testing of possibilities and technology
for successful vine protection via irrigation system; comparison
of traditional methods with different methods of plant
protective spraying and with the use of helicopters.

7925 Eichhorn, K.W.; Lorenz, D.H.   DE 509153/77/0003
Untersuchungen über Möglichkeiten zur Verringerung der
Anwendung chemischer Pflanzenschutzmittel im Weinbau.
Chances of reduction in using chemical protectives in
viticulture.

7926 Flegg; McNamara   GB 011001/00/0012
Control of nematode vectors of virus diseases.

7927 Egger, E.; Borgo, M.; Liuni, C.   IT 021300/70/0001 R
Effetti collaterali dei trattamenti in genere sul metabolismo
della vite. Collateral effects of treatments on vine general
metabolism. Publications.

7928 Brandolini, V.   IT 040211/77/0477 R
Analisi dei residui e dei metaboliti di fitofarmaci e

fitoregolatori. The persistence of pesticides and growth
regulators in grapes and wine–making products.

Edible nut fruits (B 3660)

See also 4482

Ornamentals and ornamental products in general (B 3700)

See also 7878, 7879, 7880, 7881

7929 Hand   GB 011102/00/0009
Air pollution and the effects of air pollutants on crops.

7930 Kempton; Maw   GB 011103/00/0008
The effect of methyl bromide as a soil fumigant on glasshouse
crops.

7931 Scopes   GB 011108/00/0005
Integration of chemical and biological control of glasshouse
pests.

7932 Morgan   GB 011109/00/0026
Behaviour of systemic pesticides in soils and plants.

7933 Researcher not indicated   GB 050102/00/0014 R
Pests and diseases of protected crops and their control.

7934 Researcher not indicated   GB 050142/00/0010
Pests and diseases of protected crops and their control.

7935 Foster   GB 060309/00/0004
The implementation of integrated control of glasshouse pests.

7936 Goidanich, G.   IT 040211/77/0201 R
Patologia delle piante ortensi. Pathology of garden plants.

Bulbs (B 3710)

See also 4560, 4565, 4566, 4567, 4568, 4570, 4572, 4573

7937 Researcher not indicated   GB 050104/00/0005 R
Pests and diseases of bulbs and allied crops and their control.

7938 Rec; Bb35   GB 050144/00/0001
Pests and diseases of bulbs and allied crops and their control 3
8.

7939 Rooy, M. de   NL 010205/67/1510
Chemische bestrijding van ziekten en plagen in bol–en
knolgewassen. Chemical control of diseases and pests in
bulbous crops. Publications.

Flowers and pot plants (B 3720)

See also 7552, 7883, 7884, 7940

Ornamental shrubs (B 3730)

7940 Welvaert, W.; Roos, A.   BE 030011/77/0006
Rhododendron simsii ; invloed van de voorbehandeling van
moerplanten en stekmateriaal met systemische fungiciden op
beworteling en ziekte resistentie. Rhododendron simsii ;
influence of pre–treatment of mother –plants and cuttings with
systemic fungicides on rooting and disease protection.

D 2400 – Plant protection

7941 Scotto La Massèse, C.; Berge, J.B. FR 010504/64/1713
Etude des facteurs spécifiques de résistance chez les
porte–greffes d'essences fruitières et des rosiers. **Research on
specific factors of resistance in rootstocks of fruit–trees and
roses.** Publications.

7942 Kelly, J.; Lamb, J.G.D. IE 060300/67/0110 R
Herbicides and soil sterilants in nursery stock production.
Publications.

7943 Fogliani, G. IT 040404/77/1053 N
Malattie delle piante arboree ornamentali nella problematica
del verde malato. **The diseases of ornamental trees with
reference to the general problems of "diseased green".**

7944 Slavekoorde, S.M. NL 010203/66/2573
Gebruiksmogelijkheden van grondontsmettingsmiddelen in de
boomkwekerij. **The use of chemical soil fumigation products in
the nursery.** Publications.

Forests in general (B 3800)

7945 Skatulla, U. DE 160090/74/0001
Einsatz von Bakterien, Viren und Hormonen im Forstschutz.
Use of bacteria viruses and hormones in forest protection.

Pine forests in general (B 3810)

7946 Kramer, H.; Kenneweg, H. DE 132720/75/0006
Ermittlung von Zuwachsverlusten in immissionsgeschädigten
Nadelholzbeständen aufgrund von luftbildsichtbaren
Schädigungsmerkmalen – Kronenschäden –. **Determination of
losses of increment in coniferous stands damaged by immissions
by means of damage characteristics visible on aerial
photographs – damages to crowns –.**

Spruce and fir forests (B 3812)

7947 Koltzenburg, C. DE 132780/75/0003
Holzanatomische Untersuchungen von Auswirkungen
mechanischbiologischer Schälschutzmassnahmen an Fichte.
**Wood anatomical studies of the effects of
mechanico–biological peeling protection measures on spruce.**

7948 Gardiner, J.J.; Coen, R.C. IE 120107/77/9090 N
**The isolation and identification of microorganisms from sitka
spruce litter.**

7949 Morandini, R. IT 021700/64/0001
Protezione di formazioni naturali compreso Abies
nebrodensis. **Protection of natural environments including
Abies nebrodensis.** Publications.

Leafwoods in general (B 3820)

See also 4966

Oak tree stands (B 3821)

7950 Lang, K.J. DE 160061/74/0004
Zur Kenntnis des Buchensterbens. **Causes of the dieback of
beech.**

Stimulant crops (B 3910)

7951 Thresh; Adams GB 011005/00/0009 R
Produce healthy clones.

7952 Researcher not indicated GB 050106/00/0002 R
Pests and diseases of hops and their control.

7953 Researcher not indicated GB 050146/00/0001 R
Pests and diseases of hops and their control.

Spice and seasoning plants of temporate climates (B 3930)

See also 7724

7954 Moser, E.; Locher, B. DE 144720/78/0002 N
Verbesserung der Applikationstechnik im Hopfenbau zur
Verminderung der Umweltbelastung und zur Einsparung von
Wirkstoffmitteln. **Improvement of plant protection technics in
hop–growing to reduce impact on environment and to spare
active substances.**

Drugs and medicine plants (B 3970)

See also 6003

7955 Bockstaele, L.; Derolez, J.; Maddens, K.
BE 140000/79/0062 N
Bestrijding van ziekten en parasieten bij tabak. **Control of
diseases and parasites of tobacco.** Publications.

D 2410 – Pests of plants and pest control

See also 1044, 1070, 1107, 2069, 2618, 2619, 16005, 16006,
16007, 16050, 16051, 16052, 20022, 20145

7956 Breny, R.; Magema; Baurant, R. BE 010011/69/0007
Etude des scolytidae en forêt ardennaise belge. **Study of
Scolytidae in Belgian Ardenne Forest.** Publications.

7957 Van den Bruel, W.; Quoilin, J. BE 010017/72/0003
Recherches sur les comportements de Trialeurodes
vaporarium Westn en milieu conditionné. **Research on the
behaviour of Trialeurodes vaporarium Westn in conditionned
conditions.**

7958 D'Herde, C.J.; Coolen, W.A.; Hendrickx, G.
BE 070100/76/0017 R
Methodologisch onderzoek naar de kwantitatieve
extractiemogelijkheden van nematoden uit grond en uit
plantedelen. **Methodological study of the quantitatieve
extraction posssibilities of nematodes from soil and from plant
tissues.**

7959 De Clercq, R. BE 070100/76/0018 R
Studie van de epigeische arthropodenfauna i.v.m. de
geintegreerde bestrijding van graanparasieten. **Study of the
epigean arthropode fauna in vew of the integrated control of
cereal pests.**

7960 Van Laere, O.; Maertens, D.; De Greef, M.
BE 070100/77/0019 R
Onderzoek naar de betekenis van insekten, vooral van de
honingbij bij de verspreiding van bakterievuur. **Study of the
role of insects, especially of the honeybee, in the spread of
fireblight.**

7961 D'Herde, C.; Coolen, W.   BE 070100/78/0020 R
Resistentieonderzoek van nematoden tegen systemische
nematiciden. **Study of nematode resistance against systemic
nematicides.**

7962 Bernard, J.   BE 081100/50/0001
Recherches sur l'éthologie, l'écologie et les méthodes de lutte
contre les rongeurs nuisibles. **Investigations on the ecology, the
behaviour and the control of noxious rodents.** Publications.

7963 Moens, R.   BE 081100/60/0003
Etude de la mise au point d'une technique de lutte contre le rat
musqué et dynamique des populations de ce rongeur dans le
foyer belge. **Study of up to day technique to control muskrats
and of the dynamics of that rodent populations in Belgium.**
Publications.

7964 Nøddegård, E.   DK 010116/61/0004
Insekticiders virkning mod skadedyr på korn, græs og
bælgplanter, fortrinsvis bladlus, biller og flue- og myggelarver.
**Effect of insecticides on pests of cereals, grasses and legumes,
chiefly aphids, beetles, and fly and mosquito larvae.**

7965 Lindhardt, K.   DK 010116/64/4529 R
Udbredelse og levevis af havrenematoden samt svampes
indvirkning på opformering af nematoden. **Occurrence and
habits of the cereal root eelworm, and the influence of fungi on
the propagation of the nematode.**

7966 Lindhardt, K.   DK 010116/66/0011
Undersøgelse af importerede skadedyr og usædvanlige
alvorlige angreb af insekter/nematoder. **Investigation of
imported pests and unusually severe attacks of insects and
nematodes.**

7967 Nøddegård, E.   DK 010116/68/0009
Bekæmpelsesmidler mod fluelarver i kål, gulerod og løg.
**Chemicals for the control of fly larvae in cabbage, carrot and
onion.**

7968 Jakobsen, J.   DK 010116/70/4533
Migrerende nematoders udbredelse og biologi med henblik på
forebyggelse og bekæmpelse i prydplantekulturer i væksthuse.
**Occurrence and biology of migratory nematodes with
reference to prevention and control in ornamental plant
cultures in glasshouses.**

7969 Nøddegård, E.   DK 010116/72/0005
Insekticiders virkning mod jordboende skadedyr i bederoer.
Effect of insecticides on soil–dwelling pests in beet.

7970 Nøddegaard, E.; Kirknel, E.   DK 010116/72/4527
Pesticidernes indflydelse på skadedyrenes naturlige fjender.
Influence of pesticides on natural enemies of pests.

7971 Reitzel, J.   DK 010116/72/4536
Udarbejdelse af metoder til biologisk bekæmpelse af skadedyr
på væksthusgrønsager ved hjælp af snyltehvepse og rovmider.
**Development of biological control methods for pests of
glasshouse vegetables with the aid of parasitic wasps and
predatory mites.**

7972 Esbjerg, P.   DK 010116/72/4537
Biologi og udbredelse af skadedyrenes naturlige fjender med
henblik på integreret bekæmpelse af skadedyr på
frilandsgrønsager. **Biology and occurrence of the natural**

enemies of pests with reference to integrated control of pests on
outdoor vegetables.

7973 Jakobsen, J.; Andersen, H.J.   DK 010116/73/4532
Migrerende nematodeangrebs omfang i Danmark med henblik
på tilrettelæggelse af sædskifte og andre
bekæmpelsesforanstaltninger. **Extent of migratory nematode
attacks in Denmark with reference to the planning of crop
rotations and other control measures.**

7974 Nøddegård, E.   DK 010116/74/0016
Insekticiders virkning på skadedyr, især øresnudebiller og
nematoder på jordbær. **Effect of insecticides on pests: root
weevils and nematodes on strawberries.**

7975 Nøddegård, E.   DK 010116/74/0017
Insekticiders virkning på øresnudebiller på forskellige
prydplanter i væksthus samt planteskoler. **Effect of insecticides
on root weevils on ornamentals in glasshouses and nurseries.**

7976 Reitzel, J.; Esbjerg, P.   DK 010116/74/4534
Bekæmpelses– og kulturforanstaltningers indflydelse på
kornskadedyrenes naturlige fjender. **Influence of control and
growing methods on the natural enemies of cereal pests.**

7977 Esbjerg, P.   DK 010116/74/4538
Øresnudebillens biologi, dens naturlige svampesygdomme og
disses anvendelighed ved biologisk bekæmpelse af billen.
**Biology of the root weevil, its natural fungal diseases and their
use in the biological control of the weevil.**

7978 Jakobsen, J.   DK 010116/75/4530
Udbredelse, betydning og isolering af nematoder på bederoer
og korsblomstrede planter. **Occurrence, importance and
isolation of nematodes on beet and cruciferous plants.**

7979 Nøddegård, E.   DK 010116/77/0006
Midler til bekæmpelse af knoporme i kartofler. **Chemicals for
cutworm control in potatoes.**

7980 Nøddegård, E.   DK 010116/77/0018
Nematiciders virkning på forskellige nematodearter på
prydplanter i væksthus. **Effect of nematicides on different
nematode species on glasshouse ornamentals.**

7981 Lindhardt, K.   DK 010116/78/0012
Integreret bekæmpelse af glimmerbøsser i raps. **Integrated
control of blossom beetles in rape.**

7982 Jørgensen, J.   DK 030109/63/0008
Den store kålflues biologi i Vendsyssel. **Biology of the large
cabbage fly in Vendsyssel.**

7983 Bresciani, J.   DK 030109/67/0004
Undersøgelser over parasitiske hvirvelløse dyrs integument.
Investigations into the integument of parasitic invertebrates.

7984 Bejer–Petersen, B.   DK 030109/77/0002
Elmebarkbiller, forsøg med feromonlokning samt
undersøgelse af udbredelse. **Elm bark beetles; an experiment
with pheromone attraction and an investigation of their
propagation.**

7985 Bejer–Petersen, B.   DK 030109/77/0003
Biologi af granens rodbille (Hylastes sp.). **Biology of the spruce
root beetle (Hylastes sp.).**

7986 Jørgensen, J.　　　　　　DK 030109/78/5232
Integreret bekæmpelse af skadedyr. (insektvirologi, vanding og klima, feromoner, skadetærskler og resistens). **Integrated pest control (insect virology, irrigation and climate, pheromones, damage thresholds and resistance).**

7987 Bounias, M.　　　　　　FR 010602/78/5137
Recherches sur le métabolisme des lipides et sa régulation chez divers insectes d'intérêt agronomique. **Investigations on lipids metabolism and its regulation in insects of agronomic interest.**

7988 Vey, A.; Amargier, A. Mme.　　　FR 010612/56/1769
Pathogénèse des mycoses d'insectes. **Pathogenegis of insect mycosis.** Publications.

7989 Thompson　　　　　　GB 011806/00/0002 R
Investigate factors affecting insecticidal control of insect pests of Cruciferae.

7990 Finch　　　　　　GB 011806/00/0004 R
Study biology and chemicals affecting behaviour of insect pests to develop new methods of control.

7991 Thompson　　　　　　GB 011806/00/0007 R
Investigate factors affecting insecticidal control of insect pests of Umbelliferae.

7992 Thompson; Percivall　　　　GB 011806/00/0012 R
Control of insect pests of lettuce with insecticides.

7993 Suett; Wheatley　　　　　GB 011806/00/0016 R
Persistence, degradation and behaviour of insecticides in soil and crops.

7994 Thompson; Percivall　　　　GB 011806/00/0017 R
Insecticidal control of onion fly maggot.

7995 Finch; Skinner　　　　　GB 011806/00/0020 R
Study the biology and behaviour of carrot fly.

7996 Dunn; Kempton　　　　　GB 011806/00/0021 R
Biology of pests of lettuce.

7997 Finch　　　　　　GB 011806/75/0022 R
Study of biology and behaviour of onion fly.

7998 Stephenson; Edwards　　　　GB 012004/78/0038 R
Behaviour physiology ecology and control of slugs.

7999 Bowden; Sherlock　　　　GB 012004/78/0041 R
Biology ecology and control of cutworms and Tipulids.

8000 Taylor; French　　　　　GB 012004/78/0042 R
Monitoring and analysing aerial populations and the dispersal of insects.

8001 Lewis; Wall　　　　　GB 012004/78/0043 R
Identification development and evaluation of behaviour–controlling chemicals for pest monitoring and control.

8002 Turner　　　　　　GB 012004/78/0050 R
Nature causes and spread of resistance to pesticides in aphids and other pests.

8003 Williams; Wall　　　　　GB 012004/78/0051 R
Behavioural and physiological aspects of insect communication.

8004 Turner　　　　　　GB 012004/78/0052 R
Population dynamics and genetics of invertebrates.

8005 Bailey; Wilding　　　　　GB 012004/78/0055 R
Infection processes of insect viruses and microorganisms and methods of insect resistance.

8006 Bailey; Wilding　　　　　GB 012004/78/0056 R
Characterisation distribution and ecology of insect viruses and microorganisms.

8007 Phillips; Etheridge　　　　GB 012005/00/0036 R
Chemicals and formulations for control of leaf–cutting ants.

8008 Lord　　　　　　GB 012005/00/0044 R
Factors influencing uptake of pesticides from soils by slugs and worms.

8009 Jones; F.G.W.　　　　　GB 012006/00/0002 R
Population studies, ecology and integrated control of nematodes.

8010 Hooper; Stone　　　　　GB 012006/00/0005 R
Taxonomy of economically important plant parasitic nematodes, host ranges, life cycles, general biology.

8011 Stone; Williams　　　　　GB 012006/00/0006 R
Pathotypes of cyst–nematodes.

8012 Shepherd; Clark　　　　　GB 012006/00/0008 R
Detailed morphology and fine structure of economically important genera of plant parasitic nematodes.

8013 Clarke; Greet　　　　　GB 012006/00/0009 R
Sexual, physiological and behavioural chemicals in nematode–plant relationships.

8014 Doncaster; Seymour　　　　GB 012006/00/0010 R
Feeding, behaviour, behavioural physiology, morphology and general biology of plant parasitic nematodes.

8015 Oswald　　　　　　GB 012006/75/0021 R
Nematode parasites of insects.

8016 Turl　　　　　　GB 030902/69/0309 R
Analysis of aphid–parasitic Hymenoptera from 40 foot (12 metre) suction trap at East Craigs.

8017 Woof　　　　　　GB 030903/00/0305 R
Field evaluation of control of stage pests through adjustment of the physical environment.

8018 Cuthbert　　　　　　GB 030903/00/0410 R
Investigation of poisons and stupefacients as avicides.

8019 Bell　　　　　　GB 030903/74/0301 R
Detection and monitoring of insect populations for insecticidal resistance mainly to malathion.

8020 Brodie　　　　　　GB 030903/76/0405 R
Ecology of rural rats.

8021 Courtney　　　　　　GB 040301/77/0021 R

D 2410 – Pests of plants and pest control

Weed control in minor water courses and allied areas.

8022 MacNaeidhe, F. IE 060201/71/0422 R
The absorbtion, persistence and movement of herbicides in different peat types. Publications.

8023 Springhetti, A. IT 040239/74/0637
Effetti letali e morfogenetici dei derivati del farnesolo sulla differenziazione embrionale e castale degli insetti. **Lethal and morphogenetic effects of farnesol based products on insect embryonal and taxonomic differentiation.**

Banks, shores, dikes and their vegetation (B 1530)

8024 Galichet, P.F.; Bonfils, J. FR 010602/75/5102
Conséquences écologiques de la culture de la Canne de Provence dans le Delta du Rhône. **Ecological consequences of the Donax–reed culture in the Rhône Delta.**

Parks, gardens, urban greenspaces, plantations (B 1610)

See also 8429

8025 Vite, J.P. DE 126150/78/0001 N
Ulmensterben: Aggregationsverhalten der Ulmenborkenkäfer. **Dutch elm disease: aggregation behavior of the elm bark beetle.**

8026 Kraan, C. van der NL 010301/75/6111
Gebruik van attractantia bij de signalering van iepespintkevers (Scolytus multistnatus en Scolytus scolytus). **Use of attractants to detect Scolytus beetles.** Publications.

8027 Doom, D. NL 010601/42/6265
Jaarlijkse inventarisatie van het optreden van insekten en mijten in boom – en struikbeplantingen. **Annual survey of insect and mite pests in forest trees and shrubs.** Publications.

8028 Grijpma, P. NL 010601/78/7625
Onderzoek naar de vatbaarheid van eiken (Quercus spp.) voor aantasting door de bastaardsatijnvlinder (Euproctis chrysorrhoea L.). **Susceptibility of oak (Quercus spp.) to attacks of the browntail moth (Euproctis chrysorrhoea L.).**

Sportfields, play and camping grounds (B 1630)

See also 8597

Plants and parts of plants in general (B 2100)

8029 Larsen, L.M.; Sørensen, H. DK 030106/75/0015
Undersøgelser over monofagi/oligofagi hos visse biller i forbindelse med forekomsten af attraktanter/deterrenter i glucosinoltholdige planter. **Investigations into monophagy/oligophagy in certain beetles in connection with the occurrence of attractants/deterents in plants with glucosinolates.**

8030 Vidano, C. IT 041206/78/1146 N
Screening di prodotti juvenoidi su strati giovanili di insetti. **Screening juvenoids on layers of young insects.**

8031 Lamberti, F. IT 060600/77/0947
Fisiologia, citopatologia e biochimismo di mematodi fitoparassiti di piante da essi attaccate. **Physiology,**

cytopathology and biochemistry of phytoparasitic nematodes.

Plant communities as ecological systems (B 2200)

See also 8070

8032 Rodolphe, F. FR 010214/74/8260
Dynamique population: équilibre hôte x ravageur aleurode (3). **Population dynamic = prey–predator equilibrium aleurode.**

Animal and plant communities as ecological systems (B 2500)

8033 Moreau, J.P.; Boulay, C. Melle FR 010106/63/1771
Relations plantes–insectes vecteurs. **Plants–vector insects relationships.** Publications.

Animal – plant interrelationships (B 2600)

See also 2291, 8240, 8241, 8311, 8794

8034 Breny, R.; Baurant, R. BE 010011/67/0001 R
Etude des relations insectes–hôtes pour les chermes de l'Epicéa. **Study of the Cherme Sprure–gall–aphids : relation between the insects and host–plant.** Publications.

Crops in general (B 3000)

See also 319, 794, 1040, 2243, 2291, 2791, 5174, 5319, 7958, 7961, 7973, 7988, 8987, 9036, 9074, 9162, 9186, 9198, 9203, 9220, 10142, 20067

8035 Meyer, J.; Decallonne, J.; Rouchaud, J. BE 020302/72/0003 R
Etude de la métabolisation des pesticides systématiques dans les plantes cultivées et effets secondaires. **Study of metabolisation of systematic pesticides in cultivated crops and side effects.** Publications.

8036 Gillard, A.; De Grisse, A. BE 030024/69/0001 R
Ultrastructuur van enkele plantenparasiterende nematoden geslachten met het oog op hun bestrijding. **Study of the ultrastructure of some phytoparasite nematode species viewing control.** Publications.

8037 Gillard, A.; Pelerents, G.; Buysse, G.; Van Keymeulen, M. BE 030024/71/0006 R
Studie van een bestrijdingsmethode van schadelijke insekten gesteund op de induktie van letale genen bij de mannetjes bij middel van radioactieve straling. **Study on a control method of harmful insects by induction of letal male genes bij irradiation.** Publications.

8038 Van Assche, C.; Coosemans, J. BE 040203/70/0010 R
Relatie tussen plantparasitaire nematoden en fungi als oorzaak van complex diseases. **Relation between plant parasitic nematodes and fungi as a base of complex diseases.** Publications.

8039 Van Assche, C.; Grauwels, G. BE 040203/78/0023 N
Studie van de relatie van bromide in de grond met het bromide in de plant na gebruik van methylbromide. **Relation of bromide in soil with bromide in plant after use of methylbromide.**

8040 D'Herde, C.; Coolen, W. BE 070100/49/0008 R

Nematologisch diagnostisch onderzoek. **Nematological diagnostic examination.**

8041 D'Herde, C.; Coolen, W.; Hendrickx, G.
BE 070100/68/0005 R
Screening van nieuw ontwikkelde nematiciden. **Screening of new developed nematicides.** Publications.

8042 Klingauf, F.; Schwarz–Andersch, E.; Krafft, L.
DE 111352/78/0001 N
Die Wirtswahl von phytophagenen Insekten. **Host selection of pest insects.**

8043 Klingauf, F.; Blaeser, M.; Wachendorff, U.; Sayampol, B.; Salem, I.
DE 111352/78/0002 N
Wirksamkeit und Nebenwirkungen von Pflanzenschutzmitteln – besonders Insektizide und Herbizide. **Effectiveness and side effects of pesticides especially of insecticides and herbicides.** Publications.

8044 Schmutterer, H.; Lefevre, M.
DE 129180/73/0001
Wirkungsweisen von Juvenilhormon– und Ecdyson–Analogen auf Piesma quadratum. **The effects of juvenile hormone and ecdysone analoga on Piesma quadratum.**

8045 Researcher not indicated
DE 129180/77/0001
Wirkung von Insektenwachstumsreglern – IWR – auf virusübertragende Blattläuse und ihre natürlichen Feinde unter Gewächshaus– und Freilandbedingungen. **Effects of insect growth regulators on virus–transmitting aphids and their natural enemies under greenhouse and field conditions.**

8046 Schmutterer, H.; Hendi, A.
DE 129184/73/0001
Inter– und intraspezifische Konkurrenzwirkungen bei Blattläusen. **Effects of inter– and intraspecific competition on aphids.**

8047 Schmutterer, H.; Steets, R.
DE 129184/73/0002
Untersuchungen von Pflanzenextrakten auf ihre insektiziden Wirkungen. **Investigations on the insecticidal effects of vegetable extracts.**

8048 Schmutterer, H.; Özgür, F.
DE 129184/73/0003
Wirkungen niedrig dosierter systemischer Insektizide mit Zusatz von Synergisten auf Blattläuse und Coccinelliden. **Effects of low–dosed systemic insecticides with additive synergists on aphids and coccinellidae.**

8049 Schmutterer, H.; Klein–Koch, C.
DE 129184/74/0002
Wirkungen von Insektenwachstumsregulatoren auf Spinnmilben und deren natürliche Feinde. **Effects of insect growth regulators on spider mites and their natural enemies.**

8050 Schmutterer, H.; Weiss, M.
DE 129184/74/0003
Strahlenbiologische Untersuchungen an Piesma quadrata. **Radiobiological effects on Piesma quadrata.**

8051 Schmutterer, H.; Holst, H.
DE 129184/74/0005
Mikrobielle Metabolite mit insektizider Wirkung. **Insecticidal effects of microbial metabolites.**

8052 Rössner, J.; Dibs, S.
DE 129185/73/0001
Nebenwirkungen verschiedener Fungizide auf wandernde Wurzelnematoden und nematodenfangende Pilze. **Secondary effects of divers fungicides on migratory root nematodes and nematode–catching fungi.**

8053 Rössner, J.
DE 129185/74/0002
Umweltschonende Verfahren zur Verhütung von Nematodenschäden. **Environment preserving treatments to prevent damage by nematodes.**

8054 Kranz, J.; Hau, B.
DE 129554/73/0001
Analytische Untersuchungen –System Analyse– von Wirt/Pathogen/ Umwelt–Beziehungen, die Quantifizierung von Ereignissen und Beziehungen und die Anwendung mathematischer Modelle zur Simulation und Voraussage dieser Beziehungen. **Analytic investigations –systems analysis– of host/pathogen/ environment relations, the quantification of processes and relations and the applications of mathematical models to simulation and prediction of these phenomena.**

8055 Kranz, J.; Hau, B.
DE 129554/75/0001
Epidemiologische Modelle. **Models in epidemiology.**

8056 Wilbert, H.
DE 132211/72/0001
Die Suchfähigkeit der Larven von Aphidoletes aphidimyza. **Searching ability of the larvae of Aphidoletes aphidimyza.**

8057 Wilbert, H.; Schüler, C.; Akel, O.
DE 132211/73/0002
Einfluss verschiedener Faktoren auf die Wirksamkeit von Blattlausfeinden. **Influence of different factors on the efficiency of natural enemies of aphids.**

8058 Wilbert, H.; Pfaue
DE 132211/77/0001
Einfluss der Heterogenität von Pflanzenbeständen auf den Befall durch tierische Schädlinge. **Effect of crop heterogeneity on animal pests.**

8059 Wilbert, H.; Osmers
DE 132211/77/0003
Nebenwirkungen von Herbiziden auf Blattläuse. **Secondary effects of herbicides on aphids.**

8060 Wilbert, H.; Mpkagiannis
DE 132211/77/0004
Lebensweise und Beutesuche von Anthocoris nemorum. **Biology and searching for prey in Anthocoris nemorum.**

8061 Wilbert, H.; Wohlers
DE 132211/77/0005
Anwendung des Alarm–Pheromons der Blattläuse. **Application of the alarm pheromone of aphids.**

8062 Heyns, K.; Francke, W.
DE 135151/73/0001
Sexualpheromone bei Hylecoetus dermestoides L. – Coleoptera: Lymexglidae –. **Sex pheromones of Hylecoetus dermestoides L. – Coleoptera: Lymexglidae –.**

8063 Meyer, E.; Schliesske, J.
DE 138330/73/0002
Untersuchungen zur Biologie und Gradologie der Gallmilbe Vasates fockeui NAL et TRT. **Studies of biology and population dynamics of the eriophyid gall mite Vasates fockeui NAL et TRT.**

8064 Wyss, U.
DE 138330/74/0002
Reaktion von Wurzelzellen auf den Parasitismus von Trichodorus similis. **Cellular response of roots to parasitism by Trichodorus similis.**

8065 Wyss, U.
DE 138330/74/0004
Histologische Untersuchungen an Saugstellen wandernder Wurzelnematoden. **Investigations on the histology of feeding sites of migratory root nematodes.**

8066 Wyss, U.   DE 138330/74/0005
Embryonalentwicklung und Schlüpfvorgang bei Trichodorus und Longidorus spp.. **Embryology and hatching of Trichodorus and Longidorus spp.**.

8067 Kraus, W.; Grimminger, W.; Cramer, R.; Buhlert, J.; Frenger, W.; Michel, H.   DE 144165/77/0001
Natürliche Insektizide. **Naturally occurring insecticides.**

8068 Langenscheidt, M.   DE 144550/70/0001
Wirkung von Chemosterilantien auf Spinnmilben, Wirkungsmechanismus. **Effects of chemosterilants on spider mites, mechanism of action.**

8069 Richter, I.   DE 213020/75/0014
Lipide zur Bekämpfung von Schädlingen. **Lipids for pest control.**

8070 Schuphan, I.; Ebing, W.   DE 215010/78/0001 N
Entwicklung eines geschlossenen Kulturpflanzenökosystems als Modell zur quantitativen Ermittlung des Verbleibs applizierter Pflanzenschutzmittel. **Development of a closed ecosystem of cultivated plants as a model for quantitative determination of the fate of pesticides applied.**

8071 Rassmann, W.   DE 215030/75/0001
Untersuchungen über die Verbreitung und den Grad der Resistenz gegen verschiedene Insektizide bei Vorratsschädlingen in einheimischen Lagern und Lebensmittelbetrieben. **Studies on distribution and rate of resistance of pests of stored products in native storehouses and in food trade to diverse insecticides.**

8072 Stüben, M.   DE 215070/77/0007
Möglichkeiten zur Bekämpfung der weissen Fliege Trialeurodes vaporariorum mit Juvenilhormonanalogen und Chemosterilantien. **Chances of control of white fly Trialeurodes vaporariorum by juvenile hormone analogues and chemosterilants.**

8073 Basedow, T.   DE 215120/78/0006 N
Untersuchungen zur Prognose und Bekämpfung der Erdraupen - Scotia segetum -. **Studies on forecasting and control of cut worms - Scotia segetum -.**

8074 Krieg, A.; Gröner, A.; Huber, J.   DE 215170/73/5003
Biotechnologische Forschung zur Charakterisierung und Produktion insektenpathogener Viren. **Biotechnological research on the characterization and production of insect pathogenic viruses.**

8075 Hassan, S.A.   DE 215170/75/0002
Rationelle Verfahren zur Massenproduktion räuberischer und parasitischer Insekten und Milben zur biologischen Bekämpfung von Gewächshausschädlingen. **Scientific methods for mass production of predatory and parasitic insects and mites for biological control of pests in greenhouses.**

8076 Huger, A.M.   DE 215170/77/0004
Diagnostische Untersuchungen über das Auftreten von Krankheiten in Freilandpopulationen wichtiger Schadinsekten. **Diagnostic studies on the occurrence of diseases in field populations of important insect pests.**

8077 Krieg, A.   DE 215170/77/0006
Grundlagen der Wirkung von Bacillus thuringiensis gegen Insekten. **Basis of the effect of Bacillus thuringiensis on insects.**

8078 Hassan, S.A.   DE 215170/77/0007
Nebenwirkungen von Pathogenpräparaten auf Nutzinsekten. **Side-effects of pathogenic preparations on useful insects.**

8079 Krieg, A.   DE 215170/77/0008
Untersuchungen des Titers potentiell pathogener Bakterien in Präparaten zur mikrobiologischen Schädlingsbekämpfung. **Analysis of titer of potentially pathogenic bacteria in preparations for microbiological pest control.**

8080 Huber, J.   DE 215170/77/0009
Erprobung von insektenpathogenen Viren zur praktischen Bekämpfung von Schadinsekten. **Testing of insect-pathogenic virus for practical control of pest insects.**

8081 Zimmermann, G.   DE 215170/77/0010
Erprobung von insektenpathogenen Pilzen zur praktischen Bekämpfung von Schadinsekten. **Testing of insect-pathogenic fungi for practical control of pest insects.**

8082 Langenbruch, G.A.   DE 215170/77/0011
Entwicklung und Erprobung verbesserter Applikationsverfahren von Insektenpathogenen einschliesslich der spezifischen Erfolgskontrolle. **Development and testing of improved application of insect pathogens including specific success control.**

8083 Hassan, S.A.   DE 215170/78/0001 N
Einsatz von räuberischen und parasitischen Insekten und Milben im Pflanzenschutz. **Use of predatory and parasitic insects and mites for plant protection.**

8084 Schütte, F.; Hauss, R.   DE 215180/74/5001
Untersuchungen zur Populationsdynamik und Entwicklung einer integrierten Bekämpfung des Maikäfers. **Studies on the population dynamics and development of integrated control of cockchafer.**

8085 Basedow, T.   DE 215180/75/0001
Zur Siedlungsdichte räuberischer Arthropoden der Bodenoberfläche im landwirtschaftlichen Bereich. **Settling density of predatory arthropodes in topsoil in agriculture.**

8086 Schütte, F.   DE 215180/77/0008 N
Allgemeine und epidemiologische Untersuchungen. **General and epidemiological studies.**

8087 Sturhan, D.   DE 215190/70/4009
Untersuchungen über Vorkommen und Verbreitung pflanzenparasitärer Nematoden in der Bundesrepublik Deutschland. **Studies on occurrence and distribution of plant-parasitic nematodes in the Federal Republic of Germany.**

8088 Weischer, B.   DE 215190/77/0001
Untersuchungen über den Einfluss pflanzenverträglicher Nematizide auf das Verhalten von Nematoden. **Investigations on effects of plant tolerant nematocides on the behaviour of nematodes.**

8089 Sturhan, D.   DE 215190/77/0002
Parasiten und Feinde pflanzenparasitärer Nematoden und ihr Einfluss auf die Populationsentwicklung. **Parasites and enemies of plant parasitic nematodes and their influence on the growth of population.**

8090 Thielemann, R.   DE 215190/77/0005
Versuche zur Bekämpfung von Heterodera–Arten mit systemisch wirksamen pflanzenverträglichen Substanzen. **Experiments on Heterodera control with systemic agents tolerated by plants.**

8091 Rumpenhorst, H.J.   DE 215190/77/0006
Prüfung von Kulturpflanzen auf Resistenz gegenüber pflanzenparasitären Nematoden. **Testing of cultivated plants for resistance to plant–parasitic nematodes.**

8092 Thielemann, R.   DE 215190/77/0016
Untersuchungen zur Populationsdynamik des Rübennematoden Heterodera schachtii unter modernen Anbaubedingungen. **Studies on population dynamics of Heterodera schachtii under modern cultivation conditions.**

8093 Rumpenhorst, H.J.   DE 215190/78/0003 N
Physiologie der Wirt–Parasit–Beziehungen bei pflanzenschädigenden Nematoden. **Physiology of host–parasit–relationships in plantparasitic nematodes.**

8094 Dickler, E.; Hassan, S.A.   DE 215210/74/5005
Untersuchungen über die Wirkung von Pflanzenschutzmitteln auf Nutzarthropoden im Freiland. **Investigations on the effect of plant protectives on pest destroying arthropods in open fields.**

8095 Dickler, E.   DE 215210/78/0005 N
Untersuchungen zur Verbreitung der Pfirsichmotte, Anarsia lineatella und des Pfirsichwicklers, Grapholitha molesta in der Bundesrepublik Deutschland. **Investigations on the geographical distribution of Anarsia lineatella and Grapholitha molesta in the Federal Republic of Germany.**

8096 Scholz, M.   DE 215240/77/0001
Durchführung des Meldedienstes über das Auftreten von Schaderregern an Kulturpflanzen in der Bundesrepublik Deutschland. **Working of warning service on the incidence of pests on cultivated plants in the Federal Republic of Germany.**

8097 Quantz, L.   DE 215240/77/0002
Entwicklung und Erprobung von quantitativen Meldeverfahren zur Schaderregerüberwachung. **Development and acquisition of quantitative warning methods for pest control.**

8098 Lüders, W.; Schletzer, B.; Schuler, H.   DE 501053/78/0001 N
Applikationsversuche mit Pflanzenschutzgeräten. **Application experiments with sprayers.** Publications.

8099 Wallnöfer, P.R.; Rast, H.G.; Engelhardt, G.   DE 502051/77/0002
Mikrobieller Abbau von Thiomethyl–substituierten Phenolen – Insektizidmetaboliten –. **Microbial degradation of Thiomethyl–substitued phenols – insecticide metabolites –.** Publications.

8100 Bauchhenss, J.   DE 502055/73/0008
Untersuchungen zur Biologie der Onychiuridae – Collembola – Apterygota –. **Studies on the biology of Onychiuridae – Collembola – Apterygota –.**

8101 Bauchhenss, J.   DE 502055/73/0009
Verteilungsmuster von Collembolen – Apterygota – und Oribatiden – Acari – in Wiesenböden. **Distribution of Collemboles and Oribatid mites in meadow soils.**

8102 Bauchhenss, J.   DE 502055/73/0010
Experimentell–ökologische Untersuchungen an Collembolen – Apterygota –. **Studies on experimental ecology of Collemboles – Apterygota –.**

8103 Rosopulo, A.; Zach, G.   DE 502055/73/0019
Die Bestimmung von Pestizidrückständen in den verschiedenen Pflanzen. **Analysis of pesticide residues in different plants.**

8104 Keil, W.   DE 506200/72/0002 R
Untersuchungen zur Siedlungsdichte, Ernährungsbiologie und Ökologie insektenfressender Vogelarten zwecks Einsatz zur Bekämpfung von Eichenwickler und Lärchenminiermotte. **Investigations on the population density, biology and ecology of insectivorous birds in connection with the biological control of Tortrix viridana and Coleophora laricella.**

8105 Lauenstein, G.   DE 507250/75/0001
Untersuchungen über die Verbreitung und die Bekämpfung von Heterodera pallida und Pathotypen von Heterodera rostochiensis. **Investigations into occurrence and control of Heterodera pallida and of pathotypes of Heterodera rostochiensis.**

8106 Vagt, W.; Hauschildt, H.   DE 507304/70/0001
Pflanzenschutzmittel und ihre Wirkungen auf Blatt und Frucht. **Plant protectives and the effect on leaf and fruit.**

8107 Vagt, W.; Hauschildt, H.   DE 507304/75/0001
Nebenwirkung auf andere Schädlinge und Krankheiten als durch eigentliche Indikationen erwartet. **Secondary effects on other pests and diseases other than expected by the actual indication.**

8108 Nøddegaard, E.; Rasmussen, A.N.   DK 010116/78/0029 N
Jorddesinfektion af frøbede i planteskoler. **Sterilization of soil in nurseries.**

8109 Coutin, R.   FR 010106/43/5241
Biologie des Cécidomyies et méthodes de lutte. **Biology and control of gall–midges.**

8110 d'Aguilar, J.   FR 010106/50/5242
Etude systématique et biologique des Diptères Tachinidae. **Systématique and biological studies on Diptera Tachinidae.**

8111 Chevin, H.   FR 010106/65/5243
Etude systématique et biologique des Hyménoptères symphytes. **Systematical and biological studies of Hymenoptera Symphyta.**

8112 Chambon, J.P.   FR 010106/71/5247
Biologie Systématique des Tortricidae Paléarctiques. **Biology, Taxonomy of the Paloerarctic Tortricidae.**

8113 Della Giustina, W.; Stagiaires; Meunier, M.   FR 010106/73/5249
Expérimentation de produits destinés à combattre les ravageurs aériens des cultures protégées. **Pesticide experimentation for protected cultivation aerial pest control.**

8114 Rapilly, F.; Foucault, M.　　　FR 010107/71/1018
Biologie du Septoria Nodorum. Epidémiologie de la maladie.
Biology of Septoria Nodorum. Epidemiology of the disease.

8115 Millier, C.　　　FR 010312/75/8286
Description des interactions climat x hôte x ravageur (3).
Description of interactions climate x host x defoliator.

8116 El Shishiny, H.; Millier, C.　　　FR 010312/76/8285
Dynamique de populations d'aleurodes des serres (3).
Dynamics of population of glasshouse aleurodes.

8117 Mars, S.　　　FR 010502/73/6010
Absorption du brome par la plante sur terrains traités au
bromure de méthyle et en cultures sans sol. **Bromine
absorption by plant, on methyl bromide treated soils and soilless
cultures.**

8118 Iperti, G.; Brun, P.　　　FR 010503/63/1789
Connaissances biologiques des principales espèces de
coccinelles aphidiphages afin d'évaluer leur efficacité
prédatrice potentielle. **Biological knowledge of main species of
aphido phagous coccinellideae to Assess their predaceous
efficiency.** Publications.

8119 Iperti, G.; Ferran, A.; Brun, J.; Euverte, G.
　　　FR 010503/63/5030
Bioécologie des Coccinelles aphidiphages indigènes.
Bioecology of the native aphidophagous Coccinellids.

8120 Lyon, J.P.; Leclant, F.; Lavergne, A.M.; Iperti, G.
　　　FR 010503/63/5031
Bioécologie des Syrphides. **Bioecology of the Syrphids.**

8121 Iperti, G.; Brun, J.　　　FR 010503/66/5032
Elevage de coccinelles coccidiphages exotiques. **Rearing of
exotic coccidophagous coccinellids.**

8122 Lyon, J.P.; Onillon, J.C.; Lavergne, A.M.; Rabasse,
J.M.; Ferran, A.; Pralavorio, M.　　　FR 010503/68/5033
Lutte biologique contre les pucerons en serre. **Biological
control of aphids in glasshouses.**

8123 Arambourg, Y.; Lyon, J.P.; Pralavorio, R. Mme; Iperti,
G.　　　FR 010503/68/5034
Lutte biologique contre Prays oleae. **Biological control of Prays
oleae.**

8124 Pralavorio, M.; Jourdheuil, P.　　　FR 010503/69/1792
Lutte biologique contre l' araignée rouge en serre. **Biological
control of red spider in glasshouses.** Publications.

8125 Benassy, C.; Bianchi, H.　　　FR 010503/70/1790
Etude de l'utilisation pratique de divers parasites dans la lutte
contre les cochenilles diaspines. **Study on practical use of
various parasites in control of scale–insect diaspididae.**
Publications.

8126 Panis, A.; Marro, J.P.　　　FR 010503/70/5036
Lutte biologique contre la cochenille noire Saissetia oleae.
Biological control of the black–scale, Saissetia oleae.

8127 Panis, A.; Marro, J.P.　　　FR 010503/70/5037
Démécologie de la cochenille noire de l'Olivier, Saissetia
oleae. **Demecology of the black scale, Saissetia oleae.**

8128 Pralavorio, M.; Lyon, J.P.; Onillon, J.C.; Rabasse, J.M.
　　　FR 010503/70/5038
Lutte biologique contre les Acariens en serres. **Biological
control of Mites in Glasshouses.**

8129 Ferran, A.　　　FR 010503/71/1796
Alimentation artificielle des entomophages. **Artificial feeding
of entomophagous insects.** Publications.

8130 Rabasse, J.M.; Brunel, E.; Robert, Y.; Dedryver, Ch.
　　　FR 010503/71/5039
Rôle des parasites et des champignons dans les populations de
pucerons de quelques cultures types. **Part played by aphid
parasites (and fungi) in the natural control of aphids in the
fields.**

8131 Iperti, G.; Brun, J.　　　FR 010503/72/5040
Etude de la migration des coccinelles aphidiphages. **Study of
the aphidophagous coccinellids migration.**

8132 Ferran, A.; Lyon, J.P.; Laroque, Melle; Bonnot, G.;
Iperti, G.; Bounias, M.　　　FR 010503/73/5041
Elevage des coccinelles en milieu artificiel. **Rearing of the
Coccinellids on artificial meida.**

8133 Voegele, J.; Bonnot, G.; Daumal, J.; Grenier, S.;
Pizzol, J.　　　FR 010503/73/5042
Production et stockage des Trichogrammes. **Mass production
and conservation of the Trichogramma.**

8134 Voegele, J.; Pintureau, B.; Berge, J.B.; Pointel, J.G.
　　　FR 010503/74/5047
Caractérisation des Trichogrammes. **Characterization of the
trichogramma.**

8135 Onillon, J.C.; Franco, E.; Onillon, J.; Lyon, J.P.;
Pralavorio, M. Mme; Coulon, J.　　　FR 010503/74/5049
Lutte biologique contre les Aleurodes en serres. **Biological
control of whiteflies in glass–houses.**

8136 Voegele, J.; Martouret, D.; Galichet, P.F.; Poitout, S.
　　　FR 010503/75/5050
Utilisation des Trichogrammes en lutte biologique. **The use of
Trichogramma in biological control.**

8137 Ferran, A.; Larroque, M.M.　　　FR 010503/75/5051
Etude des rapports hôte–prédateur chez une coccinelle
aphidiphage, Semiadalia undecimnotata. **Study of the
host–prèdator relationships in an aphidiphagous coccinellid,
Semiadalia undecimnotata.**

8138 Onillon, J.C.; Onillon, J.; Rodolphe, F.; Franco, E.;
Tomassonne, R.　　　FR 010503/75/5052
Modélisation du système hôte–parasite (Aleurothrixus
floccosus – Cales noacki). **Simulation models for the
host–parasite complex Aleurothrixus floccosus – Cales noacki.**

8139 Blanc, D. Mme; Pralavorio, M. Mme; Triboi, A.M.;
Rabasse, J.M.; Onillon, J.C.　　　FR 010503/75/5054
Action des fertilisants sur l'association plante–phytophage.
Effects of the fertilizers on the plant–phytophaga association.

8140 Onillon, J.C.; Rodolphe, F.; Rabasse, J.M.;
Tomassonne, R.　　　FR 010503/76/5055
Dynamique des populations d'Homoptères en serre. **Dynamics**

of Homoptera populations in glasshouses.

8141 Rabasse, J.M.; Tomassone, R.; Rodolphe, F.
FR 010503/76/5056
Dynamique des populations d'Aphides en serre et lutte
biologique à l'aide de parasites. **Populations dynamics and
biological control of aphid pests in glasshouses.**

8142 Scotto La Massese, C.; Cayrol, J.C.; Ritter, M.;
Dalmasso, A.; Berge, J.B.   FR 010504/60/5150
Recherches sur la systématique des nématodes. **Researches on
systematics of plant nematodes.**

8143 Scotto la Massese, C.; dalmasso, A.; berge, J.B.
FR 010504/64/5153
Lutte intégrée contre les nématodes – Recherche de
porte–greffe résistant aux nématodes nuisibles – Conséquences
de leur emploi en pratique agricole. **Integrated control of
nematodes : Research on roostocks resistance – Consequencies
of their use in agriculture.**

8144 Laumond, C.   FR 010504/67/5154
Etude générale des nématodes parasites d'insectes. **General
survey of insect parasite nematodes.**

8145 Cayrol, J.C.; Frankowski, J.P.; Combettes, S. Mme;
Couderc, C.   FR 010504/68/5155
Utilisation des champignons nématophages comme agents de
lutte biologique contre les nématodes en agriculture. **Use of
nematodes trapping fungi as a method of biological control
against nematodes in agriculture.**

8146 Laumond, C.   FR 010504/70/5156
Taxonomie, biologie et écologie des Allantonematidae.
Utilisation en lutte biologique. **Taxonomy, biology and ecology
of Allantonematidae – Use in biological control.**

8147 Dalmasso, A.; Berge, J.B.; Laumond, C.
FR 010504/70/5157
Diversité génétique chez les nématodes. **Genetic polymorphism
in nematodes.**

8148 Scotto la Massese, C.   FR 010504/70/5158
Influence des techniques culturales sur la nématofaune des
cultures. **Cultural practice effects on the crop nematofauna.**

8149 Laumond, C.   FR 010504/72/1714
Multiplication semi–industrielle des nématodes
Néoplectanidae. **Semi–industrial multiplication of
Neoplectanidae nematodes.**

8150 Combettes, S. Mme; Dalmasso, A.; Scotto la Massese,
C.   FR 010504/72/5162
Techniques d'élevage monoxénique des nématodes.
Techniques for monoxenic rearing of Nematodes.

8151 Laumond, C.   FR 010504/72/5163
Multiplication de masse de Neoaplectana. **Mass culture of
Neoaplectana.**

8152 Cuany, A.; Arvieu, J.C.; Scotto la Massese, C.; Pistre,
R.; Nourrisseau, J.G.; Leclair, Ph.   FR 010504/72/5164
Utilité et limites d'emploi des substances nématicides. **Utility
and use limits of nematicides.**

8153 Berge, J.B.; Cardin, M. C. Mme; Dalmasso, A.; Ritter,

M.   FR 010504/73/5161
Virulence et agressivité des Meloîdogynes en conditions
naturelles. Conséquences agronomiques. **Virulence and
aggressivity of the Meloïdogyne species.**

8154 Scotto la Massese, C.   FR 010504/74/5166
Pathogénie des nématodes. **Pathogenicity of nematodes.**

8155 Scotto la Massese, C.   FR 010504/74/5167
Monographie du genre Rotylenchus. **Revision of the genus
Rotylenchus.**

8156 Cuany, A.; Berge, J.B.; Arvieu, J.C.   FR 010504/75/5169
Etude des mécanismes d'action des nématicides. **Mode of
action of nematicides.**

8157 Laumond, C.; Doucet, M.M.   FR 010504/75/5170
Recherches sur les Mermithidae. **Researchs on Mermithidae.**

8158 Causse, R.   FR 010602/66/5088
Rythmes biologiques liés au cycle photopériodique chez
Ceratitis capitata. **Biological rhythms related to the
photoperiodic cycle in Ceratitis capitata.**

8159 Cayrol, R.; Anglade, P.; Poitout, S.; Causse, R.
FR 010602/70/5089
Etude des migrations de Noctuelles sur le territoire français.
Study of the Noctuids migrations throughout France.

8160 Cayrol, R.   FR 010602/70/5090
Etude de la diapause chez les larves et les chrysalides de
quelques espèces de Noctuidae. **Study of diapause in larvae and
pupae of some Noctuidae species.**

8161 Bounias, M.; Bonnot, G.   FR 010602/73/5134
Rôle de l'acide ascorbique dans la nutrition artificielle de
lépidoptères phytophages. **Role of ascorbic acid in artificial
nutrition of phytophagous Lepidoptera.**

8162 Bues, R.; Poitout, S.   FR 010602/74/5092
Hibernation et cycle évolutif de Arctia caja L. (Lépidoptère
Arctiidae). **Overwintering and life cycle of Arctia caja L.
(Lepidoptera : Arctiidae).**

8163 Poitout, S.; Bues, R.   FR 010602/75/5093
Etude de génétique écologique sur plusieurs espèces de
Noctuidae. **Study of bioecology and genetics of some Noctuidae
species.**

8164 Guennelon, G. Mme; Atger, P.   FR 010602/75/5098
Amélioration de l'état sanitaire des élevages de Lépidoptères
en milieux artificiels. **Control of Microorganisms in
Lepidoptera breedings on artificial media.**

8165 Guennelon, G. Mme   FR 010602/75/5099
Mise au point d'un élevage en masse de Laspeyresia pomonella
sur milieu artificiel. **Techniques leading to Laspeyresia
pomonella mass rearing on artificial media.**

8166 Galichet, P.F.   FR 010602/75/5100
Lutte biologique contre deux espèces de Diatraea et Chilo.
Biological control of two species of Diatraea and Chilo.

8167 Galichet, P.F.   FR 010602/75/5101
Etude d'un parasite, Lydella thompsoni (Tachinidae) dans
différentes biocoenoses. **Study of a parasite, Lydella thompsoni**

(Tachinidae) in several biocoenoses.

8168 Bues, R.; Poitout, S. FR 010602/76/5105
Etudes écophysiologiques des espèces françaises de Noctuelles des genres Euxoa et Scotia. **Ecophysiological studies on the French Noctuidae species of the genera Euxoa and Scotia.**

8169 Poitout, S.; Laumond, C.; Bues, R.; Voegele, J.; Burgerjon, A. FR 010602/76/5106
Etude de Scotia segetum Schiff. pour l'amélioration des méthodes de lutte. **Study of Scotia segetum Schiff., for better control of this species.**

8170 Causse, R.; Poitout, S. FR 010602/76/5107
Caractéristiques éthologiques du mouvement migratoire au niveau de l'individu, chez les Noctuidae. **Ethological characteristics of the individual flight behaviour.**

8171 Poitout, S.; Causse, R. FR 010602/77/5108
Etude éthologique du vol de migration au niveau des populations, spécialement en montagne, chez les Noctuidae. **Ethological study of population – flight during migrations especially in mountains.**

8172 Meynadier, G. FR 010612/50/5176
Recherche des maladies spontanées et de germes pathogènes chez les insectes nuisibles en agriculture. **Screening and study of spontaneous diseases and microbiol agents in insects harmful to agriculture.**

8173 Bergoin, M.; Meynadier, G.; Guelpa, B.; Duthoit, J. L.; Croizier, G.; Ricou, G. FR 010612/63/5182
Etude du Baculovirus du Diptère Tipula. **Study of the Baculovirus of Tipula.**

8174 Louis, C. FR 010612/65/1767
Rapports physiologiques entre les microorganismes symbiotiques intracellulaires et leurs cellules – hôtes. **Physiological relationship between intracellular symbiotic microorganisms and their host– cells.** Publications.

8175 Vey, A.; Quiot, J.M. FR 010612/66/1766
Etude in vitro des réactions hémocytaires des invertébrés. **In vitro study of haemocytic reactions in invertebrates.** Publications.

8176 Monsarrat, P.; Duthoit, J. L.; Vago, C.; Meynadier, G.; Croizier, G.; Hurpin, B. FR 010612/73/5202
Etude du Baculovirus du Coléoptère Oryctes rhinoceros et de sa pathogénèse. **Characterization and pathogenesis of Oryctes Baculovirus.**

8177 Monsarrat, P.; Robert, P.; Croizier, G.; Hurpin, B. FR 010612/74/5203
Tests d'inocuité du Baculovirus d'Oryctes rhinoceros. **Safety tests on Oryctes Baculovirus.**

8178 Croizier, G.; Amargier, A.; Veyrunes, J.C.; Quiot, J.B.; Meynadier, G.; Burgerjon FR 010612/74/5204
Caractérisation de Baculovirus inféodés aux Lépidoptères nuisibles aux cultures. **Identification of Baculovirus pathogen for Lepidoptera.**

8179 Roehrich, R.; Carles, J.P. FR 010707/75/5219
"Méthode de confusion" par phéromone appliquée à l'Eudemis. **The mating disruption by pheromon for Labesia**

botrana.

8180 Stengel, M.; Schubert, G. FR 010904/66/1746
Régulation neuroendocrine du comportement migratoire de la femelle du Hanneton commun. **Neuroendocrinal regulation of migratory behaviour of cockchafer female.** Publications.

8181 Robert, P.; Blaisinger, P. FR 010904/70/1748
Lutte chimique contre l'adulte du hanneton commun. **Chemical control of cochchafer adult.** Publications.

8182 Stengel, M.; Schubert, G. FR 010904/74/5024
Etude du comportement migratoire du Hanneton commun. **Study of the migratory behaviour of the Cockchafer.**

8183 Bouchery, Y.; Jacky, F.; Wiss, L. FR 010904/75/5025
Etude des populations de pucerons ailés en vol par piégeage. **Study of flying aphids with traps.**

8184 Catroux, G. FR 011002/70/6295
Ecologie de beauveria tenella introduit dans le sol (lutte biologique contre le vers blanc du hanneton). **Ecology of beauveria tenella inserted in soil (biological control against cockchafer grub).**

8185 Bessard, A.; Stawiecki, Mme. FR 011007/66/1740
Faunistique écologique des oribates (Acariens). **Ecological faunistic of oribatidae (Acarians).**

8186 Bonnel, L. FR 011007/70/1741
Etude des populations d'Heterodera Avenae en Champagne. **Study of Heterodera Avenae populations in Champagne.**

8187 de Guiran, G.; Bonnel, L. FR 011007/71/1744
Phase sol du cycle des Meloidogyne. **Soil phase of Meloidogyne cycle.** Publications.

8188 Abiracned, M.; Dalmasso, A.; Martin, J. FR 011007/76/5018
Etude ultrastructurale et biochimique des inter–relations plante – meloidogyne. **Ultrastructural and biochemical studies of interrelationships plant – meloidogyne.**

8189 Fargues, J.; Rougier, M.; Robert, P.; De Conchard, H. FR 011102/70/5063
Ecopathologie des Hyphomycètes entomopathogènes. **Ecopathology of the Entomogenous Hyphomycetes.**

8190 Ferron, P.; Marchal, M. FR 011102/71/1722
Lutte biologique avec Beauveria Tenella contre Melolontha Melolontha D. Dans la région d'Etalans (Doubs). **Biological control of Melolontha Melolontha with Beauveria Tenella in Etalans area (Doubs).**

8191 Fargues, J.; Picquot, Mme FR 011102/72/1724
Etude des effets secondaires des interventions cryptogamiques sur les ravageurs qui survivent à la Maladie. **Study of secondary effects of the use of cryptogamic germs on pests surviving to disease.**

8192 Robert, P.; Vergara, S.; Meynadier, G. FR 011102/72/5064
Enquêtes épizootiologiques dans les populations d'insectes ravageurs. **Investigations on epizootic diseases in pest insect populations.**

8193 Ferron, P.; Glandard, A.; Marchal, M.; Robert, P.
FR 011102/73/5065
Lutte biologique contre Melolontha melolontha à l'aide de Beauveria tenella. **Biological control of the common Cockchafer with Beauveria tenella.**

8194 Fargues, J.; De Conchard, H. FR 011102/73/5066
Utilisation des hyphomycètes entomopathogènes contre le doryphore et les noctuelles. **The use of entomogenous Hyphomycetes in control of the Colorado Beetle and the Noctuids.**

8195 Fargues, J.; Robert, P.; De Conchard, H.
FR 011102/73/5067
Spécificité des Hyphomycètes entomopathogènes. **Specificity of the entomogenous Hyphomycetes.**

8196 Fargues, J.; De Conchard, H.; Robert, P.; Vergara, S.
FR 011102/73/5068
Caractérisation des pathotypes de champignons entomopathogènes. **Characterization of the pathotypes of some entomogenous fungi.**

8197 Burgerjon, A.; Poitout, S.; Biache, G.; Bues, R.; Croizier, G. FR 011102/73/5069
Lutte biologique contre les Noctuelles à l'aide de virus. **Microbial control of Noctuids with viruses.**

8198 Burgerjon, A.; Biache, G. FR 011102/74/5070
Lutte biologique contre Lymantria dispar à l'aide de Baculovirus. **Biological control of Lymantria dispar with Baculoviruses.**

8199 Ferron, P.; Marchal, M. FR 011102/74/5071
Lutte contre Otiorrhynchus sp. à l'aide d'un champignon entomopathogène. **Microbial control of Otiorrhynchus sp. with an entomogenous fungus.**

8200 Ferron, P.; Vincent, J.J.; Marchal, M.
FR 011102/74/5072
Lutte microbiologique contre le Carpocapse à l'aide de Beauveria bassiana. **Microbial control of the Codling Moth with Beauveria bassiana.**

8201 Ferron, P.; Glandard, A.; Robert, P. FR 011102/74/5073
Production et titrage des préparations de champignons entomopathogènes. **Mass production and titrating of the Entomogenous fungi preparations.**

8202 Riba, G. FR 011102/77/5077
Amélioration génétique d'Hyphomycètes entomopathogènes. **Genetic improvment of some entomogenous Hyphomycetes.**

8203 Riba, G.; Poitout, S. FR 011102/77/5078
Etude des relations génétiques hôte–pathogène chez les mycoses d'insectes. **Study of the host–parasite genetic relationships in the insect mycoses.**

8204 Riba, G. FR 011102/77/5079
Identification des recombinés mitotiques d'Hyphomycètes et étude de leur pouvoir pathogène. **Identification of the mitotic recombinants of Hyphomycetes and study of their virulence.**

8205 De Luca, Y. FR 011210/49/5231
Etude des coléoptères bruchides. **Study of the Coleoptera Bruchidae.**

8206 Bournier, A.; Pivot, Y. FR 011210/50/5232
Etude des Thysanoptères nuisibles aux plantes cultivées. **Study of the Thysanoptera harmful to cultivated plants.**

8207 Leclant, F. FR 011210/63/1733
Etude faunistique des aphidés de France. **Faunistic study of Aphidae in France.** Publications.

8208 Leclant, F. FR 011210/63/5233
Relations des pucerons avec leur milieu. **Aphids'relationships with their environment.**

8209 Fauvel, G.; Rambier, A. FR 011210/71/1730
Etude expérimentale de la régulation de populations d'acariens phytophage. **Experimental study on regulation of mite populations by a predaceous Anthocoridae.**

8210 Leclant, F.; Renoust, M. FR 011210/71/1732
Epidémiologie de la sharka. **Sharka epidemiology.** Publications.

8211 Missonnier, J.; Cayrol, R. FR 011306/57/5110
Déterminisme des arrêts de développement (diapause) et de l'élimination de la diapause chez les Diptères–Muscidés. **Determinism of diapause and of its suppression in the Diptera–Muscidae.**

8212 Missonnier, J.; Brunel, E. FR 011306/60/1736
Remplacement des insecticides organo–chlorés en traitement des sols. **Substitution of organochlorine insecticides in soil treatment.** Publications.

8213 Missonnier, J. FR 011306/60/1737
Déterminisme des arrêts de développement (Diapause et quiescence) chez les diptères muscidés. **Determinism of arrest of development (Diapause and quiescence) in diptera muscidae.** Publications.

8214 Brunel, E.; Missonnier, J. FR 011306/60/5111
Etude des conditions d'application des traitements insecticides du sol contre les diptères souterrains et le taupin. **Study of the best conditions for treating the soil with insecticides against underground diptera and the wireworms.**

8215 Robert, Y.; Rabasse, J.M.; Brunel, E.
FR 011306/67/5114
Méthodologie d'étude des vols de pucerons par piégeage. **Methods for studying Aphids flight by trapping.**

8216 Robert, Y. FR 011306/67/5115
Rythmes d'activité saisonnière de vol des pucerons en Bretagne ; influence des facteurs climatiques. **Rythms of seasonal flight activity of Aphids in Brittany : Effects of the climatic factors.**

8217 Rahn, R.; Descoins, C. FR 011306/68/5117
Influence de la plante–hôte sur l'attraction sexuelle des femelles d'Acrolepia assectella et sur le comportement de ponte. **Influence of the host–plant on the sex attraction of Acrolepia assectella females and on the oviposition behaviour.**

8218 Brunel, E. FR 011306/70/5119
Comportement des adultes de Psila rosae: influence de la plante–hôte, rythme d'activité. **Adult behaviour of Psila rosae: influence of the host–plant, rate of activity.**

8219 Robert, Y.; Rabasse, J.M.; Dedryver, Ch.
FR 011306/75/5128
Etude des processus des épizooties à Entomophthora chez Aphis fabae. **Epizootics of Entomophthora in Aphis fabae, study of the processes.**

8220 Rahn, R.; Ajjundi, A.; Poitout, S.　FR 011306/75/5130
Etude des noctuelles de l'Ouest de la France. **Study of the Noctuids in the West of France.**

8221 Person, F.; Rivoal, R.; Caubel, G.; Mugniery, D.
FR 011306/76/5133
Etude génétique des races d'Heterodera sp. et de Ditylenchus dipsaci, facteurs d'isolement des races et recherche de variétés résistantes. **Genetical study of the "races" of Heterodera sp. and Ditylenchus dipsaci; isolation of the "races" and search for resistant varieties.**

8222 Kermarrec, A.; Anais, G.　FR 011605/72/1726
Rotations et population de nématodes. **Crop rotations and nematode populations.**

8223 Kermarrec, A.; Plaza, G.; Malato, G.; Coleno, A.
FR 011605/73/5271
Etude de la fourmi manioc en Guadeloupe : pathologie et physiologie de Acromyrmex octospinosus. **Pathological and physiological study of Acromyrmex octospinosus.**

8224 Malausa, J.C.　FR 011605/77/5274
Etude bioécologique des noctuelles en Guadeloupe. **Bioecological study of Guadeloupe Noctuidae.**

8225 Pesson, P.; Foldi, I.　FR 011800/74/5292
Etude ultrastructurale et cytochimique des glandes tégumentaires des Diaspines. **Ultrastructural and cytochemical study of diaspidid tegumentary glands.**

8226 Chansigaud, J.　FR 011800/74/5293
Nouvelles méthodes de lutte contre les Acridiens. **New methods for the control of Acrididae.**

8227 Strebler, G.　FR 011800/76/5295
Etude de la digestion chez des insectes phytophages. **Study of digestion in phytophagous insects.**

8228 Bonnot, G.; Grenier, S.; Delobel, B.　FR 012217/69/5058
Elevage in vitro, de Phryxe caudata et autres Tachinaires. Détermination des besoins nutritionnels. **In vitro rearing of Phryxe caudata and other Tachinids.**

8229 Grenier, S.; Bonnot, G.; Delobel, B.　FR 012217/71/5059
Méthode d'élevage in vitro de Phryxe caudata et d'autres Tachinaires. Réalisation et mise au point de l'élevage. **In vitro rearing methods of Phryxe caudata and other Tachinids.**

8230 Delobel, B.; Grenier, S.; Bonnot, G.　FR 012217/74/5060
Méthodes d'élevage in vitro de Phryxe caudata et d'autres Tachinaires. Etude des exigences métaboliques et comportementales des larves du troisième stade larvaire. **In vitro rearing methods of Phryxe caudata and other Tachinids.**

8231 Grenier, S.　FR 012217/77/5061
Elevage des Trichogrammes en milieux artificiels – Développement in vitro. **In vitro rearing methods of Trichogrammatidae.**

8232 Bonnot, G.; Delobel, B.　FR 012217/77/5062 N
Nutrition de Trichogramma. **Nutrition of trichoframma.**

8233 Maillard, J.　FR 012221/73/3019
Analyse des agrocoenoses dans des exploitations agricoles utilisant des pesticides. **Study of insect populations in the soil of farms using pesticides.**

8234 Maillard, J.; Dejardin, D.; Voreux, H.
FR 012221/73/9049
Etude des agrocénoses animales et de l'influence qu'exercent sur elles les techniques culturales (relations, emplois des pesticides, etc..). **Study of agricultural pest communities as influenced by farming practices (mutual relations, pesticide use, etc..).**

8235 Carre, S.; Tasei, J.N.　FR 012223/74/5083
Biologie et piégeage des Trichodes ravageurs des nids de Megachile pacifica. **Biology and trapping of Trichodes, destroyers of Megachile pacifica nests.**

8236 Correia, M.; Tasei, J.N.　FR 012223/74/5084
Biologie d'Heriades truncorum (Megachilidae). **Biology of Heriades truncorum.**

8237 Cantot, P.; Bournoville, R.　FR 012223/76/5086
Etude biologique de Sitona Humeralis. **Biology of Sitona humeralis.**

8238 Descoins, C.; Lettere, M.; Lalanne–Cassou, B.; Gallois, M.　FR 012225/73/5173
Etude des phéromones sexuelles des lépidoptères ravageurs des cultures. Isolement, identification et synthèse. Rôle éthologique. **Study of sex pheromones of lepidoptera harmful to cultivated plants. Extraction, identification and synthesis. Ethological function.**

8239 Payne　GB 011108/00/0009 R
The use of baculoviruses in pest control.

8240 Burges; Hall　GB 011108/00/0011 R
Potential of fungi for control of arthropod pests.

8241 Burges; Jarrett　GB 011108/00/0012 R
Potential of bacterial pathogens in control of arthropod pests.

8242 Richardson,　GB 011108/77/0020 R
The role of entophilic nematodes in the natural control of insects.

8243 Stringer; Lyons　GB 011514/00/0006 R
Behaviour and control of snails and slugs.

8244 Green　GB 011811/00/0003 R
Incidence and dynamics of migratory plant parasitic nematodes.

8245 Lowe　GB 011906/00/0021 R
Investigate resistance to aphids in arable crops.

8246 Lowe　GB 011906/00/0024 R
Value of intermediate levels of resistance in controlling aphids.

8247 Edwards; Lofty　GB 012004/78/0037 R
Effect of pesticides on pests and beneficial soil fauna.

8248 Bardner; Fletcher GB 012004/78/0045 R
Effect of insect damage on growth and yield of crop plants.

8249 Wilding GB 012004/78/0047 R
Integrated control of pests using fungal pathogens.

8250 Elliott; Janes GB 012005/00/0024 R
Relation between structure and activity of insecticides,
especially pyrethroids.

8251 Burt; Gregory GB 012005/00/0026 R
Neurophysiological and histological studies on the mode of
action of insecticides.

8252 Sawicki; Devonshire GB 012005/00/0027 R
Physiology, biochemistry and genetics of resistance of insects
(housefly, aphid) to insecticides.

8253 Sawicki; Needham GB 012005/00/0028 R
Monitoring for resistance of insects to insecticides in the field.

8254 Phillips; Jeffs GB 012005/00/0030 R
Influence of environmental factors and formulations on
persistance and effectiveness of pesticides.

8255 Graham–Bryce; Briggs GB 012005/00/0032 R
Principles determining the biological availability of pesticides in
soil.

8256 Greenway; Mudd GB 012005/00/0035 R
New methods of pest control and management of beneficial
insects using chemicals that control behaviour.

8257 Whitehead; William GB 012006/00/0001 R
Control of plant parasitic nematodes in arable crops with
nematicides.

8258 Evans; Kerry GB 012006/00/0003 R
Effect of nematodes on plant growth, physiology and nutrition.

8259 Day; Parkinson GB 012008/00/0004 R
Physiological factors in water use by crops in field.

8260 Parkinson; Leach GB 012008/00/0005 R
Physiological factors in water use by plants in the laboratory.

8261 Cockbain GB 012009/00/0006 R
Biology of virus vectors.

8262 Dunning; Thornhill GB 012301/00/0019 R
Biology of soil insects and seedling pests.

8263 Meech GB 012701/00/0001 R
Mechanism of control of neuronal membrane permeability.

8264 Moreton GB 012701/00/0002 R
Electrophysiological study of the action of molluscicides on the
nervous system.

8265 Lane; Schofield GB 012701/00/0003 R
Structure and physiology of the insect blood–brain barrier, in
relation to penetration by chemicals.

8266 Moreton; Trherne GB 012701/00/0004 R
Subcellular distribution of inorganic ions in insect nerves.

8267 Sattelle GB 012701/00/0005 R
Synaptic transmission in insects.

8268 Heslop; Howes GB 012701/00/0006 R
Axoplasmic transport.

8269 Bridges; Dwivedy GB 012701/00/0007 R
Aspects of lipid metabolism essential for the growth and
development of insects.

8270 Maddrell; Gardiner GB 012701/00/0009 R
Control mechanisms and ionic basis of insect excretion.

8271 Berridge GB 012701/00/0010 R
The role of cyclic amp and calcium in the action of insect
hormones.

8272 Lane; Skaer GB 012701/00/0011 R
Electron microscope studies of invertebrate tissues, especially
those of insect nervous systems.

8273 Sattelle GB 012701/77/0018 R
Laser light scattering studies of cellular and molecular
hydrodynamics.

8274 Chipchase GB 012701/79/0020 N
The role of insosital in calcium–gatting mechanisms in the insect
salivary gland.

8275 Chipchase GB 012701/79/0021 N
Ultrastructural studies of cholinergic systems in insect nerudus
tissue.

8276 Heslop GB 012701/79/0022 N
Mechanisms of hormonal control in insect secretory cells.

8277 Pratt; White GB 012801/00/0001 R
Isolation, identification and mode of action of insect hormones.

8278 Grove GB 012801/00/0003 R
Chemistry and mode of action of low molecular weight toxins
from entomogenous fungi.

8279 Blight; Wadhams GB 012801/00/0004 R
Isolation and identification of chemicals influencing the
behaviour of insect pests of protected crops.

8280 Brooks; Lewis GB 012801/00/0005 R
Nature, role and inhibition of enzymes involved in the
metabolism of pesticides and hormones.

8281 Bullard GB 012901/00/0002 R
Biochemistry of contractile proteins from insect and other
muscles.

8282 Harrison GB 030704/00/0014 R
Identification of viruses in relation to diseases of other crop
plants.

8283 Alphey GB 030705/00/0003 R
Chemical control of virus vector and other plant parasitic
nematodes.

8284 Trudgill GB 030705/00/0004 R
Feeding of Longidorus and Xiphinema spp. in relation to plant

response and virus transmission.

8285 Trudgill; Brown GB 030705/00/0007 R
Ecology of Longidorus and Xiphinema spp in relation to their role as plant pathogens.

8286 Woodford; Gordon GB 030705/00/0008 R
Ecology of aphids infesting potatoes, raspberry and bulbous ornamentals.

8287 Woodford; Robertson GB 030705/00/0009 R
Aphid feeding behaviour and ultrastructure in relation to virus transmission.

8288 Woodford; Aveyard GB 030705/77/0013 R
Control of aphids and virus diseases of potatoes, raspberries and bulbous ornamentals.

8289 Turl GB 030902/00/0321 R
Operation of aphid suction traps and identification of catches.

8290 Turl GB 030902/00/0401 R
Assessment of grasses as an overwintering source of aphids.

8291 Tucker; Cutler GB 030903/75/0103 R
Survey of pesticides usage against leatherjackets in SW Scotland.

8292 Cutler; Tucker GB 030903/76/0110 R
Survey of pesticide usage on glasshouse crops and hardy nursery stock.

8293 Henderson GB 030903/76/0406 R
Factors regulating rabbit populations.

8294 Henderson GB 030903/76/0407 R
Economic assessment of rabbit damage.

8295 Seymour GB 050310/74/0004 R
Taxonomy of noctuid larvae.

8296 Stroyan; Prior GB 050311/47/0001 R
Biotaxonomy of aphids.

8297 George; Tuppen GB 050312/76/0003 R
Pest monitoring scheme.

8298 Cotten; Roberts GB 050313/69/0001 R
Biology and ecology of Longidorus.

8299 Cotten; Roberts GB 050313/69/0002 R
Host range and population dynamics of Xiphinema.

8300 Mackintosh GB 060218/00/0004 R
Plant parasitic nematodes in relation to crop production.

8301 Foster GB 060309/00/0006 R
Seasonal and regional abundance of aphids.

8302 Stewart GB 060309/00/0007 R
Assessment of the effectiveness of seed dressings, seed bed and transplant treatments against crop pests.

8303 Feeney, A. IE 060500/76/1239 R
Monitoring of overwintering and spring sources of aphids together with summer populations and the relationship of these

to the occurence of virus yellows. Publications.

8304 Feeney, A.; McCafferty, C. IE 060500/76/1302 R
The biology and population dynamics of collembola, millepedes, symphylids and atomaria spp under the influence of insecticides. Herbicides. Publications.

8305 Whelan, J.; MacDonald, R.A. IE 120101/79/9174 N
The potential and realized impact of the rook (corvus frugil egus) on agriculture in county dublin, ireland.

8306 De Bellis, E. IT 011801/76/0005
Prove di lotta con nuovi insetticidi chimici a bassa tossicità contro le larve del Crittorrinco (Cryptorrhynchus lapathilarvae). Control trials with new low–toxicity insecticides against Cryptorrhynchus lapathi larvae. Publications.

8307 De Bellis, E. IT 011801/76/0006
Prove di lotta con nuovi insetticidi chimici a bassa tossicità contro le larve di Saperda maggiore (Saperda carcharias). Control trials with new low–toxicity insecticides against Saperda carcharias larvae. Publications.

8308 De Bellis, E. IT 011801/76/0012
Indagini sull'impiego di attrattivi sessuali nella lotta contro Rhyacionia buoliana. Investigations on the use of sex attraction in the control of Rhyacionia buoliana.

8309 Cavalcaselle, B.; De Bellis, E. IT 011801/76/0013
Valutazione dell'efficacia e della persistenza d'azione di alcuni preparati biologici contro le larve di Rhyacionia buoliana. Evaluation of the effectiveness and persistence of some biological insecticides against Rhyacionia buoliana larvae.

8310 Niccoli, A. IT 020400/72/0004
Ricerche sull'attività di feromoni e ormoni nei confronti di alcune specie fitofaghe di importanza agraria. Pheromones and hormones against some phytophagous insects of economic importance to agriculture. Publications.

8311 Zocchi, R. IT 020400/72/0585
Lotta integrata contro i nemici animali delle piante. Integrated control of animal pests of plants. Publications.

8312 Del Bene, G.; Melis, G. IT 020400/76/0005 N
Ricerche istologiche ed istochimiche sulla modalità di azione dell'insetticida Difluron. Histological and histochemical researches on the mode of action of Difluron insecticide..

8313 Del Bene, G. IT 020400/76/0006 N
Prove di lotta contro Aleyrodidi. Tests for white–flies control.

8314 Marinari Palmisano, A. IT 020400/77/0228 R
Dinamica delle popolazioni e caratterizzazione di patotipi di heterodera rostochiensis ed heterodera carotae in funzione di applicazioni di lotta integrata. Population dynamics and characterization of heterodera rostochiensis and heterodera carotae pathotypes in view of integrated control.

8315 Fenili, G.A. IT 020400/78/0005
Ricerche faunistiche e sistematiche sugli Imenotteri Sinfiti. Faunistic and systematic research of Hymenoptera Symphyta.

8316 Marinari Palmisano, A.; Ambrogioni, L.
IT 020400/79/0002 N
Rapporti ospite–parassita e dinamica delle popolazioni di

Heterodera carotae Jones,1950. **Host–parasite relations and population dynamics of Heterodera carotae Jones, 1950.**

8317 Tacconi, R.; Olimpieri, R.   IT 021100/75/0015 N
Prova di lotta contro heterodera schachtii schm., chaetocnema tibialis ill. e temnorrhinus mendicus gyll con geodisinfettanti di recente formulazione. **Control trial against heterodera schachtii schm., chaetocnema tibialis ill. and temnorrhinus mendicus gyll. with new soil disinfectants.**

8318 Lupo, V.   IT 023200/77/0220 R
Lotta integrata contro i nemici animali delle piante. **Integrated suppression of the animals pests of plants.**

8319 Martelli, G.   IT 040106/72/0107
Ricerche sulle virosi delle piante con particolare riguardo agli aspetti epidemiologici e fisiopatologici. **Research on plant viruses with specific reference to epidemiological and physiopathological aspects.** Publications.

8320 Roberti, D.   IT 040113/72/0546
Studio dei parassiti di insetti fitofagi viventi su piante coltivate e spontanee al fine della formulazione di programmi di applicazione lotta integrata. **Study on parasites of phytophagous insects living on cultivated and wild plants in view of formulating integrated control programmes.** Publications.

8321 Nuzzaci, G.   IT 040113/74/0001
Biologia degli Eriofidi (Acarina– Eriophyidea). **Eriophyd Biology (Acarina – Eriophyidea).** Publications.

8322 Parenzan, P.   IT 040113/75/0002
Biogeografia e dynamica delle popolazioni dei Lepidotteri. **Biogeography and Lepidoptera dinamic of population.** Publications.

8323 De Marzo, L.   IT 040113/77/0001
Studi sulle larve dei Coleotteri Ditiscidi. **Coleoptera Ditiscidae larvae studies.** Publications.

8324 Roberti, D.   IT 040113/77/0283 R
Studi di parassiti di insetti fitofagi e di interventi di lotta artificiale ai fini della formulazione di programmi di lotta integrata. **A study of the parasites of phytophagous insects and of artificial methods of suppression in view of integrated pest suppression programs.**

8325 Solinas, M.   IT 040113/77/0300 R
Ricerche sulla fecondità dei nemici animali, insetti ed acari, delle piante e dei prodotti agricoli conservati, al fine di formulare programmi di lotta biologica avanzata. **Research on the fecundity of animal, insect and acarian pests attacking plants and stocked agricultural produce in view of programming advanced biological control.**

8326 Principi, M.M.   IT 040205/72/0121
Lotta integrata contro i nemici animali delle piante. **Integrated control of animal pests of plants.** Publications.

8327 Mellini, E.   IT 040205/73/2147
Problemi del parassitismo negli insetti entomofagi utili. **Problems of parasitism in useful enthomophagous insects.**

8328 Mellini, E.   IT 040205/77/0238 R
Problemi del parassitismo negli insetti entomofagi utili. **The**

problems of parasitism in beneficial entomophagous insects.

8329 Principi, M.M.   IT 040205/77/0268 R
Problemi ecologici e tecnici della lotta integrata contro i nemici animali delle piante. **Ecological and technical problems of the integrated suppression of the animal pests of plants.**

8330 Domenichini, G.   IT 040402/72/0095
Lotta integrata contro i nemici animali delle piante. **Integrated control of animal pests of plants.** Publications.

8331 Domenichini, G.   IT 040402/77/0168 R
Problemi ecologici e tecnici della lotta integrata contro i nemici animali delle piante. **Ecological and technical problems of the integrated suppression of the animal pests of plants.**

8332 Domenichini, G.; Cravedi, P.; Ottolini, P.; Molinari, F.
          IT 040402/79/0001 N
Mezzi alternativi alla lotta chimica contro Aleyrodidae (Rhynchota Homoptera). **Possible alternatives substituting for chemicals in controlling Aleyrodidae (Rhynchota Homoptera).**

8333 Canonica, L.   IT 040604/77/0481 R
Derivati strutturali e funzionali degli ecdisoni, fotochimica degli ecdisoni e inibitori della loro biosintesi. **Structural and functional ecdysone derivates, photochemistry and bio–synthesis inhibitors of ecdysones.**

8334 Martelli, M.   IT 040607/72/0508
Lotta integrata contro i nemici animali delle piante. **Integrated control of animal pests of plants.** Publications.

8335 Martelli, M.   IT 040607/77/0233 R
Problemi ecologici e tecnici della lotta integrata contro i nemici animali delle piante coltivate. **Ecological and technical problems of the integrated control of cultivated plants animal pests.**

8336 Tremblay, E.   IT 040703/70/0001
Ricerche morfobiologiche sui braconidi parassiti di afidi. **Morphobiological researches on braconid wasp parasites oen–Morphobiological researches on braconid wasp parasites of aphids. f aphids.** Publications.

8337 Trembley, E.   IT 040703/72/0574
Lotta integrata contro i nemici animali delle piante. **Integrated control of animal pests of plants.** Publications.

8338 Iaccarino, F.M.   IT 040703/74/0002
Ricerche di sistematica e biologia sugli Homoptera aleyrodidae. **Taxonomic and biological researches on Homoptera aleyrodidae.**

8339 Scaramella, D.   IT 040703/76/0002
Ricerche sui roditori d'interesse agrario. **Researches on agriculturally harmful rodents.** Publications.

8340 Tremblay, E.   IT 040703/77/0308 R
Immunodiagnosi di insetti di interesse agrario. **Immunodiagnosis in insects of agronomic interest.**

8341 Viggiani, G.   IT 040703/77/0312 R
Lotta integrata contro i nemici animali delle piante. **Integrated control of animal plant pests.**

8342 Viggiani, G.; Mazzone, P.   IT 040703/79/0001 N

Introduzione di nemici naturali di Trialeurodes Vaporaziorum (Westw). **Introductions of natural enemies of Trialeurodes Vaporaziorum (Westw).**

8343 Mangoni, L.   IT 040746/77/0512
Sintesi di ecdisteroli e di sostanze modello ad azione ecdisonica. **Synthesis of ecdysterols and of ecdysoids.**

8344 Servadei, A.   IT 040805/72/0561
Lotta integrata contro i nemici animali delle piante. **Integrated control of animal pests of plants.** Publications.

8345 Servadei, A.   IT 040805/77/0299 R
Lotta integrata contro i nemici animali delle piante. **Integrated suppression of animal plant pests.**

8346 Zangheri, S.   IT 040805/77/0813
Controllo su basi ecologiche degli insetti dannosi in campo agricolo e forestale. **Ecological control of insects harmful to agriculture and forestry.**

8347 Ragusa, S.   IT 040903/77/0278 R
Controllo biologico degli acari fitofagi con artropodi utili. **Biological control of phytophagous acarids by beneficial arthropods.**

8348 Gendusio, P.   IT 040903/78/1060 N
Ricerche sui ditteri "fasiini" parassitoidi di "rincoti" eterotteri. **Research on "fasiini" Diptera parasitoids of "rincoti" Heteroptera.**

8349 Fiori, G.   IT 041008/77/0182 R
Problemi ecologici e tecnici della lotta integrata contro i nemici animali delle piante. **Technical and ecological problems in the integrated suppression of the animal pests of plants.**

8350 Lotti, G.   IT 041103/77/0219 R
Ricerche sull'impiego di bromuro di metile. **Research on the utilization of methyl bromide.**

8351 Rossi, R.   IT 041103/77/0289 N
Sintesi chimica e studio dell'attività biologica e delle possibilità di impiego di feromoni e di paraferomoni nella lotta contro gli insetti fitofagi e infestanti prodotti alimentari immagazzinati. **Chemical synthesis of pheromones and parapheromones. A study of their biological activity and their possible use in the suppression of phytophagous insects and stored food attacking insects.**

8352 Bibolini, C.   IT 041106/74/0518
Studi sulla fauna parassitologica del Dialeurodes citri ashm. hom. Leyr, diffusione di polynema striaticorne gir. hym. miramidae, parassita dofago di ceresa bubalus f. Ricerche sulla lotta integrata contro la saissetia oleae bern. hom. cocc., sull'olivo. Ricerca sulla fauna parassitologica delle specie di casside dannose alle piante coltivate. **Studies on parasitological fauna of dialeurodes citri ashm. hom. leyr, diffusion of polynema striaticorne gir. hym. miramidae, dophagous parasite of ceresa bubalus f. Research on integrated control of saissetia oleae bern. hom. cocc., on olive tree. Research on parasitological fauna of cassidae injurious to crops.**

8353 Crovetti, A.; Quaglia, F.   IT 041106/79/0001 N
L'impiego di feromoni di sintesi e naturali contro la Bega africana (Epichoristodes acerbella (Walk.)) e quella europea (Cacoecimorpha pronubana Hübn.) in serra. **Utilisation of natural and synthetic pheromones for carnation moths (Epichoristodes acerbella (Walk.) and Cacoecimorpha pronubana Hübn.) control in greenhouses.**

8354 Goidanich, A.   IT 041206/72/0477
Lotta integrata contro i nemici animali delle piante. **Integrated control of animal pests of plants.** Publications.

8355 Goidanich, A.   IT 041206/77/0200 R
Problemi ecologici e tecnici della lotta integrata contro i nemici animali delle piante. **Ecological and technical problems connected with the integrated control of animal pests of plants.**

8356 Vidano, C.   IT 041206/78/1119 N
Lotta integrata contro i nemici animali delle piante. **Integrated control of the animal pests of plants.**

8357 Degani, J.   IT 041234/77/0488
Sintesi di iuvenoidi e sinergici di insetticidi. **Synthesis of juvenoids and of insecticide synergetics.**

8358 Prota, R.   IT 041314/77/0269
Ricerche sui problemi ecologici e tecnici della lotta integrata contro i nemici animali delle piante. **Research on the ecological and technical problems of the integrated suppression of the animal pests of plants.**

8359 Bullini, L.   IT 041604/78/1039 N
Ricerche sull'ecologia e sull'etologia di imenotteri eumenidi parassiti di larve di lepidotteri. **Research on the ecology and etology of hymenoptera eumenides parasites of lepidopterous larvae.**

8360 Springhetti, A.   IT 042003/77/0302
Azione di ormoni, feromoni e sostanze ritardanti la crescita sulla differenziazione e la riproduzione degli insetti. **The action of hormones, pheromones and growth inhibiting substances on insect differentiation and reproduction.**

8361 Lamberti, F.   IT 060600/70/0033
Ricerche sulla sistematica, morfologia e distribuzione geografica dei nematodi fitoparassiti. Distribuzione geografica dei nematodi appartenenti alla famiglia Longidoridae. **Systematics, morphology and geographical distribution of plant parasitic nematodes. Geographical distribution of nematodes belonging to the family Longidoridae.** Publications.

8362 Lamberti, F.   IT 060600/70/0034
Ricerche sulla sistematica, morfologia e distribuzione geografica dei nematodi fitoparassiti. Sistematica dei generi Longidorus, Paralongidorus e Xiphinema. **Systematics, morphology and geographical distribution of plant parasitic nematodes. Systematics of the genera Longidorus, Paralongidorus and Xiphinema.** Publications.

8363 Inserra, R.   IT 060600/70/0038
Ricerche su biologia e patogenicità di nematodi fitoparassiti. Fluttuazioni stagionali delle popolazioni di Tylenchulus semipenetrans nell' Italia meridionale e insulare. **Biology and pathogenicity of plant parasitic nematodes. Seasonal fluctuations of populations of Tylenchulus semipenetrans in Southern and Insular Italy.** Publications.

8364 Vovlas, N.   IT 060600/71/0036
Ricerche sulla sistematica, morfologia e distribuzione geografica dei nematodi fitoparassiti. Morfologia di specie dei

generi Heterodera e Meloidogyne. **Systematics, morphology and geographical distribution of plant parasitic nematodes. Morphology of species of the genera Heterodera and Meloidogyne.** Publications.

8365 Di Vito, M.　　　　　　　IT 060600/72/0039
Ricerche su biologia e patogenicità di nematodi fitoparassiti. Effetto del clima sulla biologia di specie del genere Xiphinema. **Biology and pathogenicity of plant parasitic nematodes. Effect of climate on the biology of species of the genus Xiphinema.**

8366 Coiro, M.　　　　　　　　IT 060600/72/0043
Ricerche su biologia e patogenicità di nematodi fitoparassiti. Effetto delle rotazioni sulle popolazioni di nematodi fitoparassiti. **Biology and pathogenicity of plant parasitic nematodes. Effect of rotations on the populations of plant parasitic nematodes.** Publications.

8367 Renzoni, G.　　　　　　　IT 060600/72/0058
Nematocidi. Effetti collaterali dei nematocidi sulle piante. **Nematicides. Side effects of nematicides on plants.** Publications.

8368 Bleve Zacheo, T.　　　　　IT 060600/73/0035
Ricerche sulla sistematica, morfologia e distribuzione geografica dei nematodi fitoparassiti. Morfologia e ultrastruttura di specie dei generi Longidorus e Xiphinema. **Systematics, morphology and geographical distribution of plant parasitic nematodes. Morphology and ultrastructure of species of the genera Longidorus and Xiphinema.** Publications.

8369 Coiro, M.　　　　　　　　IT 060600/73/0040
Ricerche su biologia e patogenicità di nematodi fitoparassiti. Influenza del tipo di terreno sulla riproduzione di Xiphinema index. **Biology and pathogenicity of plant parasitic nematodes. Influence of the soil type on reproduction of Xiphinema index.**

8370 Zacheo, G.　　　　　　　IT 060600/73/0045
Fisiologia e biochimismo di nematodi fitoparassiti e di piante da essi attaccate. Studi di chemiotassonomia sui generi Meloidogyne e Heterodera. **Physiology and biochemistry of plant parasitic nematodes and the plants they attack. Chemiotaxonomy of the genera Meloidogyne and Heretodera.**

8371 Basile, M.　　　　　　　IT 060600/73/0057
Nematocidi. Dinamica e persistenza dei nematocidi nel terreno e in parti eduli di piante. **Nematicides. Dynamics and persistence of nematicides in soil and edible parts of plants.**

8372 Lamberti, F.　　　　　　IT 060600/74/0037
Ricerche sulla sistematica, morfologia e distribuzione geografica dei nematodi fitoparassiti. Indagini citologiche su alcune popolazioni mediterranee di Ditylenchus dipsaci. **Systematics, morphology and geographical distribution of plant parasitic nematodes. Cytological investigations on Mediterranean populations of Ditylenchus dipsaci.** Publications.

8373 Greco, N.　　　　　　　IT 060600/74/0042
Ricerche su biologia e patogenicità di nematodi fitoparassiti. Comportamento biologico di diversi isolati geografici di Meloidogyne spp. **Biology and pathogenicity of plant parasitic nematodes. Biological characters of different geographical isolates of Meloidogyne spp.**

8374 Lamberti, F.　　　　　　IT 060600/77/0908

Prove di laboratorio e di campo per la lotta chimica contro i nematodi parassiti delle colture. **Laboratory and field tests in view of the chemical control of nematodes infesting cultures.**

8375 Lamberti, F.　　　　　　IT 060600/77/0911
Indagini sulla distribuzione geografica e studi sulla sistematica e morfologia dei nematodi fitoparassiti. **Research on the geographical distribution, studies on the behaviour and morphology of plant infesting nematodes.**

8376 Lamberti, F.　　　　　　IT 060600/77/0912
Biologia e patogenicità dei nematodi trovati associati con le colture di interesse agrario. **Biology and pathogenecity of nematodes found in agricultural crops.**

8377 Foschi, S.　　　　　　　IT 061800/77/1010
Prove di lotta contro la Zeuzera pyrina L. (Lepidopt. cossidae) in base alla sua etologia. **Trials of Zeuzera pyrina L. (Lepidopt. cossidae) control based on its ethology.**

8378 Ferrero, A.　　　　　　IT 121400/77/0494
Impiego di diserbanti in post–emergenza, impiego di fungicidi ed insetticidi, concia dei semi e disinfestazione del terreno. **Use of weed–killers in post–emergency, use of fungicides and insecticides, seed treatment and soil pests control.**

8379 Harten, A. van　　　　　NL 010108/29/7678
Waarnemen en interpreteren van bladluizenvluchten in verband met de epidemiologie van schadelijke plantevirussen. **Monitoring of aphid flights in relation to the epidemiology of plant virus diseases.** Publications.

8380 Nijveldt, W.C.　　　　　NL 010108/49/7680
Identificatie en bestrijding van galmuggen, schadelijk in landen tuinbouwgewassen. **Identification and control of gall–midges, noxious to agricultural and horticultural crops.**

8381 Seinhorst, J.W.　　　　NL 010108/59/1051
Dynamica van het systeem aaltje – plant. **Dynamics of the nematode – plant system.** Publications.

8382 Mosch, W.H.M.　　　　　NL 010108/71/3107
Ontwikkeling en toepassing van chromatografie en electroforese ten behoeve van fytopathologisch en entomologisch onderzoek. **Development and application of chromatography and electroforese for phytopathological and entomological research.** Publications.

8383 Nijveldt, W.　　　　　　NL 010108/75/6392
De betekenis van roofvijanden voor de bestrijding van schadelijke insekten in akkerbouwgewassen. **The inportance of predators for the control of noxious insects in arable crops.** Publications.

8384 Feldmann, A.M.　　　　　NL 010110/71/3589
Genetische invloeden van straling in de spintmijt Tetranychus urticae C.L. Koch. **Genetic control of the two–spotted spidermite Tetranychus urticae Koch.** Publications.

8385 Driest, J.Ph. van　　　　NL 010207/78/9007 N
Bestrijding van aardrupsen, Agrotis segetum (Denis and Schiff). **Control of cutworms, Agrotis segetum (Denis and Schiff).**

8386 Rossem, G. van　　　　　NL 010300/44/9081 N
Taxonomisch en nomenclatorisch onderzoek over economisch

belangrijke insekten, mijten en andere Arthropoden.
Taxonomy and nomenclature of economically important insects, mites and other Arthropods.

8387 Maas, P.W.Th. NL 010300/51/8609 N
Taxonomisch onderzoek bij planteparasitaire nematoden.
Taxonomic research on plant parasitic nematodes.

8388 Maas, P.W.Th. NL 010300/51/8610 N
Ontwikkeling nematologische onderzoektechnieken.
Development nematological research technics.

8389 Maas, P.W.Th. NL 010300/51/9093 N
Inventarisatie van in Nederland voorkomende
planteparasitaire nematoden. **Inventory of plant–parasitic nematodes in the Netherlands.**

8390 Rossem, G. van NL 010300/62/9095 N
Nevenwerkingen van bestrijdingsmiddelen in de grond van
akkerbouwland op de micro–arthropoden. **Side effects of soil pesticides on micro–arthropods on arable land.**

8391 Bruin, Th. de NL 010300/75/9079 N
Langjarige grondontsmettingproeven; de invloed nagaan van
jaarlijkse grondontsmetting op bodemorganismen, grond en
gewas. **Long–term experiments on soil disinfectation; effect on soil–borne organisms, soil and crop.**

8392 Smissaert, H.R. NL 010301/71/3456
Mechanisme van remming en substraathydrolyse van
acetylcholinesterase als basis voor rationele ontwikkeling van
cholinesteraseremmers. **Mechanism of inhibition and substratehydrolysis of acetylcholinesterase as a base for rational development of cholinesterase–inhibitors.** Publications.

8393 Oppenoorth, F.J. NL 010301/73/3760
Onderzoek over selectieve toxiciteitsfactoren bij insekten.
Research on factors of selective toxicity in insects.
Publications.

8394 Voerman, S. NL 010301/73/8876 N
Synthese en onderzoek van sex–feromonen en attractantia van
insekten. **Synthesis and investigation of insect sex pheromones and attractants.** Publications.

8395 Smissaert, H.R. NL 010301/76/6871
Onderzoek naar het werkingsmechanisme van PH 60–42.
Research in the mode of action of PH 60–42.

8396 Kraan, C. van der NL 010301/76/6872
Onderzoek over het gedrag van insekten bij toepassing van sex
feromonen. **Research on the behaviour of insects as reaction upon the application of sex pheromones.**

8397 Oppenoorth, F.J. NL 010301/77/7887
Onderzoek over het werkingsmechanisme van diflubenzuron
en verwante verbindingen. **Research on the mode of action of diflubenzuron and related compounds.** Publications.

8398 Nollen, H.M. NL 010302/69/2859
Grondbewerkingen na grondontsmetting. **Tillage after soil disinfection.** Publications.

8399 Nollen, H.M. NL 010302/70/1382
Langjarige DD grondontsmettingsproeven. **Long–term experiments on soil disinfestation with DD.**

8400 Stortenbeker, C.W.; Schaeffner, B. NL 010602/64/4556
Voorkomen of beperken van de schade door vogels.
Prevention and restriction of damage by birds (damage to crops and risks for aeroplanes).

8401 Kuchlein, J.H. NL 020013/75/6784
Oecologie en classificatie van N.W.–Europese Pyraliden.
Ecology and classification of the Pyralid moths occuring in N.W.–Europe. Publications.

8402 Dinther, J. van NL 020014/75/6362
Voedingsgedrag van de zuidamerikaanse roofwants Podisus
sagitta. **Feeding behaviour in the predacious bug Podisus sagitta, a neotropical species.**

8403 Schooneveld, H. NL 020014/77/8642 N
Insektenbestrijding door induktie van hormonale afwijkingen.
Insect control by hormonally induced abnormalities.

8404 Dinther, J. van NL 020014/77/8839 N
Evaluatie van effekten van ontginning op onkruiden
bodemorganismen (aaltjes) en fytofage insekten bij de teelt
van 1–jarige gewassen op niet–bevloeibare gronden.
Evaluation of effects of defnestation on wild flora (weeds) and fauna (eelworms, insects) in connection with the permanent cultivation of annual crops on non–irrigated soil.

8405 Visser, J.H. NL 020014/78/8640 N
De olfactorische oriëntatie van insekten. **Olfactory orientation of insects.**

8406 Wilde, J. de NL 020014/78/8840 N
Geografische differentiatie in de fotoperiodiciteit en
saisonaliteit bij de Coloradokever. **Geographic differentiation in photoperiodism and seasonality in the Colorado potato beetle.**

8407 Dieleman, F.L. NL 020014/78/8841 N
Voedselopname en verteringssysteem van galmuglarven in
verband met abnormale groeiverschijnselen van planten: 1.
structuur en functie van larvale speekselklieren. **Feeding en digestive system of gallmidgelarvae in relation to abnormal growth in plants: 1. structure and function of larval salivary glands.**

8408 Schooneveld, H. NL 020014/78/8842 N
Neuro–endocriene regulatie processen in de Coloradokever.
Control of neuroendocrine processes in the Colorado potato beetle.

8409 Schoonhoven, L.M. NL 020017/76/7308
De rol van chemische factoren bij het eileggedrag van vlinders.
The role of chemical factors in oviposition behaviour of butterflies. Publications.

8410 Schoonhoven, L.M. NL 020017/76/8508
Voedselplantbinding bij bladluizen. **Host–plant relationships in aphids.** Publications.

8411 Schoonhoven, L.M. NL 020017/77/8510
Evolutie van smaakzintuigen bij insekten. **Evolution of contact chemoreceptors in insects.** Publications.

8412 Schoonhoven, L.M. NL 020017/77/8511
Voedselkeuze–gedrag in relatie tot zintuigveranderingen en

nutritieve aanpassingen bij insekten. **Food selection behaviour in relation to sensory adaptation and metabolic adaptation in insects.** Publications.

8413 Gommers, F.J.   NL 020037/71/4503
Chemisch en biochemisch onderzoek aan nematode/plant relaties. **Chemistry and biochemistry research of nematode–plant relationships.** Publications.

8414 Hoekstra, H.   NL 020037/74/4871
Relaties tussen nematoden en andere organismen: complexe ziekten, biologische bestrijding van en door aaltjes en andere correlaties. **Relationships between nematodes and other organisms: complex diseases, biological control of and by nematodes and other interrelationships.** Publications.

8415 Simons, W.   NL 020037/75/6234
Bestrijding van insekten met entomofage nematoden. **Biological control of insects by nematodes.** Publications.

8416 Bunt, J.A.   NL 020037/76/6813
Phytofarmaceutische aspecten van de toepassing van nematiciden. **Fytopharmaceutical aspects of the application of nematicides.** Publications.

8417 Loof, P.A.A.   NL 020037/78/8533
Taxonomisch en morfologisch onderzoek aan Diplogasteridae. **Taxonomy and morphology of Diplogasteridae.**

8418 Loof, P.A.A.   NL 020037/78/8534
Taxonomisch en morfologisch onderzoek aan virus–overdragende nematoden. **Taxonomy and morphology of virus transmitting nematodes.**

8419 Loof, P.A.A.   NL 020037/78/8535
Surveywerk van planteparasitaire nematoden ten behoeve van ontwikkelingslanden. **Survey of plant parasitic nematodes for developing countries.**

8420 Loof, P.A.A.   NL 020037/78/8536
Taxonomisch en morfologisch onderzoek aan Criconematoidea. **Taxonomy and morphology of Criconematoidea.**

8421 Sanders, H.   NL 020037/78/8837 N
Taxonomisch onderzoek aan het geslacht Meloidogyne. **Taxonomy of Meloidogyne.**

8422 Gommers, F.J.   NL 020037/78/8838 N
Fotochemie van natuurlijk voorkomende nematiciden. **Photochemistry of naturally occurring nematicides.**

8423 Beek, C.P. van der   NL 020061/75/6868
Isolatie en karakterisering van het messenger–RNA dat codeert voor polyedereiwit. **Isolation and characterization of the messenger RNA wich codes for polyhedral protein.** Publications.

8424 Vlak, J.M.   NL 020061/76/6870
De vermenigvuldiging van kernpolyedervirussen in insektencellen. **Multiplication of nuclear polyhedrosis viruses in cultured insect cells.** Publications.

8425 Helle, W.   NL 040001/61/6729 R
Genetische variabiliteit van spintmijten, Tetranychidae. **Genetic variability in spider mites, Tetranychidae.**

Publications.

8426 Veerman, A.   NL 040001/73/3789 R
Inductie van dia–pauze en reactivatie bij de spintmijt Tetranychus urticae. **Induction of diapause and reactivation in the spidermite Tetranychus urticae.** Publications.

8427 Helle, W.   NL 040001/73/3790
Ontwikkeling van resistentie tegen organische fosforverbindingen bij de roofmijt Phytoseiulus persimilis. **The induction of resistance against organic phosphorus compounds in the predacious mite Phytoseiulus persimilis.** Publications.

8428 Geest, L.P.S. van der   NL 040001/74/7895
Toepassingsmogelijkheden van Entomophtora soorten voor de bestrijding van bladluizen. **Possibilities of Entomophtora species for the control of aphids.**

8429 Duym, J.   NL 040007/70/8430
Onderzoek naar de bestrijding van plagen (vnl. insekten) met chemische middelen. **Chemical control of insects and rodents in crops and plantations.**

8430 Schulten, G.G.M.   NL 040012/73/3790
Ontwikkeling van resistentie tegen organische fosforverbindingen bij de roofmijt Phytoseiulus persimilis. **The induction of resistance against organic phosphorous compounds in the predacious mite Phytoseiulus persimilis.** Publications.

8431 Ritter, F.J.; Persoons, C.J.   NL 050210/70/3659
Ontwikkeling van signaalstoffen voor landbouwinsekten. **Development of pheromones for agricultural pests.** Publications.

Cereals in general (B 3100)

See also 7959, 7964, 7976, 8594, 9305, 9325, 9528, 15855

8432 Schmutterer, H.; Sagenmüller, A.   DE 129184/73/0004
Wirkungen systemischer Fungizide auf die Reproduktivität und Mortalität von Getreideblattläusen. **Effects of systemic fungicides on reproductivity and mortality of cereal aphids.**

8433 Kohsiek, H.; Rohlfing, H.   DE 215110/75/0001
Filterwirkung von Pflanzen des Getreidebaus bei der Applikation von flüssigen Pflanzenschutzmitteln. **Filter effect exercised by cereal plants during the application of pesticide sprays.**

8434 Basedow, T.   DE 215120/78/0007 N
Untersuchungen zur Anfälligkeit von Getreidesorten gegenüber Schadinsekten. **Studies on the susceptibility of cereal varieties to the attack by insect pests.**

8435 Bode, E.   DE 215170/75/0003
Untersuchungen zur Populationsdynamik von Getreideblattläusen mit besonderer Berücksichtigung der Wirkung natürlicher Blattlausvertilger. **Investigation on population dynamics of cereal aphids with special regard to the effect of natural aphicides.**

8436 Basedow, T.   DE 215180/77/0007
Untersuchungen zur Prognose des Auftretens der Getreideblattläuse. **Studies on forecasting of incidence of cereal aphids.**

8437 Rumpenhorst, H.J.; Steudel, W.   DE 215190/71/4004
Versuche zur Populationsdynamik und zum Auftreten
biologischer Rassen von Heterodera avenae bei Daueranbau
von anfälligen und resistenten Getreidesorten. **Experiments on
the population dynamics and the occurrence of biological races
of Heterodera avenae when susceptible and resistant cereals are
grown continuously.**

8438 Weischer, B.   DE 215190/78/0002 N
Populationsdynamik pflanzenparasitärer Nematoden in
verschiedenen Getreidefruchtfolgen. **Population dynamics of
plant–parasitic nematodes in various rotations of cereals.**

8439 Neuffer, G.   DE 501055/78/0001 N
Biologische Bekämpfung des Maiszünslers – Ostrinia nubilalis
– mit dem Eiparasiten Trichogramma evanescens – Hym.,
Chalcidoidea –. **Biological control of the corn moth Ostrinia
nubilalis with the egg–parasite Trichogramma evanescens –
Hym., Chalcidoidea –.** Publications.

8440 Diercks, R.; Obst, A.   DE 502051/74/0001
Pflanzenschutzprobleme in Getreidemonokultur. **Crop
protection in cereal monocultures.**

8441 Lauenstein, G.   DE 507250/75/0002
Untersuchungen über die Verbreitung von
Getreidenematoden – Heterodera avenae – Pathotypen von
Heterodera avenae und anderen zystenbildenden Nematoden.
**Investigations into the spread of Heterodera avenae and its
pathotypes and of other cyst nematodes injurious to cereals.**

8442 Nøddegaard, E.; Hansen, K.E.   DK 010116/78/0024 N
Insekticiders virkning mod bladlus i korn. **Effect of insecticides
on aphids in cereals.**

8443 Moreau, J.P.; Warin, S.; Boulay, Ch.   FR 010106/65/5244
Relations céréales (y compris maïs) et pucerons. **Small grains
(corn included) – Aphides Relationships.**

8444 Chambon, J.P.; d'Aguilar, J.; Genestier, G.
FR 010106/74/5252
Epidémiologie des Agromyza mineurs de feuilles de céréales –
Lutte. **Epidemiology on Agromyza leafmineurs on cereales –
Control.**

8445 Della, Giustina, W.   FR 010106/74/5253
Etude sur les cicadelles des céréales. **Study on Cereals
Auchenorhyncha.**

8446 Caubel, G.; Rivoal, R.; Mugniery, D.   FR 011306/66/5113
Caractérisation des races de nématodes des céréales et de la
pomme de terre. **Characterisation of "races" of Nematodes
harmful to cereals and potato.**

8447 Brunel, E.   FR 011306/71/5122
Etude des diptères des céréales dans l'Ouest de la France.
Study of the diptera harmful to cereals in the west of France.

8448 Caubel, G.; Rivoal, R.   FR 011306/74/5126
Influence des rotations céréalières intensives sur les
nématodes. **Effects of intensive cereal rotations on harmful
Nematodes.**

8449 Brunel, E.; Rivoal, R.; Caubel, G.; Dedryver, C.
FR 011306/74/5127
Conséquences de la monoculture céréalière sur les

dépredateurs animaux. **Conséquences of the cereal monoculture
on animal pests.**

8450 Dedryver, C.; Robert, Y.   FR 011306/75/5129
Conditions d'implantation et de développement des pucerons
des céréales. **Factors conditioning the establishment and the
development of aphids harmful to cereals.**

8451 Wyatt; Sunderland   GB 011108/00/0019 R
**Effect of modern farming practice on natural control of cereal
aphids.**

8452 Smith   GB 011514/76/0008 R
Cereal aphids and spread of barley yellow dwarf virus.

8453 Griffiths; Scott   GB 012005/00/0033 R
**Chemicals and seed treatments for control of cereal pests
(especially wheat bulb fly).**

8454 Cook   GB 012101/00/0012 R
**Expression and genetic control of nematode resistance in
cereals.**

8455 Turl   GB 030902/69/0402 R
Analysis of parasitic Hymenoptera from cereal aphids.

8456 Researcher not indicated   GB 050121/00/0005 R
Cereals: pest control.

8457 Researcher not indicated   GB 050161/00/0005 R
Cereals: pest control.

8458 Blasdale   GB 060218/00/0006 R
Assessment of crop losses due to rabbits.

8459 Minks, A.K.   NL 010108/73/3952
Geïntegreerde bestrijding van bladluizen in
akkerbouwgewassen, in het bijzonder in granen. **Integrated
control of aphids in arable crops, particular cereals.**

8460 Kort, J.   NL 010300/73/1393 N
Toetsing van graanrassen op resistentie tegen het
graancystenaaltje (Heterodera avenae). **Testing of cereals for
resistance to the cereal cyst nematode.**

8461 Ankersmit, G.W.   NL 020014/75/6363
Populatie dynamica bij graanbladluizen (Sitobion avenae,
Rhopalosiphum padi en Metopolophium dirhodum).
**Population dynamics in cereal aphides (Stitobion avenae,
Rhopalosiphum padi and Metopolophium dirhodum).**

8462 Rabbinge, R.   NL 020054/76/6859
Populatiedynamica van bladluizen in granen. **Population
dynamics of cereal aphids.** Publications.

Barley (B 3110)

See also 15856

8463 McKelvie   GB 060216/00/0004 R
Barley stunt disorder.

Maize (B 3120)

See also 8480, 9420

8464 Ohnesorge, B.; Schäufele, D.; Weinbuch, H.
DE 144550/77/0002
Einfluss der Witterung, insbesondere der Temperatur, auf das Schlüpfen und den Flug des Maiszünslers. **Effects of weather – especially of temperature – on hatching and flight of the European corn borer.**

8465 Schütte, F.
DE 215180/75/0002
Untersuchungen zur Populationsdynamik der Fritfliege und zur Auswirkung der Frassschäden an Maispflanzen. **Studies on population dynamics of the frit fly and on effects of feeding damages on maize plants.**

8466 Anglade, P.; Voegele, J.; Stengel, P. FR 010106/73/5250
Mise au point de la lutte intégrée contre la pyrale du Maïs. **Integrated control for Corn Borer.**

8467 Voegele, J.; Durand, Y.; Anglade, P.; Stockel, J.
FR 010503/74/5048
Utilisation des Trichogrammes contre la pyrale du maïs. **The use of Trichogramma in biological control of the European corn borer.**

8468 Stengel, M.; Schubert, G. FR 010904/75/5026
Lutte biologique contre la pyrale du maïs à l'aide de Trichogrammes. **Biological control of the European corn borer with Trichogramma.**

8469 Stengel, M.; Schubert, G. FR 010904/77/5027
Dynamique des populations de la Pyrale du Maïs et prévisions de dégâts. **Dynamics of European Corn Borer populations and crop loss forecasting.**

8470 Bruin, T. de NL 010302/72/3642
Bestrijding van de fritvlieg in maïs. **Control of Oscinella frit in maize.**

Oats (B 3130)

See also 7965

8471 Ohnesorge, B.; Harmuth, P. DE 144550/75/0001
Mechanismen der Resistenz von Kulturhafersorten gegen das Getreidezystenälchen Heterodera avenae Wollenweber 1924. **Resistance of oat varieties to Heterodera avenae Woll.: mechanism of resistance.**

8472 Rivoal, R. FR 011306/73/5125
Déterminisme de l'adaptation des cycles biologiques des races géographiques d'Heterodera avenae à leur milieu. **Determinism of the adaptation of the life cycles of the Heterodera avenae geographical races to their environment.**

Rice (B 3140)

8473 Marie, R.; Grillard, M. FR 011201/74/0288
Lutte biologique contre la pyrale du riz (chilo suppressalis). **Biological control of the rice stem borer.**

8474 Moletti, M.; Villa, B. IT 011400/79/0009 N
Controllo dell'afide vettore del virus del giallume del riso, Rhopalosiphum padi, mediante l'impiego di insetticidi granulari sistemici. **Control of the aphid Rhopalosiphum padi vector of the "yellowing" rice virus by granular systemic insecticides.**

Sorghum (B 3160)

See also 8480, 9420

Wheat (B 3170)

See also 2115, 9420

8475 Basedow, T. DE 215180/70/5001
Untersuchungen zur Populationsdynamik der Weizengallmücken. **Studies on population dynamics of the wheat blossom midges.**

8476 Chevin, H. FR 010106/75/5261
Etude biologique du Cèphe du blé. **Biological study of european wheat–stern sawfly.**

8477 Cooper GB 030903/74/0101 R
Assessment of population dynamics of wheat bulb fly to evaluate need for chemical control.

8478 George; Tuppen GB 050312/76/0001 R
Cereal aphids and crop losses.

8479 McKinlay GB 060107/00/0003 R
Wheat bulb fly in Scotland; control by optimal pesticide usage, safeguarding the environment.

8480 Suss, L. IT 040607/77/0527 R
Censimento per fitofagi di frumento, mais, sorgo, valutazione della loro dannosità e ricerca di nuovi fitofarmaci non inquinanti per limitare l'attività. **A census of the phytophagous population in wheat, maize, sorgo, evaluation of their noxiousness; finding new non polluting pesticides to control their activity.**

8481 Genduso, P. IT 040903/77/0192 R
Ricerche ecologiche sui pentatomidi del frumento e loro simbionti. **Ecological research on wheat pentatomids and their symbionts.**

Olive (B 3220)

See also 8690, 8777, 8778

8482 Arambourg, Y.; Pralavorio, R. Mme FR 010503/73/5043
Parasites de Dacus oleae. **Dacus oleae parasites.**

8483 Onillon, J.C.; Arambourg, Y. FR 010503/75/5053
Bioécologie du Psylle de l'Olivier (Euphyllura olivina C.). **Bioecology of Euphyllura olivina C.**

8484 Poitout, S.; Bues, R. FR 010602/69/1753
Etude du développement estival de Mamestra Oleracea et M. Brassicae. Diapause–Variabilité du caractère–Existence d'écotypes. **Study of estival development of Mamestra Oleracea and M. Brassicae–Diapause variability, ecotypes.**

8485 Pegazzano, F.; Castagnoli, M. IT 020400/70/0005
Ricerche sulla acarofauna dell'olivo. **Mite fauna associated with olive trees in Italy.** Publications.

8486 Castagnoli, M. IT 020400/76/0001
Dinamica delle popolazioni di Acari Eriofidi dell'olivo. **Population dynamics of Eriophyids mites of olive–trees.**

8487 Zocchi, R.   IT 020400/78/1037 N
Indagini sulla Saissetia oleae olivi ed i suoi entomofagi.
Indagini sui nematodi dei generi Globodera ed Heterodera.
**Research on Saissetia oleae olive trees and their
entomophagous hosts. Research on nematodes of Gobodera
and Heterodera genera.**

8488 Arras, A.   IT 024000/76/0002 N
Studio delle fluttuazioni delle popolazioni di Saissetia oleae
Bern e dei suoi parassiti.. **Studies of Saissetia oleae Bern and its
parasites populations, fluctuations.**

8489 Arras, G.   IT 024000/76/0003 N
Studio delle fluttuazioni delle popolazioni di Dacus oleae
Gmelin e dei suoi parassiti.. **Studies of Dacus oleae Gmelin and
its parasites populations, fluctuations.**

8490 Solinas, M.; Monaco, R.; Nuzzaci, G.   IT 040113/75/0001
Studi bio–ecologici sul Dacus oleae e sulla Rhagoletis cerasi.
Bio–ecological research on Dacus oleae and Rhagoletis cerasi.

8491 Monaco, R.   IT 040113/77/0814 R
Ricerche biologiche sul Dacus oleae. Ricerche biologiche sulla
mosca delle ciliegie. Comparazione insetticidi più attrattivi,
individuazione di nuove miscele attrattive. **Biological research
on Dacus oleae. Biological research on the cherry fly.
Comparison of attractants, discovery of new mixtures of
attractants.**

8492 Jommi, G.C.   IT 040604/77/0504 R
Feromone sessuale del dacus oliae, estrazione identificazione
natura chimica, messa a punto della sintesi. **Sexual pheromone
of dacus oleae, extraction and chemical identification,
perfecting synthesis.**

8493 Fimiani, P.   IT 040703/70/0002
Ricerche bioecologiche sulla mosca delle olive (Dacus oleae) e
i suoi nemici naturali. **Bioecological researches on olive fly
(Dacus oleae) and its natural enemies.** Publications.

8494 Girolami, V.   IT 040805/78/1138 N
Indagine sulle sostanze chimiche naturali che influiscono sul
comportamento del Dacus oleae GMELIN e ricerche
sull'allevamento del dittero. **Research on the natural chemical
compounds influencing the behaviour of Dacus oleae GMELIN
and research on diptera breeding.**

8495 Fiori, G.   IT 041008/77/0496 R
Campionamento adulti dacus con cartelle cromotropiche per
stimare dinamica e livello numerico popolazione di campo.
Campionamento olive per valutare livello infestazione,
determinazione soglia economica. **Chromotropic chart
sampling of adult dacus to assess the dynamics and density of
the field population. Sampling of olives to assess the grade of
infestation, definition of rentability.**

8496 Quaglia, F.   IT 041106/77/0274 R
Studio morfo–biologico della P.oleae costa e dell'E. olivina
berle. et silv. rhynchota–cocc. in Toscana e Liguria e ricerche
sul complesso dei loro predatori e parassiti, al fine del controllo
biologico e integrato. **Morphological and biological study of P.
oleae costa and E. olivina berle. et silv. rhynchota–cocc. in
Tuscany and Liguria, research on all their predators and
parasites in view of an integrated biological control.**

8497 Quaglia, F.   IT 041106/78/1092 N

Ricerche sull'entomofauna dell'olivo in Liguria e Toscana.
Ricerche sull'artropodofauna delle colture protette in Liguria,
Toscana e Lazio. **Research on the insect population of the olive
tree in Liguria and Tuscany. Research on the arthropod
population of the protected cultures in Liguria, Tuscany and
Latium.**

8498 Crovetti, A.   IT 041106/78/1131 N
Ricerche in pieno campo ed in laboratorio su alcuni fattori
abiotici limitanti lo sviluppo ed il pullulamento della mosca
delle olive. **Openfield and laboratory research on certain
abiotic factors interfering with the growth and multiplication of
the olive fly.**

8499 Rossi, R.   IT 041130/78/1143 N
Sintesi di attraenti del Dacus oleae e della Ceratitis capitata.
Synthesis of Dacus oleae and Ceratitis capitata attractants.

8500 Delrio, G.   IT 041301/78/1134 N
Definizione della soglia economica di intervento contro il
Dacus oleae GMEL. Prove di lotta con esche proteiche
avvelenate contro il Dacus oleae GMEL. **Definition of the
economic threshhold for the control of Dacus oleae GMEL.
Control tests with poisoned protein baits for Dacus oleae
GMEL.**

8501 Lamberti, F.   IT 060600/70/0052
Lotta chimica contro i nematodi fitoparassiti. Meloidogyne
spp. su olivo, tabacco e colture ortive e floreali. **Chemical
control of plant parasitic nematodes. Meloidogyne spp. on olive,
tobacco and vegetable and flower crops.** Publications.

8502 Fabbri, L.   IT 121200/77/0492
Esame di esche attrattive proteiche verso il dacus oleae. Rilievi
dell'effetto di trattamenti dachicidi sull'entomofauna con
raccolta degli insetti colpiti, elaborazione di una trappola con
impiego di sostanze coloranti applicate con bombola spray. **A
study on protein attractants for dacus oleae. The action of
anti–dacus treatment on the insect population as documented by
the collection of exposed insects, creating a trap using glue
compounds to be sprayed from a spraying bottle.**

8503 Fantini, R.   IT 121300/77/0493
Sperimentazione di campo con insetticidi più esche proteiche
per la lotta contro la mosca dell'olivo dacus oleae, al fine di
verificarne l'efficacia in confronto a trattamenti con solo
insetticida e di mettere a punto le relative tecniche di lotta.
**Field trials on insecticides and protein baits combined for the
suppression of the olive fly, dacus oleae, with a view to
assessing their efficacy compared with the use of insecticides
alone so as to establish the relevant pest suppression
techniques.**

8504 Fusé, M.   IT 121400/77/0498
Impiego esche proteiche ed insetticidi contro dacus oleae,
ceratitis capitata, rhagoletis cerasi. **Use of protein baits and
insecticides against dacus oleae, ceratitis capitata, rhagoletis
cerasi.**

Rape (B 3230)

See also 7978, 7981, 9432, 9543

8505 Weinmann, W.   DE 215080/78/0008 N
Untersuchung über das Rückstandsverhalten von
Parathion–äthyl auf/in Futterraps und Stoppelrüben nach einer

Behandlung gegen Rübenblattwespe und Blattlaus. **Investigation on the residur–behaviour of Parathion–ethyl on/in rape and turnips after a treatment against turnip sawfly and aphid.**

8506 Schütte, F.; Brüggemann, H.   DE 215120/78/0010 N
Regulatoren des Pflanzenwachstums zur Bekämpfung von Rapsschädlingen. **Plant growth regulators for control of pests of rape.**

8507 Bauers   DE 511100/78/0001 N
Entwicklung einer integrierten Bekämpfung der wichtigsten Rapsschädlinge unter Berücksichtigung der am jeweiligen Standort gegebenen Phase der Abundanzdynamik. **Development of an integrated control of most important rape pests in consideration of abundance dynamics on a specific site.**

8508 Laumond, C.; Jourdheuil, P.   FR 010503/68/1791
Essai d'utilisation de nématodes contre les ravageurs de cultures annuelles et notamment les insectes du colza. **Biological control of annual crop pests in particular rape pests, using entomophagous nematodes.** Publications.

8509 Free   GB 012004/78/0046 R
Ecology behaviour and control of oil seed rape pests.

Other fibre plants and oil crops (B 3290)

8510 Calabuca, E.; Micieli, De Biase, L.   IT 040703/78/0001 N
Biologia e Distribuzione degli Afidi del Cotone in Grecia. **Biology and Distribution of Aphids of Cotton in Grece.** Publications.

8511 Bruin, Th. de   NL 010300/79/9102 N
Bestrijding van de blauwmaanzaadgalwesp (Timaspis papaveris). **Control of the poppyseed gallwasp (Timaspis papaveris).**

Sugarbeets and starch producing plants in general (B 3300)

8512 Caubel, G.; Mugniery, D.   FR 011306/67/1739
Caractérisation des races, pathotypes——des groupes D. dipsaci, H. avenae et H. rostochiensis. **Characterisation of strains, pathotypes——of Ditylenchus dipsaci, Heterodera avenae, H. rostochiensis.** Publications.

Potatoes (B 3310)

See also 7979, 8446, 9500

8513 Meyer, E.; Härig, R.   DE 138330/73/0003
Untersuchungen über Wechselwirkungen zwischen Heterodera rostochiensis und verschiedenen Bodenpilzen an Kartoffeln. **Investigations into the interactions between Heterodera rostochiensis and divers soil fungi on potatoes.**

8514 Rumpenhorst, H.J.   DE 215190/74/0003
Untersuchungen über die Ursachen der Resistenz bei Kartoffeln gegenüber verschiedenen Pathotypen von Globodera rostochiensis. **Studies on causes of resistance in potato to different pathotypes of Globodera rostochiensis.**

8515 Rumpenhorst, H.J.   DE 215190/77/0013
Charakterisierung der in der Bundesrepublik vorkommenden Pathotypen der Kartoffelnematoden. **Characterization of pathotypes of potato nematodes occurring in the Federal Republic of Germany.**

8516 Hunnius, W.   DE 502058/73/0007
Züchtung auf Resistenz gegen Biotyp A und weitere Biotypen des Kartoffelnematoden 1973. **Breeding for resistance to biotype A and other biotypes of potato nematode.** Publications.

8517 Moreau, J.P.; Derridj, S.   FR 010106/71/5246
Relations pomme de terre – doryphore. **Colorado potato beetle relationships with its host.**

8518 Robert, Y.   FR 011306/66/1738
Ecologie des pucerons de la pomme de terre. **Ecology of potato aphids in Western France.** Publications.

8519 Bromilow   GB 012005/00/0043 R
Improvements of chemical control of potato cyst nematode principally using granular formulations.

8520 Alphey   GB 030705/00/0002 R
Biology and ecology of trichodorid species and their role as virus vectors.

8521 Trudgill; Brown   GB 030705/00/0010 R
Assessment of the damage caused by potato cyst and other plant parasitic nematodes in Scotland.

8522 Wastie   GB 030805/78/0010 R
Study biology of potato cyst eelworm including host parasite relationships and the nature of resistance.

8523 Wastie   GB 030805/78/0011 R
Assess potato breeding material for resistance to potato cyst eelworm. Improve screening techniques.

8524 Wastie   GB 030805/78/0013 R
Assess potato breeding material for resistance to soil–borne viruses. Improve screening techniques.

8525 Turl   GB 030902/00/0306 R
Location and potential of overwintering potato aphid populations in Scotland.

8526 Turl   GB 030902/00/0307 R
Development of aphid populations on potato at selected sites in Scotland.

8527 Howell   GB 030902/00/0308 R
Relationship between aphid borne potato viruses and weather.

8528 Hamilton; Mackenzie   GB 030902/73/0302 R
Development of improved methods for identifying nematodes in the genera Globodera and Heterodera.

8529 Mabbott   GB 030902/73/0305 R
Distribution of potato cyst nematode species and pathotypes in scotland as shown by statutory soil tests.

8530 Mabbott   GB 030902/77/0319 R
Relevance to potato cropping of a nacobbus–like nematode found recently in Scotland.

8531 Winslow   GB 040701/76/0002 R
Chemical control of potato cyst nematode in soil.

8532 Winslow; McKenna   GB 040701/77/0004 R
Varietal resistance to potato cyst eelworms(PCE).

8533 Edwards; Bell   GB 040702/76/0006 R
Aphid vectors of potato virus diseases.

8534 Bartlett   GB 050310/74/0005 R
Biology and control of Colorado beetle in relation to phytosanitary controls.

8535 Osborne   GB 060107/00/0007 R
Control of potata cyst eelworm.

8536 McKinlay   GB 060107/00/0011 R
Potato aphids(aphid vectors of potato viruses).

8537 Stewart   GB 060309/00/0003 R
Integrated control of potato cyst eelworm.

8538 Cremaschi, D.   IT 021100/76/0001 N
Lotta chimica contro gli afidi della patata. **Chemical control of the potato aphides..**

8539 Cremaschi, D.; Govoni, I.; Affatato, E.; Pollini, A.
     IT 021100/76/0003 N
Indagini sulla popolazione afidica in coltivazioni di patata da seme.. **Surveys on the population of aphides in seed potato growings..**

8540 Seinhorst, J.W.   NL 010108/56/8708 N
De bestrijding van aardappelcystenaaltjes (Globodera rostochiensis, G. pallida) met systemische nematiciden. **Control of potato cyst nematodes (Globodera rostochiensis, G. Pallida) with systemic nematicides.** Publications.

8541 Harrewijn, P.   NL 010108/62/1043
Invloed van de fysiologische toestand van de waard–plant op het populatieverloop van zuigende insekten, met name bladluizen. **Investigations on the influence of the physiological condition of the host plant on the development and population growth of sucking insects viz.aphids.** Publications.

8542 Wal, A.F. van der   NL 010120/76/7434
Methodologie van het onderzoek naar resistentie tegen aardappelmoeheid. **Development of methods for studies on resistance to the potato cyst nematode.**

8543 Wal, A.F. van der   NL 010120/78/8884 N
Ontwikkeling van een methodiek om aardappelen te toetsen op tolerantie tegen het aardappelcystenaaltje. **Development of a method to screen potatoes for tolerance to the potato cyst nematode.**

8544 Wal, A.F. van der   NL 010120/78/8885 N
Aanpassing van populaties van aardappelcystenaaltjes aan resistente aardappelen. **Adaption of potato cyst nematode populations to resistant potatoes.**

8545 Kort, J.   NL 010300/57/1390 N
Toetsing van nieuwe aardappelrassen op resistentie tegen aardappelcystenaaltjes (Globodera rostochiensis en G. pallida). **Testing of new potato varieties for resistance to potato cyst nematodes.**

8546 Kort, J.   NL 010300/58/1397 N

Onderzoek naar het vóórkomen van biotypen van de aardappelcystenaaltjes (Globodera rostochiensis en G. pallida). **The occurrence of pathotypes of the potato cyst nematodes.**

8547 Kort, J.   NL 010300/60/1396 N
Biologisch onderzoek bij biotypen van aardappelcystenaaltjes (Globodera rostochiensis en G. pallida). **Biology of pathotypes of the potato cyst nematodes.**

8548 Kort, J.   NL 010300/63/8611 N
Ontwikkeling van toetstechnieken voor de bepaling van resistentie tegen het aardappelcystenaaltje bij nieuwe aardappelrassen. **Development of techniques for testing new potato cultivars for resistance to potato cyst nematode.**

8549 Bruin, Th. de   NL 010300/69/9112 N
Langjarige grondontsmetting; onderzoek naar de bestrijdingsmogelijkheden van het aardappelcystenaaltje met diverse nematiciden. **Long term experiments on soil disinfection; research on control of the potato cyst nematode with various nematicides.**

8550 Bruin, T. de   NL 010302/71/3641
Bestrijding van ritnaalden bij aardappelen. **Control of larvae from Elateridae in potatoes.**

8551 Wilde, J.de   NL 020014/73/3657
Vergelijking van de fotoperiodieke reaktie bij de laboratoriumstam en bij een wildstam van de Coloradokever. Vaststellen van de plaats van perceptie van de fotoperiode. **Comparison of photoperiodic response of laboratory and a wild strain of the Coloradobeetle. Site of perception of the photoperiod.** Publications.

8552 Kort, C.A.D. de; Dortland, J.F.; Schooneveld, H.; Kramer, S.J.; Wilde, J. de   NL 020014/73/5046
Betekenis van neurohormonen en Juveniel hormoon bij regulatie van daglengteafhankelijke processen in de Coloradokever; in het bijzonder de reproduktie en de diapauze. **Contribution of neuro–hormones and juvenile hormone on the control of daylength–dependent processes in the Colorado potato beetle, with special reference to reproduction and diapause.** Publications.

8553 Kort, C.A.D.   NL 020014/77/8643 N
Substraatvoorziening en substraatutilisatie in de vliegspier van de Coloradokever. **Substrate–supply and substrate utilization of flight muscles of the Colorado beetle.**

Sugarbeets and other sugar crops (B 3320)

See also 5502, 7969, 8630, 9528

8554 Heyland, K.-U.; Kochs, H.-J.; Vogler, B.
     DE 111252/74/0001
Nematizid– und Insektizideinsatz bei einer Fruchtfolge mit einem Zuckerrübenanteil von 50 %. **Application of nematicides and insecticides in a 50 per cent sugar beet rotation.**

8555 Weltzien, H.C.; Kraus, R.; Szopka, U.; Wachendorff, U.; Sikora, R.A.; Klingauf, F.   DE 111353/78/0004 N
Nebenwirkungen von Herbiziden auf Schaderreger bei Zuckerrüben. **Side–effects of herbicides on pests and diseases of sugar beets.**

8556 Ulber, W.　　　　　　DE 132211/77/0002
Einfluss von Herbiziden auf Collembola in
Zuckerrübenbeständen. **Influence of herbicides on Collembola in sugar beets.**

8557 Müller, J.　　　　　　DE 215190/77/0017
Aphanomyces cochlioides als Fruchtfolgeschädling bei
Zuckerrüben und Wechselbeziehungen zu pflanzenparasitären
Nematoden. **Aphanomyces cochlioides as crop rotation pest in sugar–beets and correlations to plant–parasitic nematodes.**

8558 Diercks, R.; König, K.　　　　DE 502051/75/0005
Untersuchungen über die Auswirkungen der Anwendung
verschiedener insektizider Wirkstoffe und
Applikationstechniken auf die Fauna eines
Zuckerrübenfeldes. **Studies on the effects of various insecticides applied with different techniques on the fauna of a sugar beet field.**

8559 Robert, P.; Simonis, M.T.　　　　FR 010904/77/5028
Action des signaux chimiques des plantes hôtes et non hôtes
sur le comportement et la reproduction d'un insecte
phytophage : la Teigne de la Betterave Scrobipalpa ocellatella
Boyd. **Influence of chemical signals from host and non host plants on the behaviour and the reproduction of the sugar beet moth Scrobipalpa ocellatella Boyd.**

8560 Robert, Y.　　　　　　FR 011306/67/5116
Epidémiologie de l'enroulement de la pomme de terre transmis
par pucerons dans l'Ouest de la France. **Epidemiology of the aphid–transmitted potato leaf roll in the west of France.**

8561 Mugniery, D.　　　　　FR 011306/71/5123
Lutte intégrée contre les nématodes en culture de pomme de
terre primeur. **Integrated control of potato Nematodes.**

8562 Lowe　　　　　　GB 011906/00/0022 R
Investigate resistance to aphids in sugar beet.

8563 Cooke　　　　　　GB 012301/00/0016 R
Population dynamics and control of Longidorus and Trichodorus.

8564 Cooke　　　　　　GB 012301/00/0017 R
Control of beet cyst nematode Heterodera schachtii.

8565 Dunning　　　　　GB 012301/00/0018 R
Seed treatments: pesticides in pelleted seed.

8566 Dunning　　　　　GB 012301/00/0020 R
Seedling pest control.

8567 Dunning　　　　　GB 012301/00/0022 R
Epidemology and control of aphids and viruses.

8568 Duggan, J.J.　　　　　IE 060300/00/1430 N
Beet cyst eelworm survey. Publications.

8569 Duggan, J.J.　　　　　IE 060300/00/1431 N
Studies on population trends, host–range and bionomics of beet cyst eelworm (heterodera schacatii) and on the pathogenicity of other species of eelworm. Publications.

8570 Foschi, S.　　　　　IT 061800/77/1013
Ricerche sulla difesa antiparassitaria e sul diserbo della
barbabietola. **Beetroots : research on pest control and on weeding.**

8571 Kaai, C.　　　　　NL 010108/73/7662 N
Bietecystenaaltjes (Heterodera schachtti en Heterodera sp.) in
akkerbouw– en tuinbouwgewassen. **Beet cyst nematodes (Heterodera schachtti and Heterodera sp.) in agricultural and horticultural crops.** Publications.

8572 Kaai, C.　　　　　NL 010207/73/7662 N
Bietecystenaaltjes (Heterodera schachtti en Heterodera sp.) in
akkerbouw en tuinbouwgewassen. **Beet cyst nematodes (Heterodera schachtti and Heterodera sp.) in agricultural and horticultural crops.** Publications.

8573 Maenhout, C.A.A.　　　　NL 010207/74/7632 R
Onderzoek naar de betekenis van vruchtwisselingsgebonden
bietenpathogenen. **The influence of soilborne pathogenes on sugar beet.** Publications.

8574 Bruin, Th. de　　　　NL 010300/74/8716 N
Onderzoek naar de invloed van intensieve teelt van
suikerbieten op het populatieverloop van Heterodera
schachtii, het bietencystenaaltje. **Investigations on the influence of intensive cropping of sugar beets on the populations of Heterodera schachtii.**

8575 Dinther, J. van　　　　NL 020017/75/8639 N
Seizoensritmiek van de suikerrietschuimcicade Aeneolamia
flavilatera in Suriname. **Seasonal rhythmicity of the froghopper Aeneolamia flavilatera in Surinam.**

8576 Heijbroek, W.　　　　NL 060003/38/7720 R
Chemische bestrijding van bietecystenaaltjes bij suikerbieten.
Chemical control of beet cystnematodes in sugar beet.
Publications.

8577 Heijbroek, W.　　　　NL 060003/39/7722 R
Populatiedynamica van nematoden ter bepaling van de
maximaal haalbare frequentie van suikerbieten in
verschillende gebieden. **Population dynamics of nematodes to determine the maximum frequency of sugar beet in different areas.** Publications.

8578 Heijbroek, W.　　　　NL 060003/53/3669
Zaad– en kiemplantbescherming bij suikerbieten tegen
bodeminsekten o.a. geïntegreerde bestrijding van Onychiurus
armatus en Blaniulus guttulatus. **The protection of sugar beet seed and seedlings against soil insects a.o. integrated control of Onychiurus armatus and Blaniulus guttulatus.** Publications.

8579 Heijbroek, W.　　　　NL 060003/73/7724 R
Bestrijding van de vergelingsziekte in suikerbieten door
chemische bestrijding van bladluizen. **Control of virus yellows in sugar beet by chemical control of aphids.**

8580 Heijbroek, W.　　　　NL 060003/76/7721 R
Vaststelling van de tolerantiegrens voor bietecystenaaltjes bij
suikerbieten. **Determination of the tolerance limit for beet cystnematodes in sugar beet.**

Grasses and forage crops in general (B 3400)

8581 Cook; York　　　　　GB 012101/78/0027 R
Host nematode relationships in grasses and forage legumes.

8582 Winslow; McKenna　　　　GB 040701/77/0003 R

Nematode pests of forage crops.

Grasses (B 3410)

See also 7964, 8290, 8593

8583 Robert, P.; Blaisinger, P.; Kienlen, J.C.
FR 010904/70/5020
Protection des prairies permanentes contre les vers blancs du Hanneton commun : lutte chimique contre les adultes. **Control of the white grubs of the cockchafer Melolontha melolontha in permanent grass–land : chemical control of the adults.**

8584 Jewiss   GB 011202/00/0011 R
Reasons for differences in yield and persistancy between forage grass varieties.

8585 Clements   GB 011202/79/0713 N
Control of stem boring diptera in perennial and Italian ryegrasses (with RES and WPBS).

8586 Henderson   GB 012004/78/0040 R
Effects of insects and other invertebrates on the productivity of grassland and forage legumes.

8587 Graves   GB 030704/75/0020 R
Viruses infecting grasses.

8588 Willis   GB 040702/00/0003 R
Economic significance of shoot fly infested grasses.

8589 Lovato, A.   IT 040201/72/0495
Ricerche sull'impiego del diserbo chimico e delle sostanze regolatrici della crescita in colture erbacee da pieno campo. **Research on the use of chemical weeding and of growth regulating substances in grass cultivation in the field.** Publications.

Pastures, grassland (B 3420)

See also 19555

8590 Weinmann, W.   DE 215080/75/0007
Untersuchungen über die Höhe der Rückstände an Parathionäthyl in Weidegras nach einer Tipulabekämpfung im Frühjahr. **Investigation of the amount of parathion ethyl residues in pasture grass after Tipula control in spring.**

8591 Weinmann, W.   DE 215080/78/0002 N
Untersuchung über das Rückstandsverhalten von Parathion auf/in Gras nach einer Spritzbehandlung von Weiden gegen Tipula. **Investigation on the residue–behaviour of Parathion on/in grass after spray treatment on meadows against leather jacket.**

8592 Lauenstein, G.   DE 507250/78/0001 N
Untersuchungen zur Populationsdynamik und Bekämpfung der Feldmaus – Microtus arvalis Pallas – und pflanzenschädlicher Tipuliden – Diptera: Tipulidae – auf Grünland. **Investigations on population dynamics and control of the common vole – Microtus arvalis Pallas – and on plant injurious leatherjackets – Diptera: Tipulidae – on grassland.** Publications.

8593 Ricou, G. Mme; Vago, C.   FR 012220/66/5283
Dynamique des populations de Tipules dans les prairies et gazons. **Population dynamics of crane–flies in the meadows and grasses.**

8594 D'aguilar, J.; Chevin, H.; Maillard, J.; Chambon, J.P.
FR 012221/77/9049
Etude agrocénotique sur plantes de grande culture (prairie, céréales). **Agrocoenotic faunistic study of major field crops (pasture, cereals).**

8595 Blasdale   GB 060218/00/0002 R
Effect of insecticide treatment on yield of grassland.

8596 Newbold   GB 060309/00/0008 R
Forecasting and control of leatherjacket(Tipulid) populations (including other minor pests of grassland).

8597 Bruin, Th. de   NL 010300/78/9113 N
Bepaling van de schadedrempel en de bestrijding van rouwvlieglarven in graszoden. **Determination of damage criteria and control of Dilopus febrilis in turfs.**

8598 Bunt, J.A.   NL 020037/76/6814
De invloed van nematoden in het herinzaaiprobleem van grasland. **The influence of nematodes in the resowing problem of pasture.**

Mangolds (B 3430)

See also 7978, 8568, 8569

Legumes in general (B 3440)

See also 7964

Grassland legumes (B 3441)

8599 Bournoville, R.; Descoins, C.; Cantot, P.
FR 012223/71/5080
Insectes ravageurs de la production de semences de luzerne. Ecologie et méthode de lutte. **Insect pests of seed production of lucerne ecology and control.**

8600 Bournoville, R.; Cantot, P.   FR 012223/72/5081
Etude sur les insectes nuisibles à la luzerne fourrage en rapport avec le végétal. **Studies about insect pests of lucerne grown for forage in relation to the host plant.**

8601 Tasei, J.N.; Carre, S.   FR 012223/72/5082
Elevage et multiplication en champ de Megachile pacifica Pr. **Field rearing and multiplication of Megachile pacifica.**

8602 Bournoville, R.; Guy, P.; Cantot, P.   FR 012223/77/5087
Etude de la résistance variétale de la luzerne envers le puceron du pois. **Varietal resistance of lucerne to pea aphid.**

8603 Scott, W.   GB 012107/77/0014 R
Host resistance to clover rot in red and white clover and lucerne and the epidemiology of the pathogen.

8604 Scott, W.   GB 012107/77/0015 R
Resistance of clovers to powdery mildew.

8605 Tafuri, F.   IT 061600/74/0028
Persistenza nel terreno e residui nell'erba medica degli isomeri alpha e gamma del clordano. **Persistence of alpha and gamma isomers of chlordane in soil and residues in lucerne.**

Publications.

Cereals used for forage (B 3450)

See also 8594

8606 Maenhout, C.A.A.A.　　　NL 010207/78/9008 N
De betekenis van wortelaaltjes bij maïs. **Inportance of root nematodes in maize.**

8607 Jacobs, J.J.　　　NL 020037/78/8537 R
Betekenis van vrijlevende wortelaaltjes bij mais. **Importance of free–living root nematodes in maize.**

Turnips (B 3460)

See also 8505

Other forage crops (B 3490)

See also 8484

Vegetables in general (B 3500)

See also 7957, 7971, 7972, 8501, 9597, 9601

8608 Gillard, A.; Pelerents, C.; Van de Steene, F.
　　　BE 030024/63/0003 R
Onderzoek over de biologie en de chemische bestrijding van insekten parasieten op groentegewassen. **Research on the biology and the chemical control of insects parasite on vegetables.** Publications.

8609 Researcher not indicated　　　DE 138330/77/0001
Untersuchungen zur biologischen Bekämpfung von Schadmilben, insbesondere Tetranychus urticae Koch mit der Raubmilbe Phytoseiulus persimilis A.–H und einheimischen Raubmilben im Pflanzenbau unter Glas. **Studies on biological control of injurious mites, especially Tetranychus urticae Koch, with the predatory mite Phytoseiulus persimilis A.–H and domestic predatory mites in plant culture under glass.**

8610 Ohnesorge, B.　　　DE 144550/77/0003
Analyse des Systems Spinnmilbe – Raubmilbe – Tetranychus urticae – Phytoseiulus persimilis – unter den Bedingungen des Gewächshauses. **Analysis of the system Tetranychus urticae – prey – Phytoseiulus persimilis – predator – under greenhouse conditions.**

8611 Stüben, M.　　　DE 215070/78/0001 N
Untersuchungen über die Biologie der Weissen Fliege. **Studies on the bionomics of the greenhouse white fly.**

8612 Crüger, G.　　　DE 215200/77/0001
Erarbeitung der biologischen Grundlagen zur Festsetzung von Schadensschwellen bei Gemüsefliegenbefall. **Development of biological fundamentals for determination of injury threshold in attack of vegetable fly.**

8613 Häfner, M.　　　DE 501054/78/0001 N
Untersuchungen von Fehlermöglichkeiten bei Rückstandsanalysen, bedingt durch die Probenahme. **Studies on mistakes in the results of residue analysis in vegetables and fruits based on sampling.** Publications.

8614 Nøddegaard, E.; Rasmussen, A.N.

DK 010116/78/0026 N
Insekticiders virkning over for insekter i væksthus. **Effect of insecticides on insects in glasshouse cultures.**

8615 Maude; Presley　　　GB 011807/00/0015 R
Biology and control of Botrytis diseases of vegetables.

8616 Green　　　GB 011811/00/0002 R
General survey of nematodes associated with vegetable crops.

8617 Boag　　　GB 030705/75/0011 R
Migratory plant parasitic nematodes associated with vegetable crops in Scotland.

8618 Mowat; Martin　　　GB 040702/00/0004 R
Control of vegetable root flies.

8619 Dunne, R.M.　　　IE 060300/73/0872 N
Control of vegetable pests. Publications.

8620 Marinari Palmisano, A.; Ambrogioni, L.
　　　IT 020400/71/0003
Nematodi Tylenchida associati con piante ortensi, floricole e arboree. **Tylenchid nematodes associated with horticultural and flower crops and with fruit trees.** Publications.

8621 Fiume, F.; Ferrari, V.; Santoro, R.　IT 021000/72/0004 R
Lotta integrata contro il Trialeurodes vaporariorum. **Integrated control against Trialeurodes vaporariorum.** Publications.

8622 Loi, G.　　　IT 041106/78/1067 N
Gli insetti terricoli dannosi alle colture orticole in piena aria nella Toscana litoranea. **Soil insects harmful to open field cultivation of market garden plants in the Tuscan littoral region.**

8623 Kaai, C.　　　NL 010108/64/1057
Stengelaaltjes in groentegewassen. **Stem nematodes in vegetable crops.** Publications.

8624 Theunissen, J.A.B.M.　　　NL 010108/76/7677
Onderzoek over geleide en geïntegreerde bestrijding van schadelijke rupsen en luizen in groentegewassen in de vollegrond. **Research on supervised and integrated control of noxious caterpillars and aphids in outdoor vegetables.**

8625 Ouden, H. den　　　NL 010108/77/7679
Onderzoek over geleide en geïntegreerde bestrijding van schadelijke vliegen en kevers in groentegewassen in de vollegrond. **Research on supervised and integrated control methods of noxious flies and beetles in outdoor vegetables.**

8626 Theune, D.　　　NL 010206/72/4672
Biologie en bestrijding van dierlijke parasieten (insekten, mijten) bij kasgewassen. **Biology and control of pests in glasshouse vegetables.**

8627 Woets, J.　　　NL 010206/72/4674
Biologische bestrijding van dierlijke parasieten in kasgewassen. **Biological control of glasshouse pests.** Publications.

Root, tuber and bulb vegetables (B 3510)

See also 7967, 8571, 8572, 8655

8628 Sol, R.   DE 135052/72/0001
Zur Biologie und Bekämpfung der an Meerrettich schädlichen Fliegenarten. **Biology and control of flies damaging horse radish.**

8629 Meyer, E.; Wohanka, W.   DE 138330/73/0001
Der Einfluss der Düngung auf die Wirt–Parasit–Beziehung Pflanze – Nematode. **The influence of fertilization on the host–parasiterelationship carrot – root–knot nematode.**

8630 Weinmann, W.   DE 215080/77/0006
Analysenmethode zur Bestimmung von Dimethoat auf und in Möhren, Zuckerrüben, Schnittlauch, Salat, Blumenkohl. **Analytical method for determination of dimethoate on and in carrots, sugar–beets, chives, lettuce and cauliflower.**

8631 Overbeck, H.   DE 215200/74/0001
Entwicklung von Verfahren des integrierten Pflanzenschutzes zur Bekämpfung der Möhrenfliege – Psila rosae –. **Development of methods for integrated control of carrot fly – Psila rosae –.**

8632 Brunel, E.   FR 011306/64/5112
Dynamique des populations de la mouche de la carotte Psila rosae. **Population dynamics of the carrot rust fly, Psila rosae.**

8633 Brunel, R.; Rabasse, J.M.; Dedryver, Ch.   FR 011306/71/5121
Dynamique des populations des pucerons de la carotte (Cavariella aegopodii). **Dynamics of the carrot–willow aphids'populations (Cavariella aegopodii).**

8634 Ellis; Hardman   GB 011806/00/0008 R
Investigate resistance of carrots to insect pests.

8635 Green   GB 011811/00/0001 R
Control of nematode attack on onions.

8636 Greco, N.   IT 060600/72/0054
Lotta chimica contro i nematodi fitoparassiti. Ditylenchus dipsaci su cipolla e fragola. **Chemical control of plant parasitic nematodes. Ditylen–chus dipsaci on onion and strawberry.** Publications.

8637 Greco, N.   IT 060600/72/0055
Lotta chimica contro i nematodi fitoparassiti. Heterodera carotae su carota. **Chemical control of plant parasitic nematodes. Heterodera carotae on carrot.** Publications.

8638 Ticheler, J.   NL 010108/76/7532
Ontwikkelingsfase genetische bestrijding van de uievlieg, Delia antiqua (Meigen). **Developmental phase genetic control of the onion fly, Delia antiqua (Meigen).**

8639 Robinson, A.S.   NL 010110/72/5902
Genetische bestrijding van de uievlieg (Hylemya antiqua Meigen) door structurele chromosoomafwijkingen. **Genetical control of Hylemya antiqua Meigen by structural chromosome rearrangements.**

Greens and leafy vegetables (B 3520)

See also 8484, 8571, 8572, 8630, 9644

8640 De Clercq, R.   BE 070100/78/0021 R

Studie van de biologie en de bestrijding van de witloofvlieg Napomyza chicorei. **Study of the biologie and control of Napomyza chicorii the leafminer of the witloofchicory.**

8641 Weinmann, W.   DE 215080/77/0023
Untersuchung über das Rückstandsverhalten von Chorfenvinfos nach einer Anwendung gegen Möhrenfliege an Petersilie. **Investigations on residual action of chlorfenvinfos after application against carrot fly in parsley.**

8642 Weinmann, W.   DE 215080/77/0027
Untersuchung über das Rückstandsverhalten von Dimethoat nach einer Anwendung gegen Zwiebelfliege an Schnittlauch. **Investigations on residual action of dimethoate after application against onion fly in chives.**

8643 Weinmann, W.   DE 215080/77/0034
Untersuchung über das Rückstandsverhalten von Parathion–äthyl auf und in Kopfsalat nach einer Giessbehandlung gegen Drahtwürmer. **Investigations on residual action of parathion–ethyl on and in lettuce after watering treatment against witeworms.**

8644 Weinmann, W.   DE 215080/78/0007 N
Untersuchung über das Rückstandsverhalten von Azinphosäthyl auf/in Spinat und Bohnen nach einer Behandlung gegen beissende und saugende Insekten. **Investigation on the residue–behaviour of Azinphos–ethyl on/in spinach and beans after a treatment against biting and sucking insects.**

8645 Weinmann, W.   DE 215080/78/0009 N
Untersuchung über das Rückstandsverhalten von Parathion–äthyl und Oxydemeton–methyl auf/in Salat und Wirsingkohl nach einer Behandlung gegen beissende und saugende Insekten. **Investigation on the residue–behaviour of Parathion–ethyl and Oxydemeton–methyl on/in lettuce and savoy cabbage after a treatment against biting and sucking insects.**

8646 Weinmann, W.   DE 215080/78/0011 N
Untersuchung über das Rückstandsverhalten von Methomyl auf/in Spinat und Wirsingkohl nach einer Behandlung gegen beissende und saugende Insekten. **Investigation on the residue–behaviour of Methomyl on/in spinach and savoy cabbage after a treatment against biting and sucking insects.**

8647 Weinmann, W.   DE 215080/78/0015 N
Untersuchung über das Rückstandsverhalten von Propoxur auf/in Kopfsalat nach einer Behandlung gegen beissende und saugende Insekten. **Investigation on the residue–behaviour of Propoxur on/in lettuce after a treatment against biting and sucking insects.**

8648 Poitout, S.; Fargues, J.; Bues, R.; Laumond, C.; Burgerjon, A.; Voegele, J.   FR 010602/75/5094
Etude de la biocoenose de la culture du chou, particulièrement de Mamestra brassicae, mise au point de la lutte intégrée. **Study of the cabbage culture biocoenosis, especially of Mamestra brassicae, development of integrated control.**

8649 Dunn; Ellis   GB 011806/00/0005 R
Study resistance of cruciferae to insect pests.

8650 Osborne   GB 060107/00/0009 R
Cabbage root fly control.

8651 Tremblay, E.   IT 040703/74/0001
Biologia di nuove specie dannose di insetti. **Biological researches on new insect pest outbreaks.** Publications.

8652 Driest, J.Ph. van   NL 010207/67/1569
Bestrijding van de koolvlieg (Chortophila brassicae Bouché). **Control of the cabbage root fly (Chortophila brassicae Bouché).** Publications.

8653 Driest, J.Ph. van   NL 010207/74/6506
Bestrijding van rupsen in koolgewassen. **Control of caterpillars in Brassica crops.**

8654 Dinther, B.M.J. van   NL 020014/70/4414
Betekenis Carabiden als natuurlijke vijanden van de koolvlieg. **Importance of Carabia as predators of the cabbage root fly.**

Leguminous vegetables (B 3531)

See also 8644

8655 Weinmann, W.   DE 215080/77/0003
Analysenmethode zur Bestimmung von Dicofol auf und in Bohnen, Erdbeeren, Zwiebeln, Äpfeln, Gurken, Johannisbeeren, Kirschen, Möhren und Erde. **Analytical method for determination of dicofol on and in beans, strawberries, onions, apples, cucumbers, currants, cherries, carrots and earth.**

8656 Weinmann, W.   DE 215080/77/0026
Untersuchung über das Rückstandsverhalten von Parathion–äthyl nach einer Anwendung gegen Erbsengallmücke an Erbsen. **Investigations on residual action of parathion–ethyl after application against pea midge.**

8657 Weinmann, W.   DE 215080/77/0029
Untersuchung des Rückstandsverhaltens von Dinobuton nach einer Anwendung gegen Spinnmilben an Stangen– und Buschbohnen im Freiland und unter Glas. **Investigations on residual action of dinobuton after application against red spider in pole beans and dwarf beans in fields and under glass.**

8658 Stockel, J.; Berjon, J.   FR 010707/71/1783
Etude de la bruche de la Féverole. **Study of european broadbean beetle.**

8659 Bouchery, Y.; Jacky, F.   FR 010904/72/5022
Limitation des populations du puceron noir de la fève (Aphis fabae) sur la Féverole. **Control of black bean Aphid (Aphis fabae) on field beans.**

8660 Robert, Y.; Rabasse, J.M.; Dedryver, Ch.   FR 011306/69/5118
Epidémiologie des mycoses à entomophthorales : cas d'Aphis fabae sur hôte primaire et sur féverole. **Epidemiology of mycoses by Entomophthorales.**

8661 Hooper   GB 012006/00/0020 R
Beans: incidence, persistence, bionomics and control of stem nematodes in soil and seeds.

8662 Di Vito, M.   IT 060600/72/0056
Lotta chimica contro i nematodi fitoparassiti. Heterodera gottingiana su pisello. **Chemical control of plant parasitic nematodes. Heterodera gottingiana on pea.** Publications.

Other vegetable fruits (B 3539)

8663 Weinmann, W.   DE 215080/77/0025
Untersuchung über das Rückstandsverhalten von Methomyl nach einer Anwendung gegen beissende und saugende Insekten auf Paprika unter Glas. **Investigations on residual action of methomyl after application against biting and sucking insects on Capsicum under glass.**

8664 Weinmann, W.   DE 215080/77/0028
Untersuchung des Rückstandsverhaltens von Dinobuton nach einer Anwendung gegen Spinnmilben an Paprika. **Investigations on residual action of dinobuton after application against red spider in Capsicum.**

8665 Rast, A.Th.B.   NL 010108/75/8705 N
Virusziekten bij paprika. **Virus diseases of sweet pepper.**

Mushrooms and other edible fungi (B 3540)

8666 Weinmann, W.   DE 215080/75/0009
Untersuchung über das Rückstandsverhalten von Dichlorvos auf und in Champignons nach einer Milbenbekämpfung. **Analysis of residual Dichlorvos on and in mushrooms after mite control.**

8667 Weinmann, W.   DE 215080/77/0048
Untersuchung über das Rückstandsverhalten von Dichlorvos + Pyrethrum + Piperonylbutoxid auf und in Champignons nach einer Milbenbekämpfung. **Investigations on residual action of dichlorvos + pyrethrum + piperonylbutoxide on and in mushrooms after mite control.**

8668 Binns; E White P   GB 011108/00/0002 R
Biology and control of mushroom pests.

8669 Hesling   GB 011108/00/0017 R
Role of saprobic nematodes as vectors of mushroom pathogens.

8670 Wood   GB 011109/00/0018 R
Microbiology and chemistry of mushroom compost.

8671 Wood   GB 011109/00/0019 R
Stimulation of mushroom fruiting.

8672 Gandy   GB 011109/00/0023 R
Microclimate of mushroom sporophores in relation to disease.

8673 Zaayen, A. van   NL 010204/74/5931 R
Preventie en bestrijding van plagen en bij champignonsziekten met onbekende oorzaak. **Prevention and control of pests and unknown diseases of the cultivated mushroom.** Publications.

Fruits in general (B 3600)

See also 8611, 8613, 8614, 8620, 8622, 9597, 9722

8674 Coutin, R.   FR 010106/70/1777
Dynamique des populations de carpocapse. **Population dynamics of carpocapsa.** Publications.

8675 Bonnemaison, L.   FR 010106/71/1779
Biologie et méthode de lutte contre la capua des arbres fruitiers. **Biology and control of capua in fruit trees.** Publications.

8676 Milaire, H.; Chambon, J.P.   FR 010106/77/5269
Tordeuse de la pelure et lutte intégrée en vergers. **Leafrollers in integrated control in orchards.**

8677 Gardan, L.; Luisetti, J.   FR 010402/66/1005
Les pseudomonas dans la phyllosphère des arbres fruitiers. Ecologie–épidémiologie. **Ecology. Epidemiology of pseudomonas of fruit – trees.** Publications.

8678 Causse, R.   FR 010602/65/1754
Action de la photopériode sur le comportement de la mouche méditerranéenne des fruits Ceratitis capitata. **Action of photoperiod on the behaviour of mediterranean fruit fly Ceratitis capitata.** Publications.

8679 Audemard, H.   FR 010602/69/1758
Recherches de méthodes d'évaluation des niveaux de populations de carpocapses et de leur potentiel d'attaque sur Fruits. **Studies on population level of carpocapsa and their incidence on fruit crop.** Publications.

8680 Audemard, H.   FR 010602/70/1759
Réduction de la lutte chimique contre la carpocapse et ses conséquences pour la protection phytosanitaire (chimique et biologique) des vergers. **Reduction of chemical control of carpocapsa and result in phytosanitary protection (chemical and biological) of orchards.** Publications.

8681 Audemard, H.; Milaire, H.; Causse, R.; Roehrich, R.; Guennelon, G. Mme; Descoins, C.   FR 010602/75/5096
Dynamique des populations de la Tordeuse orientale. **Dynamics of the oriental fruit moth populations.**

8682 Arnoux, M.; Marboutie, C.   FR 010610/70/3005
Etude des possibilités de lutte intégrée en vergers. **Studies on the possible use of integrated control in orchards.** Publications.

8683 Cranham   GB 011001/00/0003 R
Fruit pests.

8684 Cranham   GB 011001/00/0004 R
Control of tetranychid mites on apple.

8685 Flegg; McNamara   GB 011001/00/0013 R
Ecology of nematode vectors of virus diseases.

8686 Solomon   GB 011001/00/0017 R
Ecology and control of pests.

8687 Flegg; McNamara   GB 011001/76/0019 R
Ecology and host–parasite relations of nematodes attacking roots of perennial plants.

8688 Flegg; McNamara   GB 011001/76/0021 R
Ecology of nematode vectors of virus diseases.

8689 Umpelby   GB 050314/68/0002 R
Red spider mite control on hops and in orchards.

8690 Barbagallo, S.   IT 040301/78/1130 N
Dinamica delle popolazioni e interventi di lotta con l'ausilio degli attrattivi contro Ceratitis capitata Wied, Dacus oleae GMEL e Rhagoletis cerasi L. **Ceratitis capitata Wied, Dacus oleae GMEL and Rhagoletis cerasi L. : population dynamics and pest control using attractants.**

8691 Micieli, L.; Calabuca, E.   IT 040703/78/0002 N
Biologia degli Afidi dannosi a Fruttiferi ed Agrumi in Campania. **Biology of Aphid pest on Fruit and Citrus trees.** Publications.

8692 Tremblay, E.   IT 040703/78/1095 N
Individuazione di biotipi in due specie di cocciniglie–Planococcus citri risso e Planococcus sign. **Identification of biotypes in two species of coccinella – Planococcus citri risso and Planococcus sign.**

8693 Jong, D.J. de   NL 010108/77/7673
Biologie en bestrijding van secundaire (of incidenteel voorkomende) insektenplagen in fruitgewassen. **Biology and control of secondary insect pests in fruit.**

8694 Jong, D.J. de   NL 010212/77/7673
Biologie en bestrijding van secundaire (of incidenteel voorkomende) insektenplagen in fruitgewassen. **Biology and control of secondary insect pests in fruit.**

8695 Kraan, C. van der; Deventer, P. van   NL 010301/73/3761
Gebruik van sex–feromonen bij de bestrijding van Tortriciden (bladrollers). **Use of sex–pheromons in controlling leafrollers.** Publications.

8696 Rabbinge, R.   NL 020054/72/4415
Het opbouwen van een simulatiemodel dat de populatieontwikkeling van de fruitspintmijt en haar roofmijten beschrijft ter ondersteuning van de biologische bestrijding. **Simulation of the development of a population of spider mites by means of a digital computer in order to support the biological control.** Publications.

8697 Gruys, P.; Freriks, J.M.   NL 050101/76/6748
Bijeenbrengen en toetsen van de vatbaarheid voor ziekten en plagen van (vnl. oude Nederlandse) fruitrassen. **Collecting and assessment of the susceptibitity to diseases and pests of (mainly old Dutch) fruit varieties.**

Top fruit in general (B 3610)

8698 Soenen, A.; Marcelle, R.; Simon, P.; Pittevils, J.; Gilles, G.; Porreye, W.   BE 140000/72/0006 R
Studie van entomologische problemen in de fruitteelt. **Study of entomological problems in orcharding.** Publications.

8699 Weinmann, W.   DE 215080/78/0010 N
Untersuchung über das Rückstandsverhalten von Benzomat auf/in Äpfeln nach einer Behandlung gegen Spinnmilben. **Investigation on the residue–behaviour of Benzomat on/in apples after a treatment against mites.**

8700 Wenzel, G.; Jacobsen, E.   DE 402000/78/0002 N
Erstellung von Zuchteltern mit Resistenz gegen Heterodera pallida durch Züchtung über Dihaploide. **Breeding of parental stock with resistance to Heterodera pallida by breeding via dihaploids.**

8701 Nøddegaard, E.; Hansen, T.   DK 010116/61/0011 N
Insekticiders virkning mod skadedyr på kerne– og stenfrugt. **Effect of insecticides on pests on kernel and stone fruit.**

8702 Lyth; Easterbrook   GB 011001/00/0008 R
Resistance of hosts to pests (aid to breeding).

8703 Jong, D.J. de   NL 010108/58/8707 N
Geleide bestrijding van plagen in boomgaarden. **Supervised pest control in orchards.** Publications.

8704 Jong, D.J. de   NL 010108/73/3796
Geintegreerde bestrijding in boomgaarden, waaronder toepassing van geintegreerde bestrijding in praktijkboomgaarden. **Integrated pest control in orchards a.o. implementation of integrated control.** Publications.

8705 Jong, D.J. de   NL 010212/58/8707 N
Geleide bestrijding van plagen in boomgaarden. **Supervised pest control in orchards.** Publications.

8706 Jong, D.J. de   NL 010212/73/3796
Geintegreerde bestrijding in boomgaarden, waaronder toepassing van geintegreerde bestrijding in praktijkboomgaarden. **Integrated pest control in orchards a.o. implementation of integrated control.** Publications.

8707 Ponsen, M.B.   NL 020061/73/6866
De infectie van polyedervirus in Adoxophyes orana. **Infection of polyhedral virus in A. orana.**

8708 Gruys, P.   NL 050101/73/3796
Geintegreerde bestrijding in boomgaarden, waaronder toepassing van geintegreerde bestrijding in praktijkboomgaarden. **Integrated pest control in orchards a.o. implementation of integrated control.** Publications.

8709 Gruys, P.   NL 050101/73/8973 N
Toepasbaarheid van de geintegreerde bestrijding in de praktijk van de fruitteelt. **Implementation of integrated control in orchards.** Publications.

Apple (B 3611)

See also 5817, 9769

8710 Soenen, A.; Marcelle, R.; Simon, P.; Pittevils, J.; Gilles, G.; Porreye, W.   BE 140000/72/0002 R
Studie over de bestrijding en ziekten van appel, peer en kers. **Research on the control of fungi and diseases of apple, pear and cherry.** Publications.

8711 Weinmann, W.   DE 215080/77/0040
Untersuchung über das Rückstandsverhalten von Chlorthiophos auf und in Äpfeln nach einer Behandlung gegen beissende und saugende Insekten, Obstmade. **Investigations on residual action of chlorothiophos on and in apples after application against biting and sucking insects, against codling moth.**

8712 Weinmann, W.   DE 215080/77/0041
Untersuchung über das Rückstandsverhalten von Tricyclotin auf und in Äpfeln nach einer Bekämpfung von Spinnmilben. **Investigations on residual action of tricyclotine on and in apples after red spider control.**

8713 Weinmann, W.   DE 215080/77/0044
Untersuchung über die Auswirkungen unterschiedlicher Spritzbrühhund Aufwandmengen auf die Rückstände von Diazinon auf und in Äpfeln nach einer Bekämpfung gegen Obstmade. **Investigations on effects of different spray mixtures and dosages on diazinon residues on and in apples after codling**

moth control.

8714 Weinmann, W.   DE 215080/78/0006 N
Untersuchung über das Rückstandsverhalten von Diazinon und Fenvalerate auf/in Äpfeln nach einer Behandlung gegen beissende und saugende Insekten, Obstmade, Blutlaus und Spinnmilbe. **Investigation on the residue–behaviour of Diazinon and Fenvalerate on/in apples after a treatment against biting and sucking insects, codling moth, wolly aphid and mites.**

8715 Weinmann, W.   DE 215080/78/0019 N
Untersuchung über das Rückstandsverhalten von Chlorthiophos auf/in Äpfeln nach einer Behandlung gegen beissende und saugende Insekten. **Investigation on the residue–behaviour of Chlorthiophos on/in apples after a treatment against biting and sucking insects.**

8716 Dickler, E.   DE 215210/73/4006
Untersuchungen zur Biologie und Populationsdynamik des Apfelbaumglasflüglers Synanthedon myopaeformis. **Investigations on the biology and population dynamics of the apple clearwing moth Synanthedon myopaeformis.**

8717 Dickler, E.; Huber, J.   DE 215210/74/5006
Freilandversuch zur Bekämpfung des Apfelwicklers Laspeyresia Pomonella mit Granuloseviren. **Field experiments on codling moth Laspeyresia pomonella control using granulosis virus.**

8718 Dickler, E.   DE 215210/77/0005
Untersuchungen zur Biologie und Populationsdynamik des Apfelwicklers Laspeyresia pomonella. **Studies on biology and population dynamics of Laspeyresia pomonella.**

8719 Milaire, H.; Arnoux, M.; Audemard, H.; Marboutie; Rambier, A.   FR 010106/75/5258
Lutte intégrée en vergers de pommiers. **Integrated control in apple orchards.**

8720 Audemard, H.; Voegele, J.; Causse, R.; Ferron, P.; Guennelon, G. Mme   FR 010602/75/5095
Lutte intégrée en verger : étude de nouvelles méthodes de lutte contre le Carpocapse. **Integrated control of Codling Moth in apple orchard.**

8721 Massonie, G.; Meymerit, J.C.   FR 010707/76/5227
Résistance du pommier au puceron cendre du pommier Dysaphis Plantaginea Pass. **Resistance to Rosy apple aphid Dysaphis Plantaginea in apple.**

8722 Martouret, D.; D'Aguilar, J.   FR 011102/66/1720
Entomofaune d'un verger de pommier. **Entomocoenosis in apple orchards in Northern France.** Publications.

8723 Martouret, D.; Goujet, M.   FR 011102/71/1721
Etude des effets des interventions pesticides sur l'entomofaune du verger de pommier et de son environnement. **Effect of pesticides on entomofauna of an apple orchard and its environment.**

8724 Martouret, D.; Goujet, R.; Guilbot, R.   FR 011102/74/5074
Etude des Tordeuses des bourgeons de pommiers. **Study of the apple Budworms.**

8725 Martouret, D.; Goujet, R.; Voegele, J.

FR 011102/75/5075

Lutte biologique à l'aide de Trichogrammes contre le Carpocapse des pommes. **Biological control of the Codling Moth with Trichogramma.**

8726 Martouret, D.; Guilbot, R.　　　FR 011102/76/5076
Influence de l'époque de taille sur les populations de Tordeuses des bourgeons en verger de pommiers. **Influence of the time of pruning on the apple–Budworms populations in apple orchards.**

8727 Rambier, A.; Cotton, M.　　　FR 011210/66/1731
Synécologie des tétranyques du pommier. **Synecology of apple tetranychus–mites.** Publications.

8728 Cranham　　　GB 011001/00/0005 R
Resistance to pesticides of Tetranychid mites of apple and hops.

8729 Cranham; Solomon　　　GB 011001/00/0006 R
Integrate control of pests.

8730 Easterbrook; Lyth　　　GB 011001/00/0007 R
Assess damage by apple pests.

8731 Glen; Solomon　　　GB 011514/00/0001 R
Ecology and intergrated control of moth pests.

8732 Edwards　　　GB 040702/00/0005 R
Apple orchard pests.

8733 Bergamini, A.; Cappellini, P.; Sacco, M.
　　　IT 021500/79/0003 N
Prove di lotta integrata su melo, pero e ciliegio e studio sull'influenza della concimazione azotata sulle popolazioni di parassiti del melo. **Study on the nitrogen fertilization influence on the apple parasite population and integrated control program on apple, pear and cherry trees.**

8734 Gruys, P.　　　NL 050101/65/3795 R
Basisonderzoek t.b.v. een geïntegreerd bestrijdingsschema voor appelboomgaarden. **Development of an integrated control programme for orchards in the Netherlands.** Publications.

Pear (B 3612)

See also 7960, 8710, 8733

8735 Onillon, J.C.; Atger, P.; Franco, E.; Onillon, J.
　　　FR 010503/73/5045
Bioécologie du Psylle du poirier et du Psylle du mimosa. **Bioecology of the pear psylla and of the mimosa psylla.**

8736 Atger, P.　　　FR 010602/76/5103
Etude de la biocoenose des vergers de poirier et de la lutte contre le psylle. **Study of the pear–orchards biocoenosis and control of the Psyllid.**

8737 Audemard, H.; Martouret, D.; Arambourg, Y.;
Descoins, C.　　　FR 010602/77/5109
Lutte intégrée en verger : étude de nouvelles méthodes de lutte contre la Zeuzère (Zeuzera pyrina L.). **Integrated control in orchard : New approaches in leopard moth control.**

8738 Fideghelli, C.　　　IT 021500/75/0013
Miglioramento genetico del pero. **Pear breeding.**

Other top fruit (B 3619)

See also 8490, 8491, 8504, 8710, 8733

8739 Jacob, H.; Rössner, J.　　　DE 129163/75/0004
Nematodenübertragbare Viren bei Kirschen. **Nematode–transmitted viruses in cherries.**

8740 Haisch, A.　　　DE 502055/70/0001
Genetische Bekämpfung der Kirschenfliege durch Strahlensterilisation einschliesslich deren Massenaufzucht und Untersuchung der Populationsdynamik. **Genetic control of the European cherry fruit fly by radio sterilisation inclusively its mass rearing and investigation of its population dynamics.**

8741 Milaire, H.; Rambier, A.; Leclant, F.; Arnoux, M.;
Audemard, A.; Marboutie – Iperti, G.　　　FR 010106/75/5259
Lutte intégrée en vergers de pêchers. **Integrated control in peach orchards.**

8742 Milaire, H.; Marboutie; Audemard, H.; Arnoux, M.
　　　FR 010106/75/5260
Aménagement de la lutte contre la Tordeuse orientale du pêcher. **Supervised control on oriental fruit moth.**

8743 Audemard, H.; Guennelon, G. Mme; Causse, R.
　　　FR 010602/75/5097
Dynamique des populations de la petite mineuse du pêcher. **Dynamics of the peach twig borer moth populations.**

8744 Guennelon, G. Mme　　　FR 010602/76/5104
Mise au point d'un élevage permanent d'Anarsia lineatella sur milieu artificiel. **Development of a permanent rearing of the Peach twig borer, Anarsia lineatella, on artificial media.**

8745 Massonie, G.; Meymerit, J.C.; Maison, P.
　　　FR 010707/74/5218
Résistance du pêcher au Puceron vert du Pêcher Myzus persicae Sulzer. **Resistance to peach aphid Myzus persicae Sulz. in Peach tree.**

8746 Massonie, G.; Maison, P.; Meymerit, J.C.
　　　FR 010707/78/5230
Lutte culturale dirigée contre les pucerons du pêcher et du prunier. **Control of peach and plum aphids by agricultural methods.**

8747 Blaisinger, P.; Simonis, M.T.　　　FR 010904/70/5021
Lutte intégrée en vergers de pruniers. **Integrated control in plum orchards.**

8748 Baker; Miller　　　GB 050310/74/0012 R
Survival of cherry fruit fly.

8749 Ferrari, R.　　　IT 020400/79/0001 N
Prove di lotta guidata contro i fitofagi del pesco. **Supervised control against peach pests.**

8750 Ricciardi, P.; Manzo, P.　　　IT 021500/73/0033 N
Indagini sulla stanchezza del terreno coltivato a pescheto. **Study on the peach replant disease.** Publications.

8751 Frilli, F.　　　IT 040402/78/1136 N
Ricerche su Rhagoletis cerasi, studio sulla dinamica delle popolazioni nel piacentino e sull'andamento delle infestazioni anche in rapporto alle diverse cultivar. **Research on Rhagoletis**

cerasi, a study on population dynamics in the Piacenza region and on the infection progress also in relation to different cultivars.

8752 Fimiani, P.   IT 040703/74/0006
Ricerche biologiche sulla mosca delle ciliege (Rhagoletis Cerasi). **Biological researches on the European cherry fruit fly (Rhagoletis cerasi).** Publications.

8753 Foschi, S.   IT 061800/77/1009
Ricerche per un adeguamento della difesa contro Psylla ps. su pero. **Research aiming to improve Psylla ps. control in the peach tree.**

Soft fruit (berries and cane fruits) (B 3620)

See also 7974, 8627, 8636, 8655

8754 Soenen, A.; Marcelle, R.; Simon, P.; Pittevils, J.; Gilles, G.; Porreye, W.   BE 140000/72/0005 R
Parasieten– en ziektebestrijding in de aardbeienteelt. **Control of parasites and diseases in strawberry culture.** Publications.

8755 Weinmann, W.   DE 215080/78/0021 N
Untersuchung über das Rückstandsverhalten von Endosulfan auf/in Johannisbeeren nach einer Behandlung gegen Johannisbeergallmilbe. **Investigation on the residue–behaviour of Endosulfan on/in black currant after a treatment against black currant gall mite.**

8756 Seemüller, E.; Grünwald, J.   DE 215210/74/5003
Untersuchungen über die Zerstörung des Periderms der Himbeerrute durch die Himbeerrutengallmücke und den Pilz Leptosphaeria coniothyrium. **Investigations on the destruction of the raspberry spur periderm by the raspberry gall midge and by Leptosphaeria coniothyrium.**

8757 Warmbrunn, K.; Bosch, J.   DE 501050/73/0001
Entwicklung integrierter Bekämpfungsverfahren im Beerenobstbau. **Development of integrated control methods in soft fruit growing.**

8758 Bouchery, Y.; Jacky, F.   FR 010904/71/1749
Etude physiologique de la transmission de virus par Aphis Fabae. **Physiological study on transmission by Aphis Fabae.**

8759 Lyth; Easterbrook   GB 011001/76/0020 R
Resistance of hosts to pests (aid to breeding).

8760 Stacey   GB 011108/00/0015 R
Integrated control of strawberry pests.

8761 Trudgill; Brown   GB 030705/75/0012 R
Ecology and control of Pratylenchus spp. associated with soft fruit.

8762 Cutler; Symonds   GB 030903/75/0105 R
Survey of pesticide usage on soft fruit.

8763 Chapman; Cotten   GB 050314/74/0004 R
Control of root nematodes Xiphinema and Longidorus in relation to fruit certification schemes.

8764 Dunne, R.M.   IE 060300/76/1324 R
Control of wingless weevils (ceutorrhynchus sp) on strawberries. Publications.

8765 Miotto, F.   IT 020400/78/0007
Ricerche sull'acarofauna del Gelso (Artropoda – Arachnida). **Researches on Mites of Mulberry–tree (Artropoda – Arachnida).**

Citrus fruit (B 3630)

8766 Benassy, C.; Franco, E.; Bianchi, H.; Euverte, G.; Pinet, C.   FR 010503/63/5029
Lutte biologique contre les cochenilles diaspines des agrumes. **Biological control of the diaspipid citrus–scales.**

8767 Onillon, J.C.   FR 010503/68/1793
Lutte biologique contre l'aleurode des agrumes. **Biological control of the citrus woolly white fly.** Publications.

8768 Onillon, J.C.; Brun, A.; Brun, P.; Rodolphe, R.; Franco, E.; Benassy, C.   FR 010503/69/5035
Lutte biologique contre les Aleurodes des Citrus. **Biological control of the citrus whiteflies.**

8769 Brun, P.   FR 012202/73/0107
Etude des ravageurs des agrumes et de leurs rapports avec la physiologie de la plante et les techniques culturales. **Citrus pests and their relationships with plant physiology and cultivation methods.**

8770 Di Martino, E.   IT 021600/69/0002
Ricerche sui problemi dei residui di antiparassitari negli agrumi. **Studies on pesticides residues in citrus fruits.**

8771 Di Martino, E.; Benfatto, D.   IT 021600/79/0006 N
Indagine sulla etologia delle cocciniglie degli agrumi e dell'entomofauna utile. **Studies on biology of citrus scales and useful Arthropods.**

8772 Di Martino, E.; Benfatto, D.; Caruso, A.   IT 021600/79/0007 N
Studi sulla lotta contro i fitofagi degli agrumi. **Studies on control against citrus pests.**

8773 Genduso, P.   IT 040903/74/0559
Ricerche sulle tignole degli agrumi e sui loro simbionti. **Research on citrus fruit moths and their symbionts.**

8774 Liotta, G.   IT 040903/77/0215 R
Gli entomofagi delle cocciniglie e degli aleurodidi degli agrumi. **Predaters of coccinellidae and aleyrodidae of citrus fruits.**

8775 Mineo, G.   IT 040903/77/0243 R
Ricerche sul Prays Citri Mill.– tignola degli agrumi– e sui suoi simbionti. **Research on Prays Citri Mill. – citrus fruit weevil – and on its symbionts.**

8776 Inserra, R.   IT 060600/70/0053
Lotta chimica contro i nematodi fitoparassiti. Tylenchulus semipenetrans su agrumi. **Chemical control of plant parasitic nematodes. Tylenchulus semipenetrans on citrus.** Publications.

8777 Inserra, R.   IT 060600/74/0041
Ricerche su biologia e patogenicità di nematodi fitoparassiti. Patogenicità di Pratylenchus vulnus su agrumi ed olivo. **Biology and pathogenicity of Pratylenchus vulnus on citrus and olive trees.**

8778 Pessina, F.　　　　　　　　IT 121100/78/1140 N
Prove di laboratorio di adulti di mosca degli agrumi con
attrattivi e con insetticidi addizionati di attrattivi e prove di
campo con miscele di insetticidi e di attrattivi contro la mosca
dell'olivo. **Laboratory tests on citrus fruit adult flies with
attractants and insecticides containing attractants; field tests
with mixtures of insecticides and attractants to control the olive
fly.**

Grapes (B 3650)

See also 9883, 15348

8779 Alleweldt, G.; Martin, C.　　　DE 144450/74/0003
Kohlenhydraternährung der Reblaus in Beziehung zu
Resistenzfragen. **Nutrition of Phylloxera with carbohydrates in
relation to resistance.**

8780 Steche, W.; Müller, F.　　　　DE 144541/74/0002
Ursachen der Vergiftung von Bienenvölkern durch im
Weinbau eingesetzte Insektizide. **The causes of bee colonies
intoxication by insecticides applied in vine–growing.**

8781 Rilling, G.　　　　　　　　DE 204000/78/0005 N
Einfluss der Reblausnodositäten auf die Einlagerung von
Reservestoffen in die Rebwurzeln. **Influence of phylloxera root
galls on accumulation of reserve substances in grape roots.**

8782 Rilling, G.　　　　　　　　DE 204000/78/0006 N
Die genetische Potenz der Rebe zur Abkapselungsreaktion in
den Wurzelgallen der Reblaus. **The genetic potential of grape
vine for isolating necrotic tissues in phylloxera root galls.**

8783 Rilling, G.　　　　　　　　DE 204000/78/0007 N
Histologische Untersuchungen über die Abkapselungsreaktion
in den Wurzelgallen der Reblaus. **Histological investigations on
the isolation of necrotic tissues in phylloxera root galls.**

8784 Rilling, G.　　　　　　　　DE 204000/78/0008 N
Die genetische Potenz der Rebe zur Induktion geflügelter
Rebläuse. **The genetic potential of grape vine for induction of
winged phylloxera.**

8785 Rilling, G.　　　　　　　　DE 204000/78/0009 N
Die Bedeutung der Stickstoffversorgung der Rebe für die
Induktion geflügelter Rebläuse. **Effects of nitrogen nutrition of
grape vine on development of winged phylloxera.**

8786 Weinmann, W.　　　　　　　DE 215080/75/0008
Untersuchungen über das Rückstandsverhalten von Schwefel
auf und in der Weintraube nach der Bekämpfung gegen
Mehltau. **Investigation of the residue behaviour of sulphur on
and in grapes after mildew control.**

8787 Weischer, B.　　　　　　　DE 215190/77/0007
Untersuchungen über die Übertragung von Rebenvirosen
durch Arten der Gattungen Longidorus und Xiphinema.
**Studies on transmission of vine viroses by Longidorus and
Xiphinema.**

8788 Weischer, B.　　　　　　　DE 215190/77/0008
Untersuchungen zur Biologie und Pathogenität
virusübertragender Nematoden bei Reben. **Studies on biology
and pathogenicity of virus–transmitting nematodes in vine.**

8789 Englert, W.D.　　　　　　　DE 215220/77/0007
Untersuchungen über den Zusammenhang zwischen Düngung
der Reben und starkem Spinnmilbenbefall. **Studies on
correlations between fertilization of vines and strong attack of
red spider.**

8790 Englert, W.D.　　　　　　　DE 215220/77/0014
Untersuchungen zur Biologie und Bekämpfung von tierischen
Rebschädlingen. **Investigations on biology and control of vine
pests.**

8791 Englert, W.D.　　　　　　　DE 215220/77/0015
Ausarbeitung von Verfahren zur Prüfung von
Schädlingsbekämpfungsmitteln im Weinbau. **Development of
methods for testing of pesticides in viticulture.**

8792 Englert, W.D.　　　　　　　DE 215220/78/0001 N
Untersuchungen zum Einsatz von Insektiziden im Weinbau mit
dem Hubschrauber. **Investigations on pest–control in vineyards
by helicopter.**

8793 Englert, W.D.　　　　　　　DE 215220/78/0006 N
Die Bekämpfung des Gefurchten Dickmaulrüsslers –
Brachyrhinus sulcatus F. – ohne Aldrin. **Control of black vine ·
weevil – Brachyrhinus sulcatus – without aldrin.**

8794 Englert, W.D.　　　　　　　DE 215220/78/0007 N
Biologie von nützlichen Insekten und Milben im Weinbau und
deren Einfluss auf Schädlingspopulationen. **Biology of useful
insects and mites in viticulture and their influence on pest
populations.**

8795 Schruft, G.　　　　　　　　DE 501100/74/0001
Untersuchungen über Nebenwirkungen von
Pflanzenschutzmitteln im Weinbau. **Investigations on
side–effects of pesticides in viticulture.**

8796 Schruft, G.　　　　　　　　DE 501100/74/0002
Integrierte Bekämpfungsmassnahmen gegenüber tierischen
Schädlingen im Weinbau. **Integrated methods of pest control in
viticulture.**

8797 Staudt, G.　　　　　　　　DE 501100/78/0001 N
Die Einwirkung von Pflanzenschutzmitteln auf Pollenkeimung
und Pollenschlauchwachstum. **The influence of pesticides on
pollen germination and pollen tube growth.**

8798 Lemperle, E.　　　　　　　DE 501105/74/0003
Wirkstoffrückstände von Fungiziden und Insektiziden in
Weintrauben und Wein. **Active residues of fungicides and
insecticides in grapes and wine.**

8799 Schruft, G.　　　　　　　　DE 501107/75/0001
Untersuchungen zum Auftreten und zur Bekämpfung von
Erdraupen im Weinbau. **Studies on the occurrence and control
of cutworms in viticulture.**

8800 Dieter, A.; Geiger, K.　　　　DE 502153/77/0001
Vergleichende Untersuchungen über die Intensität der
Infektion mit Pilzkrankheiten und Schädlingen in
flurbereinigten und nicht flurbereinigten Weinbergen.
**Comparative investigations on the intensity of infection by
fungi and pests in consolidated and non–consolidated
vineyards.**

8801 Dieter, A.　　　　　　　　DE 502154/74/0001

D 2410 – Pests of plants and pest control

Untersuchungen zur Populationsdynamik und Bekämpfung ektoparasitärer Wurzelnematoden in Rebschulen und Junganlagen. **Examination of population dynamics of ectoparasitic root nematodes and their control in vine nurseries and new–plant vineyards.**

8802 Dieter, A.　　　　　　　　DE 502154/74/0003
Untersuchungen über die Wirkungsweise von Juvenilhormonpräparaten und Metamorphosehemmern bei weinbaulich schädlichen Lepidopteren. **Experiments on the efficacy of juvenile hormone preparations and growth inhibitors to parasitic Lepidopteres on grapes.**

8803 Rühling, W.; Bäcker, G.; Brendel, G.
　　　　　　　　　　　　　　DE 506113/78/0002 N
Eignungsuntersuchung zu Applikationsverfahren mit geringen Aufwandmengen für den Weinbau. **Studies on the possibilities of using low volumes for pest control in vineyards.**

8804 Eichhorn, K.W.　　　　　　DE 509153/77/0005
Untersuchungen zur Nebenwirkung von Fungiziden auf Spinnmilben. **Secondary effects of fungicides on Tetranychidae.**

8805 Schumann, F.　　　　　　　DE 509154/73/0003
Untersuchungen an Vitis–Arten und –Kreuzungen bezüglich ihrer Verwendung als Unterlage zur Reblausbekämpfung Untersuchungen über Wildreben – V.v.L.var.Silvestris Gmel – und ihre Erhaltung am Oberrhein. **Studies on vitis–species and –crossings in regard to their use as rootstocks Studies on Vitis v.L.var.silvestris Gmel. and on their conservation in the forests of the Rhine valley.**

8806 Schaefer, H.　　　　　　　DE 509154/73/0005
Untersuchungen über den Stoffwechsel von gesunden und reblausbefallenen Rebenblättern. Untersuchungen über den Einfluss der Bodenart und der Unterlage auf den Stoffwechsel der Rebenblätter. **Studies on the metabolism of healthy and phylloxera–galled grape leaves Studies on the influence of soil types and of the rootstocks on the metabolism of grape leaves.**

8807 Rüdel, M.　　　　　　　　DE 509157/77/0001 N
Wirtspflanzenkreis – einschlieblich Rebsorten – der Nematodenüberträger von Rebviren. **Host plants – incl. vines – as vectors of vine nematodes.**

8808 Dalmasso, A.; Cardin, M.C: Mme; Cuany, A.
　　　　　　　　　　　　　　FR 010504/62/5151
Les nématodes du vignoble – Ecologie et lutte. **Nematodes in vineyard – Ecology and control.**

8809 Roehrich, R.; Maison　　　FR 010707/69/1786
Nuisibilité de l'Eudémis. **Harmfulness of Eudemis in vines.**

8810 Roehrich, R.; Carles, J.P.; Saint Luc, L.
　　　　　　　　　　　　　　FR 010707/74/5217
Utilisation des phéromones en vignoble comme indications de présence ou d'abondance. **Use of pheromons in vineyard for prognosis and population estimate.**

8811 Bournier, A.; Pivot, Y.　　FR 011210/57/1734
Etude d'un thrips nuisible à la vigne. **Study of a noxious thrips of vine.** Publications.

8812 Rambier, A.　　　　　　　FR 011210/74/5238
Effets secondaires des pesticides sur les Acariens de la vigne.

Secondary effects of pesticides on mites harmful to vine yards.

8813 Pegazzano, F.　　　　　　IT 020400/73/0004
Ricerche sulla acarofauna della vite. **Mite fauna associated with grape vine in Italy.**

8814 Egger, E.; Borgo, M.; Corino, L.　　IT 021300/65/0001 R
Prove di lotta contro particolari parassiti della vite. **Trials of struggle against some vine parasites.** Publications.

8815 Egger, E.; Borgo, M.; Costacurta, A.; Calò, A.
　　　　　　　　　　　　　　IT 021300/78/0001 R
Studio sul comportamento di viti ammalate, risanate con la termoterapia, rispetto a viti risultate sane ai saggi biologici nei riguardi della sensibilità ai nematodi ed ai virus. **Study on the behaviour of thermotherapy healed vines in comparison with healthy ones.**

Edible nut fruits (B 3660)

8816 Rambier, A.　　　　　　　FR 011210/75/5239
Protection du noyer contre les Acariens. **Protection of the walnut–tree against mites.**

8817 Fauvel, G.; Rambier, A.; Audemard, H.
　　　　　　　　　　　　　　FR 011210/75/5240
Rôle de l'environnement du verger sur les insectes et acariens prédateurs. **Role of the orchard environment on the predatory insects and mites.**

8818 Genduso, P.　　　　　　　IT 040903/73/0220
Ricerche sugli oofagi degli eterotteri del nocciolo. **Research on the oophores of hazelnut Heteroptera.** Publications.

Ornamentals and ornamental products in general (B 3700)

See also 7975, 7980, 8501, 8609, 8611, 8614, 8620, 9597

8819 Gillard, A.; Pelerents, G.; Van Daele, E.; Heungens, A.
　　　　　　　　　　　　　　BE 030024/59/0002 R
Bescherming van sierplanten tegen hun dierlijke parasieten inzonderheid mijten, nematoden en insekten. **Protecting ornamentals against specific animal parasites as acari, nematodes and insects.** Publications.

8820 D'Herde, C.; Coolen, W.; Hendrickx, G.
　　　　　　　　　　　　　　BE 070100/68/0004 R
Studie van de efficaciteit van systemische nematiciden in de sierteelt; knolbegonia's potplanten en boomteelt. **Study of the effectiveness of systemic nematicides in ornamental plantgrowing: tuberous begonia's potplants and tree nurseries.** Publications.

8821 D'Herde, C.J.; Coolen, W.A.; Hendrickx, G.
　　　　　　　　　　　　　　BE 070100/76/0016 R
Nematologisch survey onderzoek in de boomteelt en in de kasrozenteelt. **Nematological survey in tree nurseries and in glasshouse roses.**

8822 Nøddegaard, E.; Rasmussen, A.N.
　　　　　　　　　　　　　　DK 010116/78/0027 N
Nematiciders virkning på nematodearter på prydplanter, friland. **Effect of nematicides on nematodes on outdoor ornamentals.**

8823 Nøddegaard, E.; Rasmussen, A.N.
DK 010116/78/0028 N
Insekticiders virkning på bladlus i prydplanter, væksthus. **Effect of insekticides on aphids on glasshouse ornamentals.**

8824 Panis, A. FR 010503/70/1794
Lutte biologique et chimique contre les cochenilles des plantes ornementales de serre et de plein champ (sauf diaspididae). **Biological and chemical control of scale–insects of ornemental plants under glass house and field conditions (except diaspididae).** Publications.

8825 Stacey GB 011108/00/0004 R
Control of red spider mite on glasshouse ornamentals by the predator Phytoseiulus persimilis.

8826 Stacey GB 011108/00/0006 R
Control of whitefly in glasshouse by the parasite Encarsia formosa.

8827 Wyatt GB 011108/00/0007 R
Control of aphids on glasshouse crops by parasites.

8828 Scopes GB 011108/00/0008 R
Chemical control of glasshouse pests.

8829 Burges GB 011108/00/0010 R
Ecology of protozoa and their potential for control of arthropod pests.

8830 Payne GB 011108/00/0013 R
Virus infections of invertebrate pests.

8831 Burges; Payne GB 011108/00/0014 R
Biological control of secondary glasshouse pests.

8832 Turl GB 030902/68/0403 R
Identification and taxonomy of immature stages of Spodoptera litura and S. littoralis.

8833 Foster; Tones GB 060309/00/0005 R
The biology and control of the glasshouse symphilid.

Bulbs (B 3710)

8834 Powell GB 050310/76/0010 R
Fumigation of narcissus bulbs to control bulb scale mite.

8835 Windrich, W.A. NL 010108/66/1059
Stengelaaltjes in bolgewassen. **Stem eel worms in bulbous crops.** Publications.

8836 Windrich, W.A. NL 010205/66/1059
Stengelaaltjes in bolgewassen. **Stem eel worms in bulbous crops.** Publications.

Flowers and pot plants (B 3720)

See also 280, 7968, 8610, 8627, 9601

8837 Köllner, V. DE 215070/75/0002
Untersuchungen über die Biologie und die chemische Bekämpfung des Südafrikanischen Nelkenwicklers. **Studies on biology and chemical control of the South African carnation Tortrix.**

8838 Pralavorio, R. Mme; Desportes; Descoins, Ch.
FR 010503/73/5044
Biologie de la tordeuse de l'oeillet et méthodes de lutte. **Biology and control of the Carnation tortricid.**

8839 Powell GB 050310/75/0009 R
Alternatives to cold storage against Lepidoptera eggs and larvae in imported plant propagation material.

8840 Baker; Miller GB 050310/76/0006 R
Survival of carnation Tortricids.

8841 Duggan, J.J. IE 060300/73/0878 R
Control of nematodes occurring on pot plants. Publications.

8842 Mazzucchi, U. IT 040211/72/0511
Ricerche sulla caratterizzazione fitopatologica e biomolecolare dei batteri coliformi del marciume molle con particolare riguardo alle forme del gruppo chrysanthemi. **Research on phytopathological and biomolecular characterization of coliform bacteria of Marciume Molle with specific reference to the forms of chrysanthemum groups.** Publications.

8843 Evenhuis, H.H. NL 010108/72/3620
Economische betekenis en bestrijding van de gegroefde lapsnuittor, Otiorrhynchus sulcatus L. **Economic significance and control of the black vine weevil, Otiorrhynchus sulcatus L.**

8844 Vrie, M. van de NL 010108/75/6390
Onderzoek over bestrijding van schade door tripsaantasting in de sierteelt onder glas. **Research on control of damage by trips in floriculture.** Publications.

8845 Vrie, M. van de NL 010108/75/6391
Onderzoek over geintegreerde bestrijding van de spintmijt Tetranychus urticae Koch in de sierteelt onder glas. **Research on integrated control of Tetranychus urticae Koch in floriculture.** Publications.

8846 Vrie, M. van de NL 010108/76/7675
Levenswijze en bestrijding van de mineervlieg, Liriomyza sonchi (Hendel) op Gerbera. **Biology and control of the leaf miner, Liriomyza sonchi (Hendel) on Gerbera.**

8847 Vrie, M. van de NL 010108/76/7676
Levenswijze en bestrijdingsmogelijkheden van de bladroller, Clepsis spectrana (Treitschke), op sierteeltgewassen onder glas. **Biology and possibilities for control of the leafroller, Clepsis spectrana in ornamentals in glasshouses.**

8848 Vrie, M. van de NL 010108/77/7674
Levenswijze en bestrijdingsmogelijkheden van de witte vlieg, Trialeurodes vaporariorum op sierteeltgewassen onder glas. **Biology and control of the white fly, Trialeurodes vaporariorum, in ornamentals in glasshouses.**

8849 Vrie, M. van de NL 010108/78/7976 R
Spodoptera exigua op sietteeltgewassen in kasteelten. **Spodoptera exigua on ornamentals in glasshouse cultures.**

8850 Vrie, M. van de NL 010201/75/6390
Onderzoek over bestrijding van schade door tripsaantasting in de sierteelt onder glas. **Research on control of damage by trips in floriculture.**

8851 Vrie, M. van de NL 010201/75/6391

Onderzoek over geïntegreerde bestrijding van de spintmijt Tetranychus urticae Koch in de sierteelt onder glas. **Research on integrated control of Tetranychus urticae Koch in floriculture.**

8852 Vrie, M. van de   NL 010201/76/7675
Levenswijze en bestrijding van de mineervlieg, Liriomyza sonchi (Hendel) op Gerbera. **Biology and control of the leaf miner, Liriomyza sonchi (Hendel) on Gerbera.**

8853 Vrie, M. van de   NL 010201/76/7676
Levenswijze en bestrijdingsmogelijkheden van de bladroller, Clepsis spectrana (Treitschke), op sierteeltgewassen onder glas. **Biology and possibilities for control of the leaf roller, Clepsis spectrana in ornamentals in glasshouses.**

8854 Vrie, M. van de   NL 010201/77/7674
Levenswijze en bestrijdingsmogelijkheden van de witte vlieg, Trialeurodes vaporariorum op sierteeltgewassen onder glas. **Biology and control of the white fly, Trialeurodes vaporariorum in ornamentals in glasshouses.**

8855 Vrie, M. van de   NL 010201/78/7976
Spodoptera exigua op sierteeltgewassen in kasteelten. **Spodoptera exigua on ornamentals in glasshouse cultures.**

8856 Rabbinge, R.; Sabelis, M.E.; Wit, C.T. de
                                            NL 020054/75/6293
Geïntegreerde bestrijding van kasspint, tetranychus urticae (koch), in de sierteelt onder glas. **Integrated control of the two spotted spidermite, tetranychus urticae (koch), in roses in the glasshouse.** Publications.

Ornamental shrubs (B 3730)

See also 8597, 8735, 8843

8857 Börner, H.; Rienow, W.   DE 148200/78/0002 N
Chemische Bekämpfung von Meloidogyne hapla in Rosenkulturen. **Chemical control of Meloidogyne hapla in roses.**

8858 Caron, J.E.A.   NL 010203/71/8665 N
Middelen beproeving ter bestrijding van larve en kever van de Taxuskever. **Reserach on new insecticides for control of larve and beetle of the black vine weevil (Otiorrhynchus sulcatus L.).**

8859 Bruin, Th. de   NL 010300/78/9115 N
Bestrijding mestkevers (Aphodius sp.) bij de graszodenteelt. **Control of Aphodius sp. in sod production.**

8860 Ankersmit, G.W.   NL 020014/77/8641 N
Het isolerend effect van kassen op kasinsecten i.h.b. bij het "Rozen–ijltje" Clepsis spectrana. **The isolating effect of green houses on greenhouse insects especially with Clepsis spectrana.**

Other ornamentals and ornamental products (B 3790)

8861 Kohler, A.; Zeltner, G.-H.; Ottow, J.C.G.
                                            DE 144990/75/0002
Der Einfluss bakteriellen Aufwuchses auf submerse Makrophyten insbesondere auf Potamogeton lucens u. P. crispus bei unterschiedlicher Belastung mit NH+4 und PO3–4. **The influence of bacterial growth on submerged macrophytes, especially on Potamogeton lucens and P. crispus during**

differing pollution with NH+4 and PO3–4.

8862 Willis; Gallaher   GB 040702/77/0007 R
Biology and control of the vine weevil as a pest of ornamental trees and shrubs.

Forests in general (B 3800)

See also 1826, 2072, 2206, 7956, 8027

8863 Huygh, A.; Nef, L.   BE 140000/72/0013
Plagenbestrijding in bossen en hun ecologische nevenwerkingen. **Pest control in forests and their ecological side effects.** Publications.

8864 Vite, J.P.   DE 126150/77/0001
Untersuchungen von Borkenkäferlockstoffen und ihre mögliche Nutzanwendung. **Studies on bark beetle pheromones and their potential application.**

8865 Speidel, G.; Hoesch, M.   DE 126450/75/0003
Die Wildschäden im Wald aus der Sicht der Forstökonomie. **Damages by game in the view of forest economy.**

8866 Bombosch, S.; Sanders, W.   DE 132660/70/0006
Untersuchungen zur Lenkung freilebender Blattlausräuber durch optische und chemische Marken. **Investigations into the directing of free–living Aphidophagous predators by use of optical and chemical signals.**

8867 Führer, E.   DE 132660/70/0007
Der Einfluss des Parasitismus von Apanteles glomeratus L. – Hym., Braconidae – auf den Stoffwechsel der Wirtslarve Pieris brassicae L. – Lep., Pieridae –. **The effect of the parasitism of Apanteles glomeratus L. – Hym., Braconidae – on the metabolism of the host larva, Pieris brassicae L. – Lep., Pieridae –.**

8868 Bombosch, S.   DE 132660/70/0008
Versuche zur Rationalisierung der Bekämpfung des Buchdruckers – Ips typographus – durch ein schnell arbeitendes Injektionsverfahren. **Attempts for rationalizing the control of Ips typographous by using a rapid injection method.**

8869 Führer, E.   DE 132660/74/0001
Untersuchungen über die Toleranz von Schmetterlingspuppen gegenüber Puppenparasiten und über ihre Ursachen. **Investigations on the tolerance of Lepidoptera pupae of pupal parasites and on the causes.**

8870 Kerck, K.   DE 132660/74/0004
Möglichkeiten der Anwendung von Primärlockstoffen in biotechnischen Bekämpfungsverfahren gegen forstlich schädliche Käfer. **Applicability of primary baits to biotechnical methods of control of noxious beetles in forests.**

8871 Sanders, W.   DE 132660/75/0003
Das Fortpflanzungsverhalten des Borkenkäfers Pityogenes chalcographus. Der Einfluss von Aussenreizen auf Suchaktivität und Brutbaumwahl. **The reproductive behaviour of the bark beetle Pityogenes chalcographus. The influence of external stimuli on the searching activity and the selection of breeding places.**

8872 Führer, E.   DE 132660/75/0004
Untersuchungen über fortpflanzungsbiologische

Unverträglichkeit von Populationen des Kupferstechers–
Pityogenes chalcographus – unterschiedlicher Herkunft.
**Studies on reproductive incompatibility between allopatric
populations of the scolytid beetle – Pityogenes chalcographus
–. Publications.**

8873 Schneider, I.   DE 132660/77/0002
Untersuchungen zur biotechnischen Bekämpfung des
Kiefernknospentriebwicklers Rhyacionia buoliana mit Hilfe
von Sexualpheromonen. **Investigations of a possible
biotechnical control of Rhyacionia buoliana by sex pheromones.**

8874 Führer, E.; Elsufty, R.   DE 132660/77/0003
Die Bedeutung physiologischer Effekte von Endoparasiten für
den Aufbau der Wirtskutikula und für die Empfindlichkeit der
Wirte gegenüber pathogenen Pilzen. **The significance of
physiological effects, caused by endoparasites, for the
deposition on the host's cuticle and for the susceptibility of the
hosts to pathogenic fungi.**

8875 Führer, E.; Willers, D.   DE 132660/77/0004
Untersuchungen über die Tauglichkeit verschiedener
Wirtsarten für die Larvenentwicklung einiger Puppenparasiten
der Unterfamilie Pimplinae. **Studies on the suitability of
different host species for the larval development of some pupal
parasitoids, subfamily Pimplinae.**

8876 Führer, E.; Vyplel, G.; Meyerweissflog, C.; Sturies,
H.–J.   DE 132660/77/0005
Untersuchungen über die cytologischen und physiologischen
Ursachen der reproduktiven Inkompatibilität allopatrischer
Populationen des Borkenkäfers Pityogenes chalcographus.
**Studies on the cytological and physiological reasons of
reproductive incompatibility of allopatric Pityogenes
chalcographus populations. Publications.**

8877 Lunderstädt, J.; Schopf, R.   DE 132660/77/0006
Physiologische Grundlagen der Auswertbarkeit von
Pflanzeninhaltsstoffen, insbesondere Proteinen, durch
Schadinsekten. **Physiological basis of the utilizability of plant
born compounds, especially proteins, by forest damaging
insects.**

8878 Bombosch, S.; Göttsche, D.   DE 132660/77/0007
Untersuchungen über Einsatzmöglichkeiten biotechnischer
Verfahren gegen Samenschädlinge der Gattung Megastigmus.
**Investigations into using biotechnical methods against seed
chalcids of the genus Megastigmus.**

8879 Führer, E.; Wulf, A.   DE 132660/77/0008
Untersuchungen über die Anfälligkeit allopatrischer
Populationen und aus ihnen hervorgegangener Hybriden des
Borkenkäfers Pityogenes chalcographus für den pathogenen
Pilz Beauveria bassiana. **Studies on the susceptibility of
allopatric populations and of their F1–hybrids of the Scolytid
beetle Pityogenes chalcographus for the pathogenic fungus
Beauveria bassiana.**

8880 Heyns, K.; Heemann, V.   DE 135151/72/0001
Nachweis und Identifizierung von Substanzen, die den
Primärund Sekundärbefall von Xyloterus lineatus Oliv. –
Coleoptera: Scolytidae – auslösen. **Identification of substances
inducing the attack of Xyloterus lineatus Oliv. – Coleoptera:
Scolytidae –.**

8881 Heyns, K.; Francke, W.   DE 135151/72/0002
Nachweis und Identifizierung von Aggregationssubstanzen in
dem Ambrosiakäfer Xyloterus domesticus L. – Coleoptera:
Scolytidae –. **Aggregating–pheromones in the ambrosia beetle
Xyloterus domesticus L. – Coleoptera: Scolytidae –.**

8882 Francke, W.; Bühring, M.   DE 135151/73/0002
Flüchtige Inhaltsstoffe der grossen roten Waldameise Formica
rufa. **Volatile substances in Formica rufa.**

8883 Skatulla, U.   DE 160091/72/0001
Feststellung und Untersuchung von Krankheiten
forstschädlicher Insekten 1971. **english title not indicated.**

8884 Postner, M.   DE 160091/72/0004 R
Untersuchungen über die Borkenkäferfauna des im Walde
verbleibenden Holzes und ihre wirtschaftliche Bedeutung
1971–1978. **Investigations on the bark beetle fauna in residual
wood in forest and economic importance.**

8885 Haeselbarth, E.   DE 160091/72/0005 R
Beziehungen zwischen Schlupfwespen als Parasiten und
forstschädlichen Insekten als Wirte in bayerischen
Nadelwäldern 1971. **Correlations between Ichneumonidae as
parasites and injurious forest insects as hosts in coniferous in
Bavaria.**

8886 Skatulla, U.   DE 160092/72/0001 R
Labor- und Freiland–Untersuchungen über die Wirkungsweise
von Bacillus thuringiensis Berl. auf forstschädliche
Lepidopteren 1971. **Experiments in laboratory and open field
on the action of Bacillus thuringiensis Berl. on injurious
Lepidoptera.**

8887 Krump, A.   DE 160092/72/0004 R
Über die Wirkung neuer chemischer Insektizide und
Ausbringungsverfahren bei der Bekämpfung von
Forstschädlingen 1970. **On the action and methods of spreading
of novel chemical insecticides for the control of injurious forest
pests.**

8888 Plochmann, R.; Mergner, W.   DE 160301/78/0001 N
Einfluss des Schalenwildes auf die bäuerliche
Waldbewirtschaftung. **The influence of ungulates on the
management of small farm woodlots.**

8889 König, E.; Bogenschütz, H.   DE 501508/77/0001 N
Erarbeitung eines Verfahrens zur Prüfung der
Nebenwirkungen von Pflanzenschutzmitteln auf
Coccygomimus turionellae unter Freilandbedingungen.
**Development of a guideline for testing side–effects of
pesticides on Coccygomimus turionellae under field
conditions.**

8890 König, E.; Korsch, J.   DE 501508/77/0003
Ermittlung von Kriterien zur Beurteilung der waldbaulich
tragbaren Schalenwilddichte. **Investigation concerning deer
population in relation to silvicultural criterions.**

8891 König, E.; Bogenschütz, H.   DE 501508/77/0006
Erarbeitung und Prüfung von Methoden zur
Schädlingskontrolle mittels Pheromonen. **Research and tests of
methods to control noxious animals by means of pheromones.**

8892 König, E.; Bogenschütz, H.   DE 501508/78/0002 N
Untersuchungen über den Zusammenhang zwischen

D 2410 – Pests of plants and pest control

Schädlingsbefall und Schaden an Waldbäumen. **Studies on the correlation between population density of pest insects and damage of forest trees.**

8893 Keil, W. DE 506200/70/0001 R
Biologische Bekämpfung des Eichenwicklers durch die Erhöhung der Siedlungsdichte von Höhlenbrütern 1966. **Biological control of Tortrix viridana by increasing the population density of cave–breeders.**

8894 Siebert, H.; Dimitri, L. DE 506453/72/0003
Untersuchungen über die Möglichkeiten der Wildschadensminderung durch Anbau von Verbissgehölzen zur Äsungsverbesserung. **Research on the possibilities of preventing damages by game through improvement of fodder supply.**

8895 Altenkirch, W. DE 507652/70/0004
Untersuchungen zur Ökologie des kleinen Frostspanners, Operophthera brumata L.. **Investigations on the ecology of the winter moth Operophthera brumata L..**

8896 Niemeyer, H. DE 507652/71/0001 N
Prüfung der praktischen Anwendung von Ameisenansiedlungen einschliesslich Erprobung verschiedener Nestschutzmethoden als Forstschutzmassnahme gegen bestimmte Schädlinge. **Investigations on propagation of ants– including methods of protection – for control of certain forest pests.**

8897 Niemeyer, H. DE 507652/71/0003
Untersuchungen über Biologie, Massenwechsel und Bekämpfung forstschädlicher Mäuse. **Investigations on biology, mass–appearance and control of rodents harmful in forestry – mice and voles –.**

8898 Niemeyer, H.; Winter, K. DE 507652/75/0001
Verbesserung der Bekämpfung rindenbrütender Borkenkäfer. **Investigations for improving the control methods of bark breeding Scolytidae.**

8899 Kolbe, H. DE 507652/75/0006
Entwicklung eines Prognoseverfahrens für den Befall durch Hylobius abietis. **Development of a method of prognosis of Hylobius abietis attack.**

8900 Kolbe, H. DE 507652/75/0007
Untersuchungen zur Biologie von Brachyderes incanus. **Investigations on the biology of Brachyderes incanus.**

8901 Winter, K. DE 507652/75/0008
Biologie und Verhalten forstlich wichtiger Bockkäfer unter besonderer Berücksichtigung holzzerstörender Arten. **Biology and behaviour of wood destroying longhorn beetles – Cerambycidae –.**

8902 Du Merle, P. FR 010609/64/4124
Biologie et écologie des Diptères Bombylides. **Biology and ecology of Dipterae Bombylidae.**

8903 Raymond, H. FR 011800/68/5288
Ecologie des Tabanidae de haute montagne. **Ecology of Tabanidae in high mountains.**

8904 Lieutier FR 011800/73/5290
Nématodes parasites des insectes xylophages, en particulier Scolytides. **Nematodes parasite of xylophagous insects, especially the Scolytids.**

8905 McAree, D.; Ward, D. IE 050100/57/7202 N
Evaluation and control of forest insect pests. Publications.

8906 Luitjes, L.; Frenken, G. NL 010601/77/7542
De ontwikkeling van schadelijke bastkevers in niet–marktwaardig dunningshout in relatie tot het tijdstip van velling en de diameter. **Development of harmful bark beetles in non–marketable thinning material in relation the time of felling and diameter.**

Pine forests in general (B 3810)

See also 1915

8907 Vite, J.P. DE 126150/77/0002
Molekulare Grundlagen des Aggregationsverhaltens forstschädlicher Kieferninsekten. **Molecular bases of the aggregation behavior of insect pests on pine.**

8908 Schneider, I. DE 132660/74/0002
Untersuchungen zur Populationsdynamik verschiedener Schädlinge an Koniferenzapfen. **Investigations into the population dynamics of several insect pests on cones of conifers.**

8909 Postner, M. DE 160091/72/0003 R
Wirkungen systemischer Insektizide gegen Forstschädlinge Gallmilben –Eriophyidae– als Schädlinge an Nadelbäumen 1971. **Identification and analysis of diseases of injurious foren–Effects of systemic insecticides on forest pest. Eriophest insects. yidae as pests in coniferous trees.**

8910 Altenkirch, W.; Kolbe, H. DE 507652/78/0001 N
Untersuchungen aus Anlass der Massenvermehrung von Kiefern–Grossschädlingen in Niedersachsen: Massenwechsel, Schaden und Schadensfolgen, neue Möglichkeiten zur Prognose und Bekämpfung. **Investigations on occasion of the outbreak of pine insect pests in Lower Saxony: population cycles, damage and consequences of damage, new possibilities of prognosis and control.**

8911 du Merle, P. FR 010609/64/4113
Insectes entomophages ennemis de la phase hypogée de la Processionnaire du Pin. **Entomophagous insects ennemies of Thaumetopoea Pityocampa (underground phase).**

8912 Dusaussoy, G.; Millet, A. FR 010609/71/4119
Impact des Tenthrèdes défoliatrices (Diprionidae) sur les reboisements de Pins. **Influence of defoliator saw–flies (Diprionidae) on Pine young plantations.**

8913 Carle, P.; Descoins, G. FR 010609/74/4102
Attractifs naturels et synthétiques de Scolytides des résineux. **Natural and synthetic attractants for Bark–Beetles of coniferous trees.**

8914 Carle, P. FR 010609/75/4103
Attaques du Scolytide Dendroctonus Micans sur les peuplements résineux du Massif Central. **Attacks of Dendroctonus Micans (Scolytideae) on coniferous stands in the Massif Central.**

8915 Demolin, G.; Dusaussoy, G. FR 010609/75/4112
Protection des jeunes reboisements en résineux contre leurs

principaux défoliateurs, notamment Thaumetopoea Pityocampa. **Protection of young coniferous stands against the defoliator insects, mainly Thaumetopoea Pityocampa.**

8916 Binazzi, A. IT 020400/76/0003 N
L'Afidofauna delle Conifere in Italia.. **Conifer aphid fauna of Italy.**

Fir forests (B 3811)

See also 8925

8917 Bejer, B.; Annila, E. DK 030109/79/0001 N
Skadeinsekter på Contortafyr (Pinus contorta). **Pests on Pinus contorta.**

8918 Scotto la Massese, C.; Baujard, P.; Laumond, C.; Boulbria, A. FR 010504/74/5168
Rôle des nématodes dans le dépérissement du Pin maritime. **Nematodes as possible agents of Pinus pinaster decline.**

8919 Fabre, J.P.; Gerbinot, B.; Riom, J. FR 010609/69/4110
Impact du prédateur Elatophilus Nigricornis sur les populations de Matsucoccus Feytaudi dans la forêt de Pin maritime en Provence. **Predator effect of Elatophilus Nigrocornis on populations of Matsucoccus feytaudi in the forest of Pinus Pinaster in Provence.**

8920 Schvester, D.; Toth, J.; Fabre, J.P.; Bonneau, M.; Carle, P. FR 010609/72/4109
Impact des insectes, notamment de Matsucoccus Feytaudi et possibilités d'avenir des régénérations naturelles de Pin maritime en Provence. **Role of insects, especially Matsucoccus Feytaudi, and chances of survival for Pinus Pinaster young stands in Provence.**

8921 De Bellis, R.; Tarsia, N.; Eccher, A.; Valenziano, S.; Lubrano, L. IT 011801/76/0014
Indagini sull'influenza dell'ambiente e dello stato vegetativo della pianta sull'intensità degli attacchi di Rhyacionia buoliana al Pinus radiata. **Investigations on the influence of environmental and tree vegetative conditions on the intensity of Rhyacionia buoliana attacks to Pinus radiata.**

8922 Binazzi, A. IT 020400/73/0003
Ricerche sull'entomofauna del Gen. Pinus: 2. Le successioni di insetti xilofagi sul Pinus pinaster in rapporto al deperimento delle pinete litoranee in Toscana. **Insect fauna associated with Pinus spp.: 2. The successions of xylophagous insects on Pinus pinaster in relation to decay of pine forests along the sea coast of Tuscany.**

8923 Doom, D.; Tol, G. van NL 010601/77/7534
Ontwikkeling van de grote dennesnuitkever, Hylobius abietis, in naaldhoutstobben in relatie tot het vellingstijdstip. **Development of the large pine weevil, Hylobius abietis, in coniferous stumps in relation to felling date.**

Spruce and fir forests (B 3812)

See also 8034

8924 Bombosch, S.; Lunderstädt, J.; Ramakers, P.M.H.
DE 132660/70/0004
Untersuchungen über die Nahrungsqualität von Fichtennadeln für forstliche Schadinsekten. **Investigations into the food**

quality of spruce needles for forest insect pests.

8925 Bombosch, S. DE 132660/75/0002
Versuche mit Photosyntheseblockern zur Verhinderung der Befallsdisposition nach chemischer Läuterung – Bu, Fi, Kie–. **Experiments with photosynthetic blocks in order to prevent bark beetle infestations after chemical cleaning – beech, spruce, pine –.**

8926 Schwenke, W.; Pausch, H. DE 160090/78/0001 N
Beiträge zur Populationsdynamik von Fichtenborkenkäfern. **Contributions to the population dynamics of spruce bark beetles.**

8927 Reemtsma, J.B. DE 507651/74/0002
Untersuchungen über Möglichkeiten zur Erhöhung der Frostresistenz und des Schädlingsbefalls von Freiflächenkulturen mit Eiche und Douglasie. **Investigations on the possible increase in frost resistance and on infestation of open–space oak and Douglas fir cultures by pests.**

8928 Brennan, P.A. IE 120101/78/9075 N
Investigations on the biology, population dynamics and economic importance of the green spruce aphid (elatobium abietinum).

8929 Covassi, M.; Tiberi, R. IT 020400/76/0004 N
Ricerche sulla biologia ed ecologia della Pristiphora abietina Christ. nell'Appennino settentrionale (Hym. Tenthredinidae). **On the Biology and Ecology of Pristiphora abietina Christ. in Northern Apennines (Hym. Tenthredinidae)..**

Larch forests (B 3813)

8930 Altenkirch, W. DE 507652/72/0001 R
Populationsdynamik und Bekämpfungsmöglichkeiten der Lärchenminiermotte – Coleophora laricella – im Emsland. **Population dynamics and control of the larch case bearer – Coleophora laricella – in Northwestern Germany.**

Other pine forests (B 3819)

8931 Rabasse, J.M. FR 010503/77/5057
Introduction de parasites des pucerons du cèdre. **Introduction of cedar aphids parasites.**

8932 Fabre, J.P.; Foing, J.J.; Barault, M. FR 010609/74/4115
Principaux insectes ravageurs du Cèdre en France. **Main insect pests of Cedrus in France.**

Oak tree stands (B 3821)

See also 8925

8933 Bombosch, S.; Kerck, K. DE 132660/70/0001
Die Populationsdynamik von Xyloterus domesticus in chemisch geläuterten Buchenbeständen. **The population dynamics of Xyloterus domesticus in chemically thinned beech stands.**

8934 Bombosch, S.; Kerck, K. DE 132660/70/0002
Kriterien und Veränderung der Befallsdisposition chemisch abgetöteter Buchen für Stammholzschädlinge. **Characteristics and changes of the disposition of chemically killed beech trees for the infestation by woodbreeding insect pests.**

8935 Horstmann, K.   DE 176051/71/0005
Untersuchungen über die Parasiten der an Eichen lebenden Tortriciden, insbesondere von Tortrix viridana L..
Investigations into the parasites of Tortricidae on oak specially Tortrix viridana L..

8936 König, E.; Bogenschütz, H.; Altherr, E.
        DE 501508/78/0001 N
Der Einfluss der Buchenwollschildlaus – BWS – auf Gesundheitszustand und Zuwachs der Buchen. **Influence of Cryptococcus fagi on the physiology and growth of beech.**

8937 Demolin, G.; Hames, R.   FR 010609/75/4123
Ecobiologie d'un ravageur de la chênaie des garrigues: Lymantria Dispar (Lep., Lymantriidae). **Ecobiology of an insect pest of oaks in the "garrigues", Lymantria Dispar.**

Ash tree stands (B 3822)

See also 8927

Other leafwoods (B 3829)

8938 De Clercq, R.   BE 070100/78/0022 R
Onderzoek naar de bestrijding van de beukenwolluis Phyllaphis fagi L. en van de taxus kever Otiorrhynchus sulcatus (F.). **Study of the control of the beech aphid Phyllaphis fagi L. and of the vine weevil Otiorrhynchus sulcatus (F.).**

8939 Blight; Wadhams   GB 012801/76/0006 R
Isolation and identification of chemicals which affect the behaviour of elm–bark beetles.

Other forests (B 3890)

8940 Cornic, J.F.; Kermarrec, A.; de Crecy, J.
        FR 011605/76/5272
Etude d'Hexacolus Guyanensis: ravageur des plantations d'Acajous du Honduras (Swietenia macrophylla). **Bioecology of Hexacolus Guyanensis in the rain forest.**

Stimulant crops (B 3910)

See also 10039

8941 Muir   GB 011001/00/0001 R
Control of damson–hop aphid.

8942 Cranham; Muir   GB 011001/00/0002 R
Resistance to pesticides of damson–hop aphid.

8943 Cranham   GB 011001/00/0018 R
Resistance to pesticides of tetranychid mites of hops.

8944 Neve   GB 012201/00/0006 R
Selection for resistance to, and biological control of, Phorodon humuli.

8945 Neve   GB 012201/00/0008 R
Effect of virus infection on yield and brewing value.

Spice and seasoning plants of temporate climates (B 3930)

8946 Weinmann, W.   DE 215080/78/0005 N
Untersuchung über das Rückstandsverhalten von Methomyl auf/in Hopfen nach einer Bekämpfung mit Methomyl gegen Liebstöckelrüssler, Kartoffelbohrer und Blattläuse. **Investigation on the residue–behaviour of Methomyl on/in hops after a treatment with Methomyl against alfalfa snout beetle, potatoe stem borer and hop aphid.**

Drugs and medicine plants (B 3970)

See also 8501

8947 Avigliano, M.; Sannino, L.   IT 022300/79/0014 N
Prove di lotta ai nematodi galligeni (Meloidogyne sp.) del tabacco con prodotti sperimentali nuovi e tradizionali, con indagini sui residui su tabacco secco. **Control of root–knot nematodes (Meloidogyne sp.) with new and old chemicals with study on its residues on dry tobacco.**

8948 Zacheo, G.   IT 060600/72/0044
Fisiologia e biochimismo di nematodi fitoparassiti e di piante da essi attaccate. Influenza dei nematodi sul contenuto in nicotina di foglie di tabacco. **Physiology and biochemistry of plant parasitic nematodes and the plants they attack. Influence of nematodes on the nicotine content of tobacco leaves.** Publications.

Other crops (B 3990)

8949 Cappellozza, L.; Miotto, F.   IT 020400/75/0009
Studi sul Tisanottero dannoso alle foglie del gelso Pseudodendrothrips mori. **Studies on Pseudodendrothrips mori (Ins.Thysanoptera) damaging mulberry' leaves.** Publications.

D 2420 – Plant diseases and disease control

See also 6006, 13785, 19554

8950 Fraselle, J.   BE 010021/76/0010
Etude des luttes chimiques contre les maladies des céréales (blés d'hiver et de printemps, escourgeon et orge de printemps). . **Chemical control of diseases of cereals (wheat, barley).**

8951 Semal, J.; Kummert, J.   BE 010022/65/0002
Recherches sur les virus à ARN des végétaux supérieurs. **Studies on RNA viruses of higher plants.** Publications.

8952 Van Assche, C.; Uyttebroeck, P.   BE 040203/79/0022 N
Zaaizaadontsmetting. **Seed–treatments.**

8953 Detroux, L.; Seutin, E.   BE 080700/64/0018 R
Etude des insecticides et acaricides applicables en arboriculture fruitière. **Study of insecticides and acaricides applicable in orchards.** Publications.

8954 Detroux, L.; Seutin, E.   BE 080700/65/0017
Etude de l'efficacité des insecticides de substitution aux organochlorés en culture maraîchère. **Efficacity study of insecticides to substituate the organochlorated in vegetables crops.**

8955 Detroux, L.; Seutin, E.   BE 080700/67/0021
Etude des insecticides et acaricides applicables en culture florale. **Study of insecticides and acaricides applicable in ornemental plants.**

8956 Detroux, L.; Seutin, E.   BE 080700/68/0015

Etude de mise au point de programme de lutte contre les déprédateurs entomologiques des céréales. **Study of control schemes against entomological depredators in cereals.**

8957 Detroux, L.; Seutin, E.   BE 080700/68/0019
Etude des insecticides et nématicides en culture betteravière. **Study of insecticides and nematicides used in the sugarbeets culture.**

8958 Detroux, L.; Meeus, P.   BE 080700/69/0009
Lutte chimique contre les maladies tardives des céréales (Oïdium, rouille, septoriose, fusariose, rhynchosporiose,...). **Chemical control of the late diseases on cereals (Oidium, rust, septoriose, fusariose, rynchosporiose, ...).** Publications.

8959 Detroux, L.; Meeus, P.   BE 080700/69/0014
Lutte contre quelques maladies des plantes ornementales telles les rouilles, l'oïdium et le botrytis. **Control of some diseases of ornemental plants chiefly rusts, oidium and botrytis.**

8960 Detroux, L.; Meeus, P.   BE 080700/70/0011
Lutte contre les maladies dans les cultures maraîchères principalement de concombres et scorsonères (Oidium), céleris (Septoriose), fraises, salades, haricots (Botrytis), pois (Peronospora). **Control of the diseases in vegetables crops chiefly cucumber and scorzonera (Oidium), celery (Septoriose), strawberries, salad, haricot beans (Botrytis), peas (Peronospora).** Publications.

8961 Detroux, L.; Meeus, P.   BE 080700/71/0010
Lutte contre les maladies de semences des céréales (Tilletia caries et Ustilago nuda). **Control of the seed–diseases in cereals (Tilletia caries and Ustilgo nuda).**

8962 Detroux, L.; Seutin, E.   BE 080700/71/0016
Etude de la formulation des insecticides en culture maraîchère pour la lutte contre les diptères. **Study on the formulation of the insecticides in vegetable crops to control disease caused by diptera.**

8963 Fouarge, G.   BE 080900/74/0004
Etude de la lutte contre les pucerons de la pomme de terre. **Study of aphid control of potatoes.**

8964 Verhoyen, M.   BE 140000/79/0072 N
Etudes des techniques d'identifications sérologiques ou bactériologiques des phytobactérioses. **Study of technics for identification by serologic or bacteriophagic tests against principals phytobacterial diseases.**

8965 Nøddegård, E.   DK 010116/61/0001
Bejdsemidler mod frøbårne sygdomme på korn og græsser. **Seed dressings against seedborne diseases in cereals and grasses.**

8966 Nøddegård, E.   DK 010116/61/0014 N
Fungiciders virkning på svampesygdomme, især filtrust og skivesvamp på frugtbuske. **Effect of fungicides on fungal diseases: currant rust and currant leaf spot on fruit bushes.**

8967 Nøddegård, E.   DK 010116/61/0015
Fungiciders virkning på svampesygdomme, især meldug og gråskimmel på jordbær. **Effect of fungicides on fungal diseases: mildew and grey mould on strawberries.**

8968 Nøddegård, E.   DK 010116/65/0013 N

Undersøgelser af diverse fungiciders virkning på lagersygdomme på kerne– og stenfrugt, fortrinsvis æbler. **Investigations on the effect of diverse fungicides on storage diseases of pomes and stone fruits, chiefly apples.**

8969 Nøddegård, E.   DK 010116/71/0003
Fungiciders virkning mod jordbårne sygdomme i korn og græsarter. **Effect of fungicides on soilborne diseases in cereals and grasses.**

8970 Nøddegård, E.   DK 010116/72/0002
Fungiciders virkning mod meldug, rust og andre bladpletsygdomme på korn og græsser. **Effect of fungicides on mildew, rust and other leafspot diseases on cereals and grasses.**

8971 Nøddegård, E.   DK 010116/74/0007
Post–harvest midler til behandling af kartofler. **Post–harvest chemicals for treatment of potatoes.**

8972 Nøddegård, E.   DK 010116/74/0008
Midler til bekæmpelse af sygdomme med jordsmitte i kartofler. **Chemicals for the control of soilborne diseases in potatoes.**

8973 Kristensen, H.R.   DK 010116/77/0009
Undersøgelser over linier af virusgulsot. **Investigations of lines of virus yellows.**

8974 Nøddegård, E.   DK 010116/77/0010
Fungiciders virkning mod svampesygdomme på spiselige afgrøder i væksthus. **Effect of fungicides on fungal diseases of edible glasshouse crops.**

8975 Kristensen, H.R.   DK 010116/77/0012
Diagnostik af plantevira og mykoplasmata ved anvendelse af suspensions– og snitpræparater. **Diagnosis of plant vira and mycoplasmata by means of suspension and sectioning techniques.**

8976 Kristensen, H.R.   DK 010116/77/0013
Fremstilling af antisera og serodiagnostik af plantevira. **Production of antisera and serodiagnosis of plant vira.**

8977 Kristensen, H.R.   DK 010116/77/0014
Diagnosticering af virus– og mykoplasmaangreb hos indsendt plantemateriale samt orienterende forsøg vedrørende nye forekomster af vira og mykoplasmata. **Diagnosis of virus and mycoplasm attacks in plant material, and preliminary experiments on new occurrences of vira and mycoplasmata.**

8978 Nøddegård, E.   DK 010116/77/0019
Fungiciders virkning på svampesygdomme, især Phytium og Phytophthora i væksthus og på friland. **Effect of fungicides on fungal diseases: Phytium and Phytophthora on glasshouse and outdoor crops.**

8979 Bach, E.   DK 030106/71/0014
Struktur af phytotoksiner fra Pyrenophora teres. **The structure of phytotoxins from Pyrenophora teres.**

8980 Koch, J.; Wagn, O.   DK 030147/55/0004
Fomes annosus i hegn. Undersøgelse over infektionsforløbet i levende hegn ud fra inficerede pæle. **Fomes annosus in hedges. Investigation of the course of infection in hedges from infested stakes.**

8981 Hockenhull, J.   DK 030147/75/0003

Undersøgelse af ildsotangrebne tjørn. 1. Placering samt in vivo identifikation af patogenet, Erwinia amylovora i vævsnit gennemført ved anvendelse af fluorescent antibody teknikken. 2. Symptomløse infektioner. **Investigation of thorn attacked by fire blight. 1. Localisation and in vivo identification of the pathogen, Erwinia amylovora, in tissue sections using the fluorescent antibody technique. 2. Symptom–free infections.**

8982 Koch, J.   DK 030147/76/0005
Undersøgelse over forekomster af marine svampe på forarbejdet træ ved den jyske vestkyst. **Investigation of the occurrence of marine fungi on manufactured wood on the west coast of Jutland.**

8983 Lundsgaard, T.   DK 030147/76/0008
Undersøgelse af et rhabdovirus i Festuca gigantea med stribemosaik. **Investigation of a rhabdovirus in Festuca gigantea with stripe mosaic.**

8984 Bech, K.; Kovacs, G.   DK 030147/77/0007
Undersøgelse over smittekilder, smitteveje og smittebetingelser for mycogone perniciosa på den dyrkede champignon. **Investigation of infection sources, infection pathways and infection conditions for Mycogone perniciosa in the cultivated mushroom.**

8985 Smedegaard–Petersen, V.; Stölen, O.
                                                        DK 030147/77/0010
Nedarving af resistens i byg mod stribesygesvampen Pyrenophora graminea. **Inheritance of resistance to the stripe disease fungus, Pyrenophora graminea, in barley.**

8986 Løschenkohl, B.   DK 030147/79/0004 N
Latente infektioner af Phoma exigua var. foveata på kartoffel, deres etablering, betydning som smittekilde og bekæmpelse. **The establishment and control of latent infection of Phoma exigua var. foveata on potato tubers.**

8987 Markham; Townsend   FR 011210/74/5237
Ecologie des cicadelles et des orthoptéres du maquis. **Mycoplasma genetics. Diseases and structure.**

8988 Watkins; Clark   GB 011504/00/0002 R
Photochemical breakdown of fungicides.

8989 Tomlinson   GB 011807/00/0004 R
Studies on virus technology.

8990 Walker; Neely   GB 011807/00/0006 R
Factors affecting virus development and symptom expression.

8991 Maude; Presly   GB 011807/00/0007 R
Biology and control of seed–borne fungal diseases.

8992 Hughes   GB 012011/76/0009 R
Crystal structures of fungicides (Avenaciolide).

8993 Kavanagh, J.A.; Harrington, T.   IE 120106/77/9045 N
Investigations on the biology and control of phytophthora syringae kleb. in Irish orchards. Publications.

Parks, gardens, urban greenspaces, plantations (B 1610)

See also 4759, 6009, 9341, 9550, 9781, 10011, 10036

8994 Butin, H.   DE 215230/77/0006
Prüfung systemischer Schutzmittel zur Bekämpfung der "Holländischen Ulmenkrankheit". **Testing of systemic preservatives for the control of the "Dutch elm disease".**

8995 Gremmen, J.   NL 010601/67/1871
Onderzoek naar het voorkomen en de aard van ziekten bij schietwilgen (Salix alba). **Research on the incidence and cause of willow diseases (Salix alba).** Publications.

8996 Heybroek, H.M.   NL 010601/74/6008
Inventarisatie iepziekte. **Surveying Dutch Elm Disease.** Publications.

Sportfields, play and camping grounds (B 1630)

See also 9559

Plants and parts of plants in general (B 2100)

See also 6021

8997 Rassel, A.; Maroquin, C.   BE 080100/74/0017 R
Etude de certains problèmes de phytopathologie par microscopie électronique (transmission et balayage). **The use of electron microscopy (transmission and scanning) for the study of some problems of phytopathology.** Publications.

8998 Fiers, W.; Vandendriessche, L.; Gillis, E.; Alam Zulfagar Khan.; Van Emmelo, J.   BE 140000/74/0040 R
Nieuwe detectiemethoden van plantenvirosen en mycoplasmeninfecties. **New detection methods of plant viroses and mycoplasmosis.** Publications.

8999 Knösel, D.   DE 135052/74/0002
Beziehungen zwischen Enzym– bzw. Toxin– Produktion und Virulenz bei phytopathogenen Bakterien. **Correlations of enzyme and toxin production to virulency by phytopathogenic bacteria.**

9000 Marani, F.   IT 040211/78/1073 N
Virus e virosi delle piante. **Virus and virus diseases of plants.**

9001 Betto, E.   IT 040612/78/1034 N
Virus e virosi delle piante. **Virus and virus diseases in plants.**

9002 Marte, M.   IT 041015/78/1077 N
Ricerche sulla microscopia ottica ed elettronica su piante malate con particolare riferimento a quelle colpite da virosi. **Research on optic and electronic microscopy applied to diseased plants, especially those affected by virus diseases.**

9003 Scaramuzzi, G.   IT 041110/74/0627
Studio sulla interferenza di sostanze nel processo di infezione virale nei vegetali. **Study on the interference of substances in the process of virus infection of plants.**

9004 Bottalico, A.   IT 061700/77/0922
Diffusione nelle piante e nei vettori di alcuni virus fitopatogeni. **The spreading in plants and vectors of certain pathogenous plant viruses.**

9005 Bottalico, A.   IT 061700/77/0923
Ricerche sulla produzione, la natura, e l'attività biologica di metaboliti prodotti da batteri fitopatogeni. **Research on the production, nature and biological activity of metabolites**

produced by pathogenous plant bacteria.

9006 Bottalico, A.   IT 061700/77/0925
Ricerche sulla produzione, la natura e l'attività biologica di
tossine prodotte da funghi fitopatogeni. **Research on the
production, nature and biological activity of the toxines
produced by pathogenous plant fungi.**

9007 Verduin, B.J.M.   NL 020061/71/4276
De assemblage van een bolvormig plantevirus. **The assembly of
a spherical plant virus.** Publications.

Other subjects related to plants and animals in general (B 2900)

See also 8997

Crops in general (B 3000)

See also 547, 892, 1067, 2798, 6033, 8038, 8054, 8055, 8086,
8098, 8106, 8107, 8108, 8261, 8284, 8287, 8378, 8382, 8391,
8951, 8973, 8975, 8976, 8977, 8978, 8987, 8987, 9454, 10106,
10154, 13784, 20152

9008 Meyer, J.; Maraite; Perreaux, D.   BE 020302/70/0002 R
Recherche sur champignons et bactéries phytopathogenes –
pathogénéité – relations hôte parasite – adaptation aux
fongicides. **Research on plant pathogenic fungi and bacteria –
pathogeneity – host pest relation. Tolerance to fungicides.**
Publications.

9009 Ledoux, L.; Thiry–Braipson, J.; Mergeay, P.
                                          BE 140000/76/0047 R
Utilisation de DNA plasmidien pour induire une résistance aux
bactéries phytopathogènes. **Utilization of DNA plasmodium for
induced bacterious resistance.** Publications.

9010 Nienhaus, F.; Gliem, G.; Wegen, H.W.; Wienhold, W.;
Streit, C.   DE 111351/78/0003 N
Histologisch–cytologische und biochemische Veränderungen
in Tumorgeweben nach Infektion durch Tabaktumorvirus.
**Histological, cytological and biochemical changes in tumor
tissue infected by tobacco tumor virus.** Publications.

9011 Sänger, H.L.   DE 129182/74/0002
Isolation und Charakterisierung von Viroiden. **Isolation and
characterization of viroids.**

9012 Sänger, H.L.; Singh, A.   DE 129182/74/0003
Das chromatographische Verhalten von Viroid–RNA. **The
chromatographic behaviour of viroid RNA.**

9013 Schlösser, E.; Rashid, T.   DE 129183/74/0001
Wirkungsmechanismus systemischer Fungizide, sowie
Resistenzbildung phytopathogener Pilze. **Mode of action of
systemic fungizides and development of resistant
phytopathogenic fungal strains.**

9014 Stein, W.   DE 129200/70/0003
Biologie und Verhalten von Apion virens. **Biology and
behaviour of A. virens.**

9015 Kranz, J.   DE 129554/70/0002
Untersuchungen über Grundlagen der Entwicklung von
Epidemien und der vergleichenden Epidemiologie. **Studies on
principles governing the development of epidemics and the**

comparative epidemiology.

9016 Kranz, J.   DE 129554/71/0001
Untersuchungen über den Rostparasiten Eudarluca caricis.
Studies on the rust parasite Eudarluca caricis.

9017 Kranz, J.; Hindorf, H.   DE 129554/71/0003
Systematische Untersuchungen über Möglichkeiten zur
Langzeitprognose von Pflanzenkrankheiten auf
phänologischer Grundlage. **Systematic investigations of
long–term prognosis of plant diseases based on phenological
observations.**

9018 Bashi, E.; Aust, H.–J.; Kranz, J.; Hau, B.; Palti, J.
                                          DE 129554/74/0005
Kompensationsphänomene bei ökologischen Faktoren und
ihre Auswirkung auf die Epidemiologie von
Pflanzenkrankheiten. **The epidemiology of plant diseases in
dependence on compensational phenomena of ecological factors.**

9019 Kranz, J.; Koch, H.   DE 129554/75/0002
Untersuchung zur Methodik des Schätzens von
Pflanzenkrankheiten im Felde. **Methods of diseases assessment
in crops.**

9020 Kranz, J.; Eckehardt, H.   DE 129554/78/0001 N
Entwicklung von Testverfahren für systemanalytische
Simulatoren von Epidemien – Pflanzenkrankheiten –.
**Development of tests for simulators of plant disease epidemics
based on system analysis.**

9021 Kickuth, R.; Klages, F.W.   DE 132034/70/0001
Untersuchungen über die Ursache einer wirtschaftlich
bedeutungsvollen regelmässig auftretenden Schädigung –
Stein– felskrankheit – landwirtschaftlicher Kulturpflanzen im
nördlichen Vorharzgebiet 1969. **Investigations on the cause of a
disease of cultivated plants occurring regularly and of economic
importance – Steinfels–krankheit – in the northern Lower Harz
Mountains.**

9022 Rudolph, K.   DE 132210/70/0008
Chemische Charakterisierung des durch Pseudomonas
phaseolicola gebildeten Chlorose–induzierenden Toxins und
Aufklärung des Wirkungsmechanismus 1970. **Chemical
identification of chlorose–inducing toxin formed by
Pseudomonas phaseolicola and clarification of the mechanism of
action.**

9023 Mendgen, K.   DE 132210/78/0001 N
Parasitierung des Gelbrosts durch Verticillium. **Verticillium a
hyperparasite of stripe rust.**

9024 Wolf, G.   DE 132210/78/0002 N
Die Rolle von Proteinen für die Resistenz von Pflanzen gegen
pathogene Pilze. **The role of proteins in the resistance of plants
to pathogenic fungi.**

9025 Fehrmann, H.; Scheepens, P.C.; Mieskes, G.
                                          DE 132212/77/0003
Stoffwechseldefekte bei obligat parasitischen phytopathogenen
Pilzen – Peronosporales –. **Metabolic defects of obligately
parasitic phytopathogenic fungi – Peronosporales –.**

9026 Fehrmann, H.; Speakman, J.   DE 132212/78/0001 N
Genetische Nebenwirkungen von MBC. **Genetical side–effects**

of MBC.

9027 Rehm, S.; El–Deepah, H. DE 132240/77/0002
Wirkung vesikulär–arbuskulärer Mykorrhiza bei
salzresistenten und salzempfindlichen Arten auf Böden
verschiedenen Salzgehaltes. **Effect of vesicular–arbuscular
mycorrhiza in salt–resistant and salt–sensitive plants on soils
with different salt content.**

9028 Schickedanz, F.; Lichte, H.–F. DE 135052/70/0001
Wirkungsweise und Anwendung systemischer Fungizide. **Way
of action and application of systemic fungicides.**

9029 Bünemann, G.; Bremer, H. DE 138240/77/0001
Nebenwirkungen von Fungiziden. **Secondary effects of
fungicides.**

9030 Weil, B.; Pahlow, G. DE 138331/73/0002
Untersuchungen über Wechselbeziehungen zwischen Arten
der Pilzgattung Pythium und bodengebundenen Virusarten
unter spezieller Berücksichtigung der Verhältnisse im
Gartenbau. **Investigations into interactions between species of
the genus Pythium and soil–bound viruses, especially
considering circumstances in horticulture.**

9031 Hein, A. DE 144540/72/0001
Identifizierung von Virus–Isolaten aus Statice. **Identification of
virus isolates from Statice.**

9032 Buchenauer, H. DE 144540/72/0005
Umwandlung systemischer Fungizide in und auf der Pflanze
sowie durch Mikroorganismen. **Conversion of systemic
fungicides in and on plants as well as by micro–organisms.**

9033 Buchenauer, H. DE 144540/72/0008
Untersuchungen über Aufnahme und Transport systemischer
Fungizide sowie ihre Wirkung gegenüber
Pflanzenkrankheiten. **Uptake and transport of systemic
fungicides and their effect against plant diseases.**

9034 Buchenauer, H. DE 144540/75/0001
Wirkungsmechanismen neuerer Fungizide in Pilzen. **Mode of
action of newer fungicides in fungi.**

9035 Resz, A. DE 144540/75/0006
Überdauern phytopathogener Bakterien im Gartenkompost.
Survival of phytopathogenic bacteria in garden compost.

9036 Hein, A. DE 144540/75/0008
Untersuchungen zur Wirkung von Öl bei der Übertragung
nicht–persistenter Viren durch Blattläuse. **Investigations on the
effect of oil on the transmission of non–persistent viruses by
aphids.**

9037 Grossmann, F.; Redlhammer, S. DE 144540/77/0001
Bildung extrazellulärer Enzyme durch Cercosporella
herpotrichoides Fron in vitro und in vivo. **Production of
extracellular enzymes by Cercosporella herpotrichoides Fron in
vitro and in vivo.**

9038 Sarkar, S. DE 144540/78/0008 N
Isolation von pflanzlichen Protoplasten und ihre Infektion mit
phytopathogenen Viren. **Isolation and infection of plant cell
protoplasts with phytopathogenic viruses.** Publications.

9039 Müller, F.; Kälberer, R.; Grossmann, F.
 DE 144541/74/0001
Einfluss der Infektion auf die Verteilung systemischer
Fungizide in der Pflanze. **The influence of infection on the
distribution of systemic fungizides in plants.**

9040 Hein, A.; Sydow, B. von DE 144542/77/0001
Untersuchungen zum Wirkungsmechanismus von
Virus–Hemmstoffen. **Investigations on the mode of action of
virus inhibitors.**

9041 Schuck, H.J. DE 160061/75/0002 N
Einfluss flüchtiger Terpene auf die vegetative Entwicklung von
Fomes annosus. **Influence of volatile terpenes on the vegetative
development of Fomes annosus.**

9042 Schwenke, W.; Timans, U. DE 160090/78/0002 N
Untersuchungen über die Wirkungen von UV–Licht auf
Insektenviren. **Studies on the effects of UV–light on insect
viruses.**

9043 Hoffmann, G.M.; Kiebacher, H. DE 161050/77/0006
Untersuchungen zur Resistenz gegen systemische Fungizide
aus der Gruppe der Benzimidazolderivate bei Venturia
inaequalis – Spilocaea pomi –. **Resistance in Venturia
inaequalis against systemic fungicides of benzimidazol
derivatives.** Publications.

9044 Oberwinkler, F.; Sebald, F.R. DE 173051/78/0003 N
Feinstrukturstudien zur Ontogenie von Arten der Uredinales
und verwandter Basidiomyceten. **Fine structural studies on the
ontogeny of Uredinales strains and related Basidiomycetes.**

9045 Blaich, R.; Welker, R. DE 204000/75/0006
Art– und Rassenabgrenzung von Schadpilzen durch die
Analyse von Hydrolasen und Phenoloxidasen. **Classification of
species and races of parasitic fungi by the analysis of hydrolases
and phenoloxidases.**

9046 Marwitz, R. DE 215040/77/0005
Erforschung von mykoplasmaähnlichen Organismen in
wirtschaftlich genutzten und nicht genutzten
–Mykoplasmenreservoir– Wirtspflanzen, sowie Versuche zu
ihrer Ausbreitung und Bekämpfung. **Research on
mycoplasm–like organisms in utilized and non–utilized –
mycoplasm reservoir – host plants and trials on distribution and
control.**

9047 Petzold, H. DE 215040/77/0006
Physiologische, histologische und zytologische
Untersuchungen von gesunden und mit mykoplasmaähnlichen
Organismen befallenen Wirtspflanzen. **Physiological,
histological and cytological investigations on intact host plants
and on those infested with mycoplasm–like organisms.**

9048 Marwitz, R. DE 215040/77/0007
Erforschung der möglichen Rolle von rickettsienähnlichen
Bakterien als Krankheitserreger von Kulturpflanzen. **Research
on potential action of Rickettsia–like bacteria as pathogenic
agent in cultivated plants.**

9049 Petzold, H. DE 215040/77/0008
Rasterelektronenmikroskopische Untersuchungen an
Bakterienkolonien und an von mykoplasmaähnlichen
Organismen befallenen Wirtspflanzen. **Scanning electron
microscopic studies on bacterial colonies and on host plants
infested with mycoplasm–like organisms.**

9050 Köhn, S. DE 215040/77/0009
Diagnose von Pflanzenkrankheiten mit Verdacht auf
Bakteriosen. **Diagnosis of plant diseases under suspicion of
bacteriosis.**

9051 Gerlach, W. DE 215040/77/0010
Diagnose und Erforschung ätiologisch unklarer oder neuer
Pflanzenkrankheiten mit Verdacht auf Mykosen. **Diagnosis of
and research on etiologically unclear or new plant diseases
under suspicion of mycosis.**

9052 Gerlach, W. DE 215040/77/0011
Erforschung der Biologie und Taxonomie von Arten der
Gattung Fusarium. **Research on biology and taxonomy of
strains of Fusarium.**

9053 Schneider, R. DE 215040/77/0012
Untersuchungen über Arten wichtiger
Pyknidienpilzgattungen. **Studies on strains of important
Pycnidia fungi.**

9054 Kröber, H. DE 215040/77/0013
Forschungen über Arten der Gattungen Phytophthora und
Pythium und von ihnen verursachte Pflanzenkrankheiten.
**Research on strains of Phytophthora and Pythium as agents of
plant diseases.**

9055 Schneider, R. DE 215040/77/0014
Forschungen zur Methodik der Isolierung, Kultur,
Konservierung und Differenzierung von phytopathogenen
Pilzen. **Research on methods for isolation, culture,
preservation, and differentiation of phytopathogenic fungi.**

9056 Kröber, H.; Oezel DE 215040/77/0015
Zytologische Untersuchungen über Reaktionen anfälliger und
resistenter Sorten verschiedener Kulturpflanzenarten auf
Infektion durch verschiedene Rassen falscher Mehltau–Arten.
**Cytological studies on reactions of susceptible and resistant
varieties of diverse species of cultivated plants to infection by
different strains of downy mildew.**

9057 Kloke, A.; Leh, H.–O.; Schoenhard, G.
DE 215060/77/0001
Diagnose von nichtparasitären Pflanzenkrankheiten. **Diagnosis
of non–parasitic plant diseases.**

9058 Weinmann, W. DE 215080/77/0008
Entwicklung einer Analysenmethode zur Bestimmung der
Wirkstoffe in flüssigen Präparaten von Dichlorprop–Salz.
**Development of an analytical method for determination of
active substances in liquid preparations of dichloroprop–salt.**

9059 Weinmann, W. DE 215080/77/0009
Entwicklung einer Analysenmethode zur Bestimmung der
Wirkstoffe in festen Kombinationspräparaten von Dichlorprop
und 2.4.5–T–Salzen. **Development of an analytical method for
determination of active substances in solid combination
preparations of dichloroprop– and 2.4.5–T–salts.**

9060 Lyre, H.; Ehle, H.; Heidler, G.; Martin, J.
DE 215090/77/0002
Entwicklung von Methoden – Richtlinien – für die Prüfung von
Fungiziden und Herbiziden auf Wirksamkeit und
Phytotoxizität für neue Anwendungsbereiche im Rahmen des
Zulassungsverfahrens. **Development of methods – guidelines –
for testing of fungicides and herbicides for efficiency and
phytotoxicity for new ranges of application in admittance
process.**

9061 Lerch, H. DE 215140/75/0009
Entwicklung chemotherapeutischer Verfahren gegen
pflanzenpathogene Viren zur Sanierung von
Vermehrungsmaterial. **Development of chemotherapeutical
treatments against plant pathogenic viruses for the restoration
of propagating material.**

9062 Lerch, B. DE 215140/77/0009
Pflanzliche Abwehrmechanismen gegen Virusinfektionen.
Defence mechanisms in plants against virus diseases.

9063 Huth, W. DE 215150/77/0002
Bestandsaufnahme über das Vorkommen von Virosen bei
landwirtschaftlichen Kulturen. Ausarbeitung von
empfindlichen Diagnoseverfahren. **Inventory on the
occurrence of viroses in crops. Development of sensitive
diagnostic methods.**

9064 Paul, H.L. DE 215150/78/0001 N
Ausarbeitung von Reindarstellungsverfahren für Pflanzenviren
für deren Charakterisierung und Klassifizierung mit
physikalischen und chemischen Methoden. **Investigations on
the purification of plant viruses for their characterization and
classification by means of physical and chemical methods.**

9065 Huth, W. DE 215150/78/0003 N
Eliminierung von Viren aus Kulturpflanzen mittels
Meristemkultur. **Elimination of viruses from different
cultivated plants by meristem culture.**

9066 Rohloff, H. DE 215150/78/0004 N
Untersuchungen zur Chemotherapie pflanzenpathogener
Viren. **Investigations on chemotherapy against plant viruses.**

9067 Casper, R. DE 215150/78/0009 N
Entwicklung und Verbesserung serologischer und
biochemischer Diagnoseverfahren für Viren. **Development and
improvement of serological and biochemical methods for virus
assay.**

9068 Casper, R. DE 215150/78/0012 N
Entwicklung eines serologischen Tests zum Routinenachweis
eines Stammes des Tristeza Virus. **Development of a serological
test for routine assay of a strain of tristeza virus.**

9069 Lesemann, D.; Koenig, R.; Weidemann, H.–L.
DE 215150/78/0025 N
Differenzierung und Diagnose von Potyviren anhand der
Zytologie infizierter Zellen. **Differentiation and diagnosis of
potyviruses using the cytology of infected cells.**

9070 Lesemann, D.; Koenig, R. DE 215150/78/0027 N
Weiterentwicklung der Immunelektronenmikroskopie als
spezifisches, schnelles und hochempfindliches
Nachweisverfahren für Pflanzenviren. **Further development of
immune electron microscopy as a specific, time–saving, and
highly sensitive method of detection of plant viruses.**

9071 Rohloff, H. DE 215150/78/0031 N
Epidemiologische Untersuchungen über die Dynamik der
Virusausbreitung in landwirtschaftlichen Kulturen.
Epidemiological investigations on the dynamics of virus spread

in agricultural crops.

9072 Krieg, A. DE 215170/74/0003 R
Entwicklung von UV–Schutzstoffen für Präparate auf der
Basis von insektenpathogenen Mikroorganismen. **Development
of UV protection for preparations based on insect pathogenic
microorganisms.** Publications.

9073 Huger, A.M. DE 215170/74/0005 R
Versuche zur künstlichen Infektion von gesunden Freiland–
Populationen wichtiger Schadinsekten, vor allem
Mikrosporidien. **Trials on artificial infection of intact field
populations of important insect pests using microsporidia.**

9074 Zimmermann, G. DE 215170/74/5001
Untersuchungen über die Nebenwirkung von Fungiziden auf
insektenpathogene Pilze. **Investigations on side–effect of
systemic fungicides on insect pathogenic fungi.**

9075 Huger, A.M. DE 215170/77/0005
Diagnose eingesandter Insekten auf Krankheiten durch Viren,
Bakterien, Pilze, Rickettsien oder Protozoen. **Diagnosis of
transmitted insects for diseases by virus, bacteria, fungi,
rickettsia, or protozoa.**

9076 Zeller, W. DE 215180/77/0012
Biochemische Untersuchung zum Wirkungsmechanismus des
Toxins von Erwinia amylovora, dem Erreger des
Feuerbrandes. **Biochemical studies on action of the toxin of
Erwinia amylovora, the agent of fire blight.**

9077 Zeller, W. DE 215180/77/0013
Erforschung der Feuerbrandkrankheit unter besonderer
Berücksichtigung seiner Bekämpfung. **Research on fire blight
with special regard to control.**

9078 Müller, J. DE 215190/77/0003
Untersuchungen zum Einfluss mykophager Nematodenarten
auf pilzliche Krankheitserreger. **Investigations on the influence
of mycophagous nematode species on fungal disease agents.**

9079 Müller, J. DE 215190/77/0019
Einfluss pflanzenparasitärer Nematoden auf die Resistenz
verschiedener Kulturpflanzen gegenüber Welkepilzen.
**Influence of plant–parasitic nematodes on the resistance of
different cultivated plants to wilt fungi.**

9080 Weinhold, E.; Triemer, B. DE 305030/73/0001
Prüfung von Desinfektionsmitteln auf Viruzidie. **Testing of
disinfectants against virus.**

9081 Hopp, H. DE 501101/74/0001
Untersuchungen zur Biologie von Botrytis–Stämmen. **Studies
on the biology of Botrytis strains.**

9082 Lohweg, E. DE 502102/78/0002 N
Prüfung von 3 Holzschutzmitteln auf Pflanzenverträglichkeit.
The compatibility of three wood preservatives to plants.

9083 Eichhorn, K.W.; Lorenz, D.H. DE 509153/78/0001 N
Untersuchungen über mögliche Resistenzbildung von Botrytis
cinerea gegen Ronilan. **Studies on the possible development of
resistance of Botrytis cinerea to Ronilan.**

9084 Eichhorn, K.W.; Lorenz, D.H. DE 509153/78/0002 N
Untersuchungen über mögliche Resistenzbildung von Botrytis

cinerea gegen Rovral. **Studies on the possible development of
resistance of Botrytis cinerea to Rovral.**

9085 Kristensen, H.R.; Thomsen, A.; Begtrup, J.W.
DK 010116/74/4519
Forekomst, forebyggelse og bekæmpelse af
mykoplasmalignende organismer. **Occurrence, prevention and
control of mycoplasma–like organisms.**

9086 Nissen, T.V. DK 010117/78/0001 N
Kløverbakteriers forekomst i landbrugsjorder. **The distribution
of clover rhizobia (bacteria) in agricultural soils.**

9087 Olesen, J. DK 010701/76/0010
Udvikling af ny sprøjteteknik for marksprøjter til reduktion af
kemikalie– og væskemængder til landbrugsafgrøder.
**Development of new spraying technique for field sprayers to
reduce the quantities of chemical and liquid carrier applied to
agricultural crops.**

9088 Moreau, J.P.; Boulay, C.; Warin, S. FR 010106/63/5245
Transmission des maladies à virus et à mycoplasmes par
insectes piqueurs. **Virus and Mycoplasms Transmission by
Aphides and Leafhoppers.**

9089 Hawlitzky, N. Melle; Boulay, C. Melle
FR 010106/71/1778
Mise en place et évolution des réserves d'un parasite
entomophage. **Evolution of nutrient reserves of an
entomophagous parasite.** Publications.

9090 Spire, D. FR 010107/70/1022
Virus de penicillium. Etude des caractéristiques des particules
virales. **Penicillium viruses. Studies on viral particles
characters.** Publications.

9091 Boistard, P.; Macquaire, M. FR 010107/70/1029
Etude génétique des relations entre replication virale et
métabolisme cellulaire pour le couple E. Coli phage à RNA.
**Genetic study of viral replication and cellular metabolism
relationships for E. Coli and RNA phage.**

9092 Cousin, M.T. Mme FR 010107/71/1016
Etude des interactions mycoplasme–plante hôte. **Studies on
mycoplasma host–plant relation–ships.** Publications.

9093 Maury, Y.; Laquerriére, Fr. FR 010107/71/1027
Infection de protoplastes. **Protoplast infection.** Publications.

9094 Joannes, H.; Millier, C. FR 010312/76/8289
Mécanismes de compétition dans la colonisation de pythium
sp. (3). **Competition patterns in colonization by a fungus
(pythium sp.).**

9095 Digat, B.; Poutier, F. FR 010402/71/1000
Méthodologie de la détection des bactéries phytopathogènes.
Methods of detection of phytopathogenic bacteria.

9096 Cayrol, J.C.; Caubel, G.; Combettes, S. Mme
FR 010504/63/5152
Utilisation des nématodes mycophages comme agents de lutte
biologique contre certains champignons phytopathogènes. **Use
of mycophagous nematodes as a method of biological control
against some phytopathogenic fungi.**

9097 Scotto La Massése, C.; Berge, J.B. FR 010504/67/1706 R

Influence des porte–greffes ou des variétés résistantes à certains nématodes sur la nématofaune globale. **Effect of rootstocks and nematode–resistant varieties on the total nematofauna.**

9098 Cuany, A.; Scotto La Massése, C.   FR 010504/68/1707
Conditions et limites d'emploi des substances nématicides en champ. **Conditions and limits of field use of nematicides.** Publications.

9099 Dalmasso, A.; Scotto La Massése, C.   FR 010504/68/1710
Transmissions de virus par nématodes. **Virus transmissions by nematodes.** Publications.

9100 Cayrol, J.C.   FR 010504/68/1717
Lutte biologique contre les nématodes par utilisation de champignons antagonistes. **Biological control of nematodes by antagonistic fungi.** Publications.

9101 Laumond, C.   FR 010504/69/1715
Utilisation des Néoplectana en lutte biologique. **Use of Neoplectana in biological control.** Publications.

9102 Scotto La Massèse, C.   FR 010504/70/1704
Influence des méthodes culturales sur la nématofaune. **Influence of cultural methods on nematofauna.** Publications.

9103 Scotto La Massèse, C.   FR 010504/71/1705
Rôle des nématodes dans l'équilibre biologique des sols et les complexes pathologiques. **Role of nematodes in soil biological balance and pathological complexes.** Publications.

9104 Cayrol, J.C.; Combettes, S. Mme   FR 010504/71/1716
Etude des relations entre nématodes libres et bactéries dans les composts. **Studies on the relations between saprophytic nematodes and bacteria in composts.** Publications.

9105 Poitout, S.; Causse, R.   FR 010602/71/1752
Etude éthologique des migrations chez les noctuelles. **Ethological study of noctuid migrations.** Publications.

9106 Marchoux, G.; Giannotti, J.   FR 010604/68/1009
Etude des maladies à mycoplasmes affectant les plantes cultivées dans le Sud–Est de la France. **Studies of mycoplasma diseases on crops in South–East France.** Publications.

9107 Croizier, G.; Meynadier, G.   FR 010612/51/1764
Pathogénèse des Baculovirus. **Pathogenesis of Baculoviruses.** Publications.

9108 Quiot, J.B.; Bergoin, M.   FR 010612/55/1770
Etude de la pathogénèse des virus et des rickettsies des invertébrés sur cultures cellulaires. **Study of pathogenesis of invertebrate viruses and rickettsies in cell cultures.** Publications.

9109 Bergoin, M.; Vago, C.   FR 010612/63/1763
Etude des Entomopoxvirus. **Studies on Entomopox viruses.** Publications.

9110 Meynadier, G.; Croizier, G.   FR 010612/64/1768
Etude de la spécificité et des relations sérologiques des rickettsies. **Study of specificity and serologic relations of rickettsies.** Publications.

9111 Giannotti, J.   FR 010612/67/1762

Etiologie et transmission des jaunisses. **Etiology and transmission of yellow diseases.** Publications.

9112 Giannotti, J.; Leclant, F.; Dollet, M.; Marchoux, G.; Ghosh, S.   FR 010612/69/5190
Etude des propriétés biologiques des mycoplasmes des plantes. **Study of biological properties of plant mycoplasma.**

9113 Giannotti, J.; Louis, C.; Benhamou, N.; Vago, C.; Ghosh, S.; Leclant, F.   FR 010612/69/5191
Etude des rickettsoïdes des plantes. **Study of plant rickettsoids.**

9114 Meynadier, G.; Croizier, G.   FR 010612/70/1765
Recherche et étude des Baculovirus de Lépidoptères. **Research and study of Baculoviruses of Lepidoptera.** Publications.

9115 Giannotti, J.; Vago, C.; Louis, C.; Leclant, F.; Benhamou, N.; Signoret   FR 010612/74/5207
Maladies végétales complexes à transmission biologique. **Vectorborn complex plant diseases.**

9116 Massonie, G.; Maison, P.   FR 010707/73/5215
Transmission de la Sharka par Myzus Persicae Sulzer. **Transmission of Sharka by Myzus Persicae.**

9117 Fos, A.   FR 010707/77/5229
Etude de la spécificité de la transmission de maladies à mycoplasmes par des Cicadelles vectrices. **Study of specificity of transmission of mycoplasm diseases by leafhoppers.**

9118 Ferron, P.; Deotte, A.   FR 011102/68/1723
Etude des conditions favorables au développement de la mycose due au Beauveria II. Sensibilité de l'hôte à l'infection. **Study of favourable conditions to mycosis development (Beauveria) II. Host sensitivity to infection.** Publications.

9119 Burgerjon, A.   FR 011102/72/1725
Essais biologiques sur les virus des noctuelles. **Biological tests on viruses of noctuid species.** Publications.

9120 Leclant, F.; Renoust, M.   FR 011210/70/5234
Aphides vecteurs de la Sharka. **Aphids, vectors of the Sharka virus.**

9121 Dennis; Davies   GB 010204/79/0034 N
Dicarboximide resistance in Botytis cinerea; frequency and characteristics of resistant strains.

9122 Kirkham; Hignett   GB 011005/00/0015 R
Host–parasite relations in fungal diseases.

9123 Talboys   GB 011005/00/0027 R
Physiology of infection of Verticillium wilt diseases of fruit and hops.

9124 Kirkham   GB 011005/00/0033 R
Effect of light on plant disease and on plant infection in disease assay systems.

9125 Crosse; Hignett   GB 011005/75/0035 R
Factors responsible for host specificity and virulence in the phytopathogenic Pseudomonas Species.

9126 Hollings   GB 011110/00/0011 R
Characterization and identification of viruses.

9127 Stone; Hollings — GB 011110/00/0013 R
Production of virus–free horticultural plants.

9128 Hollings — GB 011110/00/0020 R
Propagation and distribution of virus–free nuclear stocks.

9130 Horne; Hobart — GB 011403/76/0004 R
Plant virus ultrastructure.

9131 Watts; Burgess — GB 011403/76/0006 R
Plant cell protoplasts relating to virology.

9132 Roberts; Phillips — GB 011403/76/0007 R
Virus infection of Chlamydomonas.

9133 Wells — GB 011403/76/0009 R
Plants–plant virus nucleic acid, protein ultrastructure.

9134 Trim; Dickerson — GB 011404/00/0002 R
Analytical studies of plant virus nucleic acid.

9135 Rees; Short — GB 011404/76/0005 R
Chemical relationships between tobacco mosaic virus strains.

9136 Rees; Short — GB 011404/76/0006 R
A comparative study of the structures of plant viruses and their proteins.

9137 Rees; Short — GB 011404/76/0007 R
Sequence determination of viral proteins.

9138 Dawson — GB 011404/76/0008 R
Comparative studies of tomato mosaic viruses.

9139 Dawson — GB 011404/76/0009 R
A study of plant cell protoplasts infected by multicomponent viruses.

9140 Hull — GB 011404/76/0010 R
Biochemistry and biophysics of plant viruses.

9141 Markham; Townsend — GB 011404/76/0011 R
Procaryotic plant pathogens.

9142 Markham; Townsend — GB 011404/76/0012 R
Insects as vectors of plant diseases.

9143 Woodcock — GB 011504/00/0001 R
Synthesis of compounds for chemical structure – fungicidal activity studies.

9144 Watkins; Clark — GB 011504/00/0003 R
Lipid components of cell walls and membranes of fungal spores.

9145 Clifford; Hislop — GB 011504/00/0004 R
Role of components of the fungicide/pathogen/host plant system in fungitoxicity.

9146 Richmond — GB 011508/00/0006 R
Mode of action of fungicides used in agriculture.

9147 Tomlinson — GB 011807/00/0019 R
Plant virus disease chemotherapy.

9148 Miflin; Thomson — GB 012001/00/0019 R
Comparative biochemistry of plants and their pathogens.

9149 Pierpoint; Antoniw — GB 012001/76/0024 R
Novel proteins formed in leaves with induced resistance to viruses.

9150 Kassanis — GB 012009/00/0001 R
Characterisation of plant viruses.

9151 Bainbridge — GB 012009/00/0010 R
Aerobiology and epidemiology of pathogens spread by air and water.

9152 Kassanis; White — GB 012009/00/0029 R
Direct methods of controlling plant viruses.

9153 Byford — GB 012301/00/0024 R
Leaf diseases (powdery mildew, downy mildew).

9154 Harrison; Perry — GB 030702/00/0010 R
Soil microbiology and root disease; seed quality–soil interactions and their effects on seedling growth.

9155 Perry; Harrison — GB 030702/00/0012 R
Plant and pathogen physiology; seed quality, causes of its variation and its effect on yield.

9156 Williamson; Wilson — GB 030702/00/0020 R
Phytopathological methods: development of histological and histochemical techniques.

9157 Wilson; Williamson — GB 030702/00/0021 R
Phytopathological methods: immunofluorescent and fluorescent techniques in histology.

9158 Duncan — GB 030702/00/0025 R
Soil microbiology and root disease: rhizosphere and allied phenomena affecting plant health.

9159 Lyons — GB 030702/00/0026 R
Plant and pathogen physiology: the nature and implications of quiescent fungal and bacterial infections.

9160 Dashwood — GB 030702/00/0027 R
Plant and pathogen physiology: studies of plant pathogens.

9161 Harrison — GB 030704/00/0002 R
Viruses with nematode vectors and/or multipartite genomes.

9162 Robertson — GB 030705/00/0005 R
Ultrastructure of nematode vectors of plant viruses with reference to their feeding apparatus.

9163 Rennie — GB 030901/00/0504 R
Evaluation of laboratory tests for seed–borne disease.
Epidemiology of seed–borne diseases.

9164 Richardson; Whittle — GB 030902/76/0406 R
Seed pathology of newly introduced or newly promoted crops.

9165 Swinburne — GB 041504/00/0002 R
Disease development in the presence of benzimidazole–tolerant pathogens.

9166 Swinburne — GB 041504/76/0003 R

D 2420 – Plant diseases and disease control

Physical and chemical factors at the surfaces of plants and their effect on fungal infection structures.

9167 Swinburne   GB 041504/76/0004 R
Mechanisms involved in latency in several fungal diseases in plants.

9168 Dickens; Sharp   GB 050323/73/0031 R
International collaborative work on seed health testing.

9169 Gray   GB 060217/00/0003 R
Monitoring of crop diseases.

9170 Shipton   GB 060217/00/0004 R
Influence of water potential in soil and plant tissues on crop disease.

9171 Shaw   GB 060218/00/0003 R
Monitoring crop pests.

9172 Dunne, B.   IE 060500/72/0244
Studies on the epidomology and control of septoria spp. and other foliar pathogens. Publications.

9173 Kavanagh, J.A.   IE 120106/77/9047 N
Studies on the isolation and control of phytophthora cinnamomi rands.

9174 Ercolani, G.L.   IT 040106/74/0547
Studi ecologici e di lotta integrata contro batteri fitopatogeni. Ecological and integrated control studies on phytopathogenic bacteria.

9175 Ciccarone, A.   IT 040106/77/0148 R
Parassiti fungini in Italia meridionale. Fungal parasites in Southern Italy.

9176 Martelli, G.   IT 040106/77/0232 R
Virus e virosi delle piante. Viruses and virus diseases of plants.

9177 Rana, G.L.   IT 040106/77/0524 R
Studio biologia botrytis, lotta contro botrytis con anticrittogamici noti e meno noti. A biological study of botrytis, botrytis control using well–known and less well–known fungicides.

9178 Rana, G.L.   IT 040106/78/1156 N
Studi chimico–fisici e sterologici comparativi tra nepovirus. Comparative chemical, physical and sterological studies on nepovirus.

9179 Canova, A.   IT 040211/73/0130
Ricerche sui virus e le virosi nelle piante con particolare riguardo alle caratteristiche patogenetiche ed epidemiologiche. Research on viruses and diseases of plants with specific reference to pathogenetic and epidemiological characteristics. Publications.

9180 Mazzucchi, U.   IT 040211/77/0237 R
Interazioni piante batteri e fitopatogeni con particolare riguardo alla prevenzione della reazione di ipersensibilità ed ai meccanismi patogenetici coinvolti nei marciumi molli. Plant, bacterium and phytopathogen interaction with particular attention to the prevention of the hypersensitivity reaction and the pathogenetical processes involved in soft rot.

9181 Mezzetti, A.   IT 040211/77/0240 R
Azione di antimetaboliti nei rapporti fra pianta ospite e agenti patogeni. The action of antimetabolites in the host plant–pathogenic agents relationship.

9182 Brunelli, A.   IT 040211/77/0479 R
Fungicidi sistemici, secondo screening e prove di pre–campo. Fungicide systems, second screening and pre–field trials.

9183 D'Ercole, N.   IT 040211/78/1132 N
Biologia ed epidemiologia della Botrytis cinerea, rapporti condizioni intrinseche ed agronomiche sull'infezione. Biology and epidemiology of Botrytis cinerea, the influence of specific conditions and agronomic factors on the infection.

9184 Todesco, P.E.   IT 040244/77/0528
Nuovi eterociclici ad azione sistemica. New heterocyclics with a systemic action.

9185 Refatti, E.   IT 040302/73/0291
Studi sull'interferenza di sostanze chimiche nel processo d'infezione da virus nei vegetali e loro caratterizzazione. Study on the interference of chemical substances in the process of virus infection of plants, and their determination. Publications.

9186 Fontana, P.   IT 040406/72/0460
Studio su ditiocarbammati, contenenti il gruppo solfossido ad azione antiparassitaria. Study on dithiocarbamates containing the sulphoxide group having an anti–parasitic action. Publications.

9187 Bolchi Serini, G.   IT 040607/77/0476 R
Studio della biologia della Botrytis, interventi fitoiatrici, ricerca dei residui tossici. A study on Botrytis biology, the use of pesticides, research on toxic residues.

9188 Battistotti, B.   IT 040611/73/0141
Distribuzione, caratteri fenotipici e genotipici del genere pedio–coccus. Distribution and phenotypical and genotypical characteristics of the Pediococcus genus. Publications.

9189 Bisiach, M.   IT 040612/73/0151
Indagini su nouvi fungicidi sistematici e protettivi nella terapia delle malattie delle piante. Studies on new systematic and protective fungicides used in the treatment of plant diseases. Publications.

9190 Belli, G.   IT 040612/73/0910
Virus e virosi delle piante. Plants viruses and virus diseases.

9191 Betto, E.   IT 040612/74/0515
Ricerche sulla struttura e sulla fisiologia degli austori fungini nelle malattie trofiche. Research on the structure and physiology of austori fungin in trophic diseases.

9192 Betto, E.   IT 040612/77/0474 R
Nuovi fungicidi sistemici, primo secondo screening, studio del meccanismo d'azione e prove pre–campo. New systemic fungicides, first and second screening, a study of their mechanism of action, pre–field tests.

9193 Bisiach, M.   IT 040612/77/0475 R
Studio biologia botrytis, interventi fitoiatrici, ricerca residui tossici. A biological study of Botrytis, treatments, detecting toxic residues.

9194 Locci, R. IT 040612/77/0507 R
Studio dell'importanza dei danni provocati da patogeni e batterici ed indagini sulle alterazioni tra microrganismi parassiti e saprofiti. **A study of the extent of damage caused by pathogens and bacteria, research on changes among micro–organisms, parasites and saprophytes.**

9195 Merlini, L.; Maroni, G.; Galli, R.; Gerali, G.
IT 040620/75/0006
Sintesi di nuovi fungicidi. **Sinthesis of new fungicides.**

9196 Merlini, L. IT 040620/77/0513 R
Sintesi di fungicidi. **Synthesis of fungicides.**

9197 Alghisi, P.; Di Lenna, P.; Magro, P. IT 040808/77/0001
Ricerche sulla patogenicità di Botrytis cinerea. **Researches on the Botrytis cinerea pathogenicity.** Publications.

9198 Raggi, V. IT 041015/73/0289
Ricerche sul metabolismo azotato, fotorespiratorio, fissazione di CO_2 al buio in piante affette da parassiti obbligati principalmente con radioisotopi etc.. **Research on the nitrogen, photorespiratory metabolism, CO_2 fixation in the dark by plants affected by parasites marked primarily with radioisotopes, etc..** Publications.

9199 Raggi, V. IT 041015/77/0277 R
Variazioni dell'attività fotorespiratoria, del punto di compensazione di diverse coppie ospite–parassita e con diversi gradi d'infezione. **Variations in the photorespiratory activity, variations of the compensation point in various host–parasite pairs with varying degrees of infection.**

9200 Scaramuzzi, G. IT 041110/72/0128
Virologia vegetale. **Plant virology.** Publications.

9201 Gambogi, P. IT 041110/72/0466
Patologia dei semi delle piante di interesse agrario. **Pathology of seed crops.** Publications.

9202 Scaramuzzi, G. IT 041110/77/0294 R
Virosi e virus delle piante. **Plant viruses and virus diseases.**

9203 Vidano, C. IT 041206/77/0311 R
Lotta antiparassitaria in agricoltura e difesa dell'entomofauna utile. **Parasite suppression in agriculture and the protection of beneficial insects.**

9204 Marras, F. IT 041307/77/0230 R
Virus e virosi delle piante. **Viruses and virus diseases in plants.**

9205 Guarnieri, M. IT 042002/77/0502
Sintesi di antifungini a struttura pirazolica. **Synthesis of antifungicides with a pyrazol structure.**

9206 Vivarelli, P. IT 042301/77/0531
Nuovi fungicidi sistemici, sintesi di composti eterociclici aza–attivati. **New systemic fungicides, synthesis of aza–activated heterocyclic compounds.**

9207 Pennazio, S. IT 060100/70/0102
Coltura di apici meristematici e termoterapia di piante virosate. **Meristem tip culture and thermotherapy of virus–infected plants.** Publications.

9208 Redolfi, P. IT 060100/74/0103
Ipersensibilità alla infezione virale nelle piante. **Hypersensitivity to virus infection in plants.**

9209 Lovisolo, O. IT 060100/77/0976
Caratterizzazione dei virus nelle piante. **Characterisation of viruses in plants.**

9210 Lovisolo, O. IT 060100/77/0977
Vettori ed epidemiologia dei virus. **Vectors and virus epidemiology.**

9211 Lovisolo, O. IT 060100/77/0978
Indagini sierologiche dei virus. **Serological studies on viruses.**

9212 Lovisolo, O. IT 060100/77/0979
Microscopia elettronica dei virus. **Electronic microscopy of viruses.**

9213 Lovisolo, O. IT 060100/77/0980
Terapia e fisiopatologia dei virus. **Therapy and physio–pathology of viruses.**

9214 Turchetti, T. IT 061100/72/0212
Indagine sulla moria dei semenzali. **Research on damping off.** Publications.

9215 Panconesi, A. IT 061100/74/0213
Indagini sulla Ceratocystis fimbriata (Ell. and Halst.) Davidson f. platani Walter. **Investigations on Ceratocystis fimbriata (Ell. and Halst.) Davidson f. platani Walter.**

9216 Mittempergher, L. IT 061100/74/0214
Comportamento di alcuni isolati di Endothia parasitica. **Behaviour of some isolates of Endothia parasitica.**

9217 Bottalico, A. IT 061700/77/0924
Studio su alcune specie di Fusarium parassite di piante agrarie e su alcuni aspetti fisiologici del loro parassitismo. **A study of certain Fusarium species infesting cultivated plants and of certain physiological aspects of their parasitism.**

9218 Avancini, D. IT 102101/77/0469 R
Analisi dei residui di fungicidi e sperimentazione nuovi anti–botritici. **Analysis of fungicide residues and trials with new anti–botrytis compounds.**

9219 Domenichini, P. IT 121100/77/0490
Studio programmi lotta chimica, biologia botrytis, calendari intervento, fungicidi da usare. **A study of chemical pest suppression programs, the biology of Botrytis, intervention calendar, fungicides to be used.**

9220 Kovacs, A. IT 121200/77/0505
Messa a punto di un metodo per la distribuzione di antiparassitari polverulenti, esame dell'influenza degli erbicidi e fitoregolatori sulle malattie e sugli afidi, del seme con fungicidi e fitoregolatori. **A new method of applying powder pesticides, analysis of the action of herbicides and growth–regulators on diseases and aphids, seed treatment with fungicides and growth regulators.**

9221 Michieli, G.A. IT 121300/77/0514
Studio rapporti tra resistenza sviluppo infezioni, biologia ospite, calendario di lotta del fungo. **A study of the relationship between resistance to infection, host biology and the fungus control calendar.**

9222 Quak, F. NL 010108/57/1012
Het virusvrijmaken van volledig met virus besmette rassen van cultuurgewassen. **Freeing fully infected varieties of cultivated plants from virus.** Publications.

9223 Maas Geesteranus, H.P. NL 010108/58/1000
Bacterieziekten algemeen; diagnostiek, ziekteverloop en bestrijding. **Bacterial diseases in general; diagnostics, pathogenesis and control.** Publications.

9224 Hoof, H.A. van NL 010108/59/1020
Onderzoek van grondvirussen. **Soil–borne viruses.** Publications.

9225 Maat, D.Z. NL 010108/59/1021
Serologie van virussen, als ziekteverwekkers in land– en tuinbouwgewassen. **Serology of plant viruses.** Publications.

9226 Huttinga, H. NL 010108/66/1025
Het ontwikkelen en toepassen van voornamelijk biofysische methodieken ter karakterisering van plantvirussen. **Development and application of biophysical methods for characterization of plant viruses.** Publications.

9227 Vruggink, H. NL 010108/69/2466
Onderzoek naar de actinomycetenflora in de grond in samenhang met de daarop voorkomende gewassen i.v.m. de mogelijkheid tot harmonische bestrijding van planteparasitaire bodemschimmels. **Study of the actinomycetes flora in the soil with regard to the possibility of integrated control of plantparasitic soil fungi.** Publications.

9228 Vruggink, H. NL 010108/73/3947
Serologie van voor plant pathogene bacteriën en schimmels. **Serology of plant pathogenic bacteria and fungi.** Publications.

9229 Slogteren, D.H.M. van NL 010205/72/3581
Serologische diagnostiek en antiserum–productie ten behoeve van de bestrijding van virusziekten in land– en tuinbouwgewassen. **Serological diagnosis and production of antisera for the control of virus diseases in agricultural and horticultural crops.** Publications.

9230 Langerak, C.J. NL 010210/72/3628
Bestudering van met het zaad overgaande ziekten. **Study of seed–borne diseases.**

9231 Boerema, G.H. NL 010300/54/9103 N
Oriënterend en aanvullend onderzoek naar onbekende resp. nieuwe schimmelziekten. **Orientating and supplementary research to unknown or new fungal diseases.**

9232 Boerema, G.H. NL 010300/59/9108 N
Differentiërende kenmerken van hyalien–sporige pycniden–vormende schimmels. **Taxonomic research on Sphaeropsidale fungi.**

9233 Boerema, G.H. NL 010300/59/9109 N
Evaluatie en toetsing van moderne technieken t.b.v. de identificatie van plantepathogene schimmels. **Evaluation and testing of modern diagnostic methods for the identification of plant pathogenic fungi.**

9234 Miller, H.J. NL 010300/70/9084 N
Taxonomisch en nomenclatorisch onderzoek van plantepathogene bacteriën. **The taxonomy and nomenclature of plant pathogenic bacteria.**

9235 Miller, H.J. NL 010300/75/9086 N
Aanvullend onderzoek naar onbekende, minder bekende en/of gevaarlijke bacterieziekten die zich in ons land zouden kunnen vestigen. **Supplementary research to unknown, less known or dangerous bacterial diseases that could become established in the Netherlands.**

9236 Kooistra, T. NL 010300/75/9096 N
Onderzoek naar de effektiviteit van preparaten die in de "alternatieve" landbouw gebruikt worden m.b.t. schimmelziekten bij planten. **Effectiveness of preparations for fungal disease control in "alternative" agriculture.**

9237 Miller, H.J. NL 010300/76/9083 N
Ontwikkelen en toepassen van snelle en nauwkeurige diagnostische methoden voor bacterieziekten. **Development and application of rapid and accurate diagnostic methods for bacterial plant diseases.**

9238 Miller, H.J. NL 010300/78/9088 N
Biologie van het pathogeen ten behoeve van de epidemiologie en bestrijding van bacteriële plantenziekten. **Biology of the pathogen on behalf of the epidemiology and control of bacterial diseases.**

9239 Bos, C.J. NL 020015/69/4421
Heterokaryose en mitotische recombinatie bij fungi. **Heterocaryosis and mitotic recombinations in fungi.**

9240 Dekker, J. NL 020018/62/5078
Optreden van resistentie in schimmels tegen fungiciden. **Development of resistance in fungi against fungicides.** Publications.

9241 Bollen, G.J.; Hoeven, E.P. van der; Pol–Luiten, B. van der; Verf, M.M. NL 020018/65/4433
Ecologische aspecten van selectieve bestrijding van pathogene wortelschimmels door middel van grondpasteurisatie en toepassing van systemische fungiciden. **Ecological aspects of selective control of root–infecting fungi by means of soil pasteurization and systemic fungicides.** Publications.

9242 Limonard, T. NL 020018/70/4434
Onderzoek van factoren, welke de aanwezigheid, en pathogeniteit van pathogenen in de rhizosfeer beïnvloeden. **Factors which affekt the presence, activity and pathogenity of pathogens in the rhizosphere.**

9243 Fuchs, A.; Vries, F.W. de NL 020018/71/4425
De rol van fytoalexinen in waardplant–parasiet relaties. **The role of phytoalexins in host–parasite relations.** Publications.

9244 Davidse, L.C.; Flach, W. NL 020018/72/5077
Werkingsmechanisme van benzimidazool–derivaten en fysiologischbiochemische aspecten van de ontwikkeling van resistentie tegen deze fungiciden. **Mode of action of benzimidazole derivatives and physiologicalbiochemical aspects of resistance against these fungicides.** Publications.

9245 Bollen, G.J. NL 020018/75/4257
Biologische bestrijding van kiemplanteziekten. **Biological control of seedling diseases caused by fungi.**

9246 Waard, M.A. de   NL 020018/77/8437
Werking van fungiciden tegen schimmels behorende tot de
Oömyceten. **Activity of fungicides on fungi belonging to
Oömycetes.** Publications.

9247 Waard, M.A. de   NL 020018/78/8438
Resistentie van schimmels tegen fungiciden die interfereren
met de sterol–biosynthese. **Resistance in fungi to fungicides
which interfere with sterolbiosyntheses.** Publications.

9248 Limonard, Th.   NL 020018/78/8794 N
Effect van verschillende teeltsystemen en teeltmaatregelen op
de ontwikkeling van vesiculaire mycorrhiza (VAM) bij
landbouwgewassen. **Effect of various farming systems and
cultural practices on the development of vesicular mycorrhiza in
agricultural crops.**

9249 Schaafsma, T.J.   NL 020034/73/4925 R
Molecuul–fysische studies van eiwit–eiwit en eiwit–RNA
interactie in Tabakmozaïek virus met behulp van
NMR–spectroscopie. **Molecular physics of protein–protein and
protein–RNA interactions of Tobacco mosaic as studied by
NMR–spectroscopy.**

9250 Schaafsma, T.J.   NL 020034/78/8628 N
Magnetische resonantie van plantevirussen: structuur en
dynamica. **Magnetic resonance of plant viruses: structure and
dynamics.**

9251 Bruinsma, J.   NL 020041/78/8846 N
Relatie tussen genen die hypersensitiviteit reguleren en het
optreden van verhoogde locale ethyleensynthese in op
virusinfectie hypersensitief reagerende tabak. **Relationship
between genes regulating hypersensitivity and the occurrence of
increased, local synthesis of ethylene in tabacco reacting
hypersensitively to virus infection.**

9252 Jager, C.P. de   NL 020061/68/4273
Genetica van het cowpea–mozaïek virus. **Genetics of cowpea
mozaic Virus.**

9253 Ie, T.S.   NL 020061/71/4275
De cellulaire biologie en de structuur van het
tomatenbronsvlekkenvirus (TSWV). **Cellular biology and
structure of tomato spotted with virus.**

9254 Peters, D.   NL 020061/71/4864
De karakterisering en overdracht van plantevirussen die door
bladluizen worden verspreid. **Characterization and
transmission of plant viruses, spread by aphides.** Publications.

9255 Vlak, J.M.   NL 020061/76/6869
Membraanvirussen uit planten. **Membrane viruses infecting
plants.** Publications.

9256 Dijkstra, J.   NL 020061/78/8862 N
Identificatie van virussen en diagnose van virusziekten in
planten. **Identification of viruses and diagnosis of virus diseases
of plants.**

9257 Kammen, A. van   NL 020068/78/8631 N
Mechanisme en regulering van de virusspecifieke
eiwitsynthese. **Mechanism and regulation of virus specific
protein synthesis.** Publications.

9258 Kammen, A. van   NL 020068/78/8632 N
Zuivering en werkingsmechanisme van cowpea mozaiek virus
RNA replicase. **Purification of cowpea mosaic virus RNA
replicase and its mechanism of action.** Publications.

9259 Kammen, A. van   NL 020068/78/8633 N
Onderzoek naar de structuur van de RNAs van cowpea
mozaiekvirus. **The structure of RNAs of cowpea mosaic virus.**
Publications.

9260 Verhoeff, K.   NL 040004/65/7892
Mechanismen van resistentie en vatbaarheid van planten t.a.v.
pathogene schimmels. **Mechanisms of disease resistance and
susceptibility of plants to pathogenic fungi.** Publications.

9261 Schippers, B.   NL 040004/65/7893
Oecologie van pathogene en saprofytische mikro–organismen.
Ecology of plant pathogenic and saprophytic micro–organisms.

9262 Wieringa–Brants, D.H.   NL 040004/72/7894
Resistentie mechanismen van planten tegen virusinfekties.
Mechanisms of resistance of plants to virus infections.

9263 Kerkenaar, A.   NL 050104/51/3769
Onderzoek naar het werkingsmechanisme van (systemische)
fungiciden en bacericiden. **Investigations on the mode of action
of (systemic) fungicides and bactericides.** Publications.

9264 Overeem, J.C.   NL 050104/56/3775
De isolatie en structuuropheldering van natuurlijke fungiciden
en van stoffen die in planten ontstaan na infectie met
schimmels. **Isolation and structuredetermination of natural
fungicides and of compounds accumulating in plants during
fungal infections.** Publications.

9265 Overeem, J.C.; Vries, L. de   NL 050104/70/3766
Ontwikkeling van chemische en/of biologisch afbreekbare
systemische fungiciden. **Development of degradable systemic
fungicides.** Publications.

Cereals in general (B 3100)

See also 2898, 2996, 3007, 5394, 5396, 5398, 6099, 8440, 8452,
8956, 8958, 8961, 8965, 8969, 8970, 9445, 9528, 13519

9266 Briquet, M.; Dutrecq, A.   BE 020105/78/0004
Isolement, purification et mode d'action de la toxine produite
par"Helminthosporium sativum", agent responsable de
l'helminthosporiose des céréales. **Isolation, purification and
mode of action of the toxin of "Helminthosporium sativum",
agent of seedling blight and root rot of cereals.**

9267 Heitefuss, R.; Wiedemann, A.   DE 132210/72/0001
Nebenwirkungen von Herbiziden auf Pflanzenkrankheiten in
Getreide und auf Faktoren der Bodenfruchtbarkeit. **Secondary
effects of herbicides on plant diseases in small grains and on soil
fertility.**

9268 Fehrmann, H.; Reinecke, P.; Duben, J.
     DE 132212/73/0001
Ökologische Untersuchungen zur Epidemiologie von
Getreidekrankheiten, Entwicklung von Warndienstsystemen
für den standort– und zeitgerechten Einsatz von Fungiziden.
**Ecological investigations on the epidemiology of cereal
diseases, development of warning service systems for the right
timing of fungicidal control.**

9269 Fehrmann, H.; Horsten, J.; Niklahs, V.
DE 132212/77/0001
Fungizidresistenz von getreidepathogenen Pilzen, speziell bei Septoria nodorum, Cercosporella herpotrichoides und Mehltau. **Fungicidal resistance in cereal pathogenic fungi, especially with Septoria nodorum, Cercosporella herpotrichoides and powdery mildew.**

9270 Fehrmann, H.; Reinecke, P.; Duben, J.; Nirenberg, H.
DE 132212/77/0004
Untersuchungen zum Auftreten von Getreide–Fusskrankheiten, Weiterentwicklung von einem speziellen Warndienstsystem. **Investigations on the occurrence of cereal foot–rots, further improvement of a special prognosis system.**

9271 Buchenauer, H.; Förster, H.; Röhner, E.
DE 144540/78/0004 N
Einfluss von Triadimefon und Triadimenol auf den Stoffwechsel von Getreidepflanzen. **Effect of triadimefon and triadimenol on the metabolism of cereal plants.** Publications.

9272 Buchenauer, H.; Konstantinidou, S.
DE 144540/78/0006 N
Einfluss von Nuarimol und Imazalil auf den Stoffwechsel der Getreidepflanzen. **Effect of nuarimol and imazalil on the metabolism of cereal plants.**

9273 Hoffmann, G.M.
DE 161050/77/0001
Biologie von Rhizoctonia an Getreide. **Biology of Rhizoctonia on cereals.** Publications.

9274 Krüger, J.
DE 161050/77/0002 N
Biologie von Septoria–Arten an Getreide. **Biology of Septoria on cereals.** Publications.

9275 Ehle, H.
DE 215090/73/4002
Wirkung von quecksilber–freien Beizmitteln auf samenbürtige Pilze bei Getreidesaatgut. **Effects of non–mercurial dressings on seed–borne fungi in cereal seed.**

9276 Fuchs, E.
DE 215120/75/0004 R
Bestimmung und Charakterisierung der Pathotypen des Getreide–Gelbrostes– Puccinia striiformis –. **Determination and characterization of pathotyps of yellow rust in cereals– Puccinia striiformis –.**

9277 Zeller, W.
DE 215120/78/0008 N
Untersuchungen zur Epidemiologie des Feuerbrandes unter besonderer Berücksichtigung der Physiologie. **Epidemiological studies of fireblight with special regard to physiology.**

9278 Mielke, H.
DE 215120/78/0009 N
Auffinden von resistentem Getreidematerial gegen Fuss–, Blatt– und Ährenkrankheiten bei Getreidemutanten. **Discovery of lines resistant to foot–, leaf blotch– and ear–diseases in mutants of cereals.**

9279 Huth, W.
DE 215150/78/0022 N
Untersuchungen über des Resistenzverhalten von Getreide gegenüber Viren. **Investigation on the resistance of cereals to viruses.**

9280 Huth, W.; Paul, H.L.; Lesemann, D.
DE 215150/78/0024 N

Bestandsaufnahme über das Vorkommen von Virosen bei Getreide und Futtergräsern. **Survey on the occurrence of virus diseases in cereals and fodder grasses.**

9281 Mielke, H.
DE 215180/77/0003
Bodenentseuchung durch hygienische Fruchtfolgemassnahmen, insbesondere von Getreidefusskrankheiten im Vergleich zur Anwendung von neuen Fungiziden. **Soil decontamination by hygienic crop rotation measures, esp. of cereal eyespot, in comparison with the use of new fungicides.**

9282 Mielke, H.
DE 215180/77/0004
Untersuchungen über die Wirkung verschiedener Fungizide auf die Ähren–, Blatt– und Fusskrankheiten des Getreides. **Studies on effects of different fungicides on spike, leaf, and root diseases of cereals.**

9283 Schiff, H.; Schrödter, H.
DE 301030/72/0004
Entwicklung eines agrarmeteorologisch gesteuerten Warndienstes gegen Getreidemehltau – Erysiphe graminis – auf der Grundlage meteorologisch–epidemiologischer Beziehungen. **Elaboration of an agrometeorological warning system against Erysiphe graminis on cereals based on meteorologicalepidemiological relations.**

9284 Schiff, H.; Schrödter, H.
DE 301030/72/0005
Entwicklung eines agrarmeteorologisch gesteuerten Warndienstes gegen die Spelzenbräune des Getreides– Septoria nodorum – auf der Grundlage meteorologisch–epidemiologischer Beziehungen. **Elaboration of an agrometeorological warning system against Septoria nodorum on cereals based on meteorological– epidemiological relations.**

9285 Hecht, H.
DE 502051/75/0003
Der qualitative und quantitative Nachweis von Getreideviren in Bayern. **The qualitative and quantitative detection of cereal viruses in Bavaria.**

9286 Kristensen, H.R.; Engsbro, B.; Christensen, M.; Begtrup, J.W.
DK 010116/60/4512
Forekomst, forebyggelse og bekæmpelse af viroser på korn, græsser og bælgplanter hovedsagelig: havre–rødsot, byg–stribemosaik. rajgræs–mosaik, hvidkløver–mosaik, rødkløver–mosaik og hundegræs–mosaik. **Occurrence, prevention and control of viruses on cereals, grasses and legumes, chiefly oat rootblight, barley stripe mosaic, ryegrass mosaic, cocksfoot mosaic, white clover mosaic, red clover mosaic and orchard grass.**

9287 Schulz, H.
DK 010116/69/4501
Forskellige jordtypers og sædskifters betydning for forekomsten af fodsyge i korn samt metoder til bekæmpelse af fodsyge. **Influence of different soil types and rotations on the occurrence of take–all in cereals, and methods for the control of take–all.**

9288 Welling, B.
DK 010116/74/4503
Forekomst af bladsygdomme på kornblade samt sygdommenes bekæmpelse. **Occurrence of leaf diseases on cereal leaves and their control.**

9289 Cassini, R.
FR 010107/70/1017
Méthode d'évaluation de la résistance des céréales au Fusarium Roseum. **Methods of assessing cereal resistance to**

Fusarium Roseum.

9290 Rapilly, F.; Pauvert, P.   FR 010107/71/1019
Etude de l'Oïdium des céréales (Érysiphe graminis). **Study of powdery mildew on cereals (Erysiphe graminis).**

9291 Clarkson; Sanderson   GB 010502/00/0011 R
Effect of soil borne pathogens on root growth and function.

9292 Butler; Watson   GB 011503/78/0006 R
Effects of diseases on physiologocal processes, growth and development of cereals.

9293 Jordan   GB 011508/76/0013 R
Epidemiology of cereal diseases in the south west.

9294 Russell   GB 011906/00/0002 R
Nature of resistance to powdery mildews of cereals.

9295 Wolfe; Bennett   GB 011906/00/0006 R
UK cereal pathogen virulence survey to determine the occurrence of virulence genes in Erysiphe graminis.

9296 Wolfe   GB 011906/00/0007 R
Variation and dynamics of changes in populations of Erysiphe graminis f.sp. tritici and f.sp. hordei.

9297 Johnson; Taylor   GB 011906/00/0012 R
Assessment and genetic control of resistance to rust diseases in cereal varieties.

9298 Johnson; Taylor   GB 011906/00/0014 R
Variability of rust pathogens in cereals.

9299 Scott; Hollins   GB 011906/00/0017 R
Search for and study of inherited resistance to Gaeumannomyces graminis.

9300 Scott; Benedikz   GB 011906/00/0019 R
Improve resistance to Septoria spp.; resistance mechanisms and inheritance.

9301 Scott; Hollins   GB 011906/00/0020 R
Improve resistance to Cercosporella herpotrichoides; resistance mechanisms and inheritance.

9302 Mattingly; Slope   GB 012003/00/0010 R
Effect of phosphate fertilisers on root diseases.

9303 McIntosh; Bateman   GB 012005/00/0040 R
Fungicidal seed treatments for controlling cereal diseases.

9304 Plumb   GB 012009/00/0004 R
Epidemiology and control of virus diseases of cereals and maize.

9305 Slope; Hornby   GB 012009/00/0011 R
Effects of crop rotations and nutrition on soil–borne diseases of cereals.

9306 Slope; Hornby   GB 012009/00/0012 R
Biology and epidemiology of fungi associated with take–all disease of wheat and barley.

9307 Salt; MacFarlane   GB 012009/00/0014 R
Survey and effects of lesser known pathogens on roots.

9308 Salt; Hornby   GB 012009/00/0015 R
Physiological effects and assessment of root disease fungi.

9309 Jenkyn   GB 012009/00/0016 R
Fungicides for cereal crops.

9310 Jenkyn   GB 012009/00/0019 R
Fungal diseases of cereal foliage and their effects.

9311 Rawlinson; Muthyalo   GB 012009/00/0031 R
Viruses of fungal pathogens.

9312 Clifford   GB 012101/00/0008 R
Expression and genetic control of resistance to rusts of barley, wheat and oats.

9313 Jones   GB 012101/00/0010 R
Expression and genetic control of resistance to powdery mildew in oats and barley.

9314 Clifford.; Clothie   GB 012101/00/0018 R
National physiologic race survey of barley crown rust and rhynchosporium; oat mildew and brown rust and wheat brown rust.

9315 Clifford   GB 012101/76/0024 R
Pathogen variability and epidemic potential of cereal pathogens.

9316 Carr   GB 012107/00/0006 R
Cereal host/pathogen relationships at cell level.

9317 Abrook   GB 012107/00/0007 R
Epidemiology of cereal and grass viruses.

9318 Abrook; Catherall   GB 012107/00/0008 R
Isolation, identification, effects and properties of cereal and grass viruses.

9319 Whittle; Richardson   GB 030902/67/0503 R
Assessment of the importance of seed treatment, seed– and soil–borne pathogens in cereal crop performance.

9320 Symonds   GB 030902/75/0504 R
Cost benefit analysis of broad spectrum fungicide use on cereals in South East Scotland.

9321 Malone   GB 041501/00/0002 R
Assessment and control of seed–borne diseases of cereals.

9322 Malone   GB 041501/00/0006 R
Fungicidal control of foliar diseases of cereals.

9323 Researcher not indicated   GB 050121/00/0006 R
Cereals: foliar diseases and effect of fungicides.

9324 Researcher not indicated   GB 050161/00/0006 R
Cereals: foliar diseases and effects of fungicides.

9325 Greaves   GB 050309/00/0003 R
Cereal yellow dwalf virus and its vectors.

9326 Greaves; George   GB 050314/73/0003 R
Studies of infection of cereal crops with barley yellow dwarf virus.

See also 3008, 5414, 6169, 8950, 8985, 9408

9327 Greaves　　　　　　　GB 050314/76/0009 R
Barley yellow dwarf virus on oats and maize and its status in volunteer cereals as a source of innoculum.

9328 King; Polley　　　　　GB 050320/67/0071 R
Surveys of diseases of wheat and barley.

9329 Lennard　　　　　　　GB 060105/00/0003 R
Physiological and biochemical aspects of fungal attack on cereals.

9330 Gilmour　　　　　　　GB 060107/00/0002 R
Incidence and control of and resistance of varieties to cereal leaf diseases.

9331 Shipton　　　　　　　GB 060217/00/0001 R
Biology and control of foot rotting diseases of cereals and their effects on crop production.

9332 Shipton　　　　　　　GB 060217/00/0002 R
The appraisal of fungicides for control of diseases in cereals.

9333 Channon; Potts　　　　GB 060310/00/0002 R
Studies of cereal leaf diseases and methods of control.

9334 Cunningham, P.C.　　　IE 060500/60/0235 N
Studies of disease levels as affected by various practices in continuous cereal systems. Publications.

9335 Cunningham, P.C.; Dunne, B.　　IE 060500/69/0237 N
Screening of systemic fungicides for the control of powdery mildew of cereals. Publications.

9336 Saponaro, A.; Porta–Puglia, A.; Montorsi, F.
　　　　　　　　　　　　IT 020300/78/0001
Ricerche sulla trasmissibilità delle malattie portate dal seme.
Seed borne diseases and mechanism of seed transmission.

9337 Appiano, A.　　　　　IT 060100/69/0101
Indagini citologiche al microscopio elettronico su graminacee affette da virus. **Electron microscopy cytological studies on virus infected graminaceous plants.** Publications.

9338 Spek, J. van der　　　　NL 010108/69/2465
Voetziekten van granen. **Footrots of cereals.** Publications.

9339 Beemster, A.B.R.　　　NL 010108/69/2772
Virusziekten in Gramineaën. **Virus diseases in gramineous crops.** Publications.

9340 Zadoks, J.C.; Blokland, A.J.; Hoogkamer, W.; Wal, A.F. van der; Geerds, C.F.; Frinking, H.D.
　　　　　　　　　　　　NL 020018/61/4431
Epidemiologie van phytopathogene schimmels bij granen.
Epidemiology of phytopathogenic fungi on cereals. Publications.

9341 Duym, J.　　　　　　NL 040007/73/4172
Onderzoek naar de bestrijding van ziekten (vnl. schimmels) met chemische middelen in granen, koolzaad en houtgewassen. **Chemical control of fungi in cereals, rape-seed and tree species.**

Barley (B 3110)

9342 Aust, H.-J.　　　　　DE 129554/74/0001
Untersuchungen von biophysikalischen Einflussgrössen in der Epidemiologie am Wirt–Parasit–Paar Gerste/echter Mehltau unter besonderer Berücksichtigung von Kompensationsphänomenen. **The role of biophysical factors in the epidemiology of the host parasite pair relationship between barley and powdery mildew with special regard to compensational phenomena.**

9343 Kranz, J.; Forche, S.　　DE 129554/74/0004
Einfluss der Resistenz auf die Dynamik von Epidemien des Gerstenmehltaues. **Influence of resistence on the dynamics of barley mildew epidemics.**

9344 Heitefuss, R.; Hwang, B.K.; Ordonez, M.T.
　　　　　　　　　　　　DE 132210/71/0003
Untersuchungen zur Resistenz von Gerste gegenüber Erysiphe graminis. **Investigations into the resistance of barley to Erysiphe graminis.**

9345 Buchenauer, H.; Hippe, S.　　DE 144540/78/0005 N
Ultrastrukturelle Veränderungen in Sporidien von Ustilago avenae, Gerstenmehltau und Weizenbraunrost nach Behandlung mit Triadimefon, Fluotrimazol, Nuarimol und Imazalil. **Ultrastructural changes in sporidia of Ustilago avenae, barley mildew and leaf wheat rust after treatment with triadimefon, fluotrimazol, nuarimol and imazalil.**

9346 Hoffmann, G.M.; Kiessling, U.　　DE 161050/77/0004
Zum Problem der Koexistenz und der Interaktionen bei pilzlichen Blattparasiten an Gerste. **Coexistence and interaction of leaf parasites on barley.**

9347 Fischbeck, G.; Kendlbacher, R.　　DE 161250/77/0001
Differenzierung früher Stadien der Keimung von Konidien verschiedener Kulturen von Gerstenmehltau auf den Blättern älterer und neuerer Zuchtsorten. **Differentiation of early stages of germination of conidia in different cultures of mildew on the leaves of older and younger varieties of barley.**

9348 Fuchs, E.　　　　　　DE 215120/77/0004
Analyse des Resistenzverhaltens von Weizen– und Gerstensorten gegenüber Gelbrost. **Analysis of resistance of wheat and barley varieties to yellow rust.**

9349 Paul, H.L.; Huth, W.　　DE 215150/78/0005 N
Ausarbeitung von serologischen Diagnosemethoden für barley yellow dwarf– und barley yellow mosaic–Virus für Testserien. **Development of serological diagnostic methods for the test screening of barley yellow dwarf and barley mosaic virus.**

9350 Huth, W.　　　　　　DE 215150/78/0023 N
Untersuchungen zur biologischen Charakterisierung mehrerer Stämme des barley yellow dwarf Virus. **Investigations on biological properties of strains of barley yellow dwarf virus.**

9351 Rintelen, J.　　　　　DE 502051/74/0004
Physiologische Rassen des Zwergrostes der Gerste – Puccinia hordei –. **Physiologic races of brown rust of barley – Puccinia hordei –.**

9352 Dalgaard, L.　　　　　DK 030106/78/0006 N
Kemiske undersøgelser af parasit–værtsplante forhold mellem meldug og byg. **Chemical investigations of parasite–host**

relationship between mildew and barley.

9353 Stølen, O.; Smedegaard–Petersen, V.
DK 030145/77/0027 N
Nedarvning af resistens i byg mod stribesygesvampen Pyrenophora graminea. **Inheritance in barley of Pyrenophora graminea.**

9354 Stølen, O.; Smedegaard–Petersen, V.
DK 030145/77/0028 N
Meldugresistensens indflydelse på bygplanters udbytte. **The effect of mildew on the yield of barley.**

9355 Skou, J.P. DK 030146/77/0019 N
Meldugresistensens fysiske og fysiologiske baggrund. **The physical and physiological basis for the resistance to barley powdery mildew.**

9356 Skou, J.P. DK 030146/77/0020 N
Resistensbiologi for byggens stribesyge. **Biology of resistance to barley leaf stripe (Drechslera graminea).**

9357 Haahr, V.; Jensen, H.P.; Jørgensen, J.H.
DK 030146/77/0021 N
Kulde– og sygdomsresistens i vinterbyg. **Winterhardiness and disease resistance in winter barley.**

9358 Giese, H.; Jensen, H.P.; Jørgensen, J.H.
DK 030146/78/0001 N
Meldugresistensgener i M1–a regionen på byggens kromosom 5. **Powdery mildew resistance genes in the M1–a region of barley chromosome 5.**

9359 Klug–Andersen, S.; Jensen, H.P.; Jørgensen, J.H.
DK 030146/79/0001 N
Selektionsværdi af "unødvendige" virulensgener i bygmeldug. **Selection value of "unnecessary" virulence genes in the barley powdery mildew fungus.**

9360 Jørgensen, J.H.; Østergaard, H.; Christiansen, F.B.
DK 030146/79/0002 N
Matematiske modeller for værtplante–patogen systemer. **Mathematical models of host–parasite systems (barley powdery mildew).**

9361 Hollomon GB 012005/00/0041 R
Systemic fungicides used to control foliar pathogens of cereals: mode of action and mechanisms of tolerance.

9362 Byford; Dunning GB 012301/00/0025 R
Root and seedling diseases.

9363 Owen GB 023902/79/0002 N
Initiation of resistance in barley to Rhynchosporium secalis.

9364 Polley; King GB 050320/00/0022 R
Factors favouring development of brown rust of barley and development of evaluation methods.

9365 Waples; Dickens GB 050320/00/0023 R
Factors determining infection, overwintering symptoms and sporulation by Rechslera graminis on barley.

9366 Dickens; Sharp GB 050323/69/0011 R
Seed–borne aspects of barley leaf stripe.

9367 Shipton GB 060217/79/0005 N
Foliar diseases of barley and their effect on the quality of grain for malting.

9368 Clark GB 060310/00/0003 R
Barley root and stem base diseases.

9369 Cooke, B.M.; Pigott, J. IE 120106/79/9171 N
Preliminary investigations into the epidemiology of net blotch disease of barley caused by pyrenophora teres drechsler.

9370 Conti, M. IT 060100/74/0094
Caratterizzazione di isolati Italiani del virus del mosaico striato dell'orzo. **Characterization of Italian isolates of barley stripe mosaic virus.**

9371 Stubbs, R.W. NL 010108/62/1086
Gele roest in gerst (Puscinia striiformis Westend.). **Yellow rust in barley (Puccinia striiformis Westend.). Publications.**

Maize (B 3120)

See also 6236, 9412, 9414, 9416, 9419, 9420, 9579

9372 Höfner, W.; Grieb, R. DE 129044/75/0004
Veränderungen des Gehalts organischer Säuren in Mais– und Sonnenblumenpflanzen mit genetisch bedingter, durch absoluten oder durch induzierten Eisenmangel bedingter Chlorose in Abhängigkeit vom Fe– und Mo– Versorgungsgrad der Pflanzen. **Changes of organic acid concentration in maize and sunflower plants, due to genetically or nutritionally caused chlorosis.**

9373 Kranz, J.; Hamelink, J. DE 129554/74/0003
Experimentelle Untersuchungen zur Ermittlung der Befallsverlustrelationen von Maiskrankheiten in Togo. **Experimental investigations on the relationship between attack intensity and loss rate in maize diseases in Togo.**

9374 Grossmann, F.; May–Hacker, M. DE 144540/78/0003 N
Entwicklung von Methoden zur Beurteilung der Stengelfäule–Resistenz bei Mais. **Development of methods for assessment of stalk rot resistance in maize.**

9375 Börner, H.; Starck, S. DE 148200/77/0003
Das Wirt–Parasit–Verhältnis von Mais und Ustilago maydis– Maisbeulenbrand. **Host–parasite–relationship of maize and Ustilago maydis.**

9376 Krüger, W.; Rogdaki–Papadaki, C. DE 215180/74/5004
Epidemiologische Studien über Pilze, die Stengel– und Wurzelfäule beim Mais verursachen. **Epidemiological studies on fungi causing stalk and root rot on maize.**

9377 Krüger, W. DE 215180/77/0005
Untersuchungen über die Wurzel– und Stengelfäule des Maises. **Studies on root and stem rot of maize.**

9378 Saponaro, A. IT 020300/74/0001
Ricerche sul marciume da Fusarium del culme e delle radici del mais. **Maize. Studies on Fusarium root and stalk rot. Publications.**

9379 Saponaro, A.; Montorsi, F. IT 020300/79/0004 N
Ricerche su alcune specie di Acremonium (Cephalosporium) presenti nei semi di Mais. **Studies on some seed–borne species of**

D 2420 – Plant diseases and disease control

Acremonium (Cephalosporium) of Maize.

9380 Mariani, G.; Desiderio, E.; Saponaro, A.
IT 020800/79/0012 N
Mais – Effetto di differenti fattori ambientali sulle manifestazioni di marciume del culmo. **Maize – Stalk rot as affected by different environmental factors.**

9381 D'Ercole, N.   IT 040211/77/0486 R
Epidemiologia e biologia delle malattie crittogamiche del mais e del sorgo. **Epidemiology and biology of cryptogamic diseases of maize and sorghum.**

9382 Conti, M.   IT 060100/72/0098
Epidemiologia e relazione con i vettori del virus del nanismo ruvido del mais. **Epidemiology and vector relationship of maize rough dwarf virus.** Publications.

9383 Spek, J. van der   NL 010108/75/6383
Voet – en stengel rot van mais. **(Fusarium) stalk rot.**

Oats (B 3130)

See also 9345

9384 Saponaro, A.; Porta–Puglia, A.   IT 020300/79/0003 N
Ricerche su alcune specie di Drechslera presenti nei semi di Avena. **Studies on some seed–borne species of Drechslera on oat seeds.**

Rice (B 3140)

See also 8474

Sorghum (B 3160)

See also 9381, 9414, 9419, 9420, 9579

Wheat (B 3170)

See also 3127, 3129, 6340, 6341, 8950, 9345, 9348

9385 Heitefuss, R.; Schreiber   DE 132210/74/0001
Untersuchungen zur Resistenz von Weizen gegenüber Gelbrost – Puccinia striiformis –. **Investigations on the resistance of wheat to yellow rust – Puccinia striiformis –.**

9386 Fehrmann, H.; Bartels, G.   DE 132212/77/0002
Feldapplikation systemischer Fungizide gegen Erysiphe graminis in Winterweizen – zeitlich optimal terminierte Spritzung –. **Field application of systemic fungicides against Erysiphe graminis in winter wheat – right timing of spraying.**

9387 Fehrmann, H.   DE 132212/78/0002 N
Entwicklung eines lokal durchführbaren Warnsystems zur termingerechten Bekämpfung der Halmbruchkrankheit in Winterweizen mit dem Ziel der Reduzierung von Fungiziden. **Development of warning system realizable locally for the control of stem break in winter wheat in due time for the purpose of reduction of fungicides.**

9388 Reiner, L.; Mangstl, A.; Englert, G.; Brummer, A.
DE 161250/78/0005 N
Einflüsse des Standortes, der Witterung und der Anbaumassnahmen auf den Befall von Spelzenbräune – Septoria nodorum – beim Weizen. **Influences of location,** climatological conditions and cultivation on the attack of wheat with Septoria nodorum.

9389 Harms, H.   DE 201010/77/0002
Phenolische Verbindungen als Resistenzfaktoren des Weizens gegen Gelbrost: Untersuchungen der WirtParasit–Stoffwechselbeziehungen an Weizen– Kalluskulturen und aseptischen Bedingungen. **Phenolic compounds as resistance factors of wheat against yellow rust: studies on metabolic interrelations between host and parasite in wheat callus cultivation and on aseptic conditions.**

9390 Weinmann, W.   DE 215080/77/0037
Untersuchung über das Rückstandsverhalten von Maneb auf und in Winterweizen nach einer Behandlung gegen Halmbruchund Ährenkrankheiten. **Investigations on residual action of maneb on and in winter wheat after application against eyespot and spike diseases.**

9391 Weinmann, W.   DE 215080/77/0038
Untersuchung über das Rückstandsverhalten von Captafol auf und in Winterweizen nach einer Behandlung gegen Halmbruchund Ährenkrankheiten. **Investigations on residual action of captafol on and in winter wheat after application against eyespot and spike diseases.**

9392 Weinmann, W.   DE 215080/77/0039
Untersuchung über das Rückstandsverhalten von Mancozeb auf und in Winterweizen nach einer Behandlung gegen Ährenkrankheiten. **Investigations on residual action of mancozeb on and in winter wheat after application against spike diseases.**

9393 Weinmann, W.   DE 215080/78/0001 N
Untersuchung über das Rückstandsverhalten von Maneb und Captafol auf/in Winterweizen nach einer Behandlung gegen Halmbruch– und Ährenkrankheiten. **Investigation on the residue–behaviour of Maneb and Captafol on/in winter–wheat after a treatment against eye spot and ear–diseases.**

9394 Bartels, G.   DE 215120/73/4003
Untersuchungen zur Epidemiologie und Bekämpfung von Blatt– und Ährenkrankheiten des Getreides. **Investigations on the epidemiology and control of leaf and spike diseases in wheat.**

9395 Bartels, G.   DE 215120/77/0005
Untersuchungen und Prüfungen von Weizensorten und –zuchtstämmen bezüglich ihres unspezifischen Resistenzverhaltens gegenüber dem Getreidemehltau – Erysiphe graminis –. **Investigations on and testing of varieties and seed–stock of wheat regarding unspecific resistance to Erysiphe graminis.**

9396 Bartels, G.   DE 215120/78/0004 N
Untersuchungen über den Einfluss des Getreidemehltaus auf die Ertragsbildung von Winter– und Sommerweizen unter dem Aspekt einer Schadensprognose. **Investigations on the effect of mildew on the yield formation of winter– and summer wheat under the aspect of damage forecast.**

9397 Bartels, G.   DE 215120/78/0005 N
Untersuchungen zum Auftreten und zur Verbreitung von Mehltau – Erysiphe graminis – an Weizen in der Bundesrepublik Deutschland. **Investigations on the occurrence and spread of mildew in wheat in the Federal Republic of**

Germany.

9398 Diercks, R.; Obst, A.   DE 502051/70/0004
Untersuchungen zur Epidemiologie und Prognose der
Spelzenbräune – Septoria nodorum – des Weizens.
**Investigations on the epidemiology and prognosis of glume
blotch – Septoria nodorum – of wheat.**

9399 Obst, A.; Montag   DE 502051/73/0002 N
Gezielte Bekämpfung der Halmbruchkrankheit des Weizens.
Systematic control of eyespot in wheat.

9400 Diercks, R.; Klein, W.   DE 502051/75/0001
Untersuchungen über den Einfluss des Getreidemehltaues –
Erysiphe graminis – auf die Ertragsbildung von Winterund
Sommerweizen unter dem Aspekt einer Befalls– und
Schadensprognose. **Studies on the influence of powdery mildew
of cereals – Erysiphe graminis – on yield formation of winter
and spring wheat with respect to a prognosis of infestation and
yield loss.**

9401 Obst, A.   DE 502051/78/0001 N
Entwicklung eines lokal durchführbaren Warnsystems zur
termingerechten Bekämpfung der Halmbruchkrankheit in
Winterweizen mit dem Ziel der Reduzierung von Fungiziden.
**Development of warning system realizable locally for the
control of stem break in winter wheat in due time for the
purpose of reduction of fungicides.**

9402 Reschke, M.   DE 507250/78/0002 N
Entwicklung eines lokal durchführbaren Warnsystems zur
termingerechten Bekämpfung der Halmbruchkrankheit in
Winterweizen mit dem Ziel der Reduzierung des Einsatzes von
Fungiziden. **Development of a monitoring system on a local
scale for the timely control of stembreak in winter wheat with
the aim of reducing the amount of applied fungicides.**

9403 Hanuss   DE 509050/78/0001 N
Entwicklung eines durchführbaren Warnsystems zur
termingerechten Bekämpfung der Halmbruchkrankheit in
Winterweizen mit dem Ziel der Reduzierung von Fungiziden.
**Development of warning system realizable locally for the
control of stem break in winter wheat in due time for the
purpose of reduction of fungicides.**

9404 Holden; Festenstein   GB 012001/00/0021 R
Biochemical studies on take–all infection of wheat.

9405 King; Polley   GB 050320/69/0011 R
**Development of methods for evaluating losses and assessing
disease in wheat caused by Septoria species.**

9406 Polley; King   GB 050320/72/0012 R
**Development of methods for evaluating losses in wheat caused
by Puccinia striformis.**

9407 Polley; Clarkson   GB 050320/74/0013 R
Feasibility of aerial surveys for wheat take–all.

9408 Cooke, B.M.; Fitzgerald, W.   IE 120106/77/9083 N
**The effect on yield of cross–infection by septoria nodorum (
berk.) berk. and br. (leptosphaeria nodorum muller) on wheat
and barley.** Publications.

9409 Basile, R.; Corazza, L.   IT 020300/53/0001
Identificazione delle razze fisiologiche di Puccinia graminis

var. tritici e di Puccinia recondita var. tritici in Italia, su
frumente, ospiti intermedi e graminacee spontanee.
**Identification of physiologic races of Puccinia graminis var.
tritici and Puccinia recondita var. tritici in Italy on wheat,
intermediate hosts and wild grasses.** Publications.

9410 Agnello, A.V.   IT 020300/75/0002
Lotta chimica contro le malattie fungine del grano. **Chemical
control of fungal diseases of wheat.**

9411 Troccoli, C.   IT 040101/77/0530 R
Azione di trattamento anticrittogamico su varietà di grano
duro, interazione tra concimazione e trattamenti di difesa
contro le malattie, influenza della successione colturale sulla
risposta di varietà di frumento a trattamenti di difesa. **The
effect of pesticides on certain varieties of Durum wheat,
interaction of fertilizers and the pharmacological prevention of
diseases, the influence of cultural rotation on the reaction of
certain varieties of wheat to prevention.**

9412 Piglionica, V.   IT 040106/77/0522 R
Verifica di applicazione di fungicidi al frumento per prevenire
gli attacchi dei più importanti agenti patogeni, prevenzione e
lotta contro le fusariosi del mais derivanti dallo impiego di
semente infetta. **Trials of fungicide application to wheat to
prevent attacks from the most important pathogenic agents,
prevention and suppression of seed–borne Fusarium in maize.**

9413 Ciccarone, A.   IT 040106/77/0815
Studio dei parassiti crittogamici del grano duro e di specie
compatibili analisi delle popolazioni e del loro evolversi in
rapporto agli ospiti e all'ambiente ricerche di punti di
resistenza e della loro natura valutazione fitopatologica dei
frumenti in fase di più avanzata costituzione. **A study of Durum
wheat cryptogamic parasites and other compatible species,
analysis of the populations and of their evolution in function of
hosts and environment, identification of resistance bridges,
their nature; phytopathological evaluation of new lines.**

9414 Baldoni, R.   IT 040201/77/0470 R
Impiego di fungicidi ed erbicidi su frumento sorgo e mais. **The
use of fungicides and herbicides on wheat, sorghum and maize.**

9415 Govi, G.   IT 040211/77/0501 R
Rilevamento epidemiologico e geografico delle malattie
fungine del frumento. **Epidemiological and geographic survey
of the fungus diseases of wheat.**

9416 Polelli, M.   IT 040606/78/1144 N
Valutazione dei costi dei trattamenti antiparassitari al
frumento e al mais contro le malattie crittogamiche e contro
erbe infestanti. **The cost of pesticide treatments used to control
the fungus diseases and weeds infesting wheat and maize.**

9417 Sisto, D.; Milia, M.; Marras, G.F.   IT 041302/74/0001
Comportamento dei frumenti verso puccinia e Erysiphe
graminis tritici. **Reaction of wheats to common chick weed and
Erysiphe graminis.** Publications.

9418 Foschi, S.   IT 061800/77/0816
Ricerca sull'attività e epoca di applicazione dei fungicidi contro
Oidio e Ruggini del frumento; effetti collaterali dei diserbanti
contro le malattie fungine e sulla sensibilità varietale di
frumento tenero e duro all'impiego dei diserbanti. **Research on
fungicide action against wheat mildew and rust, on the timing of
application; collateral effects of herbicides : on fungus diseases,**

on the sensitivity to herbicides of soft and durum wheat varieties.

9419 Patuzzo, L.   IT 121100/77/0519
Applicazione di anticrittogamici al frumento, dalla concia del seme, alla spigatura e loro interazione con trattamenti erbicidi o insetticidi da eseguire durante la vegetazione, Lotta contro le erbe infestanti piu difficili a combattersi su mais e sorgo. **The application of fungicides to wheat from the seed dressing stage to earing, interaction with herbicide or insecticide treatments during growth. Weed control of the most troublesome weeds in maize and sorghum.**

9420 Vandoni, G.   IT 121100/78/1145 N
Studio delle applicazioni di anticrittogamici al frumento dalla concia del seme alla spigatura e loro interazione con trattamenti erbicidi ed insetticidi da effettuarsi in vegetazione, lotta contro le erbe infestanti più difficili a combattersi su mais e sorgo. **A study on fungicides applied to wheat from the seed stage to ear forming, their interaction with herbicide and insecticide treatment during vegetation; the control of the most resistant weeds infesting maize and sorghum.**

9421 Stubbs, R.W.   NL 010108/62/1085
Gele roest in tarwe (Puccinia striifernis Westend). **Yellow rust in wheat (Puccinia striifernis Westend.).** Publications.

9422 Ubels, E.   NL 010108/73/3960 R
Bald– en aarziekten bij tarwe. **Leaf and ear diseases of wheat.**

9423 Stubbs, R.W.; Silfhout, C.H. van   NL 010108/74/6384
Bestrijding van schimmelziekten in tarwe in ontwikkelingslanden. **Control of fungal diseases of wheat in developing countries.**

9424 Wal, A.F. van der   NL 010120/78/8562
De bruine roest van tarwe (Puccinia recondita). **Leaf rust of wheat (Puccinia recondita).**

9425 Hag, B.A. ten   NL 010207/78/8416
Optimale ziektebestrijding in tarwe. **Optimum disease control in wheat.**

9426 Zadoks, J.C.   NL 020018/77/7316
Het ontwikkelen van een waarschuwingssysteem voor tarweziekten. **Development of a disease warning system in wheat.** Publications.

9427 Zadoks, J.C.   NL 020018/78/8440
Model–onderzoek naar de eco–fysiologische aspecten van schade aan tarwe veroorzaakt door fytopathogene schimmels en stressfactoren–EPIDAM. **Simulation of the ecophysiological aspects of crop loss in wheat caused by phytopathologic fungi and stress (EPIDAM).**

Fibre plants and oil crops in general (B 3200)

See also 3135, 9341

Olive (B 3220)

9428 Motta, E.   IT 020300/79/0002 N
Diffusione dell'infezione da Verticillium spp. sull'Olivo. **Distribution of Verticillium infection on olive trees.**

9429 Ercolani, G.L.   IT 040106/77/0169 R

Biologia ed epidemiologia di "pseudomonas savastanoi" in relazione alla sensibilità dell'olivo al freddo e alla rogna. **Biology and epidemiology of "pseudomonas savastanoi" in relation to the sensitivity of olive trees to cold and to olive knot.**

9430 Prota, U.   IT 041307/78/1091 N
Ricerche sulla patologia dell'olivo in Sardegna, con particolare riferimento all'occhio di Pavone da Spilocaea oleagina hughes. **Research on the pathology of the olive tree in Sardinia, in particular on "occhio di Pavone" due to Spilocaea oleagina hughes.**

Rape (B 3230)

9431 Krüger, W.   DE 215180/70/5006
Untersuchungen über den Erreger des Rapskrebses und seine Bekämpfung. **Studies on stem rot – Sclerotinia sclerotiorum– on rape – Brassica napus L. – and its control.**

9432 Krüger, W.   DE 215180/77/0001
Phomabefall und integrierte Bekämpfung der Rapsschädlinge. **Phoma attack and integrated control of pests in rape.**

9433 Naton, E.   DE 502051/72/0002
Erarbeitung von Grundlagen zum integrierten Pflanzenschutz im Ölfruchtanbau. **Search for principles of integrated control of rape.**

9434 Reseachers not yet known   GB 011910/79/0022 N
Resistance to oilseed rape diseases.

9435 Spek, J. van der   NL 010108/76/6747
Vallers en kankerstronken van koolzaad. **Blackleg of colza.**

Sunflower (B 3250)

See also 9372

9436 Zazzerini, A.   IT 041015/74/0647
Ricerche biologiche ed epidemiologiche sulla plasmopara Helianthi del girasole in Umbria. **Biological and epidemiological studies on plasmopara Helianthi of sunflower in Umbria.**

9437 Zazzerini, A.   IT 041015/77/0319 R
Ricerche biologiche ed epidemiologiche sulla peronosporadel girasole –Plasmopara Helianthi– in Umbria. **Biological and epidemiological research on the sunflower peronospora – Plasmopara Helianthi – in Umbria.**

Sugarbeets and starch producing plants in general (B 3300)

9438 D'Ambra, V.   IT 062200/74/0131
Ricerche su isolati di Cercospora beticola Sacc. resistenti al Benomyl. **Researches on isolates of Cercospora beticola Sacc. resistant to Benomyl.**

9439 Mutto, S.   IT 062200/74/0132
Caratterizzazioni di isolati di Cercospora beticola Sacc. provenienti da diverse località italiane. **Investigations on Italian isolates of Cercospora beticola Sacc.**

Potatoes (B 3310)

D 2420 – Plant diseases and disease control

See also 2898, 8520, 8527, 8533, 8536, 8963, 8971, 8972, 8986, 9671, 9719, 9943, 16073, 20281

9440 Sarkar, S.   DE 144540/78/0007 N
Isolation und Charakterisierung des Kartoffelblattrollvirus.
Isolation and characterization of the potato leafroll virus.
Publications.

9441 Schöber, B.   DE 215120/75/0006
Untersuchungen über die Bildung von Toxinen und Phytoalexinen in Kartoffeln nach Infektion mit dem Erreger der Braunfäule, Phytophthora infestans – Mont. – de Bary.
Investigations into the production of toxins and phytoalexins in potato tubers after infection with late blight fungus, Phytophthora infestans – Mont. – de Bary.

9442 Schöber, B.   DE 215120/77/0001
Physiologische Untersuchungen über die Resistenz von Kartoffelknollen gegenüber Fäuleerregern. **Physiological investigations on the resistance of potato tubers to rot agents.**

9443 Langerfeld, E.   DE 215120/77/0003
Untersuchungen über die pathogenen Eigenschaften von Pathotypen des Kartoffelkrebserregers Synchytrium endobioticum. **Investigations on pathogenic properties of pathotypes of Synchytrium endobioticum.**

9444 Schöber, B.   DE 215120/78/0003 N
Vergleich der europäischen Methoden zur Erfassung der Resistenz von Kartoffelsorten und Zuchtstämmen gegen Braunfäule – Phytophthora infestans – und gewöhnlichen Schorf – Streptomyces scabies –. **Comparison of the European methods for determining the resistance of potato varieties and breeding lines against late blight of tubers – Phytophthora infestans – and Streptomyces scabies –.**

9445 Bode, O.   DE 215150/77/0010
Untersuchungen zur Resistenz gegen Viren bei Kartoffeln, Bohnen, Erbsen, Salat, Futtergräsern und Getreide.
Investigations on resistance in potatoes, beans, peas, lettuce, forage grasses and cereals to virus.

9446 Bartels, R.   DE 215150/78/0008 N
Herstellung hochtitriger Antiseren zum Nachweis von Kartoffelviren. **Production of antisera with high titres for the identification of potato viruses.**

9447 Casper, R.   DE 215150/78/0010 N
Routinetest auf Kartoffelblattrollvirus – potato leafroll virus – mit dem ELISA–Verfahren – enzyme–linked immuno–sorbent assay–. **Routine testing for potato leafroll virus by ELISA.**

9448 Weidemann, H.–L.   DE 215150/78/0013 N
Untersuchung der Virusresistenzen bei Kartoffelsorten.
Investigation of virus resistance in potato varieties.

9449 Weidemann, H.–L.   DE 215150/78/0017 N
Untersuchungen von latent bleibenden Kartoffel–Y–Viren – PVY – in Pflanzkartoffelbeständen. **Investigations on latent potato virus Y–strains – PVY – in seed potatoes.**

9450 Weidemann, H.–L.   DE 215150/78/0018 N
Verbreitung von Kartoffel–M–Virus – PVM – im Kartoffelsortiment. **Distribution of potato virus M – PVM – in potato varieties.**

9451 Koenig, R.   DE 215150/78/0019 N
Entwicklung und Verbesserung von serologischen Routinenachweisverfahren für Viren, die unter EG–QuarantäneRichtlinien fallen, in Kartoffelknollen.
Development and improvement of routine serological assay procedures for EC quarantine viruses in potato tubers.

9452 Mygind, H.   DK 010116/41/4505
Afprøvning af nye kartoffelsorters resistens mod kartoffelbrok samt bekæmpelse af sygdommen ved kemisk jorddesinfektion.
Testing the resistance of new potato varieties to wart disease and the control of this disease by chemical soil disinfection.

9453 Kristensen, H.R.; Christensen, M.; Engsbro, B.   DK 010116/61/4513
Forekomst, forebyggelse og bekæmpelse af viroser på kartofler hovedsagelig: bladrullesyge, Y–virose, X–virose, S–virose og rattle. **Occurrence, prevention and control of viruses on potato, chiefly leaf roll, Y–virus, X–virus, S–virus and rattle.**

9454 Kristensen, H.R.   DK 010116/77/0008
Meristemterapi i kartoffel og andre planter. **Meristem therapy in potato and other plants.**

9455 Lyshede, O.B.   DK 030104/75/0002
Anatomiske undersøgelser over kartoflens hudvæv med særligt henblik på de infektionsbiologiske forhold. **Anatomical investigations of potatoe peal with particular reference to the biological state of infection.**

9456 Løschenkohl, B.; Lyshede, O.B.   DK 030104/78/0004 N
Infektionsveje for Phytophthora infestans i kartoffelknoldens primordier. **Anatomical examination of epidermal tissue of the potato tuber with special reference to infection by pathogens.**

9457 Løschenkohl, B.; Lyshede, O.B.   DK 030147/78/0002 N
Indfaldsveje for Phytophthora infestans i kartoffelknoldens primordier. **Infecting pathways for Phytophthora infestans in potato primordia.**

9458 Boistard, P.; Cornuet, P.   FR 010107/72/1030
Recherche de mutants avirulents du virus Y de la pomme de terre. **Research of non virulent mutants of Y potato virus.**

9459 Clifford; Cooke   GB 011504/78/0010 R
Factors affecting the chemical control of fungal diseases of potatoes.

9460 Jellis   GB 011910/00/0015 R
Resistance to tuber diseases.

9461 Fuller; Gray   GB 011910/00/0017 R
Potato virus Y and leaf roll testing for SPBS.

9462 Pierpoint; Ireland   GB 012001/00/0012 R
Potato virus and its modification by O–Quinones.

9463 McIntosh   GB 012005/00/0039 R
Chemical control of soil–borne and other diseases of potatoes.

9464 Lapwood   GB 012009/00/0021 R
Diseases of potatoes:seed tubers and growing crops.

9465 Fox; Dashwood   GB 030702/00/0016 R
Plant protection: chemical and cultural control of potato gangrene.

9466 Fox   GB 030702/00/0017 R
Soil microbiology and root disease: biology of potato gangrene.

9467 Peromberlon   GB 030702/00/0018 R
Plant and pathogen physiology: diseases of potato tubers.

9468 Peromberlon   GB 030702/00/0019 R
Epidemiology and etiology: gangrene, black leg and soft rot and
recontamination of VTSC seed potato stocks.

9469 Harrison; Barker   GB 030704/00/0001 R
Potato viruses, especially soil–borne viruses.

9470 Wastie   GB 030805/78/0012 R
Assess potato breeding material for resistance to and infection
with viruses X Y leaf roll and spindle tuber.

9471 Wastie   GB 030805/78/0014 R
Study the biology of common scab,gangrene, skin spot and dry
rot.

9472 Wastie   GB 030805/78/0015 R
Assess potato breeding material for resistance to fungal
diseases. Improve screening techniques.

9473 Macer   GB 030805/78/0028 R
Study mechanisms of genetic variability in Phytophthora
infestans and the evolution of new pathogenic types.

9474 Macer   GB 030805/78/0029 R
Study mechanisms of quantitative resistance to potato late
blight identify resistant parental material.

9475 Laidlaw   GB 030902/00/0104 R
Improved means of testing varietal reaction to potato mop top
virus.

9476 Jones; Harris   GB 030902/00/0313 R
Development of tomato test method for detection of potato
spindle tuber viroid.

9477 Laidlaw   GB 030902/00/0316 R
Development of methods of sampling soil for wart disease spores
and for extraction and viability testing.

9478 MacDonald; Cameron   GB 030902/64/0301 R
Resusciation of virus–infected varieties by meristem culture.

9479 Graham; Quinn   GB 030902/68/0208 R
Characterisation of the blackleg pathogen and other members
of the genus Erwinia.

9480 Macdonald   GB 030902/70/0206 R
Monitoring of VTSC–derived stocks for new latent and overt
infection with skinspot (Polyscytalum pustulans).

9481 Quinn   GB 030902/72/0211 R
New fumigants to control potato tuber diseases.

9482 Graham; Quinn   GB 030902/73/0207 R
Spread of soft rot coliform bacteria by flies originating from
potato dumps.

9483 Graham; Quinn   GB 030902/73/0209 R
Aerial dispersal of potato bacteria.

9484 Laidlaw   GB 030902/73/0310 R
Rapid serological diagnosis of tobacco veinal necrosis virus from
potato leaf samples.

9485 Laidlaw   GB 030902/73/0311 R
Faster diagnosis of tobacco veinal necrosis virus from potato
leaf samples using indicator plants.

9486 Hamilton   GB 030902/74/0210 R
Bulk fumigation with 2–aminobutane to control gangrene in
stored potatoes.

9487 Laidlaw; Sharma   GB 030902/74/0315 R
Taxonomy of synchytrium species to help check persistence of
wart disease in old outbreak sites.

9488 Macdonald   GB 030902/75/0204 R
Monitoring of vtsc–derived stocks for new latent and overt
infection with gangrene (Phoma exigua var foveata.

9489 Hamilton   GB 030902/75/0303 R
Morphology and ecology of a Globodera species confusable with
potato cyst nematode.

9490 Harris; Jones   GB 030902/75/0314 R
Alternative test methods for detection of potato spindle tuber
viroid.

9491 Carnegie   GB 030902/76/0102 R
Reassessment of methods of testing varietal reaction to late
blight.

9492 Tierney; Adam   GB 030902/76/0103 R
Alternative methods of testing varietal reaction to potato wart
disease.

9493 Adam   GB 030902/76/0205 R
Detailed consideration of nature and origins of latent infection
with gangrene in stocks of VTSC origin.

9494 Laidlaw; Tierney   GB 030902/76/0317 R
Evaluation of less resistant rg2 potato varieties as potential
sources of viable wart disease resting spores.

9495 Mabbott   GB 030902/76/0318 R
Development of a standard test for partial resistance to potato
cyst nematodes in potato varieties.

9496 Hall; Ali   GB 030902/77/0203 R
Monitoring vtsc stocks for leaf roll virus infection.

9497 Logan   GB 041503/00/0001 R
Potato black leg.

9498 Logan   GB 041503/00/0002 R
Potato gangrene.

9499 Calvert   GB 041505/00/0001 R
Latent virus diseases of potatoes in Northern Ireland.

9500 Cheshire; Sly   GB 050312/75/0005 R
Aphid trap data in relation to virus incidence in seed potato
production areas.

9501 Pratt   GB 050321/72/0012 R

Effect of edaphic factors on wart disease infections developing in cultivars of varying susceptibility.

9502 Boyd   GB 060107/00/0005 R
Control of tuber diseases of potatoes other than blight.

9503 Boyd   GB 060107/00/0006 R
Potato blight: incidence, sources, forecasting and control.

9504 Paton   GB 060220/79/0010 N
Infection of potatoes with blackleg its source extension and control.

9505 Holmes   GB 060310/00/0004 R
Chemical control of potato blight.

9506 Dowley, L.J.   IE 060500/64/0238 R
Disease resistance testing of potato seedlings. Publications.

9507 Dowley, L.J.   IE 060500/75/1088 R
Chemical, biological and cultural control of potato gangrene. Publications.

9508 Kavanagh, J.A.; Wilson, U.E.; Goulding, H.
           IE 120106/76/9046 N
Studies on obligate parasitism : host–pathogen interactions in wart disease of potato caused by synchytrium endobioticum (schilb.) perc. Publications.

9509 Pratella, G.C.; Menniti, A.M.   IT 040216/79/0002 N
Prevenzione e lotta antifusarium su patate. **Fusarium prevention and control on potatoes.**

9510 Turkensteen, L.J.   NL 010108/49/3961 N
Resistentie van aardappelen tegen Phytophthora infestans. **Resistance of potatoes against Phytophthora infestans.** Publications.

9511 Kliffen, C.   NL 010108/69/2470
Onderzoek naar de biochemische achtergrond van veldresistentie van aardappel t.a.v. Phytophthora infestans. **Research on the biochemical aspects of mature plant resistance to Phytophthora infestans.** Publications.

9512 Maas Geesteranus, H.P.   NL 010108/73/3943
Zwartbenigheid bij aardappel veroorzaakt door Erwinia carotovora var. atroseptica. **Black leg of potato caused by Erwinia carotovora var. atroseptica.**

9513 Bokx, J.A. de   NL 010108/73/3949
Virusziekten bij aardappelen. **Virus diseases of potatoes.** Publications.

9514 Turkensteen, L.J.; Tichelaar, G.M.
           NL 010108/73/3959 R
Identificatie methoden en epidemiologie van gangreen bij aardappel. **Identification methods and epidemiology of potato gangrene.** Publications.

9515 Hoof, H.A. van   NL 010108/76/6745
Onderzoek aangaande overdracht van non–persistente virussen door bladluizen. **Transmission of non–persistent viruses by aphids.** Publications.

9516 Turkensteen, L.J.   NL 010108/78/8700 N
Schimmel– en bacterieziekten van de aardappel in

ontwikkelingslanden. **Fungal and bacterial potato diseases in developing countries.**

9517 Quak, F.   NL 010108/78/8701 N
Linkage–project between the International Potato Center CIP, Lima, Peru and IPO, to free the Mexican Germplasm Bank of virus by meristem culture.

9518 Schepers, A.; Bus, C.B.   NL 010207/62/3243
Onderzoek naar de mogelijkheden van bestrijding van virusziekten bij aardappelen. **Investigations of the possibilities of control of virus diseases in potatoes.**

9519 Bus, C.B.   NL 010207/74/5420 R
Landbouwkundige aspecten van het optreden en de bestrijding van knolziekten bij aardappelen. **Agricultural aspects of the appearance and the control of tuber diseases of potatoes.**

9520 Loon, C.D. van   NL 010207/78/9017 N
Onderzoek naar het effect van landbouwkundige maatregelen op het optreden van Verticillium dahliae bij aardappelen. **Investigation on the effect of cultural practices on the infection of potatoes by Verticillium dahliae.**

9521 Kort, J.   NL 010300/31/1389 N
Toetsing van nieuwe aardappelrassen op resistentie tegen wratziekte (Synchytrium endobioticum) (Schilb.) Perc. **Testing of new potato varieties for resistance to wart disease.**

9522 Bruin, Th. de   NL 010300/70/9105 N
Onderzoek m.b.t. de knolziekte veroorzaakt door Phoma e.v. foveata. **Tuber disease, caused by Phoma e.v. foveata in potato.**

9523 Bruin, Th. de   NL 010300/78/9104 N
Verticillium aantasting in aardappel. **Wilting disease, caused by Verticillium dahliae in potato.**

9524 Bruin, Th. de   NL 010300/78/9114 N
Het verloop van kwikconcentraties in dompelbaden en bestrijdingseffekten hiervan bij centraal continu ontsmetten van pootaardappelen. **The influence of the mercury concentration on the effect of the central disinfection of seed–potatoes.**

9525 Langerak, C.J.; Haanstra–Verbeek, J.
           NL 020018/73/4435
De vervanging van kwikhoudende fungiciden bij de ontsmetting van plantgoed van aardappelen en bloembollen en van zaaizaden. **The replacement of organic mercurial fungicides in the disinfection of planting material (potatoes, bulbs, sowing seed).**

Sugarbeets and other sugar crops (B 3320)

See also 3263, 3286, 7855, 8555, 8560, 8567, 8570, 8573, 8957, 9567, 9840

9526 Detroux, L.; Meeus, P.   BE 080700/78/0025 R
Incidence de l'oïdium de la betterave et lutte chimique. **Incidence of sugar beet oïdium and chemical struggle.** Publications.

9527 Weltzien, H.C.; Ahrens, W.   DE 111353/78/0003 N
Untersuchungen zur Resistenz von Zuckerrüben gegenüber dem Echten Mehltau, Erysiphe betae. **Studies on sugar beet**

resistance against powdery mildew, Erysiphe betae 1977–1980 (Diss.. Publications.

9528 Maas, G.   DE 215130/78/0002 N
Beeinflussung des Umsatzes der organischen Substanz und des mikrobiellen Herbizid–Abbaus durch fungizide/insektizide Zweitkomponenten in Zuckerrüben–Getreide– und GetreideFruchtfolgen. **Influence of additional fungicide/insecticide applications on the mineralization of organic matter and the herbicide breakdown by micro–organisms in crop rotations with sugar–beets and cereals.**

9529 Lesemann, D.; Koenig, R.   DE 215150/78/0029 N
Untersuchungen über die Rizomania–Krankheit der Zuckerrüben – beet necrotic yellow vein virus –. **Rizomania disease of sugar–beets – beet necrotic yellow vein virus –.**

9530 Casarini, B.; Cerato, C.; Alessandrini, L.
                                         IT 021100/79/0011 N
Ricerche sulla epidemiologia di CERCOSPORA beticola. **Researches on the epidemiology of Cercospora beticola.**

9531 Casarini, B.; Cerato, C.; Alessandrini, L.; Ranalli, P.; De Biaggi, M.   IT 021100/79/0012 N
Indagine sulla infeziosità del virus del giallume della barbabietola in Val Padana e sulla tolleranza di alcune linee in fase di selezione. **Study of infection capacity of the sugar beet virus yellow in the Po Valley and on the tolerance of some lines with selection in progress.**

9532 D'Ambra, V.   IT 062200/77/0859
Ricerche sull'ultrastruttura di Polymyxa betae Keskim. **Research on Polymyxa betae Keskim ultrastructure.**

9533 D'Ambra, V.   IT 062200/77/0861
Ricerche sull'eziologia della rizomania. **Research on the etiology of rhizomania.**

9534 D'Ambra, V.   IT 062200/77/0862
Ricerche sull'alterazione del trasporto ionico i radici di bietola indotta dalla rizomania. **Research on ionic transfer unbalance induced by rhizomania in sugarbeet.**

9535 D'Ambra, V.   IT 062200/77/0863
Studio dell'aromatizzazione dei carboidrati e del metabolismo dei composti fenolici in bietole rizomani. **Study on carbohydrate aromatisation and on phenolic compound metabolism in sugarbeets affected by rhizomania.**

9536 D'Ambra, V.   IT 062200/77/0864
Ricerche sull'incorporazione di derivati nucleotidici del glucosio e fruttosio in bietole sane e rizomani. **Research on the assimilation of glucose and fructose nucleotide derivates in healthy beetroots and in beetroots affected by rhizomania.**

9537 Tichelaar, G.M.   NL 010108/76/6746
Pathogene bodemschimmels bij suikerbieten. **Pathogenic soil fungi on sugarbeets.**

Other starch producing plants (B 3390)

See also 9441, 9444, 9446, 9448, 9449, 9450, 9451, 9456, 9508, 9718, 16073, 20281

9538 Nienhaus, F.; Hoffmann, S.; Reckhaus, P.

                                              DE 111351/78/0002 N
Untersuchungen zur Ätiologie von Maniok– und Yammosaik in Westafrika. **Investigations on the etiology of mosaic diseases in cassava and yams in West Africa.**

9539 Kranz, J.; Dengel, H.J.   DE 129554/73/0003
Untersuchungen über das Maniok–Mosaik in Togo. **Investigations into the cassava mosaic in Togo.**

9540 Hecht, H.; Hunnius, W.; Heinlein, R.; Kell, R.
                                         DE 502051/75/0002
Antivirale Agentien, Virus– und Vektorinhibitoren zur Bekämpfung von Viruskrankheiten der Kartoffel. **Antiviral agents, virus and vector inhibitors for the control of potato virus diseases.**

9541 Nøddegaard, E.; Hansen, K.E.   DK 010116/78/0025 N
Midler til bekæmpelse af kartoffelskimmel. **Fungicides for control of Late Blight in potatoes.**

9542 Luisoni, E.   IT 060100/74/0097
Virosi della manioca africana. **Virus diseases of african cassava.**

Grasses and forage crops in general (B 3400)

9543 Jenkyn; MacFarlane   GB 012009/00/0020 R
Foliage fungal diseases of grasses and forage legumes.

Grasses (B 3410)

See also 3452, 8965, 8969, 8970, 8983, 9280, 9286, 9337, 9339, 9445, 9564

9544 Fischbeck, G.; Schwarzbach, E.   DE 161240/72/0002 N
Untersuchungen über Auftreten, Ursachen und Vererbung rassenunspezifischer Resistenz gegen Mehltau in natürlichen Populationen der zweizeiligen Wildgerste – Hordeum spontaneum –. **Occurrence, causes and inheritance of race–unspecific mildew resistance in Hordeum spontaneum.**

9545 Ullrich, J.   DE 215120/78/0001 N
Untersuchungen über Gräserkrankheiten, insbesondere Roste. **Studies on diseases of grasses, especially rusts.**

9546 Huth, W.   DE 215150/77/0016
Untersuchungen über das Resistenzverhalten von Futtergräsern und –leguminosen gegenüber Viren und den Einfluss der Virosen auf Samenertrag und Grünmasse. **Studies on resistance of forage grasses and leguminosae to viruses and influence of viroses on seed yield and green material.**

9547 Paul, H.L.   DE 215150/78/0006 N
Untersuchung der Verwandtschaft und Einordnung von Viren aus Futtergräsern mit serologischen Methoden. **Investigations on the relationship and classification of viruses of fodder grasses by means of serology.**

9548 Lesemann, D.; Huth, W.   DE 215150/78/0028 N
Vergleichende Untersuchung der Zytologie von virusinfizierten Gramineen. **Comparative study on the cytology of virus–infected gramineae.**

9549 Teuteberg, A.   DE 215180/75/0005
Untersuchungen zur Epidemiologie und Bekämpfung pilzlicher Krankheitserreger im Samenbau von Lolium–Arten.

D 2420 – Plant diseases and disease control

Studies on epidemiology and control of fungal pathogens in seed production of Lolium species.

9550 Teuteberg, A. DE 215180/77/0006
Untersuchungen über Fusarium–Arten an Futter– und Rasengräsern. **Studies on Fusarium strains in forage grasses and lawn grasses.**

9551 Nitzsche, W. DE 402000/78/0001 N
Untersuchungen zur Genetik der Virusresistenz bei Futtergräsern. **Investigations on genesis of virus resistance in forage grasses.**

9552 Ziegenbein, G.; Morgner, F. DE 506155/75/0004
Untersuchungen über Fusskrankheiten bei Kulturgräsern im Samenbau. **Studies on footrots of cultivated grasses in seed production.**

9553 Welling, B. DK 010116/70/4504
Fysiogene forholds inflydelse på svampesygdomme hos græsser, beskrivelse af patogenerne og mulighederne for resistensforædling, frøpatologiske undersøgelser. **Influence of physiogenic factors on fungal diseases of grasses, description of the causal pathogens, the possibilities of resistance breeding, and seed pathological investigations.**

9554 Heard GB 011202/00/0005 R
Epidemiology and control of grass viruses (with RES and WPBS).

9555 Catherall GB 012107/00/0005 R
Screen and select for resistance to barley yellow dwarf and ryegrass mosaic viruses.

9556 Stoddart; Jones TWA GB 012110/00/0004 R
Effects of phyllody infection on enzyme activity in white clover.

9557 Cooper GB 041505/76/0002 R
Studies on virus diseases of grasses and cereals in Norrthern Ireland.

9558 Holmes GB 060310/00/0001 R
Grass diseases.

9559 Labruyère, R.E. NL 010108/73/3942
Schimmelziekten bij grassen. **Fungal diseases of grasses.** Publications.

Pastures, grassland (B 3420)

9560 de Guiran, G.; Dalmasso, A. FR 010504/70/1708
Ecologie de la nématofaune prairiale. **Ecology of grassland nematofauna.** Publications.

9561 Heard GB 011202/00/0006 R
The effect of fungal pathogens on grassland productivity (with RES).

9562 Wilkins; Carr GB 012107/00/0004 R
Screen and select for resistance to fungal pathogens.

9563 Malone GB 041501/00/0003 R
Assessment of losses caused by fungi in grassland.

9564 O'Rourke, C. IE 060500/66/0254 R
Effects of selected grass and clover diseases on forage yields,

persistence, quality and other factors. Publications.

Mangolds (B 3430)

9565 Weinmann, W. DE 215080/77/0035
Untersuchung über das Rückstandsverhalten von Fentinhydroxid auf und in Futterrüben und –blättern nach einer CercosporaBekämpfung. **Investigations on residual action of fentine hydroxide on and in fodder beet and beet leaves after Cercospora control.**

9566 Bagger, O. DK 010116/54/4511
Kortlægning af virusgulsots udbredelse og opformering i bederoemarkerne med henblik på varslingstjeneste. **Survey of the occurrence and spread of virus yellows in beet fields with reference to warning service.**

9567 D'Ambra, V. IT 062200/74/0133
Ricerche al microscopio elettronico su Polymyxa betae Keskin. **Electron microscopy investigations on Polymyxa betae Keskin.**

Legumes in general (B 3440)

See also 9286, 9336, 9546, 9578

9568 Rohloff, H. DE 215150/78/0032 N
Bestandsaufnahme über das Vorkommen von Virosen in Leguminosen und Determinierung von Resistenzen in Sorten. **Survey on virus diseases in leguminous plants and determination of resistances in cultivated varieties.**

9569 Davidse, L.C. NL 020018/77/7315
Fysiologie en biochemie van de waardplant–parasiet relatie luzerne – Phytophthora megasperma var. sojae. **Physiology and biochemistry of the host–parasite combination alfalfa – Phytophthora megasperma var. sojae.**

Grassland legumes (B 3441)

See also 9086, 9564

9570 König, K. DE 502051/70/0003
Untersuchungen zur Ökologie und Bekämpfung des Kleekrebses. **Studies on ecology and control of clover stem rot and crown rot.**

9571 Carr GB 012107/00/0001 R
Effect of phyllody and other mycoplasma on herbage legumes.

9572 Carr GB 012107/00/0002 R
Resistance of lucerne to Verticillium wilt.

9573 Scott Abrook, W. GB 012107/77/0016 R
Effects properties and epidemiology of viruses infecting clovers and prospects for resistance breeding.

9574 Malone GB 041501/00/0019 R
Fungal diseases of red and white clover.

9575 Ribaldi, M. IT 041002/72/0543
Ricerche sulle fitopatie dell'erba medica in Umbria ed in altre regioni dell'Italia centrale. **Research on lucerne diseases in Umbria and in other regions of central Italy.** Publications.

Other legumes (B 3449)

D 2420 – Plant diseases and disease control

9576 Schmutterer, H.; Gaudchau, M.   DE 129184/74/0004
Beziehungen zwischen dem Ackerbohnenmosaikvirus und den
Vektoren und möglichen Beeinflussungen der
Virusübertragung durch Juvenilhormonanaloga. **Correlations
between horse bean mosaic virus and vectors and possible
effects of juvenile hormone analogues on virus transmission.**

9577 Teuteberg, A.   DE 215180/74/5002
Untersuchungen über Blattkrankheiten der Ackerbohne.
Studies on leaf diseases of broad bean.

Cereals used for forage (B 3450)

See also 9336, 9376, 9383

9578 Nøddegaard, E.; Kirknel, E.   DK 010116/78/0022 N
Indhold af pesticidrester i grønafgrøder og ensilage af
grønafgrøder. **Content of triadimefon– and parathion–residues
in green barley and silage.**

9579 Pancaldi, D.   IT 040211/77/0518 R
Ricerche applicative sui fitofarmaci, difesa delle colture
foraggere di mais e sorgo a maturazione cerosa. **Applied
research on pesticides, protection of maize and sorghum fodder
crops in waxy maturation.**

Turnips (B 3460)

See also 9641

9580 Mattusch, P.   DE 215200/78/0001 N
Untersuchungen zum Einfluss des Silierprozesses bei
Herbstrüben auf die Lebensfähigkeit der Dauersporen von
Plasmodiophora brassicae. **Investigations on the influence of
the ensiling process of stubble turnips on the viability of resting
spores of Plasmodiophora brassicae.**

9581 Benetti, M.P.; Kaswalder, F.   IT 020300/79/0001 N
Ricerche sul virus del mosaico giallo della rapa (TYMV) in
Italia e su altre virosi delle Crocifere. **Studies on turnip yellow
mosaic virus (TYMV) in Italy and other virus diseases of
cruciferous plants.**

Other forage crops (B 3490)

See also 9641

9582 King; Polley   GB 050320/77/0042 R
Disease assessment of forage and fodder crops other than grass.

Vegetables in general (B 3500)

See also 8613, 8954, 8960, 8962, 8974, 9719

9583 Weidemann, H.–L.   DE 215150/77/0003
Bestandsaufnahme über das Vorkommen von Viren und
Viruskrankheiten in gartenbaulichen Kulturen. Ausarbeitung
von empfindlichen Diagnoseverfahren. **Inventory on the
occurrence of virus and virus diseases in horticulture.
Development of sensitive diagnostic methods.**

9584 Weidemann, H.–L.; Koenig, R.; Lesemann, D.; Paul,
H.L.; Casper, R.   DE 215150/78/0016 N
Analyse von Viruskrankheiten in Gemüsekulturen auf ihre
Erregerviren. **Identification of unknown viruses in
virus–diseased vegetable crops.**

9585 Mattusch, P.   DE 215200/77/0002
Studien zur Bedeutung und Verbreitung der Erreger von
Keimlings– und Auflaufkrankheiten bei Gemüse. **Studies on
importance and distribution of the agents of damping off of
seedlings and of coming up of vegetables.**

9586 Crüger, G.   DE 215200/77/0003
Entwicklung von Verfahren zur Bekämpfung samenbürtiger
Krankheitserreger bei Gemüse. **Development of methods for
control of seed–born pathogenic agents in vegetables.**

9587 Crüger, G.   DE 215200/77/0005
Untersuchungen zum Auftreten und zur Verbreitung von
Pathotypen der falschen Mehltaupilze im Gemüsebau. **Studies
on occurrence and distribution of pathotypes of downy mildew
fungi in vegetable growing.**

9588 Mattusch, P.   DE 215200/77/0006
Prüfung und Sorten verschiedener Gemüsearten auf Resistenz
gegenüber Krankheitserregern. **Testing of varieties of diverse
species of vegetables for their resistance to pathogenic agents.**

9589 Fischer, H.   DE 215200/77/0007
Zum Parasitismus von Ampelomyces quisqualis auf echten
Mehltaupilzen an Gemüsepflanzen. **Parasitism of
Ampelomyces quisqualis on mildew fungi in vegetable plants.**

9590 Fischer, H.   DE 215200/77/0008
Untersuchungen über Fungizide und Fungizid–Kombinationen
für Anwendungsbereiche bei Gemüse, in denen bisher
Quintozen–Präparate eingesetzt wurden. **Investigations on
fungicides and fungicide combinations for application to
vegetables treated so far with Quintozen preparations.**

9591 Kristensen, H.R.; Paludan, N.   DK 010116/71/4516 R
Forekomst, forebyggelse og bekæmpelse af viroser hos
væksthus– og frilandsgrønsager, hovedsagelig: agurkmosaik,
agurkgrønmosaik, løgmosaik, tobakmosaik og tobaknekrose.
**Occurrence, prevention and control of virósis in vegetables
grown in greenhouses and outdoors– especially cucumber
mosaic, cucumber green mottle mosaic, onion yellow dwarf,
tobacco mosaic and tobacco necrosis.**

9592 Mygind, H.; Jørgensen, H.A.   DK 010116/73/4507
Diagnosticering, forebyggelse og bekæmpelse af forskellige
svampe– og bakteriesygdomme på prydplanter og spiselige
afgrøder i væksthus. **Diagnosis, prevention and control of
different fungal and bacterial diseases of ornamentals and
edible glasshouse crops.**

9593 Fraser; Whenham   GB 011803/00/0004 R
**Studies on the biochemical basis of plant resistance to pests and
diseases.**

9594 Taylor; Dudley   GB 011807/00/0011 R
**Biology, epidemiology and control of bacterial diseases of
vegetables.**

9595 White   GB 011807/00/0016 R
**Effectiveness of soil fumigation for the control of soil–borne
vegetable diseases.**

9596 Murrant   GB 030704/00/0007 R
Viruses infecting umbelliferous crop plants.

D 2420 - Plant diseases and disease control

9597 Cirulli, M.   IT 040106/77/0149 R
Studi sulla virulenza dei parassiti e sulla resistenza delle piante ortensi, con applicazioni di miglioramento genetico per la resistenza verso le malattie. **Studies on parasite virulence and on the resistance of market garden plants to disease. Applied biological and integrated pest control.**

9598 Bos, L.   NL 010108/73/4070
Virusziekten bij groentegewassen in de volle grond. **Virus diseases of vegetable crops in the open.** Publications.

9599 Dorst, H.J.M. van   NL 010206/72/4670
Indentificatie en bestrijding van virusziekten bij glasgewassen. **Indentification and control of virus diseases in glasshouse crops.** Publications.

9600 Theune, D.   NL 010206/72/4671
Indentificatie, biologie en bestrijding van schimmelziekten bij glasgewassen. **Identification, biology and control of fungal diseases in glasshouse crops.** Publications.

9601 Nederpel, L.J.   NL 010206/72/4673
Bestrijding van bodemparasieten onder glas. **Control of soilborn diseases in glasshouse crops.** Publications.

Root, tuber and bulb vegetables (B 3510)

See also 9438, 9439, 9591, 9641, 9699

9602 Schickedanz, F.   DE 135052/77/0001
Virulenzverhalten von Phytophthora megasperma Drechsl. gegenüber Möhren. **Virulence behaviour of Phytophthora megasperma Drechsl. towards carrots.**

9603 Weinmann, W.   DE 215080/77/0013
Untersuchung über das Rückstandsverhalten von Mancozeb auf und in Radies und Kopfsalat unter Glas nach einer Anwendung gegen falschen Mehltau. **Investigations on residual action of mancozeb on and in little radish and lettuce under glass after application against downy mildew.**

9604 Weinmann, W.   DE 215080/77/0015
Untersuchung über das Rückstandsverhalten von Captan nach einer Anwendung gegen Auflaufkrankheiten auf und in Radies unter Glas. **Investigations on residual action of captan after application against sprouting diseases on and in little radish under glas.**

9605 Weinmann, W.   DE 215080/77/0019
Untersuchung des Rückstandsverhaltens von Metiram nach einer Anwendung gegen die Papierflecken- und Purpurfleckenkrankheit des Porree. **Investigations on residual action of metiram after application against paper and purple blotch of leek.**

9606 Weinmann, W.   DE 215080/77/0021
Untersuchung über das Rückstandsverhalten von Captafol auf und in Radies nach einer Anwendung gegen pilzliche Krankheiten unter Glas. **Investigations on residual action of captafol on and in little radish after application against fungal diseases under glass.**

9607 Rintelen, J.; Brielmaier, U.   DE 502051/77/0001
Der Einfluss von Vorfrüchten auf den Befall von Rettich mit Aphanomyces raphani – Rettichschwärze –. **Influence of the preceding crops on the infestation of radish by Aphanomyces raphani.**

9608 Jensen, A.   DK 010116/74/4506
Diagnosticering, forebyggelse og bekæmpelse af forskellige svampe– og bakteriesygdomme på frilandsgrønsager– gulerødder, kål mv. **Diagnosis, prevention and control of different fungal and bacterial diseases of outdoor vegetables– carrots, cabbages, etc.**

9609 Tomlimson   GB 011807/00/0001 R
Biology, epidemiology and control of beet western yellows virus disease of lettuce.

9610 Maude   GB 011807/00/0008 R
Biology, epidemiology and control of neck rot disease of onions.

9611 Lapwood; Adams   GB 012009/00/0028 R
Scab on red beet.

9612 Perry; Harrison   GB 030702/00/0015 R
Plant and pathogen physiology; disorders of carrots.

9613 Dickens; Sharp   GB 050323/74/0061 R
Control of Sclerotinia sclerotiorum with Coniothyrium minitans.

9614 Ryan, E.W.; Cassidy, J.C.   IE 060600/74/0868 R
Onion diseases – neck rot control. Publications.

9615 Kavanagh, J.A.; Hall, K.   IE 120106/78/9173 N
Investigations of leaf spot of onions caused by cladosporium allii–cepae.

9616 Bakel, J.M.M. van   NL 010108/60/7531
Witrot bij ui. **White rot of onions.**

9617 Bakel, J.M.M. van   NL 010108/74/6728
Witrot bij uien (Sclerotium cepivorum (Berk). **White rot of onions (Sclerotium cepivorum (Berk).**

9618 Bakel, J.M.M. van   NL 010207/60/7531
Witrot bij ui. **White rot of onions.**

Greens and leafy vegetables (B 3520)

See also 9445, 9603, 9608, 9889, 9926

9619 Van Assche, C.; Vanachter, A.; Van Wambeke, E.
     BE 040203/72/0011 R
Studie over de invloed van culturale bodemfungiciden in de bestrijding van sclerotiniarot bij witloof. **Research on the influence of cultural soilfungicides in the control of Sclerotinia on witloof chicory.** Publications.

9620 Weil, B.; Gröschel, H.   DE 138331/73/0001
Untersuchungen zu Differenzierung und serologischen Eigenschaften von Viren an Asparagus officinalis L.. **Studies of the differentiation and serological properties of viruses in Asparagus officinalis L..**

9621 Zinkernagel, V.   DE 161050/77/0003 N
Resistenzmechanismen bei Salat gegen Bremia lactucae. **Mechanisms of resistance on lettuce against Bremia lactucae.**

9622 Fritz, D.; Weichmann, J.   DE 161260/77/0001
Ursachen der Innenblattnekrose von Chinakohl. **Causes of**

internal leaf necrosis of Chinese cabbage.

9623 Reimann–Philipp, R.; Schum, A.; Renaud, M.

DE 206000/78/0008 N

Untersuchungen über Konkurrenzerscheinungen zwischen den drei physiologischen Rassen A, B und C von Peronospora spinaciae auf Spinatsorten unterschiedlicher Anfälligkeitsgrade als Grundlage für die Rassen–Eliminierung durch Sorten–Anbausteuerung und zur Klärung der Voraussetzungen für die mutative Neuentstehung von Pathotypen. **Studies on the competition of 3 physiological strains A, B and C of Peronospora spinaciae with respect to their occurrence and survival on host plants of differently resistant spinach varieties in order to investigate possibilities of "disruptive selections" and conditions for the mutative origination of new pathogenic strains.**

9624 Köhn, S.; Leh, H.–O.   DE 215040/72/4003

Untersuchungen über die Zementfäule an Kohlsamenträgern. **Studies on a stem rot of seed–bearing cabbage plants.**

9625 Köhn, S.; Krüger, G.   DE 215040/74/0001

Untersuchungen zur unterschiedlichen Anfälligkeit von Chinakohlarten gegen Erwinia carotovora var. carotovora. **Studies on varying susceptibility of Chinese cabbage species to soft rot by Erwinia carotovora var. carotovora.**

9626 Weinmann, W.   DE 215080/77/0014

Untersuchung des Rückstandsverhaltens von Captan nach einer Anwendung gegen Sclerotinia und Pythium auf und an Schnittlauch. **Investigations on residual action of captan after application against Sclerotinia and Phytium on and in chives.**

9627 Weinmann, W.   DE 215080/77/0016

Untersuchung über das Rückstandsverhalten von Dichlofluanid auf und in Feldsalat nach einer Anwendung gegen falschen Mehltau und Botrytis im Freiland und unter Glas. **Investigations on residual action of dichlofluanid on and in field salad after application against downy mildew and Botrytis in fields and under glass.**

9628 Weinmann, W.   DE 215080/77/0020

Untersuchung über das Rückstandsverhalten von Folpet auf und in Feldsalat nach einer Anwendung gegen falschen Mehltau und Botrytis im Freiland und unter Glas. **Investigations on residual action of folpet on and in field salad after application against downy mildew and Botrytis in fields and under glass.**

9629 Weidemann, H.L.; Rohloff, H.   DE 215150/73/4004 R

Untersuchungen zur Epidemiologie der Viruskrankheiten in Salat–, Bohnen– und Erbsenbeständen als Grundlage für die Beurteilung von Viruskontrollmassnahmen. **Studies in the epidemiology of virus diseases in lettuce, bean and pea as basis for estimation of virus control measures.**

9630 Weidemann, H.–L.; Lesemann, D.

DE 215150/78/0015 N

Virusbedingte Lagerschäden an Weisskohl. **Virus–induced damages in stored white cabbage.**

9631 Mattusch, P.   DE 215200/77/0004

Untersuchungen zur Verbreitung des Erregers der Kohlhernie und seiner Pathotypen sowie zur Anfälligkeit der Wirtspflanzen. **Studies on distribution of the agent of Plasmodiophora brassicae and pathotypes as well as on**

susceptibility of host plants.

9632 Researcher not indicated   DE 215200/78/0002 N

Resistenzmechanismen bei Wurzel– und Gefässkrankheiten an Brassica–Arten, speziell zur Umfallkrankheit des Kohls. **Mechanisms for resistance to root and vascular diseases of Brassica spp., especially to blackleg of cabbage.**

9633 Tomlinson   GB 011807/00/0002 R

Seed transmission of lettuce mosaic disease and study of resistant cultivars.

9634 Buczacki   GB 011807/00/0010 R

Biology, epidemiology and control of clubroot.

9635 Crute; Norwood   GB 011807/00/0020 R

Identification and utilisation of chemical and host resistance factors for control of lettuce downy mildew.

9636 Tomlinson; Ward   GB 011807/75/0025 R

Studies on Brassica virus diseases.

9637 Rawlinson; MacFarlane   GB 012009/00/0032 R

Diseases of brassica crops.

9638 Brokenshire   GB 060107/00/0010 R

Brassica club root investigation.

9639 Channon   GB 060310/00/0005 R

Clubroot control.

9640 Channon   GB 060310/00/0007 R

Glasshouse lettuce: winter production disease control.

9641 Staunton, W.P.; Ryan, E.W.   IE 060300/71/0953 R

Diseases of crucifers. Publications.

9642 Cristinzio, M.   IT 040707/77/0155 R

Virosi della lattuga, delle cucurbitacee, dello albicocco, del pesco e del nocciolo. **Virus diseases in lettuces, cucumbers, apricot trees, peach trees and hazel–hut trees.**

9643 Cristinzio, M.   IT 040707/77/0829

Virosi della lattuga e del peperone. **Virus diseases of the lettuce and Capsicum.**

9644 Roca, F.   IT 060600/73/0051

Nematodi vettori di virus vegetali. Trasmissione attraverso nematodi di alcuni virus del carciofo. **Nematodes vectors of plant viruses. Transmission of some artichokes virus by nematodes.** Publications.

9645 Bakel, J.M.M. van   NL 010108/65/1566

Ziekten en bodemmoeheid in asperge. **Decline and replanting problems on Asparagus officinalis L.**

9646 Bakel, J.M.M. van; Blok, I.   NL 010108/73/3948

Valse meeldauw (Bremia lactucae Regel) in sla. **Downy mildew (Bremia lactucae Regel) in lettuce.** Publications.

9647 Blok, I.   NL 010108/73/3958 R

Wolf (valse meeldauw) bij spinazie. **Downy mildew of spinach.**

9648 Tichelaar, G.M.   NL 010108/75/6718

Inventarisatie van schimmels die smet bij sla veroorzaken. **Survey of fungi causing blackrot of lettuce.**

D 2420 – Plant diseases and disease control

9649 Kooistra, T.   NL 010300/77/9100 N
Onderzoek naar de mogelijkheden van gerichte bestrijding van
Rhizoctonia in (kas)sla op basis van planning teeltperiode en
grondmonsteronderzoek. **Control of Rhizoctonia in
(glasshouse) lettuce based on planning of the growing period
and soil sample research.**

Vegetable fruits in general (B 3530)

See also 9926

Leguminous vegetables (B 3531)

See also 8952, 9336, 9445, 9568, 9629, 9699

9650 Rudolph, K.   DE 132210/70/0006
Bakteriostatische Effekte in der Interzellularflüssigkeit aus
Bohnenblättern in Korrelation zur Resistenz gegen
Bohnen–Bakteriosen 1970. **Bacteriostatic effects of the
intercellular fluid of bean leaves in correlation with resistence to
bacterial diseases of beans.**

9651 Rudolph, K.   DE 132210/70/0007
Vergleichende Untersuchungen über die
Krankheitsausbreitung auf dem Felde bei 3 bohnenpathogenen
Bakterienarten – Pseudomonas phaseolicola, Pseudomonas
syringae, Xanthomonas phaseoli var. fusceans – in
Abhängigkeit von der Witterung unter Berücksichtigung
verschieden virulenter Rassen 1970. **Comparative studies on
the spread of disease in bean cultures by three pathogenic
bacterial diseases: Pseudomonas phaseolicola, Pseudomonas
syringae, Xanthomonas phaseoli var. fasceans in dependence on
climatic conditions in consideration of varying virulent species.**

9652 Mendgen, K.   DE 132210/74/0004
Infektionsvorgang des Bohnenrosts bei anfälligen und
resistenten Bohnen. **Bean rust infection of resistant and
susceptible varieties of bean.**

9653 Mendgen, K.   DE 132210/74/0005
Elektronenmikroskopische Untersuchung von
Resistenzmechanismen des Bohnenrosts. **Electron microscopic
studies on mechanisms of resistance of the bean rust fungus.**

9654 Heitefuss, R.; Hoppe, H.H.; Hümme, B.
            DE 132210/74/0006
Physiologische Untersuchungen zum Wirt–Parasit–Verhältnis
von Phaseolus vulgaris und Uromyces phaseoli. **Physiological
investigations on host–parasite relations of Phaseolus vulgaris
and Uromyces phaseoli.**

9655 Rudolph, K.; El–Banouby, F.E.   DE 132213/77/0001
Mechanismus der durch Pseudomonas phaseolicola im
Bohnenblattgewebe induzierten wasserdurchtränkten Flecken.
**Mechanism of the induction of water–congested areas in bean
leaves by Pseudomonas phaseolicola.**

9656 Rudolph, K.; Lehmann, H.   DE 132213/77/0004
Untersuchungen zur Resistenz von Phaseolus vulgaris gegen
bohnenpathogene Bakterien. **Investigations on the resistance of
Phaseolus vulgaris towards bean–pathogenic bacteria.**

9657 Persiel, F.; Rockstroh, K.   DE 206000/71/4011 R
Untersuchungen zur Resistenz gegen Pseudomonas
phaseolicola bei Bohnen – Phaseolus vulgaris – 1971.
**Investigations on resistance of bean – Phaseolus vulgaris – to
Pseudomonas phaseolicola.**

9658 Persiel, F.; Rockstroh, K.   DE 206000/71/4012 R
Untersuchungen zur Resistenz gegen das
Bohnen–Mosaik–Virus 1 bei Bohnen – Phaseolus vulgaris –
1970. **Investigations on resistance of bean – Phaseolus vulgaris –
to bean mosaic virus 1.**

9659 Persiel, F.; Anders, S.   DE 206000/71/5005 R
Untersuchungen zur Resistenz von Erbsensorten und
Wilderbsen gegen Pythium ultimum 1970–1980. **Investigations
on resistance of pea varieties and wild peas to Pythium ultimum.**

9660 Junge, H.; Mattiesch, L.   DE 206000/75/0004
Chemische Untersuchungen über die Resistenz von Erbsen
gegen Pythium ultimum. **Chemical studies on the resistance of
peas to Pythium ultimum.**

9661 Walkey; Dance   GB 011807/75/0026 R
**The production of navy bean cultivars resistant to bean common
and bean yellow mosaic virus.**

9662 Cockbain; Bowen   GB 012009/00/0003 R
Epidemiology and control of virus diseases of Vicia.

9663 Bainbridge   GB 012009/00/0018 R
Fungal diseases of field bean and the effect of chemicals.

9664 Cockbain; Bowen   GB 012009/00/0030 R
Epidemiology and control of virus diseases of legumes.

9665 Perry; Harrison   GB 030702/00/0013 R
**Soil microbiology and root disease; biology of root diseases in
field peas and beans.**

9666 Ryan, E.W.; Murphy, R.F.; McDonnell, M.
            IE 060600/74/0867 R
Control of diseases in peas. Publications.

9667 Prendeville, G.N.; Crowley, J.; O'Brien, M.C.
            IE 110202/76/9148 N
**Effects of herbicides on cell membrane permeability in
phaseolus and lemna minor and evaluation of their usefulness as
a means of detecting herbicide residues in terrestial and aquatic
species.**

9668 Gerlagh, M.   NL 010108/53/1070 R
Resistentie–onderzoek erwterassen m.b.t. schimmelziekten.
Investigation into resistance of pea varieties to fungus diseases.
Publications.

9669 Gerlagh, M.   NL 010108/53/1071 R
Resistentie–onderzoek bonerassen m.b.t. schimmelziekten.
**Investigation into resistance of bean varieties to fungus
diseases.** Publications.

9670 Maas Geesteranus, H.P.   NL 010108/73/3944
Bacterieziekten bij bonen. **Bacterial diseases of Phaseolus
beans.**

Tomatoes (B 3532)

See also 9591

9671 Welvaert, W.   BE 030011/69/0002 R
Fytopathologische problemen bij Middelandse zee gewassen, veroorzaakt door virus, mycoplasmen, bacteries en fungi, vooral bij Artisjok, Tomaat, Peper, Cucurbitacaea en Aardappel. **Phytopathological problems of Mediterranean crops, caused by virus mycoplasma, bacteria and fungi, especially on Artichoke, Tomato, Pepper, Cucurbitaceae and Potato.**

9672 Rudolph, K.; Mavridis, A.   DE 132213/78/0002 N
Ökonomisch wichtige Bakterienkrankheiten an Tomaten– Lycopersicon esculentum – und Paprika – Capsicum annuum– in Griechenland und Möglichkeiten ihrer Bekämpfung. **Economically important bacterial diseases of tomato– Lycopersicon esculentum – and red pepper – Capsicum annuum – in Greece and possibilities of control.**

9673 Weil, B.; Krebs, E.K.   DE 138331/73/0003
Wechselbeziehungen zwischen Pilzen und Viren am Beispiel des Krankheitskomplexes Verticillium – Tabakmosaikvirus– Tomate. **Interactions between fungi and viruses demonstrated in the disease complex Verticillium – Tobacco mosaic virus– tomato.**

9674 Grossmann, F.; Menke, G.   DE 144540/70/0003
Trennung, Reinigung und Charakterisierung pektolytischer und zellulolytischer Enzyme von Fusarium oxysporum f. lycopersici, dem Erreger der Tomatenwelke. **Separation, purification and characterization of pectolytic and cellulolytic enzymes of Fusarium oxysporum f. lycopersici, the causal organism of tomato wilt.**

9675 Boistard, P.; Denarié. J.   FR 010107/00/1031
Déterminisme génétique de laviruleuce et recherche de mutants avirulents de Ps. Solanacearum responsable du flétrissement de la tomate. **Genetic determinism of virulence, and search of non virulent mutants of Pseudomonas Solanocearum causing shrivelling of tomato.**

9676 Laterrot, H.; Thomas, H.   FR 010603/76/0470
Etude raciale des souches de Cladosporium Fulvum des cultures françaises de tomate sous serre. **Racial study of the Cladosporium Fulvum from glasshouse french tomato.**

9677 Ebben   GB 011109/00/0005 R
Brown root rot of tomato.

9678 Channon   GB 060310/00/0006 R
Control of several tomato diseases.

9679 Canova, A.   IT 040211/77/0827
Virosi di piante ortensi ed ornamentali (pomodoro, peperone, crisantemo, filodendro e rosa). **Virus diseases of market garden and ornamental plants (tomato, Capsicum, chrysanthemum, philodendron, rose).**

9680 Canova, A.   IT 040211/77/0848
Terapia antivirale, premunità e lotta contro i vettori in pomodoro e peperone. **Antivirus therapy, prevention and control of vectors in tomatoes and Capsicum Grossum.**

9681 Lisa, V.   IT 060100/69/0092
Indagini sulla distribuzione in natura e sulla identificazione dei virus del pomodoro. **Tomato virus identification and field distribution studies.** Publications.

9682 Rast, A.Th.B.   NL 010108/62/8704 N
Virusziekten van tomaat. **Virus diseases of tomato.** Publications.

9683 Gerlagh, M.   NL 010108/66/1074 R
Resistentie–onderzoek t.a.v. schimmelziekten van tomaat. **Investigation into resistance of tomato to fungous diseases.** Publications.

9684 Steekelenburg, N.A.M. van   NL 010108/70/3422
Fusarium verwelkingsziekte in tomaat. **Fusarium wilt disease of tomato.** Publications.

9685 Steekelenburg, N.A.M. van   NL 010108/73/3946
Didymella–achtige schimmels bij tuinbouwgewassen onder glas. **Didymella–like fungi on glasshouse crops.** Publications.

9686 Rast, A.T.B.   NL 010206/62/8704 N
Virusziekten van tomaat. **Virus diseases of tomato.** Publications.

9687 Steekelenburg, N.A.M. van   NL 010206/70/3422
Fusarium verwelkingsziekte in tomaat. **Fusarium wilt disease of tomato.** Publications.

9688 Steekelenburg, N.A.M. van   NL 010206/73/3946
Didymella–achtige schimmels bij tuinbouwgewassen onder glas. **Didymella–like fungi on glasshouse crops.**

9689 Wit, P.J.G.M., de; Schaft, N. van der NL 020018/74/6343
De fysiologische en biochemische achtergrond van de waardplantparasiet relatie tomaat–Cladosporium fulvum. **Physiology and biochemistry of the host–parasite combination tomato–Cladosporium fulvum.** Publications.

9690 Verkley, F.N.   NL 020061/78/8654 N
Een analyse van een deficiente stam van het bronsvlekkenvirus van de tomaat, dat geen complete deeltjes vormt. **Analysis of an isolate of tomato spotted wilt virus which is defect in the formation of complete particles.**

Cucumbers (B 3533)

See also 9591, 9642, 9671, 9685, 9688, 9699, 9889

9691 Kabsch, U.   DE 148200/77/0004
Das Wirt–Parasit–Verhältnis von Gurke und Sphaerotheca fuliginea – Gurkenmehltau. **Host–parasite–interactions between cucumber and Sphaerotheca fuliginea.** Publications.

9692 Leclant, F.; Labonne, G.   FR 011210/74/5235
Etude de la variabilité du virus de la mosaïque du concombre en conditions naturelles et contrôlées. **Study of the cucumber mosaic virus variability under natural and controlled conditions.**

9693 Ebben; Spencer   GB 011109/00/0007 R
Epidemiology and control of stem rot, powdery mildew and black root rot of cucumber.

9694 Gerlagh, M.   NL 010108/66/1083 R
Resistentie tegen schimmelziekten van komkommer en augurk. **Resistance to fungal diseases of cucumber and gherkins.** Publications.

9695 Blok, I.   NL 010108/74/5927

D 2420 – Plant diseases and disease control

Onderzoek over valse meeldauw (Pseudo pero nospera cubensis) van komkommer en augurk ten behoeve van de resistentieveredeling. **Downey mildew of cucumber and gherkin.** Publications.

Other vegetable fruits (B 3539)

See also 9642, 9643, 9671, 9672, 9679, 9680, 9694, 9695, 9889

9696 Kühne, H.   DE 135052/77/0002
Untersuchungen von Viruskrankheiten an Zucchini–Cucurbita pepo L. –. **Investigations on viroses of zucchini–Cucurbita pepo L. –.**

9697 Weinmann, W.   DE 215080/77/0017
Untersuchung über das Rückstandsverhalten von Dichlofluanid auf und in Paprika nach einer Botrytisbekämpfung unter Glas. **Investigations on residual action of dichlofluanid on and in Capsicum after Botrytis control under glass.**

9698 Porta–Puglia, A.   IT 020300/79/0005 N
Prove di lotta in campo contro la Phytophthora capsici del peperone e osservazioni circa la trasmissibilità del fungo per seme. **Field control against Phytophthora capsici on pepper (Capsicum sp.). Studies on the possible seed transmission of the pathogen.**

9699 Noviello, C.   IT 040730/72/0524
Patologia piante ortensi: malattie da funghi parassiti della fragola, anguria, cipolla, pomodoro, fagiolo. **Pathology horticultural plants: disease caused by parasitical fungi on strawberries, water–melon, onions, tomatoes, beans.** Publications.

9700 Conti, M.   IT 060100/72/0091
Indagini sulla distribuzione in natura e sulla identificazione dei virus del peperone. **Pepper virus identification and field distribution studies.** Publications.

9701 Conti, M.   IT 060100/74/0099
Virus non–persistenti del peperone. **Stylet–borne viruses of pepper.**

9702 Rast, A.Th.B.   NL 010206/75/8705 N
Virusziekten bij paprika. **Virus diseases of sweet pepper.**

Mushrooms and other edible fungi (B 3540)

See also 3998, 8673, 8984

9703 Weltzien, H.C.; Heuel, B.   DE 111353/78/0002 N
Pilzkrankheiten an kultivierten Speisepilzen. **Fungal diseases of cultivated mushrooms.**

9704 Weinmann, W.   DE 215080/77/0022
Untersuchung über das Rückstandsverhalten von Benomyl nach einer Anwendung gegen Mycogone und anderen Pilzkrankheiten in Champignonkulturen. **Investigations on residual action of benomyl after application against Mycogone and other fungal diseases in mushroom cultivation.**

9705 Weinmann, W.   DE 215080/78/0020 N
Untersuchung über das Rückstandsverhalten von Benomyl auf/in Champignon nach einer Behandlung gegen Mycogone. **Investigation on the residue–behaviour of Benomyl on/in mushrooms after a treatment against Mycogone.**

9706 Koenig, R.   DE 215150/77/0007
Untersuchungen über eine neue gefährliche Viruskrankheit des Kulturchampignons. **Investigations on a new dangerous virus disease of cultivated mushroom.**

9707 Lesemann, D.; Koenig, R.   DE 215150/78/0026 N
Viruskrankheiten des Kulturchampignons, Diagnose und Bestandsaufnahme. **Virus diseases of cultivated mushrooms, diagnosis and survey.**

9708 Bech, K.; Kovács, G.   DK 030147/79/0001 N
Undersøgelse over patogenese af den af Verticillium fungicola fremkaldte sygdom på den dyrkede champignon med særligt henblik på bekæmpelsesmuligheder. **The pathogenesis of the disease on cultivated mushroom caused by Verticillium fungicola with special reference to control methods.**

9709 Bech, K.; Kovacs, G.   DK 030171/79/0001 N
Undersøgelse over patogenese af den af Verticillium fungicola fremkaldte sygdom på dyrkede champignons. **Investigations on the pathogenesis of the disease by Verticillium fungicola on the cultivated mushroom.**

9710 Cornuet, P.; Lapierre, H.   FR 010107/71/1023
Infection et replication des virus des champignons. **Host. Infection and replication of mycoviruses.** Publications.

9711 Cayrol, J.C.   FR 010504/63/1712
Recherche sur les nématodes du champignon de couche. **Research on mushroom nematodes.** Publications.

9712 Gandy   GB 011109/00/0017 R
Etiology, epidemiology and control of mushroom diseases and competitors.

9713 Hollings   GB 011110/00/0014 R
Virus diseases of mushrooms.

9714 Zaayen, A. van   NL 010204/74/5928 R
Onderzoek naar de afstervingsziekte, een virusziekte van champignons. **Research on virus disease of cultivated mushrooms.** Publications.

9715 Zaayen, A. van   NL 010204/74/5930 R
Onderzoek naar gedrag en bestrijding van pathogene en concurrerende schimmels in de champignonteelt. **Research on behaviour and control of pathogenic and competitive fungi in mushroom culture.** Publications.

Fruits in general (B 3600)

See also 8613, 8684, 8685, 8688, 8697, 8953, 8968, 9583, 9597

9716 Geenen, J.; Veldeman, R.   BE 070500/72/0002 R
Studie van Erwinia amylovora – bacterievuur diagnosemethodiek epidemiologie waardplantengevoeligheid. **Research on "Fire Blight" caused by Erwinia amylovora diagnostic methods, epidemiology susceptibility of the host plants.** Publications.

9717 Naumann, G.   DE 111300/77/0001
Obstvirosen. **english title not indicated. en–Fruit viruses.** Publications.

9718 Jacob, H. DE 129163/75/0001
Zum Vorkommen von Viren des Types des Kartoffel–Y–Virus in Obstgewächsen. **Potato–Y–viruses in fruit plants.**

9719 Frank, H.K.; Langerfeld, E. DE 215120/75/0005
Fusarium–Toxine bei Obst, Gemüse und Kartoffeln. **Fusarium–toxins on fruits, vegetables and potatoes.**

9720 Casper, R.; Rohloff, H. DE 215150/78/0002 N
Untersuchung labiler und latenter Obstviren. **Investigation of labile and latent fruit viruses.**

9721 Krczal, H.; Kunze, L. DE 215210/77/0002
Versuche zur Wärmetherapie viruskranker Obstgewächse. **Trials on heat therapy of virus–infested fruit plants.**

9722 Berling, R. DE 502051/70/0007
Untersuchungen über integrierten Pflanzenschutz im Obstbau. **Research for integrated control of orchards.**

9723 Graf, H. DE 507303/75/0001
Pilzliche, viröse, bakterielle und mykoplasmatische Erkrankungen im Obstbau. **Fungous, virous, bacterial and mycoplasmatical pests in fruit growing.**

9724 Luisetti, J.; Gaignard, G. FR 010402/69/1003
Les relations hôte–parasite (bactériose des arbres fruitiers). **Host–parasite relationships (bacterial diseases of fruit–trees).** Publications.

9725 Prunier, J.P.; Gaignard, J.L. FR 010402/69/1004
Etude bactériologique des pseudomonas saprophytes et pathogènes des arbres fruitiers. **Bacteriological studies on pathogenic and saprophytic pseudomonas bacteria on fruit–trees.** Publications.

9726 Bernhard, R.; Lansac, M.; Marenaud, Cl.; Mazy, K. FR 010701/72/0106
Influence d'hôtes particuliers transitoires sur la virulence ultérieure d'une souche virale et les symptômes qu'elle peut alors induire chez les arbres fruitiers et leurs indicateurs. **Influence of special transitory hosts on the ulterior virulence of a strain and the symptoms it may then induce on fruit–trees and their indicators.**

9727 Marenaud, C.; Bernhard, R. FR 010701/76/0319
Recherches concernant les mécanismes de défense in vivo des plantes fruitiéres à l'égard des agents pathogènes de type mollicute. **Search for in–vivo protection–systems of fruit–plants against pathogenic agents.**

9728 Fos, A. FR 010707/75/5220
Etude des Cicadelles vectrices de maladies à mycoplasmes des arbres fruitiers. **Studies of Leafhoppers mycoplasm diseases vectors on fruit trees.**

9729 Clark; Flegg GB 011005/00/0010 R
Taxonomy, serology, purification and microstructure of viruses.

9730 Crosse; Flegg GB 011005/76/0038 R
Etiology and biology of diseases attributed to mycoplasmas and related organisms.

9731 Byrde GB 011508/00/0001 R
Factors affecting pathogenicity of Sclerotinia fructigena, with
special reference to extracellular enzymes.

9732 Byrde; Corke GB 011508/00/0002 R
Chemical control of fruit diseases.

9733 Harper GB 040403/76/0011 R
The role of anthranilic acid in appressorium formation by fungal spores.

9734 Scheer, H.A.Th. van der NL 010212/68/2240
Toetsing van fungiciden en spuitmethoden in de fruitteelt. **Testing of fungicides and spray methods in fruitgrowing.** Publications.

9735 Scheer, H.A.Th. van der NL 010212/75/6113
Vaststellen van de economische schadedrempel voor appelmeeldauw. **Determination of the economic injury threshold for apple mildew.**

9736 Scheer, H.A.Th. van der NL 010212/75/6114
Bastaantastingen bij appel en peer. **Bark infection of apple and pear.**

Top fruit in general (B 3610)

9737 Soenen, A.; Gilles, G. BE 140000/74/0028
Studie van virussen op vruchtbomen en aardbeien. **Research of viruses on fruittrees and strawberries.** Publications.

9738 Stösser, R.; Buchloh, G. DE 144445/77/0023
Gummosis beim Steinobst. **Gummosis in stone fruit trees.**

9739 Zeller, W. DE 215180/77/0011
Prüfung von Kernobst– und Ziergehölzarten sowie Cotoneastersämlingen auf Feuerbrandresistenz. **Testing of pome fruit and ornamental trees and cotoneaster seedlings for their resistance to fire blight.**

9740 Schmidle, A. DE 215210/77/0007
Anatomisch–histologische Untersuchungen über die Infektionswege von Nectria galligena beim Kernobst und über die Ausbreitung des Erregers im Baum in Abhängigkeit von klimatischen Einflüssen. **Anatomic–histological studies on infection course of Nectria galligena in pome fruit and on distribution of the agent in trees in dependence on climatic conditions.**

9741 Kunze, L. DE 215210/78/0001 N
Versuche mit latenten Kernobstviren. **Experiments with latent viruses in pome fruit.**

9742 Poulsen, E. DK 010102/76/0008
Teknik ved sprøjtning af frugttræer med henblik på at reducere kemikalie– og væskemængden. **Spraying technique for fruit trees in order to reduce quantities of chemical and liquid carrier.**

9743 Nøddegaard, E.; Hansen, T. DK 010116/61/0012 N
Fungiciders virkning mod svampesygdomme på kernefrugt. **Effect of fungicides on fungal diseases in kernel fruit.**

9744 Kristensen, H.R.; Thomsen, A. DK 010116/64/4514
Forekomst, forebyggelse og bekæmpelse af viroser på frugttræer hovedsagelig: mosaik, hestesko–ar, stjernerevner, sten i pære, splitbark, dværgsyge og ringplet. **Occurrence, prevention and control of viruses on fruit trees, chiefly mosaic,**

horseshoe scar, star crack, split bark, stony pit in pears, dwarfing and ring spot.

9745 Billing GB 011005/00/0001 R
Epidemiology and control of fireblight of apples and pears.

9746 Crosse; Garrett GB 011005/00/0002 R
Epidemiology and control of bacterial cankers.

9747 Cross; Billing GB 011005/00/0003 R
Biology and ecology of bacterial plant pathogens and bacterial epiphytes.

9748 Garrett GB 011005/00/0004 R
Epidemiology and control of crown gall.

9749 Thresh; Adams GB 011005/00/0005 R
Epidemiology of virus and mycoplasma diseases.

9750 Thresh GB 011005/00/0007 R
Epidemiology of virus and mycoplasma diseases.

9751 Butt GB 011005/00/0011 R
Epidemiology of mildew and scab of apple and pear.

9752 Butt GB 011005/00/0012 R
Control of mildew and scab of apple and pear.

9753 Edney; Harris GB 011005/00/0013 R
Epidemiology of fungi causing storage rotting of apple and pear.

9754 Edney; Harris GB 011005/00/0014 R
Control of storage rotting by fungi of apple and pear.

9755 Bennett GB 011005/00/0019 R
Epidemiology of silver leaf and control by chemotherapy.

9756 Sewell; Harris GB 011005/00/0021 R
Epidemiology and host reaction of collar rot.

9757 Sewell; Wilson GB 011005/00/0022 R
Specific replant diseases.

9758 Corke GB 011508/00/0009 R
Biological or integrated control of fruit diseases.

9759 Campbell; Sparks GB 011510/00/0010 R
Improve performance of clones by elimination of virus.

9760 Researcher not indicated GB 050325/78/0011 R
Development of serological methods for monitoring top fruit viruses.

9761 Tonini, G.; Menniti, A.M. IT 040216/79/0004 N
Prevenzione e lotta della Monilia laxa nelle drupacee. **Monilia laxa prevention and control on stony fruits.**

9762 Meer, F.A. van der NL 010108/64/1015
Virusziekten bij fruitgewassen. **Virus diseases of fruit crops.** Publications.

Apple (B 3611)

See also 8710, 8968

9763 Schulz, F.A. DE 148200/72/0001
Das Resistenzverhalten des Apfels gegenüber Pezicula malicorticis – Gloeosporium perennans –. **Resistance of apple against Pezicula malicorticis – Gloeosporium perennans –.** Publications.

9764 Schoenhard, G. DE 215060/70/4004
Untersuchungen über die Stippigkeit des Apfels. **Investigations on bitter pit of apples.**

9765 Weinmann, W. DE 215080/77/0042
Untersuchung über die Auswirkungen unterschiedlicher Spritzbrüh– und Aufwandmengen auf die Rückstände von Benomyl auf und in Äpfeln nach Bekämpfung von Schorf. **Investigation on effects of different spray mixtures and dosages on benomyl residues on and in apples after speckle control.**

9766 Weinmann, W. DE 215080/77/0043
Untersuchung über die Auswirkungen unterschiedlicher Spritzbrühund Aufwandmengen auf die Rückstände von Mancozeb auf und in Äpfeln nach Bekämpfung von Schorf. **Investigations on effects of different spray mixtures and dosages on mancozeb residues on and in apples after speckle control.**

9767 Loeschcke, V. DE 215140/75/0012
Isolierung und Charakterisierung des Toxins von Erwinia amylovora – Feuerbrand von Kernobst –. **Isolation and characterization of the toxin of Erwinia amylovora – fire blight of pome fruit –.**

9768 Schmidle, A.; Alt, D. DE 215210/74/0010 R
Untersuchungen über die Ursachen und Formen der Resistenz von Arten und Sorten des Apfels gegen Phytophora cactorum. **Investigations on causes and forms of resistance of varieties and species of apple to Phytophtora cactorum – collar rot.**

9769 Warmbrunn, K.; Steiner, H. DE 501050/70/0001
Entwicklung des integrierten Pflanzenschutzes im Apfelanbau und seine Einführung in die Praxis. **Development of integrated control in apple orchards and its practical application.**

9770 Graf, H. DE 507303/77/0001
Versuche zur Bekämpfung des Obstbaumkrebses. **Trials for control of apple canker.**

9771 Marenaud, Cl. FR 010701/74/0103
Recherche sur l'étiologie d'une forme de dépérissement du cerisier en vue de déterminer les méthodes de lutte à envisager. **Researches on the etiology of a cherry–tree withering form to develop control methods.** Publications.

9772 White GB 011003/00/0012 R
Control of Cox disease.

9773 Bennett GB 011005/00/0017 R
Epidemiology of apple canker.

9774 Bennett GB 011005/00/0018 R
Control of apple canker.

9775 White; Parry GB 011008/00/0049 R
Control of Cox disease.

9776 Hislop GB 011508/00/0011 R
The eradication of over–wintering powdery mildew fungi by

dormant season spraying with surfactants.

9777 Ross GB 041205/00/0001 R
Apple scab control.

9778 Ross GB 041205/00/0002 R
Effect of viruses on apple production.

9779 Cartwright GB 041205/00/0005 R
Apple mildew control.

9780 Swinburne GB 041504/00/0001 R
Control of apple canker.

9781 Maas Geesteranus, H.P. NL 010108/73/3945 R
Bacterievuur bij Pomacaea veroorzaakt door Erwinia amylovora. **Fire blight of Pomaceae, caused by Erwinia amylovora.**

Pear (B 3612)

See also 8710, 9767, 9781, 9985

9782 Seemüller, E.; Kunze, L.; Petzold, H.
DE 215210/70/4015
Untersuchungen zum licht– und fluoreszenzoptischen Nachweis des Birnenverfalls und zur Einflussnahme des Erregers auf den Wirt. **Investigations into light– and fluorescence-optical proof of pear decline and influence of agent on host.**

9783 Graf, H. DE 507303/77/0002
Untersuchungen zum Birnbaumsterben an der Niederelbe. **Research on pear decline in the Lower Elbe region.** Publications.

9784 Talboys; Davies GB 011005/00/0026 R
Verticillium wilt – epidemiology and control – pear and quince.

9785 Ponti, I. IT 023500/79/0004 N
Difesa del pero da Alternaria S.P. **Pear–tree protection against Alternaria S.P. attacks.**

Other top fruit (B 3619)

See also 4258, 5869, 8710, 8739, 9642, 9889

9786 Jacob, H. DE 129163/75/0002
Latente Viruskrankheiten beim Steinobst. **Latent virus diseases in stone–fruit.**

9787 Casper, R.; Rohloff, H. DE 215150/77/0011
Untersuchungen zur Scharka–Anfälligkeit von Pflaumensorten bei Einzel– und Misch–Infektionen. **Investigations on susceptibility of plum varities to Scharka virus in single and mixed infections.**

9788 Zeller, W.; Schmidle, A. DE 215180/73/5015 N
Prüfungen von Sauerkirschsorten auf Resistenz gegen den Erreger des Bakterienbrandes. **Testing of varieties of sour cherry for resistance to bacterial blight.**

9789 Schmidle, A.; Seemüller, E.; Zeller, W.
DE 215210/71/5007
Untersuchungen über pilzliche und bakterielle Rindenund Holzschäden an Süss– und Sauerkirschen. **Experiments about bark and timber diseases of sweet and sour cherry caused by bacteria.**

9790 Schmidle, A.; Zeller, W. DE 215210/72/4002
Untersuchungen über den Einfluss von Temperatur und Feuchtigkeit auf die Infektion durch Pseudomonas–Arten bei Sauerkirschen. **Investigations into the influence of temperature and humidity on the infection of sour cherries by Pseudomonas.**

9791 Krczal, H.; Kock, T.; Kunze, L. DE 215210/74/0009
Untersuchungen über die Scharka–Krankheit bei Pflaume und Pfirsisch. **Investigations on the plum and peach pox by Scharka virus.**

9792 Kunze, L.; Clark, M.F.; Coles, C.L. DE 215210/75/0002
Untersuchung von Steinobstviren. **Investigation of stone fruit viruses.**

9793 Seemüller, E.; Schmidle, A. DE 215210/75/0005
Untersuchungen über die Sortenspezifität von Pseudomonas syringae an Sauerkirschen. **Investigations into strain species of Pseudomonas syringae in sour cherries.**

9794 Milaire, H. FR 010106/72/1773
Establissement des bases de la lutte intégrée en vergers de péchers. **Ecological basis of integrated control in peach orchards.** Publications.

9795 Marenaud, Cl. FR 010701/58/0281
Polymorphie symptomatologique du virus du chlorotic leaf spot chez les prunus. **Symptomatologic polymorphy of chlorotic leaf spot virus in prunus.**

9796 Leclant, F.; Bonfils, J.; Lauriaul, F. FR 011210/74/5236
Etude de la dissémination de l'enroulement chlorotique de l'abricotier. **Study of the dissemination of the apricot tree mycoplasms.**

9797 Clark; Adams GB 011005/76/0039 R
Detection and epidemiology of plum pox virus.

9798 Paton GB 060220/00/0004 R
Bacterial canker of stone fruit trees.

9799 Gualaccini, F. IT 020300/69/0001
Trasmissione a piante erbacee di alcune virosi del Ciliegio e del Cotogno. **Transmission of some virus diseases of Cherry and Quince to herbaceous hosts.** Publications.

9800 Ponti, I. IT 023500/79/0002 N
Lotta guidata contro le malattie del pesco. **Chemical control against peach–trees deseases.**

Soft fruit (berries and cane fruits) (B 3620)

See also 8754, 8756, 8757, 8966, 8967, 9699, 9737

9801 Jacob, H. DE 129163/75/0003
Identifizierung von virus– und virusähnlichen Krankheiten bei der Gallmilbe. **Identification of virus and viruslike diseases on Ribes.**

9802 Hoffmann, G.M.; Melzer, R. DE 161050/77/0005
Zur Biologie und Bekämpfung des Erregers eines Triebsterbens – Godronia cassandrae – an Kulturheidelbeere – Vaccinium corymbosum L. –. **Biology and control of Godronia**

cassandrae on highbush blueberry.

9803 Weinmann, W.   DE 215080/78/0012 N
Untersuchung über das Rückstandsverhalten von Benomyl auf/in Johannisbeeren nach einer Behandlung gegen Blattfallkrankheit. **Investigation on the residue–behaviour of Benomyl on/in currants after a treatment against leaf spots.**

9804 Krczal, H.   DE 215210/77/0001
Untersuchungen über die Brennesselblättrigkeit der schwarzen Johannisbeere. **Investigations on nettle–leafiness of black currant.**

9805 Krczal, H.   DE 215210/78/0002 N
Untersuchungen über Viruskrankheiten der Erdbeere. **Investigations on virus diseases of strawberry.**

9806 Krczal, H.; Seemüller, E.   DE 215210/78/0003 N
Untersuchungen über die Viruskrankheiten der Himbeere. **Investigations on virus diseases of raspberry.**

9807 Seemüller, E.   DE 215210/78/0004 N
Bekämpfung der Rhizomfäule der Erdbeere. **Control of rhizome rot of strawberry.**

9808 Kristensen, H.R.; Thomsen, A.   DK 010116/62/4515
Forekomst, forebyggelse og bekæmpelse af viroser på frugtbuske og jordbær hovedsagelig: hindbær–ringplet. **Occurrence, prevention and control of viruses on fruit bushes and strawberries, chiefly raspberry ring spot.**

9809 Thresh; Adams   GB 011005/00/0006 R
Epidemiology of virus and mycoplasma diseases.

9810 Bennett   GB 011005/00/0020 R
Control of grey mould (Botrytis rot).

9811 Talboys; Davies   GB 011005/00/0025 R
Verticillium wilt – epidemiology and control – strawberry.

9812 Butt   GB 011005/00/0028 R
Epidemiology and control of spur blight (Didymella).

9813 Butt   GB 011005/00/0029 R
Control of American gooseberry mildew and black currant leaf spot.

9814 Talboys; Harris   GB 011005/19/0040 N
Epidemiology and control of soil borne fungus diseases of soft fruit.

9815 Garrett   GB 011005/76/0036 R
Epidemiology and control of crown gall.

9816 Thresh; Stickels   GB 011005/76/0037 R
Produce healthy clones.

9817 Williamson; Hargreaves   GB 030702/00/0001 R
Plant protection: chemical and cultural control and economic importance of diseases of cane and bush fruits.

9818 Williamson   GB 030702/00/0003 R
Epidemiology and etiology: shoot disorders of cane and bush fruits.

9819 Montgomory; Kennedy   GB 030702/00/0004 R

Plant protection: chemical and cultural control and economic importance of strawberry red core.

9820 Montgomory; Kennedy   GB 030702/00/0005 R
Epidemiology and etiology: analysis of, and screening for resistance to diseases of soft fruit.

9821 Duncan   GB 030702/00/0024 R
Soil microbiology and root disease: autecology of the strawberry red core fungus.

9822 Jones; Murrant   GB 030704/00/0003 R
Viruses infecting raspberry.

9823 Chambers   GB 030704/00/0004 R
Production of virus–tested raspberry stocks.

9824 Foxe, M.; Hennerty, M.; Hannon, M.; Hunter, A.   IE 120108/78/9102 N
Production of virus free strawberry plants by meristem culture.

9825 Miotto, F.   IT 020400/78/0008
Studi sul Fusarium lateritium (Nees) (Deuteromycetes–Moniliales) che danneggia i germogli di Gelso. **Studies on Fusarium lateritium (Nees) (Deuteromycetes – Moniliales) damaging Mulberry's buds.** Publications.

9826 Scheer, H.A.Th. van der   NL 010212/70/2239
Onderzoek naar infecties via de wortels en het rhizoom van aardbei–planten, in het bijzonder van roodwortelrot, stengelbasisrot en verwelkingsziekte. **Research on infection of the roots and the rhizoom of strawberry plants, particulary of Phytophthora fragariae, P. cactorum and Verticillium dalliae.**

Citrus fruit (B 3630)

See also 4358

9827 Sänger, H.L.   DE 129182/70/0001
Untersuchungen am Exocortis viroid von Citrus. **Investigations into Exocortis viroid of citrus.**

9828 Vogel, R.   FR 012202/70/0218
Lutte contre les maladies bactériennes et cryptogamiques des citrus. **Citrus protection against bacterial and cryptogamic deseases.** Publications.

9829 Spina, P.   IT 021600/73/0312
Indagini sul mal secco degli agrumi ai fini della lotta. **Studies on "mal secco" of citrus fruit in view of its control.**

9830 Lanza, G.   IT 021600/77/0002 N
Prove di efficacia su vecchi e nuovi fungicidi nella prevenzione dei marciumi degli agrumi.. **Trials on old and new fungicides for the control of citrus fruit decay..**

9831 Cutuli, G.   IT 021600/77/0158 R
Indagine sui meccanismi di resistenza al mal secco degli agrumi e valutazione di talune proprietà dei prodotti chimici impiegati per la lotta. **Research on resistance mechanisms to dry rot in citrus fruits and evaluation of some of the properties of the chemical products used in disease control.**

9832 Salerno, M.   IT 040106/73/0302
Ricerche sul problema della lotta contro il mal secco degli agrumi etc. **Research on the problem of controlling "mal secco"**

of citrus fruit, etc. Publications.

9833 Salerno, M.   IT 040106/78/1106 N
Ricerche sulla lotta contro il "mal secco" degli agrumi a mezzo di prodotti antisporulanti e nuovi fungicidi sistemici. **Research on the control of citrus fruit "dry rot" by means of antispore substances and new systemic fungicides.**

9834 Somma, V.   IT 040905/74/0632
Ricerche su aspetti epidemiologici e fisiopatologici del mal secco degli agrumi in Sicilia e prove di lotta contro la malattia. **Research on epidemiological and phytopathological aspects of the mal secco of citrus in Sicily and disease control trials.**

9835 Marras, F.   IT 041307/72/0131
Ricerche sulle virosi delle piante. Agrumi, vite. **Research on plant virosis. Citrus, grapes.** Publications.

9836 Crescimanno, F.G.   IT 060900/72/0164
Prosecuzione delle indagini sul patrimonio varietale agrumicolo italiano, con particolare riguardo per la ricerca di cultivar di limone resistente al "mal secco". **Investigations on Italian citrus varieties, with particular interest for the c.v. research of lemon resisting to the "mal secco".** Publications.

9837 Crescimanno, F.G.   IT 060900/74/0169
Indagini sulla resistenza al "mal secco" mediante l'introduzione dell'intermediario arancio dolce. **Enquirements on the resistance on the "mal secco" through the introduction of the sweet orange rootstocks.**

Tropical and sub–tropical fruits (B 3640)

9838 Molot, P.; Mas, P.   FR 010604/70/1007
Fusariose du melon – étude des mécanismes de résistance du melon à Fusarium Oxysporum f. sp. Melonis. **Studies on melon resistance to Fusarium Oxysporum f. sp. melonis.** Publications.

9839 Gerlagh, M.   NL 010108/73/3957 R
Verwelkingsziekte van meloen, veroorzaakt door Fusarium oxysporum f. melonis. **Vascular disease of muskmelon caused by Fusarium oxysporum f. melonis.**

Grapes (B 3650)

See also 7417, 8798, 8800, 8806, 8814, 8815, 9835, 15348

9840 Nienhaus, F.; Rumbos, I.; Green, S.; Brelie, D. von–der
DE 111351/78/0001 N
Rickettsienähnliche Organismen in vergilbungskranken Reben, in Zuckerrüben mit Rosettenkrankheit und Lärchen mit Hexenbesen: Untersuchungen zur Pathogenität, Kultivierung und Nachweis von RLO. **Rickettsia–like organisms in yellows–diseased grape vines, in sugarbeet with rosette disease and larch trees affected by witches broom: investigations on pathogenicity, cultivation and indexing of the RLO.** Publications.

9841 Jacob, H.   DE 129163/75/0005
Identifikation, Übertragung und Testung von Rebvirosen. **Identification, transmission and indexing of virus diseases on grapes.**

9842 Sänger, H.L.   DE 129182/75/0001
Viroid–Nachweis bei der Blattrollkrankheit der Weinrebe. **Identification of viroids in grapevine leaf roll.**

9843 Alleweldt, G.; Hill, B.   DE 144450/74/0002
Untersuchungen über die Botrytisresistenz bei Reben. **Investigations on the resistance of grape vine to botrytis disease.**

9844 Alleweldt, G.; Hahn, H.   DE 204000/71/4020
Thermotherapeutische Behandlung von Reben zur Erzeugung von virusfreiem Pflanzengut. **Thermotherapeutical treatment of vines to obtain plants free from virus.**

9845 Hahn, H.   DE 204000/78/0004 N
Prüfung auf Resistenz gegen Botrytis cinerea. **Testing for resistance against Botrytis cinerea.**

9846 Herwig, K.   DE 204000/78/0016 N
Der Einfluss von stauender Nässe und des pH im Kultursubstrat auf die Chlorosebildung von Rebenneuzuchten. **The influence of water surplus caused by insufficient draining of gravitational water and of the pH of the growth media on the induction of chlorosis in grapes.**

9847 Blaich, R.   DE 204000/78/0018 N
Die Anwendung von Botrytis–Toxin bei der Frühdiagnose von Botrytis–Resistenz bei Rebensämlingen. **Application of Botrytis toxin for the early diagnosis of Botrytis resistance in vine seedlings.**

9848 Weinmann, W.   DE 215080/78/0003 N
Untersuchung über das Rückstandsverhalten von Methylmetiram, Triziman D und Captan auf/in Weintrauben nach einer Spritzbehandlung gegen Peronospora und Botrytis. **Investigation on the residue–behaviour of Methylmetiram, Triziman D and Captan on/in grapes after a spray treatment against downy mildew of grape and grey mould.**

9849 Paul, H.L.; Stellmach, G.   DE 215150/77/0006
Untersuchungen über die Reisigkrankheit der Rebe. **Investigations on fanleaf of vine.**

9850 Paul, H.L.   DE 215150/78/0007 N
Untersuchungen von Reben mit virusähnlichen Symptomen unbekannter Ätiologie. **Investigations on grapevines showing virus–like symptoms of still unknown origin.**

9851 Stellmach, G.   DE 215220/77/0008
Untersuchungen über Verfahren zur Prüfung des Virusverhaltens von Reben entsprechend den Forderungen der Rebenpflanzgutverordnung. **Studies on methods for testing of virus defence of vines according to the conditions of "Rebenpflanzgutverordnung" – vine plants regulation –.**

9852 Stellmach, G.   DE 215220/77/0009
Prüfung im Freiland von Rebenklonen, die einer Hitzebehandlung zwecks Viruseliminierung – Thermotherapie – unterworfen waren. **Field testing of vine clones treated with heat for elimination of viruses.**

9853 Stellmach, G.   DE 215220/77/0010
Untersuchungen über die durch einzelne pathogene Viren und Kombinationen mehrerer Viren ausgelösten Symptome – Erstellung eines Symptomkataloges zur visuellen Bonitur in Vermehrungsanlagen. **Investigations on symptoms caused by single pathogenic virus and combinations of different viruses. – Establishment of a catalogue of symptoms for visual classification of propagation.**

9854 Holz, B.   DE 215220/77/0011
Untersuchungen über Mykoplasmen und Rickettsien als
Ursache der Rebenvergilbung an Mosel und Rhein.
**Investigations on mycoplasms and rickettsia as causal agents of
yellow virosis of vines by Moselle river and Rhine river.**

9855 Holz, B.   DE 215220/77/0012
Untersuchungen über Spätfolgen der durch Agrobacterium
tumefaciens verursachten Mauke unter besonderer
Berücksichtigung des sektorialen Kümmerwuchses bei
Ertragsreben – Verfahren zur Vorbeugung und Behebung.
**Investigations on late effects of malanders caused by
Agrobacterium tumefaciens with special regard to sectoral
dwarfness in vines – methods for prophylaxis and control.**

9856 Holz, B.   DE 215220/77/0013
Untersuchungen über die Oosporenbildung in Rebblättern
nach Infektionen durch Plasmopara viticola – Möglichkeiten
ihrer Schädigung vor Beginn der Vegetation, um den Aufwand
von Fungiziden zu vermindern. **Investigations on the formation
of oospores in vine leaves after infections with Plasmopara
viticola – possibilities of elimination before initial vegetation in
order to reduce the dosage of fungicides.**

9857 Stellmach, G.   DE 215220/78/0002 N
Auswirkung einer Virus–Reinfektion von Reben, die durch
sanitäre Selektion, insbesondere aber duch Thermo–Therapie
von pathogenen Viren befreit worden sind. **Effects of
virus–reinfection of grapes released from pathogenic viruses by
means of sanitary selection, in particular by thermo–therapy.**

9858 Holz, B.   DE 215220/78/0005 N
Taxonomie, Pathogenität, Bekämpfung der an Fruchtruten der
Weinrebe vorkommenden Pilze. **Taxonomy, pathogenicity and
control of fungi growing on canes of grapes.**

9859 Eichhorn, K. W.; Lorenz, D. H.   DE 509153/77/0006
Prüfung der biologischen Wirksamkeit neuer Fungizide gegen
Plasmopara viticola unter erschwerten Bedingungen im
Freiland. **Testing of biological efficiency of new fungicides on
Plasmopara viticola under difficult field conditions.**

9860 Rüdel, M.   DE 509157/77/0002
Bedeutung der Unkräuter als Nematodenwirte für die
Epidemiologie von Rebvirosen. **Significance of weeds as hosts
of nematodes to epidemiology of vine viroses.**

9861 Rüdel, M.   DE 509157/77/0003
Untersuchungen zur Übertragung von Rebviren durch
Nematoden. **Nematodes as vectors of vine viruses.** Publications.

9862 Brückbauer, H.   DE 509157/77/0004
Untersuchungen über den natürlichen Wirtspflanzenkreis der
Rebenviren. **Natural plant hosts of vine viruses.**

9863 Brückbauer, H.   DE 509157/77/0005
Untersuchungen zur Wärmetherapie viruskranker Reben.
Thermal therapy of virus–infected vines.

9864 Brückbauer, H.   DE 509157/77/0006
Klärung der Symptomatologie durch Rückübertragung
definierten Virusmaterials auf gesunde Reben.
Symptomatology of viruses defined by transfer to sound vines.

9865 Brückbauer, H.   DE 509157/77/0007

Untersuchungen über das Verhalten von Rebenneuzüchtungen
nach Virusinfektion. **Behaviour of new vine breeds after virus
infection.**

9866 Dalmasso, A.   FR 010504/62/1711
Nématodes des vignobles. **Vineyard nematodes.** Publications.

9867 Rives, M.; Doazan   FR 010702/72/0086
Etude de la transmission chez la vigne de deux virus
(court–noué et marbrure) par la graine et le pollen. **Studies on
the transmission of two virus (grape fanleaf virus and mottle
virus) by seed and pollen in vine.**

9868 Hevin; Doazan, J.P.; Leclair, Ph.; Ottenwaelter, M.
    FR 010702/72/0091
Etude de la transmission chez la vigne de deux viroses
(court–noue et marbrure) par la graine et le pollen.
**Investigations on the transmission of two virus deseases
through the seed and pollen (fan–leaf and marbrure) in the
grape–vine.**

9869 Delas, J.; Molot, P.; Pathologie, Bordeaux
    FR 010704/71/6072
Relations entre la nutrition minérale et la sensibilité de la vigne
aux parasites et aux accidents de végétation. **Relations between
mineral nutrition and the vine susceptibility to parasites and to
vegetative accidents.**

9870 Moutous, G. Mme; Fos, A.   FR 010707/73/5214
Etude de Scaphoideus littoralis, Cicadelle vectrice de la
flavescence dorée Maladie à mycoplasme de la vigne. **Study of
Scaphoideus littoralis, a leafhopper vector of mycoplasm disease
on grapes.**

9871 Pistre, R.; Boubals, D.   FR 011204/75/0458
Lutte contre le pourridiè de la vigne. **Armillaria root rot of
grape.**

9872 Boubals, D.   FR 011204/76/0455
Etudes des causes de la mort de greffes soudées provenant de
pépinières de vigne du Midi de la France. **Dying of grafted
material of grape after the nursery.**

9873 Royle   GB 012201/00/0025 R
Resistance mechanisms in downy and powdery mildews.

9874 Ebbels   GB 050324/76/0011 R
**Significance of Fusarium canker in production of certified hop
cuttings.**

9875 De Sanctis, F.; Barba, M.   IT 020300/76/0001 N
Ricerche su alcune virosi della vite nell'Italia Centrale.
**Researches on some virus diseases of grapevine in Central
Italy.**

9876 Agnello, A.V.   IT 020300/79/0006 N
Lotta antibotritica sulla vite nel Lazio. **Control of grey–mould
on grapevine in Latium.** Publications.

9877 Santoro, G.; Di Jorio, N.; Ferri, E.   IT 020600/79/0001 N
Prove sperimentali con atomizzatori ed irroratori nella lotta
alla Botritis cinerea (Muffa grigia) della vite. **Experimental
tests using atomizers and sprayers in the fight against vine's
Botritis cinerea.**

9878 Egger, E.; Borgo, M.; Provasi, C.; Ottaviano, E.

D 2420 – Plant diseases and disease control

IT 021300/77/0001 R
Ricerche sull'ereditabilità e stabilità della resistenza della vite ad alcune crittogame. **Investigation on hereditability and stability of vine resistence to some diseases.** Publications.

9879 Egger, E.; Borgo, M.; Vuittenez, IT 021300/77/0002 R
Indagine sulla corrispondenza tra sintomi macroscopici di virosi sul legno e foglie di viti europee e portinnesti e risultati della sierodiagnosi. **Study on correspondence between viral macroscopic symptoms on leaves and wood of european vines and of rootstocks and serological diagnosis results.**

9880 Ponti, I.; Flori, P. IT 023500/79/0001 N
Difesa della vite dalla "Botrytis Cinerea". **Vine protection against Botrytis Cinerea attacks.**

9881 Corte, A. IT 024000/76/0001
Ricerche sulla biologia e lotta della Guignardia bidwellii (Ellis) Vialà et Ravaz., su vite. **Investigations on the biology and control of Guignardia bidwellii (Ellis) Vialà et Ravaz., on grapevine.**

9882 Garibaldi, A. IT 041215/77/0499 R
Biologia della Botrytis cinerea nei vigneti piemontesi. Interventi fitoiatrici contro la muffa grigia. **Botrytis cinerea biology in Piedmont vines. The use of pesticides to control grey.**

9883 Roca, F. IT 060600/74/0050
Nematodi vettori di virus vegetali. Efficienza vettrice di specie di Xiphinema nella trasmissione di virus della vite. **Nematodes vectors of plant viruses. Vector efficiency of species of Xiphinema in the transmission of grapevine virus.**

9884 Zambelli, P. IT 121200/77/0532
Trattamenti sperimentali antioidici, andamento fermentazione in mosti trattati con antioidici, esame nuovi antibotritici. **Experimental treatments with anti–mildew substances, fermentation processes in anti–mildew treated musts, investigation of new anti–botrytis compounds.**

Edible nut fruits (B 3660)

See also 5869, 9642

9885 Marenaud, Cl. FR 010701/66/0105
Etude de quelques aspects agronomiques de la transmission par pollen et par semence de diverses maladies à virus chez les arbres fruitiers. **Study of some agronomic aspects of pollen and seed transmission of various virus diseases in stone fruit–trees.** Publications.

9886 Marenaud, C. FR 010701/66/0278
Etude des relations "Etat sanitaire" – Physiologie de l'arbre fruitier. **Relationships between "sanitary state" and fruit–tree physiology.** Publications.

9887 Marenaud, C. FR 010701/70/0280
Etiologie de la mosaïque du noisetier. **Etiology of the haizel–tree mosaic.**

9888 Marenaud, Cl. FR 010701/73/0104
Etude des problèmes de détection, d'épidémiologie, et des méthodes de lutte posées par la présence du virus de la sharka en France. **Study of detection, epidemiology and control methods of the sharka virus in France (stone fruit–trees).**

9889 Cristinzio, M. IT 040707/73/0921
Virosi nocciuolo–virosi albicocco–virosi cucurbitacee–virosi lattuga–virosi peperone–chemioterapia virosi peperone e lattuga–censimento virosi delle piante legnose da frutto e delle piante ortensi. **Virus diseases of hazel, apricot tree, cucurbits, lettuce and pepper –chemical control of pepper and lettuce virus diseases– inventory of virus diseases of woody fruit plants and of horticultural plants.**

9890 Lovisolo, O. IT 060100/70/0095
Seccume del nocciolo gentile delle Langhe. **Hazelnut "seccume" in Piedmont.** Publications.

9891 Fassi, B. IT 120100/73/2236
Indagine sul deperimento del nocciolo gentile delle Langhe, nell'interesse del Laboratorio di fitovirologia applicata. **Survey on decline of the "gentile" hazelnut in the Langhe. In the interest of the Laboratory for applied phytovirology.** Publications.

Ornamentals and ornamental products in general (B 3700)

See also 8959, 9583, 9592, 9597

9892 Koenig, R.; Lesemann, D. DE 215150/77/0004
Bestandsaufnahme über das Vorkommen von Virosen und Viren bei Zierpflanzen. Ausarbeitung von empfindlichen Diagnoseverfahren. **Inventory on the occurrence of viroses and virus in ornamental plants. Development of sensitive diagnostic methods.**

9893 Dinesen, I.; Jørgensen, H.A.; Mygind, H.; Jensen, A.
DK 010116/75/4509
Specialstudier vedrørende bakteriesygdomme især hos havebrugsplanter, samt registrering og diagnosearbejde vedrørende enkelte land– og havebrugsplanter. **Specialized studies of bacterial diseases, particularly in horticultural crops, and the registration and diagnosis for certain agricultural and horticultural plants.**

9894 Spencer GB 011109/00/0001 R
Evaluation of fungicides for glasshouse crops.

9895 Ebben GB 011109/00/0002 R
Effect of environmental and cultural factors on infection by pathogens.

9896 Ebben; Spencer GB 011109/00/0003 R
Control of disease by antagonistic microorganisms, especially control of cucumber black root rot.

9897 Spencer; Gandy GB 011109/00/0010 R
Etiology and control of fungal and bacterial diseases of glasshouse ornamentals.

9898 Price GB 011109/00/0011 R
Diseases of ornamental bulbs, corms and tubers.

9899 Hollings; Stone GB 011110/00/0001 R
Virus diseases of glasshouse flower crops (carnation, chrysanthemum, orchids and Pelargonium).

9900 Hollings GB 011110/00/0010 R
Virus diseases of glasshouse vegetable crops (cucumber, lettuce, tomato, pepper).

9901 Dickens; Knight   GB 050323/74/0051 R
Biology of juniper–pear rust Gymnosporangium asiaticum.

9902 Cartia, G.   IT 040302/77/0828 R
Virosi delle piante ortensi ed ornamentali e caratterizzazione
dei rispettivi agenti causali. Epidemiologia dei virus
economicamente importanti. Virus diseases in market–garden
and ornamental plants, characterization of their respective
causal agents. Epidemiology of economically relevant viruses.

9903 Galbiati, C.   IT 040612/78/1164 N
Virosi delle piante ortensi ed ornamentali. Virus diseases in
market–garden and ornamental plants.

9904 Castellani, E.   IT 041215/72/0087
Malattie delle piante da fiore ed ornamentali. Diseases of
flowering and ornamental plants. Publications.

9905 Castellani, E.   IT 041215/77/0135 R
Malattie delle piante da fiore e ornamentali. Diseases of
ornamental plants and flowers.

Bulbs (B 3710)

See also 9525

9906 D'Herde, C.; Coolen, W.; Hendrickx, G.
     BE 070100/73/0007 R
Studie van complexe ziekten te wijten aan interacties tussen
Fytonematoden en andere micro–organismen. Research on
complex diseases due to the interaction between nematodes and
some micro–organismes.

9907 Kristensen, H.R.; Paludan, N.; Thomsen, A.
     DK 010116/61/4517
Forekomst, forebyggelse og bekæmpelse af viroser på
prydplanter hovedsagelig: dværgsyge, klorotisk spætning,
ætsning, stregsyge, tomat–ringplet, mosaik, tobak–nekrose,
rattle, gul stregsyge og gulmosaik hos chrysanthemum, nellike,
pelargonie, tulipan, narcis og rose. Occurrence, prevention and
control of viruses on ornamentals, chiefly dwarfing, chlorotic
mottle, etching, stripe, tomato ringspot, mosaic, tobacco
necrosis, rattle, yellow stripe, and yellow mosaic on
chrysanthemum, carnation, pelargonium, tulip, narcissus and
rose.

9908 Brunt   GB 011110/00/0004 R
Virus diseases of bulbs, corms and tubers (dahlia, Iridaceae,
Liliaceae, narcissus).

9909 Lyons   GB 030702/00/0007 R
Epidemiology and etiology: biology of diseases of ornamental
bulbs.

9910 Mowat; Chambers   GB 030704/00/0010 R
Viruses infecting bulbous ornamentals.

9911 Mowat; Chambers   GB 030704/00/0011 R
Production of virus–tested bulb stocks.

9912 Dickens; Sharp   GB 050321/76/0021 R
Aetiology of neck rot of narcissus.

9913 Labruyère, R.E.   NL 010108/70/2770
Stromatinia gladioli (Drayton) Whetzel bij gladiool.

Stromatinia gladioli in gladiolus. Publications.

9914 Kamerman, W.   NL 010205/65/1462
Het geelziek van de hyacint (Xanthomonas hyacinthi). Yellow
disease of hyacinth (Xanthomonas hyacinthi). Publications.

9915 Weststeijn, G.; Vink, P.   NL 010205/65/1463
Wortelrotverschijnselen bij bolgewassen. Root rot phenomena
in bulbous crops.

9916 Bergman, B.H.H.   NL 010205/65/1468
Fusarium oxysporum in narcissen (bolrot). Fusarium
oxysporum in narcissus (bulb rot). Publications.

9917 Bergman, B.H.H.   NL 010205/65/1469
Fusarium oxysporum in tulpen (zuur). Fusarium oxysporum in
tulips. Publications.

9918 Slogteren, D.H.M. van   NL 010205/67/1508
Mycoplasmaziekten van bol– en knolgewassen. Mycoplasma
diseases of bulbous crops. Publications.

9919 Aartrijk, J. van   NL 010205/68/1505
Weefselcultuur van bolgewassen met het oog op de produktie
van virusvrij materiaal en snelle vegetatieve vermeerdering.
Meristem culture of bulbous plants to obtain virus free plant
material and to achieve rapid vegetative propagation.
Publications.

9920 Kamerman, W.   NL 010205/71/3221
Diagnostiek en bestrijding van ziekten in bol– en
knolgewassen, veroorzaakt door bacteriën. Diagnosis and
control of bacterial diseases in bulbous crops. Publications.

9921 Doornik, A.W.   NL 010205/73/3938
Rhizoctonia species in bol– en knolgewassen. Rhizoctonia in
bulbous crops. Publications.

9922 Beijersbergen, J.C.M.; Duineveld, T.J.
     NL 010205/75/6255
Begeleidend onderzoek bij de toepassing van systemische
fungiciden in de bloembollencultuur. Research on problems
resulting from the use of systemic fungicides for disease control
in bulbous crops. Publications.

9923 Asjes, C.J.; Derks, A.F.L.M.   NL 010205/76/6994
Diagnostiek van virusziekten en identifikatie van virussen in
bloembolgewassen. Diagnosis of virus diseases and
identification of viruses in flower bulb crops. Publications.

9924 Derks, A.F.L.M.   NL 010205/76/6995
Ontwikkeling van methoden t.b.v. de identifikatie van virussen
en de diagnostiek van virusziekten in bloembolgewassen.
Development of methods for the identification of viruses and
the diagnosis of virus diseases in flower bulb crops.
Publications.

9925 Asjes, C.J.   NL 010205/76/6996
Bestrijding van virusziekten in bloembolgewassen. Control of
virus diseases in flower bulb crops. Publications.

Flowers and pot plants (B 3720)

See also 5884, 8955, 9599, 9600, 9601, 9679, 9685, 9688, 9907,
9966, 9971

D 2420 – Plant diseases and disease control

9926 Meyer, J.; Verhoyen, M.; Matthieu, J.
BE 020302/67/0001 R
Etude des viroses d'espèces ornementales et maraîchères entre autre oeillets, chrysanthèmes – pruniers, pommiers, poiriers–céleris, laitues, poireaux. **Study of virus diseaeses of ornemental and vegetable species, especially carnation, chrysanthemum, cherries, appels and pears – celery, lettuce and leek.** Publications.

9927 Welvaert, W.; Roos, A.
BE 030011/63/0004
Fytopathologisch onderzoek van Cylindrocladium scoparium Morgan op Azalea indica ; zwam– en bakterieaantastingen op Calathea spp. en het uittesten van nieuwe systemische fungiciden. **Phytopathological research on Cylindrocladium scoparium Morgan on Azalea indica ; fungal and bacterial diseases in Calathea spp. and testing new systemic fungicides.** Publications.

9928 Welvaert, W.; Samyn, G.
BE 030011/71/0001 R
Onderzoek over en bestrijding van virusziekten in sierplanten, meer speciaal bij Begonia, Hydrangea, Pelargonium, Orchidee, cactussen en groene kamerplanten. **Research on virus diseases and control in ornamentals, especially Begonia, Hydrangea, Pelargonium, Orchids, Cacti and the so called green plants.** Publications.

9929 Welvaert, W.; Samyn, G.
BE 030011/72/0003 R
Studie van mycoplasmaziekten in de sierplantenteelt, meer in het bijzonder bij Hydrangea. **Study on mycoplasmadiseases of ornamentals especially in Hydrangea.**

9930 Bosmans, P.
BE 070500/68/0008 R
Uitwerken van bedrijfsmethoden en testen van verschillende fungiciden tegen de voornaamste ziekten van Begonia tuberosa. **To work out of exploitation methodes and control the principal diseases of Begonia tuberosa with different fungicides.** Publications.

9931 Geenen, J.; Veldeman, R.
BE 070500/69/0004 R
Studie van de infektiebiologie, de chemische– en biologische bestrijding van de voornaamste schimmels op "Azalea indica". **Research on the infection biology, the chemical and biological control of the principal fungi "Azalea indica".** Publications.

9932 Kamoen, O.; Jamart, G.
BE 070500/72/0003 R
Studie van systemische fungiciden bij ziekten in de sierplantenteelt. **Research on the systemic fungicides on diseases on ornemental plants.** Publications.

9933 Sauthoff, W.
DE 215070/75/0004
Einfluss der Kulturbedingungen auf den Befall von Pelargonien durch Xanthomonas pelargonii. **Influence of conditions of cultivation on infestation of Pelargonium by Xanthomonas pelargonii.**

9934 Koenig, R.; Lesemann, D.
DE 215150/78/0020 N
Entwicklung von empfindlichen serologischen Nachweisverfahren für Viren in Chrysanthemen. **Development of sensitive serological assay procedures for viruses in chrysanthemums.**

9935 Koenig, R.; Lesemann, D.
DE 215150/78/0021 N
Entwicklung von Verfahren zum Nachweis von Viren in Pelargonien mit besonderer Berücksichtigung des unter EG–Quarantäne–Richtlinien fallenden Tomatenringfleckenvirus. **Development of procedures for the detection of viruses in pelargonium with special reference to tomato ringspot virus – an EC quarantine virus.**

9936 Lesemann, D.; Koenig, R.
DE 215150/78/0030 N
Viruskrankheiten in Orchideenkulturen. Nachweis, Verbreitung, Ätiologie, Epidemiologie. **Virus diseases in orchids, diagnosis, distribution, etiology, epidemiology.**

9937 Poupet, A.
FR 010501/70/1015
Etude de la résistance des oeillets méditerranéns à l'infection par le virus de la marbrure. **Studies of mediterranean carnation resistance to carnation mottle virus.** Publications.

9938 Ponchet, J.; Andreoli, C.
FR 010501/71/1012
Utilisation de mutants du Phytophthora Nicotianae de l'oeillet à l'étude du pouvoir pathogène et en lutte biologique. **Use of carnation Phytophthora Nicotianae mutants for the study of pathogenic power and biological control.**

9939 Ebben; Spencer
GB 011109/00/0008 R
Control and host–parasite relations of carnation wilt diseases.

9940 Dickens; Sellar
GB 050323/76/0071 R
Biology and control of white rust of Chrysthemum (Puccinia horiana).

9941 Pergola, G.; Dalla Guda, C.; Garibaldi, A.; D'Aquila, F.
IT 021200/77/0001 N
Ricerche sulla possibilità di lotta contro la fusariosi vascolare del garofano. **Research on the possibility of controlling fusarium wilt in carnation.**

9942 Pergola, G.; Dalla Guda, C.
IT 021200/77/0004 N
Prove risanamento portainnesti virosati della rosa.. **Attempts at the recovering virus infected rose root stock.**

9943 Belli, G.
IT 040612/77/0825
Virosi di piante ortensi ed ornamentali (crisantemo e patata). **Virus diseases of market garden and ornamental plants (chrysanthemum and potato).**

9944 Lisa, V.
IT 060100/64/0093
Indagini sulla distribuzione in natura e sulla identificazione dei virus del garofano. **Carnation virus identification and field distribution studies.** Publications.

9945 Hakkaart, F.A.
NL 010108/62/0965
Inventarisatie en onderzoek van virusziekten van anjers. **Virus diseases of carnations.** Publications.

9946 Rattink, H.
NL 010108/69/2771
Verwelkingsziekten in anjers. **Wilt diseases in carnations.** Publications.

9947 Rattink, H.
NL 010108/71/3611
Bacterieziekte in Begonia (Xanthomonas begoniae). **Bacterial disease of Begonia (Xanthomonas begoniae).**

9948 Hakkaart, F.A.
NL 010108/74/5900
Virussen van Pelargonium. **Virus diseases of Pelargonium.** Publications.

9949 Hakkaart, F.A.
NL 010108/77/7975
Incidenteel virologisch onderzoek bij bloemisterijgewassen. **Incidental virological research on flower crops.**

9950 Rattink, H.   NL 010108/78/8696 N
Phytophtora bij potplanten. **Phytophthora diseases of potplants.**

9951 Rattink, H.   NL 010108/78/8697 N
Alternaria–bladvlekkenziekte bij anjers. **Alternaria leafspot of carnations.**

9952 Rattink, H.   NL 010108/78/8698 N
Incidenteel onderzoek inzake schimmel– en bacterieziekten bij bloemisterijgewassen. **Incidental research on fungal and bacterial diseases in Horticulture.**

9953 Vruggink, H.   NL 010108/78/8699 N
Bacterieziekte van Pelargonium. **Bacterial diseases of Pelargonium.**

9954 Hakkaart, F.A.   NL 010108/78/8702 N
Virusziekten van Begonia. **Virus diseases of Begonia.**

9955 Hakkaart, F.A.   NL 010108/78/8703 N
Virusziekten van Alstroemeria. **Virus diseases of Alstroemeria.**

9956 Hakkaart, F.A.   NL 010201/62/0965
Inventarisatie en onderzoek van virusziekten van anjers. **Virus diseases of carnations.** Publications.

9957 Rattink, H.   NL 010201/69/2771
Verwelkingsziekten in anjers. **Wilting diseases in carnations.** Publications.

9958 Rattink, H.   NL 010201/71/3611
Bacterieziekte in Begonia (Xanthomonas begoniae). **Bacterial disease of begonia (Xanthomonas begoniae).**

9959 Hakkaart, F.A.   NL 010201/74/5900
Virussen van Pelargonium. **Virus diseases of Pelargonium.**

9960 Hakkaart, F.A.   NL 010201/77/7975
Incidenteel virologisch onderzoek bij bloemisterijgewassen. **Incidental virological research on flower crops.**

9961 Rattink, H.   NL 010201/78/8696 N
Phytophtora bij potplanten. **Phytophthora diseases of potplants.**

9962 Rattink, H.   NL 010201/78/8698 N
Incidenteel onderzoek inzake schimmel– en bacterieziekten bij bloemisterijgewassen. **Incidental research on fungal and bacterial diseases in floriculture.**

9963 Hakkaart, F.A.   NL 010201/78/8702 N
Virusziekten in Begonia. **Virus diseases of Begonia.**

9964 Hakkaart, F.A.   NL 010201/78/8708 N
Virusziekten van Alstroemeria. **Virus diseases of Alstroemeria.**

9965 Dirkse, F.B.   NL 010201/78/9006 N
Wortelrotbestrijding bij potplanten. **Root–rot control in pot–plants.**

Ornamental shrubs (B 3730)

See also 9739, 9781

9966 Welvaert, W.; Roos, A.   BE 140000/77/0031 R
Rhododendron simsii:invloed van de voorbehandeling van moerplanten en stekmateriaal met systemische fungiciden op beworteling en ziekteresistentie. **Rhododendron simsii:influence of pre–treatment of mother–plants and cuttings with systemic fungicides on rooting and disease protection.**

9967 Schulz, F.A.; Hanella, A.   DE 148200/78/0003 N
Vorkommen und Bedeutung epiphytischer Bakterien für die Feuerbrand–Krankheit – Erwinia amylovora – an Ziersträuchern. **Occurrence and importance of epiphytic bacteria for fire blight – Erwinia amylovora – of ornamentals.** Publications.

9968 Persiel, F.; Anders, S.   DE 206000/71/5006 R
Untersuchungen zur Resistenz von Rosenunterlagen – Rosa inermis und Rosa multiflora – gegen Sphaerotheca pannosa und Diplocarpon rosae 1970. **Investigations on resistance of rose rootstocks – Rosa inermis and Rosa multiflora – to Sphaerotheca pannosa and Diplocarpon rosae.**

9969 Dimitri, L.; Butin, H.   DE 215230/75/0007
Untersuchungen über die Entstehung und Verhütung von Wundfäulen bei Nadelbäumen. **Investigations on causes and control of wound rots in coniferous trees.**

9970 Jørgensen, H.A.; Mygind, H.; Jensen, A.   DK 010116/70/4508
Diagnosticering, forebyggelse og bekæmpelse af forskellige svampe– og bakteriesygdomme – herunder ildsot – på planteskolekulturer. **Diagnosis, prevention and control of different fungal and bacterial diseases, including fireblight, on nursery garden cultures.**

9971 Devergne, J.C.; Cardin, C.   FR 010501/71/1013
Maladies à virus du rosier: mosaïques, anomalies de croissance. **Virus diseases of rose–tree: mosaic, growth abnormalities.** Publications.

9972 Berge, J.B.; Cuany, A.   FR 010504/69/1709
Nématofaune des rosiers. **Nematofauna of roses.** Publications.

9973 Thresh; Manwell   GB 011005/00/0030 R
Produce healthy clones of woody ornamentals.

9974 Talboys   GB 011005/00/0032 R
Control of verticillium wilt.

9975 Billing   GB 011005/00/0034 R
Fireblight – epidemiology and control.

9976 Smith, P.   GB 011109/00/0009 R
Epidemiology and control of Phytophthora cinnamomi and other pathogens of hardy nursery stock.

9977 Thomas   GB 011110/00/0009 R
Virus diseases of roses and hardy nursery stock.

9978 Campbell; Sweet   GB 011510/00/0011 R
Improve quality and virus status of hardy ornamental trees and shrubs.

9979 Maude; Miller   GB 011807/75/0023 R
Biology, epidemiology and control of diseases of bedding plants.

9980 Swinburne   GB 041504/76/0005 R
A study of the biology and control of root diseases incited by the genus Phytophthora.

9981 Clancy, K.J.   IE 120106/73/9081 N
Studies on a disease of seedlings of chamaecyparis lawsoniana, caused by phytophthora eriugena sp. nov. Publications.

9982 Meer, F.A. van der   NL 010108/73/4071
Virusziekten van houtige siergewassen. Virus diseases of woody ornamentals. Publications.

9983 Elk, B.C.M. van   NL 010203/60/3751
Het gebruik van schimmelbestrijdingsmiddelen bij het enten van boomkwekerijgewassen. Treatment of grafts with fungicides in the nursery. Publications.

9984 Slavekoorde, S.M.   NL 010203/66/2568
Verwelkingsziekte in coniferen en ericaceeën. Wilt diseases of conifers and ericaceous ornamentals. Publications.

9985 Kooistra, T.   NL 010300/75/9099 N
Onderzoek naar de mogelijkheden van chemische bestrijding van bacterievuur (Erwinia amylovora). Possibilities of chemical control of fire blight caused by Erwinia amylovora.

9986 Miller, H.J.   NL 010300/78/9089 N
Aantasting van es (Fraxinus excelsior) door Pseudomonas savastanoi. Infection of ash by Pseudomonas savastanoi.

9987 Frinking, H.D.   NL 020018/74/6342
Invloed milieufactoren op de infectiecyclus van meeldauw (Sphaerotheca pannosa) op roos. Influence of environmental factors on the development of mildew (Spaerotheca pannosa) on roses.

Forests in general (B 3800)

See also 2072, 4759, 8874, 9341, 9970

9988 Hüttermann, V.   DE 132630/72/0001
Hydrolytische und phosphorolytische Enzyme bei Fomes annosus und deren biologische Bedeutung im System Pilz–Baum. Hydrolytic and phosphorolytic enzymes of Fomes annosus and their role in the system fungus–tree.

9989 Volger, C.; Ahnert, G.; Finke, E.; Ganser, H.K.; Noelle, A.   DE 132751/78/0003 N
Untersuchungen zur physiologischen und genetischen Variabilität sowie zur Inkompatibilität von Fomes annosus–Isolaten. Studies on the physiological and genetical variability of Fomes annosus isolates and their incompatibility. Publications.

9990 Butin, H.   DE 215230/77/0005
Verbesserung der "Mündener Scheibenmethode" zur Prüfung wässeriger Bläueschutzmittel. Improvement of the "disk method of Muenden" for testing of watery preservatives against blue stain.

9991 Dimitri, L.; Bonnemann, I.   DE 506453/75/0001
Wundfäule forstlicher Baumarten. Wound rot of forest tree species.

9992 Hartmann, G.   DE 507652/71/0004

Untersuchungen zur Biologie des Hallimasch – Armillaria mellea – in Forstkulturen. Biology of Armillaria mellea in forest plantations.

9993 Delatour, C.; Bachacou; Perrin, R.; Bouchon, J.; Guinot, G.   FR 010305/71/4131
Ecologie du Fomes Annosus. Ecology of F. annosus.

9994 Seaby; Malone   GB 041502/76/0001 R
Biological control of Fomes annosus.

9995 McAree, D.   IE 050100/57/7201 N
Evaluation and control of forest diseases. Publications.

9996 Grasso, V.   IT 061100/77/0853
Comportamenti di alcuni isolati di Endothia parasitica. Behaviour of some Endothia parasitica colonies.

9997 Grasso, V.   IT 061100/77/0855
Indagine sulla moria dei semenzali. Research on seed death.

Pine forests in general (B 3810)

9998 Hildebrandt, G.; Cagirici; Mahmut   DE 126451/74/0003
Untersuchungen zur Luftbildinterpretation von Waldkrankheiten in Kiefern– und Eichenwäldern. Air–photo interpretation of forest diseases in pine and oak stands.

9999 Siepmann, R.   DE 215230/78/0002 N
Untersuchungen über Wurzel– und Stammfäulen an Koniferen, verursacht durch den Hallimasch – Armillaria mellea –. Studies on root– and stem–rot of conifers, caused by Armillaria mellea.

10000 Delatour, C.; Pinon, J.; Birot, Y.   FR 010305/75/4132
Rouille vésiculeuse des pins à 5 feuilles. White pine Blister Rust.

Fir forests (B 3811)

10001 Kloke, A.   DE 215060/77/0003 N
Untersuchungen zur Ätiologie und Abwehr des Omorika–Sterbens. Investigations on etiology and control of the dieback of Omorika.

10002 Siepmann, R.   DE 215230/70/4002
Anfälligkeit verschiedener Schwarzkiefernherkünfte gegenüber Scleroderris lagerbergii. Susceptibility of different proveniences of Austrian pine to Scleroderris lagerbergii.

10003 Rack, K.   DE 215230/77/0001
Infektionsbiologische Untersuchungen an Sporen von Lophodermium–pinastri–Kiefernschütte– und Naemacyclus–niveus. Infection biological studies on spores of Lophodermium–pinastri and Naemacyclus–niveus.

10004 Rack, K.   DE 215230/77/0003
Untersuchungen über die Wirkung systemischer Fungizide auf den Erreger der Kiefernschütte–Lophodermium–pinastri. Investigations on the effects of systemic fungicides on Lophodermium–pinastri.

10005 Wachter, A.   DE 501508/75/0001
Untersuchungen zur Tannenerkrankung. Investigations into the disease of fir–trees.

10006 Stuart, M.R.; Donnelly, D.M.X.; Heslin, M.C.; O'Morchu, P.   IE 120401/77/9163 N
Biological role of fomannoxin, a phytotoxic metabolite of fomes annosus, cause of root and butt rot of conifers. Publications.

10007 Moriondo, F.   IT 061100/72/0207
Aspetti particolari della biologia del Cronartium flaccidum, agente patogeno della ruggine vescicolosa dei pini a due aghi. Some biological aspects of Cronartium flaccidum, causal organism of blister rust in two needle pines. Publications.

10008 Naldini, B.   IT 061100/72/0209
Agenti particolari della biologia della Melampsora pinitorqua, agente patogeno della ruggine curvatrice dei rametti di pino. Some biological aspects of Melampsora pinitorqua, causal organism of pine twist rust. Publications.

10009 Grasso, V.   IT 061100/77/0856
Aspetti particolari della biologia della Melampsora pinitorqua, agente patogeno della rugine curvatrice dei rametti di Pino. Aspects of the biology of Melampsora pinitorqua, the pathogenic agent of pine twist rust.

10010 Grasso, V.   IT 061100/77/0857
Aspetti particolari della biologia del Cronartium flaccidum, agente patogeno della rugine vescicolosa dei pini a 2 aghi. Miglioramento genetico di alcune specie di Pino a 2 aghi al Cronartium flaccidum. Aspects of the biology of Cronartium flaccidum, the pathogenic agent of blister rust in two-needle pines. Genetic improvement of certain two-needle Pine species as regards its resistance to C ronartium flaccidum.

10011 Tol, G. van   NL 010601/72/3532
Topsterfte van Corsicaanse den. Dieback in Corsican Pine. Publications.

Spruce and fir forests (B 3812)

See also 7678

10012 Volger, C.; Haniel, J.; Murach, D.
DE 132751/78/0002 N
Chemische und biologische Behandlung von Fichtenstubben zur Verhütung der Primärinfektion in Aufforstungsbeständen durch Fomes annosus. Chemical and biological treatment of freshly cut stumps to avoid the primary infection in spruce afforestations by Fomes annosus. Publications.

10013 Rehfuess, K.E.; Alcubilla, M.   DE 160030/72/0001
Pilzhemmwirkung des Fichtenholzes in Abhängigkeit vom Standort, vom Ernährungszustand und vom Genotyp. Spruce wood–inhibition of Fomes annosus in relation to stand, nutrition and genotype.

10014 Siepmann, R.   DE 215230/72/0003
Infektionsverlauf des Stammfäuleerregers Fomes annosus bei der Fichte. Course of infection of Fomes annosus in spruce stands.

10015 Siepmann, R.   DE 215230/75/0001
Ursache und Ausmass der Stammfäule bei der Douglasie in der Bundesrepublik Deutschland. Causes and extent of stem rot of Douglas fir in the Federal Republic of Germany. Publications.

10016 Schlenker, G.; Schönhar, S.; Evers, F.H.
DE 501502/72/0003
Mykorrhiza–Untersuchungen in Zusammenhang mit Feinwurzelfäule und Rotfäule der Fichte 1972. Studies of mykorrhiza in relation to fine–root rot and red rot of spruce.

10017 Schönhar, S.   DE 501508/77/0005
Untersuchungen über die Rotfäule der Fichte. Investigations on root and stem rot of spruce. Publications.

10018 Dimitri, L.   DE 506453/71/0007
Untersuchungen über die Rotfäuleresistenz von Klonen, Provenienzen und Arten der Gattung Picea. Research on the resistance of clones, provenances and species of the genus Picea against Fomes annosus.

10019 Hartmann, G.   DE 507652/73/0002
Befallsvoraussetzungen und Bekämpfung von Phomopsis pseudotsugae in Douglasienkulturen. Predisposition to infestation by and control of Phomopsis pseudotsugae in Douglas fir plantations. Publications.

10020 Hartmann, G.; Kleinschmit, J.   DE 507652/78/0002 N
Resistenzprüfung von Douglasienherkünften gegen rostige Douglasienschütte – Rhabdocline pseudotsugae –. Testing of resistance of Pseudotsugae menziesii to Rhabdocline pseudotsugae.

10021 Benetti, M.P.; Motta, E.   IT 020300/76/0004 N
Ricerche su alcuni ospiti naturali di Seiridium cardinale.. Researches on some natural hosts of Seiridium cardinale..

10022 Intini, I.   IT 061100/72/0210
Cancro del cipresso da Coryneum cardinale. Cypress canker by Coryneum cardinale. Publications.

10023 Grasso, V.   IT 061100/77/0858
Cancro del Cipresso da Coryneum cardinale. Miglioramento genetico del Cupressus sempervirens per la resistenza al Coryneum cardinale. Cypress cancer induced by Coryneum cardinale. Genetic improvement of Cupressus sempervirens as regards its resistance to Coryneum cardinale.

Larch forests (B 3813)

See also 9840

Leafwoods in general (B 3820)

10024 Kloke, A.; Bau; Leh, H.–O.   DE 215060/75/0007
Untersuchungen zur Physiologie von Nähr– und Schadelementen bei Laubgehölzen unter besonderer Berücksichtigung der Disposition für nichtparasitäre Krankheiten. Studies on the physiology of nutritional and noxious elements in deciduous trees with special reference to the disposition for nonparasitical diseases.

Oak tree stands (B 3821)

See also 9998

10025 Butin, H.; Parameswaran, N.   DE 215230/78/0001 N
Untersuchungen über den "Buchrindenschorf", eine neue Krankheit auf Fagus sylvatica. Studies on beech bark scab, a new disease of Fagus sylvatica.

10026 König, E.; Bogenschütz, H.   DE 501508/77/0004
Untersuchungen zum Buchen–Rindensterben. Investigations

on beech bark disease.

10027 Hartmann, G.; Waragai, A.   DE 507652/77/0001
Befallsvoraussetzungen für und Schäden durch Eichenmehltau
an Traubeneichen. **Predisposing factors for and damages by**
powdery mildew on sessile oak.

10028 Perrin, R.   FR 010305/73/4127
Le chancre du hêtre à Nectria Ditissima. **Nectria ditissima**
Beech Canker.

Ash tree stands (B 3822)

10029 Pinon, J.   FR 010305/74/4128
Maladie hollandaise de l'orme. **Dutch elm disease.**

Poplar tree stands (B 3823)

10030 Veldeman, R.   BE 070500/67/0001 R
Biologische studie en bestrijding van de voornaamste
schimmels op populier, beuk en olm. **Investigations into the**
biology and control of the most important fungi on poplar,
beech and elm. Publications.

10031 Kechel, H.   DE 912000/74/0001
Untersuchungen an Pappeln zur Erhöhung der Resistenz
gegen Krankheiten. **Studies on poplars for raising the**
resistance to diseases.

10032 Delabraze, P.; Keller, R.; Frochot, H.; Garbaye, J.;
Lemoine, B.   FR 010301/69/4073
Peupliers et saules: effets de la concurrence des mauvaises
herbes, désherbage des pépinières et des plantations, contrôle
chimique des gourmands. **Poplar and willow: weed competition**
study, weed control in nursery or plantations (in commercial
willow beds), chemical control of the epicormie buds.

10033 Pinon, J.   FR 010305/71/4126
Marssonina Brunnea : facteurs influençant l'infection des
peupliers cultivés. **M. Brunnea : factors influencing the**
infection of cultivated poplars.

10034 Pinon, J.; Morelet, M.; Lemoine, M.
          FR 010305/73/4125
Leuce et étude des principaux parasites cryptogamiques des
peupliers de la section en France. **Survey and study of the main**
fungi parasites of Leuce poplars in France.

10035 Boccardo, G.   IT 060100/70/0096
Indagini sul virus del mosaico del pioppo. **Poplar mosaic virus**
studies. Publications.

10036 Gremmen, J.   NL 010601/63/1870
Onderzoek naar de bacteriekanker van de populier
veroorzaakt door Aplanobacter populiRidé. **Bacterial canker**
of poplar caused by Aplanobacter populi Ridé. Publications.

Other leafwoods (B 3829)

See also 10030

10037 D'Ambra, V.; Ferrata, M.T.   IT 040808/74/0001
Ricerche sul "canker stain" del platano (Ceratocystis
fimbriata). **Researches on canker stain of plane–tree**
(Ceratocystis fimbriata). Publications.

10038 Grasso, V.   IT 061100/77/0854
Indagine sulla Ceratocystis fimbriata(Ell. and Halst) Davidson
f. platani Walter. **Research on Ceratocystis fimbriata (Ell. and**
Halst) Davidson f. platani Walter.

Stimulant crops (B 3910)

10039 Weinmann, W.   DE 215080/78/0014 N
Untersuchung über das Rückstandsverhalten von
Oxydemetonmethyl und Propineb auf/in Tabak nach einer
Behandlung gegen Blattlaus und Blauschimmel. **Investigation**
on the residue–behaviour of Oxydemeton–methyl and Propineb
on/in tobacco after a treatment against aphid and downy
mildew.

10040 Sewell; Talboys   GB 011005/00/0023 R
Epidemiology of verticillium wilt.

10041 Wilson; Sewell   GB 011005/00/0024 R
Control of verticillium wilt.

10042 Royle   GB 012201/00/0024 R
Biology of Fusarium hop canker.

10043 Royle   GB 012201/00/0026 R
Epidemiology and control of downy and powdery mildews.

Spice and seasoning plants of temperate climates (B 3930)

See also 5032

10044 Weinmann, W.   DE 215080/78/0004 N
Untersuchung über das Rückstandsverhalten von
Dichlofluanid, Propineb und Kupferoxychlorid auf/in Hopfen
nach einer Spritzbehandlung gegen Botrytis cinerea und
Peronospora. **Investigation on the residue–behaviour of**
Dichlofluanid, Propineb, Copperoxychloride on/in hops after a
spray treatment against Botrytis diseases and downy mildew.

10045 Rohloff, H.   DE 215150/77/0008
Untersuchungen zur Neuinfektionsrate virusfreier
Hopfenbestände. **Investigations on rate of new infection in**
virus–free hopgardens.

10046 Rohloff, H.   DE 215150/77/0013
Viruseliminierung bei Hopfensorten und virusfreier Anbau
von Hopfenkulturen. **Elimination of virus in hops varieties and**
virus–free cultivation of hops.

10047 Warmbrunn, K.; Weidner, G.   DE 501052/78/0001 N
Entwicklung elektronischer Warngeräte für die Bekämpfung
des Apfelmehltaus – Podosphaera leucotricha – und des
Falschen Mehltaus des Hopfens – Pseudoperonospora humuli
–. **Development of electronic warning–apparatus for the control**
of apple mildew – Podosphaera leucotricha – and downy mildew
– Pseudoperonospora humuli – in hops.

10048 Rintelen, J.; Poschenrieder, G.   DE 502051/73/0005
Untersuchungen zur Verticillium–Welke des Hopfens.
Research on Verticillium wilt of hops.

10049 Kohlmann, J.; Heindl, M.   DE 502059/73/0002 R
Quantität und Qualität des Spritzbelages bei verschiedenen
Hopfensorten und unterschiedlichen Ausbringungsverfahren
1972. **Quantity and quality of spraying layer on different species**

D 2420 – Plant diseases and disease control

of hops and in dependence on different methods of spraying.

10050 Kremheller, H.T.; Breitner, G.　　DE 502059/73/0004
Untersuchungen zur Abhängigkeit des epidemischen
Auftretens der Hopfenperonospora unter besonderer
Berücksichtigung von Problemen des integrierten
Pflanzenschutzes. **Investigation of the conditions of epidemic
occurrence of Pseudoperonospora humili with special reference
to problems of integrated plant protection.**

10051 Maier, J.; Pichelmaier, K.　　DE 502059/74/0002
Untersuchungen zum Einfluss der Stickstofformulierung auf
die Krankheitsdisposition des Hopfens gegenüber der
Welkekrankheit. **Studies on the influence of nitrogen
formulation on susceptibility of hops to wilt disease.**

Drugs and medicine plants (B 3970)

10052 Avigliano, M.　　IT 022300/75/0018
Indagine sull'evoluzione della Peronospora e dell'Oidio del
tabacco in Italia. **Experience on Blue–mould and
powdery–mildew evolution in Italy.** Publications.

10053 Caponigro, V.　　IT 022300/79/0004 N
I microrganismi della fillosfera del tabacco. Specie fungine
responsabili dell'accumulo di prodotti tossici. **The parasites of
tobacco leaves: fungi producing toxic products.** Publications.

10054 Avigliano, M.; Sannino, L.　　IT 022300/79/0013 N
Studio sulla trasmissibilità dei virus a mezzo insetti con
impostazione di prove di lotta contro alcuni parassiti del
tabacco in campo con prodotti fosforganici nuovi e tradizionali.
**Study on the virus transmissibility by insects with control of
some tobacco parasites in the field with organophosphorus
products.** Publications.

10055 Loon, L.C. van　　NL 020041/72/4510
De hormonale regulatie van de enzymsynthese in met TMV
geïnfecteerde tabaksplanten. **Hormonal regulation of enzyme
synthesis in TMV infected tobacco plants.** Publications.

Other crops (B 3990)

10056 Hollings　　GB 011110/00/0017 R
Viruses of tropical crops.

10057 Plumb　　GB 012009/00/0002 R
Virus diseases of tropical crops.

D 2430 – Weeds and weed control

See also 545, 12609, 16279

10058 Antoine, A.; Fraselle, J.; Rondia, G.; Dekker, A.
　　BE 010003/69/0009 R
Etude technique et économique dans le Nord Tunisien sur
l'emploi des herbicides sélectifs en grandes cultures. **Technical
and economic study in North Tunisia on the use of selective
desherbing products in large scale production.** Publications.

10059 Stryckers, J.; Van Himme, M.; Bulcke, R.
　　BE 030018/58/0001
Biologisch onderzoek en bestrijding van onkruid. **Biological
and control research on weeds.** Publications.

10060 Detroux, L.; Haquenne, W.　　BE 080700/58/0006 R

Etude du désherbage en cultures maraîchères et fruitières.
Study of weed control in vegetables and fruit crops.

10061 Detroux, L.; Salembier, J.　　BE 080700/60/0007 R
Etude du désherbage dans la culture de la pomme de terre.
Study on weed control in potato.

10062 Detroux, L.; Salembier, J.　　BE 080700/62/0005 R
Etude de la lutte contre les plantes adventices dans les
céréales. **Study on weed control in cereals.** Publications.

10063 Detroux, L.; Salembier, J.　　BE 080700/67/0008 R
Etude de la technique de désherbage en culture betteravière.
Study on weed control technics in sugar beets. Publications.

10064 Koch, W.; Dissogi, L.G.　　DE 144540/75/0005
Biologie und Bekämpfung einiger Wasserpflanzen im Sudan.
Biology and control of some aquatic plant species in the Sudan.

10065 Thorup, S.; Odgaard, P.　　DK 010119/63/4814
Kemisk–analytisk bestemmelse af herbicidrester i jord– og
plantemateriale. **Chemical determination of herbicide residues
in soil and plant material.**

10066 Streibig, J.C.　　DK 030145/76/0019
Undersøgelse af danske jordbundstypers indflydelse på
jordherbicidernes biologiske effekt, især med henblik på
doseringens tilpasning til jordbundstype og ukrudtsbestand.
**Investigation of the influence of Danish soil types on the
biological effect of soil–acting herbicides, with special
reference to the regulation of dosage according to soil type and
weed population.**

10067 Streibig, J.C.; Dennis, B.; Haas, H.　　DK 030145/76/0020
Ukrudtsarternes biologi, sociologi og økologi. **The biology,
sociology and ecology of weed species.**

10068 Robson; Barrett　　GB 010807/00/0001 R
Develop methods for the control of emergent weeds.

10069 Robson; Barrett　　GB 010807/00/0002 R
**Develop chemical methods of controlling submerged and
floating vascular plants and algae.**

10070 Robson; Fowler　　GB 010807/00/0003 R
Assess potential of grass carp for the control of aquatic weeds.

10071 Robson; Barret　　GB 010807/00/0004 R
Herbicidal control of weeds in flowing water.

Banks, shores, dikes and their vegetation (B 1530)

See also 10165, 10179

Man–made recreational resources (B 1600)

See also 10178

Parks, gardens, urban greenspaces, plantations (B 1610)

See also 2031, 2646, 10165, 10179, 10283, 10311

10072 Thorup, S.; Bakkendrup–Hansen, G.; Noye, G.
　　DK 010119/52/4809

D 2430 – Weeds and weed control

Herbiciders indflydelse på ukrudt og på planteskolekulturer. **Influence of herbicides on weeds and nursery garden crops.**

10073 Tol, G. van   NL 010601/72/8672 N
De invloed van grondbewerking en onkruidbestrijding op de ontwikkeling van landschappelijke beplantingen. **The influence of soil preparation and weed control on the development of roadside and amenity plantings.**

Plants and parts of plants in general (B 2100)

See also 10059

10074 Decleire, M.; Van Roey, V.; De Cat, W.
     BE 100000/70/0017 R
Influence des herbicides sélectifs sur les protéines et l'activité enzymatique et le métabolisme des plantes. **Effect of selective herbicides on the proteins and the enzymatic activity and the metabolism of plants.** Publications.

10075 Süss, A.; Siegmund, H.   DE 502050/72/0017
Adsorption von Herbiziden an Boden und deren Verfügbarkeit für Pflanzen. **Adsorption of herbicides to soil and their availability to plants.**

10076 Helweg, A.   DK 010117/50/4607
Pesticidernes inaktivering og nedbrydning i jordbunden og pesticidernes indflydelse og jordbundens mikroorganismer. **Inactivation and degradation of pesticides in the soil and the influence of pesticides on soil microorganisms.**

10077 Thorup, S.; Petersen, E.J.; Rubow, T.W.
     DK 010119/67/4817
Totalbekæmpelse af plantevækst, selektiv bekæmpelse på sportspladser, strandenge og flerårige græsområder. **Total control of vegetation, selective control on sport fields, seashore meadows and permanenf grassland.**

10078 Haas, H.; Dennis, B.   DK 030145/66/0004
Morfologiske undersøgelser af ukrudtsarter på unge udviklingsstadier ("bekæmpelsesstadier"). **Morphological investigations of weed seedlings.**

10079 Pizzolongo, P.   IT 040704/77/0263 R
Ricerche ultrastrutturali su piante parassite e saprofite. Ricerche sulla origine e sulla evoluzione degli organuli cellulari. **Research on the ultrastructure of parasitic and saprophytic plants. Research on cellular organelles origin and evolution.**

10080 Rensen, J.J.S. van   NL 020042/65/4716
Onderzoek naar werkingsmechanismen van herbiciden die op de fotosynthese aangrijpen. **Mode of action of photosynthesis inhibiting herbicides.** Publications.

Plant communities as ecological systems (B 2200)

See also 10114, 10185, 10239, 10250

10081 Eggers, T.   DE 215130/72/4005
Biologie und Ökologie wirtschaftlich bedeutender Unkrautarten. **Biology and ecology of weeds of economic importance.**

10082 Thorup, S.; Permin, O.; Rubow, T.W.; Jensen, P.E.
     DK 010119/46/4815

Undersøgelser over ukrudtets biologi og konkurrenceevne. **Studies of the biology and competitive ability of weeds.**

10083 Thorup, S.; Thonke, K.E.; Jensen, P.E.
     DK 010119/66/4811
Udarbejdelse af hensigtsmæssige behandlingsmetoder med bladherbicider. **Development of suitable treatment methods for foliage–acting herbicides.**

10084 Thorup, S.; Røyrvik, H.J.   DK 010119/67/4812
Jordherbiciders virkning, transport og omsætning i jorden. **Effect, transport and degradation in the soil of soilacting herbicides.**

Crops in general (B 3000)

See also 188, 322, 546, 627, 799, 904, 1067, 1543, 2761, 2762, 2763, 2877, 2881, 5319, 8043, 8378, 8404, 9060, 9199, 9220, 10058, 10065, 13550, 19814, 20009

10085 Müller, H.; Böhm, H.H.; Kümmerlin, R.
     DE 126801/75/0001
Experimentelle Untersuchungen über Aufnahmekinetik, Wirkung und Anreicherung von Herbiziden auf der Stufe der Primärproduktion – Algen –. **Experimental investigations concerning uptake kinetics, action and accumulation of herbicides in primary producers – algae –.**

10086 Baeumer, K.; Böttger, W.   DE 132181/75/0001
Bekämpfung der Quecke in Ackerbausystemen mit reduzierter Bearbeitungsintensität – zugleich ein Beitrag zur Schätzung der Schadensschwelle im Getreide –. **Control of Agropyron repens in cropping systems with reduced tillage intensity – Assessment of the economic injury level –.**

10087 Hoppe, H.H.   DE 132213/78/0001 N
Untersuchungen zum Wirkungsmechanismus herbizider Diphenoxypropionsäurederivate. **Investigations on the mode of action of herbicidal diphenoxypropionic acid derivatives.**

10088 Heitefuss, R.; Zacher, H.   DE 132213/78/0003 N
Ursachen der Selektivität herbizider Diphenoxypropionsäurederivate. **Causes of selectivity of herbicidal diphenoxypropionic acid derivatives.**

10089 Koch, W.   DE 144540/70/0005
Veränderungen in der Unkrautflora 1960. **Changes in weed flora in consequence of application of herbicides.**

10090 Koch, W.; Rauber, R.; Schuler, E.; Röttele, M.
     DE 144540/73/0003
Möglichkeiten der kurz– und langfristigen Prognose von Unkrautproblemen auf der Grundlage populationsdynamischer Studien. **Possibilities of short– and longterm prognosis of weed problems on the basis of studies in population dynamics of weeds.**

10091 Bischof, F.   DE 144540/73/0006
Unkräuter Nordafrikas und ihre Keimpflanzen. **Weeds in North Africa and their seedlings.**

10092 Hurle, K.   DE 144540/75/0003
Phytotoxizität von Herbiziden in Abhängigkeit von der Nährstoffversorgung der Pflanze. **Interactions between the phytotoxicity of herbicides and plant nutrition.**

10093 Hurle, K.　　　　　　　DE 144540/78/0001 N
Einfluss des Strohverbrennens auf Aktivität, Sorption und
Abbau von Herbiziden im Boden. **Effect of straw burning on
the activity, adsorption and degradation of herbicides in soil.**

10094 Grossmann, F.; Philipp, W.–D.　　DE 144540/78/0002 N
Verhalten und Nebenwirkungen von Herbiziden unter
besonderer Berücksichtigung ökologischer Zusammenhänge.
**Behaviour and side–effects of herbicides in special
consideration of ecological aspects.**

10095 Bischof, F.　　　　　　　DE 144540/78/0009 N
Biologie und Bekämpfung von Striga hermonthica im Sudan.
Biology and control of Striga hermonthica in the Sudan.

10096 Koch, W.; Unterladstätter, R.　　DE 144540/78/0010 N
Biologie und Bekämpfung von Rottboellia exaltata. **Biology
and control of Rottboellia exaltata.**

10097 Müller, F.　　　　　　　DE 144541/70/0001
Transport von Herbiziden in verschiedenen
Entwicklungsstadien von ausdauernden Unkräutern.
**Translocation of herbicides in different stages of shoot
development of perennial weeds.**

10098 Müller, F.; Vassiliou, G.　　　DE 144541/77/0001
Untersuchungen über die unterschiedliche Empfindlichkeit
von Kulturumbelliferen gegen Metoxuron. **Investigations on
the different sensitivity of cultivated Umbelliferae to
metoxuron.**

10099 Zach, M.　　　　　　　DE 201040/78/0008 N
Erarbeitung von Schwellenwerten für den vertretbaren
Unkrautbesatz in Pflanzenbeständen durch unterschiedliche
Bodenbearbeitungsmassnahmen. **Elaboration of threshold
values for a justifiable weed population using different soil
tillage.**

10100 Weinmann, W.　　　　　DE 215080/77/0010
Entwicklung einer Analysenmethode zur Bestimmung der
Wirkstoffe in festen und flüssigen Kombinationspräparaten
von 2.4.5–T– und Mecoprop–Salzen. **Development of an
analytical method for determination of active substances in solid
and liquid combination preparations of 2.4.5–T– and
mecoprop–salts.**

10101 Weinmann, W.　　　　　DE 215080/77/0011
Entwicklung einer Analysenmethode zur Bestimmung des
Phenolgehaltes in 2.4.5–T–, 2.4–D– und MCPA–Säure.
**Development of an analytical method for determination of
phenol content in 2.4.5–T–, 2.4–D– and MCPA–acid.**

10102 Weinmann, W.　　　　　DE 215080/77/0012
Entwicklung einer Analysenmethode zur Bestimmung von
Methabenzthiazuron. **Development of an analytical method for
determination of methabenzthiazuron.**

10103 Malkomes, H.–P.　　　　DE 215130/73/4005
Untersuchungen über die Populationsdynamik und die
physiologische Leistungsfähigkeit von Mikroorganismen des
Bodens nach Herbizidanwendung. **Studies in the population
dynamics and physiological performance of soil microorganisms
after application of herbicides.**

10104 Eggers, T.　　　　　　DE 215130/73/4007
Unkrautbekämpfung in der Landschaftspflege. **Weed control**
in landscape management.

10105 Maas, G.; Niemann, P.　　　DE 215130/73/4014
Erarbeitung von Grundlagen für die standortgerechte
Dosierung von Herbiziden. **Development of principles for
locally adequate dosages of herbicides.**

10106 Malkomes, H.–P.　　　　DE 215130/74/0002
Untersuchungen über den Einfluss von Herbiziden auf das
Verhalten von Antagonisten gegen bodenbürtige
Pflanzenkrankheiten. **Investigations on the effect of herbicides
upon micro–organisms antagonistic to soil–borne plant
diseases.**

10107 Pestemer, W.　　　　　DE 215130/74/0005
Entwicklung und Prüfung von Biotestmethoden zur Ermittlung
von Rückständen und Pflanzenverfügbarkeit wichtiger
Herbizide. **Development and testing of bioassay methods for
determination of residues and availability of important
herbicides in plants.**

10108 Maas, G.　　　　　　　DE 215130/75/0002
Sorten– und Stadienempfindlichkeit von Kulturpflanzen
gegenüber Herbiziden. **Susceptibility of certain species and
growth stages of cultivated plants to herbicides.**

10109 Eggers, T.　　　　　　DE 215130/77/0001
Bedeutung der Unkräuter und Auswirkungen der
Unkrautbekämpfung in der Agrozönose. **Importance of weeds
and effects of weed control on agrocoenosis.**

10110 Niemann, P.　　　　　　DE 215130/77/0002
Erarbeitung von Grundlagen zur Prognose der Verunkrautung
und deren Anwendung im integrierten Pflanzenschutz des
Ackerbaues. **Development of guidelines for forecasting of weed
infestation and their use in integrated plant protection in
agriculture.**

10111 Eggers, T.　　　　　　DE 215130/77/0003
Pflanzensoziologische und anbautechnische Erhebungen über
die Anpassungsfähigkeit wirtschaftlich bedeutender
Unkrautarten. **Plant sociological and cultivating observations
on the adaptability of weeds of economic importance.**

10112 Pestemer, W.　　　　　DE 215130/77/0005
Erarbeitung von Serienanalysen zur Bestimmung von
Herbizidrückständen in Boden, Pflanzenmaterial und Wasser.
**Development of serial analyses for determination of herbicide
residues in soil, plants and water.**

10113 Malkomes, H.–P.　　　　DE 215130/77/0008
Erarbeitung eines einheitlichen Rahmens zur Beurteilung von
Herbizidnebenwirkungen auf Bodenmikroorganismen und
deren Funktionen. **Development of standard framework for
determination of herbicidal side–effects on soil
micro–organisms and their functions.**

10114 Eggers, T.　　　　　　DE 215130/78/0003 N
Biologische Bekämpfung wirtschaftlich bedeutender
Unkrautarten. **Biological control of weeds of economic
importance.**

10115 Weischer, B.; Bembenek, M.　　DE 215190/78/0001 N
Ökotoxikologische Nebenwirkungen von Herbiziden im
Hinblick auf pflanzenschädigende Nematoden.
Ecotoxicological side–effects of herbicides on plantparasitic

nematodes.

10116 Olberg–Kallfass, R.　　　DE 501508/75/0003
Vorversuche zur integrierten Unkrautbekämpfung bei der
Pflanzenanzucht. **Preliminary experiments for integrated weed
control in plant nurseries.**

10117 Diercks, R.; Reuss, H.U.　　　DE 502051/75/0004
Untersuchung des Einflusses produktionstechnischer und
ökologischer Faktoren auf die quantitative und qualitative
Veränderung der standörtlichen Unkrautflora auf Ackerland.
**Studies on the influence of production technical and of
ecological factors on the quantitative and qualitative alteration
of the locational weed flora on tilled land.**

10118 Süss, A.　　　DE 502055/73/0011
Reaktionen von verschiedenen Bodenherbiziden mit einzelnen
Bodenkomponenten. **Reactions of different soil herbicides with
soil components.**

10119 Süss, A.; Fuchsbichler, G.　　　DE 502055/74/0002
Verhalten von verschiedenen Herbizid–Metaboliten in Boden,
deren Aufnahme durch Pflanzen und Mikroorganismen.
**Behaviour of different herbicide metabolites in soil and their
uptake by plants and microorganisms.** Publications.

10120 Thorup, S.; Ravn, K.　　　DK 010119/46/4801
Herbiciders indflydelse på ukrudt og på frøafgrøder. **Influence
of herbicides on weeds and seed crops.**

10121 Hancock, M.　　　DK 030106/77/0002 N
Syntese og reaktivitet af rhodium (III) komplekser. **Synthesis
and reactivity of rhodium (III) complexes.**

10122 Hancock, M.; Springborg, J.　　　DK 030106/77/0003 N
Flerkernede forbindelser af rhodium (III). **Polynuclear
compounds of rhodium (III).**

10123 Mortensen, G.; Dennis, B.　　　DK 030145/77/0021
Konkurrence mellem afgrøde og ukrudt. **Competition between
crop and weeds.**

10124 Lefebvre, E.　　　FR 010103/74/6336
Mesure de la bioactivité des herbicides. **Measurement of
herbicide bioactivity.**

10125 Soulas, G.　　　FR 011002/72/6304
Etude de l'influence de l'adsorption des herbicides sur leur
dégradation par les microorganismes. **Study on influence of
herbicide adsorption their dégradation by microorganisms.**

10126 Blanchard, M.; Champion, R.　　　FR 011103/48/0190
Etude des déprédations causées par les orobanches sur les
plantes de grande culture; lutte contre ces parasites. **Study of
the damages involved by broomrape on large–scale cultivated
plants; control of these pests.**

10127 Descoins, C.　　　FR 012225/77/5174
Etude des phénomènes d'allélopathie entre adventices et
plantes cultivées. Obtention d'herbicides et de fongicides
naturels. **Allelopathic effects between weeds and cultivated
plants. Isolation of natural herbicides and fongicides.**

10128 Mercer; Hill　　　GB 010504/00/0001 R
Behaviour of herbicides in soil and their uptake by plants.

10129 Richardson; Blair　　　GB 010801/00/0001 R
**Evaluate biological activity, selectivity and soil persistence of
new herbicides.**

10130 Hance　　　GB 010802/00/0005 R
Factors affecting the performance of soil–applied herbicides.

10131 Chancellor; Peters　　　GB 010804/00/0001 R
**Periodicity of germination of weed seeds. Evaluate chemicals for
breaking seed dormancy.**

10132 Parker　　　GB 010809/00/0001 R
**New herbicide treatments for use in tropical crops against
annual and established perennial weeds.**

10133 Elliot　　　GB 010813/00/0001 R
**Survey and analysis of information about weeds and weed
control in agriculture.**

10134 Osborne; Sargent　　　GB 010814/77/0001 R
Dormancy and variability of weed seeds.

10135 Osborne; Sargent　　　GB 010814/77/0002 R
**Importance of stress conditions in germination and seedling
establishment.**

10136 Osborne; Wright　　　GB 010814/77/0003 R
Factors regulating perennation and regeneration of plant parts.

10137 Osborne; Sargent　　　GB 010814/77/0004 R
Control of seed shedding in weed species.

10138 Roberts; Lockett　　　GB 011808/00/0001 R
**Weed seed populations in soil and their germination
requirements.**

10139 Walker　　　GB 011808/00/0004 R
Uptake of herbicides from soil by plants.

10140 Walker; Bond　　　GB 011808/00/0007 R
Persistence of herbicides in soils.

10141 Thurston; Williams　　　GB 012002/00/0001 R
**Effect on weed population of Rothamsted "classical"
experiments.**

10142 Walker; Spokes　　　GB 012010/00/0002 R
Herbicide and pesticide decomposition.

10143 Cooper; Warwick　　　GB 030901/73/0302 R
**Survey of buried seeds from soil samples examined for potato
cyst eelworm.**

10144 Erskine　　　GB 060107/00/0012 R
Weed control (general).

10145 Scragg　　　GB 060216/00/0006 R
Weed control in arable crops.

10146 Potts　　　GB 060314/00/0017 R
Development of weed control systems.

10147 Prendeville, G.N.; O'Leary, N.F.　　　IE 110202/79/9182 N
Effects of herbicides on selected foliose lichens.

10148 Ialongo, M.　　　IT 020300/72/0001

Le specie di Puccinia parassite delle Centaurea (Compositae) infestanti, in Italia. **The species of Puccinia parasites on Centaurea (Compositae) weeds in Italy.** Publications.

10149 Ialongo, M.   IT 020300/76/0003 N
Indagini preliminari sulla specificità di alcune ruggini delle piante infestanti in vista di un loro possibile impiego nella lotta biologica contro le loro matrici.. **Preliminary studies on host specificity of some rusts on weeds for their possible use on biological control of this plants..**

10150 Poli, E.   IT 040317/73/0284
Ricerche sulla flora e vegetazione infestante le colture siciliane per fini pratico–applicativi. **Research on the flora and vegetation infesting crops in Sicily in view of practical application.** Publications.

10151 Poli, E.   IT 040317/77/0265 R
Ricerche geobotaniche ed ecologiche sulle infestanti le colture siciliane allo scopo di fornire indicazioni valide in campo applicativo. **Geo–botanical and ecological research on pests affecting Sicilian cultures with a view to practical suggestions.**

10152 Vazzana, C.   IT 040502/78/1117 N
Fotosintesi di alcune specie infestanti in relazione a luce, temperatura ed umidità. **Photosynthesis of certain species of pests in relation to light, temperature and humidity.**

10153 Pizzolongo, P.; Tucci, G.F.; Ponzi, R.; Melchionna, M.
                    IT 040704/78/0001 N
L'ecologia della germinazione in alcune piante parassite. **The ecology of seed germination in some parasitic plants.** Publications.

10154 Tafuri, F.   IT 061600/77/0874
Sintesi e proprietà erbicide e fungicide di alcuni derivati dell'acido canfosolfonico. **Synthesis of certain derivates of canphor sulphonic acid, their herbicide and fungicide action.**

10155 Magherini, R.   IT 062800/73/0191
Problemi di tecnica vivaistica. Diserbo. **Nursery technical problems. Weed control.** Publications.

10156 Ziliotto, U.   IT 062900/73/0139
Studio sulla biologia del Taraxacum officinale. **Study on the biology of Taraxacum officinale.**

10157 Giardini, L.   IT 062900/73/0140
Studio sulla biologia del Sorghum halepense. **Study on the biology of Sorghum halepense.**

10158 Giovanardi, R.   IT 062900/73/0141
Ricerche sulla lotta chimica ed agronomica contro il Sorghum halepense. **Research on the chemical and agronomic control of Sorghum halepense.**

10159 Toniolo, L.   IT 062900/77/0877
Minimum tillage e zero tillage, "lavorazione terreno". **Minimum tillage and zero tillage, "soil preparation".**

10160 Toniolo, L.   IT 062900/77/0878
Persistenza diserbanti. **Persistence of herbicides.**

10161 Toniolo, L.   IT 062900/77/0879
Biologia delle malerbe: lotta alla sorgagna, influenza degli avvicendamenti sull'evoluzione della flora infestante. **Weed biology: control of sorgagna. Influence of rotations on the evolution of infesting weeds.**

10162 Oorschot, J.L.P. van   NL 010102/67/7929 R
Werking en selektiviteit van herbiciden en groeiregulatoren op fotosynthese, ademhaling en verdamping. **Action and selectivity of herbicides and growth regulators on photosynthesis, respiration and transpiration.** Publications.

10163 Reisler, A.   NL 010102/72/7999
Kasonderzoek met stoffen die de groei en ontwikkeling van planten beïnvloeden. **Research in greenhouse with substances influencing growth and development of plants.** Publications.

10164 Staas–Ebregt, E.M.   NL 010102/75/7300
De biologie en oecologie van akkeronkruiden in relatie tot teeltsystemen en teeltmaatregelen. **The biology and ecology of weeds in relation to the cultivation systems.**

10165 Hoogerkamp, M.   NL 010102/77/8929 N
Niet agrarisch gebruikte landschapselementen als onkruidinfectiebron van cultuurland. **Non agricultural landscape element, as a source of weed infestation in cultivated areas.**

10166 Oorschoot, J.L.P. van   NL 010102/78/7983 R
Invloed van herbiciden op celdeling, kieming en ontwikkeling van kiemplanten. **Influence of herbicides on cell division, germination and early seedling development of plants.**

10167 Oorschot, J.L.P. van   NL 010102/78/7985
De betekenis van fotosynthese en verdamping voor de onkruidontwikkeling. **The significance of photosynthesis and transpiration for weed growth.**

10168 Zandvoort, R.   NL 010102/78/8300
Gevolgen van interacties van biociden voor het gedrag van herbiciden in de grond en voor de groei van planten. **Consequences of biocid interaction on the fate of herbicides in the soil and on plant growth.**

10169 Dorschot, J.P.L. van   NL 010102/78/8875 N
Fysisch–chemisch gedrag van herbiciden in het bodem–plant systeem. **Physico–chemical behaviour of herbicides in the soil–plant system.**

10170 Scheepens, P.C.   NL 010102/78/8926 N
Selectieve bestrijding van enige probleemonkruiden met pathogene micro–organismen in een systeem van geïntegreerde gewasbescherming. **Selective control of some weeds with pathogenic micro–organisms within a system of integrated plant protection.**

10171 Zon, J.C.j. van   NL 010102/79/8924 N
Onkruidproblemen, probleemonkruiden en onkruidbestrijding in de akkerbouw bij diverse vormen van beheer. **Weed problems, problem weeds and weed control in different agricultural systems.**

10172 Zandvoort, R.; Reisler, A.   NL 010104/74/5382
Bodemkundige aspecten van de chemische onkruidbestrijding. **Soil aspects of chemical weed control.** Publications.

10173 Klooster, J.J.   NL 010106/75/6415
Onderzoek naar de mogelijkheden van een overwegend mechanische onkruidbestrijding. **Research on the possibilities**

D 2430 – Weeds and weed control

of a for the most part mechanical weed control.

10174 Aarts, H.F.M.   NL 010207/77/6623 N
De invloed van gewas, teelttechniek en vruchtwisseling op
onkruidpopulaties. **The influence of crop, growing system and
crop rotation on weed populations.**

10175 Cevaal, P.K.   NL 010207/78/8418
Het bepalen van een economisch verantwoorde
onkruidbestrijding door afweging van mechanische en
chemische bestrijdingsmogelijkheden. **To estimate an economic
system of weed control by comparison of mechanical and
chemical weed control possibilities.**

10176 Leistra, M.   NL 010301/78/8875 N
Fysisch–chemisch gedrag van herbiciden in het bodem–plant
systeem. **Physico–chemical behaviour of herbicides in the
soil–plant system.**

10177 Hiele, F.J.H. van; Klei, N.M. van der
                                          NL 020025/70/4472
Reacties van enige cultuurgewassen en hun rassen, alsmede
van enige onkruiden, op toepassingsvarianten van herbiciden.
**Behaviour of plant species and varieties under herbicides; the
effect of herbicides in relation to environmental conditions.**

10178 Karssen, C.M.   NL 020041/75/4516
Ecofysiologie van onkruidzaden. **Ecophysiology of weed seeds.**

10179 Duym, J.   NL 040007/73/8429
Onderzoek naar de bestrijding van onkruiden met chemische
middelen in landbouwgewassen, houtsoorten en watergangen.
Chemical weed control in crops, tree species and waterways.

10180 Rijn, P.J. van   NL 040012/70/3803
Ontkieming en opkomst van onkruiden. **Germination and
emergence of weeds.**

Cereals in general (B 3100)

See also 1146, 8440, 9267, 10062

10181 Heitefuss, R.; Beer   DE 132210/77/0001
Untersuchungen zur Ermittlung der Schadensschwelle von
Unkräutern im Getreide. **Investigations on the determination of
the economic injury level of weeds in small grain.**

10182 Müller, F.   DE 144541/72/0001
Aufnahme, Translokation und Umwandlung von
Bodenherbiziden in Getreide und Mais. **Uptake, translocation
and conversion of soil herbicides in cereals and maize.**

10183 Müller, F.   DE 144541/77/0003
Metabolismus von Phenylharnstoff–Herbiziden als Ursache für
Empfindlichkeitsunterschiede bei Getreidesorten sowie für die
Herbizidwirkung auf Unkräuter. **Metabolism of phenylurea
herbicides as cause of differences in the sensitivity of cereal
varieties as well as of the herbicidal effect on weeds.**

10184 Leh, H.–O.   DE 215060/77/0006
Untersuchungen über Nebenwirkungen von Herbiziden auf
den Mineralstoffhaushalt von Getreide. **Studies on side–effects
of herbicides on mineral balance in cereals.**

10185 Niemann, P.   DE 215130/78/0001 N
Auswirkungen des Unkrautbesatzes im Getreidebestand auf

Arbeitsqualität und Leistung von Mähdreschern. **Effect of
different weed densities in cereal crops on harvesting losses and
harvesting efficiency of combine harvesters.**

10186 Steiner, H.; Bosch, J.; El–Titi, A.; Richter, J.; Weng,
W.   DE 501051/78/0001 N
Ökologische und ökonomische Auswirkungen der
Unkrautbekämpfung mit Herbiziden und mit nichtchemischen
Methoden im Ackerbau – Getreide, Zuckerrüben, Feldgemüse
–. **Ecological and economic effects of weed control using
herbicides and non–chemical measures in agriculture – cereals,
sugar beets and vegetables –.**

10187 Kees, H.   DE 502051/73/0001
Sortenspezifische Reaktion von Getreide auf Herbizide. **The
tolerance of varieties of cereals to a treatment with herbicides.**

10188 Thorup, S.; Permin, O.   DK 010119/46/4802
Herbiciders og jordbearbejdningens indflydelse på ukrudt og
på kornafgrøder. **Influence of herbicides and soil operations on
weeds and cereal crops.**

10189 Lutman; Thornton   GB 010811/00/0001 R
**Herbicide treatments for the control of wild oat and blackgrass
in cereals.**

10190 Wilson; Cussans   GB 010811/00/0003 R
**Development of economic long term systems for the control of
wild oats and blackgrass in cereals.**

10191 Cussans; Ayres   GB 010811/00/0005 R
**Growth and control of Agropyron repens and Agrostis gigantea
in cereal and other cropping systems.**

10192 Courtney   GB 040301/00/0031 R
Control of corn marigold in cereals.

10193 Courtney   GB 040301/00/0032 R
Eradication of wild oats in Northern Ireland.

10194 Researcher not indicated   GB 050121/00/0004 R
Cereals: weed control.

10195 Researcher not indicated   GB 050161/00/0004 R
Cereals: weed control.

10196 Erskine   GB 060107/00/0004 R
Control of weeds in cereals.

10197 Sijtsma, R.   NL 010102/60/7998
Onkruidbestrijding in de akkerbouw. **Weed control in field
crops.** Publications.

Barley (B 3110)

See also 3033

10198 Dennis, B.; Mortensen, G.   DK 030145/78/0001 N
Bygsorters konkurrenceevne over for ukrudt. **Competitive
ability of barley varieties to weeds.**

Maize (B 3120)

See also 6246, 9414, 9416, 9419, 9420, 10209, 10364

10199 Weinmann, W.   DE 215080/78/0016 N

D 2430 – Weeds and weed control

Untersuchung über das Rückstandsverhalten von Atrazin auf/in Gemüsemais nach einer Unkrautbekämpfung. **Investigation on the residue–behaviour of Atrazin on/in maize after a treatment against weeds.**

Rice (B 3140)

10200 Moletti, M.; Baldi, G.; Villa, B.; Mazzini, F.
IT 011400/79/0006 N
Studio sul controllo delle infestanti del riso coltivato in condizioni di irrigazione turnata. **Studies on weed control of rice grown in conditions of periodical irrigation.**

10201 Moletti, M.; Villa, B.   IT 011400/79/0007 N
Prove sull'impiego di erbicidi nel controllo delle infestanti della risaia Heteranthera limosa e H. reniformis. **Trials on the control of the rice weeds Heteranthera limosa and H. reniformis by chemicals.**

10202 Moletti, M.; Villa, B.   IT 011400/79/0008 N
Ricerche sull'impiego di prodotti chimici per il controllo del riso crodo nelle coltivazioni risicole. **Studies on the control of red rice by chemicals in rice fields.**

10203 Russo, S.; Lupotto, E.; Venturo, R.; Caresana, C.
IT 020800/79/0009 N
Riso – Ricerche sui mezzi di controllo dell'infestazione con granella a pericarpio rosso. **Rice – Researches on the possibility to control the infestation of red pericarp grain.**

Sorghum (B 3160)

See also 3083, 9414, 9419, 9420

Wheat (B 3170)

See also 9414, 9416, 9418, 9419, 9420

10204 Rixhon, L.; Crohain, A.; Delhaye, R.; Guiot, J.; Frankinet, M.; Couvreur, L.   BE 080800/75/0010 R
Techniques modernes de production de froment d'hiver. **Modern crop husbandry for winter wheat.**

10205 Heyland, K.–U.; Braun, H.; Goldhammer, T.
DE 111252/73/0022
Nebenwirkung von Bodenherbiziden auf Winterweizensorten bei unterschiedlicher Applikationstechnik. **Response of winter wheat varieties to secondary effects of soil herbicides at different techniques of application.**

10206 Agnello, A. V.; Leandri, A.; Imbroglini, G.C.
IT 020300/79/0007 N
Prove di diserbo contro le avene del grano. **Trials of chemical weeding against wild oats in wheat fields.**

10207 De Robertis, A.; De Giorgio, D.   IT 020500/76/0004 N
Diserbo grano duro a taglia bassa. **The weeding of low durum wheat..**

10208 Vecchio, V.   IT 040502/77/0309 R
Studio dell'accrescimento di alcune specie infestanti del grano con particolare riguardo alla competizione inter e intraspecifica. **A study on increased wheat infestation by certain species with particular reference to inter and intraspecific competition.**

10209 Cantele, A.   IT 062900/74/0135
Prove di aggiornamento sul diserbo delle principali colture agrarie (frumento, mais, barbabietola da zucchero) e della soia. **Field trials on new commercial herbicides for the most important crops (wheat, corn, sugar beet) and for soy–bean.**

Other cereals (B 3190)

See also 3083

10210 Rixhon, L.; Crohain, A.; Couvreur, L.
BE 080800/71/0016 N
Techniques modernes de production d'épeautre. **Modern crop husbandry for Triticum spelta.**

Fibre plants and oil crops in general (B 3200)

See also 10197

Rape (B 3230)

See also 9433

10211 Heitefuss, R.; Korpraditskul, V.   DE 132210/78/0003 N
Nebenwirkungen von Herbiziden auf Phyllosphäre von Raps. **Side effects of herbicides on phyllosphere of rape.**

10212 Börner, H.; Winkler, K.   DE 148200/77/0002
Nebenwirkungen von Herbiziden an Raps. **Side–effects of herbicides on rape.**

Soyabean (B 3240)

See also 10209

Sunflower (B 3250)

10213 Pirani, V.   IT 021100/79/0025 N
Diserbo del girasole. **Sunflower weeding.**

Sugarbeets and starch producing plants in general (B 3300)

See also 10197, 10209

Potatoes (B 3310)

See also 10061, 10275

10214 Frost, M.C.   IE 060500/75/1067 R
Potatoes as weeds (control). Publications.

10215 Cremaschi, D.   IT 021100/76/0002 N
Diserbo chimico della patata.. **Potato chemical weed control..**

10216 Foschi, S.   IT 061800/77/1012
Studio della selettività dei prodotti impiegati nella lotta contro le cuscute delle colture di patata, cipolla, carota e melone. **Study on the selectivity of the substances employed to control dodder in potato, onion, carrot and melon crops.**

Sugarbeets and other sugar crops (B 3320)

See also 7855, 8570, 10063, 10186, 10364

10217 Cussans   GB 010811/79/0009 N

D 2430 - Weeds and weed control

Factors affecting the success of weed beet in agricultural land.

10218 Jaggard; Webb GB 012301/00/0014 R
Weed control.

10219 Longden; Johnson GB 012301/00/0033 R
Methods of controlling annually flowering sugar beet plants in the root crop.

10220 Mitchell, B. IE 060500/70/0194 N
Weed control in sugar beet. Publications.

10221 Barry, P.; Conroy, N.J. IE 120105/68/9041
An evaluation of herbicide systems for weed control in sugar beet.

10222 Giordano, I.; D'Amato, A. IT 021100/75/0001 R
Diserbo chimico della barbabietola da zucchero. **Sugar beet chemical weeding.** Publications.

10223 Pirani, V. IT 021100/79/0008 N
Trattamenti erbicidi in associazione con concimi azotati liquidi distribuiti in presemina normale di barbabietola da zucchero. **Treatments with herbicides to the sugar beet in association with nitrogenous fluid fertilizers spread earlier at pre-sowing or at pre-sowing as usual.**

10224 Catizone, P.; Rubboli, P.; Viggiani, P.
IT 040201/79/0004 N
Studio della selettività del Phenmediphan e del Lenacil nei confronti della bietola da zucchero. **Selectivity of Phenmediphan and Lenacil toward sugar beet.**

10225 Jorritsma, J. NL 060003/57/7716 R
Onkruidbestrijding in suikerbieten. **Weed control in sugar beet.**

Grasses and forage crops in general (B 3400)

See also 10197

10226 Detroux, L.; Haquenne, W. BE 080700/79/0028 N
Etude du désherbage en culture fourragère. **Study of weed control in fodder crops.**

10227 Ziegenbein, G. DE 506155/75/0002 N
Untersuchungen auf Einsatzmöglichkeiten von Herbiziden im Futterpflanzen–Samenbau. **Studies on the possibilities of employing herbicides in forage plant seed production.**

10228 Thorup, S.; Ravn, K. DK 010119/46/4803
Herbiciders indflydelse på ukrudt og på rodfrugter. **Influence of herbicides on weeds and root crops.**

10229 Thorup, S.; Permin, O. DK 010119/46/4804
Herbiciders indflydelse på ukrudt og på grovfoderafgrøder. **Influence of herbicides on weeds and forage crops.**

Grasses (B 3410)

See also 3509, 5533

10230 Hayes GB 040301/00/0022 R
Competition of sown grasses with volunteer Gramineae.

10231 Courtney GB 040301/00/0029 R

Control of Agropyron repens and Agrostis gigantea.

10232 Courtney GB 040301/00/0030 R
Docks (Rumex spp) in Northern Ireland.

10233 Courtney GB 040301/00/0034 R
Agricultural significance of noxious weeds in grassland.

10234 Sijtsma, R. NL 010104/75/6222
Onderzoek naar de mogelijkheden ter bestrijding van onkruiden in de graszaadteelt. **Research on weed control in grass seed crops.** Publications.

Pastures, grassland (B 3420)

See also 3509, 3547, 3620

10235 Andries, A.; Carlier, L. BE 070400/76/0026 R
Kweekbestrijding in grasland. **Destruction of coach grass in grassland.**

10236 Schulz, H. DE 144500/73/0001
Unkrautbekämpfung auf Grünland mit Hilfe von Herbiziden und verschiedener Düngerformen. **Weed control on grassland by way of herbicides and various fertilizers.** Publications.

10237 Voigtländer, G.; Imhoff, H.; Kühbauch, W.
DE 161255/72/0002 R
Untersuchungen über Entwicklung und Reservestoffwechsel besonders persistenter Unkräuter des Dauergrünlandes als Grundlage für Bekämpfungsmassnahmen 1971–1978. **english title not indicated.**

10238 Schöllhorn, J.; Müller, A. DE 501202/77/0002
Prüfung verschiedener Verfahren und Mittel zur Bekämpfung der grossblättrigen Ampferarten – Rumex crispus und Rumex obtusifolius – im Dauergrünland. **Testing of different methods and herbicides for controlling great-leaved docks – Rumex crispus and Rumex obtusifolius – in permanent grassland.**

10239 Wasshausen, W. DE 507350/78/0007 N
Untersuchungen zur Ausbreitung der Quecke auf dem Grünland. **Studies on spread of Agropyron repens on grassland.**

10240 Ernst, P. DE 508301/75/0012
Ampferbekämpfung auf Grünland mit Asulam und Mecoprop. **The control of dock on grassland by means of Asulam and Mecoprop.**

10241 Williams; Chancelor GB 010804/74/0003 R
Grassland weed ecology.

10242 Lewis GB 012109/00/0008 R
Chemical control of weeds in grass and clover seed crops.

10243 Davies GB 030304/00/0004 R
Determination of effect of bracken control on herbage production and pasture formation.

10244 Erskine GB 060107/00/0001 R
Grass and legumes weed control.

10245 Scragg GB 060216/00/0003 R
Weed control in grassland.

10246 Williams GB 060307/00/0002 R
Chemical control of bracken and after treatments.

10247 Corte, A. IT 024000/75/0001
Efficacia del trattamento con Tordon e delle concimazioni azoto–fosfatiche sul miglioramento di pascoli montani infestati da piante cespugliose, principalmente Genista radiata Scop. e Ononis spinosa L. **Effect of Tordon treatment and nitrogen–phosphorus fertilizer on the improvement of upland pastures infested by shrubby plants, chiefly Genista radiata Scop. and Ononis spinosa L.**

10248 Hoogerkamp, M. NL 010102/74/8930 N
Oorzaken en gevolgen van veronkruiding van grasland. **Causes en effects of weed infestation in grassland.**

10249 Luten, W.; Roozeboom, L. NL 010208/73/3881 R
Bestrijding van ongewenste grassoorten in grasland. **Control of undesired grasses in grassland.** Publications.

Mangolds (B 3430)

10250 Bürcky, K. DE 907010/78/0002 N
Rübenschosser und Unkrautrüben. **Bolters and weed beets.**

Grassland legumes (B 3441)

10251 Weinmann, W. DE 215080/78/0017 N
Untersuchung über das Rückstandsverhalten von Metobromuron auf/in Feldsalat nach einer Behandlung zur Unkrautbekämpfung. **Investigation on the residue–behaviour of Metobromuron on/in corn salad after a treatment against weeds.**

Other legumes (B 3449)

See also 3659

Cereals used for forage (B 3450)

See also 10199

10252 Haan, G.H. de; Hag, B.A. ten NL 010207/77/7644
Invloed van mechanische onkruidbestrijding op de opbrengst van mais. **Mechanical weed control in forage maize.**

Turnips (B 3460)

10253 Weinmann, W. DE 215080/78/0018 N
Untersuchung über das Rückstandsverhalten von Trifluralin auf/in Stoppelrüben nach einer Behandlung zur Unkrautbekämpfung. **Investigation on the residue–behaviour of Trifluralin on/in turnips after a treatment against weeds.**

10254 Erskine GB 060107/00/0008 R
Control of weeds in turnips and swedes.

Vegetables in general (B 3500)

See also 10060, 10186

10255 Pestemer, W. DE 215130/73/4011
Verhalten und Wirkung von Herbiziden im Boden bei unterschiedlicher Dosierung. **Behaviour and action of herbicides in soils depending on different dosage.**

10256 Maas, G. DE 215130/77/0004
Methoden der Unkrautbekämpfung in gesäten Gemüsekulturen. **Methods of weed control in cultivation of sown vegetables.**

10257 Maas, G. DE 215130/78/0004 N
Einfluss von Herbiziden auf wertgebende Inhaltsstoffe einiger Gemüsearten. **Influence of herbicides on quality–determining constituents of some vegetables.**

10258 Süss, A.; Stärk, H. DE 502055/73/0004
Aufnahme verschiedener Bromidverbindungen aus Boden durch Gemüsepflanzen nach einer Bodenentseuchung mit Methylbromid. **Uptake of different bromide compounds from soil by vegetable plants after soil treatment with methyl bromide.**

10259 Thorup, S.; Bakkendrup–Hansen, G.; Noye, G. DK 010119/46/4805
Herbiciders indflydelse på ukrudt og på frilandsgrønsager. **Influence of herbicides on weeds and outdoor vegetables.**

10260 Thorup, S.; Røyrvik, H.J.; Thonke, K.E. DK 010119/72/4816
Herbicidernes virkning under væksthusforhold. **Effect of herbicides under glasshouse conditions.**

10261 Roberts; Bond GB 011808/00/0006 R
Evaluation of herbicides and herbicide programmes.

10262 Lawson; Wiseman GB 030701/00/0024 R
Weed ecology and control in vegetables.

10263 Dawson GB 041206/00/0001 R
Vegetable weed control by herbicides.

10264 Researcher not indicated GB 050103/00/0019 R
Use of herbicides in vegetable crops.

10265 Researcher not indicated GB 050143/00/0016 R
Use of herbicides in vegetable crops.

10266 Restaino, F.; D'Amore, R.; Petralia, S. IT 021000/79/0015 N
Diserbo chimico selettivo in orticoltura. **Selective chemical weeding in horticulture.**

10267 Foschi, S. IT 061800/77/1011
Ricerche sull'attività e selettività di erbicidi nelle colture orticole protette. **Research on the action and selectivity of herbicides used in protected market garden cultures.**

10268 Staalduine, D. van NL 010104/75/6224
Onderzoek naar onkruidbestrijdingsmogelijkheden in tuinbouwgewassen. **Research on weed control in horticultural crops.** Publications.

10269 Boer, W. den NL 010206/50/0923
Onkruidbestrijding bij glasgewassen. **Chemical weed control in glasshouse crops.**

10270 Jonkers, J.; Aarts, H.F.M. NL 010207/55/1575
Onderzoek naar de gebruikswaarde van herbiciden voor de vollegrondsgroenteteelt. **Evaluation of herbicides for outdoor vegetable crops.**

D 2430 – Weeds and weed control

Root, tuber and bulb vegetables (B 3510)

See also 10216, 10288

10271 Weinmann, W.   DE 215080/77/0032 N
Untersuchung über das Rückstandsverhalten von Propachlor
nach einer Unkrautbekämpfung in Radies unter Glas.
**Investigations on residual action of propachlor after control in
cultivation of little radish under glass.**

10272 Weinmann, W.   DE 215080/77/0033
Untersuchung über das Rückstandsverhalten von Propachlor
nach einer Unkrautbekämpfung in Kohlrabi unter Glas.
**Investigations on residual action of propachlor after weed
control in cultivation of kohlrabi under glass.**

10273 Weinmann, W.   DE 215080/78/0013 N
Untersuchung über das Rückstandsverhalten von Chloroxuron
auf/in Möhren, Steckzwiebeln, Petersilie und Dill nach einer
Behandlung zur Unkrautbekämpfung. **Investigation on the
residue–behaviour of Chloroxuron on/in carrots, seed–onions,
parsley and dill after a treatment against weeds.**

10274 Friis, E.; Madsen, A.   DK 030104/78/0006 N
Undersøgelse af klorofyl– og betacyaninindholdet i
bederoeblade behandlet med ethofumesat
(2–ethoxy–2,3–dihydro–3,3–dimethylbenzofuran–5–yl
methansulfonat). **Chlorophyll and betacyan in beet leaves
treated with ethofumesat
(2–ethoxy–2,3–dihydro–3,3–dimethylbenzofuran–5–yl
methanesulphonate).**

10275 Blanco, V.V.   IT 062700/77/0943
Ricerche sul diserbo chimico (carota,
cavolo–broccolo–cetriolo, fagiolino, patata). **Research on
chemical weeding (carrot, broccoli, cucumber, bean, potato).**

10276 Pimpini, F.   IT 062900/74/0137
Studio sul diserbo chimico della cipolla nell'ambiente di
Chioggia. **Study on chemical weeding of onions in the
horticoltural area of Chioggia (Venice).**

Greens and leafy vegetables (B 3520)

See also 10275, 10285, 10288

10277 Weinmann, W.   DE 215080/77/0024
Untersuchung über das Rückstandsverhalten von Carbetamid
nach einer Anwendung gegen Unkräuter in Kopfsalat unter
Glas. **Investigations on residual action of carbetamide after
application against weeds in lettuce under glass.**

10278 Weinmann, W.   DE 215080/77/0031
Untersuchung über das Rückstandsverhalten von Propachlor
nach einer Unkrautbekämpfung in Schnittlauchkulturen.
**Investigations on residual action of propachlor after weed
control in cultivation of chives.**

Leguminous vegetables (B 3531)

See also 5747, 10275

Tomatoes (B 3532)

10279 Toniolo, L.   IT 062900/77/0876
Diserbo colture orticole (pomodoro). **Weed–killing in market**

garden cultures (tomato).

Other vegetable fruits (B 3539)

See also 10275

Fruits in general (B 3600)

See also 4066, 10060, 10267, 10268

10280 Tiemann, K.–H.; Dammann, H.–J.   DE 507306/75/0001
Herbizideinsatz im Obstbau. **Herbicides in fruit growing.**

10281 Thorup, S.; Bakkendrup–Hansen, G.; Noye, G.
       DK 010119/56/4807
Herbiciders indflydelse på ukrudt og på frugttræer og–buske.
Influence of herbicides on weeds and fruit trees and bushes.

10282 Davison; Bailey   GB 010806/00/0002 R
Effect of important weeds on fruit production.

Top fruit in general (B 3610)

10283 Groeneveld, R.M.W.; Hoogerkamp, M.; Staalduine,
D. van   NL 010102/75/7301
Onderzoek naar het effect van bodembedekkers op de
onkruidbezetting in en op de groei en ontwikkeling van
meerjarige houtige gewassen. **Research on soil covering plants
in relation to weed control in perennial woody crops.**

10284 Bolding, P.   NL 010212/59/1802
Chemische onkruidbestrijding bij groot fruit. **Chemical weed
control in topfruit.**

Apple (B 3611)

10285 Weinmann, W.   DE 215080/77/0005
Analysenmethode zur Bestimmung von Omethoat auf und in
Salat, Äpfeln, Hopfen und Erde. **Analytical method for
determination of omethoate on and in lettuce, apples, hops, and
earth.**

10286 O'Kennedy, N.D.   IE 060302/62/0503 N
Chemical weed control (apple plantations). Publications.

Other top fruit (B 3619)

10287 Karnatz, A.; Sprenger, F.   DE 105204/77/0002
Auswirkung zehnjähriger Simazinbehandlung des Bodens auf
Obstkulturen– Sauerkirschen und Schwarze Johannisbeeren–.
**Effect of ten years soil treatment with Simazin on sour cherries
and black currants.**

Soft fruit (berries and cane fruits) (B 3620)

See also 10281, 10287

10288 Weinmann, W.   DE 215080/77/0004
Analysenmethode zur Bestimmung von Diuron auf und in
Erdbeeren, Möhren, Erde und Spargel. **Analytical method for
determination of diuron on and in strawberries, carrots, earth
and asparagus.**

10289 Lawson; Wiseman   GB 030701/00/0021 R
Weed ecology and control in soft fruit.

10290 Ross   GB 041204/00/0001 R
Control of weeds in soft fruits by herbicides.

Citrus fruit (B 3630)

10291 Lo Giudice, V.   IT 021600/79/0005 N
Applicazioni e fisiologia degli erbicidi. **Physiology of herbicides and weed control.**

Tropical and sub–tropical fruits (B 3640)

See also 10216

Grapes (B 3650)

See also 193, 294

10292 Julliard, B.; Ancel, J.   FR 010902/55/0275
Utilisation optimale des désherbants chimiques dans les vignobles. **Optimal utilisation of chemical weed–killers in vineyards.** Publications.

10293 Cantele, A.   IT 062900/74/0136
Studio sul diserbo chimico della vite. **Study on chemical weeding of grapes.**

Edible nut fruits (B 3660)

10294 Rizzo, V.; De Giorgio, D.   IT 020500/73/0001 R
Agrotecnica del mandorlo in aridocoltura (diserbo, concimazione azotata, potatura). **Almond–tree cultivation techniques in dry–farming (weed control, nitrogen fertilization and pruning).**

Ornamentals and ornamental products in general (B 3700)

See also 10260, 10268

Bulbs (B 3710)

10295 Thorup, S.; Bakkendrup–Hansen, G.; Noye, G.
     DK 010119/56/4806
Herbiciders indflydelse på ukrudt og på blomsterløg. **Influence of herbicides on weeds and flower bulbs.**

10296 De Ranieri, M.; De vita, M.; Cirrito, M.; Zizzo, G.; Grassotti, A.   IT 021200/77/0014 N
Prove di diserbo chimico su bulbose. **Chemical weed control trials on bulbouses.**

10297 Rooy, M. de   NL 010205/67/1509
Chemische onkruidbestrijding in bol- en knolgewassen. **Chemical weed control in bulbous crops.** Publications.

Flowers and pot plants (B 3720)

See also 10269

10298 Koch, W.; Hansen–del–Orbe, R.   DE 144540/75/0004
Vergleichende Untersuchungen zur Biologie und Bekämpfung von Amaranthus–Arten aus der Dominikanischen Republik und aus Mitteleuropa. **Comparative studies on the biology and control of Amaranthus species from the Dominican Republic and from Central Europe.**

Ornamental shrubs (B 3730)

10299 Lohweg, E.   DE 502102/78/0001 N
Verträglichkeitsprüfung von 9 Herbiziden in verschiedenen bodendeckenden Gehölzen. **The compatibility of nine herbicides to different cultures of soil covering woody plants.**

10300 Caron, J.E.A.   NL 010203/50/1516
Chemische onkruidbestrijding in de boomkwekerij. **Chemical weed control in the nursery.** Publications.

Forests in general (B 3800)

See also 2031

10301 Sukopp, H.; Rijpert, J.M.S.   DE 105060/77/0003
Zur spontanen Ansiedlung und Verbreitung des Neophyten Prunus serotina in europäischen Laub- und Nadelwäldern. **On the colonisation and dissemination of the neophyte Prunus serotina in Europaean deciduous and coniferous forests.**

10302 Abetz, P.; Spiecker, H.   DE 126600/78/0002 N
Untersuchungen über die Entstehung und Verhinderung von Kleba sten – Wasserreiser – bei Tanne, Douglasie und Fichte. **Investigations on formation and prevention of watersprouts in fir, douglas fir and oak.**

10303 Volger, C.; Stolzenburg   DE 132751/75/0001
Neue Herbizide für die Bekämpfung von Pteridium aquilinum. **New herbicides for the control of Pteridium aquilinum.**

10304 Volger, C.   DE 132751/78/0001 N
Die Reaktion forstlicher Kulturpflanzen auf die Bekämpfung von Adlerfarnbeständen mit Asulam. **The reaction of forest trees to bracken control with Asulam.** Publications.

10305 Schütt, P.; Schuck, H.J.   DE 160060/72/0008 R
Allelopathische Wirkung von Unkräutern auf Forstpflanzen 1972. **Allelopathic effect of weeds on forest plants.**

10306 Grasblum, M.   DE 215100/77/0002
Untersuchungen zur Bekämpfbarkeit der Himbeere im Forst mit Herbiziden im Hinblick auf die Rückstandssituation. **Investigations on the control of raspberry in forests with herbicides regarding residues.**

10307 Gussone, H.A.; Hartmann, J.   DE 507651/77/0003
Untersuchungen über den Einfluss konkurrierender Vegetationen auf die Wirtschaftsbaumarten in Forstkulturen und Naturverjüngungen mit dem Ziel einer deutlichen Abgrenzung der Notwendigkeit einer Herbizidanwendung im Walde. **Investigations on the influence of competitive vegetations on productive tree species in forest plantations and natural regeneration with the aim of clear delimitations of necessity of herbicide application in forests.**

10308 Thorup, S.; Rubow, T.W.   DK 010119/50/4810
Herbiciders indflydelse på ukrudt og på skovkulturer og læplanter. **Influence of herbicides on weeds and forest nursery stock and shelter plants.**

10309 Delabraze, P.; Frochot, H.   FR 010301/56/4075
Emploi des phytocides dans les régénérations naturelles et artificielles d'espèces forestières. **Use of herbicides for natural and artificial regeneration in forest.**

D 2430 – Weeds and weed control

10310 Robinson, J.D.; Brosnan, J.   IE 050100/72/7206 N
Classification of forest weed problems and evaluation of herbicides. Publications.

10311 Tol, G. van   NL 010601/76/6876
Onderzoek naar de uitbreiding van de Amerikaanse vogelkers (Prunus serotina Ehrh.). **Research on spreading of Black Cherry (Prunus serotina Ehrh.).** Publications.

Pine forests in general (B 3810)

10312 Detroux, L.; Haquenne, W.   BE 080700/79/0029 N
Etude du désherbage sélectif dans les jeunes plantations forestières de résineux. **Selective brush control in young pine–forest plantations.**

10313 Tol, G. van   NL 010601/53/2396
Onkruidbestrijding in jonge bosbeplantingen. **Weed control in young forest plantations.** Publications.

Fir forests (B 3811)

See also 5978

Spruce and fir forests (B 3812)

10314 König, E.; Olberg–Kallfass, R.   DE 501508/77/0002
Untersuchungen über den Einfluss starker Unkrautkonkurrenz und Wildverbiss auf die Entwicklung von Fichtenkulturen. **Investigations on the influence of strong weed concurrence and of browsing on young stands of Norway spruce.**

Stimulant crops (B 3910)

See also 10136

10315 Grossmann, F.; Janicke, R.; Müller, F.
    DE 144541/77/0002
Metabolismus von Pyracarbolid in Kaffee und einer empfindlichen Pflanzenart, sowie Wirkung auf pflanzlichen Stoffwechsel. **Metabolism of pyracarbolid in coffee and in a sensitive plant as well as its effect on plant metabolism.**

Spice and seasoning plants of temporate climates (B 3930)

See also 10273, 10285

Drugs and medicine plants (B 3970)

See also 17021

10316 Maas, G.   DE 215130/73/4001
Rationelle Unkrautbekämpfungsverfahren in ArzneipflanzenKulturen. **Scientific methods of weed control in medicinal herb cultures.**

Other crops (B 3990)

10317 Meneghini, A.; Cappellozza, L.   IT 020400/70/0002
Diserbo e interventi fitosanitari in gelsicoltura. **Pest– and weed–control in mulberry–tree cultures.** Publications.

D 2490 – Miscellaneous plant disorders

10318 Impens, R.; Nangniot, P.; Delcarte, E.;

Deroanne–Bauvin, J.   BE 010021/71/0007 R
Contamination de l'environnement par le plomb en relation avec la circulation automobile. **Contamination of the environment by lead near highways.** Publications.

10319 Kirchmann.; Impens, R.; Delcarte, E.; Nangniot, R.
    BE 010021/73/0004 R
Transfert du plomb dans la chaîne alimentaire. **Transfert of lead in the food–chain.** Publications.

10320 Schmidt–Vogt, H.; Lessel, A.   DE 126300/75/0001
Erfassung und Analyse des Kiefernschneebruchs 1968 und 1975 im Pfälzer Wald. **Inventory and analysis of the pine snow break in 1968 and 1975 in the Palatinate Forest.**

10321 Baumgartner, A.; Mayer, H.   DE 160120/75/0001
Sturmgefährdung des Waldes. **Storm damage risk of forests.**

10322 Nøddegård, E.   DK 010116/61/0022
Pesticidforsøg med jord og planter med henblik på analysering for rester. **Pesticide experiments with soil and plants in relation to residue analysis.**

10323 Løschenkohl, B.; Allerup, S.   DK 030147/79/0003 N
Cytologiske forandringer som følge af beskadigelser af kartofler. **Cytological changes resulting from mechanical injuries to potato tubers.**

10324 Dragsted, J.   DK 030181/77/0011
Vejtræer og salt. **Roadside trees and salt.**

10325 Tjell, J.C.   DK 030202/77/8039
Kilder for uønsket tilførsel af bly, cadmium og kviksølv i landbrugsplanter. **Sources of undesirable supplies of lead, cadmium and mercury in agricultural plants.**

10326 Boleij, J.S.M.   NL 020070/78/8396
De effecten van luchtverontreiniging op planten en ecosystemen. **The effects of air pollution on plants and eco–systems.**

Man–made recreational resources (B 1600)

10327 Impens, R.; Nangniot, P.; Leduc, A.; Delcarte, E.
    BE 010021/72/0005 R
Contamination de l'environnement par les métaux lourds en zone urbaine et industrielle. **Contamination of industrial and urban areas by heavy metals.** Publications.

Parks, gardens, urban greenspaces, plantations (B 1610)

See also 10392

10328 Knabe, W.   DE 508303/77/0001
Resistenzprüfung von Forst– und Parkgehölzen unter Immissionsbelastung. **Investigations on the tolerance of woody plants in forests and parks under the stress of air pollution.** Publications.

10329 Mooi, J.   NL 010108/70/3424
Onderzoek naar de gevoeligheid van houtige gewassen voor luchtverontreiniging. **Investigation on the susceptibility of woody plants to air pollution.** Publications.

10330 Dissen, H.D.   NL 010601/67/2384

D 2490 – Miscellaneous plant disorders

Problemen betreffende beplantingen in steden. **Problems with tree plantations in urban agglomerations.** Publications.

Plants and parts of plants in general (B 2100)

See also 10318

10331 Impens, R.; Paul, R.; Lacroix, J.   BE 010021/70/0006 R
Métabolisme des plantes en présence d'anhydride sulfureux.(SO_2). **Metabolism of plants in the present of sulfuric anhydrid (SO_2).** Publications.

10332 Istas, J.; Termonia, M.; Alaerts, G.; Raekelboom, E.; De Temmerman, L.   BE 100000/72/0008 R
Studie van de uitwerking van gasvormige luchtvervuilers op planten. **Study of the effects of gaseous pollutants on plants.** Publications.

10333 Istas, J.; De Temmerman, L.; Baeten, H.; De Borger, R.   BE 100000/78/0038 N
Onderzoek over de invloed van het gebruik van dooizout en pekeloplossingen in de winterbehandeling van de wegen op de bodem en de plantengroei. **Study on the effects of the winter treatments of the roads by salt solutions on the soil and the vegetation.**

10334 Baumeister, W.; Austenfeld, F.–A.   DE 164050/75/0004
Untersuchungen über die physiologische Wirkung von Schwermetallen. **Studies on physiological effects of heavy metals.**

10335 Rossi, N.   IT 040202/78/1097 N
Detersivi nel suolo. Interazione con i componenti della frazione argillosa e risultanti effetti sulla cresciuta delle piante. **Detergents in the soil. Interaction with the clay fraction components and risulting action on plant growth.**

10336 Silva, S.; Carini, F.; Beghi, B.; Rastelli, A.; Anguissola Scotti, I.   IT 040406/78/0003
Determinazione dei fattori di trasferimento di ^{131}J dalla acqua ai vegetali. **Determination of the transfer coefficients of ^{131}J from water to vegetable.**

10337 Posthumus, A.C.; Floor, H.   NL 010108/72/3956
Onderzoek fotochemische en andere luchtverontreinigingen in Nederland, im verband mat beschadigingen van planten. **Photochemical and other air pollutions in The Netherlands in relation with damage to plants.** Publications.

Plant communities as ecological systems (B 2200)

10338 De Temmerman, L.; Baeten, H.; Raekelboom, E.   BE 100000/78/0037 N
Biologisch onderzoek van luchtverontreiniging met behulp van hogere planten. **Biological research on air pollution by aid of vascular plants.**

Crops in general (B 3000)

See also 5239, 10322, 10325, 10326, 13870

10339 Impens, R.; Mathy, P.; Nangniot, P.   BE 010021/76/0008 R
Pénétration – translocation et transfert du Cd chez les végétaux. **Penetration translocation and transfer of Cd by plants.**

10340 Van Assche, C.   BE 040203/76/0017 R
Fytiatrie ten opzichte van de zware metalen met kationenwisselaars. **Plant diseases caused by heavy metals and their phytiatry with cation exchanges.** Publications.

10341 Bornkamm, R.; Faensen, A.; Overdieck, D.   DE 105100/77/0001
Reaktionen von höheren Pflanzen auf die Einwirkung von Luftverunreinigung. Pflanzenökologischer Teil. **Reaction of higher plants to the immission of air pollutants. Plant ecological part.** Publications.

10342 Daunicht, H.–J.; Noffke, K.   DE 105203/74/0003
Akkumulation und Wirkung phytogener Schadgasemanationen in Kulturräumen. **Accumulation and effects of phytogenic emanations of toxic gases in protected cultivation.**

10343 Stegemann, H.; Roeb, L.   DE 215140/75/0008
Enzymanalysen zur Frühdiagnose negativer Umwelteinflüsse. **Enzyme analysis for early diagnosis of harmful ecological influences.**

10344 Brandtner, E.   DE 301050/75/0006
Luftbildinterpretation für phänologische Zwecke. **Remote sensing for phenological purposes.**

10345 Kofoed, A.D.; Larsen, S.D.; Klausen, P.S.   DK 010101/74/3003
Landbrugets anvendelse af slam fra rensningsanlæg og andre affaldsstoffer fra byer og industrier. **Utilization of sewage slurry and other wastes from towns and industries for agricultural purposes.**

10346 Dumortier, B.   FR 010106/72/1774
Inhibition de la diapause. "Créneaux de lumière". **Diapause inhibition by night short lighting. "Battlements light".**

10347 Allemand, P.; Augé, P.   FR 010501/69/1010
Etude de la résistance au froid. **Studies on frost resistance.**

10348 Del Re, A.   IT 040406/72/0448
Alterazioni del metabolismo vegetale provocate da residui di anti–parassitari sistematici. **Alterations of the plant metabolism due to residues of systema–tic pesticides.** Publications.

10349 Goldberg Federico, L.   IT 040603/78/1063 N
Tossicità di metalli inquinanti e loro effetti sul metabolismo di diverse specie di vegetali cresciute in soluzioni idroponiche. **Toxicity of polluting metals and their effect on the metabolism of different species of plants cultivated in hydroponic solutions.**

10350 Marte, M.   IT 041015/77/0231 R
Ricerche di isto–citologia e istochimica in piante colpite da malattie non parassitarie. **Histo–cytological and histo–chemical research on plants affected by non–parasitic diseases.**

10351 Clementi, F.   IT 063800/77/0825
Tossicità ed analisi dei residui e dei metaboliti di fungicidi e tossicità di nuovi fitofarmaci. **Toxicity and analysis of fungicide residues and metabolites; toxicity of new pesticides.**

10352 Posthumus, A.C.   NL 010108/59/1060 R
Gevoeligheid van gewassen voor luchtverontreiniging.

D 2490 – Miscellaneous plant disorders

Susceptibility of crops to air pollution. Publications.

10353 Raay, A. van — NL 010108/60/1063
Onderzoek naar het voorkomen van plaatselijke luchtverontreiniging van geringe omvang. **Research on the investigation of local air pollution of a small extent.** Publications.

10354 Raay, A. van — NL 010108/64/1064 R
Monitoring van effecten van luchtverontreiniging rond een aluminium fabriek te Heveskes (Gr.). **Monotoring of air pollution effects around an aluminium factory near Heveskes (Gr.).**

10355 Raay, A. van — NL 010108/66/3097 R
Onderzoek naar het voorkomen van luchtverontreiniging in Zeeland. **Research on the occurrence of air pollution on Sealand.**

10356 Eerden, L.J.M. van der — NL 010108/78/8396 N
De effekten van luchtverontreiniging op planten en ecosystemen. **The effects of air pollution on plants and ecosystems.**

10357 Eerden, L.J.M. van der — NL 010108/78/8709 N
Invloed van ammoniak houdende lucht van itensieve veehouderijbedrijven op planten. **Influence of ammonia containing air from intensive livestock farmings on plants.**

Cereals in general (B 3100)

10358 Heitefuss, R.; Ibenthal, W.–D. — DE 132210/71/0001
Physiologische Ursachen der Nebenwirkung von Herbiziden auf Blattkrankheiten des Getreides. **Physiological causes of side–effects of herbicides on leaf diseases of cereales.**

10359 Ellis; Barnes — GB 010503/79/0020 N
Effects of straw residues on growth of cereals and populations of earthworms.

10360 Ellis; Christian — GB 010503/79/0021 N
Agronomic effects of the presence of straw residues on cereal production.

Maize (B 3120)

See also 10364

10361 Soyer, J.P.; Chignon; Juste, C.; Lubet — FR 010704/73/6079
Intoxication ammonique du maïs. **Ammonia toxicity on corn.**

Wheat (B 3170)

10362 Häckel, H. — DE 301100/77/0003
Zusammenhang zwischen der Höhe eines Weizenbestandes und seiner Anfälligkeit auf Pilzkrankheiten. **Correlations between height of wheat crop and sensitivity to fungal diseases.**

Potatoes (B 3310)

See also 10323

10363 Pratella, G.C.; Menniti, A.M. — IT 040216/79/0003 N
Prevenzione dei danni da freddo su patate. **Chilling injury prevention on potatoes.**

Sugarbeets and other sugar crops (B 3320)

10364 Casarini, B.; Olimpieri, R. — IT 021100/79/0013 N
Saggio biologico dell'effetto dei residui dei trattamenti erbicidi sulla precedente coltura di Mais, su barbabietola da zucchero. **Sugar beet biological test of the effect of residues of treatments with herbicides on the previous maize growing.**

Other starch producing plants (B 3390)

10365 Heilinger, F. — DE 201040/75/0028
Chemische Ursachen der "Blauverfärbung" von unverletztem Gewebe der Kartoffelknolle. **Chemical causes of the 'blueing' of undamaged tissue of the potato tuber.**

Vegetables in general (B 3500)

10366 Hansen, H.; Bohling, H. — DE 211010/77/0001 R
Klärung der Ursachen von physiologischen Erkrankungen bei Obst und Gemüse. **Elucidation of the causes of physiological disorders in fruits and vegetables.**

10367 Schmid, G.; Rosopulo, A. — DE 502055/74/0007
Untersuchungen über den Blei– und Cadmiumgehalt im Gemüsebau an stark befahrenen Verkehrswegen und in Einflug– schneisen von Flugplätzen. **Analysis of Pb and Cd content in vegetables along greatly frequented highways and in flying lanes of airports.**

10368 Steenbergen, P. — NL 010206/51/0924
Niet parasitaire beschadiging, inclusief luchtverontreiniging bij glasgewassen. **Non parasitic damages, including air pollution in glasshouse crops.**

Root, tuber and bulb vegetables (B 3510)

10369 Häckel, H. — DE 301100/77/0001
Wurzelfrostschutz in Pflanzcontainern. **Frost protection of roots in plant containers.**

Leguminous vegetables (B 3531)

10370 Persiel, F.; Rockstroh, K. — DE 206000/75/0009
Untersuchungen zur Resistenz gegen Auflaufschäden bei Phaseolus vulgaris. **Studies on the resistance of beans– Phaseolus vulgaris – to impairs during emergence.**

Mushrooms and other edible fungi (B 3540)

10371 Couvy, J. Mme; Crouau, Y. — FR 010703/75/0385
Etude des champignons déformés. **Study sporophore malformations in the cultivated mushrooms.**

Fruits in general (B 3600)

See also 10366

10372 Wagner, C. — DE 301100/77/0005
Abhängigkeit der Frostresistenzschwelle verschiedener Obstgehölze von der meteorologischen Vorgeschichte. **Threshold of frost resistance in different fruit trees in dependence on meteorological antecedents.**

10373 Quast, P. — DE 507301/75/0001 N
Diagnose und Beka mpfung physiologischer Fruchtscha den.

D 2490 – Miscellaneous plant disorders

Diagnosis and control of physiological fruit damages.

Top fruit in general (B 3610)

10374 Pratella, G.C.; Biondi, G.; Brigati, S.
IT 040216/79/0001 N
Prevenzione dei danni da freddo sulle drupacee. **Prevention of chilling injury on stony fruits.**

Apple (B 3611)

10375 Beinhauer, R.; Todt, T.   DE 301060/75/0002
Frostschutz durch Wassernebel zur Obstblütezeit. **Frost protection by fog at apple blossom time.**

10376 Lalatta, F.   IT 040605/78/0001
Indagini sul deperimento del melo "Delicious rosse" in Valtellina. **Researches on the decay of Delicious red apple trees in Valtellina.**

Other top fruit (B 3619)

See also 4258

10377 Wagner, C.   DE 301100/77/0004
Frostresistenzprüfung von Kirschunterlagen und UnterlagenSorten–Kombination. **Testing of frost resistance on cherry stocks and in combined stock varieties.**

10378 Monet, R.   FR 010701/74/0193
Etude de la fente de noyaux chez les variétés de pêches précoces. **Study of stone splitting in early peach varieties.**

10379 Monet, R.   FR 010701/74/0297
Etude de la résistance au gel des bourgeons floraux du pêcher. **Frost resistance of peach–tree flower–buds.**

Soft fruit (berries and cane fruits) (B 3620)

10380 Lefebvre, J.M.; Lavielle, G.   FR 011001/73/6139
Etude du dépérissement du framboisier. **Study of raspberry decay.**

Citrus fruit (B 3630)

10381 Averna, V.   IT 040901/77/0108 R
Danni da gas nelle piante, assorbimento del fluoruro atmosferico in piante di limone ed effetti sul metabolismo. **Gas damage to plants, absorption of atmospheric fluoride in lemon plants and effects on their metabolism.**

Grapes (B 3650)

10382 Gärtel, W.   DE 215220/78/0004 N
Untersuchungen über Pigmentstörungen, Deformationen und Nekrosen, die durch Pflanzenbehandlungsmittel an Rebblättern ausgelöst werden. **Investigations on pigment defects, deformations and necroses induced by pesticides on grapevine leaves.**

Ornamentals and ornamental products in general (B 3700)

10383 Van Onsem, J.G.; Verdonck, O.   BE 070600/79/0082 N
Studie van de remmende faktoren bij optimale groei van sierplanten. **Study of factors that inhibit optimal growth of** ornamentals.

Bulbs (B 3710)

10384 Wolting, H.G.   NL 010108/73/3954
Onderzoek naar de gevoeligheid van monocotyle siergewassen voor chronische HF–inwerking. **Investigations to the susceptibility of monocotyledonous ornamental crops to chronical HF–influence.** Publications.

10385 Wolting, H.G.   NL 010108/73/3955
Onderzoek naar het voorkomen van industriële luchtverontreiniging in het tuinbouwgebied IJmond. **Investigations to the appearance of industrial air–pollution in the horticultural area IJmond.**

Flowers and pot plants (B 3720)

See also 10368, 10384, 10385

Ornamental shrubs (B 3730)

See also 10329

Forests in general (B 3800)

See also 2074, 10321, 10329

10386 Burschel, P.; Julio, G.   DE 160151/78/0002 N
Diagnose der Waldbrandprobleme in der Bundesrepublik Deutschland. **Analysis of forest fire problems in the Federal Republic of Germany.**

10387 Delabraze, P.; Valette, J.C.   FR 010301/66/4076
Entretien chimique des pare–feu arborés en région méditerranéenne. **Chemical control in mediterranean woody fire breaks.**

10388 Delabraze, P.; Frochot, H.   FR 010301/71/4074
Migration des phytocides et leurs conséquences en sylviculture. **The movement of herbicides and their consequence in forestry.**

10389 Decourt, N.; Bedeneau; Romary   FR 010301/73/4072
Effets de la pollution par le fluor et le dioxyde de soufre sur un massif forestier. **Effects of fluor and sulphur dioxide pollution on a forest.**

10390 Delabraze, P.; Valette, J.C.   FR 010301/73/4077
Inflammabilité et combustibilité de la végétation forestière méditerranéenne. **Inflammability and combustibility of the mediterranean forest vegetation.**

10391 Sulli, M.; Buresti, E.   IT 021700/72/0001
Ricerche sugli incendi in foresta. **Forest fire researches and control.**

10392 Kopinga, J.   NL 010601/77/7540
Behandeling van boomwonden. **Treatment of tree wounds.** Publications.

Pine forests in general (B 3810)

10393 Kechel, H.; Weisgerber, H.   DE 506451/71/0008
Untersuchungen über die Resistenzeigenschaften verschiedener Laub– und Nadelbaumarten gegenüber biotischen und abiotischen Einflüssen. **Research on the**

resistance of coniferous and broadleaved tree species against biotic damages.

Fir forests (B 3811)

10394 Carle, P.; Fabre, J.P.; Toth, J.   FR 010609/65/4108
Dépérissement du Pin maritime dans le Sud Est méditerranéen. **Dieback of Pinus Pinaster in french mediterranean region.**

Spruce and fir forests (B 3812)

See also 8927

10395 Schütt, P.; Lo, H.C.   DE 160061/74/0006
Anatomische Untersuchungen an verletzten Fichtenrinden. **Anatomy of injured barks of Picea abies.**

10396 Schönborn, A.von; Schindlbeck, W.
                  DE 160180/75/0001
Untersuchungen über die Schadwirkung von Schwefeldioxyd auf Picea abies. **Investigations of the noxious influence of sulfur dioxide on Picea abies.**

10397 Dimitri, L.; Vaupel, O.   DE 506453/72/0004
Läuterung und Durchforstung von Fichten–Jungbeständen auf gefährenden Standorten. **Research on tending and thinning of young spruce stands on sites susceptible to wind blow and other abiotic damages.**

Leafwoods in general (B 3820)

See also 10393

Oak tree stands (B 3821)

10398 König, E.; Korsch, J.   DE 501508/78/0003 N
Untersuchungen über das Auftreten von Buchenrotkern in Abhängigkeit von Standortsunterschieden. **Studies on the correlations between red heartwood of beech and differences of site–conditions.**

Ash tree stands (B 3822)

See also 8927